Nonlinear System Identification

Oliver Nelles

Nonlinear System Identification

From Classical Approaches to Neural Networks, Fuzzy Models, and Gaussian Processes

Second Edition

Oliver Nelles
University of Siegen
Netphen, Germany

ISBN 978-3-030-47438-6 ISBN 978-3-030-47439-3 (eBook)
https://doi.org/10.1007/978-3-030-47439-3

This Springer imprint is published by the registered company Springer Nature Switzerland AG.
The registered company address is: Gewerbestrasse 11, 6330 Cham, Switzerland

Preface to the Second Edition

The second edition of this book contains novel chapters on the following topics:

- Input selection with local model networks, Chap. 15.
- Gaussian process models, Chap. 16.
- Design of experiments applications, Chap. 26.
- Input selection applications, Chap. 27.
- Local model network toolbox, Chap. 29.

This book focuses mainly on the key benefits that local model networks offer, compared to alternative model architectures. These features are presented with significant width and depth. Additionally, extensively treated new topics are:

- Axis-oblique partitioning strategies, in particular, HILOMOT.
- Design of experiment strategies utilizing the special structure of local model networks.
- Input selection strategies exploiting the possibility for distinct input spaces in local model networks, namely the input space where the local models live, and the input space where the partitioning takes place.
- Nonlinear finite impulse response (NFIR) local model networks that take advantage of the input space separation mentioned above and the local estimation approach for the linear models.
- Bagging, boosting, regularization, and knowledge discovery in the context of local model networks.

Furthermore, besides a couple of small corrections, the following issues are addressed which were newly discovered or gained higher significance since the first edition of this book:

- Deep learning was discovered/invented since the first edition of this book. Due to its overwhelming success, whole monographs have been written on this topic. Section 11.2.9 touches the key ideas and major ingredients of deep learning.
- Long short-term memory (LSTM) networks are extremely promising neural network architectures with internal dynamics. They are very successfully applied

to machine learning problems with complex dynamics, and currently their properties in an engineering context for system identification are an open question of future research. Refer to Sect. 23.5 for a short explanation of LSTM networks in the terminology and language of this book.
- Excitation signal design for nonlinear dynamic processes is a very important but neglected field in research on system identification. A new approach based rather on an optimal signal generator than on a certain signal type is proposed and analyzed. Refer to Sects. 19.8 and 26.4.
- A more elaborate discussion of finite impulse response (FIR) models addresses the important progress made in this field. A detailed comparison between FIR and ARX models is carried out methodically. Due to the new and very promising developments in *regularized* FIR model estimation, the pros and cons of (N)ARX vs. (N)FIR becomes much more important. Refer to Sects. 18.6.1 and 18.6.2 for the linear case and to Sect. 22.7 for the nonlinear, local model network case.
- L_1-norm regularization has significantly gained popularity as it is capable of producing sparse models. Therefore, some basics and possibilities are discussed in this book. Refer to Sects. 3.1.6 and 13.2.5.

On the real-world applications' side, two new chapters are included in this edition on the following topics:

- Design of experiments for two *static* applications (in structural health monitoring and combustion engine modeling) with an active learning strategy exploiting local model network characteristics. In addition, a new nonlinear *dynamic* excitation signal generator is applied to a common rail injection system.
- The power of input selection is demonstrated on a couple of processes of different complexity. In particular, the feature of distinct input spaces for local model networks is analyzed and exploited. Furthermore, metamodeling for the efficiency modeling of fans depending on their geometrical characteristics is discussed in detail. Data is generated by computational fluid dynamics (CFD) simulations, which are orders of magnitude too slow for a fan efficiency optimization in the industry calling for fast metamodels.

Finally, the reader is additionally pointed to more than 200 novel references, mostly detailing the newly discussed topics and refreshing the state-of-the-art on the mature material.

I would like to thank my dear colleagues Claus-Peter Fritzen and Thomas Carolus for contributing to the applications part of this book via successful joint projects. Furthermore, many thanks for numerous fruitful discussions to my former and current research assistants Benjamin Hartmann, Tobias Ebert, Julian Belz, Geritt Kampmann, Mark Schillinger, Tim Oliver Heinz, Tobias Münker, and Max Schüssler.

Siegen, Germany
March 2020

Oliver Nelles

Preface to the First Edition

The goal of this book is to provide engineers and scientists in academia and industry with a thorough understanding of the underlying principles of nonlinear system identification. The reader will be able to apply the discussed models and methods to real problems with the necessary confidence and the awareness of potential difficulties that may arise in practice. This book is self-contained in the sense that it requires merely basic knowledge of matrix algebra, signals and systems, and statistics. Therefore, it also serves as an introduction to linear system identification and gives a practical overview on the major optimization methods used in engineering. The emphasis of this book is on an intuitive understanding of the subject and the practical application of the discussed techniques. It is not written in a theorem/proof style; rather the mathematics is kept to a minimum and the pursued ideas are illustrated by numerous figures, examples, and real-world applications.

Fifteen years ago, nonlinear system identification was a field of several ad-hoc approaches, each applicable only to a very restricted class of systems. With the advent of neural networks, fuzzy models, and modern structure optimization techniques a much wider class of systems can be handled. Although one major characteristic of nonlinear systems is that almost every nonlinear system is unique, tools have been developed that allow the use of the same approach for a broad variety of systems. Certainly, a more problem-specific procedure typically promises superior performance, but from an industrial point of view a good tradeoff between development effort and performance is the decisive criterion for success. This book presents neural networks and fuzzy models together with major classical approaches in a unified framework. The strict distinction between the model architectures on the one hand (Part II) and the techniques for fitting these models to data on the other hand (Part I) tries to overcome the confusing mixture between both that is frequently encountered in the neuro and fuzzy literature. Nonlinear system identification is currently a field of very active research; many new methods will be developed and old methods will be refined. Nevertheless, I am confident that the underlying principles will continue to be valuable in the future.

This book offers enough material for a two-semester course on optimization and nonlinear system identification. A higher level one-semester graduate course can focus on Chap. 7, the complete Part II, and Chaps. 19, 20, 21, 22, and 23 in Part III. For a more elementary one-semester course without prerequisites in optimization and linear system identification Chaps. 1, 2, 3, 4, and 18 might be covered, while Chaps. 10, 14, 20, and 23 can be skipped. Alternatively, a course might omit the dynamic systems in Part III and instead emphasize the optimization techniques and nonlinear static modeling treated in Parts I and II. The applications presented in Part IV focus on a certain model architecture and will convince the user of the practical usefulness of the discussed models and techniques. It is recommended that the reader should complement theses applications with personal experiences and individual projects.

Many people supported me while I was writing this book. First of all, I would like to express my sincerest gratitude to my Ph.D. advisor, Professor Rolf Isermann, Darmstadt University of Technology, for his constant help, advice, and encouragement. He gave me the possibility of experiencing the great freedom and pleasure of independent research. During my wonderful but much too short stay as a postdoc at the University of California in Berkeley, Professor Masayoshi Tomizuka gave me the privilege of taking utmost advantage of all that Berkeley has to offer. It has been an amazing atmosphere of inspiration. The last 6 years in Darmstadt and Berkeley have been the most rewarding learning experience in my life. I am very grateful to Professor Isermann and Professor Tomizuka for giving me the chance to teach a graduate course on neural networks for nonlinear system identification. Earlier versions of this book served as a basis for these courses, and the feedback form the students contributed much to its improvement.

I highly appreciate the help of my colleagues in Darmstadt and Berkeley. Their collaboration and kindness made this book possible. Big thanks go to Dr. Martin Fischer, Alexander Fink, Susanne Töpfer, Michael Hafner, Matthias Schüler, Martin Schmidt, Domink Füssel, Peter Ballé, Christoph Halfmann, Henning Holzmann, Dr. Stefan Sinsel, Jochen Schaffnit, Dr. Ralf Schwarz, Norbert Müller, Dr. Thorsten Ullrich, Oliver Hecker, Dr. Martin Brown, Carlo Cloet, Craig Smith, Ryan White, and Brigitte Hoppe.

Finally, I want to deeply thank my dear friends Martin and Alex and my wonderful family for being there, whenever I needed them. I didn't take it for granted. This book is dedicated to you!

Kronberg, Germany
June 2000

Oliver Nelles

Contents

Notation

Abbreviation	Full Name
4SID	state space system identification
AIC	Akaike information criterion
AIC_C	corrected Akaike information criterion
ANN	artificial neural network
AR	autoregressive
ARMAX	autoregressive moving average with exogenous input
ARARX	autoregressive autoregressive with exogenous input
ART	adaptive resonance theory
ARX	autoregressive with exogenous input
ASA	adaptive simulated annealing
ASCMO	advanced simulation for calibration, modeling, and optimization
ASMOD	adaptive spline modeling of observation data
B&B	branch-and-bound
BA	Boltzmann annealing
Bagging	Bootstrap aggregating
BIC	Bayesian information criterion
BFGS	Broyden-Fletcher-Goldfarb-Shanno
BLUE	best linear unbiased estimator
BJ	Box-Jenkins
CMAC	cerebellar model articulation controller
COR-LS	correlation and least squares
DFP	Davidon-Fletcher-Powell
DSFI	discrete square roots filtering in information form
DoE	design of experiments
EA	evolutionary algorithm
ECU	engine control unit
ELS	extended least squares
ES	evolution strategy
FA	fast annealing
FIR	finite impulse response

FL	fuzzy logic
FUREGA	fuzzy rule extraction by a genetic algorithm
HILOMOT	hierarchical local model tree
GA	genetic algorithm
GCV	generalized cross validation
GLS	generalized least squares
GP	genetic programming
GP	Gaussian process
GPM	Gaussian process model
GPR	Gaussian process regression
GRNN	general regression neural network
HPFS	high pressure fuel supply system
IIR	infinite impulse response
IV	instrumental variable
LH	Latin hypercube
LLM	local linear model
LMN	local model network
LOO	leave-one-out
LMS	least mean square
LOLIMOT	local linear model tree
LS	least squares
LSTM	long short-term memory
MA	moving average
MAP	maximum a-posteriori
MIMO	multiple input multiple output
MISO	multiple input single output
MLP	multilayer perceptron
MSF	membership function
N	nonlinear
NARMAX	nonlinear autoregressive moving average with exogenous input
NARX	nonlinear autoregressive with exogenous input
NDE	nonlinear differential equation
NF	neuro-fuzzy
NFIR	nonlinear finite impulse response
NIIR	nonlinear infinite impulse response
NLS	nonlinear least squares
NN	neural network
NOBF	nonlinear orthonormal basis function
NRBF	normalized radial basis function
OBF	orthonormal basis function
OE	output error
OLS	orthogonal least squares
OMNIPUS	optimized nonlinear input signal
PCA	principle component analysis
PD	proportional differential

pdf	probability density function
PI	proportional integral
PID	proportional differential integral
PLS	partial least squares
PPL	projection pursuit learning
RBF	radial basis function
RELS	recursive extended least squares
RPEM	recursive prediction error method
RPLR	recursive pseudo-linear regression
RIV	recursive instrumental variable
RLS	recursive least squares
SA	simulated annealing
SHM	structural health monitoring
SISO	single input single output
SIMO	single input multiple output
SQP	sequential quadratic programming
SVD	singular value decomposition
TS	Takagi-Sugeno (fuzzy system) or tabu search
VFSR	very fast simulated reannealing
WLS	weighted least squares
XOR	exclusive or

Variable Name	**Description**
$f(\cdot)$	function to be approximated
k	discrete time
m	dynamic order
n	number of parameters
n	noise
n_{eff}	effective number of parameters
nx	number of inputs for the consequents
nz	number of inputs for the premises
p	number of inputs
q	number of outputs
q^{-1}	delay operator
$\underline{u}$	inputs
$\underline{x}$	inputs for the consequents
y	process output (potentially with noise)
y_u	process output (noise-free)
$\hat{y}$	model output
$\underline{z}$	inputs for the premises
λ	regularization strength
$\mu(\cdot)$	membership functions
σ	standard deviation
$\underline{\theta}$	parameter vector
$\underline{K}$	kernel matrix

M	number of local models
N	number of data points
$\Psi(\cdot)$	splitting functions
$\Phi(\cdot)$	validity functions
$\underline{\Sigma}$	covariance matrix

Chapter 1
Introduction

Abstract This chapter gives an outline of the whole book and explains what will be discussed in its four similarly comprehensive parts, dedicated to optimization, static models, dynamic models, and applications. The whole model development path in system identification from choice of the model inputs up to model validation is introduced. The recurring topic of "fiddle parameters" is addressed, which often is hidden but plays a crucial role in the real application of any method or algorithm. The distinction of modeling approaches into white box, black box, and gray box models is explained, and the different gray shades are discussed. Finally, the terminology is introduced that consistently will be used in all chapters of this book, making it much easier for the reader to focus on the content rather than on formal issues.

Keywords System identification · Nonlinear · Dynamic · Model · Terminology · White box · Black box · Gray box

In this chapter the relevance of nonlinear system identification is discussed. Standard problems whose solutions typically rely heavily on the use of models are introduced in Sect. 1.1. Section 1.3 presents and analyzes the various tasks that have to be carried out in order to solve a nonlinear system identification problem. Several modeling paradigms are reviewed in Sect. 1.4. Section 1.5 characterizes the purpose of this book and gives an outline with some reading suggestions. Finally, a few terminological issues are addressed in Sect. 1.6.

1.1 Relevance of Nonlinear System Identification

Models of real systems are of fundamental importance in virtually all disciplines. Models can be useful for system analysis, i.e., for gaining a better understanding of the system. Models make it possible to predict or simulate a system's behavior. In engineering, models are required for the design of new processes and for the

O. Nelles, *Nonlinear System Identification*,
https://doi.org/10.1007/978-3-030-47439-3_1

analysis of existing processes. Advanced techniques for the design of controllers, optimization, supervision, fault detection, and diagnosis components are also based on models of processes.

Since the quality of the model typically determines an upper bound on the quality of the final problem solution, modeling is often the bottleneck in the development of the whole system. As a consequence, a strong demand for advanced modeling and identification schemes arises.

1.1.1 Linear or Nonlinear?

Before thinking about the development of a nonlinear model, a linear model should be considered first. If the linear model does not yield satisfactory performance, one possible explanation besides many others is a significantly nonlinear behavior of the process. Whether this is indeed the case can be investigated by a careful comparison between the process and the linear model (e.g., step responses can make operating point dependent behavior quite obvious) and/or by carrying out a *nonlinearity test*. A good overview of nonlinearity tests is given in [199], where time domain- and correlation-based tests are recommended.

When changing from a linear to a nonlinear model, it may occur that the nonlinear model, if it is not chosen flexible enough, performs worse than the linear one. A good strategy for avoiding this undesirable effect is to use a nonlinear model architecture that contains a linear model as a special case. Examples for such models are polynomials (which simplify to a linear model for degree) or local linear neuro-fuzzy models (which simplify to a linear model when the number of neurons/rules is equal to one). For other nonlinear model architectures, it can be advantageous to establish a linear model in parallel: so the overall model output is the sum of the linear and the nonlinear model parts. This strategy is very appealing because it ensures that the overall nonlinear model performs better than the linear model.

1.1.2 Prediction

Figure 1.1a illustrates the task of one-step prediction.[1] The previous inputs $u(k-i)$ and outputs $y(k-i)$ of a process are given, and the process output $y(k)$ at the next sampling instant is predicted. Most real processes have no direct feedthrough: that is, their output $y(k)$ cannot be influenced immediately by input $u(k)$. This property is guaranteed to hold if sampling is not done exactly concurrently. For example, sampling is usually implemented in control systems as follows. The process output y

[1]In the literature this is frequently called one-step *ahead* prediction. Here, the term "ahead" is omitted since this book does not deal with backward predictions.

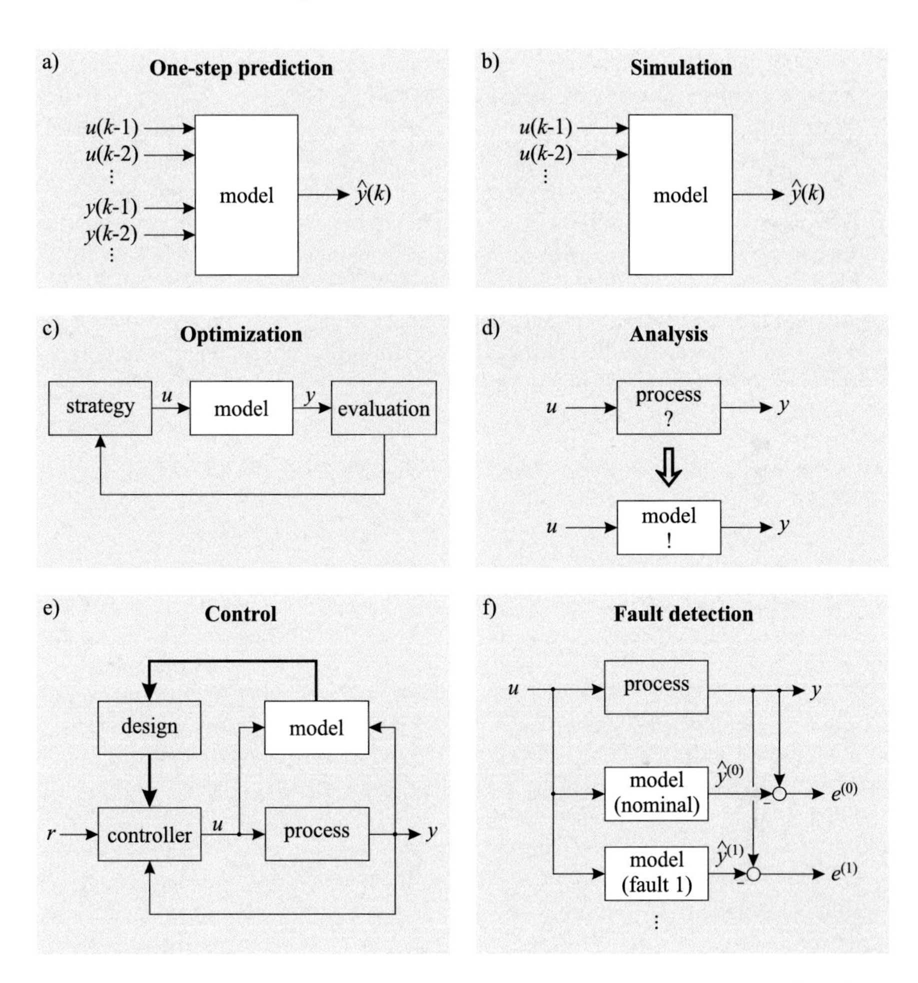

Fig. 1.1 Models can be utilized for (**a**) prediction, (**b**) simulation, (**c**) optimization, (**d**) analysis, (**e**) control, and (**f**) fault detection

(controlled variable) is sampled, and then the controller output u (actuation signal) is computed and applied. Throughout this book it is assumed for the sake of simplicity that $u(k)$ does not affect $y(k)$ instantaneously, and thus $u(k)$ is not included in the model. It can happen, however, that models which include $u(k)$ as input perform better because the effect of u on y can be much faster than one sampling time interval.

Typical examples for one-step prediction tasks are short-term stock market or weather forecasts. They can also arise in control problems. For a one-step prediction problem, it is necessary to measure the process outputs (or possibly the states).

If more than one step is to be predicted into the future, this is called a *multi-step* or *l-step prediction* task. The number of steps predicted into the future l is called the *prediction horizon*. Two alternative solutions exist for multi-step prediction tasks:

- A model can be built that directly predicts l steps into the future. Of course, such a model must additionally include the inputs $u(k)$, $u(k+1)$, ..., $u(k+l-1)$. The drawbacks of this approach are that (i) the input space dimensionality and thus the model complexity grows with l and (ii) the model can only predict exactly l steps ahead, while often a prediction of $1, 2, \ldots, l$ steps ahead is required.
- The model shown in Fig. 1.1a can be used to predict one step into the future. Next, the same model is used to predict a further step ahead by replacing $k \rightarrow k+1$ and utilizing the result $\hat{y}(k)$ from the previous prediction step. This procedure can be repeated l times to predict l steps ahead altogether. The drawback of this approach is that the prediction relies on previous predictions, and thus prediction errors may accumulate. In fact, this is a system with feedback that can become unstable.

The latter alternative is equivalent to simulation when the prediction horizon approaches infinity ($l \rightarrow \infty$).

1.1.3 Simulation

Figure 1.1b shows that, in contrast to prediction, for simulation only the process inputs are available. Simulation tasks occur very frequently. They are utilized for optimization, control, and fault detection. Another important application area is the design of software sensors, that is, the replacement of a real (hardware) sensor by a model that is capable of describing the quantity measured by the sensor with the required accuracy. Software sensors are especially important whenever the real sensor is not robust with respect to the environment conditions, too large, too heavy, or simply too expensive. They may allow the realization of (virtual) feedback control based on a software sensor "signal" where before only feedforward control was achievable since the controlled variable could not be measured.

The fundamental difference between simulation and prediction is that simulation typically requires feedback components within the model. This makes the modeling and identification phase harder and requires additional care in order to ensure the stability of the model.

1.1.4 Optimization

Figure 1.1c illustrates one possible way in which a prediction or simulation model can be utilized for optimization. The model can be used to find an optimal operating point or an optimal input profile. The advantages of using the model instead of the real process are that the optimization procedure may take a long time and may involve input signals or operating conditions that must not be applied during the normal operation of the process. Both issues are not problematic when the optimization is performed with a model on a computer instead of carrying it out in the real world. However, it is important to assess whether the model is accurate enough under all operating conditions encountered during optimization; otherwise the achieved optimum for the model may be very different from the optimum for the process.

1.1.5 Analysis

Figure 1.1d illustrates the very ambitious idea of exploiting the model of a process in order to improve understanding of the functioning of the process. In the simplest case, some insights into the process behavior can be gained by playing around with the model and observing its responses to certain input signals. Depending on the particular model architecture employed, it may be possible to infer some process properties by analyzing the model structure, e.g., by extracting fuzzy rules.

1.1.6 Control

Figure 1.1e shows how a model can be utilized for controller design. Most advanced control design schemes rely on a model of the process, and many even require more information such as the uncertainties of the model (an estimate of the model accuracy). The controller design may utilize various information from the model, not just its output. For example, the controller design may be based on a linearization of the model about the current operating point (u, y). The information transfer is indicated by the thick line from the model to the design block and similarly from the design to the controller block.[2]

It is important to understand that in Fig. 1.1e the model just serves as a tool for controller design. What finally matters is the performance of the controller, not that of the model. Although it can be expected that the controller performance and the model performance will be closely linked (the best controller certainly results from the "perfect" model), this is no monotonic relationship: refer to [251, 574] for more details. For this reason, the research area of *identification for control* originated, which tries to understand what makes a model well suited for control and how to identify it from data.

As a rule of thumb in the linear case, the model should be most accurate in the medium frequency range around the crossover frequency of the closed-loop

[2]Note that Fig. 1.1e does not cover all model-based controller design methods. For example, in nonlinear model-based predictive control, the design block would use the model block for prediction or simulation of potential actuation sequences. This would require an arrow from the design towards the model block. The purpose of Fig. 1.1e is just to explain the main ideas of model-based controller design.

transfer function (bandwidth), since this is decisive for stability [276, 574]. For reference tracking, the model additionally should be accurate in the frequency range of the reference signal, which is usually at low frequencies. For a regulator problem (setpoint control), one can rely on the integral action of the controller to compensate for process/model mismatch at low frequencies.

1.1.7 Fault Detection

Figure 1.1f shows one possible strategy for using models in fault detection. Many alternative model-based strategies are available, but this one has the advantage that it can be pursued in a black box manner. A model is built describing the process under normal conditions (nominal case), and one model is built for each fault, describing the process behavior when this fault has occurred (fault $1, 2, \ldots$). By comparing the outputs of these models ($\hat{y}^{(0)}$ for the nominal case and $\hat{y}^{(i)}$ for fault i) with the process output y, a fault i can be detected if any of the $e^{(i)}$ with $i > 0$ is larger than $e^{(0)}$. This comparison is usually based on filtered versions of these errors or carried out over a specified time period that trades off the detection speed and the robustness against modeling errors and noise.

Typically, models utilized for fault detection are required to be more accurate in the lower frequency range. The reason for this is that high frequencies are usually encountered during fast transient phases of the process, which do not last very long. In these short phases, however, detection is hardly possible if the measurements are disturbed because filtering must be applied to avoid false alarms. Of course, detection speed and sensitivity can be improved if the model describes the process accurately over a larger frequency range but for most applications low frequency models may be sufficient.

One characteristic feature of fault detection systems is that the cost associated with a missed fault (no detection when a fault occurs) is several orders of magnitude higher than the cost of a false alarm (detection when no fault occurs). This implies that the system should be made extremely sensitive, tolerating frequent false alarms. However, the probability of a fault is usually (hopefully) several orders of magnitude lower than the probability of the nominal case. Thus, this compensates for the abovementioned effect. Too many false alarms will certainly result in the fault detection system being switched off. A good adjustment of its sensitivity is the trickiest part in the development of a fault detection system.

1.2 Views on Nonlinear System Identification

This book deals with nonlinear system identification. Figure 1.2 visualizes how this field can be categorized with respect to linear/nonlinear and static/dynamic systems.

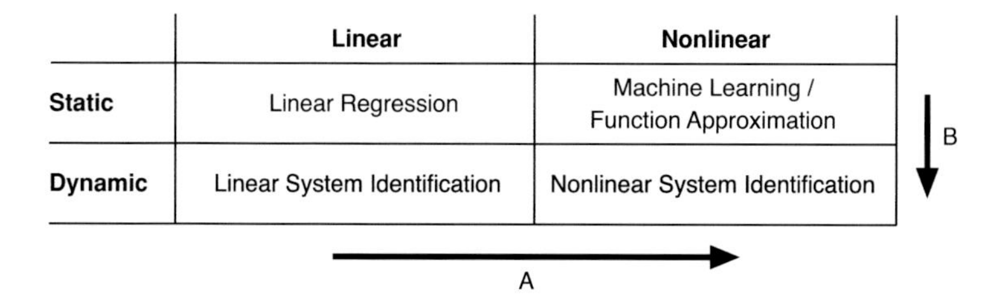

Fig. 1.2 Views on nonlinear system identification

Building static linear models from data typically (in the simplest case) corresponds to the linear regression problem of estimation a line (1D), plane (2D), or hyperplane (higher D). It is discussed in Sect. 10.1.

Building static nonlinear models from data is the classical problem of function approximation dealt with in most machine learning approaches and also in nonlinear statistics. These communities mainly focus on different model architectures (neural networks, deep learning, Gaussian process models, tree-based approaches, ...); meta-algorithms (boosting, bagging); and design of experiments. Some of these issues are discussed in Part II.

Building dynamic linear models from data is the topic of linear system identification. It is discussed in Chap. 18. This community mainly focuses on control-relevant issues: dynamics realizations (ARX, OE, FIR, OBF, state space, ...); noise descriptions; correlations; frequency domain point of view; and excitation signal design.

Of main interest to both communities are issues like optimization, regularization and local model networks, and piecewise affine systems.

Building dynamic nonlinear models from data is the most challenging task shown in the table of Fig. 1.2. It needs merging of the expertise of both communities mentioned above: control engineers and researchers (A) and machine learning people and statisticians (B). While the former focuses a lot on structurally simpler settings like on block-oriented structures, the latter prefers extremely flexible and powerful approaches like long short-term memory (LSTM) networks. Some of these issues are discussed in Part III.

At the time of writing this book, both communities are quite separated. Applications and research call for engineers and computer scientists to combine forces in education, industry, and research.

1.3 Tasks in Nonlinear System Identification

Modeling and identification of nonlinear dynamic systems is a challenging task because nonlinear processes are unique in the sense that they do not share many properties. A major goal for any nonlinear system modeling and identification scheme is universality: that is, the capability of describing a wide class of structurally different systems.

Figure 1.3 illustrates the task of system identification. For the sake of simplicity, it is assumed that the process possesses only a single output. An extension to the case of multiple outputs is straightforward. A model should represent the behavior of a process as closely as possible. The model quality is typically measured in terms of a function of the error between the (disturbed) process output and the model output. This error is utilized to adjust the parameters of the model.

This section examines the major steps that have to be performed for a successful system identification. Figure 1.4 illustrates the order in which the following steps have to be carried out. The parentheses indicate steps that are necessary only when dealing with dynamic systems.

1. Choice of the model inputs.
2. Choice of the excitation signals.
3. Choice of the model architecture.
4. (Choice of the dynamics representation.)
5. (Choice of the model order.)
6. Choice of the model structure and complexity.
7. Choice of the model parameters.
8. Model validation.

These eight steps are discussed further in Sects. 1.3.1–1.3.8. The complexity of and the need for prior knowledge in these steps typically decreases from Step 1 to Step 7. In the following, the term "training data" is used to characterize the measurement data that is utilized for carrying out Steps 1–7. Finally, some general remarks on the role of so-called fiddle parameters are given in Sect. 1.3.9.

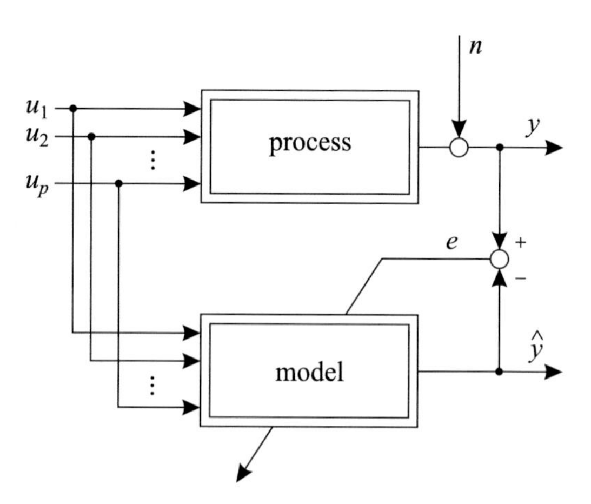

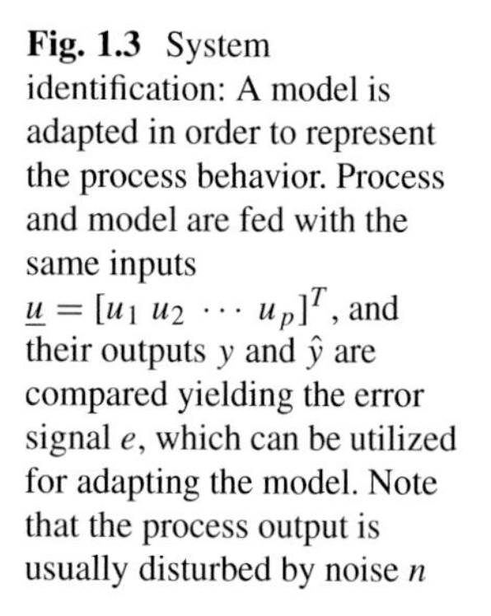
Fig. 1.3 System identification: A model is adapted in order to represent the process behavior. Process and model are fed with the same inputs $\underline{u} = [u_1\ u_2\ \cdots\ u_p]^T$, and their outputs y and $\hat{y}$ are compared yielding the error signal e, which can be utilized for adapting the model. Note that the process output is usually disturbed by noise n

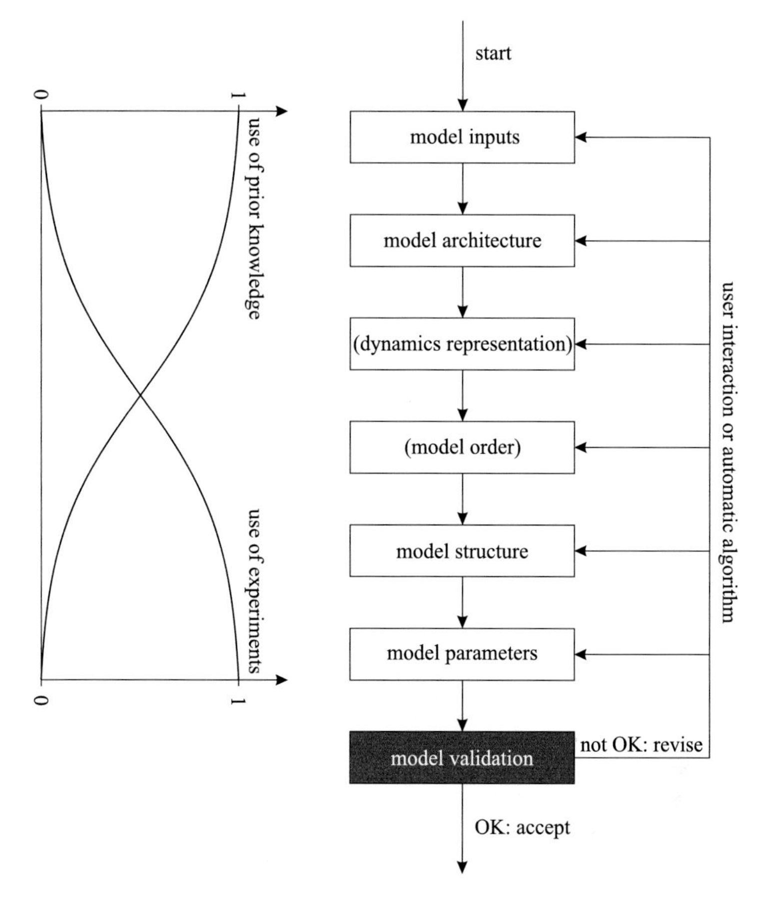

Fig. 1.4 The system identification loop (parentheses indicate steps that are necessary only when dealing with dynamic systems)

1.3.1 Choice of the Model Inputs

Step 1 is typically realized in a trial-and-error approach with the help of prior knowledge. In mechanical processes the influence of the different variables is usually quite clear, and the relevant model inputs can be chosen by available insights into the physics of the process. However, as the process categories become more complex, changing from physical laws to chemical, biological, economic, or even social relationships [280], the number of potential model inputs typically increases, and the insight into the influence of the different variables decreases. Tools for data-driven input selection might be very helpful in order to reduce the model development time by reducing the required user interaction. Basically, the following four different strategies can be distinguished:

- *Use all inputs*: This approach leads to extremely high-dimensional approximation problems, which implies the need for a huge number of parameters, a vast amount of data, and long training times. Although in connection with regularization methods this approach might be practical if the number of potential inputs is small, it becomes infeasible if the number of potential inputs is large.
- *Try all input combinations*: Although this approach, in principle, yields the best combination of inputs, it is usually infeasible in practice since the number of input combinations grows combinatorically with the number of potential inputs. Even for a moderate number of potential inputs, the computational demand would be too high.
- *Unsupervised input selection*: The typical tool for unsupervised input selection is principal component analysis (PCA). It makes it possible to discard non-relevant inputs with a low computational demand; see Sect. 6.1. PCA or similar techniques are very common, especially if the number of potential inputs is huge. The main drawback is that the PCA criterion for relevance of an input is based on the input data distribution only. Consequently, in some cases a PCA might discard inputs that are highly relevant for the model performance. Thus, the expected quality of unsupervised input selection schemes is relatively low.
- *Supervised input selection*: The inputs are selected with respect to the highest possible model accuracy. For linear models the standard tool is correlation analysis. For nonlinear models supervised input selection is a complex optimization problem. It can either be solved by general purpose structure optimization techniques such as evolutionary algorithms (EAs), which require a huge computational demand, or model-specific algorithms can be developed, which usually operate much more efficiently. Supervised input selection schemes represent the most powerful but also computationally most demanding approaches.

The LOLIMOT algorithm discussed in Chaps. 13, 14, and 22 belongs to the last category. Alternative approaches, which are also based on a specific model architecture, are briefly discussed in the following.

For neural networks, especially MLPs, numerous pruning techniques are available that discard nonsignificant parameters or whole neurons from the network; for a good survey, refer to [479]. Specifically tailored for neural network input and structure selection are the evolution strategy (ES) proposed in [72] and the genetic algorithm (GA) presented in [599]; see also Chap. 5. For fuzzy models, evolutionary algorithms are applied for input and rule structure selection as well [4, 115, 210, 244, 258, 285, 353, 410, 425, 444, 556]; see also Sect. 12.4.4. Another common strategy is the adaptive spline modeling of observation data (ASMOD) algorithm and related approaches, which are pruning and growing strategies for additive fuzzy models [61–63, 83, 316]; see also Sect. 12.4.5. In [327] an input and structure selection approach for polynomial models based on stepwise regression with an orthogonal least squares (OLS) algorithm is proposed; see Sect. 3.4. A tree construction approach for input selection is presented in [495, 496] for basis function networks. Such tree construction strategies are well known from the classification and regression trees (CART) [74] and the multivariate adaptive regression splines (MARS) [172]. Finally, genetic programming is on its way to becoming a very powerful but extremely computationally expensive tool for general structure search [330, 364, 365]; see also Sect. 5.2.3.

1.3.2 Choice of the Excitation Signals

Step 2 requires prior knowledge about the process and the purpose of the model. For black box modeling (see Sect. 1.4), the measurements are the most important source of information. Process behavior that is not represented within the data set cannot be described by the model unless prior knowledge is explicitly incorporated. Consequently, this step limits the achievable model quality and is more important than the subsequent steps. Unfortunately very few tools exist and little research is devoted to this subject. Probably this neglect is due to the highly application-specific nature of the problem. Nevertheless, some general guidelines for the design of excitation signals are given in Chap. 19. This step is probably the one that involves the highest engineering expertise.

If the process under consideration cannot be actively excited, a training data set still has to be designed by selecting a data set from the gathered measurements that is as representative as possible. This requires similar knowledge as the explicit design of an excitation signal.

It is important to store the information about the operating conditions that are covered by the training data set in order to be able to assess the reliability of the model's predictions and possibly to avoid or at least limit extrapolation.

1.3.3 Choice of the Model Architecture

Step 3 is possibly the hardest and most subjectively influenced decision. The following (certainly incomplete) list gives an idea about the criteria that are important for the choice of an appropriate model architecture.

- *Problem type*: Classification, approximation of static systems, or identification of dynamic systems.
- *Intended use*: The desired model properties differ according to whether the model is to be used for simulation, optimization, control, fault detection, etc.
- *Problem dimensionality*: The number of relevant inputs (and outputs) restricts the class of suitable model architectures.
- *Available amount and quality of data*: If data is very sparse and noisy, a global approach might be more successful than local ones because they have a higher tendency to average out disturbances.
- *Constraints on development, training, and evaluation time*: The development time depends strongly on the training time and the automation level of an iden-

tification technique. The automation level determines how much user interaction is required. This often relates to the number and interpretability of the fiddle parameters.

Training and evaluation time are typically in conflict. Long training times enable a high level of information compression (i.e., relatively small structures) by the utilization of complex optimization techniques. This allows a fast model evaluation. An extreme case is the multilayer perceptron (MLP) network; see Sect. 11.2. Short training times often imply a low level of information compression (i.e., relatively large structures) by the application of simpler optimization techniques. This requires longer model evaluation times. An extreme case is the general regression neural network (GRNN), in which no optimization is carried out at all and training just consists of storing all data in the memory; see Sect. 11.4.1.

- *Memory restrictions*: In areas with a high number of produced units such as consumer products or the automotive industry, memory restrictions are still an important issue. In the future, however, these restrictions will lose significance since memory costs decrease even faster than computation costs.
- *Offline or online learning*: While all architectures are suitable for offline training, online learning in a reliable and robust manner is a question for future research. From the current point of view only local modeling approaches seem to be appropriate for this task.
- *Experience of the user*: Users will always prefer those models and methods that they have been trained on and they are used to, even if an alternative approach offers objective advantages. Experience of the work with a modeling and identification scheme is required in order to develop confidence in its properties and to acquire some "feeling" for its application. Consequently, it is very important that a modeling technique is a universal tool because then it can be utilized for a large number of different applications. The generality of *neural network* based approaches is one of the major reasons for their success.
- *Availability of tools*: As the complexity of software grows exponentially with time like the performance increase of hardware, tools become increasingly important in order to keep software manageable and development times short.
- *Customer acceptance*: A key issue in acceptance of a new modeling and identification approach is usually at least some understanding about the operation of the system. This implies that the implemented models should be interpretable. Pure black box solutions typically are not convincing.

1.3.4 Choice of the Dynamics Representation

Step 4 is mainly determined by the intended use of the model. If it is to be utilized for one-step prediction, a NARX or NARMAX representation may be a good choice; see Chap. 18. For simulation the chosen model architecture and the available prior knowledge about the process also influence the decision. If very little is known, an internal dynamics approach as discussed in Chap. 23 might be a good first attempt. Otherwise, the external dynamics approach is the much more common choice. Currently, the advantages and drawbacks of different nonlinear dynamics representations are still under investigation. Thus, subjective factors such as the user's previous experience with a specific approach play an important role. The fundamental properties of different dynamics representations are analyzed in Chap. 19.

1.3.5 Choice of the Model Order

Step 5 is typically carried out by a combination of prior knowledge and trial and error. With the external dynamics approach, the choice for a higher dynamic order of the model increases the dimensionality of the problem and hence its complexity. Compared with linear system identification, when dealing for nonlinear systems, the user often may be forced to accept significant unmodeled dynamics by choosing too low a model order. The reason for this is that a tradeoff exists between the errors introduced by neglected dynamics and the static approximation error. Since the overall model complexity is limited by the bias/variance dilemma, a high dynamic order of the model (increasing its dimensionality) may have to be paid for by a loss in static approximation accuracy. It is the experience of the author that low-order models often suffice because the model error is dominated by the approximation error caused by an inaccurate description of the process nonlinearity.

1.3.6 Choice of the Model Structure and Complexity

Step 6 is much harder than parameter optimization. It can be carried out automatically if structure optimization techniques such as orthogonal least squares (OLS) for linear parameterized models or evolutionary algorithms (EAs) for nonlinear parameterized models are applied. An alternative to these general approaches is model-specific growing and/or pruning algorithms such as LOLIMOT for local linear neuro-fuzzy models (Chaps. 13, 14, and 22), ASMOD for additive singleton neuro-fuzzy systems (Sect. 12.4.5 and [63]), or the wide variety of algorithms available for multilayer perceptron networks [479]. Furthermore, regularization methods can be utilized for complexity determination. An overview of the different approaches for structure and complexity optimization is given in Chap. 7, especially in Sects. 7.4 and 7.5. Although quite a large number of sophisticated algorithms have been developed for these tasks, many problems are still solved with non-automated trial-and-error approaches, especially in industry. The main reasons for the dominance of manual structure selection seem to be related to the following shortcomings of the automatic algorithms:

- long training times;
- large number of (partly nontransparent) fiddle parameters to determine;
- no or limited suitability for dynamic models (particularly suited for simulation, not just one-step prediction);
- limited software support in the form of toolboxes.

With the local linear model tree (LOLIMOT) algorithm presented in Chaps. 13, 14 and 22, these handicaps can be significantly weakened or even overcome. Nevertheless, many problems of structure and complexity optimization are still open for future research. Step 6 seems to be the most promising candidate for an advantageous integration of different information sources such as measurement data, qualitative knowledge (e.g., in the form of fuzzy rules), equations obtained by first principles, etc. Step 7 is dominantly measurement data driven since efficient and accurate optimization techniques are available, while Steps 1 to 3 are primarily based on prior knowledge. Figure 1.4 illustrates the importance of prior knowledge and measurement data for the six steps.

1.3.7 Choice of the Model Parameters

Step 7 is the simplest and easiest to automate. It is usually solved by the application of linear and nonlinear optimization techniques; see Part I. While linear optimization techniques operate fully automatically, nonlinear optimization typically requires some user interaction. Depending on the chosen method, initial estimates of the parameters and optimization technique specific parameters have to be determined by the user. Nevertheless, parameter optimization techniques are in a mature state. They are easy to use (almost in a black box fashion) and many excellent toolboxes are available; see, e.g., [70]. Current research focuses on problems with a huge number of parameters, with constraints, with multiple objectives, or with a global search goal.

1.3.8 Model Validation

Step 8 checks whether all preceding steps have been carried out successfully or not. The specific criteria utilized for making this decision are highly problem dependent. In many cases, the model is not the ultimate goal but rather serves as a basis for further steps, such as the design of a controller or a fault detection system. Then the proper criteria may be the performance of the closed-loop control or the sensitivity of the fault detection. Typically, however, even in these cases, before the model is used as a basis for further steps, the model quality is investigated directly in order to decouple the modeling and the subsequent design steps as far as possible.

The easiest type of validation is to check the model quality on the training data. If this does not yield satisfactory results, the model certainly cannot be accepted. In this case, the problem is known to *not* lie in insufficient training data or poor model generalization behavior. Rather it can be concluded that either information is missing (an additional input may be needed) or the model is not flexible enough to describe the underlying relationships.

If the performance achieved on the training data is acceptable, it is necessary (or at least desirable) to test the model on fresh test data. The need for this separate test data set increases the smaller and more noisy the training data set and the more complex the model are. The test data set should excite the process and the model particularly at those operating conditions that are considered important for the subsequent use of the model. For example, if the model is to be utilized for controller design, the test data should excite especially at frequencies around the (expected) closed-loop bandwidth, while the steady-state characteristics of a model utilized for fault detection may be more relevant.

If no separate test data set is available or in addition to the use of test data, other validation methods exist that are based on correlation considerations or information criteria. These approaches are discussed in Chap. 7. Furthermore, it is advisable to investigate the model's response to some simple input signals such as steps with different heights and directions at different operating points or sinusoidal signals of different amplitudes and frequencies. Even if no measurement data is available for a comparison between process and model, the analysis of such model responses makes it possible to gain insights into the model's behavior. Characteristic features such as gains, time constants, and minimum phase behavior for different operating points can be inspected. Often enough prior knowledge about the process is available to identify unrealistic model responses and to revise the model. Many details may be missed by just examining the model on training and test data.

1.3.9 The Role of Fiddle Parameters

Fiddle parameters are parameters that are not automatically adjusted by the identification algorithm that is applied to the model but rather have to be chosen by the user. Typical examples of fiddle parameters are the initialization of model parameters that have to be optimized, the learning rate in a training method, the number of clusters, regressors, neurons, rules, etc. of a model, the error threshold that terminates an algorithm, etc. Virtually *all* modeling schemes possess such fiddle parameters. Basically, the user performs a manual optimization while playing with them and seeking their best values. Often when a model is revised in the system identification loop (see Fig. 1.4), the user changes a fiddle parameter trying to improve the design. The reasons for doing this manually can be twofold. Either the identification algorithm is not sophisticated enough to optimize these parameters automatically (because this not easily possible or too time-consuming) or the objective(s) of optimization (loss function) cannot be properly expressed. The latter issue occurs quite often since typically many objectives have to be taken into account (multi-objective optimization) but their tradeoffs are not very clear in the beginning. Thus, the user goes through a learning process or exploits his or her experience (implicit knowledge) by adjusting the fiddle parameters. Positively speaking, fiddle parameters enable the user to play with the model in order to adjust it to the specific problem. Negatively speaking, fiddle parameters force the user into a tedious trial-and-error tuning approach, which slows down the model development procedure and introduces subjectiveness.

Which one of the above-formulated points of view is correct? The answer depends on the specific properties of the fiddle parameters. The following properties are desirable:

- *Decoupled fiddle parameters*: A human cannot really handle more than two fiddle parameters at the same time.[3] Therefore, the influence of the different fiddle parameters should be mainly decoupled, which allows separate tuning.
- *Small number of fiddle parameters*: Since the model development time is limited, the number of fiddle parameters should be small in order to allow the user to find a good adjustment under time restrictions.
- *High level interpretation of the fiddle parameters*: An expert of the domain to which the process under investigation belongs should be able to interpret the fiddle parameters without knowing the details about the model and the identification algorithm. Otherwise, two experts would be required for modeling (the domain expert and the model and identification tool expert), which is unrealistic in practice. Such a high level interpretation implies that the fiddle parameters affect mainly the upper levels of the system identification loop in Fig. 1.4. An example of a high level fiddle parameter is the model smoothness; examples of low level fiddle parameters are learning rate and parameter initialization.
- *Reasonable problem-independent default values for the fiddle parameters*: The user needs an orientation about the order of magnitude of the fiddle parameters that is problem independent. This is particularly important for lower-level parameters with limited interpretation. Otherwise, a tedious trial-and-error procedure would be necessary. Furthermore, it should be possible to obtain a reasonably performing model with the default settings to give the user a quick first guess (at least an order of magnitude) about the model quality that can be expected.
- *Steady sensitivity and unimodal influence of the fiddle parameters*: If the sensitivity of the fiddle parameters varies strongly, or several locally optimal values exist, a user would be unlikely to find a reasonable adjustment.

When all these demands are met, the user should be capable of utilizing the model together with the identification algorithm in an efficient manner.

[3]This fact is partly responsible for the success of PI controllers, which have just two knobs, one for the proportional and one for the integral component. The manual tuning of PID controllers, which additionally possess a derivative component, is a much more difficult problem and cannot really be carried out without automatic support.

1.4 White Box, Black Box, and Gray Box Models

Many combinations and nuances of theoretical modeling from first principles and empirical modeling based on measurement data can be pursued. Basically, the following three different modeling approaches can be distinguished; see Fig. 1.5:

- *White box models* are fully derived by first principles, i.e., physical, chemical, biological, economical, etc. laws. All equations and parameters can be determined by theoretical modeling. Typically, models whose structure is completely derived from first principles are also subsumed under the category white box models even if some parameters are estimated from data. Characteristic features of white box models are that they do not (or only to a minor degree) depend on data and that their parameters possess a direct interpretation in first principles.
- *Black box models* are based solely on measurement data. Both model structure and parameters are determined from experimental modeling. For building black box models, no or very little prior knowledge is exploited. The model parameters have no direct relationship to first principles.
- *Gray box models* represent a compromise or combination between white and black box models. Almost arbitrary nuances are possible. Besides the knowledge from first principles and the information contained in the measurement data, other knowledge sources such as qualitative knowledge formulated in rules may also be utilized in gray box models. Gray box models are characterized by an integration of various kinds of information that are easily available. Typically, the determination of the model structure relies strongly on prior knowledge, while the model parameters are mainly determined by measurement data.

	White box	Gray box	Black box
Information sources	first principles insights	qualitative knowledge (rules) some insights + some data	experiments data
Features	good extrapolation good understanding high reliability scalable		short development time little domain expertise req. can be used in addition also for not understood proc.
Drawbacks	time consuming detailed domain expertise req. knowledge restricts accuracy only for well understood proc.		no reliable extrapolation not scalable data restricts accuracy little understanding
Application areas	planning, construction, design rather simple processes		only for existing processes rather complex processes

Fig. 1.5 White box, black box, and gray box models. The blank fields for gray box models can be almost any combination of white and black box models

Figure 1.5 gives an overview of the differences between these modeling approaches. Most entries for gray box models are left blank because it is impossible to characterize the overwhelming number of different gray box modeling approaches or their combinations. The advantages and drawbacks of gray box models lie somewhere between the two extremes (white and black box), making them a good compromise in practice. Typically, when utilizing gray box models, one tries to overcome some of the most restrictive factors of the white and black box approaches for a specific application. For example, some prior knowledge may be incorporated into a black box model in order to ensure reasonable extrapolation behavior.

Note that in reality pure white or black box approaches rarely exist. Nothing is black or white, everything is gray. Often the model structure may be determined by first principles but the model parameters may be estimated from data (light gray box), or a neural network may be used but the data acquisition procedure, e.g., design of excitation signals, requires prior knowledge (dark gray box). Thus, the transition from white through gray to black box approaches is fuzzy, not crisp. Usually, if prior knowledge is clearly the dominating factor, one speaks of white box models and if experimental data is the major basis for modeling, one speaks of black box models. If both factors are balanced, one speaks of gray box models. This book focuses on black box modeling, with some outlooks towards gray box modeling.

1.5 Outline of the Book and Some Reading Suggestions

This books emphasizes intuitive explanations, tries to keep things as simple as possible, and stresses the existing links between the different approaches. Unnecessary formalism and equations are avoided, and the basic principles are illustrated with many figures and examples. The book is written from an engineering perspective. Implementation details are skipped whenever they are not necessary to motivate or understand the approach. Rather the goals of this book are to enable the reader

- to understand the advantages, the drawbacks, and the areas of application of the different models and algorithms;
- to choose a suitable approach for a given problem;
- to adjust all fiddle parameters properly;
- to interpret and to comprehend the obtained results; and
- to assess the reliability and limitations of the identified models.

The book is structured in four parts. They cover optimization issues, identification of static and dynamic models, and some real-world application examples. An elementary understanding of systems and signals is sufficient to easily follow most of this book. Some previous experience with function approximation and identification of linear dynamic systems is helpful but not required. Thus, this book is tailored for engineers in industry and for final-year undergraduate or first-year graduate courses.

Part I introduces the basic principles and methods for optimization that are a prerequisite for an appropriate understanding of the subsequent chapters. Although many excellent books on optimization techniques are available, they usually focus on particular methods. Part I gives a broad overview of all optimization issues related or helpful to nonlinear system identification. Chapter 3 deals with *linear optimization* methods and focuses on the least squares approach for parameter and structure optimization. *Nonlinear local optimization* techniques are summarized in Chap. 4 which concentrates on gradient-based approaches but also covers direct search methods. The incorporation of constraints is briefly treated as well. Chapter 5 covers *nonlinear global optimization* techniques and can be skipped for a first reading, although some issues addressed there are helpful for the understanding of structure optimization problems. The class of so-called *unsupervised learning* methods is treated in Chap. 6, which can also be omitted for a first reading. A crucial preparation for the remaining parts of the book, however, is Chap. 7, which deals with *model complexity optimization* and thus gives a general, independent of model type, treatment of all issues concerning the question: How complex should a model be? Very important and fundamental topics such as the bias/variance dilemma, regularization, and the curse of dimensionality are introduced and thoroughly explained. Dependent on the previous experience of the reader, Chap. 7 can be quite abstract. In this case, it might be better to just browse through at a first reading and to look up the details when addressed in the following chapters.

Part II introduces model architectures for approximation of static systems. These architectures also serve as a basis for the dynamic models: thus the reading of at least some selected chapters is essential for proceeding further to Part III. Chapter 10 introduces the very widely applied classical architectures based on *linear, polynomial, and look-up table models*. In particular, look-up table models are examined in greater detail since they are extremely popular in industry. *Neural networks* are discussed in Chap. 11. The emphasis is on the widely applied multilayer perceptron and radial basis function networks, but some other promising architectures are covered as well. An introduction to fuzzy logic and an analysis of the link between neural networks and *fuzzy models* towards *neuro-fuzzy models* are given in Chap. 12. Chapters 13 and 14 deal with *local linear neuro-fuzzy models*, which represent a particularly promising model architecture. Therefore, they are discussed in greater detail, and two full chapters are devoted to them, although from an organization point of view they belong in Chap. 12.

Part III extends the static models to dynamic ones. An introduction to the foundations of *linear dynamic system identification* is given in Chap. 18. In Chap. 19 these concepts are generalized to *nonlinear dynamic system identification* in a way that is independent of the specific model architecture. The classical approaches based on *polynomials* are discussed in Chap. 20, and the *neural network* and *fuzzy model* architectures are treated in Chap. 21. Similar to the organization of Part II, a special chapter is devoted to *local linear neuro-fuzzy models* (Chap. 22). Finally, a different kind of dynamics realization for neural networks, the so-called *internal dynamics* approach, is addressed in Chap. 23

Part IV illustrates the features of some model architectures and algorithms discussed in this book, with a strong focus on local linear neuro-fuzzy models. Chapter 24 presents several *static* function approximation and optimization problems arising in the automotive electronics and control area. Nonlinear *dynamic* system identification of a cooling blast, a Diesel engine turbocharger, and several subprocesses of a thermal pilot plant are discussed in Chap. 25. Finally, an outlook to *online adaptation*, *control*, *fault detection*, and *reconfiguration* applied to a heat exchanger is given in Chap. 28.

The two appendices summarize some definitions of matrix and vector derivatives (Appendix A) and give an introduction to some basic statistical relationships and properties (Appendix B).

1.6 Terminology

The terminology used in this book follows the standard system identification and optimization literature rather than the neural network language. For the sake of brevity, the following expressions are often used although some of them are strictly speaking not correct.

- *Process*: Used as a synonym for the plant or system under study. It can be static or dynamic.
- *Linear parameters*: Parameters that influence the model error in a linear way.
- *Nonlinear parameters*: Parameters that influence the model error in a nonlinear way.
- *Training*: Optimization of the model structure and/or parameters in order to minimize a given loss function for the training data. When the emphasis is on structure optimization, the expression *learning* is also used. When the emphasis is on parameter optimization, the expression *estimation* is also used.
- *Generalization*: Evaluation of the model output for an input data sample that is not contained in the training data set. Generalization can be *interpolation* when the input is within the range covered by the training data; otherwise, it is *extrapolation*.
- *Neural network (NN)*: Short for artificial neural network.

Part I
Optimization

Chapter 2
Introduction to Optimization

Abstract This chapter introduces the topic of optimization. It outlines the key ideas in an intuitive fashion. Furthermore, the distinction between supervised and unsupervised learning is explained, and their strengths and weaknesses are discussed. For both learning approaches, the choice of the loss function and their implications are discussed in detail.

Keywords Optimization · Loss function · Supervised · Unsupervised · Maximum likelihood · Bayes

This chapter gives an introduction to optimization from the viewpoint of modeling and identification. The whole of Part I deals with different optimization approaches that allow one to determine the model parameters and possibly the model structure from measurement data for a given model architecture. Suitable model architectures for static and dynamic processes are treated in Parts II and III, respectively. Although there exist close links (some of the historical character) between special model architectures and special optimization techniques, it is very important to distinguish carefully between the model on the one hand and the optimization technique used for parameter and structure determination on the other hand.

Three different approaches to optimization can be distinguished. They differ in the amount of information required about the desired model behavior. These three approaches are:

- supervised learning,
- reinforcement learning,
- unsupervised learning.

The so-called supervised learning methods are based on knowledge about the input and output data of a process. This means that for each input, a desired model output, namely, the measured process output, is known. In supervised learning the objective is to minimize some error measure between the process and the model behavior in order to obtain the "best" model. An exact mathematical formulation of the error measure in the form of a loss function and thus a definition of the

O. Nelles, *Nonlinear System Identification*,
https://doi.org/10.1007/978-3-030-47439-3_2

term "best" is given in Sect. 2.3. Since for most problems addressed in this book the output can be measured, supervised learning techniques play a dominant role. Chapters 3, 4, 5, and 7 analyze different supervised learning approaches.

In reinforcement learning some information about the quality of the model is available. However, no desired output value is known for each input. Typical application examples are games where it is not possible to evaluate the quality of each move but the final result of the game (win, loss, draw) contains information about the quality of the applied strategy. Any supervised learning problem can be artificially transferred into a reinforcement learning problem by discarding information. For example, a pole balancing problem, on the one hand, can be treated as a supervised learning problem if the information about the deviation (error) of the pole from the desired upright position is utilized. On the other hand, if only the information on whether the run was successful or a failure is exploited, a reinforcement problem arises. Clearly, it is always advantageous to exploit all available information. Therefore, reinforcement learning techniques are not addressed in this book. They mainly become interesting for strategy learning where no desired output and, consequently, no error signal is available for each step.

For unsupervised learning methods, only input data is utilized. The objective of unsupervised methods is grouping or clustering of data. The exploited information is the input data distribution. Unsupervised learning techniques are primarily applied to data preprocessing. They are discussed in Chap. 6.

This chapter gives an introduction and a brief overview of the most important optimization techniques. The focus is on the application of these techniques to modeling and identification. Most of the methods described are parameter optimization techniques. In Fig. 2.1 the basic concept is depicted from a modeling point of view. A model $f(\cdot)$ maps the inputs gathered in the input vector $\underline{u}$ to the scalar output y. The model is parameterized by a set of n parameters gathered in the parameter vector $\underline{\theta}$ such that $\hat{y} = f(\underline{u}, \underline{\theta})$. The goal of a parameter optimization technique is to find the "best" approximation $\hat{y}$ of the measured output y, which may be spoiled with noise n, by adapting the parameter vector $\underline{\theta}$. A more precise definition of "best" will be given in Sect. 2.3. It is helpful to look at this problem

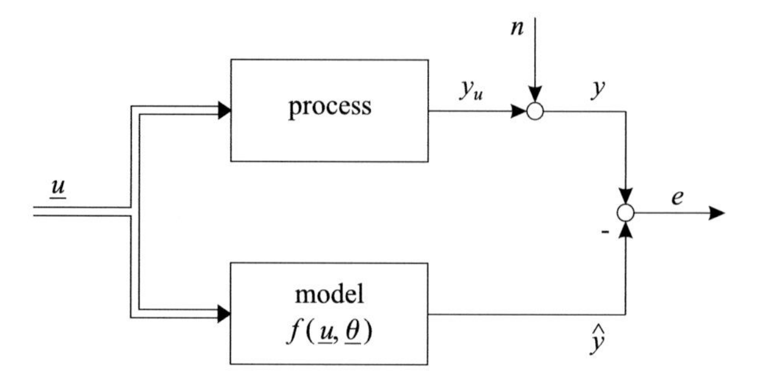

Fig. 2.1 Process and model

as a search for the optimal point in an n-dimensional parameter space spanned by the parameter vector $\underline{\theta}$. The Chaps. 3, 4, and 5 address such parameter optimization techniques.

Besides these parameter optimization techniques, the so-called structure optimization techniques represent another category of methods. They deal with the problem of searching an optimal model structure, e.g., the optimal kind of function $f(\cdot)$ and the optimal number of parameters. These issues are discussed in Chap. 7 and partly addressed also in Chap. 5.

This chapter is organized as follows. First, a brief overview on the supervised learning methods is given. Section 2.2 gives an illustration of these techniques by means of a humorous analogy. In Sects. 2.3 and 2.4, some definitions of loss functions for supervised and unsupervised learning are given.

2.1 Overview of Optimization Techniques

The supervised learning techniques can be divided into three classes: the linear, the nonlinear local, and the nonlinear global optimization methods. Out of these three, linear optimization techniques are the most mature and most straightforward to apply. They are discussed in Chap. 3. Nonlinear local optimization techniques, summarized in Chap. 4, are a well-understood field of mathematics, although active research still takes place, especially in the area of constrained optimization. By contrast, many questions remain unresolved for nonlinear global optimization techniques, and therefore, this is quite an active research field at the moment; see Chap. 5. Figure 2.2 illustrates the relationship between the discussed supervised optimization techniques.

2.2 Kangaroos

Plate and Sarle give the following wonderful description of the most common nonlinear optimization algorithms with respect to neural networks (NN) in [460] (comments in brackets by the author are related to Chap. 4):

Training a NN is a form of numerical optimization, which can be linked to a kangaroo searching the top of Mt. Everest. Everest is the global optimum, *the highest mountain in the world, but the top of any other really tall mountain such as K2 (a good* local optimum*) would be satisfactory. On the other hand, the top of a small hill like Chapel Hill, NC, (a bad local optimum) would not be acceptable.*

This analogy is framed in terms of maximization, while neural networks are usually discussed in terms of minimization of an error measure such as the least squares criterion, but if you multiply the error measure by -1*, it works out the same. So in this analogy, the higher the altitude, the smaller the error.*

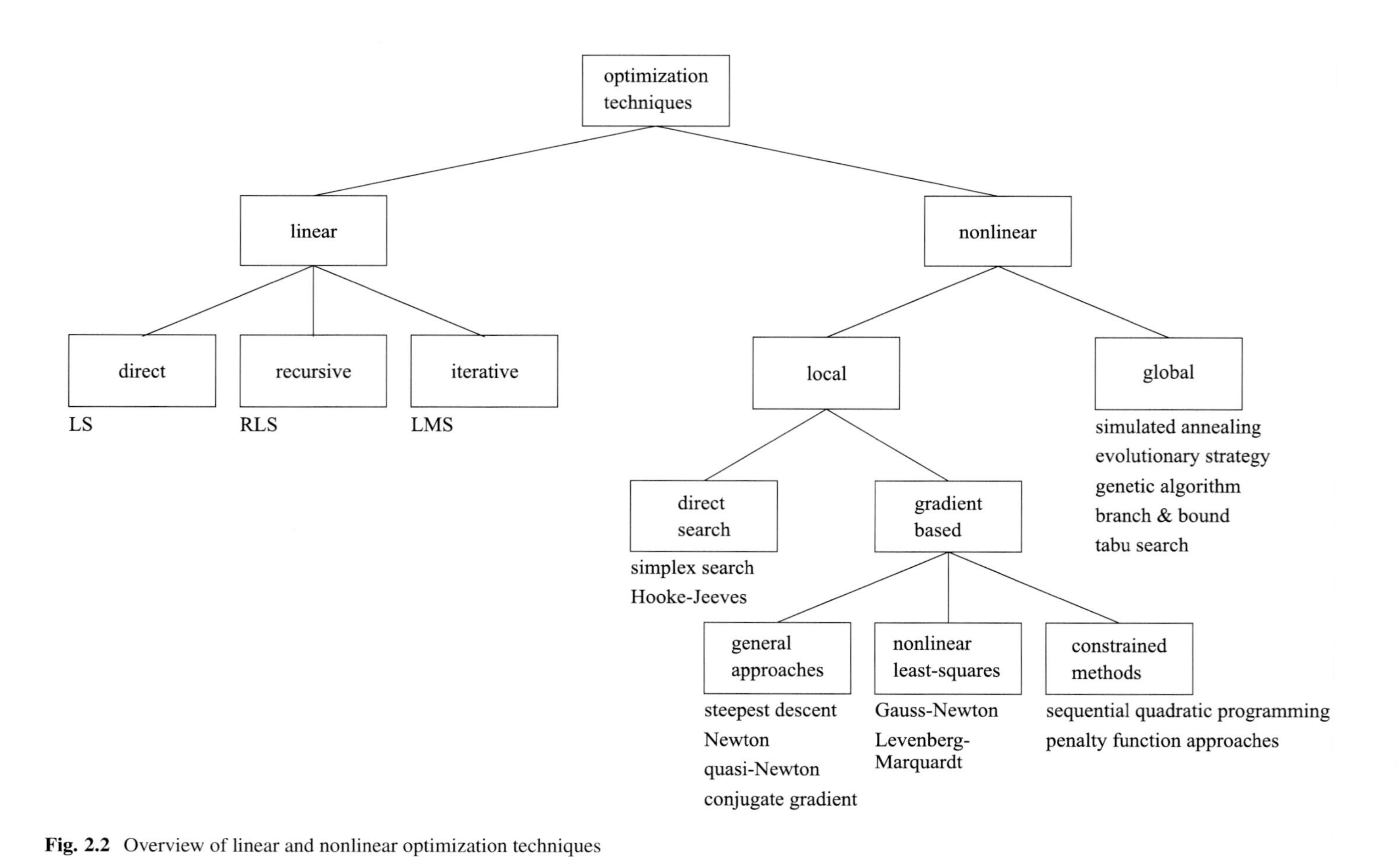

Fig. 2.2 Overview of linear and nonlinear optimization techniques

The compass directions represent the values of synaptic weights [parameters] *in the network. The north/south direction represents one weight, while the east/west direction represents another weight. Most networks have more than two weights, but representing additional weights would require a multidimensional landscape, which is difficult to visualize. Keep in mind that when you are training a network with more than two weights, everything gets more complicated.*

Initial weights are usually chosen randomly, which means that the kangaroo is dropped by parachute somewhere over Asia by a pilot who has lost the map. If you know something about the scales of input, you may be able to give the pilot adequate instructions to get the kangaroo to land near the Himalayas. However, if you make a really bad choice of distributions for the initial weights, the kangaroo may plummet into the Indian Ocean and drown.

With Newton-type (second order) algorithm [with fixed step size $\eta = 1$], *the Himalayas are covered with fog, and the kangaroo can only see a little way around her location* [first- and second-order derivative information]. *Judging from the local terrain, the kangaroo makes a guess about where the top of the mountain is, assuming that the mountain has a nice, smooth, quadratic shape. The kangaroo then tries to leap all the way to the top in one jump.*

Since most mountains do not have a perfect quadratic shape, the kangaroo will rarely reach the top in one jump. Hence, the kangaroo must iterate, *i.e., jump repeatedly as previously described until she finds the top of the mountain. Unfortunately, there is no assurance that this mountain will be Everest.*

In a stabilized Newton algorithm [with variable step size η], *the kangaroo has an altimeter, and if the jump takes her to a lower point, she backs up to where she was and takes a shorter jump. If ridge stabilization* [the Levenberg-Marquardt idea] *is used, the kangaroo also adjusts the direction of her jump to go up a steeper slope. If the algorithm isn't stabilized, the kangaroo may mistakenly jump to Shanghai and get served for dinner in a Chinese restaurant* [divergence].

In steepest ascent with line search, the fog is very *dense, and the kangaroo can only tell, which direction leads up most steeply* [only first-order derivative information]. *The kangaroo hops in this direction until the terrain starts going down. Then the kangaroo looks around again for the new steepest ascent direction and iterates.*

Using an ODE (ordinary differential equation) solver is similar to steepest ascent, except that kangaroo crawls on all fives to the top of the nearest mountain, being sure to crawl in steepest direction at all times.

The following description of conjugate gradient methods was written by Tony Plate (1993):

> *The environment for conjugate gradient search is just like that for steepest ascent with line search – the fog is dense and the kangaroo can only tell, which direction leads up. The difference is that the kangaroo has some memory of the direction it has hopped in before and the kangaroo assumes that the ridges are straight (i.e., the surface is quadratic). The kangaroo chooses a direction to hop that is upwards, but that does not result in it going downwards in the previous directions it has hopped in. That is, it chooses an upwards*

direction, moving along which will not undo the work of previous steps. It hops upwards until the terrain starts going down again, then chooses another direction.

In standard backprop, the most common NN training method, the kangaroo is blind and has to feel around on the grounds to make a guess about which way is up. If the kangaroo ever gets near the peak, she may jump back and forth across the peak without ever landing on it. If you use a decaying step size, the kangaroo gets tired and makes smaller and smaller hops, so if she ever gets near the peak she has a better chance to actually landing on it before the Himalayas erode away. In backprop with momentum, the kangaroo has poor traction and can't make sharp turns. With online training, there are frequent earthquakes, and mountains constantly appear and disappear. This makes it difficult for the blind kangaroo to tell whether she has ever reached the top of a mountain, and she has to take small hops to avoid falling into the gaping chasms that can open up at any moment.

Notice that in all the methods discussed so far, the kangaroo can hope at best to find the top of a mountain close to where she starts. In other words these are local ascent *methods. There's no guarantee that this mountain will be Everest, or even a very high mountain. Many methods exist to try to find the global optimum.*

In simulated annealing, the kangaroo is drunk and hops around randomly for a long time. However, she gradually sobers up and the more sober she is, the more likely she is to hop up hill [temperature decreases according to the annealing schedule].

In a random multi-start method, lots of kangaroos are parachuted into the Himalayas at random places. You hope at least one of them will find Everest.

A genetic algorithm begins like random multi-start. However, these kangaroos do not know that they are supposed to be looking for the top of a mountain. Every few years, you shoot the kangaroos at low altitudes and hope that the ones that are left will be fruitful, multiply, and ascend. Current research suggests that fleas may be more effective than kangaroos in genetic algorithms, since their faster rate of reproduction more than compensates for their shorter hops [crossover is more important than mutation].

A tunneling algorithm can be applied in combination with any local ascent method but requires divine intervention and a jet ski. The kangaroo first finds the top of any nearby mountain. Then the kangaroo calls upon her deity to flood the earth to the point that the waters just reach the top of the current mountain. She gets on her ski, goes off in search of a higher mountain, and repeats the process until no higher mountains can be found.

2.3 Loss Functions for Supervised Methods

Before starting with any optimization algorithm, a criterion needs to be defined that is the exact mathematical description of what has to be optimized. In supervised learning the error $e(i)$ is usually computed as the difference between the measured

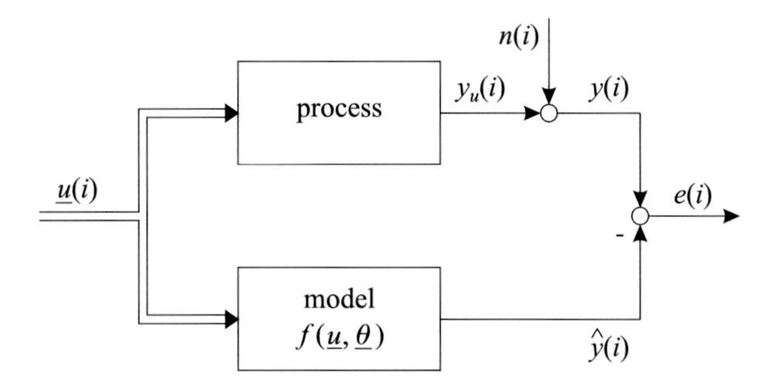

Fig. 2.3 Process and model for data sample i

process output $y(i)$ and the model output $\hat{y}(i)$ for a given number N of training data samples $i = 1, \ldots, N$: the so-called training data set (see Fig. 2.3). Usually the measured process output $y(i)$ is corrupted with noise $n(i)$. So the actually desired output $y_u(i)$ is unknown. The most common choice for a criterion is the sum of squared errors or its square root,

$$I(\underline{\theta}) = \sum_{i=1}^{N} e^2(i) \qquad \text{with} \quad e(i) = y(i) - \hat{y}(i)\,. \tag{2.1}$$

Since the objective is to find the minimum of this function $I(\underline{\theta})$, it is called a *loss function*. Linear and nonlinear optimization problems applying this special kind of loss function are called *least squares (LS)* and *nonlinear least squares (NLS)* problems. The reasons for the popularity of this loss function are the following. To achieve the minimum of a loss function, its gradient has to be equal to zero. With (2.1) this leads to a linear equation system if the error itself is linear in the unknown parameters. Thus, for linear parameters, the error sum of squares leads to an easy-to-solve linear optimization problem; see Chap. 3. Another property of this loss function is the quadratic scaling of the errors, which favors many small errors over a few larger ones. This property can often be seen as an advantage. Note, however, that this property makes the sum of squared errors sensitive to outliers.

The sum of squared errors loss function can be extended by weighting the contribution of each squared error with a factor, say q_i,

$$I(\underline{\theta}) = \sum_{i=1}^{N} q_i\, e^2(i)\,. \tag{2.2}$$

This offers the additional advantage that knowledge about the relevance of or confidence in each data sample i can be incorporated in (2.2) by selecting the q_i appropriately. Problems applying this type of criterion are called *weighted least squares (WLS)* problems.

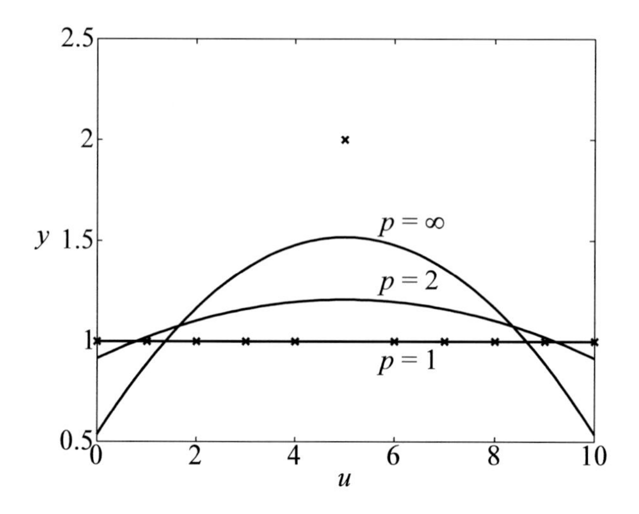

Fig. 2.4 Fit of second-order polynomials through 11 data samples for loss function (2.3) with $p = 1, 2$, and ∞. Higher values of p yield a higher sensitivity with respect to the outlier at (5,2)

Even more general is the following loss function definition:

$$I(\underline{\theta}) = \left(\sum_{i=1}^{N} q_i \, \|e(i)\|^p \right)^{1/p} . \tag{2.3}$$

Besides $p = 2$ common choices for p are $p = 1$ (i.e., the sum of absolute errors) and $p = \infty$ (i.e., the maximum error).[1] Figure 2.4 shows the fit of a second-order polynomial through 11 data samples by minimizing 3 different loss functions of type (2.3) with $p = 1, 2$, and ∞, respectively.

2.3.1 Maximum Likelihood Method

It seems as if the loss functions introduced above do not make any explicit assumptions about the properties of the corrupting noise n. However, it can be shown that a sum of squared errors loss function implicitly assumes uncorrelated noise $n(i)$ (i.e., $E\{n(i)n(j)\} = 0$ for $i \neq j$ when $E\{n(i)\} = 0$) with a Gaussian distribution that has zero mean and constant variance σ^2. Therefore, since $e(i)$ follows the same distribution as $n(i)$, the probability density function (pdf) of the error $e(i)$ is assumed to be

$$p\left(e(i)\right) = \frac{1}{\sqrt{2\pi}\sigma(i)} \exp\left(-\frac{1}{2} \frac{e^2(i)}{\sigma^2(i)} \right) . \tag{2.4}$$

[1]Note that taking the pth root in (2.3) just scales the absolute value of the loss function. It is important, however, to let (2.3) converge to the maximum error for $p \to \infty$.

Because the errors are assumed to be independent, the maximum likelihood function is equal to the product of the pdfs for each sample:

$$L\,(e(1), e(2), \ldots, e(N)) = p\,(e(1)) \cdot p(e(2)) \cdot \ \ldots \ \cdot p(e(N))\,. \tag{2.5}$$

The optimal parameters are obtained by maximizing (2.5). Utilizing the negative (since the minimum, not the maximum, of the loss function is sought) logarithm of the maximum likelihood function in (2.5) as a loss function leads to

$$I\,(\underline{\theta}) = -\ln\left(p(e(1))\right) - \ln\left(p(e(2))\right) - \ldots - \ln\left(p(e(N))\right)\,. \tag{2.6}$$

It is easy to see that for the Gaussian noise distribution with zero mean in (2.4), (2.6) results in the weighted sum of squared errors loss function (by ignoring constant factors and offsets)

$$I\,(\underline{\theta}) = \frac{1}{\sigma^2(1)}\,e^2(1) + \frac{1}{\sigma^2(2)}\,e^2(2) + \ldots + \frac{1}{\sigma^2(N)}\,e^2(N)\,. \tag{2.7}$$

Therefore, it is optimal in the sense of the maximum likelihood to weight each squared error term with the inverse noise variance $\sigma^2(i)$. If furthermore equal noise variances $\sigma^2(i)$ are assumed for each data sample, the standard sum of squared errors loss function is recovered:

$$I\,(\underline{\theta}) = e^2(1) + e^2(2) + \ldots + e^2(N)\,. \tag{2.8}$$

In most applications, the noise pdf $p(n)$ and therefore the error pdf $p(e)$ are unknown, and the maximum likelihood method is not directly applicable. However, the assumption of a normal distribution is very reasonable and often at least approximately valid in practice, because noise is usually composed of a number of many different sources. The central limit theorem states that the pdf of a sum of arbitrary[2] distributed random variables approaches a Gaussian distribution as the number of random variables increases. Thus, the assumption of Gaussian distributed noise can be justified, and the error sum of squares loss function is of major practical importance. Then, the least squares estimate that minimizes the error sum of squares, and the maximum likelihood estimate are equivalent. For different noise distributions, other loss functions are obtained by the maximum likelihood principle. For example, assume uncorrelated noise $n(i)$ that follows the double exponential distribution [125]

$$p\,(e(i)) = \frac{1}{2\sigma}\,\exp\left(-\frac{|e(i)|}{\sigma}\right)\,. \tag{2.9}$$

[2]Some "exotic" distributions (e.g., the Cauchy distribution) exist that do not meet the so-called Lindeberg condition, and therefore their sum is not asymptotically Gaussian distributed. But these exceptions are without practical significance in the context of this book.

Then the loss function obtained by the maximum likelihood principle is the sum of absolute errors

$$I(\underline{\theta}) = |e(1)| + |e(2)| + \ldots + |e(N)| . \tag{2.10}$$

It can also be shown that for equally distributed noise

$$p(e(i)) = \begin{cases} 1/2c & \text{for } |e(i)| \le c \\ 0 & \text{for } |e(i)| > c \end{cases} \tag{2.11}$$

the optimal loss function in the sense of maximum likelihood is

$$I(\underline{\theta}) = \max(e(1), e(2), \ldots, e(N)) . \tag{2.12}$$

It is intuitively clear that for noise distributions with a high probability of large positive or negative values ("fat tail" distributions), the optimal loss function should have a low sensitivity to outliers and vice versa. So for equally distributed noise as in (2.11), an error $e(i)$ larger than c cannot be caused by noise, and therefore no further outliers exist (if the noise assumption is correct).

2.3.2 Maximum A Posteriori and Bayes Method

It is of mainly theoretical interest that the maximum likelihood method can be derived as a special case from the maximum a posteriori or from the Bayes method. This is because the Bayesian approach requires even more knowledge (which usually is not available) about the process than the maximum likelihood method. The idea of the maximum a posteriori and the Bayes methods is to model the observed process outputs $\underline{y} = [y_1 \; y_2 \; \cdots \; y_N]^T$ *and* the parameters $\underline{\theta} = [\theta_1 \; \theta_2 \; \cdots \; \theta_n]^T$ by a *joint* probability density function (pdf) [141]. Thus, the parameters are also treated as random variables. The observations $\underline{y}$ are known from the measurement data. The maximum a posteriori estimate is given by the parameters that maximize the conditional pdf

$$p(\underline{\theta}|\underline{y}) \longrightarrow \max_{\underline{\theta}} . \tag{2.13}$$

This pdf is known as a posteriori pdf because it describes the distribution of the parameters after taking measurements $\underline{y}$. Therefore, this estimate is called the maximum a posteriori (MAP) estimate. It can be calculated via the Bayes theorem if the a priori pdf $p(\underline{\theta})$ is given. This a priori pdf has to be chosen by the user on the basis of prior knowledge about the parameters. Furthermore, the conditional pdf $p(\underline{y}|\underline{\theta})$ has to be known. It describes how the measurements depend on the parameters. This conditional pdf is "sharp" if little noise is present because one can conclude with little uncertainty from the parameters $\underline{\theta}$ to the output $\underline{y}$.

The a posterior pdf $p(\underline{\theta}|\underline{y})$ can be calculated via the Bayes theorem

$$p(\underline{\theta}|\underline{y})\, p(\underline{y}) = p(\underline{y}, \underline{\theta}) = p(\underline{y}|\underline{\theta})\, p(\underline{\theta}) \tag{2.14}$$

to be equal to

$$p(\underline{\theta}|\underline{y}) = \frac{p(\underline{y}|\underline{\theta})\, p(\underline{\theta})}{p(\underline{y})} = \frac{p(\underline{y}|\underline{\theta})\, p(\underline{\theta})}{\int p(\underline{y}, \underline{\theta}) d\underline{\theta}} \,. \tag{2.15}$$

With this formula the a priori probability $p(\underline{\theta})$ of the parameters, that is, the knowledge before taking measurements $\underline{y}$, is converted to the a posteriori probability $p(\underline{\theta}|\underline{y})$ by exploiting the information contained in the measurements. The more measurements are taken, the "sharper" the a posteriori pdf becomes, i.e., the more accurately the parameters $\underline{\theta}$ are described.

The Bayes method extends the MAP approach by incorporating a cost function $C(\underline{\theta})$. This cost function allows one to take into account different benefits and risks associated with the solutions. Note that contrary to the loss function $I(\underline{\theta})$, the cost function $C(\underline{\theta})$ operates *directly* on the *parameters* and not on the process and model *outputs*. Thus, with the Bayes method, the a posteriori pdf is not maximized. Instead the a posteriori pdf is weighted with the cost function, and the Bayes estimate is obtained by

$$\int C(\underline{\theta})\, p(\underline{\theta}|\underline{y})\, d\underline{\theta} \longrightarrow \min_{\underline{\theta}} \,. \tag{2.16}$$

If no knowledge about the cost function $C(\underline{\theta})$ is available, it is reasonable to choose the cost function constant. Then (2.16) simplifies to the MAP estimate.

In some cases, the a priori pdf $p(\underline{\theta})$ can be chosen on the basis of previous modeling attempts or prior knowledge about the process. Some model parameters might be known more accurately than others. This can be expressed at least qualitatively in differently "sharp" a priori pdfs. The MAP or Bayes estimates then preserve this prior knowledge in the sense that uncertain parameters are influenced more strongly by the measurements than more certain parameters. Therefore the MAP or Bayes estimates are methods for *regularization*, a framework explained in Sect. 7.5. Unfortunately, for many problems, no or very little prior knowledge about the probability distribution of the parameters is available. Consequently, it is often reasonable to assume a constant a priori pdf $p(\underline{\theta})$, i.e., all parameters are assumed to be equally likely before measurements are taken. It can be shown that in this case the MAP and Bayesian approach reduce to the maximum likelihood method. For more details, see [453].

The relationships between the Bayes, maximum a posteriori, maximum likelihood, weighted least squares, and least squares methods are shown in Fig. 2.5 [279].

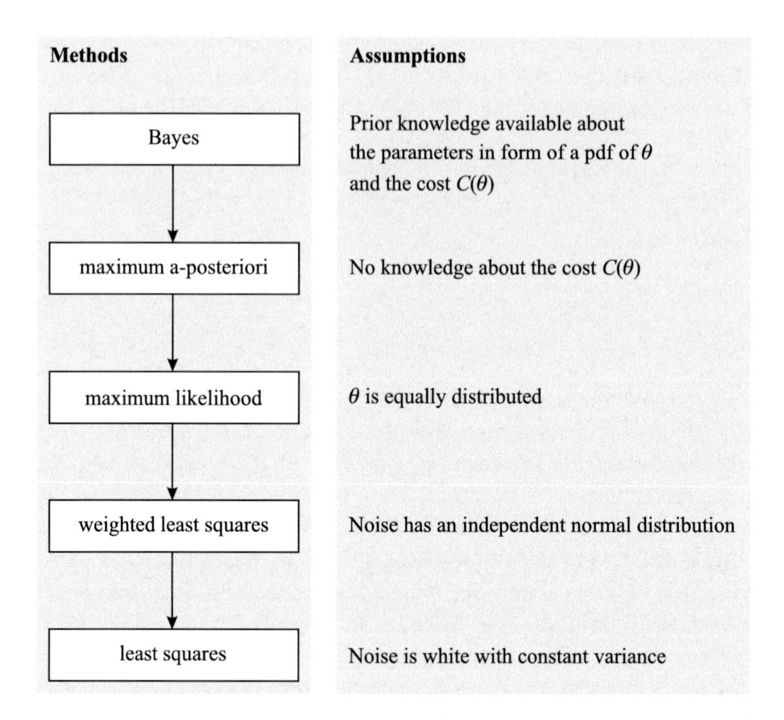

Fig. 2.5 The Bayes method is the most general approach but requires detailed knowledge about the probability density distributions (pdfs). The maximum a posteriori, maximum likelihood, weighted least squares, and least squares methods follow from the Bayes method by making special assumptions [141]

If not explicitly stated otherwise, in this book, the loss function is always chosen as the most commonly applied (possibly weighted) sum of squared errors.

2.4 Loss Functions for Unsupervised Methods

For unsupervised learning, the loss functions presented above are not suitable since no measured outputs $y(i)$ are available. Therefore, the optimization criterion can evaluate the inputs $\underline{u}(i)$ only. Usually, the criterion is based on neighborhood relations. For example, in a two-dimensional input space, a clustering algorithm may search for circles, ellipses, or rectangulars of given or variable proportion and size. Thus, the loss function is usually a distance measure of the training data input samples to some geometric form, e.g., in the simplest case the center of their nearest cluster (i.e., group of data samples).

Chapter 3
Linear Optimization

Abstract This chapter deals with linear regression in detail. Modeling problems where the model output is linear in the model's parameters are studied. The method of "least squares" which minimizes the sum of squared model errors is derived, and its properties are analyzed. Furthermore, various extensions are introduced like weighting and regularization. The concept of the smoothing or hat matrix, which maps the measured output values to the model output values, is introduced. It will be required to understand the leave-one-out error in linear regression and the local behavior of kernel methods in later chapters. Also, the important aspect of the "effective number of parameters" is discussed as it will be a key topic throughout the whole book. In addition, methods for recursive updating are introduced that are capable of dealing with data streams. Finally, the more advanced issue of linear subset selection is discussed in detail. It allows to select regressors incrementally, thereby carrying out structure optimization. These approaches are also a recurrent theme throughout the book.

Keywords Optimization · Least squares · Regressors · Regularization · Weighted · Recursive · Subset selection

If the error between process and model output is linear in the parameters, and the error sum of squares is applied as a loss function, a linear optimization problem arises. Also, a linear optimization problem can be artificially generated if the error is a nonlinear function $g(\cdot)$ of the parameters, but the loss function is chosen as a sum of those inverted nonlinearities $g(\cdot)^{-1}$ of the errors. Note, however, that this loss function may not be suitable for the underlying problem. This approach will be discussed in more detail in Chap. 4. Linear optimization techniques have the following important properties:

- A unique optimum exists, which hence is the global optimum.
- The surface of the loss function is a hyperparabola (Fig. 3.1) of the form $\frac{1}{2}\underline{\theta}^T\underline{H}\,\underline{\theta} + \underline{h}^T\underline{\theta} + h_0$ with the n-dimensional parameter vector $\underline{\theta}$, the $n \times n$-dimensional matrix $\underline{H}$, the n-dimensional vector $\underline{h}$, and the scalar h_0.

O. Nelles, *Nonlinear System Identification*,
https://doi.org/10.1007/978-3-030-47439-3_3

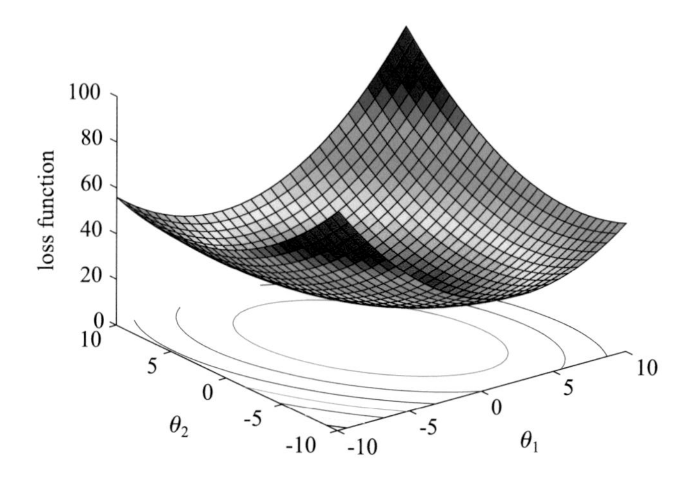

Fig. 3.1 Loss function surface of a linear optimization problem with two parameters

- A one-shot solution can be computed analytically.
- Many numerically stable and fast algorithms exist.
- A recursive formulation is possible.
- The techniques can be applied online.
- The leave-on-out error can be computed at negligible additional cost in closed form.

Those features make linear optimization very attractive, and it is a good idea to first attack all problems with linear methods before applying more complex alternatives. For a linear optimization problem with an error sum of squares loss function, the model output $\hat{y}$ depends linearly on the n parameters θ_i:

$$\hat{y} = \theta_1 x_1 + \theta_2 x_2 + \ldots + \theta_n x_n = \sum_{i=1}^{n} \theta_i x_i \qquad \text{with} \quad x_i = g_i(\underline{u}) \, . \tag{3.1}$$

In statistics the x_i are called *regressors* or *independent variables*, the θ_i are called *regression coefficients*, y is called the *dependent variable*, and the whole problem is called *linear regression*. Note that there are as many regressors as parameters. The parameters θ_i in (3.1) will be called *linear parameters* to emphasize that they can be estimated by linear optimization.

As can be seen in (3.1), only the parameters θ_i have to enter linearly; the regressors x_i can depend in any nonlinear way on the measured inputs $\underline{u}$. Therefore, with some a priori knowledge, many nonlinear optimization problems can be transformed into linear ones. For example, with voltages and currents as measured inputs u_1, u_2 and the knowledge that the output depends only on the electrical power, a linear optimization problem arises by taking the product of voltage and current as regressor $x = u_1 u_2$. Intelligent preprocessing can often reduce the complexity by simplifying nonlinear to linear optimization problems.

In this chapter, first the standard least squares solution and some extensions are extensively discussed. Then the recursive least squares algorithm is introduced, and the issue of constraints is briefly addressed. Finally, the problem of selecting the important regressors is treated in detail.

3.1 Least Squares (LS)

The well-known least squares method was first developed by Gauss in 1795. It is the most widely applied solution for linear optimization problems. In the following, it will be assumed that $i = 1, \ldots, N$ data samples $\{\underline{u}(i), y(i)\}$ have been measured; see Fig. 2.3. The process output y may be disturbed by white noise n. For a detailed discussion of the assumed noise properties, refer to Sect. 3.1.7. The number of parameters to be optimized will be called n; the corresponding n regressors $x_1, \ldots, x_n$ can be calculated for the data. The goal is to find the model output $\hat{y}$ that best approximates the process output y in the least squares sense, i.e., with the minimal sum of squared error loss function value. According to (3.1), this is equivalent to finding the best linear combination of the regressors by optimizing the parameters $\theta_1, \ldots, \theta_n$. Following this, an expression for the optimal parameters will be derived.

First, some vector and matrix definitions are introduced for a least squares problem with n parameters and N training data samples:

$$\underline{e} = \begin{bmatrix} e(1) \\ e(2) \\ \vdots \\ e(N) \end{bmatrix} \quad \underline{y} = \begin{bmatrix} y(1) \\ y(2) \\ \vdots \\ y(N) \end{bmatrix} \quad \underline{\hat{y}} = \begin{bmatrix} \hat{y}(1) \\ \hat{y}(2) \\ \vdots \\ \hat{y}(N) \end{bmatrix} \quad \underline{n} = \begin{bmatrix} n(1) \\ n(2) \\ \vdots \\ n(N) \end{bmatrix}, \tag{3.2}$$

$$\underline{X} = \begin{bmatrix} x_1(1) & x_2(1) & \cdots & x_n(1) \\ x_1(2) & x_2(2) & \cdots & x_n(2) \\ \vdots & \vdots & & \vdots \\ x_1(N) & x_2(N) & \cdots & x_n(N) \end{bmatrix} \quad \underline{\theta} = \begin{bmatrix} \theta_1 \\ \theta_2 \\ \vdots \\ \theta_n \end{bmatrix}. \tag{3.3}$$

Note that the columns of the *regression matrix* $\underline{X}$ are the *regression vectors*

$$\underline{x}_i = \begin{bmatrix} x_i(1) \\ x_i(2) \\ \vdots \\ x_i(N) \end{bmatrix} \quad \text{for } i = 1, \ldots, n. \tag{3.4}$$

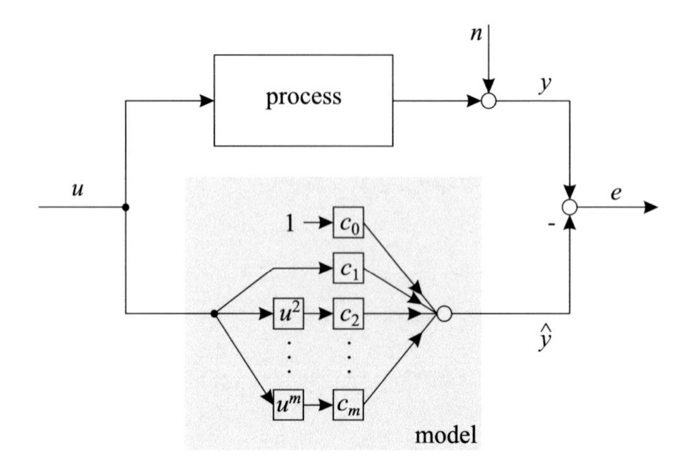

Fig. 3.2 Polynomial model with $\hat{y} = c_0 + c_1 u + c_2 u^2 + \ldots + c_m u^m$

Consequently, the regression matrix can be written as

$$\underline{X} = [\underline{x}_1 \; \underline{x}_2 \; \cdots \; \underline{x}_n] \,. \tag{3.5}$$

The following examples illustrate how different problems can be formulated in this way.

Example 3.1.1 Least Squares for Polynomials
The common problem of fitting a polynomial of order m to data is discussed as a simple least squares example. The model output $\hat{y}$ is (see Fig. 3.2)

$$\hat{y} = c_0 + c_1 \, u + c_2 \, u^2 + \ldots + c_m \, u^m = \sum_{i=0}^{m} c_i \, u^i \,. \tag{3.6}$$

Accordingly, the error $e(k) = y(k) - \hat{y}(k)$ becomes (see Fig. 3.3)

$$e(k) = y(k) - c_0 - c_1 \, u - c_2 \, u^2 - \ldots - c_m \, u^m \,. \tag{3.7}$$

The regression matrix $\underline{X}$ for N measurements and the parameter vector $\underline{\theta}$ are

$$\underline{X} = \begin{bmatrix} 1 & u(1) & u^2(1) & \cdots & u^m(1) \\ 1 & u(2) & u^2(2) & \cdots & u^m(2) \\ \vdots & \vdots & \vdots & & \vdots \\ 1 & u(N) & u^2(N) & \cdots & u^m(N) \end{bmatrix} \qquad \underline{\theta} = \begin{bmatrix} c_0 \\ c_1 \\ c_2 \\ \vdots \\ c_m \end{bmatrix} . \tag{3.8}$$

□

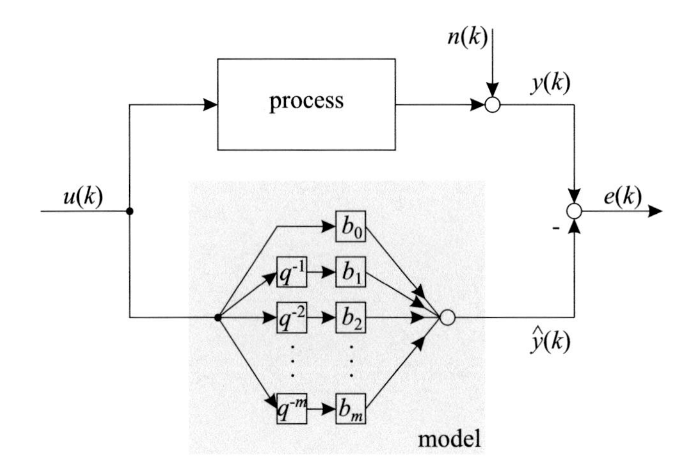

Fig. 3.3 Linear finite impulse response filter with $\hat{y}(k) = b_0u(k) + b_1u(k-1) + \ldots + b_mu(k-m)$. q^{-1} is the delay operator, i.e., $q^{-1}x(k) = x(k-1)$

Example 3.1.2 Least Squares for Linear FIR Filters
Another important basic LS problem is the identification of a time-discrete linear dynamic process by a finite impulse response (FIR) filter. The model output for a one-step predictor (Sect. 18.6.1) of dynamic order m is

$$\hat{y}(k) = b_0u(k) + b_1u(k-1) + \ldots + b_mu(k-m)\,. \tag{3.9}$$

Accordingly, the error $e(k) = y(k) - \hat{y}(k)$ becomes (see Fig. 3.3)

$$e(k) = y(k) - b_0u(k) - b_1u(k-1) - \ldots - b_mu(k-m)\,. \tag{3.10}$$

The regression matrix $\underline{X}$ for N measurements and the parameter vector $\underline{\theta}$ are

$$\underline{X} = \begin{bmatrix} u(m+1) & u(m) & \cdots & u(1) \\ u(m+2) & u(m+1) & \cdots & u(2) \\ \vdots & \vdots & & \vdots \\ u(N) & u(N-1) & \cdots & u(N-m) \end{bmatrix} \qquad \underline{\theta} = \begin{bmatrix} b_0 \\ b_1 \\ \vdots \\ b_m \end{bmatrix}. \tag{3.11}$$

This regression matrix is only $(N-m) \times n$, *not* $N \times n$ as in the previous example. The reason for this is that the regressors have to be constructed from the inputs with time delays between 0 and m. An additional row in the regression matrix would require measurements of $u(k)$ either for $k > N$ (lower left entry of $\underline{X}$) or for $k < 1$ (upper right entry of $\underline{X}$), but the data set contains only $u(k)$ for $k = 1, \ldots, N$. □

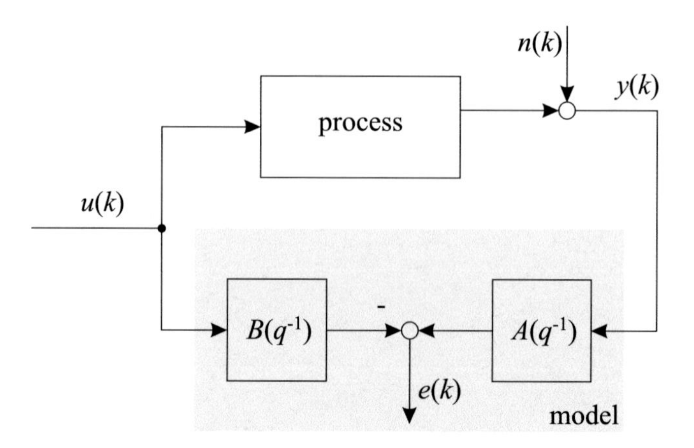

Fig. 3.4 Linear infinite impulse response filter for one-step prediction with $B(q^{-1}) = b_1q^{-1} + b_2q^{-2} + \ldots + b_mq^{-m}$ and $A(q^{-1}) = 1 + a_1q^{-1} + a_2q^{-2} + \ldots + a_mq^{-m}$

Example 3.1.3 Least Squares for Linear IIR Filters
A more complex LS problem is the identification of a time-discrete linear dynamic process by an infinite impulse response (IIR) filter. The model output for a one-step predictor (Sect. 18.5.1) of dynamic order m is

$$\hat{y}(k) = b_1u(k-1)+\ldots+b_mu(k-m) - a_1y(k-1)-\ldots-a_my(k-m). \tag{3.12}$$

Obviously, in contrast to the previous examples, some regressors (the $y(k-i)$, $i = 1, \ldots, m$) depend on the process output. Therefore, it violates the assumption (3.1) that the regressors depend only on the inputs. This fact has important consequences; see the warning below. Accordingly, the error $e(k) = y(k) - \hat{y}(k)$ becomes (see Fig. 3.4)[1]

$$e(k) = y(k) + \ldots + a_my(k-m) - b_1u(k-1) - \ldots - b_mu(k-m). \tag{3.13}$$

The regression matrix $\underline{X}$ for N measurements and the parameter vector $\underline{\theta}$ are

$$\underline{X} = \begin{bmatrix} u(m) & \cdots & u(1) & -y(m) & \cdots & -y(1) \\ u(m+1) & \cdots & u(2) & -y(m+1) & \cdots & -y(2) \\ \vdots & & \vdots & \vdots & & \vdots \\ u(N-1) & \cdots & u(N-m) & -y(N-1) & \cdots & -y(N-m) \end{bmatrix}, \tag{3.14}$$

[1]This error is called the *equation* error and is different from the *output* error (difference between process and model output) used in the other examples. The reason for using the equation error here is that for IIR filters the output error would *not* be *linear* in the parameters; see Chap. 18.

$$\underline{\theta} = [b_1 \ \cdots \ b_m \ a_1 \ \cdots \ a_m]^T \, . \tag{3.15}$$

For the same reasons as in the FIR filter example above, the regression matrix is only $(N - m) \times n$.

Warning: Although it seems at first sight as if the above IIR filter example is a standard least squares problem, it possesses a special property that makes analysis much harder. The regression matrix $\underline{X}$ contains measured process outputs, which usually are disturbed by noise. Therefore, the regression matrix contains random variables and cannot be treated as deterministic anymore. All proofs relying on a deterministic regression matrix, such as those leading to (3.34) and (3.35), cannot be applied since $\mathrm{E}\{\underline{X}\} \neq \underline{X}$. For a thorough discussion of this subject, refer to Chap. 18. □

In the vector/matrix notation, the model output can be written as $\underline{\hat{y}} = \underline{X}\,\underline{\theta}$, and the least squares problem becomes

$$I(\underline{\theta}) = \frac{1}{2}\,\underline{e}^T\underline{e} \longrightarrow \min_{\underline{\theta}} \qquad \text{with} \quad \underline{e} = \underline{y} - \underline{\hat{y}} = \underline{y} - \underline{X}\,\underline{\theta}\, . \tag{3.16}$$

Note that for convenience, the loss function is multiplied by 1/2 in order to get rid of the factor 2 in the gradient. In some books the loss function is defined as $\frac{1}{N}\,\underline{e}^T\underline{e}$ or $\frac{1}{2N}\,\underline{e}^T\underline{e}$, which simply realizes a normalization by the number of data samples and makes the loss function equal to (half) the error variance.

The loss function (3.16) is a parabolic function in the parameter vector $\underline{\theta}$:

$$I(\underline{\theta}) = \frac{1}{2}\,\underline{\theta}^T\underline{H}\,\underline{\theta} + \underline{h}^T\underline{\theta} + \frac{1}{2}\,h_0 \tag{3.17}$$

with the quadratic term

$$\underline{H} = \underline{X}^T\underline{X}\, , \tag{3.18}$$

the linear term

$$\underline{h} = -\underline{X}^T\underline{y}\, , \tag{3.19}$$

and the constant term

$$h_0 = \underline{y}^T\underline{y}\, . \tag{3.20}$$

The quadratic term $\underline{H}$ is the Hessian, i.e., the second derivative of the loss function (see below).

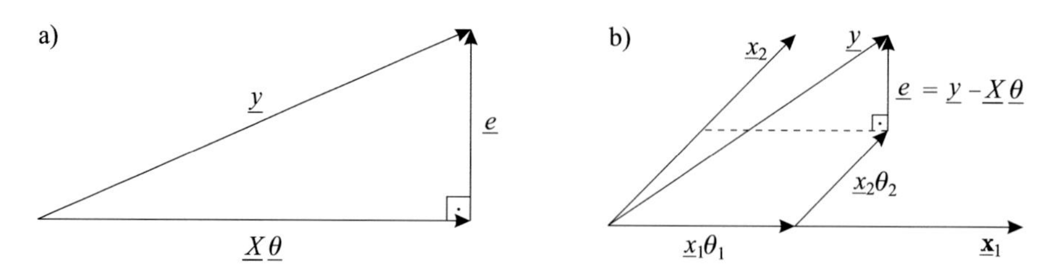

Fig. 3.5 At the optimum $\underline{X}\,\underline{\theta}$ is closest to $\underline{y}$ and therefore the error $\underline{e}$ is orthogonal to all columns (regressors) in $\underline{X}$: (**a**) projection of an n-regressor problem, (**b**) a two regressor problem

Considering (3.16), the gradient of $I(\underline{\theta})$ with respect to the parameter vector $\underline{\theta}$ has to be equal to zero. This leads to the famous orthogonal equations, which express that at the optimum, the error $\underline{e}$ is orthogonal to all regressors $\underline{x}_i$ (columns of $\underline{X}$):

$$\frac{\partial I(\underline{\theta})}{\partial \underline{\theta}} = \underline{g} = -\underline{X}^T \underline{e} = -\underline{X}^T \left(\underline{y} - \underline{X}\,\underline{\theta}\right) = \underline{0}\,. \tag{3.21}$$

Figure 3.5 illustrates the orthogonal equations as a projection for the general n-regressor problem (a) and in more detail for a two regressor problem (b).

The orthogonal equations lead to the least squares estimate

$$\hat{\underline{\theta}} = \left(\underline{X}^T \underline{X}\right)^{-1} \underline{X}^T \underline{y}\,. \tag{3.22}$$

The difference $\underline{e}$ between the measured output $\underline{y}$ and the LS estimates output $\hat{\underline{y}} = \underline{X}\,\hat{\underline{\theta}}$ is called the *residual*. In the ideal case (perfect model structure, optimal parameters, no noise), the residuals should be zero. In practice, examination of the residuals can reveal many details about the estimation quality. Residuals close to white noise indicate a good model since then all information is exploited by the model; that is, the model fully explains the output. For a more detailed discussion, refer to Chap. 7.

It is interesting to note that the LS estimate in (3.22) can be formulated in terms of correlation functions. By computing $\underline{X}^T \underline{X} =$

$$N \begin{bmatrix} \frac{1}{N}\sum_{i=1}^{N} x_1^2(i) & \frac{1}{N}\sum_{i=1}^{N} x_1(i)x_2(i) & \cdots & \frac{1}{N}\sum_{i=1}^{N} x_1(i)x_n(i) \\ \frac{1}{N}\sum_{i=1}^{N} x_2(i)x_1(i) & \frac{1}{N}\sum_{i=1}^{N} x_2^2(i) & \cdots & \frac{1}{N}\sum_{i=1}^{N} x_2(i)x_n(i) \\ \vdots & \vdots & \ddots & \vdots \\ \frac{1}{N}\sum_{i=1}^{N} x_n(i)x_1(i) & \frac{1}{N}\sum_{i=1}^{N} x_n(i)x_2(i) & \cdots & \frac{1}{N}\sum_{i=1}^{N} x_n^2(i) \end{bmatrix} \tag{3.23}$$

and computing $\underline{X}^T \underline{y} =$

$$N \begin{bmatrix} \frac{1}{N} \sum_{i=1}^{N} x_1(i) y(i) \\ \frac{1}{N} \sum_{i=1}^{N} x_2(i) y(i) \\ \vdots \\ \frac{1}{N} \sum_{i=1}^{N} x_n(i) y(i) \end{bmatrix}, \tag{3.24}$$

the LS estimate can be written as

$$\hat{\underline{\theta}} = \text{corr}\{\underline{x}, \underline{x}\}^{-1} \cdot \text{corr}\{\underline{x}, y\}, \tag{3.25}$$

where $\text{corr}\{\underline{x}, \underline{x}\}$ denotes the autocorrelation matrix, which is composed of the auto- and cross-correlations between all regressor combinations x_i and x_j,

$$\text{corr}\{\underline{x}, \underline{x}\} = \begin{bmatrix} \text{corr}\{x_1, x_1\} & \text{corr}\{x_1, x_2\} & \cdots & \text{corr}\{x_1, x_n\} \\ \text{corr}\{x_2, x_1\} & \text{corr}\{x_2, x_2\} & \cdots & \text{corr}\{x_2, x_n\} \\ \vdots & \vdots & \ddots & \vdots \\ \text{corr}\{x_n, x_1\} & \text{corr}\{x_n, x_2\} & \cdots & \text{corr}\{x_n, x_n\} \end{bmatrix}, \tag{3.26}$$

and $\text{corr}\{\underline{x}, y\}$ denotes the cross-correlation vector, which is composed of the cross-correlations between all regressors x_i and the output y,

$$\text{corr}\{\underline{x}, y\} = \begin{bmatrix} \text{corr}\{x_1, y\} \\ \text{corr}\{x_2, y\} \\ \vdots \\ \text{corr}\{x_n, y\} \end{bmatrix}. \tag{3.27}$$

Therefore, the LS estimate can be interpreted as the cross-correlation of input and output divided by the auto-correlation of the input. Note that if the cross-correlation vector $\text{corr}\{\underline{x}, y\}$ is equal to zero ($\underline{0}$), the parameter estimate is zero ($\underline{0}$) independent of the autocorrelation matrix. Such a case occurs if all regressors $\underline{x}_i$ are orthogonal to the output vector $\underline{y}$. This property is obvious, because the output $\underline{y}$ is approximated by a linear combination of all regressors $\underline{X}\,\underline{\theta}$, and if all these regressors are orthogonal to the output, the smallest possible modeling error is $\underline{e} = \underline{y}$ (see Fig. 3.5) which can simply be achieved by $\hat{\underline{\theta}} = \underline{0}$.

In the least squares solution (3.22), the expression $\left(\underline{X}^T \underline{X}\right)^{-1} \underline{X}^T$ is called the *pseudoinverse* of the matrix $\underline{X}$ and is sometimes written as $\underline{X}^+$. If $\underline{X}$ has full rank n, which of course requires at least as many measurements as unknown parameters ($N \geq n$),[2] the matrix inversion can be performed. In practice, the data is noisy, and

[2]Strictly speaking, the regression matrix must have at least as many rows as columns. (Recall that for the FIR and IIR filter examples, the number of rows is smaller than N.) Moreover, this condition is not sufficient since additionally the columns must be linearly independent.

thus the regression matrix is (almost) never exactly singular. Rather the condition number of the matrix $\underline{X}^T \underline{X}$ is decisive for the accuracy of a numerical inversion; see Example 3.1.4. Usually, the matrix inversion is not carried out in the form of (3.22). Rather one of the following, more sophisticated, approaches is used to avoid the bad numerical properties of a direct matrix inversion:

- Solving the normal equations $\underline{X}^T \underline{X}\, \underline{\theta} = \underline{X}^T \underline{y}$ by Gaussian elimination or by forming the Cholesky decomposition of $\underline{X}^T \underline{X}$
- Forming an orthogonal decomposition of $\underline{X}$ by Gram-Schmidt, modified Gram-Schmidt, Householder, or Givens transformations,
- Forming a singular value decomposition of $\underline{X}$

For more details on these numerically advanced algorithms, refer to [193].

It is important to note that the matrix $\underline{X}^T \underline{X}$ is identical to the Hessian (see Appendix A) of the loss function

$$\underline{H} = \frac{\partial^2 I(\underline{\theta})}{\partial \underline{\theta}^2} = \underline{X}^T \underline{X} . \tag{3.28}$$

Thus, the Hessian has to be well conditioned in order to obtain accurate parameter estimates. The condition of a matrix, χ, can be defined by the eigenvalue spread of the matrix, that is, the ratio of the largest to the smallest eigenvalue of $\underline{H}$[3]:

$$\chi = \frac{\lambda_{\text{max}}}{\lambda_{\text{min}}} . \tag{3.29}$$

Example 3.1.4 Loss Function, Contour Lines and Hessian
Figure 3.6 shows the contour lines of loss functions with diagonal Hessians with the eigenvalue spreads $\chi_{\text{a}} = 2$ and $\chi_{\text{b}} = 5$. The corresponding loss functions are $I_{\text{a}}(\underline{\theta}) = 10\,\theta_1^2 + 5\,\theta_2^2$ and $I_{\text{b}}(\underline{\theta}) = 10\,\theta_1^2 + 2\,\theta_2^2$, respectively.

Note that the loss functions $I_{\text{a}}(\underline{\theta})$ and $I_{\text{b}}(\underline{\theta})$ are very special and simple cases, but valid realizations of the general loss function form $\underline{\theta}^T \underline{A}\,\underline{\theta} + \underline{b}^T \underline{\theta} + c$ with a diagonal $\underline{A}$ and $\underline{b} = \underline{0}$, and $c = 0$.

Owing to numerical errors, the optimal parameter value $\underline{\theta}_{\text{opt}}$ (here at (0,0)) can not be expected to be reached exactly. However, some loss function value close to the minimum will be reached, e.g., the most inner contour line. While θ_1 can be determined quite accurately for any point within the most inner contour line (gray-shaded area), the parameter θ_2 is estimated $\sqrt{2}$ or $\sqrt{5}$ times more inaccurately. The

[3]Note that the Hessian $\underline{H}$ is symmetric and therefore all eigenvalues are real. Furthermore, the eigenvalues are non-negative because the Hessian is positive semi-definite since $\underline{H} = \underline{X}^T \underline{X}$. If $\underline{X}$ and thus $\underline{H}$ are not singular (i.e., have full rank), the eigenvalues are strictly positive.

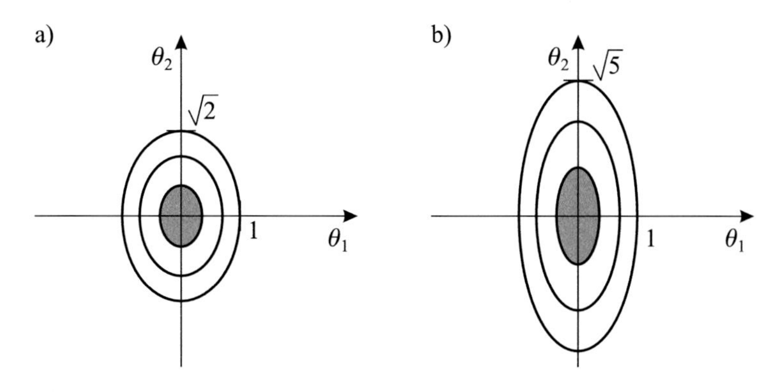

Fig. 3.6 Contour lines of loss functions with diagonal Hessians. For a Hessian with equal eigenvalues ($\chi = 1$), the contours are perfect circles. For an eigenvalue spread of (**a**) $\chi = 2$ or (**b**) $\chi = 5$, the contours become elliptic. The gray-shaded areas are the regions of parameter uncertainty for an optimization accuracy represented by the innermost contour line

ratio of the achieved accuracy for θ_2 and θ_1 is given by $\sqrt{\chi}$. A redundant parameter could not be determined at all, since the corresponding ellipse axis expands to infinity and the Hessian is rank deficient (has rank $n - 1$), i.e., is singular and therefore not invertible. In that case the condition number would be $\chi = \infty$ because the smallest eigenvalue $\lambda_{\min} = 0$. The solution would be a line instead of a point in the parameter space. □

3.1.1 Covariance Matrix of the Parameter Estimate

As shown above, the parameters may be estimated with different accuracy. It is possible to describe the accuracy of the estimated parameters by its covariance matrix

$$\mathrm{cov}\{\underline{\hat{\theta}}\} = \mathrm{E}\left\{\left(\underline{\hat{\theta}} - \mathrm{E}\{\underline{\hat{\theta}}\}\right)\left(\underline{\hat{\theta}} - \mathrm{E}\{\underline{\hat{\theta}}\}\right)^T\right\}, \tag{3.30}$$

where

$$\underline{\hat{\theta}} = \left(\underline{X}^T\underline{X}\right)^{-1}\underline{X}^T\underline{y} \quad \text{and} \quad \mathrm{E}\{\underline{\hat{\theta}}\} = \left(\underline{X}^T\underline{X}\right)^{-1}\underline{X}^T\mathrm{E}\{\underline{y}\} \tag{3.31}$$

when $\mathrm{E}\{\underline{X}\} = \underline{X}$, i.e., $\underline{X}$ is deterministic.

Therefore

$$\underline{\hat{\theta}} - \mathrm{E}\{\underline{\hat{\theta}}\} = \left(\underline{X}^T\underline{X}\right)^{-1}\underline{X}^T\left(\underline{y} - \mathrm{E}\{\underline{y}\}\right). \tag{3.32}$$

Since from Fig. 3.14 $\underline{y} - \mathrm{E}\{\underline{y}\} = \underline{y} - \mathrm{E}\{\underline{y}_u + \underline{n}\} = \underline{y} - \underline{y}_u = \underline{n}$, the parameter estimate is related to the noise properties. Thus, (3.30) becomes

$$\begin{aligned} \mathrm{cov}\{\hat{\underline{\theta}}\} &= \mathrm{E}\left\{\left(\left(\underline{X}^T\underline{X}\right)^{-1}\underline{X}^T\underline{n}\right)\left(\left(\underline{X}^T\underline{X}\right)^{-1}\underline{X}^T\underline{n}\right)^T\right\} \\ &= \left(\underline{X}^T\underline{X}\right)^{-1}\underline{X}^T\mathrm{E}\{\underline{n}\,\underline{n}^T\}\underline{X}\left(\underline{X}^T\underline{X}\right)^{-1} \end{aligned} \tag{3.33}$$

because the Hessian is symmetric: $\left(\underline{X}^T\underline{X}\right)^T = \underline{X}^T\underline{X}$. If n is white noise with variance σ^2, then $\mathrm{E}\{\underline{n}\,\underline{n}^T\} = \sigma^2\underline{I}$ ($\mathrm{E}\{n(i)n(i)\} = \sigma^2$ and $\mathrm{E}\{n(i)n(j)\} = 0$ for $i \neq j$). Finally, the covariance matrix of the estimated parameters becomes

$$\mathrm{cov}\{\hat{\underline{\theta}}\} = \sigma^2\left(\underline{X}^T\underline{X}\right)^{-1} = \sigma^2\underline{H}^{-1}. \tag{3.34}$$

Note that the diagonal entries of this symmetric matrix give the variances of the parameter estimates. Hence, good models are obtained if the regressors $\underline{x}_i$ are large, and the noise variance σ is small; that is, the signal-to-noise ratio is large. Note that the covariance matrix of the parameters is proportional to $1/N$ if the entries in $\underline{X}^T\underline{X}$ increase linearly with N. Thus, by collecting "enough" data, any noise level can be compensated. Furthermore, $-\underline{X}$ is the derivative of the error $\underline{e}$ with respect to the parameters $\underline{\theta}$ representing the sensitivity with respect to $\underline{\theta}$. Consequently, the smaller the variance of the parameters, the more sensitive the error is to these parameters. If the error tends to be independent of some parameter, this parameter cannot be estimated, that is, its variance will tend to infinity. The practical benefit of (3.34) lies more in the obtained feeling for the relative accuracy of the parameter estimates than in the absolute values within $\mathrm{cov}\{\hat{\underline{\theta}}\}$. It allows one to compare the accuracy of different parameters of one estimation or of different estimations.

If one is interested in the absolute values of $\mathrm{cov}\{\hat{\underline{\theta}}\}$, an estimate of σ^2 is required. Because the noise variance is usually unknown, it must be estimated from the residuals with the following unbiased estimator of σ^2 [540]:

$$\hat{\sigma}^2 = \frac{\underline{e}^T\underline{e}}{N-n} = \frac{2\,I(\hat{\underline{\theta}})}{N-n}. \tag{3.35}$$

The denominator in the above formula represents the degrees of freedom of the residuals, that is, the number of data samples minus the number of parameters. It is wise to use these estimates carefully since they are based on the assumption of additive white measurement noise and a correctly assumed model structure. These assumptions can be quite unrealistic. Especially, considerable errors due to a structural mismatch between process and model can be expected for almost any application.

Example 3.1.5 Parameter Variances
In Example 3.1.4 presented above, the Hessians were chosen to

$$\underline{H}_{\mathrm{a}} = \begin{bmatrix} 10 & 0 \\ 0 & 5 \end{bmatrix} \quad \text{and} \quad \underline{H}_{\mathrm{b}} = \begin{bmatrix} 10 & 0 \\ 0 & 2 \end{bmatrix} .$$

With (3.34) this leads to parameter variance estimates of $\mathrm{var}_{\mathrm{a}}\{\hat{\theta}_1\} = \frac{1}{10}\sigma^2$, $\mathrm{var}_{\mathrm{b}}\{\hat{\theta}_1\} = \frac{1}{10}\sigma^2$ and $\mathrm{var}_{\mathrm{a}}\{\hat{\theta}_2\} = \frac{1}{5}\sigma^2$, $\mathrm{var}_{\mathrm{b}}\{\hat{\theta}_2\} = \frac{1}{2}\sigma^2$, respectively. Due to the diagonal structure of the Hessians (indicating orthogonal regressors), the covariances $\mathrm{cov}\{\hat{\theta}_1, \hat{\theta}_2\}$ and $\mathrm{cov}\{\hat{\theta}_2, \hat{\theta}_1\}$ are equal to zero. □

3.1.2 Errorbars

The concept of errorbars is very important because it allows one to estimate the accuracy of a linear parameterized model. From a practical point of view, a model is virtually useless without an estimate of its accuracy. An indicator that qualifies the model's precision for a given input is highly desirable. Such information can be exploited in various ways. For example, a controller can be designed in such a way that it acts strongly in operating regimes and frequency ranges where the model is good and acts weakly (carefully) where the model is inaccurate. The prediction of a model can be discarded if it is too inaccurate. For predictive control, an optimal prediction horizon may be determined in dependency on the model quality. For each input, the most accurate model can be chosen from a bank of models with different architectures or complexities. Many active learning algorithms that add further training data in order to gain a maximum amount of new information on the process are based on the estimated model quality.

The concept of errorbars takes the input data applied for training into account. Generally speaking, a model that was estimated from data can be expected to be good in regions where the data was dense and to be poor in regions where the data was sparse. As described by (3.34), the linear parameters of a model can be estimated only with a certain variance, given finite and noisy data. Obviously, the parameter covariance matrix $\text{cov}\{\underline{\hat{\theta}}\}$ determines the accuracy of the model output for a given input:

$$\begin{aligned}\text{cov}\{\underline{\hat{y}}\} &= \text{E}\left\{\left(\underline{\hat{y}} - \text{E}\{\underline{\hat{y}}\}\right)\left(\underline{\hat{y}} - \text{E}\{\underline{\hat{y}}\}\right)^T\right\} \\ &= E\left\{\left(X\left(\underline{\hat{\theta}} - \text{E}\{\underline{\hat{\theta}}\}\right)\right)\left(X\left(\underline{\hat{\theta}} - \text{E}\{\underline{\hat{\theta}}\}\right)\right)^T\right\} \\ &= \underline{X}\,\text{E}\left\{\left(\underline{\hat{\theta}} - \text{E}\{\underline{\hat{\theta}}\}\right)\left(\underline{\hat{\theta}} - \text{E}\{\underline{\hat{\theta}}\}\right)^T\right\}\underline{X}^T . \end{aligned} \quad (3.36)$$

Thus, the covariance matrix of the model output $\underline{\hat{y}}$ is

$$\text{cov}\{\underline{\hat{y}}\} = \underline{X}\,\text{cov}\{\underline{\hat{\theta}}\}\,\underline{X}^T . \quad (3.37)$$

Since the diagonal entries of $\text{cov}\{\underline{\hat{y}}\}$ represent the variances of the model output $\text{E}\{y^2(i)\}$ for each data sample in $\underline{X}$, the errorbars can be defined as $\underline{\hat{y}}$ plus and minus the standard deviation of the estimated output, that is,

$$\underline{\hat{y}} \pm \sqrt{\text{diag}\left(\text{cov}\{\underline{\hat{y}}\}\right)} . \quad (3.38)$$

These errorbars allow the user to compare the expected model accuracy for different input data. Therefore, they can serve as a tool for designing a good training data set. The errorbars represents just a qualitative measure of the parameter and model accuracy. In order to compute quantitative *confidence intervals*, i.e., intervals in which the parameters and outputs lie with some given *probability*, the noise distribution must be known. Usually, a Gaussian pdf is assumed, and then (3.38) would represent the one-sigma interval, which covers the model output with 52% probability. Typically, a 1.96 times wider interval is considered, which covers the model output with about 95% probability; see [125].

Example 3.1.6 Errorbars and Missing Data
In order to illustrate the effect of missing data on the errorbars, the following function

$$y = 0.5u^2 + 5u + 2 \quad (3.39)$$

will be approximated by a second-order polynomial from 1000 noisy data samples equally distributed in [0, 10]. Figures 3.7 and 3.8 demonstrate three different cases with all data available (Fig. 3.7), missing data in the middle region (Fig. 3.8a), and missing data at the boundary (Fig. 3.8b). Obviously missing data in the middle region is not so problematic as close to the boundary, since global approximators

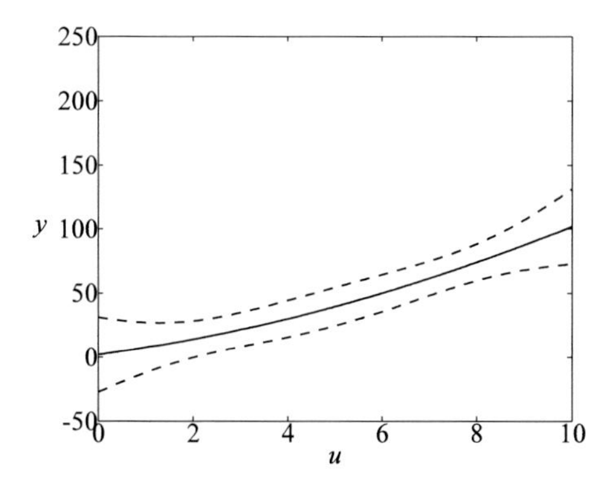

Fig. 3.7 Approximation of the function $y = 0.5u^2 + 5u + 2$ by a second-order polynomial. The training data is equally distributed in [0, 10]. The dotted curves represent the errorbars. Note that the estimated model error increases close to the boundaries of the training data. This effect is due to missing training data left from 0 and right from 10. For extrapolation, these errorbars increase further

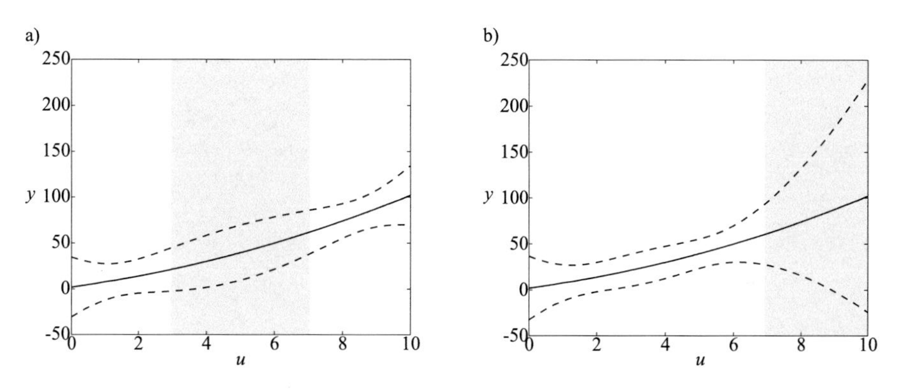

Fig. 3.8 (**a**) No training data is available in the interval (3, 7). (**b**) No training data is available in the interval (7, 10]. The estimated model error increases in those regions where no training data is available; see Fig. 3.7. Note that the errorbars in (**b**) are much larger than in a although more data is missing in a. The reason for this is that extrapolation occurs in (**b**) for $u > 7$

(such as polynomials) are able to fill these data holes by interpolation. Note that local approximators (such as fuzzy systems) are much more sensible in this respect, and their errorbars in Fig. 3.8a would be much larger. □

The errorbars allow the estimation of the model accuracy for a given input. It is important to note that these estimations are calculated under the assumption of a correct model structure. The model errors are solely due to the fact that the estimated parameters do not correspond to their (theoretically) optimal values, which could be estimated from infinite noise-free data. In practice, however, large modeling errors may also occur due to a structural mismatch between process and model. These errors are not (and cannot be) included in the errorbars. They have to be taken into consideration by the user. Essentially, only knowledge about the process allows a reasonable assessment of the model quality.

3.1.3 Orthogonal Regressors

An important special case of the least squares solution (3.22) results for mutually orthogonal regressors, i.e., $\underline{x}_i^T \underline{x}_j = 0$ for $i \neq j$. Then the Hessian $\underline{H}$ becomes

$$\underline{H} = \underline{X}^T \underline{X} = \text{diag}\left(\underline{x}_1^T \underline{x}_1,\ \underline{x}_2^T \underline{x}_2,\ \cdots,\ \underline{x}_n^T \underline{x}_n\right) \tag{3.40}$$

because in the matrix $\underline{H} = \underline{X}^T \underline{X}$, all off-diagonal terms are inner products between two distinct regressors $\underline{x}_i^T \underline{x}_j$ $(i \neq j)$, which are equal to zero when the regressors are mutually orthogonal; see also (3.23).

Therefore, the inversion of the Hessian is trivial:

$$\underline{H}^{-1} = \left(\underline{X}^T \underline{X}\right)^{-1} = \text{diag}\left(\frac{1}{\underline{x}_1^T \underline{x}_1},\ \frac{1}{\underline{x}_2^T \underline{x}_2},\ \cdots,\ \frac{1}{\underline{x}_n^T \underline{x}_n}\right). \tag{3.41}$$

For orthonormal regressors (i.e., $\underline{x}_i^T \underline{x}_i = 1$), (3.40) and (3.41) even simplify to the identity matrix $\underline{H} = \underline{H}^{-1} = \underline{I}$. From (3.41) it is obvious that for orthogonal regressors, no matrix inversion is necessary. This property is used quite often (see Sect. 3.4); it is illustrated in Fig. 3.9. The optimal parameter estimate then becomes

$$\underline{\hat{\theta}} = \left(\underline{X}^T \underline{X}\right)^{-1} \underline{X}^T \underline{y} = \text{diag}\left(\frac{1}{\underline{x}_1^T \underline{x}_1},\ \frac{1}{\underline{x}_2^T \underline{x}_2},\ \cdots,\ \frac{1}{\underline{x}_n^T \underline{x}_n}\right) \underline{X}^T \underline{y}. \tag{3.42}$$

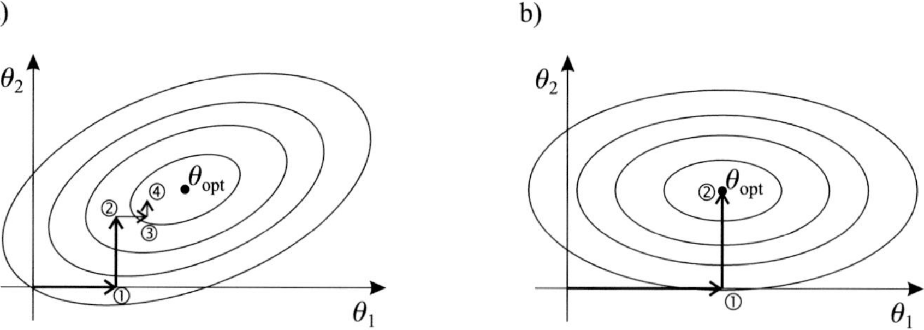

Fig. 3.9 Contour lines for loss functions with (**a**) non-orthogonal and (**b**) orthogonal regressors. If the regressors are orthogonal (i.e., the Hessian has orthogonal columns), the minimum can be reached by performing subsequent single parameter optimizations. This significantly reduces the complexity of the problem. For non-orthogonal regressors, subsequent single parameter optimization steps approach the minimum but do not reach it. The closer to orthogonality the regressors are, the better conditioned the Hessian is, and the faster such a staggered parameter optimization procedure will converge to the minimum

Consequently, each parameter θ_i in $\underline{\theta}$ can be determined separately by

$$\hat{\theta}_i = \frac{\underline{x}_i^T \underline{y}}{\underline{x}_i^T \underline{x}_i} . \tag{3.43}$$

The optimal parameter estimate $\hat{\theta}_i$ depends only on the single regressor $\underline{x}_i$ and the measured output $\underline{y}$. There is no correlation between the different regressors. Therefore, it can be concluded that for non-orthogonal regressors, the inverse Hessian $\underline{H}^{-1}$ decorrelates the correlated regressors.

Besides the obvious computational advantages of orthogonal regressors, another benefit emerges. Since each parameter can be estimated separately, one can include or remove regressors without affecting the other parameter estimates. This allows one to incrementally build up complex models with many parameters from simple models with only a few parameters in a very efficient manner; see Sect. 3.4.

From local basis function approaches (see Sect. 11.3) often nearly orthogonal regressors arise; that is, the scalar product of two regressors is close but not identical to zero. Although such problems do not simplify as shown above, at least the Hessian be can expected to be well conditioned owing to the approximate orthogonality. It is apparent from Fig. 3.9 that subsequent single parameter optimization steps approach the minimum rapidly if the regressors are close to orthogonal. In these cases, the standard least squares solution may be computationally more expensive than iterative single parameter optimization steps, in particular when it is sufficient to reach the minimum roughly. Also, other iterative linear optimization schemes such as the LMS benefit from almost orthogonal regressors by increased convergence speed.

3.1.4 Regularization/Ridge Regression

The problem of poorly conditioned Hessians often becomes severe if the number of regressors is large. It is well known that the probability of poor conditioning increases with the matrix dimension. Therefore the variance of the worst estimated parameters increases with increasing model flexibility. There exists a fundamental tradeoff between the benefits of additional regressors due to higher model flexibility and the drawbacks due to increasing estimation variance; see Chap. 7. Also, poorly conditioned Hessians may arise if the process is not properly excited. It is obvious that the input data must be chosen in such a way that the influence of all regressors shows up in the output; otherwise they cannot be separated from each other.

Since neural networks and fuzzy systems usually utilize a large number of parameters, serious variance problems must be tackled. Methods for controlling the variance are called *regularization techniques*. Some of them are discussed in

Chap. 7. In the context of least squares approaches, the following regularization method is very common. The loss function is extended by a penalty term to

$$I(\underline{\theta}, \lambda) = \frac{1}{2}\left(\underline{e}^T\underline{e} + \lambda\,\underline{\theta}^T\underline{\theta}\right) = \frac{1}{2}\left(\underline{e}^T\underline{e} + \lambda\sum_{i=1}^{n}\theta_i^2\right) \longrightarrow \min_{\underline{\theta}}\,. \tag{3.44}$$

Equivalently, the non-penalized objective can be constrained as follows:

$$I(\underline{\theta}, \lambda) = \frac{1}{2}\underline{e}^T\underline{e} \longrightarrow \min_{\underline{\theta}}$$

$$\text{subject to} \quad \sum_{i=1}^{n}\theta_i^2 \leq t^2\,. \tag{3.45}$$

Solving this optimization problem by the method of Lagrange multipliers exactly yields the above unconstrained problem. For every constraint limit t, a corresponding regularization parameter λ exists.

The idea behind the additional penalty term $\lambda|\underline{\theta}|^2$ is remarkably simple. Those parameters that are not important for solving the least squares problem are driven toward zero in order to decrease the penalty term. Therefore only the significant parameters effectively will be used since their error reduction effect is larger than their penalty term. The level of significance is adjusted by the choice of λ. For $\lambda \to 0$ the standard least squares problem is recovered, while $\lambda \to \infty$ forces all parameters to zero. The loss function (3.44) will (for positive λ) always lead to a biased solution with a smaller variance than in the LS case (3.22). This approach is called *ridge regression* in statistics. Following a comparable derivation as in (3.16)–(3.22), the regularized LS problem leads to the following parameter estimate:

$$\underline{\hat{\theta}} = \left(\underline{X}^T\underline{X} + \lambda\underline{I}\right)^{-1}\underline{X}^T\underline{y}\,. \tag{3.46}$$

3.1.4.1 Efficient Computation

Since the parameter in (3.44) show up quadratically, they can be interpreted as "pseudo"-errors weighted with $\sqrt{\lambda}$. Thus, the problem in (3.44) is identical to solving the following linear equation system in the least squares sense

$$\underline{\widetilde{e}} = \underline{\widetilde{y}} - \underline{\widetilde{X}}\,\underline{\theta} \tag{3.47}$$

with the augmented output vector and regression matrix:

$$\underline{\widetilde{y}} = \begin{bmatrix} y(1) \\ y(2) \\ \vdots \\ y(N) \\ 0 \\ 0 \\ \vdots \\ 0 \end{bmatrix} \quad \text{and} \quad \underline{\widetilde{X}} = \begin{bmatrix} x_1(1) & x_2(1) & \cdots & x_n(1) \\ x_1(2) & x_2(2) & \cdots & x_n(2) \\ \vdots & \vdots & & \vdots \\ x_1(N) & x_2(N) & \cdots & x_n(N) \\ \sqrt{\lambda} & 0 & \cdots & 0 \\ 0 & \sqrt{\lambda} & \cdots & 0 \\ \vdots & \vdots & \ddots & \vdots \\ 0 & 0 & \cdots & \sqrt{\lambda} \end{bmatrix} . \tag{3.48}$$

With this definition, the first N equations measure the model error, the last n equations measure θ_i^2, $i = 1, \ldots, n$, weighted with λ after squaring the augmented error by calculating $\underline{\widetilde{e}}^T \underline{\widetilde{e}}$. This approach allows to utilize standard approaches for calculating linear regression like QR factorization with the Matlab operator "\" which would not be possible directly with (3.46).

Since in (3.46) λ is added to all diagonal entries of the Hessian $\underline{X}^T \underline{X}$, its eigenvalues λ_i, $i = 1, \ldots, n$, are changed. While the addition of λ influences the significant eigenvalues ($\lambda_i \gg \lambda$) only negligibly, the small eigenvalues ($\lambda_i \ll \lambda$) are virtually set to λ. Therefore, the condition of the Hessian can be directly controlled by λ via

$$\chi_{\text{reg}} = \frac{\lambda_{\text{max}}}{\lambda_{\text{min}}} \approx \frac{\lambda_{\text{max}}}{\lambda} . \tag{3.49}$$

A parameter is said to be *significant* or *effective* if its corresponding eigenvalue is larger than λ, and it is *non-significant* or *spurious* for eigenvalues smaller than λ.

3.1.4.2 Covariances for Ridge Regression

For LS the covariance matrix of the estimated parameters is proportional to $\left(\underline{X}^T \underline{X}\right)^{-1}$. For ridge regression, it becomes smaller since $\lambda \underline{I}$ is added before the inverse is taken. It also becomes less critical because the inverse always exists for $\lambda > 0$ even if $\underline{X}^T \underline{X}$ is singular. The covariance matrix for ridge regression is given by[4]:

$$\text{cov}\{\hat{\underline{\theta}}\} = \sigma^2 \left(\underline{X}^T \underline{X} + \lambda \underline{I}\right)^{-1} \underline{X}^T \underline{X} \left(\underline{X}^T \underline{X} + \lambda \underline{I}\right)^{-1} . \tag{3.50}$$

[4]Found under https://onlinecourses.science.psu.edu/stat857/node/155.

Example 3.1.7 Ridge Regression: Matrix Condition
The eigenvalue spread of the Hessian

$$\underline{H} = \begin{bmatrix} 100 & 0 \\ 0 & 0.01 \end{bmatrix}$$

is $\chi = 10{,}000$. With $\lambda = 1$ the modified Hessian becomes

$$\underline{H}_{\text{reg}} = \begin{bmatrix} 100 & 0 \\ 0 & 0.01 \end{bmatrix} + 1 \begin{bmatrix} 1 & 0 \\ 0 & 1 \end{bmatrix} = \begin{bmatrix} 101 & 0 \\ 0 & 1.01 \end{bmatrix}$$

with an eigenvalue spread of only $\chi_{\text{reg}} \approx 100$. □

Example 3.1.8 Ridge Regression: Contour Lines
This example illustrates the influence of ridge regression on the shape of the loss function. The following loss function is considered:

$$I(\underline{\theta}) = (\theta_1 - 1)^2 + 2\,(\theta_2 - 0.5)^2 + \theta_1\theta_2\,. \tag{3.51}$$

The contour lines of this loss function are depicted in Fig. 3.10a. The minimum is at $\underline{\theta} \approx [0.86\ 0.29]^T$. Ridge regression with the regularization parameter λ leads to the following loss function:

$$I(\underline{\theta}, \lambda) = (\theta_1 - 1)^2 + 2\,(\theta_2 - 0.5)^2 + \theta_1\theta_2 + \lambda\,(\theta_1^2 + \theta_2^2)\,. \tag{3.52}$$

Figure 3.10b–d shows the effect that λ has on the shape of the contour lines. As λ increases, the minimum of the loss function tends to [0 0], and the lines of constant loss function values change from ellipses to circles, that is, the Hessian becomes better conditioned.

□

Since for problems with many regressors, the Hessian is usually poorly conditioned, ridge regression is a popular method for the reduction of the estimation variance. It is a very simple and easy to use alternative to the subset selection methods described in Sect. 3.4. The price to be paid for this approach is an increasing estimation bias with increasing λ (similar to subset selection with a decreasing number of regressors) and the necessity for an iterative approach in order to find good values for λ. Various studies discuss the determination of an optimal λ [125]. Note, however, that it will be computationally prohibitive to handle very large problems (i.e., with very many regressors) with ridge regression, since the matrix inversion has cubic complexity. In these cases, subset selection techniques have to be applied.

Example 3.1.9 Ridge Regression for Polynomial Modeling
In order to illustrate the benefits gained by ridge regression, the approximation of the following function by a polynomial from a small, noisy data set is considered:

$$y = 1 + u^3 + u^4\,. \tag{3.53}$$

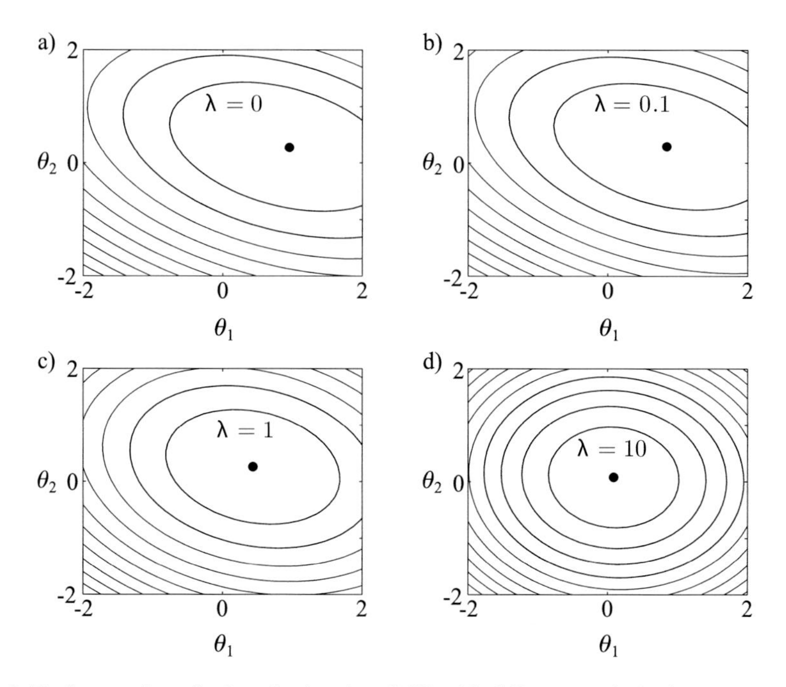

Fig. 3.10 Contour lines for loss the function (3.52) with different regularization parameters λ. Higher regularization parameter values lead to more "circle-like" contour plots with a minimum closer to the origin. (**a**) No. (**b**) Low. (**c**) Medium. (**d**) High regularization

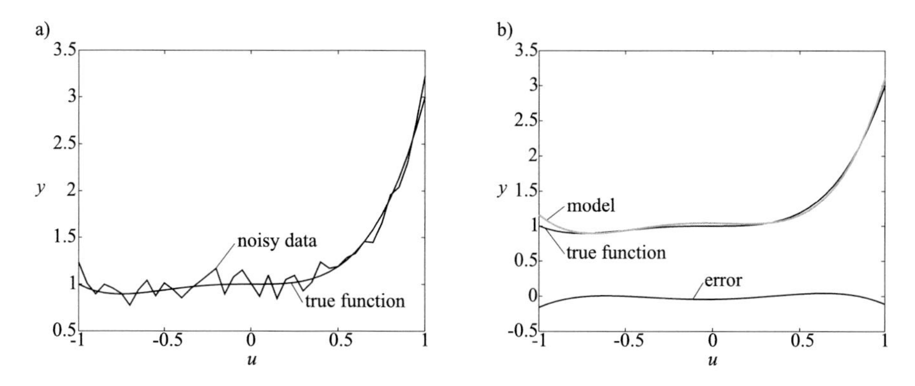

Fig. 3.11 (**a**) True function (3.53) and noisy data used for training. (**b**) Approximation of (3.53) by a fourth-order polynomial estimated by standard least squares

Figure 3.11a shows the true function and the equally distributed 40 training data samples corrupted by white Gaussian noise with variance 0.1. Under the assumption that no knowledge about the special structure of (3.53) is available, it is reasonable to start with a first-order polynomial and then to increase the polynomial order step by step. Finally, a fourth-order polynomial leads to good results; see Fig. 3.11b. The

Table 3.1 Comparison of standard least squares and ridge regression

	True value	Least squares	Ridge regression with $\lambda = 0.1$
c_0	1.0000	1.0416	0.9998
c_1	0.0000	−0.0340	0.0182
c_2	0.0000	−0.3443	0.0122
c_3	1.0000	1.0105	0.9275
c_4	1.0000	1.4353	1.0475
Error sum of squares	—	0.0854	0.0380

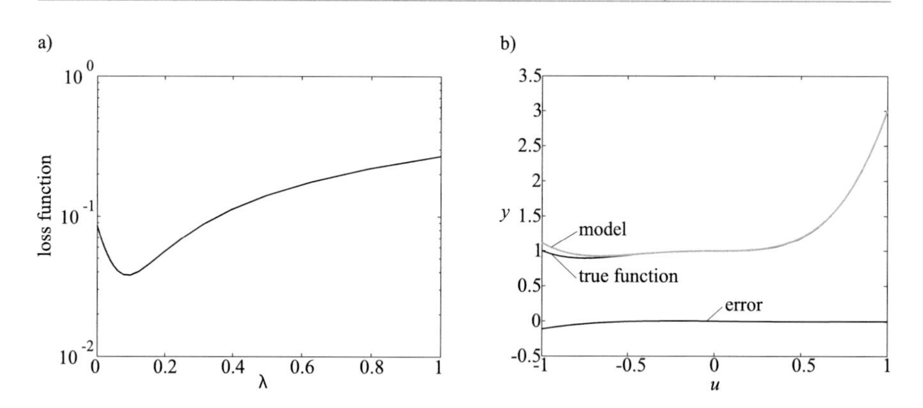

Fig. 3.12 (**a**) Loss function obtained by ridge regression in dependency of the regularization parameter λ. The optimal value 0.1 represents the best bias/variance tradeoff. (**b**) Approximation of (3.53) by a fourth-order polynomial estimated by ridge regression with the optimal regularization parameter $\lambda = 0.1$

estimated coefficients are summarized in Table 3.1. Since the data set is small and quite noisy, the estimated coefficients deviate considerably from their true values. It might be possible to detect that the original function does not depend on u^1; however, the estimation of the quadratic term $\hat{c}_2 = -0.3443$ certainly does not reveal the independence from u^2.

Applying ridge regression to this approximation problem leads to better or worse solutions depending on the regularization parameter λ. Figure 3.12a shows the obtained error sum of squares for different values of λ. Starting from the least squares solution with $\lambda = 0$ first, for increasing λ, the variance decrease overcompensates the bias increase. The minimum is realized for $\lambda \approx 0.1$ and represents the best bias/variance tradeoff; see Sect. 7.2. A further increase of the regularization parameter degrades the approximation quality again. Table 3.1 summarizes the estimated coefficients for the optimal value of λ. They are much closer to their true values than in the least squares case, and consequently, the obtained loss function is much smaller; see also Fig. 3.12b. Moreover, the structure of the original function can be guessed because $\hat{c}_1 \approx \hat{c}_2 \approx 0$. Therefore, the model complexity (i.e., number of regressors) could be reduced by a second standard least squares estimation procedure that takes only the three significant regressors $[1\ u^3\ u^4]$ into account. Such an approach is an alternative method for regressor selection. A comparison with a subset selection technique is presented in Sect. 3.4.

For a more thorough discussion about ridge regression for polynomial, refer to Sect. 10.2.1. □

Regularization with ridge regression is a special case of the so-called *Tikhonov regularization* where instead of the identity matrix $\underline{I}$, a general matrix $\underline{L}$ is used in (3.44) and (3.46) [564]. As discussed above, the choice $\underline{L} = \underline{I}$ drives the non-significant parameter toward zero. Often $\underline{L}$ is chosen as a discrete approximation of the first or the second derivative of the model. Then the Tikhonov regularization drives the parameters toward values that represent a constant (zero gradient/first derivative) or linear model (zero curvature/second derivative). By this strategy, it is thus possible to incorporate prior knowledge on the model into the optimization procedure [299].

3.1.4.3 Prior Parameters for Ridge Regression

The standard ridge regression problem in (3.44) can be extended in order to pull the parameters to values $\underline{\theta}_0$ known a priori instead toward $\underline{0}$. Then the penalty term penalizes the deviation of the estimated parameters from $\underline{\theta}_0$:

$$I(\underline{\theta}, \lambda) = \frac{1}{2}\left(\underline{e}^T\underline{e} + \lambda(\underline{\theta} - \underline{\theta}_0)^T(\underline{\theta} - \underline{\theta}_0)\right) \longrightarrow \min_{\underline{\theta}} . \tag{3.54}$$

This leads to the following solution:

$$\hat{\underline{\theta}} = \left(\underline{X}^T\underline{X} + \lambda\underline{I}\right)^{-1}\left(\underline{X}^T\underline{y} + \lambda\,\underline{\theta}_0\right) . \tag{3.55}$$

Obviously, as $\lambda \to \infty$, $\hat{\underline{\theta}} \to \underline{\theta}_0$. This is a simple and easy way to introduce prior knowledge establishing a kind of gray box model.

3.1.5 Ridge Regression: Alternative Formulation

The typical scenario for least squares and ridge regression is many data points N and only a few parameters n, i.e., $N \geq n$. But what happens in the case $N < n$? The system of linear equations is underdetermined. Least squares in its original form fails because $\underline{X}^T\underline{X}$ then is singular. All data points can be exactly met and additional degrees of freedom remain; infinitely many different parameter vectors can yield a zero error. Thus, the LS problem is usually recast in finding the "smallest" parameter vector (in the sense $\underline{\theta}^T \underline{\theta} \rightarrow \min$) that solves the equations exactly. This results in the following LS solution for $N < n$:

$$\hat{\underline{\theta}} = \underline{X}^T \left(\underline{X}\,\underline{X}^T\right)^{-1} \underline{y}\,. \tag{3.56}$$

Note that $\underline{\theta}^T \underline{\theta}$ is exactly the penalty in ridge regression in (3.44). Thus, ridge regression in (3.44) balances these two objectives: $\underline{e}^T \underline{e}$ and $\underline{\theta}^T \underline{\theta}$ with the tradeoff tuning factor λ, while LS optimizes either $\underline{e}^T \underline{e}$ for $N \geq n$ or $\underline{\theta}^T \underline{\theta}$ for $N < n$.

Comparing the original LS solution in (3.22) and the alternative form in (3.56), one notices that always the smaller matrix of dimension $\min(n, N) \times \min(n, N)$ is inverted, i.e., $n \times n$ in (3.22) and $N \times N$ in (3.56), respectively.

For ridge regression, such an alternative formulation exists as well:

$$\hat{\underline{\theta}} = \underline{X}^T \left(\underline{X}\,\underline{X}^T + \lambda \underline{I}\right)^{-1} \underline{y}\,. \tag{3.57}$$

Note that the size of $\underline{I}$ here is $N \times N$ while in (3.46) it is $n \times n$.

It is just a matter of computational demand and practicability which formulation to choose. The added identity matrix (times λ) makes the matrices in both cases invertible due to the regularization. For more details on this topic, refer to [348]. The alternative formulation of ridge regression in (3.57) often comes up in the context of kernel methods; see Sect. 16.3 and [501]. Also compare with the next box that summarizes the primal and dual form of ridge regression.

In the original (primal) ridge regression form, the objective is $\underline{e}^T \underline{e}$, and the penalty $\lambda \underline{\theta}^T \underline{\theta}$ can also be written as a constraint $\underline{\theta}^T \underline{\theta} \leq t$ where t has a 1:1 relationship with λ. The estimation yields an n-dimensional *parameter* vector.

In the alternative (dual) ridge regression form, objective and constraint swap places. Instead of an N-dimensional vector in the objective and an n-dimensional vector in the constraint, now $\frac{1}{\lambda} \underline{\theta}^T \underline{\theta}$ is the objective, and $\underline{e}^T \underline{e} \leq t$ is the constraint. The estimation yields an N-dimensional *weight* vector.

Ridge Regression
N data points
n parameters
Loss function:

$$I(\underline{\theta}, \lambda) = \underline{e}^T \cdot \underbrace{\underline{e}}_{N \times 1} + \lambda \underline{\theta}^T \cdot \underbrace{\underline{\theta}}_{n \times 1} \longrightarrow \min_{\underline{\theta}}. \tag{3.58}$$

Primal Form
Optimal parameters:

$$\underbrace{\underline{\hat{\theta}}}_{n \times 1} = \underbrace{\left(\underline{X}^T \underline{X} + \lambda \underline{I}\right)^{-1}}_{n \times n} \underbrace{\underline{X}^T \underline{y}}_{n \times 1} \tag{3.59}$$

Model output:

$$\underline{\hat{y}} = \underbrace{\underline{X}}_{N \times n} \cdot \underbrace{\underline{\hat{\theta}}}_{n \times 1} = \underbrace{\underline{X} \left(\underline{X}^T \underline{X} + \lambda \underline{I}\right)^{-1} \underline{X}^T}_{\underline{S}} \underline{y} \tag{3.60}$$

Dual Form
Optimal dual "parameters":

$$\underbrace{\underline{\hat{w}}}_{N \times 1} = \underbrace{\left(\underline{X}\,\underline{X}^T + \lambda \underline{I}\right)^{-1}}_{N \times N} \underbrace{\underline{y}}_{N \times 1} \tag{3.61}$$

Model output:

$$\underline{\hat{y}} = \underbrace{\underline{X}\,\underline{X}^T}_{N \times N} \cdot \underbrace{\underline{\hat{w}}}_{N \times 1} = \underbrace{\underline{X}\,\underline{X}^T \left(\underline{X}\,\underline{X}^T + \lambda \underline{I}\right)^{-1}}_{\underline{S}} \underline{y} \tag{3.62}$$

Kernel Methods

$$\underline{X}\,\underline{X}^T \rightarrow \underline{K} \tag{3.63}$$

3.1.6 L_1 Regularization

In recent years, regularizing a quadratic objective by the L1 norm of the parameter vector instead of the L2 norm as in ridge regression has proven very powerful. It is known under the name lasso (least absolute shrinkage and selection operator) [560]. The penalty is proportional to the sum of the *absolute* parameter *values*:

$$I(\underline{\theta}, \lambda) = \frac{1}{2}\left(\underline{e}^T\underline{e} + \lambda\,||\underline{\theta}||_1\right) = \frac{1}{2}\left(\underline{e}^T\underline{e} + \lambda\sum_{i=1}^{n}|\theta_i|\right) \longrightarrow \min_{\underline{\theta}}. \tag{3.64}$$

Equivalently, the non-penalized objective can be constrained as follows:

$$I(\underline{\theta}, \lambda) = \frac{1}{2}\underline{e}^T\underline{e} \longrightarrow \min_{\underline{\theta}}$$

$$\text{subject to} \sum_{i=1}^{n}|\theta_i| \le t. \tag{3.65}$$

Solving this optimization problem by the method of Lagrange multipliers exactly yields the above unconstrained problem. For every constraint limit t, a corresponding regularization parameter λ exists.

Figure 3.13 compares the non-regularized least squares solution to the ridge and lasso estimate. One very appealing property of the lasso can already be observed in this figure: Lasso drives many parameters toward zero. Thus, it produces *sparse* parameter vectors, i.e., many parameters values are exactly zero. This is not the case for ridge regression because the square lightens the penalty; the closer one parameter value approaches zero. Therefore in the ridge regression case, there exists no mechanism to drive it *exactly* toward zeros. If the penalty is linear, as in the lasso case, the pressure on parameters toward zero remains intact.

Due to this *sparsity* property, many significant advances were made in many application areas, most notably in the field known as compressed (or compressive)

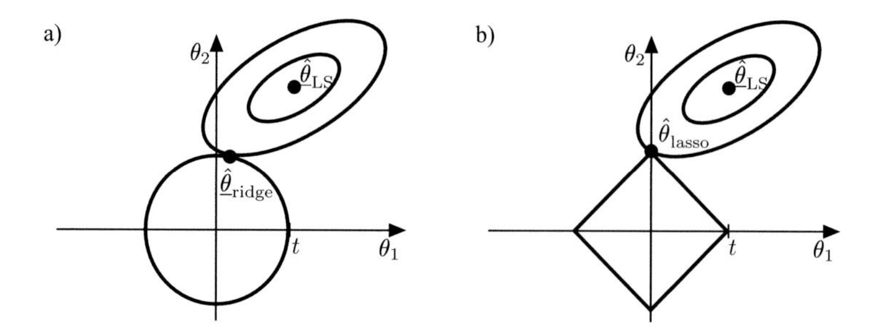

Fig. 3.13 (**a**) L2 and (**b**) L1 regularization for a quadratic objective

sensing [123]. If the amount of information is not sufficient to estimate all parameters, the assumption that many parameters are zero seems to be a very successful idea.

The computation of the lasso is more demanding than simple least squares because with the L1 regularization no analytic solution for the parameter vector exists. Luckily, it turns out that with LARS (least angle regression), a very powerful and extremely efficient algorithm is available [133]. LARS computes not only a solution of (3.64) for one value λ; rather it evaluates the whole *regularization path* giving the optimal parameter values for all possible values of λ. Because more and more parameters become zero as λ increases, the corresponding regressors drop out. Therefore, the lasso approach, together with the LARS algorithm is an extremely attractive (and often preferable) alternative to classical subset selection algorithms as described in Sect. 3.4.

3.1.7 Noise Assumptions

Up to now the role of the noise n in Fig. 2.3 has not been thoroughly discussed. It is obvious that the optimal estimate $\underline{\hat{\theta}}$ is dependent on the noise properties. It can be shown that the least squares estimate in (3.22) is the best linear unbiased estimate (BLUE) of the true parameters $\underline{\theta}_{\text{opt}}$ if the noise is white. This means that the covariance matrix of the estimated parameters (3.34) is the smallest of all possible linear unbiased estimators. If the white noise is also Gaussian, then there does not even exist a better nonlinear unbiased estimator [540]. Note, however, that generally there may exist better nonlinear or biased estimators.

If the noise is not white, it is intuitively clear that the noise properties have to be considered in the estimation. Indeed, the LS estimator in (3.22) is only the BLUE if the noise signal contains no exploitable information, i.e., n is white noise, or the process output can be measured undisturbed ($n = 0$). Otherwise the BLUE is called the *Markov* estimate

$$\underline{\hat{\theta}} = \left(\underline{X}^T \underline{\Omega}^{-1} \underline{X}\right)^{-1} \underline{X}^T \underline{\Omega}^{-1} \underline{y} \tag{3.66}$$

with the covariance matrix of the noise $\underline{\Omega} = \text{cov}\{\underline{n}\}$. For white noise with variance σ^2, the covariance matrix of the noise signal is $\sigma^2 \underline{I}$ and therefore (3.66) equals the linear least squares approach (3.22). By (3.66) information about the noise signal is exploited for the parameter estimation. Intuitively (3.66) can be interpreted as follows. Data corrupted with large noise levels, i.e., high values in the covariance matrix, is regarded as less reliable than data corrupted with small noise levels. Therefore, the more disturbed the data is, the less it contributes to the estimation due to the inversion of $\underline{\Omega}$ in (3.66). This concept becomes clearer under the assumption of a diagonal structure of the noise covariance matrix:

$$\underline{\Omega} = \begin{bmatrix} \Omega_{11} & 0 & \cdots & 0 \\ 0 & \Omega_{22} & \cdots & 0 \\ \vdots & \vdots & \ddots & \vdots \\ 0 & 0 & \cdots & \Omega_{NN} \end{bmatrix}. \tag{3.67}$$

Then each measurement i ($i = 1, \ldots, N$) is weighted with $1/\Omega_{ii}$.

Although (3.66) can improve the estimation quality considerably if the noise is highly correlated, the major problem in practice is to determine the noise covariance matrix $\underline{\Omega}$. Very often no a priori knowledge about noise properties is available. Usually either simply white noise is assumed, which leads to (3.22), or the noise is modeled by a linear dynamic filter with unknown filter coefficients driven by a white noise signal; see Fig. 3.14. The filter parameters may be estimated by the LS residuals, and subsequently the noise covariance matrix can be computed. The extra effort with this approach, however, is usually justified only for highly disturbed measurements.

Example 3.1.10 Noise Assumptions
This short example will illustrate which benefits can be obtained by taking knowledge about noise properties into account. The task is to estimate the coefficients of a second-order polynomial $y = c_0 + c_1 u + c_2 u^2$ based on 21 disturbed, equally distributed measurements between $u = 0$ and $u = 10$. The disturbance is a white noise signal with a standard deviation proportional to u; see Fig. 3.15.

Table 3.2 compares the estimated coefficients obtained by three estimators. The first one is a standard least squares estimator that ignores the noise distribution. Consequently, all data samples are weighted equally. The second estimator follows (3.66), where $\underline{\Omega}$ is a diagonal matrix that reflects the correct standard

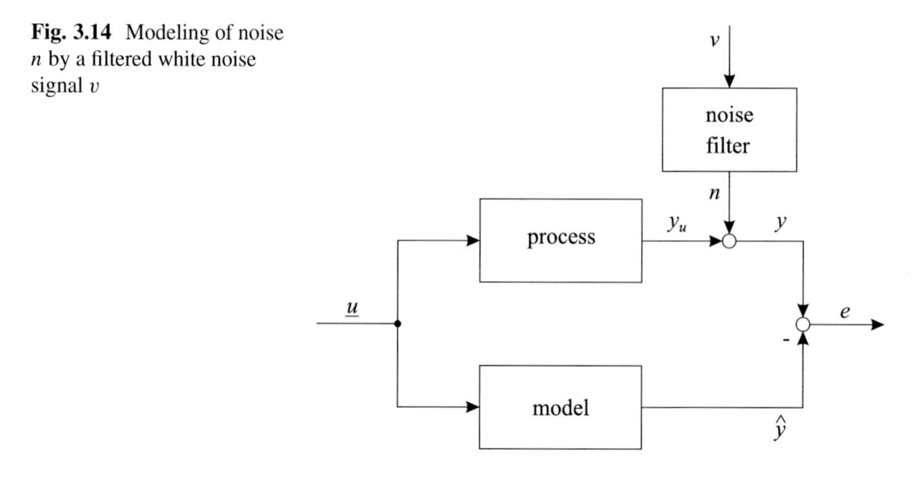

Fig. 3.14 Modeling of noise n by a filtered white noise signal v

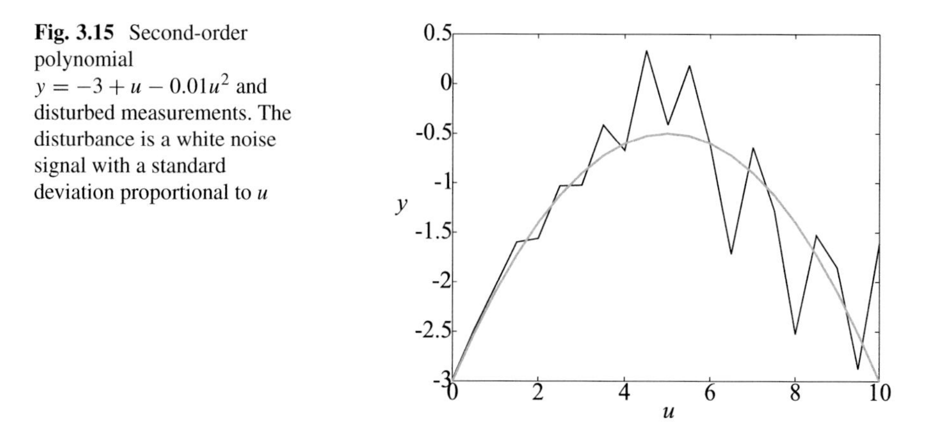

Fig. 3.15 Second-order polynomial $y = -3 + u - 0.01u^2$ and disturbed measurements. The disturbance is a white noise signal with a standard deviation proportional to u

Table 3.2 Comparison of least squares estimates with correct, wrong, and no noise assumptions

Coefficients	True values	Without noise assumptions	With correct noise assumptions	With wrong noise assumptions
c_0	−3.0000	−2.8331	−2.9804	−2.0944
c_1	1.0000	0.9208	1.0138	0.6283
c_2	−0.0100	−0.0913	−0.0106	−0.0674

deviations of the disturbance. Therefore, the data samples with small values of u are weighted higher than those with large values of u. As Table 3.2 reveals, this leads to a significant improvement in parameter estimation quality. The last column in the table demonstrates what happens in the worst case, that is, the noise assumptions are totally wrong (the assumed matrix $\underline{\Omega}$ is the inverse of the true matrix). Obviously, the parameter estimate is very poor because the estimation is based primarily on the highly distributed measurements.

This example clearly has a very artificial nature. In practice, good estimates of $\underline{\Omega}$ will be hard to obtain. However, any assumption on the noise properties that is closer to reality than the assumption that all measurements are equally disturbed will improve the estimation quality. □

3.1.8 *Weighted Least Squares (WLS)*

As mentioned in Sect. 2.3, a weighted least squares (WLS) criterion can be applied. The most general sum of weighted squared errors loss function is

$$I(\underline{\theta}) = \frac{1}{2} \underline{e}^T \underline{Q}\, \underline{e} \qquad (3.68)$$

with the weighting matrix $\underline{Q}$. In most cases the weighting matrix has a diagonal structure

$$\underline{Q} = \begin{bmatrix} Q_{11} & 0 & \cdots & 0 \\ 0 & Q_{22} & \cdots & 0 \\ \vdots & \vdots & \ddots & \vdots \\ 0 & 0 & \cdots & Q_{NN} \end{bmatrix}, \tag{3.69}$$

and (3.68) becomes equal to (2.2). Then each single squared error value $e(i)$ is weighted by the corresponding matrix entry Q_{ii} in $\underline{Q}$. The general solution to the weighted least squares optimization problem is

$$\underline{\hat{\theta}} = \left(\underline{X}^T \underline{Q}\, \underline{X}\right)^{-1} \underline{X}^T \underline{Q}\, \underline{y}\,. \tag{3.70}$$

For $\underline{Q} = q\underline{I}$, (3.70) recovers the standard least squares solution (3.22). The weighted least squares approach is used in all situations where the data samples and therefore the corresponding errors have different relevance or importance for the estimation. Assume, for example, a function approximation problem with small tolerances in one region and lower precision requirements in some other region. This task can be solved with WLS by choosing high weighting factors for all data samples in the small tolerance region and small weights in the other region.

Since many program packages provide sophisticated algorithms for numerical calculation of the pseudoinverse $\left(\underline{X}^T \underline{X}\right)^{-1} \underline{X}^T$, it is useful to note how (3.70) can be expressed in this form. If $\underline{Q}$ is a diagonal matrix, its square root can be computed by taking the square root of all entries, and the matrix can be written as $\underline{Q} = \sqrt{\underline{Q}^T}\sqrt{\underline{Q}}$, since $\underline{Q}^T = \underline{Q}$. Then with the transformation $\underline{\tilde{X}} = \sqrt{\underline{Q}}\underline{X}$ and therefore $\underline{\tilde{X}}^T = \underline{X}^T\sqrt{\underline{Q}^T}$ and with $\underline{\tilde{y}} = \sqrt{\underline{Q}}\,\underline{y}$, (3.70) can be written as $\underline{\hat{\theta}} = \left(\underline{\tilde{X}}^T \underline{\tilde{X}}\right)^{-1} \underline{\tilde{X}}^T \underline{\tilde{y}}$.

The covariance matrix of the estimated parameters cannot be obtained directly by replacing $\underline{X}$ in (3.34) with $\underline{\tilde{X}}$ because the weighting in $\underline{\tilde{y}}$ has to be considered in the derivation (3.31)–(3.33); see Sect. 3.1.1. Rather the covariance matrix of parameters estimated with WLS becomes

$$\text{cov}\{\underline{\hat{\theta}}\} = \sigma_n^2 \left(\underline{X}^T \underline{Q}\underline{X}\right)^{-1}. \tag{3.71}$$

Comparing (3.66) with (3.70) leads to the conclusion that a weighted least squares approach with white noise assumption is identical to a standard (non-weighted) least squares approach with colored noise assumptions if $\underline{Q} = \underline{\Omega}^{-1}$, that is, the weighting matrix is equal to the inverse of the noise covariance matrix. This relationship becomes intuitively clear by realizing that it must be reasonable to give highly disturbed data (large values in $\underline{\Omega}$) small weights and vice versa. Thus, the relevance of data (how important is it?) and the reliability of data (how noisy is it?) are treated in exactly the same manner by the LS estimation.

3.1.9 Robust Regression

As discussed in Sect. 2.3.1, the LS approach is optimal in the maximum likelihood sense if the output noise is following a Gaussian distribution. Large errors ($|e(i)| > 1$) are progressively penalized, but this is balanced by the fact that extreme values are very unlikely in case of a Gaussian distribution. In contrast, the double exponential distribution allows for more outliers (*fat tail distribution*) and yields the sum of *abolute*, not squared, errors as optimal loss function, compare Sect. 2.3.1. Figure 3.16a compares both approaches. The term *robust regression* usually refers to all approaches that are robust with respect to outliers by not penalizing them disproportionately high. Although minimizing the sum of absolute errors is significantly more demanding in terms of computation than LS, this is a *convex optimization* problem with a unique optimum that can be solved relatively effectively. Also popular for robust regression is to use a quadratic loss function for small errors but a linear one for large errors. This is realized by the *Huber* loss function shown in Fig. 3.16b. Sophisticated algorithms exist for properly determining the transition point where quadratic behavior crosses over to linear. Huber loss can be minimized very efficiently by iterated re-weighted LS (compare Sect. 3.1.8) with just a couple of times the computational demand of ordinary LS.

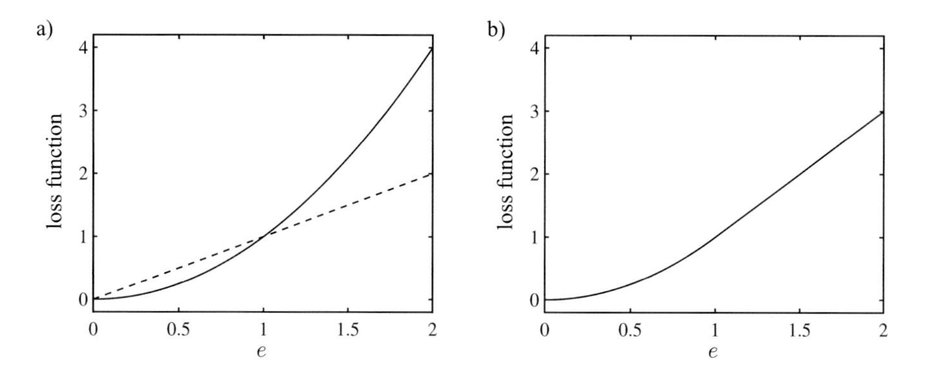

Fig. 3.16 Different loss functions: (**a**) quadratic (solid) and absolute value (dashed), (**b**) Huber (combination of both)

3.1.10 Least Squares with Equality Constraints

Sometimes, some of the parameters of a linear optimization problem are known to be dependent on each other. Then the parameters cannot be estimated directly by LS because the optimal parameters obtained by (3.22) do not necessarily meet these constraints. If inequality constraints are considered, the optimization problem becomes more difficult; see Sect. 3.3. However, for equality constraints, it can be solved easily. Equality constraints are, for example, $\theta_1 = \theta_2$ or $\theta_1 = \theta_2 + \theta_3$ or $\theta_1 - \theta_2 = 5$ or combinations of such equations. Each of these equations reduces the degrees of freedom of the model. Thus, the number of free parameters is equal to the number of nominal parameters minus the number of constraints. Thus, in order to be able to solve such a linearly constrained linear optimization problem, the number of constraints must be smaller than the number of nominal parameters n. If the number of constraints is equal to the number of nominal parameters, all parameters are fully determined by the constraints, and actually no optimization problem remains to be solved.

A linear optimization problem with linear equality constraints can be formulated as the following linear optimization problem [540]:

$$\underline{y} = \underline{X}\,\underline{\theta} + \underline{e} \qquad \text{with} \quad \frac{1}{2}\underline{e}^T\underline{e} \longrightarrow \min_{\underline{\theta}} \tag{3.72}$$

with the linear equality constraints

$$\underline{A}\,\underline{\theta} = \underline{b}\,. \tag{3.73}$$

Each equation in the linear equation system (3.73) represents one constraint. Thus, the number of rows of $\underline{A}$ and the dimension of $\underline{b}$ must be smaller than n. There exist two alternative solution strategies for this kind of problem.

The indirect way is to substitute the linear dependent parameters in (3.72) by the equations (3.73). Then an unconstrained linear optimization problem with $n - \text{rank}\{A\}$ parameters arises, which can be solved directly by LS.

The identical result can be obtained with the direct solution by performing the constrained optimization with Lagrange multipliers. This leads to the following parameter estimate [540]:

$$\hat{\underline{\theta}}_{\text{constr}} = \hat{\underline{\theta}}_{\text{unconstr}} - \underline{H}^{-1}\underline{A}^T\left(\underline{A}\,\underline{H}^{-1}\underline{A}^T\right)^{-1}\left(\underline{A}\,\hat{\underline{\theta}}_{\text{unconstr}} - \underline{b}\right), \tag{3.74}$$

where $\hat{\underline{\theta}}_{\text{unconstr}}$ is the unconstrained LS estimate in (3.22) and $\underline{H}^{-1} = \left(\underline{X}^T\underline{X}\right)^{-1}$ is the inverse Hessian. For a linearly constrained weighted least squares, $\hat{\underline{\theta}}_{\text{unconstr}}$ is the weighted LS estimate in (3.70) and correspondingly $\underline{H}^{-1} = \left(\underline{X}^T\underline{Q}\,\underline{X}\right)^{-1}$.

3.1.11 Smoothing Kernels

So far the model has been described as a linear combination of the regressors. The regressors are the columns in the regression matrix $\underline{X}$; they are weighted with the parameters in $\underline{\theta}$ and finally summed up to compute the model output. In matrix/vector formulation, the model output is thus

$$\underline{\hat{y}} = \underline{X}\,\underline{\hat{\theta}}\,. \tag{3.75}$$

It is interesting to eliminate the parameter vector in (3.75) with the optimal LS solution $\underline{\hat{\theta}} = \left(\underline{X}^T \underline{X}\right)^{-1} \underline{X}^T \underline{y}$ in order to obtain the direct relationship between the measured process outputs and the model outputs where $\underline{S}$ is called the smoothing matrix:

$$\underline{\hat{y}} = \underline{X}\left(\underline{X}^T \underline{X}\right)^{-1} \underline{X}^T \underline{y} = \underline{S}\,\underline{y}\,. \tag{3.76}$$

For some advanced aspects, it is convenient to noticed that the model error can be expressed in terms of the smoothing matrix as

$$\underline{e} = \underline{y} - \underline{\hat{y}} = \underline{y} - \underline{S}\,\underline{y} = \left(\underline{I} - \underline{S}\right)\underline{y}. \tag{3.77}$$

While in (3.75) the measured process outputs enter only indirectly via the optimal parameter vector $\underline{\hat{\theta}}$, their influence is more obvious in (3.76). The relationship between $\hat{y}$ and y is linear since both the model and the estimator are linear. The $N \times N$ matrix $\underline{S}$ determines the contribution of each measured output value to each model output value. To clarify this relationship, it is helpful to analyze a single row of (3.76):

$$\hat{y}(j) = s_{j1}y(1) + s_{j2}y(2) + \ldots + s_{jN}y(N) = \underline{s}_j^T \underline{y}\,, \tag{3.78}$$

where s_{ji} denote the entries of $\underline{S}$, and the vector $\underline{s}_j^T$ is the row in $\underline{S}$ representing the jth model output. Obviously, the model output can be interpreted as a filtered or smoothed version of the process output measurements. For this reason the $\underline{s}_j^T$ (rows of $\underline{S}$) are called the *smoothing kernels*. The matrix $\underline{S}$ is often called the smoothing matrix or hat matrix. The model output is a smoothed version of the measured output because usually, the number of parameters is smaller than the number of measurements ($n < N$), and thus the degrees of freedom are reduced.

The smoothing kernels are of considerable interest because they allow one to analyze the effect of each measurement on the model. An often desired property of the smoothness kernels is that they are *local*, i.e., the influence of two measurements decreases with increasing distance of the two corresponding inputs. The following example illustrates the insights that can be gained by an analysis of the smoothing kernels.

Example 3.1.11 Global and Local Smoothing Kernels
Two different one-dimensional models $\hat{y} = f(u)$ will be considered for 31 input values equally distributed in [−5, 5]. The task is to approximate the simple linear function

$$y = u/10 + 0.5\,, \tag{3.79}$$

which transforms the input in [−5, 5] to output values between 0 and 1; see Fig. 3.17.

Example (a) is a simple linear model,

$$\hat{y} = \theta_1 + \theta_2 u\,, \tag{3.80}$$

with the regressors 1 and u shown in Fig. 3.18a. Example (b) is a Gaussian radial basis function (RBF) network with ten neurons placed equidistantly in the input space (see Sect. 11.3)

$$\hat{y} = \sum_{i=1}^{10} \theta_i \exp\left(-(u - i + 5.5)^2\right) \tag{3.81}$$

with the regressors shown in Fig. 3.18b.

The most important difference that can be observed from Fig. 3.19 is that the smoothing kernels of the linear model possess global character, while the RBF network's smoothing kernels are local in the sense that the influence of measure-

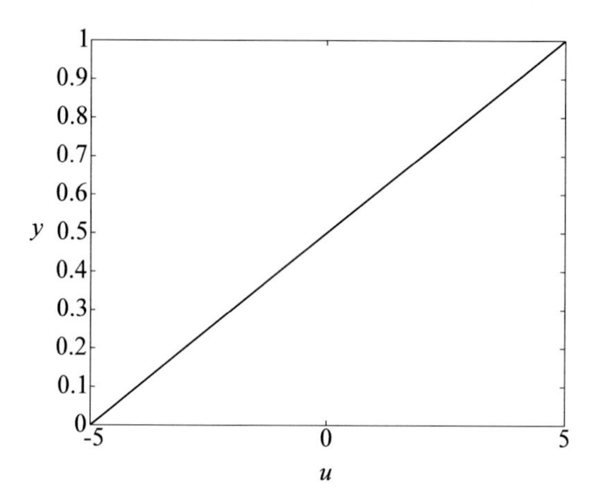

Fig. 3.17 The linear function to be approximated by the models (**a**) and (**b**)

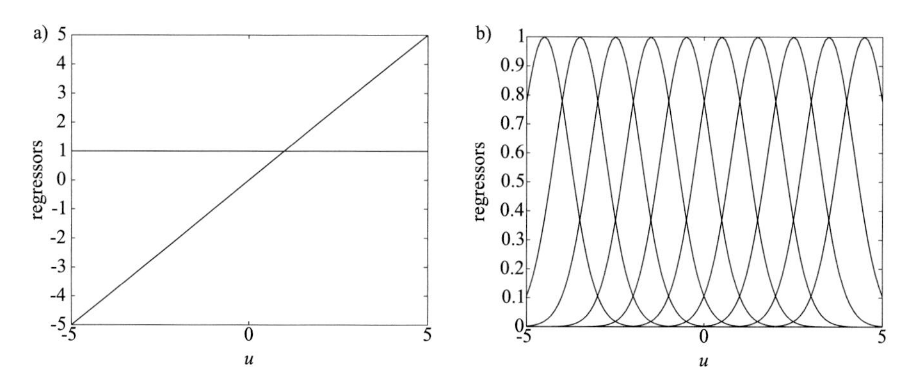

Fig. 3.18 The regressors for (**a**) the linear model and (**b**) the RBF network model

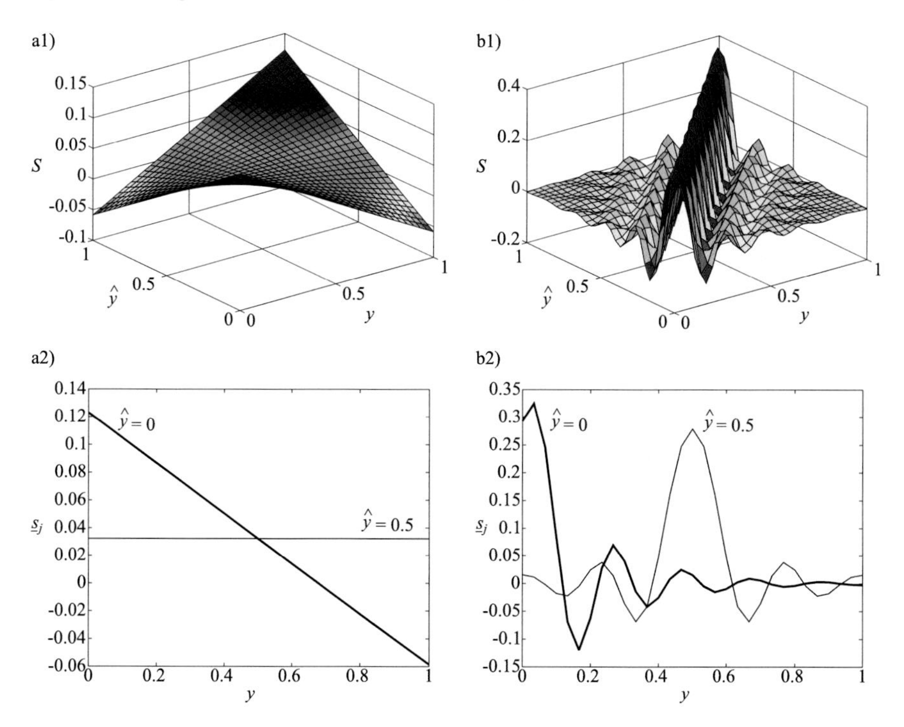

Fig. 3.19 The smoothing kernels for (**a**) the linear model and (**b**) the RBF network model. The upper plots (**a1** and **b1**) show the entries of the smoothing matrix $\underline{S}$. The lower plots (**a2** and **b2**) show two smoothing kernels (rows of $\underline{S}$) corresponding to different model outputs

ments decreases with increasing distance between y and $\hat{y}$. The interpretation of the smoothing kernels is as follows.

The thick line in Fig. 3.19a2 represents the influence of measurements of y between 0 and 1 (corresponding to $u = -5 \ldots 5$) on the model output $\hat{y} = 0$ (corresponding to $u = -5$). If the measurement is made in the neighborhood of

$u = -5$, it affects the model positively, i.e., a larger measurement value for y increases the model output $\hat{y}$, too. However, right of $y \approx 0.65$, the effect becomes negative, i.e., a larger measurement value for y decreases the model output $\hat{y}$. This is because a larger measurement far to the right tends to increase the slope of the line, and such an increased slope decreases the model output far to the left. Because the smoothing kernels are global, all measurements influence the model output everywhere. For online adaptive systems, such a property is usually not desirable, and local smoothing kernels as for the RBF network are better suited. □

3.1.11.1 Ridge Regression

In the case of ridge regression instead of least squares, the direct effect from $\underline{y}(i)$ to $\hat{\underline{y}}(i)$ becomes weaker due to the regularization which results in lower overfitting. The smoothing matrix then becomes

$$\hat{\underline{y}} = \underline{X}\left(\underline{X}^T\underline{X} + \lambda\underline{I}\right)^{-1}\underline{X}^T\underline{y} = \underline{S}\,\underline{y}\,. \tag{3.82}$$

3.1.12 *Effective Number of Parameters*

It can be shown that the trace (sum of diagonal entries) of the smoothing matrix gives the effective degrees of freedom or *effective number of parameters*; see also Sect. 7.5.1:

$$n_{\text{eff}} = \text{tr}\{\underline{S}\}\,. \tag{3.83}$$

This is the standard formula used in statistics, compare Section 7.6 in [230] for theoretical arguments. Sometimes the slightly different expression $n_{\text{eff}} = \text{tr}\{\underline{S}\,\underline{S}^T\}$ is used in the literature; see, e.g. [402].

In case of least squares, (3.83) always is equal to the nominal number of parameters of the model, i.e., $n_{\text{eff}} = n$. In case of ridge regression, it is smaller due to the regularization effect, i.e., $n_{\text{eff}} < n$. In the extreme case $\lambda \to \infty$, $n_{\text{eff}} \to 0$.

The effective number of parameters is important to assess the flexibility of the model and shall be used in the information criteria discussed in Sect. 7.3.3 that trade off model training error with model flexibility/complexity. The diagonal entries of $\underline{S}$ or their average value is also used for calculating the leave-one-out (LOO) error or the generalized cross-validation (GCV) criterion, respectively, compare Sect. 7.3.2.

3.1.13 L_2 Boosting

Boosting was first introduced by Schapire in the context of classification [505] where *weak* classifiers ("weak" means simple and not very discriminatory) are iteratively combined while increasing the weight of wrongly classified data points. Later these ideas were extended to regression problems by Breiman [79] and Friedman [171] in the framework of *stochastic gradient boosting*. The simplest form originates from sum of squared errors loss functions and is called *L_2 boosting* because it minimizes the L_2 norm of the error.

L_2 boosting iteratively trains a model $\hat{y}^{(1)}$, evaluates the model error $e^{(1)} = y - \hat{y}^{(1)}$, trains a model $\hat{y}^{(2)}$ on this error as desired output values, evaluates the model error $e^{(2)} = y - \left(\hat{y}^{(1)} + \hat{y}^{(2)}\right)$, and so on. Thus, the final model corresponds to the sum of the original model and all error models $\hat{y} = \hat{y}^{(1)} + \hat{y}^{(2)} + \ldots$.

It is shown in [86] that after m iterations of L_2 boosting, the overall smoothing matrix of the boosted model becomes

$$\underline{S}^{(m)} = \underline{I} - (\underline{I} - \underline{S})^m \tag{3.84}$$

where $\underline{S}$ is the original model's smoothing matrix. This means that if the model and its training method has an appropriate "learning capacity" (is sufficiently flexible, i.e., $||\underline{I} - \underline{S}|| < 1$ for some matrix norm) $\underline{S}^{(m)} \to \underline{I}$ as $m \to \infty$ which corresponds to the perfect overfitting case $\hat{y}(i) = y(i)$, $i = 1, \ldots, N$. Of course, this is highly undesirable and boosting needs to be stopped before. This overfitting does not happen in the next example because the utilized model is not a non-parametric smoother as in [86] but a local model network with just ten local models.

3.1.13.1 Shrinkage

It has been proven to be very successful in shrinking the model update by multiplying the error model with a shrinkage factor $0 < \nu \leq 1$ or equivalently to not try to fit the complete error from the previous iteration but rather only some fraction ν of it. Thus, more iterations are necessary to build the overall model, which results in increased robustness and lower model variance but in additional computational effort. Typical shrinkage values are $\nu = 10^{-1}, 10^{-2}, 10^{-3}$. Similar to bagging, boosting belongs to the category of ensemble methods where many models are cleverly combined.

Example 3.1.12 L_2 Boosting
This example illustrates the steps in the algorithm. Figure 3.20a shows the original process to be fit (gray) and the first model (black). For all iterations, the model is a local model network with ten LMs constructed with the HILOMOT algorithm that optimizes the partitioning explained in Sect. 14.8.

Next the model error is evaluated. It is shown in Fig. 3.20b as gray line and represents the target function for the model in iteration 2. This model $\hat{y}^{(2)}$ is shown

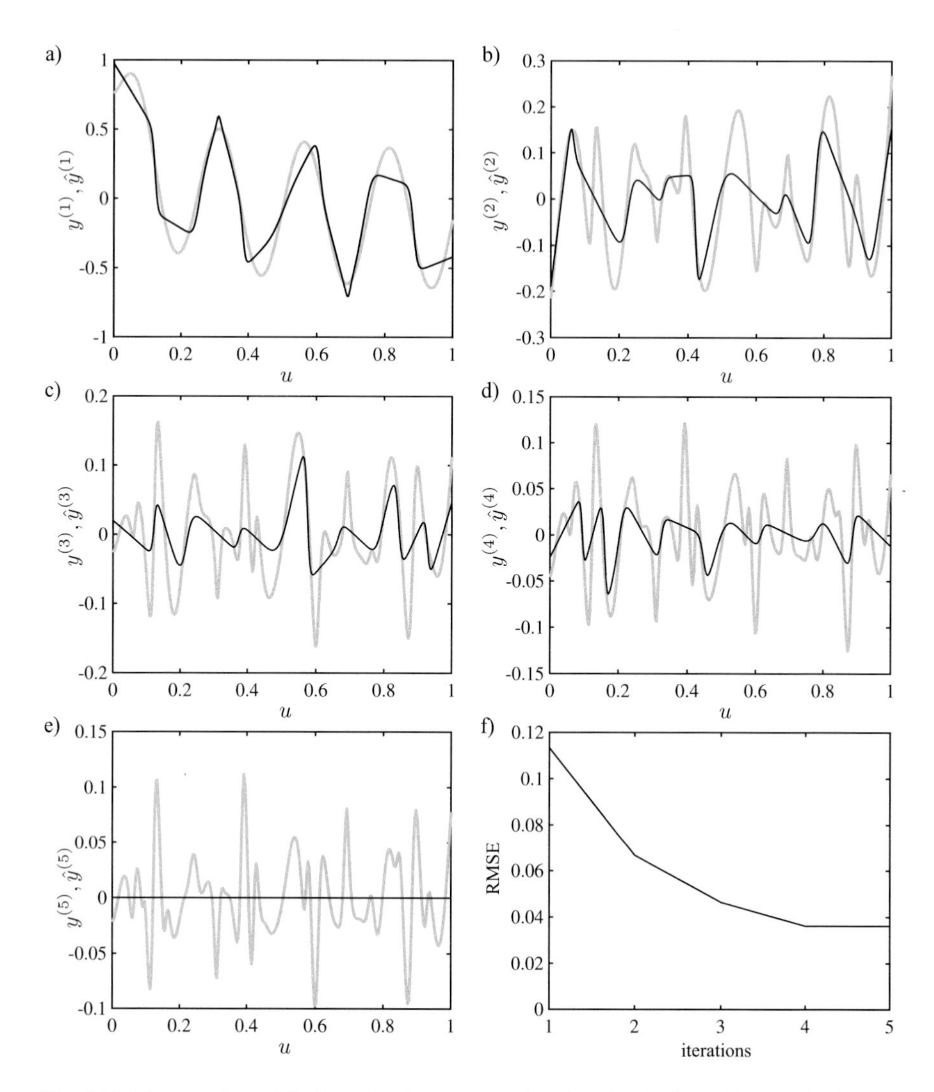

Fig. 3.20 Five iterations of L_2 boosting for a process (gray) and a local model network with ten LMs trained with HILOMOT from Sect. 14.8 (black) and the boosted model error in (**f**). Note the different y-axis scales in (**a**)–(**e**)

in (b) as black line. The procedures continue in (c) with $\hat{y}^{(3)}$ and in (d) with $\hat{y}^{(4)}$. Notice that the shape of the error becomes increasingly rough; high frequencies are more and more dominating. Also, note that the characteristics of the model tend to alternate between any two iterations. Observe, for example, that the model's slope changes for $u \approx 0$ from (a) negative, (b) positive, (c) negative, and (d) positive. This is due to the fact that the model tries to compensate for *all* of the error in each

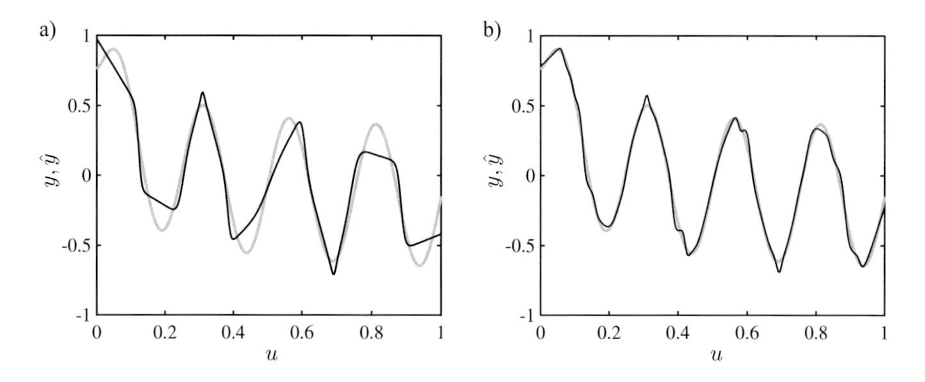

Fig. 3.21 L_2 boosted model. Models are networks of ten local linear models with incremental tree learning algorithm (HILOMOT): (**a**) iteration 1, (**b**) iteration 4. No further progress can be made in subsequent iterations

iteration. Shrinkage would overcome this effect by approaching the error in a slower fashion, similar to any algorithm using a smaller step size.

As illustrated in Fig. 3.20e, in iteration 5, no further progress can be made with local linear model networks, at least in combination with the applied training algorithm. Therefore the error model is zero and boosting converges. The root mean squared error is shown in Fig. 3.20f.

A comparison of the model performance between the first and the last iteration of boosting is displayed in Fig. 3.21. The improvement is very significant.

It is very important to understand that this improvement highly depends on the exact model architecture and training algorithm. A standard procedure would have been to train a local model network with HILOMOT up to a good bias/variance tradeoff. This can be realized by stopping the incremental partitioning and local model splitting at the best AIC_C value. Refer to Chap. 7 for an explanation about the model complexity optimization and to Sect. 14.8 for a detailed discussion of the HILOMOT algorithm. In this standard training approach, it is hard to achieve significant performance improvements with L_2 boosting. HILOMOT builds up models from simple to complex. With high frequency (in space) errors that are typical for boosting and also displayed in Fig. 3.20a–e, boosting terminates the algorithm early and unsuccessfully. Therefore the number of local models was fixed to ten in order to keep significant low-frequency error components to illustrate the operation of boosting. For comparison, HILOMOT with automatic AIC_C termination yields a network with 33 local models and an RMSE of 0.009 which is significantly better than the achieved value of 0.036 in Fig. 3.20f. The boosting results may improve; however, if shrinkage and/or more flexible model architectures are applied. □

3.2 Recursive Least Squares (RLS)

In the previous section, it was assumed for derivation of the least squares solution that all the data samples $\{\underline{u}(i), y(i)\}$ entering $\underline{X}$ and $\underline{y}$ had been previously recorded. When the LS method is required to run online in real time, a new algorithm needs to be developed, since the computational effort of the LS method grows with the number of data samples collected. A recursive formulation of the LS method, the so-called *recursive least squares (RLS)*, calculates a new update for the parameter vector $\hat{\underline{\theta}}$ each time new data comes in. The RLS requires a constant computation time for each parameter update, and therefore it is perfectly suited for online use in real-time applications.

The basic idea of the RLS algorithm is to compute the new parameter estimate $\hat{\underline{\theta}}(k)$ at time instant k by adding some correction vector to the previous parameter estimate $\hat{\underline{\theta}}(k-1)$ at time instant $k-1$. This correction vector depends on the new incoming measurement of the regressors $\underline{x}(k) = [x_1(k)\ x_2(k)\ \cdots\ x_n(k)]^T$ and the process output $y(k)$. Note that in the previous and following text, $\underline{x}_i$ denotes an N-dimensional vector that represents the ith regressor (ith column in the regression matrix $\underline{X}$). In contrast, here $\underline{x}(k)$ denotes the n-dimensional vector of *all* regressors at time instant k.

It can be easily shown (see, e.g. [278]) that the RLS update is

$$\hat{\underline{\theta}}(k) = \hat{\underline{\theta}}(k-1) + \underline{P}(k)\underline{x}(k)e(k) \qquad (3.85)$$

$$\text{with} \quad \underline{P}^{-1}(k) = \underline{P}^{-1}(k-1) + \underline{x}(k)\underline{x}^T(k)\,,$$

where $\underline{P}^{-1}(k) = \underline{X}^T(k)\underline{X}(k)$ is an approximation of the Hessian $\hat{\underline{H}}$ based on the already processed data, and thus $\underline{P}(k)$ is proportional to the covariance matrix of the parameter estimates; see (3.34). For this reason $\underline{P}(k)$ often is sloppily called the *covariance matrix*. The vector $\underline{x}(k)$ is the new measurement for all regressors, and $e(k) = y(k) - \underline{x}^T(k)\hat{\underline{\theta}}(k-1)$ is the one-step prediction error, that is, the difference between the measured process output $y(k)$ and the predicted output $\hat{y}(k|k-1) = \underline{x}^T(k)\hat{\underline{\theta}}(k-1)$ with a parameter vector based on all old measurements. Thus, the amount of correction is proportional to the prediction error. After starting the RLS algorithm, usually the initial parameter vector $\hat{\underline{\theta}}(0)$ is a poor guess of the optimal one, $\underline{\theta}_{\text{opt}}$, and a large parameter correction takes place. The algorithm then converges to the optimal parameter values. The initial value of $\underline{P}(k)$ is usually chosen as $\underline{P}(0) = \lambda\underline{I}$ with large values for λ (say 100 or 1000), because this leads to high correction vectors and therefore to fast convergence. Another interpretation of $\underline{P}(0)$ is that it represents the uncertainty of the initial parameter values gathered in $\hat{\underline{\theta}}(0)$, because $\underline{P}$ is proportional to the parameter covariance matrix. Consequently, λ should be chosen large if little prior knowledge about the initial parameters is available, and $\hat{\underline{\theta}}(0)$ is assumed to be considerably different from $\underline{\theta}_{\text{opt}}$.

There is another intuitive way of examining the RLS algorithm in (3.85). As will be seen in Chap. 4, (3.85) has the same form as any gradient-based nonlinear optimization technique. The term $-\underline{x}(k)\,e(k)$ is the gradient for the new

Table 3.3 Relationship between linear recursive and nonlinear optimization techniques

Derivative information	Linear optimization	Nonlinear optimization
First order	LMS	Steepest descent method
Second order	RLS	Newton's method

measurement; it is equivalent to the gradient expression in (3.21) for a single data sample. Therefore, the RLS correction vector is the negative of the (approximated) inverse Hessian $\underline{P}(k) = \underline{\hat{H}}^{-1}(k)$ times the gradient. As will be seen in Chap. 4, this is equivalent to the Newton optimization method. Since the loss function surface is a perfect hyperparabola, the Newton method converges in one single step. Thus, the RLS can be seen as the application of Newton's method in each time instant, and it reaches the global minimum of the loss function within each iteration step (provided that the influence of the initial values is negligible).

While the RLS is a recursive version of Newton's method applied to a linear optimization problem, what corresponds to the famous steepest descent method (see Chap. 4)? Since steepest descent follows the opposite gradient direction, it can be obtained by simply replacing the (approximated) inverse Hessian in (3.85) by a step length η (strictly speaking by $\eta \underline{I}$). This results in Widrow's least mean squares (LMS) method:

$$\underline{\hat{\theta}}(k) = \underline{\hat{\theta}}(k-1) + \eta\, \underline{x}(k)\, e(k)\,. \tag{3.86}$$

The relationship between the linear optimization algorithms RLS, LMS on the one hand and the nonlinear optimization techniques of Newton and the steepest descent type, on the other hand, are summarized in Table 3.3. First-order methods use only gradient (first derivative) information, while second-order methods also use curvature (second derivative) information. The LMS algorithm (3.86) converges much slower than the RLS, since all information about the surface curvature available in $\underline{\hat{H}}$ is not exploited. A similar statement holds in the nonlinear optimization case for the steepest descent and Newton methods; see Sect. 4.4. However, it is of much lower computational complexity and therefore can be applied for faster processes in real time. Sometimes even $\underline{x}(k)$ is replaced by $\text{sign}\left(\underline{x}(k)\right)$ to save the multiplication operation in (3.86).

3.2.1 *Reducing the Computational Complexity*

The RLS in (3.85) requires the inversion of the Hessian $\underline{\hat{H}}$ or $\underline{P}$, respectively. Therefore, the complexity of this algorithm is $\mathcal{O}(n^3)$, with n being the number of parameters. Thus, it is not feasible for fast processes or many parameters in real time. By applying a matrix lemma (see, e.g., [278]), the RLS in (3.85) can be replaced by the following algorithm:

$$\underline{\hat{\theta}}(k) = \underline{\hat{\theta}}(k-1) + \underline{\gamma}(k)e(k)\,, \quad e(k) = y(k) - \underline{x}^T(k)\underline{\hat{\theta}}(k-1) \tag{3.87a}$$

$$\underline{\gamma}(k) = \frac{1}{\underline{x}^T(k)\underline{P}(k-1)\underline{x}(k)+1}\,\underline{P}(k-1)\underline{x}(k) \tag{3.87b}$$

$$\underline{P}(k) = \left(\underline{I} - \underline{\gamma}(k)\,\underline{x}^T(k)\right)\underline{P}(k-1)\,. \tag{3.87c}$$

The recursive weighted least squares (RWLS) where the weighting of data $\underline{x}(k)$ is denoted as $q(k)$ becomes

$$\underline{\hat{\theta}}(k) = \underline{\hat{\theta}}(k-1) + \underline{\gamma}(k)e(k)\,, \quad e(k) = y(k) - \underline{x}^T(k)\underline{\hat{\theta}}(k-1) \tag{3.88a}$$

$$\underline{\gamma}(k) = \frac{1}{\underline{x}^T(k)\underline{P}(k-1)\underline{x}(k)+1/q(k)}\,\underline{P}(k-1)\underline{x}(k) \tag{3.88b}$$

$$\underline{P}(k) = \left(\underline{I} - \underline{\gamma}(k)\,\underline{x}^T(k)\right)\underline{P}(k-1)\,. \tag{3.88c}$$

This algorithm is of complexity $\mathcal{O}(n^2)$ and is most widely applied in control engineering applications for online system identification. Some numerically improved versions of (3.87a)–(3.87c) such as UD-composition-based algorithms or discrete square root filtering in information form (DSFI) are also popular; see, e.g. [282]. Note that in all those algorithms, $\underline{P}$ does not need to be inverted explicitly since it is computed directly in (3.87c). However, because the complexity of (3.87a)–(3.87c) depends on the square of the number of parameters, it is still out of reach for fast signal processing applications in telecommunications, such as adaptive filters in mobile digital phones with sampling times around 1 ms and large moving average filters with about 1000 parameters. In recent years, extensive research has been carried out on so-called *fast* RLS algorithms. An RLS algorithm is called fast if it has a linear complexity in n, i.e., $\mathcal{O}(n)$. It is very impressive that even the $\mathcal{O}(n^2)$ algorithm in (3.87a)–(3.87c) contains enough redundancy to allow such a speed up. The problem with fast RLS algorithms is that by removing all (implicitly existing) redundant information from (3.87a)–(3.87c), numerical instabilities increase dramatically. Most fast RLS algorithms are numerically unstable. But recently, robust, fast lattice and QR-decomposition-based RLS algorithms have been developed that exploit internal feedback for numerical stabilization; see [232] for a detailed description of fast RLS algorithms. Nevertheless, these fast RLS algorithms are 6–7 times computationally more demanding than the simple LMS algorithm. This is the price to be paid for much faster convergence.

3.2.2 Tracking Time-Variant Processes

The discussed RLS algorithm generally converges to the optimal parameter vector. For long times ($k \to \infty$), the rate of convergence will slow down as $\underline{P}$ approaches zero. This is no problem when dealing with stationary environments, i.e., time-invariant processes. However, the RLS is often applied to non-stationary systems, as, e.g., in adaptive control. Then not convergence to (constant) optimal parameters is of interest but rather tracking of time-varying parameters. Good tracking capability can be ensured by preventing that $\underline{P}$ becomes too small. This is done by the introduction of a forgetting factor $\lambda \leq 1$. Data, j samples ago, is weighted by λ^j, i.e., exponential forgetting is applied, by changing the RLS algorithm to

$$\underline{\hat{\theta}}(k) = \underline{\hat{\theta}}(k-1) + \underline{\gamma}(k)e(k)\,, \quad e(k) = y(k) - \underline{x}^T(k)\underline{\hat{\theta}}(k-1) \tag{3.89a}$$

$$\underline{\gamma}(k) = \frac{1}{\underline{x}^T(k)\underline{P}(k-1)\underline{x}(k) + \lambda}\,\underline{P}(k-1)\underline{x}(k) \tag{3.89b}$$

$$\underline{P}(k) = \frac{1}{\lambda}\left(\underline{I} - \underline{\gamma}(k)\,\underline{x}^T(k)\right)\underline{P}(k-1)\,. \tag{3.89c}$$

The recursive weighted least squares with exponential forgetting where the data $\underline{x}(k)$ is weighted with $q(k)$ becomes

$$\underline{\hat{\theta}}(k) = \underline{\hat{\theta}}(k-1) + \underline{\gamma}(k)e(k)\,, \quad e(k) = y(k) - \underline{x}^T(k)\underline{\hat{\theta}}(k-1) \tag{3.90a}$$

$$\underline{\gamma}(k) = \frac{1}{\underline{x}^T(k)\underline{P}(k-1)\underline{x}(k) + \lambda/q(k)}\,\underline{P}(k-1)\underline{x}(k) \tag{3.90b}$$

$$\underline{P}(k) = \frac{1}{\lambda}\left(\underline{I} - \underline{\gamma}(k)\,\underline{x}^T(k)\right)\underline{P}(k-1)\,. \tag{3.90c}$$

The forgetting factor λ is usually set to some value between 0.9 and 1. While $\lambda = 1$ recovers the original RLS without forgetting where all data is weighted equally no matter how far back in the past, for $\lambda = 0.9$ new data is 3, 8, 24, and 68 times more significant than old data 10, 20, 30, and 40 samples back, respectively. The adjustment of λ is a tradeoff between high robustness against disturbances (large λ) and fast tracking capability (small λ). This tradeoff can be made dynamically dependent on the quality of the excitation. If the excitation is "rich," that is provides significant new information, λ should be decreased and otherwise increased. For more details on adaptive tuning of the forgetting factor, refer to [147, 153, 165, 323]. By this procedure, a "blow-up" of the $\underline{P}$ matrix can also be prevented as it may happen for constant λ and low excitation, since then $\underline{P}(k-1)\underline{x}(k)$ approaches zero and therefore in (3.89c) $\underline{P}(k) \approx \frac{1}{\lambda}\underline{P}(k-1)$.

3.2.3 *Relationship Between the RLS and the Kalman Filter*

The Kalman filter is very closely related to the RLS algorithm with forgetting. Usually, the Kalman filter is applied as an observer for the estimation of states, not parameters. However, formally the parameter estimation problem can be stated in the following state space form [540]:

$$\underline{\theta}(k+1) = \underline{\theta}(k) \tag{3.91a}$$

$$y(k) = \underline{x}^T(k)\underline{\theta}(k) + v(k) \tag{3.91b}$$

where the *measurement noise* $v(k)$ is a white noise signal with variance σ_v^2 disturbing the model output.

The standard state-space equations for a state vector $\underline{x}(k)$ and a single output $y(k)$ are usually written as:

$$\underline{x}(k+1) = \underline{A}\,\underline{x}(k) + \underline{b}\,u \tag{3.92a}$$

$$y(k) = \underline{c}^T\underline{x}(k) + v(k) \tag{3.92b}$$

By interpreting the parameter vector as the state vector $\underline{x}(k) \rightarrow \underline{\theta}(k)$ and setting $\underline{A} \rightarrow \underline{I}$, $\underline{b} \rightarrow \underline{0}$, and $\underline{c}^T \rightarrow \underline{x}^T(k)$, these equations are formally equal.

Thus, (3.91a) is only a dummy equation generated in order to formally treat the parameters as states. Since the parameters are assumed to be time variant, this property has also to be expressed in some way. This can be done by incorporation of a noise term in (3.91a) [540]:

$$\underline{\theta}(k+1) = \underline{\theta}(k) + \underline{w}(k) \tag{3.93a}$$

$$y(k) = \underline{x}^T(k)\underline{\theta}(k) + v(k) \tag{3.93b}$$

where the *process noise* $\underline{w}(k)$ is an n-dimensional vector representing white noise with an $n \times n$-dimensional covariance matrix $\underline{W}$ (n is the number of parameters). Thus, the time variance of the parameters is modeled as a random walk or drift [540]. The covariance matrix $\underline{W}$ is typically chosen diagonal. The diagonal entries can be interpreted as the strength of time variance of the individual parameters. Thus, if a parameter is known to vary rapidly, the corresponding entry in $\underline{W}$ should be chosen to be large and vice versa. This procedure allows one to control the forgetting individually for each parameter, which is a significant advantage over the RLS, where only a single forgetting factor λ can be chosen for the complete model. If no knowledge about the speed of the time-variant behavior is available, the covariance matrix $\underline{W}$ can be simply set to $\zeta \underline{I}$. A forgetting factor $\lambda = 1$ is equivalent to $\underline{W} = \underline{0}$ (no time-variant behavior). Generally, small values for λ correspond to large entries in $\underline{W}$ and vice versa. This relationship becomes obvious from the Kalman filter algorithm as well, where for simplicity the equations are normalized such that the measurement noise variance $\sigma_v^2 = 1$ [540]:

$$\underline{\hat{\theta}}(k) = \underline{\hat{\theta}}(k-1) + \underline{\gamma}(k)e(k)\,, \quad e(k) = y(k) - \underline{x}^T(k)\underline{\hat{\theta}}(k-1) \tag{3.94a}$$

$$\underline{\gamma}(k) = \frac{1}{\underline{x}^T(k)\underline{P}(k-1)\underline{x}(k) + 1}\underline{P}(k-1)\underline{x}(k) \tag{3.94b}$$

$$\underline{P}(k) = \left(\underline{I} - \underline{\gamma}(k)\underline{x}^T(k)\right)\underline{P}(k-1) + \underline{W}\,. \tag{3.94c}$$

With weighting the Kalman filter is

$$\underline{\hat{\theta}}(k) = \underline{\hat{\theta}}(k-1) + \underline{\gamma}(k)e(k)\,, \quad e(k) = y(k) - \underline{x}^T(k)\underline{\hat{\theta}}(k-1) \tag{3.95a}$$

$$\underline{\gamma}(k) = \frac{1}{\underline{x}^T(k)\underline{P}(k-1)\underline{x}(k) + 1/q(k)}\underline{P}(k-1)\underline{x}(k) \tag{3.95b}$$

$$\underline{P}(k) = \left(\underline{I} - \underline{\gamma}(k)\underline{x}^T(k)\right)\underline{P}(k-1) + \underline{W}\,. \tag{3.95c}$$

Indeed for $\lambda = 1$ and $\underline{W} = \underline{0}$, the RLS and the Kalman filter are equivalent. In the Kalman filter algorithm, the adaptation vector $\underline{\gamma}(k)$, which determines the amount of parameter adjustment, is called the *Kalman gain*. In the Kalman filter algorithm, the $\underline{P}$ matrix does not "blow up" exponentially as with the RLS but according to (3.94c) only linearly $\underline{P}(k) \approx \underline{P}(k-1) + \underline{W}$ in the case of non-persistent excitation (i.e., $\underline{x}(k) = \underline{0}$).

3.3 Linear Optimization with Inequality Constraints

Linear optimization problems with linear inequality constraints can be formulated as

$$\frac{1}{2}\underline{\theta}^T\underline{H}\,\underline{\theta} + \underline{h}^T\underline{\theta} + h_0 \longrightarrow \min_{\underline{\theta}} \tag{3.96}$$

with the linear inequality constraints

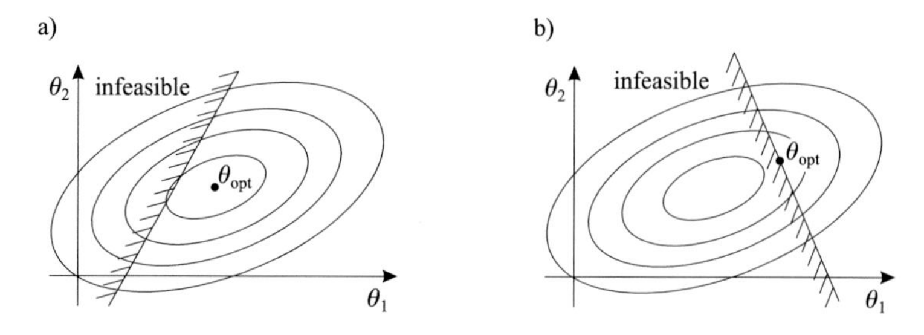

Fig. 3.22 Linear optimization with linear inequality constraint. (**a**) Constraint is not active since the minimum of the unconstrained optimization problem can be realized. (**b**) Constraint is active and the minimum lies on it

$$\underline{A}\,\underline{\theta} \geq \underline{b}\,. \tag{3.97}$$

Note that this is the same type of problem as considered in Sect. 3.1.10 except that in (3.97) inequality, not equality constraints have to be satisfied. Such problems arise, for example, in linear predictive control, where the loss function represents the control error and possibly the actuation signal power. The constraints may represent bounds on the actuation signal and its derivative and possibly operating regime restrictions of the process. Such linear optimization problems can be efficiently solved with *quadratic programming (QP)* algorithms [185].

For the optimization of (3.96), (3.97), two phases can be distinguished [70]. First, a feasible point must be found, i.e., a point that meets all constraints. Second, this point is taken as the initial value in an iterative search procedure for the minimum. Most quadratic programming algorithms are so-called active set methods. This means that they estimate the active constraints at the minimum. Figure 3.22 shows one example for an active and inactive constraint. If more constraints are involved, usually some of them are active at the feasible minimum, and some are not. In each iteration of the QP algorithm, the parameter vector and the estimate of the active constraints are updated. See [70, 185] for more details. Linear optimization problems with linear inequality constraints can be solved so robustly and efficiently that they are utilized in real-world predictive control applications in the chemical and process industry.

3.4 Subset Selection

Up to now linear least squares problems have been discussed in the context of parameter optimization. The n regressors, which are the columns in matrix $\underline{X}$, were assumed to be known a priori, and only the associated n parameters were unknown. This section deals with the harder problem of structure or subset selection, that is, the determination of the proper n_s regressors out of a set of n given regressors. The monograph [375] and the article [99] treat this subject extensively.

The measured output $\underline{y}$ can be described as the sum of the predicted output $\underline{\hat{y}} = \underline{X}\,\underline{\hat{\theta}}$ and the prediction error $\underline{e}$ (see Sect. 3.1)

$$\underline{y} = \underline{X}\,\underline{\hat{\theta}} + \underline{e} \tag{3.98}$$

or in expanded form

$$\begin{bmatrix} y(1) \\ y(2) \\ \vdots \\ y(N) \end{bmatrix} = \begin{bmatrix} x_1(1) & x_2(1) & \cdots & x_n(1) \\ x_1(2) & x_2(2) & \cdots & x_n(2) \\ \vdots & \vdots & & \vdots \\ x_1(N) & x_2(N) & \cdots & x_n(N) \end{bmatrix} \begin{bmatrix} \hat{\theta}_1 \\ \hat{\theta}_2 \\ \vdots \\ \hat{\theta}_n \end{bmatrix} + \begin{bmatrix} e(1) \\ e(2) \\ \vdots \\ e(N) \end{bmatrix}. \tag{3.99}$$

Now, (3.99) will be examined more closely. The task is to model the measured output $\underline{y}$ by a linear combination of regressors $\underline{x}_i$ ($\underline{X} = [\underline{x}_1\ \underline{x}_2\ \cdots\ \underline{x}_n]$). The output $\underline{y}$ can be seen as a point in an N-dimensional space and the regressors $\underline{x}_i$ are vectors in this space that have to be combined to approach $\underline{y}$ as closely as possible; see Fig. 3.5. In general, it will require $n = N$ (i.e., one regressor for each measurement) linear independent regressors to reach $\underline{y}$ *exactly*. Such an interpolation case, where the number of parameters equals the number of measurements, is not desirable for various reasons; see Chap. 7. Rather the goal of modeling is usually to find a set of regressors that allows one to approximate $\underline{y}$ to a desired accuracy with as few regressors as possible. This means that the most important or most significant regressors out of a set of given regressors are searched. These types of problems arise, e.g., for finding the time delays in a linear dynamic system, finding the degree of a polynomial or the basis functions in an RBF network. From these examples, the importance and wide applicability of subset selection methods becomes obvious.

There are only two restrictions to the application of the subset selection techniques described below. First, the model has to be linear in the parameters. Second, the set of regressors from which the significant ones will be chosen must be precomputed. This means that all regressors are fixed during the selection procedure. There exists an important case, namely, the search for fuzzy rules, where this second restriction is violated; see Chap. 12. If these restrictions are not met, a linear subset selection technique cannot be applied. Nevertheless, other approaches (e.g., genetic algorithms) can be utilized; refer to Sect. 5.2. However, these other techniques do not reach by far the computational efficiency of the linear regression-based methods discussed below.

3.4.1 *Methods for Subset Selection*

The subset selection problem is to select n_s significant regressors out of a set of n given regressors. The most obvious solution is to examine all candidate regressor combinations. However, this requires one to estimate $2^n - 1$ different models, which is such a huge number for most applications that this brute force approach is not feasible in practice except for very simple problems. In the following, the three main strategies for efficient subset selection are discussed:

- Forward selection
- Backward elimination
- Stepwise selection

The most common approach to subset selection is the so-called *forward selection*. First, every single regressor out of all n possible ones are selected, and the performance with each of these regressors is evaluated by optimizing the associated parameters. Then the regressor that approximated $\underline{y}$ best, i.e., the most significant one, is selected. This regressor and its associated parameter will be denoted as $\underline{x}_A$ and $\hat{\theta}_A$, respectively. Second, the part of $\underline{y}$ not explained by $\underline{x}_A$ can be calculated as $\underline{y}_A = \underline{y} - \underline{x}_A \hat{\theta}_A$. Next, each of the remaining (not selected) $n - 1$ regressors are evaluated for explaining $\underline{y}_A$. Again this is done by optimizing the associated parameters. This second selected regressor and its associated parameter will be denoted as $\underline{x}_B$ and $\hat{\theta}_B$, respectively. Now, $\underline{y}_B = \underline{y}_A - \underline{x}_B\hat{\theta}_B$ has to be explained by the non-selected regressors. This procedure can be performed until n_s regressors have been selected. It is very fast, since only $n - i + 1$ times a one-parameter estimation is required at step i. However, the major drawback of this approach is that no interaction between the regressors is taken into account. So the parameters of the selected regressors are estimated by subsequent one-parameter optimizations, while the correct solution would require to optimize all parameters simultaneously; see Figs. 3.9 and 3.23. Only if all regressors are orthogonal, no interactions will take place, and this approach would yield good results. Since orthogonality of the

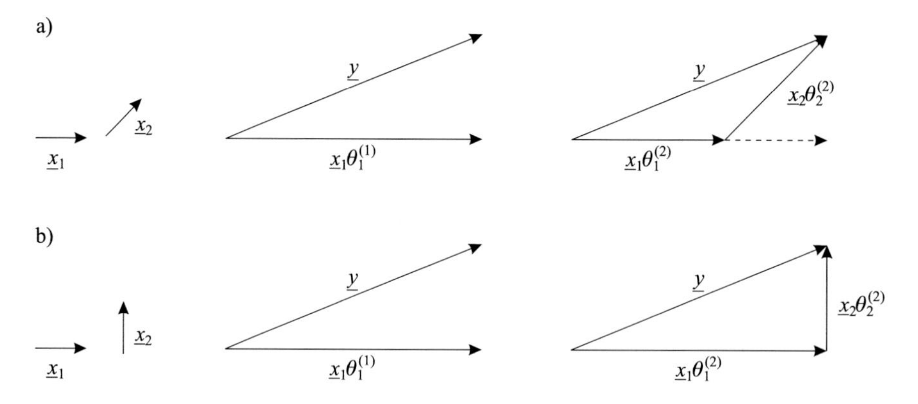

Fig. 3.23 Non-orthogonal and orthogonal regressors $\underline{x}_1$ and $\underline{x}_2$: (**a**) If the regressors are non-orthogonal, the optimal parameter $\hat{\theta}_1^{(1)}$ for $\underline{x}_1$ as the only regressor and $\hat{\theta}_1^{(2)}$ for $\underline{x}_1$ and $\underline{x}_2$ as joint regressors are not identical. This implies that selecting a new regressor requires the recomputation of all parameters. (**b**) If the regressors are orthogonal, the optimal parameters $\hat{\theta}_1^{(1)}$ and $\hat{\theta}_1^{(2)}$ are identical, that is, the regressors do not interact with each other, and therefore the parameters of all regressors can be calculated independently

regressors cannot be expected (not even approximately) in most applications, the above algorithm usually yields poor results. A solution to these difficulties would be to estimate all parameters simultaneously. However, this would require n one-parameter estimations for the first step, $n-1$ two-parameter estimations for the second step, $n-2$ three-parameter estimations for the third step, and so on. The required computational effort for this approach becomes unacceptable for a large number of selected regressors n_s.

A logical consequence of the difficulties discussed above is making the regressors orthogonal to eliminate their interaction. This approach is called forward selection with *orthogonalization*. However, since the number of given regressors n is usually quite large and the number of selected regressors n_s is small, it would be highly inefficient to orthogonalize the full regression matrix $\underline{X}$. An efficient procedure is as follows. First, the most significant regressor is selected. Next, all other (not selected) $n-1$ regressors are made orthogonal to the selected one. In the second step of the algorithm, the most significant of the remaining $n-1$ regressors is again selected, and all $n-2$ non-selected regressors are made orthogonal with respect to the selected one, and so on. In contrast to the approach described above, owing to the orthogonalization, the regressors no longer interact. This means that the optimal parameter values associated with each selected regressor can be determined easily; see Sect. 3.1. Since all remaining regressors are made orthogonal to all selected ones in each step of the algorithm, the improvement of each selectable regressor is isolated. This orthogonal least squares algorithm is the most common strategy for subset selection. Therefore, a mathematical formulation is given in the following paragraph.

Example 3.4.1 Illustration of the Forward Subset Selection
This simple example will illustrate the structure selection problem. Given are $N=3$ data samples and a three column regression matrix $\underline{X}$:

$$\begin{bmatrix} y(1) \\ y(2) \\ y(3) \end{bmatrix} = \begin{bmatrix} x_1(1) \; x_2(1) \; x_3(1) \\ x_1(2) \; x_2(2) \; x_3(2) \\ x_1(2) \; x_2(2) \; x_3(2) \end{bmatrix} \cdot \begin{bmatrix} \theta_1 \\ \theta_2 \\ \theta_3 \end{bmatrix} + \begin{bmatrix} e(1) \\ e(2) \\ e(3) \end{bmatrix} . \tag{3.100}$$

Because the desired output $\underline{y}$ can be reached exactly with the three independent regressors in $\underline{X}$, the error $\underline{e}$ is equal to $\underline{0}$. For $\underline{y}$ and $\underline{X}$, the following numerical values are assumed:

$$\begin{bmatrix} 1 \\ 2 \\ 3 \end{bmatrix} = \begin{bmatrix} 1\;0\;0 \\ 0\;1\;0 \\ 0\;0\;1 \end{bmatrix} \cdot \begin{bmatrix} \theta_1 \\ \theta_2 \\ \theta_3 \end{bmatrix} . \tag{3.101}$$

Figure 3.24 depicts the vector of desired outputs $\underline{y}$. Now, the task is to select one regressor in $\underline{X}$ and to optimize the corresponding parameter θ_i in order to approximate $\underline{y}$ as closely as possible. As Fig. 3.24b shows, regressor 3 with $\theta_3=3$ is the most relevant for modeling $\underline{y}$. The second important regressor is given by

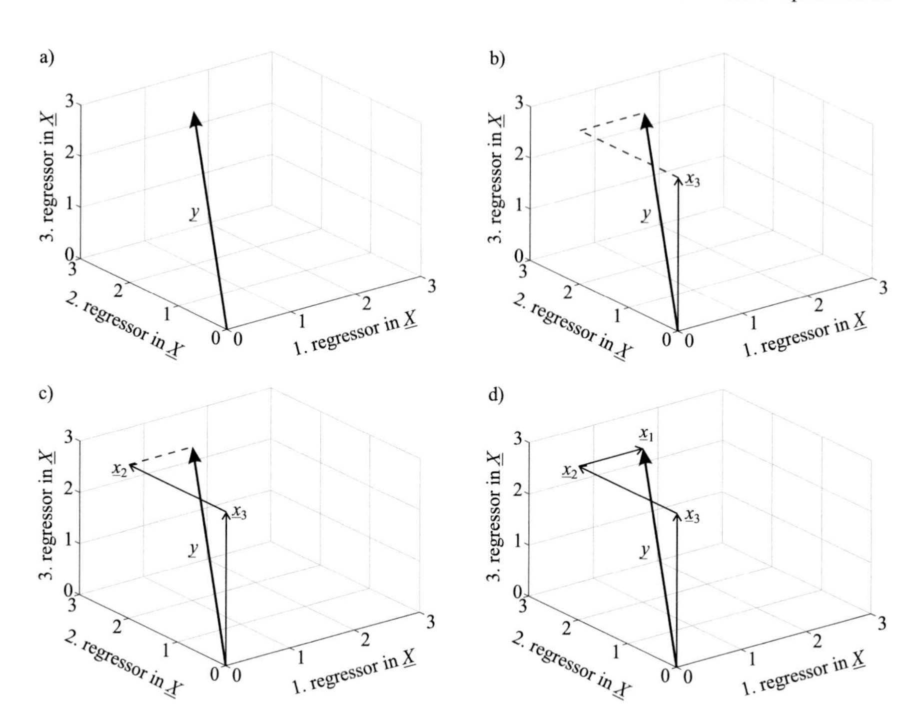

Fig. 3.24 Illustration of the subset selection problem for $N = 3$ data samples and three orthogonal regressors

column 2 with $\theta_2 = 2$; see Fig. 3.24c. Finally, the least relevant regressor is number 1 with $\theta_1 = 1$; see Fig. 3.24d. This analysis is so simple because the regressors are orthogonal to each other. Therefore, it is possible to determine the influence of each regressor and its corresponding optimal parameter separately. Because, in practice, usually the regressors are not orthogonal, most linear structure selection techniques are based on an orthogonalization of the columns in $\underline{X}$. In this toy example, the number of data samples was chosen as $N = 3$ to allow visualization. In practice, however, the number of data samples is usually much larger than the number of selected regressors. Then an overdetermined linear equation system has to be solved, and the desired output $\underline{y}$ cannot be reached exactly by the linear combination of the selected regressors. □

An alternative to forward selection is the so-called *backward elimination*. Instead of increasing the number of selected regressors step by step, in backward elimination, the algorithm starts with all n regressors and removes the least significant regressor in each step. Such an approach may be favorable for problems where $n_s \approx n$. It is, however, not reasonable if $n_s \ll n$, as is the case in most applications.

Another possibility is to combine forward selection and backward elimination into the so-called *stepwise selection*. At each iteration before a new regressor is selected, all already selected regressors undergo some statistical significance test, and those regarded as insignificant are removed from the model. Of course, stepwise selection is much more advanced and complex than simple forward selection or backward elimination. However, in [125], it is the recommended subset selection technique owing to its superior performance. Note that precaution must be taken in order to avoid cycling, that is, the selection and elimination of the same regressors over and over again.

The forward selection method (the same is valid for backward elimination) is based on the selection of the most significant regressor in each step. This means that in each step, the performance improvement is maximized. It is important to note that such an approach does not necessarily lead to the optimal choice of regressors. In general, the solution will be suboptimal. The only way to guarantee the global optimum is by exhaustive search, which has already been ruled out owing to its excessive computational demand. There exist two common extensions to the discussed selection methods that address this suboptimality problem. One possibility is to check at each step in the algorithm the significance of all previously selected regressors by some statistical test. A very simple idea is to remove a previously selected regressor if it degrades the performance less than the improvement by the newest selected regressor. Another alternative that extends these ideas is the use of sequential replacement algorithms. At each step, all regressors are systematically checked for replacement with another (not already selected) regressor. Such a replacement is carried out if it improves the performance. At the point where no further improving replacement can be found, a new regressor is added, and the procedure starts again. Although these algorithms often select a superior set of regressors, in many applications the simple orthogonal least squares for forward selection will be a good choice with a reasonable tradeoff between performance in terms of model accuracy and computational complexity, in particular for large n and n_s.

3.4.2 *Orthogonal Least Squares (OLS) for Forward Selection*

The following description of the orthogonal least squares (OLS) approach for forward selection is taken mainly from [99, 100]. Note that in the literature, OLS is sometimes the abbreviation for *ordinary* least squares (in contrast to weighted least squares etc.), but here it will be used only as *orthogonal* least squares. The starting point is

$$\underline{y} = \underline{X}\,\underline{\theta} + \underline{e}\,. \tag{3.102}$$

The OLS method involves the transformation of the set of regressors $\underline{x}_i$ into a set of orthogonal basis vectors and thus makes it possible to calculate the individual contribution to the desired output variance from each basis vector. The regression matrix $\underline{X}$ can be decomposed into $\underline{X} = \underline{V}\,\underline{R}$, where $\underline{R}$ is an $n \times n$ triangular matrix of the following structure:

$$\underline{R} = \begin{bmatrix} 1 & r_{12} & r_{13} & \cdots & r_{1n} \\ 0 & 1 & r_{23} & \cdots & r_{2n} \\ 0 & 0 & \ddots & & \vdots \\ \vdots & & & & r_{n-1n} \\ 0 & \cdots & & 0 & 1 \end{bmatrix} \tag{3.103}$$

and $\underline{V}$ is an $N \times n$ matrix with orthogonal columns $\underline{v}_i$ ($\underline{V} = [\underline{v}_1\ \underline{v}_2\ \cdots\ \underline{v}_n]$) such that $\underline{V}^T\underline{V} = \underline{S}$ where $\underline{S}$ is diagonal with entry $s_i = \underline{v}_i^T\,\underline{v}_i$.

The space spanned by the set of orthogonal basis vectors $\underline{v}_i$ is the same as the space spanned by the set of $\underline{x}_i$. Therefore, (3.102) can be written as

$$\underline{y} = \underline{V}\,\underline{\vartheta} + \underline{e} \tag{3.104}$$

with a transformed parameter vector $\underline{\vartheta}$ that is related to the original parameter vector $\underline{\theta}$ by satisfying the following triangular system:

$$\underline{R}\,\underline{\theta} = \underline{\vartheta}\,. \tag{3.105}$$

The solution of (3.104) is given by

$$\underline{\vartheta} = \left(\underline{V}^T\underline{V}\right)^{-1}\underline{V}^T\underline{y} = \underline{S}^{-1}\underline{V}^T\underline{y} \tag{3.106}$$

or

$$\vartheta_i = \frac{\underline{v}_i^T\,\underline{y}}{\underline{v}_i^T\,\underline{v}_i} \quad \text{with} \quad i = 1, \ldots, n\,. \tag{3.107}$$

Any orthogonalization method like Gram-Schmidt, modified Gram-Schmidt, Householder, or Givens transformations [193] can be used to derive (3.105). The simplest but numerically least sophisticated Gram-Schmidt orthogonalization method is applied in the following. The numerical robustness of the implemented algorithm is not of major importance for this problem, since only the significant n_s regressors should be selected, which usually leads to well-conditioned matrices. Solving the full n-dimensional system (3.105), however, would generally require numerically more robust techniques. Before the explicit algorithm is introduced, note that the output variance (assuming zero mean) is given by

$$\frac{1}{N}\underline{y}^T\underline{y} = \frac{1}{N}\sum_{i=1}^{n}\vartheta_i^2\,\underline{v}_i^T\,\underline{v}_i + \frac{1}{N}\underline{e}^T\underline{e}\,, \tag{3.108}$$

since $\underline{v}_i$ and $\underline{v}_j$ are orthogonal (and certainly $\underline{v}_i$ and $\underline{e}$ are orthogonal). The above equation for the output variance can be derived by multiplying (3.104) with itself and dividing by the number of measurements N. It can be seen from (3.108) that $\frac{1}{N}\,\vartheta_i^2\,\underline{v}_i^T\,\underline{v}_i$ is the part of the output variance explained by regressor $\underline{v}_i$. A regressor is significant (important for modeling the output) if this amount is large. Therefore, the error $\underline{e}$ is reduced by regressor $\underline{v}_i$ by the following error reduction ratio:

$$err_i = \frac{\vartheta_i^2\,\underline{v}_i^T\,\underline{v}_i}{\underline{y}^T\,\underline{y}}\,. \tag{3.109}$$

Utilizing the conventional Gram-Schmidt orthogonalization, the forward subset selection procedure can be summarized as follows. In the first step, for $i = 1, \ldots, n$ compute (the iteration index is denoted as $(\cdot)^{(i)}$)

$$\underline{v}_1^{(i)} = \underline{x}_i\,, \tag{3.110}$$

$$\vartheta_1^{(i)} = \frac{\underline{v}_1^{(i)\,T}\,\underline{y}}{\underline{v}_1^{(i)\,T}\,\underline{v}_1^{(i)}}\,, \tag{3.111}$$

$$err_1^{(i)} = \frac{\left(\vartheta_1^{(i)}\right)^2\,\underline{v}_1^{(i)\,T}\,\underline{v}_1^{(i)}}{\underline{y}^T\,\underline{y}}\,. \tag{3.112}$$

Find the largest error reduction ratio

$$err_1^{(i_1)} = \max_i\left(err_1^{(i)}\right) \qquad \text{with} \quad i = 1, \ldots, n \tag{3.113}$$

and select the regressor associated with the number i_1

$$\underline{v}_1 = \underline{v}_1^{(i_1)} = \underline{x}_{i_1}\,. \tag{3.114}$$

At the kth step ($k = 2, \ldots, n_s$), for $i = 1, \ldots, n$ with $i \neq i_1, \ldots, i_{k-1}$ compute

$$r_{jk}^{(i)} = \frac{\underline{v}_j^T\,\underline{x}_i}{\underline{v}_j^T\,\underline{v}_j} \qquad \text{with} \quad j = 1, \ldots, k-1\,, \tag{3.115}$$

$$\underline{v}_k^{(i)} = \underline{x}_i - \sum_{j=1}^{k-1} r_{jk}^{(i)}\,\underline{v}_j\,, \tag{3.116}$$

$$\vartheta_k^{(i)} = \frac{\underline{v}_k^{(i)\,T}\,\underline{y}}{\underline{v}_k^{(i)\,T}\,\underline{v}_k^{(i)}}\,, \tag{3.117}$$

$$err_k^{(i)} = \frac{\left(\vartheta_k^{(i)}\right)^2 \underline{v}_k^{(i)\,T} \underline{v}_k^{(i)}}{\underline{y}^T \underline{y}} \,. \tag{3.118}$$

Find the largest error reduction ratio

$$err_k^{(i_k)} = \max_i \left(err_k^{(i)}\right) \quad \text{with } i = 1, \ldots, n \text{ and } i \neq i_1, \ldots, i_{k-1} \tag{3.119}$$

and select the regressor associated with the number i_k:

$$\underline{v}_k = \underline{v}_k^{(i_k)} = \underline{x}_{i_k} - \sum_{j=1}^{k-1} r_{jk}^{(i_j)} \underline{v}_j \,. \tag{3.120}$$

The basis vectors are orthogonalized in (3.116), where the original regressors $\underline{x}_i$ are transformed to the new ones $\underline{v}_i$. This is done according to the Gram-Schmidt method by subtracting a linear combination of all previously orthogonalized regressors. The same idea is utilized in the so-called *innovation* algorithms known from statistics and system identification. For every newly incoming piece of data, the actual information content (the innovation) is calculated: innovation = new data − prediction of the new data based on the existing data. For example, an innovation is equal to zero if the newly incoming data can be perfectly predicted by the already existing data, that is, the new data contains no useful information. Dealing with innovations simplifies the algorithms since all innovations are mutually uncorrelated (orthogonal when interpreted as vectors).

There exist several ways to terminate this subset selection algorithm. Either the number of regressors to be selected (n_s) can be predetermined or the algorithm can be stopped if the amount of unexplained output variance drops below some limit ε:

$$1 - \sum_{j=1}^{n_s} err_j^{(i_j)} < \varepsilon \,. \tag{3.121}$$

Another alternative is to stop when the improvement for selecting a new regressor is below some threshold. Then it must be taken into account that the error reduction ratios for $k = 1, \ldots, n_s$ are not necessarily monotonically decreasing. Furthermore, some information criterion (Chap. 7) can be chosen to terminate the algorithm.

Example 3.4.2 OLS for Polynomial Modeling
For a simple demonstration of the OLS algorithm, the ridge regression Example 3.1.9 can be utilized. The following function has to be approximated by a fourth-order polynomial (see Fig. 3.11)

$$y = 1 + u^3 + u^4 \,. \tag{3.122}$$

The OLS has to select the significant regressors out of the set of possible regressors $[1\ u\ u^2\ u^3\ u^4]$. This also includes the task of deciding how many regressors are

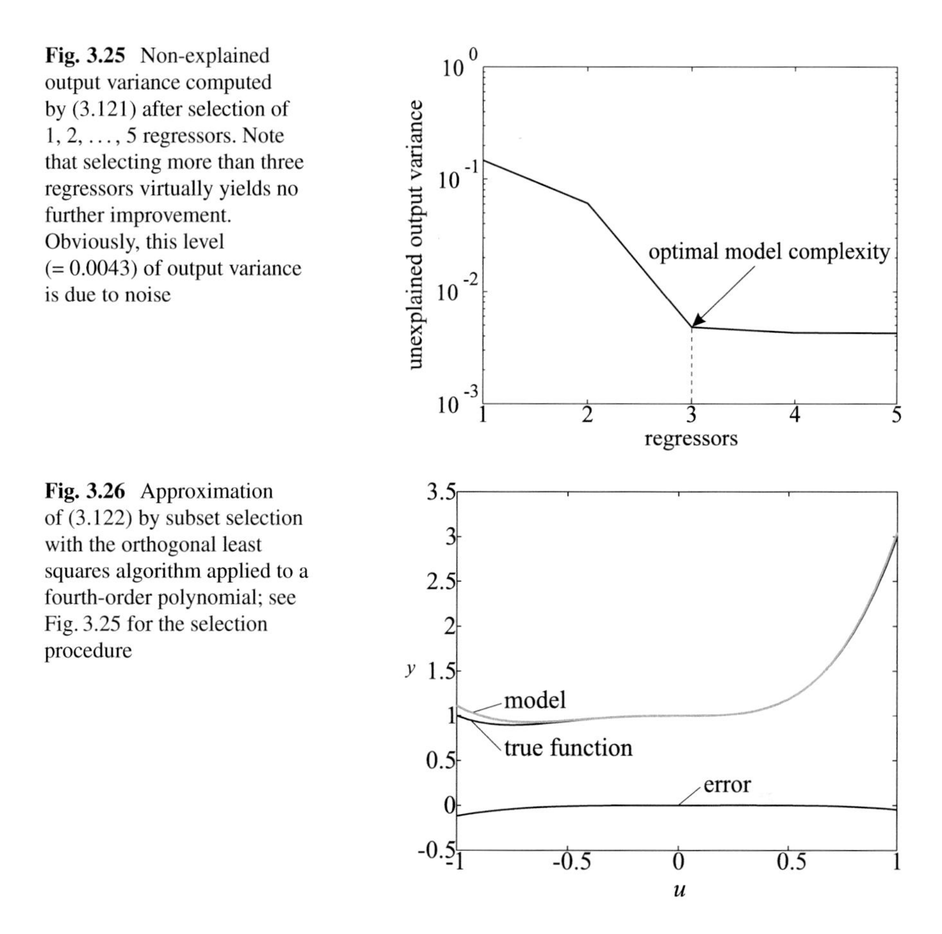

Fig. 3.25 Non-explained output variance computed by (3.121) after selection of $1, 2, \ldots, 5$ regressors. Note that selecting more than three regressors virtually yields no further improvement. Obviously, this level ($= 0.0043$) of output variance is due to noise

Fig. 3.26 Approximation of (3.122) by subset selection with the orthogonal least squares algorithm applied to a fourth-order polynomial; see Fig. 3.25 for the selection procedure

significant, i.e., n_s is not given a priori. Figure 3.25 depicts the convergence of the unexplained output variance in dependency on the number of selected regressors. The five regressors were selected in the sequence $1,\ u^3,\ u^4,\ u,\ u^2$. Obviously, selecting more than three regressors cannot further improve the model quality significantly. Therefore, only the first three regressors were selected. Figure 3.26 shows the accuracy of the function approximation. Table 3.4 summarizes the estimated coefficients and the obtained loss function values. These results are compared with the standard least squares and the ridge regression approach from Example 3.1.9. □

3.4.3 Ridge Regression or Subset Selection?

How does subset selection compare with ridge regression? As shown in Example 3.1.9, ridge regression can also be utilized to select the significant regressors by pursuing the following procedure. First, a ridge regression is performed (possibly

Table 3.4 Comparison between subset selection, standard least squares, and ridge regression

Coefficients	True value	Least squares	Ridge regression	Subset selection
c_0	1.0000	1.0416	0.9998	0.9987
c_1	0.0000	−0.0340	0.0182	0.0000
c_2	0.0000	−0.3443	0.0122	0.0000
c_3	1.0000	1.0105	0.9275	0.9652
c_4	1.0000	1.4353	1.0475	1.0830
Error sum of squares	—	0.0854	0.0380	0.0443

for different regularization parameters λ). Second, the regressors corresponding to parameters with estimated values close to 0 are decided to be non-significant (here the user has to choose some threshold). In order to make the comparison between the parameter values, sensible, the regressors have to be normalized before optimization. Third, these non-significant regressors are discarded, and a standard least squares estimation is performed with the remaining regressors. Steps 2 and 3 are useful since the final model can be of considerably lower complexity than the original one. Although one cannot expect a better performance of the final model since the ridge regression has already realized a good bias/variance tradeoff, simpler models are generally preferred in terms of interpretation and computation time.

In many applications, this regressor selection based on ridge regression yields better results than simple OLS subset selection. This is because the ridge regression utilizes the information of all regressors simultaneously. In contrast, the simple OLS selects one regressor at each iteration and cannot discard regressors if they become insignificant. The OLS algorithm is handicapped by its step-by-step selection. However, it is exactly this step-by-step approach that makes the OLS so efficient. For problems with many potential regressors, ridge regression becomes infeasible owing to the high computational effort (a matrix inversion is of $\mathcal{O}(n^3)$). Subset selection methods, as discussed in this section, can then be the only practical alternative.

3.5 Summary

The linear optimization techniques discussed in this chapter are very mature and thoroughly analyzed methods. When dealing with a specific modeling problem, it is advisable to search for linear parameterized models first in order to exploit the following features:

- The unique global optimum can be found analytically in one step.
- The estimation variance of the parameters and errorbars for the model output can be calculated easily.
- Robust and fast recursive formulations exist.

- Ridge regression and subset selection techniques are powerful tools for bias/variance tradeoff.
- Subset selection techniques allow an efficient model structure determination.

All these advantageous properties vanish for the nonlinear optimization problems that are treated in the next chapter. However, some basic results and insights gained in this chapter are very useful in the context of nonlinear optimization schemes as well. This comes from the fact that any *smooth* nonlinear function can be locally approximated by a second-order Taylor series expansion and consequently can be described by the gradient and the Hessian in a local neighborhood around any point.

3.6 Problems

1. Which of the following models are linear in their parameters?

   (a) $y = \theta_1 + \theta_2 e^{-2u} + \theta_3 e^{-3u}$.
   (b) $y = \sum_{i=1}^{10} \theta_i \sin(2iu)$.
   (c) $y = \theta_1 + \theta_2 \theta_3 u$.
   (d) $y = e^{-\theta_1 u}$.
   (e) $y = \theta_1 u_1 + \theta_2 u_2 + \theta_3 u_1 u_2$.

2. Write down the regression matrix $\underline{X}$ for those models from Problem 1 which are linear in their parameters.
3. Consider the simple linear model $y = \theta_1 + \theta_2 u$. For the estimation of the two parameters θ_1 and θ_1, the following three data points are given: $\{u(1),\ y(1)\} = \{1,\ 4\}$, $\{u(2),\ y(2)\} = \{2,\ 5\}$, $\{u(3),\ y(3)\} = \{3,\ 7\}$. Write down the equation system $\underline{y} = \underline{X}\,\underline{\theta}$. Calculate the loss function for the least squares problem and the Hessian $\underline{H}$. Draw the contour plot of the loss function in the θ_1-θ_2-plane. Calculate the eigenvalue spread κ.
4. The Hessian $\underline{H} = \text{diag}\,(80, 0.03, 0.1, 6)$ arises from a least squares problem. What does this imply for the solution of the problem?
5. Calculate the eigenvalues and the eigenvalue spread of the Hessian in Problem 4. Assume that ridge regression should be applied. What is the smallest value for the regularization parameter λ that makes the least significant parameter (which one is that?) spurious?
6. Explain the relationship between the weighted least squares method and least squares method that takes the noise covariance matrix into account. How can this relationship be understood intuitively?
7. What are the dimensions of the matrix $\underline{A}\,\underline{H}^{-1}\underline{A}^T$ that has to be inverted in order to solve a least squares problem with the equality constraints $\underline{A}\,\underline{\theta} = \underline{b}$? How does this compare to the dimensions of an unconstrained least squares problem with the same number of parameters?

8. Gather experience with the RLS algorithm by following these steps:
   (a) Write a subroutine that implements the RLS algorithm with forgetting. The input parameters are the "old" estimated parameter vector $\underline{\hat{\theta}}(k-1)$, the "old" matrix $\underline{P}(k-1)$, the forgetting factor λ, and the "new" incoming data $\underline{x}^T(k)$ and $y(k)$. The output parameters are the "new" estimated parameter vector $\underline{\hat{\theta}}(k)$ and the "new" matrix $\underline{P}(k)$.
   (b) Write a main program that sets the initial values $\underline{\hat{\theta}}(0) = \underline{0}$ and $\underline{P}(0) = \lambda \underline{I}$ and runs through a loop for $k = 1, 2, \ldots$ in which new data is generated, the RLS subroutine is called, and the estimated parameter vector $\underline{\hat{\theta}}(k)$ and the matrix $\underline{P}(k)$ are stored.
   (c) The data shall be generated from the following two simple models: (i) $y = \theta_1$ and (ii) $y = \theta_1 + \theta_2 u$. First, a value for the input u is chosen by a random number generator in the interval [0, 1]. Second, the associated model output is calculated according to (i) or (ii), respectively. Finally, the model output is disturbed by noise with variance σ^2 which is also generated by a random number generator.
   (d) Vary the following quantities to become familiar with the behavior of the RLS algorithm: output noise variance σ^2, initialization value λ, and forgetting factor λ. Observe the convergence behavior and the sensitivity to noise of $\underline{\hat{\theta}}(k)$ and $\underline{P}(k)$.
   (e) Make one or both parameters of the model (ii) slowly time-variant, e.g., vary θ_1 continuously from 1 to 3 over 2000 steps. Observe how fast the RLS tracks this time-variant model parameter in dependency on the forgetting factor λ.
9. Why are subset selection methods necessary, is it not better to go through all possible choices and simply choose the best one?
10. Compare subset selection with ridge regression. What are the main advantages and drawbacks of either approach?

Chapter 4
Nonlinear Local Optimization

Abstract This chapter deals with fundamental nonlinear optimization techniques that strive to find a local minimum of a possible multimodal loss function. Most of this chapter focuses on unconstrained optimization, but some basics also deal with constrained optimization. First, the exact criteria to optimize are investigated: Batch adaptation, sample adaptation, and mini-batch adaptation as a way in between are discussed. The role of the initial parameters as the starting point for a local search is explained. Existing methods for local nonlinear optimization can be separated into two classes: (i) direct search approach and (ii) gradient-based algorithms. The latter can again be subdivided into general and nonlinear least squares methods. The ideas of the most important algorithms are explained – no algorithmic details are given. The reader/user will understand which approach possesses which properties and thus might be suited for a specific problem.

Keywords Optimization · Direct search · Gradient · Newton method · Nonlinear least squares · Line search · Batch · Sample

If the gradient of the loss function $I(\underline{\theta})$ is nonlinear in the parameters $\underline{\theta}$, a nonlinear optimization technique has to be applied to search for the optimal parameters $\underline{\theta}_{\text{opt}}$. These problems are very common in all engineering disciplines. The parameters will be called *nonlinear parameters*. For example, the hidden layer weights in a neural network or the membership functions' positions and widths in fuzzy systems are nonlinear parameters. Even in the context of linear system identification nonlinear parameters can arise if, e.g., the output error of a dynamic model is minimized; see Sect. 18.5.4. It is important to understand the basic concepts of the different nonlinear optimization techniques in order to decide which one is the most suitable for a particular problem. Because most algorithms are available in common optimization toolboxes as in [70], no implementation details are addressed.

O. Nelles, *Nonlinear System Identification*,
https://doi.org/10.1007/978-3-030-47439-3_4

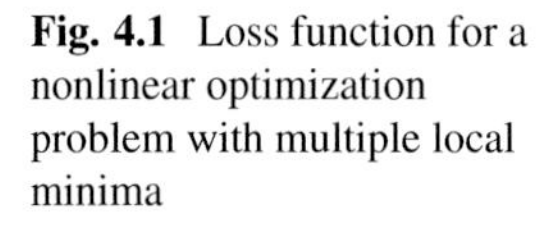

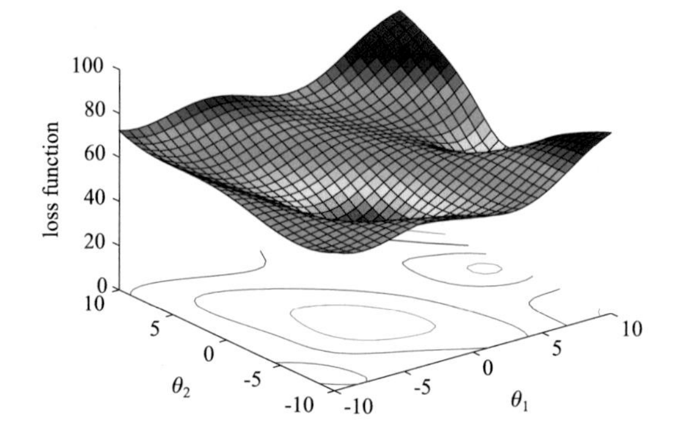

Fig. 4.1 Loss function for a nonlinear optimization problem with multiple local minima

Nonlinear optimization problems generally have the following properties:

- many local optima exist (see Fig. 4.1),
- the surface in the neighborhood of a local optimum can be approximated (using a second-order Taylor series expansion) by a hyperparabola of the form $\underline{\theta}^T \underline{A}\, \underline{\theta} + \underline{b}^T \underline{\theta} + c$,
- no analytic solution exists,
- an iterative algorithm is required,
- they can hardly be applied online.

This chapter deals with approaches for nonlinear local optimization, that is, the iterative search is started at one initial point and only the neighborhood of this point is examined. Usually, with local optimization algorithms, one of the local optima closest to the initial point is found. However, a search of global character can be constructed by restarting a local method from many different initial points (multi-start technique) and finally choosing the best local solution.

Example 4.0.1 Banana Function
The purpose of this example is to illustrate the nonlinear local optimization schemes discussed in the following for the minimization of the so-called "banana" function (also called Rosenbrock's function) depicted in Fig. 4.2. This example is partly taken from the MATLAB optimization toolbox [70]. The function (here denoted as $I(\underline{\theta})$) of two inputs (here denoted as parameters θ_1 and θ_2) is

$$I(\underline{\theta}) = 100\,(\theta_2 - \theta_1^2)^2 + (1 - \theta_1)^2 \,. \tag{4.1}$$

It is called the banana function because of the way the curvature bends around the origin. It is notorious in optimization examples because of the slow convergence that most methods exhibit when trying to solve this problem. This function has a unique minimum at the point $\underline{\theta}_{\text{opt}} = [1.0\ 1.0]^T$ with $I(\underline{\theta}_{\text{opt}}) = 0$. Figure 4.2b shows the contour lines of the banana function with its minimum and the two starting points at $[-1.9\ 2.0]^T$ and $[0.0\ 2.0]^T$ used as initial values for nonlinear optimization.

The goal of this example is to find the minimum of the banana function with a parameter accuracy of at least 0.001. The number of iterations and function evalua-

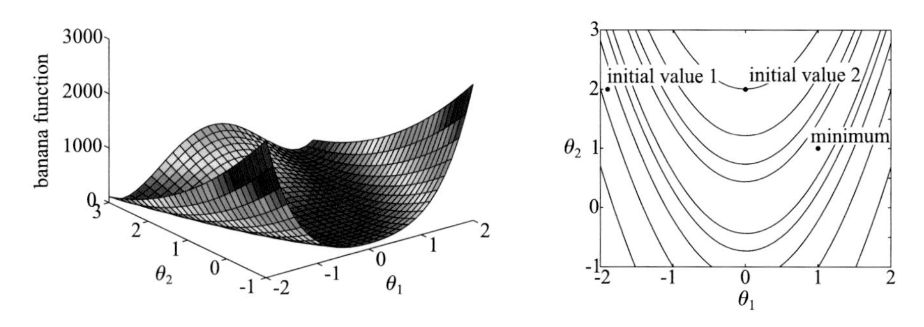

Fig. 4.2 (**a**) Banana function. (**b**) Contour lines of the banana function with its minimum at $[1.0\ 1.0]^T$ and two starting points at $[-1.9\ 2.0]^T$ and $[0.0\ 2.0]^T$, which are used as initial values for a minimum search

tions required from both initial values are listed for all nonlinear local optimization techniques. Furthermore, for the first initial value, the rate of convergence and the path followed towards the minimum is shown. Thus, some insight can be gained into how these algorithms actually behave. It should be noted that the steepest descent method is poorly suited for minimizing the banana function. No *generally valid* conclusions can be drawn based upon the performance of the algorithms on this specific function. However, some advantages and drawbacks of the algorithms become obvious. The results achieved for the different nonlinear local optimization techniques are summarized in Sect. 4.7.

The gradient of the banana function is required by all gradient-based local methods, and computes to

$$\underline{g}(\underline{\theta}) = \begin{bmatrix} 400\theta_1\,(\theta_1^2 - \theta_2) + 2\,(\theta_1 - 1) \\ 200\,(\theta_2 - \theta_1^2) \end{bmatrix}, \tag{4.2}$$

and the Hessian (see Appendix A), required by the Newton method, is

$$\underline{H}(\underline{\theta}) = \begin{bmatrix} 1200\,\theta_1^2 - 400\,\theta_2 + 2 & -400\,\theta_1 \\ -400\,\theta_1 & 200 \end{bmatrix}. \tag{4.3}$$

□

4.1 Batch and Sample Adaptation

The goal of all optimization techniques is to find the minimum of a given loss function with respect to the parameters $\underline{\theta}$; see Sect. 2.3. Thus, a natural procedure is to evaluate this loss function and possibly its derivatives for different parameter values $\underline{\theta}_k$. At each iteration k, a new parameter value can be computed from the past parameter values, the previous loss function values, and possibly its derivatives

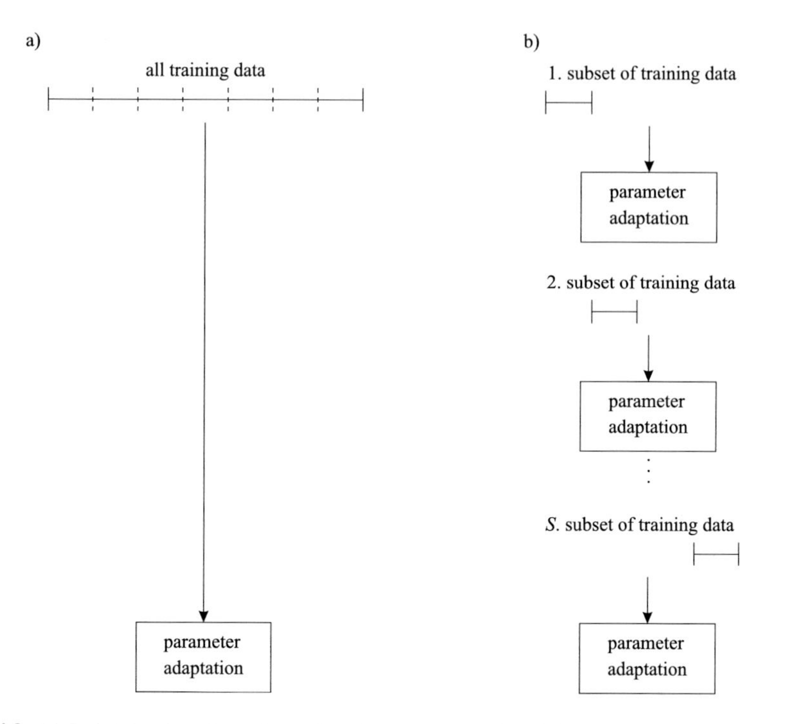

Fig. 4.3 (**a**) In batch adaptation all training data is processed and then one parameter update is performed. (**b**) For large data sets it can be reasonable to update the parameters S times with a $1/S$ part of the training data (mini-batches). If S is chosen equal to the number of training data samples N, this is called sample adaptation

$$\underline{\theta}_k = f\left(\underline{\theta}_j,\ I(\underline{\theta}_j),\ \frac{\partial}{\partial\underline{\theta}_j} I(\underline{\theta}_j),\ \ldots\right) \quad \text{with } j = k-1,\ k-2,\ \ldots,\ 0. \tag{4.4}$$

Usually in (4.4) only the previous iteration step $j = k - 1$ is utilized, guided by the idea that the computation of $\underline{\theta}_{k-1}$ already includes the information about the past $(j < k - 1)$

$$\underline{\theta}_k = f\left(\underline{\theta}_{k-1},\ I(\underline{\theta}_{k-1}),\ \frac{\partial}{\partial\underline{\theta}_{k-1}} I(\underline{\theta}_{k-1}),\ \ldots\right). \tag{4.5}$$

A direct consequence of (4.5) is that between two parameter updates, the loss function and possibly its derivatives have to be evaluated. This approach is called *batch adaptation* or somewhat misleadingly "offline" learning in neural network terminology, since each update requires a sweep through the whole training data. Hence, for problems with huge training data sets, the computational effort for each parameter update is very high, and the whole algorithm becomes prohibitively slow. A common solution to this problem is to divide the training data into S subsets or *mini-batches* and apply (4.5) successively to all these training data subsets; see Fig. 4.3. The advantage of this approach is that S times more parameter updates are computed for the same amount of data. This benefit becomes paramount in the big data context. The drawback is that S different loss functions, each based on a $1/S$ part of the data, are optimized. Hence, the quality of each update will decrease with increasing S, since smaller training data subsets are less representative than the whole data set.

4.1.1 Mini-Batch Adaptation

In big data applications, the size of the training data set can be so enormous that it does not fit the computer RAM anymore. It such cases, it is much simpler and faster to choose the subsets or mini-batches randomly instead of partitioning the whole training data set in advance. If data is scarce, care should be taken to present every data point equally often. If data is abundant, this occurs on average anyway.

The random presentation of the training data introduces additional fluctuation effects which even may be welcomed as they can help to escape local optima. Typical batch sizes are 10 to 1000, corresponding to $S = N/10$ to $N/1\,000$.

4.1.2 Sample Adaptation

An extreme realization of this idea is to choose $S = N$ subsets (with N being the number of training data samples), that is, one subset for each training data sample. Then the iteration in (4.5) simplifies to (assuming a squared error loss function)

$$\underline{\theta}_k = f\left(\underline{\theta}_{k-1},\ e^2(\underline{\theta}_{k-1}),\ \frac{\partial}{\partial \underline{\theta}_{k-1}} e^2(\underline{\theta}_{k-1}),\ \ldots\right). \tag{4.6}$$

This approach is called *sample adaptation* or *instantaneous learning*. In the neural network terminology, it is known as "online" learning. Again this terminology is somewhat misleading because it is not related to "online" in the sense of "in real time with the process." Usually, nonlinear optimization techniques are not suitable for online use owing to their iterative nature.

There is a big difference between iterative and recursive algorithms. Recursive algorithms can be used online since they represent an exact solution. They are just formulated to cope with sample-wise incoming data. A truly recursive algorithm would not gain any information by sweeping through the same data several times. By contrast, iterative algorithms necessarily require many sweeps through the data, and consequently, convergence is orders of magnitude slower. Although running an iterative algorithm in sample mode and online is (in principle) possible, it would usually exhibit poor performance, slow convergence, and non-robust behavior with regard to the ordering of the data.

Sample adaptation historically stems from the backpropagation algorithm to train multilayer perceptrons and since then has become very popular for all optimization tasks within the neural network community. Batch adaptation is the standard approach in statistics and engineering. It is the basis for all sophisticated algorithms. The major problem of the sample adaptation in (4.6) is that the actual error measure is based on a single data sample and thus gives only very vague information about the whole loss function. Therefore, each update in (4.6) is of very poor quality. However, since all data samples are presented after one sweep of (4.6) through the training data (and hence after N parameter updates), these effects will average out. The batch and sample adaptation approaches can be seen intuitively in the following way. In batch adaptation, a lot of information (N samples) is gathered to determine a good next parameter update that eventually is performed. In sample adaptation, after every single piece of new information (one sample), a parameter update is performed. In Fig. 4.4, the two approaches are compared. It is easy to see that in

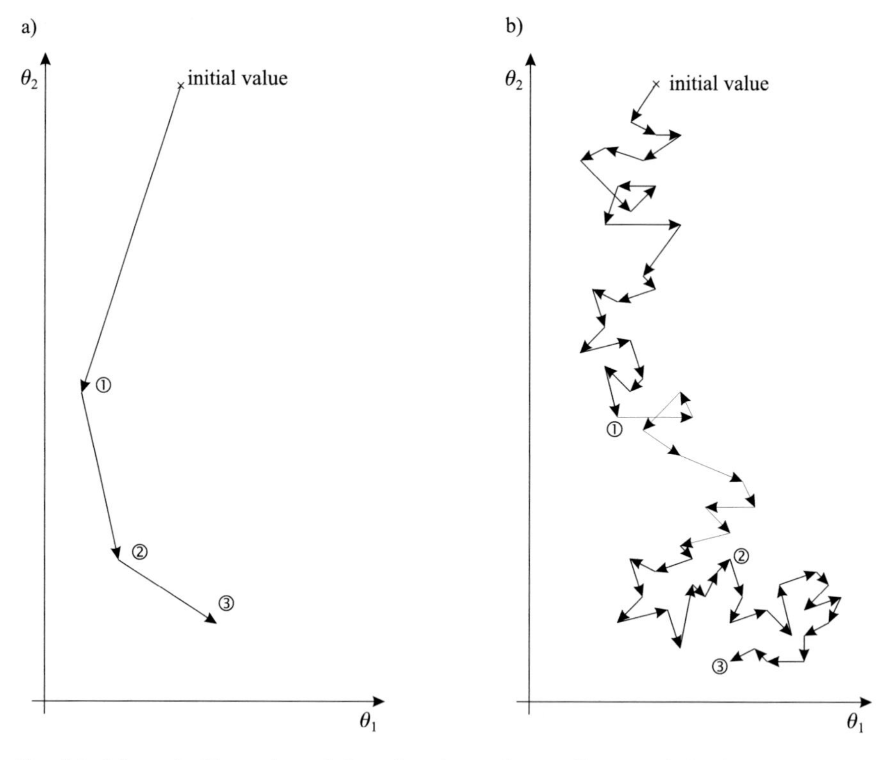

Fig. 4.4 Schematic illustration of three iterations of a nonlinear optimization algorithm in (**a**) batch and (**b**) sample mode. In batch mode, one large parameter update is performed in each iteration, while in sample mode, many small parameter adaptations are performed, each based on one single data sample. Although the average direction of the sample updates are close to the batch update direction, both approaches may end in different local optima and usually have different convergence speeds

batch adaptation one big step towards a good direction (in the sense of making the loss function smaller) is carried out, while in sample adaptation, many small steps in bad directions but a good mean direction are made.

Applying sample adaptation can be interpreted as adding noise on the parameter update. This reduces the probability of getting stuck in a local optimum, since noise can move the parameter point out. However, this also means that a sample adaptation algorithm will never converge to a single point. Remedies are to reduce the step size or to switch to batch adaptation for the last iterations. Other difficulties with sample adaptation are that all advanced nonlinear optimization schemes applying line search algorithms and exploiting second-order derivative information cannot be utilized reasonably. Furthermore, the results obtained with sample mode adaptation are highly dependent on the ordering of the incoming data. The parameters will always fit the last data samples in the training data set better than the first ones. Owing to these drawbacks, in recent times batch adaptation has become more and more popular in combination with neural networks [32, 33, 95, 207]. At least, when applying sample adaptation, the last few iterations should be performed in batch mode to get rid of the data ordering dependence.

4.2 Initial Parameters

Since nonlinear optimization techniques are iterative, at the first iteration $k = 1$, some initial parameter vector $\underline{\theta}_0$ has to be determined. It is obvious that a good initial guess $\underline{\theta}_0$ of the optimal parameters $\underline{\theta}_{\text{opt}}$ leads to fast convergence of the algorithm and a high probability of converging to the global optimum. If the parameters $\underline{\theta}$ represent physical variables or other interpretable quantities, usually, enough prior knowledge is available to choose a reasonable $\underline{\theta}_0$. Often previous experience and simple experiments can yield a good guess for $\underline{\theta}_0$ or at least give range limits for $\underline{\theta}_{\text{opt}}$. Particular model architectures induce certain initial values, e.g., for fuzzy systems typically rules developed by experts yield a good initialization. A different situation occurs for black box models such as neural networks, since the parameters have no direct physical relevance and allow no interpretation. However, even in these cases, parameter initializations that are better than random are possible, speeding up the learning procedure considerably; see Chap. 11.

Figure 4.5 shows a possible loss function dependent on one parameter θ. Generally, for a nonlinear local optimization technique only with an initial parameter value smaller than θ_{C}, convergence to the global optimum at θ_{A} can be expected. Typically, for initializations with $\theta \geq \theta_{\text{C}}$, the local optimum at θ_{D} will be found. Saddle points like B usually do not cause practical problems since it is virtually impossible to end up in θ_{B} exactly, and so the algorithm will have a good chance of escaping from saddle points.

Another approach for determination of proper initial parameters is the transformation of the nonlinear into a linear optimization problem. Figure 4.6 shows an example that may occur at the output node of a multilayer perceptron with a sigmoidal activation function (see Sect. 11.2)

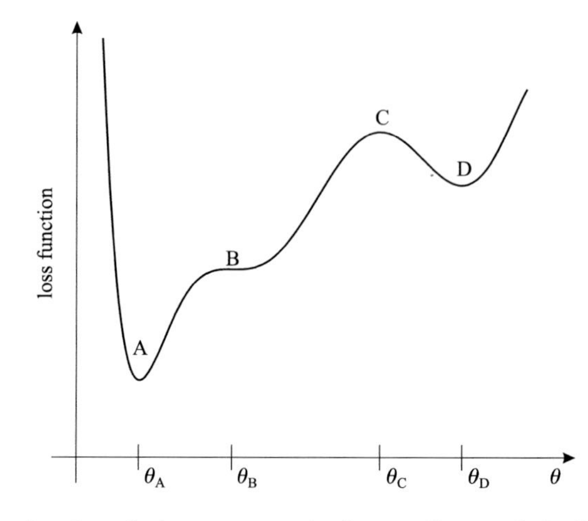

Fig. 4.5 Loss function dependent on one parameter for a nonlinear optimization problem

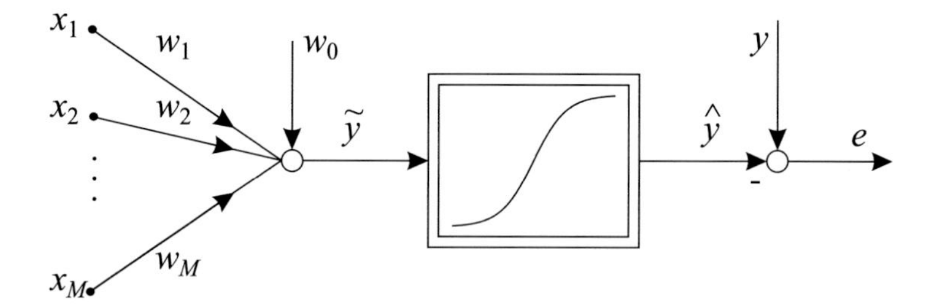

Fig. 4.6 Multilayer perceptron neuron with M inputs

$$\hat{y} = \frac{1}{1 + e^{-\tilde{y}}} \,. \tag{4.7}$$

with M hidden nodes and one bias node ($x_0 = 1$)

$$\tilde{y} = \sum_{i=0}^{M} w_i \, x_i \,. \tag{4.8}$$

Owing to the nonlinear activation function, it is a nonlinear optimization problem to adapt the parameters w_i in order to minimize the error sum of squares loss function when the errors are calculated at the output ($e = y - \hat{y}$). However, if the desired output at the input of the activation function was known, a simple linear least squares problem would arise. The idea is now to transform the measured output behind the nonlinearity y to a desired output before the nonlinearity y_{trans} by inverting (4.7):

$$y_{\text{trans}} = \ln \frac{y}{1-y}. \tag{4.9}$$

The new error before the nonlinearity becomes $e_{\text{trans}} = y_{\text{trans}} - \widetilde{y}$. The corresponding optimization problem is linear in the parameters w_i and thus can be easily solved. Note, however, that the optimal parameters of this transformed, linear problem are in general different from the optimal parameters of the original problem. In the example above, all outputs that are in the saturation of the sigmoid will be scaled by high factors. Therefore, the loss function based on e_{trans} will be dominated by those data samples. Nevertheless, the LS estimated parameters may be a reasonable initial value for a subsequent iterative nonlinear optimization technique, especially for weak nonlinearities. Note that many complex nonlinearities *cannot* be transformed to linear problems as shown above.

4.3 Direct Search Algorithms

Direct search algorithms are based on the evaluation of loss function values only. No derivatives are required. Consequently, it is not reasonable to apply these methods if the derivatives of the loss function are easily available with low computational effort. Although the direct search methods do not require the derivatives to exist, higher performance can be expected on smooth functions. The advantages of direct search methods are that they are easy to understand and to implement. Their application is recommended to problems where gradients are not available or tedious to evaluate. It should be clearly stated that the direct search methods usually have slow convergence and are popular mainly because they are the simplest choice. A more complex but also more powerful alternative for the case of unavailable gradients is to apply a gradient-based method in connection with finite difference techniques to compute the gradients numerically. The following introduction of the simplex and Hooke-Jeeves methods is a summary of the more extensive treatment in [481]. Both approaches utilize different strategies for the generation of the search directions.

4.3.1 Simplex Search Method

The direct search strategy called simplex search or S^2 method has no relationship to the simplex method for linear programming, which will not be discussed here. The goal of the search algorithm is to find the n-dimensional parameter vector that minimizes the loss function. It is based on a regular simplex, that is, in n dimensions a polyhedron composed of $n + 1$ equidistant points, which form its vertices. For example, in two dimensions, a simplex is an equidistant triangle; see Fig. 4.7a. The basic idea of simplex search is to compute the loss function at each vertex. The

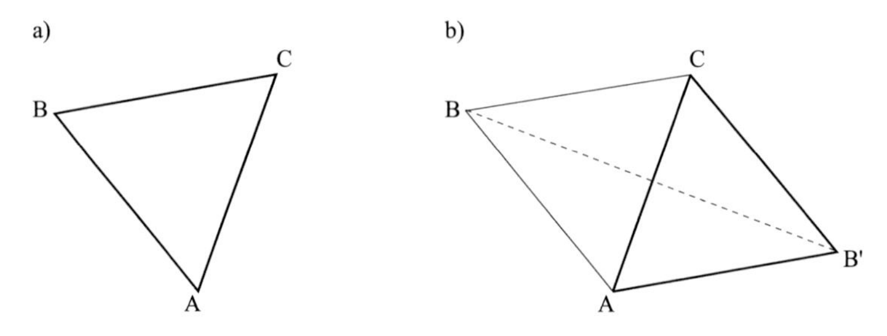

Fig. 4.7 (**a**) A simplex (equidistant triangle) in a two-dimensional parameter space. (**b**) The vertex with the highest loss function value (here B) is reflected at the centroid (becoming B') to create a new simplex

vertex with the largest loss function value is regarded as the worst point and is therefore reflected at the centroid; see Fig. 4.7b. The worst point is deleted and the new point is used to generate a new simplex. By iterating this algorithm, the simplex "roles" downhill until a local minimum is reached. Thus, each iteration of this algorithm requires only one loss function evaluation. However, several additional precautions must be taken to ensure convergence. For example, cycling between two or more identical simplices has to be avoided. This can be achieved by choosing not the worst but the second-worst vertex for reflection and by reducing the size of the simplex if cycling is detected. The size of the simplex is conceptionally close to the step size η in gradient-based techniques; see Sect. 4.4. Usually, the search is started with a large simplex to enable fast convergence, and while the algorithm progresses, the simplex is shrunk whenever cycling is detected.

The basic simplex search introduced above was extended by Nelder and Mead to partially eliminate the following drawbacks:

- All directions (i.e., parameters) are scaled by the same factor if the simplex size is reduced.
- The size of the simplex can only be reduced and therefore no acceleration of the convergence speed is possible.
- Contracting the simplex requires the recomputation of all vertices.

The regularity of the simplex (i.e., equidistant vertices) is no longer demanded by the Nelder and Mead algorithm. This enables stretching and shrinking of the simplex in each reflection procedure. First, a normal reflection is checked. If this reflection yields a loss function decrease, an expanded reflection is performed. If the normal reflection yields a loss function increase, a contracted reflection is performed; see Fig. 4.8. The expansion and contraction procedures allow a significant convergence speed-up.

Example 4.3.1 Simplex Search for Function Minimization
In this example, the banana function is minimized by applying the simplex search technique due to Nelder and Mead. This method makes no explicit use of the

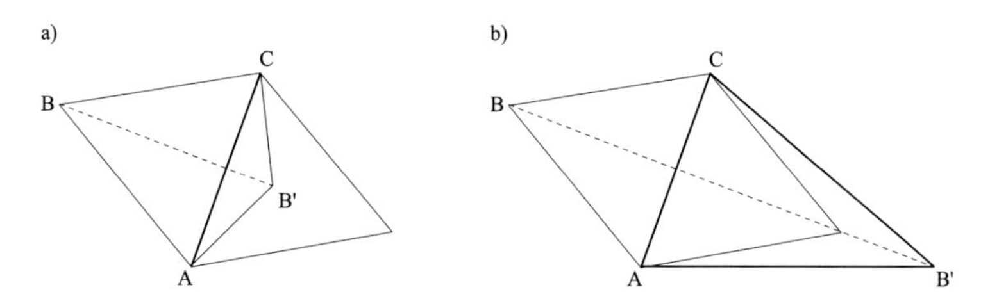

Fig. 4.8 The algorithm due to Nelder and Mead allows (**a**) contraction and (**b**) expansion of the simplex to adapt the step size of the search procedure and to speed up convergence

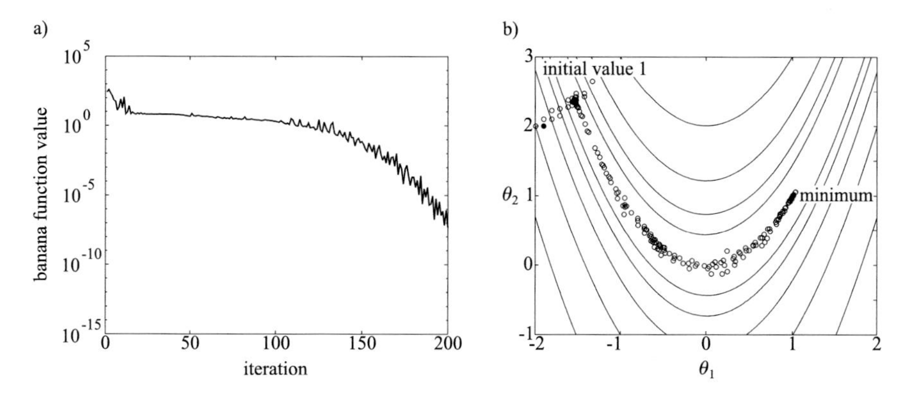

Fig. 4.9 Simplex search method for minimization of the banana function. The rate of convergence (**a**) over the iterations, (**b**) in the parameter space (the circles mark the function evaluations)

gradient and Hessian but only evaluates the loss function values. Taking this into account, the rate of convergence shown in Fig. 4.9 is surprisingly fast. Note that the convergence curve in Fig. 4.9 is not monotonically decreasing because the reflection procedure may yield a higher loss function value. The total number of function evaluations, which is equivalent to the number of iterations, is equal to 200 and 221 when starting from the first and second initial value, respectively. Although the convergence of simplex search is quite fast, it will be by far outperformed by the gradient-based methods discussed in Sect. 4.4. Consequently, simplex search is only recommended for problems with unavailable gradients.

Note that simplex search was the only method that converged prematurely when starting at initial value 2. It converged to $\underline{\theta} \approx [0\ 0]^T$ instead of the minimum at $\underline{\theta}_{\text{opt}} = [1\ 1]^T$. This effect is due to a missing line search procedure. However, convergence to the correct minimum could be achieved by setting the desired accuracy temporary from 0.001 to 0.00001. □

4.3.2 Hooke-Jeeves Method

The concept of the Hooke-Jeeves algorithm is to search the n-dimensional parameter space in n independent fixed search directions. Usually, orthogonal search directions are chosen, for example, along the coordinate axes. The idea is to start at a so-called base point $\underline{\theta}_0$ and search for the minimum in one direction with a given step size. If the loss function value increases, the opposite direction is checked. If no decrease in the loss function can be obtained, the step size is reduced. Next, the search is performed in the second search direction and so on. When all search directions have been investigated a new base point $\underline{\theta}_1$ is set. From this base point, the procedure is iterated. Note that different step sizes can be selected for each search direction.

The above algorithm can be significantly improved by searching in the direction $\underline{\theta}_k - \underline{\theta}_{k-1}$, since this vector is a good overall direction at iteration k. It approximates the opposite gradient direction. Thus, after a new base point is set, the Hooke-Jeeves algorithm searches in this direction with step size one; see Fig. 4.10. If the loss function value decreases, this new point becomes a base point. Otherwise the search along $\underline{\theta}_k - \underline{\theta}_{k-1}$ is discarded and the next search starts at the old base point.

Example 4.3.2 Hooke-Jeeves for Function Minimization
The performance of the Hooke-Jeeves method on the banana function minimization problem is depicted in Fig. 4.11. Like the simplex search, Hooke-Jeeves makes no explicit use of the gradient and Hessian but only evaluates the loss function values. The rate of convergence of Hooke-Jeeves is similar to that of the simplex search. From the two initial values, the algorithm required 230 and 175 function evaluations, respectively. Nevertheless, simplex search is generally considered to be superior to Hooke-Jeeves [481]. □

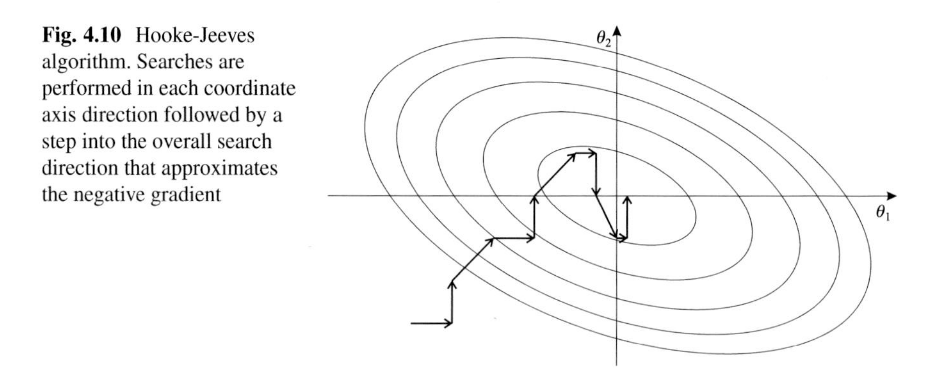

Fig. 4.10 Hooke-Jeeves algorithm. Searches are performed in each coordinate axis direction followed by a step into the overall search direction that approximates the negative gradient

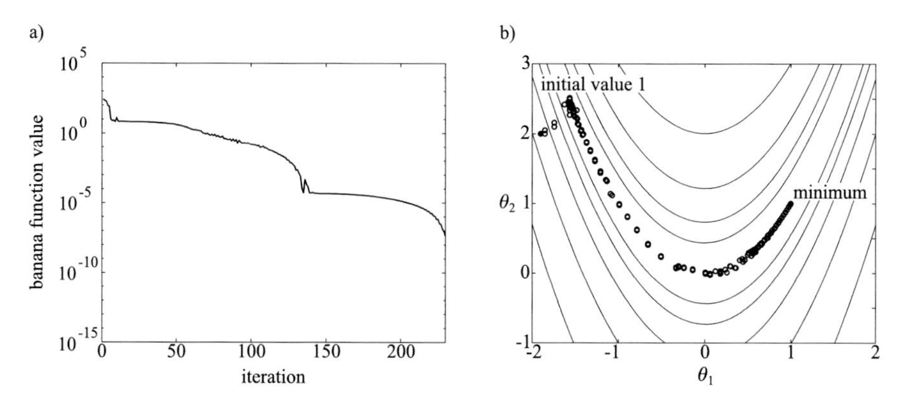

Fig. 4.11 Hooke-Jeeves method for minimization of the banana function. The rate of convergence (**a**) over the iterations, (**b**) in the parameter space (the circles mark the function evaluations)

4.4 General Gradient-Based Algorithms

Gradient-based algorithms are the most common and important nonlinear local optimization techniques. In the following, it is assumed that the gradient $\underline{g} = \partial I(\underline{\theta}) / \partial \underline{\theta}$ of the loss function $I(\underline{\theta})$ with respect to the parameter vector $\underline{\theta}$ is known by analytic calculations or approximated by finite difference techniques; see Sect. 4.4.2. The principle of all gradient-based algorithms is to change the parameter vector $\underline{\theta}_{k-1}$ proportional to some step size η_{k-1} into a direction $\underline{p}_{k-1}$ that is the gradient direction $\underline{g}_{k-1}$ rotated and scaled by some direction matrix $\underline{R}_{k-1}$:

$$\underline{\theta}_k = \underline{\theta}_{k-1} - \eta_{k-1}\, \underline{p}_{k-1} \qquad \text{with} \qquad \underline{p}_{k-1} = \underline{R}_{k-1}\, \underline{g}_{k-1}\,. \tag{4.10}$$

Clearly, the goal of optimization is that each iteration step should decrease the loss function value, i.e., $I(\underline{\theta}_k) < I(\underline{\theta}_{k-1})$. This is the case for positive definite direction matrices $\underline{R}_{k-1}$. The simplest choice $\underline{R}_{k-1} = \underline{I}$ leads into the steepest descent direction exactly opposite to the gradient $\underline{g}_{k-1}$.

The existing algorithms can be distinguished by different choices of the scaling and rotation matrix $\underline{R}$ and the step size η. In the following, first, the determination of the step size η is discussed. Then the steepest descent, Newton, quasi-Newton (or variable metric), and conjugate gradient methods are introduced. For an excellent book on these methods, refer to [504]. Although this chapter deals with nonlinear optimization techniques, it will sometimes be helpful for gaining some insights to analyze the convergence behavior of these algorithms on a linear least squares problem, i.e., for the case of a quadratic loss function. The reason for this is that close to an optimum each nonlinear function can be approximated by a hyperparabola.

4.4.1 Line Search

In order to compute the iteration (4.10), a step size η has to be chosen. Some algorithms fix the step size at a constant value. It is far more powerful, however, to search for the optimal step size η_{k-1} that minimizes the loss function in the direction $\underline{p}_{k-1}$. This is a univariate (single parameter) optimization task. It can be divided into two procedures. First, an interval that contains the loss function minimum along $\underline{p}_{k-1}$ must be found (interval location). Second, the minimum within this interval must be searched by interval reduction.

4.4.1.1 Interval Reduction

The interval reduction methods can be divided into function comparison and polynomial interpolation methods. Starting from some interval [A, B], the comparison methods compute the loss function values at points within the interval; see Fig. 4.12. The optimal method is called *Fibonacci search*; in this, at each iteration, the interval is divided by the ratio of two subsequent Fibonacci numbers. The difficulty of Fibonacci search is that the number of iterations has to be determined a priori. Almost as efficient is the *Golden Section search*, which asymptotically approaches Fibonacci search but does not suffer from its drawbacks. Hereby, the interval is divided with the ratio of the Golden Section. All function comparison methods have only first-order convergence, and they are independent of the specific shape of the loss function.

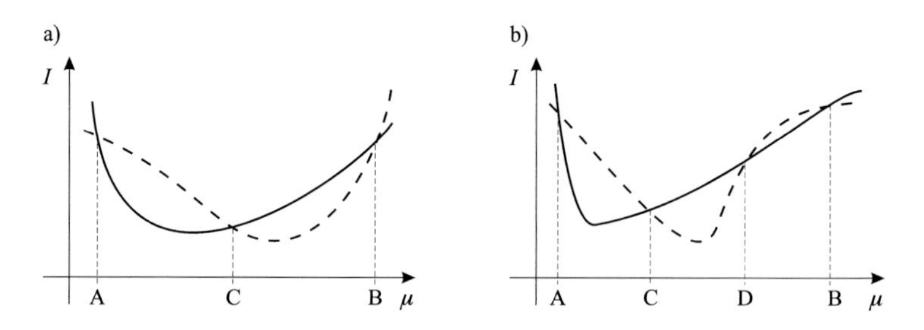

Fig. 4.12 Interval reduction [504]: (**a**) Starting from the interval [A, B] one loss function evaluation is not sufficient to reduce the size of the interval, since the local minimum can be located to the left or right side of this point C. (**b**) Two loss function evaluations, however, can reduce the interval (here to [A, D]). Note that the minimum could not be achieved in [D, B] because only a single minimum is assumed to exist in [A, B]. Intelligent search methods such as Golden Section search utilize the loss function evaluation at C in the next iteration

The polynomial interpolation methods estimate the minimum within the interval by fitting a (usually second or third order) polynomial through known loss function values and possibly the derivatives. Since these methods use absolute loss function values, and not just ratios as function comparison methods do, their convergence behavior is dependent on the loss function's shape. However, they usually achieve a higher rate of convergence than function comparison methods do.

4.4.1.2 Interval Location

The interval location methods can also be divided into function comparison and polynomial extrapolation methods. The first interval bound is simply the old parameter vector $\underline{\theta}_{k-1}$. Function comparison methods find the other interval bound usually by starting with an initial step size and then double this size in each iteration until the loss function value increases in the search direction; see Fig. 4.13. Polynomial extrapolation basically works in an equivalent way to the interpolation for the interval reduction methods. The advantages and drawbacks of both approaches are similar to those for interval reduction.

The interaction between the line search algorithm and the nonlinear optimization technique is an important but difficult topic. The desired accuracy of the interval reduction certainly depends on the chosen optimization technique. It has to be determined by a tradeoff between fewer iterations with higher computational effort each (due to more exact line search) and more iterations with lower computational effort each (due to less exact line search). For example, it is known that for quasi-Newton methods, a very rough line search is sufficient and computationally advantageous, while conjugate gradient methods require a much more accurate line search.

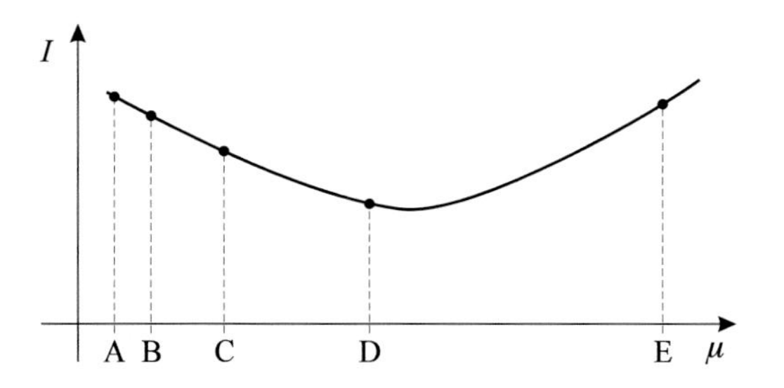

Fig. 4.13 Interval location [504]: To locate an interval that brackets a local minimum in the search direction η, the loss function $I(\underline{\theta})$ is evaluated at points (here A, B, C, D, and E) until it starts increasing. To guarantee fast speed of this algorithm, the step size is doubled at each iteration, ensuring exponentially fast progress

4.4.2 Finite Difference Techniques

For all optimization methods discussed in this and the following section, the gradient, i.e., the first derivative, of the loss function is required in order to calculate the search direction $\underline{p}$. For many problems it is possible to find an analytical expression for the gradient, as in the banana function example.

The gradient of the loss function always requires calculation of the first derivative of the model output $\hat{y}$ with respect to the parameters $\underline{\theta}$. For example, for the most commonly applied sum of squared errors loss function, the gradient can be computed as follows:

$$\underline{g} = \frac{\partial I(\underline{\theta})}{\partial \underline{\theta}} = \frac{\partial \frac{1}{2} \sum_{i=1}^{N} e^2(i)}{\partial \underline{\theta}} = \frac{1}{2} \sum_{i=1}^{N} \frac{\partial e^2(i)}{\partial \underline{\theta}} = \sum_{i=1}^{N} e(i) \frac{\partial e(i)}{\partial \underline{\theta}}$$

$$= - \sum_{i=1}^{N} e(i) \frac{\partial \hat{y}(i)}{\partial \underline{\theta}} , \tag{4.11}$$

where the errors $e(i) = y(i) - \hat{y}(i)$ are the difference between the measured outputs $y(i)$ and the parameter-dependent model outputs $\hat{y}(i)$ and N is the number of data samples. For the model structures analyzed in Parts II and III that possess nonlinear parameters, the analytical expression for the gradient of the model output with respect to its parameters is given. Note that if (4.11) is used as the gradient in (4.10), the minus signs cancel.

In some cases it might be impossible to derive an analytical expression for the gradient, or it might be computationally too expensive to evaluate [504]. Then so-called *finite difference techniques* can be used for a numerical calculation of the gradient.

The gradient component i at point $\underline{\theta}$ can be approximated by the difference quotient

$$g_i(\underline{\theta}) \approx \frac{I(\underline{\theta} + \Delta\theta_i) - I(\underline{\theta})}{\Delta\theta_i} , \tag{4.12}$$

where $\Delta\theta_i$ is a "small step" into the ith coordinate axis direction. This means that for approximation of the full gradient $\underline{g}$ at point $\underline{\theta}$, the expression (4.12) has to be evaluated for $i = 1, 2, \ldots, n$, that is, for all search space directions. Consequently, the computational effort required for gradient determination is equal to n loss function evaluations at the points $I(\underline{\theta} + \Delta\theta_i)$, where n is the number of parameters. Thus, the computation of the gradient by finite differences can easily become the dominating factor in optimization.

In principle, it is also possible to approximate the Hessian by applying finite difference techniques to the gradient. The ith column $\underline{h}_i$ of the Hessian $\underline{H}$ can be calculated by

$$\underline{h}_i(\underline{\theta}) \approx \frac{\underline{g}(\underline{\theta} + \Delta\theta_i) - \underline{g}(\underline{\theta})}{\Delta\theta_i}, \tag{4.13}$$

In practice, this is feasible only if the gradients are available analytically; otherwise the computational effort would be overwhelming.

A practical problem in the application of finite difference techniques is how large to choose the "small step" $\Delta\theta_i$. On the one hand, it should be chosen as small as possible to make the approximation error small. On the other hand, the step size has to be chosen large enough to avoid large quantization errors resulting from the subtraction of two nearly equal values on a real computer. For more details refer to [504].

4.4.3 Steepest Descent

Steepest descent is the simplest version of (4.10), since the direction matrix $\underline{R}$ is set to identity $\underline{I}$:

$$\underline{\theta}_k = \underline{\theta}_{k-1} - \eta_{k-1}\, \underline{g}_{k-1}\,. \tag{4.14}$$

Hence, the search direction is the opposite gradient direction. This guarantees a decreasing loss function value for each iteration, if the line search algorithm finds the minimum along $\underline{p} = -\underline{g}$. The strategy to follow the opposite gradient direction is quite natural and gives acceptable results for the first few iterations. When converging to the minimum, however, it faces the same difficulties as the LMS algorithm for linear least squares problems (see Sect. 3.2), if the Hessian has a large eigenvalue spread χ, i.e., is badly conditioned. Figure 4.14 shows the typical behavior of a steepest descent algorithm in such a case, the so-called zig-zagging. This behavior follows from the orthogonality of successive search directions in steepest descent. Orthogonal search directions are a consequence of the line search procedure since at each iteration the minimum in the opposite gradient direction is found. The gradient in this minimum is orthogonal to that

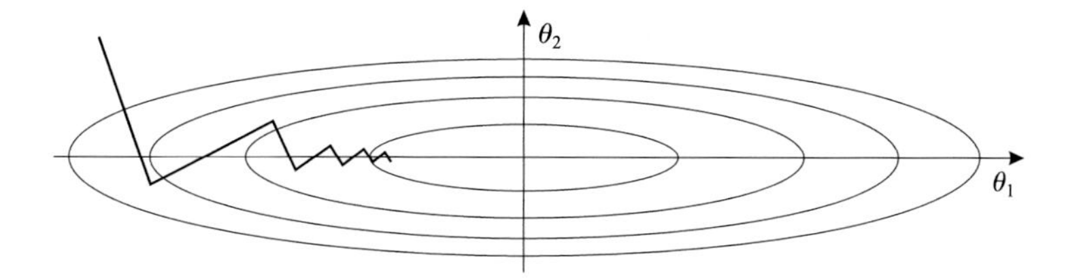

Fig. 4.14 "Zig-zagging": Typical behavior of steepest descent on a loss function with a Hessian with large eigenvalue spread. From most points in the parameter space, the gradient does not point to the minimum

in the previous iteration. Owing to this "zig-zagging" effect, a steepest descent algorithm will only optimize (in reasonable time) the parameters that correspond to dominant eigenvalues in the Hessian. Even for a linear optimization problem, the steepest descent algorithm (then equivalent to the LMS) would need an infinite number of iterations to converge. Owing to these drawbacks, steepest descent is not usually applied for nonlinear optimization. To summarize, steepest descent has the following important properties:

- no requirement for second-order derivatives;
- easy to understand;
- easy to implement;
- linear computational complexity;
- linear memory requirement complexity;
- very slow convergence;
- affected by a linear transformation of the parameters;
- generally requires an infinite number of iterations for the solution of a linear optimization problem.

It is interesting to note how the training of multilayer perceptrons fits into the concept of nonlinear optimization techniques. Werbos discovered the so-called backpropagation algorithm for multilayer perceptron training in 1974 [596], and in 1986 Rummelhard, Hinton and Williams rediscovered it [491] and started the modern boom in neural networks (together with the results of Hopfield's research in 1982 [259]). Strictly speaking, backpropagation of errors is nothing else but the calculation of the gradients by applying the chain rule for a multilayer perceptron. However, commonly the whole optimization algorithm is referred to as backpropagation. Sometimes even multilayer perceptrons (which are special neural network architectures) are called "backpropagation networks." Mixing up network architectures with training methods certainly underlines the disordered situation under which this algorithm was discovered.

The backpropagation algorithm is a simplified version of steepest descent utilizing sample adaptation. No line search is performed for determination of the step size η_{k-1} in each iteration. Rather some heuristic constant value η is chosen by the user. Such a primitive approach certainly leads to the difficulty of how to determine η. A large η may lead to oscillatory divergence, while a small η slows down the algorithm. Furthermore, the optimal value of η varies from one iteration to the next. Starting in the late 1980s and up to now, countless suggestions such as adaptive learning rates, momentum terms, etc. have been made to overcome or at least reduce the problem of choosing η. Since the early 1990s, multilayer perceptrons have been increasingly seen as a special kind of approximator with nonlinear parameters in the neural network community. Hence, backpropagation loses its importance, and more sophisticated nonlinear optimization techniques as described in this chapter are applied. In retrospect, it appears very strange that these widely known and mature algorithms were not utilized right from the beginning of the development of neural networks.

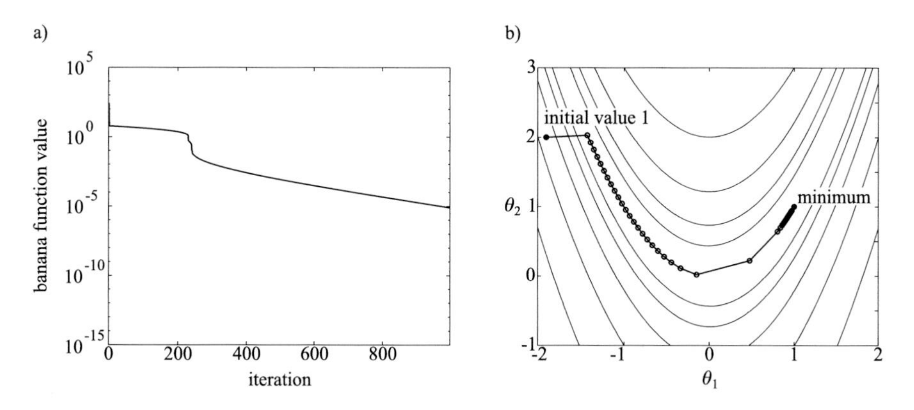

Fig. 4.15 Steepest descent method for minimization of the banana function. The rate of convergence (**a**) over the iterations, (**b**) in the parameter space (the circles mark every tenth function evaluation)

Example 4.4.1 Steepest Descent for Function Minimization
In this example, the banana function is minimized by applying the steepest descent technique. This method uses the gradients directly in order to perform a line search in this direction within each iteration. No second derivative information is exploited. The numbers of function evaluations and iterations are 3487 and 998, respectively, when starting at the first initial value, and 10956 and 3131, respectively, for the second initial value. This means that for each iteration on average about three function evaluations are required for the line search procedure. Line search in this and all following banana function examples is implemented as a mixed cubic and quadratic polynomial extrapolation/interpolation method; see Sect. 4.4.1. Figure 4.15 shows the rate of convergence. Note that in contrast to all other banana function examples, the circles in Fig. 4.15b mark only every tenth iteration because the rate of convergence is so slow. The performance of steepest descent is catastrophic on the banana function and can certainly not be generalized to all extent. However, it is quite realistic that steepest descent requires significantly more iterations than techniques that utilize second derivative information. □

4.4.4 Newton's Method

In Newton's method the direction matrix $\underline{R}$ in (4.10) is chosen as the inverse of the Hessian $\underline{H}_{k-1}^{-1}$ of the loss function at the point $\underline{\theta}_{k-1}$

$$\underline{\theta}_k = \underline{\theta}_{k-1} - \eta_{k-1}\,\underline{H}_{k-1}^{-1}\underline{g}_{k-1}\,. \tag{4.15}$$

Hence, for Newton's method, all second-order derivatives of the loss function have to be known analytically or estimated by finite difference techniques. In the classical Newton method, the step size η is set to 1 since this is the optimal choice for a linear optimization problem where the optimum would be reached after a single iteration. This follows directly from a second-order Taylor series expansion of the loss function. For nonlinear optimization problems, however, the optimum generally cannot be reached with a single iteration. Then a constant step size equal to 1 can be too small or too large owing to the non-quadratic surface of the loss function. To improve the robustness of Newton's method, the step size η is usually determined by line search (damped Newton method).

A problem with Newton's method is that (4.15) will decrease the loss function value (go "downhill") only for a positive definite Hessian $\underline{H}_{k-1}$. This is always true in the neighborhood of the optimum, but a positive definite Hessian cannot necessarily be expected for the initial point $\underline{\theta}_0$ and the first iterations. To avoid this problem a modified Newton method is often applied in which the Hessian is replaced by a matrix $\underline{\widetilde{H}}_{k-1}$ that is guaranteed to be positive definite but is close to $\underline{H}_{k-1}$.

The fundamental importance of Newton's method comes from the fact that its rate of convergence is of second order if $\underline{H}_{k-1}$ is positive definite [504]. This is the fastest rate normally encountered in nonlinear optimization. The drawbacks of Newton's method are the requirement for the second-order derivatives and the high computational effort for inverting the Hessian. This restricts the application of Newton's method to small problems.

Newton's method is widely known and probably more familiar for the numerical search of the zero of a given function $y = f(x)$. The iteration then becomes $x_k = x_{k-1} - f(x_{k-1}) / f'(x_{k-1})$. If the minimum of the function $f(x)$ is searched, the zero of the first derivative of the function has to be found by iterating $x_k = x_{k-1} - f'(x_{k-1}) / f''(x_{k-1})$. This is equivalent to (4.15) for the single parameter case and $\eta = 1$.

To summarize, Newton's method has the following important properties:

- requirement for second-order derivatives;
- cubic computational complexity owing to matrix inversion;
- quadratic memory requirement complexity, since the Hessian has to be stored;
- fastest convergence normally encountered in nonlinear optimization;
- requires a single iteration for the solution of a linear optimization problem;
- unaffected by a linear transformation of the parameters;
- suited best for small problems (order of ten parameters).

Example 4.4.2 Newton for Function Minimization

In this example, the banana function is minimized by applying the Newton method. It multiplies the inverse Hessian with the gradient in order to perform a line search in this direction within each iteration. The second derivatives are given analytically; see Example 4.0.1. The numbers of function evaluations and iterations are 54 and 17, respectively, when starting at the first initial value, and 36 and 11, respectively, for the second initial value. Figure 4.16 shows the rate of convergence from initial

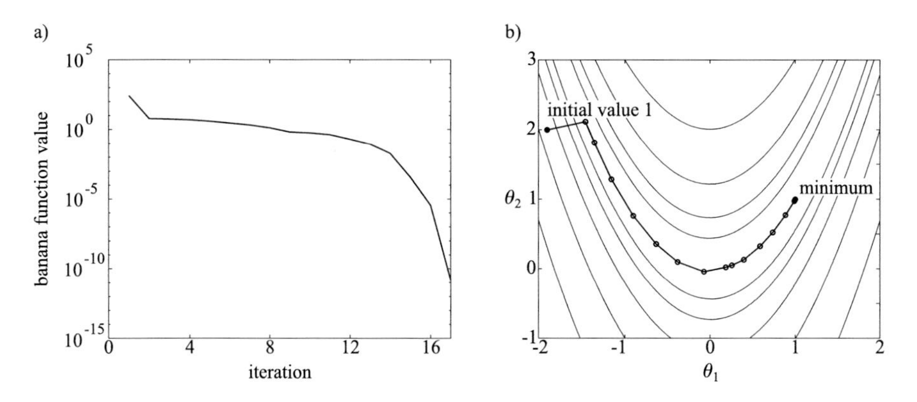

Fig. 4.16 Newton's method for minimization of the banana function. The rate of convergence (**a**) over the iterations, (**b**) in the parameter space (the circles mark the function evaluations)

value 1. The Newton method is outperformed only by the Gauss-Newton method on the banana function. Note, however, that it is the only technique that requires knowledge of the Hessian. Compared with the quasi-Newton methods that are discussed in the following subsection, the Newton algorithm has significantly faster convergence close to the minimum, because in this region a second-order Taylor expansion is very accurate. □

4.4.5 *Quasi-Newton Methods*

The main drawback of Newton's method is the requirement for second-order derivatives. If they are not available analytically, the Hessian has to be found by finite difference techniques, which requires $\mathcal{O}(n^2)$ gradient calculations. Therefore, Newton's method becomes impractical even for medium-sized problems if the Hessian is not available analytically. But even if the second-order derivatives are computationally cheap to obtain, the required inversion of the $n \times n$ Hessian strongly limits the size of problems that can be tackled. The idea of quasi-Newton methods (also known as variable metric methods) is to replace the Hessian or its inverse in (4.15) by an approximation

$$\underline{\theta}_k = \underline{\theta}_{k-1} - \eta_{k-1}\, \underline{\hat{H}}_{k-1}^{-1} \underline{g}_{k-1}\,, \tag{4.16}$$

$$\text{with either} \quad \underline{\hat{H}}_k = \underline{\hat{H}}_{k-1} + \underline{Q}_{k-1} \quad \text{or} \quad \underline{\hat{H}}_k^{-1} = \underline{\hat{H}}_{k-1}^{-1} + \underline{\widetilde{Q}}_{k-1}\,. \tag{4.17}$$

The approach of approximating the inverse Hessian directly has the important advantage that no matrix inversion has to be performed in (4.16). Hence, this approach is by far the most common one. The approximation of the Hessian or its inverse is usually started with $\underline{H}_0 = \underline{I}$, that is, the search is started in the opposite gradient direction.

The most widely known quasi-Newton methods are based on the Broyden-Fletcher-Goldfarb-Shanno (BFGS) formula with $\Delta\underline{\theta}_{k-1} = \underline{\theta}_k - \underline{\theta}_{k-1}$ and $\Delta\underline{g}_{k-1} = \underline{g}_k - \underline{g}_{k-1}$ [504],

$$\underline{\hat{H}}_k^{-1} = \left(\underline{I} - \frac{\Delta\underline{\theta}_{k-1}\,\Delta\underline{g}_{k-1}^T}{\Delta\underline{\theta}_{k-1}^T\,\Delta\underline{g}_{k-1}}\right)\underline{\hat{H}}_{k-1}^{-1}\left(\underline{I} - \frac{\Delta\underline{\theta}_{k-1}\,\Delta\underline{g}_{k-1}^T}{\Delta\underline{\theta}_{k-1}^T\,\Delta\underline{g}_{k-1}}\right)^T + \frac{\Delta\underline{\theta}_{k-1}\,\Delta\underline{\theta}_{k-1}^T}{\Delta\underline{\theta}_{k-1}^T\,\Delta\underline{g}_{k-1}}, \qquad (4.18)$$

or on the Davidon-Fletcher-Powell (DFP) formula [504]. However, the practical experience of both algorithms has made it clear that the BFGS formula is superior in almost all cases [504]. Numerous studies show that it is sufficient and even more efficient to carry out the line search for η only very roughly [504]. Note that there exist equivalent formulations of BFGS and DFP that update $\underline{\hat{H}}_k$ instead of the inverse Hessian. The kind of formula for $\underline{Q}$ or $\underline{\tilde{Q}}$ in (4.17) distinguishes all quasi-Newton methods. The most important properties of quasi-Newton methods are

- no requirement for second-order derivatives;
- quadratic computational complexity owing to matrix multiplication;
- quadratic memory requirement complexity, since the Hessian has to be stored;
- very fast convergence;
- requires at most n iterations for the solution of a linear optimization problem;
- affected by a linear transformation of the parameters;
- suited best for medium-sized problems (order of 100 parameters).

Example 4.4.3 Quasi-Newton for Function Minimization
In this example, the banana function is minimized by applying the BFGS quasi-Newton method. It exploits only gradient information and accumulates knowledge about the Hessian according to (4.18). The numbers of function evaluations and iterations are 79 and 24, respectively, when starting at the first initial value, and 55 and 17, respectively, for the second initial value. Figure 4.17 shows the rate of convergence from initial value 1. Because the information about the Hessian matrix is not used, the quasi-Newton methods converge slower than the Newton technique. For many applications, however, the Hessian may not be available analytically. In these cases, the application of Newton's methods would require finite difference techniques for the determination of the Hessian. From the number function evaluations given above, it is clear that a Newton method with these additional finite difference computations will be much slower than a quasi-Newton method. Even if the Hessian is computationally cheap to obtain, the Newton method falls behind quasi-Newton methods in terms of computational demand as the dimensionality of the problem increases. Since each iteration of Newton's method is $\mathcal{O}(n^3)$ and of quasi-Newton methods is $\mathcal{O}(n^2)$, there exist a number of parameters n at which quasi-Newton methods have a lower overall computational demand although they require more iterations. □

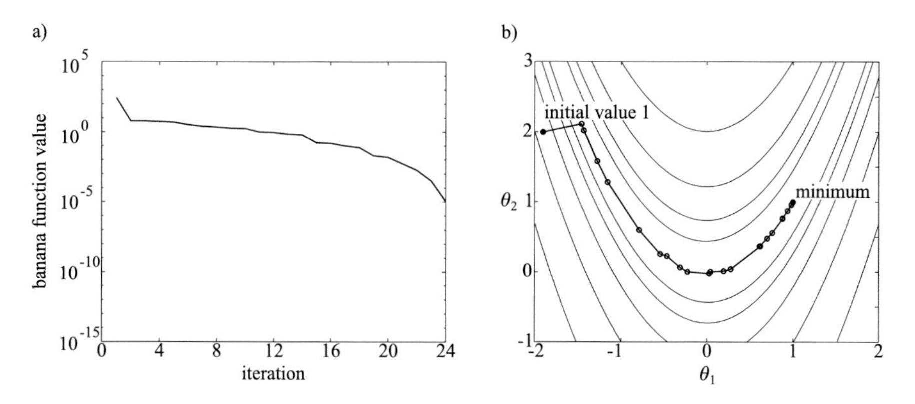

Fig. 4.17 Quasi-Newton method for minimization of the banana function. The rate of convergence (**a**) over the iterations, (**b**) in the parameter space (the circles mark the function evaluations)

4.4.6 Conjugate Gradient Methods

All quasi-Newton methods have quadratically increasing memory requirements and computational complexity with the number of parameters n. For large problems, it often does not pay off to approximate the Hessian. Conjugate gradient methods avoid a direct approximation of the Hessian and therefore are linear in their memory requirements and computational complexity with respect to n. A conjugate gradient method results if the $\underline{H}_{k-1}^{-1}$ matrix BFGS formula (4.18) is reset to $\underline{I}$ in each iteration. Then it is possible to update the search direction $\underline{p}$ directly, which makes the handling of $n \times n$ matrices superfluous. Because the conjugate gradient methods can be seen as rough approximations of the quasi-Newton methods, generally the search directions will be worse and conjugate gradient methods will require more iterations for convergence than quasi-Newton methods. However, the overall computation time of the conjugate gradient methods will be smaller for large problems, since each iteration is much less computationally expensive. Conjugate gradient algorithms can be described by

$$\underline{\theta}_k = \underline{\theta}_{k-1} - \eta_{k-1}\, \underline{p}_{k-1} \tag{4.19a}$$

$$\text{with} \quad \underline{p}_{k-1} = \underline{g}_{k-1} - \beta_{k-1}\, \underline{p}_{k-2} \tag{4.19b}$$

with a scalar β_{k-1} that distinguishes different conjugate gradient methods. The most popular choices are due to Fletcher and Reeves with

$$\beta_{k-1} = \frac{\underline{g}_{k-1}^T \underline{g}_{k-1}}{\underline{g}_{k-2}^T \underline{g}_{k-2}} \tag{4.20}$$

or due to Hestenes and Stiefel or Polak and Ribière [504]. The scalar factor β represents the knowledge carried over from the previous iterations. Thus, the conjugate gradient methods can be seen as a compromise between steepest descent, where no information about the previous iterations is exploited, and the quasi-Newton methods, where the information about the approximation of the second-order derivatives in form of the whole Hessian matrix is exploited. It can also be seen as a sophisticated batch mode version of backpropagation with momentum, where the learning rate is determined by line search and the momentum is determined by the choice of β [57]. Note that for fast convergence, in contrast to quasi-Newton methods, conjugate gradient algorithms generally require an accurate line search procedure for η [504].

On a linear optimization problem, all choices for β_{k-1} are equivalent, and the conjugate gradient converges to the optimum in at most n iterations. In fact, conjugate gradient methods are also used to solve linear least squares problems, in particular, if they are large and sparse.

In (4.19b), conjugate search directions are created. Two search directions $\underline{p}_i$ and $\underline{p}_j$ are conjugate if $\underline{p}_i^T \underline{H}\, \underline{p}_j = 0$. This implies that $\underline{p}_i$ and $\underline{p}_j$ are independent search directions. The geometric interpretation of a conjugate gradient method is as follows. At each iteration k, the search is performed along a direction that is orthogonal to all previous (maximally n) gradient changes $\Delta \underline{g}_i$, $i = 0, \ldots, k-1$. By taking into account the information about the previous search directions, the "zig-zagging" effect known from steepest descent can be avoided. As pointed out in [57], during the running of the algorithm, the conjugacy of the search directions tends to deteriorate, and so it is common practice to restart the algorithm after every n steps by resetting the search vector $\underline{p}$ to the negative gradient direction. To summarize, the most important properties of conjugate gradient methods are

- no requirement for second-order derivatives;
- linear computational complexity;
- linear memory requirement complexity, since no Hessian has to be stored;
- fast convergence;
- requires at most n iterations for the solution of a linear optimization problem;
- affected by a linear transformation of the parameters;
- suited best for large problems (order of 1000 and more parameters).

Example 4.4.4 Conjugate Gradient for Function Minimization
In this example, the banana function is minimized by applying the conjugate gradient algorithm due to Fletcher and Reeves. It exploits only gradient information but utilizes information about the previous gradient directions. In contrast to quasi-Newton methods, no complete approximation of the Hessian is performed. Therefore, the computational effort in each iteration is smaller than for quasi-Newton methods, but the total number of iterations is usually larger. For the banana function problem, the numbers of function evaluations and iterations are 203 and 36, respectively, when starting at the first initial value, and 261 and 44, respectively, for the second initial value. Figure 4.18 shows the rate of convergence from initial value 1. In comparison to the quasi-Newton method, conjugate gradient algorithms are much more sensible to inaccurate line search. For the more accurate line search

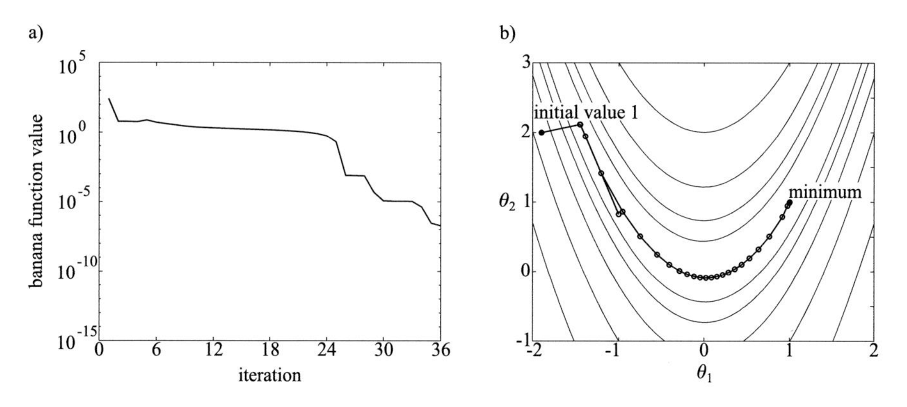

Fig. 4.18 Conjugate gradient method for minimization of the banana function. The rate of convergence (**a**) over the iterations, (**b**) in the parameter space (the circles mark the function evaluations)

procedure, more function evaluations (about six instead of three) are necessary in each iteration. If the same (inaccurate) line search technique as for the quasi-Newton method is applied in combination with conjugate gradient, 561 function evaluations and 187 iterations are required from the first initial value. □

4.5 Nonlinear Least Squares Problems

In the previous section, general gradient-based nonlinear optimization algorithms were introduced. These algorithms do not make any assumptions about the loss function except its smoothness. However, as discussed in Sect. 2.3, a (possibly weighted) quadratic loss function of the form

$$I(\underline{\theta}) = \sum_{i=1}^{N} q_i e^2(i, \underline{\theta}) \tag{4.21}$$

is by far the most common in practice. If the parameters are linear, a least squares problem arises. For nonlinear parameters, the optimization of the loss function

$$I(\underline{\theta}) = \sum_{i=1}^{N} f^2(i, \underline{\theta}) \tag{4.22}$$

is known as a nonlinear least squares problem. Note that (4.21) is a special case of (4.22). In this section, effective methods are introduced that exploit this information on the loss function's structure. In vector form (4.22) becomes

$$I(\underline{\theta}) = \underline{f}^T \underline{f} \qquad \text{with} \quad \underline{f} = [f(1, \underline{\theta})\ f(2, \underline{\theta})\ \cdots\ f(N, \underline{\theta})]^T\,, \tag{4.23}$$

where here and in the following the argument "$(\underline{\theta})$" is dropped for brevity.

In the following, the gradient and the Hessian of this loss function will be derived [504]. The jth component of the gradient is

$$g_j = 2 \frac{\partial I(\underline{\theta})}{\partial \theta_j} = 2 \sum_{i=1}^{N} f(i) \frac{\partial f(i)}{\partial \theta_j} . \tag{4.24}$$

Therefore, with the Jacobian

$$\underline{J} = \begin{bmatrix} \partial f(1) \, / \, \partial \theta_1 & \cdots & \partial f(1) \, / \, \partial \theta_n \\ \vdots & & \vdots \\ \partial f(N) \, / \, \partial \theta_1 & \cdots & \partial f(N) \, / \, \partial \theta_n \end{bmatrix} \tag{4.25}$$

the gradient can be written as

$$\underline{g} = 2 \, \underline{J}^T \underline{f} . \tag{4.26}$$

The entries of the Hessian of the loss function are obtained by calculation of the derivative of the gradient (4.24) with respect to parameter θ_l:

$$H_{lj} = \frac{\partial^2 I(\underline{\theta})}{\partial \theta_j \partial \theta_l} = 2 \sum_{i=1}^{N} \left(\frac{\partial f(i)}{\partial \theta_l} \frac{\partial f(i)}{\partial \theta_j} + f(i) \frac{\partial^2 f(i)}{\partial \theta_l \partial \theta_j} \right) . \tag{4.27}$$

The first term in the sum of (4.27) is the squared Jacobian of $\underline{f}$, and the second term is $\underline{f}(i)$ multiplied by the Hessian of $f(i)$. Denoting the entries of the Hessian of $f(i)$ as $T_{lj}(i) = \partial^2 f(i) \, / \, \partial \theta_l \partial \theta_j$, the Hessian of the loss function in (4.27) becomes

$$\underline{H} = 2 \, \underline{J}^T \underline{J} + 2 \sum_{i=1}^{N} f(i) \, \underline{T}(i) . \tag{4.28}$$

If the second term in (4.28) is denoted as $\underline{S}$, the Hessian of the loss function can be written as

$$\underline{H} = 2 \, \underline{J}^T \underline{J} + 2 \, \underline{S} . \tag{4.29}$$

The nonlinear least squares methods introduced below exploit this structure of the Hessian to derive an algorithm that is generally more efficient than the general approaches discussed in the previous section. The nonlinear least squares algorithms can be divided into two categories.

Methods of the first category neglect $\underline{S}$ in (4.29). The approximation of the Hessian $\underline{H} \approx \underline{J}^T \underline{J}$ is justified only if $\underline{S} \approx 0$. This condition is met for small $f(i)$ in (4.28). Since the $f(i)$ usually represents residuals (errors), these methods are called *small residual algorithms*. The big advantage of this approach is that the Hessian can be obtained by the evaluation of first-order derivatives (Jacobian) only.

Since for most problems the gradients of the residuals are available analytically, the approximate Hessian can be computed with low computational cost. The two most popular algorithms are the Gauss-Newton and Levenberg-Marquardt methods described below.

The approaches of the second category are called *large residual algorithms*. They do not neglect the $\underline{S}$ term but spend extra computation on either approximating $\underline{S}$ or switching between a universal Newton and the nonlinear least squares Gauss-Newton method. Note that, in principle, it is possible to compute $\underline{S}$ exactly. However, this would require the evaluation of many second-order derivative terms. As for the general Newton method, this would be practical only if the second-order derivatives were analytically available and the dimensionality of the optimization problem was low. Therefore, such an approach would lead directly to the general Newton method, annihilating the advantages of the nonlinear least squares structure. Owing to the extra effort involved, the large residual algorithms are recommended only if $\underline{S}$ is significant. For a more detailed treatment, see [504].

4.5.1 Gauss-Newton Method

The Gauss-Newton method is the nonlinear least squares version of the general Newton method in (4.15). Since the gradient can be expressed as $\underline{g} = \underline{J}^T \underline{f}$ and the Hessian is approximated by $\underline{H} \approx \underline{J}^T \underline{J}$, the Gauss-Newton algorithm becomes

$$\underline{\theta}_k = \underline{\theta}_{k-1} - \eta_{k-1} \left(\underline{J}_{k-1}^T \underline{J}_{k-1} \right)^{-1} \underline{J}_{k-1}^T \underline{f}_{k-1} . \tag{4.30}$$

It approximately (as $\underline{S} \rightarrow \underline{0}$) shares the properties of the general Newton algorithm, but no second-order derivatives are required. As with the classical Newton algorithm, in its original form, no line search is performed and η_{k-1} is set to 1. However, a line search makes the algorithm more robust, and consequently Gauss-Newton with line search, the so-called damped Gauss-Newton, is widely applied.

In practice, the matrix inversion in (4.30) is not usually performed explicitly. Instead the following n-dimensional linear equation system is solved to obtain the search direction $\underline{p}_{k-1}$:

$$\left(\underline{J}_{k-1}^T \underline{J}_{k-1} \right) \underline{p}_{k-1} = \underline{J}_{k-1}^T \underline{f}_{k-1} . \tag{4.31}$$

This system can either be solved via Cholesky factorization or it can be formulated as the linear least squares problem $||\underline{J}_{k-1} \underline{p}_{k-1} + \underline{f}_{k-1}|| \rightarrow \min$ and then be solved via orthogonal factorization of $\underline{J}_{k-1}$ [193]. However, no matter how the search direction is evaluated, problems occur if the matrix $\underline{J}_{k-1}^T \underline{J}_{k-1}$ is poorly conditioned or even singular. The smaller the least eigenvalue of $\underline{J}_{k-1}^T \underline{J}_{k-1}$ is, the slower is the rate of convergence of the Gauss-Newton method. The Levenberg-Marquardt algorithm discussed next deals with these problems.

Example 4.5.1 Gauss-Newton for Function Minimization
In this example, the banana function is minimized by applying the Gauss-Newton method. In order to apply a nonlinear least squares technique for function minimization, the function has to be a sum of squares as in (4.22). The banana function $I(\underline{\theta}) = 100\,(\theta_2 - \theta_1^2)^2 + (1-\theta_1)^2$ has this form. It consists of two quadratic terms. Consequently, $\underline{f}$ in (4.22) is given by

$$\underline{f}(\underline{\theta}) = \begin{bmatrix} 10\,(\theta_2 - \theta_1^2) \\ 1 - \theta_1 \end{bmatrix}, \tag{4.32}$$

and the Jacobian becomes

$$\underline{J}(\underline{\theta}) = \begin{bmatrix} -20\,\theta_1 & 10 \\ -1 & 0 \end{bmatrix}. \tag{4.33}$$

The Gauss-Newton method approximates the Hessian of the loss function by $2\,\underline{J}^T\underline{J}$ as explained above. Since the banana function actually is zero at its minimum, this approximation is justified. The damped Gauss-Newton method requires 40 and 12 function evaluations and iterations, respectively, when starting at the first initial value, and 21 and 7, respectively, for the second initial value. Figure 4.19 shows the rate of convergence from initial value 1. Therefore, this method is the fastest on the banana function example. On most problems, however, a Newton method will give superior performance because it utilizes exact second derivative information. If the Hessian is not available analytically, nonlinear least squares methods have an advantage over quasi-Newton approaches because they do not need to accumulate second-order information. Thus, the Gauss-Newton or the more robust Levenberg-Marquardt algorithm discussed in the next subsection is

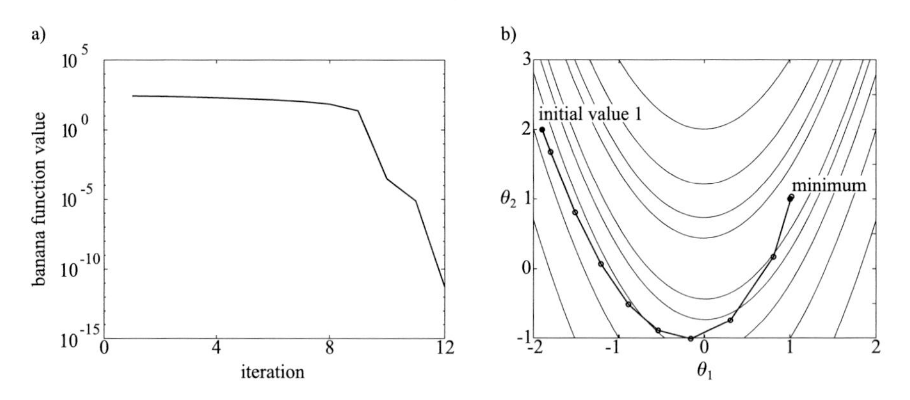

Fig. 4.19 Gauss-Newton method for minimization of the banana function. The rate of convergence (**a**) over the iterations, (**b**) in the parameter space (the circles mark the function evaluations)

recommended whenever the problem can be formulated in nonlinear least squares form. □

4.5.2 Levenberg-Marquardt Method

The Levenberg-Marquardt algorithm is an extension of the Gauss-Newton algorithm. The idea is to modify (4.30) to

$$\underline{\theta}_k = \underline{\theta}_{k-1} - \eta_{k-1} \left(\underline{J}_{k-1}^T \underline{J}_{k-1} + \alpha_{k-1} \underline{I} \right)^{-1} \underline{J}_{k-1}^T \underline{f}_{k-1}, \tag{4.34}$$

where again the matrix inversion is not performed explicitly but by solving

$$\left(\underline{J}_{k-1}^T \underline{J}_{k-1} + \alpha_{k-1} \underline{I} \right) \underline{p}_{k-1} = \underline{J}_{k-1}^T \underline{f}_{k-1}. \tag{4.35}$$

The addition of $\alpha_{k-1}\underline{I}$ in (4.34) to the approximation of the Hessian $\underline{J}_{k-1}^T \underline{J}_{k-1}$ is equivalent to the regularization technique in ridge regression for linear least squares problems; see Chap. 3.1. It solves the problems of a poorly conditioned $\underline{J}_{k-1}^T \underline{J}_{k-1}$ matrix.

The Levenberg-Marquardt algorithm can be interpreted as follows. For small values of α_{k-1}, it approaches the Gauss-Newton algorithm, while for large values of α_{k-1}, it approaches the steepest descent method. Close to the optimum the second-order approximation of the loss function performed by the (Gauss)-Newton method is very good, and a small α_{k-1} should be chosen. Far away from the optimum, the (Gauss)-Newton method may diverge, and a large α_{k-1} should be chosen. For sufficiently large values of α_{k-1}, the matrix $\underline{J}_{k-1}^T \underline{J}_{k-1} + \alpha_{k-1}\underline{I}$ is positive definite, and a descent direction is guaranteed. Therefore, a good strategy for the determination of α_{k-1} is as follows. Initially, some positive value for α_{k-1} is chosen. Then at each iteration α_{k-1} is decreased by some factor, since the parameters are assumed to approach their optimal values where the Gauss-Newton method is powerful. If the decrease of α_{k-1} leads to a bad search direction (i.e., the loss function value increases), then α_{k-1} is again increased by some factor until a downhill direction results.

Note that although the Levenberg-Marquardt algorithm is a nonlinear least squares technique, its regularization idea can also be applied to the general Newton or quasi-Newton methods.

Example 4.5.2 Levenberg-Marquardt for Function Minimization
In this example, the banana function is minimized by applying the Levenberg-Marquardt method. The advantages are basically the same as for the Gauss-Newton method. However, owing to the modified search direction, the Levenberg-Marquardt algorithm is more robust. The numbers of function evaluations and iterations are 63 and 18, respectively, when starting at the first initial value, and 17 and 6,

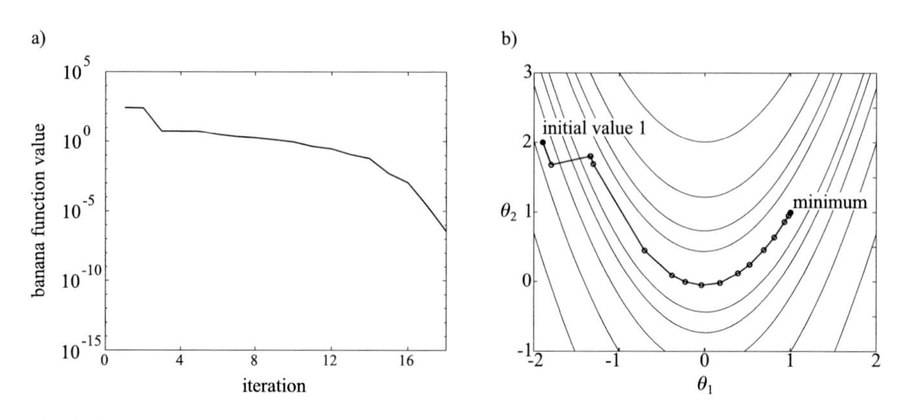

Fig. 4.20 Levenberg-Marquardt method for minimization of the banana function. The rate of convergence (**a**) over the iterations, (**b**) in the parameter space (the circles mark the function evaluations)

respectively, for the second initial value. Figure 4.20 shows the rate of convergence from initial value 1. The Levenberg-Marquardt algorithm is slower than Gauss-Newton from initial value one and slightly faster from initial value 2. Owing to its fast convergence and robustness, the Levenberg-Marquardt algorithm is applied to most nonlinear local optimization problems treated within this book. Note that a lot of expertise in determining the regularization parameter α is necessary for a good implementation [70]. □

4.6 Constrained Nonlinear Optimization

Techniques for nonlinear local optimization with constraints are much newer and mathematically more complicated than the unconstrained methods discussed above. For an extensive treatment of such techniques, refer to [577]. Constraints emerge from available knowledge or restrictions about the parameters. This implies that usually only interpretable parameters are constrained. A typical example is a control variable that is bounded by its minimum and maximum value (e.g., valve open and valve closed). Another less direct example is the position of membership functions in a fuzzy system. They are also constrained to the range of the physical variable that they represent in order to keep the fuzzy system interpretable. From these examples, it can be concluded that controller optimization and fuzzy membership function optimization are characteristic applications of constrained optimization techniques.

Obviously, knowledge about constraints gives valuable information about the parameters. It reduces the size of the search space and therefore should lead to a speed-up in optimization. But constraints are not easy to incorporate into the optimization technique and certainly require extra effort. However, if a constrained optimization technique is chosen, each additional constraint generally improves the convergence of the algorithm.

A general constrained nonlinear optimization problem may include inequality and equality constraints. Thus, the task can be formulated as follows:

$$I(\underline{\theta}) \longrightarrow \min_{\theta} \tag{4.36}$$

subject to

$$g_i(\underline{\theta}) \leq 0 \qquad i = 1, \ldots, m\,, \tag{4.37a}$$

$$h_j(\underline{\theta}) = 0 \qquad j = 1, \ldots, l\,. \tag{4.37b}$$

This leads to the following Lagrangian, which is the unconstrained loss function plus the weighted constraints:

$$L(\underline{\theta}, \underline{\lambda}) = I(\underline{\theta}) + \sum_{i=1}^{m} \lambda_i\, g_i(\underline{\theta}) + \sum_{j=1}^{l} \lambda_{j+m}\, h_j(\underline{\theta})\,. \tag{4.38}$$

The famous Kuhn-Tucker equations give the following necessary (and often sufficient) conditions for optimality [577]

$$\underline{\theta}_{\text{opt}} \quad \text{is feasible, i.e., meets the constraints}\,, \tag{4.39}$$

$$\lambda_i\, g_i(\underline{\theta}_{\text{opt}}) = 0 \qquad i = 1, \ldots, m \qquad \lambda_i \geq 0\,, \tag{4.40}$$

$$\frac{\partial I(\underline{\theta}_{\text{opt}})}{\partial \underline{\theta}} + \sum_{i=1}^{m} \lambda_i \frac{\partial g_i(\underline{\theta}_{\text{opt}})}{\partial \underline{\theta}} + \sum_{j=1}^{l} \lambda_{j+m} \frac{\partial h_j(\underline{\theta}_{\text{opt}})}{\partial \underline{\theta}} = \underline{0}\,. \tag{4.41}$$

Equation (4.39) demands that the parameters meet the constraints (4.37a) and (4.37b). Equation (4.40) imposes that the Lagrange multipliers of those inequality constraints with $g_i(\underline{\theta}_{\text{opt}}) < 0$ are zero. Equation (4.41) requires the first-order derivative of the Lagrangian to be equal to zero.

Most modern constrained nonlinear optimization algorithms take the Lagrange multipliers into account or even approximate (linear or quadratic) the Kuhn-Tucker equations. Perhaps the most powerful of these algorithms are sequential quadratic programming (SQP), which utilizes second-order derivative information. It is included in many optimization packages; see, e.g., [70]. On the other hand, there are some very simple and easy-to-use approaches, that are very widely applied, although they are regarded as being obsolete [577]. Nevertheless, they are of significant practical importance, and thus these methods are discussed briefly in the following.

The simplest and most straightforward way to deal with the constraints is to modify the loss function in order to incorporate the constraints. Then a conventional

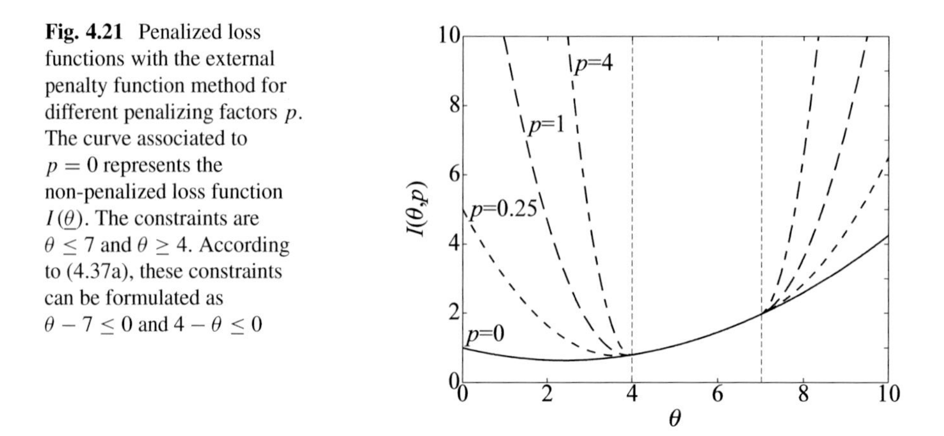

Fig. 4.21 Penalized loss functions with the external penalty function method for different penalizing factors p. The curve associated to $p = 0$ represents the non-penalized loss function $I(\underline{\theta})$. The constraints are $\theta \leq 7$ and $\theta \geq 4$. According to (4.37a), these constraints can be formulated as $\theta - 7 \leq 0$ and $4 - \theta \leq 0$

unconstrained optimization algorithm can be applied to solve the problem. The loss function is extended by an additional penalty term to limit constraint violations:

$$I_p(\underline{\theta}, p) = I(\underline{\theta}) + p\, P(\underline{\theta})\,. \tag{4.42}$$

The penalty function $P(\underline{\theta})$ should be large for constraint violations and small for feasible parameters $\underline{\theta}$. This penalty function is weighted with a factor of p. For $p = 0$ the loss function of the unconstrained problem is recovered, and constraint violations are not penalized. For $p = \infty$ constraint violations are prevented but the original loss function is extremely distorted. Consequently, large penalty values p lead to numerical ill-conditioning of the optimization procedure. Therefore, some tradeoff for the choice of p is required. Usually, this tradeoff is performed by changing p while the optimization progresses.

Two different approaches must be distinguished: the exterior and interior penalty function methods. For an exterior approach, the penalty function may be defined as (see Fig. 4.21)

$$P_{\text{exterior}}(\underline{\theta}) = \sum_{i=1}^{m} \left(\max\left(0, g_i(\underline{\theta})\right)\right)^2 + \sum_{j=1}^{l} \left(h_j(\underline{\theta})\right)^2\,. \tag{4.43}$$

Thus, if all constraints are satisfied ($g_i(\underline{\theta}) \leq 0$ and $h_j(\underline{\theta}) = 0$), the penalty function is equal to zero ($P(\underline{\theta}) = 0$). The reason for the quadratic form of (4.43) is that this guarantees a smooth transition from the regions where the constraints are satisfied to the regions where they are violated. This smooth transition is important to ensure acceptable convergence of the applied nonlinear optimization technique. Note, however, that only loss function values and first-order derivatives are continuous at the constraints boundary while all higher-order derivatives are discontinuous. This fact causes problems with all sophisticated optimization methods that utilize second-order information such as Newton, quasi-Newton, and

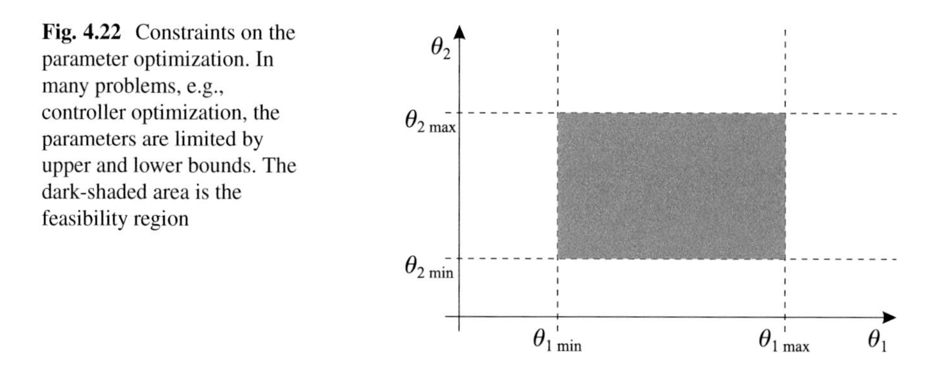

Fig. 4.22 Constraints on the parameter optimization. In many problems, e.g., controller optimization, the parameters are limited by upper and lower bounds. The dark-shaded area is the feasibility region

conjugate gradient methods. Although these penalty function methods generally have poor performance, they can be effectively applied for problems with simple constraints, such as shown in Fig. 4.22. This kind of constraint arises in control, for example, where the manipulated variable is always bounded by the upper and lower limits of the actuator. Since in Fig. 4.22 the region of feasible parameters is convex and the initial parameters typically are within this region, the degradation of the loss function by the penalty function is of minor importance. For the controller optimization example, the exterior penalty function approach usually causes no problems at all. For example, if any constraint due to the actuator is *violated* during the optimization of the manipulated variable, the process performs exactly as it would work *on* the constraints. Therefore, the loss function is constant in the infeasible regions. Thus, a small value of p (preventing ill-conditioning) is sufficient to drive the parameters (manipulated variables) into the feasible region.

As discussed above, the choice of p requires a tradeoff between minimal distortion of the original loss function and maximum constraint violation penalty. This tradeoff is usually performed by starting with small values for p in order to obtain a good conditioned optimization problem. As the optimization process progresses, p is increased in order to drive the parameters out of the constraint violation regions. Experiments are necessary to find a good schedule for increasing p.

While the exterior methods start with parameters that possibly violate the constraints and then iteratively increases the penalty, the interior methods follow the opposite philosophy. The penalty function approaches infinity as the parameters approach the constraint boundaries (see Fig. 4.23):

$$P_{\text{interior}}(\underline{\theta}) = \sum_{i=1}^{m} \frac{-1}{g_i(\underline{\theta})} + \sum_{j=1}^{l} \left(h_j(\underline{\theta})\right)^2 . \tag{4.44}$$

Therefore, if the initial parameters are satisfying the constraints, they are forced to stay feasible during the optimization procedure. Thus, the strategy for the choice of p is exactly opposite to the exterior method. The interior penalty function

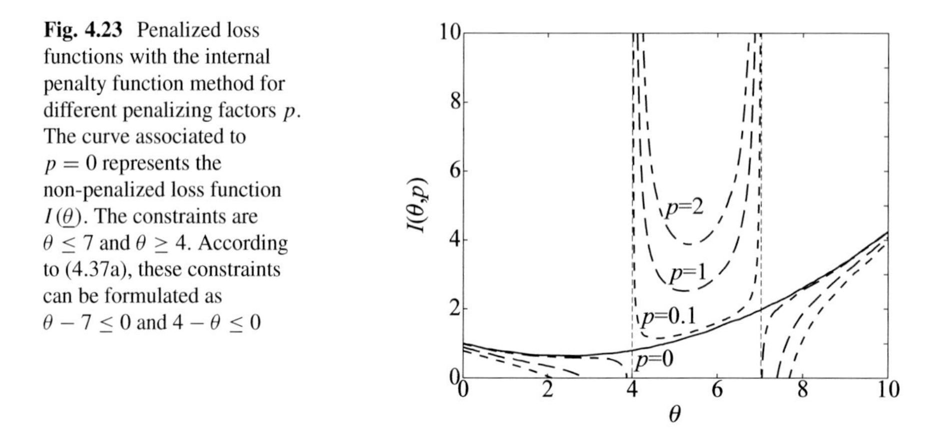

Fig. 4.23 Penalized loss functions with the internal penalty function method for different penalizing factors p. The curve associated to $p = 0$ represents the non-penalized loss function $I(\underline{\theta})$. The constraints are $\theta \leq 7$ and $\theta \geq 4$. According to (4.37a), these constraints can be formulated as $\theta - 7 \leq 0$ and $4 - \theta \leq 0$

method starts with large values of p and decreases p until zero as the optimization progresses.

4.7 Summary

To clarify the somewhat confusing naming of the unconstrained nonlinear local optimization algorithms, Table 4.1 shows the relationship between those techniques based on general loss functions and those that solve nonlinear least squares problems.

Table 4.2 gives a rough orientation to the problems which for which general nonlinear local optimization techniques may be best suited. It is important to note that in batch mode optimization, steepest descent is generally outperformed by conjugate gradient algorithms on any problem size (number of parameters). However, in sample mode optimization, the performance of steepest descent is not as catastrophic since line search techniques cannot be reasonably applied [32].

Table 4.3 summarizes the results obtained by the minimization of the banana function with the nonlinear local optimization techniques. The number of function evaluations and the number of iterations required in order to find the minimum with a parameter accuracy of at least 0.001 are shown. For the simplex and Hooke-Jeeves search, the two numbers are the same because no line search is performed. For all gradient-based methods, the number of function evaluations is larger than the number of iterations since the line search procedure performed in each iteration requires additional function evaluations. Note that the Newton, quasi-Newton, and nonlinear least squares techniques require only three to four function evaluations in each iteration because the line search is carried out roughly, while the conjugate gradient method demands accurate line search and consequently requires more function evaluations per iteration.

Table 4.1 Relationship between nonlinear local optimization techniques for general loss functions and for nonlinear least squares problems

General loss functions	Nonlinear least squares
Search methods	—
Steepest descent	—
Newton	Gauss-Newton
Regularized Newton	Levenberg-Marquardt
Quasi-Newton	—
Conjugate gradient	—

Table 4.2 Overview of general nonlinear local optimization techniques

Technique	No. of para.	Remarks
Search methods	1–10	For non-available analytic first derivative
Steepest descent	100–	Only reasonable for sample adaptation
Newton	1–10	For available analytic second derivatives
Quasi-Newton	10–100	Approximation of second derivatives
Conjugate gradient	100–	Best suited for large problems

Table 4.3 Number of function evaluations/number of iterations for the banana function minimization with nonlinear local optimization techniques. Note that both the number of function evaluations and the number of iterations required do not correspond exactly to the necessary computation time because the effort for evaluation of the search direction $\underline{p}$ is different for each technique

Technique	Initial value 1	Initial value 2
Simplex search	200/200	221/221
Hooke-Jeeves	230/230	175/175
Steepest descent	3487/998	10956/3131
Newton (with line search)	54/17	36/11
Quasi-Newton (BFGS)	79/24	55/17
Conjugate gradient (Fletcher-Reeves)	203/36	261/44
Gauss-Newton (damped)	40/12	21/7
Levenberg-Marquardt	63/18	17/6

Note that steepest descent does not always perform so poorly in comparison with the other techniques. Its performance is highly dependent on the shape of the specific function, as illustrated by the zig-zagging effect in Fig. 4.14. Especially close to the minimum, where the other (higher-order) methods exploit the almost quadratic shape of the loss function, steepest descent performs very poorly. If the goal is *not* to reach the minimum exactly but only approximately, steepest descent is more competitive than Table 4.3 indicates. In the context of neural networks or other very flexible model structures, where early stopping is usually applied as a regularization technique (see Chap. 7) and sample adaptation is very common, steepest descent (without line search) may still be a reasonable technique.

The reason why simplex and Hooke-Jeeves search perform significantly better than steepest descent and about the same as conjugate gradient on the banana function, although neither search method utilizes first derivative information, is as follows. The valley of the banana function is so flat that the gradient is almost zero. Thus, the step size for steepest descent is very small. The step size of the search methods, however, starts with an initial value and is reduced only if the algorithm is not able to go downhill. Therefore, as can be seen by comparing Fig. 4.15 with Figs. 4.9 and 4.11, the step size of the search methods is significantly larger, leading to faster convergence on this specific problem.

4.8 Problems

1. What exactly does "local" refer to in the chapter heading *Nonlinear Local Optimization*?
2. Explain the difference between sample and batch adaptation. What are the advantages and drawbacks of either approach?
3. Consider an optimization problem where the loss function is defined as a sum of squared errors and the error is nonlinear in the parameters to be optimized. Furthermore, assume this nonlinear relationship can be made linear by a transformation (inversion). Now the transformed problem can be solved by a least squares method. In what aspects is this procedure different from solving a standard linear least squares problem? Explain what difficulties may arise.
4. What type of problems should be approached with a direct search technique and for what problem type a gradient-based approach may be more suitable?
5. What are the advantages and drawbacks of applying line search within a gradient-based optimization? For which type of optimization technique shall line search by used?
6. Order the gradient-based optimization techniques according to their computational effort for one iteration step. What does this ranking imply for the speed of these techniques?
7. Explain the main difference between a simple gradient search and Newton's method. What are the consequences of this difference (positive and negative)?
8. Why do quasi-Newton methods make sense, why not simply using Newton's method?
9. What is the difference between nonlinear least squares techniques and the others? What information (input parameters for a function realizing an optimization algorithm) must be provided to a "standard" nonlinear optimization technique and a nonlinear least squares technique?
10. What is the key idea in the Levenberg-Marquardt method?

Chapter 5
Nonlinear Global Optimization

Abstract This chapter gives an overview of a topic that normally requires a complete monograph on its own: nonlinear optimization. Here the goal is to find the global minimum, not just a local one. Thus, it can only emphasize the key ideas of the most popular approaches with a focus on simulated annealing, evolution strategies, and genetic algorithms. The concept of a tradeoff between exploitation (going for the minimum with high convergence rate) and exploration (discover many regions in the parameter space) is explained and discussed. Also, hybrid approaches, like starting local searches from multiple initial parameters, are considered.

Keywords Optimization · Simulated annealing · Genetic algorithm · Evolution strategy · Local minimum · Global minimum · Exploration

The distinction between local and global techniques is only necessary in the context of nonlinear optimization, since linear problems always have a unique optimum; see Chap. 3. The nonlinear local optimization techniques discussed in the previous chapter starts from an initial point in the parameter space and search in directions obtained by neighborhood information such as first and possibly second-order derivatives. Obviously, such an approach leads to an optimum that is close to the starting point and in general, not the global one. The methods discussed in this chapter search for the global optimum. Although for some methods convergence to the global optimum can be proved, this usually cannot be claimed within finite time. Thus, one cannot expect those global methods to find the global optimum, especially for large problems. Rather the intention is to find a good local optimum. Since in most applications the loss function value of the global optimum is not known, it is difficult to assess the quality of an optimum. However, comparisons with results obtained by truly local techniques can justify the use of global approaches.

The simplest strategy for searching a good local optimum is a multi-start approach. It simply starts a conventional nonlinear local optimization technique from several different initial parameter values. Each local optimization run discovers a local optimum, and the best one is chosen as the final result. Assessing the quality and number of *different* local optima achieved in this procedure reveals some

O. Nelles, *Nonlinear System Identification*,
https://doi.org/10.1007/978-3-030-47439-3_5

information about the possible benefits that can be expected from a global search. It is advisable to start any global optimization with such a strategy in order to get a "feeling" of sensitivity on the initial parameters and the difference in the quality of different local optima. Besides their simplicity the multi-start methods have the advantage of intrinsic parallelization, that is, each local method can be run on a different computer. The big drawback of multi-start methods is the difficulty of how to choose the different initial parameters reasonably.

Another common strategy for extending local methods to a more global search is to add noise on the parameter update (4.10). The additional noise deteriorates the convergence speed of the algorithm, but it can drive the parameters out of local optima. As shown in Sect. 4.1, sample adaptation is similar to adding noise on the parameter update, and therefore it can also escape from shallow local optima. Typically, the noise level is reduced as the optimization progresses in order to speed up convergence. With sample adaptation, this corresponds to a decreasing step size. These noise addition methods, however, due to their local orientation can only search in the neighborhood of the initial parameters.

If these approaches fail to reach satisfactory results, one of the global techniques described in the following may succeed. They should be applied mainly if one (or several) of the following conditions is (are) met:

- search for a good local or even global optimum is required;
- loss function's surface or its derivative is highly non-smooth;
- some parameters are not real but integer or binary or even of mixed types;
- combinatorial optimization problems.

The main drawback of all global approaches is the huge computational demand. The most obvious way to search for the global optimum is to cover the input space of the loss function with a grid. If each parameter is discretized into $\Delta - 1$ intervals, then Δ^n loss function evaluations have to be computed (n is the number of parameters, i.e., the dimension of the parameter space). Clearly, this effort grows exponentially with the number of parameters. Thus, such a strategy is hopeless even for moderately sized problems. The exponential increase of complexity with the number of parameters is a fact that does not depend on the specific algorithm applied for global optimization. It simply follows from the problem and is a typical manifestation of the curse of dimensionality; see Sect. 7.6.1. The way most algorithms solve this dilemma is to examine more closely those regions of the parameter space that are more likely to contain good local optima. Parameter values in the other regions are evaluated with lower probability. This idea can only be successful if the underlying loss function contains some "regularity," a property that fortunately can be expected from all real-world problems.

Most nonlinear global optimization techniques incorporate some stochastic elements. In order to let the algorithm more easily escape from local optima, it must be allowed to forget the corresponding parameter values. Thus, the loss function is not guaranteed to be monotonically decreasing over the iterations. However, when the optimization procedure is finally terminated it is reasonable to choose the best overall solution, not the best solution in the final iteration. Therefore, it is advisable to store the best solution over all iterations, if this information is not transferred from one iteration to the next by the algorithm itself.

Since the global methods have to examine the whole parameter space (although with non-uniform density) their convergence to any optimum is prohibitively slow. Therefore, it is a good idea to use the estimated parameters from any global method as initial values for a subsequent local optimization procedure in order to converge to the optimum fast and accurately.

Loosely speaking, *global methods are good at finding regions, while local methods are good at finding points.*

Because nonlinear global, nonlinear local, and linear optimization techniques each have their specific advantages and drawbacks, in practice it is often effective to combine different approaches. Figure 5.1 depicts some common combinations of different optimization approaches. The already mentioned idea to first run a global search and subsequently start a local search is shown in Fig. 5.1a.

Figure 5.1b illustrates a nested optimization approach. This is a good alternative if some parameters are easy to optimize, e.g., linear parameters and others are hard to optimize, e.g., structural parameters. Typical examples are neural networks, fuzzy systems, or polynomials, where the outer loop may optimize the model structure (number of neurons, fuzzy rules, or polynomial terms), which is a combinatorial optimization problem, and the inner loop estimates the (often linear) parameters. Thus, each calculation of the loss function value by the global technique involves

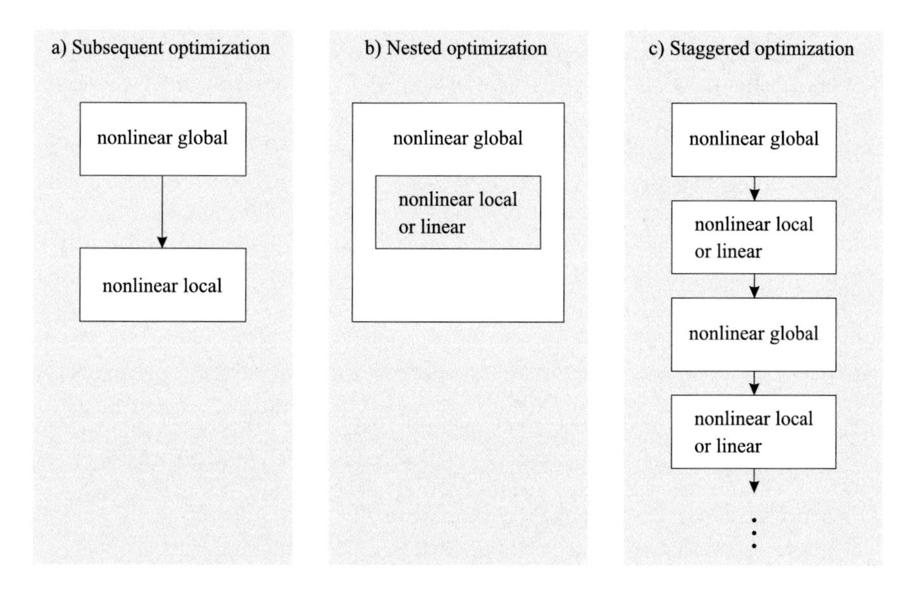

Fig. 5.1 Combinations of different optimization techniques: (**a**) global optimization of all parameters and a subsequent local optimization of all parameters, (**b**) global optimization of some parameters or structures and within (in each iteration) a nonlinear local or linear optimization of the other parameters, (**c**) repeated global optimization of some parameters and subsequently nonlinear local or linear optimization of the other parameters

the solution of a linear or nonlinear local optimization problem. Consequently, the computational effort of each iteration is considerable. However, such a strategy usually speeds up convergence compared with a nonlinear global optimization of all (even the linear) parameters because significantly fewer iterations are required.

Figure 5.1c shows an alternative approach for the same type of problem. The structural parameters may be optimized by a global technique with all other parameters fixed, and subsequently, all other parameters are optimized keeping the structure fixed. These steps are repeated until convergence. This staggered optimization approach is particularly successful if the structural parameters and all other parameters are not closely linked, i.e., they are almost orthogonal; see Sect. 3.1.3.

All these approaches try to exploit the different character of different parameters by applying individually well-suited optimization techniques. Whether the enhanced performance is worth the additional implementation effort is highly problem dependent. Nevertheless, as a rule of thumb, one should always try to optimize linear parameters with a linear optimization technique since their advantages are overwhelming.

5.1 Simulated Annealing (SA)

Simulated annealing (SA) is a Monte-Carlo (stochastic) method for global optimization. Its name stems from the following physical analogy that describes the ideas behind the algorithm: A warm particle is simulated in a potential field. Generally, the particle moves down toward lower potential energy, but since it has a non-zero temperature, i.e., kinetic energy, it moves around with some randomness and therefore occasionally jumps to higher potential energy. Thus, the particle is capable of escaping local minima and possibly finding a global one. The particle is annealed in this process, that is, its temperature decreases gradually, so the probability of moving uphill decreases with time. It is well known that the temperature must decrease slowly to end up at the global minimum energy; otherwise, the particle may get caught in a local minimum. In the context of optimization, the particle represents the parameter point in search space and the potential energy represents the loss function.

Simulated annealing was first proposed by Kirkpatrick [321] in 1983. In the following, the general form of a simulated annealing algorithm is discussed.

1. Choose an initial "large enough" temperature T_0.
2. Choose an initial parameter vector $\underline{\theta}_0$, that is, a point in search space.
3. Evaluate the loss function value for the initial parameters $I(\underline{\theta}_0)$.
4. Iterate for $k = 1, 2, \ldots$
5. Generate a new point in search space $\underline{\theta}_{\text{new}}$, which has the deviation $\Delta\underline{\theta}_k = \underline{\theta}_{\text{new}} - \underline{\theta}_{k-1}$ from the old point with the generation probability density function $g(\Delta\underline{\theta}_k, T_k)$.

6. Evaluate the loss function value for the new parameters $I(\underline{\theta}_{\text{new}})$.
7. Accept this new point with an acceptance probability $h(\Delta I_k, T_k)$ where $\Delta I_k = I(\underline{\theta}_{\text{new}}) - I(\underline{\theta}_{k-1})$, that is, set $\underline{\theta}_k = \underline{\theta}_{\text{new}}$, or otherwise keep the old point, that is, set $\underline{\theta}_k = \underline{\theta}_{k-1}$.
8. Decrease the temperature according to the annealing schedule T_k.
9. Test for the termination criterion and either go to Step 4 or stop.

The algorithm starts with an initial temperature and an initial parameter vector. A choice for T_0 is not easy, and since all nonlinear problems are different, a trial-and-error approach is required. T_0 determines the initial tradeoff between globality and locality of the search. If it is chosen too high, convergence is very slow, and if it is chosen too small, the algorithm concentrates too early on the neighborhood around the initial point. Then at Step 5 a new point in search space is generated by a generation probability density function $g(\cdot)$. This generation pdf has two properties: (i) smaller parameter changes have a higher probability than larger ones, and (ii) large parameter changes are more likely for higher temperatures than for lower ones. Thus, the algorithm starts with a very wide examination of the parameter space, and as time progresses and the temperature decreases, it concentrates to an increasingly focused region. The loss function value of the newly generated point can either be better (i.e., smaller) or worse (i.e., higher) than for the old point. A typical descent approach would accept any better point and would discard any worse point. In order to enable the escape from local minima, SA accepts the new point with a probability of $h(\cdot)$ (Step 7). This acceptance probability function has the following properties: (i) acceptance is more likely for better points than for worse ones, and (ii) acceptance of worse points has a higher probability for larger temperatures than for smaller ones. The acceptance probability in standard SA is chosen as [271]

$$\begin{aligned} h(\Delta I_k, T_k) &= \frac{\exp(-I_k/T_k)}{\exp(-I_k/T_k) + \exp(-I_{k-1}/T_k)} \\ &= \frac{1}{1 + \exp(\Delta I_k/T_k)}, \end{aligned} \tag{5.1}$$

where ΔI_k represents the difference of the loss function value between the new and the old parameters. Figure 5.2 shows this acceptance probability function for different temperatures. In Step 8 the temperature is decreased according to some predefined *annealing schedule*. It is of fundamental importance for the convergence of the algorithm to the global optimum that this annealing schedule decreases the temperature slowly enough.

The traditional choice of the generation probability density function $g(\Delta\underline{\theta}, T)$ is a Gaussian distribution; see Fig. 5.3. It can be shown that the algorithm is statistically guaranteed to find the global optimum (annealing proof), if the following annealing schedule is applied:

$$T_k = \frac{T_0}{\ln k}. \tag{5.2}$$

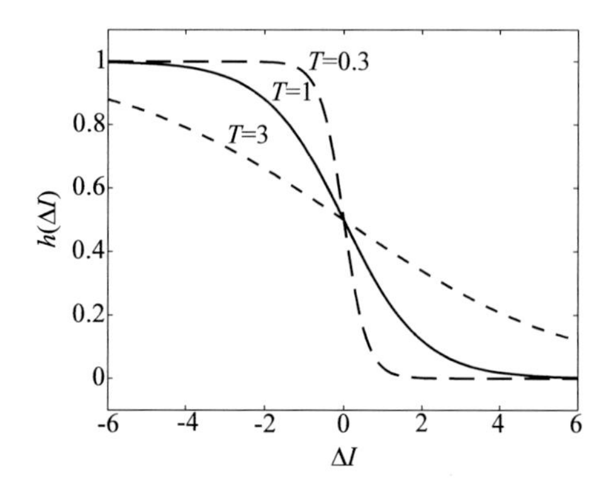

Fig. 5.2 Acceptance probability in simulated annealing for high, medium, and small temperature T. Points that decrease the loss function ($\Delta I < 0$) are always accepted with a probability higher than 0.5, while points that increase the loss function ($\Delta I > 0$) are less likely to be accepted. For very high temperatures ($T \rightarrow \infty$), all points are accepted with equal probability, while for very low temperatures ($T \rightarrow 0$), only points that improve the loss function are accepted

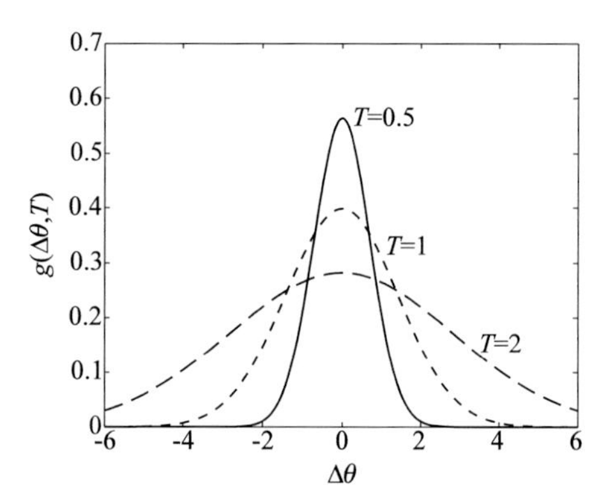

Fig. 5.3 Gaussian probability density function for different annealing temperatures T. The lower the temperature, the more likely points around "0" are to be generated. Points around "0" are close to the old point and therefore represent local search

Simulated annealing with these choices for the generation pdf and annealing schedule is known as Boltzmann annealing (BA). It is important to understand that the proof of statistical convergence to the global optimum relies on the annealing schedule (5.2). No such proof exists for BA with faster annealing. Although this proof basically states that the global optimum is found in infinite time, it is of practical significance since it guarantees that the algorithm will not get stuck in a local minimum. Also, if the parameter space is discretized (as it always is for digital representations), it may be possible to put some finite upper bounds on the search time.

The problem with BA is that its annealing schedule is very slow. Thus, many users apply faster schedules such as an exponential decrease of temperature (quenching). While this might work on some problems, it will, however, not work

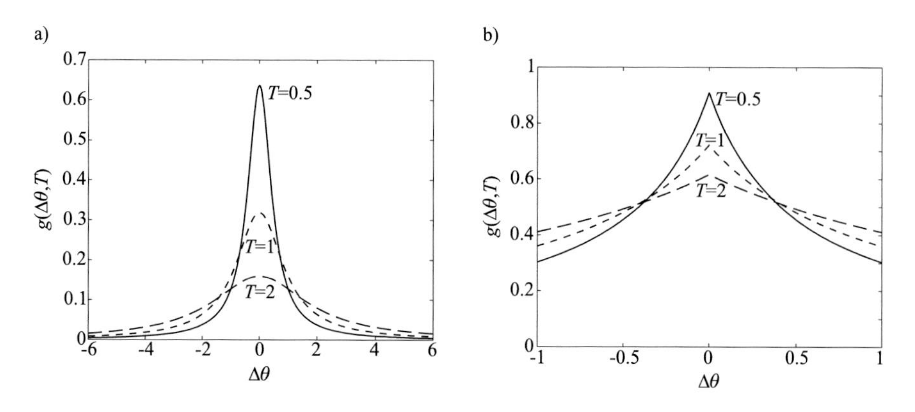

Fig. 5.4 (**a**) Cauchy probability density function for different annealing temperatures T. The Cauchy distribution has a "fatter" tail than the Gaussian in Fig. 5.3 and therefore it permits easier access to test local minima in the search for the desired global minimum. (**b**) Probability density function for VFSR and ASA. It is non-zero only for $[-1, 1]$ and can be transformed into any interval [A, B]. This pdf has a "fatter" tail than the Cauchy pdf in Fig. 5.4a, and therefore it permits easier access to test local minima in the search for the desired global minimum

on others, since the conditions of the global convergence proof are violated. In order to allow faster annealing without violating these conditions, fast annealing (FA) was developed [271]. In FA the Gaussian generation pdf is replaced by a Cauchy distribution; see Fig. 5.4a. Since it has a "fatter" tail, it leads to a more global search. Fast annealing statically finds the global optimum for the following annealing schedule:

$$T_k = \frac{T_0}{k} \,. \tag{5.3}$$

Thus, it is exponentially faster than BA. Although with the Cauchy distribution the probability of rejected (non-accepted) points is higher, the faster annealing schedule usually over-compensates this effect.

In the late 1980s, Ingber developed very fast simulated reannealing (VFSR) [270], which was continuously extended to *adaptive simulated annealing (ASA)* [271, 272]. In contrast to BA and FA, which (theoretically) sample infinite ranges with the Gaussian and Cauchy distributions, respectively, these algorithms are designed for *bounded* parameters. Figure 5.4b shows the generation pdf applied in ASA for the generation of random variables in $[-1, 1]$. They can be linearly transformed into any interval [A, B]. This approach meets the requirements of most real-world problems. Indeed, ASA has been applied in many different disciplines, ranging from physics to medicine to finance. As can be seen in Fig. 5.4b, this pdf favors global sampling over local sampling even more than the Cauchy distribution. Thus, the number of accepted points will be lower. However, the statistical convergence to the global optimum can be proven for an exponentially fast annealing schedule

$$T_k = T_0 \exp\left(-c\,k^{1/n}\right), \tag{5.4}$$

where n is the number of parameters and c is a constant that can be chosen for each parameter, that is, each parameter may have its individual annealing schedule. The term $1/n$ in (5.4) is a direct consequence of the curse of dimensionality. An increasing number of parameters has to slow down the annealing schedule exponentially in order to fulfill the global optimum proof since the problem's complexity increases exponentially with the number of parameters. Thus, for high-dimensional parameter spaces, ASA will become very slow. Experience shows that faster annealing schedules that violate the annealing proof often also lead to the global optimum. Changing the annealing schedule to $T_k = T_0 \exp\left(-c\,k^{Q/n}\right)$ with $Q > 1$ is called *quenching*, where Q is called the quenching factor. For example, the *quenching factor* can be chosen as $Q = n$ in order to make the speed of the algorithm independent of the number of parameters. Although quenching violates the annealing proof, it performs well on a number of problems [271].

Some special features such as reannealing and self-optimization justify the term "adaptive" in ASA [271]. Whenever performing a multidimensional search on real-world problems, inevitably one must deal with different changing sensitivities of the loss function with respect to the parameters in the search space. At any given annealing time, it seems reasonable to attempt to "stretch out" the range over which the relatively insensitive parameters are being searched, relative to the ranges of the more sensitive parameters. Therefore, the parameter sensitivity, that is, the gradient of the loss function, has to be known or estimated. It has been proven fruitful to accomplish this by periodically (every 100 iterations or so) rescaling the annealing temperature differently for each parameter. By this reannealing ASA adapts to the underlying loss function surface. Another feature of ASA is self-optimization. The options usually chosen by the user based on a priori knowledge and trial and error can be incorporated into the loss function. This means that ASA may optimize its own options. Although this feature is very appealing at first sight, it requires huge computational effort. In [273] ASA is compared with a genetic algorithm (GA) and is shown to perform significantly better on a number of problems.

5.2 Evolutionary Algorithms (EA)

Evolutionary algorithms (EAs) are stochastic optimization techniques based on some ideas of natural evolution, just as simulated annealing (SA) is based on a physical analogy. The motivation of EAs is the great success of natural evolution in solving complex optimization problems such as the development of new species and their adaptation to drastically changing environmental conditions. All evolutionary algorithms work with a *population* of *individuals*, where each individual represents one solution to the optimization problem, i.e., one point in the parameter space. One difference from simulated annealing is that for SA usually a single particle

Table 5.1 Comparison of terminology of evolutionary algorithms and classical optimization

Evolutionary algorithm	Classical optimization
Individual	Parameter vector
Population	Set of parameter vectors
Fitness	Inverse of loss function value
Fitness function	Inverse of loss function
Generation	Iteration
Application of genetic operators	Parameter vector update

corresponding to a single parameter point is considered, while for EAs typically a population of many (parallel) individuals are considered, which makes EAs inherently parallel algorithms.

As in nature this population of individuals evolves in *generations* (i.e., iterations) over time. This means that individuals change because of *mutation* or *recombination* or other genetic operators. By the application of these genetic operators, new individuals are generated that represent a different parameter point, consequently making it possible to search the parameter space. Each individual in the population realizes a loss function value. By *selection* the better performing individuals (smaller loss function values) are more likely to survive than the worse performing ones. The performance index is called *fitness* and represents some inverse[1] of the loss function value. Selection ensures that the population tends to evolve toward better performing individuals, thereby solving the optimization problem. Table 5.1 compares the terminology of evolutionary algorithms with the corresponding classical optimization expressions. Evolutionary algorithms work as follows [374]:

1. Choose an initial population of S individuals (parameter points) $\underline{\Theta}_0 = [\underline{\theta}_{1,0}\ \underline{\theta}_{2,0}\ \cdots\ \underline{\theta}_{S,0}]$.
2. Evaluate the fitness (some inverse of the loss function) of all individuals of the initial population $\text{Fit}(\underline{\Theta}_0)$.
3. Iterate for $k = 1, 2, \ldots$
4. Perform selection $\underline{\Theta}_{\text{new}} = \text{Select}(\underline{\Theta}_{k-1})$.
5. Apply genetic operators $\underline{\Theta}_k = \text{GenOps}(\underline{\Theta}_{\text{new}})$ (mutation, recombination, etc.).
6. Test for the termination criterion and either go to Step 3 or stop.

One feature of all evolutionary algorithms is that they are inherently parallel because a set of individuals (the population) is evaluated in each iteration. This can be easily implemented on a parallel computer. Evolution in nature itself is an example of a massive parallel optimization procedure.

The evolutionary algorithms differ in the type of selection procedure and the type of genetic operators applied. Usually, a number of *strategy parameters* such as mutation and recombination rates or step sizes have to be fixed by the user or

[1] e.g., $1/I(\underline{\theta})$ or $-I(\underline{\theta})$.

optimized by the EA itself, which again distinguishes different EA approaches. Furthermore, substantial differences lie in the coding of the parameters or structures that have to be optimized.

In engineering typically very simple selection procedures and genetic operators (mostly only mutation and recombination) are applied, trying to imitate the essential features of nature without becoming too complicated. The engineer's goal is to design a well-performing optimization technique rather than imitate nature as closely as possible. The same pragmatic approach is pursued in the context of artificial neural networks inspired by the brain, fuzzy logic inspired by rule-based thinking, and many others. All these approaches have their roots outside engineering, but as their understanding progresses they typically move away from their original motivation and tend to become mathematically analyzed and improved tools.

Evolutionary algorithms can be distinguished into *evolution strategies (ES)* (Sect. 5.2.1), *genetic algorithms (GA)* (Sect. 5.2.2), *genetic programming (GP)* (Sect. 5.2.3), *evolutionary programming*, and *classifier systems* [243]. The last two are beyond the scope of this book.

While evolution strategies and genetic algorithms are mainly used as *parameter* optimization techniques, genetic programming operates on a higher level by optimizing tree structures. Figure 5.5 compares the approaches of evolution strategies and genetic algorithms. Genetic algorithms operate on a binary level and are very similar to nature, where the information is coded in four different bases on the DNA. In order to solve a parameter optimization problem with real parameters, these parameters must be coded in a binary string. The genetic algorithm then operates on this modified coding. In contrast, evolution strategies do not imitate nature as closely but operate on real parameters, which allows them to solve parameter optimization problems more directly. A huge number of variations of all types of evolutionary algorithms exist. A phase of almost separate development of evolution strategies (mainly in Germany) and genetic algorithms (mainly in the USA) ended

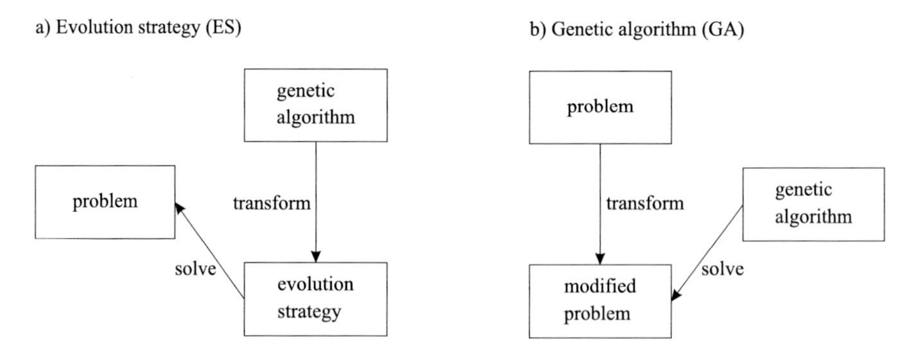

Fig. 5.5 Comparison of evolution strategies and genetic algorithms: (**a**) the genetic algorithm is modified into an evolution strategy that can solve the problem directly, (**b**) the genetic algorithm can only solve the modified problem [374]

in the 1990s, and many ideas were exchanged and merged in order to create hybrid algorithms. For the sake of clarity, this section focuses on the classical concepts of evolution strategies and genetic algorithms; almost arbitrary combinations of these ideas are possible.

In the following sections evolution strategies, genetic algorithms, and genetic programming are briefly summarized. For further details refer to [22, 117, 182, 189, 243, 256, 374, 478, 525].

5.2.1 Evolution Strategies (ES)

Evolution strategies (ESs) were developed by Rechenberg and Schwefel in Germany during the 1960s. For ESs the parameters are coded as they are. Continuous parameters are coded as real numbers, integer parameters are coded as integer numbers, etc.

In ESs *mutation* plays the dominant role of all genetic operators. The parameters are mutated according to a normal distribution. In the simplest case, all parameters are mutated with the same Gaussian distribution (see Fig. 5.6a)

$$\theta_i^{(\text{new})} = \theta_i + \Delta\theta\,, \tag{5.5}$$

where $\Delta\theta$ is distributed according to the following pdf:

$$p(\Delta\theta) = \frac{1}{\sqrt{2\pi}\sigma} \exp\left(-\frac{1}{2}\frac{\Delta\theta^2}{\sigma^2}\right). \tag{5.6}$$

Because (5.6) has zero mean, the parameters are changed with equal probabilities to smaller and larger values. Furthermore, (5.6) generates smaller mutations with

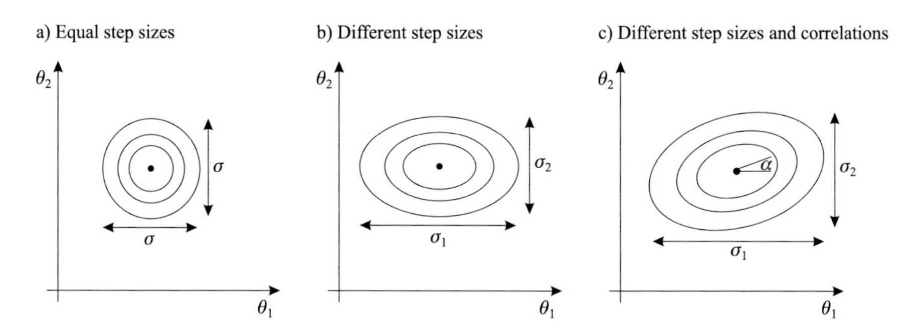

Fig. 5.6 Mutation in an evolution strategy is usually based on a normal distribution: (**a**) all parameters are mutated independently with equal standard deviations of the Gaussian distribution, (**b**) all parameters are mutated independently but with individual standard deviations, (**c**) the mutations of all parameters have individual standard deviations and are correlated (all entries of the covariance matrix in the Gaussian distribution are non-zero)

a higher probability than larger mutations. The standard deviation σ of the pdf controls the average step size $\Delta\theta$ of the mutation.

If (5.6) is not flexible enough, it can be extended in order to allow individual step sizes (standard deviations) for each parameter (see Fig. 5.6b)

$$\theta_i^{(\text{new})} = \theta_i + \Delta\theta_i \, , \tag{5.7}$$

where $\Delta\theta_i$ are distributed according to the following pdfs:

$$p(\Delta\theta_i) = \frac{1}{\sqrt{2\pi}\,\sigma_i} \exp\left(-\frac{1}{2}\frac{\Delta\theta^2}{\sigma_i^2}\right) . \tag{5.8}$$

This is the most common implementation. Furthermore, it is possible to incorporate correlations between the parameter changes. This allows one to specify search directions that are not axis-orthogonal similar to the Hooke-Jeeves algorithm; see Sect. 4.3.2. Then the joint pdf of all step sizes $\Delta\underline{\theta} = [\Delta\theta_1 \ \Delta\theta_2 \ \cdots \ \Delta\theta_n]^T$ becomes (see Fig. 5.6c)

$$p(\Delta\underline{\theta}) = \frac{1}{2\pi \ \det(\underline{\Sigma})} \exp\left(-\frac{1}{2}\Delta\underline{\theta}^T \underline{\Sigma}\Delta\underline{\theta}\right) , \tag{5.9}$$

where $\underline{\Sigma}^{-1}$ is the covariance matrix that contains the information about the standard deviations and rotations of the multidimensional distribution. The two previous mutation pdfs are special cases of (5.9) where $\underline{\Sigma}^{-1} = \text{diag}(\sigma_1^2, \sigma_2^2, \ldots, \sigma_n^2)$ or $\underline{\Sigma}^{-1} = \sigma^2 \underline{I}$, respectively.

If the parameters are not real but integer or binary, the mutation approach described above has to be altered slightly. Integers may be utilized to code structural parameters such as the number of neurons in a neural network, the number of rules in a fuzzy system, or the number of terms in a polynomial. For these parameters, the normal distributed step size $\Delta\underline{\theta}$ can be rounded to ensure that $\theta_i^{(\text{new})}$ stays integer. For binary parameters, a normal distribution of the mutation does not make any sense. Rather some fixed mutation probability can be introduced as for genetic algorithms; see the next section.

From the above discussion, it becomes clear that it is hard for the user to choose reasonable values for the step sizes (standard deviations). In local optimization techniques the step size is optimized by line search; see Sect. 4.4.1. A similar strategy is pursued here. The step sizes are coded in the individuals and thus also subject of optimization. A typical individual in an ES is depicted in Fig. 5.7, where the ϑ_i denotes the so-called *strategy parameters*, which represent the standard deviations. For example, if individual standard deviations are used, the individual contains n parameters θ_i and additionally n strategy parameters ϑ_i. The optimization of strategy parameters that influence the optimization algorithm is called *self-adaptation* or *second-level learning*.

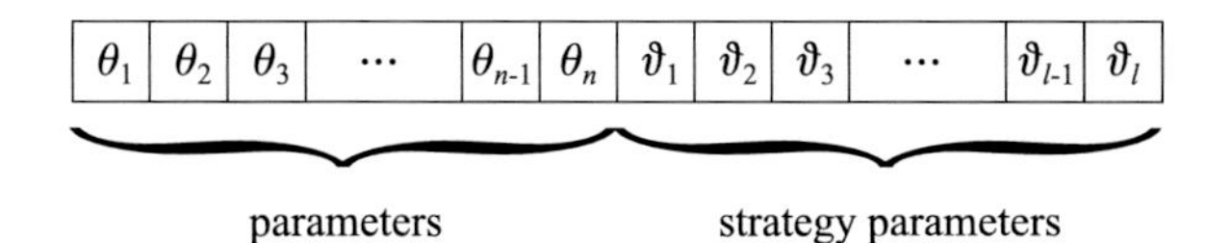

Fig. 5.7 The individual for an evolution strategy contains the parameters to be optimized and so-called strategy parameters (or meta parameters) which control the internal behavior of the evolution strategy. These strategy parameters are also subject to the optimization

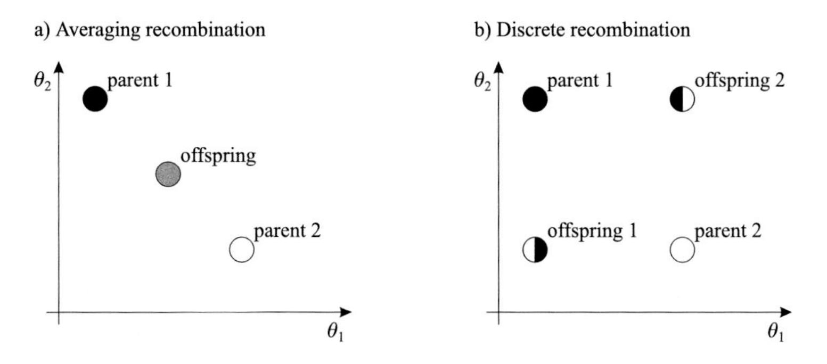

Fig. 5.8 Recombination in an evolution strategy is performed either by (**a**) averaging or (**b**) discrete combination

The idea behind this approach is that individuals with well-adapted step sizes have an advantage in evolution over the others. They are more successful because they tend to produce better offspring. Thus, not only do the parameters converge to their optimal values but so do the strategy parameters as well. Note that well-suited strategy parameters should reflect the surface of the loss function. As the optimization progresses, the strategy parameters should adapt to the changing environment. For example, assume that the optimization starts in a deep valley in the direction of one parameter. Therefore, only one parameter should be changed, and individuals with high step sizes for this parameter and small step sizes for the other parameters are very successful. At some point, as the optimization progresses, the character of the surface may change and other parameters should be changed as well in order to approach the minimum. Then the step sizes should gradually adapt to this new environment.

The mutation of the strategy parameters does not follow the normal distribution. One reason for this is that step sizes are constrained to be positive. Rather strategy parameters are mutated with a log-normal distribution, which ensures that changes by factors of F and $1/F$ are equally probable.

Besides mutation, the *recombination* plays a less important role in evolution strategies. Figure 5.8 shows two common types of recombination. In the averaging approach, the parameters (including strategy parameters) are simply averaged between two parents to generate the offspring. This follows the very natural idea that the average of two good solutions is likely to be a good (or even better) solution as well if the loss function is smooth. An alternative is discrete recombination, in which the offspring inherits some parameters from one parent and the other parameters from the second parent. Both approaches can be extended to more than two parents.

Selection is typically performed in a deterministic way. This means that the individuals are ranked with regard to their fitness (some inverse of the loss function value), and the best individuals are selected. This deterministic selection has different consequences depending upon which of the two distinct variants of evolution strategies is applied: the *"+"-strategy* or the *","-strategy*. In general, evolution strategies are denoted by $(\mu + \lambda)$ and (μ, λ), respectively, where μ is the number of parents in each generation and λ is the number of offspring they produce by mutation and recombination. The selection reduces the number of individuals to μ in each generation by picking the best individuals. If $\mu = 1$ the algorithm can hardly escape from a local optimum, while larger values for μ lead to a more global search character. A ratio of $1/7$ is recommended in [337] for μ/λ as a good compromise between local and global search.

In the "+"-strategy the parents compete with their offspring in the selection procedure. This means that the information about the best individual cannot be lost during optimization, i.e., the fitness of the best individual is monotonically increasing over the generations (iterations). This property is desirable for fast search of a local optimum. However, if one good solution is found, the whole population tends to concentrate in this area. The "+"-strategy can be applied either with self-adaptation of the strategy parameter or alternatively with the simple 1/5-rule, which says that the step size should be adjusted such that one out of five mutations is successful [243].

For a more global search, the ","-strategy is better suited. All parents die in each generation, and the selection is performed only under the offspring. This strategy sacrifices convergence speed for higher adaptivity and broader search. In a "+"-strategy the strategy parameters (step sizes) tend to converge to very small values if a local optimum is approached. So the "+"-strategy is not capable of global search anymore and cannot track possibly time-variant optimization objectives (changing fitness functions). In the ","-strategy possibly successful parents are forgotten in order to keep the ES searching and adapting its strategy parameters. The price to be paid for this is that an ES with ","-strategy never really converges, similarly to the sample mode adaptation; see Sect. 4.1.

The optimal number of parents and offspring is very problem-specific. More complex problems demand larger populations and more generations. Typical population sizes for evolution strategy are around 100; e.g., the (15, 100)-strategy is common. Fortunately, all evolutionary algorithms are quite insensitive with respect to the choice of these meta-parameters, i.e., a rough adjustment according to a rule of thumb is usually sufficient.

Table 5.2 summarizes the main properties of evolution strategies and compares them with genetic algorithms, which are discussed in the next subsection.

Table 5.2 Comparison of evolution strategies and genetic algorithms

	Evolution strategy	Genetic algorithm
Coding	Real (problem oriented)	Binary (nature oriented)
Mutation	Important	Minor
Recombination	Minor	Important
Selection	Deterministic	Probabilistic
Strategy parameters	Adapted	Fixed

5.2.2 *Genetic Algorithms (GA)*

Genetic algorithms (GAs) have been developed independently and almost concurrently with the evolution strategies by Holland in the USA [256]. In contrast to evolution strategies, GAs represent all types of parameters (real, integer, binary) with a binary coding. Thus, an individual is a bit-string; see Fig. 5.9. The number of bits used for the coding of each parameter is chosen by the user. Of course, the accuracy doubles with each additional bit that is spent for the coding of a parameter. Because the complexity of the optimization problem increases with the length of the bit-string, coding should be chosen to be as coarse as possible. In many applications, the highest reasonable accuracy for a parameter is given by the resolution of the A/D and D/A converters. Often a much rougher representation of the parameters may be sufficient, especially if the GA optimization is followed by a local search for fine-tuning. Note that some parameters may require a higher accuracy than others.

The motivation for the binary coding of GAs stems from the natural example. In nature the genetic information is stored in a code with four symbols, the "A," "G," "C," and "T" bases of the DNA. The binary code with its symbols "0" and "1" is a simplified version of the natural genetic code.

Compared with the real coding of ESs, the binary coding of GAs creates much larger individuals and seems less rational if the original parameters are real numbers. In structure optimization problems, however, where each bit may control whether some part (e.g., neuron, fuzzy rule, polynomial term) of a model is active (switched on) or not (switched off), the binary coding of GAs is very rational and straightforward.

As a consequence of the binary coding of parameters, upper and lower bounds represented by "00 . . . 0" and "11 . . . 1", respectively, have to be fixed by the user. If the parameters are easily interpretable (mass of a car, temperature inside a room, control action of a valve), these limits are typically known. Then the GA has the advantage of an automatic incorporation of minimum and maximum constraints, i.e., the parameter in the individual is guaranteed to be feasible. With clever coding, even more complicated constraints can be easily incorporated into the GA. For example, it might be desirable that a set of parameters is monotonically increasing because they represent centers of fuzzy membership functions such as very small, small, medium, large, and very large. Then each center (parameter) can be coded as the positive distance from the previous center (parameter). This ensures that

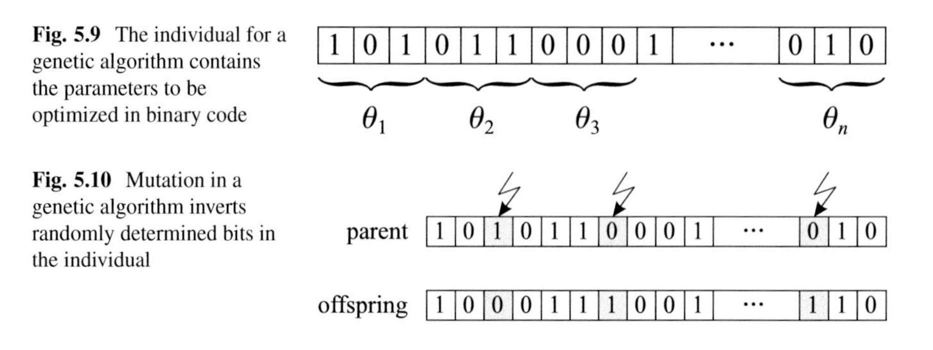

Fig. 5.9 The individual for a genetic algorithm contains the parameters to be optimized in binary code

Fig. 5.10 Mutation in a genetic algorithm inverts randomly determined bits in the individual

during the whole optimization procedure, very small is less than small, small is less than medium, etc. On the other hand, the automatic incorporation of a lower and upper bound on the parameters runs into severe problems if the numerical range of the parameters is not known beforehand. This is typically the case for non-interpretable parameters, as they occur in black box models. The only solution for this difficulty is to choose a very large interval for these parameters. Then the expected resolution and accuracy of these parameters are very low. So the binary coding of GAs can be either an advantage or a drawback depending on the type of the original parameters. For a deeper discussion about the practical experiences with binary coding, refer to [117, 182]. It seems that binary coding is usually much worse than coding that represents problem-specific relationships.

Mutation does not play the key role in GAs as it does for ESs. Figure 5.10 illustrates the operation of mutation in GAs. It simply inverts one bit in the individual. The mutation rate, a small value typically around 0.001 to 0.01, determines the probability for each bit to become mutated. Mutation imitates the introduction of (not repaired) errors into the natural genetic code as they may be caused by radiation. The main purpose of the mutation is to prevent a GA from getting stuck in certain regions of the parameter space. Mutation has a similar effect as adding noise on the parameter update, a strategy for escaping from local optima that is well known from local optimization techniques; see Sect. 4.1.

A closer look at mutation in GAs reveals a weakness of genetic algorithms compared with evolution strategies. In a standard GA, all bits are mutated with equal probability but not all bits have the same significance. For example, an integer parameter between 0 and 15 may be coded by the bit-strings "0000" (=0) to "1111" (=15). A mutation in the first (most significant) bit changes the value of the parameters by $+8$ or -8, while a mutation in the last (least significant) bit leads to changes of only $+1$ or -1. This means that huge changes are equally probable as tiny ones. In most cases, this is no realistic model of reality, and the normal distribution chosen for mutation changes in ESs seems much more reasonable. To circumvent this problem, either the mutation rate can be adjusted to the significance of the corresponding bit or a gray coding of the parameters can be implemented [374].

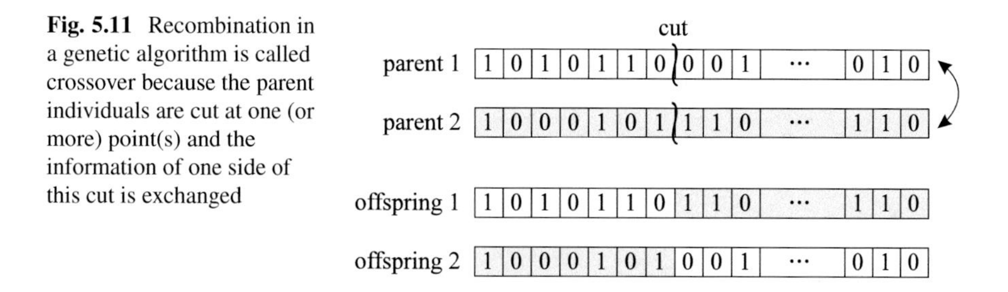

Fig. 5.11 Recombination in a genetic algorithm is called crossover because the parent individuals are cut at one (or more) point(s) and the information of one side of this cut is exchanged

Recombination is the major genetic operator in GAs. The bit-strings of two (or more) parents are cut into two (or more) pieces, and the parts of the bit-string are crossed over; see Fig. 5.11. Therefore, this type of recombination is called *crossover*. Both offspring contain one part of information from one parent and the rest of the other parent, as in the discrete recombination strategy for ESs shown in Fig. 5.8. The crossover rate, a large value typically around 0.6 to 0.8, determines the probability with which crossover is performed. The point where the parents are cut is determined randomly, with the restriction that it should not cut within one parameter but only between parameter boundaries. Crossover imitates the natural example where an offspring inherits half of its genes from each parent (at least for humans and most animals).

The standard genetic operators as described above work for simple problems. Often, however, not all parameter combinations or structures coded in the individual make sense with respect to the application. Then the standard mutation and crossover operators may probably lead to nonsense individuals, which cannot be reasonably evaluated by the fitness function. Consequently, application-specific mutation and crossover operators must be developed that are tailored to the requirements of the application. The development of an appropriate coding and the design of these operators are typically the most important and challenging steps toward the problem solution. For example, refer to [117, 182, 374].

Selection in GAs works stochastically, contrary to the deterministic ES selection. This means that no individual is guaranteed to survive. Rather, random experiments decide which individuals are carried over to the next generation. Again, this is very close to nature, where also the fittest individuals have high probabilities but no guarantee of surviving. Different GA selection schemes can be distinguished. The most common one is the so-called *roulette wheel selection*. Each individual has a probability of surviving that is proportional to its fitness. An individual that is twice as good as another one has twice the chance of surviving. Figure 5.12 shows how this selection scheme can be illustrated as a spinning roulette wheel. The probabilities $P_i^{(\text{select})}$ for each individual can be calculated by normalizing the (positive!) fitness values of all individuals:

$$P_i^{(\text{select})} = \frac{Fit_i}{\sum_{j=1}^{\text{pop. size}} Fit_j} . \tag{5.10}$$

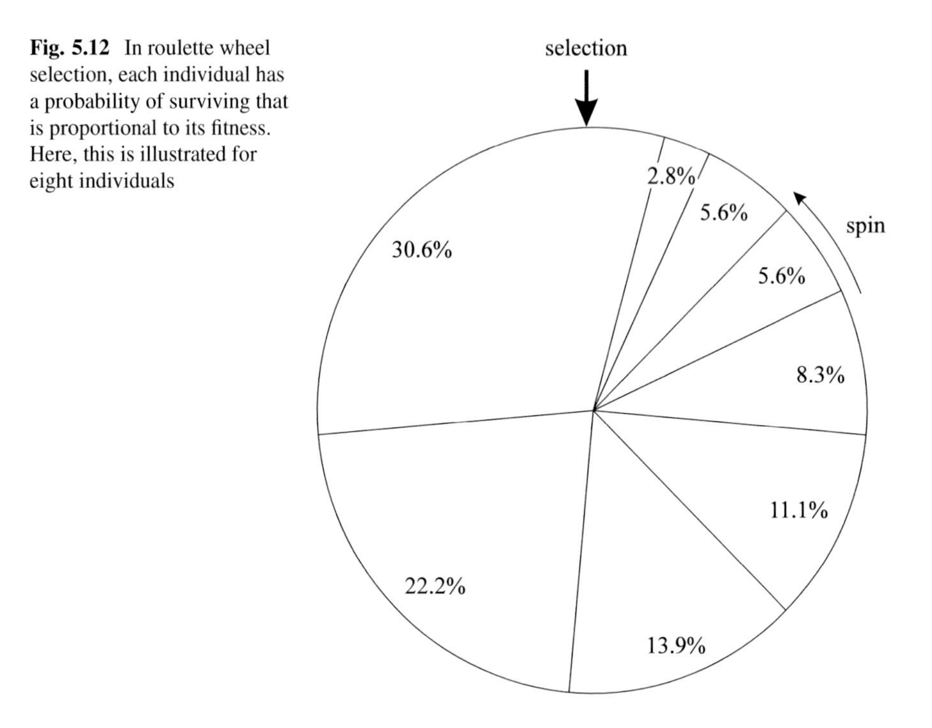

Fig. 5.12 In roulette wheel selection, each individual has a probability of surviving that is proportional to its fitness. Here, this is illustrated for eight individuals

The fitness of each individual is required to be positive in order to evaluate (5.10). This can be achieved by, e.g., inverting the loss function $1/I(\underline{\theta})$ or changing the sign of the loss function $C - I(\underline{\theta})$ with a C larger than the largest loss function value. One problem with roulette wheel selection becomes obvious here. The selection process depends on the exact definition of the fitness function and not only on the ranking of the individuals as is the case for ESs. Any nonlinear transformation of the fitness function influences the selection process. Therefore, the following problem typically arises with roulette wheel selection. During the first generations, the population is very heterogeneous, i.e., the fitness of the individuals is very different. As the GA starts to converge, the fitness of all individuals tends to be much more similar. Consequently, all individuals survive with almost the same probability; in other words, the selection pressure becomes very small; see Fig. 5.13. Then the GA degenerates to random search. In order to avoid this effect, several methods for scaling the fitness values have been proposed. A simple solution is to scale the fitness values such that the highest value is always mapped to 1 and the lowest fitness value is mapped to 0. With such a scaling, the selection pressure can be kept constant over all generations. For more details refer to [189].

Alternatives to roulette wheel selection are the ranking based selection schemes such as tournament selection. They base the selection on the ranking of each individual in the whole population rather than on its absolute fitness value. The relationship between roulette wheel selection and ranking based selection is similar to the relationship between the mean and the median. For more details about selection schemes, refer to [189].

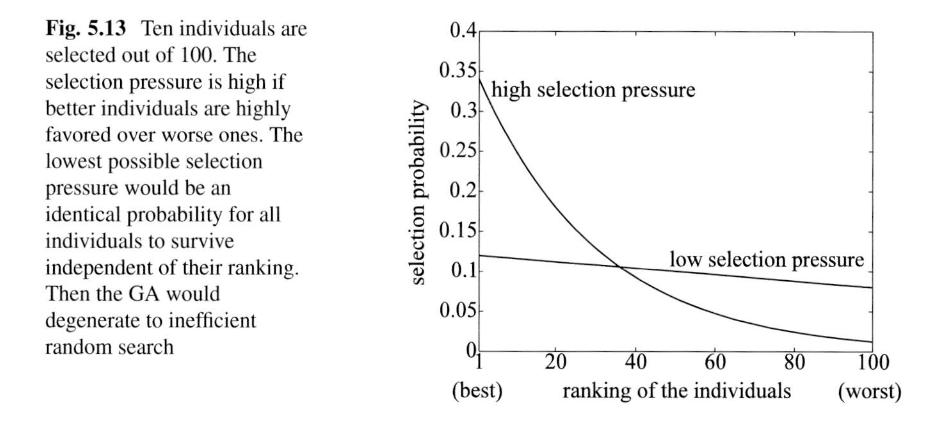

Fig. 5.13 Ten individuals are selected out of 100. The selection pressure is high if better individuals are highly favored over worse ones. The lowest possible selection pressure would be an identical probability for all individuals to survive independent of their ranking. Then the GA would degenerate to inefficient random search

Because the selection is probabilistic, GAs tend to sample the parameter space more stochastically than ESs. Since in GAs good individuals can die, they are similar to the ","-strategy in ESs. The convergence behavior is controlled by the *selection pressure*. This corresponds to the inverse of the annealing temperature in simulated annealing; see Sect. 5.1. Figure 5.13 illustrates that for very high selection pressures, the diversity of the population decreases, since only the very best individuals have a significant probability of surviving. This may lead to *premature convergence*, i.e., the GA loses its capability of exploring new regions of the parameter space, and the search becomes too locally focused. On the other hand, for selection pressures that are too low, the GA deteriorates to inefficient random search, which is too global. This means that by designing the fitness function (and possibly scaling algorithms) and choosing a selection scheme, the user decides on the selection pressure and thus carries out a problem-specific tradeoff between a global and local character of the search.

As for evolution strategies, the choice of the number of parents and offspring in GAs is very problem-specific. Typical population sizes are around 100. Table 5.2 summarizes the properties of GAs and ESs. For GAs a clever coding of the parameters utilizing prior knowledge is probably the most decisive step toward a successful solution of complex optimization problems.

In the following an extension of GAs, genetic programming is described that applies the genetic operators on a higher structural level than bit-strings.

5.2.3 Genetic Programming (GP)

Genetic programming (GP) proposed by Koza [330–332] is very similar to genetic algorithms, but it operates on tree structures rather than on binary strings. Trees are flexible structures that allow one to represent relationships efficiently. Figure 5.14 shows two examples of how logic expressions and mathematical equations can be realized with trees. The leaves of a tree typically represent variables or constants,

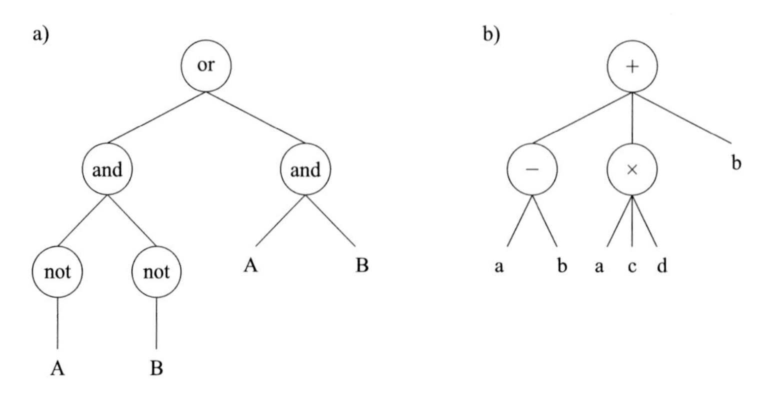

Fig. 5.14 The individual for genetic programming is represented in a tree structure. Tree structures allow one to realize (**a**) logic expressions (here: XOR), (**b**) mathematical equations (here $a + acd$ in a redundant manner), and many other structures

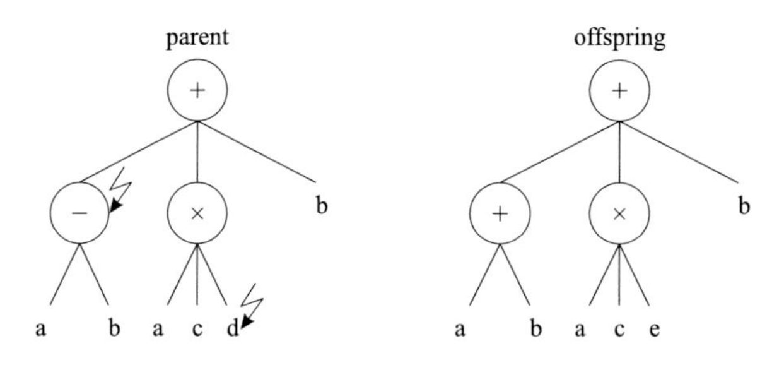

Fig. 5.15 Mutation in genetic programming changes parts of the tree structure, like operators or variables

while the other nodes implement operators. The user specifies a set of variables, constants, and operators that may be used in the tree.

Clearly, the GA mutation and crossover operators have to be modified for GPs. Figure 5.15 illustrates how mutation can be performed. Parts of the tree (variables, constants, or operators) can be randomly changed, deleted, or newly generated. Figure 5.16 shows how crossover can be realized for GPs. Parts (subtrees) of both parents are randomly cut and exchanged to generate the offspring. As the example in Fig. 5.16 shows, the trees can grow and shrink in this procedure. For selection basically, the same schemes as for GAs can be applied.

One important issue in genetic programming is how to deal with real parameters. If, for example, the function $y = 2u_1 + 3.5u_2^3 + 1.75$ is to be searched by GP, the task is to find not only the correct (or a good approximate) structure of the problem but also the numerical values 2, 3.5, and 1.75. One approach is to code and optimize the parameters as bit-strings, as is done in GAs. A much more promising alternative, however, is to optimize all parameters within each fitness function call by local (if possible linear) optimization techniques [364, 365]. This corresponds to the nested optimization strategy, as illustrated in Fig. 5.1b. Such a decomposition of

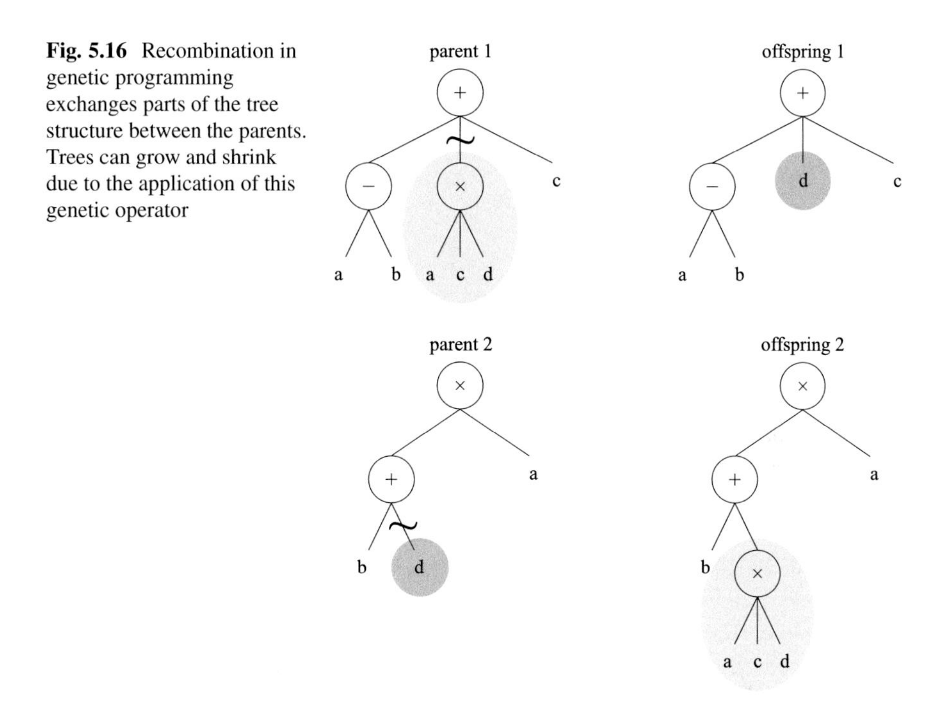

Fig. 5.16 Recombination in genetic programming exchanges parts of the tree structure between the parents. Trees can grow and shrink due to the application of this genetic operator

the problem into a structural part solved by GP and a parameter optimization part solved by local search or linear regression reduces the complexity significantly and speeds up convergence.

5.3 Branch and Bound (B&B)

Branch and bound (B&B) is a tree-based search technique that is very popular for the solution of combinatorial optimization problems. For a restricted class of problems, it can be applied to parameter optimization as well. Then the parameters have to be discretized, i.e., real parameters are mapped to, say, B different values similar to GAs where $B = 2^L$ with L being the length of the binary code. The idea of branch and bound is to build a tree that contains all possible parameter combinations (see Fig. 5.17) and to search only the necessary part of this tree. Branch and bound employs tests at each node of the tree, which allows one to cut parts of the tree and thus save computations compared with an exhaustive search.

In B&B it is assumed that upper and lower bounds for the loss function are known (or can be easily derived). This information can be utilized to prune away the whole subtrees. For example, in a traveling salesman problem, the optimal route through a number of given cities is sought. The best existing solution to this problem represents an upper bound on the loss function, i.e., the length of the route or its

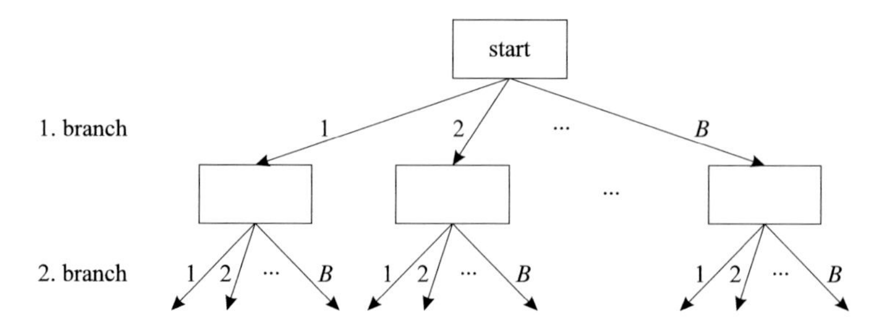

Fig. 5.17 Illustration of the branch and bound method with B alternatives

required time. Only solutions that can be better than the existing ones are relevant. In the B&B tree, each node would represent one city. If the B&B algorithm branches several times, i.e., subsequently visits several cities, the current loss function value for this route may exceed the best existing solution. If this happens the whole subtree can be pruned since all routes in this subtree are known to be worse than the best existing solution. If during the search procedure a better solution is discovered, its loss function value becomes the new upper bound. The more restrictive the bounds are, the more subtrees can be pruned and the faster the search becomes.

It is very easy to incorporate various kinds of constraints into a B&B technique. At each node, the constraints can be checked. If they are violated, this part of the tree can be deleted.

For the practical application of branch and bound techniques, a number of characteristics have to be specified that influence the performance of the algorithm significantly. For example, the search can be performed in depth, broadly, or based on the best bounds. For complex problems, various heuristics and rules can be incorporated to further prune the tree, although this gives up the guarantee of finding the globally best solution.

The difficulty for an extension of the B&B ideas for optimization of real parameters is that it is not easy to make a reasonable comparison with the lower and upper bounds during the tree search. For example, a model with three parameters is to be optimized. Each parameter is discretized into B values. Then the tree has three levels and B^3 final nodes (leaves), which represent the final models. The difficulty now is that it is usually not possible to evaluate the model performance when only one (after the first branch) or two (after the second branch) parameters out of three are determined. The third parameter (which is unknown in the first two levels of the tree) may have fundamental importance for the model performance. Thus, it may be not possible to prune subtrees, and B&B becomes senseless or at least ineffective. In [16] a strategy is proposed that demonstrates how B&B can be applied for optimization in nonlinear predictive control. This strategy is based on the fact that each parameter (here: control action) causes a minimum fixed contribution to the loss function (here: control error). Thus, the upper bound may be exceeded during the B&B search, and subtrees can be pruned. The key to a successful application of branch and bound techniques is the decomposition of the whole problem into subproblems whose performance can be evaluated individually. For a deeper analysis of branch and bound methods, refer to [103, 341, 381].

5.4 Tabu Search (TS)

Tabu search (TS) is a promising new alternative to the established simulated annealing and evolutionary algorithms. It is especially applied to combinatorial optimization problems in the operations research area (traveling salesman, routing, scheduling, etc. problems). The philosophy of tabu search is to utilize the efficiency of local search techniques in the global search strategy. The simplest and ad hoc version is the multi-start approach in which local search is performed, starting from different (typically randomly chosen) initial values; compare the introduction to this chapter. This approach is not very efficient because it may happen that the same (or similar) initial values are investigated several times. In contrast, tabu search is memory-based, i.e., it stores the history of the investigated points and tries to avoid considering them twice (already investigated points are tabu). In order to search the parameter space globally while avoiding getting stuck in local minima, it is necessary to temporarily allow a deterioration (a step back) in performance. In simulated annealing and evolutionary algorithms, the decision whether a new parameter vector is accepted or not is made in a probabilistic manner. In contrast, tabu search typically accepts a performance deterioration (increase of the loss function) only if further local search is not expected to yield better results and the subsequent search region to investigate has not been visited before (for a given number of iterations). Tabu search is currently the subject of active research, and as for all global search methods, a large number of different variants exist. For more detailed information, refer to [34].

5.5 Summary

Nonlinear global optimization techniques are suited for problems that cannot be satisfactorily solved by nonlinear local optimization methods. These problems typically fulfill one or more of the following properties: multi-modal, non-continuous derivatives of the loss function, non-continuous parameters, and structure optimization problems. All nonlinear global optimization methods require some tradeoff between the extent of the global character and the convergence speed. No method can guarantee to find the global solution for complex practical applications (in finite time). However, in most practical cases, the user will be satisfied if the obtained solution meets the specifications even without knowing whether a better solution exists or not. It is highly recommended to combine global search methods with nonlinear local or linear optimization approaches, as described in the introduction of this chapter. It is very hard to give general guidelines, in which a global optimization method is expected to work best on which class of problems. Probably for practical considerations, the experience of the user with the (not few!) fiddle parameters of each algorithm such as annealing schedule, quenching, mutation, or crossover rates is more important than the actual used method. The careful application of global search techniques requires more experience and trial and error than other optimization methods. It has often been reported in the literature that the incorporation of prior knowledge into the algorithm is decisive not only for good performance but also for practical feasibility of the solution [117, 374].

5.6 Problems

1. What is the difference between a local and a global optimization method? What are their key advantages and drawbacks independent of the specific algorithm?
2. Explain some ideas on how the strength of global and local optimization methods can be combined.
3. Write a subroutine that implements simulated annealing with a logarithmic annealing schedule. Input parameters are the initial annealing temperature, the initial parameter vector, and the maximum number of iterations. Output parameters are the optimized parameter vector.
4. Explain the key differences between evolution strategies and genetic algorithms. What type of problem is each variant best suited for?
5. Explain the key ingredients of genetic programming in comparison to the ordinary genetic algorithms.

Chapter 6
Unsupervised Learning Techniques

Abstract This chapter introduces unsupervised learning approaches. In the context of this book, unsupervised approaches are "only" used as a tool that may be helpful in solving a supervised learning problem efficiently. Typically they are considered not for their own sake but as a means for achieving something else. Therefore, it is sufficient to cover the most important approaches briefly. Two categories are discussed: (i) principal component analysis, which transforms the coordinate system and can be used for dimensionality reduction, and (ii) clustering techniques which allow to group data. One particularly important approach is the Gustafson-Kessel clustering algorithm as it can be applied to solve supervised problems in the context of local model networks, which are extensively covered in this book. The idea behind using this unsupervised method for a supervised learning task is that it can be carried out in product space rather than in input space in a meaningful manner. Besides this specialty, the algorithms in this chapter are usually applied in a preprocessing phase of a supervised learning scheme.

Keywords Optimization · Clustering · Principal component analysis · Self-organizing map · Input data · Data distribution

In so-called unsupervised learning, the desired model output y is not known or is assumed to be not known. The goal of unsupervised learning methods is to process or extract information with the knowledge about the input data $\{\underline{u}(i)\}$, $i = 1, \ldots, N$, only. In all problems addressed in this book the desired output is known. However, unsupervised learning techniques can be very interesting and helpful for data preprocessing; see Fig. 6.1. Preprocessing transforms the data into another form, which hopefully can be better processed by the subsequent model. In this context, it is important to keep in mind that the desired output is actually available, and there may exist some efficient way to include this knowledge even into the preprocessing phase.

The following example illustrates the typical use of unsupervised learning techniques for a simple classification problem. Figure 6.2 shows the distribution of the input data in the u_1-u_2-input space. Assume that u_1 and u_2 represent two features

O. Nelles, *Nonlinear System Identification*,
https://doi.org/10.1007/978-3-030-47439-3_6

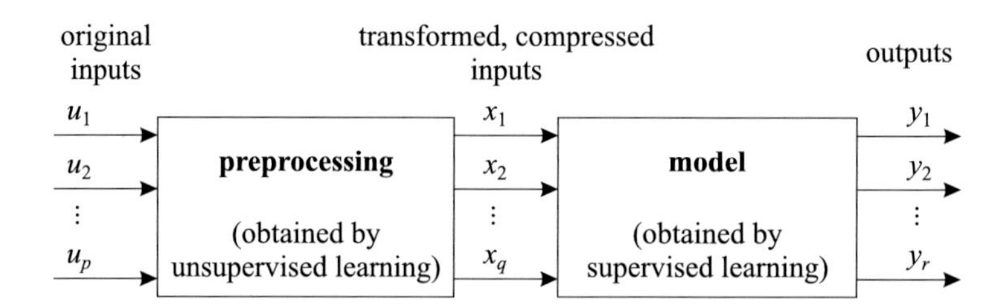

Fig. 6.1 Data preprocessing units are often trained by unsupervised learning. The original p inputs $u_1, u_2, \ldots, u_p$ are not mapped directly to the outputs by the model. Rather, in a preprocessing step, the inputs are transformed to $x_1, x_2, \ldots, x_q$. The number of transformed inputs q is typically smaller than the number of original inputs p since they usually represent the input space in some compressed form. The goal of preprocessing is to reduce the required complexity of the model

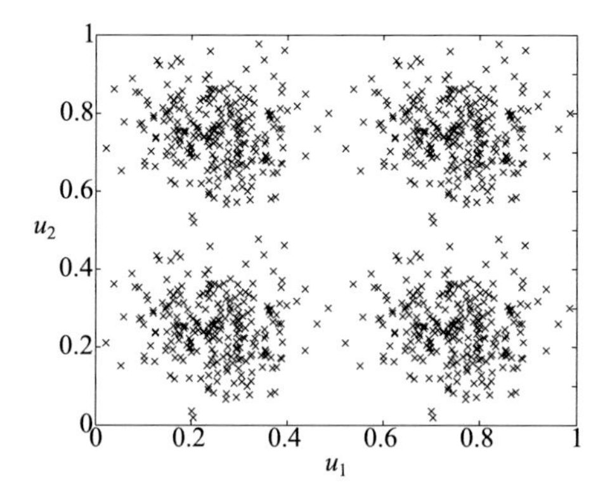

Fig. 6.2 Four clusters in a two-dimensional input space. Without information about the associated outputs, an unsupervised method would discover four groups of data

that have to be mapped to classes represented by integer values of the output y. For example, the correct classification for an XOR-like problem is to assign the upper left and lower right groups of data (clusters) to class 1 and the other two clusters to class 0. In a supervised learning problem, each training data sample would consist of both the input values u_1 and u_2 and the associated output value y, being either 1 or 0. With such training data information, this classification problem can be solved easily. For an unsupervised learning problem, the outputs are unknown, that is, the training data consists only of the input data without any information about the associated classes. Therefore, the best an unsupervised learning technique can do is to group or cluster the input data somehow. For example, an algorithm may find four clusters with their centers at approximately (0.25, 0.25), (0.75, 0.25), (0.25, 0.75), and (0.75, 0.75) in the form of circles with an approximate radius of 0.25 each. Next, in a second step, these four clusters can be mapped by a supervised learning technique to the associated two classes. Thus, the clustering performed by the unsupervised learning technique has transformed the problem of mapping a vast amount of input data samples to their associated class to the problem of mapping four clusters into two classes. Thereby the complexity of the second mapping step is considerably reduced.

The above example is typical in that a single supervised learning approach can be replaced by a two-step procedure consisting of an unsupervised learning preprocessing phase followed by a supervised learning phase. Often the two-step approach is computationally much less demanding than the original single supervised learning problem. The first phase in such a two-step strategy can be considered as information compression or dimensionality reduction. It naturally is most promising and regularly applied for problems with a huge amount of data and/or high-dimensional input spaces.

The above example is somewhat misleading about the performance that can be expected from such unsupervised learning techniques. There is no guarantee that the input data distribution is related to the corresponding output values of the underlying problem. Although unsupervised learning methods are standard tools for preprocessing, their benefits are highly problem-specific.

In the following section, principal component analysis is introduced as a tool for coordinate axis transformation and dimensionality reduction. Section 6.2 addresses popular clustering techniques, starting from the classical k-means and its fuzzy version through the Gustafson-Kessel algorithm to neural network inspired methods such as the famous self-organizing map or neural gas network. Finally, Sect. 6.2.7 discusses possible strategies for incorporating the information about the output values into the clustering framework.

6.1 Principal Component Analysis (PCA)

Principal component analysis (PCA) is a tool for coordinate axis transformation and dimensionality reduction [487]. With a PCA a new set of orthogonal coordinate axes is calculated by maximizing the sample variance of the given data points along these axes. The idea behind PCA is that the direction where the data variance is highest is the most important one and the direction with the smallest data variance is the least significant.

Figure 6.3 shows a simple two-dimensional example. By looking at the data distribution, one can discover that both inputs are highly correlated. It seems very likely that input u_2 is a linear function of input u_1 ($u_2 = c\,u_1$) and the deviations from the exact linear dependency are due to noise. However, with certainty, such a conclusion can never be drawn solely from data without process knowledge. If the assumption of the linear dependency of u_1 and u_2 is true, the input data can properly be described by just one new input along $\underline{x}_1 = \underline{u}_1 + c\,\underline{u}_2$, where $\underline{u}_1 = [u_1\ 0]^T$ and $\underline{u}_2 = [0\ u_2]^{\mathrm{T}}$. This direction $\underline{x}_1$ is the one with the highest data variance. The (orthogonal) second new input axis then is $\underline{x}_2 = \underline{u}_2 - c\,\underline{u}_1$, which is the direction with the lowest data variance. If the axes $\underline{u}_1$ and $\underline{u}_2$ are transformed to the new ones $\underline{x}_1$ and $\underline{x}_2$, the underlying problem is simplified considerably. Since $\underline{x}_1$ contains most information on the data and $\underline{x}_2$ basically describes the noise, a two-

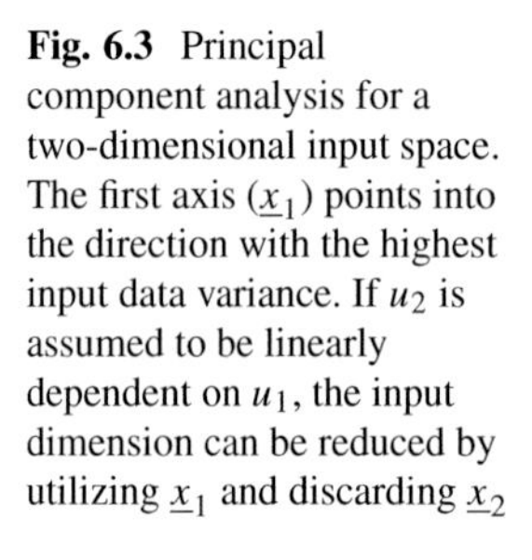

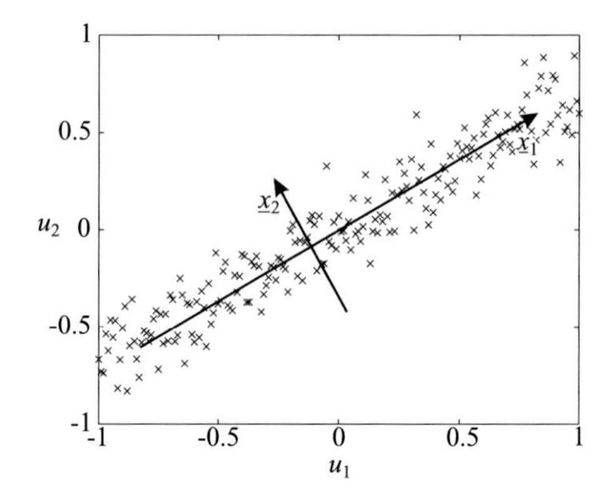

Fig. 6.3 Principal component analysis for a two-dimensional input space. The first axis ($\underline{x}_1$) points into the direction with the highest input data variance. If u_2 is assumed to be linearly dependent on u_1, the input dimension can be reduced by utilizing $\underline{x}_1$ and discarding $\underline{x}_2$

dimensional problem formulated in the $\underline{u}_1$–$\underline{u}_2$-input space can be transformed into a one-dimensional problem in the $\underline{x}_1$-input space by discarding $\underline{x}_2$. Again, note that discarding $\underline{x}_2$ relies on the linear dependency of u_1 and u_2, which is only assumed, not known.

For the general p-dimensional case, the PCA can be formulated as follows [487]. The goal is to maximize the variance along the new axes $\underline{x}_i = [x_{i1}\ x_{i2}\ \cdots\ x_{ip}]^T$, $i = 1, \ldots, p$, where each axis is a linear combination of the original axes. Since there exists an infinite number of vectors pointing in equal directions, in order to obtain a unique solution the variance maximization is constrained by normalizing the vectors, i.e., $\underline{x}^T\underline{x} = 1$. The data can be written in an $N \times p$ matrix $\underline{U}$, where N is the number of data points and p is the dimension of the input space. It is assumed that the data has zero mean (this can be enforced by subtracting the mean). Then the variance of the data along direction $\underline{x}$ is found by calculating the square of the projection of the data on the new axis $\underline{U}\,\underline{x}$. Therefore, the following constrained maximization problem arises:

$$\left(\underline{U}\,\underline{x}\right)^T \left(\underline{U}\,\underline{x}\right) + \lambda \left(1 - \underline{x}^T\underline{x}\right) \longrightarrow \max_{\underline{x}}\,, \tag{6.1}$$

where λ is the Lagrangian multiplier. The first term represents the variance of the data along direction $\underline{x}$, and the second term represents the constraint $\underline{x}^T\underline{x} = 1$. The solution of (6.1) leads to the following eigenvalue problem:

$$\left(\underline{U}^T\underline{U}\right)\underline{x} = \lambda\underline{x}\,. \tag{6.2}$$

The eigenvectors $\underline{x}_i$ are the optimal axes, and the corresponding eigenvalues λ_i are the data variances along these directions ($i = 1, \ldots, p$). Note that the eigenvectors and eigenvalues of $\underline{U}^T\underline{U}$ can be computed by the singular value decomposition (SVD) of $\underline{U}$ [193]. Thus, if the eigenvalues of $\underline{U}^T\underline{U}$ are arranged in descending order, a dimensionality reduction can be performed by selecting the first p_{sel} axes, that is, the directions with the highest data variance.

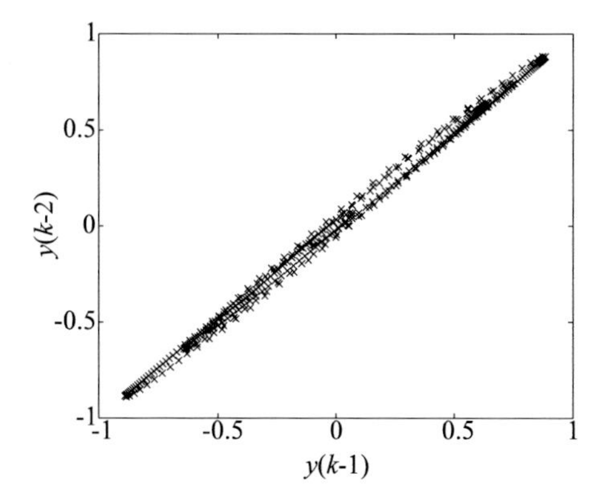

Fig. 6.4 Correlation between the two subsequent process outputs $y(k-1)$ and $y(k-2)$ of a dynamic process with $y(k) = f(u(k-1), u(k-2), y(k-1), y(k-2))$. Although a PCA suggests eliminating one of these variables, the information of both is required for an appropriate process description. Note that a PCA without any dimensionality reduction that performs only a transformation of the axes may lead to a better process description or may not, depending on the specific shape of the function $f(\cdot)$ and the model architecture applied for approximation of $f(\cdot)$

The PCA is an unsupervised learning technique. Only input data variances are evaluated for the axes' transformations. Thus, extreme care must be taken when a PCA is applied for dimensionality reduction. There is generally no reason why low input data variance should imply low significance of the corresponding input. This can be clearly seen in Fig. 6.4, which depicts a typical input data distribution for a dynamic system identification problem. It shows the process outputs delayed by one and two time instants, respectively. Figure 6.4 thus represents the measured data with input $u(k)$ and output $y(k)$ in the input $y(k-1) - y(k-2)$-subspace for an identification of a second-order process, e.g., $y(k) = f(u(k-1), u(k-2), y(k-1), y(k-2))$. Since the sampling time must be chosen in order to cover all significant dynamics of the process, two subsequent process outputs ($y(k-1)$ and $y(k-2)$) necessarily are highly correlated. Nevertheless, the information of *two* previous process outputs is required for building an appropriate model. In this example, a dimensionality reduction would discard a considerable amount of information.

Often therefore a PCA is applied only to coordinate axis transformation and not for dimensionality reduction. But even in this case, it is by no means guaranteed that the transformed axes are a better description with respect to the output values than the original ones.

A better measure for the significance of each input is the complexity of the output y. Methods that take the output into account belong to the class of supervised learning techniques. Although this chapter does not deal with supervised approaches, the basic ideas will be discussed briefly in the following to point out the connection and relationship between these methods.

It will be assumed that the output y can be written as a function of the inputs u_i, $i = 1, \ldots, p$, that is, $y = f(u_1, u_2, \ldots, u_p)$. Obviously, if $\partial f / \partial u_i = 0$ for all u_i, the output y does not depend on input u_i. This redundant input can then be discarded without any information loss. The problem, however, is more complicated, since not only redundant inputs but also their redundant linear combinations are searched. If the function $f(\cdot)$ is assumed to be linear, this problem can be solved analogously to a PCA by a partial least squares regression (PLS). Instead of the data matrix $\underline{U}$, the covariance (cross-correlation) matrix $\left(\underline{U}^T \underline{y}\right)^T$ has to be used in (6.1) and (6.2), where $\underline{y}$ is the N-dimensional output data vector. However, for nonlinear $f(\cdot)$ the PLS does not work properly, since the covariance matrix can only represent linear dependencies. Finding the optimal directions in nonlinear problems is a very complicated task. It obviously becomes more important, the higher dimensional the input space is. Since the linear combinations (directions) of inputs u_i, $i = 1, \ldots, p$ influence the output in a nonlinear way, necessarily nonlinear optimization techniques are required to solve this problem. Common algorithms are projection pursuit learning (PPL) [269], originated from statistics, and the multilayer perceptron (MLP) network in the context of neural networks. It can be shown that the MLP is a special kind of PPL. Section 11.2.7 addresses this matter more extensively.

6.2 Clustering Techniques

Clustering techniques search for groups of data. A cluster can be defined as a group of data that are more similar to each other than data belonging to other clusters [51]. Figure 6.5 shows four examples. The user has to define what kinds of clusters will be sought by defining a *similarity measure*. The most common cluster shapes are filled circles or (in higher dimensions) spheres, respectively. Then the similarity measure may be the distance of all data samples within a cluster to the cluster center. In this case, the cluster center represents the cluster, and thus it is called the *prototype* of the cluster. For other similarity measures, the cluster prototype can be different, e.g., a line in Fig. 6.5b or a circle's/ellipses' center and its radius in Fig. 6.5c, d.

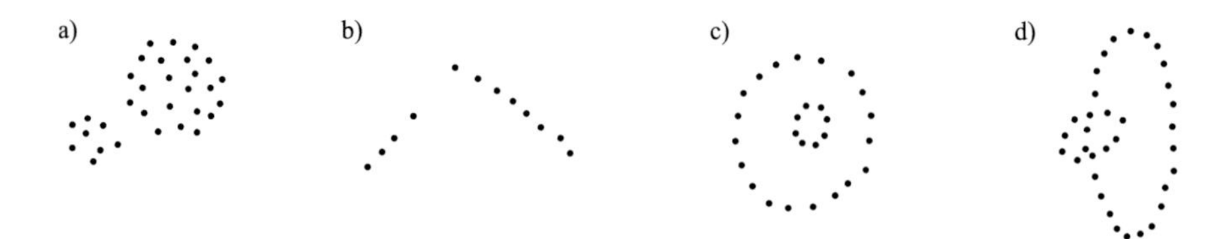

Fig. 6.5 Examples for different shapes of clusters: (**a**) filled circles, (**b**) lines, (**c**) hollow circles, (**d**) hollow ellipses [16]

The user specifies what the clusters will look like and chooses a suitable clustering algorithm for this task. The clustering techniques treated in this section require the user to choose the number of clusters a priori. More advanced methods can determine the number of clusters automatically in dependency on a granularity measure given by the user. Only non-hierarchical clustering approaches are considered here, i.e., methods in which all clusters have a comparable interpretation. Hierarchical clustering techniques generate clusters that can again include clusters and so on.

Classical clustering methods such as k-means (Sect. 6.2.1) assign each data sample fully to one cluster (hard or crisp partition). Modern clustering techniques generate a fuzzy partition. This means that each data sample is assigned to each cluster with a certain *degree of membership* $\mu(\underline{u})$. For each data sample, all degrees of membership sum up to 1. For example, one data sample $\underline{u}(i)$ may be assigned to cluster 1 with $\mu(\underline{u}(i)) = 0.7$, to cluster 2 with $\mu(\underline{u}(i)) = 0.2$, to cluster with $\mu(\underline{u}(i)) = 0.1$, and to all other clusters with $\mu(\underline{u}(i)) = 0$.

Since the loss functions minimized by clustering techniques are typically nonlinear, the algorithms operate iteratively starting from initially chosen clusters. Generally, convergence to the global optimum cannot be guaranteed. The sensitivity on the initial values depends on the specific algorithm. If the initial clusters are chosen reasonably by prior knowledge instead of randomly, convergence to poor local optima can usually be avoided.

Another important issue related to the choice of the similarity measure is the normalization of the data. Most clustering algorithms are very sensitive to the scale of the data. For example, a clustering technique that looks for filled circles in the data is strongly influenced by the scaling of the axes, since that changes circles to ellipses. Non-normalized data, say two inputs in the ranges $0 < u_1 < 1$ and $0 < u_2 < 1000$, can degrade the performance of a clustering technique because for the calculation of distances, the value of u_1 is almost irrelevant compared with u_2. Thus, data should be normalized or standardized before applying clustering techniques. An exception to this rule is methods with adaptive similarity measures that automatically rescale the data, such as the Gustafson-Kessel algorithm described in Sect. 6.2.3. But even for these rescaling algorithms, normalized data yields numerically better-conditioned problems.

In order to summarize, clustering techniques can be distinguished according to the following properties:

- Type of the variables they can be applied to (continuous, integer, binary)
- Similarity measure
- Hierarchical or non-hierarchical
- Fixed or self-adaptive number of clusters
- Hard or fuzzy partitions

For a more detailed treatment of clustering methods, refer to [9, 16, 260, 488].

6.2.1 k-Means Algorithm

The k-means algorithm discussed in this section is the most common and simple clustering method. It can be seen as the basis for the more advanced approaches described below. Therefore, some illustrative examples are presented to demonstrate the properties of k-means, while the other algorithms are treated more briefly. In the name k-means clustering, also known as c-means clustering [9], the "k" or "c" stands for the fixed number of clusters, which is specified by the user a priori.

k-means clustering minimizes the following loss function:

$$I = \sum_{j=1}^{C} \sum_{i \in \mathcal{S}_j} ||\underline{u}(i) - \underline{c}_j||^2 \longrightarrow \min_{\underline{c}_j}, \tag{6.3}$$

where the index i runs over all elements of the sets $\mathcal{S}_j$, C is the number of clusters, and $\underline{c}_j$ are the cluster centers (prototypes). The sets $\mathcal{S}_j$ contain all indices of those data samples (out of all N) that belong to the cluster j, i.e., which are nearest to the cluster center $\underline{c}_j$. The cluster centers $\underline{c}_j$ are the parameters that the clustering technique varies in order to minimize (6.3). Therefore, the loss function (6.3) sums up all quadratic distances from each cluster center to its associated data samples. It can also be written as

$$I = \sum_{j=1}^{C} \sum_{i=1}^{N} \mu_{ji} ||\underline{u}(i) - \underline{c}_j||^2, \tag{6.4}$$

where $\mu_{ji} = 1$ if the data sample $\underline{u}(i)$ is associated (belongs) with the cluster j and $\mu_{ji} = 0$ otherwise.

The k-means algorithm by MacQueen to minimize (6.4) works as follows [9]:

1. Choose initial values for the C cluster centers $\underline{c}_j$, $j = 1, \ldots, C$. This can be done by picking randomly C different data samples.
2. Assign all data samples to their nearest cluster center.
3. Compute the centroid (mean) of each cluster. Set each cluster center to the centroid of its cluster, that is,

$$\underline{c}_j = \frac{\sum\limits_{i \in \mathcal{S}_j} \underline{u}(i)}{N_j}, \tag{6.5}$$

   where i runs over those N_j data samples that belong to cluster j, i.e., are in the set $\mathcal{S}_j$, and N_j is the number of the elements in the set $\mathcal{S}_j$ ($\sum_{j=1}^{C} N_j = N$).
4. If any cluster center has been moved in the previous step, go to Step 2; otherwise stop.

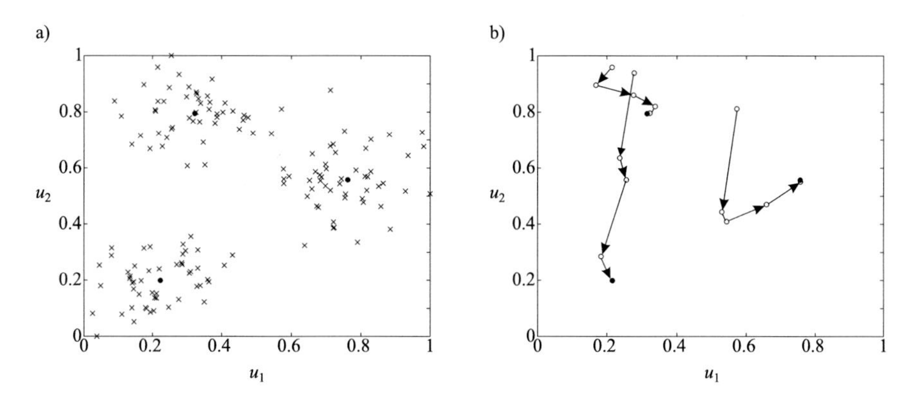

Fig. 6.6 (**a**) Clustering of the data with k-means leads (**b**) to convergence in five iterations. The black filled circles represent the final cluster centers

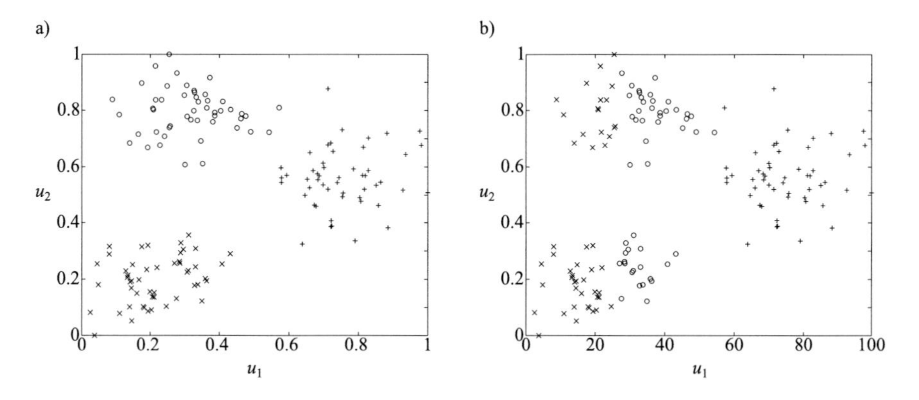

Fig. 6.7 (**a**) Comparison of normalized data and (**b**) non-normalized data and its effect on k-means clustering. Although the cluster centers in (**b**) are still reasonably placed, the data is assigned almost solely to the cluster that is closest in the u_1-dimension because this coordinate dominates the distance measure

Figure 6.6 illustrates the convergence behavior of k-means with $C = 3$ clusters. Figure 6.6a shows the two-dimensional data set. Figure 6.6b depicts the three cluster centers for the five iterations required for convergence. The initial cluster centers were chosen randomly from the data set. Although these initial values lie quite poorly, k-means converges fast to the final cluster centers.

Figure 6.7 illustrates the importance of normalization of the data. Figure 6.7a shows which data samples belong to which cluster for the normalized data from Fig. 6.6a. Figure 6.7b shows the clusters for the non-normalized data set where u_1 lies between 0 and 100 and u_2 lies between 0 and 1. The distance in (6.4) is dominated by the u_1-dimension, and the cluster boundaries depend almost solely on u_1.

An alternative to normalization of the data is changing the distance metric used in (6.4). The quadratic Euclidean norm

$$D_{ij}^2 = ||\underline{u}(i) - \underline{c}_j||^2 = (\underline{u}(i) - \underline{c}_j)^T (\underline{u}(i) - \underline{c}_j) \tag{6.6}$$

can be extended to the quadratic general Mahalanobis norm

$$D_{ij,\underline{\Sigma}}^2 = ||\underline{u}(i) - \underline{c}_j||_{\underline{\Sigma}}^2 = (\underline{u}(i) - \underline{c}_j)^T \underline{\Sigma}(\underline{u}(i) - \underline{c}_j) \,. \tag{6.7}$$

The norm matrix $\underline{\Sigma}$ scales and rotates the axes. For the special case where the covariance matrix is equal to the identity matrix ($\underline{\Sigma} = \underline{I}$), the Mahalanobis norm is equal to the Euclidean norm. For

$$\underline{\Sigma} = \begin{bmatrix} 1/\sigma_1^2 & 0 & \cdots & 0 \\ 0 & 1/\sigma_2^2 & \cdots & 0 \\ \vdots & \vdots & \ddots & \vdots \\ 0 & 0 & \cdots & 1/\sigma_p^2 \end{bmatrix} , \tag{6.8}$$

where p denotes the input space dimension, the Mahalanobis norm is equal to the Euclidean norm with the scaled inputs $u_l^{(\text{scaled})} = u_l/\sigma_l$. In the most general case, the norm matrix scales and rotates the input axes. Figure 6.8 summarizes these distance measures. Note that an choice in (6.4) is equivalent to the Euclidean norm with transformed input axes. One restriction of k-means is that the chosen norm is fixed for the whole input space and thus for all clusters. This restriction is overcome with the Gustafson-Kessel algorithm, which employs individual adaptive distance measures for each cluster; see Sect. 6.2.3.

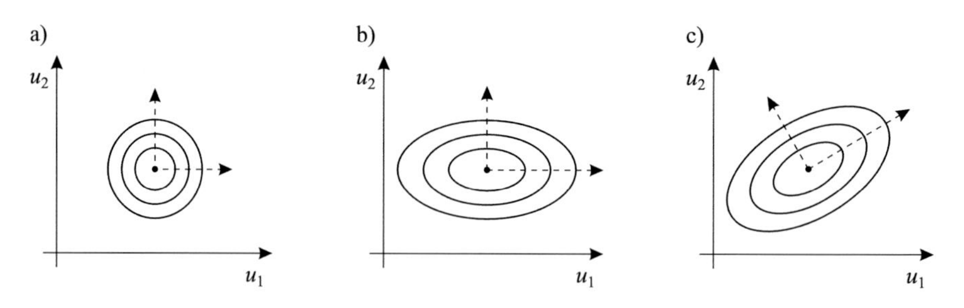

Fig. 6.8 Lines with equal distance for different norms: (**a**) Euclidean ($\underline{\Sigma} = \underline{I}$), (**b**) diagonal ($\underline{\Sigma}$ = diagonal), and (**c**) Mahalanobis norm ($\underline{\Sigma}$ = general)

6.2.2 Fuzzy C-Means (FCM) Algorithm

The fuzzy c-means (FCM) algorithm is a fuzzified version of the classical k-means algorithm described above. The minimized loss function is almost identical:

$$I = \sum_{j=1}^{C} \sum_{i=1}^{N} \mu_{ji}^{\nu} ||\underline{u}(i) - \underline{c}_j||_{\underline{\Sigma}}^2 \qquad \text{with} \quad \sum_{j=1}^{C} \mu_{ji} = 1\,. \tag{6.9}$$

So, the degree of membership μ_{ji} of one data sample $\underline{u}(i)$ to cluster j is not required to be equal to one for one cluster and zero for all others. Rather each data sample may have *some* degree of membership between 0 and 1 for each cluster under the constraint that all these degrees of membership sum up to 1, i.e., 100%. The degree of membership is raised to the power of ν that determines the fuzziness of the clusters. This weighting exponent ν lies in the interval $(1, \infty)$, and without prior knowledge, the typically chosen value is $\nu = 2$. If the clusters are expected to be easily separable, then low values close to 1 are recommended for ν since this reduces the fuzziness and drives the degrees of membership close to 0 or 1. On the other hand, if the clusters are expected to be almost not distinguishable, high values for ν should be chosen.

The degree of membership μ_{ji} of one data sample $\underline{u}(i)$ to cluster j is defined by [16, 51, 260]

$$\mu_{ij} = \frac{1}{\sum\limits_{l=1}^{C} \left(D_{ij,\underline{\Sigma}}^2 \,/\, D_{il,\underline{\Sigma}}^2 \right)^{\frac{1}{\nu-1}}} \tag{6.10}$$

with

$$D_{ij,\underline{\Sigma}}^2 = ||\underline{u}(i) - \underline{c}_j||_{\underline{\Sigma}}^2 = (\underline{u}(i) - \underline{c}_j)^T \,\underline{\Sigma}(\underline{u}(i) - \underline{c}_j)\,. \tag{6.11}$$

Figure 6.9 illustrates the definition of the distances for one data sample and two cluster centers and the Euclidean distance measure $\underline{\Sigma} = \underline{I}$. Then (6.10) becomes

$$\mu_{ij} = \frac{1}{\left(D_{i1}^2/D_{i1}^2 + D_{i1}^2/D_{i2}^2\right)^{\frac{1}{\nu-1}}} = \frac{1}{\left(1 + D_{i1}^2/D_{i2}^2\right)^{\frac{1}{\nu-1}}} \tag{6.12}$$

with the distances in Fig. 6.9. Obviously, as the data sample approaches the cluster center ($D_{ij,\underline{\Sigma}}^2 \to 0$), the degree of membership to this cluster approaches 1 ($\mu_{ij} \to 1$), and as $D_{ij,\underline{\Sigma}}^2 \to \infty$, $\mu_{ij} \to 0$. Clearly, (6.10) automatically fulfills the constraint that the sum of μ_{ij} over all clusters is equal to 1 for each data sample.

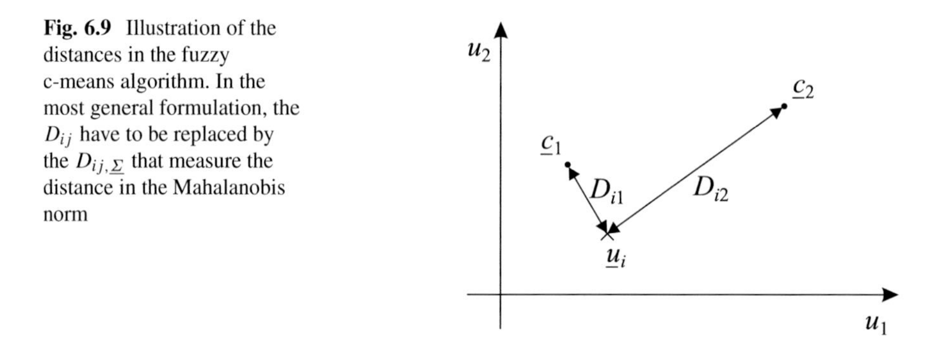

Fig. 6.9 Illustration of the distances in the fuzzy c-means algorithm. In the most general formulation, the D_{ij} have to be replaced by the $D_{ij,\underline{\Sigma}}$ that measure the distance in the Mahalanobis norm

When calculating the degree of membership according to (6.10), the following two special cases must be taken into account:

- If in (6.10), (6.11) the data sample $\underline{u}(i)$ lies exactly on a cluster center $\underline{c}_l$ that is not cluster j ($l \neq j$), then one denominator in (6.10) ($D^2_{il,\underline{\Sigma}}$) becomes zero and $\mu_{ij} = 0$.
- If in (6.10), (6.11) the data sample $\underline{u}(i)$ lies exactly on the cluster center $\underline{c}_j$, then μ_{ij} can be chosen arbitrarily if the constraint $\sum_{j=1}^{C} \mu_{ij} = 1$ is met.

The fuzzy c-means algorithm that minimizes (6.9) works as follows [16, 51, 260]:

1. Choose initial values for the C cluster centers $\underline{c}_j$, $j = 1, \ldots, C$. This can be done by picking randomly C different data samples.
2. Calculate the distances $D^2_{ij,\underline{\Sigma}}$ of all data samples to $\underline{u}(i)$ to each cluster center $\underline{c}_j$ according to (6.11).
3. Compute the degrees of membership for each data sample $\underline{u}(i)$ to each cluster j according to (6.10).
4. Compute the centroid (mean) of each cluster. Set each cluster center to the centroid of its cluster, that is,

$$\underline{c}_j = \frac{\sum\limits_{i=1}^{N} \mu_{ij}^{\nu}\, \underline{u}_i}{\sum\limits_{i=1}^{N} \mu_{ij}^{\nu}} . \tag{6.13}$$

   In an extension to (6.5) in the classical k-means algorithm, the data samples are weighted with their corresponding degrees of membership (weighted sum).
5. If any cluster center has been moved significantly, say more than ϵ, in the previous step then go to Step 3; otherwise stop.

Like the k-means algorithm, this fuzzified version searches for filled circles ($\underline{\Sigma} = \underline{I}$), axis-orthogonal ellipses ($\underline{\Sigma}$ = diagonal), or arbitrarily oriented ellipses ($\underline{\Sigma}$ = general). The shape of the cluster has to be fixed by the user a priori. It can neither be adapted to the data nor be different for each individual cluster.

6.2.3 Gustafson-Kessel Algorithm

The Gustafson-Kessel clustering algorithm [16, 260] is an extended version of the fuzzy c-means algorithm. Each cluster j possesses its individual distance measure $\underline{\Sigma}_j$:

$$D^2_{ij,\underline{\Sigma}_j} = ||\underline{u}(i) - \underline{c}_j||^2_{\underline{\Sigma}_j} = (\underline{u}(i) - \underline{c}_j)^T \underline{\Sigma}_j (\underline{u}(i) - \underline{c}_j) \,. \tag{6.14}$$

Furthermore, not only the cluster centers $\underline{c}_j$ but also the norm matrices $\underline{\Sigma}_j$ are subject to the minimization of the loss function (6.9). The $p \times p$ norm matrices $\underline{\Sigma}_j$ can be calculated as the inverse *fuzzy covariance matrix* of each cluster

$$\underline{\Sigma}_j = \underline{F}_j^{-1} \tag{6.15}$$

with

$$\underline{F}_j = \frac{\sum\limits_{i=1}^{N} \mu_{ij}^{\nu} (\underline{u}(i) - \underline{c}_j)(\underline{u}(i) - \underline{c}_j)^T}{\sum\limits_{i=1}^{N} \mu_{ij}^{\nu}} \,. \tag{6.16}$$

One difficulty is that the distance (6.14) and consequently the loss function (6.9) can be reduced by simply making the determinant of $\underline{\Sigma}_j$ small. To prevent this effect, the norm matrices or the fuzzy covariance matrices have to be normalized. Typically, the determinant of the norm matrices is constrained to a user-defined constant:

$$\det(\underline{\Sigma}_j) = \nu_j \,. \tag{6.17}$$

Thus, the norm matrices are defined by

$$\underline{\Sigma}_j = \underline{F}_j^{-1} \cdot \left(\nu_j \det(\underline{F}_j)\right)^{1/p} \,, \tag{6.18}$$

with p being the dimensionality of the input space. By this normalization, the volume of the clusters is restricted to ν_j. Thus, the Gustafson-Kessel algorithm searches for clusters with given and equal volume. If prior knowledge about the expected cluster volumes is available, the ν_j can be chosen individually for each cluster. However, in practice, typically all cluster volumes are chosen as $\nu_j = 1$.

Figure 6.10 illustrates the relationship between the distance measure (6.14) and the eigenvectors and eigenvalues of the fuzzy covariance matrix $\underline{F}_j$. A large eigenvalue spread $\lambda_{\max}/\lambda_{\min}$ indicates a widely stretched ellipse, while for an eigenvalue spread close to 1, the ellipse has spherical character. In [16] the Gustafson-Kessel algorithm is applied for detecting linear functions (hyperplanes),

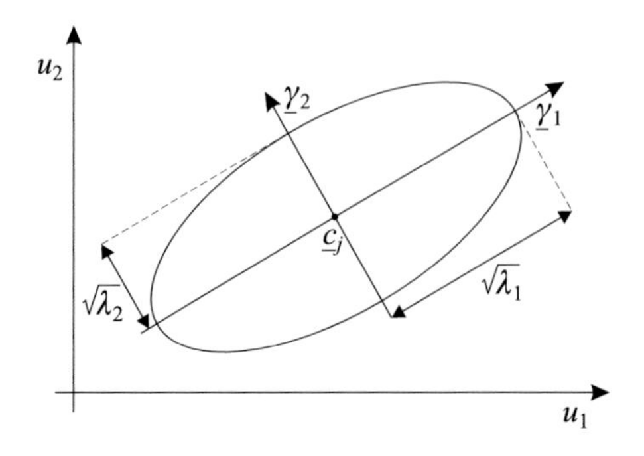

Fig. 6.10 Interpretation of the eigenvalues λ_l and eigenvectors $\underline{\gamma}_l$, $l = 1, \ldots, p$, of the fuzzy covariance matrix $\underline{F}_j$. The ellipse is a contour line of $(\underline{u}(i) - \underline{c}_j)\underline{\Sigma}(\underline{u}(i) - \underline{c}_j)^T = 1$ with $\underline{\Sigma}$ according to (6.18). The area (for higher dimensions, the volume) of the cluster is fixed, but its orientation and shape are determined by the eigenvectors and eigenvalues of $\underline{F}_j$, which is adapted by the algorithm

which results in one very small eigenvalue that is orthogonal to the corresponding hyperplane.

The Gustafson-Kessel algorithm is basically identical to the fuzzy c-means algorithm described in the previous section. However, for the calculation of the distance $D^2_{ij,\underline{\Sigma}_j}$ in Step 2, the equations (6.16), (6.18), and (6.14) must be evaluated subsequently. Thus, each iteration involves the inversion of C $p \times p$ matrices according to (6.18). Since a matrix inversion has quadratic complexity for symmetric matrices, this restricts the applicability of Gustafson-Kessel clustering to relatively low-dimensional problems.

Besides the computational demand, the only major drawback of the Gustafson-Kessel algorithm is the restriction to clusters of constant volume. This restriction is overcome in *fuzzy maximum likelihood estimates clustering* (also known as *Gath-Geva algorithm* [260]). The price to be paid is a much higher sensitivity on the initial clusters and thus a higher probability of convergence to poor local minima of the loss function [16].

6.2.4 Kohonen's Self-Organizing Map (SOM)

Kohonen's self-organizing map (SOM) [326] is the most popular neural network approach to clustering. It is an extension of the *vector quantization (VQ)* technique, which has also been developed by Kohonen. Vector quantization is basically a simplified version of k-means clustering (Sect. 6.2.1) in sample adaptation, i.e., it updates the parameters, not after one sweep through the whole data set (batch adaptation) as k-means does, but after each incoming data sample; see Sect. 4.1. Vector quantization operates with C "neurons," which correspond to the clusters in k-means. Each neuron has p (the input dimension) parameters or "weights" corresponding to the p components of each cluster center vector $\underline{c}$. Since the distinction between neurons and cluster is solely terminological, the parameters of each neuron will also be denoted as $\underline{c}$.

The algorithm for vector quantization is as follows:

1. Choose initial values for the C neuron vectors $\underline{c}_j$, $j = 1, \ldots, C$. This can be done by picking randomly C different data samples.
2. Choose one sample for the data set. This can be done either randomly or by systematically going through the whole data set (cyclic order).
3. Calculate the distance of the selected data sample to all neuron vectors. Typically, the Euclidean distance measure is used. The neuron with the vector closest to the data sample is called the *winner neuron*.
4. Update the vector of the winner neuron in a way that moves it toward the selected data sample $\underline{u}$:

$$\underline{c}_{\text{win}}^{(\text{new})} = \underline{c}_{\text{win}}^{(\text{old})} + \eta \left(\underline{u} - \underline{c}_{\text{win}}^{(\text{old})} \right) . \tag{6.19}$$

5. If any neuron vector has been moved significantly, say more than ϵ, in the previous step then go to Step 2; otherwise stop.

In Step 4 the step size (learning rate) η must be chosen appropriately. The step size $\eta = 1$ would move each neuron vector exactly on the actual data samples and the algorithm would not converge. In fact, as for all sample adaptation approaches, the step size η is required to approach zero for convergence. Thus, for fast convergence speed, it is recommended to start with a large step size, say 0.5, which is decreased in each iteration of the algorithm. Learning vector quantization is essentially the same as a sample adaptation version of k-means clustering [9, 499]. In k-means, however, the step size is normalized by the number of data samples that belong to the winner neuron. This ensures that the vector of the winner neuron converges to the centroid (mean) of these data samples.

The neural network approaches to clustering are often referred to as *competitive learning* because in Step 3 all neurons of the network compete for the selected data sample. The strategy in VQ is called *winner takes it all* since only the neuron that best matches the data sample is updated.

Kohonen's *self-organizing map (SOM)* is an extension of the vector quantization algorithm described above [179, 326, 488, 499]. In an SOM the neurons are not just abstract structures that represent the cluster center. Rather, neurons are organized in one-, two-, and sometimes higher-dimensional topologies; see Fig. 6.11. For most applications a two-dimensional topology with hexagonal or rectangular structure (as in Fig. 6.11c) is used. The neurons' weight vectors (cluster centers) are p-dimensional as for VQ or k-means independent of the chosen topology of the SOM.

The idea of the self-organizing map is that neurons that are neighbors in the network topology should possess similar weight vectors (cluster centers). Thus, the distance of the different cluster centers in the p-dimensional space is represented in a lower (typically two-) dimensional space. Of course, this projection of a

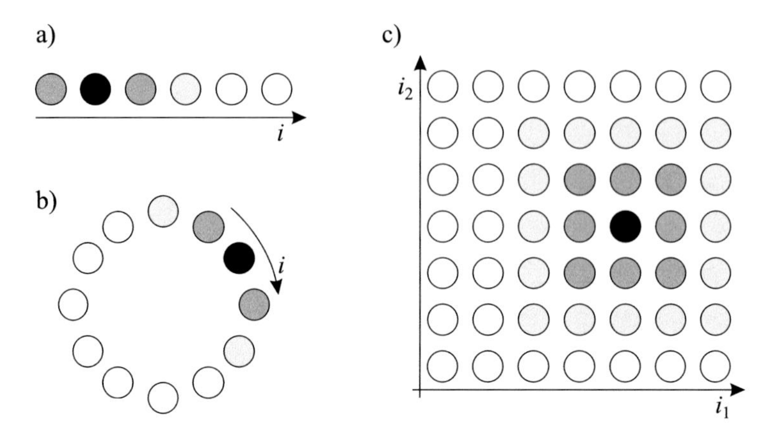

Fig. 6.11 Different topologies of self-organizing maps: (**a**) linear, (**b**) circular, (**c**) two-dimensional grid. The black neuron is the winner for an incoming data sample, and the gray neurons represent its neighborhood

high-dimensional onto a low-dimensional space cannot be performed perfectly and involves information compression. The two-dimensional SOM is an excellent tool for the visualization of high-dimensional data distributions.

In order to ensure that neighbored neurons represent similar regions in the p-dimensional input space, the VQ algorithm is extended as follows. A *neighborhood function* is introduced that defines the activity of those neurons that are neighbors of the winner. In contrast to VQ, not only is the winner neuron updated as in (6.19) but also its neighbors. The neighborhood function $h(\underline{i})$ usually is equal to 1 for the winner neuron and decreases with the distance of the neurons from the winner. The neighborhood function is defined in the topology of the SOM. For example, the SOMs in Fig. 6.11a, b have one-dimensional neighborhood functions, e.g.,

$$h(i) = \exp\left(-\frac{1}{2}\frac{(i^{(\mathrm{win})} - i)^2}{\sigma^2}\right), \tag{6.20}$$

where $i^{(\mathrm{win})}$ denotes the index of the winner neuron and i denotes the index of any neuron. The SOM in Fig. 6.11c has a two-dimensional neighborhood function, e.g.,

$$h(i_1, i_2) = \exp\left(-\frac{1}{2}\frac{(i_1^{(\mathrm{win})} - i_1)^2 + (i_2^{(\mathrm{win})} - i_2)^2}{\sigma^2}\right), \tag{6.21}$$

where $i_1^{(\mathrm{win})}$ and $i_2^{(\mathrm{win})}$ denote the indices of the winner neuron and i_1 and i_2 denote the indices of any neuron. The black, dark gray, and light gray neurons in Fig. 6.11 indicate the value (i.e., the neuron activity) of such a neighborhood function. As long as the neighborhood function has local character, its exact shape is not crucial. For the SOM learning algorithm, (6.19) in Step 4 of the VQ algorithm is extended to

$$\underline{c}_j^{(\text{new})} = \underline{c}_j^{(\text{old})} + \eta\, h(\underline{i}) \left(\underline{u} - \underline{c}_j^{(\text{old})} \right) , \tag{6.22}$$

where $h(\underline{i})$ is the neighborhood function of the dimension of the SOM's topology. Note that (6.22) is evaluated for all active neurons $j = 1, \ldots, C$, not just the winner neuron. By (6.22) a whole group of neighboring neurons is moved toward the incoming data sample. The closer to the winner a neuron is, the larger is $h(\underline{i})$ and thus the step size.

To illustrate the effect of the neighborhood function, it is helpful to consider the two cases of an extremely sharp ($\sigma \to 0$ in (6.20) or (6.21)) and wide ($\sigma \to \infty$) neighborhood function. In the first case, the SOM network reduces to VQ, and no relationship between neighbored neurons is generated. In the second case, all neurons identically learn the centroid of the whole data set and the SOM is useless. The SOM learning algorithm starts with a broad neighborhood function and shrinks it in each iteration. By this strategy, in the first iterations, the network learns a rough representation of the data distribution and refines it as the neighborhood function becomes more and more local. This is the same strategy as taken for the step size η. If the shrinking process is implemented slowly, the danger of convergence to a local minimum reduces.

6.2.5 *Neural Gas Network*

The strength of the SOM for visualization of high-dimensional data in a one- or two-dimensional network topology is also its weakness. There is no reason for the low-dimensional configuration of neurons other than the limited capabilities of humans to see and think in high-dimensional spaces. The neural gas network gives up the easy interpretability of the SOM for better performance. In [368] the neural gas network is proposed, and it is shown that it compares favorably with other clustering techniques such as k-means, Kohonen's SOM, and maximum entropy clustering.

For the neural gas network, the neighborhood function is defined in the p-dimensional input space. First, all neurons $\underline{c}_j$, $j = 1, \ldots, C$, are ranked according to the distance of their cluster centers to the incoming data sample $\underline{u}$. Then the neighborhood function is defined as

$$h(j) = \exp\left(-\frac{\text{ranking}(\underline{c}_j) - 1}{\sigma} \right) . \tag{6.23}$$

For the neuron with the cluster center $\underline{c}_{\text{closest}}$ that is closest to $\underline{u}$, $\text{ranking}(\underline{c}_{\text{closest}}) = 1$, for the neuron with the most distant cluster center $\text{ranking}(\underline{c}_{\text{farthest}}) = C$. Thus, the neighborhood function is equal to 1 for the most active neuron and decays for less active neurons to (almost) zero. How fast the neighborhood function decreases with the ranking of the neurons is determined by σ. As for the SOM, the neural gas network starts with a large value for σ and shrinks it in each iteration. This approach is closely related to simulated annealing (see Sect. 5.1), in which a slowly decreasing temperature ensures convergence to the global optimum. By analogy, the neural gas network converges to the global minimum if σ is decreased slowly enough. In practice, no guarantee for convergence to the global minimum can be given in finite time. However, the neural gas network is significantly less sensitive to the initialization of the cluster centers. As $\sigma \to 0$, basically only the most active neuron is updated, and the neural gas network approaches vector quantization.

6.2.6 Adaptive Resonance Theory (ART) Network

Adaptive resonance theory (ART) networks have been developed by Carpenter and Grossberg [91, 92]. ART is a series of network architectures for unsupervised learning, from which ART2 is suitable for clustering continuous data and thus will be discussed here. So everything that follows in this section is focused on the ART2 architecture, even if the abbreviation ART is used for simplicity. The main motivation of ART is to imitate cognitive phenomena in humans and animals. This is the main reason why most publications on the ART architecture use totally different terminology than in engineering and mathematics. Here, ART is discussed in the clustering framework following the analysis in [179, 499].

ART networks have been developed to overcome the so-called *stability/plasticity dilemma* coined by Carpenter and Grossberg. In this context "stability" does not refer to dynamic systems. Rather, "stability" is used as a synonym for convergence. The stability/plasticity dilemma addresses the following issue. A learning system has to accomplish at least two tasks: (i) it has to adapt to new information, (ii) it has to converge to some optimal solution. The dilemma is how to design a system that is able to adapt to new information (being plastic) without forgetting or overwriting already learned relationships (being stable). In ART networks, this conflict is solved by the introduction of a *vigilance* parameter. This vigilance controls whether already learned relationships should be adapted (and thus partly forgotten) by the new incoming information or whether a new cluster should be generated that represents the new data sample.

The ART algorithm is similar to vector quantization. However, an ART network starts without any neurons and builds them up as the learning progresses. In each iteration, the winner neuron is determined and the distance between the cluster center of the winner and the considered data sample is calculated. If this distance is smaller than the vigilance parameter ρ, i.e.,

$$||\underline{u} - \underline{c}_{\text{win}}|| < \rho\,, \tag{6.24}$$

then the winner neuron is updated as in the vector quantization algorithm in Step 4; see Sect. 6.2.4. If (6.24) is not fulfilled (or no neuron exists at all, as is the case in the first iteration), a new neuron is generated with its cluster center on the data sample:

$$\underline{c}_{\text{new}} = \underline{u} \,. \tag{6.25}$$

Clearly, the behavior of ART networks depends strongly on the choice of the vigilance parameter ρ. Large values of ρ lead to very few clusters, while too small values for ρ may create one cluster for each data sample. Since this choice of ρ is so crucial, the user must determine a good vigilance parameter for each specific application by a trial-and-error approach. Furthermore, in [499, 500] ART networks are criticized: they do not yield consistent estimates of the optimal cluster centers, they are very sensitive to noise, and the algorithm depends strongly on the order in which the training data is presented. All these are undesirable properties of adaptive resonance theory networks.

6.2.7 Incorporating Information About the Output

The main restriction of all unsupervised learning approaches for modeling problems is that the available information about the process output is not taken into account. At least two strategies exist to incorporate this information into the clustering procedure without explicitly optimizing a loss function depending on the output error (which would lead to a supervised learning approach):

- A scheme consisting of an unsupervised preprocessing phase is followed by a model obtained by supervised learning (see Fig. 6.1) can be applied iteratively. In the first iteration, the unsupervised learning technique does not utilize any information about the output. In all following iterations, it can exploit the information about the model error obtained in the previous iteration. For example, the error can be used to drive the clusters into those regions of the input space where the error is large. This error is expected to be large in regions where the process behavior is complex, and the error is probably small where the process behavior is simple. Consequently, the clustering does not rely solely on the input data distribution but also takes the complexity of the considered process into account. For applications of this idea, see Sect. 11.3.3 and [206, 433].
- Product space clustering is an alternative way to incorporate the output, which is extensively applied in the area of fuzzy modeling [16]. Instead of applying clustering to the *input space* spanned by $\underline{u} = [u_1\ u_2\ \cdots\ u_p]^T$, it is performed in the *product space* consisting of the inputs and output

$$\underline{p} = [u_1\ u_2\ \cdots\ u_p\ \ y]^T \,. \tag{6.26}$$

All clustering techniques can be applied to the product space by simply replacing $\underline{u}$ with $\underline{p}$. In [16] it is demonstrated how product space clustering with the Gustafson-Kessel algorithm (Sect. 6.2.3) can be used for discovering hyperplanes $y = w_0 + w_1u_1 + w_2u_2 + \ldots + w_pu_p$ in the data. Furthermore, it is shown how the parameters w_i of the hyperplane can be extracted from the corresponding covariance matrix $\underline{F}_j$ and that these parameters are equivalent to those obtained by a total least squares (TLS) optimization; see Sect. 13.3.

Some of the unsupervised learning techniques can be extended to supervised learning methods. Examples are the learning vector quantization (LVQ) based on the VQ approach discussed in Sect. 6.2.4 and ARTMAP based on the ART network (Sect. 6.2.6).

6.3 Summary

Unsupervised learning techniques extract compressed information about the input data distribution. They are typically used as preprocessing tools for a subsequent supervised learning approach that maps the extracted features to the model output. Since the only source of information utilized by unsupervised learning techniques is the input data distribution (for exceptions and extensions see Sect. 6.2.7), they are not optimal with respect to the final goal of modeling of a process. They do not take the complexity of the process into account. Nevertheless, since most unsupervised methods are computationally inexpensive compared with supervised learning techniques, their application can improve performance and reduce the overall computational demand. The unsupervised learning techniques can be distinguished into the following two categories.

- Principal component analysis (PCA) can be used for the transformation of the input axes, which may be better suited than the original ones for representing the data. Often PCA is utilized for dimensionality reduction by discarding those axes that appear to contain the least information about the data. However, for some applications, this dimensionality reduction step may lose significant information because it is based solely on the input data distribution. Nevertheless, for very high-dimensional problems, a PCA with dimensionality reduction is a promising and perhaps the only feasible strategy.
- Clustering techniques find groups of similar data samples. A user-chosen similarity measure defines the shape of the clusters that a clustering method searches for. The most popular techniques are the classical k-means, the fuzzy c-means, and the Gustafson-Kessel algorithms. Also, several neural networks are used for clustering tasks, such as vector quantization, Kohonen's self-organizing map, the neural gas network, and adaptive resonance theory networks. Clustering techniques allow the incorporation of information about the output, which in many cases can make them more powerful.

6.4 Problems

1. What is the difference between supervised and unsupervised methods? Give examples for some application areas of both approaches.
2. What is the output of a PCA and why is this useful?
3. Calculate a PCA for two-dimensional data generated in the following way: Draw 1000 samples from a random number generator and store them in the vector $\underline{u}_1$. Calculate $\underline{u}_2 = 3\underline{u}_1 + \underline{n}$, where $\underline{n}$ represents noise and the elements of $\underline{n}$ are drawn from a random number generator and scaled by a factor of $1/10$.
4. Write a subroutine that implements an arbitrary dimensional k-means clustering algorithm. Input parameters are the initial coordinates of the clusters and the data; both can be represented in matrix form. Output parameters are the cluster centers after k-means has converged. Play around with different data distributions, different initializations, and different number of clusters.
5. Explain the differences between self-organizing maps and neural gas networks.

Chapter 7
Model Complexity Optimization

Abstract This chapter deals with various very different approaches to model complexity optimization. This is probably the most important prerequisite for all the machine learning approaches discussed in the remainder of this book. Finding a good model complexity is a crucial issue for all data-driven modeling, and many very distinct approaches exist for doing that. First, the fundamental tradeoff between bias and variance is illustrated and discussed from multiple points of view. Then four main strategies for dealing with the explained dilemma are discussed: (i) different data sets and statistical approaches, (ii) explicit model structure optimization, (iii) regularization, and (iv) model structuring. For each model architecture, only some of the discussed approaches are suitable or preferable, but it is important to understand the many alternative possibilities. Finding the right model complexity is often more important than finding the right model architecture. Thus, this chapter can be considered as the most crucial in Part A.

Keywords Optimization · Bias/variance tradeoff · Model complexity · Cross-validation · Information criteria · Regularization · Structure optimization

This chapter does not focus on special optimization techniques. Rather it is a general discussion about the fundamental importance to all kinds of data-driven modeling approaches, independent of their specific properties, such as whether the models are linear or nonlinear parameterized, etc. This chapter deals with questions of the following type: "How complex should the model be?" The terms "overfitting" and "underfitting" are very well known in this context and describe the use of too complex and too simple a model, respectively. Surprisingly, part of this question can be analyzed and answered independently of the particular type of model used. Thus, understanding of the following sections is important when dealing with identification tasks independent of whether the models are linear or nonlinear, classical or modern, and neuro or fuzzy models.

After a brief introduction into the basic ideas of model complexity optimization, the bias/variance dilemma is explained in detail. It gives some insight into the effect that model complexity has on the model performance. Next, the importance of

O. Nelles, *Nonlinear System Identification*,
https://doi.org/10.1007/978-3-030-47439-3_7

different data sets for identification and validation is analyzed, and some statistical approaches for measuring the model performance are introduced. The subsequent two chapters describe explicit and implicit strategies for model complexity optimization. While the explicit approaches influence the model complexity by, e.g., increasing or decreasing the number of neurons, rules, etc. of the model, the implicit approaches control the model complexity by regularization, e.g., by restricting the degrees of freedom in the model. Finally, several modeling approaches are presented that reduce the complexity of the modeling problem by making assumptions about the structure of the process.

7.1 Introduction

What is *model complexity*? There is no need for a strict definition. The term "model complexity" certainly can hardly be defined uniquely. There might be a type of model that looks simple from an engineering point of view but complex in the eyes of a biologist, or vice versa. What is meant here with the term "model complexity" is not related to such subjective assessments. It is also not equivalent to the computation time that might be required for evaluating the model or the length of its mathematical formula. (Although these would be very reasonable definitions for model complexity in another context!) Rather, here, the model complexity will be related to the number of parameters that the model possesses. A model becomes more complex if additional parameters are added, and it becomes simpler if some parameters are removed. Model complexity consequently expresses the flexibility of the model. Here the terms *complexity* and *flexibility* are used as synonyms. This definition does *not* imply a one-to-one relationship between the model complexity and the number of parameters since each parameter is not necessarily equally important; refer to Sect. 7.5. Thus it is not necessarily correct to say that of two models the one with more parameters is more complex (although this is very often the case). Nevertheless, the number of parameters is tentatively used as a measure for model complexity until this relationship is formulated more precisely in Sect. 7.5.

The fundamental idea of this chapter can be briefly summarized as follows. A model should not be too simple, because then it would not be capable of capturing the process behavior with a reasonable degree of accuracy. On the other hand, a model should not be too complex because then it would possess too many parameters to be estimated with the available finite data set. Thus, it is clear that somewhere in between there must exist something like an *optimal model complexity*. Of course, the optimal model complexity depends on the available data, the specific model, etc. But some very helpful and surprisingly general analysis can be carried out.

7.2 Bias/Variance Tradeoff

This section analyzes the influence of the number of parameters on the model's performance. It is shown that the model error can be decomposed into two parts: the *bias error* and the *variance error*. This decomposition helps us to understand the influence of the number of parameters on the model. First, a mathematical derivation of this decomposition is presented. Then, in the subsequent sections, this expression is intuitively explained in greater detail.

Consider Fig. 7.1, which depicts a process with its true output y_u disturbed by the noise n, resulting in the measurable process output y. The model with output $\hat{y}$ will describe the process. This is achieved by minimizing some loss function depending on the error e with respect to the model parameters.

In a probabilistic framework, the expectation of the squared error may be used as a loss function. This is analogous to the sum of squared errors that is used in a deterministic setting. Initially, the loss function will be composed into two parts,

$$\mathrm{E}\{e^2\} = \mathrm{E}\{(y - \hat{y})^2\} = \mathrm{E}\{(y_u + n - \hat{y})^2\} = \mathrm{E}\{(y_u - \hat{y})^2\} + \mathrm{E}\{n^2\}, \tag{7.1}$$

since the cross terms $\mathrm{E}\{(y_u - \hat{y})\, n\}$ vanish because the noise is uncorrelated with the process and model outputs. (Note that actually e, y_u, and $\hat{y}$ are functions of the input u; this argument is omitted here for better readability.) The first term on the right side of (7.1) represents the model error between the true (unmeasurable) process output and the model output, and the second term represents the noise variance. The loss function is minimal if the model describes the process perfectly, i.e., $\hat{y} = y_u$. In this case, the first term vanishes and the minimal loss function value is equal to the noise variance. Since the noise variance cannot be influenced by the model, only the first term is considered in the following.

The model error $y_u - \hat{y}$ can be further decomposed as follows:

$$\begin{aligned}
\mathrm{E}\{(y_u - \hat{y})^2\} &= \mathrm{E}\{[y_u - \mathrm{E}\{\hat{y}\} - (\hat{y} - \mathrm{E}\{\hat{y}\})]^2\} \\
&= \mathrm{E}\{[y_u - \mathrm{E}\{\hat{y}\}]^2\} + \mathrm{E}\{[\hat{y} - \mathrm{E}\{\hat{y}\}]^2\} \\
&= [y_u - \mathrm{E}\{\hat{y}\}]^2 + \mathrm{E}\{[\hat{y} - \mathrm{E}\{\hat{y}\}]^2\}
\end{aligned} \tag{7.2}$$

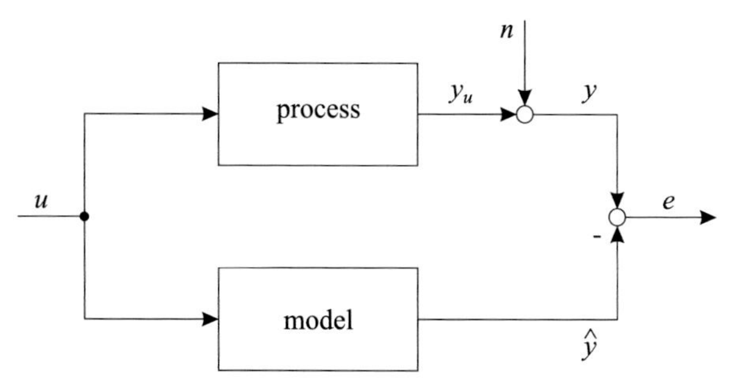

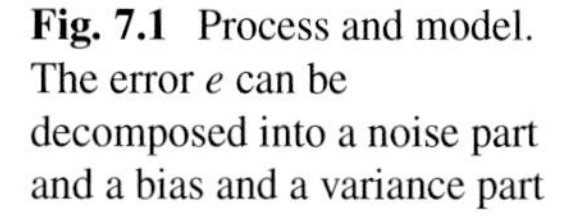
Fig. 7.1 Process and model. The error e can be decomposed into a noise part and a bias and a variance part

since the first term on the right-hand side is deterministic and all cross terms vanish. This decomposition can also be expressed as

$$(\text{model error})^2 = (\text{bias error})^2 + \text{variance error}\,. \tag{7.3}$$

The intuitive meanings of the bias error and variance error are explained in the following sections. The number of parameters decisively influences this decomposition. As will be shown, the bias and variance error are in conflict. A compromise – the so-called *bias/variance tradeoff* – has to be realized in order to find the optimal model complexity. Good references concerning this bias/variance tradeoff are [57, 181, 593].

For a better understanding of the following, it is important to distinguish between two types of data: the *training data* and the *test data*. The training data is used for training the model, i.e., estimating or optimizing its parameters. The test data is used for measuring the performance of the model, i.e., evaluating the quality of the model. In practice, often two distinct data sets are used for training and testing. The reason for this becomes obvious from the following discussion. It is important to realize that one is primarily interested in a good model performance on new, fresh data: that is, on the test data. For a more detailed discussion refer to Sect. 7.3.1.

7.2.1 Bias Error

The bias error is that part of the model error which is due to the restricted flexibility of the model. In reality, most processes are quite complex, and the class of models typically applied are not capable of representing the process *exactly*. Even if the model's parameters were set to their optimal values (which are not known in practice, of course, but instead have to be estimated from data) and no noise was present, an error between process and model would typically occur. This error is called the *bias error*. It is due purely to the structural inflexibility of the model.

As an example, assume a linear fifth-order process that will be modeled by a linear third-order model. Then even if the parameters of the model were set to their optimal values the model would not be capable of an exact description of the process, and the error due to this process/model mismatch is called the bias error. If, however, the model was of fifth-order as well, the bias error would be zero. This means that for flexible models a small bias error can be expected. However, if the real process was nonlinear (which in reality all processes are more or less) and of fifth order, then with any linear model, there would be a non-zero bias error since no linear model is capable of an exact description of a nonlinear process. Nevertheless, for weak nonlinear processes, the bias error may be negligible.

From the above discussion, it is clear that linear processes can be described by linear models without any bias error if the model order is "high enough." In the literature, this assumption is often made for proofing theorems. A *nonlinear* process usually cannot be modeled without a bias error. The only exception occurs when the true nonlinear structure of the process is known, e.g., from first principles modeling. More commonly, nonlinear structures are only approximated by some universal approximator (polynomial, neural network, fuzzy system, etc.). In this case, an approximation error can always be expected, which makes the bias error non-zero. Typically, the best one can hope for is that with increasing model complexity (degree of the polynomial, number of neurons, number of rules, etc.), the bias error *approaches* zero. If this property is fulfilled by an approximator for all (smooth) processes, it is called a *universal approximator*.

From the above discussion, it is clear that the bias error describes the systematic deviation between the process and the model that in principle exists due to the model structure. The model error is always at least as big as the bias error. In a probabilistic framework, the bias error can be expressed as (see Sects. B.7 and (B.27))

$$\text{bias error} = y_u - \mathrm{E}\{\hat{y}\}\,, \tag{7.4}$$

where y_u is the noise-free process output and $\hat{y}$ is the model output. The model output can be seen as a random variable owing to the stochastic character of the data with which the model's parameters were estimated. From (B.27) the reason for the term bias error in (7.4) is obvious.

The bias error is large for inflexible models and decreases as the model complexity grows. Since the model complexity is related to the number of parameters, the bias error qualitatively depends on the number of parameters of the model, as shown in Fig. 7.2. It is typical that the bias error decreases strongly for simple models (few parameters) and saturates for complex models as the number of parameters increases. For the example mentioned above with the linear fifth-order process, the bias error curve would be similar to the one in Fig. 7.2, but a linear model with ten parameters (fifth-order model) would be capable of an exact process description. Consequently, the bias error would drop to zero at $n = 10$ and stay at zero for all even more complex models.

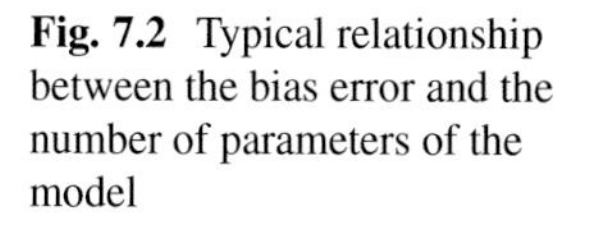
Fig. 7.2 Typical relationship between the bias error and the number of parameters of the model

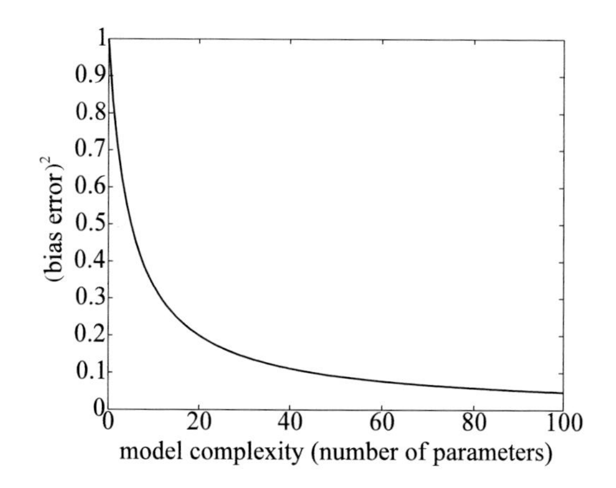

One goal of modeling is to make the bias error small, which implies making the model very flexible. As the number of parameters increases, however, the benefit of an incorporation of additional parameters reduces. Nevertheless, if the model error was dominated by the bias error, the model should be made as flexible as possible, that is, the number of parameters should be chosen as high as the available computational possibilities allow. Thus, the tradeoff would simply be between computational demand and model quality. This, however, is usually not the case because the model error is decisively determined by a second part, the variance error.

7.2.2 Variance Error

The variance error is caused by the sensitivity of the model with respect to the training data. Training data $\{\underline{u}(i), y(i)\}, i = 1, \ldots, N$ can vary in its

- output $y(i)$ due to measurement noise and in its
- inputs $\underline{u}(i)$ due to their distribution as a consequence of the design of experiments or process operation strategy during data collection.

Figure 7.5 shows an example of the first issue, Example 7.2.1 illustrated the second issue.

The variance error can also be seen as that part of the model error that is due to a deviation of the estimated parameters from their (unknown) perfect values. Since, in practice, the model parameters are estimated from a finite and noisy data set, these parameters usually deviate from their perfect values. This introduces an error, which is called the *variance error*. In other words, the variance error describes that part of the model error that is due to uncertainties in the estimated parameters. For linear parameterized models, these parameter uncertainties can be calculated explicitly by (3.34) (see Sect. 3.1.1), and the variance error is represented by the size of the confidence intervals or the error bars in (3.38); see Sect. 3.1.2.

For an infinitely sized training data set, the variance error will be equal to zero if a consistent estimator is used, i.e., the estimated parameters (in the mean) are equal to the perfect ones. In contrast, if the training data contains only as many data samples as there are parameters in the model, the variance error reaches its maximum. In such a case, the degrees of freedom in the model allows one to fit the model perfectly to the training data, which of course means that the parameters precisely represent the noise contained in the training data. Consequently, such a model can be expected to perform much worse on the test data set which contains another noise realization and input locations. For even smaller training data sets, the degrees of freedom in the model exceeds the number of data samples, and the model parameters cannot be determined uniquely, which may result in an arbitrarily large variance error. From this discussion, it becomes clear that the number of parameters in the model should always be smaller than the number of training data samples.[1]

The variance error can be expressed as

$$\text{variance error} = \mathrm{E}\{[\hat{y} - \mathrm{E}\{\hat{y}\}]^2\}\,. \tag{7.5}$$

This expression can be interpreted as follows. Assume that the identical input sequence is applied to the same process several times. Then, different data sets will be gathered owing to the stochastic effects caused by the noise n. If models are estimated on these different data sets, then (7.5) measures the mean squared deviation of these different model outputs from the mean model output. This corresponds to the first component of the variance error and is illustrated in Fig. 7.5. Obviously, without any noise and a purely deterministic process behavior, all data sets and consequently all models would be identical, and thus this component of the variance error would be zero.

A second component comes from a possible variation in the training data *locations*. This is illustrated in Example 7.2.1. In practice usually, both components contribute to the variance error although in quite some cases the first component is zero or negligible. One example is metamodeling where data comes from simulations, not measurement, and therefore are noise-free. Another example is stationary measurements as they are carried out, e.g., on combustion engine test stands where the data samples are typically not single but averaged measurement with negligible noise corruption.

The fewer parameters the model possesses, the more accurately they can be estimated from the training data. Thus, the variance error increases with the number of parameters in the model. Directly from this fact follows the *parsimony principle* or *Occam's razor* which states that from all models that can describe a process accurately, the simplest one is the best [356]. This statement can be generalized by saying that in any context the simplest of comparably performing solutions shall be preferred. It can be shown [230] that for large training data sets and linear models, the variance error increases linearly with the number of parameters in the model:

$$\begin{aligned} \text{variance error} &\sim \sigma^2 \frac{n}{N} \qquad (7.6)\\ &\sim \text{noise variance} \cdot \frac{\text{number of parameters}}{\text{number of training data samples}}\,. \end{aligned}$$

This expression holds only asymptotically for a linear model! Nevertheless, in many practical cases, (7.6) is a good guideline unless the number of parameters in the model approaches the number of training data samples. So some care is recommended when dealing with very small data sets or very complex models.

[1] If no regularization technique is employed in order to reduce the effective number of parameters, see Sect. 7.5.

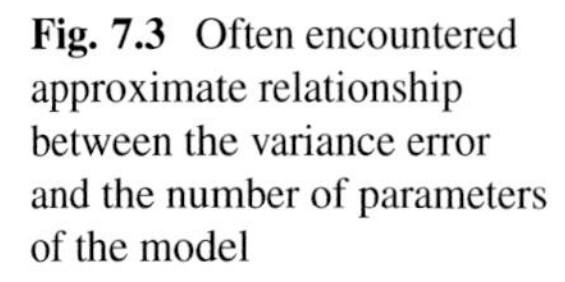
Fig. 7.3 Often encountered approximate relationship between the variance error and the number of parameters of the model

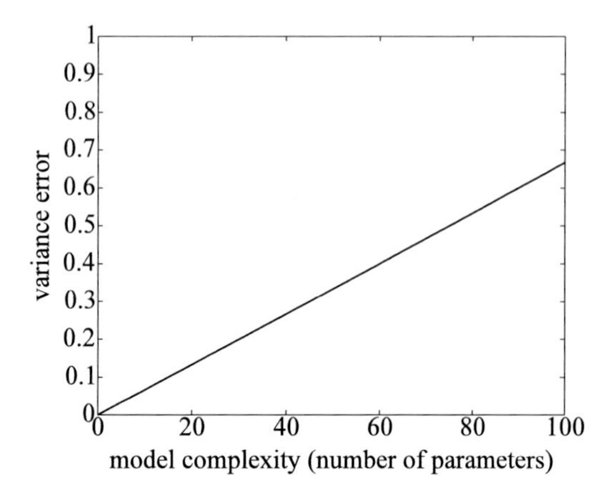

Figure 7.3 shows how the variance error depends on the number of parameters in the model according to (7.6). Note that the slope of this line is determined by the noise variance σ^2 and the number of training data samples N. Higher noise levels lead to higher variance errors. Larger training data sets lead to smaller variance errors. Thus, by collecting huge amounts of data, any noise level can be "compensated." Intuitively, more data allows the estimator to average out the noise in the data better. Note that the noise variance σ^2 cannot be so easily and accurately estimated in practice. Formulas such as (3.35) in Sect. 3.1.1, which allow one to estimate the noise variance σ^2 for linear parameterized models, cannot generally be applied directly for nonlinear processes. The reason for this is that the residuals are often dominated by the bias error, while (3.35) implies that the residuals are solely due to the variance error.

For very flexible models, the bias error can be neglected, and the total model error is dominated by the variance part. Then the squared model error is about equal to the variance error; see (7.2). Thus, the squared model error is proportional to $1/N$, and the model error is proportional to $1/\sqrt{N}$. The generality of this relationship is remarkable. To summarize, for *flexible enough* models, the model error decreases with the inverse square root of the number of training data samples:

$$\text{model error} \approx \sqrt{\text{variance error}} \sim \sigma \frac{\sqrt{n}}{\sqrt{N}}\,. \tag{7.7}$$

Even in the more realistic case where the model error is significantly influenced by the bias error, the above expression underlines the importance of the amount of data. It clearly shows the fundamental limitations imposed on the model performance by the available amount (N) and quality (σ^2) of data.

7.2.3 *Tradeoff*

Figure 7.4a summarizes the effect of the bias and variance error on the model error. Obviously, a very simple model has a high bias but a low variance error, while a very complex model has a low bias but a high variance error. Somewhere in between lies the optimal model complexity. Figure 7.4a clearly shows that models, that are too simple, can be improved by the incorporation of additional parameters because the increase in the variance error is overcompensated by the decrease in the bias error. Contrary, a model, that is too complex, can be improved by discarding parameters because the increase in the bias error is overcompensated by the decrease in the variance error. The fact that the bias and variance error are in conflict (it is not possible to minimize both simultaneously) is often called the *bias/variance dilemma*.

One important goal in modeling is to realize the optimal model complexity, or at least to get close. Sometimes the optimal model complexity may be so huge that it is impossible in practice to estimate such a model. Then computational restrictions do enforce the choice of a model that is too simple. However, in most real-world applications, the data set is so small and noisy that the increase in variance error restricts the model complexity rather than the computational aspects.

Figure 7.4b depicts the bias/variance tradeoff for a training data set that is two times larger than in Fig. 7.4a. This leads to variance error that is two times smaller. Consequently, the optimal model complexity is higher than in Fig. 7.4a. Intuitively, one might express this relationship by saying that more data allows one to estimate more parameters. Note that the same effect is caused by a lower noise variance instead of an increased amount of data or a combination of both.

It is important to understand that the points of optimal model complexity in Fig. 7.4 do not represent the best *overall* solution. They just give the best bias/variance tradeoff for this *specific* model class. There might exist other model architectures that are better suited for describing the process. Another model

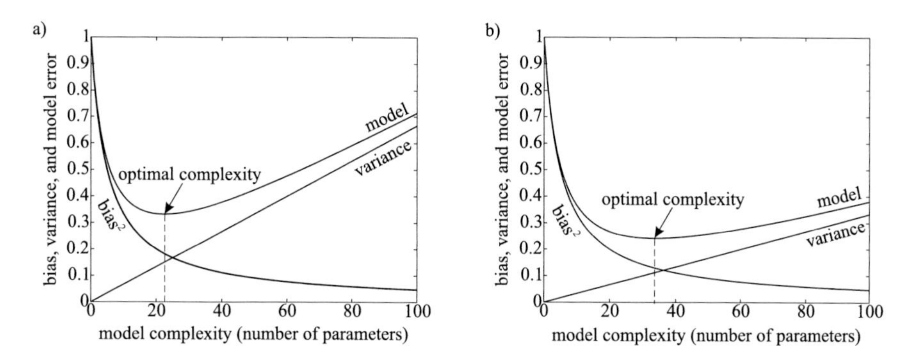

Fig. 7.4 Bias/variance tradeoff. The model error can be decomposed into a bias and variance part: (**a**) relatively high variance error, (**b**) lower variance error due to more data or less noise than in a

architecture might be structurally closer to the process, which would mean that the bias error could be significantly reduced without increasing the number of parameters.

The difficulty with the bias/variance tradeoff in practice is that the bias and variance error are unknown. The most straightforward approach would be to estimate many models of different complexity and to compare the resulting errors evaluated on the test data set. This approach works fine if the estimation of a model requires little computational effort, as is usually the case for linear models. If the estimation of a model is computationally expensive, other less time-consuming strategies must be applied. These other strategies can be divided in two categories: explicit and implicit structure optimization, which are discussed in Sects. 7.4 and 7.5, respectively.

Finally, it is important to make some comments on the use of training data and test data. If the training data were used for measuring the performance of the model, the variance error could not be detected at all. The error on the training data only overoptimistically measures the bias part. The variance error is detected only if a data set with a different noise realization is used. The reason for this is as follows. The variance error is due to the uncertainties in the model parameters, which are caused by the particular noise realization and input location distribution in the training data. If the same data set is used for training and evaluation of the model performance, the parameter uncertainties cannot be discovered, since the parameters represent exactly this noise realization in the training data.

This means that the error on the training data (which is approximately equal to the bias error) decreases with the model complexity, while the error on the test data (which is equal to the bias error plus the variance error) starts to increase again beyond the point of optimal complexity. If this effect is ignored, one typically ends up with overly complex models, which perform well on the training data but poorly on the test data. This effect is often called *overfitting* (low bias, high variance). In contrast, *underfitting* characterizes the use of too simple a model (high bias, low variance).

Figure 7.5 illustrates the effect of underfitting and overfitting. All the figures on the left-hand side show the training data, while all the figures on the right-hand side show the test data. Figure 7.5a1 depicts the true function to be modeled and 11 training data samples, which are distributed equally over the input space and disturbed by noise. Figure 7.5a2 shows the same function and the test data set, which has the same inputs as the training data but a different noise realization.

Figure 7.5b1, b2 shows a first-order polynomial whose coefficients have been fitted to the training data. Obviously, the model is not flexible enough to represent the underlying function with reasonable accuracy. This is called *underfitting*. The error on the test data is slightly higher than on the training data because it includes a (small) variance error. Nevertheless, the model error is dominated by the systematic deviation and thus the variance error is almost negligible.

Figure 7.5c1, c2 show a tenth order polynomial whose coefficients have been fitted to the training data. On the training data, the error is zero. Obviously, the bias error is zero since the model is flexible enough to describe the underlying function

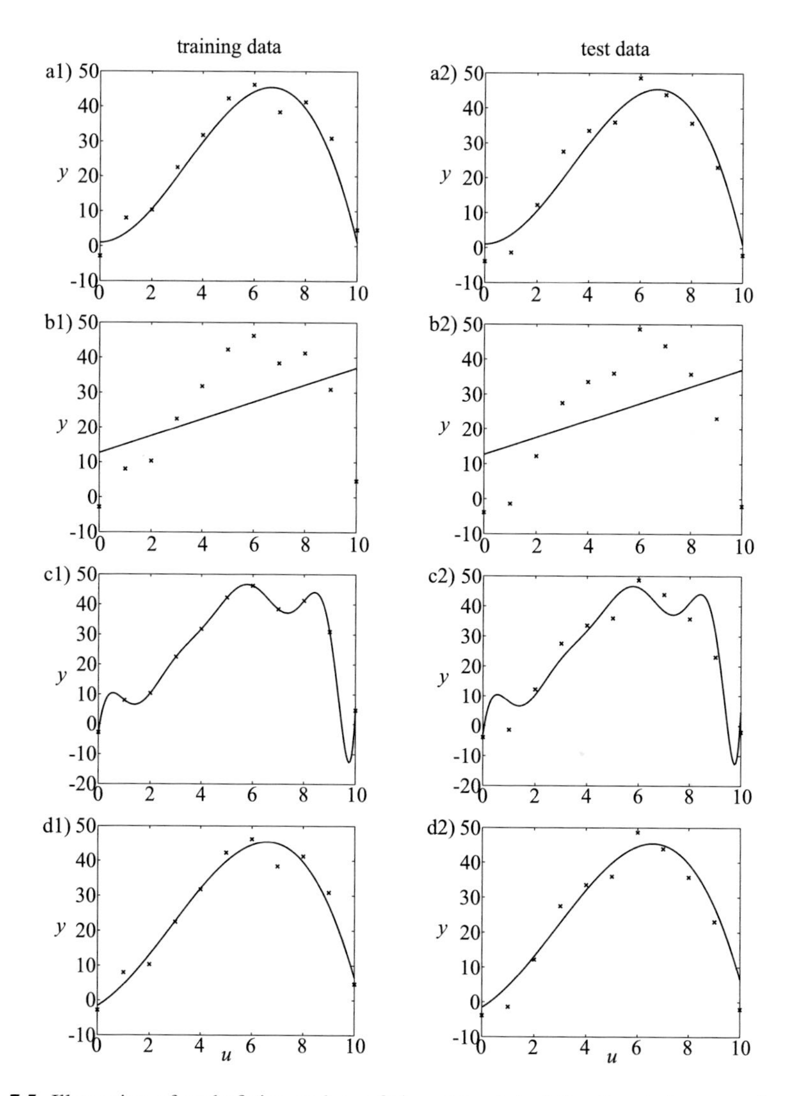

Fig. 7.5 Illustration of underfitting and overfitting: (**a**) original function to be approximated, (**b**) first-order polynomial model (underfitting: large bias, small variance), (**c**) tenth-order polynomial model (overfitting: small (here: zero) bias, large variance), (**d**) fourth-order polynomial model, good bias/variance tradeoff: medium bias and variance – very close to unknown original function in (**a**)

exactly. Nevertheless, there exists a large deviation of the model from the underlying function. In practice, the underlying function is unknown; however, this effect can be discovered by analyzing the model's performance on the test data set. A look at the test data reveals a significant variance error, which cannot be detected on the training data. Such a behavior is called *overfitting*. It is more dangerous than

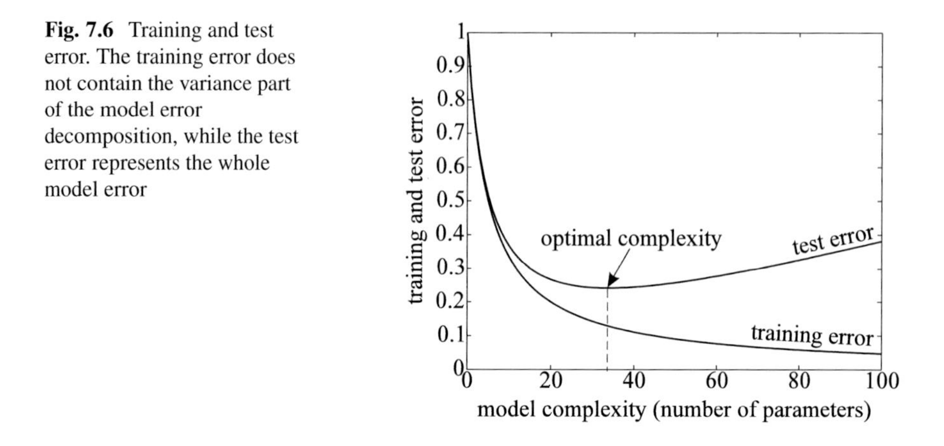

Fig. 7.6 Training and test error. The training error does not contain the variance part of the model error decomposition, while the test error represents the whole model error

underfitting since underfitting is always obvious to the user while overfitting cannot be observed on the training data alone but only on the test data.

Figure 7.5b1, b2 represents too simple a model and Fig. 7.5c1, c2 represents too complex a model. Figure 7.5d1, d2 depicts the best bias/variance tradeoff, here given by a fourth-order polynomial. Figure 7.6 summarizes the behavior of the training and test error. Obviously, the distinction between training and test data is of fundamental importance. Up to this point, it has been assumed that these data sets are somehow given. Section 7.3 discusses the choice of training and test data sets by the user.

7.2.3.1 Dependency on the Amount of Data

Instead of plotting the error over the model flexibility, i.e., number of parameters, new insights are provided by investigating the dependency on the number of training data points. The following plots and the subsequent example are taken from an excellent and intuitive online course on machine learning.[2]

Figure 7.7a shows the typical behavior of the training and test error over the number of data points. They approach the same limit as $N \to \infty$. Obviously, the curves are idealized and just represent the principle behavior. More accurately the training error is zero for $N \leq n$ if no regularization is involved because the model perfectly fits all data points (interpolation case). Thus, the plots start from $N > n$. From that point on, the training error increases since it is increasingly difficult to fit the data as N grows. The overfitting effect weakens as N grows and would vanish for $N = \infty$. On the other hand, the quality of the model improves as N grows and therefore the test error decreases. The gap between test and training error tightens and represents the amount of overfitting that occurs. Figure 7.8 illustrates how these curves change from simpler (a) to more complex (b) models. Clearly, the

[2]Machine Learning Course – CS 156 at Caltech, http://www.work.caltech.edu/teaching.html.

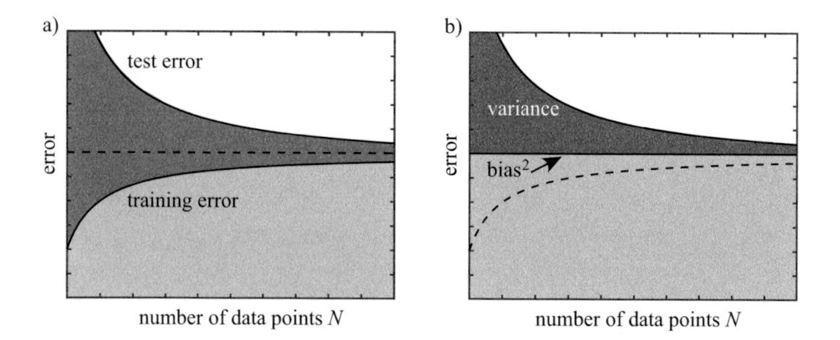

Fig. 7.7 (**a**) Typical training and test error curves. (**b**) Decomposition into bias^2 and variance

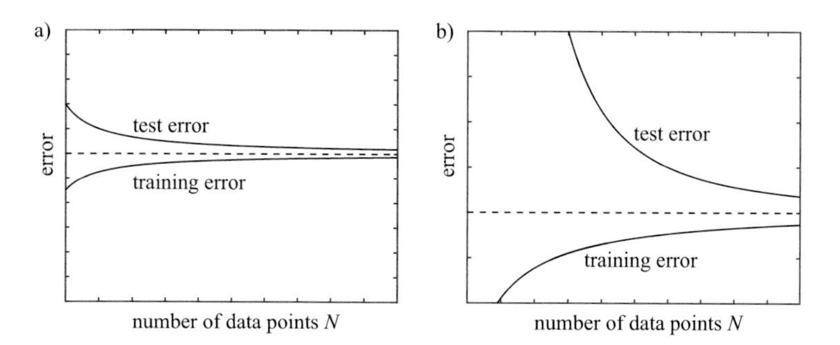

Fig. 7.8 Typical training and test error curves: (**a**) Simple model. (**b**) Complex model

gap between test and training error is bigger for the more complex model as it is more sensitive with respect to overfitting. Also, the curves saturate much later since the ratio N/n governs the strength of the overfitting effect.

Figure 7.7b clarifies the relationship to bias and variance. The bias is the error of an average model on infinitely many training data sets of size N. It is independent of the training data size N. This can also be seen from Example 7.2.1 in Figs. 7.11 and 7.12 where 100 model realizations for $N = 2$ and $N = 20$ are plotted whose average values approach the models in Fig. 7.10.

7.2.3.2 Optimism

In [230] it is explained that the degrees of freedom of a model which is called the effective number of parameters in this book are proportional to the difference of two mean squared errors. One is the training error, and the other is the *repeated error* which occurs when measuring the output at the same input locations $\underline{u}(i)$ several times (with different noise realizations) and then calculating the mean squared error.

$$n_{\text{eff}} = \frac{1}{2\sigma^2} \underbrace{(\text{ repeated error} - \text{training error})}_{\text{optimism}} \tag{7.8}$$

The difference on the right-hand side is called *optimism* as it expresses how much better the model is assumed to be by looking at the training error alone. This is what makes very flexible models so dangerous. Very inflexible models yield bad training errors and are insufficient at first sight. However, very flexible models can only be assessed correctly if the bias/variance dilemma or at least the necessity for a test data set is known to the user.

Note that there is an important distinction between the repeated error and the test error. Both are disturbed by a new noise realization. But only the test data is measured at new/independent locations compared to the training data. Thus, the test data fully reveals the quality of interpolation and extrapolation of the model while the repeated data only shows the distortion towards a certain noise realization in the training data.

Example 7.2.1 To illustrate the bias/variance issue with a very simple example,[3] the following function shall be approximated in the interval $[-1, 1]$:

$$y = \sin(\pi u) . \tag{7.9}$$

The training data consists of two noise-free measurements ($N = 2$). Thus, this example excellently demonstrates the second root cause for the variance error which is due to the data distribution. Two models are compared:

- Constant: $y = \theta_0$,
- Linear: $y = \theta_0 + \theta_1 u$.

Figure 7.9 illustrates two estimated models of both types with randomly chosen input data. It is obvious that the linear model is capable, in principle, of approximat-

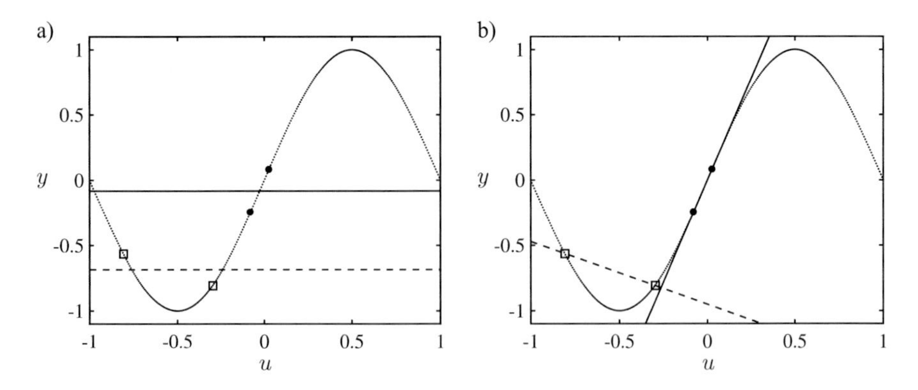

Fig. 7.9 Two models (dashed and solid lines) estimated on $N = 2$ data samples each (squares and dots): (**a**) Constant models. (**b**) Linear models

[3]This example is taken from the Machine Learning Course – CS 156 at Caltech, http://www.work.caltech.edu/teaching.html

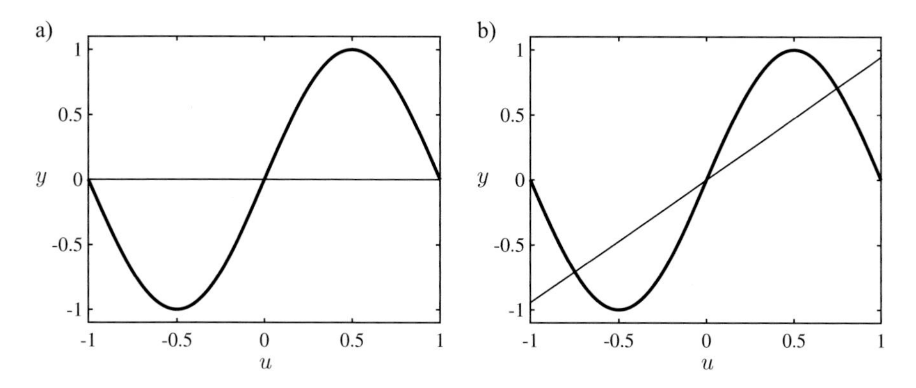

Fig. 7.10 Best models of (**a**) constant and (**b**) linear type in the least squares sense

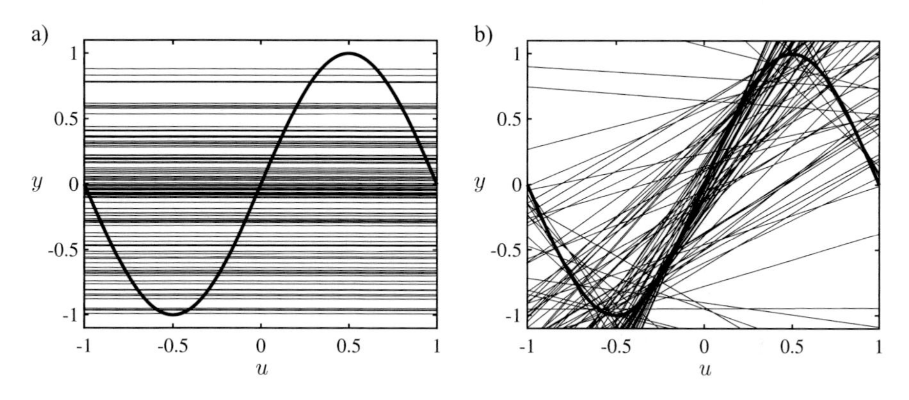

Fig. 7.11 100 (**a**) constant and (**b**) linear models for $N = 2$ data points

ing the sine function much better. However, it depends very much on the location of the data points.

Figure 7.10 shows the best case for both models. These are the results of optimally approximating the sine function in the interval $[-1, 1]$ with a constant and a line, respectively. The same outcome can be achieved with equally distributed training data over this interval. The mean squared errors for both models are $\text{MSE}_{\text{const}} = 0.5$ and $\text{MSE}_{\text{linear}} = 0.2$, respectively.

However, in data-driven modeling, the function to be approximated is unknown, and all information is contained in the data. Obviously, $N = 2$ data points carry very little and not representative information about the whole sine function. The linear model is much more sensitive with respect to the location of the data than the constant model. This is illustrated by Fig. 7.11 where 100 constant and linear models are shown based on two data points each. The inputs are uniformly distributed in $[-1, 1]$; the outputs are noise-free as in Fig. 7.9. This illustrates very clearly that although the bias error of the linear model is much smaller than for the constant, the variance error is even higher. Extremely bad lines are estimated with a non-

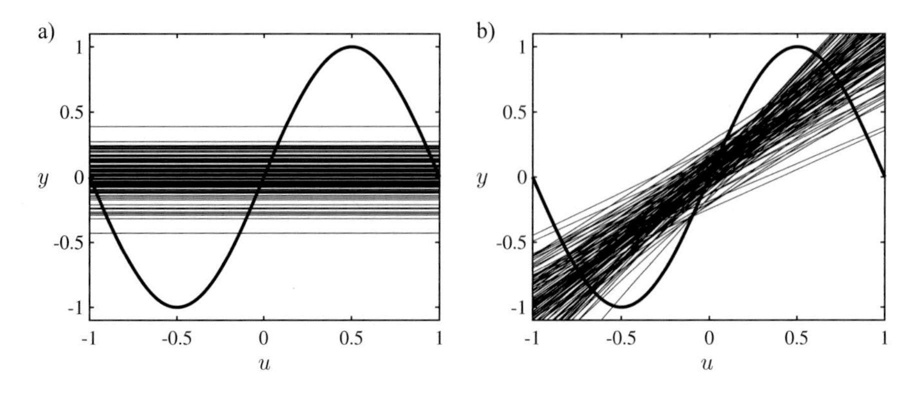

Fig. 7.12 100 (**a**) constant and (**b**) linear models for $N = 20$ data points

negligible probability. Summing up the bias and variance components, overall the line is the worse model. This result will change if more data points are measured. The variance error decreases with the number of data points N significantly, while the bias error keeps unchanged, compare Fig. 7.12 for $N = 20$ data points. In between Fig. 7.11 ($N = 2$) and Fig. 7.12 ($N = 20$), the optimal model complexity changes from a constant to a line. □

7.3 Evaluating the Test Error and Alternatives

As mentioned in the previous section, the observed performance of a model on the training data does not contain the variance part of the error. Thus, the performance of the model is overestimated by evaluating it on the training data. The goal of modeling is to build a model that performs well on fresh, previously unseen, data. The most straightforward way to determine the expected model performance on fresh data is to evaluate the model on a separate test data set that has not been used for training, see Sect. 7.3.1. Cross-validation is a more sophisticated strategy that refines this idea; see Sect. 7.3.2.

Alternatives to the direct evaluation of the test error are treated in the second part of this section. In Sect. 7.3.3, instead of evaluating the model on test data, the test error is approximated by the training error plus a complexity penalty term. Section 7.3.4 discusses the alternative approach of multi-objective optimization strategies for complexity optimization. Finally, in Sects. 7.3.5 and 7.3.6, some statistical tests and correlation-based methods are introduced that help us to determine the optimal model complexity without explicitly considering the test error.

7.3.1 Training, Validation, and Test Data

The simplest way to estimate the quality of a model on fresh data is to train it on a training data set and evaluate its performance on a different test data set. The assignment to the training and test data sets is typically done randomly, and this is called the *holdout method*. In order to realize this approach, the available data has to be split up into separate training and test data sets. If the amount of available data is huge, this causes no difficulties and is the most straightforward approach. Care must be taken that both the training and the test data are representative, i.e., cover all considered operating regimes of the process equally well. This requirement becomes increasingly difficult to fulfill as the amount of available data becomes smaller. If the training set lacks data from some regimes, the model cannot be expected to perform well in these regimes. On the other hand, if important data is missing in the test set, the evaluation of the model performance becomes unreliable. While too small a training data set inevitably leads to a low-quality model, too small a test data set may allow good models, but there is no way to prove this because the performance assessment is unreliable. From this discussion, it is obvious that more sophisticated approaches must be pursued for model performance evaluation if the amount of available data is small; see Sect. 7.3.2.

For the remaining section, it is assumed that the amount of available data is large enough to allow splitting it up into different sets. For determination of the optimal model complexity, the following strategy seems to be simple and effective: Train, say T, differently complex models with the training data and evaluate their performance on the test data. Finally, choose the model with the lowest error on the test data (test error). This strategy, however, itself leads to an optimistic estimation of the performance of the chosen model since the data utilized for selection is identical to the data used for performance evaluation. Thus, to be exact, a third data set, the *validation data*, has to be utilized for model selection, as shown in Fig. 7.13.

Why, in addition to training and test data, is the validation data necessary? The answer to this question becomes intuitively clear if a very large number T of investigated models are considered. Obviously, the larger T is, the higher is the probability that just by chance one of these models performs well on a separate validation data set. In an extreme (hypothetical) example, the training may do almost nothing at all (because the number of iterations of the optimization algorithm used may be too small). Then the randomly initialized parameters of the model decide the quality of the model. In such a case the validation error solely determines which of the T models is selected. Thus, the validation data cannot give a realistic estimate of the model performance on fresh data, and a third separate data set, the test data set, is required.

In practice, splitting up the data into three parts is rarely realized because it seems to waste the available data. If the number of investigated models T is small and the size of the validation data set is large, the performance estimate on the validation data may be reasonably, realistic and an additional test data set is not absolutely necessary. Nevertheless, it should be kept in mind that using the same data for validation and test tends to yield overly optimistic estimates of the model quality.

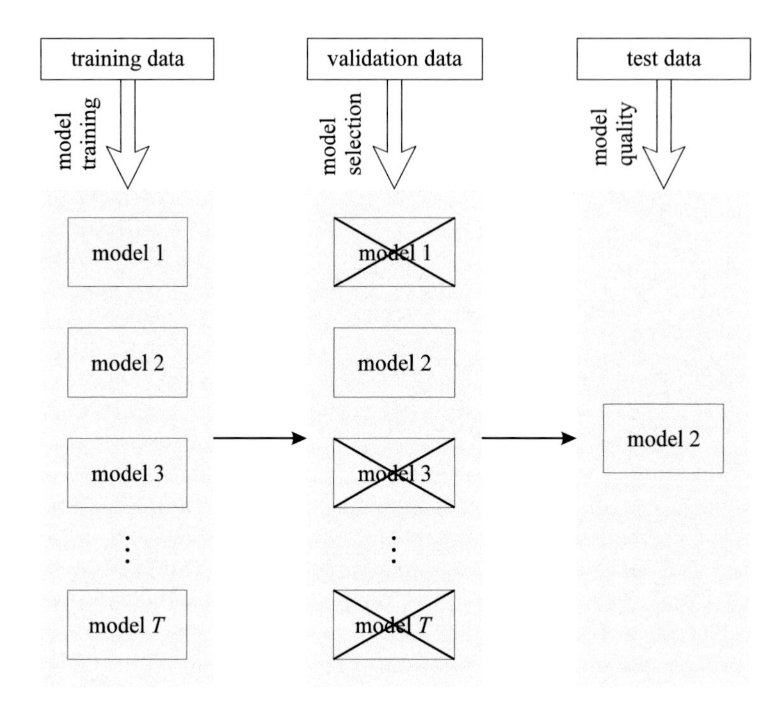

Fig. 7.13 Model training, validation, and test. Different models are trained with the training data. Next, the quality of these models is investigated with a separate validation data set, and the best model is selected. Finally, the quality of this model is evaluated with a separate test data set

Note that the T investigated models in Fig. 7.13 can be models with increasing complexity, e.g., neural networks or fuzzy systems with an increasing number of neurons or rules, respectively. They can also represent the same neural network structure that is trained for, say 100, 200, 300, etc., iterations, since the number of iterations of the optimization algorithm can also be related to the *effective* network complexity. Of course, the number of iterations does not influence the *nominal* network complexity. This relationship is explained further in Sect. 7.5.

7.3.2 Cross-Validation (CV)

In the previous section, it was assumed that the available amount of data would be large enough to allow a split into training, validation, and test data sets. This situation rarely occurs in practice. Usually, data is scarce, and the user may be willing to spend higher computational resources for better exploitation of the available data rather than a simple split into separate sets.

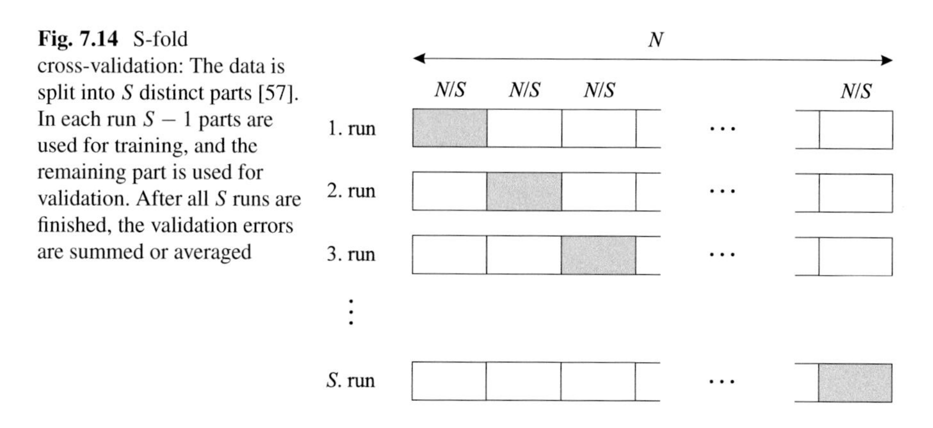

Fig. 7.14 S-fold cross-validation: The data is split into S distinct parts [57]. In each run $S - 1$ parts are used for training, and the remaining part is used for validation. After all S runs are finished, the validation errors are summed or averaged

7.3.2.1 S-Fold Cross-Validation

A commonly applied improvement is to split the data set containing N samples into S parts. Then $S - 1$ of these parts are used for training, and the single remaining part is used for validation. This procedure is repeated with all S different possible combinations of these data set parts; see Fig. 7.14. This approach is called *S-fold cross-validation*. With this strategy, it is possible to exploit a much larger fraction of the original data set for training while the validation data set is very small. However, since training and validation is repeated S times with different parts of the original data set, it is possible to sum up or average all test errors in order to obtain a reliable estimate of the model performance on new data. This technique is called *cross-validation*. Note that cross validation is just used to obtain a good estimate of the expected model performance on fresh data. This is typically utilized to adjust hyperparameters (e.g., regularization strength) or to make a structural decision (e.g., number of neurons, model order). For the final use, the model can be trained on the whole available data set in order to exploit all information.

A typical value for S is 5 or 10. Compared with an equal split into training and validation data, $S = 10$ allows one to utilize $9/10$ instead of $1/2$ of the available data for training and requires ten times the computational effort. Note that cross-validation with $S = 2$ is still better than the simple approach with training on one half and validation on the other half of the data set, because in cross-validation the complete data set is always utilized for validation since all S runs are taken into account.

7.3.2.2 Leave-One-Out Error

If data is very scarce, the most extreme case of cross-validation with $S = N$ can be applied. This is called the *leave-one-out* method because in each run only one sample is used for validation and thus left out for training. Clearly, in the general case, the leave-one-out method is feasible only for small data set, because it requires

N times the computational effort. However, for linear regression problems, a simple closed formulation exist for the leave-one-out error. For data point i, it can be computed by

$$e_{\text{LOO}}(i) = \frac{e(i)}{1 - s_{ii}} \tag{7.10}$$

where $e(i) = y(i) - \hat{y}(i)$ is the training error and s_{ii} is the ith diagonal element of the $N \times N$-smoothing matrix $\underline{S}$; see (3.76 in Sect. 3.1.11). For a pure look-up table of the data without any generalization capabilities, $s_{ii} \to 1$ since $\hat{y}(i) \approx y(i)$ and e_{LOO} become very large since no information from neighboring (or any other) points is utilized. The sum of squared LOO errors (7.10) is often called the prediction residual error sum of squares (PRESS).

As (7.10) is trivial to calculate, the only effort is the calculation of the diagonal of $\underline{S} = \underline{X}\left(\underline{X}^T\underline{X}\right)^{-1}\underline{X}^T$. Note that $\left(\underline{X}^T\underline{X}\right)^{-1}\underline{X}^T$ is already known from the LS estimation. Therefore almost no additional computational cost is required. This is another huge advantage of linear regression structures.

In [191] an enhanced LOO cross-validation is proposed known as *generalized cross-validation (GCV)* that gives different weightings to different data points which take into account the individual influences of each data point. The GCV does not depend on the orientation of the measurement coordinate system as the LOO does, i.e., the GCV is rotation-invariant. The diagonal elements of the smoothing matrix are replaced by their average value $\text{tr}[\underline{S}]/N$:

$$e_{\text{GCV}}(i) = \frac{e(i)}{1 - \text{tr}[\underline{S}]/N} . \tag{7.11}$$

7.3.2.3 Leave-One-Out Versus S-Fold CV

Since computing the leave-one-out error for linear regression is so cheap due to the analytic expressions given above, it is the main validation method if the model is linear in its parameters. At first sight, it also seems to be the most sophisticated and accurate way of validation since the amount of data is maximally exploited.

However, it can be observed that the models selected or tuned by S-fold cross-validation possess a higher variance and lower bias as S increases. Thus, high values for S tend to yield (slightly) overfitting models. For leave-one-out CV, $S = N$ and values like $S = 5$ or $S = 10$ are considered to deliver more stable and robust models. Nevertheless, for relatively simple models that tend to exhibit a relatively high bias like low-order splines or local linear models, leave-one-out CV is an excellent (and very efficient) choice.

An intuitive reason why large S values (like LOO) favor higher model variances while low S values favor a high model bias is the following. If a significant fraction of the data is missing for training as it is the case for small S, bigger data holes and wider extrapolation ranges exist. Therefore models tend to be simpler and more robust in order to perform well on the validation data fraction. In opposition, if just one data point is missing in the training data (as for LOO), the model will typically be supported by many surrounding points within the training data fraction making bad performance on the validation data point unlikely.

7.3.2.4 Bootstrapping

Even more powerful and computationally expensive alternatives to cross-validation are the jackknife and bootstrapping. For details refer to [132]. Particularly, bootstrapping has become extremely popular in recent years. Therefore this idea is explained in the following. It is a general-purpose approach to approximate statistical properties (often the variance) of an estimate. However, its ideas can also be utilized to decrease the variance of models, i.e., making them more robust by creating an *ensemble of models*.

In bootstrapping artificial data sets called *bootstrap samples* are generated from the original data. These bootstrap samples have the same number of data points N as the original data. They are created by drawing N data points from the original data with replacement; see Fig. 7.15. Thus, some data points will be in the bootstrap sample once, twice, three times, or even more often and others not at all. It can be shown that the number of left-out data points amounts to approximately $\frac{1}{e}$ fraction of the data. Each bootstrap sample is used to estimate a model, and the variation between them gives the model variance. Typically, 1,000 to 10,000 bootstrap samples are drawn for statistical assessment.

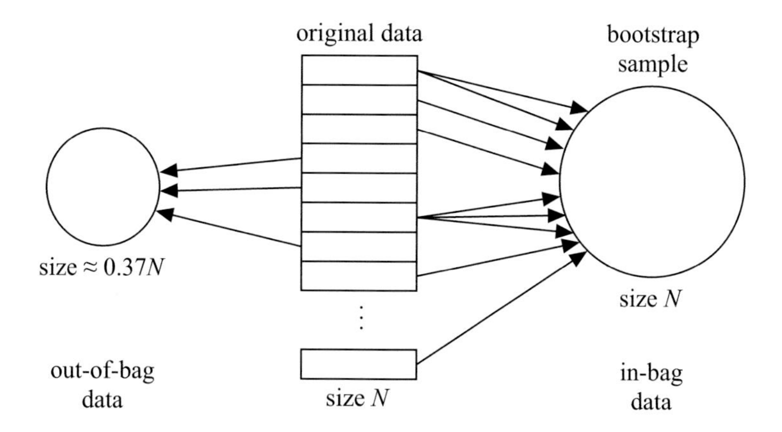

Fig. 7.15 Bootstrap sample

Models which are very sensitive with respect to changes in the training data set are often called unstable[4] models [230]. Typical examples for such models are trees that partition the input space like CART [230] or LOLIMOT (Chaps. 13 and 14). For these types of models, the outcomes for each bootstrap sample vary a lot. A very intuitive idea therefore is to average these models in order to reduce the overall variance. This approach proposed by Breiman in 1996 is called *bagging* for "bootstrap aggregation" [230]. Section 14.11 discusses how bagging can be utilized for local model trees.

The introduction of randomness by bootstrapping and the reduction of variance by averaging is a very powerful idea that has been extremely successful. Typically, such an ensemble consists of around 30 to 3,000 models. It has been carried further by artificially introducing randomness into the partitioning process going from bagged CART to random forests in 2001 also by Breiman [230]. Together with support vector machines, these are the most powerful classifiers to date. The key advantages of ensemble methods are

- Conceptionally simple.
- Robust.
- Perfectly parallel, that is a big plus in times where computers become mainly faster due to parallelization. For processors with very many cores, training and evaluation times of ensembles are not longer than for single models.

The main drawbacks are

- Difficult to interpret.
- High computational demand for training and for evaluation on a serial or only moderately parallel computer.

It is interesting to note that the so-called out-of-bag data left out of the bootstrap sample (see Fig. 7.15) can be exploited as validation data [77]. This is very similar to S-fold cross-validation.

7.3.2.5 Why Ensemble Methods Work

A nice and simple explanation is given in [194]. If K regression models $\hat{y}_i, i = 1, \ldots, K$ are averaged, then their expected squared error is

$$\mathrm{E}\left\{\left(\frac{1}{K}\sum_{i=1}^{K} e_i\right)^2\right\} = \frac{1}{K^2}\mathrm{E}\left\{\sum_{i=1}^{K}\left(e_i^2 + \sum_{\substack{i=1 \\ i \neq j}}^{K} e_i e_j\right)\right\} . \tag{7.12}$$

With identical variances of the models $v = \mathrm{E}\{e_i^2\}$ and identical covariances between each pair of models $c = \mathrm{E}\{e_i e_j\}, i \neq j$, this becomes

[4] Meant colloquially; not in the sense of *dynamic* stability in control theory.

$$\mathrm{E}\left\{\left(\frac{1}{K}\sum_{i=1}^{K} e_i\right)^2\right\} = \frac{1}{K}v + \frac{K-1}{K}c\,. \tag{7.13}$$

In the case where the model errors are perfectly correlated and $v = c$, the expected error variance in (7.13) is exactly the one of a single model, namely, v. But in the case that the model errors of two different models i and j are perfectly uncorrelated, i.e., $c = 0$, (7.13) is $\frac{v}{K}$. In the first case ($v = c$), ensemble methods achieve no benefit; in the second case ($c = 0$), they work perfectly. Therefore ensemble methods call for very distinct models. Often additional randomness is artificially induced to meet this requirement well; see, e.g., random forests [78].

7.3.3 Information Criteria

A widely applied alternative to the computationally expensive cross-validation is the use of information criteria. The data is not split up in different parts. Rather training is performed on the whole data set. In order to avoid overfitting, a complexity penalty is introduced, which grows with an increasing number of model parameters. An information criterion is introduced, that reflects the loss function value and the model complexity:

$$\text{information criterion} = \text{IC}\,(\text{loss function, model complexity})\,. \tag{7.14}$$

Then the "best" model is defined as the model with the lowest information criterion (7.14). By taking into account the model complexity in (7.14), the variance part of the error will be considered. Since the variance part is proportional to $1/N$ only for an infinite data set, no "correct" function (7.14) can be determined for the general case. All reasonable complexity penalties should increase with the number of model parameters n and should decrease with an increasing amount of data N. In the limit $N \to \infty$, the complexity penalty should tend to zero because the variance error vanishes.

Starting from different statistical assumptions, a number of proposals for the complexity term have been made. The most prominent are briefly described in the following. All of them are monotonically increasing with the number of parameters in the model. Thus, all models that realize the optimal criterion (7.14) have a finite number of parameters. It is important to note that the parameters of the models are still determined by minimizing the sum of squared errors. The criterion (7.14) is just utilized for model comparison instead of the validation on a separate data set. Since it is not clear which information criterion is the "best" one (no "best" term exists for all kinds of problems), the model complexity yielded by this approach must be more carefully supervised by the user than for the approaches in the previous two sections. Furthermore, some criteria contain a tuning parameter, ρ, which cannot be easily determined.

With the loss function defined as mean squares error

$$I(\underline{\theta}) = \frac{1}{N} \sum_{i=1}^{N} e^2(i) \tag{7.15}$$

typical choices for the information criterion in (7.14) are [3, 201]

- *Akaike's information criterion (AIC)* in a general form

$$\mathrm{AIC}(\rho) = N \ln(I(\underline{\theta})) + \rho\, n\,. \tag{7.16}$$

The most commonly used and original AIC choice is $\rho = 2$. It results from an asymptotical (i.e., $N \rightarrow \infty$) statistical derivation starting from the likelihood function minimizing the Kullback-Leibler divergence (information loss).

The AIC assumes that no true model exists; rather, it makes sense in view of the bias/variance tradeoff to estimate a more complex model if more data is available.

For small data sets (low N), a bias corrected version exists that is recommended in the literature [87]:

$$\mathrm{AIC}_\mathrm{C}(\rho) = \mathrm{AIC}(\rho) + \frac{2n(n+1)}{N-n-1}\,. \tag{7.17}$$

An important extension is the *multivariable* version, i.e., an AIC for multiple outputs [87]:

$$\mathrm{AIC}(\rho) = N \ln\left(\det \underline{\Sigma}(\underline{\theta})\right) + \rho\, n\,. \tag{7.18}$$

Here the loss function matrix for a model with q outputs is defined as

$$\underline{\Sigma}(\underline{\theta}) = \frac{1}{N} \begin{bmatrix} \sum_{i=1}^{N} e_1^2(i) & \sum_{i=1}^{N} e_1(i)e_2(i) & \cdots & \sum_{i=1}^{N} e_1(i)e_q(i) \\ \sum_{i=1}^{N} e_2(i)e_1(i) & \sum_{i=1}^{N} e_2^2(i) & \cdots & \sum_{i=1}^{N} e_2(i)e_q(i) \\ \vdots & \vdots & \ddots & \vdots \\ \sum_{i=1}^{N} e_q(i)e_1(i) & \sum_{i=1}^{N} e_q(i)e_2(i) & \cdots & \sum_{i=1}^{N} e_q^2(i) \end{bmatrix} \tag{7.19}$$

with the error at output j for data point i being $e_j(i) = y_j(i) - \hat{y}_j(i)$.

Note that it depends on the correlation between the output errors. If they are roughly uncorrelated, the loss function matrix approaches diagonal structure, and the determinant is just the product (taking the logarithm, it becomes a sum of ln-terms) of the diagonal terms, i.e., the loss functions of all individual outputs.

However, if they are all correlated, this value is corrected by subtracting off-diagonal entries in the determinant calculation. This reflects the fact that the outputs should be easier to model (i.e., with fewer parameters) if they are similar.

If the number of outputs is huge, then evaluating (7.19) can be tedious. It is then common to assume uncorrelatedness and just consider the diagonal terms avoiding the calculation of the determinate.

- *Bayesian information criterion (BIC)*:

$$\mathrm{BIC} = N \ln(I(\underline{\theta})) + \ln(N) n\,. \tag{7.20}$$

The BIC is identical to the general AIC with $\rho = \ln(N)$. For typical data set sizes around $N = 1,000 - 100,000$ data points, this yields penalty terms about five times bigger than the AIC leading to less complex models.

In contrast to the AIC, the BIC assumes that a true model exists.

- *Khinchin's law of iterated logarithm criterion (LILC)*:

$$\mathrm{LILC}(\rho) = N \ln(I(\underline{\theta})) + 2\rho \ln(\ln(N))\, n\,. \tag{7.21}$$

- *Final prediction error criterion (FPE)*:

$$\mathrm{FPE} = N \ln(I(\underline{\theta})) + N \ln\left(\frac{N+n}{N-n}\right). \tag{7.22}$$

- *Structural risk minimization (SRM)*:

$$\mathrm{SRM}(\rho_1, \rho_2) = I(\underline{\theta}) \Bigg/ \left(\rho_1 \sqrt{\frac{n \ln(2N) - \ln(n!) + \rho_2}{N}}\right). \tag{7.23}$$

If SRM is smaller than zero, it is set to ∞, i.e., the maximum possible model complexity is exceeded.

7.3.3.1 Effective Number of Parameters and Effective Amount of Data

In the above discussion, N is the number of data samples, and n is the number of parameters.

Note that in the literature, there is an agreement that in all these criteria the number of parameters n must be replaced by the effective number of parameters n_{eff} if any regularization technique is applied. This is the number of parameters that determines the model flexibility and therefore the sensitivity with respect to overfitting. For more details about the effective number of parameters, refer to Sect. 7.5.

However, the role of the number of data samples N is hardly discussed in literature. It is clear that n_{eff} should measure the degrees of freedom the model

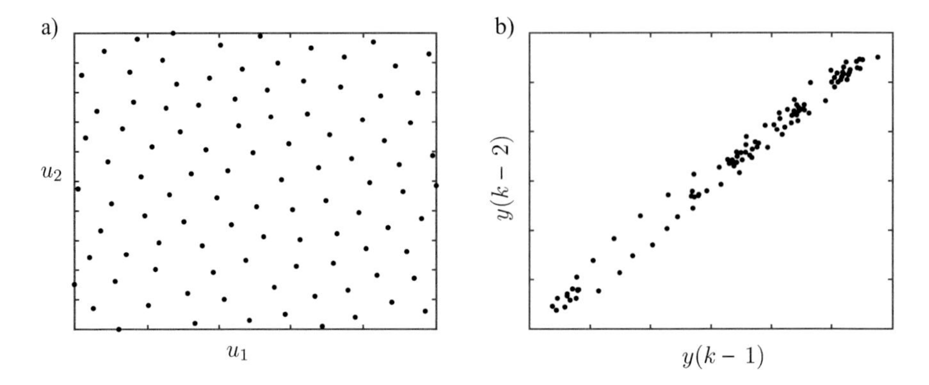

Fig. 7.16 Input space data distribution for 100 data points for **(a)** $y = f(u_1, u_2)$ with a maximin Latin hypercube design: $N_{\text{eff}} \approx N$ and **(b)** $y(k) = f(y(k-1), y(k-2))$, a second-order AR time series: $N_{\text{eff}} \ll N$

possesses. But the information content of the data is rarely a topic of discussion. In the opinion of the author, the ignorance with respect to the quality of the data distribution in practice is a significant shortcoming for the application of information criteria.

If the data is uniformly distributed in the input space like in space-filling designs (see Fig. 7.16a), the information content per data point certainly is very high. However, in many applications, the data is agglomerated. Sometimes this happens because no proper care was taken during the experimental design phase. Sometimes a bad data point distribution follows from intrinsic correlations, e.g., for dynamic processes where most inputs cannot be controlled independently, compare Fig. 7.16b. It is quite obvious that the two cases in Fig. 7.16 represent totally different amounts of information per data point. Thus, in the opinion of the author, something like *effective number of samples* N_{eff} should be defined where $N_{\text{eff}} \rightarrow N$ for situations similar to Fig. 7.16a and significantly smaller for cases like the one in Fig. 7.16b. Such an approach certainly could improve the bias/variance tradeoff carried out with information criteria.

7.3.4 Multi-Objective Optimization

For practical application of the information criteria, one important issue is the proper choice of the user-defined parameter(s). This problem emerges in almost all engineering applications. At a certain point in the design procedure, one has to make a decision on a tradeoff between model performance and model complexity. Often this can be reformulated as

$$\text{criterion}(\alpha) = f(\text{model error}) + \alpha \cdot g(\text{complexity}), \tag{7.24}$$

where the penalty factor α controls the tradeoff between (some function of) the model error and complexity. The information criteria take this or a similar form as well as ridge regression (Sects. 3.1.4 and 7.5.2.2) and other regularization approaches. Furthermore, optimization problems similar to the type in (7.24) emerge for the tradeoff between control performance and control action in controller design, between fast-tracking performance and good noise attenuation in recursive estimator design, between the efficiency of a combustion engine and its amount of NO_X exhaust gas, between product quality and production cost, between expected profit and risk of an investment, etc. The additive combination of the model error term and complexity penalty in (7.24) is the most common but not the only possible realization. For example, the product of both terms represents an alternative [50].

In modeling, the complexity term can represent an approximation of the expected variance error as it is realized by the information criteria. Then (in the ideal case) the model complexity with the best bias/variance tradeoff minimizes the criterion. Often, however, additional restrictions to the variance error increase force the user to implement simpler models:

- computation speed,
- computer memory,
- development time and cost,
- model interpretation and transparency,
- industrial acceptance.

All these restrictions may be incorporated in the penalty term.

Two alternative approaches exist for optimization of (7.24). On the one hand, (7.24) can be optimized several times for different penalty factors α, and finally the user chooses the solution which realizes the most appealing tradeoff. The difficulty with this approach is that it is usually hard to get some feeling for reasonable penalty factor values α, and thus a lot of trial and error is necessary to find a good tradeoff.

On the other hand, problem (7.24) can be solved by a multi-objective optimization method. In particular, evolutionary algorithms are popular for this approach; see Sect. 5.2 and [162, 163]. Figure 7.17 illustrates that the set of pareto-optimal solutions represents all possible tradeoffs between model error and model complexity. Each solution on this curve represents one specific penalty value of α. Multi-objective optimization techniques do not try to perform this tradeoff. Rather they generate a set of pareto-optimal solutions. Then the user can compare the models that are represented by these solutions and can choose one. Note that the curve of pareto-optimal solutions is not necessarily as smooth as depicted in Fig. 7.17, and thus the number of generated solutions should not be too small; otherwise the user may not have enough information to perform a good tradeoff.

Note that the main motivation for such a tedious tradeoff between different objectives arises from the fact that users are usually not able to exactly quantify the priorities for these objectives, at least not before they have gained some experience by considering alternative models. Essentially, multi-objective optimization techniques present a number of alternative models at the same time and thus reduce user interaction compared with a trial-and-error approach.

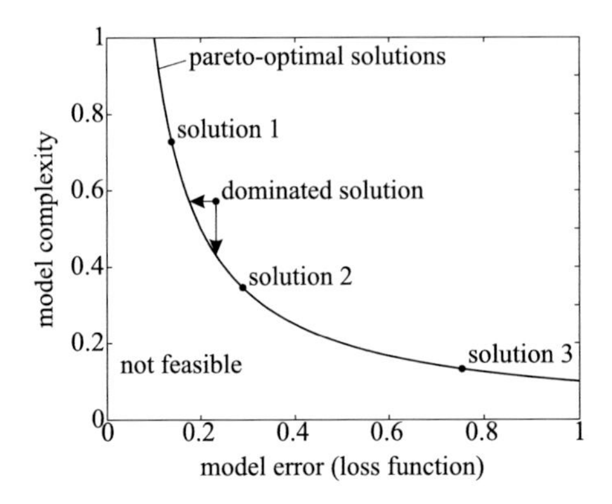

Fig. 7.17 Multi-objective optimization for two criteria: model error and model complexity. Each solution of the multi-objective optimization represents a model with a certain error and complexity. The solutions 1, 2, and 3 are all pareto-optimal, i.e., they are not dominated by other solutions. In contrast, for a dominated solution, solutions with smaller model errors and identical complexity or solutions with smaller complexity and identical model error exist. A dominated solution is always suboptimally independent of the tradeoff between the criteria

7.3.5 Statistical Tests

An alternative to the application of information criteria is the use of statistical tests. Close relationships exist between some of the information criteria mentioned above and the chi-squared test and F-test, which are described in this subsection. The following discussion is based on the dissertations of Sjöberg [535] and Kortmann [327]. For more details refer also to [125].

The main idea for the application of statistical tests to model complexity selection is to assume two models with different complexity and decide whether the more complex model makes significant use of its additional parameters. The first, simple model has n_{simple} parameters, and the second, more complex model possesses n_{complex} parameters. It is assumed that the simple model is contained in the complex model as a special case. For the sake of simplicity, it is assumed that the complex model is identical to the simple model if the additional $n_{\text{complex}} - n_{\text{simple}}$ parameters are set equal to zero.

Now, the following hypothesis is formulated:

$$\begin{array}{llll} \mathrm{H}_0: & \underline{\theta}_{\text{add}} = 0 & \text{simple model OK} \\ \mathrm{H}_1: & \underline{\theta}_{\text{add}} \neq 0 & \text{simple model not OK} \end{array} \tag{7.25}$$

where $\underline{\theta}_{\text{add}}$ denotes the vector of the additional parameters in the complex model compared with the simple one. The so-called *null hypothesis* H_0 expresses the fact that the additional parameters are zero. If this is correct, the complex model does not make any use of its additional parameters, and thus the simple model should be selected. A statistical test will decide whether the null hypothesis should be accepted or rejected.

It is necessary for the user to decide upon the significance level α, also called the first type of risk, which determines the probability of rejecting H_0 when it is actually true. Typical values for α are 0.01 to 0.1. The first type of risk should not be made too small (especially for small training data sets) because then the second type of risk increases, namely, the probability of accepting H_0 when it is false. The selection of α is the critical issue in all statistical tests and corresponds to the choice of the parameter in information criteria such as the AIC in (7.16).

This decision can be based on the performance comparison between both models measured by the difference of their loss function values I_{simple} and I_{complex} defined as sum of squared errors. The higher the improvement due to the additional parameters is, the larger is this difference. It can be shown that, if H_0 holds and $N \rightarrow \infty$ [535]

$$T = N \frac{I_{\text{simple}} - I_{\text{complex}}}{I_{\text{complex}}} \xrightarrow{\text{dist.}} \chi^2(n_{\text{complex}} - n_{\text{simple}}) \,, \tag{7.26}$$

where N is the number of training data samples and $\chi^2(n_{\text{complex}} - n_{\text{simple}})$ is the chi-squared distribution with $n_{\text{complex}} - n_{\text{simple}}$ degrees of freedom. The null hypothesis H_0 is accepted with significance α if [535]

$$T \leq \chi^2_\alpha(n_{\text{complex}} - n_{\text{simple}}) \,, \tag{7.27}$$

where $\chi^2_\alpha(\cdot)$ denotes the α-percentile (see Fig. 7.18) of the chi-squared distribution. This is the case if T is small "enough," i.e., the more complex model performs only insignificantly better than the simple one. If the significance level α is chosen to be very small, almost any T will fulfill (7.27), and thus H_0 will almost always be accepted. However, then the second type of risk increases dramatically. So the

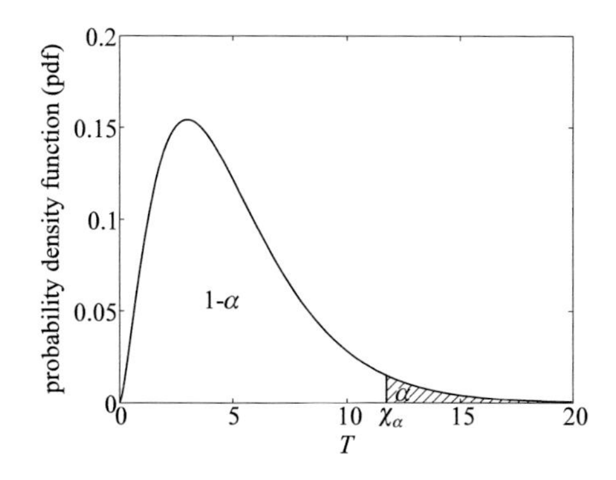

Fig. 7.18 Chi-squared distribution with α-percentile. If the null hypothesis holds, T is asymptotically chi-squared distributed. The pdf shows that very large values for T have a low probability. If the value of T is larger than the α-percentile, it is unlikely (with a probability of $1 - \alpha$) that T indeed follows this distribution. As $\alpha \rightarrow \infty$, the α-percentile tends to infinity, and almost all values of T are below this threshold: consequently H_0 is accepted

first type of risk must be chosen reasonably. Since T increases with the number of training data samples N, the first type of risk can be chosen the smaller the more data is available.

Note that the above chi-squared test is based on the assumption that N is large because only then T approximately is chi-squared distributed. In the case of linear regression, an exact distribution valid for any number of training data samples N can be derived. If H_0 holds and the noise is Gaussian distributed, then [535]

$$T = \frac{N - n_{\text{complex}}}{n_{\text{complex}} - n_{\text{simple}}} \frac{I_{\text{simple}} - I_{\text{complex}}}{I_{\text{complex}}} \tag{7.28}$$

is $F(n_{\text{complex}} - n_{\text{simple}}, N - n_{\text{complex}})$-distributed with the two degrees of freedom $n_{\text{complex}} - n_{\text{simple}}$ and $N - n_{\text{complex}}$. Corresponding to the chi-squared test, the null hypothesis is accepted with the significance α if [535]

$$T \leq F_\alpha(n_{\text{complex}} - n_{\text{simple}}, N - n_{\text{complex}}) \tag{7.29}$$

holds. Equation (7.29) is called the *F-test*.

The chi-squared test can be applied for any type of model under the assumption that the number of training data samples N is large enough (the accuracy increases as N grows). The F-test can be applied for any N, but the models have to be linearly parameterized. For both tests, the first type of risk α should not be chosen too small, especially if N is small.

Note again that in all these tests, the number of parameters n must be replaced by the effective number of parameters n_{eff} if any regularization technique is applied. For more details about the effective number of parameters, refer to Sect. 7.5.

7.3.6 Correlation-Based Methods

Another strategy for testing whether a model is appropriate is to check whether it captures all the information contained in the data. The optimal model extracts all relationships within the data. Consequently, the model error e should not be correlated with the model inputs $u_i, i = 1, \ldots, p$:

$$\text{corr}\{u_i, e\} = 0\,. \tag{7.30}$$

In practice, the correlation is not exactly zero, and a statistical test can reveal whether (7.30) holds with a user-specified probability [54, 56, 541]. If the model is linear and static, (7.30) is sufficient. However, nonlinear static, linear dynamic, or even nonlinear dynamic models require additional correlation tests. For nonlinear models, correlations between powers (or other nonlinear transformations) of the inputs and the error should be used as well, i.e., high-order correlations between u_i^2, u_i^3, etc. and e. For dynamic models, different time lags should be checked as well, e.g., correlations between $u_i(k-1)$, $u_i(k-2)$, etc. and $e(k)$ and additionally correlations between $y(k-1)$, $y(k-2)$, etc. and $e(k)$ if the previous process outputs $y(k-l)$ are utilized by the model's prediction. In [54, 56], a number of correlation tests are proposed for nonlinear dynamic models that combine both extensions. One drawback of correlation tests is that the required maximum power and time lag are unknown. The major disadvantage of all approaches based on statistical tests and many other validation strategies is that they are not constructive. They tell the user whether a given model is adequate, but not *how* to change the model in order to make it adequate.

7.4 Explicit Structure Optimization

This section discusses some common strategies for *explicit* model complexity optimization. The term "explicit" means that the bias/variance tradeoff is carried out by examining models with different numbers of parameters. By contrast, Sect. 7.5 focuses on *implicit* structure optimization where the nominal number of model parameters does not change but, nevertheless, the model complexity varies.

Structure optimization can operate directly at the parameter level, i.e., it can compare models with $1, 2, \ldots, n$ parameters, e.g., polynomials of zeroth, first, $\ldots, n-1$ order. Alternatively, structure optimization can operate at a higher structural level, such as the number of neurons in a neural network, the number of rules in a fuzzy system, or the order of a linear dynamic system. Each neuron, rule, or dynamic order is typically associated with more than one parameter. In black box models, it may be reasonable to operate at a parameter level. In gray and white box models, it is often not possible to remove or add a single parameter because the model interpretation would be lost. Then it is more reasonable to operate on whole substructures. Another reason for performing the structure optimization at a higher structural level is that the number of possible models is significantly reduced. For example, it may be feasible to determine whether the optimal neural network complexity possesses $1, 2, \ldots, 10$ neurons, but it may be computationally too expensive to compare all networks with $1, 2, \ldots, 50$ parameters (assuming that each neuron is associated with five parameters). Clearly, the lower computational effort of the first alternative results from a coarser coding, but often in practice, a rough determination of the optimal model complexity is sufficient.

The explicit structure optimization methods can be distinguished into the following four categories (see Sect. 3.4):

- *General methods*: Models with different complexity are compared. There has to be no specific relationship between these models. Usually a combinatorial optimization problem of finding the globally best solution arises. Consequently, these general methods are computationally demanding. Common techniques are as follows:

  - A comparison of all models from a minimal to a maximum complexity can be carried out. Comparing all possible models leads to the best solution but requires huge computational effort even for small problems.
  - A genetic algorithm (GA) or other structure search methods can be applied; see Sect. 5.2.
  - Genetic programming (GP) or similar strategies can be followed; see Sect. 5.2.3. In comparison with GAs, GPs structure the problem in a tree that structures the search space by a proper coding and may yield a more focused search.

- *Forward selection*: Starting with a very simple model, in each iteration, the model's complexity is increased by adding either parameters or whole substructures. In many cases the initial, very simple model is empty. These approaches follow the philosophy "try simple things first!" They are called *incremental training* or *construction* because they increase the complexity by one unit (parameter or substructure) within each iteration. This has the advantage that unnecessarily complex models do not have to be computed since the algorithm can be stopped if an increase in model complexity does not yield better performance. This feature makes the forward selection approaches the most popular and widely applied. Typical representatives are as follows:

  - The orthogonal least squares (OLS) algorithm for forward selection is one of the best approaches if the model is linearly parameterized; see Sect. 3.4.2. It exploits the linear parameter structure of the model.
  - The additive spline modeling (ASMOD) algorithm constructs additive fuzzy models of the singleton type. It can make the fuzzy model more complex by (i) increasing the number of membership functions for one input, (ii) adding new inputs to the model, or (iii) combining inputs in order to model their interaction. This algorithm also benefits from the linear parameters in singleton fuzzy systems. For more details refer to Chap. 12.
  - The local linear model tree (LOLIMOT) algorithm trains fuzzy models of Takagi-Sugeno type. It operates at the rule level, i.e., it adds one rule in each iteration. This approach exploits the linear and local properties of Takagi-Sugeno fuzzy models. For more details refer to Chap. 13.
  - The projection pursuit (PP) algorithm builds up multilayer perceptron neural networks. It also works at the neuron level, i.e., it adds one neuron in each iteration. For more details refer to Chap. 11.
  - *Growing* is the generic term for all kinds of neural network training techniques that increase the network complexity by adding either parameters or neurons. Mostly, growing methods are applied to multilayer perceptron networks. A well-known approach is the so-called *cascade-correlation* network [143]. Often growing is combined with a regularization technique; see Sect. 7.5.

- *Backward elimination*: In opposition to forward selection, backward elimination starts with a very complex model and removes parameters or substructures in each iteration. Since this approach starts with very complex models, it is usually more time-consuming than forward selection. Typical applications are as follows:

- An OLS can be used if the model possesses linear parameters.
- After a forward selection method is used, a backward elimination may be performed in order to discover and discard redundant parameters or substructures. Thus, a backward elimination component can be appended to ASMOD, LOLIMOT, PP, etc.
- *Pruning* is the generic term for all kinds of neural network training techniques that decrease the network complexity by removing either parameters or neurons. Like growing, pruning is most often applied to multilayer perceptron networks [315, 479]. Typically, it is combined with a regularization technique; see Sect. 7.5.

- *Stepwise selection*: If in each iteration forward selection and backward elimination steps are considered, this is called stepwise selection. Because unimportant parameters or substructures can be discarded in each iteration, stepwise selection usually leads to better results than pure forward selection or backward elimination. On the other hand, it requires a significantly higher computational effort. The computational demand can be reduced by splitting the training procedure into separate forward selection and backward elimination phases instead of considering both in each iteration. Typical algorithms that implement stepwise selection are as follows:
  - The classification and regression tree (CART) is proposed in [74], which incrementally builds up and prunes back a tree structure.
  - Multivariate adaptive regression splines (MARS) are proposed in [172].

7.5 Regularization: Implicit Structure Optimization

Regularization techniques allow one to influence the complexity of a model although the nominal number of parameters does not change. When regularization techniques are applied, a model is not as flexible as it might appear from considering the number of parameters alone. Thus, regularization makes a model behave as though it possesses fewer parameters than it really has. Consequently, in the bias/variance tradeoff, regularization increases the bias error (less flexibility) and decreases the variance error (fewer degrees of freedom). Obviously, the application of regularization techniques is reasonable only if the model complexity is high before regularization is applied, i.e., to the right-hand side of the optimal model complexity in Fig. 7.4a. This is the reason why regularization techniques are most frequently applied to neural networks, fuzzy systems, or other models with many parameters as they are typically used for modeling nonlinear systems.

Regularization ideas, especially the curvature penalty approaches described in Sect. 7.5.2, can be applied also in contexts other than dealing with overparameterized models. In Sect. 11.3.6, regularization theory is briefly discussed. By using the calculus of variations, it allows one to determine which model *architecture* is the best one under given smoothness assumptions. Another application of regularization ideas can be found in ridge regression for linear optimization (Sect. 3.1.4) and the Levenberg-Marquardt algorithm for nonlinear optimization (Sect. 4.5.2).

7.5.1 Effective Parameters

Loosely speaking, regularization works as follows. Not all parameters of the model are optimized in order to reach the minimal loss function, e.g., the sum of squared errors. Rather some other criteria or constraints are taken into account. Because some degrees of freedom of the model must be spent on these other criteria or constraints, the model flexibility for solving the original problem reduces. Those parameters that are still used for minimizing the original loss function are called the *effective parameters* since only they have an effect on the original loss function. The parameters that have only an insignificant influence on the original loss function are called *spurious parameters*. These spurious parameters are utilized to fulfill other criteria or constraints. Since a regularized model behaves similarly to a nonregularized model that possesses only the effective number of parameters, the approximate expression for the variance error in (7.6) must be replaced by

$$\text{model error} \approx \sqrt{\text{variance error}} \sim \sigma \frac{\sqrt{n_{\text{eff}}}}{\sqrt{N}}, \tag{7.31}$$

where n_{eff} represents the effective number of parameters. If the regularization effect is weak, n_{eff} approaches the nominal number of model parameters n. If the regularization effect is strong, n_{eff} approaches zero. Depending on the specific regularization technique, it may not be possible to draw a clear line between the effective and spurious parameters. The transition from spurious to effective parameters can be fuzzy, or parameters may be exchangeable. For example, consider the model $y = \theta_1 + \theta_2 u + \theta_3 u^2$ with the constraint $\theta_1 + \theta_2 + \theta_3 = 1$. Obviously, this model possesses three nominal parameters, but since the constraint determines one of them completely by the other two, the number of spurious parameters is one, and the effective number of parameters is two. Nevertheless, it is impossible to say *which* of the three parameters is the spurious one. Because the model structure is very simple, one parameter can be directly substituted; this is not usually as easy for complex nonlinear models.

In the following, the most common regularization techniques are explained.

7.5.2 *Regularization by Non-Smoothness Penalties*

Since regularization reduces the effective number of model parameters, all types of regularization smooth the model output. Smoothness is a property that is usually desirable because almost no real-world phenomena lead to steps, instantaneous changes, or nondifferentiable relationships.

One regularization method explicitly penalizes nonsmooth behavior and therefore forces the model to be smooth. Instead of minimizing the original loss function, e.g., the sum of squared errors, the following objective function is optimized:

$$\text{criterion} = \text{sum of squared errors} + \text{nonsmoothness penalty}\,. \tag{7.32}$$

7.5.2.1 Curvature Penalty

If the model is $\hat{y} = f(\underline{u}, \underline{\theta})$, a typical choice for the nonsmoothness penalty would be the second derivative of the model output with respect to the model inputs. With $i = 1, \ldots, N$ samples in the training data set, this becomes

$$\text{nonsmoothness penalty} = \alpha \sum_{i=1}^{N} \left| \frac{\partial^2 f(\underline{u}(i), \underline{\theta})}{\partial \underline{u}(i)^2} \right|^2 . \tag{7.33}$$

A tradeoff has to be performed between the original loss function ($\alpha = 0$) and the pure nonsmoothness penalty without taking the model performance into account ($\alpha \to \infty$). The value for the regularization parameter α must either be chosen by the user or it can be roughly estimated by some probability considerations based on Bayesian statistics [57, 64]. The penalty factor in (7.33) drives the model toward linear behavior since the second derivative of a linear model is zero everywhere.

On the one hand, (7.33) can be seen as a *curvature penalty*, i.e., smoothness is defined as low curvature (many different definitions are possible). On the other hand, (7.33) can be understood as the incorporation of the *prior knowledge* that linear models are preferable over others. With any definition of smoothness, a special type of model that matches these smoothness properties is favored over all others. For example, the (very unusual) nonsmoothness penalty $\alpha \cdot \partial f(\underline{u}, \underline{\theta})/\partial \underline{u}$, i.e., the first derivative of the model, prefers constant models because they possess a zero gradient. Owing to this relationship between the nonsmoothness penalty and the incorporation of prior knowledge, it is sometimes called the *prior*.

7.5.2.2 Ridge Regression

In practice, often an approximation of the nonsmoothness penalty is used because the evaluation of the exact derivatives is computationally expensive and can make the optimization problem more complex. For example, for linear parameterized models, the penalty term is typically chosen such that the regularized problem stays linear in the parameters.

The simplest form of nonsmoothness penalties is the ridge regression for linear optimization problems (see (3.44) in Sect. 3.1.4):

$$I(\underline{\theta}, \alpha) = \underline{e}^T \underline{e} + \alpha |\underline{\theta}|^2 . \tag{7.34}$$

This penalty drives all parameters in the direction of zero. Thus, the preferred model is a constant. If prior knowledge is available that the parameters should not be close to zero but close to $\underline{\theta}_{\text{prior}}$, (7.34) can be changed to

$$I(\underline{\theta}, \alpha) = \underline{e}^T \underline{e} + \alpha |\underline{\theta} - \underline{\theta}_{\text{prior}}|^2 . \tag{7.35}$$

7.5.2.3 Weight Decay

The idea of ridge regression can be extended to nonlinear optimization problems. In the context of neural networks, especially multilayer perceptrons, the parameters $\underline{\theta}$ are called *weights*, and the ridge regression type of regularization is called *weight decay*. Intuitively, the operation of weight decay can be understood as follows (see Sect. 3.1.4). The penalty term tries to push the network weights toward zero, while the error term $\underline{e}^T \underline{e}$ tries to move the weights toward their optimal values for the nonregularized neural network. Some compromise is found, depending on the choice of the regularization parameter α. Clearly, those network parameters that are very important for the reduction of the error term will be scarcely influenced by the penalty term because the decrease in $\underline{e}^T \underline{e}$ overcompensates for $\alpha |\underline{\theta}|^2$. In contrast, those network parameters that are not very important for the network performance will be driven close to zero by the penalty term. Thus, after training, the unimportant parameters are close to zero. In a second step, these parameters (or even whole neurons) can be removed from the neural network. This is a typical combination of a regularization technique (weight decay) with an explicit structure optimization technique (pruning).

The ridge regression or weight decay regularization can be analyzed by evaluating the eigenvalues of the Hessian matrix (the second derivative of the model output with respect to its parameters). This is illustrated in Sect. 3.1.4 for linear parameterized problems but can be extended to nonlinear optimization problems too; see [57]. It turns out that the regularization parameter α marks a threshold that allows one to determine which model parameters are effective. Each eigenvalue in the Hessian corresponds to one parameter. All parameters with eigenvalues larger than the regularization parameter α are effective, while the others are not. The larger the eigenvalue, the higher is the influence of the associated parameter on the loss function and the more accurately this parameter can be estimated. Parameters with very small corresponding eigenvalues are insignificant and cannot be estimated accurately. These parameters contribute very little to the decrease in the error term and are affected most by the penalty term. Regularization virtually sets their eigenvalues to α and consequently improves the conditioning of the Hessian. This leads to less stretched contour lines of the loss function and thus to faster convergence of training with the regularized loss function compared to the original one.

7.5.3 Regularization by Early Stopping

Another important regularization technique is *early stopping*. It can be applied when iterative optimization methods are used. Training is not performed until the model parameters have converged to their optimal values. Rather during the iterative training algorithm, the model performance on a validation data set is monitored. Training is stopped when the validation error reaches its minimum. Typically, the convergence curves on training and validation data behave as shown in Fig. 7.19. The relationship of early stopping to weighted decay is formally shown in [535]. Here, just an informal explanation will be given.

At the minimum of the validation error, the best bias/variance tradeoff is realized. At the left-hand side of this minimum, the model would underfit, while to the right-hand side, it would overfit the data. During the iterations, the effective number of model parameter increases. If the training continued until convergence, all model parameters would become effective, resulting in a large variance error.

During the training procedure, more and more model parameters converge to their optimal values. The more important a parameter is, the faster it moves toward its optimum. Thus, it can be concluded that early stopping allows all important parameters to converge and to become the effective parameters of the model, while others basically stay close to their initial values. Consequently, if the initial values are zero (or close to zero), early stopping has the same effect as weight decay.

The parameter importance is represented by their associated eigenvalue of the Hessian matrix. The loss function is very sensitive in the direction of important parameters (large eigenvalues) and flat in the direction of less relevant parameters

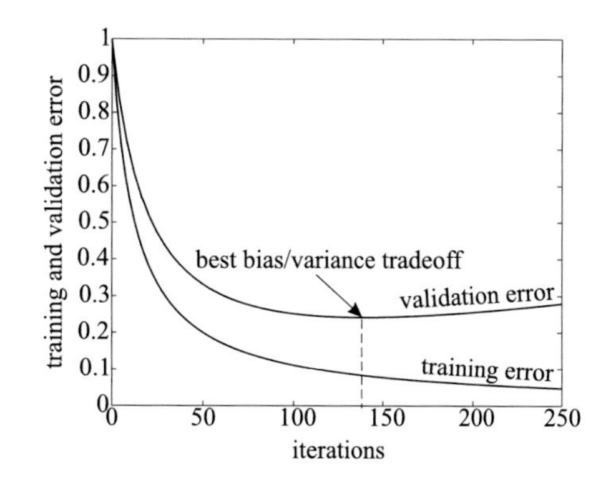

Fig. 7.19 With the early stopping regularization technique, training does not continue until all model parameters have converged. Rather, it is stopped early when the error on the validation data reaches its minimum. As the training proceeds, in each iteration the effective number of model parameters increases slightly

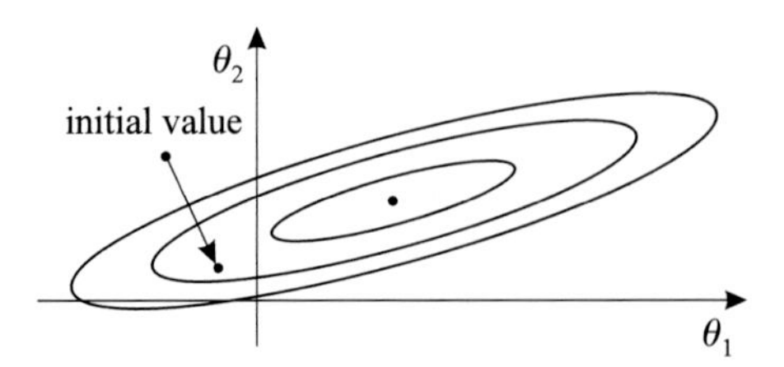

Fig. 7.20 The convergence speed of different parameters depends on the strength of their influence on the loss function. During the first iterations of a gradient-based optimization technique, mainly the important parameters (here θ_2) are driven close to their optimal values, while less relevant parameters (here θ_1) are hardly changed

Table 7.1 Comparison of early stopping and pruning

	Early stopping	Pruning
Training speed	Fast	Slow
Type of training	Parameter opt.	Structure and parameter opt.
Final model size	Large n	Small n
No. of parameters	$n_{\mathrm{eff}} < n$	$n_{\mathrm{eff}} \approx n$

(small eigenvalues). This explains why the importance of parameters determines their convergence speed; see Fig. 7.20.

The main reason for the popularity of early stopping is its simplicity. Furthermore, it reduces the computational demand since training does not have to be completed. It is important to understand that for very flexible models, convergence of all parameters is not desired! This is the reason why poorly converging optimization algorithms such as steepest descent (Sect. 4.4.3) can still work reasonably well in connection with large neural networks and early stopping.

Note, however, that during the previous discussion of this topic, an overparameterized model has been assumed, i.e., a model with too many parameters. Only then is the best bias/variance tradeoff reached before all parameters have converged. If the considered model is close to the optimal complexity, the minimum of the validation error in Fig. 7.19 moves far to the right since no overfitting is possible. Then convergence of all parameters is necessary for the best model performance, and the application of more sophisticated optimization algorithms (Sect. 4.4) is strongly recommended.

Table 7.1 compares early stopping regularization with pruning. Pruning methods try to find the optimal model structure explicitly. One major drawback of this approach compared with regularization is the higher computational demand for training. One major benefit is that the resulting model is simpler. This reduces the model evaluation time, lowers the memory requirements, and makes interpretation easier. Note, however, that although the nominal number of parameters may be much larger for the regularized model than for the pruned model, the effective number of parameters and consequently the bias and variance errors may be comparable.

Training with noise and *parameter (weight) sharing* are other regularization techniques that are frequently applied in connection with neural networks; for more details refer to [57].

7.5.4 Regularization by Constraints

Constraints restrict the flexibility of a model. Therefore, they increase the bias error and reduce the variance error without changing the nominal number of parameters. Thus, they exhibit a regularization effect. Different types of constraints can be distinguished:

- *Hard constraints* must be met. If not all hard constraints can be met simultaneously, no solution to the parameter optimization problem exists. With hard constraints, exact prior knowledge can be incorporated into the model. The following two categories of hard constraints can be distinguished:
  - Equality constraints reduce the flexibility of the model by one parameter. Examples are $\theta_1 = 4$ or $\theta_1 + \theta_2 = 1$. In these cases, one parameter of the model is fully determined by all others and can be removed from the model. With equality constraints, e.g., knowledge about the gain or a time constant of a linear dynamic process can be incorporated into the model.

    A typical application is the estimation of a transfer function $(b_0 + b_1 q^{-1} + \ldots + b_m q^{-m})/(1 + a_1 q^{-1} + \ldots + a_m q^{-1})$ with a fixed gain equal to 1, i.e., the constraint is $b_0 + b_1 + \ldots + b_m = 1 + a_1 + \ldots + a_m$. Clearly, a solution to the parameter optimization problem can be obtained only if the number of equality constraints is smaller than the number of model parameters.
  - Inequality constraints reduce the flexibility of the model only if they are active. They typically take the form $\theta_1 > 0$ or $\theta_1 + \theta_2 > 5$. It can happen that the optimal parameters for the unconstrained problem meet all these constraints, i.e., no constraint is active. Then the flexibility of the model is not affected at all (no regularization). If, however, the optimum of the unconstrained problem lies in an infeasible region of the parameter space, the inequality constraints prevent some parameters from realizing their optimal values, see Fig. 7.21. This effect regularizes the model.

    Inequality constraints can be derived from prior knowledge, e.g., the gain of a linear process must always be positive. This knowledge about the process under consideration can be transferred to the following inequality constraint: $b_0 + b_1 + \ldots + b_m > 1 + a_1 + \ldots + a_m$. Inequality constraints can also be used to ensure such properties as the monotonicity of some nonlinear function. Guaranteed monotonic behavior of a model can be very important in the context of feedback control since it determines the sign of the process gain [352].

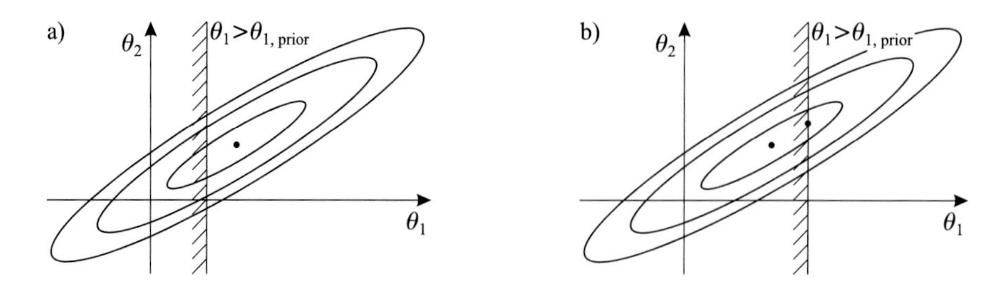

Fig. 7.21 Regularization effect of inequality constraints: (**a**) The constraint $\theta_1 > \theta_{1,\text{prior}}$ is not active at the optimum. (**b**) The constraint prevents the parameters from taking their optimal values, and the best solution is realized on the constraint boundary $\theta_1 = \theta_{1,\text{prior}}$. This results in a regularization effect because the constraint reduces the model's flexibility

- *Soft constraints* allow one to incorporate qualitative, non-exact knowledge into the model. Since soft constraints must not be met exactly, in principle, an arbitrary number of them can be implemented. Usually, soft constraints are realized by incorporation of a penalty function in the loss function (see Sect. 4.6):

$$I(\underline{\theta}, \alpha) = \underline{e}^T \underline{e} + \alpha f_{\text{sc}}\left(\underline{\theta}, \underline{\theta}_{\text{prior}}\right) . \tag{7.36}$$

In the simplest case, the soft constraints penalty function $f_{\text{sc}}(\cdot)$ can be chosen as the Euclidean distance between the parameters $\underline{\theta}$ and the prior knowledge about the parameters $\underline{\theta}_{\text{prior}}$, as is done in the ridge regression or weight decay approach in (7.35), Sect. 7.5.2. Because the soft constraints on different parameters may be of different importance, in extension to (7.35) an additional weighting of the constraints can be introduced, e.g.,

$$f_{\text{sc}}\left(\underline{\theta}, \underline{\theta}_{\text{prior}}\right) = \left(\underline{\theta} - \underline{\theta}_{\text{prior}}\right)^T \underline{Q} \left(\underline{\theta} - \underline{\theta}_{\text{prior}}\right) . \tag{7.37}$$

Thus, the tradeoff between the model performance $\underline{e}^T \underline{e}$ and the soft constraints $f_{\text{sc}}(\cdot)$ is controlled by α, while the tradeoff between the soft constraints themselves is controlled by the entries in the matrix $\underline{Q}$. The penalty factor α allows one to balance the data and the expert knowledge [352].

In [352] it is proposed to express the soft constraints $f_{\text{sc}}(\cdot)$ by a linguistic fuzzy system in order to fully exploit the qualitative character of the available prior knowledge. Soft constraints offer a powerful tool for incorporating prior knowledge in the model. They can be combined with virtually every type of model.

7.5.5 Regularization by Staggered Optimization

Usually, all model parameters are optimized simultaneously. In staggered optimization, this is not the case. Rather the parameters are divided into different subsets, and only the parameters within one subset are optimized concurrently. Then in succession, all subsets are subject to optimization. Figure 7.22 illustrates this approach for a model with two parameters, which are divided into two subsets containing one parameter each. If the influence on the loss function of the parameters of different subsets is only weakly coupled, one cycle of staggered optimization can approach the optimum closely. If the parameter subsets are fully decoupled, as is the case for orthogonal regressors (see Fig. 3.9 in Sect. 3.1.3), then one cycle of staggered optimization is sufficient for convergence to the optimum. In practice, usually all parameters are more or less coupled, and thus staggered optimization can be more or less efficient depending on the conditioning of the problem. If only one or a few cycles are carried out, a regularization effect similar to the one for early stopping occurs.

The application of staggered optimization is reasonable whenever the model parameters can be grouped naturally into different subsets. For example, many neural networks possess one or more hidden layers with nonlinear parameters and an output layer with linear parameters. Consequently, the nonlinear parameters can be collected in one subset and the linear ones in another. In this case, staggered optimization offers the advantage that the linear parameter subset can be optimized very efficiently by a least squares technique. This can speed up the neural network training. If all parameters are optimized simultaneously, a nonlinear optimization technique would have to be applied to all parameters instead of only the nonlinear ones. Clearly, in order to obtain convergence, the number of required iterations for staggered optimization is larger. However, if special properties of the

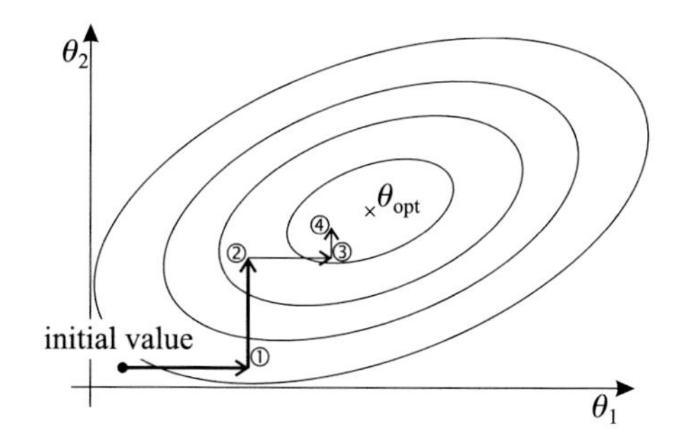

Fig. 7.22 Regularization by staggered optimization. Starting from an initial value, the parameters θ_1 and θ_2 are not optimized simultaneously. Rather, each parameter is optimized separately. Usually, after separate optimization of each parameter (here steps 1 and 2), the optimum is not reached, and a regularization effect comparable to early stopping is obtained. However, this procedure can be iterated until convergence is accomplished and the regularization effect diminishes

different parameter subsets are exploited, each iteration may be computationally less expensive, leading to a reduced overall computational demand.

Another example of staggered optimization is the so-called *backfitting* algorithm for additive model structures (Sect. 7.6.4), which is proposed in [173] for projection pursuit regression; see also Sect. 11.2.7. For backfitting the parameter subsets correspond to the parameters of each additive submodel. In each iteration one parameter subset, i.e., one additive submodel, is optimized, while all others are kept fixed. The optimization of the jth parameter subset operates on a loss function that is based on the following error:

$$e^{(j)} = y - \sum_{i=1,\, i \neq j}^{S} \hat{y}_i^{(\text{submodel})}\,, \tag{7.38}$$

where S is the number of additive submodels. Obviously, one difficulty with backfitting is that the error in (7.38) becomes more nonlinear and irregular the more submodels are utilized; so the (relatively simple) submodels are decreasingly able to describe the unmodeled part of the process.

7.5.6 Regularization by Local Optimization

A special quite common form of staggered optimization is local optimization. In this context, “local” refers to the effect of the parameters on the model; there is no relation to the expression “nonlinear local” optimization, where “local” refers to the parameter search space. Local optimization is utilized, for example, in the LOLIMOT algorithm described in Chap. 13. Several types of nonlinear models generate their output by a combination of locally valid submodels. Important examples are fuzzy systems and basis function networks. All model parameters can be divided into subsets, each containing the parameters of a locally valid submodel. Following the staggered optimization approach, the parameters of each locally valid submodel are optimized separately.

The extent of the regularization effect with local optimization depends on the coupling between the local submodels. If the submodels are strictly separated, no regularization effect occurs since then local and global optimization are identical. The larger the coupling between different local submodels is, the higher is the regularization effect.

Example 7.5.1 Local Versus Global Optimization
A data set consisting of only two samples will be approximated by the following model:

$$y = \theta_1 \Phi_1(u) + \theta_2 \Phi_2(u)\,. \tag{7.39}$$

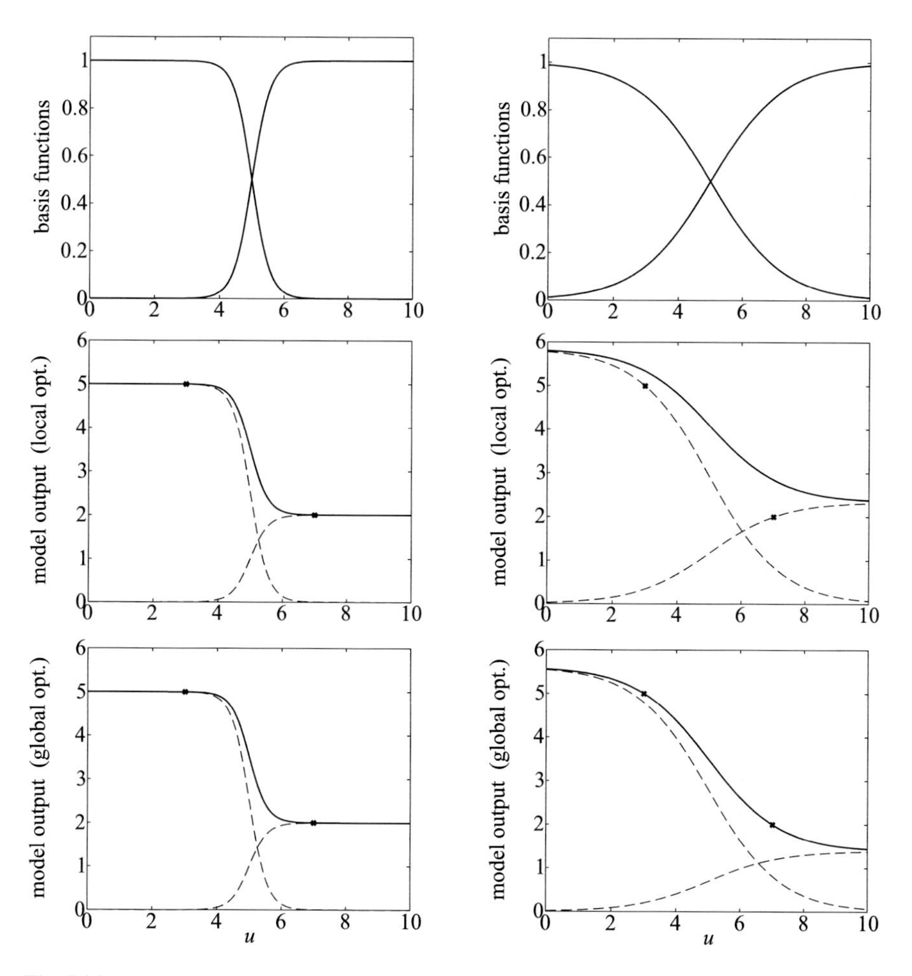

Fig. 7.23 Regularization by local optimization. A model $y = \theta_1\Phi_1(u) + \theta_2\Phi_2(u)$ approximates the data $\{(y_i, u_i)\} = \{(3, 5), (7, 2)\}$. The first and second column represent the basis functions $\Phi_1(\cdot)$, $\Phi_2(\cdot)$ with small and large overlap, respectively. They are depicted in the first row. The second row shows the model output (solid line) and the contribution of both basis functions (dashed lines) if the parameters are optimized locally. The results shown in the third row are obtained with global optimization. The crosses mark the data samples that should be approximated by the model

The shapes of the basis functions $\Phi_1(\cdot)$, $\Phi_2(\cdot)$ and the two data samples are shown in Fig. 7.23. Two different cases are analyzed. On the left of Fig. 7.23, the results for barely overlapping basis functions are shown, while on the right, the overlap is much larger.

If the parameters of this model are optimized locally, the overlap between the basis functions is neglected. This means that the parameters θ_1 and θ_2 are optimized separately without taking the effect of the other basis function into account. For local optimization of the parameters, only local data is considered, i.e., for the basis

Table 7.2 Comparison of the parameters obtained by local and global optimization

Parameters θ_1/θ_2	Small overlap	Large overlap
Local optimization	5.0041/2.0016	5.8451/2.3380
Global optimization	5.0024/1.9976	5.6102/1.3898

function $\Phi_1(u)$, the data sample at $(3, 5)$ is relevant, while $(7, 2)$ is assigned to $\Phi_2(u)$. The parameter θ_1 of the first basis function is adjusted such that $\theta_1\Phi_1(u)$ lies on the data sample $(3, 5)$, and θ_2 is determined such that $\theta_2\Phi_2(u)$ lies on the data sample $(7, 2)$. For global optimization, both parameters are estimated simultaneously with a least squares algorithm. The optimized parameter values are summarized in Table 7.2.

As can be observed from Fig. 7.23 (left), the local and global optimization yield similar results for barely overlapping basis functions. The reason for this lies in the fact that the basis functions are almost orthogonal. Indeed, orthogonal basis functions would yield identical results for local and global optimization. Nevertheless, in this example, with global optimization, the model fits both data samples perfectly, while this is not the case for local optimization owing to the error introduced by neglecting the basis function's overlap. The right part of Fig. 7.23 demonstrates that for basis functions with larger overlap, this error becomes significant. Again, locally both model components $\theta_1\Phi_1(u)$ and $\theta_2\Phi_2(u)$ describe the local data. In contrast to Fig. 7.23 (left), the resulting model performs poorly. Owing to the larger overlap, the regularization effect is larger and model flexibility decreases, which leads to a higher bias error. □

7.6 Structured Models for Complexity Reduction

This section discusses strategies for model complexity reduction which try to reduce or overcome the so-called *curse of dimensionality*. The term "curse of dimensionality" has been introduced by Bellman [37]. Basically, it expresses the intuitively clear fact that in general problems become harder to solve as the dimensionality of the input space increases.

In the next section, the curse of dimensionality is analyzed in greater detail. The subsequent sections are concerned with different strategies that scale up moderately with an increasing input dimension.

7.6.1 Curse of Dimensionality

The curse of dimensionality can be explained by considering the approximation of the following process:

$$y = f\left(u_1, u_2, \ldots, u_p\right) . \tag{7.40}$$

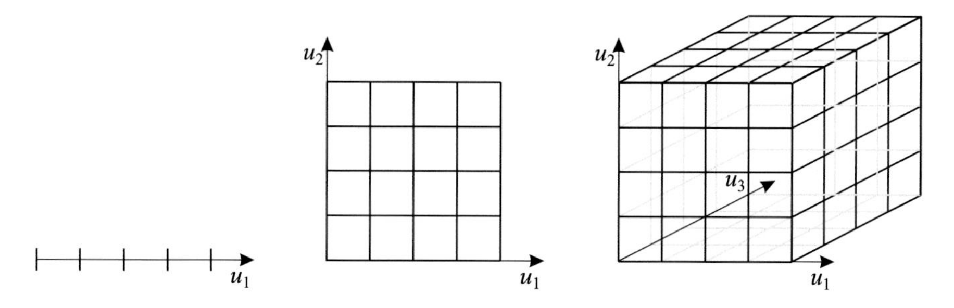

Fig. 7.24 Illustration of the curse of dimensionality. The necessary amount of data grows exponentially with the input space dimensionality

Many tasks from classification to system identification can be transformed into such a function approximation problem. In order to approximate the unknown function $f(\cdot)$ from data, the whole p-dimensional input space spanned by $[u_1\ u_2\ \cdots\ u_p]$ must be covered with data samples. The smoother the function $f(\cdot)$ is, the fewer data samples are required for its approximation to a given degree of accuracy.

Example 7.6.1 Curse of Dimensionality
If it can be assumed that the function $f(\cdot)$ in (7.40) is very smooth, it may be sufficient to cover each input with only four data samples. Thus, if the function is one dimensional ($p = 1$), the required number of data samples is four. For two- and three-dimensional input, spaces 16 and 64 data samples are required, respectively. This is illustrated in Fig. 7.24. Obviously, the necessary amount of data grows exponentially with the input space dimensionality. For approximation of a ten-dimensional function, already over 1 million data samples would be required.

□

It is important to notice that the curse of dimensionality is an intrinsic property of the problem, and is independent of the specific model employed. At first sight, the situation looks hopeless. Indeed, it is almost impossible to approximate a *general* nonlinear function in high-dimensional space (say $p > 10$). Luckily, in practice, the curse of dimensionality is much less severe because in real-world problems typically some of the following statements are fulfilled, thereby restricting the complexity of the problem.

- Non-reachable regions in the input space exist. Very often not all input combinations are feasible, e.g., high pressure and low temperature may be contradictory in a physical system, or high-interest rates and high stock market indices may not be realistic at the same time. Especially in dynamic systems typically large regions of the input space cannot be reached with the available power or energy. Of course, all non-reachable regions of the input space do not have to (and indeed cannot) be covered with data. This can reduce the necessary amount of data considerably.

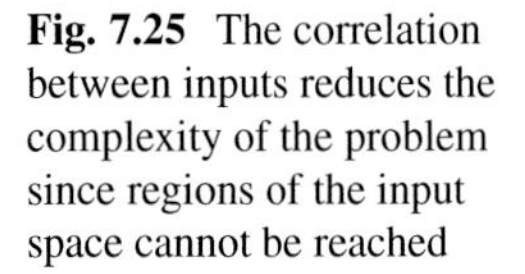

Fig. 7.25 The correlation between inputs reduces the complexity of the problem since regions of the input space cannot be reached

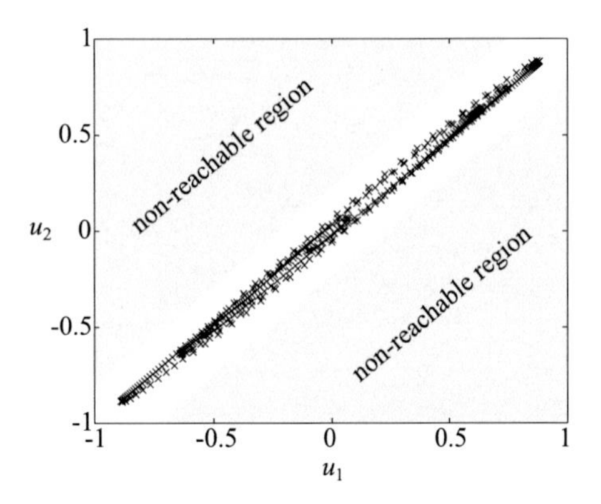

- Inputs are correlated or redundant. Redundant inputs can be removed from the model without the loss of any information. Thereby the dimensionality of the problem is reduced. Correlated inputs usually lead to non-reachable regions in the input space, as in Fig. 7.25 (see Sect. 6.1).
- The behavior is very smooth in some regions of the input space. The nonlinear function $f(\cdot)$ may be very simple, e.g., almost constant or linear, in some regions of the input space. In these regions, a reasonable model can be obtained even for extremely sparse data. This can reduce the required amount of data significantly.
- Although the function might be very complex in some regions, it depends on the specifications whether the model must possess a similar complexity in all these regions. In practice, a quite inaccurate model behavior might be acceptable in some regions. Each specific application demands different model accuracies in different operating conditions.

Owing to the properties listed above, the amount of required data for the solution of real-world problems typically do not increase exponentially with the input space dimensionality.

Another important issue is how the complexity of a *model* increases with the input space dimensionality. This issue is independent of the complexity of the real problem. Several types of models suffer from the curse of dimensionality, i.e., their complexity scales up exponentially with the number of inputs. Clearly, such a behavior strongly restricts the application of these model architectures to very low-dimensional problems. Typical models that suffer from the curse of dimensionality are conventional look-up tables and fuzzy models. Both are so-called *lattice-based* approaches, i.e., they cover the input space with a regular grid.

Lattice-based methods cannot cope with moderate to high-dimensional input spaces because they are unable to exploit the relationships that lead to complexity reduction (see the list above). A lattice covers all regions of the input space equally well. Non-reachable areas and areas where the function $f(\cdot)$ is very smooth are all covered with the same resolution as the important regions in the input space, as shown for the data in Fig. 7.24.

In the following sections, different strategies are briefly introduced that allow one to exploit the reduced complexity of a problem. None of these strategies works well on all types of problems. Therefore, it is important to incorporate as much prior knowledge in the model as possible, since this is the most efficient way to reduce the complexity of the problem; see Sect. 7.6.2. Prior knowledge does not only help to design the appropriate model structure; it also allows the user to generate a good data distribution that gathers as much information about the function $f(\cdot)$ as possible.

Finally, an important aspect of the relationship between the curse of dimensionality and the bias/variance tradeoff will be discussed. Each additional input makes the model more complex. Although each additional input may provide the model with more information about the process, this does not necessarily improve the model performance. Only if the benefit of the additional information exceeds the variance error caused by the additional model parameters will the overall effect of this input be positive. Thus, discarding inputs *can* improve model performance.

7.6.2 Hybrid Structures

Hybrid structures are composed of different (at least two) submodels that are of different types. Typically, one submodel is based on theoretical modeling by first principles (physical, chemical, biological, etc. laws) or is the currently used state-of-the-art model obtained by any combination of modeling and identification techniques. Often this prior model does not fully reflect all properties of the process. In many cases, the prior model may be linear, or its nonlinear structure may be inflexible. Then the prior model can be improved by combining it with a new, more general submodel such as a fuzzy system, neural network, etc. which may be generated from data.

The major advantage of a combination of a prior model and another data-driven model compared with the solely data-driven model is that the already available model quality is exploited and improved instead of starting from scratch and throwing away all knowledge. Typically, the incorporation of prior knowledge improves the extrapolation capabilities (without prior knowledge extrapolation is highly dangerous) and the robustness with respect to missing or low-quality data. Furthermore, industrial acceptance and confidence are usually much higher if an already existing model is the basis for the new model.

7.6.2.1 Parallel Model

The prior model can be combined with an additional data-driven model by different strategies. Figure 7.26 shows the parallel configuration. This is a reasonable approach if the prior model describes the whole process under consideration from its inputs to its output. By far the most common approach is to supplement the prior model by an *additive* data-driven model; see Fig. 7.26a. An alternative is a

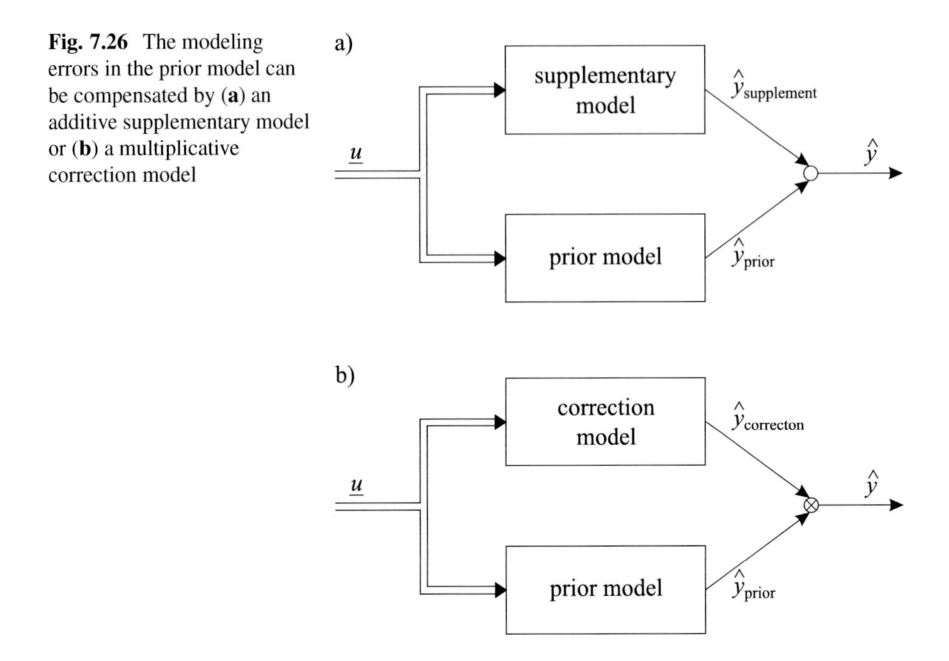

Fig. 7.26 The modeling errors in the prior model can be compensated by (**a**) an additive supplementary model or (**b**) a multiplicative correction model

multiplicative correction model as shown in Fig. 7.26b. Clearly, it depends on the process and model structure whether the additive or multiplicative approach is better suited.

When an additive supplementary model is used, it can be trained very simply as follows. The overall model output is

$$\hat{y} = \hat{y}_{\text{supplement}} + \hat{y}_{\text{prior}} , \tag{7.41}$$

which leads the model error

$$e = y - \hat{y} = y - (\hat{y}_{\text{supplement}} + \hat{y}_{\text{prior}}) = (y - \hat{y}_{\text{prior}}) - \hat{y}_{\text{supplement}} , \tag{7.42}$$

where y denotes the process output. Thus, the desired output for the supplementary model is

$$\hat{y}_{\text{supplement}}^{(\text{desired})} = y - \hat{y}_{\text{prior}} . \tag{7.43}$$

According to (7.43), the supplementary model is trained to compensate the error of the prior model $y - \hat{y}_{\text{prior}}$.

When a multiplicative correction model is used, it can be trained with the following procedure. The overall model output is

$$\hat{y} = \hat{y}_{\text{correction}} \cdot \hat{y}_{\text{prior}} \,, \tag{7.44}$$

which leads the model error

$$e = y - \hat{y} = y - \hat{y}_{\text{correction}} \cdot \hat{y}_{\text{prior}} = \hat{y}_{\text{prior}} \cdot \left(\frac{y}{\hat{y}_{\text{prior}}} - \hat{y}_{\text{correction}} \right) . \tag{7.45}$$

Thus, the desired output for the correction model is

$$\boxed{\hat{y}_{\text{correction}}^{(\text{desired})} = \frac{y}{\hat{y}_{\text{prior}}} \,.} \tag{7.46}$$

So the multiplicative model is trained to realize the correction factor $y/\hat{y}_{\text{prior}}$. Since the error in (7.45) is weighted with the term $\hat{y}_{\text{prior}}$, a suitable loss function should reflect this property. The most straightforward approach is a weighted least squares optimization with the weights

$$q_i = \hat{y}_{\text{prior}}^2(i) \,, \tag{7.47}$$

where q_i denotes the ith weight; see (2.2) in Sect. 2.3. Such a weighted least squares approach also reduces the possible problems that would occur when $\hat{y}_{\text{prior}}$ approaches zero. Nevertheless, owing to the division in (7.46), the multiplicative approach is much more sensitive to noise than the additive alternative. Thus, if there exists no indication that the parallel model should be multiplicative, an additive structure is generally preferable.

7.6.2.2 Series Model

Figure 7.27 shows a configuration in which the data-driven submodel is in series with the prior submodel. Such a series configuration can be applied if the prior model is incomplete. Then the data-driven submodel can describe those effects that are unmodeled in the prior model. A very simple example may be a prior model

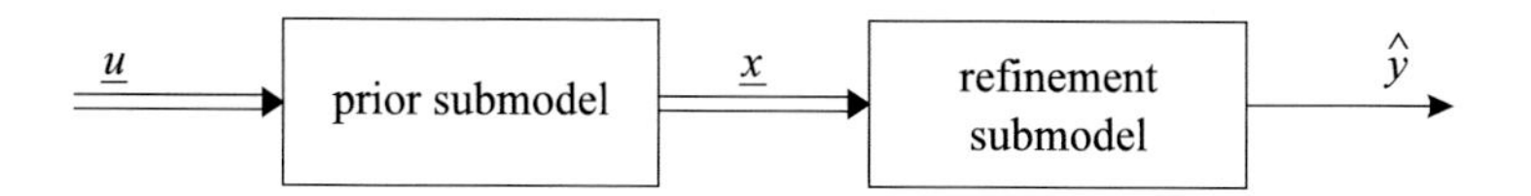

Fig. 7.27 A refinement model in series with the prior model can compensate for modeling errors

that computes the electrical power $P = U \cdot I$ from the input voltage U and current I, because it is known that the process behavior depends directly on the power rather than on voltage or current. By supplying the refinement model with this quantity, the difficulty of the modeling problem may be significantly reduced. On the one hand, the configuration in Fig. 7.27 covers applications where the prior model may be almost trivial (as in the example above) and can be seen as a kind of data preprocessing. On the other hand, Fig. 7.27 also represents applications where the prior model is just marginally refined, e.g., by a filter realizing unmodeled dynamics. Training of the refinement submodel is straightforward since its output is the overall model output.

7.6.2.3 Parameter Scheduling Model

Figure 7.28 shows a third category of combinations of prior and data-driven models. The data-driven model schedules some parameters within the prior model. As for the parallel configuration, the prior model cannot be partial; it must describe the process from the input to the output. In contrast to the parallel and serial configurations, the outputs of the data-driven model are not signals but the parameters of the prior model. These prior model parameters are denoted as $\underline{\rho}$ in Fig. 7.28 to distinguish them from the internal parameters of the experimental model $\underline{\theta}$. This approach is often applied in the chemical process industry. It is very successful if the structure of the prior model matches the true process structure well. Since only the parameters are influenced by the data-driven model, the overall model behavior can be well understood, and some model characteristics can be ensured. This advantage of the parameter scheduling approach turns into a drawback if the structure of the process is not well understood. Structural mismatches in the prior model cannot be compensated by the data-driven model at all. Another difficulty with the parameter scheduling approach is that the desired values for the outputs of the data-driven model (the prior model parameters) are not known during training. Thus a nonlinear optimization with gradient calculations is necessary in order to train the data-driven model.

The parallel, serial, and parameter scheduling configurations shown in Figs. 7.26, 7.27, and 7.28, respectively, represent general concepts that can be realized with any

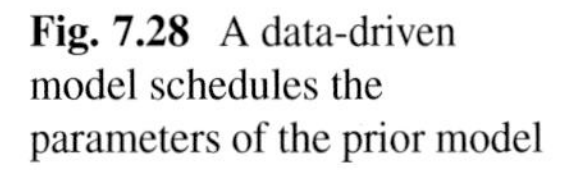
Fig. 7.28 A data-driven model schedules the parameters of the prior model

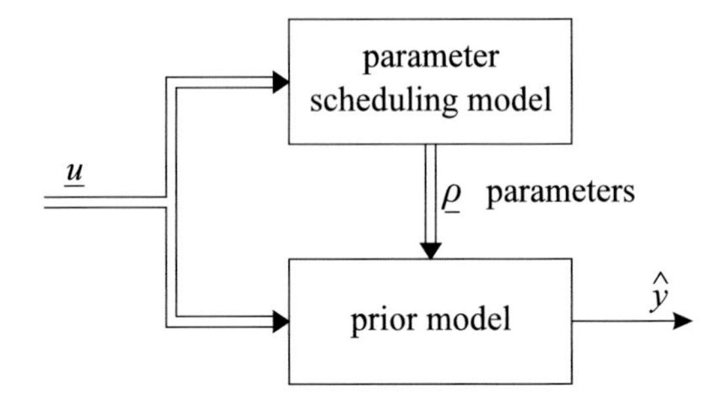

type of data-driven model. For special model architectures, it is possible to directly build in prior knowledge (or parts of it), which usually is more easily interpretable and more effective than the standard parallel and serial configurations.

7.6.3 Projection-Based Structures

Projection-based structures are the most radical way of overcoming the curse of dimensionality. The input space is projected onto some axes that represent the important information within the data. This will be illustrated in the following example.

Example 7.6.2 Projection-Based Approach
The following process is considered:

$$y = f(u_1, u_2) \,. \tag{7.48}$$

If both inputs u_1 and u_2 are highly correlated, it might be sufficiently accurate to construct a model based on a single input:

$$\hat{y} = \hat{f}(\xi) \qquad \text{with} \quad \xi = w_1 u_1 + w_2 u_2 \,, \tag{7.49}$$

where the "weights" w_1 and w_2 must be determined appropriately. The argument $\xi = w_1 u_1 + w_2 u_2$ can be interpreted as the projection (scalar product) of the input vector $\underline{u} = [u_1\ u_2]^T$ on the vector of weights $\underline{w} = [w_1\ w_2]^T$. If the inputs are not only correlated but redundant, the model in (7.49) can even be exact if the weights are chosen correctly. Thus, a high-dimensional problem can often be approximated by low-dimensional ones. This is extensively discussed in the context of multilayer perceptron networks; see Sect. 11.2. □

Projection-based approaches compute a set of relevant directions by

$$\xi_j = \sum_{i=1}^{p} w_{ji}\, u_i \,, \tag{7.50}$$

and process these directions further; see Fig. 7.29.

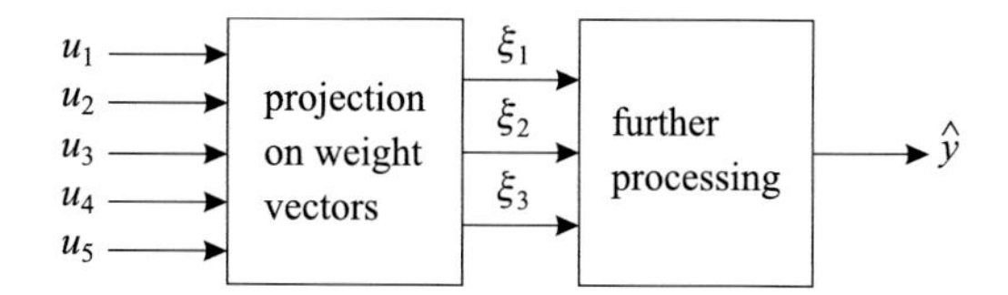

Fig. 7.29 A projection-based structure. The inputs are projections on some weight vectors. These projections are processed further

Several ways exist to determine these weights. Principle component analysis is the most common unsupervised learning technique; see Sect. 6.1. Projection pursuit learning and multilayer perceptron networks are widely applied supervised techniques; see Sect. 11.2. These methods have been very successfully applied to high-dimensional problems, and indeed there exist almost no generally working alternatives if the number of inputs is huge (say more than 50).

7.6.4 Additive Structures

Additive approaches split the high-dimensional problem into a sum of lower-dimensional problems; see Fig. 7.30. The justification of the additive structure can be drawn from a Taylor series expansion of the process:

$$\hat{y} = c_0 + c_1 u_1 + \ldots + c_p u_p + c_{11} u_1^2 + c_{12} u_1 u_2 + \ldots + c_{pp} u_p^2 + c_{111} u_1^3 + \ldots . \quad (7.51)$$

This is an additive structure. Thus, any (smooth) process can be approximated by an additive model structure. The important issue in practice is, of course, how fast the additive approximation converges to the true process behavior if the model complexity increases. This depends on the usually unknown structure of the process and the particular construction algorithm applied for building the additive model.

The major advantage of additive structures is that they can be quite easily constructed from data. Typically, simple submodels are chosen that are linear in the parameters, since then the overall model still is linear in all the submodel parameters. Considering the Taylor series expansion in (7.51), the most straight-forward way to generate an additive model is to select the relevant polynomial terms in (7.51) by a subset selection technique; see Sect. 3.4. One of the most prominent representatives of additive construction algorithms is ASMOD (additive spline modeling) as proposed by Kavli [316]. It can also be interpreted in terms of fuzzy models and is one standard approach to overcome the curse of dimensionality in fuzzy systems [61].

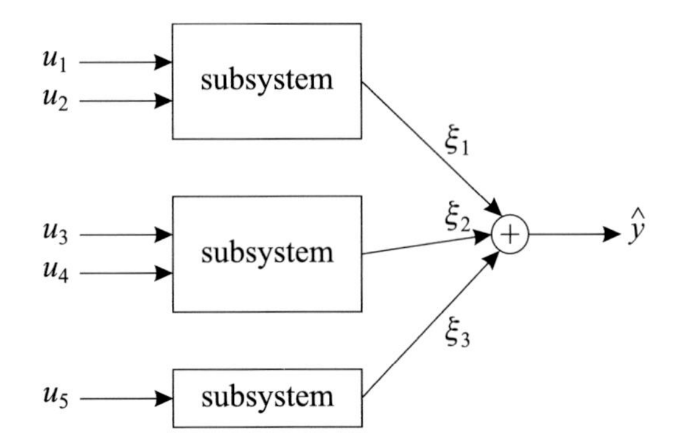

Fig. 7.30 An additive structure. The process $y = f(u_1, u_2, u_3, u_4, u_5)$ is modeled by the sum of lower-dimensional submodels: $\hat{y} = \xi_1 + \xi_2 + \xi_3$ with $\xi_1 = f_1(u_1, u_2)$, $\xi_2 = f_2(u_3, u_4)$, and $\xi_3 = f_3(u_5)$

The interpretation of the additively structured model is certainly easier than of a single black box structure. In particular, each submodel can be expected to be easy to understand because they are relatively simple. However, the additive structure can introduce unexpected effects since it is usually not possible for humans to understand the superimposition of submodels that have different input spaces. By analyzing each submodel, the overall effect of all submodels can hardly be grasped.

Instead of summing up the contributions of all submodels, they can be multiplied (or any other operator may be used). Such an approach can be reasonable if it matches the structure of the process (which assumes that structural process knowledge is available). Since this introduces strongly nonlinear behavior and consequently requires nonlinear optimization techniques, these approaches are rarely applied in practice.

7.6.5 Hierarchical Structures

Hierarchically built models as shown in Fig. 7.31 can describe the inner structure of the process. Typically, modeling with first principles leads to a set of coupled equations. Often, however, different subsystems (e.g., a chemical, mechanical, and electrical subsystem) are organized hierarchically. The subsystems' outputs correspond to states of the process. The interpretability of hierarchical models is excellent because each submodel is low dimensional and hierarchical structures are close to human reasoning. It is possible to build up complex models step by step and utilizing already available submodels. A hierarchical organization has proven useful in almost any context ranging from software engineering to user interfaces.

The big drawback of hierarchically structured models is that no satisfactory algorithms are available for constructing such models from data. First, the optimization of the hierarchical structure is a complex structure optimization task. Second, the parameters of the submodels influence the overall model output in a nonlinear way. Finally, it is usually not possible to guarantee that the submodel

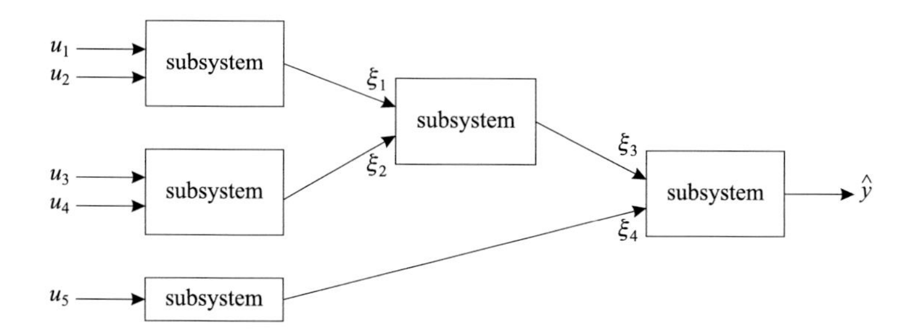

Fig. 7.31 A hierarchical structure. The process $y = f(u_1, u_2, u_3, u_4, u_5)$ is modeled by a hierarchy of lower-dimensional submodels $\hat{y} = f_5(\xi_3, \xi_4)$ with $\xi_3 = f_3(\xi_1, \xi_2)$ with $\xi_1 = f_1(u_1, u_2)$ and $\xi_2 = f_2(u_3, u_4)$ and $\xi_4 = f_4(u_5)$

outputs (ξ_i's in Fig. 7.31) correspond to a state in the real-world process (or anything else meaningful). The latter disadvantage is the most severe one because it destroys the model interpretation completely. Nevertheless, some algorithms exist that try to build hierarchical models. Typically, they rely on genetic algorithms or genetic programming in order to solve the complex structure optimization problem; see [353, 475] for an application to fuzzy systems. A promising approach based on the integration of various sources of prior process knowledge with genetic programming is proposed in [365].

7.6.6 Input Space Decomposition with Tree Structures

Input space decomposition concepts accept the high dimensionality of the problem as the projection-based methods do. Their concept is to decompose or partition the input space according to the complexity of the process. No low-dimensional submodels are generated as for additive or hierarchical structures. Rather, the complexity is decreased by splitting up the whole problem into smaller subproblems. Note that here "smaller" refers to the size of the regions in the input space and not to its dimensionality. Typically, the input space decomposition is implemented in a tree structure. Thus, most input space decomposition strategies are hierarchically organized. In contrast to the hierarchical structures discussed in the previous subsection, the characteristic feature here is the type of the input space decomposition. Whether the algorithm and the model are implemented as a hierarchical tree structure or otherwise is not decisive. Figure 7.32 compares a lattice partitioning with two commonly applied tree partitioning approaches for input space decomposition.

The key idea is to decompose the input space in such a manner that each local region can be described with a simple submodel. Often these simple submodels are chosen to be constant or linear or at least linearly parameterized. In regimes where the process is strongly nonlinear, many different regions must be generated by the

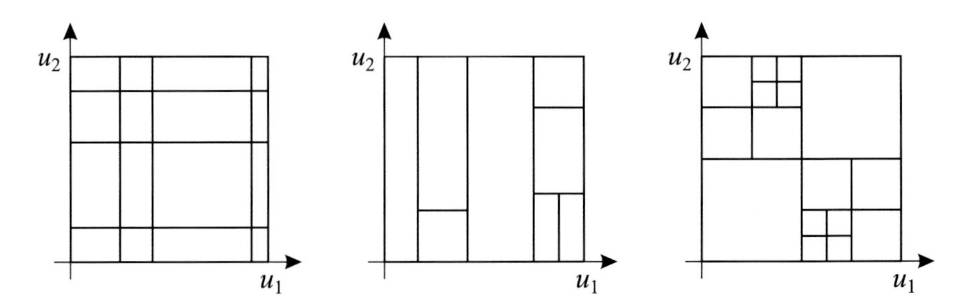

Fig. 7.32 A lattice structure (left) compared with a k-d tree (center) and a quad tree (right). Compared with the lattice, tree structures allow one to partition the input space with different resolutions (granularity) in different regions

decomposition algorithm in order to make the model sufficiently accurate. However, in regimes where the process behavior is very smooth, only a few partitions are required. By this strategy, the curse of dimensionality can be reduced.

The k-d tree and quad tree structures illustrated in Fig. 7.32 partition the input space orthogonally with respect to the input axes. Most decomposition algorithms do so because this reduces the complexity of the structure optimization problem significantly. Note, however, that axis-orthogonal splits tend to become less effective as the input space dimensionality increases. Alternative, more flexible decomposition algorithms are usually based on axis-oblique partitioning or clustering strategies (Sect. 6.2). Refer to Chaps. 13 and 14 for a typical axis-orthogonal strategy and to Sect. 14.7 for a comparison with an axis-oblique strategy.

Basically, two different strategies exist to combine all submodels together. Either, depending on the incoming data sample, the output of the corresponding submodel is used or the neighboring submodels are also taken into account. The first method switches between the submodels, which results in noncontinuous behavior, and therefore it is reasonable only for classification problems or in other cases where smoothness is not required. The second method interpolates (weights) between submodels, which can lead to smooth behavior and consequently is usually preferred for function approximation.

As in additive structures, linear parameterized submodels are usually chosen for decomposition algorithms, because the overall model preserves this advantageous property. In addition to the possible exploitation of linear relationships, the locality properties can be utilized. Since each submodel represents a local region of the input space, it may be optimized with local data almost independently of all other submodels. Such an approach is called *local optimization* and can reduce the complexity of the parameter optimization problem significantly.

Commonly applied algorithms for input space decomposition based on tree structures are classification and regression trees (CART) [74] and ID3 and C4.5 from Quinlan [472], which all switch between the submodels. Basis function trees [495, 496], multivariate adaptive regression splines (MARS) [172], and local linear model trees (LOLIMOT) [432] implement interpolation between the submodels; see also the Chaps. 13, 14, and 22.

7.7 Summary

The characteristics of the process, the amount and quality of the available data, the prior knowledge, and the type of model imply an optimal model complexity. Each additional parameter on the one hand makes the model more flexible; on the other hand, it makes it harder to accurately estimate the optimal parameter values. The model error can be split into two parts: the bias and the variance error. The bias error is due to insufficient model flexibility and decreases with a growing number of parameters. The variance error is due to inaccurately estimated parameters and increases with a growing number of parameters. The best bias/variance tradeoff marks the point of optimal model complexity. At this point, the decrease of the bias error is balanced with the increase in variance error.

In order to discover the best bias/variance tradeoff, the expected model quality on fresh data must be determined. This cannot be done by examining the model error on the training data set. Otherwise, overfitting cannot be detected, i.e., an overly complex model, which adapts well to the training data (and the noise contained in it) but generalizes much worse. Rather a separate test set must be used, or some complexity penalty must be introduced.

One way to realize a good bias/variance tradeoff is explicit structure optimization. The model complexity is varied by varying the number of parameters or whole substructures in the model such as neurons, rules, etc. Explicit structure optimization is usually computationally expensive but yields small models where each parameter is significant. This offers advantages in terms of model evaluation times, memory requirements, and model interpretation.

An alternative is implicit structure optimization by regularization. Instead of varying the model complexity by the number of nominal parameters, this is done by choosing a relatively complex model in which not all degrees of freedom are really used. Although the model may look complex and may possess a large number of parameters, the effective complexity and the effective number of parameters can be significantly smaller than the nominal ones. This is achieved by regularization techniques such as the introduction of nonsmoothness penalties or constraints, training with early stopping, and staggered or local optimization.

The required model complexity is strongly influenced by the dimension of the input space of the problem. Often the complexity of the problem or the model increases exponentially with the number of inputs. This is called the curse of dimensionality. One has to distinguish between the complexity of the problem and that of the model. Luckily, restrictions prevent many real-word problems from becoming too complex. However, some model architectures scale up exponentially with the number of inputs. All lattice-based approaches belong to this category. So it is no coincidence that many of the existing algorithms for overcoming or reducing the curse of dimensionality deal with fuzzy systems. The construction of additive, hierarchical, or tree structures are the most common approaches. The utilization of prior knowledge, e.g., in the form of hybrid models, almost always reduces the model complexity and consequently the curse of dimensionality. Finally, projection-based structures are the most radical and consequent approach dealing with high-dimensional spaces. Their excellent properties for handling large input spaces come at the price of nonlinear parameters and black box characteristics.

7.8 Problems

1. Explain the bias/variance decomposition of the model error. Find examples where the bias error is dominant. Find examples where the variance error is dominant.

2. How does the variance error typically depend on the flexibility of the model?
3. Why is it necessary to distinguish between data used for training and validation? What happens if training and validation is carried out with the same data set?
4. How can it be proceeded if only one data set is available?
5. What is the difference between validation and test data?
6. Show how an optimization problem with two objectives can be reformulated into a single-objective problem with an additional parameter. Explain what the set of pareto-optimal solutions is and how this set can be searched through.
7. What is the difference between explicit and implicit structure optimization? Give some examples of explicit structure optimization.
8. Explain the expression “effective number of parameters” and give some examples.
9. What is the effect of all regularization techniques? Why and in which cases is this effect desirable?
10. Explain the “curse of dimensionality.” Give examples for models that suffer more and less severely from it.

Chapter 8
Summary of Part I

Abstract This chapter gives a summary of Part A: Optimization.

Keywords Summary optimization

The goal of this part on optimization techniques was to give the reader a "feeling" for the most important optimization algorithms. No implementation details have been discussed. Rather the focus was on the basic properties of the different categories of techniques and the advantages and drawbacks of the particular algorithms. These insights are of fundamental importance when trying to choose an appropriate algorithm from a provided toolbox. To give a brief review, the optimization techniques can be divided into the following categories:

- *Linear optimization*: These are the most thoroughly understood and best-analyzed techniques. Linear optimization offers a number of highly desirable features such as an analytic one-shot solution, a unique optimum, and a recursive formulation that allows an online application. Furthermore, the accuracy of the estimated parameters and the model output can be assessed by the covariance matrix and errorbars. Powerful and very efficient structure optimization techniques are available, such as the orthogonal least squares (OLS) forward selection method.

  Linear optimization techniques are very fast, simple, and easy to apply; many numerically robust implementations are available in toolboxes. Consequently, a problem should be addressed with linear parameterized approaches first, and only if the obtained solutions are not satisfactory may one turn to more complex approaches requiring nonlinear optimization methods.
- *Nonlinear local optimization*: Nonlinear optimization problems often have multiple local optima. The nonlinear local techniques start from an initial value in the search space and find a local optimum by evaluation of loss function values (direct search methods) and possibly first and second derivatives of the loss function. The most popular general-purpose algorithms are the conjugate gradient (>100 parameters) and the quasi-Newton (<100 parameters) methods.

O. Nelles, *Nonlinear System Identification*,
https://doi.org/10.1007/978-3-030-47439-3_8

The simple steepest descent algorithm usually converges very slowly, and consequently cannot be recommended. However, for sample adaptation where no line search can be performed, this algorithm is a reasonable choice.

In many problems, the loss function is a sum of squares. Then nonlinear least squares methods such as the Gauss-Newton or Levenberg-Marquardt algorithm can be applied. They effectively exploit the special form of the loss function and approximate second derivatives from gradient information only. Generally, these algorithms are recommended if the loss function is of the required type.

Nonlinear local optimization techniques are also thoroughly studied and well analyzed. Many toolboxes make robust and efficient implementations of these algorithms easily available. The main restriction is that these techniques perform a local search. No attempt is made to escape from local optima in order to search for the global one. For many problems, local search is sufficient, especially if good initial parameter values are available. One may start from different initial parameters to get a "feeling" for the quality and number of the different local optima. However, if it is not possible to achieve a satisfactory solution, a global method can be applied.

- *Nonlinear global optimization*: If local search methods do not yield a satisfactory solution to a nonlinear optimization task, one reason can be convergence to a poor local optimum. The simplest possible remedy for this problem is to start a local search several times from different initial values (multi-start technique). The difficulty with this approach, however, is how to choose the initial values reasonably.

  Nonlinear global optimization techniques try to find the global optimum or (more realistically speaking) at least a good local optimum. Typically, random components are introduced into global search algorithms in order to allow them to escape from local optima. The incorporation of randomness makes the search robust against particular properties of the optimization problems (i.e., the algorithm is very universal and not tailored to a special class of problems) at the price of low performance. The most prominent global search techniques are simulated annealing and evolutionary algorithms, of which the latter can be further classified into the categories evolutionary strategies, genetic algorithms, and genetic programming.

  A fundamental dilemma that has to be solved for any nonlinear global optimization technique is the tradeoff between the degree of global character and convergence speed or in other words between diversification (exploration of new regions) and intensification (convergence in local regions).
- *Unsupervised learning*: The goal of unsupervised learning approaches is to analyze or compress the information contained in a data set. Typically, only the distribution of the input data is considered. Unsupervised learning techniques are often utilized for preprocessing of the data and for extracting features that can be used as inputs for a subsequent supervised learning technique. The most common unsupervised learning approaches are the principal component analysis (PCA) and clustering methods. PCA can be used for transformation of the data onto new axes and for dimensionality reduction. Clustering techniques allow one

to search for similar groups or objects in the data set. For clustering, classical, fuzzy, and neural network-based approaches can be used.

- *Complexity optimization*: Finding the optimal model complexity requires optimization of the model structure, and not just of its parameters. For each application, an optimal model complexity exists that is characterized by a tradeoff between the bias and the variance error. The overall model error can be decomposed into a bias and a variance part. The bias part describes the systematical error due to the restricted model flexibility. The variance part describes the stochastic error due to inaccurately estimated parameters. The bias error decreases and the variance error increases for growing model complexity. Thus, both error parts are in conflict and a compromise must be found. This is called the bias/variance dilemma.

  To measure the performance of a model, it should not be evaluated on the same data used for parameter estimation (training data) because it does not reveal the variance part of the error. The error on the training data can be arbitrarily decreased by making the model more and more complex. In the extreme case, when a model possesses as many parameters as the number of available training data samples, the training error would be zero (interpolation). Thus, an objective evaluation of the model performance must be measured on a test data set that was not used for training. Alternative strategies to the use of a separate test data set are the application of complexity penalties in information criteria or statistical tests.

  The optimal model complexity can be found by either explicit or implicit structure optimization. Explicit strategies approach the problem directly as a structure optimization task, which may be solved by a nonlinear global search technique. Implicit structure optimization or regularization does not influence the model structure. Rather, not all nominal parameters of the model are really utilized. Only a smaller number, the so-called effective parameters, are really estimated from data. All others are kept at their initial values or are constrained in some way. Regularization techniques reduce the flexibility of a model and thus reduce the variance error at the price of a higher bias error. So an adjustment of the strength of the regularization effect allows one to control the bias/variance tradeoff.

  One fundamental issue in model complexity optimization is the curse of dimensionality. It describes the fact that the amount of data and the required model complexity grow strongly with the input dimensionality of the problem. For all lattice-based approaches, the complexity increases exponentially, which makes high-dimensional models infeasible in practice. Different strategies for reduction of the model's sensitivity to the input space dimension are available, such as projection-based approaches, hybrid, additive, hierarchical structures, and input space decomposition by trees.

For a more extensive study of linear optimization techniques [125] is recommended, while a well-written treatment about the required linear algebra can be found in [193]. Excellent books for nonlinear local optimization are [481, 504, 577].

References for the other techniques can be found in the corresponding chapters. A lot of literature exists for special methods. However, a lack of overview literature can be observed addressing non-experts in optimization. This was one major motivation for this part. Another reason is the strict distinction between the type of model and the optimization method applied to determine or adapt its parameters or structure. In all other chapters of this book, references are made to special classes of optimization techniques that may be applied to a particular problem. It is then always clear from the context which algorithms, in principle, can be utilized. This part contains only a general discussion of the optimization techniques. The application-specific advantages and drawbacks are discussed in the corresponding chapters.

Part II
Static Models

Chapter 9
Introduction to Static Models

Abstract This chapter introduces the basis function formulation, which most subsequent chapters relate to. It systematizes most approaches in one general setting and discusses some very general ideas that are independent of the specific model architecture chosen. These are the notions of linear vs. nonlinear parameters, of local vs. global behavior, and of multiple inputs and/or multiple outputs. Furthermore, the basis function formulation is extended to conveniently include local model networks, which represent a core topic throughout this book. Finally, some artificial test processes are defined, and the different evaluation criteria are discussed.

Keywords SISO · MISO · MIMO · Basis function · Linear parameters · Nonlinear parameters

This part deals with the most frequently used static model architectures. It is organized as follows. In Chap. 10 common classical model architectures are reviewed. These are linear, polynomial, and look-up table models. Owing to their large industrial significance look-up tables, in particular, are discussed thoroughly. The most common neural networks are analyzed in Chap. 11. Chapter 12 treats three different categories of fuzzy and neuro-fuzzy approaches. Chapters 13 and 14 introduce and extend the local linear neuro-fuzzy model architectures and in particular the local linear model tree (LOLIMOT) training algorithm. Finally, the main results of this part are summarized in Chap. 17.

This section gives an introduction to some foundations of nonlinear static models. Section 9.1 analyzes the handling of multivariable systems. A basis function formulation of static models is given in Sects. 9.2 and 9.3. A simple static nonlinear test processes is introduced in Sect. 9.4. It is used to illustrate the behavior of the different static model architectures discussed in Part II. For comparison, a few criteria for evaluation of the properties of the different model architectures are given in Sect. 9.5.

O. Nelles, *Nonlinear System Identification*,
https://doi.org/10.1007/978-3-030-47439-3_9

9.1 Multivariable Systems

Nonlinear static models perform a mapping from p inputs u_i gathered in a p-dimensional input vector $\underline{u} = [u_1\ u_2\ \cdots\ u_p]^T$ to r outputs y_j gathered in an r-dimensional output vector $\underline{y} = [y_1\ y_2\ \cdots\ y_r]^T$. Such a general model is called a *multiple-input multiple-output (MIMO)* model; see Fig. 9.1. Typically, such a MIMO model is decomposed into r different *multiple input single output (MISO)* models (see Fig. 9.2) for the following reasons:

- Each MISO model is simpler than an overall MIMO model and thus easier to understand, to validate, and to apply in practice.
- The required accuracy of each of the r model outputs can be adjusted separately. There is no need for a single loss function that weights the r output errors and thus performs an accuracy tradeoff between the different model outputs.
- Different model architectures, structures, and optimization techniques can be applied to each MISO subproblem, which makes the modeling and identification approaches more appropriate, flexible, and powerful.

In opposition to these advantages, a MIMO model usually offers faster evaluation speed, i.e., the time required to calculate to model outputs for given model inputs. Even though the MIMO model can be expected to be significantly more complex than each of the MISO models, its complexity is usually less than r times higher. Several parts of the structure and parameters of the MIMO model are typically useful for modeling of more than one output. These common structures and parameters cannot be exploited by the separate MISO models. Nevertheless, the advantages of a MIMO model decomposition according to Fig. 9.2 are significant in most real-world situations. Thus in all that follows only MISO models and *single input single output (SISO)* models are addressed. So a static MISO model can

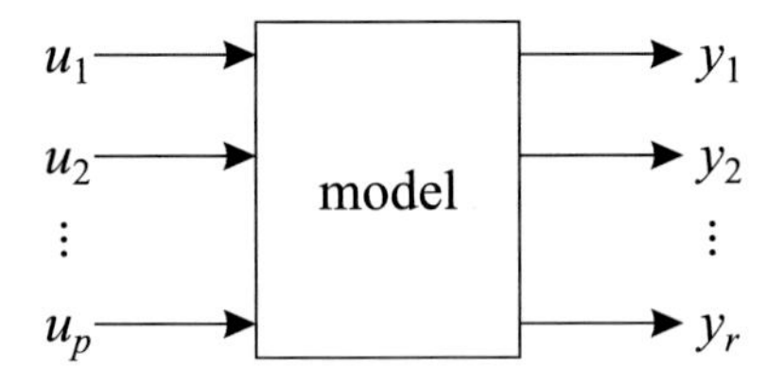

Fig. 9.1 A general MIMO model

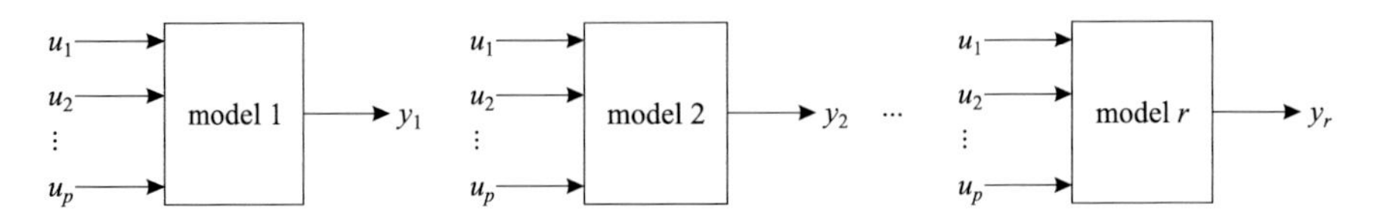

Fig. 9.2 A MIMO model can be decomposed into MISO models

be described by the following mapping from the p-dimensional input to the one-dimensional output:

$$\hat{y} = f(\underline{u}) . \tag{9.1}$$

9.2 Basis Function Formulation

From all possible realizations of this function $f(\cdot)$, almost all alternatives of practical interest can be written in the following basis function formulation:

$$\hat{y} = \sum_{i=1}^{M} \theta_i^{(\mathrm{l})} \Phi_i \left(\underline{u}, \underline{\theta}_i^{(\mathrm{nl})} \right) . \tag{9.2}$$

The output y is modeled as a weighted sum of M basis functions $\Phi_i(\cdot)$. The basis functions are weighted with the *linear* parameters $\theta_i^{(\mathrm{l})}$, and they depend on the inputs $\underline{u}$ and a set of *nonlinear* parameters gathered in $\underline{\theta}_i^{(\mathrm{nl})}$. In order to realize a nonlinear mapping, the basis functions have to be nonlinear. Thus, the parameters $\underline{\theta}_i^{(\mathrm{nl})}$ on which the basis functions depend are necessarily nonlinear.

Often models incorporate an offset parameter (sometimes called "bias") that adjusts the operating point. Such an offset can be included in the basis function formulation by the introduction of a "dummy" basis function $\Phi_0(\cdot)$, which is always equal to 1. Its corresponding linear parameter $\theta_0^{(\mathrm{l})}$ implements the offset

$$\hat{y} = \sum_{i=0}^{M} \theta_i^{(\mathrm{l})} \Phi_i \left(\underline{u}, \underline{\theta}_i^{(\mathrm{nl})} \right) \qquad \text{with} \quad \Phi_0(\cdot) = 1 . \tag{9.3}$$

The basis function formulations (9.2) and (9.3) can be illustrated as the network shown in Fig. 9.3. Generally, the basis functions $\Phi_i(\cdot)$ can be of different type for each node. If all basis functions are of the same type and differ only in their parameters, the network is called an *artificial neural network (ANN)* or, for short, a *neural network (NN)* (since no biological issues are addressed in this book). Then the nodes of the network in Fig. 9.3 are called *neurons*. This class of model architectures is discussed in Chap. 11.

9.2.1 Global and Local Basis Functions

In general, the basis functions $\Phi_i(\cdot)$ can take any form. In many cases, however, especially for fuzzy systems and neural networks, they are chosen as elementary functions or are constructed by elementary functions; see Sect. 11.1. Common

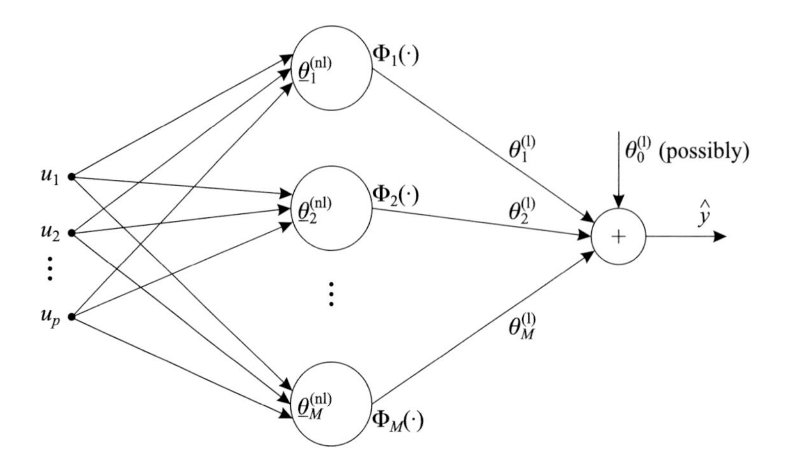

Fig. 9.3 A network of basis functions. Each node represents one basis function that depends on its nonlinear parameter vector $\underline{\theta}_i^{(\mathrm{nl})}$. Depending on the specific model, the offset $\theta_0^{(\mathrm{l})}$ may exist or not

one-dimensional basis functions are depicted in Figs. 9.4 and 9.5. They can be distinguished into the following:

- *Global* basis functions significantly contribute to the model output globally, i.e., in an infinitely sized region of the input space; see Fig. 9.4. Global behavior exists if a change in the associated linear parameter $\theta_i^{(\mathrm{l})}$ of the basis function formulation significantly influences the model output over a wide operating regime. *Strictly* global basis functions additionally possess a global derivative, while non-strictly global basis functions have a local derivative. Models with non-strictly global basis functions can construct true nonlinear behavior only in those regions where the derivative of the basis functions varies significantly; outside these regions, the basis functions are approximately constant. So, in fact, models with (non-strictly) global basis functions operate virtually locally, although changes in the linear parameters result in global effects.
- *Local* basis functions significantly contribute to the model output locally, i.e., in a finitely sized region of the input space; see Fig. 9.5. Local behavior exists if a change in the associated linear parameter $\theta_i^{(\mathrm{l})}$ of the basis function formulation significantly influences the model output only in a small region of the input space. *Strictly* local basis functions are exactly equal to zero outside their activation region (they are said to have *compact support*), while (non-strictly) local basis functions possess an insignificant contribution.

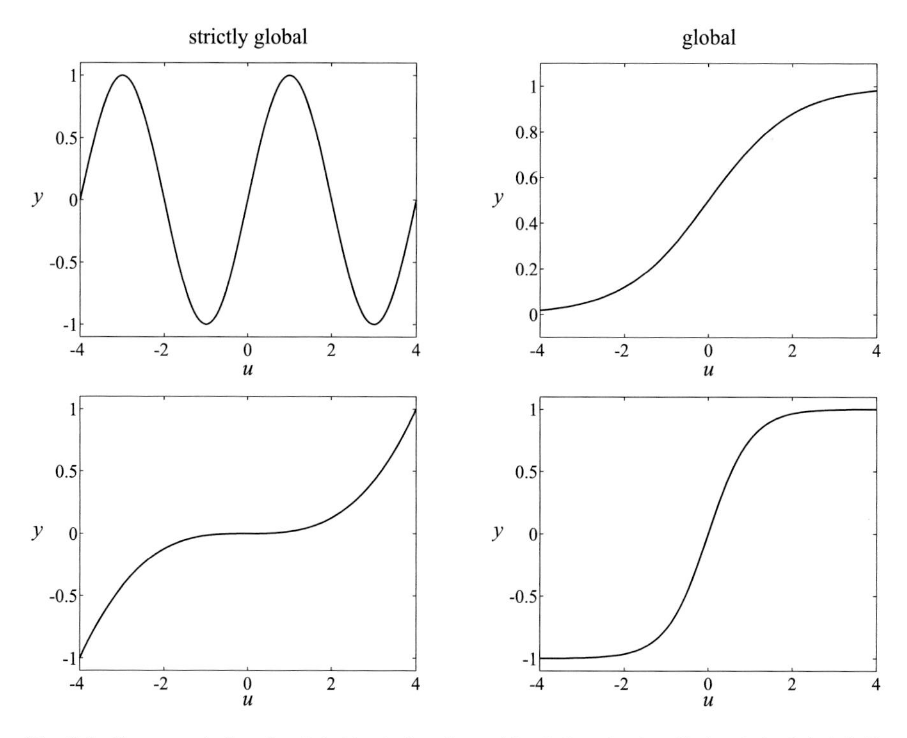

Fig. 9.4 Common choices for global basis functions. A basis function is called *strictly* global (left) if its derivative is global too, e.g., a sine function (top) or a polynomial (bottom). Otherwise it is called (non-strictly) global, e.g., a sigmoid function (top) or a tanh-function (bottom)

9.2.2 Linear and Nonlinear Parameters

Independent of the specific model architecture, the basis function formulation allows one to draw some conclusions about the fundamental properties of this wide model class.

- The model is linear in its weighting parameters $\theta_i^{(l)}$. If the basis functions are fully specified, i.e., the nonlinear parameters are determined somehow, the linear parameters can be estimated by an efficient linear optimization technique, e.g., least squares; see Chap. 3. The regression matrix $\underline{X}$ and parameter vector $\underline{\theta}^{(l)}$ are

$$\underline{X} = \begin{bmatrix} 1 & \Phi_1(\underline{u}(1)) & \Phi_2(\underline{u}(1)) & \cdots & \Phi_M(\underline{u}(1)) \\ 1 & \Phi_1(\underline{u}(2)) & \Phi_2(\underline{u}(2)) & \cdots & \Phi_M(\underline{u}(2)) \\ \vdots & \vdots & \vdots & & \vdots \\ 1 & \Phi_1(\underline{u}(N)) & \Phi_2(\underline{u}(N)) & \cdots & \Phi_M(\underline{u}(N)) \end{bmatrix} \qquad \underline{\theta}^{(l)} = \begin{bmatrix} \theta_0^{(l)} \\ \theta_1^{(l)} \\ \theta_2^{(l)} \\ \vdots \\ \theta_M^{(l)} \end{bmatrix}, \qquad (9.4)$$

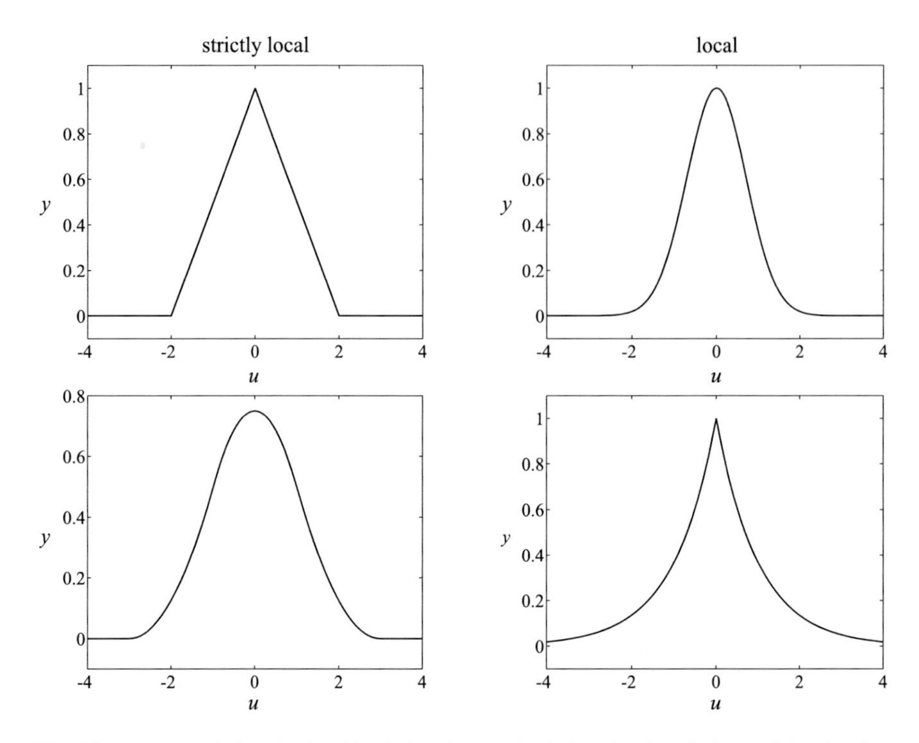

Fig. 9.5 Common choices for local basis functions. A basis function is called *strictly* local or has *compact support* (left) if it is exactly equal to zero outside its activity region, e.g., a triangular function (top) or third order B-spline (bottom). Otherwise it is called (non-strictly) local, e.g., a Gaussian (top) or a double exponential function (bottom)

where N is the number of data samples utilized for training, and the dependency of the basis functions on the nonlinear parameters is omitted for briefness. The first column in $\underline{X}$ and the first entry in $\underline{\theta}^{(l)}$ is optional for those approaches that implement an explicit offset value. Furthermore, fast linear structure selection techniques, e.g., the orthogonal least squares method, can be utilized to optimize the model structure.

- The optimization of the nonlinear parameters $\underline{\theta}_i^{(nl)}$ is much more difficult since it requires nonlinear local or global optimization schemes; see Chaps. 4 and 5. Therefore, it can be reasonable not to optimize these nonlinear parameters by supervised learning but to determine them differently. So the following approaches for the determination of the nonlinear parameters are common:

  - nonlinear optimization techniques,
  - unsupervised learning techniques, e.g., clustering,
  - exploitation of prior knowledge.

  The nonlinear parameters influence the basis functions. Typically, they specify the positions of the basis functions in the input space spanned by $\underline{u}$, and possibly some nonlinear parameters determine the smoothness or the widths of the basis functions. Thus, the nonlinear parameters allow an adjustment of the basis functions' positions and shapes in order to make the model more flexible. If nonlinear optimization is to be applied the gradient of the model output with respect to the parameters is of fundamental importance. This can be seen by considering as an example the sum of squared errors loss function:

$$J = \frac{1}{2} \sum_{i=1}^{N} e(i)^2 = \frac{1}{2} \sum_{i=1}^{N} (y(i) - \hat{y}(i))^2 \tag{9.5}$$

where N denotes the number of measurements, e is the model error, and y and $\hat{y}$ are the process and model output, respectively. For the application of any gradient-based optimization technique the gradient of the loss function with respect to each parameter θ is required:

$$\frac{\partial J}{\partial \theta} = \sum_{i=1}^{N} e(i) \frac{\partial e(i)}{\partial \theta} = - \sum_{i=1}^{N} e(i) \frac{\partial \hat{y}(i)}{\partial \theta} \,. \tag{9.6}$$

Thus, the gradient of the model output with respect to the parameters $\partial \hat{y}(i)/\partial \theta$ is required. Consequently, in the following, these derivatives are given for those model structures whose nonlinear parameters are usually optimized.

The advantages and drawbacks of different model architectures are often strongly related to the suitable methods for parameter optimization or determination. An especially interesting issue is whether it is better to (i) use models without or with fixed nonlinear parameters and to optimize only linear parameters, or (ii) to optimize models with nonlinear parameters. In the author's experience low-dimensional problems, say $p \leq 4$, can usually be solved more efficiently by the first alternative, while high-dimensional problems generally require the explicit optimization of the basis function's positions and widths in order to cope appropriately with sparsely covered input spaces and correlated inputs.

9.3 Extended Basis Function Formulation

The basis function formulation in (9.2) can be extended to a more flexible structure by replacing each linear parameter by a (typically linear parameterized) function $L_i(\cdot)$:

$$\hat{y} = \sum_{i=1}^{M} L_i \left(\underline{u}, \underline{\theta}_i^{(\mathrm{l})} \right) \Phi_i \left(\underline{u}, \underline{\theta}_i^{(\mathrm{nl})} \right) \,. \tag{9.7}$$

As long as $L_i(\cdot)$ is linear parameterized, the parameters $\underline{\theta}_i^{(\mathrm{l})}$ can be estimated with linear optimization techniques if the basis functions $\Phi_i(\cdot)$ are known. This extended basis function formulation is the foundation of models discussed in Sect. 12.2.3 and Chap. 13. In the simplest case, $L_i = \theta_i^{(\mathrm{l})}$ and the original basis function formulation is recovered. Another common alternative is to choose $L_i(\cdot)$ as a linear function of $\underline{u}$. This leads to

$$\hat{y} = \sum_{i=1}^{M} \left(w_{i0} + w_{i1}u_1 + w_{i2}u_2 + \ldots + w_{ip}u_p\right) \Phi_i\left(\underline{u}, \underline{\theta}_i^{(\mathrm{nl})}\right) , \tag{9.8}$$

where

$$\underline{\theta}_i^{(\mathrm{l})} = [w_{i0}\ w_{i1}\ w_{i2}\ \cdots\ w_{ip}]^T . \tag{9.9}$$

So the basis functions are not weighted with constants but with linear submodels. In principle, the submodels $L_i(\cdot)$ can be chosen arbitrarily complex. This issue is discussed in detail in Chap. 13.

Any extended basis function formulation with linear parameterized $L_i(\cdot)$ can be rewritten in the standard basis function form. For example, (9.8) can be written as

$$\hat{y} = \sum_{i=1}^{M\cdot(p+1)} \widetilde{\theta}_i^{(\mathrm{l})} \widetilde{\Phi}_i\left(\underline{u}, \underline{\theta}_i^{(\mathrm{nl})}\right) , \tag{9.10}$$

where

$$\widetilde{\Phi}_i(\cdot) = \Phi_i(\cdot) \qquad \text{for } i = 1, \ldots, p \tag{9.11a}$$

$$\widetilde{\Phi}_i(\cdot) = u_1\Phi_i(\cdot) \qquad \text{for } i = p+1, \ldots, 2p \tag{9.11b}$$

$$\widetilde{\Phi}_i(\cdot) = u_2\Phi_i(\cdot) \qquad \text{for } i = 2p+1, \ldots, 3p \tag{9.11c}$$

$$\vdots$$

$$\widetilde{\Phi}_i(\cdot) = u_p\Phi_i(\cdot) \qquad \text{for } i = (M-1)\cdot p+1, \ldots, M\cdot p . \tag{9.11d}$$

With these newly defined basis functions $\widetilde{\Phi}_i(\cdot)$, the standard formulation can be recovered. However, an interpretation of these model architectures is based on the extended formulation in (9.7), and thus for easier understanding the transformations in (9.11a–9.11d) are generally not carried out.

9.4 Static Test Process

In order to illustrate the functioning of the different nonlinear model architectures introduced in this Part II, the following simple static SISO process will be utilized:

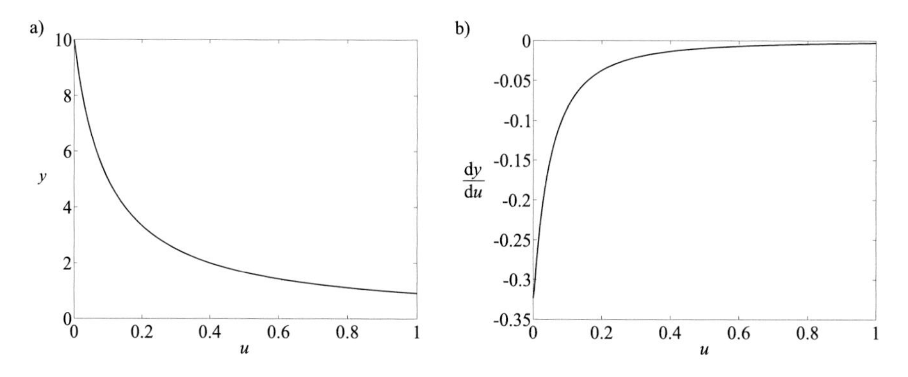

Fig. 9.6 (**a**) Static test process and (**b**) its derivative

$$y = \frac{1}{0.1 + u} \,. \tag{9.12}$$

This function is depicted in Fig. 9.6a. As can be clearly observed from its derivative in Fig. 9.6b, its curvature increases strongly (in absolute value) for $u \to 0$. It can be expected that most model architectures will require more parameters to describe the region around $u \approx 0$ than the region around $u \approx 1$. For an approximation of this function, 100 equally distributed data samples in the interval [0, 1] are generated. If not explicitly stated differently the data is not disturbed by noise.

Note that this simple example is not sufficient to assess and compare the properties of the different model architectures. Its sole purpose is to gain some insights into *how* these models approximate a function.

9.5 Evaluation Criteria

The introduced models will be evaluated according to the following properties:

- *Interpolation behavior*: What is the character of the model output between training data samples?
- *Extrapolation behavior*: What is the character of the model output outside the region where the training data lies?
- *Locality*: Are the basis functions global, strictly global, local, or strictly local?
- *Accuracy*: How accurate is the model with a given number of parameters?
- *Smoothness*: How smooth is the model output?
- *Sensitivity to noise*: Noise causes a variance error, i.e., the model parameters cannot be estimated to their (theoretically) optimal values. How does noise affect the model behavior?

- *Parameter optimization*: How can the linear and nonlinear model parameters be estimated?
- *Structure optimization*: How can the structure and complexity of the model be optimized?
- *Online adaptation*: How can the model be adapted online, and how reliable is on-line adaptation?
- *Training speed*: How quickly can the model parameters and possibly the model structure be trained from data?
- *Evaluation speed*: How quickly can the model be evaluated, i.e., what is the computational demand for an evaluation of the model for a given input?
- *Curse of dimensionality*: How does the model scale up to higher input space dimensions?
- *Interpretation*: Can the model parameters and possibly the model structure be interpreted in a way that is related to the properties of the process?
- *Incorporation of constraints*: How easily can constraints be incorporated into the model?
- *Usage*: How widespread is the model architecture?

At the end of each chapter, a table summarizes the advantages and drawbacks of each model architecture. Note that this simplified summary of the models' properties cannot reflect all details, and necessarily is quite crude. Furthermore, many properties depend on the combination of model and training strategy rather than solely on the model itself. Nevertheless, these tables may be helpful in roughly assessing the strengths and weaknesses of the different model architectures. The character of all major model architectures is illustrated with some simple static examples. Clearly, not all features of the models can be demonstrated by the means of some simple examples. Their purpose is just to give the user some "feeling" for the behavior of the models.

Chapter 10
Linear, Polynomial, and Look-Up Table Models

Abstract This chapter covers the most traditional, basic, and most widely applied model architectures for static modeling: (i) linear/affine models, (ii) polynomial models, and (iii) look-up tables. Although all of them are not very well suited to deal with general complex nonlinear problems, it is very important to understand how they work, what their characteristics are, and when and why they typically fail. In particular, in the case of look-up tables, which are the standard way to represent nonlinearities in industry, few engineers seem to understand that they can be seen as parameterized models and how their parameters can be learned from data. This topic is extensively analyzed. This allows to refrain from grid-based measurements and nevertheless establish a grid-based look-up table. This idea alone allows boosting the dimensionality (number of input variables) without resorting to sophisticated and resource-consuming modern machine learning approaches.

Keywords Linear model · Polynomial model · Optimization · Regularization · Look-up table · Constraints

This chapter analyzes some classical or traditional model architectures for nonlinear static processes. These model architectures are widely used in theory and practice. The simplest approach pursued in Sect. 10.1 is to approximate the nonlinear process behavior with a linear model. In the subsequent section, the polynomial approximator is discussed. Finally, in Sect. 10.3 the standard grid-based look-up table with linear interpolation is analyzed.

10.1 Linear Models

A linear model may be able to approximate a nonlinear process with reasonable accuracy if its nonlinear characteristic is weak. A linear model is simple and possesses a small number of parameters. Especially if only very few, noisy measurements are available, a linear model may be a good description of the

O. Nelles, *Nonlinear System Identification*,
https://doi.org/10.1007/978-3-030-47439-3_10

nonlinear process behavior compared with other more complex nonlinear models that have a much higher variance error. In other words, if the available data is sparse and noisy and the input dimensionality is high, the data may be not informative enough to estimate models that are more complex than linear.

A linear model can be written as

$$\hat{y} = w_0 + w_1 u_1 + w_2 u_2 + \ldots + w_p u_p \tag{10.1}$$

or

$$\hat{y} = \sum_{i=0}^{p} w_i u_i \qquad \text{with} \quad u_0 = 1\,. \tag{10.2}$$

Figure 10.1 shows that a linear model for two inputs represents a plane. For higher dimensions a linear model represents a hyperplane where the offset parameter w_0 determines the ordinate value at $\underline{u} = \underline{0}$ and the parameters w_i, $i > 0$, determine the slope of the hyperplane in the direction of u_i. In the basis function formulation, the inputs u_i are the basis functions, the coefficients w_i are the linear parameters and no nonlinear parameters exist.

The parameters of a linear model can be estimated by least squares (Sect. 3.1) with the following regression matrix $\underline{X}$ and parameter vector $\underline{\theta}$:

$$\underline{X} = \begin{bmatrix} 1 & u_1(1) & u_2(1) & \cdots & u_p(1) \\ 1 & u_1(2) & u_2(2) & \cdots & u_p(2) \\ \vdots & \vdots & \vdots & & \vdots \\ 1 & u_1(N) & u_2(N) & \cdots & u_p(N) \end{bmatrix} \qquad \underline{\theta} = \begin{bmatrix} w_0 \\ w_1 \\ w_2 \\ \vdots \\ w_p \end{bmatrix}. \tag{10.3}$$

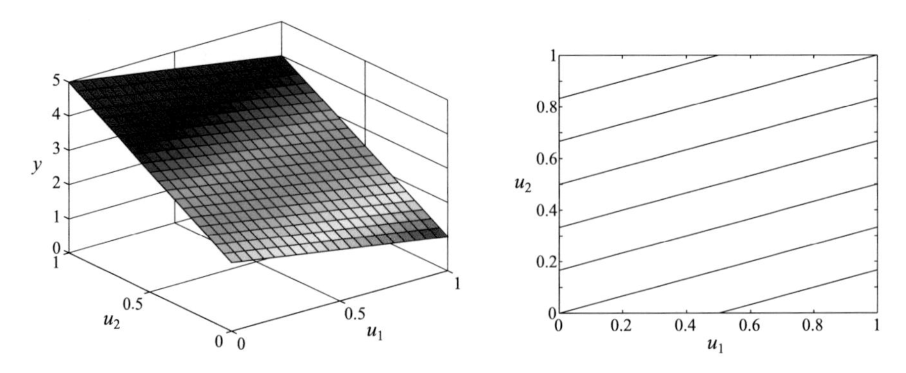

Fig. 10.1 A linear model for two inputs and its contour lines. The model is $y = 2 - u_1 + 3u_2$

The most important properties of linear models are as follows:

- *Interpolation behavior* is linear.
- *Extrapolation behavior* is linear, i.e., the slopes stay constant.
- *Locality* does not exist. A linear model possesses a fully global characteristic.
- *Accuracy* is typically low. The accuracy is the lower, the stronger the nonlinear characteristics of the process is.
- *Smoothness* is high. The derivative of the model output stays constant over the whole operating range.
- *Sensitivity to noise* is very low since all training data samples are exploited to estimate the few model parameters (global approximation characteristics).
- *Parameter optimization* can be performed very rapidly by a least squares algorithm; see Sect. 3.1.
- *Structure optimization* can be performed efficiently by a linear subset selection technique such as the orthogonal least squares (OLS) algorithm; see Sect. 3.4.
- *Online adaptation* can be realized efficiently with a recursive least squares algorithm; see Sect. 3.2.
- *Training speed* is fast. It increases with cubic complexity or even only with quadratic complexity if the Toeplitz structure of the Hessian $\underline{X}^T \underline{X}$ is exploited.
- *Evaluation speed* is fast since only p multiplications and additions are required.
- *Curse of dimensionality* is low because the number of parameters increases only linearly with the input dimensionality.
- *Interpretation* is possible if insights can be drawn from the offset and slope parameters.
- *Incorporation of constraints* for the model output and the parameters is possible if a quadratic programming algorithm is used instead of the least squares; see Sect. 3.3.
- *Incorporation of prior knowledge* about the expected parameter values is possible in the form of the regularization technique ridge regression; see Sect. 3.1.4.
- *Usage* is very high. Linear models are *the* standard models.

10.2 Polynomial Models

Polynomials are the straightforward classical extension to linear models. The higher the degree of the polynomial, the more flexible the model becomes. A p-dimensional polynomial of degree l is given by

$$\hat{y} = w_0 + \sum_{i=1}^{p} w_i u_i + \sum_{i_1=1}^{p} \sum_{i_2=i_1}^{p} w_{i_1 i_2} u_{i_1} u_{i_2} + \ldots$$
$$+ \sum_{i_1=1}^{p} \cdots \sum_{i_l=i_{l-1}}^{p} w_{i_1 \cdots i_l} u_{i_1} \cdots u_{i_l} . \tag{10.4}$$

The offset and the first sum describe a linear model; the second sum describes the second-order terms like u_1^2, $u_1 u_2$, etc.; and the last sum describes the lth-order terms like u_1^l, $u_1^{l-1} u_2$, etc. It can be shown that the p-dimensional polynomial of degree l possesses

$$M = \frac{(l+p)!}{l!\,p!} - 1 \tag{10.5}$$

terms excluding the offset (M is the number of basis functions; $M+1$ is the number of parameters) [327]. Thus, this polynomial can also be expressed as

$$\hat{y} = \sum_{i=0}^{M} \theta_i x_i \qquad \text{with} \quad x_0 = 1\,, \tag{10.6}$$

where the $\theta_i x_i$, $i = 0, \ldots, M$ correspond to the ith term in (10.4). In the formulation (9.2), the basis functions $\Phi_i(\cdot)$ correspond to the x_i, the linear parameters correspond to the θ_i, and no nonlinear parameters exist. For example, a second-degree polynomial of three inputs becomes

$$\begin{aligned} \hat{y} = \theta_0 + \theta_1 u_1 + \theta_2 u_2 + \theta_3 u_3 \\ + \theta_4 u_1^2 + \theta_5 u_1 u_2 + \theta_6 u_1 u_3 + \theta_7 u_2^2 + \theta_8 u_2 u_3 + \theta_9 u_3^2\,. \end{aligned} \tag{10.7}$$

Figure 10.2 illustrates the characteristics of this polynomial.

The polynomial model parameters can be estimated by least squares. The regression matrix $\underline{X}$ and the parameter vector $\underline{\theta}$ for a polynomial model of degree l for p inputs become

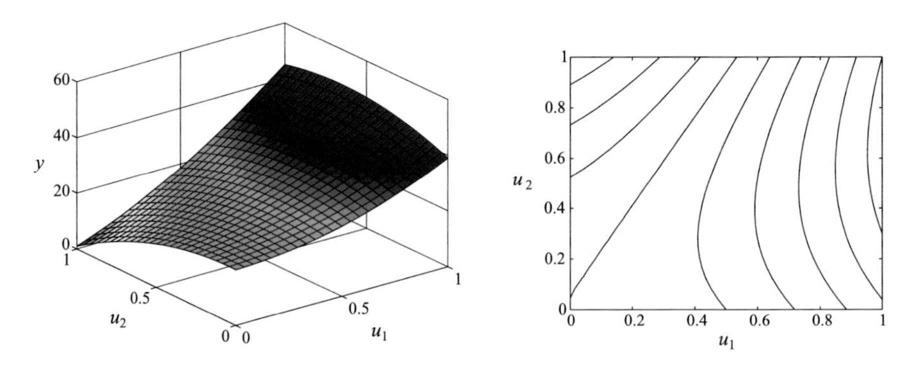

Fig. 10.2 A second-degree polynomial model for two inputs and its contour lines. The model is $y = 20 + u_1 + u_2 + 15u_1^2 + 20u_1u_2 - 20u_2^2$

$$\underline{X} = \begin{bmatrix} 1 & u_1(1) & \cdots & u_p(1) & u_1^2(1) & \cdots & u_p^l(1) \\ 1 & u_1(2) & \cdots & u_p(2) & u_1^2(2) & \cdots & u_p^l(2) \\ \vdots & \vdots & & \vdots & \vdots & & \vdots \\ 1 & u_1(N) & \cdots & u_p(N) & u_1^2(N) & \cdots & u_p^l(N) \end{bmatrix} \qquad \underline{\theta} = \begin{bmatrix} \theta_0 \\ \theta_1 \\ \theta_2 \\ \vdots \\ \theta_M \end{bmatrix} \tag{10.8}$$

with M as the number of basis functions according to (10.5).

Considering (10.5), it is obvious that the number of parameters and thus the model complexity grow strongly with an increasing number of inputs p and/or polynomial degree l. Therefore, even for moderately sized problems, the estimation of a full polynomial model is beyond practical feasibility, with respect to both the huge variance error and the high computational demand. Consequently, for almost all non-trivial problems, polynomial models can only be applied in combination with a structure selection technique. These structure selection schemes can automatically select the relevant terms from a full polynomial and lead to a reduced polynomial model with significantly fewer parameters. Efficient algorithms such as the orthogonal least squares can be utilized for structure selection since polynomials are linearly parameterized. Although the application of structure selection techniques makes polynomial models powerful, the huge number of terms of full polynomials is a severe drawback because it makes the structure selection computationally demanding.

Another drawback of polynomial models is their tendency to oscillatory interpolation and extrapolation behavior, especially when high-degree polynomials are used. Figure 10.3a and b illustrates this effect for a polynomial with degree 5 and 20. The strange interpolation behavior is unrealistic for most applications. Furthermore, the extrapolation of polynomials tends to $+\infty$ or $-\infty$ as for linear models but with a much faster rate. The extrapolation behavior is basically determined by the highest-

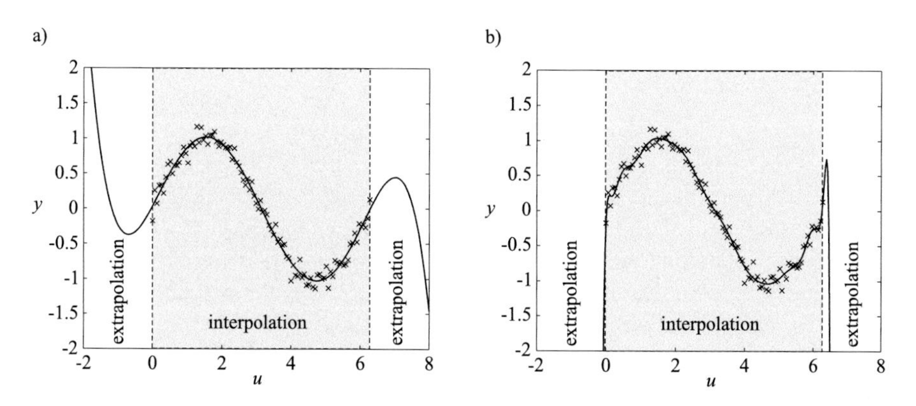

Fig. 10.3 (**a**) A polynomial model of degree 5 approximating a sine wave. (**b**) A polynomial model of degree 20 approximating the same sine wave. The 100 noisy data samples are marked as "×." The dotted lines represent the interval $[0,\ 2\pi]$ within which the training data is distributed

order terms, which results in strongly increasing or decreasing model outputs with huge derivatives. This is also unrealistic for most applications. Moreover, the extrapolation behavior can be non-monotonic and is unrelated to the tendency (e.g., the slope) of the model at the extrapolation boundary. Although the interpolation and extrapolation behavior of polynomials is usually not desirable, in some cases the polynomial model can be advantageous because it may match the true (typically unknown) process structure.

10.2.1 Regularized Polynomials

As has been previously observed, higher-degree polynomials exhibit "wild" behavior. Parameters estimated by LS tend to huge positive and negative values which (multiplied with their regressors) almost cancel each other. A remedy to this problem is to estimate the parameters by ridge regression; see Sect. 3.1.4. The penalty term forces the parameters to values close to zero and thereby makes the result more robust. Even tiny regularization parameters λ already significantly dampen the polynomial. Adjusting λ is the much more sophisticated way to perform the bias/variance tradeoff compared to plainly choosing the polynomial degree. It offers the following advantages:

- Continuous compared to discrete adjustment
- Contribution of each regressor according to its relevance for the problem compared to "in" or "out"
- Parameter estimate with lower variance compared to (potentially) arbitrary large variances

The price to be paid for these benefits is the computational effort required for finding a good or even the best value for λ. This can be done by cross-validation. Since for linear regression problems, a simple closed formula exists (see Sect. 7.3.2) for calculating the leave-one-out error, a nonlinear one-dimensional search over λ is relatively fast and easy to carry out.

Example 10.2.1 To illustrate ridge regression for polynomials, we generate six data points whose input u and output y are randomly, uniformly distributed in $[0, 1]$. Figure 10.4 compares the polynomials estimated by ridge regression for six different λ-values. Each plot shows two polynomials: (i) through the original six data points and (ii) with one point at $u \approx 0.46$ shifted upward by 0.1. It can be clearly seen that the polynomials are damped by increasing λ's. Also, the polynomials for the two data sets approach each other for increasing λ's. Note that it is particularly bothersome for the weakest regularized case that the polynomial flips, i.e., instead of coming from and going to $\rightarrow -\infty$ for one data set, it comes and goes to $\rightarrow \infty$ for the other one.

In the following, some covariance matrices calculated with (3.50) are given for this ridge regression estimation problem assuming a noise variance of $\sigma^2 = 1$.

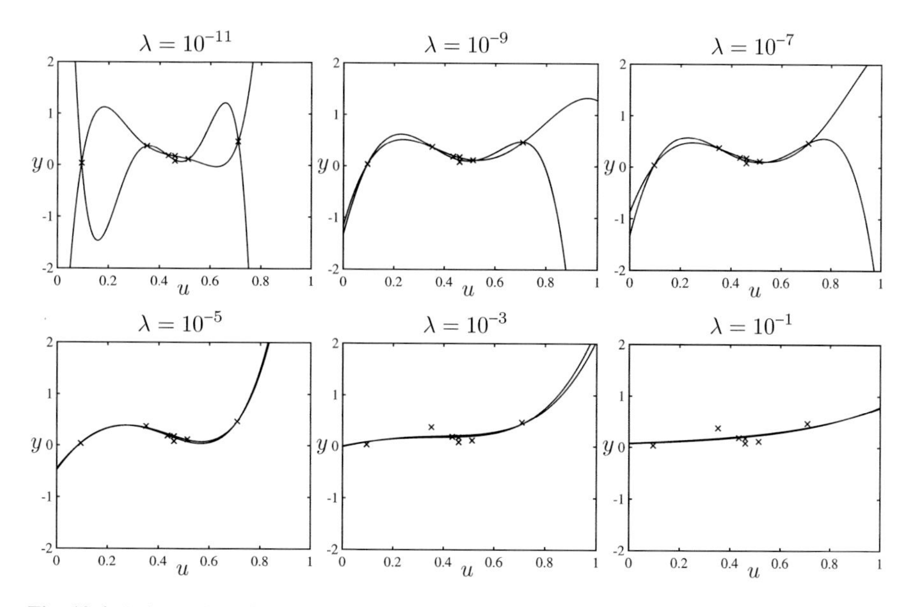

Fig. 10.4 Polynomials of degree 5 fitting 6 data points estimated with ridge regression for various regularization parameters. Each plots shows two fitted polynomials: for the original data set and for one output value increased by 0.1 at $u \approx 0.46$. The plot for $\lambda = 10^{-11}$ is very close to the interpolation polynomials

Remember that the parameter variances are the diagonal entries of the matrices sorted according to increasing powers of u.

For $\lambda = 10^{-11}$ the covariance matrix is very close to the LS solution:

$$\begin{bmatrix} 1.5 \cdot 10^5 & -2.8 \cdot 10^6 & 1.7 \cdot 10^7 & -4.6 \cdot 10^7 & 5.8 \cdot 10^7 & -2.7 \cdot 10^7 \\ -2.8 \cdot 10^6 & 5.3 \cdot 10^7 & -3.2 \cdot 10^8 & 8.7 \cdot 10^8 & -1.1 \cdot 10^9 & 5.2 \cdot 10^8 \\ 1.7 \cdot 10^7 & -3.2 \cdot 10^8 & 2.0 \cdot 10^9 & -5.3 \cdot 10^9 & 6.7 \cdot 10^9 & -3.2 \cdot 10^9 \\ -4.6 \cdot 10^7 & 8.7 \cdot 10^8 & -5.3 \cdot 10^9 & 1.4 \cdot 10^{10} & -1.8 \cdot 10^{10} & 8.5 \cdot 10^9 \\ 5.8 \cdot 10^7 & -1.1 \cdot 10^9 & 6.7 \cdot 10^9 & -1.8 \cdot 10^{10} & 2.3 \cdot 10^{10} & -1.1 \cdot 10^{10} \\ -2.7 \cdot 10^7 & 5.2 \cdot 10^8 & -3.2 \cdot 10^9 & 8.5 \cdot 10^9 & -1.1 \cdot 10^{10} & 5.1 \cdot 10^9 \end{bmatrix}.$$

For $\lambda = 10^{-5}$ it is significantly smaller:

$$\begin{bmatrix} 13.0 & -177.0 & 522.0 & -500.0 & -177.0 & 311.0 \\ -177.0 & 2.3 \cdot 10^3 & -8.0 \cdot 10^3 & 8.4 \cdot 10^3 & 1.9 \cdot 10^3 & -5.0 \cdot 10^3 \\ 522.0 & -8.0 \cdot 10^3 & 3.1 \cdot 10^4 & -3.8 \cdot 10^4 & -1.0 \cdot 10^3 & 1.9 \cdot 10^4 \\ -500.0 & 8.4 \cdot 10^3 & -3.8 \cdot 10^4 & 6.2 \cdot 10^4 & -2.5 \cdot 10^4 & -1.2 \cdot 10^4 \\ -177.0 & 1.9 \cdot 10^3 & -1.0 \cdot 10^3 & -2.5 \cdot 10^4 & 6.1 \cdot 10^4 & -4.1 \cdot 10^4 \\ 311.0 & -5.0 \cdot 10^3 & 1.9 \cdot 10^4 & -1.2 \cdot 10^4 & -4.1 \cdot 10^4 & 4.6 \cdot 10^4 \end{bmatrix}.$$

For $\lambda = 10^{-1}$ it is again significantly smaller:

$$\begin{bmatrix} 0.73 & -1.4 & -0.086 & 0.2 & 0.24 & 0.21 \\ -1.4 & 5.4 & -2.8 & -1.5 & -0.88 & -0.52 \\ -0.086 & -2.8 & 7.5 & -1.9 & -1.4 & -0.98 \\ 0.2 & -1.5 & -1.9 & 8.3 & -1.3 & -1.0 \\ 0.24 & -0.88 & -1.4 & -1.3 & 8.9 & -0.86 \\ 0.21 & -0.52 & -0.98 & -1.0 & -0.86 & 9.3 \end{bmatrix} .$$

Coupling between the parameters becomes increasingly weaker (off-diagonal entries $\to 0$). For the sake of comparison, these are the parameter vectors estimated for $\lambda = 10^{-11}, 10^{-5}, 10^{-1}$ from the offset to the coefficient for u^5:

$$\hat{\underline{\theta}} = \left[\, 14.0 \; -288.0 \; 1.7 \cdot 10^3 \; -4.9 \cdot 10^3 \; 6.2 \cdot 10^3 \; -3.0 \cdot 10^3 \,\right]^T ,$$

$$\hat{\underline{\theta}} = \left[\, -0.48 \; 6.8 \; -13.0 \; -1.6 \; 7.3 \; 9.0 \,\right]^T ,$$

$$\hat{\underline{\theta}} = \left[\, 0.07 \; 0.17 \; 0.15 \; 0.15 \; 0.13 \; 0.11 \,\right]^T .$$

It can be observed how the size of the parameter values varies in dependency on the regularization parameter λ in Fig. 10.5. They start from the LS solution for $\lambda = 0$ and tend toward zero for $\lambda \to \infty$. The zoomed plot in Fig. 10.5b illustrates how the parameters balance each other out.

Figure 10.6 shows that the sum of squared errors on the training data increases from zero for $\lambda = 0$ (interpolation case) to higher errors and how the parameter variances decrease with increasing λ-values dramatically. □

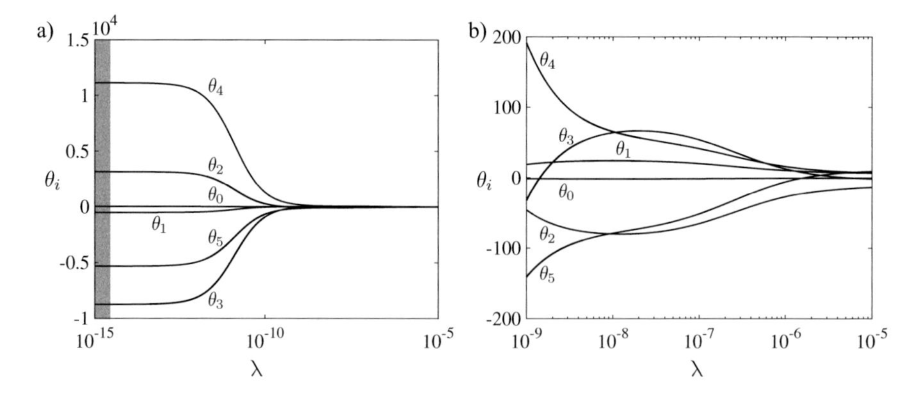

Fig. 10.5 **(a)** The regularization path of a polynomial of degree 5 for Example 10.2.1, **(b)** detailed view. θ_i is the coefficient for u^i. The gray-shaded area represents the LS solution

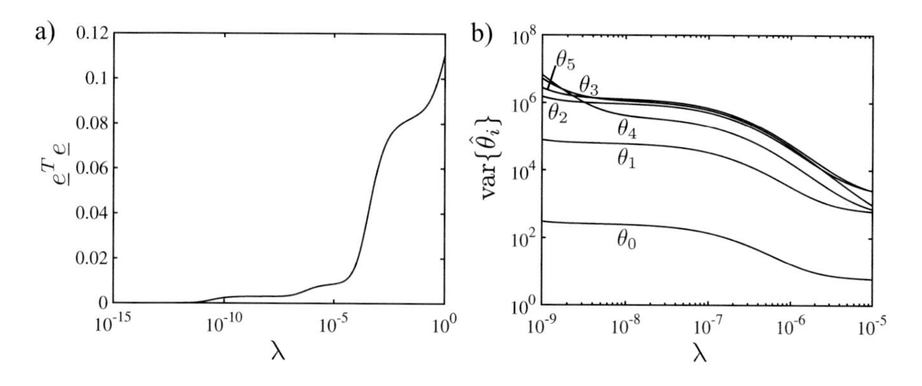

Fig. 10.6 Depending on the strength of regularization: (**a**) sum of squared errors on the training data and (**b**) parameter variances (notice the logarithmic scaling for *both* axes)

10.2.1.1 Penalization of Offset

Strictly speaking, the above approach did something wrong or at least unusual. It penalized all parameters including the *offset* (in statistics literature typically called *intercept* and in neural network literature often called *bias*). Usually, the offset is not penalized because a shift in the y-values would yield different parameter estimates and therefore different regularization results. It is reasonable to make the results independent of the position of the origin of the coordinate system.[1]

Two approaches to solve this problem typically are proposed:

- Centering the regressors: Then the model's offset does not have to be estimated. No "1"-regressor exists. $\theta_0 = \sum_{i=1}^{N} y(i)$ is simply the mean of the output values and not estimated by LS.
- Omitting the offset parameter from the penalization: In (3.44) in Sect. 3.1.4, no distinction between offset and other parameters was made. In order to distinguish the offset, it is usually denoted by θ_0. Then penalizing all parameters minimizes the following loss function:

$$I(\underline{\theta}, \lambda) = \frac{1}{2}\left(\underline{e}^T\underline{e} + \lambda\sum_{i=0}^{n}\theta_i^2\right) \longrightarrow \min_{\underline{\theta}}. \tag{10.9}$$

with the solution

$$\hat{\underline{\theta}} = \left(\underline{X}^T\underline{X} + \lambda\underline{I}\right)^{-1}\underline{X}^T\underline{y}. \tag{10.10}$$

[1] It is also reasonable to ensure that the parameters are in the same scale of magnitude by scaling the regressors. Otherwise, some parameters will be favored significantly over others.

where $\underline{I}$ is the normal unity matrix. However, if the offsets should not be penalized, it changes to

$$I(\underline{\theta}, \lambda) = \frac{1}{2}\left(\underline{e}^T \underline{e} + \lambda \sum_{i=1}^{n} \theta_i^2\right) \longrightarrow \min_{\underline{\theta}}. \tag{10.11}$$

with the solution

$$\underline{\hat{\theta}} = \left(\underline{X}^T \underline{X} + \lambda \underline{\widetilde{I}}\right)^{-1} \underline{X}^T \underline{y}. \tag{10.12}$$

with a clipped unity matrix for the zeroth parameter:

$$\underline{\widetilde{I}} = \begin{bmatrix} 0 & 0 & 0 & 0 \\ 0 & 1 & 0 & 0 \\ 0 & 0 & 1 & 0 \\ 0 & 0 & 0 & 1 \end{bmatrix} \tag{10.13}$$

10.2.2 Orthogonal Polynomials

Polynomials of different degrees can be made orthogonal to each other within a certain interval for input u. This means that the inner product between two polynomials $f(u)$ and $g(u)$ becomes zero if they have different degrees. Most often the interval $[-1, 1]$ is considered:

$$\int_{-1}^{1} f(u)g(u)du = 0. \tag{10.14}$$

If even orthonormality holds, additionally

$$\int_{-1}^{1} f(u)f(u)du = 1. \tag{10.15}$$

The polynomials that fulfill this property are called *Legendre polynomials*. This first six is shown in Fig. 10.7a and defined by the following equations[2]:

$$L_0(u) = 1$$
$$L_1(u) = u$$

[2]From https://en.wikipedia.org/wiki/Legendre_polynomials.

$$L_2(u) = \frac{1}{2}\left(3u^2 - 1\right)$$

$$L_3(u) = \frac{1}{2}\left(5u^3 - 3u\right)$$

$$L_4(u) = \frac{1}{8}\left(35u^4 - 30u^2 + 3\right)$$

$$L_5(u) = \frac{1}{8}\left(63u^5 - 70u^3 + 15u\right)$$

Other orthogonal polynomial exist where the inner product is taken with respect to a weighting function $w(u)$:

$$\int_{-1}^{1} f(u)g(u)w(u)du = 0. \tag{10.16}$$

For Legendre polynomials $w(u) = 1$. Famous are Chebyshev polynomials (first and second kind) which are orthogonal with respect to certain weightings $w(u)$ because they play an important role in polynomial interpolation theory. For Chebyshev polynomials of the first kind, the weighting approaches ∞ for $u \to -1$ and $u \to 1$. A consequence of this weighting is that the output values y of all polynomials fill the interval $[-1, 1]$; see Fig. 10.7b. Chebyshev polynomials are known to have optimal properties in minimax optimization problems.

If a continuous function shall be approximated, the above integrals appear in auto- and cross-correlation expressions for the optimization. Because of the orthogonality, the optimal parameters are completely decoupled, i.e., each optimal parameter can be calculated independently of the others. Therefore the LS fit only requires the inversion of a diagonal matrix which is trivial. Moreover, the degree

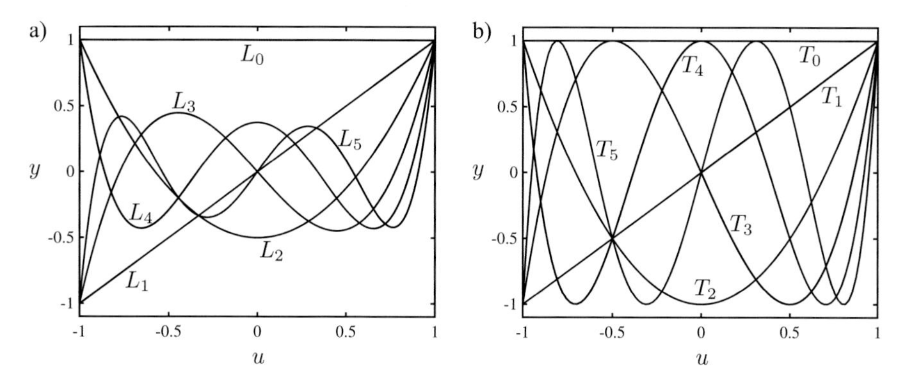

Fig. 10.7 First six orthogonal (**a**) Legendre and (**b**) Chebyshev polynomials (first kind) on $[-1, 1]$ for the degrees 0 to 5

of a Legendre polynomial can be increased or decreased without any effect on the associated optimal parameters.

However, the abovementioned major advantages for an orthogonal polynomial hold only for the continuous case where the whole input between -1 and 1 is of equal importance (or with importance according to the weighting function). In the context of data-driven modeling, the polynomial should fit data points. Then the orthogonality property is destroyed. The better the distribution of the data matches the weighting function, the better the orthogonality holds at least approximately.

For a "normal" polynomial, the regressor matrix is composed of the powers of u: $\underline{X} = [1\ u\ u^2\ u^3\ u^4\ u^5]$, while for a Legendre polynomial, it becomes $\underline{X} = [L_0\ L_1\ L_2\ L_3\ L_4\ L_5]$.

Example 10.2.2 In a perfect scenario, $\underline{X}^T\underline{X}$ would be diagonal, and the parameters do not interact at all. This is how $\underline{X}^T\underline{X}$ looks like for the estimation of a Legendre polynomial of degree 5 for three different data distributions:

1. Equidistantly distributed 100 points in $[-1, 1]$:

$$\begin{bmatrix} 100.0 & 0 & 1.0 & 7.8\cdot 10^{-16} & 1.0 & 1.6\cdot 10^{-15} \\ 0 & 34.0 & 1.8\cdot 10^{-15} & 1.0 & 7.8\cdot 10^{-16} & 1.1 \\ 1.0 & 1.8\cdot 10^{-15} & 21.0 & 1.8\cdot 10^{-15} & 1.0 & 1.8\cdot 10^{-15} \\ 7.8\cdot 10^{-16} & 1.0 & 1.8\cdot 10^{-15} & 15.0 & 8.9\cdot 10^{-16} & 1.1 \\ 1.0 & 7.8\cdot 10^{-16} & 1.0 & 8.9\cdot 10^{-16} & 12.0 & 1.8\cdot 10^{-15} \\ 1.6\cdot 10^{-15} & 1.1 & 1.8\cdot 10^{-15} & 1.1 & 1.8\cdot 10^{-15} & 10.0 \end{bmatrix}$$

   This matrix is almost diagonal. Thus, the error neglecting the off-diagonal entries and assuming perfect orthogonality may be tolerable. However, with modern resources there is no need to do that. The exact inverse can be calculated at extremely low computational cost.

2. Uniformly, randomly distributed 100 points in $[-1, 1]$:

$$\begin{bmatrix} 100.0 & 2.6 & 0.37 & -0.38 & 0.8 & 3.6 \\ 2.6 & 34.0 & 0.8 & 0.62 & 1.9 & -0.45 \\ 0.37 & 0.8 & 21.0 & 2.3 & -0.36 & 2.8 \\ -0.38 & 0.62 & 2.3 & 14.0 & 3.1 & -1.3 \\ 0.8 & 1.9 & -0.36 & 3.1 & 9.8 & 2.4 \\ 3.6 & -0.45 & 2.8 & -1.3 & 2.4 & 9.0 \end{bmatrix}$$

   It is possible to see a resemblance to a diagonal matrix but deviations are pretty high.

3. Uniformly, randomly distributed 100 points in [0, 1]:

$$\begin{bmatrix} 100.0 & 50.0 & 0.25 & -12.0 & 0.7 & 6.7 \\ 50.0 & 34.0 & 13.0 & 0.51 & -1.6 & 0.51 \\ 0.25 & 13.0 & 20.0 & 13.0 & 0.42 & -3.4 \\ -12.0 & 0.51 & 13.0 & 15.0 & 7.5 & 0.52 \\ 0.7 & -1.6 & 0.42 & 7.5 & 12.0 & 7.5 \\ 6.7 & 0.51 & -3.4 & 0.52 & 7.5 & 9.5 \end{bmatrix}$$

This matrix hardly offers any advantage compared to an estimation of a standard polynomial. □

The typical advantage the user gets in a data-driven modeling context from an orthogonal polynomial compared to a standard one is that $\underline{X}^T \underline{X}$ tends to be well-conditioned and therefore easy to invert with little numerical error.

Finally, this subsection only discussed the univariate case (one input). In principle, it can be generalized to multiple inputs by tensor products; see Sect. 12.3.1. However, this fully underlies and suffers from the curse of dimensionality.

10.2.3 Summary Polynomials

The weaknesses of polynomials have led to the development of *splines*, which are locally defined low-degree polynomials. Splines are not further discussed here since they fit into the neuro-fuzzy models with singletons addressed in Chap. 12; refer to [84] for an extensive treatment of these models.

The properties of polynomials can be summarized as follows:

- *Interpolation behavior* tends to be non-monotonic and oscillatory for polynomials of high degree. Ridge regression instead of least squares for estimation of the coefficients significantly alleviates these issues at the price of higher computational demand.
- *Extrapolation behavior* tends strongly to $+\infty$ or $-\infty$, with a rate determined by the highest-order terms.
- *Locality* does not exist. A polynomial possesses a fully global characteristic.
- *Accuracy* is limited since high-degree polynomials are not practicable.
- *Smoothness* is low. The derivative of the model output often changes sign owing to the oscillatory interpolation behavior.
- *Sensitivity to noise* is low since all training data samples are exploited to estimate the model parameters (global approximation characteristics). However, this means that training data that is locally very noisy can significantly influence the model behavior in all operating regimes.

- *Parameter optimization* can be performed very fast by a least squares algorithm; see Sect. 3.1. However, the number of parameters grows rapidly with increasing input dimensionality and/or polynomial degree.
- *Structure optimization* can be performed efficiently by a linear subset selection technique such as the orthogonal least squares (OLS) algorithm; see Sect. 3.4. However, owing to the huge number of potential regressors, structure selection can become prohibitively slow for high-dimensional problems.
- *Online adaptation* can be realized efficiently with a recursive least squares algorithm; see Sect. 3.2. However, owing to the nonlinear global characteristics of polynomials, their online adaptation is unreliable since small parameter changes in one operating regime can have a strong impact on the model behavior in all other operating regimes.
- *Training speed* is fast for parameter optimization but slows down considerably for structure optimization.
- *Evaluation speed* is medium. The higher-order terms can be computed from the low-order ones to save computations: e.g., u^6 can be computed with one multiplication only if u^5 is known.
- *Curse of dimensionality* is high. The number of parameters grows strongly with increasing input dimensionality.
- *Interpretation* is hardly possible. Only if the polynomial model matches the true structure of the process may its parameter values be meaningful in some physical sense.
- *Incorporation of constraints* for the model output is possible if a quadratic programming algorithm is used instead of the least squares; see Sect. 3.3.
- *Incorporation of prior knowledge* is hardly possible since the interpretation of the model is very limited.
- *Usage* is high. Polynomial models are commonly used since interpolation polynomials are a standard tool taught in mathematics.

10.3 Look-Up Table Models

Look-up tables[3]by far are the most common type of nonlinear static models in real-world implementations, at least for problems with one- or two-dimensional input spaces. The reason for this lies in their simplicity and their extremely low computational evaluation demand. Furthermore, in most applications where look-up tables are utilized, the "training" procedure is mere storage of the training data – no real optimization techniques are used. All these features make look-up tables the state-of-the-art models for low-dimensional static mappings. One particularly important field for the application of look-up tables is the automotive area. The

[3]For the sake of brevity, the term "look-up table" is used instead of "grid-based look-up table" which would be the more exact expression.

motor electronics and other devices in a standard passenger car of the late 1990s contain about 50 to 100 look-up tables; trucks contain even more. The reason for the vast amount of look-up tables and the immense increase in complexity over the last years is mainly a consequence of the exclusive application of one- and two-dimensional mappings. The main reasons for the restriction to these low-dimensional mappings are:

- One- and two-dimensional mappings can be visualized; higher-dimensional ones cannot.
- Low-dimensional mappings can be realized with grid-based look-up tables; higher-dimensional ones cannot, since look-up tables severely suffer from the curse of dimensionality. So, if the true relationships are higher dimensional, typically many low-dimensional look-up tables are combined in an additive or multiplicative manner.

The second point can be overcome by the use of more sophisticated models. The first issue, however, is of fundamental character and can only be weakened by the application of interpretable models with well-understood interpolation and extrapolation behavior.

10.3.1 One-Dimensional Look-Up Tables

The upper part of Fig. 10.8 shows a one-dimensional look-up table with six points (c_1, w_1) to (c_6, w_6). For six input values c_1 to c_6, the corresponding output values or heights w_1 to w_6 are stored in this look-up table. Often these values stem directly from input/output measurements of the process: that is, they represent the training data. The output of such a look-up table model output is determined by the closest look-up table points to the left and to the right of the model input. It is calculated as the linear interpolation of both corresponding heights. Thus, for a one-dimensional look-up table, the output becomes

$$\hat{y} = \frac{w_{\text{left}}\left(c_{\text{right}} - u\right) + w_{\text{right}}\left(u - c_{\text{left}}\right)}{c_{\text{right}} - c_{\text{left}}} \tag{10.17}$$

where $(c_{\text{left}}, w_{\text{left}})$ and $(c_{\text{right}}, w_{\text{right}})$ are the closest points to left and right of u, respectively. Thus, for $u = c_{\text{left}} \Longrightarrow \hat{y} = w_{\text{left}}$ and for $u = c_{\text{right}} \Longrightarrow \hat{y} = w_{\text{right}}$. This linear interpolation behavior is also shown in Fig. 10.8. For extrapolation, e.g., if u possesses either no left or no right neighbor, the output of a look-up table is not defined. However, any kind of extrapolation behavior can be artificially introduced. Usually the look-up table height is kept constant for extrapolation. For the example in Fig. 10.8, this means for $u < c_1 \Longrightarrow \hat{y} = w_1$ and for $u > c_6 \Longrightarrow \hat{y} = w_6$.

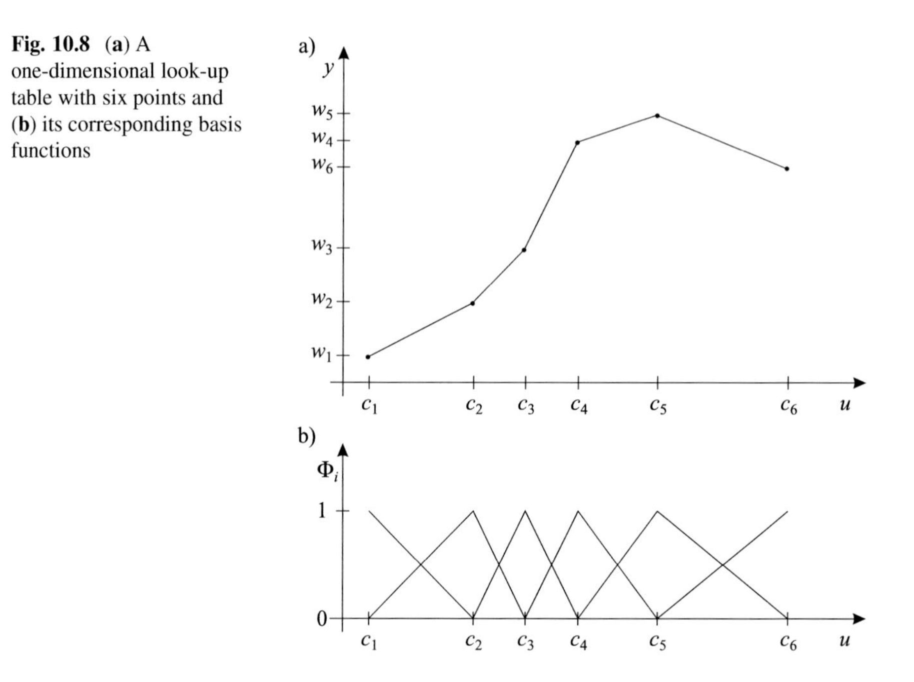

Fig. 10.8 (**a**) A one-dimensional look-up table with six points and (**b**) its corresponding basis functions

The one-dimensional look-up table model can be described in the basis function framework by the introduction of the triangular basis functions shown in Fig. 10.8b as

$$\hat{y} = \sum_{i=1}^{M} w_i \Phi_i(u, \underline{c}) \,, \tag{10.18}$$

where $\underline{c} = [c_1\ c_2\ \cdots\ c_M]^T$ contains the input values of the M look-up table points. The c_i represent the positions or centers of the basis functions. Under the assumption that these centers c_i are monotonic increasing, the basis functions can be written as

$$\Phi_i(u, \underline{c}) = \begin{cases} (u - c_{i-1})/(c_i - c_{i-1}) & \text{for } c_{i-1} \le u \le c_i \\ (u - c_{i+1})/(c_i - c_{i+1}) & \text{for } c_i < u \le c_{i+1} \\ 0 & \text{otherwise} \end{cases} . \tag{10.19}$$

Note that the ith basis function $\Phi_i(u, \underline{c})$ actually depends only on the centers c_{i-1}, c_i, and c_{i+1}, and not on the whole vector $\underline{c}$. These basis functions realize the linear interpolation behavior of the look-up table. They form a so-called *partition of unity*, which means that they sum up to 1 for any input u:

$$\sum_{i=1}^{M} \Phi_i(u, \underline{c}) = 1 \,. \tag{10.20}$$

Considering (10.18), it is obvious that the heights w_i of the points are linear parameters, while the input values c_i of the points (positions, basis function centers) are nonlinear parameters. This relationship does not matter as long as both the w_i and the c_i are directly set to measured data values. However, when look-up tables are to be optimized, this issue becomes important; see Sects. 10.3.3, 10.3.4, and 10.3.5.

If nothing else is explicitly stated, the term "look-up table" will always refer to a look-up table with linear interpolation, as shown in Fig. 10.8, or to its higher-dimensional extensions; see Sect. 10.3.2. In principle, the basis functions can be modified from simple triangular ones to higher-order splines or Gaussian functions in order to make the interpolation behavior smoother. Although there exists no fundamental difference, these approaches are computationally more involved and are discussed in the context of fuzzy models and neural networks in Chaps. 11 and 12. In fact, as demonstrated in Chap. 12, the look-up table in Fig. 10.8 can be interpreted as a fuzzy system.

For many real-world applications, the points of the look-up tables are uniformly distributed over the input. Then the basis functions are symmetrical triangles. This equidistant choice of the points allows one to access the neighboring points directly by fast address calculations. However, then the flexibility of the look-up table is restricted because the resolution (accuracy) is independent of the input. In general, the grid does not have to be equidistant.

10.3.2 Two-Dimensional Look-Up Tables

The extension of one-dimensional look-up tables to higher-dimensional input spaces is classically done by a grid-based approach. Alternatives to grid-based approaches are discussed in the context of neural networks; see Chap. 11. If not explicitly stated otherwise, look-up tables are assumed to be grid-based. Figure 10.9 shows the input values for points of an equidistant two-dimensional look-up table with 10×7 points. The number of points in each dimension can be chosen differently according to the accuracy requirements and process characteristics. In principle, this grid-based approach can be extended to arbitrary dimensional mappings. However, the number of data points for a p-dimensional look-up table is

$$M = \prod_{i=1}^{p} M_i \,, \tag{10.21}$$

where M_i is the number of points for input dimension i. Obviously, the number of data points M, which is equal to the number of basis functions, increases exponentially with the number of inputs p. Consequently, grid-based look-up tables underlie the curse of dimensionality, and thus in practice, they cannot be used for problems with more than three inputs. In [568] an extension of look-up tables based on a rectangular rather than a grid input space decomposition is proposed.

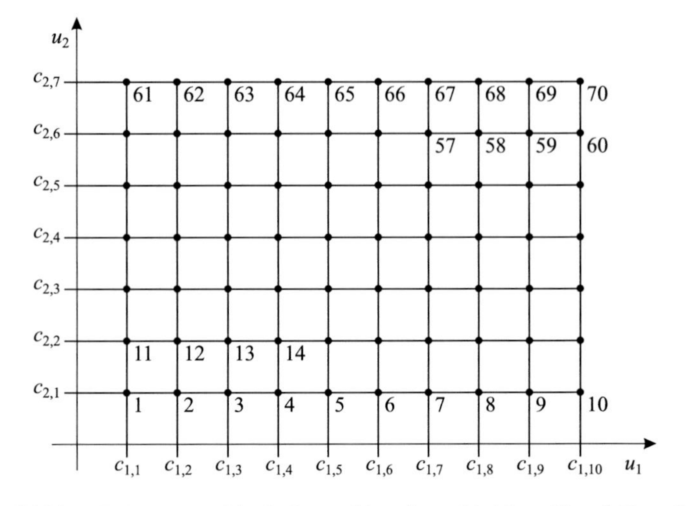

Fig. 10.9 Grid-based placement of the look-up table points with $M_1 = 10$ and $M_2 = 7$ points for the twoinputs u_1 and u_2

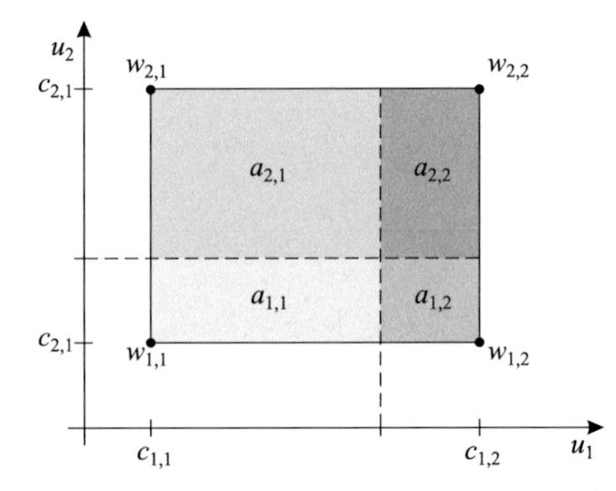

Fig. 10.10 Bilinear or area interpolation for a two-dimensional look-up table

This overcomes the exponential relationship in (10.21). Consequently, the curse of dimensionality is substantially weakened and up to four or five inputs can be handled; see also [569].

The output of a two-dimensional look-up table model is calculated as follows; see Fig. 10.10. The model output is determined by the closest points to the bottom left, bottom right, top left, and top right of the model input. Thus, for the example in Fig. 10.10, the model output is

$$\hat{y} = \frac{w_{1,1}a_{2,2} + w_{1,2}a_{2,1} + w_{2,1}a_{1,2} + w_{2,2}a_{1,1}}{a_{1,1} + a_{1,2} + a_{2,1} + a_{2,2}} \tag{10.22}$$

with the areas

$$a_{1,1} = (u_1 - c_{1,1})(u_2 - c_{2,1}), \tag{10.23a}$$

$$a_{1,2} = (c_{1,2} - u_1)(u_2 - c_{2,1}), \tag{10.23b}$$

$$a_{2,1} = (u_1 - c_{1,1})(c_{2,2} - u_2), \tag{10.23c}$$

$$a_{2,2} = (c_{1,2} - u_1)(c_{2,2} - u_2). \tag{10.23d}$$

According to (10.22)–(10.23d), each height w_{ij} is weighted with the opposite area. The equations (10.22)–(10.23d) perform a so-called *bilinear interpolation* or *area interpolation*. A pure linear interpolation cannot be realized since in general all four surrounding points cannot be guaranteed to lie on a linear function (a plane). Rather the bilinear interpolation in (10.22)–(10.23d) can be seen as the fit of the following *quadratic* function through all four surrounding points:

$$\hat{y} = \theta_0 + \theta_1 u_1 + \theta_2 u_2 + \theta_3 u_1 u_2 . \tag{10.24}$$

This is a restricted two-dimensional polynomial of degree 2 where the terms u_1^2 and u_2^2 are discarded. With appropriate parameters θ_i (10.24) is identical to (10.22)–(10.23d). This bilinear interpolation is illustrated in Fig. 10.11.

An interesting property of the bilinear interpolation is that it reduces to one-dimensional linear interpolation if one input is fixed. In other words, all axis-parallel cuts through the interpolation surface are linear functions; see Fig. 10.11. In contrast, all non-axis-parallel cuts can be described by quadratic functions.

The two-dimensional (and also any higher-dimensional) look-up table can be described in a basis function formulation similar to the one-dimensional look-up table. One basis function corresponds to each point in the look-up table and possesses the same dimensionality as the input space. Figure 10.12 illustrates the shape of these basis functions for a 5 × 5 look-up table. For the basis function

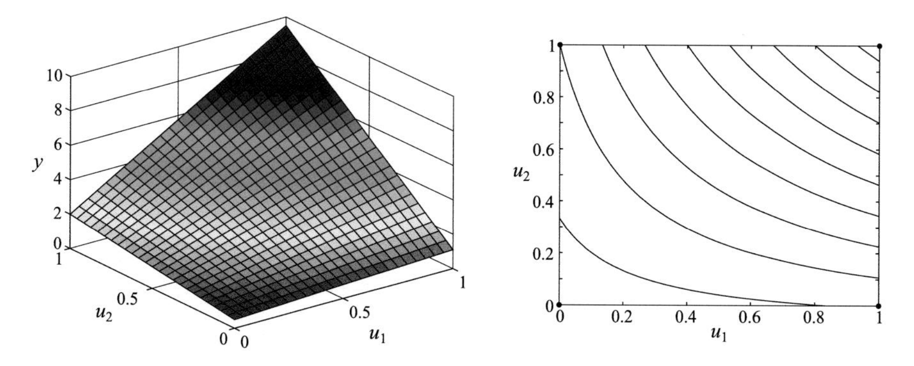

Fig. 10.11 Surface and contour lines of bilinear interpolation. The stored look-up table points are placed in the corners of the shown input space

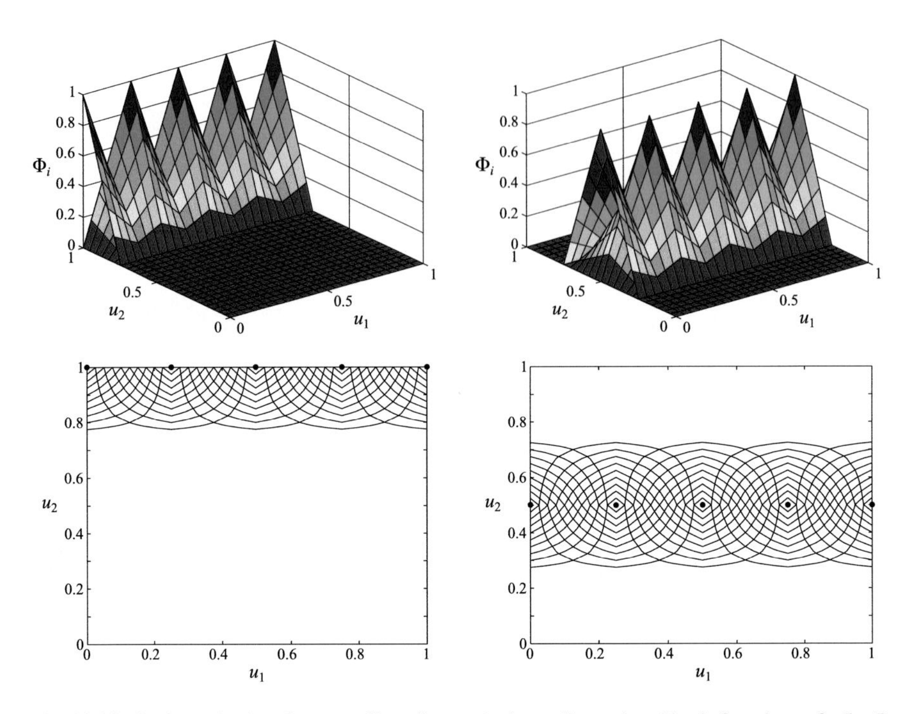

Fig. 10.12 Surfaces (top) and contour lines (bottom) of two-dimensional basis functions of a 5×5 look-up table. Only the basis functions of the fifth row (left) and third row (right) are shown for the sake of clarity

formulation, the linear parameters (heights) have to be re-indexed such that all w_{ij} ($i = 1, \ldots, M_1$, $j = 1, \ldots, M_2$) are mapped to θ_l ($l = 1, \ldots, M_1 M_2$); see Fig. 10.9.

Extension to more than two input dimensions is straightforward. The area interpolation rule can be extended to a volume and hypervolume interpolation rule. The look-up table model output is calculated by a weighted average of the 2^p heights of the surrounding points, where p is the input space dimensionality.

The subsequent subsections discuss the identification of look-up table models from data. The look-up table points are not considered as measurement data samples but as height and center parameters that can be optimized.

10.3.3 Optimization of the Heights

If the measurement data does not lie on a grid or the number of points stored in the look-up table should be smaller than the number of available data samples, then the points of the look-up table have to be estimated from data. The simplest approach to solve this task is to fix the positions of the points (the centers of the basis functions), since these are the nonlinear parameters, and to optimize the heights from data. For a p-dimensional look-up table, the number of heights is equal to the number of stored points

$$\prod_{i=1}^{p} M_i \,, \tag{10.25}$$

where M_i is the number of points for input dimension i.

The heights are linear parameters, and they consequently can be estimated by least squares. The regression matrix and parameter vector are

$$\underline{X} = \begin{bmatrix} \Phi_1(u(1), \underline{c}) & \Phi_2(u(1), \underline{c}) & \cdots & \Phi_M(u(1), \underline{c}) \\ \Phi_1(u(2), \underline{c}) & \Phi_2(u(2), \underline{c}) & \cdots & \Phi_M(u(2), \underline{c}) \\ \vdots & \vdots & & \vdots \\ \Phi_1(u(N), \underline{c}) & \Phi_2(u(N), \underline{c}) & \cdots & \Phi_M(u(N), \underline{c}) \end{bmatrix} \quad \underline{\theta} = \begin{bmatrix} w_1 \\ w_2 \\ \vdots \\ w_M \end{bmatrix}, \tag{10.26}$$

where N is the number of data samples and M is the number of look-up table points. The regression matrix $\underline{X}$ is typically sparse. Each row of $\underline{X}$ contains only 2^p non-zero entries, where p is the input space dimension. For example, in a two-dimensional look-up table, each data sample activates only four basis functions; see Figs. 10.10 and 10.12. If the measurement data does not fill the whole input space, some basis functions will not be activated at all, and $\underline{X}$ will contain columns of zeros. Thus, the regression matrix $\underline{X}$ may become singular. In order to make the least squares problem solvable, the corresponding regressors and their associated heights have to be removed from $\underline{X}$ and $\underline{\theta}$.

It may happen that an optimized height is out of the range of physically plausible values if no (or few) data samples lie close to the center of its basis function. Section 10.3.6 discusses how this effect can be prevented.

The accuracy of a look-up table with optimized heights depends decisively on how well the placement of the basis functions, e.g., the grid, matches the characteristics of the process. If no prior knowledge is available, the most natural choice would be an equidistant grid. However, such an equidistant distribution of basis function centers is suboptimal for most cases. Thus, the next subsection discusses the optimization of the grid according to the measurement data.

10.3.4 Optimization of the Grid

Optimization of the basis function centers cannot be carried out individually in order to maintain the grid structure. In fact, only

$$\sum_{i=1}^{p} M_i \tag{10.27}$$

nonlinear parameters, i.e., the sum of the number of points per input, have to be optimized. Thus for look-up tables, the number of nonlinear parameters grows only linearly with the input space dimensionality, while the number of linear parameters grows exponentially. This is due to the constraints imposed by the grid-based structure.

In principle, any nonlinear optimization technique can be applied to estimate the grid; refer to Chaps. 4 and 5. The initial values can either be determined by prior knowledge (when available) or chosen in an equidistant manner.

If linear interpolation is used, the gradients of the loss function with respect to the basis function centers are non-continuous functions and even do not exist at the centers of the basis functions. Nevertheless, gradient-based optimization schemes can be applied by explicitly setting the gradient to zero at the center points. The gradients are usually not calculated analytically because quite complex equations result. Instead finite difference techniques can be applied for gradient approximation (Sect. 4.4.2), which do not run into difficulties at the center points. The gradients for a one-dimensional look-up table can be calculated as follows:

$$\frac{\partial \hat{y}}{\partial c_i} = w_{i-1} \frac{\partial \Phi_{i-1}}{\partial c_i} + w_i \frac{\partial \Phi_i}{\partial c_i} + w_{i+1} \frac{\partial \Phi_{i+1}}{\partial c_i}, \tag{10.28}$$

since all other basis functions do not depend on i. In (10.28) it is assumed that $1 < i < M$; for the cases $i = 1$ and $i = M$, the first and the last term, respectively, of the sum in (10.28) must be discarded. The derivatives of the affected basis functions in (10.28) are (see (10.19))

$$\frac{\partial \Phi_{i-1}}{\partial c_i} = \begin{cases} 0 & \text{for } c_{i-2} \le u \le c_{i-1} \\ (u - c_{i-1})/(c_{i-1} - c_i)^2 & \text{for } c_{i-1} < u \le c_i \\ 0 & \text{otherwise} \end{cases}, \tag{10.29a}$$

$$\frac{\partial \Phi_i}{\partial c_i} = \begin{cases} -(u - c_{i-1})/(c_i - c_{i-1})^2 & \text{for } c_{i-1} \le u \le c_i \\ -(u - c_{i+1})/(c_i - c_{i+1})^2 & \text{for } c_i < u \le c_{i+1} \\ 0 & \text{otherwise} \end{cases}, \tag{10.29b}$$

$$\frac{\partial \Phi_{i+1}}{\partial c_i} = \begin{cases} (u - c_{i+1})/(c_{i+1} - c_i)^2 & \text{for } c_i \le u \le c_{i+1} \\ 0 & \text{for } c_{i+1} < u \le c_{i+2} \\ 0 & \text{otherwise} \end{cases}. \tag{10.29c}$$

For higher-dimensional look-up tables, the gradient calculation becomes even more involved since more basis functions are activated by a data sample and more different cases have to be distinguished.

In contrast to the linear optimization of the heights, the centers can be optimized with respect to loss functions other than the sum of squared errors. Common alternatives are the sum of absolute errors or the maximum error; see Sect. 2.3.

When performing nonlinear optimization of the grid, special care must be taken to avoid the following problems. During optimization, the centers may overtake each other, or they may drift outside the range of physical meaning. These complications can be avoided by imposing constraints on the optimization problems as discussed in Sect. 10.3.6.

10.3.5 Optimization of the Complete Look-Up Table

Nonlinear optimization of the grid with fixed heights is quite inefficient. Thus, it is advised to optimize both the grid and the heights simultaneously. The most straightforward way to do this is to follow the nested optimization approach depicted in Fig. 10.13; see also Fig. 5.1. This guarantees optimal values for the heights in each iteration of the nonlinear optimization technique. Compared with a nonlinear optimization of all look-up table parameters, the nested strategy is much more efficient since it exploits the linearity in the height parameters.

10.3.6 Incorporation of Constraints

Two motivations for the incorporation of constraints into the optimization task must be distinguished:

- Constraints that ensure the reliable functioning of the optimization algorithm
- Constraints that are imposed in order to guarantee certain properties of the look-up table model that are known from the true process

The first alternative affects only the nonlinear optimization of the grid and the second alternative is typically relevant only for the optimization of the heights.

nonlinear optimization of the grid

linear optimization of the heights

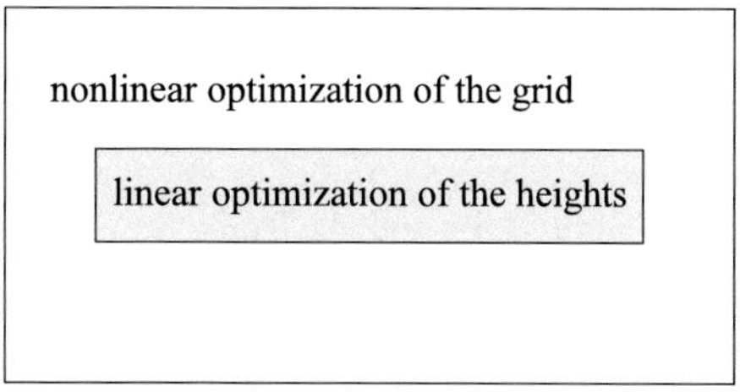

Fig. 10.13 Nested optimization of a look-up table. In an outer loop, the iterative nonlinear optimization scheme optimizes a loss function, e.g., the sum of squared errors. Each evaluation of the loss function (inner loop) by the nonlinear optimization algorithm involves a linear optimization of the heights

10.3.6.1 Constraints on the Grid

The following three difficulties can occur during nonlinear optimization of the grid:

- A center can overtake a neighboring center.
- A center can drift outside the range of physical meaning.
- A center can converge very closely to a neighboring center, which practically annihilates the effect of this center completely. This can occur only with local optimization schemes that may converge to a local optimum since a better solution must exist which utilizes all centers.

All these possible difficulties can be overcome by taking the following countermeasures. First it will be assumed that all 2^p centers in the corners of the p-dimensional input space are fixed, i.e., they are not subject to the nonlinear optimization. The center coordinates for each input are constrained to be monotonically increasing. For example, for a two-dimensional $M_1 \times M_2$ look-up table, the following constraints are applied (see Fig. 10.10):

$$c_{1,1} + \epsilon_1 < c_{1,2} \quad c_{1,2} + \epsilon_1 < c_{1,3} \quad \ldots, \; c_{1,M_1-1} + \epsilon_1 < c_{1,M_1}, \tag{10.30a}$$

$$c_{2,1} + \epsilon_2 < c_{2,2} \quad c_{2,2} + \epsilon_2 < c_{2,3} \quad \ldots, \; c_{2,M_2-1} + \epsilon_2 < c_{2,M_2}, \tag{10.30b}$$

where ϵ_i should be defined as a fraction of the range of operation $c_{i,M_i} - c_{i,1}$ of input i. This guarantees a minimum distance of ϵ_i between all center coordinates.

Second, it can be assumed that the "corner" centers $c_{1,1}$, c_{1,M_1}, $c_{2,1}$, and c_{2,M_2} are also subject to the optimization. Then the additional constraints

$$\min(u_1) < c_{1,1} \qquad c_{1,M_1} < \max(u_1) \tag{10.31a}$$

$$\min(u_2) < c_{2,1} \qquad c_{2,M_2} < \max(u_2) \tag{10.31b}$$

have to be taken into account. These constraints can be met by applying one of the following strategies:

- Addition of a penalty term to the loss function, see Sect. 4.6;
- Application of direct methods for constrained optimization such as sequential quadratic programming, see also Sect. 4.6;
- Coding of the parameters by an appropriate transformation that creates an optimization problem where the constraints are automatically met.

Independent of the method used, it is the experience of the author that the problem of optimization of the grid is usually rich in local optima. So prior knowledge utilized for a good initialization of the grid is necessary to reach a good or even the global optimum. Starting from an equidistant grid, initialization typically results in convergence to a poor local optimum.

The last of the strategies listed above for constrained optimization will be explained a little bit further since it is the easiest, most stable, and most efficient of the three alternatives. It possesses the big advantage that a constrained optimization problem is transformed into an unconstrained one that is much simpler. Furthermore, no "fiddle" parameters are introduced as with the penalty function approach. The transformed centers will be denoted by $\widetilde{c}_{i,j}$. They are subject to optimization. In the loss function, the transformed centers $\widetilde{c}_{i,j}$ first have to be converted back to the original centers $c_{i,j}$. With the original centers, the look-up table is evaluated. The transformation from $\widetilde{c}_{i,j}$ to $c_{i,j}$ is explained in the following. The development of the transformation back from $c_{i,j}$ to $\widetilde{c}_{i,j}$ is then straightforward. For the sake of simplicity, it is assumed that the data is normalized such that $\min(u_i) = 0$ and $\max(u_i) = 1$.

A sigmoid can be utilized to transform a number from the interval $(-\infty, \infty)$ to the interval $(0, 1)$. It is calculated by

$$\mathrm{sig}(v) = \frac{1}{1+\exp(-v)}\,. \tag{10.32}$$

Thus, any real number $\widetilde{v}$ within $(-\infty, \infty)$ can be transformed in a number v within (lb, ub) (lb = lower bound, ub = upper bound) by

$$v = (ub - lb) \cdot \mathrm{sig}(\widetilde{v}) + lb\,. \tag{10.33}$$

This mapping is essential for the transformation from $\widetilde{c}_{i,j}$ to $c_{i,j}$. Two cases have to be distinguished: (i) all centers are optimized; (ii) only the inner centers are optimized, while the corner centers $c_{i,1}$ and c_{i,M_i} are kept fixed.

For case (i), the following transformation is carried out (see Fig. 10.14):

- $\widetilde{c}_{i,1} \to c_{i,1}$: $lb = 0, ub = 1 - \epsilon \cdot (M_i - 1)$.
- $\widetilde{c}_{i,j} \to c_{i,j}$ for $j = 2, \ldots, M_i$: $lb = c_{i,j-1} + \epsilon, ub = 1 - \epsilon \cdot (M_i - j)$.

For case (ii), the following transformation is carried out (see Fig. 10.14):

- $\widetilde{c}_{i,1} \to c_{i,1}$: $c_{i,1} = 0$.
- $\widetilde{c}_{i,j} \to c_{i,j}$ for $j = 2, \ldots, M_i - 1$: $lb = c_{i,j-1} + \epsilon, ub = 1 - \epsilon \cdot (M_i - j)$.
- $\widetilde{c}_{i,M_i} \to c_{i,M_i}$: $c_{i,M_i} = 1$.

With these transformations, it is guaranteed that the constraints on the original centers $c_{i,j}$ are met. The lower bound lb is chosen in such a manner that the minimum distance ϵ to the adjacent centers is ensured. The upper bound ub is chosen in order to leave enough space to the 1 for all centers still to be placed while meeting the minimum distance of ϵ. Figure 10.14 illustrates this transformation strategy for $M_i = 5$.

10.3.6.2 Constraints on the Heights

Constraints on the heights are typically required if the output of a look-up table is to be guaranteed to lie in a certain interval. For example, the output may be given in

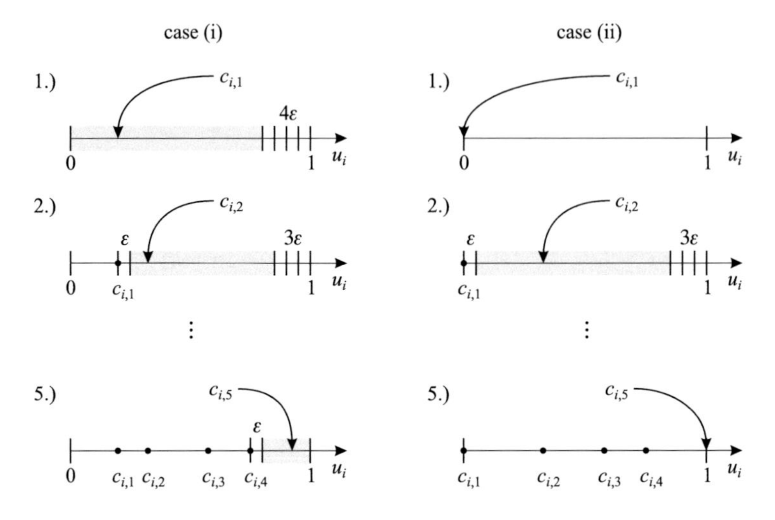

Fig. 10.14 Parameter transformation for meeting the constraints. In case (i) all centers are optimized. In case (ii) only the inner centers are optimized, while the corner centers are kept fixed. Note that the minimal and maximum values for u_i are here assumed to be 0 and 1, respectively

percent and thus lies in [0, 100], the output may be an efficiency that always is in [0, 1], or the output may be an angle that lies in [0°, 90°]. This leads to lower and upper bounds on the heights

$$lb \le w_i \le ub \tag{10.34}$$

for $i = 1, \ldots, M$. Similarly, other constraints can be incorporated in order to ensure other properties such as monotonic behavior, bounded derivatives, etc. All these kinds of linear constraints can be achieved by replacing the least squares with a quadratic programming algorithm; see Sect. 3.3. Quadratic programming is still fast enough to realize the nested optimization approach proposed in Sect. 10.3.5.

10.3.7 Properties of Look-Up Table Models

The properties of grid-based look-up table models with linear interpolation can be summarized as follows:

- *Interpolation behavior* is piecewise linear, and thus the model output cannot be differentiated at the center points. This problem can be overcome by using higher-order interpolation rules (or basis functions, respectively). The output of the model is guaranteed to stay between the minimum and maximum height, i.e., within the interval $[\min(w_i), \max(w_i)]$. This property ensures that the output of a look-up table is always in the range of physical meaning if all heights are. Furthermore, look-up tables with linear interpolation reproduce their stored points exactly. This is a drawback if the stored points are noisy measurements, but it can be seen as an advantage otherwise. In most standard applications of look-up tables, the stored points are measurements that have been averaged to reduce the noise.

- *Extrapolation behavior* does not exist. However, any kind can be defined by the user. Typically, constant extrapolation behavior is defined, i.e., the height of the nearest center determines the model output.
- *Locality* is strong. The basis functions have strictly local support, i.e., they are non-zero only in a small region of the input space. A training data sample affects only its neighboring points in the look-up table.
- *Accuracy* is medium since a large number of parameters are required to model a process to a given accuracy. This is due to the fact that the points are placed on a grid, which makes the look-up table model less flexible. It does not allow one to spend more parameters in those input regions where the process is complex and fewer parameters in the regions where the process is almost linear.
- *Smoothness* is very low since the model output is not even differentiable.
- *Sensitivity to noise* is very high since only a few training data samples are exploited to estimate the parameters owing to the strictly local basis functions. If the stored points are measurements, the noise contained in them enters the model directly without any averaging effect.
- *Parameter optimization* of the heights can be performed very fast by a least squares algorithm; see Sect. 3.1. For incorporation of constraints, quadratic programming can be used; see Sect. 3.3. The grid has to be optimized nonlinearly by either a local or a global technique; see Chaps. 4 and 5. Constraints have to be imposed in order to guarantee a meaningful grid; see Sect. 4.6.
- *Structure optimization* is very difficult to realize. No standard approaches exist to optimize the structure of a look-up table. Of course, global search techniques can be applied for structure optimization, but no specific well-suited approach exists; see Chap. 5.
- *Online adaptation* is possible. In order to avoid unpredictable difficulties with the convergence, only the linear parameters (heights) should be adapted online. This can be done by a recursive least squares (RLS) algorithm; see Sect. 3.2. Convergence is usually slow because only a few active points are adapted. Thus, the new information contained in the online measured data can be shared only by the neighboring points and does not generalize more broadly. On the other hand, the local character prevents *destructive learning effects*, also known as *unlearning* or *data interference* [12], in different operating regimes.
- *Training speed* is high if only the heights are optimized. For a complete optimization of a look-up table model, the training speed is low. Compared with other model architectures, the number of linear parameters is high, while the number of nonlinear parameters is low.
- *Evaluation speed* is very high since extremely few and cheap operations have to be carried out, especially for an equidistant grid. This is one main reason for the wide application of look-up tables in practice.

- *Curse of dimensionality* is extremely high owing to the grid-based approach. The number of inputs is thus restricted to three or four (at most).
- *Interpretation* is poor since the number of points is usually high and nothing especially meaningful is attached to them. However, a look-up table can be converted into a fuzzy model and consequently interpreted correspondingly; see Chap. 12. Note that interpretation as a fuzzy rule base is reasonably helpful only if the number of points is low! Otherwise, the huge amount of rules leads to more confusion than interpretation.
- *Incorporation of constraints* on the grid and the heights is easy; see Sect. 10.3.6.
- *Incorporation of prior knowledge* in the form of rules can be done in the fuzzy model interpretation. Typically, however, the number of points is so large that the lack of interpretability does not allow prior knowledge to be incorporated easily except with range constraints.
- *Usage* is very high for low-dimensional mappings.

10.4 Summary

The properties of linear, polynomial, and look-up table models are summarized in Table 10.1. A linear model should always be the first try. It can be successfully applied if the process is only slightly nonlinear and/or only very few data samples are available, which does not allow one to identify a more complex nonlinear model. A linear model offers advantages in almost all criteria; see Table 10.1. However, for many applications, significant performance gains can be expected from the use of nonlinear models. Nevertheless, a linear model can often be advantageously utilized in a hybrid modeling approach; see Sect. 7.6.2.

Polynomials are the classical nonlinear approximators. However, compared with other nonlinear models, they possess some important drawbacks. So their application is recommended only in some specific cases where the true structure of the process can be assumed to be approximately polynomial. The main disadvantages of polynomials are their oscillating interpolation and extrapolation behavior. However, these disadvantages fade if appropriate regularization is employed. Even with simple ridge regression for parameter estimation instead of least squares, the interpolation behavior of polynomials can improve significantly. The price to be paid is the (expensive) tuning of the regularization strength on validation data or with the help of a cross-validation technique; compare Sect. 7.3.2. The bad extrapolation behavior of polynomials also may slightly profit from regularization, but the usually undesirable tendency toward ∞ or $-\infty$ is an inherent issue.

Look-up tables represent the most widely implemented nonlinear models for low-dimensional input spaces. Their main advantage is the simple implementation and the low computational demand for evaluation. Also, constraints can be incorporated efficiently. The main drawbacks are their restriction to low-dimensional

Table 10.1 Comparison between linear, polynomial, and look-up table models

Properties	Linear	Polynomial	Look-up table
Interpolation behavior	+	−	+
Extrapolation behavior	+	−−	−
Locality	−−	−−	++
Accuracy	−−	0	0
Smoothness	++	−	−−
Sensitivity to noise	++	+	−−
Parameter optimization	++	++	++*/−−**
Structure optimization	++	0	−−
Online adaptation	+	−	0
Training speed	++	+	+*/−**
Evaluation speed	++	0	++
Curse of dimensionality	++	−	−−
Interpretation	+	0	−−
Incorporation of constraints	0	−	++
Incorporation of prior knowledge	0	−	0
Usage	++	+	++

* = linear optimization, ** = nonlinear optimization,
++ / −− = model properties are very favorable/undesirable

mappings, the non-smooth behavior for linear interpolation, and the inflexible distribution of the height parameters over the whole input space.

The severe shortcomings of the classically used models, polynomials and look-up tables, motivate the search for model architectures with better properties, resulting in neural networks and fuzzy models.

10.5 Problems

1. Write down the regression matrix for the least-squares estimation of the parameters of a polynomial of degree 3 and with two inputs.
2. Calculate the number of parameters of a polynomial of degree 3 for $1, 2, \ldots,$ and 6 inputs.
3. Why is the interpolation behavior of high-degree polynomials in many applications not desirable?
4. Explain the extrapolation of polynomials and comment on its applicability.
5. What are the main reasons for the popularity of look-up tables as nonlinear models?
6. Under which conditions does the bilinear interpolation rule for two-dimensional look-up tables yield a perfect plane between the four grid points that contribute to the model output?

7. Derive the model equation for a look-up table with linear interpolation for three inputs. (Hint: Try to generalize the one- and two-dimensional cases. How many grid points contribute to the model output?)
8. List all parameters of a look-up table. Which parameters influence the model output in a linear manner?
9. How many parameters of each kind (positions and weights) exist in dependency on the input space dimension for look-up tables with M grid points for each input?
10. What is the motivation of the nested linear/nonlinear optimization approach as proposed in Fig. 10.13?

Chapter 11
Neural Networks

Abstract This chapter introduces the two most important representatives of neural networks: (i) the radial basis function (RBF) network and (ii) the multilayer perceptron (MLP) network. Additionally, some network architectures of minor importance are also covered. A key topic of this chapter is the issue of how to map a high-dimensional input vector (or of hidden layer neurons) to a scalar quantity within each neuron of the network. The three common construction mechanism for that task are explained, analyzed, and compared: (i) ridge, (ii) radial, and (iii) tensor product construction. One main goal of this chapter is to make clear that the fundamental difference between RBF and MLP networks is not the question of locality or the shape of the activation function but rather the construction mechanism and the consequences arising from it. The third type of construction (tensor product) is treated in the next chapter.

Keywords Radial basis function network · Normalization · Multilayer perceptron · Ridge construction · Radial construction · Tensor product construction

This chapter deals with artificial neural networks for static modeling. Artificial neural networks were originally motivated by the biological structures in the brains of humans and animals, which are extremely powerful for such tasks as information processing, learning, and adaptation. Good overviews on the biological background can be found in [488, 491]. The most important characteristics of neural networks are

- large number of simple units,
- highly parallel units,
- strongly connected units,
- robustness against the failure of single units,
- learning from data.

These properties make an artificial neural network well suited for fast hardware implementations [233, 371]. Two main directions of neural network research can

O. Nelles, *Nonlinear System Identification*,
https://doi.org/10.1007/978-3-030-47439-3_11

be distinguished. On the one hand, the physician's, biologist's, and psychologist's interests are to learn more about and even model the still not well-understood fundamental properties and operation of the human and animal brain. On the other hand, the engineer's interest is to develop a universal tool for problem-solving inspired by the impressive examples of nature but without any pretension to model biological neural networks. This book addresses only the latter pursuit. In fact, most artificial neural networks used in engineering are at least as closely related to mathematics, statistics, and optimization as to the biological role model. In the following, artificial neural networks are simply called "neural networks" or "NNs" since no biological issues are addressed. Because of their biological background, many neural network publications use their own terminology. Table 11.1 gives a translation for the most important expressions into system identification terminology, which is partly taken from [499].

Sometimes it is hard to draw a clear line between neural network and non-neural network models. Here, from a pragmatic point of view, a model will be called a *neural network* if its basis functions are of the same type; see Sect. 9.2. This definition includes all neural network architectures that are addressed throughout this book. In neural network terminology, the network in Fig. 11.1 is described as follows. The node at the output is called the *output neuron*, and all output neurons together are called the *output layer* (here only a single output is considered, so the output layer consists only of one neuron). Each of the M nodes in the center that realizes a basis function is called the *hidden layer neuron*, and all these neurons

Table 11.1 Translations from neural network into system identification language

Neural network terminology	System identification and statistics terminology
Mapping or approximation	Regression
Classification	Discriminant analysis
Neural network	Model
Neuron	Basis function
Weight	Parameter
Bias or threshold	Offset or intercept
Hidden layer	Set of basis functions
Input layer	Set of inputs
Input	Independent variable
Output	Predicted value
Error	Residual
Learning or training	Estimation or optimization
Generalization	Interpolation or extrapolation
Overfitting or overtraining	High variance error
Underfitting or undertraining	High bias error
Errorbar	Confidence interval
Online learning	Sample adaptation
Offline learning	Batch adaptation

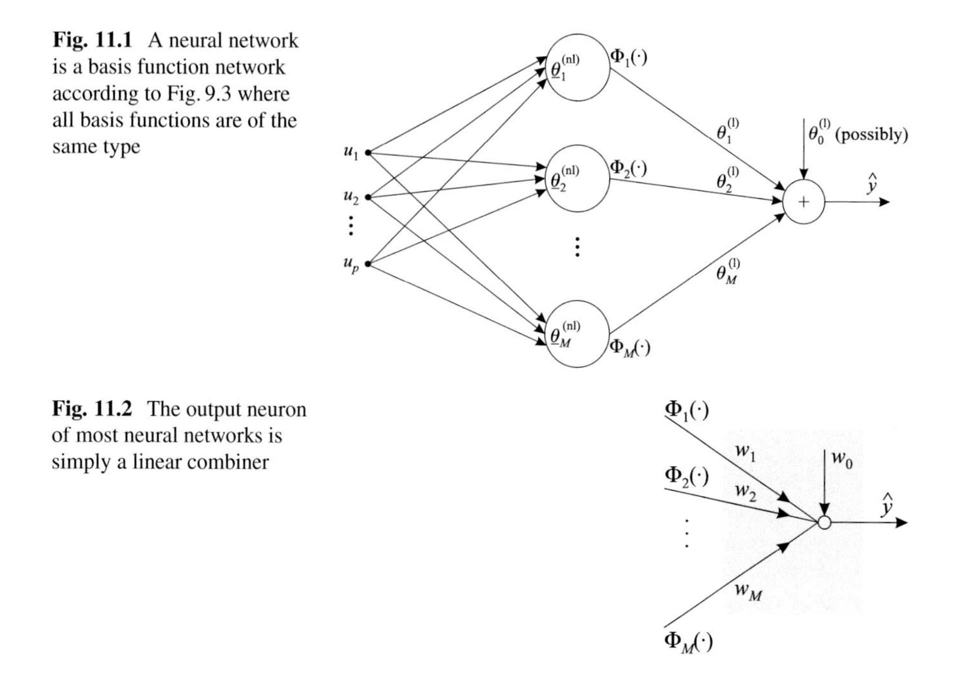

Fig. 11.1 A neural network is a basis function network according to Fig. 9.3 where all basis functions are of the same type

Fig. 11.2 The output neuron of most neural networks is simply a linear combiner

together are called the *hidden layer*. Finally, the inputs are sometimes denoted as *input neurons*, and all of them together are called the *input layer*. However, these neurons only fan out the inputs to all hidden layer neurons and do not carry out any real calculations. Because the literature is not standardized on whether the inputs should be counted as a layer or not, the network in Fig. 11.1 can be called a two- or a three-layer neural network. In order to avoid any confusion, it is sensible to call it a neural network with one hidden layer. There exist other neural network architectures that cannot be described by the scheme in Fig. 11.1. For example, the number of hidden layers may be larger, or the whole structure may be different. However, the architectures discussed here all match the standard (or extended) basis function formulation in Sects. 9.2 and 9.3 and the network structure depicted in Fig. 11.1.

For a neural network, the linear parameters associated with the output neuron(s) are called *output weights*:

$$\theta_i^{(l)} = w_i \,. \tag{11.1}$$

The output neuron is usually a linear combination of the hidden layer neurons (basis functions) $\Phi_i(\cdot)$ with an additional offset w_0, which is sometimes called "bias" or "threshold"; see Fig. 11.2. Each hidden layer neuron output is weighted with its corresponding weight. Since the hidden layer outputs lie in some interval (e.g., [0 1] or [−1 1]), the output neuron determines the scaling (amplitudes) and the operation point of the NN. The neural network architectures addressed here – in fact, all models described by the basis functions formulation (9.2) – can be distinguished solely by their specific type of hidden layer neurons.

The two most common neural network architectures – the multilayer perceptron (MLP) and the radial basis function (RBF) network – are introduced in Sects. 11.2 and 11.3. The third very frequently applied class of architectures are the neuro-fuzzy networks, which are treated in Chap. 12. Section 11.4 gives a brief overview of some other interesting but less widespread neural network approaches. Before these specific architectures are analyzed, Sect. 11.1 starts with a discussion of different construction mechanisms that allow one to generalize one-dimensional mappings to higher input space dimensionalities.

Good neural network literature for a more detailed discussion is briefly summarized in the following. Haykin [233] gives an extensive overview, ranging from the biological background to most neural network architectures and learning algorithms to hardware implementation. The book is written from a signal processing viewpoint. Bishop [57] focuses on multilayer perceptrons and radial basis function networks for the solution of pattern recognition and classification problems. He analyzes the neural networks deeply in the framework of classical Bayesian statistics. He reveals and clarifies many often unnoticed links and relationships. Similar in concept to Bishop's is Ripley's book [487] but with an even stronger statistical and mathematical perspective. A good historical review, especially from the hardware point of view, is given by Hecht-Nielsen [235]. Brown and Harris [84] have written an excellent neural network book from an engineering point of view that bridges the gap to fuzzy systems. It focuses on neuro-fuzzy models and associative memory networks; other network architectures are hardly discussed.

An excellent source for neural network literature is the answers to the frequently asked questions (FAQs) of the neural network newsgroup on the Internet [499].

11.1 Construction Mechanisms

The basis functions $\Phi_i(\cdot)$ in (9.2) are generally multidimensional, i.e., their dimensionality is defined by the number of inputs $p = \dim\{\underline{u}\}$. For all neural network approaches and many other model architectures, however, the multivariate basis functions are constructed by simple one-dimensional functions. Figure 11.3 illustrates the operation of such construction mechanisms. In the neural network context, the one-dimensional function is called the *activation function*. Note that the activation function that maps the scalar x to the neuron output y is denoted in the following by $g(\cdot)$. In contrast, the basis function $\Phi(\cdot)$ characterizes the multidimensional mapping from the neuron inputs to the neuron output and thus depends on the construction mechanism. The three most important construction mechanisms are introduced in the following subsections: ridge, radial, and tensor product construction.

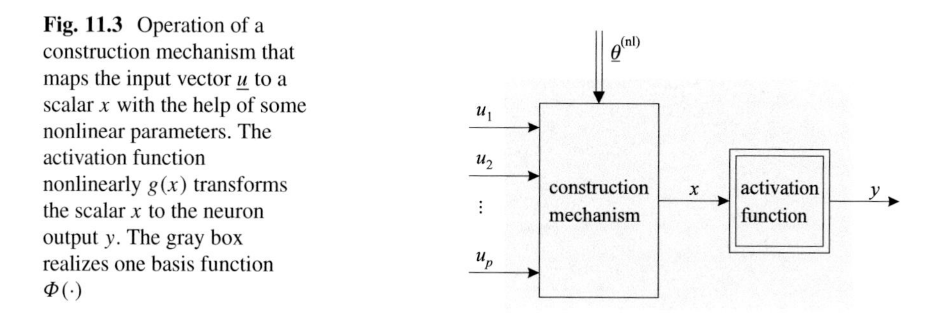

Fig. 11.3 Operation of a construction mechanism that maps the input vector $\underline{u}$ to a scalar x with the help of some nonlinear parameters. The activation function nonlinearly $g(x)$ transforms the scalar x to the neuron output y. The gray box realizes one basis function $\Phi(\cdot)$

11.1.1 Ridge Construction

Ridge construction is based on a projection mechanism as utilized for multilayer perceptrons (MLPs); see Sect. 11.2. The basis functions operate on a scalar x, which is generated by projecting the input vector on the nonlinear parameter vector; see Fig. 11.3. This can be realized by the scalar product

$$x = \underline{\theta}^{(\mathrm{nl})}\underline{\tilde{u}} = \theta_0^{(\mathrm{nl})}u_0 + \theta_1^{(\mathrm{nl})}u_1 + \ldots + \theta_p^{(\mathrm{nl})}u_p \tag{11.2}$$

with the augmented input vector

$$\underline{\tilde{u}} = \begin{bmatrix} 1 \\ \underline{u} \end{bmatrix} = \begin{bmatrix} 1 \\ u_1 \\ \vdots \\ u_p \end{bmatrix} \tag{11.3}$$

that contains the dummy input $u_0 = 1$ in order to incorporate an offset value into (11.2). So, if the nonlinear parameter vector and the augmented input vector have the same direction, the activation x is maximum; if they have opposite direction, x is minimal; and if they are orthogonal, x is zero.

This means that the multivariate basis functions constructed with the ridge mechanism possess only one direction in which they vary – namely, the direction of the nonlinear parameter vector. In all other orthogonal directions, the multivariate basis functions stay constant; see Fig. 11.4. This projection mechanism makes ridge construction well suited for high-dimensional input spaces; see Sect. 7.6.2.

11.1.2 Radial Construction

The radial construction is utilized for radial basis function networks. The scalar x is calculated as the distance between the input vector and the center of the basis function:

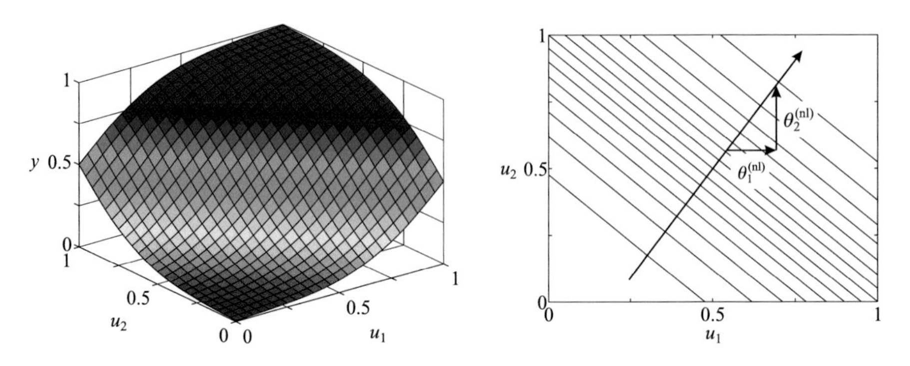

Fig. 11.4 A basis function obtained by ridge construction (left) varies only in one direction. The contour plot of the basis function (right) shows the direction of nonlinearity. The nonlinear parameters associated to the inputs determine this direction and the slope of the basis function

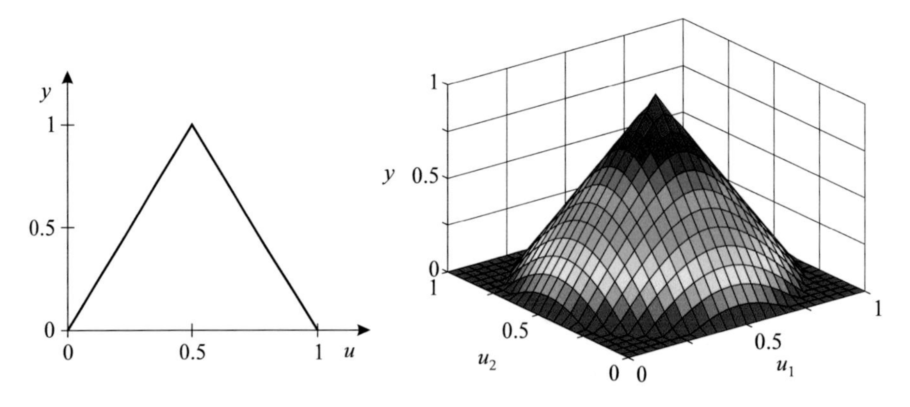

Fig. 11.5 A one- and two-dimensional radial basis function

$$x = ||\underline{u} - \underline{c}|| = \sqrt{(\underline{u} - \underline{c})^T (\underline{u} - \underline{c})}$$
$$= \sqrt{(u_1 - c_1)^2 + (u_2 - c_2)^2 + \ldots + (u_p - c_p)^2}\,, \tag{11.4}$$

where the nonlinear parameter vector contains the center vector of the basis function $\underline{c} = [c_1\ c_2\ \cdots\ c_p]^T$. Figure 11.5 illustrates a one- and two-dimensional example of a triangular radial basis function.

If, instead of the Euclidean norm in (11.4), the Mahalanobis norm is used, the distance is transformed according to the covariance matrix $\underline{\Sigma}$ (see Sect. 6.2.1):

$$x = ||\underline{u} - \underline{c}||_{\Sigma} = \sqrt{(\underline{u} - \underline{c})^T \underline{\Sigma}^{-1} (\underline{u} - \underline{c})} \tag{11.5}$$

where the nonlinear parameter vector additionally contains the parameters in the covariance matrix $\underline{\Sigma}$. The inverse covariance matrix $\underline{\Sigma}^{-1}$ scales and rotates the axes.

$\underline{\Sigma}$ characterizes the ellipses in Fig. 11.6c. For the special case where the inverse covariance matrix is equal to the identity matrix ($\underline{\Sigma}^{-1} = \underline{I}$), the Mahalanobis norm is equal to the Euclidean norm. For

$$\underline{\Sigma}^{-1} = \begin{bmatrix} 1/\sigma_1^2 & 0 & \cdots & 0 \\ 0 & 1/\sigma_2^2 & \cdots & 0 \\ \vdots & \vdots & \ddots & \vdots \\ 0 & 0 & \cdots & 1/\sigma_p^2 \end{bmatrix}, \tag{11.6}$$

where p denotes the input space dimension; the Mahalanobis norm is equal to the Euclidean norm with the scaled inputs $u_l^{(\text{scaled})} = u_l/\sigma_l$. In the most general case, the inverse covariance matrix scales and rotates the input axes. Figure 11.6 summarizes these distance measures. Note that no matter which norm is chosen, it can be replaced by the Euclidean norm by transforming the input axes. Therefore, the function in Fig. 11.7 is still called a radial basis function. Despite the fact that it is not radial (not even symmetric) with respect to the original inputs u_1 and u_2, it is radial with respect to appropriately transformed input axes.

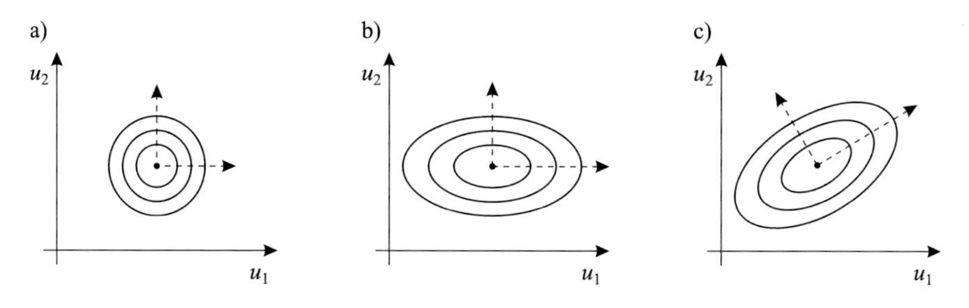

Fig. 11.6 Lines with equal distance for different norms: (**a**) Euclidean ($\underline{\Sigma}^{-1} = \underline{I}$), (**b**) diagonal ($\underline{\Sigma}^{-1}$ = diagonal), and (**c**) Mahalanobis norm ($\underline{\Sigma}^{-1}$ = symmetric)

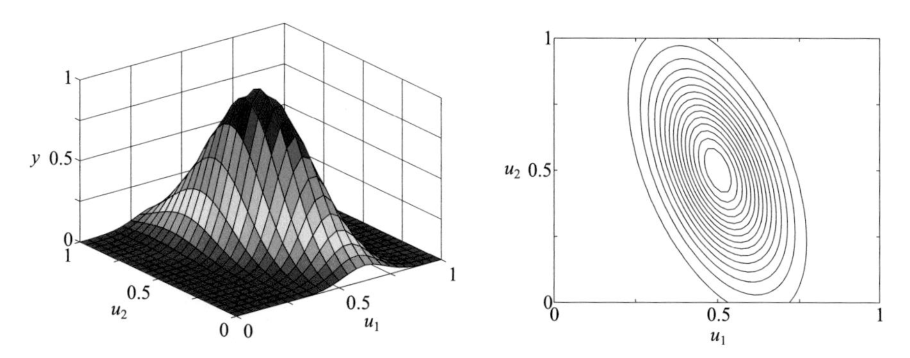

Fig. 11.7 A Gaussian radial basis function

11.1.3 Tensor Product Construction

The tensor product is the construction mechanism utilized for many types of neuro-fuzzy models, spline models, and look-up tables; see Chap. 12. It operates on a set of univariate functions that are defined for each input $u_1, \ldots, u_p$. The parameters of these univariate functions are gathered in the nonlinear parameter vector. The basis functions are calculated by forming the tensor product of these univariate functions, i.e., each function is multiplied with each function of the other dimensions. For example, M_i may denote the number of univariate functions defined for input u_i, $i = 1, \ldots, p$; then the basis functions are constructed by

$$f_1(\cdot) = g_{11}(\cdot) \cdot g_{21}(\cdot) \cdot \ldots \cdot g_{p1}(\cdot) \tag{11.7a}$$

$$f_2(\cdot) = g_{12}(\cdot) \cdot g_{21}(\cdot) \cdot \ldots \cdot g_{p1}(\cdot) \tag{11.7b}$$

$$\vdots$$

$$f_{M_1}(\cdot) = g_{1M_1}(\cdot) \cdot g_{21}(\cdot) \cdot \ldots \cdot g_{p1}(\cdot) \tag{11.7c}$$

$$f_{M_1+1}(\cdot) = g_{11}(\cdot) \cdot g_{22}(\cdot) \cdot \ldots \cdot g_{p1}(\cdot) \tag{11.7d}$$

$$\vdots$$

$$f_{2M_1}(\cdot) = g_{1M_1}(\cdot) \cdot g_{22}(\cdot) \cdot \ldots \cdot g_{p1}(\cdot) \tag{11.7e}$$

$$\vdots$$

$$f_{M_1 M_2}(\cdot) = g_{1M_1}(\cdot) \cdot g_{2M_2}(\cdot) \cdot \ldots \cdot g_{p1}(\cdot) \tag{11.7f}$$

$$\vdots$$

$$f_{M_1 M_2 \ldots M_p}(\cdot) = g_{1M_1}(\cdot) \cdot g_{2M_2}(\cdot) \cdot \ldots \cdot g_{pM_p}(\cdot)\,, \tag{11.7g}$$

where $g_{ij}(\cdot)$ denotes the jth univariate function of input u_i. Thus, the number of basis functions is (see (10.25))

$$\prod_{i=1}^{p} M_i\,. \tag{11.8}$$

The tensor product construction is suitable only for very low-dimensional mappings, say $p \leq 4$, since the number of basis functions grows exponentially with the input space dimensionality. The tensor product construction partitions the input space into a multidimensional grid and therefore fully underlies the curse of dimensionality. Figures 11.8 and 11.9 illustrate the operation of the tensor product construction for a simple two-dimensional example.

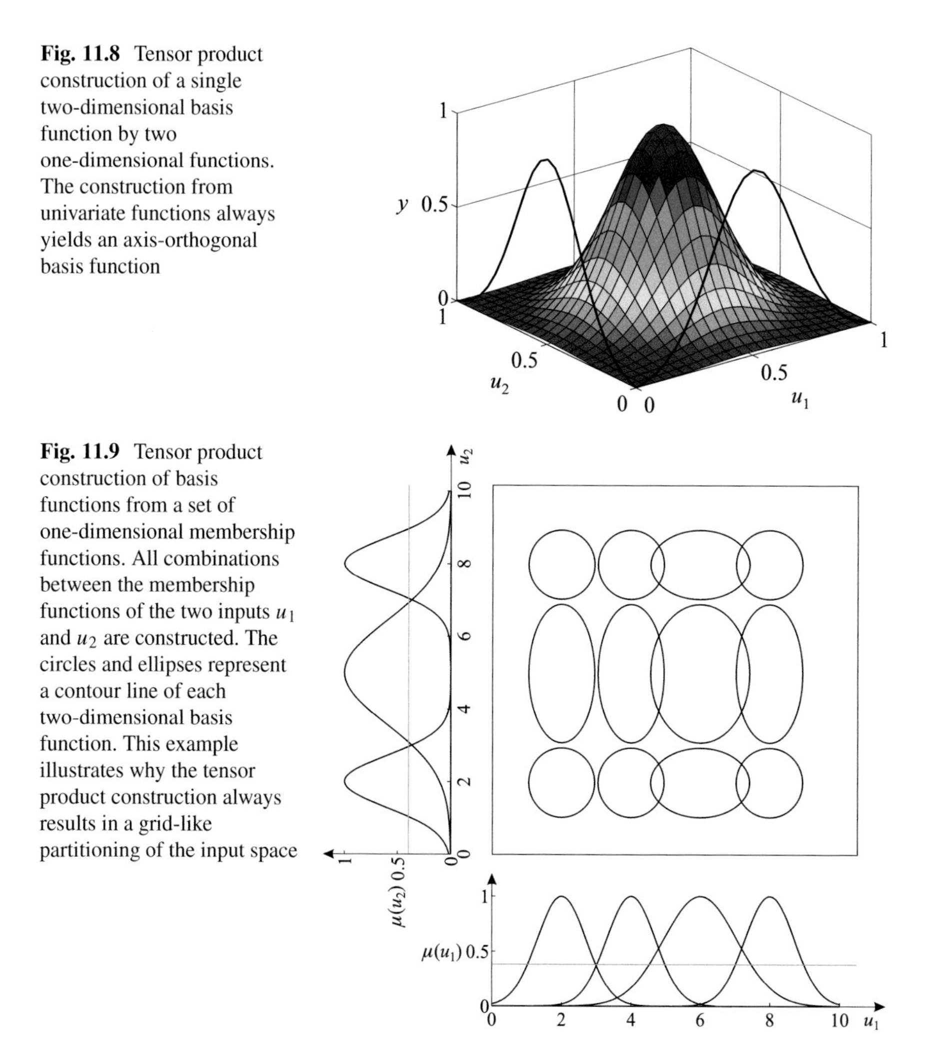

Fig. 11.8 Tensor product construction of a single two-dimensional basis function by two one-dimensional functions. The construction from univariate functions always yields an axis-orthogonal basis function

Fig. 11.9 Tensor product construction of basis functions from a set of one-dimensional membership functions. All combinations between the membership functions of the two inputs u_1 and u_2 are constructed. The circles and ellipses represent a contour line of each two-dimensional basis function. This example illustrates why the tensor product construction always results in a grid-like partitioning of the input space

11.2 Multilayer Perceptron (MLP) Network

The multilayer perceptron (MLP) is the most widely known and used neural network architecture. In many publications, the MLP is even used as a synonym for NN. The reason for this is the breakthrough in neural network research that came with the very popular book by Rumelhart, Hinton, and Williams in 1986 [491]. It contributed significantly to the start of a new neural network boom by overcoming the restrictions of perceptrons pointed out by Minsky and Papert in 1969 [378]. Although modern research has revealed considerable drawbacks of the MLP with respect to many applications, it is still the favorite general-purpose neural network.

This section is organized as follows. First, the MLP neuron is introduced and its operation is illustrated. Second, the MLP network structure is presented. In Sect. 11.2.3 the famous backpropagation algorithm is discussed, and it is extended to a more generalized approach for training of MLP networks in Sect. 11.2.4. Next, the advantages and drawbacks of the MLP are analyzed. Finally, Sects. 11.2.8 and 11.2.7 introduce some extensions of the standard MLP network architecture.

11.2.1 MLP Neuron

Figure 11.10 shows a hidden neuron of a multilayer perceptron. This single neuron is called a *perceptron*. The operation of this neuron can be split into two parts. First, ridge construction is used to project the inputs $\underline{u} = [u_1\ u_2\ \cdots\ u_p]^T$ on the weights. In the second part, the nonlinear activation function $g(x)$ transforms the projection result. Typically, the activation function is chosen to be of saturation type. Common choices are sigmoid functions such as the logistic function

$$g(x) = \text{logistic}(x) = \frac{1}{1+\exp(-x)} \tag{11.9}$$

and the hyperbolic tangent

$$g(x) = \tanh(x) = \frac{\exp(x)-\exp(-x)}{\exp(x)+\exp(-x)} = \frac{1-\exp(-2x)}{1+\exp(-2x)}\,. \tag{11.10}$$

Both functions are shown in Fig. 11.11. They can be transformed into each other by

$$\begin{aligned} \tanh(x) &= \frac{1-\exp(-2x)}{1+\exp(-2x)} = \frac{2}{1+\exp(-2x)} + \frac{-1-\exp(-2x)}{1+\exp(-2x)} \\ &= 2\,\text{logistic}(2x) - 1\,. \end{aligned} \tag{11.11}$$

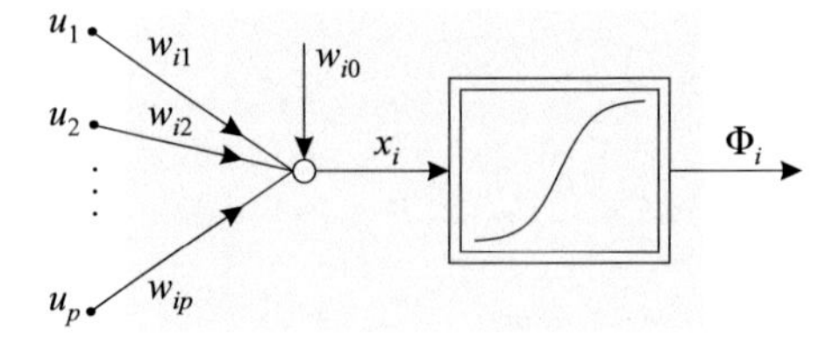

Fig. 11.10 A perceptron: the ith hidden neuron of a multilayer perceptron

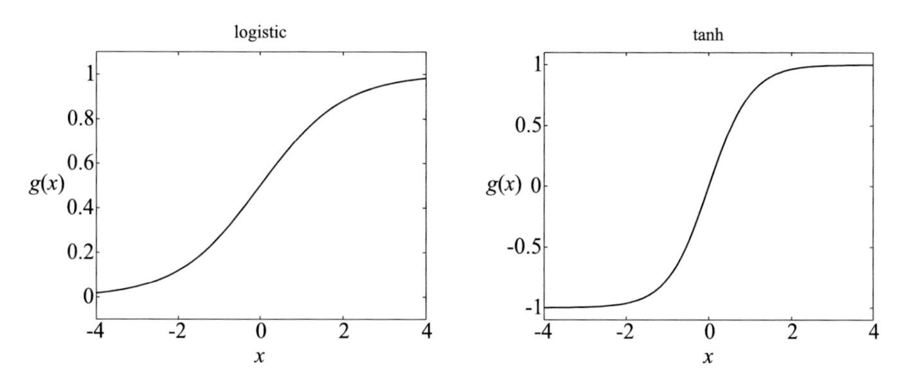

Fig. 11.11 Typical activation functions for the perceptron

The two functions share the interesting property that their derivative can be expressed as a simple function of their output:

$$\frac{d\,\Phi_i}{dx} = \frac{d\,\text{logistic}(x)}{dx} = \frac{\exp(-x)}{(1+\exp(-x))^2} = \frac{1+\exp(-x)-1}{(1+\exp(-x))^2}$$
$$= \frac{1}{1+\exp(-x)} - \frac{1}{(1+\exp(-x))^2} = \Phi_i - \Phi_i^2 = \Phi_i(1-\Phi_i)\,. \quad (11.12)$$

$$\frac{d\,\Phi_i}{dx} = \frac{d\,\tanh(x)}{dx} = \frac{1}{\cosh^2(x)} = \frac{\cosh^2(x)-\sinh^2(x)}{\cosh^2(x)}$$
$$= 1-\tanh^2(x) = 1-\Phi_i^2\,. \quad (11.13)$$

These derivatives are required in any gradient-based optimization technique applied for training of an MLP network; see Sect. 11.2.3.

The perceptron depicted in Fig. 11.10 depends on nonlinear hidden layer parameters. These parameters are called *hidden layer weights*:

$$\underline{\theta}_i^{(\text{nl})} = [w_{i0}\ w_{i1}\ w_{i2}\ \cdots\ w_{ip}]^T\,. \quad (11.14)$$

The weights w_{i0} realize an offset, and sometimes are called "bias" or "threshold". Figures 11.12 and 11.13 illustrate how these hidden layer weights determine the shape of the basis functions.

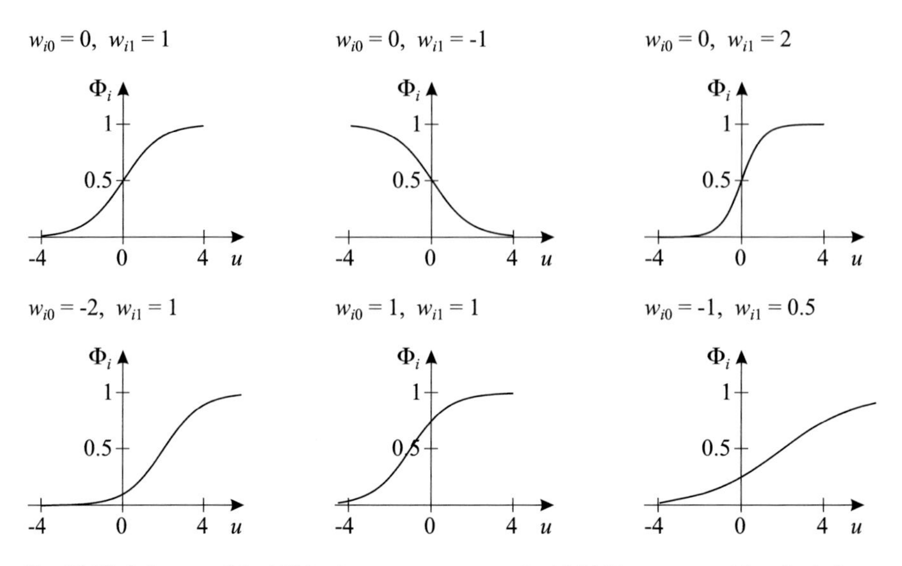

Fig. 11.12 Influence of the hidden layer parameters on the ith hidden neuron with a single input u. While the offset weight w_{i0} determines the activation function's position, the weight w_{i1} determines the activation function's slope

11.2.2 Network Structure

If several perceptron neurons are used in parallel and are connected to an output neuron, the multilayer perceptron network with one hidden layer is obtained; see Fig. 11.14. In basis function formulation, the MLP can be written as

$$\hat{y} = \sum_{i=0}^{M} w_i \Phi_i \left(\sum_{j=0}^{p} w_{ij} u_j \right) \qquad \text{with } \Phi_0(\cdot) = 1 \text{ and } u_0 = 1 \tag{11.15}$$

with the output layer weights w_i and the hidden layer weights w_{ij}. The total number of parameters of an MLP network is

$$M(p+1) + M + 1\,, \tag{11.16}$$

where M is the number of hidden layer neurons and p is the number of inputs. Since the inputs are given by the problem, the number of hidden layer neurons allows the user to control the network complexity, i.e., the number of parameters.

An MLP network is a *universal approximator* [261]. This means that an MLP can approximate *any* smooth function to an arbitrary degree of accuracy as the

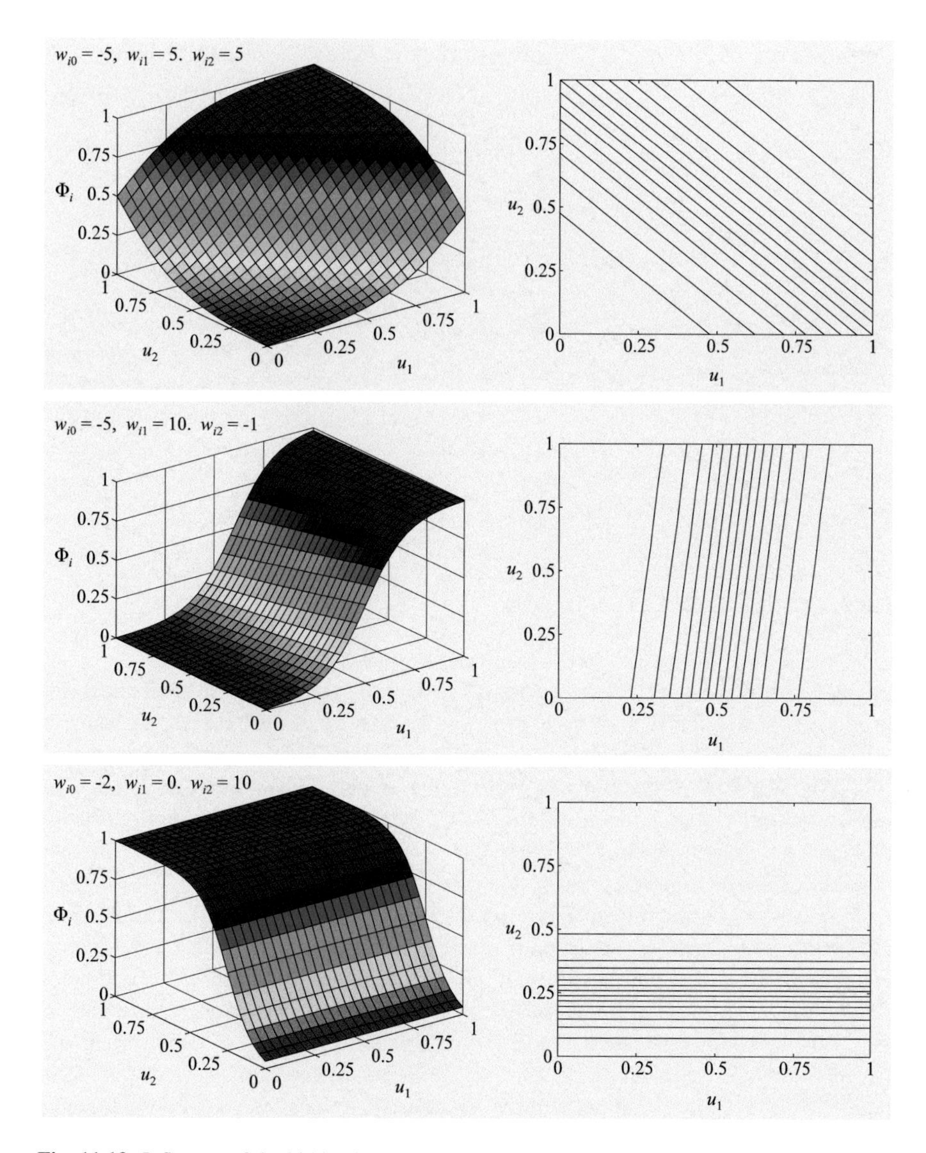

Fig. 11.13 Influence of the hidden layer parameters on the ith hidden neuron with two inputs u_1 and u_2. While the offset weight w_{i0} determines the activation function's distance to the origin, the weights w_{i1} and w_{i2} determine the slopes in the u_1- and u_2-directions. The vector $[w_{i1}\ w_{i2}]^T$ points toward the direction of nonlinearity of the basis function. Orthogonal to $[w_{i1}\ w_{i2}]^T$, the basis function stays constant

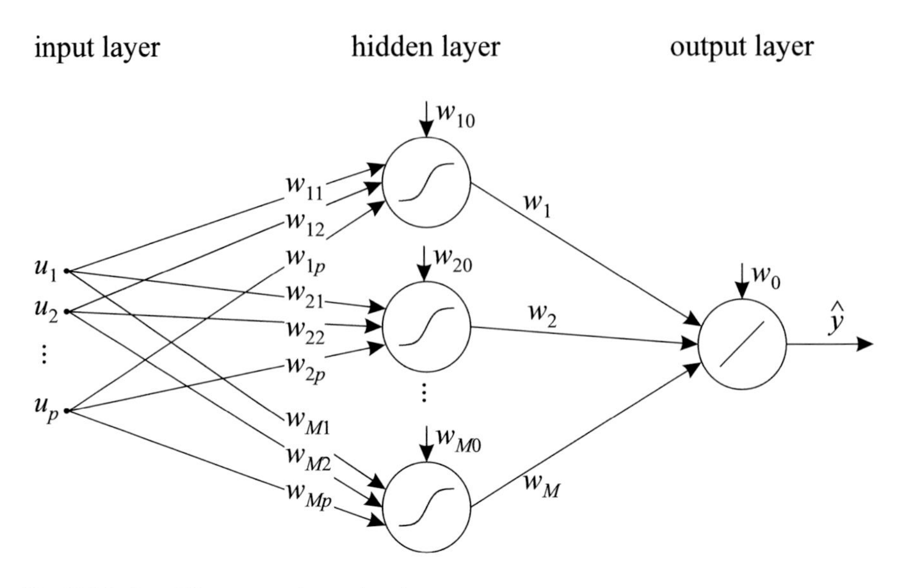

Fig. 11.14 A multilayer perceptron network

number of hidden layer neuron increases. The universal approximation capability is an important property since it justifies the application of the MLP to any function approximation problem. However, virtually all other widely used approximators (polynomials, fuzzy models, and most other neural network architectures) are universal approximators too. Furthermore, the proof for the universal approximation ability is not constructive in the sense that it tells the user *how many* hidden layer neurons would be required to achieve a given accuracy.

The reason why the universal approximation capability of the MLP has attracted such attention is that, in contrast to many other approximators, the MLP output is computed by the combination of *one*-dimensional functions (the activation functions). This is a direct consequence of the ridge construction mechanism. In fact, it is not intuitively understandable that a general function of many inputs can be approximated with arbitrary accuracy by a set of one-dimensional functions. Note that the approximation abilities of the MLP have no *direct* relationship to Kolmogorov's theorem, which states that a multidimensional function can be *exactly* represented by one-dimensional functions [187]. The famous theorem by Kolmogorov from 1957 disproved a conjecture by Hilbert in context with his 13th problem from the famous list of Hilbert's 23 unsolved mathematical problems in 1900. Girosi and Poggio [187] point out that in Kolmogorov's theorem *highly non-smooth* and *problem-specific* one-dimensional functions are required in order to *exactly* represent a multidimensional function, while an MLP consists of *smooth* and *standardized* one-dimensional functions that are sufficient to *approximate* a multidimensional function.

The MLP network as described so far represents only the most commonly applied MLP type. Different variants exist. Sometimes the output neuron is not of the pure "linear combination" type but is chosen as a complete perceptron. This means that an additional activation function at the output is used. Of course, this restricts the MLP output to the interval $[0, 1]$ or $[-1, 1]$. Furthermore, the output layer weights become nonlinear, and thus training becomes harder. Another possible extension is the use of more than one hidden layer. Multiple hidden layers make the network much more powerful and complex; see Sect. 11.2.8. In principle, an MLP can have an arbitrary number of hidden layers, but only one or two are common in practice.

An MLP network consists of two types of parameters:

- *Output layer weights* are linear parameters. They determine the amplitudes of the basis functions and the operating point.
- *Hidden layer weights* are nonlinear parameters and determine the directions, slopes, and positions of the basis functions.

An MLP network is trained by the optimization of these weights. Several MLP training strategies are discussed in Sect. 11.2.4. First, the famous backpropagation algorithm is reviewed as a foundation for the application of optimization techniques.

11.2.3 Backpropagation

Strictly speaking, the backpropagation algorithm is only a method for the computation of the gradients of an MLP network output with respect to its weights. In fact, there is nothing special about it since backpropagation is identical to the application of the well-known chain rule for derivative calculation. Nevertheless, it took a very long time until it was realized within the neural network community how simply the hidden layer weights can be optimized. One important stimulus for the neural network boom in the late 1980s and throughout the 1990s was the rediscovery of backpropagation by Rumelhart, Hinton, and Williams in 1986 [491]. It was first found by Werbos in 1974 [596]. Backpropagation solves the so-called *credit assignment problem*, that is, the question of which fraction of the overall model error should be assigned to each of the hidden layer neurons in order to optimize the hidden layer weights. Prior to the discovery of backpropagation, it was only possible to train a simple perceptron that possesses no hidden layers and thus is, of course, no universal approximator [378]. The importance of backpropagation is so large that in the literature it is often is used as a synonym for the applied training method or even for the neural network architecture (namely, the MLP). Here, the term *backpropagation* is used for what it really is, a method for gradient calculation of a neural network.

The derivatives of the MLP output with respect to the ith *output layer weight* are ($i = 0, \ldots, M$)

$$\frac{\partial \hat{y}}{\partial w_i} = \Phi_i \qquad \text{with } \Phi_0 = 1\,. \tag{11.17}$$

The derivatives of the MLP output with respect to the *hidden layer weights* are ($i = 1, \ldots, M$, $j = 0, \ldots, p$)

$$\frac{\partial \hat{y}}{\partial w_{ij}} = w_i \frac{dg(x)}{dx} u_j \qquad \text{with } u_0 = 1 \tag{11.18}$$

for the weight at the connection between the jth input and the ith hidden neuron. In (11.18) the expressions (11.12) or (11.13) for the derivatives of the activation functions $g(x)$ can be utilized. For example, with $g(x) = \tanh(x)$, (11.18) becomes

$$\frac{\partial \hat{y}}{\partial w_{ij}} = w_i (1 - \Phi_i^2) u_j \qquad \text{with } u_0 = 1\,. \tag{11.19}$$

Note that the basis functions Φ_i in the above equations depend on the network inputs and the hidden layer weights. These arguments are omitted for better readability.

Since the above gradient expressions can be thought to be constructed by propagating the model error back through the network, the algorithm is called *back-propagation*. This becomes more obvious if multiple hidden layers are considered; see Sect. 11.2.8.

11.2.4 MLP Training

Basically, three different strategies for training of an MLP network can be distinguished.

- *Regulated activation weight neural network (RAWN)*: Initialization of the hidden layer weights and subsequent least squares estimation of the output layer weights.
- *Nonlinear optimization of the MLP*: All weights are simultaneously optimized by a local or global nonlinear optimization technique.
- *Staggered training of the MLP*: A combination of both approaches.

In any case, a good method for the initialization of the hidden layer weights is required. This issue is discussed first.

11.2.4.1 Initialization

Unfortunately, the hidden layer parameters are not easily interpreted since humans are not able to think in and visualize directions in high-dimensional spaces. Thus, they can hardly be initialized by prior knowledge. Of course, the optimal hidden layer weights are highly problem-dependent but some generally valid guidelines can be given.

The simplest approach for the initialization of the hidden layer weights is to choose them randomly. This, however, will generate a number of neurons whose activation functions are in their saturation for all data because some hidden layer weights will be large. These saturated neurons give virtually the same response for all inputs contained in the data. Thus, they basically behave like constants, which is totally useless since the output layer weight w_0 already realizes an offset value. All saturated neurons are (almost) redundant, do not improve the approximation capabilities of the network, and even make the training procedure more poorly conditioned and more time-consuming. Furthermore, it is very difficult for any optimization technique to move saturated neurons in regions where they are active since the model error is highly insensitive with respect to the saturated neuron's weights. So the goal of a good initialization method must be to avoid highly saturated neurons.

In order to do so, it is important to recall how the input of the activation function of the ith hidden layer neuron is calculated:

$$x_i = w_{i0} + w_{i1}u_1 + w_{i2}u_2 + \ldots + w_{ip}u_p\,. \qquad (11.20)$$

If $x_i = 0$ then the activation function operates in its almost linear region, while $x_i \to -\infty$ or $x_i \to \infty$ drives it into saturation. Thus, a reasonable initialization should take care that the activations x_i lie in some interval around 0, say $[-5, 5]$. This can be accomplished by choosing very small weights [167, 344]. It is important, however, that all inputs are similarly scaled. Otherwise, a huge input u_j in (11.20) would dominate all others, which would make the network's output almost solely dependent on u_j. To avoid this effect, either the data should be normalized or standardized, or the initial values of the weights should reflect the amplitude range of the associated input so that the products $w_{ij}u_j$ are all within the same range for the given data. In more advanced initialization schemes it may be possible to control the distribution of the weight vector directions, slopes, and distances from the origin.

11.2.4.2 Regulated Activation Weight Neural Network (RAWN) or Extreme Learning Machine

The simplest way to train an MLP network is to initialize the nonlinear hidden layer weights and subsequently estimate the output layer weights by least squares. This approach, proposed in [558], is called *regulated activation weight neural network*

(RAWN). Since the LS estimation of the output layer weights is very efficient, the only obvious problem with the RAWN approach is that the hidden layer weights are not adapted to the specific problem. Thus the RAWN approach requires many more hidden layer neurons than the nonlinear optimization discussed in the next paragraph.

Since the importance of an appropriate choice of the projection directions becomes more severe as the input dimensionality increases, the RAWN approach becomes less efficient. For very high-dimensional problems, the RAWN idea is not feasible.

A possible extension and improvement of the RAWN approach is to replace the least squares by a linear subset selection technique such as orthogonal least squares (OLS); see Sect. 3.4. Then it is possible to initialize a large number of hidden neurons and to select only the significant ones. This strategy allows one to extend the applicability of RAWN, but nevertheless, it is limited to quite low-dimensional problems. This is due to the fact that the number of potential hidden neurons has to increase with input dimensionality in order to cover the input space at an equal density and the OLS computational demand grows strongly with an increasing number of potential regressors.

Another alternative approach with similar ideas is to replace least squares estimation with ridge regression (Sect. 3.1.4). The adjustable amount of regularization allows handling arbitrarily big MLPs (any number of hidden layer neurons) without overfitting. These ideas have become very popular recently and are known under the name *extreme learning machine*; see [264] for an overview.

11.2.4.3 Nonlinear Optimization of the MLP

By far the most common approach for training an MLP network is nonlinear local optimization of the network weights. In the early days of MLP training (late 1980s), typically simple gradient-based learning rules were used, such as

$$\underline{\theta}_k = \underline{\theta}_{k-1} - \eta \frac{\partial J_k}{\partial \underline{\theta}_{k-1}}, \tag{11.21}$$

where

$$\underline{\theta} = [w_0\ w_1\ \cdots\ w_M\ w_{10}\ w_{11}\ \cdots\ w_{1p}\ w_{20}\ w_{21}\ \cdots\ w_{Mp}]^T \tag{11.22}$$

is the weight vector containing *all* network weights, η is a *fixed* step size called the *learning rate*, and J is the loss function to be optimized. For the loss function J, the actual squared error was normally used:

$$J_k = e_k^2 = (y_k - \hat{y}_k)^2, \tag{11.23}$$

which corresponds to *sample adaptation* in contrast to *batch adaptation*

$$J = \sum_{i=1}^{N} e_i^2 = \sum_{i=1}^{N} (y_i - \hat{y}_i)^2 \,, \tag{11.24}$$

where the *sum* of squared errors over the whole training data set is used; see Sect. 4.1. In either case, backpropagation is utilized for the calculation of the loss function's gradient.

Since the weight update in (11.21) is primitive, several problems can occur during training. Owing to the first-order gradient nature of (11.21), convergence is very slow. Furthermore, the fixed step size η may be chosen too large, leading to oscillations or even divergence; or it may be chosen too small, yielding extremely slow convergence. Often it leads to the "zigzagging" effect; see Sect. 4.4.3. A large number of remedies, extensions, and improvements to (11.21) exist, such as an adaptive learning rate, weight or neuron individual learning rates, the introduction of a momentum term proportional to $\Delta\underline{\theta}_{k-1} = \underline{\theta}_{k-1} - \underline{\theta}_{k-2}$, the Rprop and Quickprop algorithms, etc. [233, 499]. In essence, all these approaches are advanced optimization techniques compared with (11.21), which is a fixed step size steepest descent algorithm; see Sect. 4.4.3.

State-of-the-art training of an MLP network is performed by the Levenberg-Marquardt nonlinear least squares (Sect. 4.5.2) [207] or a quasi-Newton (Sect. 4.4.5) optimization technique for small- and medium-sized networks and a reduced memory version of the Levenberg-Marquardt or quasi-Newton techniques or a conjugate gradient algorithm (Sect. 4.4.6) for large networks ("small," "medium," and "large" refer to the total number of network weights).

Global search techniques are rarely applied for optimization of the MLP network weights because convergence is very slow and the local optima problem is not very severe for MLP networks. This is due to the fact that in most applications MLP networks are trained with regularization, i.e., a network that is too complex is trained in order to guarantee high model flexibility, and overfitting is avoided by means of early stopping, weight decay, or the application of any other regularization technique; see Sect. 7.5. This means that the global optimum of the original loss function in (11.23) or (11.24) is *not* the goal of optimization any more, and thus difficulties with convergence to local optima are of smaller importance. Global optimization techniques are, however, frequently applied in the context of network structure optimization; see Sect. 7.4.

11.2.4.4 Combined Training Methods for the MLP

One weakness of the nonlinear optimization approach for MLP training discussed in the previous paragraph is that the linearity of the output layer weights is not explicitly exploited. This weakness can be overcome by estimating the output layer weights by least squares. Basically, the following two alternative strategies exist; see the introduction to Chap. 5.

- *Staggered training of the MLP*: The hidden layer weights and the output layer weights can be optimized subsequently, i.e., first, the output layer weights are optimized by LS, while the hidden layer weights are kept fixed; second, the hidden layer weights are optimized by a nonlinear optimization technique, while the output layer weights are kept fixed. These two steps are repeated until the termination criterion is met.
- *Nested training of the MLP*: The LS estimation of the output layer weights is incorporated into the loss function evaluation of a nonlinear optimization technique that optimizes only the hidden layer weights. Thus, in each iteration of the nonlinear optimization technique, the optimal output layer weights are computed by LS.

It depends on the specific problem whether one of these combined approaches or the nonlinear optimization of all weights converges faster.

11.2.5 Simulation Examples

The simple examples in this section will illustrate how an MLP network functions. The same examples are utilized for a demonstration of RBF and fuzzy model architectures. The results obtained are not suited for benchmarking different network architectures because they are only one-dimensional and thus may not transfer to real-world problems.

In the following example, MLP networks with logistic activation functions in the hidden layer and a linear activation function in the output layer are used. The simple backpropagation algorithm (first-order method) is applied for training the networks. All parameters are initialized randomly in the interval $[-1, 1]$, which avoids saturation of the activation functions because the input lies between 0 and 1.

The function approximation results for an MLP network with one hidden neuron shown in Fig. 11.15a, b demonstrate that even for such a simple network, two equivalent optima exist. While in Fig. 11.15a the output layer offset and weight are negative ($w_{01} = -4.4, w_{11} = -6.4$) and the hidden layer offset and weight are positive ($w_0 = 1.2, w_1 = 635$), it is vice versa for the solution in Fig. 11.15b ($w_{01} = 4.2, w_{11} = 6.4, w_0 = 552, w_1 = -551$). Furthermore, an examination of the network parameters reveals a small sensitivity with respect to some weights. In other words, the loss function possesses very flat optima, and relatively large changes in some network parameters around their optimal values affect the quality of the approximation only insignificantly. Consequently, major differences arise, depending on the optimization technique used. While second-order methods converge to the optimum, first-order methods converge only to a relatively large area around the optimum within a reasonable amount of computation time. Thus, the obtained network parameters may vary significantly depending on

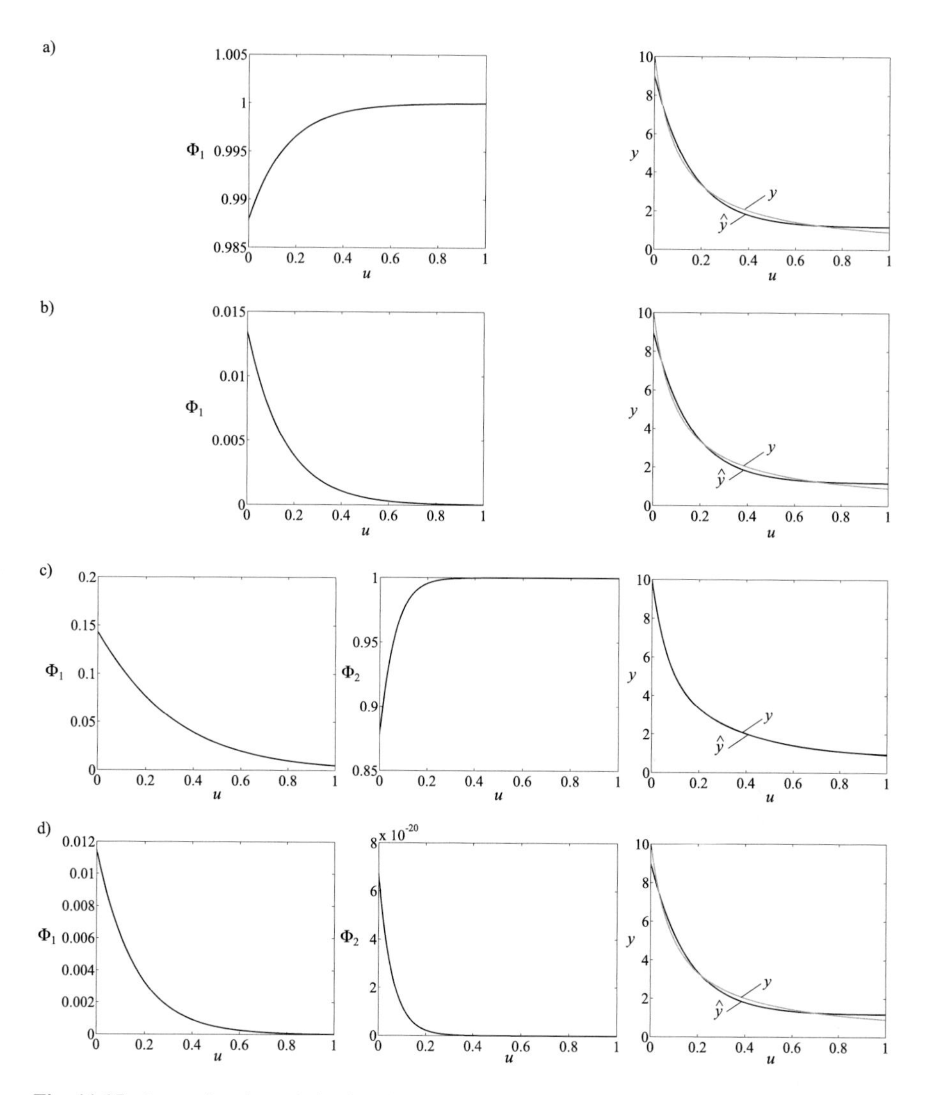

Fig. 11.15 Approximation of the function $y = 1/(u + 0.1)$ with an MLP network with one hidden neuron (upper plots) and two hidden neurons (lower plots). Two possible training results, depending on the parameter initialization for the one and two neuron(s) case, are shown in (**a**), (**b**) and (**c**), (**d**), respectively

the initialization and the utilized optimization technique, although they yield a comparable approximation accuracy. This important observation for such a simple example implies that no "meaning" can be associated with the MLP network parameters. Therefore, it makes no sense to build any kind of system that processes these values of these parameter values further.

Figure 11.15c, d shows two approximations with an MLP with two hidden neurons. In Fig. 11.15c an excellent approximation accuracy is reached where one basis function (Φ_1) has been optimized more for describing the right part of the nonlinearity and the other basis function (Φ_2), which is multiplied by a negative output layer weight is devoted mainly toward the left part of the nonlinearity. However, depending on the initialization, the reached network basis functions may also look as shown in Fig. 11.15d. The second basis function was driven to virtually zero during the optimization procedure and cannot recover since it is highly saturated, and thus the gradient of the loss function with respect to this basis function's parameters is virtually zero. Note that the situation in Fig. 11.15d does not represent a local optimum. If the optimization algorithm was run for long enough, then the network would eventually converge toward the solution shown in Fig. 11.15c. The loss function is just so insensitive with respect to one neuron that in practice it has the same effect as a local optimum.

The approximation problem shown in Fig. 11.16 seems to require more neurons for a reasonably accurate approximation than in the previous example in Fig. 11.15. Astonishingly this is not the case, as Fig. 11.16a demonstrates. This example clearly underlines the benefits obtained by the global approximation characteristics and the adjustment of the basis function shapes due to the optimization of the hidden layer parameters. However, not all trials are as successful. Rather most network trainings result in one of the three unsatisfactory solutions shown in Fig. 11.16b–d. While Fig. 11.16c represents a local optimum, the situations in Fig. 11.16b, d are similar to that in Fig. 11.15d discussed above. This example nicely illustrates the strengths and weaknesses of MLP networks. On the one hand, MLP networks possess extremely high flexibility, which allows them to generate a wide variety of basis function shapes suitable to the specific problem. On the other hand, the risk of convergence to local optima and saturation of neurons requires an extensive trial-and-error approach, which is particularly difficult and tedious when the problems become more complex and an understanding of the network (as in the trivial examples presented here) is scarcely possible.

11.2.6 MLP Properties

The most important properties of MLP networks can be summarized as follows:

- *Interpolation behavior* tends to be monotonic owing to the shape of the sigmoid functions.
- *Extrapolation behavior* is constant in the long-range owing to the saturation of the sigmoid functions. However, in the short-range extrapolation, behavior can be linear if a sigmoid function has a very small slope. A difficulty is that it is not clear to the user at which amplitude level the network extrapolates, and thus the network's extrapolation behavior is hard to predict.

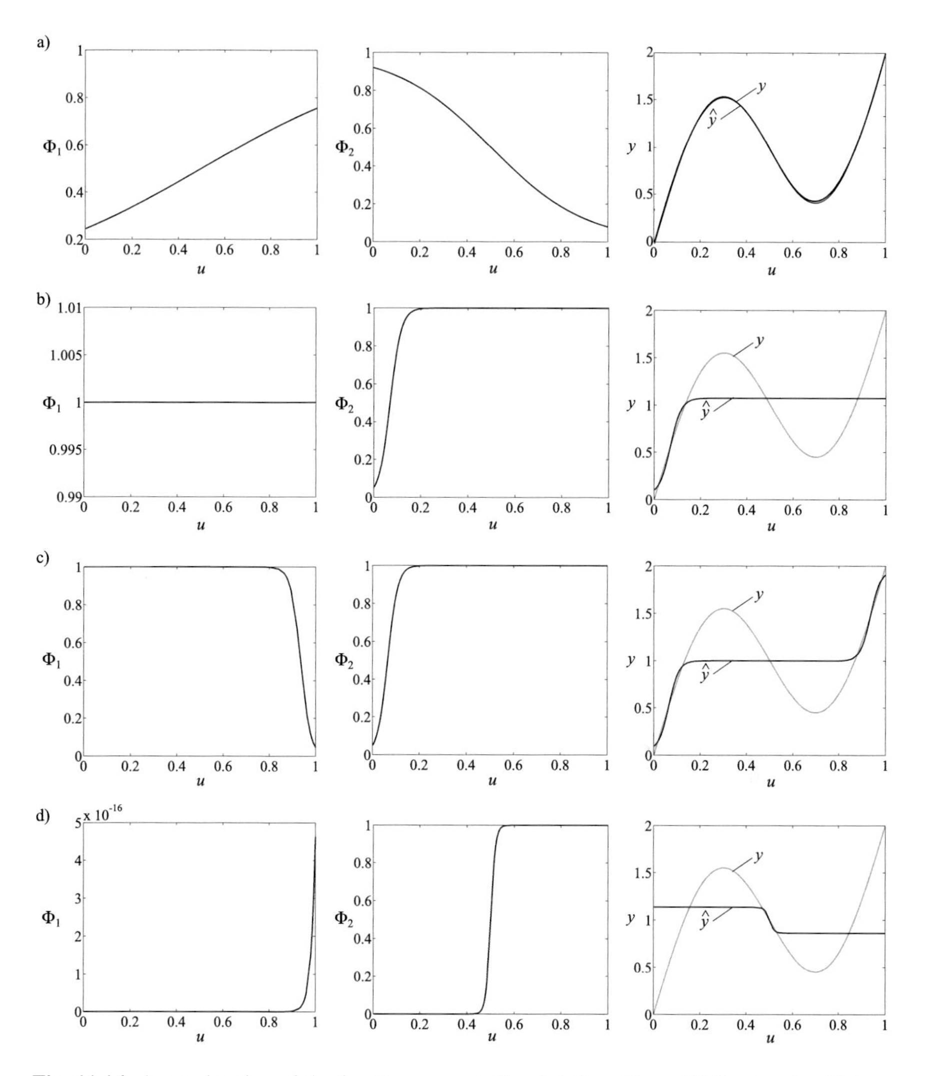

Fig. 11.16 Approximation of the function $y = \sin(2\pi u) + 2u$ with an MLP network with two hidden neurons. Four possible training results, depending on the parameter initialization, are shown in (**a**)–(**d**)

- *Locality* does not exist since a change in one output layer weight significantly influences the model output for a large region of the input space. Nevertheless, since the activation functions are not strictly global, an MLP has approximation flexibility only if the activation functions of the neurons are not saturated. So the approximation mechanism possesses some locality.
- *Accuracy* is typically very high. Owing to the optimization of the hidden layer weights, the MLP is extremely powerful and usually requires fewer neurons and fewer parameters than other model architectures to achieve a comparable

approximation accuracy. This property can be interpreted as a high information compression capability, which is paid for by long training times and the other disadvantages caused by nonlinear parameters; see below.

- *Smoothness* is very high. Owing to the tendency to monotonic interpolation behavior, the model output is typically very smooth.
- *Sensitivity to noise* is very low since, owing to the global character, almost all training data samples are exploited to estimate all model parameters.
- *Parameter optimization* generally has to be performed by a nonlinear optimization technique and thus is slow; see Chap. 4.
- *Structure optimization* requires computationally quite expensive pruning or growing methods; see Sect. 7.4.
- *Online adaptation* is slow and unreliable owing to the nonlinearity of the parameters and the global approximation characteristics.
- *Training speed* is very slow since nonlinear optimization techniques have to be applied and perhaps repeated for several network weight initializations if an unsatisfactory local optimum has been reached.
- *Evaluation speed* is fast since the number of neurons is relatively small compared with other neural network architectures.
- *Curse of dimensionality* is very low because the ridge construction mechanism utilizes projections. The MLP is *the* neural network for very high-dimensional problems.
- *Interpretation* is extremely limited since projections can be hardly interpreted by humans.
- *Incorporation of constraints* is difficult owing to the limited interpretation.
- *Incorporation of prior knowledge* is difficult owing to the limited interpretation.
- *Usage* is very high. MLP networks are still standard neural networks.

11.2.7 Projection Pursuit Regression (PPR)

In 1981 Friedman and Stuetzle [173, 265] proposed the projection pursuit regression (PPR), which is a new approach for the approximation of high-dimensional mappings. It can be seen as a generalized version of the multilayer perceptron network. Interestingly, PPR is much more complex and advanced than the standard MLP, although it was proposed 5 years earlier. This underlines the lack of idea exchange between different disciplines in the early years of the neural network era. Basically, PPR differs from an MLP in the following two aspects:

- The activation functions are flexible functions, which are optimized individually for each neuron.
- The training of the weights is done by a staggered optimization approach.

The one-dimensional activation functions are typically chosen as cubic splines or non-parametric smoothers such as a simple general regression neural network; see Sect. 11.4.1. Their parameters are also optimized from data.

PPR training is performed in three phases, which are iterated until a stop criterion is met. First, the output layer weights are estimated by LS for fixed activation functions and fixed hidden layer weights. Second, the parameters of the one-dimensional activation functions are estimated individually for each hidden layer neuron, while all network weights are kept fixed. This estimation is carried out separately for each neuron, following the staggered optimization philosophy. Since the activation functions are usually linearly parameterized, this estimation can be performed by LS, as well. Finally, the hidden layer weights are optimized for each neuron individually by a nonlinear optimization technique, e.g., the Levenberg-Marquardt algorithm. Since the activation functions and the hidden layer weights are determined for each neuron separately by staggered optimization, this procedure must be repeated several times until convergence.

Note that the additional flexibility in the activation functions of PPR compared with the MLP is a necessary feature in combination with the staggered optimization approach. The reason for this is as follows. The first few optimized neurons capture the main coarse properties of the process, and the remaining error becomes more nonlinear the more neurons have been optimized. At some point, a simple relatively inflexible neuron with sigmoid activation function will not be capable to extract any new information if its parameters are not estimated simultaneously with those of all other neurons. High-frequency components (in the space, not the time domain) can be modeled only by the interaction of several simple neurons. Thus, staggered optimization can be successful only if the activation functions are sufficiently flexible.

Since the optimization is performed cyclically, neuron by neuron, it can be easily extended to a growing and/or pruning algorithm that adds and removes neurons if necessary. More details of PPR in a neural network context can be found in [269, 274].

11.2.8 Multiple Hidden Layers

The multilayer perceptron network depicted in Fig. 11.14 can be extended by the incorporation of additional hidden layers. The use of more than one hidden layer makes the network more complex and can be seen as an alternative to the use of more hidden layer neurons. The question of which of two MLP networks with the same number of parameters, one with several hidden layers but only a few neurons in each layer, the other with a single hidden layer but more neurons, is superior cannot be answered in general; rather it is very problem dependent. Clearly, more hidden layers make the network harder to train since the gradients become more complicated and the parameters become more strongly nonlinear.

In practice, MLPs with one hidden layer are clearly most common, and sometimes two hidden layers are used. The application of more than two hidden layers is exotic. Figure 11.17 depicts an MLP network with two hidden layers. Its basis function representation is more involved since the neurons of the second hidden

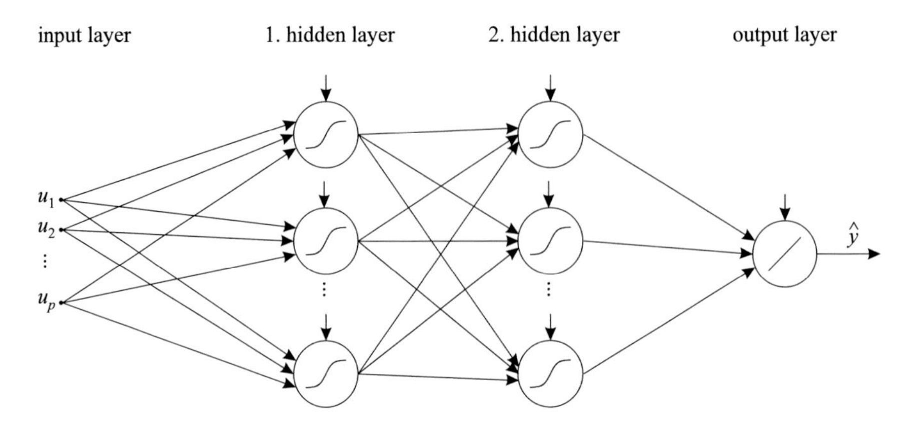

Fig. 11.17 A multilayer perceptron network with two hidden layers (the weights are omitted for simplicity)

layer are themselves composed of the neuron outputs of the first hidden layer. With M_1 and M_2 as the number of neurons in the first and second hidden layer, respectively, w_i as output layer weights, and $w_{jl}^{(1)}$ and $w_{ij}^{(2)}$ as weights of the first and second hidden layer, the basis function formulation becomes

$$\hat{y} = \sum_{i=0}^{M_2} w_i \Phi_i \left(\sum_{j=0}^{M_1} w_{ij}^{(2)} \xi_j \right) \qquad \text{with } \Phi_0(\cdot) = 1\,, \tag{11.25}$$

and with the outputs of the first hidden layer neurons

$$\xi_j = \Psi_j \left(\sum_{l=0}^{p} w_{jl}^{(1)} u_l \right) \qquad \text{with } \Psi_0(\cdot) = 1 \text{ and } u_0 = 1\,. \tag{11.26}$$

Usually the activation functions of both hidden layer Φ_i and Ψ_j are chosen to be of saturation type.

The gradient with respect to the output layer weights of an MLP with two hidden layers does not change ($i = 0, \ldots, M$) (see Sect. 11.2.3):

$$\frac{\partial \hat{y}}{\partial w_i} = \Phi_i \qquad \text{with } \Phi_0 = 1\,. \tag{11.27}$$

The gradient with respect to the weights of the second hidden layer and $\tanh(\cdot)$ activation function is similar to (11.19) ($i = 1, \ldots, M_2$, $j = 0, \ldots, M_1$):

$$\frac{\partial \hat{y}}{\partial w_{ij}^{(2)}} = w_i (1 - \Phi_i^2) \Psi_j \qquad \text{with } \Psi_0 = 1\,. \tag{11.28}$$

The gradient with respect to the weights of the first hidden layer and tanh(·) activation function is ($j = 1, \ldots, M_1, l = 0, \ldots, p$)

$$\frac{\partial \hat{y}}{\partial w_{jl}^{(1)}} = \sum_{i=0}^{M_2} w_i (1 - \Phi_i^2) w_{ij}^{(2)} (1 - \Psi_j^2) u_l \qquad \text{with } u_0 = 1\,. \tag{11.29}$$

Note that the basis functions Φ_i in the above equations depend on the network inputs and the hidden layer weights. These arguments are omitted for better readability.

The number of weights of an MLP with two hidden layers is

$$M_1(p+1) + M_2(M_1+1) + M_2 + 1\,. \tag{11.30}$$

Owing to the term $M_1 M_2$, the number of weights grows quadratically with an increasing number of hidden layer neurons.

11.2.9 Deep Learning

In previous times (before ~2010), typical MLP networks had one or two hidden layers. Deeper networks with more hidden layers usually did not yield convincing performance because learning was extremely slow and more sophisticated types of regularization than early stopping were either not invented yet or not popular. For deep networks, regularization is a key ingredient for a successful bias/variance tradeoff because of the overwhelming number of parameters. The success story of deep learning began in 2006 with Hinton's work on deep belief networks [250], was extended to all kinds of neural network architectures, and achieved a breakthrough in commercial and public interest in 2012 when a deep network impressively won the ImageNet object recognition competition [334].

This section can only briefly address some core motivations and ideas of deep learning. Due to its spectacular success deep learning has become a huge research field on its own. For a thorough up-to-date discussion, refer to [194] and the references therein. From a fundamental perspective, having many hidden layers offers many advantages. It allows the network to build a hierarchy of levels of different degrees of abstraction. For example, in the context of object detection, the network may be fed with raw data, just the pixels of an image. The neurons in the first layer might extract low-level features (edges, corners, etc.), the next layer might build on that detecting simple geometric patterns (rectangles, circles, etc.), the next level might build on that detecting more complex geometric patterns, and so on up to the final layer which might distinguish between cars, trees, and humans. Of course, the final classes could be further detailed again. This kind of hierarchical structure is probably very similar to human thinking.

In pre-deep-learning times, it was not possible to successfully train a deep network extracting the different levels of abstraction from data alone. The typical

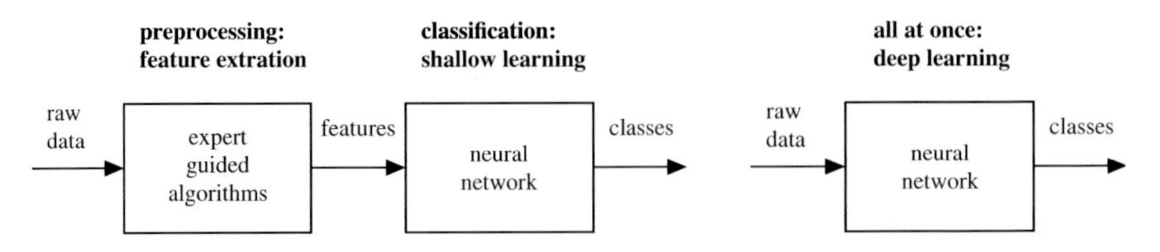

Fig. 11.18 Typical structure of shallow learning (left) and deep learning (right) classification

approach was to split the problem into two steps, compare Fig. 11.18: (i) preprocessing and (ii) classification. Feature extraction analyzes, decomposes, and constructs features from raw data with application-specific algorithms that need human expert guidance. Therefore expert knowledge typically was the key factor for success. With the possibility to train deep networks effectively, both steps merge into one, accomplished by the neural network. The overwhelming importance of domain knowledge decreases significantly because "everything" is learned from data. Of course for collecting and selecting "good" data, domain knowledge still plays an important role but not as decisive as before. Instead, expertise in the area of deep learning becomes more and more relevant.

In addition to the abovementioned very appealing properties of deep learning, theoretical and empirical evidence prove the benefits of deep network architectures. In [194] it is discussed that deep networks, in principle, can represent a function in an exponentially smaller manner than shallow networks. Furthermore, the authors state: "Empirically, greater depth does seem to result in better generalization for a wide variety of tasks" followed by a long list of investigations supporting this claim.

Significant advances in many areas made training of deep networks feasible and practical. It allowed to change in Fig. 11.18 from left to right. In the years before, the performance of neural networks or alternative AI systems had saturated. Only tiny improvements had been possible for a long period of time. Then deep learning approaches became feasible and entered the competitions. The object recognition ImageNet benchmark[1]is the most prominent benchmark example, containing over 14 million images each associated to one object. Deep-learning approaches broke the saturation boundary and even achieved superhuman performance levels (Fig. 11.19).

Some important advances in understanding and technology that made this breakthrough possible are:

- *Vanishing gradients*: During gradient-based (backpropagation) training of networks with many hidden layers with saturating (sigmoidal) activation functions, the gradients almost vanish when propagated back through the layers. This makes learning prohibitively slow for deep networks. The found solution is to use

[1] https://qz.com/1034972/the-data-that-changed-the-direction-of-ai-research-and-possibly-the-world/

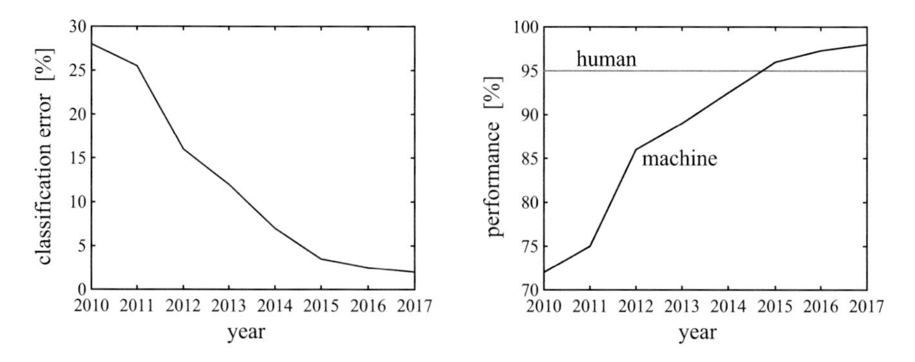

Fig. 11.19 Left: Classification performance on the ImageNet benchmark and the breakthrough occurring in 2012 with modern deep-learning approaches. Right: Best AI systems surpassed human object recognition performance. (Source: image-net.org)

non-saturating activation functions like the ReLU (rectified linear unit) which possesses fixed slope for positive inputs.

- *Big data*: In order to train deep networks with a huge number of parameters (often in the order of millions), lots of data is necessary. The rise of many internet applications and cloud-based services exactly demanded big data solutions.
- *Regularization*: The huge model/network complexities called for new regularization methods for controlling the bias/variance tradeoff in a sophisticated manner. New general and network-architecture-specific ideas prevent overfitting of deep networks.
- *Computing*: Graphics processing units (GPUs), distributed networks, and cloud-based data processing met the demand for high parallel computing power and memory resources.
- *Software*: Powerful, flexible, and freely available tools like TensorFlow^{TM} promoted the easy spread of deep-learning techniques.

Most deep learning approaches make use of the following components:

- *Mini-batches*: Typical training data sets are huge. Therefore an efficient training relies on parameter updated on a mini-batch basis. Typical mini-batches consist of 10 to 1,000 randomly drawn data samples, compare Sect. 4.1. The randomization caused by the random selection and ordering of the presented training data points makes the learning process more robust and offers bigger chances for escaping potential local optima.
- *Weight sharing*: In many network architectures, weights are shared, e.g., convolution neural networks (CNNs) reuse identical convolution masks (cross-correlation kernels) for all areas of an input image. This reduces the degrees of freedom significantly compared to the nominal number of parameters.
- *Stochastic gradient descent (SGD)*: Since the parameter updates are based on the gradients computed for the mini-batches, the performance of the network is not monotonously increasing during training but can fluctuate. The properties

of SGD seem to help to achieve well-generalizing solutions [218] in a stable manner. The key statement in [218] is: "Any model trained with stochastic gradient method in a reasonable amount of time attains small generalization error." There even is a lot of evidence that higher-order methods or alternative high-performance search schemes like AdaGrad, RMSProp, and Adam – although they converge much faster – yield optima with worse generalization performance [605]. Currently, this strange but welcome issue is a question under heavy research [612].

- *Dropout regularization*: As pointed out in [194], dropout can be seen as a clever way for performing regularization and increasing robustness for deep-learning architectures, compare Sect. 7.5. It realizes a rough approximation of bagging (see Sect. 14.11 for an example) with extremely low computational demand. The idea of dropout is to remove hidden layer neurons (typically with a probability of 0.5) and input nodes (typically with a probability of 0.2) during training for one mini-batch. For the next mini-batch, the neurons and nodes for removal are selected randomly again and so on, compare Fig. 11.20. This forces the network to become robust against these random disruptions similar to random forests; see [78]. If, for example, half of the hidden neurons have been removed during training, the trained hidden layer weights have to be divided by 2 for the final usage of the network (generalization) where the complete network with *all* neurons is employed. In consequence, a huge number of possible network structures (trained with different mini-batches) are averaged which corresponds to the basic ideas of bagging. Besides a very simple implementation, dropout has the significant advantage of being closely tied to the network architecture. Thus

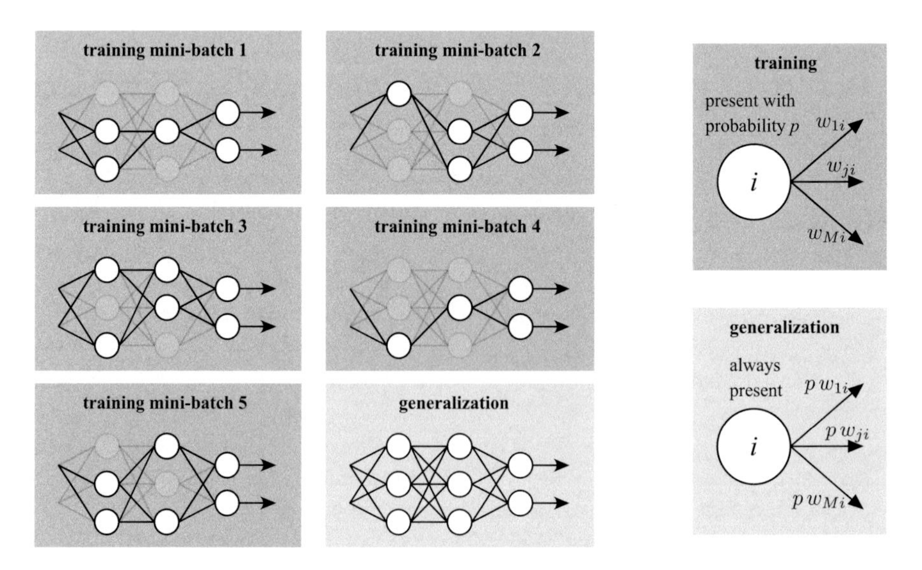

Fig. 11.20 Visualization of dropout regularization: Left: How dropout affects the network, lightly drawn neurons and connections/weights drop out. Right: How dropout affects one neuron, based on [547]

no tedious tuning of the regularization strength is necessary, just the removal probabilities have to be fixed where the abovementioned ones represent the most common values.

As discussed in [611], deep-learning approaches typically are so overparameterized, i.e., possess so many parameters, that they can be and often are trained to zero training error. According to traditional belief, this should lead to bad generalization performance (overfitting) if not appropriate; massive regularization is employed, making the effective number of parameters smaller than their nominal number. It is illustrated and discussed in [611] that these criteria obviously are not sufficient in order to explain the generalization behavior of deep-learning approaches. It is possible to train two networks with identical architectures to zero training errors, one having a small and one having a large generalization error. This challenges the current understanding about which factors determine the generalization properties.

11.3 Radial Basis Function (RBF) Networks

In contrast to the MLP network, the radial basis function (RBF) network utilizes a radial construction mechanism. This gives the hidden layer parameters of RBF networks a better interpretation than for the MLP and therefore allows new, faster training methods. Originally, radial basis functions were used in mathematics as interpolation functions without a relationship to neural networks [466]. In [81] RBFs are first discussed in the context of neural networks, and their interpolation and generalization properties are thoroughly investigated in [168, 358].

This section is organized as follows. First, the RBF neuron and the network structure are introduced in Sects. 11.3.1 and 11.3.2. Next, the most important strategies for training are analyzed. Section 11.3.5 discusses the strengths and weaknesses of RBF networks. Finally, some mathematical foundations of RBF networks are summarized in Sect. 11.3.6, and the extension to normalized RBF networks is presented in Sect. 11.3.7.

11.3.1 RBF Neuron

Figure 11.22 shows a neuron of an RBF network. Its operation can be split into two parts. In the first part, the distance of the input vector $\underline{u} = [u_1\ u_2\ \cdots\ u_p]^T$ to the center vector $\underline{c}_i = [c_{i1}\ c_{i2}\ \cdots\ c_{ip}]^T$ with respect to the covariance matrix $\underline{\Sigma}_i$ is calculated. This is the radial construction mechanism already introduced in Sect. 11.1.2. In the second part, this distance x (a scalar) is transformed by the nonlinear activation function $g(x)$. The activation function is usually chosen to possess local character and a maximum at $x = 0$. Typical choices for the activation function are the Gaussian function

$$g(x) = \exp\left(-\frac{1}{2}x^2\right) \tag{11.31}$$

and the inverse multiquadratic function (see Fig. 11.21)

$$g(x) = \frac{1}{\sqrt{x^2 + a^2}} \tag{11.32}$$

with the additional free parameter a.

Figure 11.22 shows, for a Gaussian activation function, how the hidden layer parameters influence the basis function.

The distance x_i is calculated with help of the center $\underline{c}_i$ and inverse covariance matrix $\underline{\Sigma}_i^{-1}$, which are the hidden layer parameters of the ith RBF neuron

$$x_i = ||\underline{u} - \underline{c}_i||_{\underline{\Sigma}_i} = \sqrt{\left(\underline{u} - \underline{c}_i\right)^T \underline{\Sigma}_i^{-1} \left(\underline{u} - \underline{c}_i\right)}\,. \tag{11.33}$$

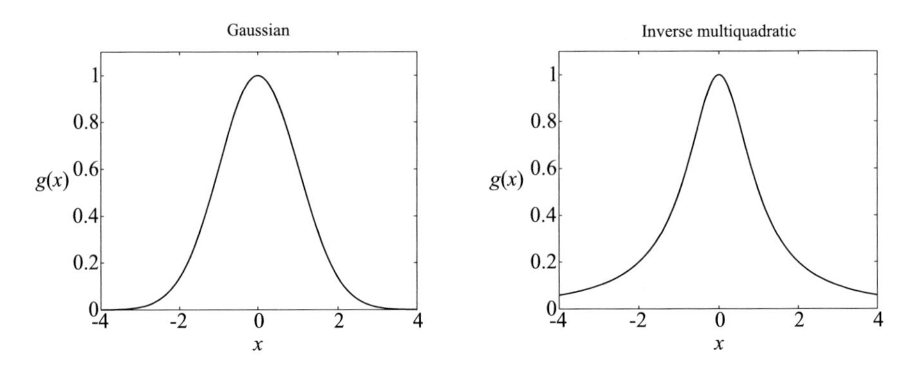

Fig. 11.21 Typical activation functions for the RBF neuron: (left) Gaussian, (right) inverse multiquadratic with $a = 1$

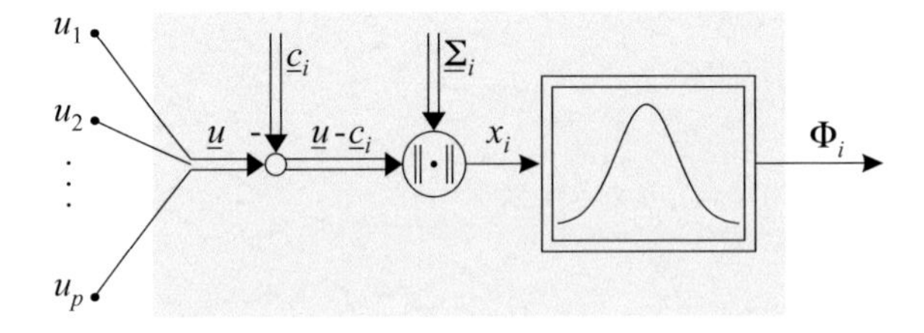

Fig. 11.22 The ith hidden neuron of an RBF network

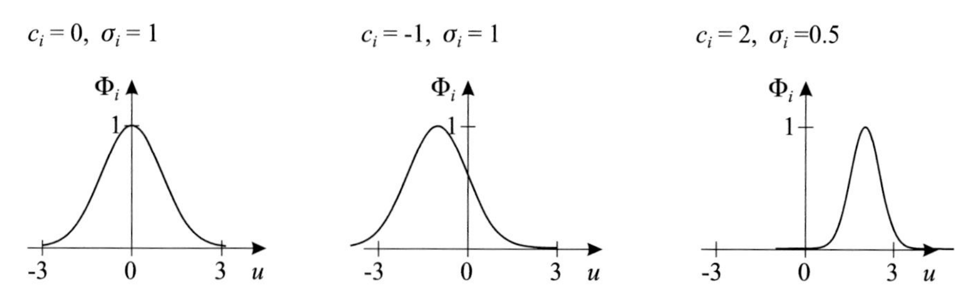

Fig. 11.23 Influence of the hidden layer parameters on the ith hidden RBF neuron with a single input u. The center determines the position and the standard deviation the width of the Gaussian

Thus, the basis functions $\Phi_i(\cdot)$ of a Gaussian RBF network are

$$\Phi_i\left(\underline{u}, \underline{\theta}_i^{(\text{nl})}\right) = \exp\left(-\frac{1}{2}||\underline{u} - \underline{c}_i||^2_{\underline{\Sigma}_i}\right), \tag{11.34}$$

where the hidden layer parameter vector $\underline{\theta}_i^{(\text{nl})}$ consists of the coordinates of the center vector $\underline{c}_i$ and the entries of the inverse covariance matrix $\underline{\Sigma}_i^{-1}$. Figure 11.23 illustrates the effect of these parameters on the basis function for the one-dimensional case. Note that, for a single input, the covariance matrix $\underline{\Sigma}_i$ simplifies to the standard deviation σ.

Often the $\underline{\Sigma}^{-1}$ matrix is chosen to be diagonal; so it contains the inverse variances for each input dimension. Then the distance calculation simplifies to

$$\begin{aligned} x_i = ||\underline{u} - \underline{c}_i||_{\underline{\Sigma}_i} &= \sqrt{\sum_{j=1}^{p}\left(\frac{u_j - c_{ij}}{\sigma_{ij}}\right)^2} \\ &= \sqrt{\left(\frac{u_1 - c_{i1}}{\sigma_{i1}}\right)^2 + \ldots + \left(\frac{u_p - c_{ip}}{\sigma_{ip}}\right)^2}, \end{aligned} \tag{11.35}$$

where

$$\underline{\Sigma}_i^{-1} = \begin{bmatrix} 1/\sigma_{i1}^2 & 0 & 0 & 0 \\ 0 & 1/\sigma_{i2}^2 & 0 & 0 \\ \vdots & \vdots & \ddots & \vdots \\ 0 & 0 & 0 & 1/\sigma_{ip}^2 \end{bmatrix}. \tag{11.36}$$

If identical variances are chosen for each input dimension, this result becomes even simpler. Figure 11.24 illustrates the effect of different covariance matrices on the shape of the basis function. The top of Fig. 11.24 shows that identical

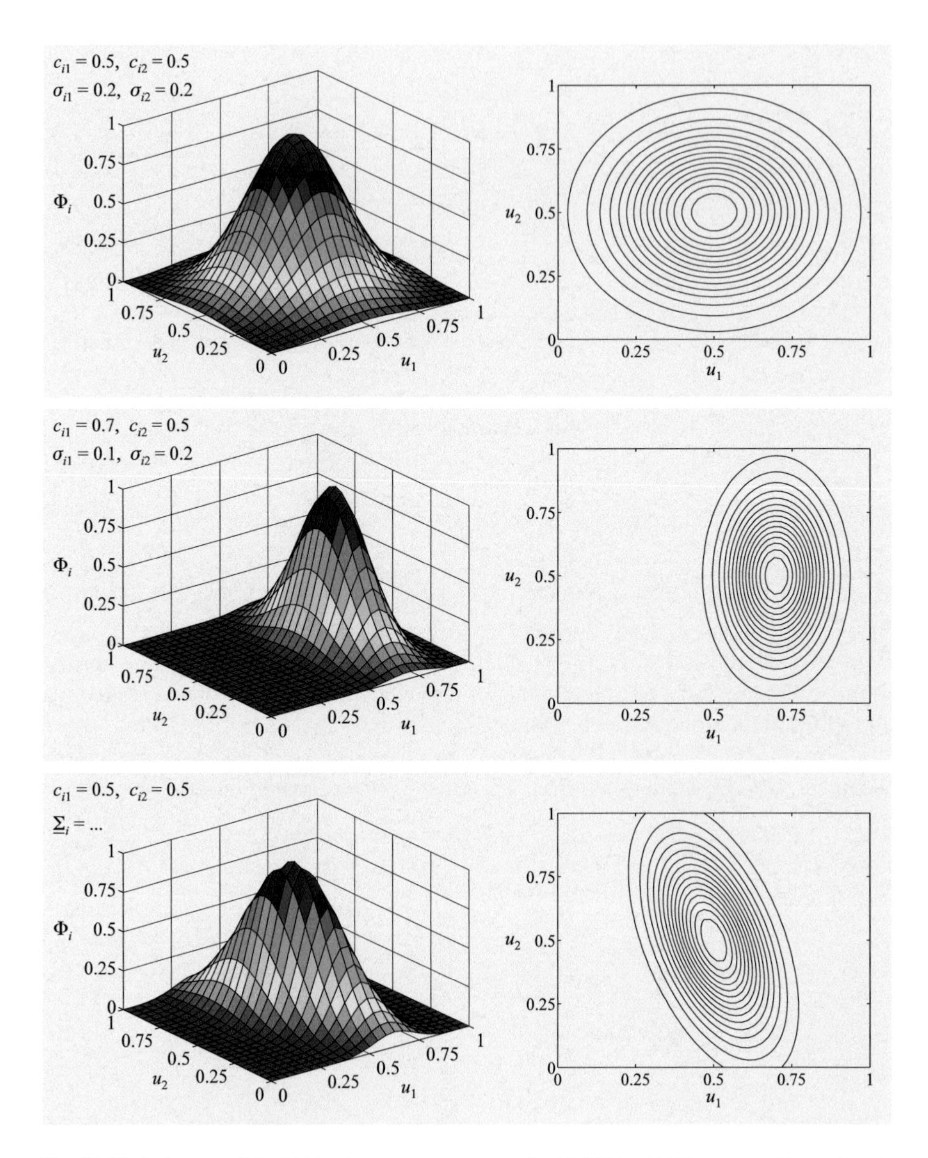

Fig. 11.24 Influence of the hidden layer parameters on the ith hidden RBF neuron with two inputs u_1 and u_2. The basis functions are shown on the left, the corresponding contour lines on the right

standard deviations for each dimension lead to a true *radial* basis function with circle contours:

$$\underline{\Sigma}^{-1} = \begin{bmatrix} (1/0.2)^2 & 0 \\ 0 & (1/0.2)^2 \end{bmatrix} . \tag{11.37}$$

The center of Fig. 11.24 shows that different standard deviations for each dimension lead to a symmetric basis function with elliptic contours:

$$\underline{\Sigma}^{-1} = \begin{bmatrix} (1/0.1)^2 & 0 \\ 0 & (1/0.2)^2 \end{bmatrix}. \tag{11.38}$$

The bottom of Fig. 11.24 illustrates the use of a complete covariance matrix, which additionally allows for rotations of the basis functions:

$$\underline{\Sigma}^{-1} = \begin{bmatrix} (1/0.1)^2 & (1/0.2)^2 \\ (1/0.2)^2 & (1/0.2)^2 \end{bmatrix}. \tag{11.39}$$

11.3.2 Network Structure

If several RBF neurons are used in parallel and are connected to an output neuron, the radial basis function network is obtained; see Fig. 11.25. In basis function formulation, the RBF network can be written as

$$\hat{y} = \sum_{i=0}^{M} w_i \Phi_i \left(||\underline{u} - \underline{c}_i||_{\underline{\Sigma}_i} \right) \qquad \text{with } \Phi_0(\cdot) = 1 \tag{11.40}$$

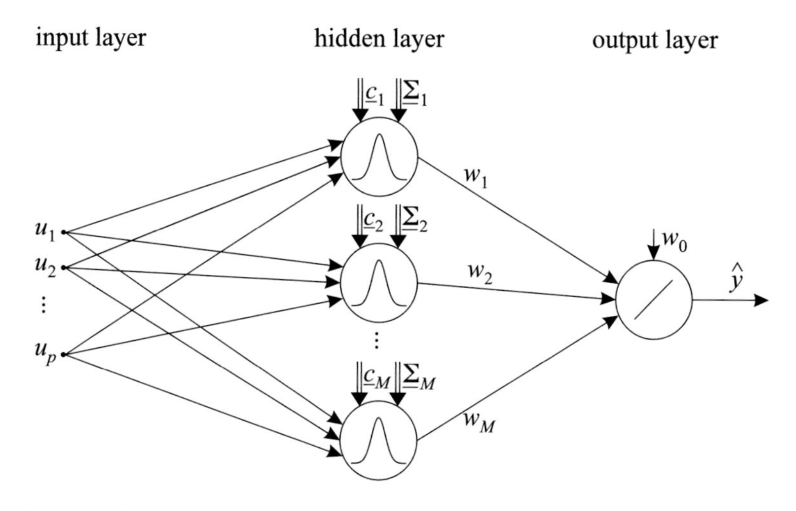

Fig. 11.25 A radial basis function network

with the output layer weights w_i. The hidden layer parameters contain the center vector $\underline{c}_i$, which represents the position of the ith basis function, and the inverse covariance matrix $\underline{\Sigma}_i^{-1}$, which represents the widths and rotations of the ith basis function. The total number of parameters of an RBF network depends on the flexibility of $\underline{\Sigma}_i^{-1}$. The number of output weights is $M + 1$, the number of center coordinates is Mp, and the number of parameters in $\underline{\Sigma}_i^{-1}$ is equal to

- M: for identical standard deviations for each input dimension, i.e., $\underline{\Sigma}_i^{-1}$ is diagonal with identical entries,
- Mp: for different standard deviations for each input dimension. i.e., $\underline{\Sigma}_i^{-1}$ is diagonal, and
- $M(p + 1)p/2$: for the general case with rotations, i.e., $\underline{\Sigma}_i^{-1}$ is symmetric.

Since the general cases require a huge number of parameters, commonly $\underline{\Sigma}_i^{-1}$ is chosen diagonal, and then the total number of parameters of an RBF network becomes

$$2Mp + M + 1\,, \tag{11.41}$$

where M is the number of hidden layer neurons and p is the number of inputs. Since the inputs are given by the problem, the number of hidden layer neurons allows the user to control the network complexity, i.e., the number of parameters.

Like the MLP, an RBF network is a *universal approximator* [220, 445, 446]. Contrary to the MLP, multiple hidden layers do not make much sense for an RBF network. The neuron outputs of a possible first hidden layer would span the input space for the second hidden layer. Since this space cannot be interpreted, the hidden layer parameters of the second hidden layer cannot be chosen by prior knowledge, which is one of the major strengths of RBF networks.

An RBF network consists of three types of parameters:

- *Output layer weights* are linear parameters. They determine the heights of the basis functions and the offset value.
- *Centers* are nonlinear parameters of the hidden layer neurons. They determine the positions of the basis functions.
- *Standard deviations (and possibly off-diagonal entries in the covariance matrices)* are nonlinear parameters of the hidden layer neurons. They determine the widths (and possibly rotations) of the basis functions.

These parameters have somehow to be determined or optimized during the training of RBF networks.

Figure 11.26 illustrates the interpolation and extrapolation behavior of RBF networks. Obviously, it is very sensitive to the choice of the basis function widths. The interpolation behavior may have "dips" if the standard deviations are too small and may "overshoot" if they are too large. The extrapolation behavior decays toward zero, the slower, the wider the basis functions are.

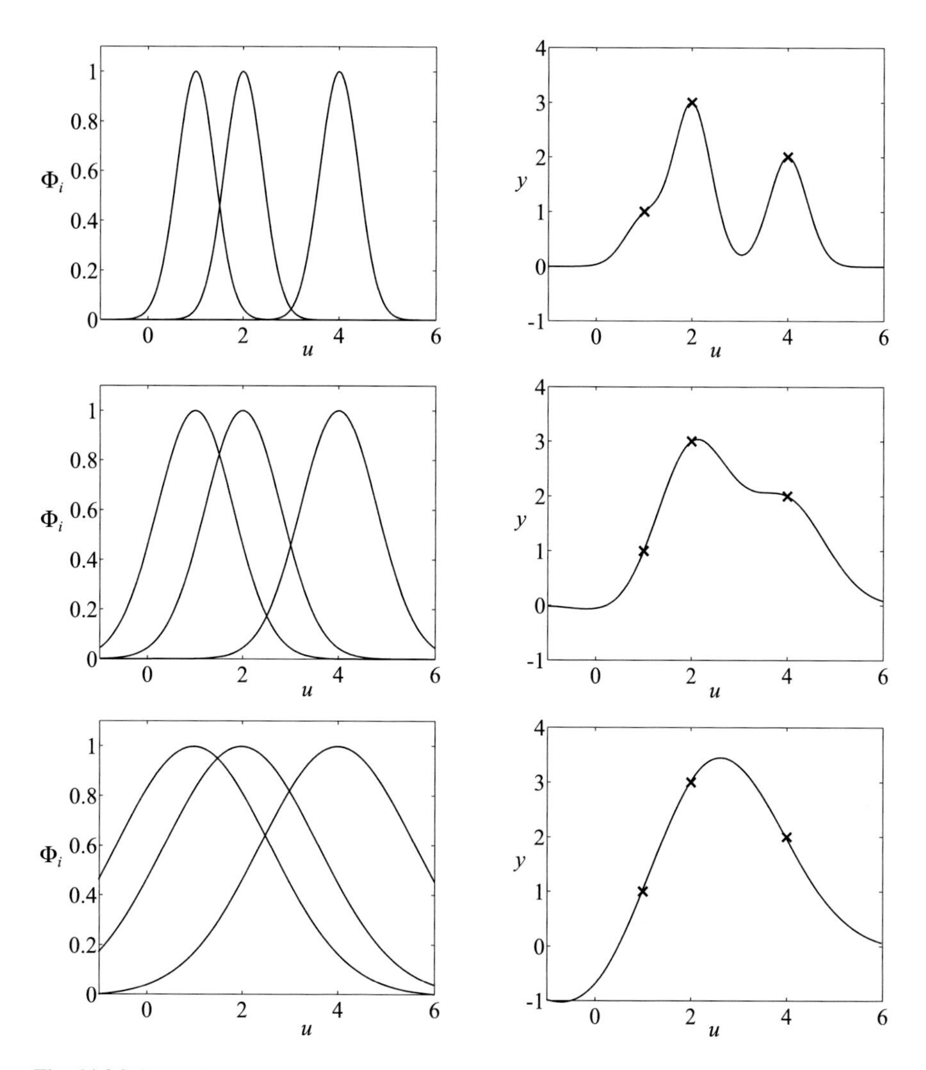

Fig. 11.26 Interpolation and extrapolation behavior of an RBF network without offset, i.e., $w_0 = 0$. The basis functions are shown on the left; the network outputs are shown on the right. The standard deviations are chosen identically for each neuron in each network as top, $\sigma = 0.4$; center, $\sigma = 0.8$; bottom, $\sigma = 1.6$. The network interpolates between the three data points marked as crosses

11.3.3 RBF Training

For training of RBF networks, different strategies exist. Typically, they try to exploit the linearity of the output layer weights and the geometric interpretability of the hidden layer parameters. Thus, most strategies determine the hidden layer parameters first, and subsequently, the output layer weights are estimated by least squares; see Sect. 9.2.2. These strategies correspond to the RAWN approach for MLP training; see Sect. 11.2.4. Alternatively, an orthogonal least squares algorithm or any other subset selection technique can be applied for combined structure and parameter optimization, or nonlinear optimization techniques can be used for the hidden layer parameters. Sample mode (online) training algorithms for RBF networks are not discussed here; refer to [169].

11.3.3.1 Random Center Placement

Selecting the centers randomly within some interval is the most primitive approach and certainly leads to inferior performance. A little more advanced is the random choice of training data samples as basis function centers since this at least guarantees that the basis functions lie in relevant regions of the input space. The first approach corresponds to the random initialization of hidden layer weights for the MLP network. The second approach corresponds to an advanced initialization technique that guarantees that all neurons are not saturated. Neither alternative is applied owing to their inferior performance, but they are reasonable initialization methods for more advanced techniques. Basically, any random choice of the basis function centers ignores the interpretability advantage of RBF over MLP hidden layer parameters.

11.3.3.2 Clustering for Center Placement

The straightforward improvement of the random selection of data samples as basis functions centers is the application of clustering techniques; see Sect. 6.2 for an overview of the most important clustering algorithms. The most commonly applied clustering technique is the k-means algorithm; see Sect. 6.2.1 and [387]. Alternatively, more complex unsupervised learning methods can be applied, such as Kohonen's self-organizing map (SOM) (Sect. 6.2.4) or other neural network-based approaches, but their superiority over simple k-means clustering with respect to basis function center determination has not been demonstrated up to now.

Clustering determines the basis function centers according to the training data distribution in the input space. Thus, many RBFs are placed in regions where data is dense, and few RBFs are placed in regions where data is sparse. On the one hand, this data representation is partly desired since it guarantees that enough data is available to estimate the heights (output layer weights) of the basis functions. So the variance error of RBF networks tends to be uniformly distributed over the input space. On the other hand, the complexity of the process that has to be approximated is not considered by the clustering technique because it operates completely unsupervised. This has the consequence that, in contrast to supervised learning of the hidden layer parameters, the RBFs are *not* automatically moved toward the regions where they are required for a good approximation of the process. Thus, the expected performance of an RBF network with clustering for center determination is relatively low, and therefore more neurons are required. Although this can be a strong drawback with respect to evaluation speed, this is not necessarily so for the training speed, since the clustering and LS algorithms are both very fast, even for relatively complex networks.

After clustering is completed, the standard deviations (or possibly the complete covariance matrices) have to be determined before the output layer weights can be computed by LS. The most common method for this is the *k-nearest neighbor* rule, which assigns each RBF a standard deviation proportional to the (possibly weighted) average distance between its center and the centers of the k nearest neighbor RBFs. The value for k is typically chosen between 1 and 5, and the proportionality factor has to be chosen in order to guarantee a desired overlap between the basis functions. With the standard k-nearest neighbor method, only a single standard deviation can be assigned to each RBF, as in (11.37). However, it can be extended to allow for different standard deviations in each input dimension by considering the distances individually for each dimension, as in (11.38). If the complete covariance matrix as in (11.39) is to be used, elliptic fuzzy clustering algorithms that generate a covariance matrix for each cluster may be utilized similar to the Gustafson-Kessel algorithm in Sect. 6.2.3; for more details refer to [51]. The last alternative, however, is rarely applied in practice since the estimation of the covariance matrix is computationally expensive and involves a large number of parameters, which makes this approach very sensitive to overfitting.

In summary, clustering is a good tool for the determination of the basis function centers. It generates a reasonable coverage of the input space with RBFs and scales up well to higher-dimensional input spaces. However, it suffers from the fundamental drawback of unsupervised methods, that is, ignorance with respect to the process complexity. Furthermore, the choice of the standard deviations is done in a very ad hoc manner. The hidden layer parameters are chosen heuristically and thus suboptimally. Therefore, typically more hidden layer neurons are required than in other approaches and network architectures where the hidden layer parameters are optimized. Nevertheless, this approach can be very successful since even networks with many hidden layer neurons can be trained rapidly. If the number of neurons and consequently the number of output layer weights become very large, efficient regularization techniques such as ridge regression can be employed in order to realize a good bias/variance tradeoff; see Sects. 3.1.4 and 7.5.2.2 and [440].

11.3.3.3 Complexity Controlled Clustering for Center Placement

The most severe drawback of clustering is that it does not take into account the complexity of the process under investigation. On the one hand, the data distribution should be reflected to avoid the generation of basis functions in regions without or with too sparse data. On the other hand, it would be desirable to generate many basis functions in regions where the process possesses complex behavior (i.e., is strongly nonlinear) and few RBFs in regions where the process is very smooth. This requires the incorporation of information about the process output into the unsupervised learning procedure. Section 6.2.7 discusses how such output information can be utilized by a clustering technique.

In [433] the following strategy is proposed to incorporate output information into the clustering technique:

1. Estimate a linear model by least squares.
2. Calculate the error of this linear model, i.e., $e_{\text{lin}} = y - \hat{y}_{\text{lin}}$.
3. Train an RBF network to approximate the error of the linear model by the following steps, i.e., use e_{lin} as the desired output for the network training; see Sect. 7.6.2.
4. For determination of the centers: Perform an extended version of k-means clustering that is sensitive to the errors; see below for details.
5. For determination of the standard deviations: Use the k-nearest neighbor method.
6. Estimate the output layer weights with least squares.

The final model has a hybrid character since it is obtained by the sum of the linear model and the RBF network. The decisive issue of this algorithm, however, is the *error-sensitive clustering (ESC)* in Step 4. It drives the RBF centers toward regions of the input space with a high approximation error of the linear model. ESC operates as follows. To each data sample a "mass" is assigned. This "mass" is equal to the squared error (or some other error measure) of the linear model for this data sample. So data samples in input regimes that are well modeled by the linear model possess small "masses," while data in regimes with high modeling errors possess large "masses." The k-means clustering is performed by taking into account these different "masses." Thus, the centroid (mean) of each cluster is calculated by the center of gravity

$$\underline{c}_j = \frac{\sum\limits_{i \in \mathcal{S}_j} \underline{u}(i) \cdot e_{\text{lin}}^2(i)}{N_j}\,, \tag{11.42}$$

where i runs over those N_j data samples that belong to cluster j, this is the set $\mathcal{S}_j$, and the following property $\sum_{j=1}^{M} N_j = N$ holds. The standard k-means clustering is extended by weighting the data samples with the "mass" $e_{\text{lin}}^2(i)$; see Sect. 6.2.1. The computation of the centroid for ESC is illustrated in Fig. 11.27.

The linear model in the above algorithm is the simplest way to generate modeling errors that can be assigned to the data samples. More sophisticated but computationally more expensive strategies may perform several iterations of RBF network training instead of using a simple linear model:

1. Train an RBF network with conventional clustering.
2. Calculate the modeling error.
3. Train an RBF network with error-sensitive clustering utilizing the modeling error from the network of the previous iteration.
4. If significant improvements could be realized, go to Step 2; otherwise stop.

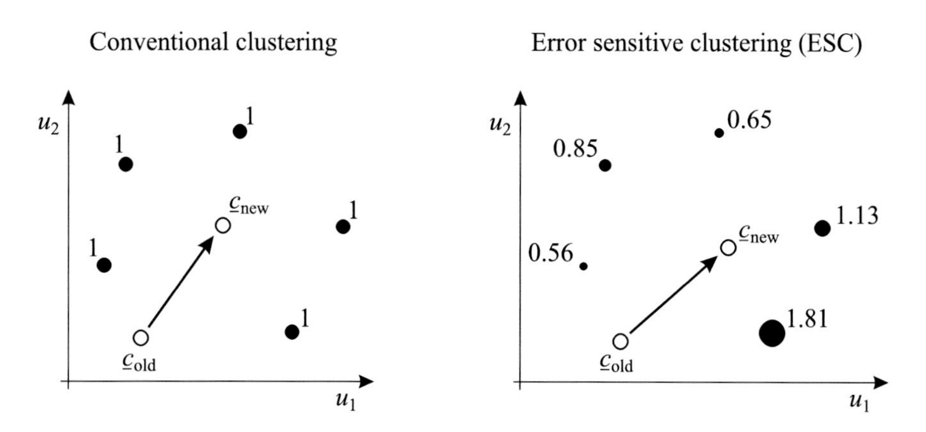

Fig. 11.27 In standard k-means clustering (left), all data samples are equivalent. In error-sensitive clustering (ESC), proposed in [433] (right), data samples that lie in poorly modeled regions of the input space are associated with a larger "mass" to drive the basis functions toward these regions

Although it is shown in [433] that error-sensitive clustering is superior to conventional clustering for basis function center determination, it still is a heuristic approach and necessarily suboptimal.

11.3.3.4 Grid-Based Center Placement

One of the major difficulties of clustering-based approaches for center placement is the determination of the standard deviations. Generally, the nearest neighbor method cannot guarantee to avoid "holes" in the space, i.e., regions where all basis functions are almost inactive. These "holes" lead to "dips" in the interpolation behavior, as demonstrated in Fig. 11.26. For low-dimensional mappings, an alternative to clustering is to place the centers on a grid. This allows one to determine the standard deviations dependent on the grid resolution in each dimension, which guarantees regular overlapping of all basis functions. Furthermore, the grid-based approach provides close links to neuro-fuzzy models with singletons; see Chap. 12.

One major difficulty of the grid-based approach is that it suffers severely from the curse of dimensionality and thus can be applied only for very low-dimensional problems; see Sect. 7.6.1. The curse of dimensionality can be weakened by exploiting the input data distribution. Those basis functions on the grid that are activated less than a given threshold by the data can be discarded [142, 311]. Figure 11.28 demonstrates that this strategy can reduce the number of neurons significantly. An alternative or additionally applied method is principal component analysis (PCA), which transforms the input axes in order to decorrelate the input data; see Sect. 6.1. Note, however, that PCA leads to a new coordinate system that typically is less interpretable.

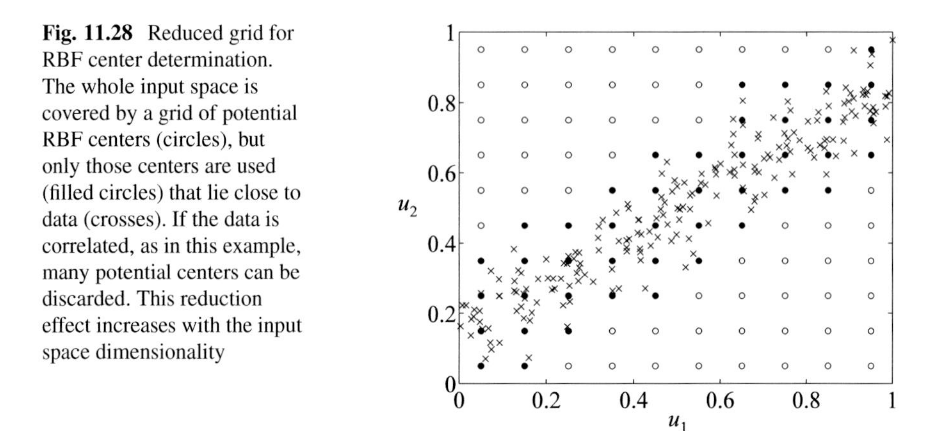

Fig. 11.28 Reduced grid for RBF center determination. The whole input space is covered by a grid of potential RBF centers (circles), but only those centers are used (filled circles) that lie close to data (crosses). If the data is correlated, as in this example, many potential centers can be discarded. This reduction effect increases with the input space dimensionality

Another difficulty of the grid-based approach is that it fully ignores the complexity of the process. Large modeling errors can be expected in regions where the process possesses complex behavior while many neurons may be wasted in regions where the process behaves smoothly (leading to an unnecessarily high variance error without any payoff in terms of bias error reduction). On the other hand, the grid-based approach has the advantage of being very simple, and it allows a good choice of the standard deviations.

11.3.3.5 Subset Selection for Center Placement

An efficient supervised learning approach for choosing the basis function centers is proposed in [100]. It is based on the orthogonal least squares (OLS) algorithm but can be extended to any other subset selection technique as well; see Sect. 3.4 and in particular Sect. 3.4.2. The subset selection strategy is as follows. First, a large number of potential basis functions are specified, i.e., their centers and standard deviations are fixed (somehow). Second, a subset selection technique is applied in order to determine the most important basis functions of all potential ones with respect to the available data. Since for the potential basis functions only the linear output layer weights are unknown, an efficient *linear* subset selection technique such as OLS can be applied. Note that the optimal weights and the relevance of each selected basis function are automatically obtained by OLS during subset selection; see Sect. 3.4.2.

In [100] one potential basis function is placed on each data sample with one fixed user-defined standard deviation for all dimensions. Thus, the number of potential basis functions is equal to the number of potential regressors for OLS. Because the computational demand for OLS grows strongly with the number of potential regressors, this strategy is feasible only for relatively small training data sets, say $N < 1000$. If more data is available, the training time can be limited by choosing only every second, third, etc. data sample as potential RBF center. As an alternative, a clustering technique with a relatively large number of clusters, say some 100, can be applied for the determination of the potential RBF centers. This would reduce the number of potential centers in order to allow additionally the definition of several potential basis functions on each center with different standard deviations. Then the subset selection technique optimizes not only the positions but also the widths of the RBFs. This is an important improvement, since the a priori choice of a standard deviation for *all* basis functions is the main drawback in the strategy originally proposed in [100].

There are several strong benefits of the subset selection approach for center selection compared with the previously introduced methods. First, it is supervised and therefore selects the basis functions in those regimes where they are most effective in terms of modeling error reduction. Second, the OLS algorithm operates in an incremental manner: that is, it starts with one RBF and incorporates a new basis function into the network in each iteration. This allows the user to fix some error threshold or desired model accuracy in advance, and the OLS algorithm automatically generates as many basis functions as are required to achieve this goal. This is a significant advantage over grid-based or clustering approaches, which require the user to specify the number of basis functions before training.

Drawbacks of OLS training for RBF networks are the heuristic and thus suboptimal choice of the standard deviations and the higher computational demand. The user can, however, easily influence the training time by choosing the number of potential basis functions correspondingly. This allows a tradeoff between model accuracy and computational effort. Note that there is no guarantee that the linear subset selection techniques do realize the *global* optimum, i.e., select the best subset of basis functions from the potential ones; see Sect. 3.4. However, in the experience of the author, this restriction is usually not severe enough to justify the application of computationally much more involved global search techniques; see Chap. 5. Rather, for better results, the application of stepwise selection is recommended instead of the simpler forward selection proposed in [100]; see Sect. 3.4. In [440] a regularized version of the OLS subset selection technique is presented in order to reduce the variance error.

For training of RBF networks, the OLS algorithm is currently one of the most effective tools. It is the standard training procedure in the MATLAB neural network toolbox [120].

In [461] another incremental approach called a *resource allocating network (RAN)* is proposed, which is extended in [312], and further analyzed in [606]. It incorporates new neurons when the error between process and model output or the distance to the nearest basis function center exceeds a given threshold. The distance threshold is decreased during training according to an exponential schedule that leads to a denser placement of the RBFs as training progresses. The RAN algorithm mixes supervised (error threshold) and unsupervised (distance threshold) criteria for center selection. However, it works only for low noise levels since it is based on the assumption that the model error is dominated by its bias part while the variance part is negligible. Otherwise, the error threshold could be exceeded by noise even in the case of a negligible systematical modeling error, and a single outlier can degrade the performance.

11.3.3.6 Nonlinear Optimization for Center Placement

All the above-introduced approaches for the determination of the hidden layer parameters in RBF networks have heuristic characteristics. They exploit the interpretability of these parameters to find good values without an explicit optimization. Therefore, the chosen centers and standard deviations are suboptimal, and the number of required neurons can be expected to be higher than for an MLP network with optimization of *all* parameters. Clearly, the nonlinear optimization of the centers and standard deviations is the most powerful but also the most computationally demanding alternative. With such an approach, however, the advantage of significantly faster training of RBF networks compared with MLP networks basically collapses. Nevertheless, the determination of good initial parameter values for a subsequent nonlinear optimization is easier for RBF networks because any of the methods described above can be applied in order to generate a good initialization.

For nonlinear optimization, the same strategies exist as for MLP networks; see Sect. 11.2.4. For efficient implementation, they require the derivatives of the network output with respect to its parameters.

The derivative with respect to the output layer weights is trivial. Note that it is required only if the staggered optimization strategy is *not* taken. The derivatives of the RBF network output with respect to the ith *output layer weight* are ($i = 0, \ldots, M$)

$$\frac{\partial \hat{y}}{\partial w_i} = \Phi_i \qquad \text{with } \Phi_0 = 1\,. \tag{11.43}$$

The derivatives with respect to the hidden layer parameters depend on the specific type of activation function. For the most common case of a Gaussian RBF network with individual standard deviations for each dimension, the hidden layer weight gradient can be calculated as follows. The basis functions are

$$\Phi_i(\cdot) = \exp\left(-\frac{1}{2}\sum_{j=1}^{p}\frac{(u_j - c_{ij})^2}{\sigma_{ij}^2}\right)\,. \tag{11.44}$$

The derivative of the RBF network output with respect to the jth coordinate of the *center* of the ith neuron is ($i = 1, \ldots, M$, $j = 1, \ldots, p$)

$$\frac{\partial \hat{y}}{\partial c_{ij}} = w_i \frac{u_i - c_{ij}}{\sigma_{ij}^2} \Phi_i(\cdot)\,. \tag{11.45}$$

The derivative of the RBF network output with respect to the *standard deviation* in the jth dimension of the ith neuron is ($i = 1, \ldots, M$, $j = 1, \ldots, p$)

$$\frac{\partial \hat{y}}{\partial \sigma_{ij}} = w_i \frac{(u_i - c_{ij})^2}{\sigma_{ij}^3} \Phi_i(\cdot) \,. \qquad (11.46)$$

It is not recommended to use nonlinear optimization techniques for an RBF network with complete covariance matrices $\underline{\Sigma}_i$ since the number of parameters grows quadratically with the input space dimensionality (see Sect. 11.3.2), implying long training times and substantial overfitting problems.

In [358] and [598], it is demonstrated that nonlinear optimization of the hidden layer parameters in RBF networks can improve the performance significantly. However, the same performance may be achieved with unsupervised learning methods for the centers and standard deviations and least squares optimization of the output layer weights if more neurons are used [233]. Often this combined unsupervised/supervised learning approach can be faster in terms of training time although more neurons are required. So it depends on the specific application whether nonlinear optimization of the hidden layer weights pays off. As a rule of thumb, nonlinear optimization is favorable if the model evaluation speed is crucial because the obtained network is much smaller. Note that in terms of the bias/variance dilemma, the number of optimized *parameters* is decisive, not the number of neurons.

11.3.4 Simulation Examples

Both approximation example problems already discussed in Sect. 11.2.5 in the context of MLP networks will be considered here again. In order to allow some comparison between the MLP and RBF networks, the same number of parameters are used. The MLP networks with one and two neurons possess four and seven parameters, respectively. As RBF network training strategies, the grid-based and the OLS-based center selection schemes are compared. The standard deviations are fixed to a reasonable value, and the output layer weights are estimated by least squares.

Figure 11.29 shows, as expected, that the OLS training procedure (b, d) yields superior results compared with the grid-based approach (a, c). The OLS selects more basis functions in the left half of the input space because the nonlinear behavior is more severe in this region. Obviously, several basis functions that lie closely together are selected. This indicates compensation effects between the neighbored RBFs, i.e., the RBFs do not reflect the underlying function locally. Indeed, an analysis of the corresponding output layer weights reveals basis function heights of opposite sign. Thus, with the application of the OLS algorithm, the local interpretation gets lost. The same is true for the grid-based approach in the case of

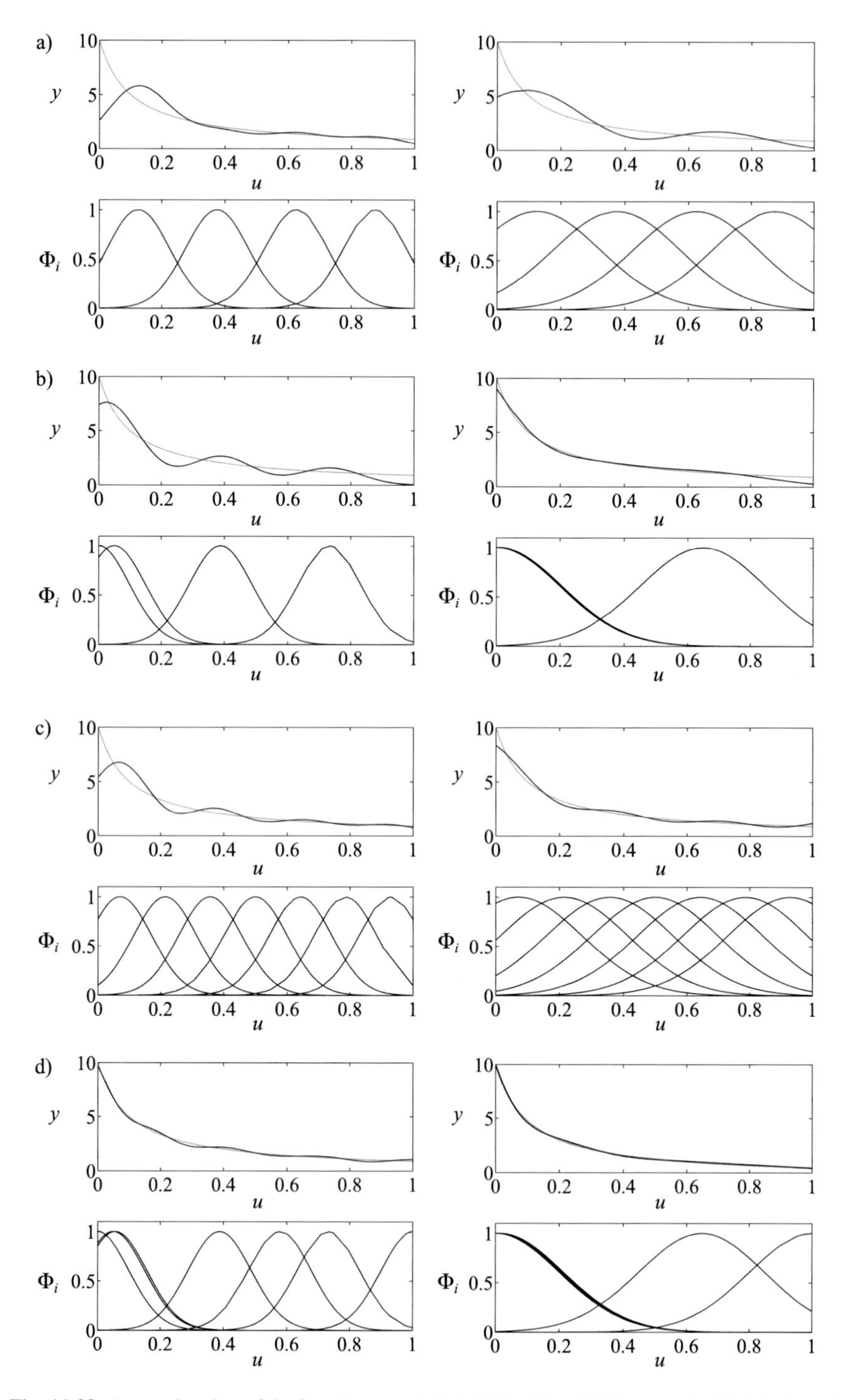

Fig. 11.29 Approximation of the function $y = 1/(u+0.1)$ with an RBF network with 4 (**a**, **b**) and 7 (**c**, **d**) neurons. The grid-based center placement is pursued in (**a**) and (**c**), while the OLS training method is used in (**b**) and (**d**). The standard deviations of the basis functions are chosen equal to 0.1 (left) and 0.2 (right), respectively

large standard deviations (right). On the other hand, small standard deviations (left) yield "dips" in the network output and realize only inferior approximation accuracy. The reason for the latter is that for large standard deviations, the optimization can balance all RBFs since all RBFs contribute significantly to the network output everywhere. For small standard deviations, however, the network is less flexible since the local behavior has to be described by the local RBFs. It is remarkable that the character of the RBF network approximation is quite different for small (left) and large (right) widths although the standard deviations differ only by a factor of 2. Consequently, the RBF network is very sensitive with respect to the basis function widths, which makes their choice difficult in practice.

Thus, with RBF networks, one has to choose between either local characteristics, good conditioned optimization problem, low accuracy, and "dips" in the interpolation behavior or global characteristics, poorly conditioned optimization problem, good accuracy, and smooth interpolation behavior. The poor conditioning in the case of large widths results from the less independent basis functions (for $\sigma \to \infty$ the RBFs become linearly dependent, while they become orthogonal for $\sigma \to 0$). It manifests itself by huge (positive and negative) optimal weights, which clearly lead to non-robust behavior with respect to small changes in the training data. Thus, for the large widths case, the parameters of RBF networks are non-interpretable, similar to those of MLP networks. For the small width case, however, the weights represent the local process characteristics.

Figure 11.30 confirms the findings from Fig. 11.29 and additionally demonstrates that wave-like shapes are well suited for Gaussian RBF networks. The four-neuron RBF networks manage to approximate the function quite well where the four-parameter (one-neuron) MLP network fails. The RBF network with seven neurons performs better and is much more reliable (without any random influences) than the seven-parameter (two-neuron) MLP network.

11.3.5 RBF Properties

The most important properties of RBF networks are as follows:

- *Interpolation behavior* tends to possess "dips" when the standard deviations of some RBFs are chosen too small, and it thus has a tendency to be non-monotonic.
- *Extrapolation behavior* tends to the DC offset value which sometimes is chosen as zero.
- *Locality* is guaranteed when local activation functions are employed.
- *Accuracy* is typically medium. Because the hidden layer parameters of the RBF network are usually not optimized but determined heuristically, many neurons are required to achieve high accuracy. If the hidden layer parameters are optimized, accuracy is comparable with that of MLP networks – perhaps slightly worse owing to the local character.

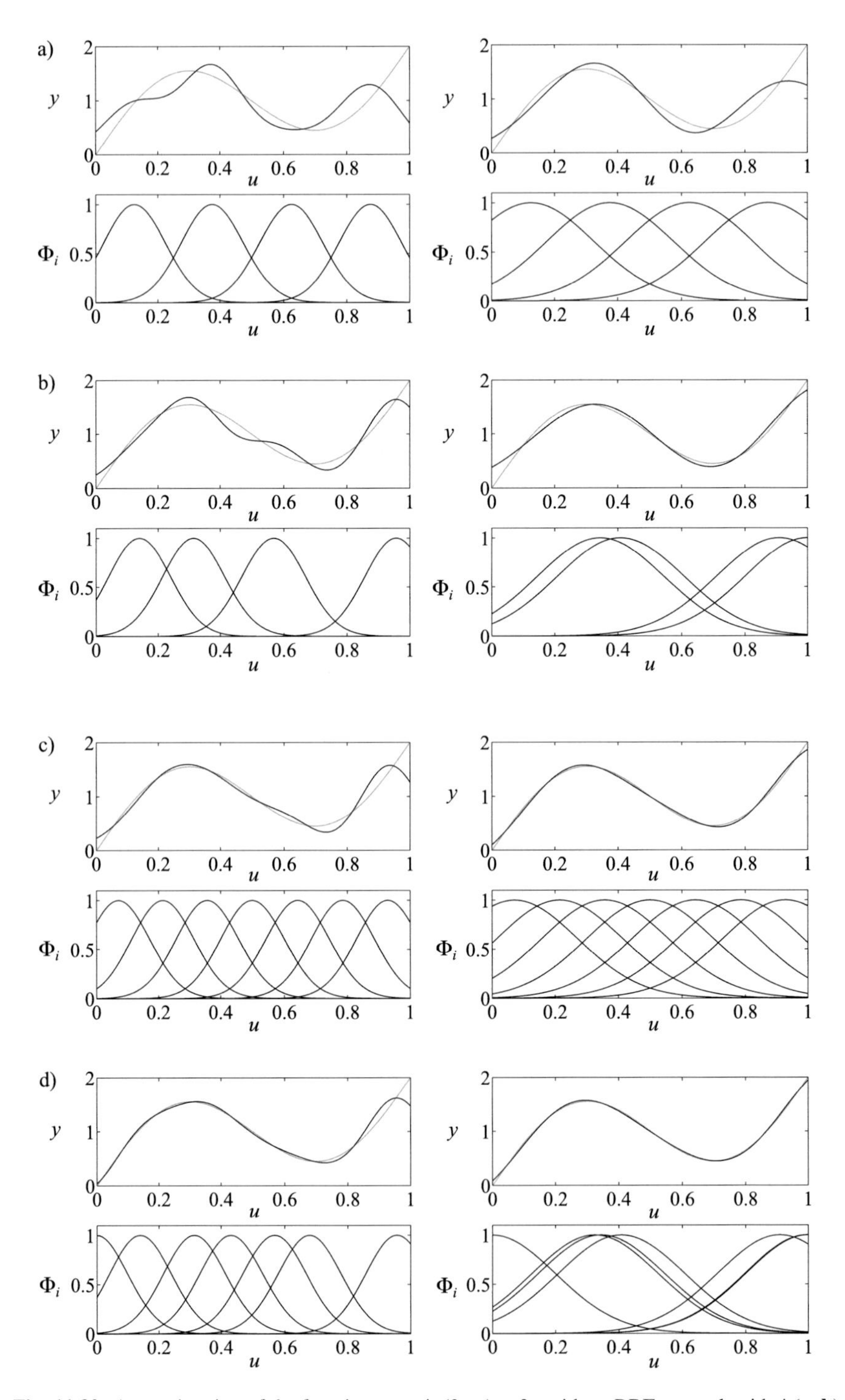

Fig. 11.30 Approximation of the function $y = \sin(2\pi u) + 2u$ with an RBF network with 4 (**a**, **b**) and 7 (**c**, **d**) neurons. The grid-based center placement is pursued in (**a**) and (**c**), while the OLS training method is used in (**b**) and (**d**). The standard deviations of the basis functions are chosen equal to 0.1 (left) and 0.2 (right), respectively

- *Smoothness* depends strongly on the chosen activation function. For the interpolation case, the definition or type of smoothness can be determined by regularization theory; see Sect. 11.3.6. Local basis functions that have too small widths lead to "dips" and thus to non-smooth behavior.
- *Sensitivity to noise* is low since basis functions are usually placed only in regions where "enough" data is available. Furthermore, small variations in the network parameters have an only local effect.
- *Parameter optimization* is fast if the combined unsupervised/supervised learning approach is taken, e.g., the centers and standard deviations are fixed by clustering and the k-nearest neighbor method and the output layer weights are optimized by least squares. In contrast, nonlinear optimization is very slow, as for MLP training. The computational demand for subset selection is medium but grows strongly with the number of potential basis functions.
- *Structure optimization* is relatively fast, since it can be performed with a linear subset selection technique like OLS; see Sect. 3.4.
- *Online adaptation* is robust and efficient if only the output layer weights are adapted. This can be done by a linear recursive least squares (RLS) algorithm; see Sect. 3.2. Owing to the locality of the basis functions, online adaptation in one operating regime does not (or only negligible) influence the others.
- *Training speed* is fast, medium, and slow for the combined unsupervised/supervised learning, the subset selection, and the nonlinear optimization method, respectively.
- *Evaluation speed* is medium because, compared with approaches where the hidden layer parameters are optimized, usually more neurons are required for the same accuracy; see the property accuracy.
- *Curse of dimensionality* is medium to very high, depending on the training strategy. The grid-based approach severely underlies the curse of dimensionality, while clustering decreases this problem significantly, and subset selection is even less sensitive in this respect. However, owing to the local character of RBFs, the curse of dimensionality is, in principle, more severe than for global approximators.
- *Interpretation* of the centers, standard deviations, and heights is possible if the basis functions are local and their widths are chosen small. However, the interpretability in high-dimensional spaces is limited, especially if the complete covariance matrix is utilized.
- *Incorporation of constraints* is possible because the parameters can be interpreted.
- *Incorporation of prior knowledge* is possible because the parameters can be interpreted, and the local character allows one to drive the network toward a desired behavior in a given operating regime.
- *Usage* is medium. RBF networks became more popular in the late 1990s.

11.3.6 Regularization Theory

Radial basis function networks have a strong foundation in mathematics. Indeed, under some assumptions, they can be derived from *regularization theory*. This subsection will illustrate this relationship without going into too much mathematical detail. For a more extensive discussion on this topic, refer to [57, 186, 233, 463, 464].

Regularization theory deals with the ill-posed problem of finding the function that fits the given data best. The interpolation case is considered, that is, the function possesses as many parameters (degrees of freedom) as there exist data samples, i.e., $n = N$. Of course, there exist an infinite number of different functions that interpolate the data with zero error. All these functions differ in the way they behave *between* the data samples, although they all go exactly through each data point. What is the best function to use? In order to be able to answer this question (transforming the ill-posed to a well-posed problem), some additional criterion has to be defined that assesses the behavior of the functions between the data points. A natural requirement for the functions is that they should be *smooth* in some sense. The criterion can express an exact definition of smoothness in a mathematical form. Thus, in regularization the following functional is minimized:

$$I(\hat{f}(\cdot)) = \sum_{i=1}^{N} \Big(y(i) - \hat{f}(\underline{u}(i)) \Big)^2 + \alpha \int |P\hat{f}(\cdot)|^2 d\underline{u} \tag{11.47}$$

where $\hat{f}(\underline{u})$ is the unknown function, $\underline{u}(i)$ and $y(i)$ are the input/output data samples, α is the regularization factor, and P is some differential operator [57]. Equation (11.47) represents a functional, which is a function of a function. So the minimum of (11.47) is an optimal function $\hat{f}^*(\cdot)$. The first term in (11.47) is the sum of squared errors. The second term is called *regularizer* or *prior* since it introduces prior knowledge about the desired smoothness properties into the functional. It penalizes non-smooth behavior of the function $\hat{f}(\cdot)$ by integrating the expression $|P\hat{f}(\cdot)|^2$ over the input space spanned by $\underline{u}$. The influence of this penalty is controlled by the regularization parameter α. For $\alpha \to 0$ the function $\hat{f}(\cdot)$ becomes an interpolation function, i.e., fits all data samples exactly. So α allows a tradeoff between the quality of fit and the degree of smoothness. Basically, (11.47) is the same formulation as in ridge regression; see Sect. 7.5.2. The only difference is that ridge regression represents a parameter optimization problem while in (11.47) a function has to be optimized.

The differential operator P mathematically defines smoothness. A typical choice of P would be the second derivative, that is, the curvature of the function $\hat{f}(\cdot)$ (∇^2 is the Laplacian operator[2]):

$$P\hat{f}(\cdot) = \nabla^2 \hat{f}(\cdot) = \sum_{i=1}^{p} \frac{\partial^2 \hat{f}(\cdot)}{\partial u_i^2} \,. \tag{11.48}$$

Equation (11.47) can be minimized by using the calculus of variations. Depending on the smoothness definition, i.e., the choice of P, different optimal functions result. For example, for the curvature penalty in (11.48), the minimization of (11.47) leads to a cubic spline. The following, more complex operator also takes higher-order derivatives into account [57]:

$$P\hat{f}(\cdot) = \sum_{i=0}^{\infty} \frac{\sigma^i}{2^i \cdot i!} D^i \hat{f}(\cdot) \tag{11.49}$$

where $D^i = (\nabla^2)^{i/2}$ for even i and $D^i = \nabla(\nabla^2)^{(i-1)/2}$ for odd i, with the gradient operator ∇ and Laplacian operator ∇^2. The optimal function for the regularizer (11.49) is a radial basis function network (without offset). Its centers are placed at all data samples, the standard deviations are equal to σ in all dimensions and identical for all neurons, and the optimal output layer weights become

$$\underline{w} = \left(\underline{X} + \alpha \underline{I}\right)^{-1} \underline{y} \,, \tag{11.50}$$

where $\underline{y} = [y(1)\ y(2)\ \cdots\ y(N)]^T$, $\underline{I}$ is an $N \times N$ identity matrix, $\Phi_i(\cdot)$ are the basis functions, and

$$\underline{X} = \begin{bmatrix} \Phi_1(\underline{u}(1)) & \Phi_2(\underline{u}(1)) & \cdots & \Phi_N(\underline{u}(1)) \\ \Phi_1(\underline{u}(2)) & \Phi_2(\underline{u}(2)) & \cdots & \Phi_N(\underline{u}(2)) \\ \vdots & \vdots & \ddots & \vdots \\ \Phi_1(\underline{u}(N)) & \Phi_2(\underline{u}(N)) & \cdots & \Phi_N(\underline{u}(N)) \end{bmatrix} \qquad \underline{w} = \begin{bmatrix} w_1 \\ w_2 \\ \vdots \\ w_N \end{bmatrix} . \tag{11.51}$$

Note that the regression matrix $\underline{X}$ here is quadratic and thus (11.50) contains the inverse matrix instead of the pseudoinverse. This is because for the interpolation case, the number of basis functions equals the number of data samples. Hence, in contrast to the approximation problem treated in Sect. 9.2.2, which is usually encountered in practice, the equation system is not over-determined. Note that

[2]The Laplacian operator sums up the second derivatives with respect to all inputs: $\nabla^2 = \partial^2/\partial u_1^2 + \partial^2/\partial u_2^2 + \ldots + \partial^2/\partial u_p^2$. The symbol is based on the gradient operator $\nabla = [\partial/\partial u_1\ \partial/\partial u_2\ \cdots\ \partial/\partial u_p]^T$.

(11.50) represents the ridge regression solution for the output layer weights; see Sect. 7.5.2. For $\alpha \to 0$ it simplifies to the least squares solution.

In summary, regularization theory gives a justification for radial basis function networks. An RBF network is the optimally interpolating function for a specific smoothness definition. Note, however, that in practice most frequently approximation, not interpolation, problems arise because the number of data samples is usually much higher than the number of neurons.

11.3.7 Normalized Radial Basis Function (NRBF) Networks

One of the undesirable properties of RBF networks is the "dips" in the interpolation behavior that occur for standard deviations that are too small. They are almost unavoidable for high-dimensional input spaces and cause unexpected non-monotonic behavior. Furthermore, the extrapolation behavior of standard RBF networks, which tends to zero, is not desirable for many applications. These drawbacks are overcome by the *normalized* RBF (NRBF) network. The NRBF network output is calculated by

$$\hat{y} = \frac{\sum_{i=1}^{M} w_i \Phi_i \left(||\underline{u} - \underline{c}_i||_{\underline{\Sigma}_i} \right)}{\sum_{i=1}^{M} \Phi_i \left(||\underline{u} - \underline{c}_i||_{\underline{\Sigma}_i} \right)} . \tag{11.52}$$

So the output of an RBF network is normalized by the sum of all (non-weighted) hidden layer neuron outputs. In the standard basis function formulation (9.2), this becomes

$$\hat{y} = \sum_{i=1}^{M} w_i \widetilde{\Phi}_i(\cdot) \qquad \text{with } \widetilde{\Phi}_i(\cdot) = \frac{\Phi_i \left(||\underline{u} - \underline{c}_i||_{\underline{\Sigma}_i} \right)}{\sum_{j=1}^{M} \Phi_j \left(||\underline{u} - \underline{c}_j||_{\underline{\Sigma}_j} \right)} . \tag{11.53}$$

Thus, the sum over all basis functions $\widetilde{\Phi}_i(\cdot)$ is equal to 1:

$$\sum_{i=1}^{M} \widetilde{\Phi}_i(\cdot) = 1 . \tag{11.54}$$

This property is called the *partition of unity*. In [597] it is shown that networks that form a partition of unity possess some advantages in function approximation. In contrast to RBF networks, typically NRBF networks are employed without offset, i.e., $w_0 = 0$, $\Phi_0(\cdot) = 0$, because the normalization allows one to fix an output level without any explicit offset value. Obviously, the basis functions $\widetilde{\Phi}_i(\cdot)$ depend on *all* neurons. This fact is of great importance for the training and interpretation of these networks because a change in one neuron (in the center or standard deviations) affects all basis functions. Figure 11.31 illustrates the interpolation and extrapolation behavior of normalized RBF networks. Comparison with the result

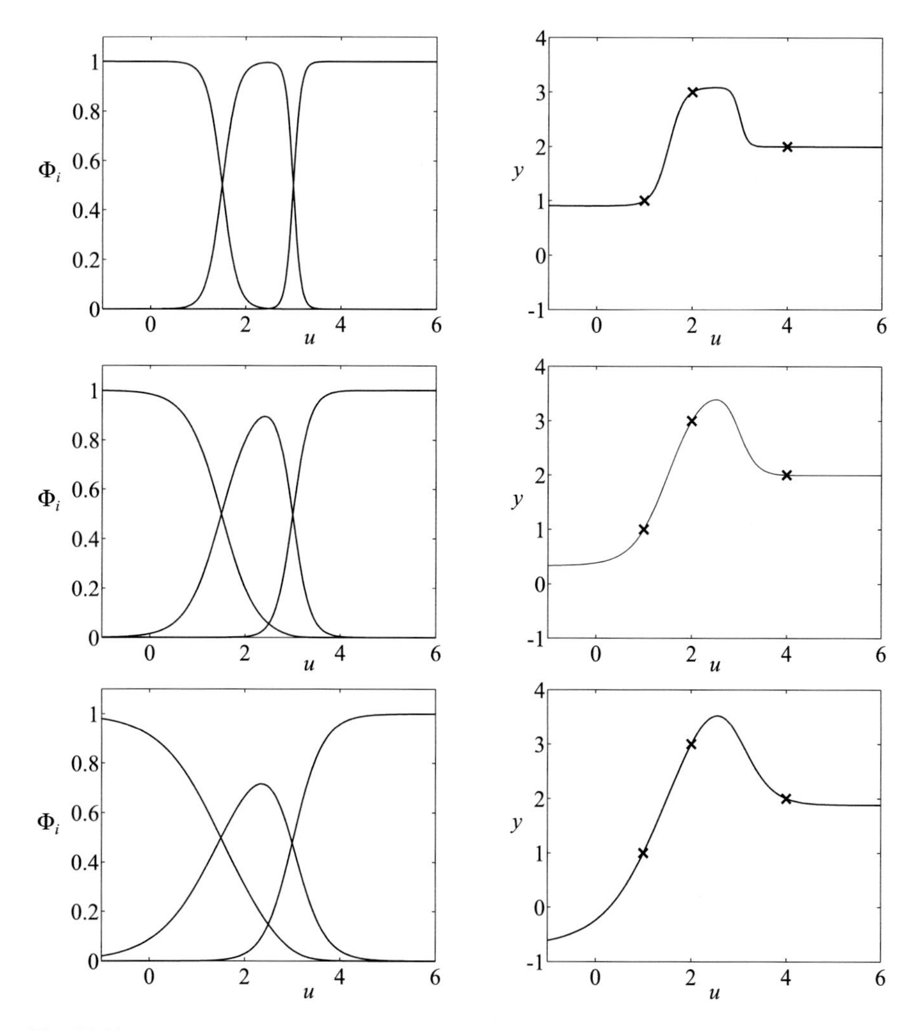

Fig. 11.31 Interpolation and extrapolation behavior of a normalized RBF network. The basis functions are shown on the left; the network outputs are shown on the right. The standard deviations are chosen identically for each neuron in each network as top, $\sigma = 0.4$; center, $\sigma = 0.6$; bottom, $\sigma = 0.8$. The network interpolates between the three data points marked as crosses. Note that "reasonable" standard deviations for NRBF networks are smaller than for RBF networks; see Fig. 11.26

of a standard RBF network in Fig. 11.26 clearly shows that the characteristics of a normalized RBF network are less sensitive with respect to the choice of the standard deviations. No "dips" or "overshoots" can occur, and the extrapolation behavior is constant. Furthermore, it can be guaranteed that the NRBF network output always lies in the interval

$$\min_i(w_i) \leq \hat{y} \leq \max_i(w_i)\,. \tag{11.55}$$

Another appealing feature of NRBF networks is that under some conditions, they are equivalent to fuzzy models with singletons [266]. All issues concerning these fuzzy models are discussed in Chap. 12. In the following, some other differences between RBF and NRBF networks are pointed out.

11.3.7.1 Training

All combined unsupervised/supervised learning strategies described in Sect. 11.3.3 can be applied for NRBF networks as well. For nonlinear optimization, the gradient calculations become more complicated owing to the normalization denominator. The only training method for RBF networks that cannot be applied directly to NRBF networks is linear subset selection techniques such as OLS. The reason for this is the dependency of each basis function on all neuron outputs. During the incremental OLS selection procedure the number of selected basis functions (regressors) increases by 1 in each iteration. At iteration i of the OLS, the selected basis functions whose indices may be gathered in the set $\mathcal{S}_i$ take the form

$$\widetilde{\Phi}_j(\cdot) = \frac{\Phi_j(\cdot)}{\sum\limits_{l \in \mathcal{S}_i} \Phi_l(\cdot)} \qquad \text{for } j = 1, 2, \ldots, i\,. \tag{11.56}$$

The normalization denominator changes in each iteration since $\mathcal{S}_i$ is supplemented by the newest selected basis function. Owing to this change in the basis functions, the orthogonalization procedure of the OLS becomes invalid. Several remedies have been proposed to overcome this problem; see Sect. 3.4.2 for more details. Nevertheless, it is fair to say that linear subset selection loses much of its advantage in both performance and training time when applied to NRBF networks. So global search strategies may be an attractive alternative. All these topics are thoroughly analyzed in the context of fuzzy models because interpretability issues play an important role too; see Sect. 3.4.2.

11.3.7.2 Side Effects of Normalization

The normalization can lead to some very unexpected and usually undesirable effects, which are discussed in detail in the context of fuzzy models in Sect. 12.3.4; refer also to [530]. These effects do *not* occur if all basis functions possess identical standard deviations for each dimension, i.e.,

$$\sigma_{1j} = \sigma_{2j} = \ldots = \sigma_{Mj} . \tag{11.57}$$

The side effects are caused by the fact that for all very distant inputs the activation of the Gaussian RBF with the largest standard deviation becomes higher than the activation of all other Gaussian RBFs. The normalization then results in a reactivation of the basis function with the largest width. This reactivation makes the basis functions non-local and multi-modal – both are properties usually not intuitively assumed in such networks. These side effects are typically not very significant for the performance of an NRBF network, but they are of fundamental importance with regard to its interpretation; therefore more details can be found in Sect. 12.3.4.

11.3.7.3 Properties

In the following, those properties of normalized RBF networks are listed that *differ* from those of non-normalized RBF networks:

- *Interpolation behavior* tends to be monotonic, similar to the MLP.
- *Extrapolation behavior* is constant, similar to the MLP. However, owing to the side effects of normalization the basis function with the largest width determines the extrapolation behavior, that is, the model output tends to the weight associated with this basis function.
- *Locality* is ensured if no strong normalization side effects occur.
- *Structure optimization* is not easily possible because linear subset selection techniques such as the OLS *cannot* be applied owing to the normalization denominator; see Sect. 3.4.
- *Interpretation* is possible as for the standard RBF network. However, care has to be taken with respect to the normalization side effects.

11.4 Other Neural Networks

This section gives a brief overview of other neural network architectures for function approximation. These networks have close relationships with conventional look-up tables (Sect. 10.3) and try to avoid some of their more severe drawbacks. Clearly, this chapter cannot treat *all* existing neural network architectures. Additionally, some architectures with links to fuzzy models are discussed in Chap. 12, and approaches based on local linear models are treated in Chaps. 13 and 14.

11.4.1 General Regression Neural Network (GRNN)

The general regression neural network (GRNN) proposed in [546] and equivalent approaches in [512, 545] are rediscoveries of the work in [403, 448, 591] on nonparametric probability density estimators and kernel regression [57, 499]. With Gaussian kernel functions, a GRNN follows the same equation as a normalized RBF network with identical standard deviations in each dimension and for each neuron:

$$\hat{y} = \frac{\sum_{i=1}^{N} y(i)\Phi_i(\cdot)}{\sum_{i=1}^{N} \Phi_i(\cdot)} \qquad \text{with } \Phi_i(\cdot) = \exp\left(-\frac{1}{2}\frac{||\underline{u} - \underline{u}(i)||^2}{\sigma^2}\right). \tag{11.58}$$

The basis functions are positioned on the data samples $\underline{u}(i)$; thus the network possesses as many neurons as training data samples. Instead of the weights w_i, the measured outputs $y(i)$ are used for the heights of the basis functions. The standard deviation that determines the smoothness of the mapping can either be fixed a priori by the user or optimized (by trial and error or automatically).

The GRNN can be derived via the following statistical description of the process. The relationship between the process output y and the process input $\underline{u}$ can be described in terms of the joint probability density function $p(\underline{u}, y)$. This pdf is the complete statistical representation of the process. If it were known, the expected output could be computed for any given input $\underline{u}$ as follows:

$$\hat{y} = \mathrm{E}\{y|\underline{u}\} = \int_{-\infty}^{\infty} y\, p(y|\underline{u})dy = \frac{\int_{-\infty}^{\infty} y\, p(\underline{u}, y)dy}{\int_{-\infty}^{\infty} p(\underline{u}, y)dy}. \tag{11.59}$$

Since $p(\underline{u}, y)$ is unknown, the task is to approximate this pdf by means of measurement data. With Gaussian kernel functions, the $(p + 1)$-dimensional pdf can be approximated as

$$\hat{p}(\underline{u}, y) = \frac{1}{N}\sum_{i=1}^{N} \frac{1}{(\sqrt{2\pi}\sigma)^{p+1}} \exp\left(-\frac{1}{2}\frac{||\underline{u} - \underline{u}(i)||^2 + (y - y(i))^2}{\sigma^2}\right). \tag{11.60}$$

Substituting (11.60) into (11.59) gives the GRNN equation in (11.58).

For a GRNN, the centers and heights can be directly determined from the available data without any training. Hence, a GRNN can be seen as a generalization of a look-up table where also the data can be stored directly without any optimization. Thus, the GRNN belongs to the class of *memory-based* networks, that is, its operation is dominated by the storage of data. In fact, the GRNN also belongs to the class of just-in-time models discussed in Sect. 11.4.4. These memory-based concepts are opposed to the *optimization-based* networks, which are characterized by high optimization effort for strong information compression like the MLP. Instead of linear interpolation as is usually used in look-up tables (Sect. 10.3), smoother normalized Gaussian interpolation functions are utilized. Furthermore, the GRNN is not grid-based but deals with arbitrarily distributed data. Owing to these properties, the GRNN is an extension of conventional look-up tables for higher-dimensional spaces.

The two major drawbacks of GRNNs are the lack of use of any optimization technique, which results in inefficient noise attenuation, and the large number of basis functions, which leads to very slow evaluation times. The latter disadvantage can be partly overcome by neglecting the contribution of all basis functions whose centers are far away from the current input vector. For such a *nearest neighbor* approach, however, sophisticated search algorithms and data structures are required in order to find the relevant basis functions in an efficient way.

The advantages of the GRNN over grid-based look-up tables with linear interpolation are the smoother model output, which can be differentiated infinitely often, and the better suitability for higher-dimensional problems. The price to be paid is a much higher computational effort, owing to the huge number of exponential functions that have to be evaluated.

In comparison with a normalized RBF network, the GRNN trains much faster (almost in zero time) since training consists only of storing the data samples for the centers and the heights of the basis functions. However, the least squares optimization of the weights in normalized RBF networks allows one to achieve the same performance with considerably fewer neurons, which results in a much faster evaluation speed. Besides these important differences, the GRNN shares most properties with the normalized RBF network; see Sect. 11.3.7.

11.4.2 Cerebellar Model Articulation Controller (CMAC)

The cerebellar model articulation controller (CMAC) network originally was inspired by the (assumed) functioning of the cerebellum, a part of the brain [84]. It was proposed by Albus in 1975 [5, 6] and thus represents one of the earliest neural network architectures. CMAC also belongs to the class of memory-based networks like the GRNN, although the optimization component is more pronounced and the memory requirements are much smaller. Owing to their good online adaptation properties, CMAC networks are mainly applied for nonlinear adaptive or learning control systems [376, 566].

The CMAC network can also be seen as an extension of conventional look-up tables; see Sect. 10.3. Like look-up tables, CMAC networks are grid-based and usually possess a strictly local characteristic. In contrast to look-up tables, the CMAC network output is calculated as the sum of a *fixed, dimensionality independent* number of basis functions. Hereby two major drawbacks of look-up tables are overcome (or at least weakened): the bad generalization capability (see the properties "sensitivity to noise" and "online adaptation" in Sect. 10.3.7) and the exponential increase of the number of basis functions, that contribute to the model output, with the input dimensionality.

The basis functions of CMAC networks are chosen to be strictly local, and typically simple *binary* basis functions are used, which have constant output when they are active and zero output otherwise [84, 566]. Each network input u_i is discretized into M_i intervals, and the whole input space is uniformly covered by these basis functions, which are placed on the $M_1 \times M_2 \times \ldots \times M_p$ *grid*. With binary basis functions for each input, only one basis function would be active. Hence, in order to achieve an averaging and generalization effect, several *overlays* are used, and each of these overlays contains a set of basis functions covering the whole input space. These overlays differ from each other in the displacement of the basis functions. The number of overlays ρ is selected by the user. Figures 11.32 and 11.33 illustrate the structure of the overlays for a one- and two-dimensional CMAC. From these figures, it becomes clear that the number of overlays is chosen equal to the support of the basis functions. Therefore, ρ is called the *generalization parameter*. It determines how many basis functions contribute to the network output and how large the contributing region in the input space is. If the generalization parameter ρ is increased, fewer basis functions are required, the averaging and generalization effect of the network is stronger, and the network output becomes smoother. Thus, for CMAC networks, the choice of ρ plays a role similar to the choice of the standard deviation σ in GRNNs; see Sect. 11.4.1. In practice, ρ is chosen as a tradeoff between model accuracy and smoothness and between memory requirements and learning speed; see [84] for more details. Typical values for ρ are 8, 16, or 32.

The CMAC output is calculated by the weighted sum of *all* M basis functions:

$$\hat{y} = \sum_{i=1}^{M} w_i \Phi_i \left(\underline{u}, \underline{\theta}_i^{(\mathrm{nl})} \right) \tag{11.61}$$

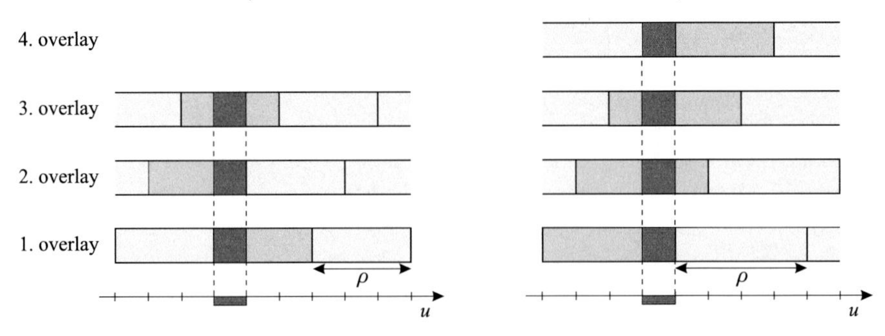

Fig. 11.32 Illustration of the overlays of a one-dimensional CMAC for the generalization parameter $\rho = 3$ (left) and $\rho = 4$ (right). The light gray cells indicate the support of the basis functions. The dark gray cells are activated by all inputs within the dark-shaded interval

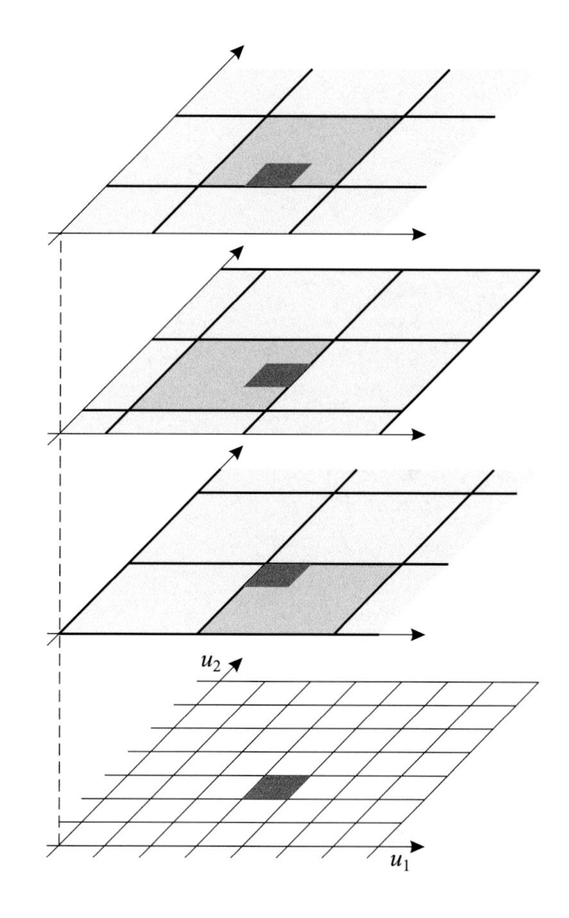

Fig. 11.33 Illustration of the overlays of a two-dimensional CMAC for the generalization parameter $\rho = 3$. The light gray cells indicate the support of the basis functions. The overlays are displaced relative to each other in both inputs. The dark gray-shaded cells are activated by all input vectors within the dark-shaded area

where $\underline{\theta}_i^{(\text{nl})}$ contains the centers of the basis functions that form a partition of unity (see Sect. 11.3.7) and the overall number of basis function and parameters is approximately [84]

$$M \approx \prod_{i=1}^{p} (M_i - 1)/\rho^{p-1} . \tag{11.62}$$

This number increases exponentially with the input space dimensionality. Thus, for high-dimensional mappings usually ρ is chosen large, and additionally *hash coding* strategies are applied to reduce the memory requirements [84, 566]. Nevertheless, CMAC networks are only well suited for low- and moderate-dimensional problems with, say, $p < 10$.

Since the basis functions are strictly local, (11.61) simplifies considerably to

$$\hat{y} = \sum_{i \in \mathcal{S}_{\text{active}}} w_i \Phi_i \left(\underline{u}, \underline{\theta}_i^{(\text{nl})} \right) , \tag{11.63}$$

where the set $\mathcal{S}_{\text{active}}$ contains the indices of all basis functions that are *activated* by the current input $\underline{u}$. Hence, in (11.63) only the weighted sum of the ρ active basis functions has to be evaluated since all other basis functions $\Phi_i(\cdot) = 0$. With binary basis functions that are equal to $1/\rho$ when they are active, (11.63) becomes simply the average of the weights:

$$\hat{y} = \frac{1}{\rho} \sum_{i \in \mathcal{S}_{\text{active}}} w_i \, . \tag{11.64}$$

In contrast to the GRNN (Sect. 11.4.1), for the CMAC, the ρ active basis functions can be easily determined owing to the grid-based coverage of the input space. Furthermore, (11.63) represents the exact network output, not just an approximation, since the CMAC's basis functions are strictly local. So the CMAC network is based on the following trick. Although the number of basis functions (11.62) becomes huge for high-dimensional mappings owing to the curse of dimensionality for grid-based approaches, the number of required basis function evaluations for a given input is equal to ρ. This means that the evaluation speed of CMAC networks depends only on ρ and is virtually independent of the input space dimensionality (neglecting the dimensionality-dependent part of effort for one basis function evaluation). In contrast, a conventional look-up table requires the evaluation of 2^p basis functions, which becomes infeasible for high-dimensional problems.

Training of CMAC networks is a linear optimization task because only the weights w_i have to be estimated. This can be performed by least squares. However, since the number of parameters is huge and the basis functions are strictly local, the Hessian (see Sect. 3.1) contains mostly zero entries. Thus, special algorithms for inversion of sparse matrices are needed. CMAC networks are typically trained in sample adaptation mode, i.e., each data sample is processed separately; see Sect. 4.1. These algorithms can be directly applied online. Usually, the least mean squares (LMS) or normalized LMS are applied because the recursive least squares (RLS) would require the storage of the huge covariance matrix; see Sect. 3.2. The strict locality of the CMAC networks' basis functions makes these approaches unusually fast since for each training data sample, only the ρ parameters associated with the active basis functions have to be updated. Locality implies that most basis functions are orthogonal, which is an important property for fast convergence of the LMS and similar first-order algorithms; see Sect. 3.1.3.

In summary, the most important advantages of CMAC networks are (i) fast training owing to the linear parameters and the strictly local characteristics, (ii) high evaluation speed owing to the small number (ρ) of active basis functions and restriction to computationally cheap operations (e.g., no exponential functions or other complex activation functions have to be evaluated), (iii) favorable features for online training, and (iv) good suitability for hardware implementation [380].

The drawbacks of CMAC networks are as follows. First, the grid-based approach results in high memory requirements. Hash coding can weaken but not overcome this disadvantage and comes at the price of increased implementation and computation effort. Furthermore, high memory compression rates decrease the network's performance owing to frequent collisions. Second, the grid-based approach implies a uniform distribution of the network flexibility over the input space. This means that the maximum approximation accuracy is fixed by the discretization of the input space (the choice of the M_i). The CMAC is unable to incorporate extra flexibility (basis functions with weights) in regions where it is required because the process behavior is complex and wastes a lot of parameters in regions with extremely smooth process behavior. Third, the discretization of the input space requires a priori knowledge about the upper and lower bounds on all inputs. An overestimation of these bounds reduces the effective network flexibility since fewer basis functions are utilized than assumed; the others are wasted in unused regions of the input space.

Fourth, the displacement strategy that guarantees that only ρ basis functions are superimposed and thus avoids the curse of dimensionality comes at a price. The CMAC network structure may cause significant approximation errors for non-additive functions [84]. Fifth, all network complexity optimization strategies discussed in Sect. 7.4 cannot really be applied to the CMAC because after the selection of ρ, no further basis functions can be added or deleted. Finally, the CMAC with binary basis functions produces a non-differentiable network output. Smoother basis functions such as B-splines of higher order can be used, but they slow down training and evaluation times.

Nevertheless, the relatively short adaptation times and the high evaluation speed make CMACs especially interesting for online identification and control applications with fast sampling rates. For a more extensive treatment of CMAC networks, refer to [84, 566].

11.4.3 Delaunay Networks

Another possible extension of conventional look-up tables is so-called *scattered data look-up tables*, which replace the grid-based partitioning of the input space by a more flexible strategy that allows an arbitrary distribution of the data points. The most common strategy is the *Delaunay* triangulation, which partitions the input space in simplices, i.e., triangles for two-dimensional input spaces; see Fig. 11.34.

Compared with a grid-based partitioning of the input space, this offers the following three major advantages. First, the arbitrary distribution of the points allows one to adapt the complexity of the model with respect to the data distribution and the process behavior and weakens the curse of dimensionality. Second, the interpolation between the data points can be realized linearly since a (hyper)plane can be fitted exactly through the $p + 1$ points of a simplex for a p-dimensional input space. Third, online adaptation capabilities of scattered data look-up tables are superior since points can be arbitrarily moved, new points can be generated, and old points can be deleted. This is not possible for grid-based look-up tables because the grid structure must be maintained.

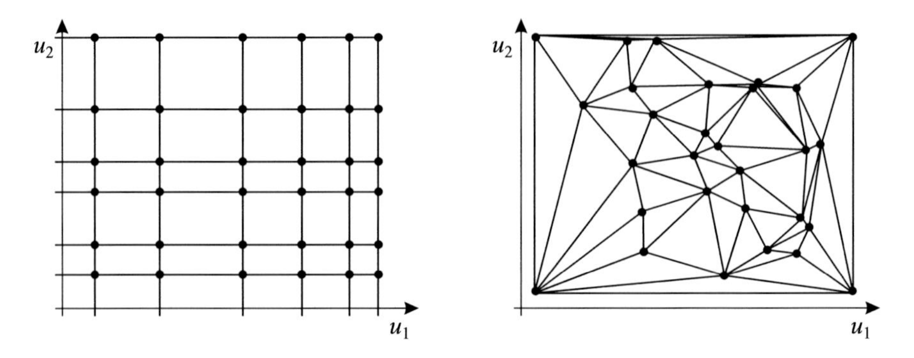

Fig. 11.34 Partitioning of the input space by a grid (left) and by Delaunay triangulation (right)

Compared with conventional look-up tables, Delaunay networks significantly reduce the memory requirements at the price of higher computational effort in training and evaluation. So the tradeoff between these two conflicting resources decides which of these two model architectures is more appropriate for a specific application. If online adaptation is required and more than two input dimensions have to be considered, Delaunay networks are clearly favorable.

These features make Delaunay networks especially attractive for automobile and all other applications where severe memory restrictions exist and cheap, low-performance micro-controllers do not allow the use of more sophisticated neural network architectures. An extensive analysis of scattered data look-up tables and especially Delaunay networks can be found in [516, 567, 573].

Although Delaunay networks and other scattered data look-up tables overcome some of the drawbacks of conventional grid-based look-up tables, they share some important restrictions. The model output is not differentiable owing to the piece-wise linear interpolation. Note that smoother interpolation rules (or basis functions) would be more computationally demanding and thus somewhat against the philosophy of look-up table models. Finally, although Delaunay networks are less sensitive to the curse of dimensionality with respect to memory requirements, the computational effort for the construction of the Delaunay segmentation increases sharply with the input dimensionality. Thus the application of these models is also limited to relatively low-dimensional problems with, say, $p \leq 4$.

11.4.4 Just-In-Time Models

For an increasing number of real-world processes, a huge amount of data is gathered in databases by the higher levels of process automation systems. Old data from other identical plants may be available, and new data is collected during the operation of the process. So there arises a need for automated methods that exploit the information contained in such a huge amount of data, an increasingly important field

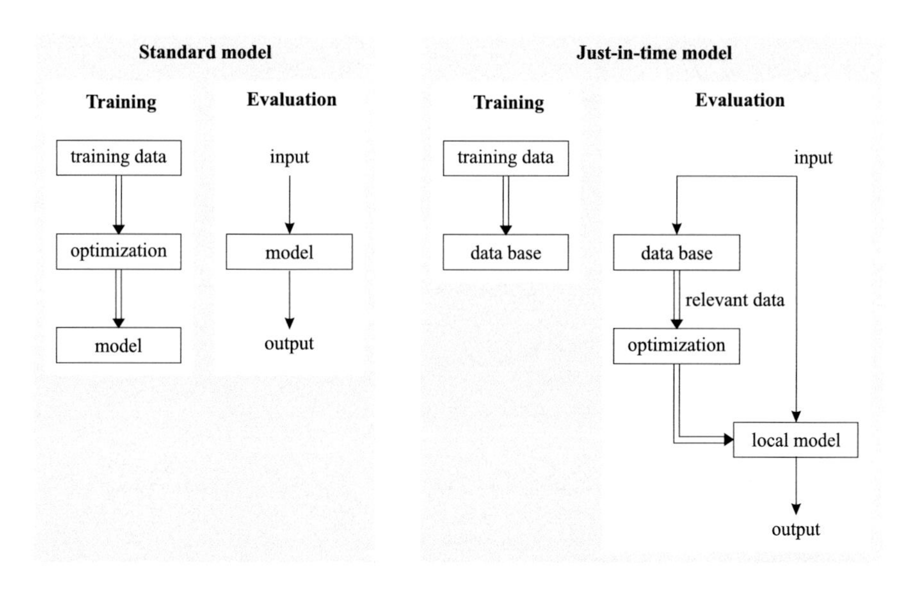

Fig. 11.35 Comparison between the training and evaluation phases in standard and just-in-time models

called *data mining*. For this purpose, so-called *just-in-time models* are a promising alternative to standard models. In the system identification context, these approaches are more commonly known as *models-on-demand*. In the computer science context, they are also called instance-based learning or *lazy learning* methods in the machine learning terminology. These names characterize their main feature: The actual model building phase is postponed until the model is needed, i.e., until the model output for a given input is requested; before that the data is simply gathered and stored.

Figure 11.35 illustrates the differences between the application of standard and just-in-time models. Standard models are typically trained offline, and for evaluation, only the fixed model is used. Thus, all training data is processed a priori in a batch-like manner. This can become computationally expensive or even impossible for huge amounts of data, and therefore data reduction techniques may have to be applied. Another consequence of this standard approach is that training is the computationally demanding step, while evaluation is very fast. Additionally, online adaptation can be implemented, which adapts the model (usually by changing its parameters) during the evaluation phase in order to track the time-variant behavior of the process.

For just-in-time models, the training phase consists merely of data gathering and effective storage in a database. The computationally involved model optimization part is performed in the evaluation phase. First, the relevant data samples that describe similar operating conditions as the incoming input data are searched in the database. This data selection is typically based on nearest neighbor ideas; see Fig. 11.36. Next, with these relevant data samples, a model is optimized. Finally,

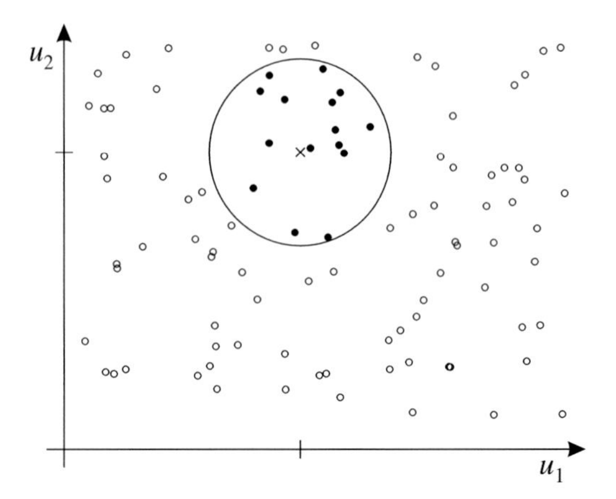

Fig. 11.36 Neighbor selection for just-in-time models. For a given input (marked by the cross), the neighboring data samples (filled circles) are used for model optimization, while all others (hollow circles) are ignored. The size of the region of relevant data is determined by a tradeoff between noise attenuation on the one hand and local approximation accuracy of the model on the other hand. It is influenced by the available amount of data and the strength of nonlinear behavior of the modeled process. Large regions are chosen if the local data density is sparse and the process behavior is locally almost linear and vice versa

this model is used to calculate the output for the given input. The GRNN described in Sect. 11.4.1 is a very simple representative of a just-in-time model since all data samples are stored in the training phase. In the evaluation phase, the GRNN computes the model output as the weighted average of the neighboring points, i.e., the GRNN is a trivial case of just-in-time models because no actual optimization is carried out.

Just-in-time models are only locally valid for the operating condition characterized by the current input. Thus, a very simple model structure can be chosen, e.g., a linear model; see Chap. 13. It is important to understand that the data selection and model optimization phase is carried out individually for each incoming input. This allows one to change the model architecture, model complexity, and the criteria for data selection online according to the current situation. It is possible to take into consideration the available amount and quality of data, the state, and condition of the plant, and the current constraints and performance goals. Obviously, just-in-time models are inherently adaptive when all online measured data is stored in the database and old data is forgotten.

An example of a just-in-time model is the McLain-type interpolating associative memory system (MIAS) [566, 567]. In MIAS a linear local model is estimated from the relevant data samples weighted with the inverse of their quadratic distance from the input. This can be seen as a just-in-time version of the local model approaches discussed in Chap. 13.

The major drawbacks of just-in-time models are their large memory requirements, the high computational effort for evaluation, and all potential difficulties that may arise from the online model optimization such as missing data and guarantee of certain response time. With increasing processing capabilities, however, just-in-time models will probably become more and more interesting alternatives in the future.

11.5 Summary

Table 11.2 summarizes the advantages and drawbacks of MLP, RBF, and normalized RBF networks. Probably the most important lesson learned in this chapter is that the curse of dimensionality can be overcome best by projection-based mechanisms as applied in MLP networks. However, this ridge construction *in principle* requires time-consuming *nonlinear* optimization techniques in order to determine the optimal projection directions. This computational effort can be avoided by combining unsupervised or heuristic learning methods with *linear* optimization techniques as is typically realized for RBF network training. Although the curse of dimensionality is stronger than in projection-based approaches, it can be weakened sufficiently in order to solve a wide class of real-world problems. The next two chapters introduce further strategies for the reduction of the curse of dimensionality without nonlinear optimization.

Figure 11.37 gives a rough overview of the memory requirements and training efforts for the neural network architectures discussed. It shows that there exists a

Table 11.2 Comparison between MLP and RBF networks

Properties	MLP	RBF	Normalized RBF
Interpolation behavior	+	−	+
Extrapolation behavior	0	−	+
Locality	−	++	+
Accuracy	++	0	0
Smoothness	++	0	+
Sensitivity to noise	++	+	+
Parameter optimization	−−	++*/−−**	++*/−−**
Structure optimization	−	+	−
Online adaptation	−−	+	+
Training speed	−−	+*/−−**	+*/−−**
Evaluation speed	+	0	0
Curse of dimensionality	++	−	−
Interpretation	−−	0	0
Incorporation of constraints	−−	0	0
Incorporation of prior knowledge	−−	0	0
Usage	++	0	−

* = linear optimization, ** = nonlinear optimization,
+ + / − − = model properties are very favorable/undesirable

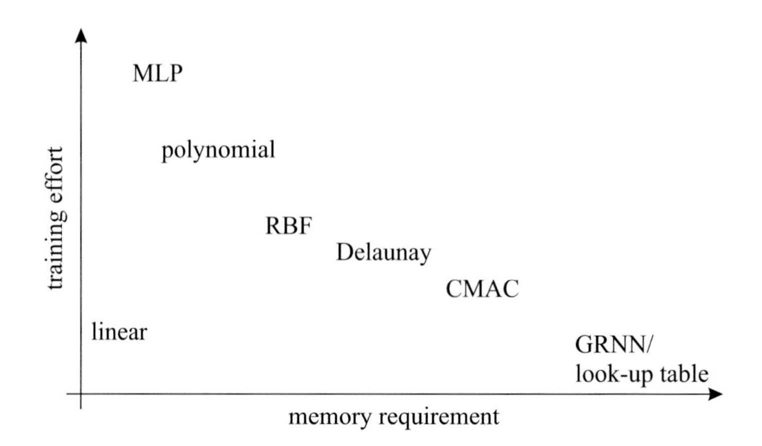

Fig. 11.37 Memory requirement versus training effort for some neural network architectures. Typically, the training effort corresponds to the amount of information compression and thus is inversely proportional to the number of parameters

tradeoff between these two properties, which in the ideal case would both be small. Linear models are the only exception from this rule, but they are not able to represent any nonlinear behavior.

Note that this diagram gives only a crude idea of the characteristics of the networks. A more detailed analysis of the specific training strategies would have to be taken into account. For example, an RBF network with grid-based center selection would be positioned closer to the lower right corner, while an RBF network trained with nonlinear optimization would be closer to the upper left corner and thereby would overtake a RAWN (MLP with fixed hidden layer weights).

11.6 Problems

1. Write down a sum of basis functions and explain how these basis functions have to be chosen in case of a linear, a polynomial, and a neural network model with two inputs u_1 and u_2.
2. What are the differences between the three main construction mechanisms (ridge, radial, and tensor product)? Conclude from these distinct properties about the advantages and drawbacks of each construction mechanism. Which construction mechanism is used in which type of neural network?
3. Sketch an MLP network with two inputs, two hidden layer neurons, and one output. The hidden layer neurons possess a logistic nonlinearity; the output neuron is linear. Identify all parameters in the network and check which parameters influence the network output in a linear way. Calculate the gradient of the network output with respect to two parameters: (i) the weight from the first hidden layer neuron to the output neuron and (ii) the weight from the first

input to the second hidden layer neuron. Explain why and how these gradients are needed for any gradient-based learning algorithm (start by calculating the gradient of the loss function).

4. Explain two different strategies for training an MLP network. How do they take into account linear and nonlinear relationships between certain parameters (weights) and the network output?
5. Calculate the total number of parameters of an MLP network with one hidden layer with M_1 neurons and an MLP network with two hidden layers each with M_2 neurons. Consider problems with p inputs and one output. Compare the total number of parameters for both networks for $M_1 = M_2$ and $M_1 = 2M_2$ and $M_1 = M_2^2$.
6. Sketch an RBF network with two inputs, two hidden layer neurons, and one output. The hidden layer neurons possess an axis-orthogonal Gaussian nonlinearity with different standard deviations in each dimension; the output neuron is linear. Identify all parameters in the network and check which parameters influence the network output in a linear way. Calculate the gradient of the network output with respect to three parameters: (i) the weight from the first hidden layer neuron to the output neuron, (ii) the second center coordinate from the Gaussian of the first hidden-layer neuron, and (iii) the standard deviation in the first input dimension from the Gaussian of the second hidden-layer neuron.
7. Explain three different training methods for RBF networks with their specific features.
8. Generate ten data points $\{x(i), y(i)\}, i = 1, \ldots, 10$, where the $x(i)$ are equally distributed in $[0, 10]$ and $y(i) = x(i)$. Choose an RBF network with ten Gaussian neurons. The centers of the Gaussians are fixed on the $x(i)$, and the standard deviations are also fixed at some value σ. Generate the regression matrix and compute the least squares solution for the output weights of the RBF network for $\sigma = 0.5, 1,$ and 2. Plot the output of the RBF network in the interval $[-5, 15]$. Are all data points met exactly? Why?
9. Repeat the steps of Problem 8 for a normalized RBF network. Compare the results with those obtained for Problem 8. How sensitive are RBF and normalized RBF networks with respect to the width of their basis functions? How does extrapolation behavior differ?
10. What is the key idea of CMAC and Delaunay networks? What major problem do these both approaches try to overcome?

Chapter 12
Fuzzy and Neuro-Fuzzy Models

Abstract This chapter introduces fuzzy systems and how they can be cast into neuro-fuzzy system, which can be learned from data. It discusses how prior knowledge can be incorporated and how knowledge can be extracted from a data-driven neuro-fuzzy model. Different kinds of fuzzy systems are analyzed where the Takagi-Sugeno variant already builds bridges to the subsequent chapters on local linear modeling approaches. Different learning schemes for neuro-fuzzy models are discussed, and their principal ideas highlighted. The role of "defuzzification" or normalization in the context of learning and interpretability is discussed in detail.

Keywords Rules · Fuzzy logic · Singleton · Takagi-Sugeno · RBF · Interpretation · Prior knowledge

This chapter first gives a short introduction to fuzzy logic and fuzzy systems and then concentrates on methods for learning fuzzy models from data. These approaches are commonly referred to as neuro-fuzzy networks since they exploit some links between fuzzy systems and neural networks. Within this chapter only one architecture of neuro-fuzzy networks is considered, the so-called singleton approach. Neuro-fuzzy networks based on local linear models are extensively treated in the next two chapters.

This chapter is organized as follows. Section 12.1 gives a brief introduction to fuzzy logic. Different types of fuzzy systems are discussed in Sect. 12.2. The links between fuzzy systems and neural networks and the optimization of such neuro-fuzzy networks from data are extensively treated in Sect. 12.3. Some advanced neuro-fuzzy learning schemes are reviewed in Sect. 12.4. Finally, Sect. 12.5 summarizes the essential results.

O. Nelles, *Nonlinear System Identification*,
https://doi.org/10.1007/978-3-030-47439-3_12

12.1 Fuzzy Logic

Fuzzy logic was invented by Zadeh in 1965 [608] as an extension of Boolean logic. While classical logic assigns to a variable either the value 1 for "true" or the value "0" for "false," fuzzy logic allows one to assign to a variable any value in the interval [0, 1]. This extension is motivated by the observation that humans often think and communicate in a vague and uncertain way – partly because of insufficient information and partly due to human nature. There exist many possible ways to deal with such imprecise statements; perhaps the most obvious one is by using probabilities. However, in order to deal with such statements in a rule-based form, new *approximate reasoning* mechanisms based on fuzzy logic had to be developed [609]. This chapter introduces fuzzy logic and approximate reasoning mechanisms only so far as they are necessary for a clear understanding of fuzzy and neuro-fuzzy models. For more information on the fundamentals of fuzzy logic, refer to [20, 65, 126, 128, 287, 313, 320, 336, 342, 343, 373, 451, 467, 615]. A collection of the fuzzy logic terminology is given in [49].

In order to illustrate how fuzzy logic works, a simple example will be introduced. Fuzzy logic allows one to express and process relationships in form of rules. For example, the following rule formulates the well-known wind-chill effect, i.e., the temperature a person senses does not depend solely on the true environment temperature but also on the wind:

$$\text{IF } temperature = \mathsf{low} \text{ AND } wind = \mathsf{strong} \text{ THEN } sensation = \mathsf{very\ cold}$$

where the components of this rule are denoted as:

$temperature, wind, sensation$	linguistic variables
low, strong, very cold	linguistic terms/labels
AND	operator; here, conjunction
$temperature =$ low, $wind =$ strong	linguistic statements
$temperature =$ low AND $wind =$ strong	premise
sensation = very cold	consequent

The linguistic variables $temperature$ and $wind$ are often simply denoted as inputs and the linguistic variable $sensation$ as output. With a *complete set of rules*, all combinations of $temperature$ and $wind$ can be covered:

$$\begin{aligned}
R_1 &: \text{IF } temp. = \mathsf{low} && \text{AND } wind = \mathsf{strong} && \text{THEN } sens. = \mathsf{very\ cold}\\
R_2 &: \text{IF } temp. = \mathsf{medium} && \text{AND } wind = \mathsf{strong} && \text{THEN } sens. = \mathsf{cold}\\
R_3 &: \text{IF } temp. = \mathsf{high} && \text{AND } wind = \mathsf{strong} && \text{THEN } sens. = \mathsf{medium}\\
R_4 &: \text{IF } temp. = \mathsf{low} && \text{AND } wind = \mathsf{weak} && \text{THEN } sens. = \mathsf{cold}\\
R_5 &: \text{IF } temp. = \mathsf{medium} && \text{AND } wind = \mathsf{weak} && \text{THEN } sens. = \mathsf{medium}\\
R_6 &: \text{IF } temp. = \mathsf{high} && \text{AND } wind = \mathsf{weak} && \text{THEN } sens. = \mathsf{warm}
\end{aligned}$$

The overall number of rules (here 6) depends on the chosen fineness or resolution of the fuzzy sets. Clearly, the accuracy of the fuzzy system depends on this property, which is called *granularity*. It can be shown that a fuzzy system can approximate any smooth input/output relationship to an arbitrary degree of accuracy if the granularity is decreased; in other words, fuzzy systems are universal approximators [329, 590].

The components of fuzzy rules are briefly discussed in the following.

12.1.1 Membership Functions

In fuzzy rules the linguistic variables are expressed in the form of fuzzy sets. In the above example, the linguistic input variables *temperature* and *wind* are labeled by the linguistic terms low, medium, and high and weak and strong. These linguistic terms are defined by their associated *membership functions (MSFs)*. Figure 12.1 shows a possible definition of these membership functions. These MSFs define the *degree of membership* of a specific *temperature* or *wind* to the fuzzy sets, e.g., the *temperature* $T = 3\,°\mathrm{C}$ is considered low with 0.7, medium with 0.3, and high with 0 degree of membership. This procedure, which calculates from a crisp input such as $T = 3\,°\mathrm{C}$ the degree of membership for the fuzzy sets, is called *fuzzification*. MSFs are usually functions of a single variable, e.g., $\mu_i(T)$ or $\mu_i(W)$, where i stands for low, medium, or high or weak or strong, respectively. So fuzzy systems typically deal with each input separately, and the inputs are combined in the rules by logic operators such as AND and OR; see Sect. 12.1.2.

In the fuzzy system, only the degrees of membership are further processed. This can be seen as a nonlinear transformation of the inputs. Often information is lost during this procedure. For example, with the MSFs in Fig. 12.1, it does not matter whether the *temperature* is $10\,°\mathrm{C}$ or $15\,°\mathrm{C}$ or some value in between, because the degrees of membership are not affected. Another interesting property of the MSFs in Fig. 12.1 is that they sum up to 1, i.e., they fulfill the property $\mu_{\mathsf{low}}(T) + \mu_{\mathsf{medium}}(T) + \mu_{\mathsf{high}}(T) = 1$, and $\mu_{\mathsf{weak}}(W) + \mu_{\mathsf{strong}}(W) = 1$, or more generally

$$\sum_{i=1}^{M} \mu_i(u) = 1 \qquad \text{for all } u\,, \tag{12.1}$$

where M denotes the number of MSFs for the linguistic variable u. Although it is not required that the MSFs are normalized, this property is often employed because it makes the interpretation easier; see Sect. 12.3.4.

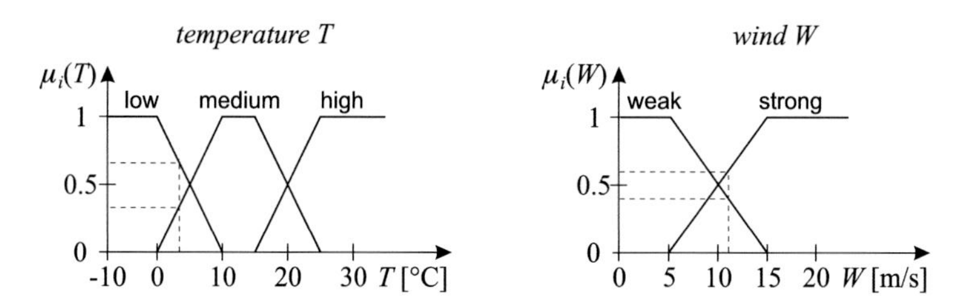

Fig. 12.1 Membership functions for the linguistic variables *temperature* T (left) and *wind* W (right)

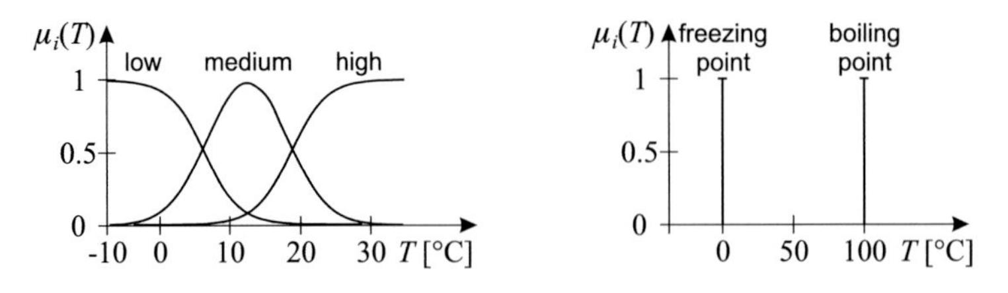

Fig. 12.2 Membership functions can be smooth (left), which reduce information loss and are better suited for learning, or they can deteriorate to singletons (right), which describes crisp fuzzy sets

In Fig. 12.1 the membership functions are of a triangular and trapezoidal type. As discussed above this leads to an information loss in regions where the slope of the MSFs is equal to zero. Furthermore, these types of MSFs are not differentiable, and thus learning from data may run into problems; see Sect. 12.3. Alternatively, smoother MSFs such as normalized Gaussians can be used, as depicted in Fig. 12.2, to avoid these difficulties.

Sometimes linguistic terms are not really fuzzy. They are either completely true $\mu = 1$ or completely false $\mu = 0$. In this case, the MSF becomes a rectangle. If it is true for just a single value and false otherwise, this deteriorates to a *singleton* as demonstrated in Fig. 12.2(right). Singletons possess the value 1 at their position (here at 0 °C and 100 °C) and are equal to 0 elsewhere.

After the degrees of membership for each linguistic statement have been evaluated, the next step is to combine these values by logic operators such as AND and OR.

12.1.2 Logic Operators

Fuzzy logic operators are an extension of the Boolean operators. This implies that fuzzy logic operators are equal to the Boolean ones for the special cases where the degrees of membership are only either zero or one. The negation operator for a linguistic statement such as $T =$ low is calculated by

$$\text{NOT}(\mu_i(T)) = 1 - \mu_i(T)\,. \tag{12.2}$$

For the *conjunction* of two linguistic statements such as $T =$ low and $W =$ strong, several alternative logic operators exist, the so-called t-norms. The most common t-norms are (see Fig. 12.3):

$$\text{Min:}\quad \mu_i(T)\ \text{AND}\ \mu_j(W) = \min[\mu_i(T), \mu_j(W)]. \tag{12.3a}$$

$$\text{Product:}\quad \mu_i(T)\ \text{AND}\ \mu_j(W) = \mu_i(T)\mu_j(W). \tag{12.3b}$$

$$\text{Bounded diff.:}\quad \mu_i(T)\ \text{AND}\ \mu_j(W) = \max[0, \mu_i(T) + \mu_j(W) - 1]. \tag{12.3c}$$

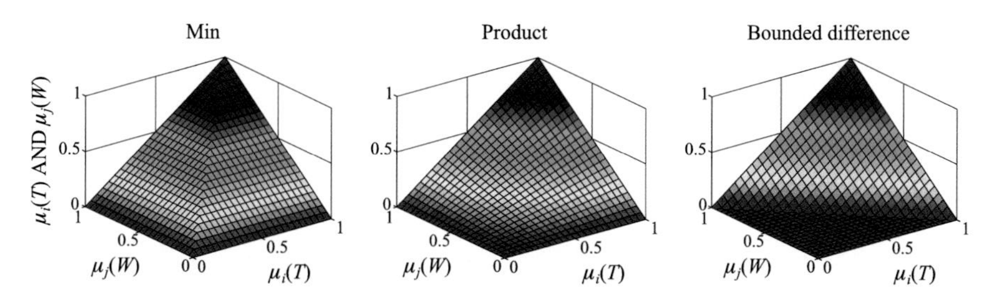

Fig. 12.3 Illustration of different t-norms (operators for conjunction)

For the *disjunction* of two linguistic statements also, several alternative logic operators exist, the so-called t-conorms. The most common t-conorms are:

$$\text{Max:} \quad \mu_i(T) \text{ OR } \mu_j(W) = \max[\mu_i(T), \mu_j(W)]. \tag{12.4a}$$

$$\text{Algebraic sum:} \quad \mu_i(T) \text{ OR } \mu_j(W) = \mu_i(T) + \mu_j(W) - \mu_i(T)\mu_j(W). \tag{12.4b}$$

$$\text{Bounded sum:} \quad \mu_i(T) \text{ OR } \mu_j(W) = \min[1, \mu_i(T) + \mu_j(W)]. \tag{12.4c}$$

For classification tasks the min and max operators are popular. For approximation and identification tasks, the product and algebraic product operators are better suited, owing to their smoothness and differentiability. For some neuro-fuzzy learning schemes, the bounded difference and sum operators offer several advantages.

All t-norms and t-conorms can be applied to an arbitrarily large number of linguistic statements by nesting the operators in either of the following ways:

$$\mathsf{A} \text{ AND } \mathsf{B} \text{ AND } \mathsf{C} = (\mathsf{A} \text{ AND } \mathsf{B}) \text{ AND } \mathsf{C} = \mathsf{A} \text{ AND } (\mathsf{B} \text{ AND } \mathsf{C})\,.$$

Often, especially for neuro-fuzzy networks, the linguistic statements are combined only by AND operators, like the rules R_1–R_6 in the above example. With this *conjunctive form* all input/output relationships can be described since disjunctive dependencies can be modeled by the introduction of additional rules. Although the conjunctive form is sufficient, it may not always be the most compact or easiest to understand representation.

12.1.3 Rule Fulfillment

With the logic operators, it is possible to combine the degrees of membership of all linguistic statements within the rule premise. For example, the *temperature* may be $T = 3\,°\text{C}$ and the *wind* $W = 11$ m/s. This gives the following degrees of membership: $\mu_{\text{low}}(T) = 0.7$, $\mu_{\text{medium}}(T) = 0.3$, $\mu_{\text{high}}(T) = 0$, $\mu_{\text{weak}}(W) = 0.4$, and $\mu_{\text{strong}}(W) = 0.6$; see Fig. 12.1. Thus, in rule R_1 the conjunction between

$\mu_{\text{low}}(T) = 0.7$ and $\mu_{\text{strong}}(W) = 0.6$ has to be calculated. This results in 0.6 for the min operator, 0.42 for the product operator, and 0.3 for the bounded difference operator. Obviously, the outcome of a fuzzy system is strongly dependent on the specific choice of operators.

The combination of the degrees of membership of all linguistic statements is called the *degree of rule fulfillment* or *rule firing strength*, since it expresses how well a rule premise matches a specific input value (here $T = 3\,^\circ\text{C}$ and $W = 11\,\text{m/s}$).

For the whole fuzzy system, only rules with a degree of fulfillment larger than zero are relevant. All others are inactive. For strictly local MSFs like the triangular ones in Fig. 12.1, only a subset of all rules is activated by an input. For example, with the input $T = 3\,^\circ\text{C}$ and $W = 11\,\text{m/s}$, the rules R_3 and R_6 have a zero degree of rule fulfillment because $\mu_{\text{high}}(T) = 0$. Care must be taken that the whole feasible input space is covered by rules in order to avoid the situation where all fuzzy rules are inactive. When non-strictly local MSFs are chosen, as in Fig. 12.2(left), automatically *all* rules are active for any input, although the degree of rule fulfillment may be arbitrarily small. So non-covered regions in the input space cannot arise even if the rule set is not complete.

12.1.4 Accumulation

After the degree of rule fulfillment has been calculated for all rules, the consequents have to be evaluated and accumulated to generate one output of the fuzzy system. Finally, for most applications this fuzzy system output, which generally is a fuzzy set, has to be *defuzzified* in order to obtain one crisp output value. Note that defuzzification is not necessary if the output of the fuzzy system is used as an input for another fuzzy system (rule chaining in hierarchical fuzzy systems) or if it is directly presented to a human. For example, the output *sensation* cannot be easily quantified because it is a subjective, qualitative measure, and thus defuzzification to a crisp value is not necessarily reasonable.

Since the exact procedure for these last steps in fuzzy inference depends on the specific type of the fuzzy rule consequents, it is described in the following section.

12.2 Types of Fuzzy Systems

This section presents three different types of fuzzy systems: linguistic, singleton, and Takagi-Sugeno fuzzy systems. For reasons given below, only the latter two are investigated further in terms of neuro-fuzzy models. The singleton fuzzy models are analyzed in this chapter, while Takagi-Sugeno fuzzy models are extensively treated in Chaps. 13 and 14. Relational fuzzy systems are not discussed; for more information refer to [451].

12.2.1 Linguistic Fuzzy Systems

Linguistic fuzzy systems, also known as *Mamdani* fuzzy systems [360], possess rules in the form

$$R_i\colon \ \text{IF } u_1 = \mathsf{A}_{i1} \text{ AND } u_2 = \mathsf{A}_{i2} \text{ AND } \ldots u_p = \mathsf{A}_{ip} \ \text{ THEN } y = \mathsf{B}_i\,,$$

where $u_1, \ldots, u_p$ are the p inputs of the fuzzy system gathered in the input vector $\underline{u}$, y is the output, the index $i = 1, \ldots, M$ runs over all M fuzzy rules, A_{ij} denotes the fuzzy set used for input u_j in rule i, and B_i is the fuzzy set used for the output in rule i. The example fuzzy system introduced in Sect. 12.1 is of this type. A comparison with this example illustrates that typically many A_{ij} are identical for different rules i (the same is valid for the B_i).

On first sight, such linguistic fuzzy systems are the most appealing because both the inputs and the output are described by linguistic variables. However, the analysis of the fuzzy inference will show that quite complex computations are necessary for the evaluation of such a linguistic fuzzy system. The following steps must be carried out:

Fuzzification → Aggregation → Activation → Accumulation → Defuzzification

The fuzzification uses the MSFs to map crisp inputs to the degrees of membership. The aggregation combines the individual linguistic statements to the degree of rule fulfillment. Both steps are identical for all types of fuzzy systems discussed here and have been explained in the previous section. The last three steps depend on the fuzzy system type considered.

In the *fuzzification* phase, the degrees of membership for all linguistic statements are calculated. They will be denoted by $\mu_{ij}(u_j)$, $i = 1, \ldots, M$, $j = 1, \ldots, p$.

In the *aggregation* phase, these degrees of membership are combined according to the fuzzy operators. When the fuzzy system is in conjunctive form and the product operator is applied as t-norm, the degree of fulfillment of rule i becomes

$$\mu_i(\underline{u}) = \mu_{i1}(u_1) \cdot \mu_{i2}(u_2) \cdot \ldots \cdot \mu_{ip}(u_p)\,. \tag{12.5}$$

In the *activation* phase, these degrees of rule fulfillment are utilized to calculate the output activations of the rules. This can, for example, be done by cutting the output MSFs at the degree of rule fulfillment, i.e.,

$$\mu_i^{\text{act}}(\underline{u}, y) = \min[\mu_i(\underline{u}), \mu_i(y)]\,, \tag{12.6}$$

where $\mu_i(y)$ is the output MSF belonging to the fuzzy set B_i and $\mu_i(\underline{u})$ is the degree of fulfillment for rule i. Figure 12.4 shows output MSFs and the output activation of rule R_1 of the example in Sect. 12.1.3, where the degree of fulfillment of the first rule was 0.42.

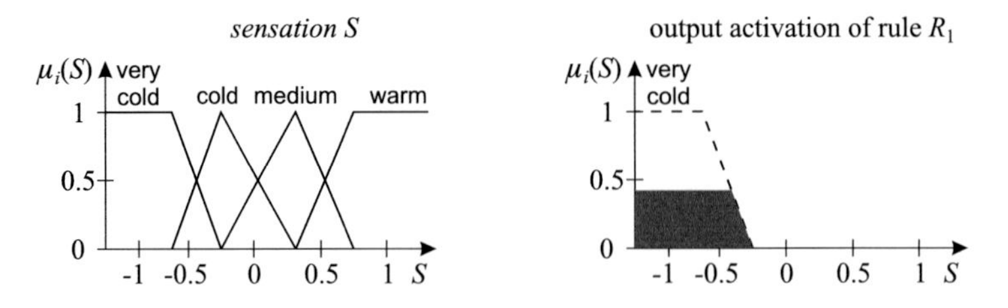

Fig. 12.4 Membership functions for the linguistic variable *sensation* S (left) and the calculation of the output activation (right). The black area represents the output activation of rule R_1 and is obtained by cutting the output MSF at the degree of rule fulfillment 0.42

In the *accumulation* phase, the output activations of all rules are joined together. This can be done by computing the maximum of all output activations, that is,

$$\mu^{\text{acc}}(\underline{u}, y) = \max_i [\mu_i^{\text{act}}(\underline{u}, y)] \,. \tag{12.7}$$

The accumulation yields one fuzzy set, which is the fuzzy output of the fuzzy system. If no crisp output is required, the inference mechanism stops here.

Otherwise, a crisp output value can be calculated by a final *defuzzification* step. The most common strategy for extraction of a crisp value from a fuzzy set is the *center of gravity* method, that is,

$$\hat{y} = \frac{\int\limits_{y_{\text{min}}}^{y_{\text{max}}} y\, \mu^{\text{acc}}(\underline{u}, y)\, dy}{\int\limits_{y_{\text{min}}}^{y_{\text{max}}} \mu^{\text{acc}}(\underline{u}, y)\, dy} \,, \tag{12.8}$$

where $\mu^{\text{acc}}(\underline{u}, y)$ is the fuzzy output set, i.e., the accumulated output activation.

Other defuzzification methods can yield substantially different results. For an extensive comparison, refer to [492].

Figure 12.5 summarizes the complete inference procedure for linguistic fuzzy systems with min and max operators for conjunction and disjunction, max operator for the accumulation, and center of gravity defuzzification.

There are reasons why linguistic fuzzy systems are not pursued further here, although they are quite popular. First, the defuzzification according to (12.8) is complicated and time-consuming, since the integrals have to be solved numerically. Second, in the general case, learning of linguistic fuzzy systems is more complex than that of singleton and Takagi-Sugeno fuzzy systems. Usually, the higher complexity does not pay off in terms of significantly improved performance. Third, better interpretability, commonly seen as a major advantage of linguistic fuzzy systems, is in some cases questionable, as the following simple example demonstrates.

When human experts are asked to construct a fuzzy rule base with the corresponding membership functions, they tend to define output MSFs with small widths

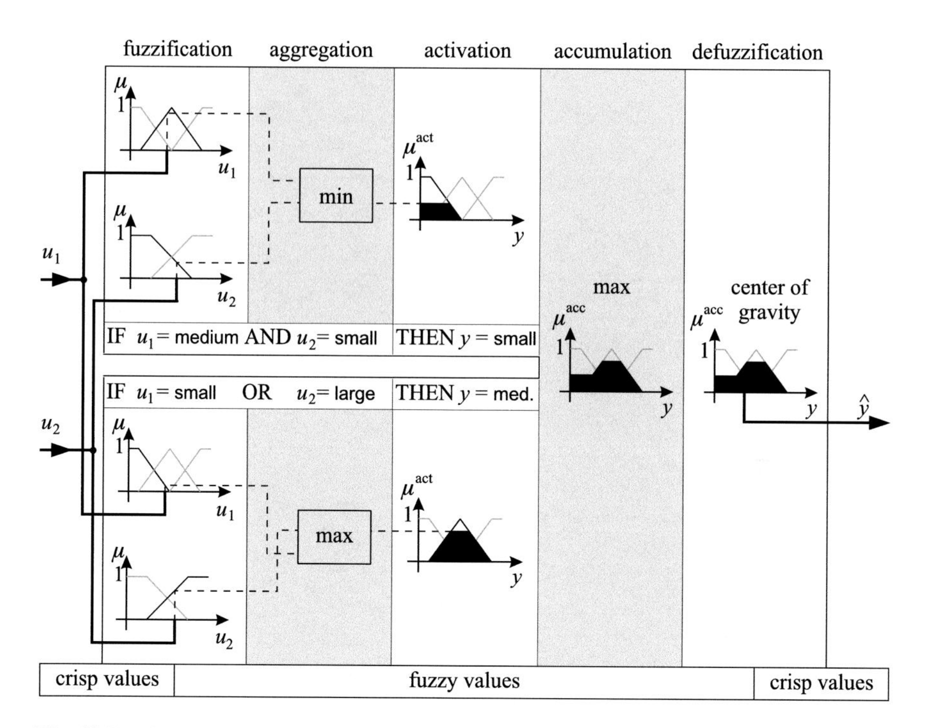

Fig. 12.5 Inference for a linguistic fuzzy system with two inputs and two rules (This figure was kindly provided by Martin Fischer, Institute of Automatic Control, TU Darmstadt.)

when they are certain about a relationship and to define very broad output MSFs when they are insecure. This is quite natural because the membership functions express the degree of uncertainty. For example, an MSF approaches a singleton as the uncertainty decreases. Unfortunately, the output MSFs' widths have just the opposite effect on the fuzzy inference to that expected. A larger width gives an MSF a higher weight in the center of gravity defuzzification and thus drives the fuzzy system output toward the most uncertain rule activation. To avoid these difficulties, additional confidences can be introduced for each rule that expresses the expert's certainty about the correctness of the rule [84].

12.2.2 Singleton Fuzzy Systems

The complex defuzzification phase in linguistic fuzzy systems can be simplified considerably by the restriction to singleton output membership functions. This also reduces the computational demand for the evaluation and learning of the fuzzy system significantly since no integration has to be carried out numerically. Therefore, singleton fuzzy systems are most widely applied in industry.

A rule of a singleton fuzzy system has the following form:

$$R_i: \text{ IF } u_1 = \mathsf{A}_{i1} \text{ AND } u_2 = \mathsf{A}_{i2} \text{ AND } \ldots u_p = \mathsf{A}_{ip} \text{ THEN } y = s_i\,,$$

where s_i is a real value called the *singleton* of rule i. It determines the position of the trivial output MSF; see Fig. 12.2. With such singleton type rules, several activation, accumulation, and defuzzification methods yield identical results. For example, it does not matter whether the singletons are cut at or multiplied with the degree of rule fulfillment in the activation calculation. In the center of gravity calculation, the integration simplifies to a weighted sum since the output MSFs do not overlap. The fuzzy system output can be calculated by

$$\hat{y} = \frac{\sum\limits_{i=1}^{M} s_i\, \mu_i(\underline{u})}{\sum\limits_{i=1}^{M} \mu_i(\underline{u})}\,, \tag{12.9}$$

where M denotes the number of rules and $\mu_i(\underline{u})$ is the degree of fulfillment of rule i according to (12.5). This means the fuzzy system output is a weighted sum of the singletons, where the "weights" are given by the degrees of rule fulfillment. The normalization denominator in (12.9) forces the fuzzy system to a *partition of unity*. This property holds for all types of fuzzy systems and is required for a proper interpretability of the rule consequents.

Note that the denominator in (12.9) is equal to 1 when the membership functions themselves sum up to 1 (see (12.1)) and the rule set is complete. The set of rules is called *complete*, if each combination of all input fuzzy sets is represented by one rule premise. Under these conditions, which are often fulfilled, the calculation of a singleton fuzzy system further simplifies to

$$\hat{y} = \sum_{i=1}^{M} s_i\, \mu_i(\underline{u})\,. \tag{12.10}$$

These equations already indicate a close relationship between singleton fuzzy systems and normalized radial basis function networks, which is discussed in more detail in Sect. 12.3. In addition, the operation of singleton fuzzy systems is identical to grid-based look-up tables; see Sect. 10.3. The basis functions in Fig. 10.8 can equivalently be interpreted as MSFs, and the weights w_i correspond to the singletons s_i. Indeed, a singleton fuzzy system is nothing else but a rule-based interpretation of grid-based look-up tables. The input MSFs determine the type of interpolation. Triangular MSFs result in linear interpolation; see Fig. 10.8. Smoother MSFs lead

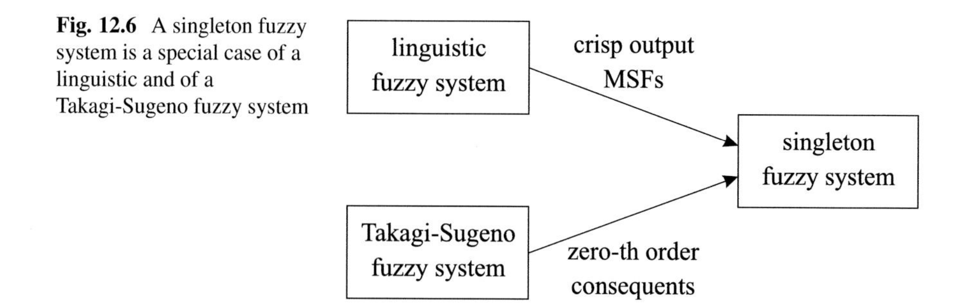

Fig. 12.6 A singleton fuzzy system is a special case of a linguistic and of a Takagi-Sugeno fuzzy system

to smoother interpolation rules. Furthermore, a singleton fuzzy system can be seen as a special case of both linguistic and Takagi-Sugeno fuzzy systems, as Fig. 12.6 illustrates; see also Sect. 12.2.3.

One drawback of singleton fuzzy systems is that generally each rule possesses its individual singleton as output MSF. This means that there exist as many singletons as rules, which makes large fuzzy systems very hard to interpret. Alternatively, the singleton fuzzy system can be transformed into a linguistic fuzzy system with fewer output MSFs but additional rule confidences. The reduced number of MSFs for the output may improve the interpretability, and the rule confidences introduce additional weights of the rules, which retain the fuzzy system's flexibility. In [84] the necessary conditions are given that allow a transformation from singleton fuzzy systems to linguistic ones with confidences.

12.2.3 *Takagi-Sugeno Fuzzy Systems*

In 1985 Takagi and Sugeno [552] proposed a new type of fuzzy system with rules in the following form:

$$R_i:\ \text{IF } u_1 = \mathsf{A}_{i1} \text{ AND } \ldots \text{ AND } u_p = \mathsf{A}_{ip} \text{ THEN } y = f_i(u_1, u_2, \ldots, u_p)\,.$$

It can be seen as an extension of singleton fuzzy systems. While singletons can still be seen as a special type of MSF, the functions $f_i(\cdot)$ are definitely not fuzzy sets. Rather, only the premise of Takagi-Sugeno fuzzy systems is really linguistically interpretable. Note, however, that this restricted interpretation holds only for *static* models. For modeling of *dynamic* processes, Takagi-Sugeno fuzzy models possess an excellent interpretation, which is superior to most if not all alternative approaches; see Chap. 22.

If the functions $f_i(\cdot)$ in the rule consequents are trivially chosen as constants (s_i), a singleton fuzzy system is recovered. This case is often called a *zeroth-order* Takagi-Sugeno fuzzy system, since a constant can be seen as a zeroth-order Taylor series expansion of a function $f_i(\cdot)$. Commonly, *first-order* Takagi-Sugeno fuzzy systems are applied. This means that the rule consequent is a *linear* function of the inputs:

$$y = w_{i0} + w_{i1}u_1 + w_{i2}u_2 + \ldots + w_{ip}u_p \, . \tag{12.11}$$

The output of a Takagi-Sugeno fuzzy system can be calculated by

$$\hat{y} = \frac{\sum_{i=1}^{M} f_i(\underline{u})\,\mu_i(\underline{u})}{\sum_{i=1}^{M} \mu_i(\underline{u})}\,, \tag{12.12}$$

which is a straightforward extension of the singleton fuzzy system output in (12.9). Figure 12.7 illustrates the operation of a one-dimensional Takagi-Sugeno fuzzy system with the following two rules:

R_1: IF $u =$ small THEN $y = u$

R_2: IF $u =$ large THEN $y = 2u - 0.2$.

Basically, all issues discussed in the context of singleton fuzzy systems in the remaining parts of this chapter can be extended to Takagi-Sugeno fuzzy systems by replacing s_i with $w_{i0}+w_{i1}u_1+w_{i2}u_2+\ldots+w_{ip}u_p$. However, a number of learning methods that are particularly tailored for Takagi-Sugeno fuzzy systems are further discussed and analyzed in Chaps. 13 and 14.

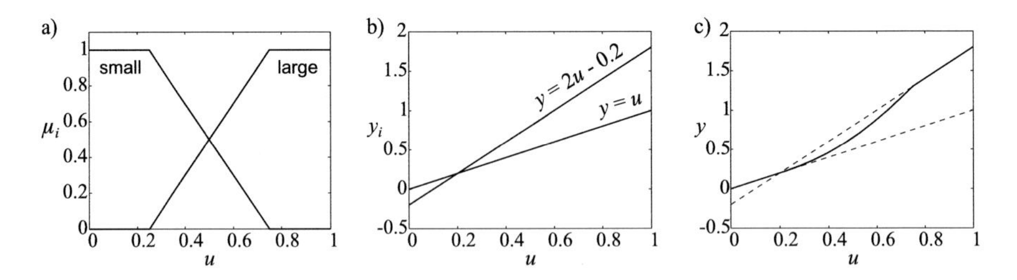

Fig. 12.7 One-dimensional Takagi-Sugeno fuzzy system with two rules: (**a**) membership functions, (**b**) linear functions of both rule consequents, (**c**) fuzzy system output (solid) and linear functions (dotted). Note that the denominator in (12.12) is equal to 1 since the MSFs already form a partition of unity

12.3 Neuro-Fuzzy (NF) Networks

This and the subsequent section discuss neuro-fuzzy networks based on singleton fuzzy models. Chapters 13 and 14 focus on Takagi-Sugeno neuro-fuzzy models. Neuro-fuzzy networks are fuzzy models that are not solely designed by expert knowledge but are at least partly learned from data. The close links between fuzzy models and neural networks motivated the first approaches for data-driven fuzzy modeling. Typically, the fuzzy model is drawn in a neural network structure, and learning methods already established in the neural network context are applied to this neuro-fuzzy network. The contemporary point of view is that fuzzy models can be *directly* optimized or learned from data without having to be drawn in a neural network structure. Many learning methods are applied that have no relationship to neural networks. Nevertheless, the original name "neuro-fuzzy networks" has survived for all types of fuzzy models that are learned from data.

This section analyzes which components of a singleton fuzzy system can be optimized from data, how efficiently this can be performed, and in which way the interpretability of the obtained model is affected. Incorporation of prior knowledge into and interpretation of fuzzy models are (or should be) primary concerns in experimental fuzzy modeling since these are the major benefits compared with other model architectures. Typically, there exists a tradeoff between interpretability and performance. In order to keep a model interpretable, it must be restricted in some way: that is, flexibility has to be sacrificed. This drawback can usually be compensated by the incorporation of prior knowledge. If no prior knowledge is available, the application of a fuzzy model does not make any sense from the model accuracy point of view. However, if accuracy is not the only ultimate goal and rather an understanding of the functioning of the process is desired, then fuzzy models are an excellent choice. In summary, the two major motivations for the use of neuro-fuzzy networks are:

- *Exploitation of prior knowledge.* This may allow one to improve the model's accuracy, to reduce the requirements on the amount and quality of data, and to define a model behavior in operating regimes or conditions where no data can be measured.
- *Improved understanding of the process.* This may allow one to utilize the model with higher confidence, especially in safety-critical applications, or even to extract information about the process that may be helpful with regard to a variety of goals.

12.3.1 Fuzzy Basis Functions

A singleton fuzzy model can be written in basis function formulation. When the fuzzy model is in conjunctive form and the product operator is used as t-norm, the degrees of rule fulfillment become

$$\mu_i(\underline{u}) = \prod_{j=1}^{p} \mu_{ij}(u_j) \,. \tag{12.13}$$

Since $\mu_i(\underline{u})$ combines the information of the one-dimensional MSFs $\mu_{ij}(u_j)$, it is also called *multidimensional* membership function. Figure 12.8 illustrates how univariate triangular and Gaussian MSFs are combined into two-dimensional MSFs by the product and min operators. As demonstrated in Fig. 12.8d, the multidimensional MSFs are only smooth if both the univariate MSFs and the logic operator are smooth.

According to (12.13), fuzzy models with complete rule sets follow the tensor product construction mechanism; see Sect. 11.1.3. This means that the rules realize the basis functions based on the conjunction of all combinations of univariate MSFs; see Fig. 11.9. Thus, fuzzy models with complete rule set are grid-based and fully underlie the curse of dimensionality. This property is obvious because complete fuzzy models with singletons are just rule-based interpretations of look-up tables; see Sect. 10.3. However, generally, the multidimensional MSFs $\mu_i(\underline{u})$ in (12.13) are

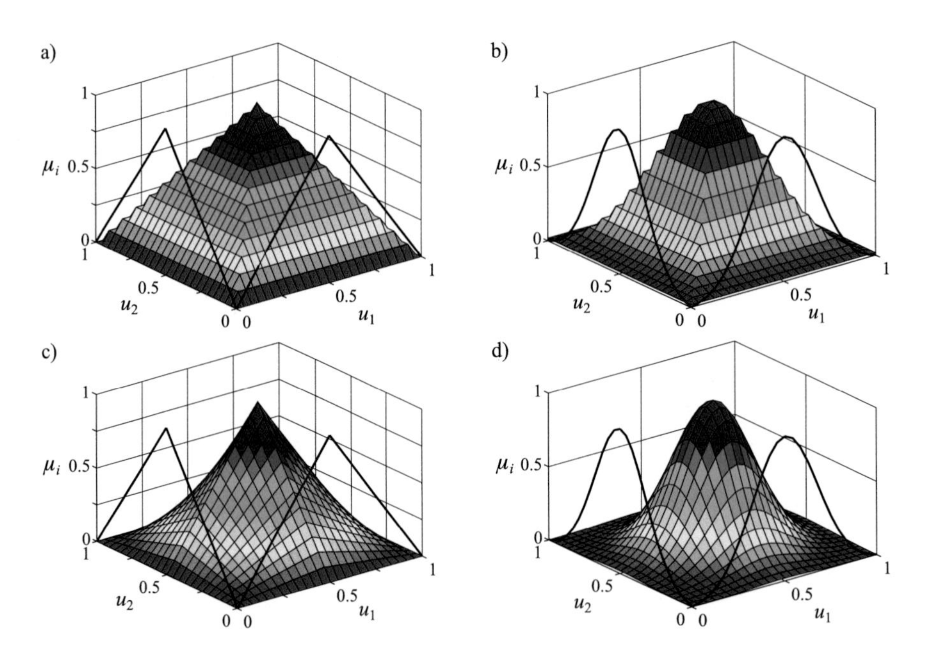

Fig. 12.8 Construction of multidimensional membership functions: (**a**) triangular MSFs with min operator, (**b**) Gaussian MSFs with min operator, (**c**) triangular MSFs with product operator, (**d**) Gaussian MSFs with product operator

not directly identical to the basis functions. Rather the basis function formulation is (see (12.9))

$$\hat{y} = \sum_{i=1}^{M} s_i \Phi_i(\underline{u}) \qquad \text{with } \Phi_i(\underline{u}) = \frac{\mu_i(\underline{u})}{\sum\limits_{j=1}^{M} \mu_j(\underline{u})} . \tag{12.14}$$

The denominator in (12.14) guarantees that the fuzzy model output is always a weighted sum of singletons, where the "weighting factors" $\Phi_i(\underline{u})$ sum up to 1 for any $\underline{u}$. When the fuzzy model has a complete rule set and the univariate MSFs form a partition of unity individually for each dimension, then the basis functions $\Phi_i(\underline{u})$ naturally form a partition of unity since $\sum_{j=1}^{M} \mu_j(\underline{u}) = 1$. Otherwise, this normalization denominator enforces basis functions $\Phi_i(\underline{u})$ that form a partition of unity, which can result in unexpected and undesired normalization side effects; see Sect. 12.3.4.

12.3.2 Equivalence Between RBF Networks and Fuzzy Models

Under some conditions singleton neuro-fuzzy systems are equivalent to normalized radial basis function networks [211, 291, 318, 468]. The singletons s_i correspond to the NRBF network output layer weights w_i. This result can be further extended to Takagi-Sugeno neuro-fuzzy systems, which are equivalent to local model networks [266]. From the basis function formulation, an NRBF network and a singleton neuro-fuzzy model are identical if Gaussian MSFs are used and the product operator is applied as t-norm. This follows since the product of several univariate Gaussian MSFs is equivalent to one multidimensional Gaussian RBF. So for each neuron i, the following identity holds:

$$\prod_{j=1}^{p} \exp\left(-\frac{1}{2}\frac{(u_j - c_{ij})^2}{\sigma_{ij}^2}\right) = \exp\left(-\frac{1}{2}\sum_{j=1}^{p}\frac{(u_j - c_{ij})^2}{\sigma_{ij}^2}\right) . \tag{12.15}$$

A neuro-fuzzy neuron with Gaussian MSFs is shown in Fig. 12.9. Compared with the RBF neuron in Fig. 11.22, the neuro-fuzzy neuron first processes each input individually and finally combines them with a t-norm. Although for other types of MSFs and logic operators no exact equivalence between NRBF and neuro-fuzzy networks holds, a strong similarity is maintained.

Two additional restrictions apply to the NRBF network in order to allow a reasonable interpretation in terms of a fuzzy system:

- *The RBFs have to be placed on a grid.* In Fig. 12.10a six RBFs are shown that do not lie on a grid. The equivalent neuro-fuzzy model consisting of six rules requires six MSFs for each input. Each MSF is used only once, and

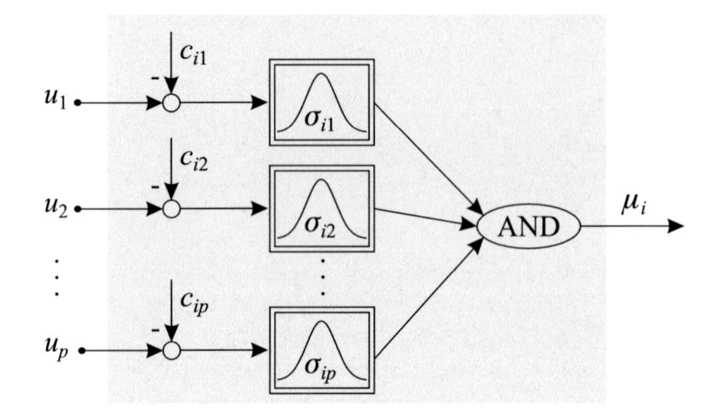

Fig. 12.9 A neuro-fuzzy neuron with Gaussian MSFs

no clear linguistic interpretation is possible because many MSFs are almost indistinguishable. Therefore, the NRBF network in Fig. 12.9a does not allow a reasonable interpretation in the form of fuzzy rules. A possible way out of this dilemma is to *merge* close MSFs; see [16] for more details. Note that the merging of MSFs reduces their number and makes an interpretation easier but, of course, loses approximation accuracy.

- *The RBFs have to be axis-orthogonal,* i.e., the norm matrix $\underline{\Sigma}$ has to be diagonal. Figure 12.10b illustrates that RBFs with non-axis-orthogonal orientation cannot be correctly projected onto the inputs. Again, an interpretable fuzzy system can be constructed by the projections at the price of reduced accuracy [16].

Both the above issues underline the general statement that a tradeoff exists between interpretability and approximation accuracy.

12.3.3 What to Optimize?

In neuro-fuzzy networks, different components can be optimized from data. As a general guideline, all components that can be determined by prior knowledge should not be optimized ("Do not estimate what you already know!" [356]). The interplay between interpretation, prior knowledge, and estimation from data is discussed in the two next sections.

12.3.3.1 Optimization of the Consequent Parameters

The rule consequent parameters correspond to the output weights in the basis function formulation and thus are relatively easy to estimate. For a singleton fuzzy model (and any Takagi-Sugeno type with linear parameterized $f_i(\cdot)$), these

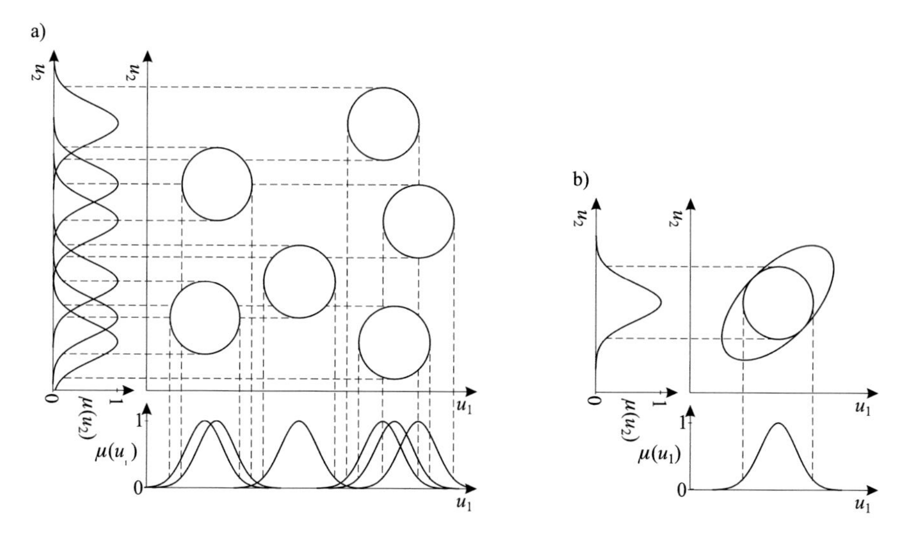

Fig. 12.10 Projection of multidimensional membership functions (RBFs) to univariate MSFs: (**a**) If the multidimensional MSFs do *not* lie on a grid, then generally the number of univariate MSFs for each input is equal to the number of rules. (**b**) If the multidimensional MSFs are *not* axis-orthogonally oriented, then their projections on the input axes cannot exactly reproduce the original multidimensional MSF (RBF)

parameters are linear. Therefore, they can simply be optimized by a least squares technique; see Chap. 3 and [84, 454, 468]. Note, however, that linear subset selection techniques such as OLS cannot be applied directly; see Sect. 12.4.3 for details.

Fuzzy models are often optimized according to a philosophy similar to that of RBF and NRBF networks. The premise parameters (hidden layer parameters) are fixed on a grid, and only the linear singletons (output layer parameters) are optimized. As demonstrated in Fig. 12.10, determination of the premise parameters with clustering techniques causes significant difficulties for the interpretation of the resulting rule base and thus is not recommended if the strengths of fuzzy systems in transparency are to be exploited.

12.3.3.2 Optimization of the Premise Parameters

The premise parameters are those of the input membership functions such as positions and widths. They correspond to the hidden layer parameters in NRBF networks and are nonlinear. They can be optimized by either nonlinear local (Sect. 12.4.1) or nonlinear global (Sect. 12.4.2) optimization techniques. For two reasons it is much more common to optimize only the rule consequent parameters instead of the premise parameters or both. First, optimization of the singletons is a linear regression problem, with all its advantages. Second, the input MSFs can be interpreted much more easily, and thus incorporation of prior knowledge is more straightforward. The number and positions of the input MSFs directly determine the local flexibility of the neuro-fuzzy model. Their widths are responsible for the smoothness of the model. So, optimization of the rule premises is recommended only if little prior knowledge is available.

The question arises whether nonlinear local or global optimization techniques should be applied. Global methods, such as genetic algorithms, prevent the optimization from becoming trapped in a poor local optimum. The price to be paid is a much higher computational demand. It seems to be reasonable to select a global method if the prior knowledge is very vague or even not available at all. If, however, the expert-developed input membership functions reflect quite accurate knowledge about the process, then a solution close to this initialization is desired by the user, and any global optimization approach is not appropriate. In such a case, a local method can start with good initial parameter values and will converge quickly. Additional care has to be taken to preserve prior knowledge during the optimization procedure; see Sect. 12.3.5. No matter how the input membership functions are chosen (by prior knowledge or optimized), special care should be taken concerning the normalization effects discussed in Sect. 12.3.4.

12.3.3.3 Optimization of the Rule Structure

Rule structure optimization is an important approach for fuzzy model learning because it allows one to determine the optimal complexity of the fuzzy model, and by that it weakens the curse of dimensionality. Only with some kind of rule structure optimization are fuzzy models feasible for higher-dimensional mappings.

Optimizing the rule structure is a combinatorial optimization problem. One can try to solve it either by modified linear subset selection schemes such as the OLS (Sect. 12.4.3) or by nonlinear global search techniques such as GAs (Sect. 12.4.4). Although linear subset selection schemes are much more efficient, nonlinear global search methods play a dominating role in rule structure selection. This is in contrast to RBF networks, where OLS algorithms are commonly applied. The reason for this is the normalization denominator in the basis function formulation, which prevents an efficient application of standard linear subset selection methods; see Sect. 12.4.3.

An alternative to direct rule selection is to impose a certain structure on the fuzzy model; see Sect. 7.6.2. Common approaches are additive and hierarchical fuzzy models; see Sect. 12.4.5 and [353, 475].

12.3.3.4 Optimization of Operators

The optimization of input and output membership functions and the rule base can be combined in various ways. Section 12.4 introduces some combinations. Depending on the specific application, a compromise must be found between the flexibility of the fuzzy system and computational complexity.

Besides the possibilities already discussed, other components of fuzzy systems can be adapted as well. For example, the defuzzification method can be optimized either by choosing the one best suited for the specific application [492] or by enhancing it with an additional parameter (inference filter) [320]. Another possibility is to optimize the fuzzy logic operators. An interesting approach is introduced in [418], where the linguistic statements are not combined by a pure conjunction or disjunction operator but by an ANDOR operator. During training the degree of AND and OR, respectively, is learned from data by a neuro-fuzzy network. This provides additional flexibility but raises questions about the interpretability.

12.3.4 Interpretation of Neuro-Fuzzy Networks

The major difference between fuzzy systems and other nonlinear approximators is the possibility of interpretation in terms of rules. Therefore, it is of major importance to discuss the circumstances under which a fuzzy system is really interpretable. Clearly, this depends on the specific application. For example, as noted above, Takagi-Sugeno fuzzy systems are easily interpretable when applied to the modeling of dynamic processes, but they are poorly interpretable when applied to static modeling. In spite of the many specific application-dependent issues, some general interpretation guidelines can be given. Certainly, good interpretation does not automatically follow from the existence of a rule structure as is sometimes claimed. When optimizing fuzzy systems, interpretation issues should always be considered. The following factors may influence the interpretability of a fuzzy system:

- *Number of rules*: If the number of rules is too large, the fuzzy system can hardly be understood by the user. Especially for systems with many inputs, the number of rules often becomes overwhelmingly large if all possible combinations of the linguistic statements are realized. Thus, it is often necessary to restrict the complexity of the fuzzy model from an interpretation point of view. For higher-dimensional mappings, this can be realized only by discarding the pure grid-based partitioning of the input space, e.g., by structural assumptions (Sect. 12.4.5) or by the use of rule premises that do not contain all linguistic statement combinations (Sect. 12.4.4).
- *Number of linguistic statements in the rule premise*: Rules with premises that possess many, say more than three or four, linguistic statements are hard to interpret. In human languages most rules include only very few linguistic statements even if the total number of inputs relevant for the problem is large. These kinds of rules can either be realized directly or can be generated by introducing *don't care* dummy MSFs. For example, for a fuzzy system with five inputs, only rules with two linguistic statements may be used:

$$\text{IF } u_1 = \mathsf{A}_{i1} \text{ AND } u_4 = \mathsf{A}_{i4} \text{ THEN } \ldots$$

or equivalently with dc ="don't care" fuzzy sets

$$\text{IF } u_1 = \mathsf{A}_{i1} \text{ AND } u_2 = \mathsf{dc} \text{ AND } u_3 = \mathsf{dc} \text{ AND } u_4 = \mathsf{A}_{i4} \text{ AND } u_5 = \mathsf{dc} \ \ldots$$

- *Dimensionality of input fuzzy sets*: One way to avoid or at least to reduce the difficulties with high-dimensional input spaces and to decrease the number of rules is to work directly with high-dimensional input fuzzy sets; see, e.g., [335]. These approaches discard the grid-based partitioning of the input space that is typical for fuzzy systems. They are equivalent to NRBF networks where the centers are not determined by a grid approach. However, it is exactly the conjunction of one-dimensional input fuzzy sets that makes a fuzzy system easy to interpret. Multidimensional input fuzzy sets with more than two inputs are certainly beyond human imagination. Since the use of multidimensional fuzzy sets does make sense only if they cannot be projected to univariate ones, almost any rule-based interpretation disappears; see Fig. 12.10. Therefore, approaches with multidimensional fuzzy sets are discussed in the context of NRBF networks.
- *Ordering of fuzzy sets*: Fuzzy sets should be ordered such that, e.g., very small is followed by small, which is followed by medium, large, etc. If a fuzzy system is developed with expert knowledge, such an ordering of fuzzy sets is intuitive. In a successive optimization procedure, however, this ordering can be lost if no precautions are taken; see Sect. 12.3.5. Although it is, in principle, possible to re-label the fuzzy sets, this will lead to difficulties in the rule interpretation, and the expert knowledge incorporated into the initial fuzzy system may get lost to a large extent.
- *Normalization of input membership functions (partition of unity)*: Often the membership functions are chosen such that they sum up to 1 for each input, e.g., a 30-year-old person may be considered young with a degree of membership of 0.2, middle age with 0.7, and old with 0.1. This property is intuitively appealing. If all membership functions sum up to 1 for each input and a complete rule base (all linguistic statement combinations) is implemented, it can be shown that the denominator in (12.14) is equal to 1. It does not hold if only a subset of the complete rule base is realized. The user of rule selection algorithms should always be aware of "strange" effects that might be caused by a modified denominator in (12.14). Thus, discarding or adding rules may change the fuzzy system in a fashion that is not easy to understand.

  There are two ways to achieve normalized input fuzzy sets. One way is to choose membership functions that naturally employ this property, such as triangles with appropriate slopes. More generally, B-splines of order m can be used [84]. Another way is to normalize arbitrary membership functions, e.g., Gaussians. Figure 12.11 shows an undesirable effect that can occur if Gaussian membership functions do not have an identical width. Owing to the normalization, the rules may have non-local influence, which can be regarded as a highly unexpected and undesirable property. Furthermore, the basis functions can become multi-modal even when the MSFs were unimodal. Note that if no explicit normalization is performed for all input fuzzy sets, this normalization is

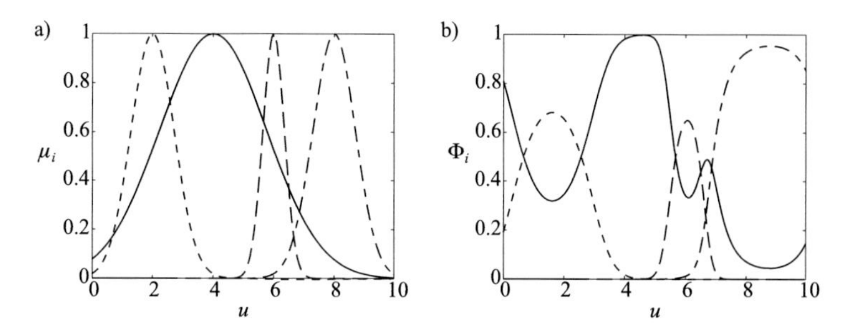

Fig. 12.11 (**a**) Gaussian membership functions with different standard deviations. (**b**) Normalized Gaussian membership functions that sum up to 1. The second membership function (solid line) has the largest standard deviation and therefore becomes dominant for $u \to -\infty$ and $u \to \infty$. Thus, the normalized membership functions become multi-modal and non-local. Rules that include the second membership function do not only influence regions around its center $u = 4$ but also have a dominant effect around $u = 0$. This behavior is usually not expected by the user. Note that owing to the normalizing denominator in (12.14), the same effects take place even if the non-normalized membership functions in (**a**) are used

automatically carried out by the denominator in (12.14). Figures 12.12 and 12.13 illustrate that the normalization side effects can occur equivalently for non-normalized and normalized MSFs. For non-normalized MSFs, they can easily go unnoticed because the multidimensional are rarely plotted (and indeed cannot be visualized for higher-dimensional input spaces).

As shown in [530], the reactivation of basis functions occurs for input u_j when

$$\frac{\sigma_{1j}}{\sigma_{2j}} < \frac{|u_j - c_{1j}|}{|u_j - c_{2j}|} \,. \tag{12.16}$$

Figure 12.14 illustrates this effect. Consequently, reactivation cannot occur if all MSFs have identical widths. Furthermore, it occurs outside the universe of discourse (those regions of the input space that are considered) and thus is harmful only for the extrapolation behavior, when the widths of the MSFs are chosen according to the distances between the MSFs.

From the above discussion, it seems as if Gaussian and other not strictly local MSFs would be inferior to B-splines or other strictly local MSFs,x which naturally fulfill the partition of unity property. In terms of normalization side effects, this is indeed the case. However, only non-strictly local MSFs allow a rule structure optimization that results in incomplete rule sets. This is due to the fact that discarded rules create holes in the grid which are undefined when strictly local MSFs are used. These holes also exist when MSFs without strictly compact support are applied, but because all MSFs are everywhere larger than zero, these holes are filled by interpolation.

All these issues discussed above impose restrictions on the fuzzy model. These restrictions are the price to be paid for the interpretability of fuzzy systems.

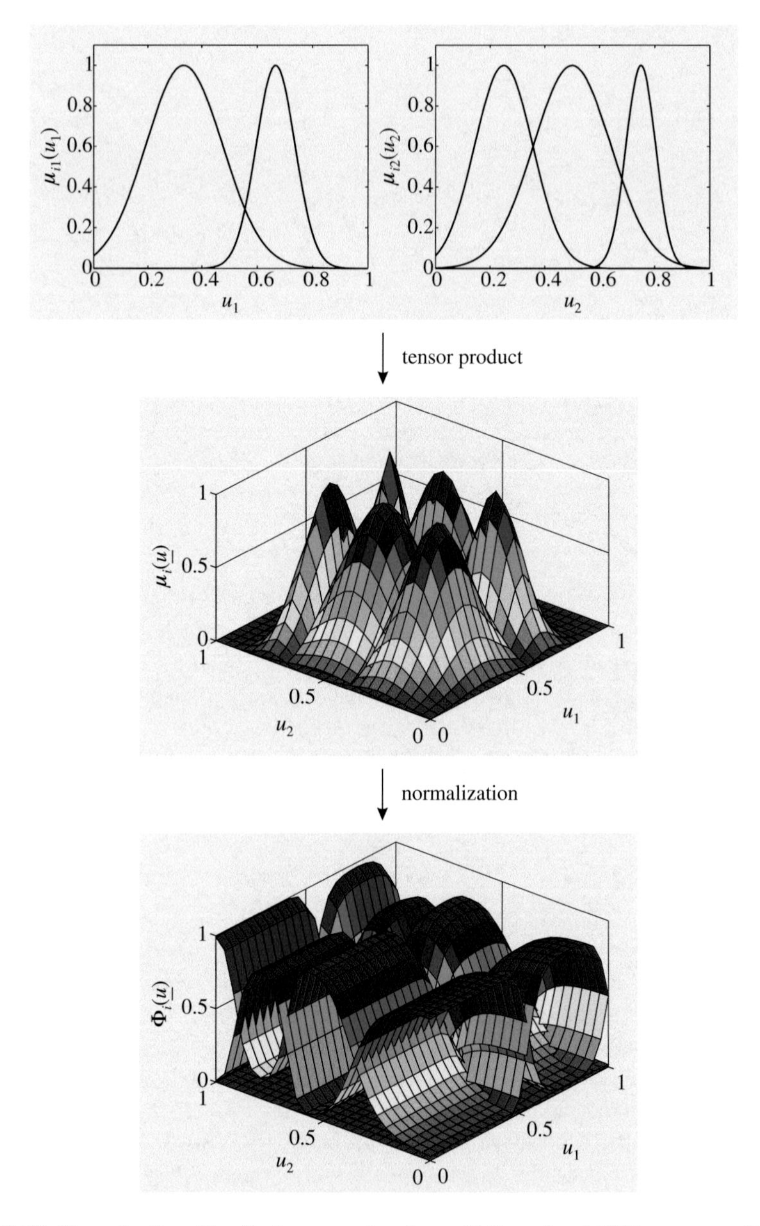

Fig. 12.12 Normalization side effects occur when the multidimensional MSFs are normalized to obtain the basis functions

If neuro-fuzzy networks are optimized from data, several constraints must be imposed to guarantee the fulfillment of the above issues (Sect. 12.3.5); otherwise, interpretability of neuro-fuzzy models can fade away the more these restrictions are violated.

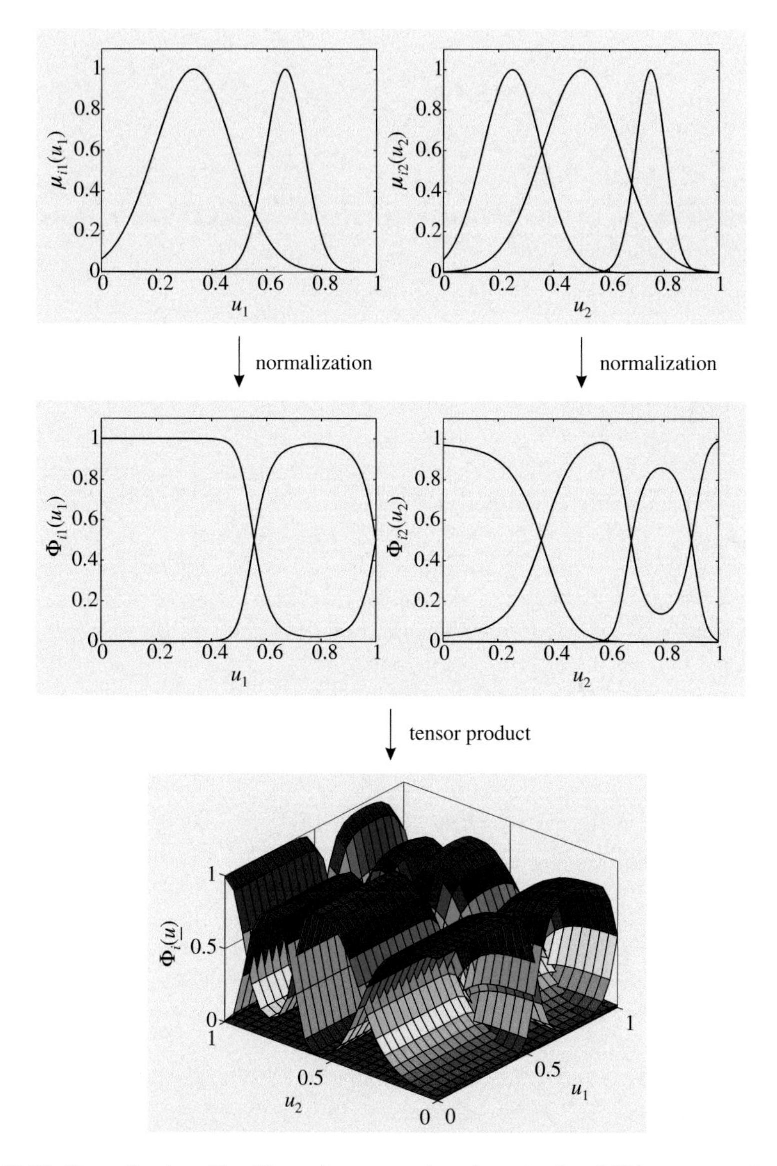

Fig. 12.13 Normalization side effects also occur when the univariate MSFs are normalized to obtain normalized univariate MSFs. These effects are maintained when the tensor product is formed. This result is identical to the one in Fig. 12.12. It does not matter in which order the normalization and tensor product operations are carried out when the input space is partitioned grid-like and a complete rule set is used

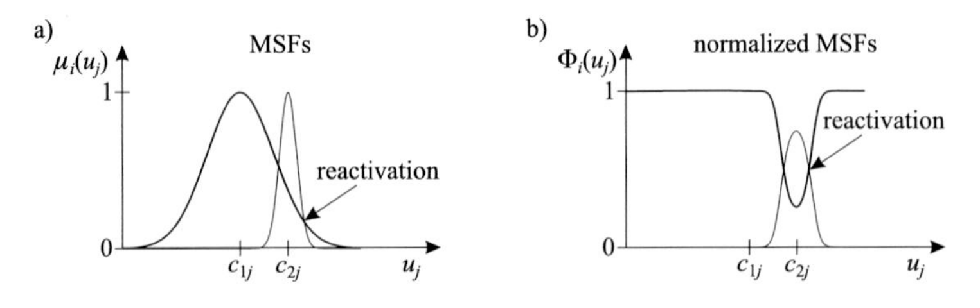

Fig. 12.14 Reactivation of basis functions occurs if (12.16) is fulfilled [530]: (**a**) Gaussian MSFs with different widths, (**b**) normalized MSFs

12.3.5 Incorporating and Preserving Prior Knowledge

One issue in the optimization of fuzzy systems that needs more attention in future research is how to incorporate prior knowledge into the fuzzy system and how to preserve it during the optimization procedure. Usually, first a fuzzy model (or components of it) is developed by expert knowledge, and subsequently, an optimization phase follows that will improve the performance of the original fuzzy model or complete the missing components based on data. The following discussion gives some ideas on this topic. A more extensive analysis can be found in [352].

The order of the input membership functions can be considered as *hard knowledge*, that is, it can either be violated or not. This order can be preserved by performing a constrained optimization subject to

$$lb_j < c_{1j} < c_{2j} < \ldots < c_{M_j j} < ub_j \tag{12.17}$$

for all inputs $j = 1, \ldots, p$, where lb_j and ub_j represent the lower and upper bound, respectively, of the universe of discourse of input j. If genetic algorithms are applied, these constraints can be elegantly incorporated by a relative coding, that is, the position of each membership function is coded not with its absolute value but as a (always positive) distance from the neighboring membership function.

Further restrictions can be considered as *soft knowledge*. The expert may like to restrict the membership functions in such a way that they do not differ "too much" from their initial values chosen by prior knowledge. "Too much" is defined in some way by the expert, e.g., by the incorporation of penalty terms in the loss function. An interesting approach is suggested in [352]. The expert specifies the degree of confidence (certainty) in the prior knowledge (assumptions) in the form of fuzzy rules. These rules are then utilized to compute the penalty function.

In [352] it is demonstrated that constrained rather than unconstrained optimization may not only lead to an easier interpretation of the fuzzy system. The performance may be higher as well because a better local optimum can be found. Furthermore, fuzzy systems are often overparameterized, i.e., have more parameters than can reasonably be estimated from the available amount of data (see the discussion of the bias/variance dilemma in Sect. 7.2). Constraining the flexibility of such a fuzzy system can be advantageous with respect to performance as well, because it has a regularization effect; see Sect. 7.5.4. However, in some applications unconstrained optimization of a fuzzy system will yield better performance since the constraints may limit the flexibility of the fuzzy model too severely; see Sect. 12.4.4 and [528].

All constraints impose restrictions on the search space, that is, they reduce its size. Therefore the rate of convergence of the applied optimization technique will usually increase with the number of constraints.

12.3.6 Simulation Examples

The approximation example problems already discussed in Sects. 11.2.5 and 11.3.4 in the context of MLP and RBF networks will be considered here again. These examples illustrates the functioning of singleton neuro-fuzzy models, which are equivalent to normalized RBF networks; see Sect. 12.3.2.

The membership functions are equally distributed over the input space according to the grid-based approach. Note that the OLS algorithm cannot be directly applied to neuro-fuzzy systems as argued in the next section. The neuro-fuzzy results shown in Fig. 12.15 are obtained for four and seven rules and with membership functions of different widths. They are clearly superior to the RBF network results obtained with the grid-based center placement strategy. The "dips" either do not exist or are at least less pronounced, and the sensitivity with respect to the basis function width is significantly smaller. Thus, the choice of the membership functions' widths becomes less decisive, and so this parameter is easier to adjust for the user. Basically, the choice of too small a width results only in an approximation behavior that is not very smooth. This may be acceptable, while for RBF networks it results in "dips" that change the whole characteristics of the approximation, which usually is unacceptable.

12.4 Neuro-Fuzzy Learning Schemes

This section discusses more sophisticated neuro-fuzzy learning schemes than a simple least squares optimization of the singletons. These more complex approaches should be applied only when the simple LS estimation with fixed input membership functions does not yield satisfactory results. This is typically the case when insufficient prior knowledge is available in order to make a good choice for the input MSFs and the rule structure. Also, higher-dimensional problems, say $p > 3$, usually require either structural knowledge or an automatic structure search to overcome the curse of dimensionality.

When combining the optimization of different fuzzy system components, two alternative strategies can be distinguished: Several components may be optimized

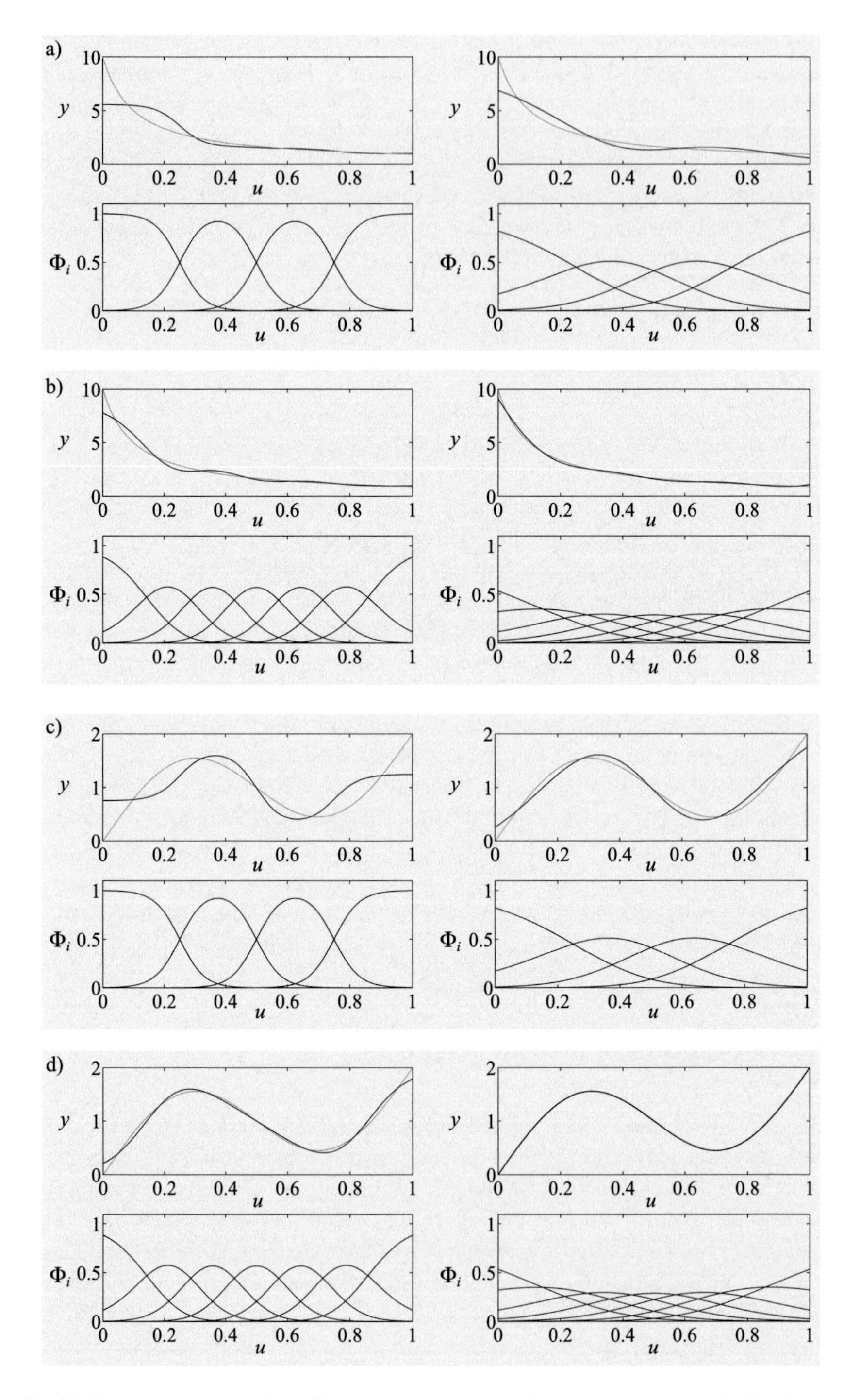

Fig. 12.15 Approximation of the functions $y = 1/(u + 0.1)$ (**a**, **b**) and $y = \sin(2\pi u) + 2u$ (**c**, **d**) with a neuro-fuzzy network with four (**a**, **c**) and seven (**b**, **d**) rules. The standard deviations of the membership functions before normalization are chosen equal to 0.1 (left) and 0.2 (right), respectively

simultaneously or separately, e.g., in a nested or staggered procedure as suggested in Sects. 12.4.4 and 12.4.5. While the first strategy offers higher flexibility, the second one is more efficient in terms of computational demand; see Chap. 5 and Sect. 7.5.5.

This section is organized as follows. First, some principal features of the nonlinear local and global optimization of input MSFs are analyzed. In Sect. 12.4.3 the application of an orthogonal least squares algorithm for rule structure selection is discussed. Finally, two structure optimization approaches are introduced; one is based on genetic algorithms for rule selection, and the other builds an additive model structure.

12.4.1 Nonlinear Local Optimization

Often it is said that gradient-based methods would require differentiable membership functions. This is not quite true, however, [352]. For example, triangle membership functions are differentiable except at a finite number of points. If the gradients at these points are artificially set to the derivative of one neighboring point or to zero, any gradient-based method can be successfully applied. The only real restriction for the use of gradient-based methods is that min or max operators should not be applied for realizing the con- and disjunction because they may lead to objective functions with whole zero-gradient areas.

The gradient for a singleton neuro-fuzzy network can be calculated as follows. The derivative of the neuro-fuzzy network output with respect to the ith *singleton* is ($i = 1, \ldots, M$)

$$\frac{\partial \hat{y}}{\partial w_i} = \Phi_i = \frac{\mu_i}{\sum\limits_{l=1}^{M} \mu_l} . \tag{12.18}$$

The derivatives of a neuro-fuzzy network output with respect to its nonlinear parameters – the positions and widths of the input MSFs – are more complex. The reason for this is that *all* basis functions depend on all nonlinear parameters through the normalization denominator, which contains all multidimensional MSFs. The derivative with respect to a *nonlinear parameter* θ_{ij} (can be a position or width) that influences the ith multidimensional MSF μ_i is ($i = 1, \ldots, M$, $j = 1, \ldots, p$)

$$\frac{\partial \hat{y}}{\partial \theta_{ij}} = \frac{\partial \mu_i / \partial \theta_{ij} \sum\limits_{l=1}^{M} (s_i - s_l) \mu_l}{\left(\sum\limits_{l=1}^{M} \mu_l \right)^2} . \tag{12.19}$$

Usually, more than just one multidimensional MSF is affected by a nonlinear parameter because they lie on a grid. Then in (12.19) the contributions of all affected multidimensional MSFs have to be summed up.

With univariate Gaussian MSFs, the multidimensional MSFs are

$$\mu_i = \prod_{j=1}^{p} \exp\left(-\frac{1}{2}\frac{(u_j - c_{ij})^2}{\sigma_{ij}^2}\right) = \exp\left(-\frac{1}{2}\sum_{j=1}^{p}\frac{(u_j - c_{ij})^2}{\sigma_{ij}^2}\right). \tag{12.20}$$

Thus, the derivative of a neuro-fuzzy network with respect to the jth coordinate of the *center* of the ith rule is ($i = 1, \ldots, M$, $j = 1, \ldots, p$)

$$\frac{\partial \mu_i}{\partial c_{ij}} = \frac{u_i - c_{ij}}{\sigma_{ij}^2}\mu_i\,. \tag{12.21}$$

The derivative of a neuro-fuzzy network with respect to the *standard deviation* in the jth dimension of the ith rule is ($i = 1, \ldots, M$, $j = 1, \ldots, p$)

$$\frac{\partial \mu_i}{\partial \sigma_{ij}} = \frac{(u_i - c_{ij})^2}{\sigma_{ij}^3}\mu_i\,. \tag{12.22}$$

Equations (12.21) and (12.22) can be inserted in (12.19) for $\theta_{ij} = c_{ij}$ and $\theta_{ij} = \sigma_{ij}$, respectively.

Since the gradient calculations are so complex, and additional case distinctions have to be made for interval-defined MSFs such as B-splines, direct search methods that do not require explicit gradients are quite popular for nonlinear local optimization of neuro-fuzzy networks; see Sect. 4.3.

As Sect. 12.4.4 points out further, the following issues should be taken into account when optimizing the input MSFs in order to maintain the interpretability of the optimized model:

- MSFs should not overtake each other during training.
- MSFs should stay in the universe of discourse.
- MSFs should be sufficiently distinct in order to be meaningful; otherwise, they should be merged.
- MSFs should stay local.
- Normalization side effects should be kept as small as possible.

12.4.2 Nonlinear Global Optimization

Nonlinear global optimization of the input membership function parameters is quite common because it is a highly multi-modal problem and the constraints can be easily incorporated. However, for many applications, the number, positions, and widths of the input MSFs can be chosen by prior knowledge. In these cases, a global optimization does not make any sense because the desired solution can be expected to be close to the prior knowledge. Rather either the chosen input MSFs should be fixed, optimizing only the output MSFs and the rule base, or a fine-tuning of the initial settings by a local optimization technique should be carried out.

Therefore, nonlinear global optimization techniques should be mainly applied when very little prior knowledge is available. They are also recommended for approaching fuzzy rule structure selection problems and for a combined optimization of rule structure and input MSFs. A more detailed overview of these approaches is given in Sect. 12.4.4.

12.4.3 Orthogonal Least Squares Learning

Linear subset selection with an orthogonal least squares (OLS) algorithm is one of the most powerful learning methods for RBF networks; see Sects. 11.3.3 and 3.4. So it is fair to assume that this also holds for fuzzy models. However, since fuzzy models are (under some conditions) equivalent to *normalized* RBF networks, this is not the case. Linear subset selection techniques *cannot* be directly applied to rule selection in fuzzy systems. The reason lies in the normalization denominator in (12.14). This denominator contains the contribution of all multidimensional MSFs computed from the corresponding rules. During a possible rule selection procedure, the number of selected rules changes. Consequently, the denominator in (12.14) changes, and thus the fuzzy basis functions change. The orthogonalization in the OLS algorithm, however, is based on the assumption that the basis functions stay constant. This assumption is not fulfilled for fuzzy models.

The only way in which the OLS algorithm can be applied directly is to completely discard any interpretability issues and to fix the denominator as the contribution of all potential (also the non-selected) rules. This approach in [589, 590] is not really a rule selection procedure since all (also the non-selected) fuzzy rules have to be evaluated when using the obtained fuzzy model. Furthermore, the normalization denominator does not match the rule structure, and indeed the obtained approximator is no true fuzzy model. These severe drawbacks are partly overcome in [255], where a second processing phase is suggested in order to restore some interpretability. An additional drawback of the approaches in [255, 589, 590] is that they are based on multidimensional membership functions without a grid-like partitioning of the input space. As Fig. 12.10 demonstrates, a fuzzy system loses interpretability with such an approach.

The following example will illustrate the difficulties encountered in fuzzy rule structure selection. A fuzzy model with two inputs $\underline{u} = [u_1\ u_2]^T$ and output y is to be constructed incrementally. Each input will be represented by three MSFs, say small, medium, and large. Thus, a complete rule set would contain nine rules corresponding to all combinations of the MSFs. The nine multidimensional MSFs and basis functions are denoted by $\mu_i(\underline{u})$ and $\Phi_i(\underline{u})$ with $i = 1, \ldots, 9$, where $i = 1$

represents the combination (u_1 = small, u_2 = small), $i = 2$ stands for (u_1 = small, u_2 = medium), and so on.

A selection procedure will be used to find the, say, three most relevant rules out of the set of all nine rules. In the first iteration, the following nine fuzzy basis functions can be selected (the arguments $\underline{u}$ are omitted for brevity):

$$\Phi_1 = \frac{\mu_1}{\mu_1} = 1\,, \quad \Phi_2 = \frac{\mu_2}{\mu_2} = 1\,, \quad \ldots\,, \quad \Phi_9 = \frac{\mu_9}{\mu_9} = 1\,. \tag{12.23}$$

This means that *all* basis functions and thus all rules are *equivalent* as long as only a fuzzy model with a single rule is considered. This is a direct consequence of the partition of unity property. Therefore, the choice of the first rule is arbitrary. It is assumed that (for whatever reason) rule 2 is selected in the first iteration. In the second iteration, the following eight basis functions are available:

$$\Phi_1 = \frac{\mu_1}{\mu_1 + \mu_2}\,, \quad \Phi_3 = \frac{\mu_3}{\mu_2 + \mu_3}\,, \quad \ldots\,, \quad \Phi_9 = \frac{\mu_9}{\mu_2 + \mu_9}\,. \tag{12.24}$$

Owing to the selection of Φ_2 in the first iteration, all denominators have changed, and thus the shape of all basis functions is altered. Now, all eight basis functions are different, and the most significant one can be chosen, say Φ_9. In the third iteration, the following seven basis functions are available:

$$\Phi_1 = \frac{\mu_1}{\mu_1 + \mu_2 + \mu_9}\,, \quad \Phi_3 = \frac{\mu_3}{\mu_2 + \mu_3 + \mu_9}\,, \quad \ldots\,,$$
$$\Phi_8 = \frac{\mu_8}{\mu_2 + \mu_8 + \mu_9}\,. \tag{12.25}$$

Again the character of all basis functions has changed. Note that the same is true for the basis functions already selected. These observations have two major consequences. First, the changes in the already selected basis functions can make previously important rules superfluous. (This can also happen with fixed basis functions, where newly selected basis functions can reduce the relevance of previously chosen ones. However, the effect is much more dominant when additionally the characteristics of the basis functions change.) Second, the change of the basis functions in each iteration prevents the direct application of the OLS algorithm as described in Sect. 3.4. It is based on an orthogonalization of all potential regressors (basis functions) with respect to the already selected ones. In a new iteration, the potential regressors are only made orthogonal to the *newly* selected regressor. Orthogonality to all *previously* selected regressors is maintained. This feature makes the OLS so computationally efficient. For fuzzy models, however, the change of the basis functions in each iteration destroys this orthogonality. Therefore, in each iteration, the potential regressors must be orthogonalized with respect to *all* selected regressors. This causes considerable additional computation effort compared with the standard OLS.

Nevertheless in [328] this approach is pursued. The first difficulty is remedied by the application of an OLS-based stepwise selection scheme that also removes insignificant rules in each iteration; see Sect. 3.4. With these techniques, it is possible to apply linear subset selection to fuzzy models successfully. The price to be paid is a much higher computational demand, which reduces the gap to genetic algorithms or other global search techniques considerably. This is the reason for the popularity of global search approaches for fuzzy model structure optimization.

12.4.4 Fuzzy Rule Extraction by a Genetic Algorithm (FUREGA)

An interesting approach for fuzzy rule extraction by a genetic algorithm (FUREGA) is proposed in [425] and extended in [410]. It is based on a genetic algorithm for selection of the fuzzy rule structure that avoids the difficulties encountered by the OLS or other linear subset selection algorithms as discussed in Sect. 12.4.3. Nested within the GA, a least squares optimization of the singletons is performed. Optionally, a subsequent nonlinear optimization of the input MSFs can be included.

First, appropriate codings of a fuzzy system for optimization with a GA are discussed. Next, a strategy for overcoming or at least weakening the curse of dimensionality is proposed. Then a combination of the GA with least squares optimization for the singletons and constrained nonlinear optimization for the input membership functions is introduced. Finally, an application example illustrates the operation of FUREGA and demonstrates its effectiveness.

12.4.4.1 Coding of the Rule Structure

Figure 12.16 shows how the rule structure of a fuzzy system is coded in an individual for the FUREGA approach. Each gene represents one rule, where "1" stands for a selected rule and "0" stands for a non-selected rule. For details on GAs, refer to Sect. 5.2.2.

A very similar coding of the rule base is applied in [285] for a fuzzy classification system. Since in [285] two classes have to be distinguished, the authors propose the following coding: "1" = class A, "-1" = class B, and "0" = non-selected. Such an approach requires some extensions of the standard GA because each gene has to take more than the standard two realizations: ("-1", "0", and "1"). In this case,

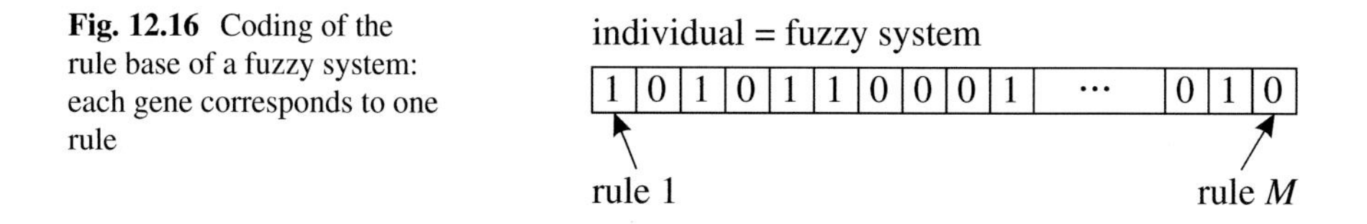

Fig. 12.16 Coding of the rule base of a fuzzy system: each gene corresponds to one rule

the mutation operator cannot be simply implemented as an inversion. However, it may be modified such that it randomly changes each gene to any permitted value. A further extension is made in [258], where each gene takes an integer value between "1" and "M_o" that codes the output fuzzy set that corresponds to each rule. Thus, both the rule structure and the output membership functions are represented by the individual. Such coding is reasonable only for linguistic fuzzy systems because it relies on a small number of output fuzzy sets M_o that are shared by several rules. In [258] no real structure selection is performed because all rules are used by the fuzzy system. This approach, however, could be easily extended to a rule selection procedure by using integer values between "0" and "M_o" where "1" to "M_o" code the output fuzzy sets and "0" switches off the rule.

All these codings share one important property. They represent one rule by one gene. This means that the individual's size is equal to the number of potential rules. Consequently, the mutation and crossover operators perform operations such as switching rules on or off and combining rules from different fuzzy systems. Since these operations are meaningful on the fuzzy rule level, GAs can be expected to be more efficient than for problems where the bits do not possess a direct interpretation; see Sect. 5.2. The choice of an appropriate coding is always a crucial step in the application of genetic algorithms.

Especially for high-dimensional input spaces, the number of potential rules M may be huge, compared with the expected number of selected rules M_s. Then the evolution process will drive the population toward very sparse individuals, i.e., individuals with genes that have a much higher probability for "0"s than for "1"s. The evolution pressure toward sparse individuals can be supported by a non-symmetric mutation rate that has a higher probability to change bits from "1" to "0" than vice versa.

An alternative coding for such sparse problems is to numerate the potential rules from 1 to M and to represent only the selected M_s rules in each individual by its integer number. Two difficulties arise with this kind of coding. First, two subsequent numbers do not necessarily represent related rules. Second, either the number of selected rules M_s must be fixed and specified a priori, or individuals of variable length have to be handled.

So far, only the coding of the rule structure and possibly the output fuzzy sets have been discussed. Several authors [244, 258, 444] have proposed optimizing the input membership functions by means of a GA as well. While in [258] the parameters of the membership functions are coded as integers, in [244] and [444] real-value coding is used for parameter optimization. These approaches support the views and comments in [117] that binary coding for parameters is often not the best choice. As pointed out in [258], the advantage of optimizing rule structure and input fuzzy sets simultaneously is that the relationship between the rule structure and the input fuzzy sets can be taken into account. Both components of a fuzzy system are highly interdependent. The price to be paid is a much higher computational effort compared with approaches that keep the input fuzzy sets fixed. Therefore, those simultaneously optimizing approaches seem to be a good choice only if little prior knowledge is available. Otherwise, a two-step procedure with separate structure optimization and fuzzy set tuning components as proposed below in the paragraph "Constrained Optimization of the Input Membership Functions" might be more efficient.

In all these methods, as much a priori knowledge as possible should be incorporated into the fuzzy system. The initial population of the GA should contain all available information about rule structure and fuzzy sets. This can be done in various ways. If, for example, information about the smoothness of the underlying function or about the accuracy or importance of the input variables is available, this can be exploited in order to guess the required number and widths of membership functions. It is shown in [444] that initializing the GA with prior knowledge leads to superior results and to much faster convergence.

Certainly, many combinations of the discussed strategies are possible. Final evaluation and comparison of all these approaches are open to future research. Furthermore, there are some interesting approaches, such as GA-based training of hierarchical fuzzy systems in [353] that have not been discussed here. A vast amount of references on combinations of GAs and fuzzy systems can be found in [4, 115].

12.4.4.2 Overcoming the Curse of Dimensionality

For the sake of simplicity, the rule extraction process will be demonstrated for a system with only two inputs u_1 and u_2, output y, and three membership functions for each input $\mathsf{A}_{11}, \ldots, \mathsf{A}_{31}$ and $\mathsf{A}_{12}, \ldots, \mathsf{A}_{32}$, respectively. The set of all possible conjunctive rules from which the extraction process will select the most significant ones contains 15 rules, i.e., three rules with u_1 in the premise only, three rules with u_2 in the premise only, and nine rules with all combinations of u_1 and u_2 in the premise. It is important to notice that rules with a reduced number of linguistic statements in their premises such as (see Fig. 12.17)

$$R_1\text{: IF } u_1 = \mathsf{A}_{21} \text{ THEN } y = ?$$

cover the same area as the three rules

$$R_2\text{: IF } u_1 = \mathsf{A}_{21} \text{ AND } u_2 = \mathsf{A}_{12} \text{ THEN } y = ?$$
$$R_3\text{: IF } u_1 = \mathsf{A}_{21} \text{ AND } u_2 = \mathsf{A}_{22} \text{ THEN } y = ?$$
$$R_4\text{: IF } u_1 = \mathsf{A}_{21} \text{ AND } u_2 = \mathsf{A}_{32} \text{ THEN } y = ?\,.$$

The singletons "?" will be optimized by a least squares technique. If the singletons of the rules R_2, R_3, and R_4 are almost equivalent, then those three rules can be approximated by R_1. This is a mechanism to overcome the curse of dimensionality. Generally speaking, one rule with just one linguistic statement in the premise can cover the same area as $M_i p - 1$ rules with p linguistic statements in the premise, where M_i is the number of membership functions for each input. Therefore, the number of rules required can be drastically reduced, and

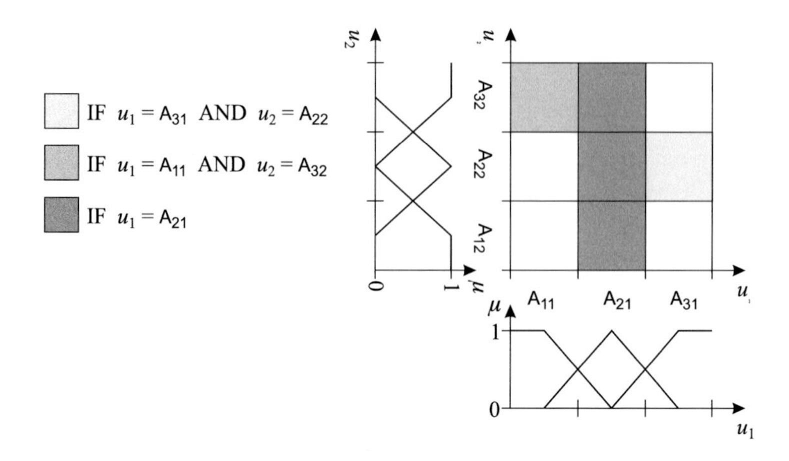

Fig. 12.17 Rules with a reduced number of linguistic statements cover larger regions of the input space and thus help to overcome the curse of dimensionality

this reduction effect increases exponentially with the input dimension p. For the case of p inputs and M_i membership functions per input (of course they are in general allowed to be different in shape and number for each input), the number of possible rules is equal to

$$M = \underbrace{\binom{p}{1} M_i}_{\text{1 ling. statement}} + \underbrace{\binom{p}{2} M_i^2}_{\text{2 ling. statements}} + \ldots + \underbrace{\binom{p}{p} M_i^p}_{p \text{ ling. statements}} . \tag{12.27}$$

Many neuro-fuzzy and GA-based fuzzy models consider only the last term of the sum in (12.27), which is dominant for $M_i > p$. This last term represents the number of rules required for full coverage of the input space with a complete rule set containing only rules with p linguistic statements. With the proposed approach, it is, in principle, possible for the GA to detect whether an input u_j is irrelevant, since then u_j will not appear in the optimal rule set.

Under some conditions, a fuzzy system is equivalent to a normalized RBF network (see Sect. 12.3.2) if all p linguistic statements appear in all premises. It is interesting to ask for an interpretation of premises with less than p linguistic statements from the neural network point of view. In [219] so-called Gaussian bar units are proposed for radial basis function networks to overcome the curse of dimensionality. Those Gaussian bar units correspond to p premises with just one linguistic statement as proposed here from the fuzzy point of view. This means that the Gaussian bar approach represents a special case of fuzzy models with reduced linguistic statements in their premises. In [219] experimental results are given to show that this semi-local approach is much less sensitive than a pure local method to the curse of dimensionality. However, the approach in [219] includes only premises with exactly one linguistic statement and is a non-normalized RBF network, i.e., gives no rule interpretation. The method presented here allows any number of linguistic statements in each rule premise ranging from 1 to p. This enables FUREGA to optimize the granularity or coarseness of the fuzzy model. The degree of locality decreases as the number of linguistic statements in the premises decreases. A rule with one linguistic statement in the premise covers $1/M_i$ part of the input space; a rule with two linguistic statements in the premise covers $1/M_i^2$, and a rule with p linguistic statements in the premise covers only $1/M_i^p$. Therefore, the GA can also control the degree of locality of the fuzzy model for all input regions separately.

12.4.4.3 Nested Least Squares Optimization of the Singletons

The philosophy behind FUREGA is as follows. It is assumed that enough knowledge about smoothness properties and the required resolution for each input variable is available in order to specify reasonable values for number, positions, and widths of the input membership functions. With these a priori defined input fuzzy sets, rules with all possible combinations of linguistic statements are computed, following the reduced linguistic statements strategy for overcoming the curse of dimensionality, which is described in the previous paragraph. The task of rule selection is performed by a GA, while within each fitness evaluation, the singletons in the rule consequents are estimated from training data by a least squares technique; see Chap. 3.

For rule extraction, all possible rules are coded in a binary string. The length of this binary string is equal to the number of all possible rules. Selected rules are represented by setting the corresponding gene to "1," while non-selected rules are symbolized by a "0". Thus, the number of "1"s in each binary string is equal to the number of selected rules. For the examples presented in the following, a population size of 30 and a crossover probability of 0.9 were used. The mutation rate was determined not for each bit, but for each individual. Each rule set on average was mutated with a probability of 0.2. The mutation probability can be calculated by dividing the individual mutation probability by the number of possible rules M. As the selection method, the roulette wheel selection was chosen.

The fitness of each individual was evaluated in the following way. First, the rules are extracted from the binary string by finding the rule numbers that are set to "1." For these rules, a least squares optimization of the singletons is performed. Then the normalized mean square error of this fuzzy system is evaluated. A penalty function that is proportional to the number of selected rules is added as well as another penalty function for singletons that have no physical meaning because they violate output constraints. The inverse of this loss function value is the fitness of the corresponding rule set. The penalty factor that determines how strongly large rule sets should be penalized is chosen by the user. Large penalty factors will lead to small rule sets, while small penalty factors will lead to large rule sets. The penalty for the singletons is calculated in the following way: A range is given for every output by the minimum and maximum values of the corresponding output data. This range is expanded by a factor determined by the user. In the following, this factor is chosen equal to 1.2, so that the singletons are allowed to exceed the output range by $\pm 20\%$. Singletons that exceed these bounds automatically introduce an additional penalty term. This penalty term is equal to the distance of the singletons from the violated bound. This procedure ensures the conformity of the learned structure with the given data.

Although the least squares optimization for each fitness evaluation is time-consuming, that approach guarantees a linear optimized fuzzy rule set for each individual and therefore leads to fast convergence. Coding the singleton positions within the GA would ignore the information about the linear dependency of the model output on these parameters. In order to further accelerate the fitness evaluation, a maximum number of rules can be determined by the user. Rule sets with a larger number of selected rules than this maximum are not optimized by least squares. Owing to the cubic complexity of the least squares optimization, this strategy saves a considerable amount of computation time. Instead, a low fitness value is returned. The GA will find a good or even the optimal rule set corresponding to the penalty value with a rule set size between one and the maximum number of rules.

12.4.4.4 Constrained Optimization of the Input Membership Functions

After selecting the significant rules out of the set of all possible rules by a GA, in a second step, the input membership functions are optimized. This is done by sequential quadratic programming (SQP), a nonlinear gradient-based constrained optimization technique, with an embedded least squares (LS) optimization of output fuzzy sets (singletons). Figure 12.18 illustrates the strategy behind FUREGA. The motivation behind this approach is to apply those optimization techniques that are most efficient for the specific task. That are (in the opinion of the author) a GA for the solution of the combinatorial rule selection problem, a nonlinear local constrained optimization technique for the optimization of the input membership functions, and a linear least squares technique for determining the output MSFs.

As input membership functions, normalized Gaussians are chosen since they guarantee a smooth and differentiable approximation behavior. In order to tune the input MSFs, a nonlinear optimization problem has to be solved. The parameters of the optimization are the centers and widths (standard deviations) of the Gaussian MSFs. If the approximation quality is the only objective of optimization, the search space of the parameters to be optimized is not limited. Although a good approximation quality can be expected for such an approach, the interpretability of the fuzzy rules may get lost. This is due to the fact that if the range of the MSFs is not restricted, often widely overlapping MSFs give good numerical results. Fuzzy membership functions as shown in Fig. 12.21b can result. They lack any physical interpretation, and the locality is lost. To avoid this, different kinds of constraints should be imposed on the optimization: equality constraints, inequality constraints, and parameter bounds. Another strategy is to add a penalty measure

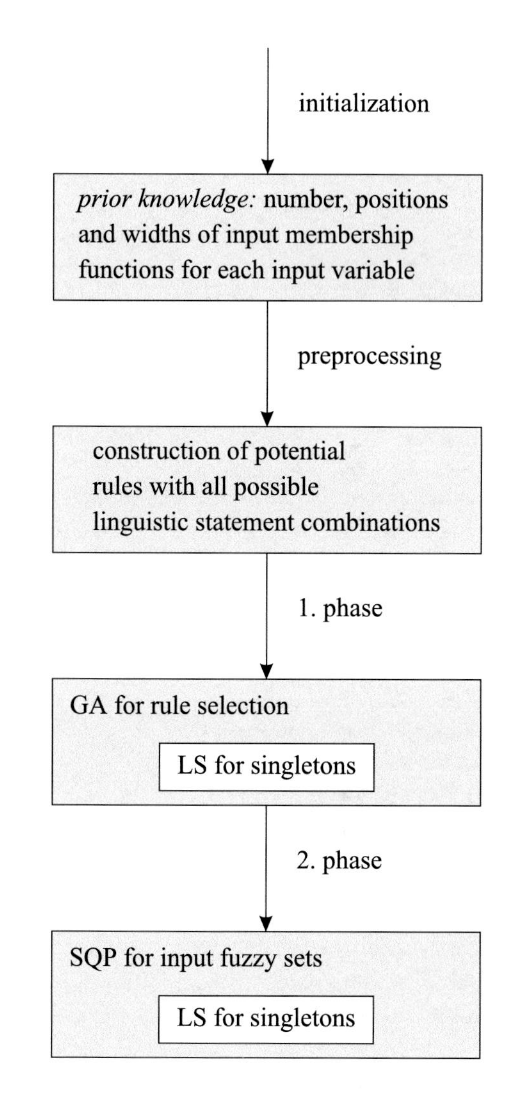

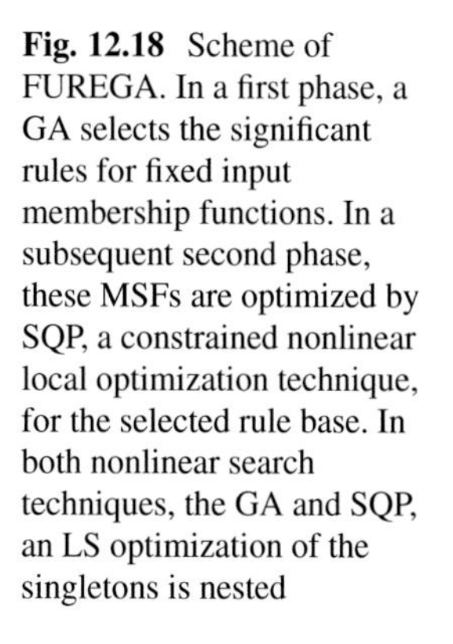

Fig. 12.18 Scheme of FUREGA. In a first phase, a GA selects the significant rules for fixed input membership functions. In a subsequent second phase, these MSFs are optimized by SQP, a constrained nonlinear local optimization technique, for the selected rule base. In both nonlinear search techniques, the GA and SQP, an LS optimization of the singletons is nested

to the objective function. However, this normally reduces the efficiency of the optimization algorithm [577].

In order to efficiently solve this constrained nonlinear optimization problem in which the loss function and the constraints may be nonlinear functions of the variables, a sequential quadratic programming (SQP) algorithm as presented in [70] is applied; see also Sect. 4.6. It iteratively solves the Kuhn-Tucker equations and builds up second-order information for fast convergence.

After the GA has selected the significant rules, the performance of the fuzzy system can further be improved by the following strategy: The input MSFs are optimized by an SQP algorithm in which a least squares optimization of the output membership functions is embedded. The loss function of the optimization is the normalized mean square error. To prevent a large overlap or even coincidental membership functions, different strategies are implemented:

1. *Minimum distance of membership functions*: The center of each membership function must have a minimum distance to the centers of the adjoining membership functions.
2. *Parameter bounds*: The center and the width of each membership function are restricted to a given range.
3. *The sum of the optimized membership functions for each input should be around 1*: The original membership functions are normalized, i.e., they sum up to 1. This is an appealing property that makes human interpretation easier. Thus, the squared difference between 1 and the sum of the optimized membership functions is integrated. This value is used as a penalty, which is weighted with a "sum penalty" factor and then added to the objective function. An alternative approach would be to keep the input MSFs normalized during optimization so that they automatically form a partition of unity. Then, however, normalization side effects become more likely.

Strategy 3 has turned out to be the most powerful one in terms of interpretation quality but also the most restrictive. To further increase the interpretation quality, a penalty factor for the singletons violating the output bounds is used; see above. In almost every combination of the different strategies, this singleton penalty factor leads to a faster convergence of the optimization algorithm.

12.4.4.5 Application Example

Figure 12.19 shows the relationship of the exhaust gas pressure to the engine speed and the injection mass for a Diesel engine. The 320 (32×10) data points have been measured at a Diesel engine test stand and are stored in a look-up table. Since the relationship seems to be quite smooth, only four (very small, small, high, very high) normalized Gaussian membership functions with considerable overlap were placed on each input axis. In order to ensure a smooth approximation behavior, the product operator is used as the t-norm. The resulting rule set contains 24 ($4 + 4 + 16$) possible rules. For a high rule penalty factor, FUREGA leads to the following rule set of only four rules:

IF $speed$ = small THEN $exhaust$ = 1.113 bar
IF $speed$ = high THEN $exhaust$ = 1.696 bar
IF $speed$ = very small AND $injec.$ = very small THEN $exhaust$ = 1.012 bar
IF $speed$ = very high AND $injec.$ = very high THEN $exhaust$ = 2.566 bar

It has to be stated clearly that this rule set is the best result obtained during a few runs of the GA and it could not always be reached. It is very easy to interpret and has the nice property that the more relevant input $engine\ speed$ is included in

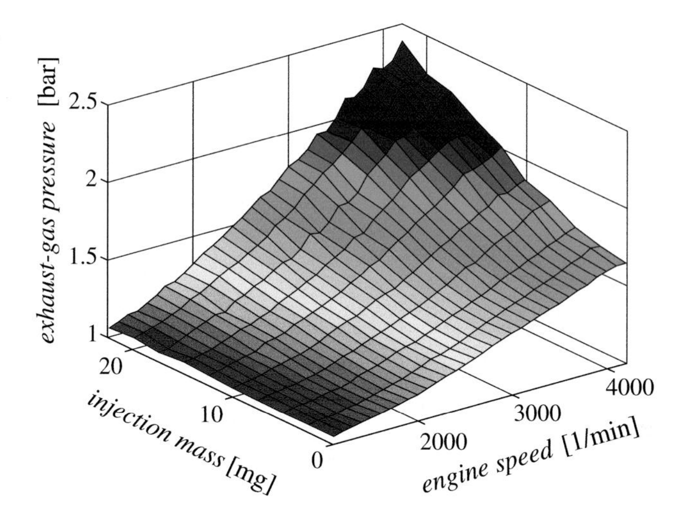

Fig. 12.19 Measured relationship between exhaust gas pressure, engine speed, and injection mass

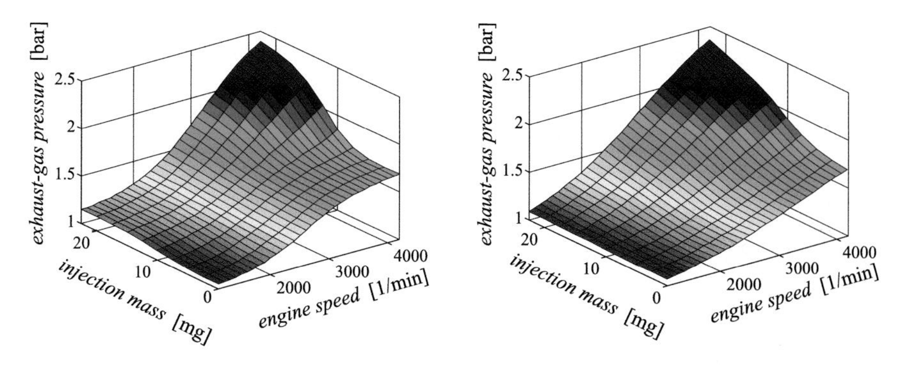

Fig. 12.20 Approximation of the characteristic map in Fig. 12.19 with four rules selected by FUREGA (left). Approximation after an unconstrained nonlinear optimization of the membership functions of these four rules (right)

all rules, while the less relevant input *injection mass* appears only in two rules in combination with the other input. Figure 12.20(left) shows the mapping generated by those four selected fuzzy rules with a normalized mean square error of 0.0243, and Fig. 12.21a depicts the corresponding fuzzy membership functions for the input *engine speed*.

The subsequent tuning of the membership functions by SQP without constraints leads to a much better normalized mean square error of 0.0018; see Fig. 12.21b. The upper curve in Fig. 12.21a–d represents the sum of all membership functions for this input. The approximation quality corresponding to the MSFs depicted in Fig. 12.21b is shown in Fig. 12.20. Obviously, with this unconstrained optimization, interpretability gets lost because the membership functions small and high lose

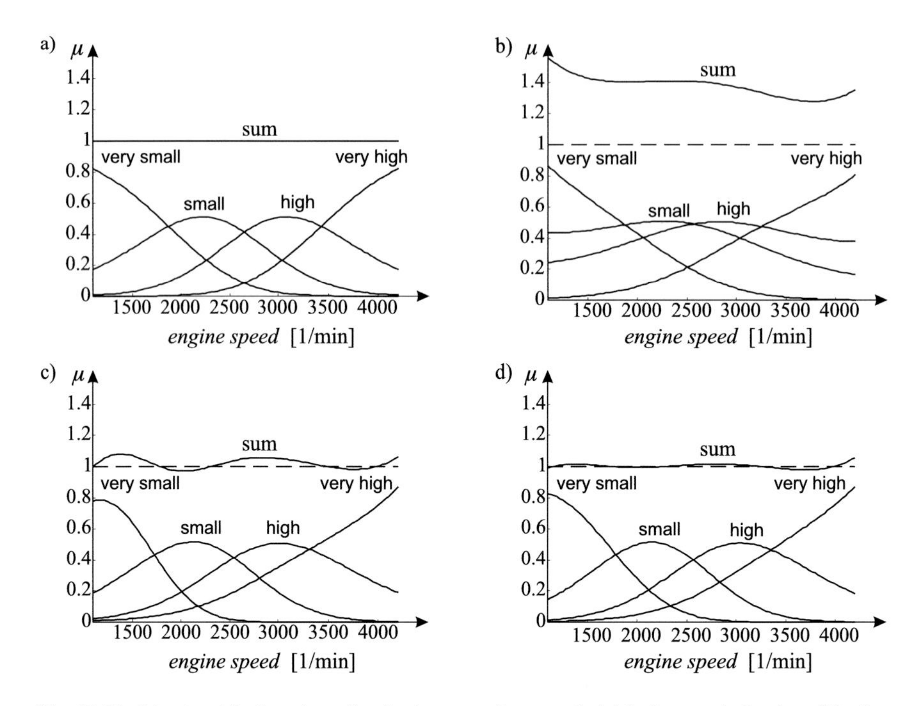

Fig. 12.21 Membership functions for the input engine speed: (**a**) before optimization, (**b**) after unconstrained nonlinear optimization, (**c**) after nonlinear optimization with constraints and a "sum penalty" factor of 0.01, (**d**) as in c but with a "sum penalty" factor of 0.1

locality. Furthermore, the optimized singleton for the first rule is at 0.390 bar. This value violates the output range of the look-up table, which varies approximately between 1 bar and 2.5 bar. Obviously, by balancing very small and large singletons with non-local input membership functions, a good overall approximation can be achieved. However, a linguistic interpretation is impossible.

Fig. 12.21c–d show examples of the tuned membership functions of the input *engine speed*. All the above-listed constraints are active. Only the "sum penalty" factor for strategy 3 is varied. Figure 12.21b shows the optimized membership functions for a "sum penalty" factor of 0.01. The singletons vary between 1.032 bar and 2.542 bar. Figure 12.21d shows the optimized membership functions for a larger "sum penalty" factor of 0.1. Here the singletons vary between 1.047 bar and 2.537 bar. The normalized mean square error is 0.0077 and 0.0144, respectively; see Table 12.1. This is a considerable improvement compared with the result obtained with the GA only. As the range of the singletons and the centers and widths of the Gaussian input MSFs show, the fuzzy models obtained by constrained optimization are easy to interpret. These results clearly demonstrate that it might be necessary to pay for increasing interpretability by decreasing performance. In this example, nonlinear optimization without constraints could improve the approximation accuracy by a factor of 3.5, while the improvement with constraints

Table 12.1 Normalized mean squared errors (NMSE) for fuzzy models trained with FUREGA; see Fig. 12.21

Optimization method	NMSE
(a) Before nonlinear optimization (GA and LS only)	0.0243
(b) Unconstrained optimization	0.0018
(c) Constrained optimization with "sum penalty" factor of 0.01	0.0077
(d) Constrained optimization with "sum penalty" factor of 0.1	0.0144

was between 1.8 and 1.3 depending on the sum penalty factor (these values correspond to the normalized *root* mean square error ratios).

12.4.5 Adaptive Spline Modeling of Observation Data (ASMOD)

The adaptive spline modeling of observation data (ASMOD) algorithm by Kavli [316] can be used for training neuro-fuzzy networks when B-splines are used as fuzzy sets. For an extensive treatment, refer to [61–63, 316]. The ASMOD algorithm constructs *additive* model structures; see Sect. 7.4. This means that the overall model is a summation of fuzzy submodels. This additive structure imposes structural assumptions on the process, and by these restrictions, the curse of dimensionality can be overcome. In order to reduce the restrictions on the model, a very flexible algorithm for model structure construction is required. ASMOD fulfills this task.

ASMOD starts from an initial model (Fig. 12.22a) that can be empty if no prior knowledge about the rule base is available. Then the following three growing and optionally pruning steps are usually considered during ASMOD training:

1. *Add/discard an input variable*: As demonstrated in Fig. 12.22b, a new input variable can be incorporated into the model.
2. *Form/split a tensor product*: In order to incorporate correlations between two or more inputs, a tensor product can be formed; see Fig. 12.22c. Note that the tensor product can also be formed between submodels that are already combined by tensor products. If too many, say more than three or four, inputs are combined by a tensor product, the curse of dimensionality becomes severe owing to the grid-like partitioning. The philosophy of ASMOD is based on the observation that such high-order tensor products are rarely necessary.
3. *Insert/remove a knot*: An additional knot can be incorporated; see Fig. 12.22d. This includes an additional MSF for an input variable and thus increases the resolution for this input.

With these steps, candidate models are built that are accepted if they pass a validation test; see Sect. 7.3.1. The exact sequence in which the growing and pruning steps are carried out can vary. Typically, first, a growing phase is performed, which

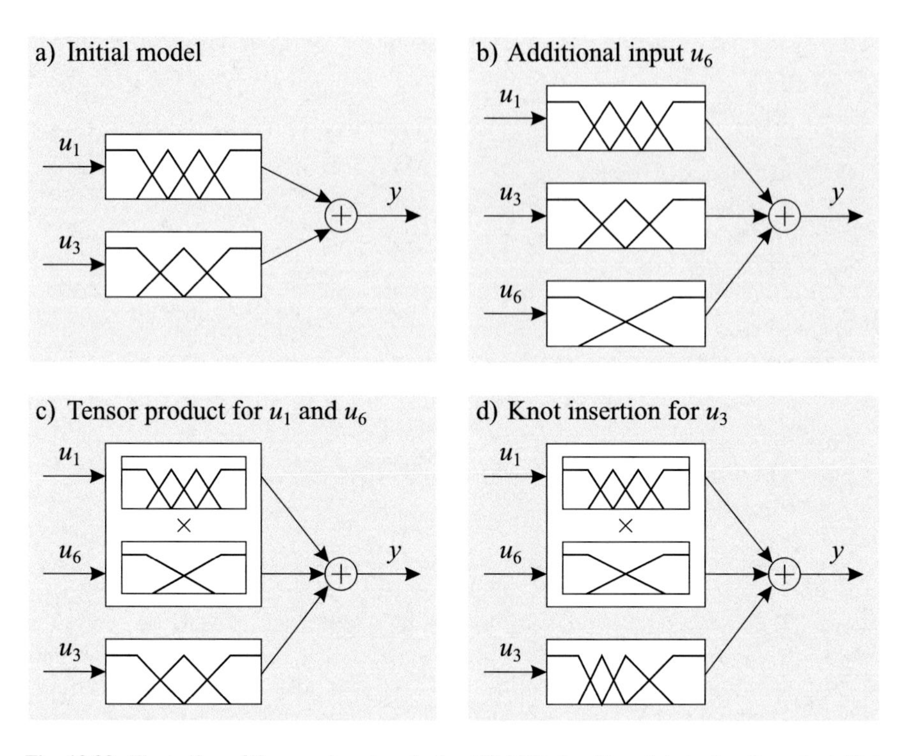

Fig. 12.22 Illustration of the growing steps in the ASMOD algorithm: (**a**) starting from the initial model, (**b**) an additional input can be incorporated into the model, (**c**) two submodels can be combined by a tensor product, and (**d**) additional knots can be inserted

subsequently carries out steps 1, 2, and 3. Then a pruning phase can be performed in the same sequence.

The steps 1, 2, and 3 possess the nice property that any more complex model obtained through a growing step can *exactly* reproduce the simpler model. This guarantees that any growing step improves the model's performance as long the bias error dominates, i.e., no overfitting occurs. A possible fourth growing and pruning step is usually not applied which changes the order of the B-splines, that is the polynomial order of the MSFs [63, 316].

An appealing property of additive models is that linear parameterized submodels can still be optimized by linear least squares techniques. Since the input MSFs and the rule structure are fixed a priori and/or manipulated by the ASMOD algorithm, only the output MSFs (the singletons) remain to be optimized. This can be carried out by LS or, for large structures, where the complexity of LS may become too high, by an iterative conjugate gradient algorithm [63, 316]. Furthermore, each candidate model generated during training can be trained on the current residuals following the staggered optimization approach instead of re-estimating all model parameters; see Sect. 7.5.5. This simplification is successful since growing steps incorporate components that are almost orthogonal to the existing submodels; see Fig. 3.9 in Sect. 3.1.3.

Finally, it is important to analyze some interpretability issues. Owing to the additive structure, ASMOD does not really generate fuzzy models. So no direct interpretation in the form of rules exists. It is possible, however, to analyze each of the submodels individually as a fuzzy model. Although this allows one to gain some insights, the overall effect of each submodel is obscure for the user. It can happen, for example, that two submodels with an almost opposite effect are generated. Then tiny differences between these two submodels may decide whether an overall effect is positive or negative. Since fuzzy models usually support only a coarse interpretation, the overall effect cannot be assessed by roughly examining the submodels. Therefore, additive models are not really interpretable in terms of fuzzy rules. However, the additive structure offers other advantages (independent of the realization of the submodels) in terms of interpretability. Typically, high-dimensional problems are broken down to lower-dimensional ones. Often only one- and two-dimensional (and possibly three-dimensional) submodels are generated by ASMOD. These simple submodels can be easily visualized and thus are very transparent to the user. This is one of the major advantages of ASMOD and other additive structure construction algorithms.

12.5 Summary

Fuzzy models allow one to utilize qualitative knowledge in the form of rules. Three major types of fuzzy systems can be distinguished. Linguistic fuzzy systems possess a transparent interface to the user because both the input and the output variables are described in terms of fuzzy sets. In contrast, singleton and Takagi-Sugeno fuzzy systems realize their rule output as equations. While in singleton fuzzy systems each rule output is a simple constant, Takagi-Sugeno type fuzzy systems use (usually linear) functions of the system inputs. Table 12.2 compares the properties of the three types of fuzzy systems.

Typically, qualitative expert knowledge is not sufficient in order to build a fuzzy model with high accuracy. Therefore, often measurement data is exploited for fuzzy model identification. Data-driven fuzzy models are commonly referred to as neuro-fuzzy models or networks. Many combinations of expert knowledge and identification methods are possible: from an (almost) black box model that is completely trained with data to a fuzzy model that is fully specified by an expert with an additional data-driven fine-tuning phase. The two main motivations for the application of neuro-fuzzy models are:

- Incorporation of prior knowledge into the model before and during identification
- Interpretation of the model obtained by identification

Only if these issues are explicitly considered can a benefit from the application of neuro-fuzzy models be expected. A list of criteria for promising employment

Table 12.2 Comparison between linguistic neuro-fuzzy models, neuro-fuzzy models with singletons, and those of Takagi-Sugeno type

Properties	Linguistic	Singleton	Takagi-Sugeno
Interpolation behavior	0	+	0
Extrapolation behavior	+	+	++
Locality	0	+	+
Accuracy	−	0	++
Smoothness	+	+	0
Sensitivity to noise	+	+	++
Parameter optimization	−	++*/−−**	++*/−−**
Structure optimization	−−	0	++
Online adaptation	−−	+	++
Training speed	−	+*/−−**	++*/−−**
Evaluation speed	−−	−	−
Curse of dimensionality	−	−	0
Interpretation	++	+	0
Incorporation of constraints	0	0	0
Incorporation of prior knowledge	+	0	0
Usage	+	++	0

* = linear optimization, ** = nonlinear optimization,
++/−− = model properties are very favorable/undesirable

of fuzzy and neuro-fuzzy models is given in [455]. If these criteria are not fulfilled, neuro-fuzzy models usually yield inferior performance compared with other nonlinear models such as neural networks. The reason for this observation is that a price has to be paid for interpretability. As discussed in Sects. 12.3.2 and 12.3.4, several restrictions and constraints must be imposed on the fuzzy model in order to keep it interpretable. The most severe example for this is probably the grid-like partitioning of the input space, which is required to obtain linguistically meaningful univariate membership functions. All restrictions and constraints reduce the approximation capabilities of the neuro-fuzzy model in comparison with a normalized RBF network.

12.6 Problems

1. Generate three-dimensional plots of the following t-conorms (OR operators): (i) max, (ii) algebraic sum, and (iii) bounded sum.
2. Explain the meaning of the term "granularity" in the context of fuzzy systems.
3. Compare the three basic types of fuzzy systems with their advantages and drawbacks.
4. What are the similarities and difference between a fuzzy system with singletons and a normalized RBF network?

5. What is difference between a fuzzy and a neuro-fuzzy system? Why is the terminology "neuro"-fuzzy system somewhat misleading?
6. Consider a fuzzy system with singletons. What are the parameters of such a fuzzy system? How many parameters exist in dependency on the number of inputs, assuming one output and M membership functions per input? What are the consequences for the interpretability of the fuzzy system if all these parameters are optimized?
7. Explain the strange effects that can occur during the normalization/defuzzification step. What are the consequences for the interpretability of the fuzzy system? How can these undesirable normalization effects be avoided?
8. What are the relationships between the curse of dimensionality and the interpretability of fuzzy system?
9. Explain how the curse of dimensionality can be (partly) overcome by using rules with a reduced number of linguistic statements. How can these rules be interpreted linguistically and geometrically?
10. Explain the major features of ASMOD.

Chapter 13
Local Linear Neuro-Fuzzy Models: Fundamentals

Abstract This chapter introduces the fundamental ideas of local linear neuro-fuzzy models. The concept is presented, and the wide variety of existing learning schemes is summarized. The equivalence to Takagi-Sugeno fuzzy systems is analyzed, and the constraints for a proper interpretation in terms of fuzzy logic are outlined. Then the problem of learning is subdivided into two parts: (i) parameters of the local models and (ii) structural parameters of the rule premises. For the relatively simple part (i), the local and global estimation approaches are detailed and compared. For the much more complex part (ii), an overview of proposed learning schemes is given, and a specific algorithm based on incremental axis-orthogonal tree construction is discussed in detail. It is called local linear model tree (LOLIMOT) and will be used extensively throughout this book. Its basic features and settings are treated in this chapter. LOLIMOT allows training a local linear neuro-fuzzy model from data deterministically and without adjusting many fiddle parameters as it is usually the case for most alternative neural network approaches.

Keywords Local model · Linear · Affine · Polynomial · Local estimation · Axis-orthogonal · Interpretation · Regularization · LOLIMOT

This chapter deals with local linear neuro-fuzzy models also referred to as Takagi-Sugeno fuzzy models and appropriate algorithms for their identification from data. The focus is on local *linear* models and on axis-*orthogonal* partitioning which allows an interpretation in terms of one-dimensional fuzzy membership functions. In Chap. 14 this will be extended to more complex local models and to more complex partitioning types. If the partitioning is not axis-orthogonal, a fuzzy interpretation becomes very difficult and unintuitive in terms of multidimensional membership functions. This loses most of the fuzzy advantages. Therefore, the more general term *local model networks* will be used in Sect. 14.8 and the following.

The local linear modeling approach is based on a divide-and-conquer strategy. A complex modeling problem is divided into a number of smaller and thus simpler subproblems, which are solved (almost) independently by identifying simple, e.g., linear, models. The most important factor for the success of such an approach is the

O. Nelles, *Nonlinear System Identification*,
https://doi.org/10.1007/978-3-030-47439-3_13

division strategy for the original complex problem. Therefore, the properties of local linear neuro-fuzzy models crucially depend on the applied construction algorithm that implements a certain division strategy. This chapter focuses on the local linear model tree (LOLIMOT) algorithm proposed by Nelles [411, 415, 432].

The basic principles of local linear neuro-fuzzy models have been more or less independently developed in different disciplines in the context of neural networks, fuzzy logic, statistics, and artificial intelligence with different names such as local model networks, Takagi-Sugeno fuzzy models, operating regime-based models, piecewise models, and local regression [303]. There are also close links to multiple models, mixtures of experts, and gain scheduling approaches. In [303] Johansen and Murray-Smith give a nice overview of the various existing approaches. The local modeling approaches can be distinguished according to the manner in which they combine the local models. Here only the soft partitioning strategies, e.g., arising from a fuzzy logic formulation, are discussed. Other strategies are hard switching between the local models [55, 248, 405, 437, 456, 457, 465], utilizing finite-state automata that generate a continuous-/discrete-state hybrid system [212, 372, 613], and probabilistic approaches [286, 309, 565].

This chapter is organized as follows. Section 13.1 introduces and illustrates the basic ideas of local linear neuro-fuzzy models. Sections 13.2 and 13.3 present and analyze parameter and structure optimization algorithms, respectively. The local linear model tree (LOLIMOT) algorithm is pursued in the remaining parts. A brief summary is given in Sect. 13.4. Chapter 14 continues this chapter with more advanced aspects. It extends the features of local linear neuro-fuzzy models and of the LOLIMOT algorithm.

13.1 Basic Ideas

The network structure of a local linear neuro-fuzzy model is depicted in Fig. 13.1. Each neuron realizes a *local linear model (LLM)* and an associated *validity function* that determines the region of validity of the LLM. The outputs of the LLMs are[1]

$$\hat{y}_i = w_{i0} + w_{i1}u_1 + w_{i2}u_2 + \ldots + w_{ip}u_p\,, \tag{13.1}$$

where w_{ij} denote the LLM parameters for neuron i.

The validity functions form a partition of unity, i.e., they are *normalized* such that

$$\sum_{i=1}^{M} \Phi_i(\underline{u}) = 1 \tag{13.2}$$

[1]Strictly speaking, owing to the existence of an offset term, the LLMs are local *affine* not local linear. Nevertheless, "local *linear* models" is the standard terminology in the literature.

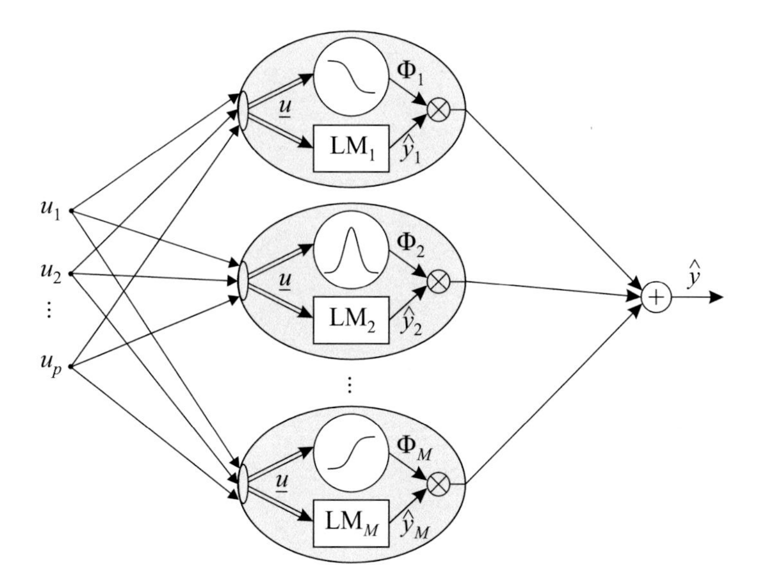

Fig. 13.1 Network structure of a static local linear neuro-fuzzy model with M neurons for p inputs

for any model input $\underline{u} = [u_1 \ u_2 \ \cdots \ u_p]^T$. This property is necessary for a proper interpretation of the $\Phi_i(\cdot)$ as validity functions because it ensures that the contributions of all local linear models sum up to 100%.

The output of a local linear neuro-fuzzy model becomes

$$\hat{y} = \sum_{i=1}^{M} \underbrace{\left(w_{i0} + w_{i1}u_1 + w_{i2}u_2 + \ldots + w_{ip}u_p\right)}_{\hat{y}_i} \Phi_i(\underline{u}) \,. \tag{13.3}$$

Thus, the network output is calculated as a weighted sum of the outputs of the local linear models where the $\Phi_i(\cdot)$ is interpreted as the operating point-dependent weighting factors. The network interpolates between different LLMs with the validity functions. The weights w_{ij} are linear network parameters; see Sects. 9.2.2 and 9.3. Their estimation from data is discussed in Sect. 13.2.

The validity functions are typically chosen as normalized Gaussians. If these Gaussians are furthermore axis-orthogonal, the validity functions are

$$\Phi_i(\underline{u}) = \frac{\mu_i(\underline{u})}{\sum\limits_{j=1}^{M} \mu_j(\underline{u})} \tag{13.4}$$

with

$$\mu_i(\underline{u}) = \exp\left(-\frac{1}{2}\left(\frac{(u_1 - c_{i1})^2}{\sigma_{i1}^2} + \ldots + \frac{(u_p - c_{ip})^2}{\sigma_{ip}^2}\right)\right)$$

$$= \exp\left(-\frac{1}{2}\frac{(u_1 - c_{i1})^2}{\sigma_{i1}^2}\right) \cdot \ldots \cdot \exp\left(-\frac{1}{2}\frac{(u_p - c_{ip})^2}{\sigma_{ip}^2}\right). \quad (13.5)$$

The normalized Gaussian validity functions $\Phi_i(\cdot)$ depend on the center coordinates c_{ij} and the dimension individual standard deviations σ_{ij}. These parameters are nonlinear; they represent the hidden layer parameters of the neural network. Their optimization from data is discussed in Sect. 13.3. The *validity functions* are also called *activation functions* since they control the activity of the LLMs or *basis functions* in consideration of the basis function formulation (9.7) in Sect. 9.3 or *interpolation functions* or *weighting functions*.

In the next section, the functioning of local linear neuro-fuzzy models is illustrated. Section 13.1.2 addresses some issues concerning the interpretation of the offset values w_{i0} in (13.3). In Sect. 13.1.3 the equivalence between local linear neuro-fuzzy models and Takagi-Sugeno fuzzy models is analyzed. Section 13.1.4 discusses the link to normalized radial basis function (NRBF) networks.

13.1.1 Illustration of Local Linear Neuro-Fuzzy Models

Figure 13.2 demonstrates the operation of a local linear neuro-fuzzy model. The nonlinear function in Fig. 13.2a(left) is to be approximated by a network with four neurons. Each neuron represents one local linear model, which is shown in Fig. 13.2a(right). Three alternative sets of validity functions are examined. All sets possess the same centers but have different standard deviations. Obviously, small standard deviations as in Fig. 13.2b lead to a relatively hard switching between the LLMs with a small transition phase. The network output becomes non-smooth (although it is still arbitrarily often differentiable). A medium value for the standard deviations as in Fig. 13.2c smooths the model output. However, if the standard deviation is chosen too high, the validity functions become very wide, and their maximum value decreases; see Fig. 13.2d. For example, the second validity function in Fig. 13.2d(right) is smaller than 0.6 for all inputs u. Consequently, the corresponding local linear model contributes always less than 60% to the overall model output. Thus, when the validity functions' overlap increases, locality and interpretability decrease.

It is interesting to consider the following two special cases:

- $\sigma \to 0$: As the standard deviation tends to zero, the validity functions become step-like. For any input $\underline{u}$, only a single local linear model is active. No smooth

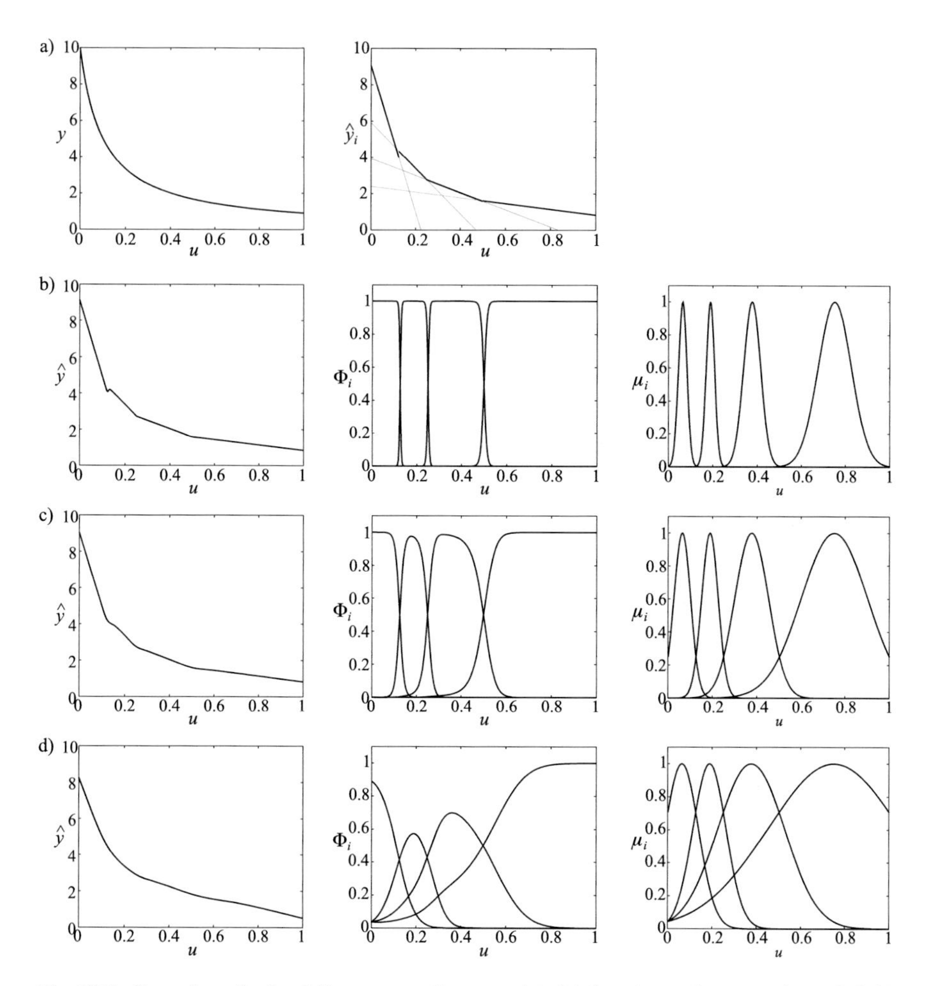

Fig. 13.2 Operation of a local linear neuro-fuzzy model: (**a**) function to be approximated (left) and local linear models (right), (**b**) small, (**c**) medium, (**d**) large standard deviations of the validity functions Φ_i

transition from one LLM to another exists. The overall model output is non-continuous and non-differentiable.

- $\sigma \rightarrow \infty$: As the standard deviation tends to infinity, the validity functions become constant with an amplitude of $1/M$. All local linear models are equally active for all inputs $\underline{u}$. The overall model output is the average of all LLMs, and thus it is globally linear.

Obviously, a compromise between these two cases must be found. Note that each validity function possesses its individual standard deviations, and therefore the tradeoff between smoothness and locality can be performed locally.

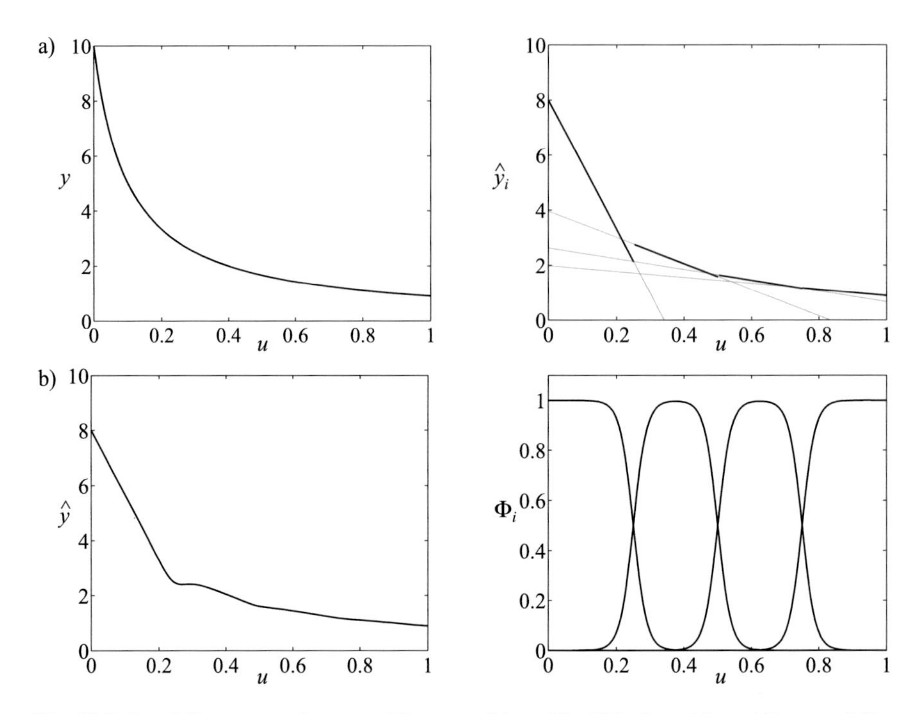

Fig. 13.3 Local linear neuro-fuzzy model comparable to Fig. 13.2c but with equidistant validity functions

The validity functions in Fig. 13.2 are well-positioned. In regions close to 0, where the curvature of the function to be approximated is high, the Φ_i is placed denser than in regions close to 1, where the function is almost linear. For comparison, Fig. 13.3 shows a model with equidistant validity functions. Its approximation performance is considerably lower than with the model in Fig. 13.2c, which possesses comparable smoothness properties. The two LLMs in Fig. 13.3 in the interval $0.5 < u < 1$ are very similar and may be merged without a significant performance loss. On the other hand, the two LLMs in the interval $0 < u < 0.5$ are very different and thus allow only a rough approximation. It is a major challenge for any identification algorithm to automatically construct well-positioned validity functions from data.

Finally, Fig. 13.4 will illustrate how local linear neuro-fuzzy models operate for two inputs. The validity functions become two-dimensional; the LLMs become planes. The extension to higher dimensions is straightforward.

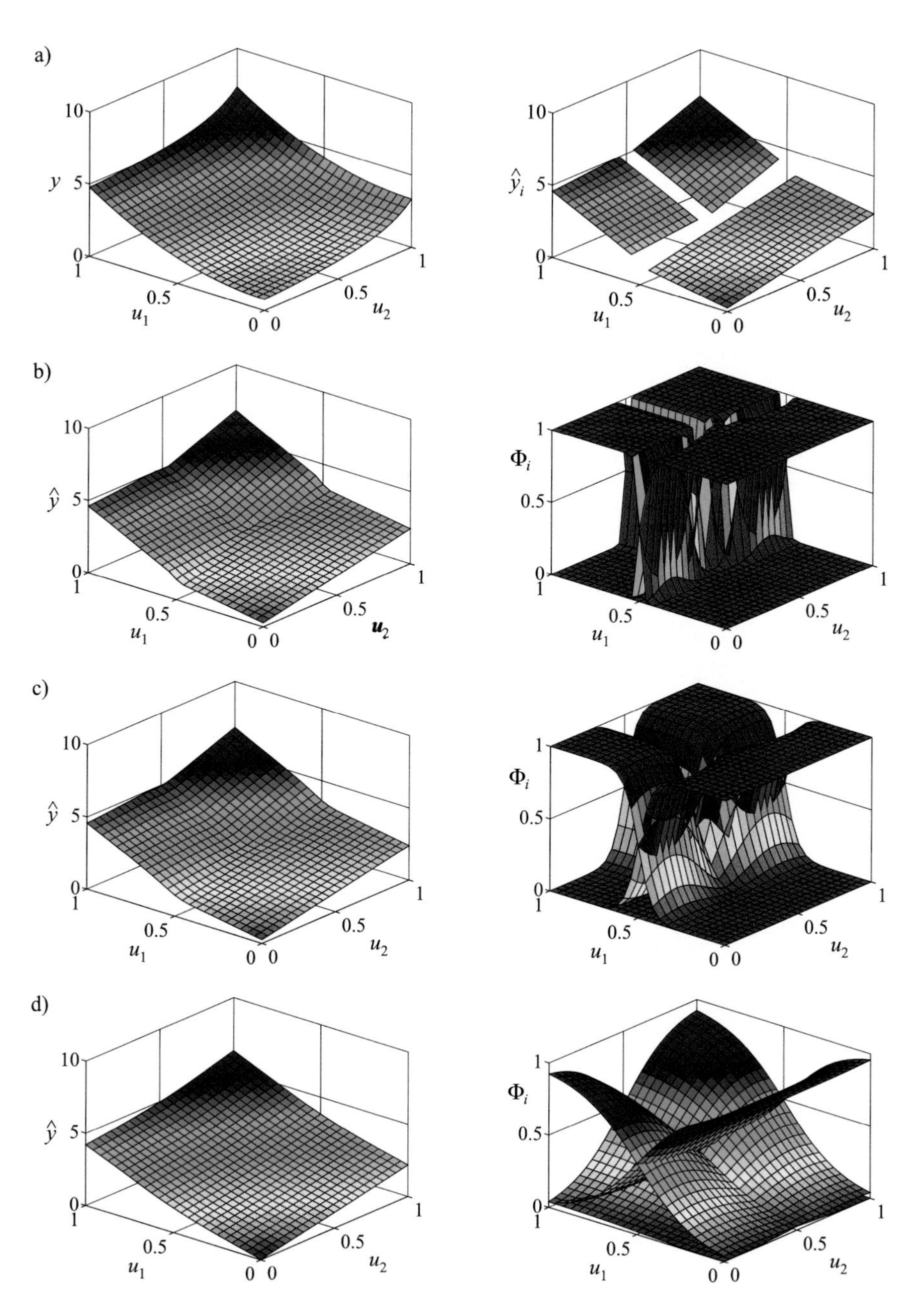

Fig. 13.4 Operation of a local linear neuro-fuzzy model: (**a**) function to be approximated (left) and local linear models (right), (**b**) small, (**c**) medium, (**d**) large standard deviations of the validity functions Φ_i

13.1.2 Interpretation of the Local Linear Model Offsets

The offsets w_{i0} in (13.3) are not directly interpretable because they are *not* equal to the output of the ith local linear model at the center $\underline{c}_i$ of the ith validity function Φ_i. The offsets w_{i0} can even be out of the range of physically reasonable values for the output y. In order to be able to interpret the offset values directly, the local linear neuro-fuzzy model in (13.3) can be rewritten in the following *local* form [335]:

$$\hat{y} = \sum_{i=1}^{M} \left(\widetilde{w}_{i0} + w_{i1}(u_1 - c_{i1}) + \ldots + w_{ip}(u_p - c_{ip})\right) \Phi_i(\underline{u}) \tag{13.6}$$

with the new, transformed offsets $\widetilde{w}_{i0}$. Figure 13.5a illustrates the interpretation of these offsets as

$$\widetilde{w}_{i0} = \hat{y}_i(\underline{u} = \underline{c}_i) \tag{13.7}$$

with $\hat{y}_i$ as the output of the ith local linear model. The relationship between $\widetilde{w}_{i0}$ in (13.6) and w_{i0} in (13.3) is as follows:

$$\widetilde{w}_{i0} = w_{i0} + w_{i1}c_{i1} + \ldots + w_{ip}c_{ip} \,. \tag{13.8}$$

13.1.2.1 Advantages of Local Description

Both formulations in (13.3) and (13.6) deliver the identical outcome, but they are not equivalent in terms of estimation. The local formulation (13.6) yields lower parameter variances and is numerically more robust since the equations become better conditioned; refer to Sect. 13.2 for details on the parameter estimation

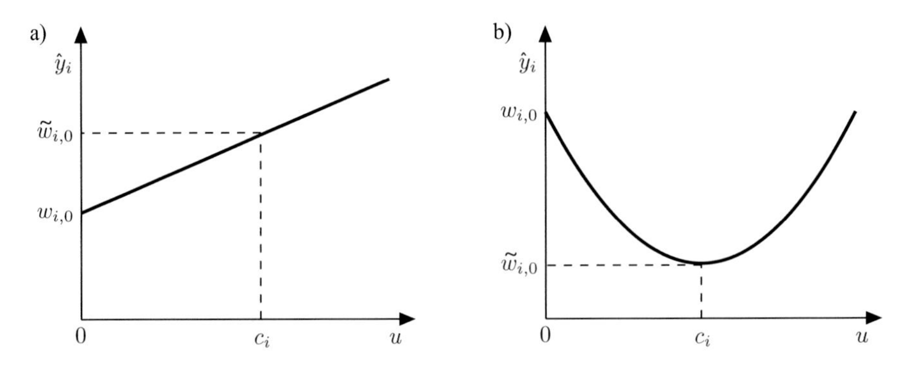

Fig. 13.5 Interpretation of the local model offsets w_{i0} and $\widetilde{w}_{i0}$ for a local (**a**) linear/affine and (**b**) quadratic model with a single input u

procedure. In addition, the version (13.6) is extremely beneficial if the local models are nonlinear, e.g., polynomials (Sect. 14.2), because then the parameters are directly equivalent with the Taylor series expansion around c_i.

In combination with *subset selection* techniques, the form (13.6) is *mandatory* as Fig. 13.5b illustrates. Only with the local description (13.6), a pure quadratic form without a linear term

$$\hat{y}_i = \widetilde{w}_{i,0} + \widetilde{w}_{i,2}(u - c_i)^2 = w_{i,0} + w_{i,1}u + w_{i,2}u^2 \tag{13.9}$$

can be described with just *two* regressors $(u - c_i)^2$ and 1 while $u - c_i$ is not required. In the original form, all three regressors u^2, u, and 1 are necessary to describe such a pure quadratic form if it is not centered at the origin. Thus, for local polynomial subset selection as mentioned in Sect. 13.3, the formulation (13.6) is *mandatory* for obtaining reasonable and sparse local models.

One drawback of formulation (13.6) is the additional implementation effort since for each local model individual regressors centered at c_i have to be built up.

13.1.3 Interpretation as Takagi-Sugeno Fuzzy System

As shown in [266] under some assumptions, the network depicted in Fig. 13.1 and given by (13.3) is equivalent to a Takagi-Sugeno fuzzy model; see Sect. 12.2.3. These assumptions are basically the same as those required for an equivalence between an RBF network and singleton fuzzy systems. The membership functions (MSFs) must be Gaussian, and the product operator has to be used for conjunction [266]. If these restrictions do not hold, then a strong similarity is still preserved. Furthermore, the validity functions should be axis-orthogonal in order to allow a projection onto the input axes for which the univariate MSFs are defined. Refer to Sect. 12.3.2 for more details.

Figure 13.6 demonstrates that a Takagi-Sugeno fuzzy system with the following three rules describes the model in Fig. 13.4c:

R_1: IF $u_1 =$ small AND $u_2 =$ don't care THEN $y = w_{10} + w_{11}u_1 + w_{12}u_2$

R_2: IF $u_1 =$ large AND $u_2 =$ small THEN $y = w_{20} + w_{21}u_1 + w_{22}u_2$

R_3: IF $u_1 =$ large AND $u_2 =$ large THEN $y = w_{30} + w_{31}u_1 + w_{32}u_2$.

This rule base covers the whole input space, although it is not complete, i.e., it does not contain a rule for all possible membership function combinations (small/small, small/ large, large/small, large/large). The first rule effectively contains only one linguistic statement since the fuzzy set don't care represents the whole universe of discourse of input u_2. This can be verified by the shape of the corresponding validity

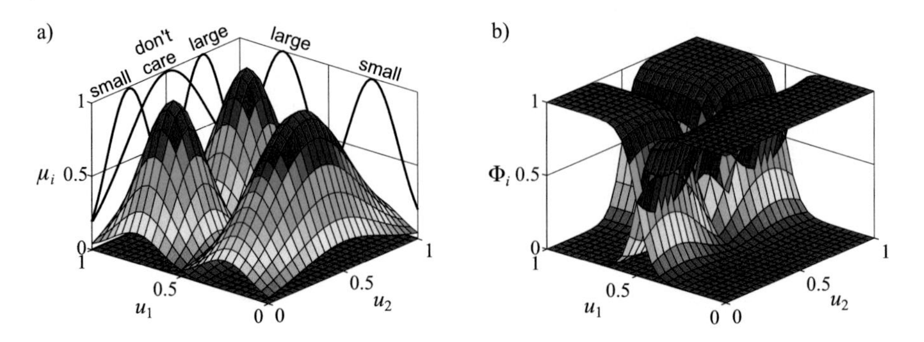

Fig. 13.6 (**a**) By a t-norm (here the product operator) univariate Gaussian MSFs μ_{ij} are combined to multidimensional MSFs μ_i. (**b**) By normalization, validity functions Φ_i are obtained from the multidimensional MSFs

function in Fig. 13.6b. It does depend only on u_1 and is insensitive to u_2. Although it is not exactly equivalent (see below) this rule can also be interpreted as

$$R_1\text{: IF } u_1 = \textsf{small} \text{ THEN } y = w_{10} + w_{11}u_1 + w_{12}u_2 .$$

With such rules that contain fewer linguistic statements in their premise than the number of inputs, the curse of dimensionality (Sect. 7.6.1) can be overcome or at least be weakened. The same strategy is pursued in the FUREGA algorithm in Sect. 12.4.4. The rule R_1 covers the same region of the input space as the following two rules in a grid partitioning approach with complete rule set:

$$R_{1a}\text{: IF } u_1 = \textsf{small} \text{ AND } u_2 = \textsf{small} \text{ THEN } y = w_{1a0} + w_{1a1}u_1 + w_{1a2}u_2$$

$$R_{1b}\text{: IF } u_1 = \textsf{small} \text{ AND } u_2 = \textsf{large} \text{ THEN } y = w_{1b0} + w_{1b1}u_1 + w_{1b2}u_2 .$$

In Fig. 13.6 it is demonstrated that a Takagi-Sugeno fuzzy model with univariate Gaussian MSFs can yield the validity functions Φ_i. This relationship is also obvious from (13.5), which builds the multidimensional MSFs from the univariate ones. Figure 13.7 illustrates that *normalized* Gaussian MSFs yield very similar but not identical results. The deviation between Figs. 13.7b and 13.6b becomes larger for less uniformly distributed MSFs. Both approaches are identical if a complete rule base (full-grid partitioning of the input space) is used; see Sect. 12.3.1. Consequently, a local linear neuro-fuzzy model, which corresponds to a fuzzy system with a complete rule base can be interpreted with either Gaussian or normalized Gaussian MSFs. In the first case, the normalization is carried out in the defuzzification step (13.4) when fuzzy basis functions are calculated; see Sect. 12.3.1. In the second case, the normalization is already incorporated into the MSFs themselves.

In the Takagi-Sugeno fuzzy model interpretation, the validity functions depend on the centers c_{ij} and standard deviations σ_{ij} of the input MSFs and on the rule

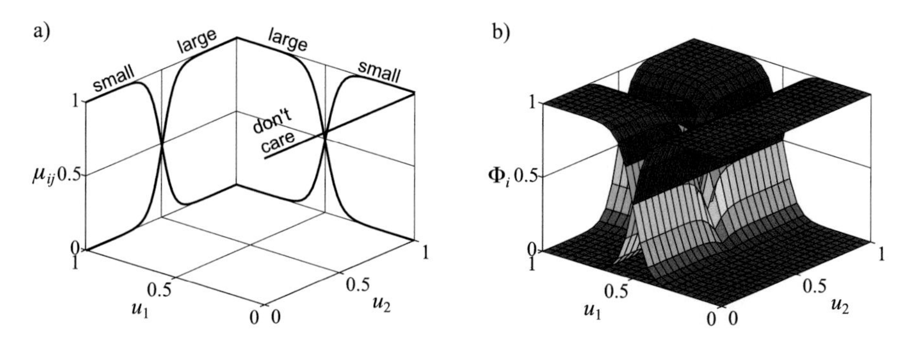

Fig. 13.7 With *normalized* Gaussian MSFs, a similar but not identical result to Fig. 13.6b is achieved: (**a**) univariate case, (**b**) multidimensional case

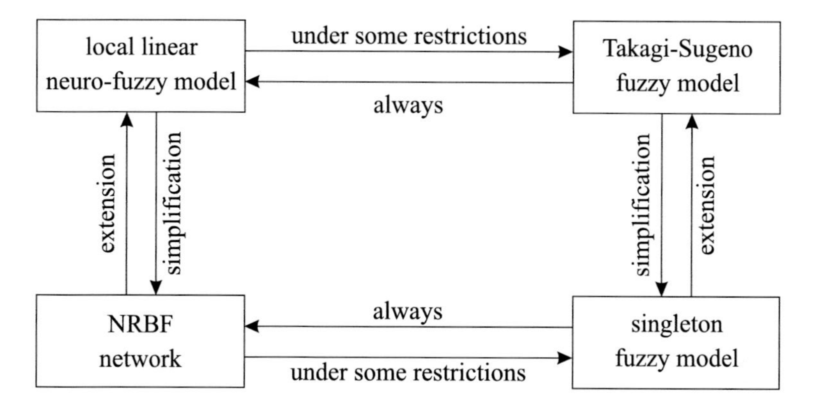

Fig. 13.8 Relationships between local neural networks and fuzzy models

structure. The optimization of the *premise structure* and the *premise parameters* is discussed in Sect. 13.3. The parameters of the local linear models w_{ij} are also called rule *consequent parameters*, and their estimation from data is addressed in Sect. 13.2.

In the following, the presented model is compared with similar modeling approaches. Local linear neuro-fuzzy models are (under some conditions) equivalent to Takagi-Sugeno fuzzy systems. Analogously, normalized RBF networks are (under some conditions) equivalent to singleton fuzzy systems. Furthermore, the local linear neuro-fuzzy and Takagi-Sugeno fuzzy models represent extensions of NRBF networks and singleton fuzzy systems, respectively. These links are depicted in Fig. 13.8.

13.1.4 Interpretation as Extended NRBF Network

Local linear neuro-fuzzy models are a straightforward extension of normalized RBF (NRBF) networks (Fig. 13.9); see Sect. 11.3.7. Herewith the normalized Gaussian basis functions are weighted with a constant w_{i0} rather than with a local linear model $w_{i0} + w_{i1}u_1 + \ldots + w_{ip}u_p$. In other words, a local linear neuro-fuzzy model in (13.3) degenerates to an NRBF network if

$$w_{i0} \neq 0 \text{ and } \quad w_{i1} = w_{i2} = \ldots = w_{ip} = 0 . \tag{13.10}$$

It is interesting to notice that the structure of a local linear neuro-fuzzy model can be interpreted in two different ways:

- *Extension of an NRBF network*: The validity functions are weighted with their local linear models.
- *Extension of a (global) linear model*: The local linear models are weighted with their validity functions.

Both interpretations are correct; they just take a different point of view.

In Fig. 13.10a the same function as in Fig. 13.4 is approximated by an NRBF network. The standard deviations are chosen comparable to those in Fig. 13.4c, which represented a good smoothness tradeoff for the local linear neuro-fuzzy model. Because one NRBF network neuron is less flexible than one local linear model, 12 instead of 3 neurons are used. Nevertheless the approximation in Fig. 13.10a is poor. It clearly reveals that the local *constant* models are a much worse local approximation than local *linear* models. The approximation performance of the NRBF network can be enhanced considerably if basis functions with wider overlap are chosen. This is illustrated in Fig. 13.10b, where the local constant regions are completely smoothed out by the basis functions with large standard deviations.

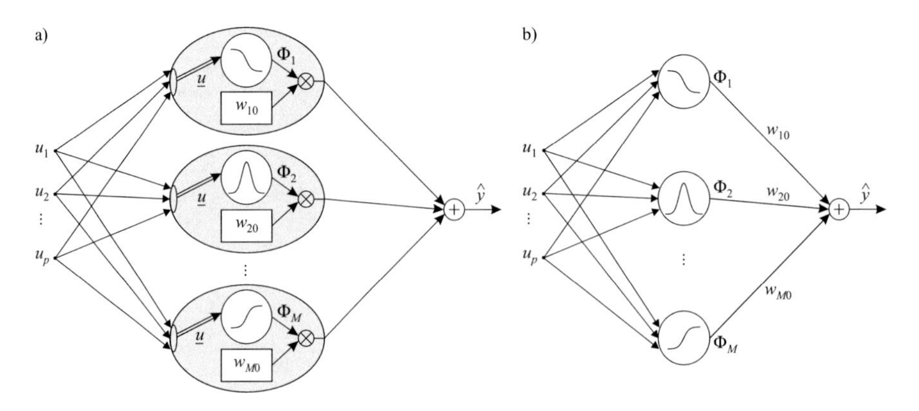

Fig. 13.9 Normalized radial basis function (NRBF) network: (**a**) in local model network representation, (**b**) in standard NRBF network representation. Note that both schemes are equivalent

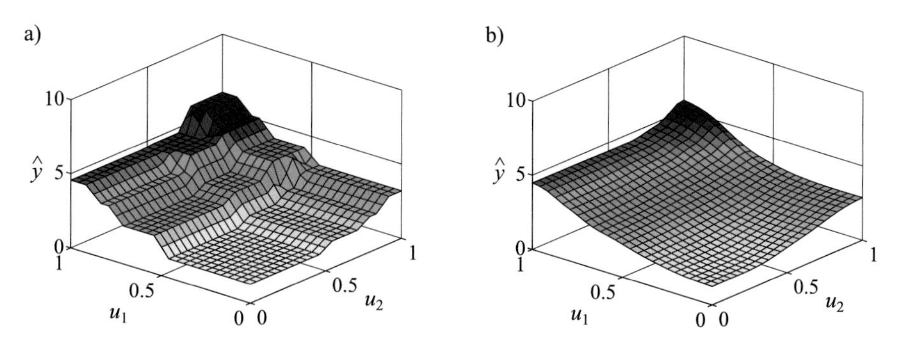

Fig. 13.10 Approximation of the function in Fig. 13.4 with an NRBF network with 12 neurons: (**a**) medium standard deviations, (**b**) large standard deviations

Obviously, the optimal basis function overlap is larger for NRBF networks than for local linear neuro-fuzzy models. Furthermore, the difference between Fig. 13.10a, b is much larger than the difference between Fig. 13.4c, d. Therefore, it can be concluded that NRBF networks are much more sensitive with respect to the chosen overlap of the basis functions than are local linear neuro-fuzzy models. This fact has important consequences for the design of an identification method; see Sect. 13.3.

13.2 Parameter Optimization of the Rule Consequents

Estimating the local linear model parameters is a linear optimization problem under the assumption that the validity functions are known. Throughout this section it is assumed that the validity functions are fully specified, i.e., their centers and standard deviations are known. In the next section, data-driven methods for determination of the validity functions are presented.

Two different approaches for optimization of the local linear model parameters can be distinguished: *global* and *local* estimation. While global estimation represents the straightforward application of the least squares algorithm, local estimation neglects the overlap between the validity functions in order to exploit the local features of the model [109, 399]. The following two subsections introduce both approaches, and Sect. 13.2.3 compares their properties. Finally, Sect. 13.2.6 extends this analysis to differently weighted data samples.

13.2.1 Global Estimation

In the global estimation approach *all* linear parameters are estimated simultaneously in a single LS optimization. The parameter vector contains all $n = M(p + 1)$ parameters of the local linear neuro-fuzzy model in (13.3) with M neurons and

p inputs:

$$\underline{w} = [w_{10}\ w_{11}\ \cdots\ w_{1p}\ w_{20}\ w_{21}\ \cdots\ w_{2p}\ \cdots\ w_{M0}\ w_{M1}\ \cdots\ w_{Mp}]^T . \tag{13.11}$$

The associated regression matrix $\underline{X}$ for N measured data samples becomes

$$\underline{X} = \left[\underline{X}_1^{(\mathrm{sub})}\ \underline{X}_2^{(\mathrm{sub})}\ \cdots\ \underline{X}_M^{(\mathrm{sub})} \right] \tag{13.12}$$

with the regression submatrices $\underline{X}_i^{(\mathrm{sub})} =$

$$\begin{bmatrix} \Phi_i(\underline{u}(1)) & u_1(1)\Phi_i(\underline{u}(1)) & u_2(1)\Phi_i(\underline{u}(1)) & \cdots & u_p(1)\Phi_i(\underline{u}(1)) \\ \Phi_i(\underline{u}(2)) & u_1(2)\Phi_i(\underline{u}(2)) & u_2(2)\Phi_i(\underline{u}(2)) & \cdots & u_p(2)\Phi_i(\underline{u}(2)) \\ \vdots & \vdots & \vdots & & \vdots \\ \Phi_i(\underline{u}(N)) & u_1(N)\Phi_i(\underline{u}(N)) & u_2(N)\Phi_i(\underline{u}(N)) & \cdots & u_p(N)\Phi_i(\underline{u}(N)) \end{bmatrix} .$$

The model output $\hat{\underline{y}} = [\hat{y}(1)\ \hat{y}(2)\ \cdots\ \hat{y}(N)]^T$ is then given by

$$\hat{\underline{y}} = \underline{X}\, \underline{w} . \tag{13.13}$$

If the parameters are to be estimated in the form (13.6), where the offsets are directly interpretable as operating points, all u_j in $\underline{X}$ have to be replaced by $u_j - c_{ij}$.

In global estimation, the following loss function is minimized with respect to the parameters:

$$I = \sum_{j=1}^{N} e^2(j) \longrightarrow \min_{\underline{w}} \tag{13.14}$$

where $e(j) = y(j) - \hat{y}(j)$ represent the model errors for data sample $\{\underline{u}(j), y(j)\}$.

The globally optimal parameters can be calculated as (for $N \geq M(p+1)$) (see Sect. 3.1):

$$\hat{\underline{w}} = \left(\underline{X}^T \underline{X} \right)^{-1} \underline{X}^T \underline{y} \tag{13.15}$$

where $\underline{y} = [y(1)\ y(2)\ \cdots\ y(N)]^T$ contains the measured process outputs.

Although the global LS estimation is a very efficient way to optimize the rule consequent parameters, its computational complexity grows cubically with the number of parameters and thus with the number of neurons M:

$$\mathcal{O}\left(M^3(p+1)^3\right) \approx \mathcal{O}\left(M^3p^3\right) . \tag{13.16}$$

The local estimation described next offers a much faster way to compute the rule consequent parameters.

13.2.2 Local Estimation

The idea behind the local estimation[2] approach is to consider the optimization of the rule consequent parameters as individual problems. The parameters for each local linear model are estimated *separately*, neglecting the interactions between the local models. As discussed below, this increases the bias error of the model. The motivation for this approach stems from the case $\sigma \to 0$ in which no interaction between the different local models takes place. The output of the model for a given input $\underline{u}$ is determined solely by a single local linear model, while all others are inactive. In this case the local estimation approach is exactly equivalent to the global one. As σ is increased, the interaction between neighbored local models grows, and the error caused by their neglect increases.

Instead of estimating all $n = M(p+1)$ parameters simultaneously, as is done in the global approach, M separate local estimations are carried out for the $p+1$ parameters of each local linear model. The parameter vector for each of these $i = 1, \ldots, M$ estimations is

$$\underline{w}_i = [w_{i0}\ w_{i1}\ \cdots\ w_{ip}]^T . \tag{13.17}$$

The corresponding regression matrices are

$$\underline{X}_i = \begin{bmatrix} 1 & u_1(1) & u_2(1) & \cdots & u_p(1) \\ 1 & u_1(2) & u_2(2) & \cdots & u_p(2) \\ \vdots & \vdots & \vdots & & \vdots \\ 1 & u_1(N) & u_2(N) & \cdots & u_p(N) \end{bmatrix} . \tag{13.18}$$

If the parameters are to be estimated in the form (13.6), where the offsets are directly interpretable as operating points, all u_j have to be replaced by $u_j - c_{ij}$.

[2]In this context, "local" refers to the effect of the parameters on the model; there is no relation to the expression "nonlinear local" optimization, where "local" refers to the parameter search space.

Note that here the regression matrices of all local linear models $i = 1, \ldots, M$ are identical since the entries of $\underline{X}_i$ do not depend on i. However, as discussed in Sect. 14.3, this can be generalized to local models of different structure; thus it is helpful to maintain the index i in $\underline{X}_i$. A local linear model with the output ($\underline{\hat{y}}_i = [\hat{y}_i(1)\ \hat{y}_i(2)\ \cdots\ \hat{y}_i(N)]^T$)

$$\underline{\hat{y}}_i = \underline{X}_i\, \underline{w}_i \tag{13.19}$$

is valid only in the region where the associated validity function $\Phi_i(\cdot)$ is close to 1. This will be the case close to the center of $\Phi_i(\cdot)$. Data in this region is highly relevant for the estimation of $\underline{w}_i$. As the validity function decreases, the data becomes less relevant for the estimation of $\underline{w}_i$ and more relevant for the estimation of the neighboring models. Consequently, it is straightforward to apply a weighted least squares optimization where the weighting factors are given by the validity function values, i.e.,

$$I_i = \sum_{j=1}^{N} \Phi_i(\underline{u}(j))\, e^2(j) \longrightarrow \min_{\underline{w}_i} \tag{13.20}$$

where $e(j) = y(j) - \hat{y}(j)$ represents the model errors for data sample $\{\underline{u}(j), y(j)\}$. For the extreme case $\sigma \to 0$, $\Phi_i(\underline{u}(j))$ is equal to either 0 or 1, that is, only a subset of the data is used for the estimation of $\underline{w}_i$. For $\sigma > 0$ all data samples are exploited for estimation, but those whose validity value is close to 0 are virtually ignored.

With the following diagonal $N \times N$ weighting matrix

$$\underline{Q}_i = \begin{bmatrix} \Phi_i(\underline{u}(1)) & 0 & \cdots & 0 \\ 0 & \Phi_i(\underline{u}(2)) & \cdots & 0 \\ \vdots & \vdots & \ddots & \vdots \\ 0 & 0 & \cdots & \Phi_i(\underline{u}(N)) \end{bmatrix} \tag{13.21}$$

the weighted LS solution for the parameters of the rule consequent i is given by (for $N \geq p + 1$)

$$\underline{\hat{w}}_i = \left(\underline{X}_i^T\, \underline{Q}_i \underline{X}_i\right)^{-1} \underline{X}_i^T\, \underline{Q}_i \underline{y}\,. \tag{13.22}$$

This estimate has to be computed successively for all $i = 1, \ldots, M$ local linear models.

The major advantage of local estimation is its low computational complexity. In contrast to global estimation, only $p + 1$ parameters must be estimated with the local LS optimization independently of the number of neurons. Since such a local estimation has to be performed individually for each local linear model, the computational complexity grows only linearly with the number of neurons, that is,

$$\mathcal{O}\left(M(p+1)^3\right) \approx \mathcal{O}\left(Mp^3\right) . \tag{13.23}$$

Consequently, the computational effort for the local estimation is significantly lower than for the global approach (see (13.16)) and this advantage increases quadratically ($\mathcal{O}(M^2)$) with the size of the neuro-fuzzy model. The price to be paid for this extremely efficient local optimization approach is the introduction of a systematic error due to the neglected interaction between the local models. It is important to understand the effect of the local estimation on the model quality in order to be able to assess the overall performance of this approach.

It is shown in [399, 401] that the local estimation approach increases the *bias error*, that is, the flexibility of the model. On the other hand, as expected from the bias/variance dilemma (Sect. 7.2), it reduces the *variance error*. Hence, local estimation reduces the number of *effective parameters* n_{eff} of the model although the nominal number of parameters n does not change. The effective number of parameters is a measure of the true model flexibility. It is smaller than the number of nominal parameters when a regularization technique is applied. For more details about the bias error, variance error, regularization, and the effective number of parameters, refer to Sects. 7.2 and 7.5. The effective number of parameters for a local linear neuro-fuzzy model with local estimation is [401]

$$\text{number of effective parameters} = n_{\text{eff}} = \text{trace}(\underline{S}^T \underline{S}) , \tag{13.24}$$

where $\underline{S}$ is the smoothing matrix of the model that relates the desired outputs y to the model outputs $\underline{\hat{y}} = \underline{S}y$; see Sect. 3.1.11. The smoothing matrix can be calculated by the sum of the local smoothing matrices

$$\underline{S} = \sum_{i=1}^{M} \underline{S}_i \tag{13.25}$$

with

$$\underline{S}_i = \underline{Q}_i \underline{X}_i \left(\underline{X}_i^T \underline{Q}_i \underline{X}_i\right)^{-1} \underline{X}_i^T \underline{Q}_i . \tag{13.26}$$

It can be shown that the effective number of parameters n_{eff} of the overall model can be decomposed in the sum of the effective parameters of the local models $n_{\text{eff},i}$ [401, 402]:

$$n_{\text{eff}} = \sum_{i=1}^{M} n_{\text{eff},i} \tag{13.27}$$

with

$$n_{\text{eff},i} = \sum_{j=1}^{M} \text{trace}\left(\underline{Q}_i \underline{X}_i \left(\underline{X}_i^T \underline{Q}_i \underline{X}_i\right)^{-1} \underline{X}_i^T \underline{Q}_i \underline{Q}_j \underline{X}_j \left(\underline{X}_j^T \underline{Q}_j \underline{X}_j\right)^{-1} \underline{X}_j^T \underline{Q}_j\right). \tag{13.28}$$

With increasing overlap the effective number of parameters decreases. The following two extreme cases can be analyzed:

- *No overlap ($\sigma \to 0$)*: The validity functions become step-like and the local estimation approach becomes equivalent to the global one (no overlap is neglected because the overlap is equal to zero). Therefore, the effective number of parameters becomes equal to the nominal number of parameters:

$$n_{\text{eff}} = n\,. \tag{13.29}$$

- *Full overlap ($\sigma \to \infty$)*: The validity functions become constant $\Phi_i(\cdot) = 1/M$ and thus all local linear models become identical. In fact, M times the same local linear models are estimated because all validity functions are identical. In (13.28) $\underline{Q}_i \to \underline{I}/M$ where $\underline{I}$ is an $N \times N$ identity matrix and consequently $n_{\text{eff}} = n/M$. Because the number of parameters for global estimation is $n = M(p+1)$, the effective number of parameters in local estimation is identical to the number of parameters of a linear model:

$$n_{\text{eff}} = p + 1\,. \tag{13.30}$$

Since local estimation decreases the effective number of parameters, it reduces the degrees of freedom of the model, and thus it is a *regularization technique*; see Sect. 7.5. Consequently, the additional bias error that is caused by the local estimation may be compensated or even over-compensated by a reduction in the variance error. Because of this effect, the overall performance of local estimation can be equal to or even higher than for global estimation, at least if no explicit regularization technique is used in combination with the global approach.

13.2.3 Global Versus Local Estimation

A simple example will illustrate the properties of the global and local parameter estimation approaches. The nonlinear function in Fig. 13.11a will be approximated

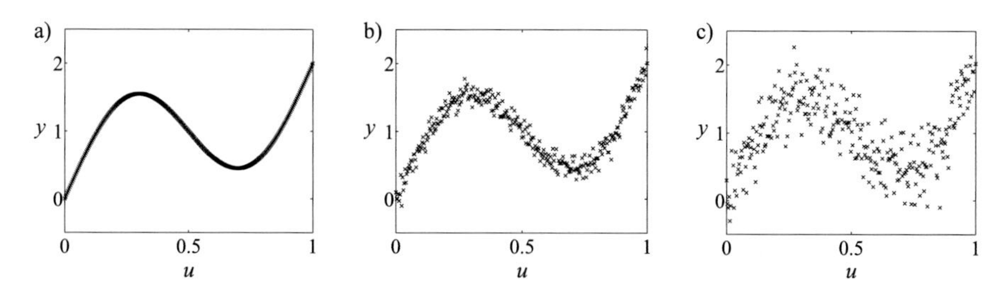

Fig. 13.11 Training data generated from the function $y = \sin(2\pi u) + 2u$ with different noise levels: (**a**) $\sigma_n = 0$, (**b**) $\sigma_n = 0.1$, (**c**) $\sigma_n = 0.3$

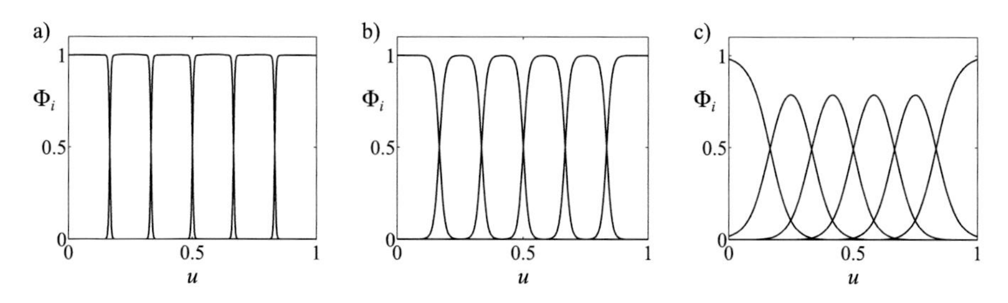

Fig. 13.12 Validity functions of the neuro-fuzzy model with different overlaps: (**a**) $\sigma = 0.125/M$, (**b**) $\sigma = 0.25/M$, (**c**) $\sigma = 0.5/M$ with $M = 6$

Table 13.1 Comparison of global and local estimation for different noise levels and different overlaps between the validity functions

100 · RMSE: global/local	$\sigma = 0.125/M$	$\sigma = 0.25/M$	$\sigma = 0.5/M$
Noise $\sigma_n = 0.0$	2.84/2.85	2.92/3.06	0.67/3.81
Noise $\sigma_n = 0.1$	3.35/3.33	3.39/3.43	2.00/3.69
Noise $\sigma_n = 0.3$	6.04/5.92	5.94/5.69	5.70/4.46
Nominal no. of parameters	12/12	12/12	12/12
Effective no. of parameters	12/10.8	12/8.4	12/5.1

RMSE = root mean square error
The mean square error can be calculated as $\frac{1}{N}\sum_{i=1}^{N} e^2(i)$ if the error $e(i) = y(i) - \hat{y}(i)$ is evaluated on the *test data*. If the error is evaluated on the *training data* $\frac{1}{N-n_{\text{eff}}}\sum_{i=1}^{N} e^2(i)$ has to be used, where n_{eff} is the effective number of parameters

by a local linear neuro-fuzzy model with $M = 6$ neurons and equidistantly positioned normalized Gaussian validity functions. As training data 300 equally distributed data samples are generated. Three different normally distributed, white disturbances with the standard deviations $\sigma_n = 0,\ 0.1,\ 0.3$ are considered; see Fig. 13.11a–c. Three different standard deviations for the validity functions $\sigma = 0.125/M, 0.25/M, 0.5/M$ with $M = 6$ are considered; see Fig. 13.12a–c.

Table 13.1 summarizes the results obtained with global and local estimation. Obviously, global and local optimization perform about equally when the validity functions have little overlap; see Fig. 13.13. Also, the estimated parameters of the

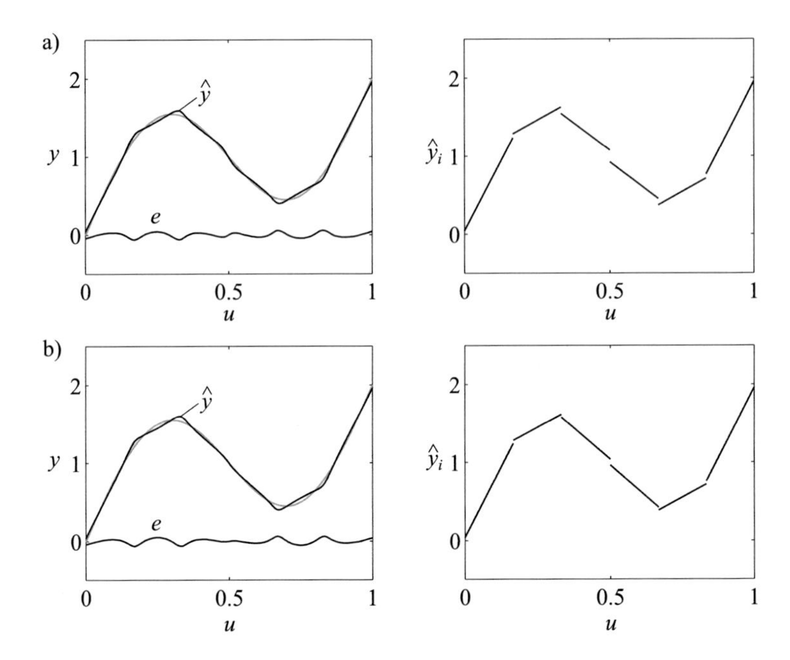

Fig. 13.13 The process and model output (left) and the local linear models for (**a**) global and (**b**) local estimation (right) with $\sigma_n = 0$ and $\sigma = 0.125/M$

local linear models are very similar. However, as the overlap of the validity functions increases, the two approaches behave increasingly distinctly.

For low noise levels, global estimation by far outperforms the local approach, as demonstrated in Fig. 13.14. This effect is intuitively clear since the interaction between the local models that is taken into account by the global estimation becomes stronger for increasing σ. Local estimation neglects this interaction and thus performs much worse. Although the approximation error with local estimation is much higher, it possesses a significant advantage in the interpretation of the model. The local linear models still represent the local behavior of the underlying function. By analyzing the LLM parameters, the user can draw conclusions about the process. This is not the case for global estimation, where the individual LLMs do not allow an interpretation. For example, the third and fourth LLMs have positive slopes, although the process possesses a large negative slope in this region. Only if the interactions with the neighboring LLMs are taken into account can information about the process behavior be extracted.

For high noise levels, global estimation may perform more poorly than local estimation owing to overfitting effects. The inherent regularization effect in local estimation makes it much less sensitive with respect to noise; see Fig. 13.15. So even for widely overlapping validity functions, two important issues are in favor of local estimation (besides the significantly lower computational demand): the better interpretability and the robustness against noise.

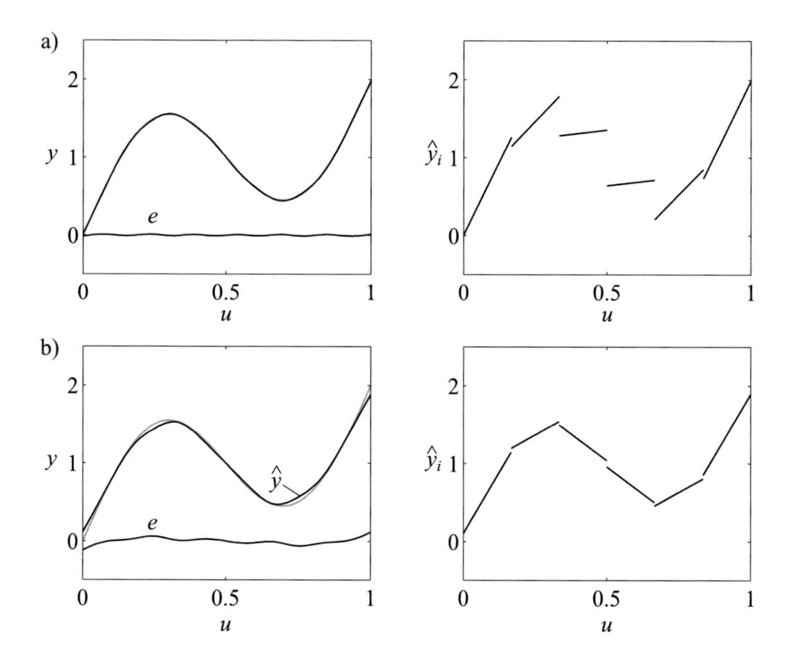

Fig. 13.14 The process and model output (left) and the local linear models for (**a**) global and (**b**) local estimation (right) with $\sigma_n = 0$ and $\sigma = 0.5/M$

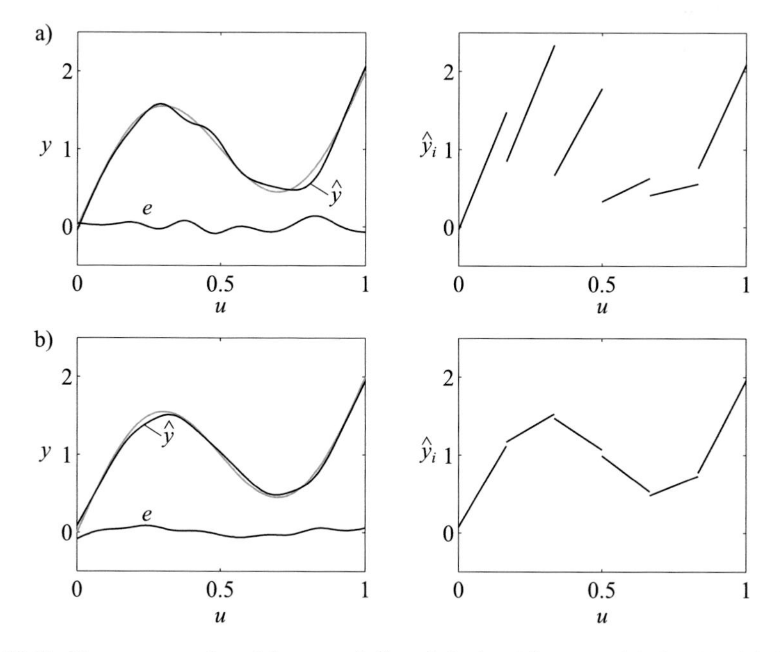

Fig. 13.15 The process and model output (left) and the local linear models for (**a**) global and (**b**) local estimation (right) with $\sigma_n = 0.3$ and $\sigma = 0.5/M$

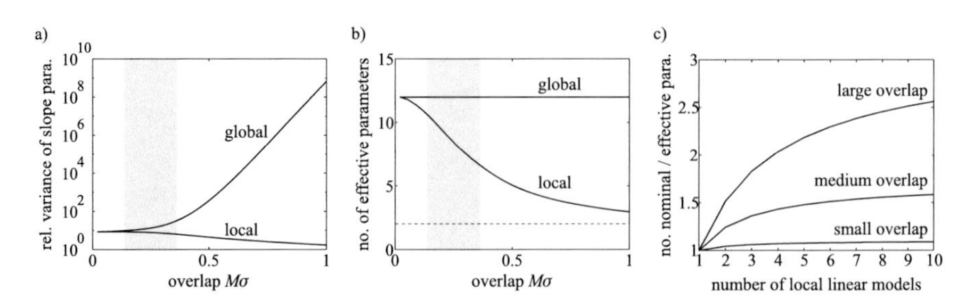

Fig. 13.16 Dependency of the different properties of global and local estimation on the overlap of the validity functions: (**a**) Parameter variance. (**b**) effective number of parameters. The gray-shaded area represents the extent of overlap that is typical for most applications. (**c**) Ratio of the nominal number of parameters to the effective number of parameters

In Fig. 13.16 the regularization effect of local estimation is illustrated. The variance of the globally estimated parameters increases sharply (note the logarithmic scaling) with the overlap of the validity functions; see Fig. 13.16a. In contrast, the locally estimated parameter variances even decrease because more data samples are considered with higher weighting factors in the weighted LS optimization (13.22). Figure 13.16b shows the effective number of parameters for different validity function overlaps. For global estimation it is constantly equal to 12, while for local estimation, it decreases from 12 ($\sigma \to 0$) to 2 ($\sigma \to \infty$).

Figure 13.16c shows the ratio of the nominal number of parameters to the effective number of parameters for local estimation in dependency of the model complexity. Obviously, the regularization effect increases with the model complexity. For most applications, medium overlaps are realistic, and thus as a rule of thumb, the effective number of parameters is 1.5 to 2 times less than their nominal number.

Clearly, the question arises: Which approach is better, global or local estimation? The answer depends on the specific problem under consideration, but some general guidelines can be given. In terms of computational demand, the local estimation approach is superior to the global one, and this advantage becomes stronger the more complex the neuro-fuzzy model is. In terms of performance, the global estimation approach is favorable with respect to the bias error. It exploits the full model flexibility. Thus, whenever a large amount of high-quality data is available, the global estimation approach should be utilized.

In many (if not most) practical situations, however, the variance error plays a dominant role since very flexible nonlinear models are actually over-parameterized. In these cases, local estimation can become superior because it possesses an inherent regularization effect. It is a nice feature of local estimation that the strength of the regularization effect is coupled with the overlap of the validity functions; see (13.28). Thus, the strength of the regularization effect is intuitively understandable. Alternatively, the global estimation approach can be combined with a user-defined regularization technique to obtain a better bias/variance tradeoff.

Finally, depending on the specific application, many different criteria besides the approximation accuracy may be important for a decision between the global and the local approach. Often the interpretability of the obtained local linear models is a highly relevant issue. As has been demonstrated in the above examples, local estimation is strongly advantageous in this respect.

In summary, local estimation seems to be superior to global estimation in most applications. The following benefits can be expected:

- *Fast training*: Owing to the significantly lower computational complexity of local estimation in (13.23) compared with (13.16), training becomes very fast. This advantage increases quadratically with the number of neurons.
- *Regularization effect*: The effective number of parameters with local estimation is decreased according to (13.28). The conditioning, i.e., the eigenvalue spread, of the Hessian of the loss function is smaller and thus the parameter variances are reduced. The bias error of the model is increased, but the variance error is decreased in comparison with global estimation. These properties are advantageous when the available training data is noisy and/or sparsely distributed, which is always the case for high-dimensional input spaces.
- *Interpretation*: The locally estimated parameters can be individually interpreted as a description of the identified process behavior in the regime represented by the corresponding validity function. The parameters of the local linear models are not sensitive with respect to the overlap of the validity functions. These properties easily allow one to gain insights into the process behavior. Globally estimated parameters can be interpreted only by taking the interaction with the neighboring models into account.
- *Online learning*: As will be shown in Sect. 14.6, the local estimation approaches offer considerable advantages when applied in a recursive algorithm for online learning. Besides the lower computational complexity and the improved numerical stability due to the better condition of the Hessian, local online learning allows one to solve the so-called stability/plasticity dilemma [91]; see Sect. 6.2.6.
- *Higher flexibility*: Local estimation permits a wide range of optimization approaches that are not feasible with the global estimation approach. For example, some local models may be linearly parameterized, while others are nonlinearly parameterized. Then local estimation allows one to apply a linear LS estimation to the first type of local model and nonlinear optimization to the latter one. Another example is the use of differently structured local linear models, as will be introduced in Sect. 14.3. Local estimation allows one to specify and realize individually desired local model complexities. Furthermore, different loss functions can be specified individually for each local model.

13.2.4 Robust Regression

As briefly discussed in Sect. 3.1.9, an outlier-insensitive alternative to LS for estimating the local models' parameters is robust regression. Instead of the "sum of squared errors" loss function, the "sum of absolute errors"

$$I = \sum_{i=1}^{N} |e(i)| \rightarrow \min . \tag{13.31}$$

or some combination of both like the Huber loss function (see Sect. 3.1.9) can be minimized. It can be distinguished whether (13.31) is utilized for optimization of:

- *Parameters of the local models*: This slows down the whole training algorithm significantly since iterative methods replace the one-shot LS solutions. However, these are convex optimization problems and can be solved reliably with around one order of magnitude higher computational demand than LS.
- *Structure of the rule premises*: Since for LOLIMOT this just involves a discrete selection of the worst LM to be split and the best split to be carried out, any loss function can be employed for these tasks without additional effort. Section 13.3.2 discussed this topic in some detail. For algorithms like HILOMOT (see Sect. 14.8) that continuously optimize the split, just the objective function for the split optimization changes – also without any significant additional effort.

An interesting and pragmatic choice would be to combine a standard LS approach to local model parameter estimation with a robust structure optimization at virtually no additional cost but with an improved robustness w.r.t. outliers.

13.2.5 Regularized Regression

As discussed in the Sects. 3.1.4 and 3.1.6, L_2-(ridge) or L_1-regularized versions of LS can be considered as sophisticated variants for local linear parameter estimation. As explained in Sect. 13.2.2, the local estimation approach already induces a significant amount of regularization. Thus, applying regularized estimation schemes on top of that may often yield too inflexible models.

However, there exist some cases where such an approach can be very beneficial:

- *Extremely high-dimensional problems*: If the number of local model variables in $\underline{x}$ is huge, i.e., $nx \approx N$ or even $nx > N$, then regularized regression is one possibility to still obtain reasonable results. Alternatively, one may think of some approach to subset selection; see Sect. 3.4.
- *Problematic data distribution*: It can happen, especially in case of strangely distributed data, that the data distribution within one or several partitions does not provide enough information to estimate some local model's slope in some direction. Such an example is illustrated in Fig. 13.17. In such a case, regularization may be necessary to stabilize some slope parameters of the local linear models. In the case of weighted LS for local model parameter estimation, as analyzed in Sect. 13.2.2, the data points outside the partition stabilize the estimate, and regularization is optional. In cases where only the data points inside

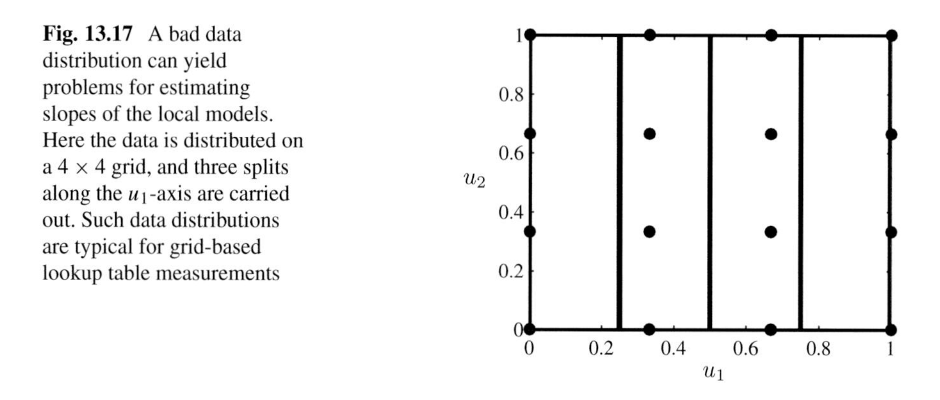

Fig. 13.17 A bad data distribution can yield problems for estimating slopes of the local models. Here the data is distributed on a 4×4 grid, and three splits along the u_1-axis are carried out. Such data distributions are typical for grid-based lookup table measurements

the partition are used for estimation, regularization is mandatory in situations as shown in Fig. 13.17.

It is possible to carry out separate bias/variance tradeoffs for each local model by adjusting individual regularization parameters. This allows to deal with different noise levels in different areas of the input space and/or different local model complexities (e.g., different polynomial degrees or dynamic orders). This is another attractive feature of the local estimation approach.

13.2.6 Data Weighting

In practice, the training data that is used for parameter optimization is often not well distributed. For example, the data may be distributed densely within a small operating region because the process has been operated at one set point for a long time. Since a sum-of-squared-errors loss function then accumulates a large number of model errors representing this operating region the parameters will be estimated in order to fit the process particularly well in this region. Thus, the model tends to specialize in the region in which the training data is most dense. Global model architectures are especially sensitive in this respect because all parameters can influence the model in the respective region and thus all parameters will be deteriorated. Local model architectures cope better with this difficulty; mainly the parameters of those local models that are significantly activated in the respective region, i.e., $\Phi_i \gg 0$, are affected.

Nevertheless, Fig. 13.18a demonstrates that some local linear models of the neuro-fuzzy model can degenerate significantly because in the white marked area the training data is very dense. The training data for this example consists of 100 equally distributed data samples in [0, 1], and an additional 100 data samples concentrated in the white marked interval. Six local linear models with equidistant validity functions are estimated. As a consequence of the unequal data distribution, the model accuracy is high in this white marked area and declines to the left and

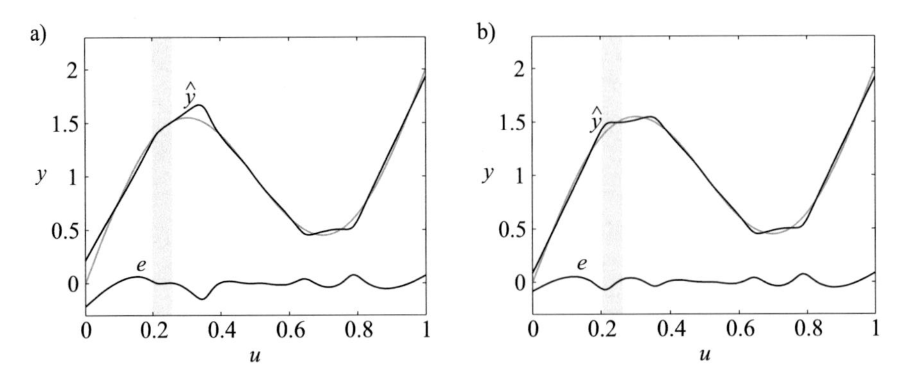

Fig. 13.18 Compensation of poorly distributed data by weighting: (**a**) many training data samples are distributed in [0.2, 0.25] (gray shaded area), (**b**) weighting of the data can compensate this effect

right. Two alternative strategies can reduce the difficulties occurring with poorly distributed data. One option is to discard so much data in the densely covered regions that the emphasis vanishes. This, of course, discards information that might be valuable in order to attenuate noise. The second and more sophisticated approach is to weight the data for the parameter estimation. This weighting can be chosen anti-proportional to the data density in order to compensate for the data distribution.[3] Figure 13.18b illustrates the effect of this weighting.

The data weighting matrix

$$\underline{R} = \begin{bmatrix} r_1 & 0 & \cdots & 0 \\ 0 & r_2 & \cdots & 0 \\ \vdots & \vdots & \ddots & \vdots \\ 0 & 0 & \cdots & r_N \end{bmatrix} \tag{13.32}$$

weights data sample i with r_i. The weighted global parameter estimate becomes

$$\underline{\hat{w}} = \left(\underline{X}^T \underline{R}\, \underline{X}\right)^{-1} \underline{X}^T \underline{R}\, \underline{y}\,. \tag{13.33}$$

In the weighted local parameter estimate, the combined weighting factor $\underline{Q}_i \underline{R}$ of the data weighting and the validity function values are taken into account:

$$\underline{\hat{w}}_i = \left(\underline{X}_i^T \underline{Q}_i \underline{R}\, \underline{X}_i\right)^{-1} \underline{X}_i^T \underline{Q}_i \underline{R}\, \underline{y}\,. \tag{13.34}$$

[3]Note that equal data distribution is not always desirable; see Sect. 14.9.

13.3 Structure Optimization of the Rule Premises

The linear parameter estimation approaches discussed in the previous section are based on the assumption that the validity functions have already been determined a priori. Basically, two strategies for determination of the validity functions can be distinguished: prior knowledge and structure identification from data. The two strategies can also be combined in various ways.

The number of validity functions and their parameters, the centers c_{ij} and the standard deviations σ_{ij}, define the *partitioning of the input space*. For example, the three validity functions in Fig. 13.6 partition the input space in the three rectangular regions depicted in Fig. 13.19. The extension to higher-dimensional input spaces is straightforward; the rectangles just become hyperrectangles. When normalized Gaussian validity functions are used, their centers c_{ij} determine the centers of the rectangles, and their standard deviations σ_{ij} determine the extensions of the rectangles in each dimension. A proportionality factor k_σ relates the standard deviations of the validity functions to the extensions of the rectangles by

$$\sigma_{ij} = k_\sigma \cdot \Delta_{ij} \,. \tag{13.35}$$

Via the link to Takagi-Sugeno fuzzy systems, the validity functions or the partitioning of the input space represents a specific rule premise structure with corresponding input membership functions; see Sect. 13.1.3. Thus, the determination of

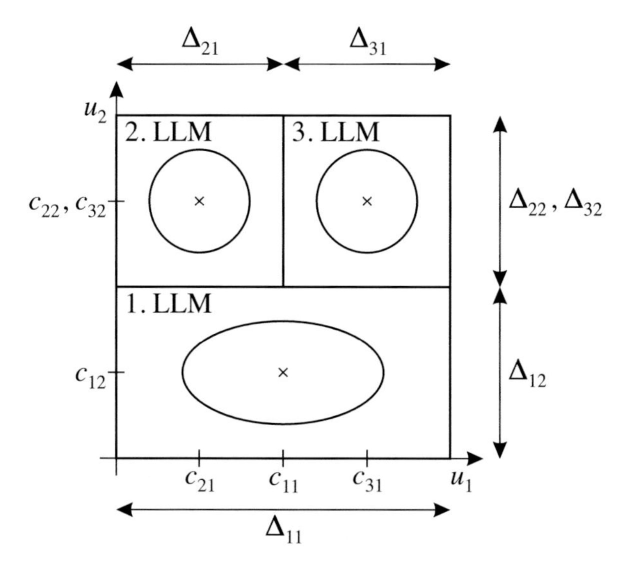

Fig. 13.19 A unique relationship between the input partitioning and the validity functions Φ_i can be defined. The circles and ellipses represent contour lines of the multidimensional membership functions μ_i (i.e., before normalization); see Fig. 13.6

the number and the parameters of the validity functions completely fix the number of fuzzy rules, their premise structures, and the input MSFs with all their parameters.

The identification of the validity functions' parameters is a nonlinear optimization problem. Basically, the following strategies exist for the determination of the parameters; see also [19]:

- *Grid partitioning*: The number of input MSFs per input is typically chosen by prior knowledge. This approach suffers severely from the curse of dimensionality. To weaken its sensitivity to the input space dimensionality, the grid can be reduced to the regions where enough data is available (see Sect. 11.3.3), or a multiresolution grid can be used [285]. All grid-based approaches are restricted to very low-dimensional problems and do not exploit the local complexity of the process.
- *Input space clustering*: The validity functions are placed according to the input data distribution [548]. Since the local process complexity is ignored, this simple approach usually does not perform well.
- *Nonlinear local optimization*: Originally, the input MSFs and the rule consequent parameters have been optimized simultaneously. The current state-of-the-art method, however, is to optimize the rule premise parameters by nonlinear local optimization and the rule consequent parameters by global least squares in a nested or staggered approach as in ANFIS (adaptive neuro-fuzzy inference system) [289, 290, 292]; see also Chap. 4. This approach is computationally expensive but typically yields very accurate results. However, a huge number of parameters are optimized, and overfitting often becomes a serious problem.
- *Orthogonal least squares*: As for singleton fuzzy systems, the OLS algorithm can be utilized for subset selection. However, the same severe restrictions apply owing to the normalization, which changes the regressors during their selection [328, 589]; see Sect. 12.4.3. Because of the normalization, the OLS cannot unfold its full efficiency, and thus this approach is computationally demanding. Furthermore, the fuzzy logic interpretation diminishes since the projection to univariate membership functions is not possible; see Sect. 12.3.4.
- *Genetic algorithms*: In order to circumvent the difficulties connected with the OLS algorithm, genetic algorithms can be applied for structure search [48, 556]. Evolutionary algorithms offer a wide spectrum of different approaches. All of them, however, suffer from extremely slow convergence compared with the alternative methods; see Chap. 5.
- *Product space clustering*: One of the most popular approaches applies the Gustafson-Kessel clustering algorithm to find hyperplanes in the product space, i.e., space spanned by $[u_1\ u_2\ \cdots\ u_p\ y]$; see Sect. 6.2.3 and [16]. More details on product space clustering approaches for the construction of local linear neuro-fuzzy models can be found in [16, 20, 21, 231, 335, 404, 447, 607]. Although product space clustering is a widely applied method, it suffers from a variety of drawbacks: (i) The computational effort grows strongly with the dimensionality of the problem. (ii) The number of clusters (i.e., rules) has to be fixed a priori. (iii) For an appropriate model interpretation in terms of fuzzy logic, the multivariate

fuzzy sets must be projected with accuracy losses onto univariate membership functions [16, 18]. (Note that the more flexible input space partitioning generated by product space clustering can also be seen as an advantage of the method as far as approximation capabilities are concerned. It can turn to a drawback only if a fuzzy logic interpretation with one-dimensional fuzzy sets is desired.) (iv) The rule premises and consequents must depend on the same variables. This is a severe restriction, which prevents many advanced concepts such as premise or consequent structure optimization or some simple ways to incorporate prior knowledge; see Sect. 14.1. (v) The local models are restricted to be linear.

- *Heuristic construction algorithms*: Probably the most widely applied class of algorithms increases the complexity of the local linear neuro-fuzzy model during training. They start with a coarse partitioning of the input space (typically with a single rule, i.e., a global linear model) and refine the model by increasing the resolution of the input space partitioning. These approaches can be distinguished into the very flexible strategies, which allow an (almost) arbitrary partitioning of the input space [399, 402] or the slightly less flexible axis-oblique decomposition strategies [137], on the one hand, and the axis-orthogonal strategies, which restrict the search to rectangular shapes [297, 298, 307, 549, 550, 552] on the other hand.

In the remaining part of this chapter, the local linear model tree (LOLIMOT) algorithm is introduced, extended, and analyzed. It is an axis-orthogoal heuristic, incremental construction algorithm and thus belongs to the last category. LOLIMOT is similar to the strategies in [297, 298, 549, 552] and utilizes some ideas from other tree-based structures as proposed in [74, 172, 275, 309, 438, 472, 495, 496]. The advantages and drawbacks of the LOLIMOT algorithm and its extensions over other approaches will be summarized in Sect. 13.4 after all features have been introduced.

13.3.1 Local Linear Model Tree (LOLIMOT) Algorithm

LOLIMOT is an incremental tree-construction algorithm that partitions the input space by axis-orthogonal splits. In each iteration, a new rule or local linear model (LLM) is added to the model. Thus, LOLIMOT belongs to the class of *incremental* or *growing* algorithms; see Sect. 7.4. It implements a heuristic search for the rule premise structure and avoids a time-consuming nonlinear optimization. In each iteration of the algorithm, the validity functions that correspond to the actual partitioning of the input space are computed, as demonstrated in Fig. 13.19, and the corresponding rule consequents are optimized by a local weighted least squares technique. The only “fiddle” parameter that has to be specified a priori by the user is the proportionality factor between the rectangles’ extension and the standard deviations. The optimal value depends on the specific application, but usually the following value gives good results:

$$k_\sigma = \frac{1}{3}\,. \tag{13.36}$$

This value is also chosen in Fig. 13.6. The standard deviations are calculated as follows:

$$\sigma_{ij} = k_\sigma \cdot \Delta_{ij} \tag{13.37}$$

where Δ_{ij} denotes the extension of the hyperrectangle of local model i in dimension u_j; see Fig. 13.19.

13.3.1.1 The LOLIMOT Algorithm

The LOLIMOT algorithm consists of an outer loop in which the rule premise structure is determined and a nested inner loop in which the rule consequent parameters are optimized by local estimation.

1. *Start with an initial model*: Construct the validity functions for the initially given input space partitioning and estimate the LLM parameters by the local weighted least squares algorithm. Set M to the initial number of LLMs. If no input space partitioning is available a priori, then set $M = 1$ and start with a single LLM, which in fact is a global linear model since its validity function covers the whole input space with $\Phi_1(\underline{u}) = 1$.
2. *Find worst LLM*: Calculate a local loss function for each of the $i = 1, \ldots, M$ LLMs. The local loss functions can be computed by weighting the squared model errors with the degree of validity of the corresponding local model according to (13.20):

$$I_i = \sum_{j=1}^{N} e^2(j)\Phi_i(\underline{u}(j))\,. \tag{13.38}$$

   Find the worst-performing LLM, that is, $\max_i(I_i)$, and denote l as the index of this worst LLM.
3. *Check all divisions*: The LLM l is considered for further refinement. The hyperrectangle of this LLM is split into two halves with an axis-orthogonal split. Divisions in all dimension are tried. For each division $dim = 1, \ldots, p$, the following steps are carried out:

   (a) construction of the multidimensional MSFs for both hyperrectangles;
   (b) construction of all validity functions[4];

[4]This step is necessary because all validity functions are changed slightly by the division as a consequence of the common normalization denominator in (13.4).

(c) local estimation of the rule consequent parameters for both newly generated LLMs;
(d) calculation of the loss function for the current overall model; see Sect. 13.3.2 for more details.

4. *Find best division*: The best of the p alternatives checked in Step 3 is selected. The validity functions constructed in Step 3 and the LLMs optimized in Step 3 are adopted for the model. The number of LLMs is incremented $M \rightarrow M + 1$.
5. *Test for convergence*: If the termination criterion is met then stop, else go to Step 2.

For the termination criterion, various options exist, e.g., a maximum model complexity, that is, a maximum number of LLMs, statistical validation tests, or information criteria. These alternatives are discussed in Sect. 7.3. Note that the number of *effective* parameters must be inserted in these termination criteria.

In Step 2 the local sum of squared errors loss function (13.38) and not their mean is utilized for the comparison between the LLMs. Consequently, LOLIMOT preferably splits LLMs that contain more data samples. Thus, the local model quality depends on the training data distribution. This consequence is desired because more data allows one to estimate more parameters.

Note that the parameter estimation in Step 3 can be performed reliably only if the number of considered data samples is equal to or higher than the number of estimated parameters. For validity functions *without* overlap, the following relationship must hold: $p + 1 \leq N^{\mathrm{loc}}$, where N^{loc} denotes the number of data samples within the activity region of the estimated local linear model. This condition represents the minimum number of data samples ensuring that the parameters can be estimated, i.e., $\underline{X}_i^T \underline{Q}_i \underline{X}_i$ in (13.22) is not singular. In the case of disturbances, more data is required to attenuate the noise. For validity functions *with* overlap, this condition can be generalized to

$$p + 1 \leq \sum_{j=1}^{N} \Phi_i(\underline{u}(j)), \tag{13.39}$$

where $\Phi_i(\underline{u})$ is the validity function associated with the estimated local linear model. It is possible to estimate the parameters with less data than in (13.39) owing to the regularization effect of the local estimation approach. Nevertheless, (13.39) gives a reasonable bound in practice.

Figure 13.20 illustrates the operation of the LOLIMOT algorithm in the first five iterations for a two-dimensional input space. In particular, two features make LOLIMOT extremely fast. First, at each iteration, not all possible LLMs are considered for division. Rather, Step 2 selects only the worst LLM whose division most likely yields the highest performance gain. For example, in iteration 4 in Fig. 13.20, only LLM 4-4 is considered for further refinement. All other LLMs are kept fixed. Second, in Step 3 the local estimation approach allows one to estimate only the parameters of those two LLMs that are newly generated by the division. For

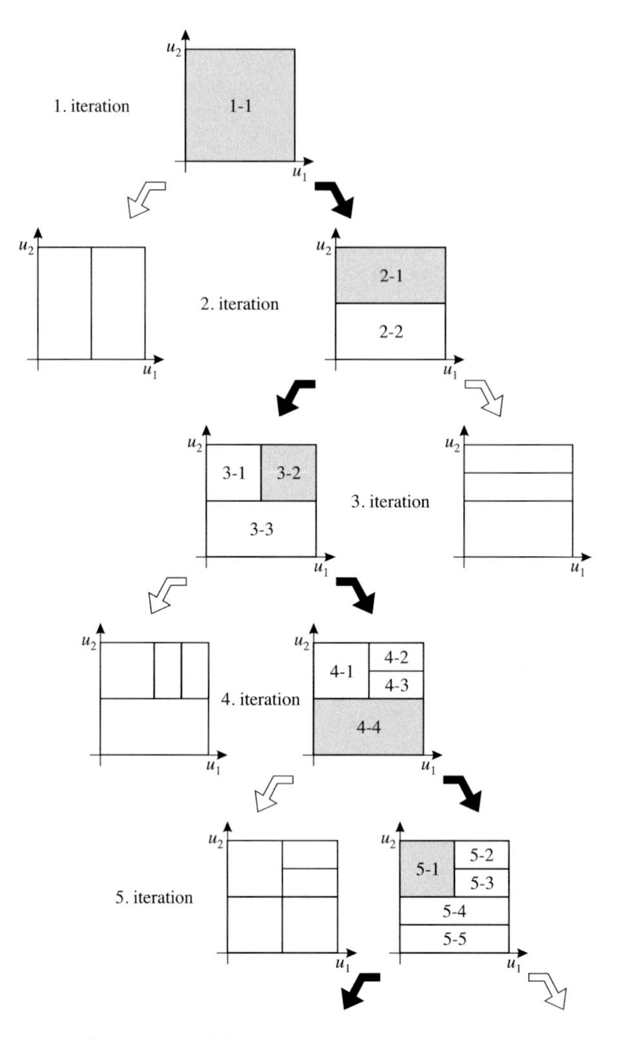

Fig. 13.20 Operation of the LOLIMOT structure search algorithm in the first five iterations for a two-dimensional input space ($p = 2$)

example, when in iteration 4 in Fig. 13.20, the LLM 4-4 is divided into LLM 5-4 and 5-5, the LLMs 4-1, 4-2, and 4-3 can be directly passed to the LLMs 5-1, 5-2, and 5-3 in the next iteration without any estimation.

13.3.1.2 Computational Complexity

The computational complexity of LOLIMOT can be assessed as follows. In each iteration the worst performing LLM is divided into two halves along p different dimensions. Thus, in each iteration the parameters of $2p$ LLMs have to be estimated, which gives the following computational complexity (see (13.23))

$$\mathcal{O}\left(2p(p+1)^3\right) \approx \mathcal{O}\left(2p^4\right) . \tag{13.40}$$

It is very remarkable that (13.40) does not depend on the iteration number which is a consequence of the local estimation approach. This means that LOLIMOT virtually[5] does not slow down during training. As a consequence, even extremely complex models can be constructed efficiently. For the whole LOLIMOT algorithm, M iterations have to be performed, i.e.,

$$\mathcal{O}\left(2Mp(p+1)^3\right) \approx \mathcal{O}\left(2Mp^4\right) . \tag{13.41}$$

Hence, the computational demand grows only linearly with the model complexity, that is, with the number of LLMs. This is exceptionally fast. Furthermore, the computational demand does not grow exponentially with the input dimensionality, which avoids the curse of dimensionality.

In [297, 298] a more general class of construction algorithms is proposed. Instead of selecting the best alternative under all divisions in Step 4, the consequences of one division for the next iteration(s) can be examined. LOLIMOT can be seen as a one-step-ahead optimal strategy, i.e., only the improvement from one iteration to the next iteration is considered. More generally, a k-step-ahead optimal strategy can be applied, where the division is carried out that yields the best overall model after the next k iterations. If the final model should possess M rules, the optimal strategy would be to look $k = M$ steps ahead. This, however, would require one to consider and optimize all possible model structures with M rules. As pointed out in [297, 298], the computational complexity grows exponentially with an increasing prediction horizon k. Therefore, only values of $k \leq 3$ are realistically feasible. The philosophy pursued here focuses on approaches that allow fast training, and therefore LOLIMOT implements $k = 1$. The price to be paid is possibly a suboptimal model structure, which implies that more local linear models are needed for the same approximation accuracy. This is an example of a general tradeoff between the computational demand and the amount of information compression.

Example 13.3.1 (Hyperbola) Figure 13.21 illustrates the operation of LOLIMOT for the test process introduced in Sect. 9.4. The function possesses a strong nonlinearity (high curvature) in regions close to $u = 0$ and becomes more linear when approaching the $u = 1$. Thus, it can be expected that more LLMs are required the smaller u is. Indeed, LOLIMOT constructs a local linear neuro-fuzzy model with

[5]Some slow-down effect can be observed because the loss function evaluations in (13.38) and Step 3 becomes more involved as the number of LLMs increases. However, for most applications, the computational demand is dominated by the parameter optimizations, and this slow-down effect can be neglected.

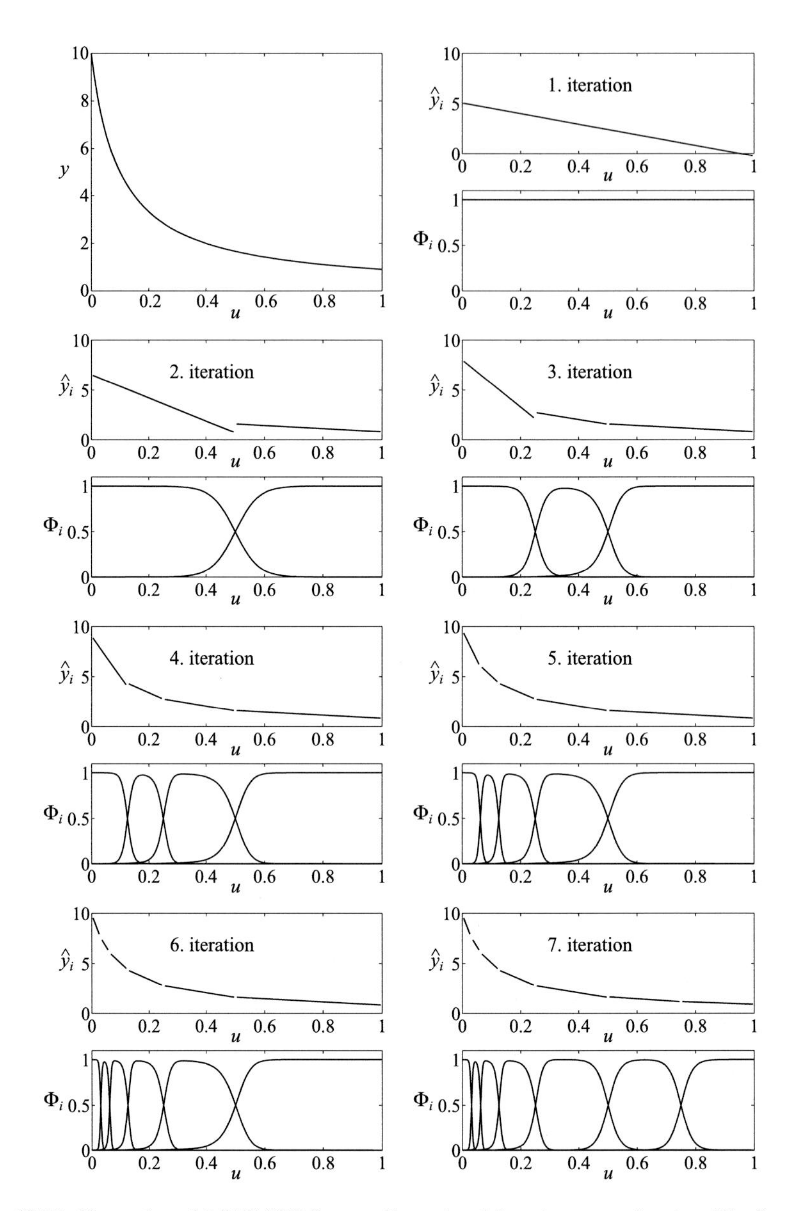

Fig. 13.21 Illustration of LOLIMOT for one-dimensional function approximation. The function (top, left) is approximated by a local linear neuro-fuzzy model constructed with LOLIMOT. The first seven iterations are shown

this property. In the first iteration, a linear model is fitted to the data. Its validity function is constantly equal to 1; so this LLM is a global linear model. In the second iteration, LOLIMOT has no alternative but to split this LLM into two halves because this is a univariate example ($p = 1$). In the third iteration, the left LLM is divided since it describes the process worse than the right LLM. Up to the sixth iteration, always the leftmost LLM is further divided since the local model error is highest in this region. However, in the seventh iteration, the approximation for small u has become so good that a division of the rightmost LLM yields the highest performance improvement.

It is important to note that in each iteration *all*, LLMs are considered for further refinement. So the algorithm is *complexity adaptive* in the sense that it always constructs new LLMs where they are needed most. In correlation with the growing density of the local models, the widths of the validity functions are reduced because they depend on the hyperrectangles' (here: intervals') extensions according to (13.37). Hence the resolution can vary from arbitrary coarse to arbitrary fine. This automatic complexity adaptation of LOLIMOT ensures a good bias/variance tradeoff since additional parameters (of newly generated LLMs) are estimated from data only if the expected reduction in the bias error is large and overcompensates the increase in the variance error. The complexity adaptivity is a good mechanism in order to overcome or at least weaken the *curse of dimensionality*. As pointed out in Sect. 7.6.1, a crucial issue for reducing the sensitivity of an algorithm with respect to the input space dimensionality is to distribute the complexity of the model according to the complexity of the process. LOLIMOT meets this requirement. □

13.3.1.3 Two Dimensions

Figure 13.22 demonstrates the first three iterations of LOLIMOT for the two-dimensional example from Fig. 13.4. It is shown how the multidimensional Gaussian MSFs μ_i develop (first row) and which normalized Gaussian validity functions Φ_i are obtained (second row). The overall model output (last row) is computed by weighting the LLMs (third row) with their validity functions and summing up their contributions.

13.3.1.4 Convergence Behavior

Since LOLIMOT is a growing algorithm, it automatically increases the number of rules steadily. Thus, if a neuro-fuzzy model with M rules is trained in M iterations, all models with $1, 2, \ldots, M - 1$ rules are also identified during the training procedure. The convergence curves as shown in Fig. 13.23a reveal useful information about the training procedure. They easily allow one to choose the optimal, or at least a good, model complexity, which is a hard task for standard neural networks. This can either be done directly by the user or it can be automated by the use of information criteria such as the corrected Akaike's information

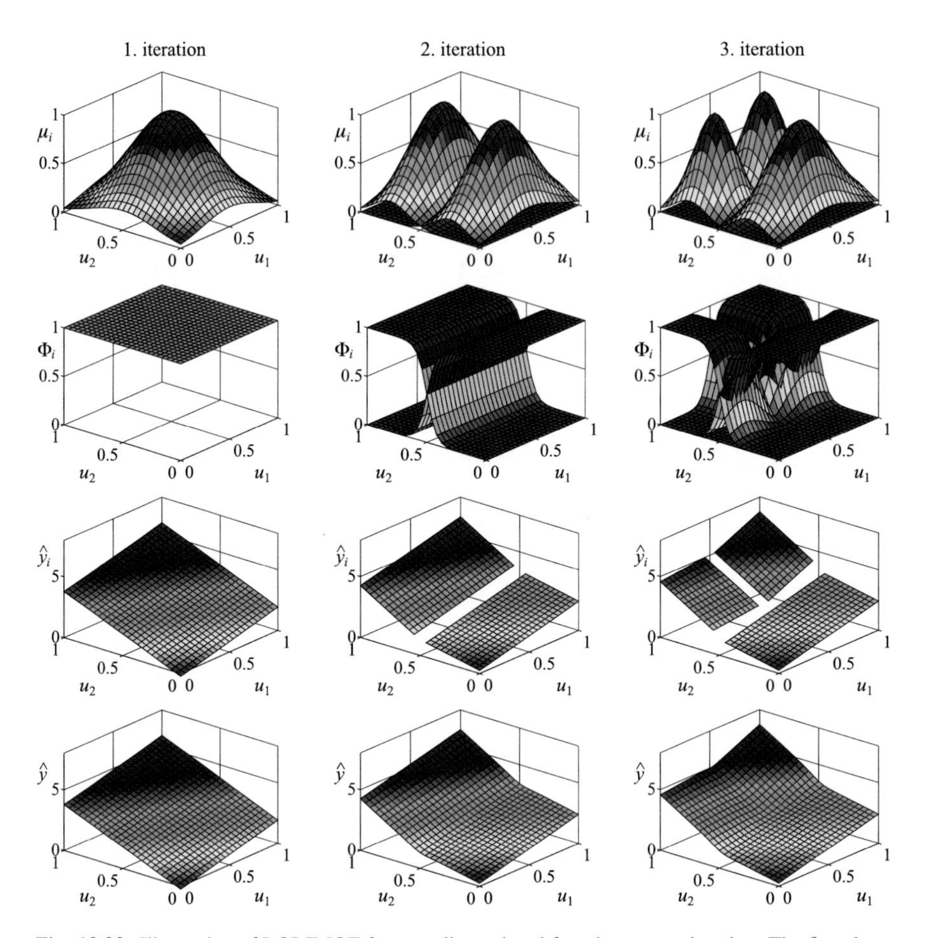

Fig. 13.22 Illustration of LOLIMOT for two-dimensional function approximation. The first three iterations are shown. First row: multidimensional Gaussian MSFs μ_i. Second row: validity functions Φ_i. Third row: local linear models. Fourth row: overall model output

criterion (AIC_C); see Sect. 7.3.3. Note, however, that then the *effective* number of parameters in (13.27) or (13.28) must be utilized in the information criteria. The convergence curves in Fig. 13.23a are obtained with the function approximation example in Fig. 13.21. The output is disturbed by additive white Gaussian noise with the standard deviations $\sigma_n = 0$, $\sigma_n = 0.01$, and $\sigma_n = 0.1$. While for undisturbed data, LOLIMOT can construct models with virtually[6] arbitrarily good approximation capabilities, for noisy data, the convergence curves saturate. For the noise level $\sigma_n = 0.01$, no significant improvement can be achieved for more than

[6] At some point the number of training data samples will not suffice to estimate the parameters. Note that because of the regularization effect of the local estimation, this point is far beyond $M = 150$ where the nominal number of parameters is equal to the number of data samples.

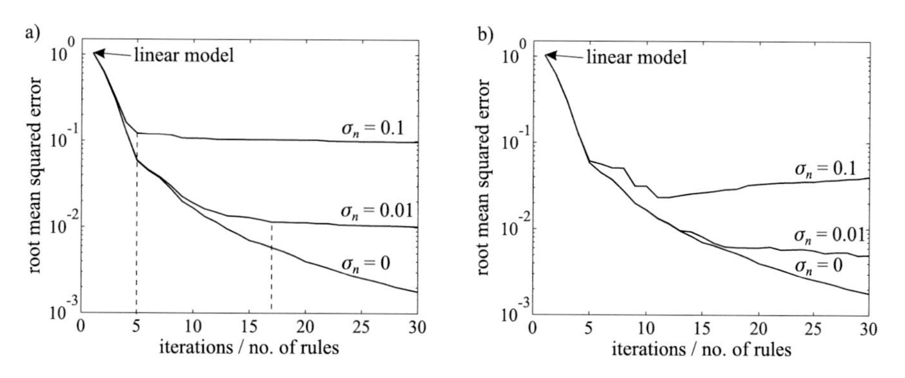

Fig. 13.23 Convergence behavior of LOLIMOT on (**a**) noisy training data and (**b**) noise-free test data. The root mean squared error is shown in dependency on the iterations of LOLIMOT, that is, the number of rules, neurons, or LLMs. The convergence curves are obtained with 300 equally distributed training data samples generated by the function in Fig. 13.21 with three different noise levels

$M = 17$ rules, and for $\sigma_n = 0.1$, more than $M = 5$ rules are not advantageous. Following the parsimony principle [356], these model complexities should be chosen.

Interestingly, the root mean squared errors achieved for the cases $\sigma_n = 0.01$ and $\sigma_n = 0.1$ are not much higher than the standard deviations of the noise. Hence, the bias error is already relatively small. Figure 13.23b depicts the convergence curves for the model error on the test data in contrast to the training data error in Fig. 13.23a. Note that the test data is chosen *noise-free* in order to represent the true behavior of the function. It reveals a small overfitting effect for the case $\sigma_n = 0.1$. Nevertheless, for all cases the test error is equal to[7] or smaller than the training data error. This highly unusual behavior is due to the obviously very advantageous regularization effect. Thus, choosing the models with $M = 5$ and $M = 17$, respectively, where the training curve optically saturates is conservative with respect to a possible overfitting effect. The actually optimal model complexities (which, of course, would not be known in a real problem) are at $M = 11$ and $M = 29$, respectively (Fig. 13.23b). The optimal model complexity for the case $\sigma_n = 0$ lies beyond $M > 150$.

13.3.1.5 AIC$_\mathrm{C}$

Figure 13.24 shows the convergence behavior on the AIC$_\mathrm{C}$ for high and low noise levels. As can be observed in Fig. 13.23a, the more disturbed the training data is, the less performance improvement can be achieved by the model. Thus, the

[7] Both curves for $\sigma_n = 0$ are, of course, identical since the training data is equal to the true process behavior.

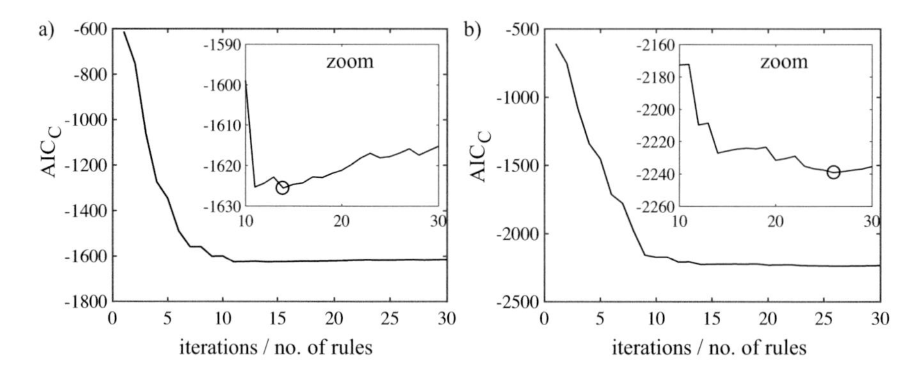

Fig. 13.24 AIC_C convergence behavior of LOLIMOT on training data with (**a**) high noise with a minimum at $M = 13$ and (**b**) low noise with a minimum at $M = 27$. The minimum is marked by a circle

penalty term for the effective number of parameters in the AIC_C lets the criterion rise at lower model complexities, here 13 local models compared to 27. However, the error and AIC_C convergence curves typically are not very smooth but rather relatively rugged. The effect is emphasized by choosing a low smoothness value. One additional reason for this phenomenon is the hard decisions taken in the partitioning process: Which local model to split? Which split is best? Therefore it is advisable to NOT stop the training iterations immediately after the AIC_C starts increasing. Rather utilizing a more robust search technique is recommended like terminating the splitting process after the AIC_C successively increased a couple of times.

13.3.2 Different Objectives for Structure and Parameter Optimization

The LOLIMOT algorithm utilizes a loss function of the type sum of weighted squared errors in order to estimate the LLM parameters. *Parameter optimization* should be based on such a loss function to be able to exploit all advantages of the linear weighted least squares method. In (13.38), which is used in Step 2 and in Step 3, however, arbitrary measures of the model's performance can be employed. These performance measures determine the criteria for *structure optimization*; special loss function types do not offer any advantage here. Consequently, the loss function utilized for structure optimization should reflect the actual objective of the user. This combination of different objectives for parameter and structure optimization is a major advantage of the LOLIMOT algorithm.

For example, constraints on the model output can be easily incorporated into the identification procedure without influencing the computational complexity (in contrast to a possible application of a constraint optimization technique). Figure 13.25a shows the function from Fig. 13.21 that is to be approximated under the constraint

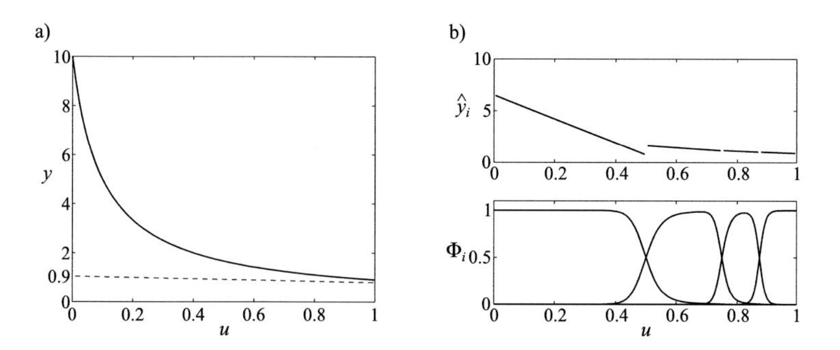

Fig. 13.25 (**a**) Function to be approximated under the constraint $\hat{y} > 0.9$. (**b**) To meet this constraint, the input space decomposition must concentrate on the regions of large inputs u first

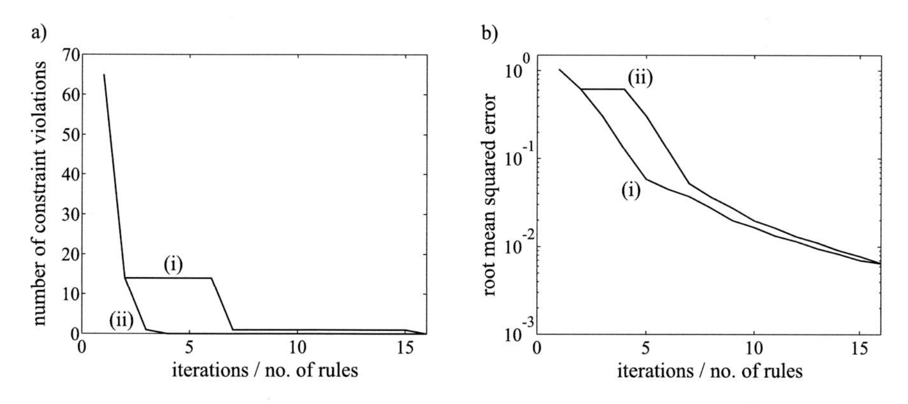

Fig. 13.26 Convergence curves for (**a**) the number of constraint violations and (**b**) root mean squared error. Algorithm (i) is identical to the one used in Fig. 13.21, that is, a sum of squared errors loss function is used for parameter and structure optimization. Algorithm (ii) utilizes the same loss function for parameter optimization but (13.43) for structure optimization. Algorithm (i) requires 16 rules to meet the constraint, while algorithm (ii) needs only 4 rules. The price to be paid is a slower improvement in the model error. Note that iteration 1 and 2 are identical for both algorithms because LOLIMOT offers no degrees of freedom in structure search for one-dimensional problems in the first two iterations

$\hat{y} > 0.9$, which is indicated by the dashed line. The loss function for structure optimization can be defined as

$$I_i^{(\text{constr})} = \sum_{j=1}^{N} \text{constr}_j \cdot \Phi_i(\underline{u}(j)) \, , \tag{13.42}$$

where $\text{constr}_j = 1$ when the model violates the constraint for data sample j and $\text{constr}_j = 0$ otherwise.

With this criterion, LOLIMOT discovers after the second iteration that the right LLM, although its modeling error is very small, is responsible for more constraint violations than the left one. Thus, in contrast to the results in Fig. 13.21, LOLIMOT divides the right LLM. The same happens in the third iteration. The local linear neuro-fuzzy model with four rules is the first that fully meets the constraint; see Fig. 13.25b. This can also be observed from the convergence curve in Fig. 13.26. If

LOLIMOT is run further, then in the subsequent iterations, only divisions of the left LLMs are carried out to improve the model quality. In order to enable LOLIMOT to discover which divisions are the best when (13.42) is equal to zero, the structure optimization criterion should be modified to

$$I_i^{(\text{structure})} = I_i^{(\text{constr})} + \alpha \sum_{j=1}^{N} e^2(j) \Phi_i(\underline{u}(j)), \tag{13.43}$$

where α is a small constant that allows a tradeoff between the importance of the constraint and model quality.

Note that the approach described above to constraint optimization is suboptimal because the rule consequent parameters are still estimated by LS ignoring the constraints. However, in many applications, the proposed approach might be sufficient. Even, because of the suboptimality, a model with higher complexity is required, it may be identified faster than with the application of constraint parameter optimization techniques.

A different loss function for structure optimization can also be utilized in order to meet a given tolerance band with the model; refer to the approximation of a driving cycle in Sect. 24.1. Another application is to prevent overfitting by the use of an information criterion objective or the use of validation data for structure optimization. Furthermore, the different objectives strategy is very useful in the context of dynamic systems; see Sect. 22.1.

13.3.3 Smoothness Optimization

The only "fiddle" parameter that has to be chosen by the user before starting the LOLIMOT algorithm is the smoothness k_σ of the model's validity functions; see (13.36). It is important to understand the influence of this parameter on the model quality. From Fig. 13.4, it can be assumed that for local linear neuro-fuzzy models, the smoothness of the validity functions is not crucial in terms of the model performance. This is a very appealing observation since it suggests that a rough choice of a reasonable value for k_σ is sufficient. Indeed, the following discussion demonstrates that an optimization of the model's smoothness is not necessary.

Figure 13.27a shows the approximation of a sine-like function by a local linear neuro-fuzzy model with $M = 8$ rules constructed by LOLIMOT. The smoothness parameter was a priori chosen as $k_\sigma = 0.33$. A subsequent univariate nonlinear optimization of k_σ with a nested re-computation of the validity functions and a local linear estimation of all rule consequent parameters yields the optimal smoothness parameter $k_\sigma^{(\text{opt})} = 0.53$. The root mean squared error of the model drops from 0.18 to 0.13; see Fig. 13.27b. The same performance gain can be achieved if a model with ten instead of eight rules is used without any nonlinear optimization. For some specific applications where only very little, noisy data is available, the smoothness

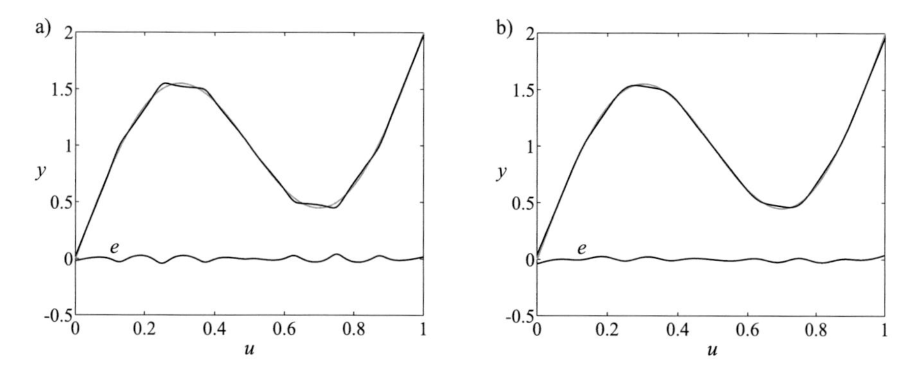

Fig. 13.27 Smoothness optimization: (**a**) before optimization, (**b**) after optimization

optimization may be worth the additional effort. In most cases, it is more efficient to accept slightly more rules and avoid the nonlinear optimization of k_σ. Of course, this tradeoff between model complexity and training time is problem-dependent.

Interestingly, when *global* estimation of the LLM parameters is applied, the optimal smoothness parameters typically tend to be much larger. In the above example shown in Fig. 13.27, the optimal k_σ is more than ten times larger than for the local estimation approach. This means that locality and interpretability are completely lost. In most applications, the optimal smoothness for global parameter estimation even tends to infinity. The reason for this strange effect is that widely overlapping validity functions make the model more flexible because all "local" (actually they are not local any more) models contribute everywhere to the model output. Thus, smoothness optimization is not recommended in connection with global estimation.

With local estimation, the smoothness optimization also does not work satisfactorily in most cases. The example in Fig. 13.27 is especially well suited because its sine-like form highly favors a certain smoothness factor. For other more monotonic functions, the optimal smoothness parameter is very close to zero when local estimation is applied. The reason is simply that a small value for k_σ also minimizes the error caused by neglecting the overlap between the validity functions. However, even if optimal performance is achieved by $k_\sigma \to 0$, such small values should not be realized in practice where the model is usually required to be smooth.

Considering the low sensitivity of the model quality on the smoothness and all the difficulties arising during nonlinear optimization of the smoothness parameter k_σ, it can be generally recommended to fix k_σ a priori by the rule of thumb (13.36).

13.3.4 Splitting Ratio Optimization

The question arises: Why should LOLIMOT split the local linear models into two equal *halves*? What improvement can be expected if various different splitting ratios such as $r_c = 1/3,\ 2/3$ or $r_c = 1/4,\ 2/4,\ 3/4$ are considered instead of checking only $r_c = 1/2$? Many tree construction algorithms such as CART even *optimize* the splitting ratio r_c [74]. One justification of the split into two equal halves is that if a sufficiently large number of local models are generated, any other splitting ratio can be approximated arbitrarily well. However, the number of required rules may be large; so a splitting ratio of 1/2 may be very suboptimal.

The optimization of r_c demands the solution of a univariate nonlinear optimization problem with two nested local LS parameter estimations of the newly generated LLMs. Typically, about ten loss function evaluations are required in order to optimize r_c. Consequently, LOLIMOT's computational demand is about ten times higher compared with the utilization of $r_c = 1/2$. Figure 13.28a demonstrates the benefit that can be expected from the splitting ratio optimization. The optimal splitting ratio in the first division is $r_c^{(\mathrm{opt})} = 0.22$. Compared with the standard LOLIMOT approach with splitting ratio $r_c = 0.5$, the root mean squared model error is reduced by a factor of 3! This improvement is remarkably large because the approximated function is strongly nonlinear. When models with more rules are applied, however, this improvement fades. Figure 13.29a depicts the convergence curves for neuro-fuzzy models with up to ten rules. This clearly demonstrates that the improvement realized by the approach with a nonlinear optimized splitting ratio vanishes as the model becomes more complex. The reason for this effect is that, as the model becomes more complex, the local linear models describe smaller operating regions of the process. The smaller these regions get,

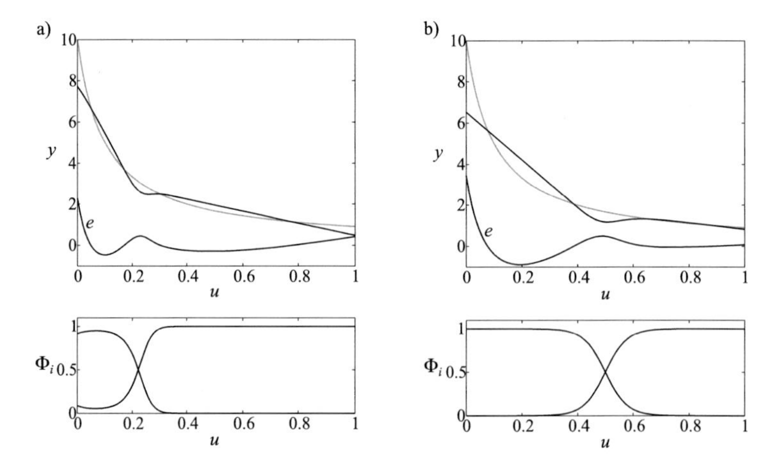

Fig. 13.28 Model quality and the corresponding validity functions with (**a**) the optimized splitting ratio $r_c^{(\mathrm{opt})} = 0.22$ and (**b**) the heuristically chosen splitting ratio $r_c = 0.5$

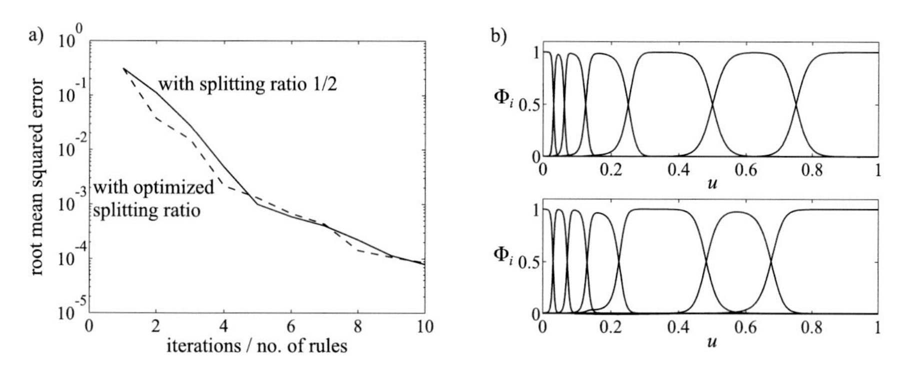

Fig. 13.29 (**a**) Convergence curves with and without splitting ratio optimization. (**b**) Validity functions without (top) and with (bottom) splitting ratio optimization

the weaker the nonlinear behavior becomes in these regions. Consequently, the expected improvement of the splitting ratio optimization reduces. Both models with and without splitting ratio optimization do not differ significantly, neither in model quality nor in the validity functions. This argumentation is underlined by the comparison of the validity functions in Fig. 13.29b, which indeed are quite similar.

The only way to achieve sustainable improvement is to optimize all splitting ratios simultaneously. This is equivalent to a nonlinear optimization of *all* input MSF centers. Only with such an approach can all splitting ratios be kept variable during the whole learning procedure and their interaction be taken into account by the optimization technique. A similar approach is proposed by Jang [289, 290, 292], but for a neuro-fuzzy model of fixed complexity. The difficulty with such an approach is the immense computational demand, which usually does not allow the application of a model complexity search in a higher level like LOLIMOT performs it.

Besides the high computational demand, a splitting ratio optimization possesses another drawback. As Fig. 13.28a shows, the left validity function starts to decrease again for $u \to 0$. This is an undesirable *normalization side effect*; see Sect. 12.3.4. It is caused by the extremely different (about a factor of 5) standard deviations of both validity functions. Thus, for small inputs $u < 0.1$, the right validity function starts to reactivate. Although this is no big problem in the above example, it can become more severe if even more extreme splitting ratios are used. If splitting ratios or centers (and possibly widths) of input MSFs are optimized, possible normalization side effects must be carefully taken into account. Note that the only reliable way to avoid normalization side effects is to choose identical widths for all validity functions. However, as Fig. 13.30a illustrates, resolution adaptive standard deviations are required in order to obtain reasonable results. Identical widths can lead to deformed validity functions with too small or too large smoothness; see Fig. 13.30b.

In summary, the larger computational effort and the difficulties with normalization side effects make the benefits of a splitting ratio optimization questionable.

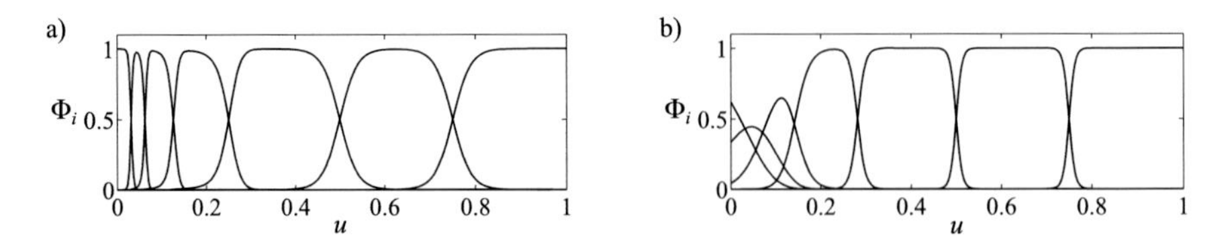

Fig. 13.30 Validity functions (**a**) with and (**b**) without resolution adaptive widths

Thus, the heuristically chosen division into two equal halves, that is, $r_c = 1/2$, is justified. Typically, a performance comparable to splitting ratio optimization can be achieved without this optimization by the use of a few more local linear models. Nevertheless, splitting ratio optimization may be beneficial in some particular cases, e.g., if one is limited to a simple model with very few rules.

13.3.5 Merging of Local Models

LOLIMOT belongs to the class of *growing* strategies (see Sect. 7.4) because it incorporates an additional rule in each iteration of the algorithm. During the training procedure, some of the formerly made divisions may become suboptimal or even superfluous. However, LOLIMOT does not allow one to undo these divisions. By extending LOLIMOT with a *pruning* strategy, which is able to merge formerly divided local linear models, this drawback can be remedied. Pruning algorithms are typically applied at the end of a growing phase or after a prior choice for an overparameterized model structure in order to remove non-significant model components [74, 172, 438, 479]. These approaches possess the disadvantage that they first build an unnecessarily complex model, which generally is quite time-consuming. In contrast, LOLIMOT favorably allows one to incorporate a possible pruning step within each iteration of the growing algorithm, thereby avoiding unnecessarily complex model structures.

Besides the more sophisticated complexity control, pruning possesses another feature. With subsequent mergers and splits, various splitting ratios r_c can be generated. Thus, similar effects as with a splitting ratio optimization can be achieved. As a consequence, however, the same difficulties can arise with respect to the normalization side effects.

It is proposed to include an additional step between Steps 4 and 5 into the LOLIMOT algorithm presented in Sect. 13.3.1. In this step the following tasks are carried out:

1. Find all LLMs that can be merged.
2. For all these LLMs, perform the following steps.

    (a) construction of the multidimensional MSF in the merged hyperrectangle;
    (b) construction of all validity functions;

   (c) local estimation of the rule consequent parameters for the merged LLM;
   (d) calculation of the loss function for the current overall model.

3. Find the best merging possibility.
4. Perform this merger if it yields an improvement compared with the model in the previous iteration with the same number of rules, i.e., one rule less than before the merger.

Step 1 is easy for a one-dimensional input space since simply all neighbored LLMs can be merged. In higher-dimensional spaces additionally, the (hyper)rectangles that two adjoined LLMs share at their common boundary must possess the same extensions in order to guarantee that a merger of the two LLMs can be described by a single (hyper)rectangle. For example, in the model obtained after the fifth iteration of LOLIMOT in Fig. 13.20, only LLMs 5-2 and 5-3 or LLMs 5-4 and 5-5 can be merged.

The computational demand of this merging procedure is low since again the local estimation approach can be exploited. However, the software implementation becomes relatively involved, especially since care must be taken in order to avoid an infinite loop caused by cycles. The easiest way to avoid cycles is to store the complete history of selected partitions and to prohibit divisions and mergers that would yield a previously selected structure. These issues require further research in the future.

Figure 13.31 compares the standard LOLIMOT algorithm (a) with the extended version (b) that contains the merging capabilities described above. The pruning strategy allows one to merge LLMs where they are not required and thus more LLMs are available in the important region around $u \approx 0$. Consequently, a performance increase of more than a factor of the two is possible.

The convergence curves in Fig. 13.31 compare both learning procedures. Five times two local linear models are merged, and new divisions improve the performance significantly. Thus 20 instead of 10 iterations are carried out (five merging and five additional division steps), which requires more than about twice the computation time. A comparable model error can be achieved when the standard LOLIMOT algorithm constructs a model with 12 rules, which requires only a 20% higher computational demand. Therefore, merging is an attractive extension only if (i) the training time is not important, (ii) the primary objective is a low model complexity, or (iii) the data is very noisy and/or sparse and thus overfitting problems are severe.

Note that this example is especially well suited for a demonstration of the merging capabilities. In the experience of the author, the merging option, similarly to the splitting ratio optimization, typically does not improve the performance significantly. For the example in Figs. 13.28 and 13.29, the behaviors of merging and splitting ratio optimization are very similar. For a more detailed investigation on merging of local models allowing for more flexible partitioning structures (e.g., L- and U-shapes), refer to [160].

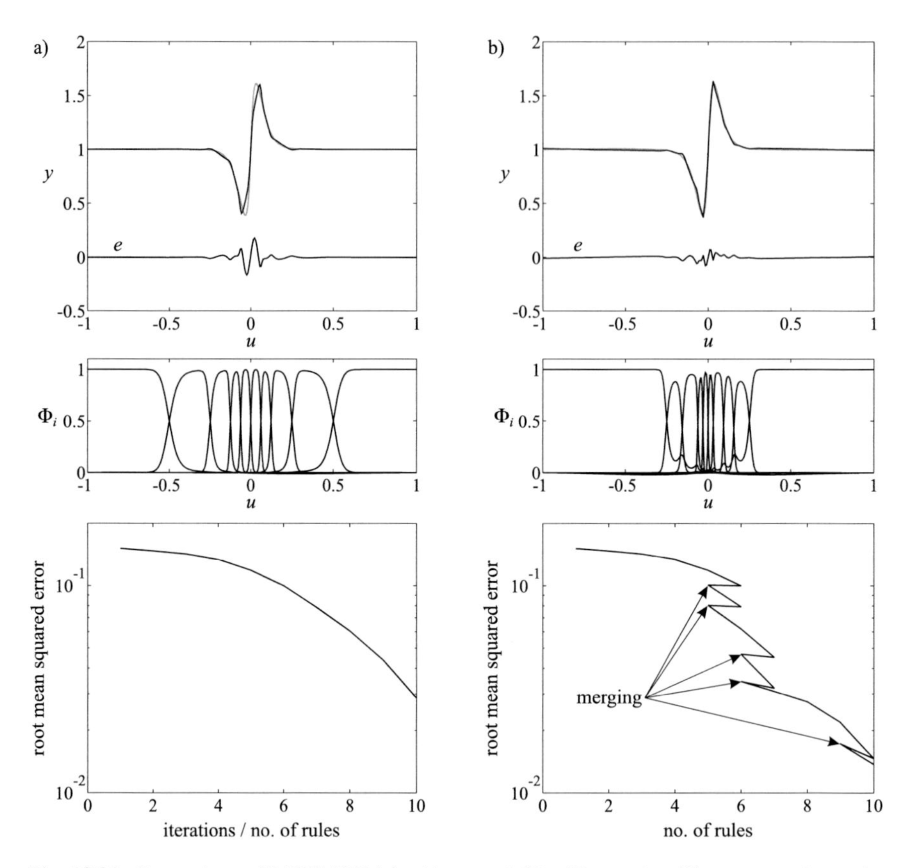

Fig. 13.31 Comparison of LOLIMOT (**a**) without and (**b**) with merging. The top row shows the function to be approximated, the model outputs, and the model errors. The middle row shows the validity functions of the final models. The bottom row shows the convergence curves

13.3.6 Principal Component Analysis for Preprocessing

One of the most severe restrictions of LOLIMOT is the axis-orthogonal partitioning of the input space. This restriction is crucial for interpretation as a Takagi-Sugeno fuzzy system and for the development of an extremely efficient construction algorithm. The limitations caused by the axis-orthogonal partitioning are often less severe than assumed at first glance because the premise input space dimensionality may be reduced significantly; see Sect. 14.1. Nevertheless, the higher dimensional the problem is, the more this restriction limits the performance of LOLIMOT.

Basically, there are two ways to solve or weaken this problem. Both alternatives diminish much of the strengths of LOLIMOT in interpretability. On the one hand, an unsupervised preprocessing phase can be included. On the other hand, an axis-oblique decomposition algorithm can be developed. The first alternative is computationally cheap and easy to implement and will be considered in this subsection. An outlook on the second alternative, which is much more universal and powerful but also computationally more expensive, is given in Sect. 14.7.

Principal component analysis (PCA) is an unsupervised learning tool for preprocessing that performs a transformation of the axes; refer to Sect. 6.1. Figure 13.32

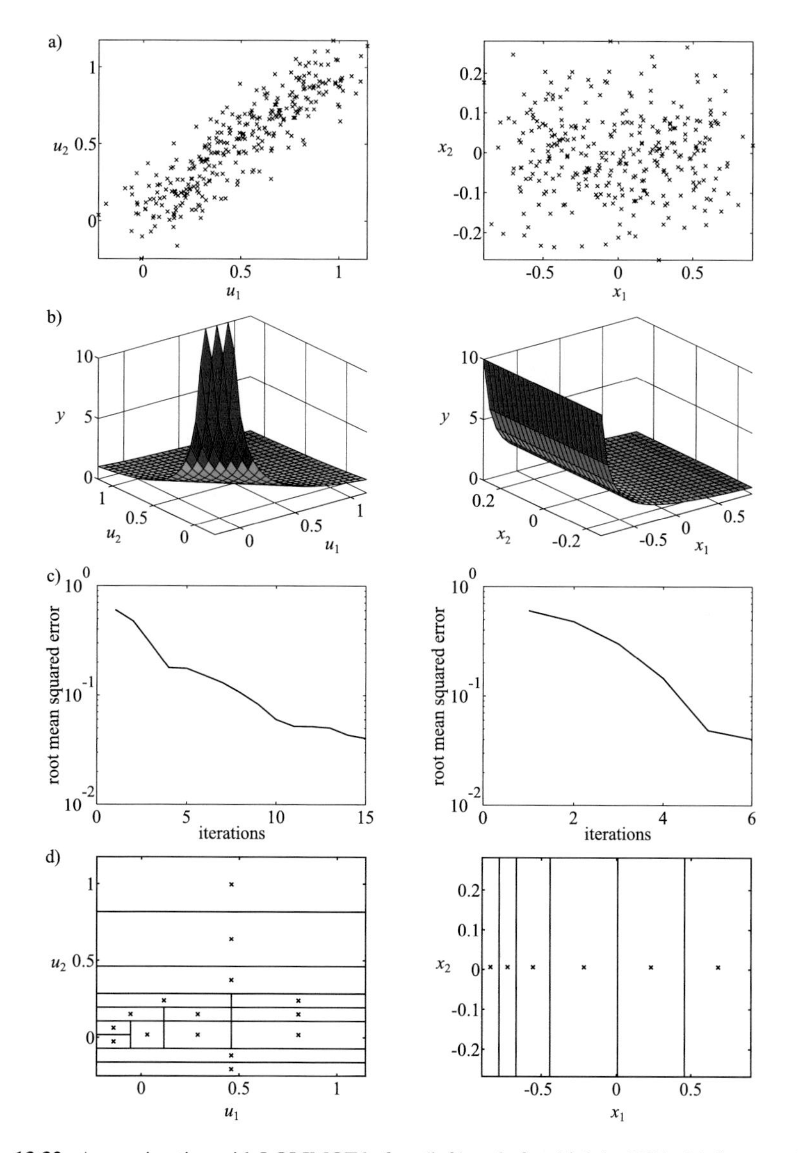

Fig. 13.32 Approximation with LOLIMOT before (left) and after (right) a PCA: (**a**) data samples, (**b**) function to be approximated, (**c**) LOLIMOT convergence curves, (**d**) partitioning of the input space performed by LOLIMOT. Note that this example represents the "best case" where a PCA yields the maximum performance gain

shows for a two-dimensional example of how a PCA can be advantageously exploited. If the inputs u_1 and u_2 are linearly correlated, the data samples are distributed along the diagonal in Fig. 13.32a(left). A PCA transforms the data as depicted in Fig. 13.32a(right). If the nonlinear behavior of the process indeed dominantly depends on one of the transformed input axes, the problem complexity can be significantly reduced. While the nonlinear behavior in the original input space (see Fig. 13.32b(left)) stretches along a diagonal direction, it becomes axis-orthogonal in the transformed input space; see Fig. 13.32b(right). In fact, in this example the second input x_2 even becomes redundant and could be discarded without any loss of information, thereby reducing the dimensionality of the problem. Note, however, that such favorable circumstances cannot usually be expected in practice.

Because of the axis-orthogonal nonlinear behavior in the transformed input space, a model with only six rules is required to achieve a satisfactory approximation performance. Figures 13.32c(right) and d(right) shows the convergence curve and the input space partitioning obtained by LOLIMOT. For a comparable performance without PCA, 15 rules are necessary, and the partitioning of the input space is carried out in both dimensions; see Fig. 13.32c(left) and d(left).

It is important to note that the above example only demonstrates how effective PCA possibly *can* be. Although highly correlated inputs such as u_1 and u_2 may sometimes arise from (almost) redundant measurements of the same physical cause (here x_1), this can definitely be concluded only from insights into the process and not from the input data distribution alone. In the above example, the nonlinearity was along the direction of x_1. If it had been along u_1, all advantages of the PCA would have turned into drawbacks (in Fig. 13.32b–d the left and right sides would have changed). Thus, PCA is a good option especially for high-dimensional problems, but its successful application cannot be guaranteed. The supervised approach discussed in Sect. 14.7 overcomes this severe limitation at the price of a much higher computational effort.

13.3.7 Models with Multiple Outputs

As explained in Sect. 9.1, systems with multiple outputs can be realized either by a single SIMO or MIMO model or by a bank of several SISO or MISO models, each representing one output. When a separate model is implemented for each output y_l ($l = 1, \ldots, r$), an individual number of neurons M_l can be chosen for each model, depending on the accuracy requirements for each output. The model structure, that is, the positions and widths of the validity functions, can also be optimized for each output individually. Consequently, a modeling problem with r outputs possesses r times the complexity of a problem with a single output ($l = 1, \ldots, r$):

$$\hat{y}_l = \sum_{i=1}^{M_l} \left(w_{i0}^{(l)} + w_{i1}^{(l)} u_1 + w_{i2}^{(l)} u_2 + \ldots + w_{ip}^{(l)} u_p \right) \Phi_i^{(l)}(\underline{u}) \,. \tag{13.44}$$

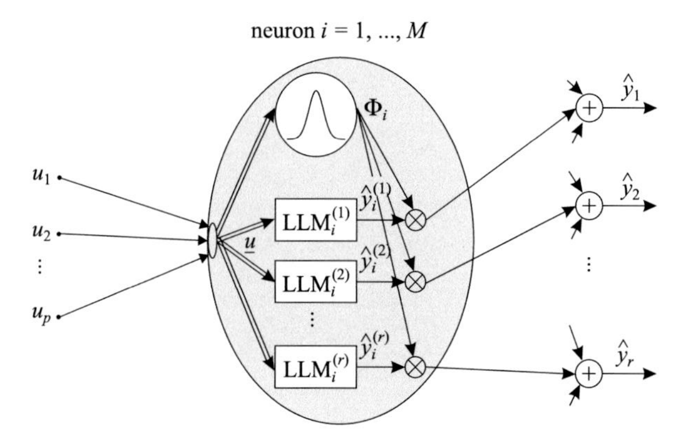

Fig. 13.33 A neuron of a local linear neuro-fuzzy model extended to the MIMO case

Alternatively, the neurons of the local linear neuro-fuzzy network can be extended to the MIMO case. Figure 13.33 shows that each neuron can realize individual local linear models (LLMs) for each output y_l. In contrast to (13.44), only one network structure exists, that is, a single set of validity functions Φ_i $(i = 1, \ldots, M)$, for the complete model $(l = 1, \ldots, r)$:

$$\hat{y}_l = \sum_{i=1}^{M} \left(w_{i0}^{(l)} + w_{i1}^{(l)} u_1 + w_{i2}^{(l)} u_2 + \ldots + w_{ip}^{(l)} u_p \right) \Phi_i(\underline{u}) . \tag{13.45}$$

In order to train the MIMO model, a loss function of the following type can be employed for global parameter estimation:

$$I = \sum_{l=1}^{r} q_l \sum_{j=1}^{N} e^2(j) \tag{13.46}$$

with individual weighting factors q_l of the outputs that reflect the desired accuracy for each y_l and compensate for possibly different scales. The loss function for local parameter estimation can be defined accordingly; see Sect. 13.2.2. For structure optimization (see Sect. 13.3.2), either (13.46) can be used or the following; different loss function can be utilized in order to ensure that LOLIMOT partitions the input space in a manner that the currently worst modeled output improves most:

$$I = \max_{l} \left\{ q_l \sum_{j=1}^{N} e^2(j) \right\} . \tag{13.47}$$

With (13.47) the structure optimization is formulated as a problem of min-max type.

Which of the two alternative ways (13.44) and (13.45) to model processes with multiple outputs is favorable? One distinct advantage of the multiple MISO model approach is its better interpretability, which also allows easier incorporation of prior knowledge. The model structures obtained reflect the nonlinear behavior of each output separately by the partitioning of the input space. This allows one to analyze and incorporate available knowledge for each output individually. In contrast, the partitioning of the input space in the MIMO approach mixes the effects of all outputs and thus is difficult to interpret.

A possible advantage of the MIMO approach is that often fewer neurons are required than in the multiple MISO model case, that is,

$$M < \sum_{l=1}^{r} M_l \,. \tag{13.48}$$

Consequently, model evaluation with the MIMO approach is faster if the calculation of the validity functions is the dominant factor. Note that the training time is usually larger for the MIMO approach since even for $M = M_l$ the number of local parameter estimations is identical for both approaches since r LLMs have to be determined for each neuron in the MIMO case; see Fig. 13.33. The extent of the reduction effect on the number of neurons in (13.48) depends on the nonlinear characteristics of the outputs. If they possess similar nonlinear behavior, all MISO models would have a similar optimal partitioning of the input space. Therefore, a MIMO model can efficiently exploit the common characteristics. In the best case $M = M_l$, and thus a MIMO model may require r times fewer neurons than the multiple MISO models. However, in the worst case, the reduction effect can be zero or even negative, especially for higher-dimensional problems (see example below).

Figure 13.34 shows examples with one input u and two outputs y_1 and y_2. In Fig. 13.34a the nonlinear characteristics of y_1 and y_2 are similar in the sense that they share regions of large curvature (small u) although the two outputs are quite different in other respects. When both outputs are approximated by LOLIMOT with separate SISO models, eight neurons are required to achieve a root mean squared error of less than 0.03, i.e., $M_1 = 8$ and $M_2 = 8$. With a SIMO model, only $M = 8$ neurons are required as well; see Fig. 13.34a(bottom). This means that 50% of the neurons can be saved. The number of local linear models is equal to 16 in both cases. If, however, the outputs possess different nonlinear characteristics, as depicted in Fig. 13.34b, a SIMO model requires 12 LLMs. The fine partitioning for large u is necessary for a sufficiently accurate description of y_1, while y_2 requires many validity function for small u. This leads to an overparameterized model and thus to an unnecessarily large variance error because too many LLMs are estimated for y_1 in the region of small u and for modeling of y_2 in regions of large u. Although the number of neurons $M = 12$ of the MIMO model is smaller than of the two SISO models with $M_1 + M_2 = 16$, the number of LLMs and thus the number of estimated parameters is 50% larger.

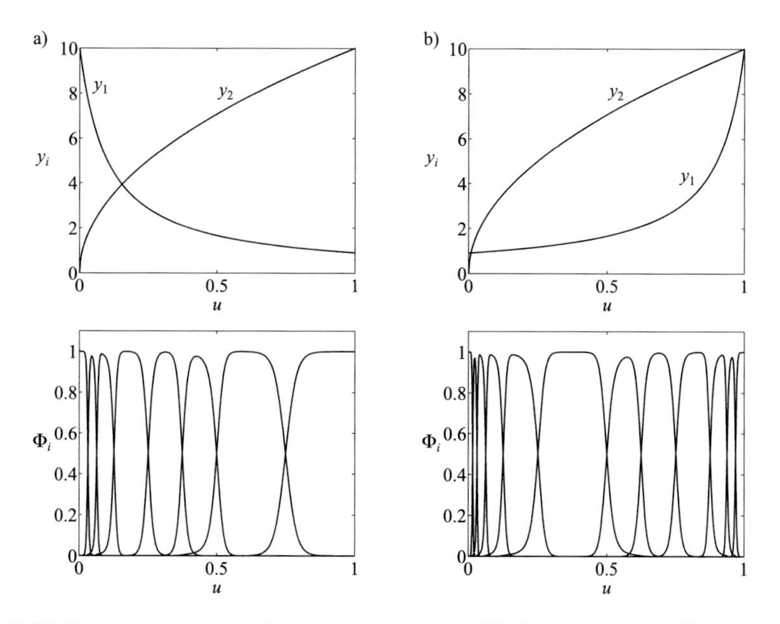

Fig. 13.34 The two outputs in (**a**) are approximated by SIMO models. The validity functions constructed by LOLIMOT are shown in (**b**)

An example of two inputs is illustrated in Fig. 13.35. As the three cases in Fig. 13.35a–c demonstrate, there are many more possible directions for the major nonlinear behavior than in the univariate example. The more highly dimensional the problem is, the less likely it is that the nonlinear behavior of the different outputs will share common characteristics. This implies that multiple MISO models become more advantageous than a single MIMO model as the input dimensionality increases. When the nonlinear characteristics of y_1 and y_2 are similar, only $M = 5$ neurons are required in order to achieve a root mean squared error of 0.1 (Fig. 13.35a). However, as the direction of the nonlinear behavior becomes more and more distinct, the number of required neurons grows sharply to $M = 9$ (Fig. 13.35b) and $M = 20$ (Fig. 13.35c), respectively. Note that this example overpronounces this effect since the original nonlinearities are axis-orthogonal.

The above examples illustrated the multiple output problems with only two outputs. The extension to more than two outputs is straightforward, and all insights can be transferred. All effects become more pronounced as the number of outputs increases.

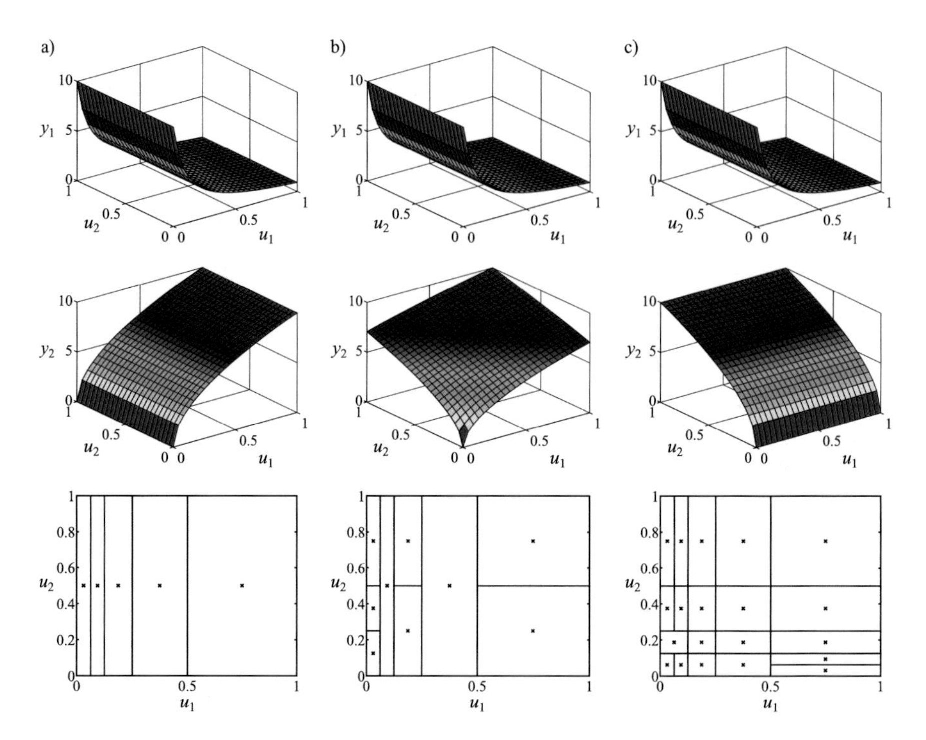

Fig. 13.35 The required number of neurons grows with the dissimilarity of the nonlinear characteristics of the outputs. The first two rows show the two process outputs y_1 and y_2. The last row shows the partitioning of the input space obtained with LOLIMOT in order to reach a root mean squared error of 0.1

13.4 Summary

Local linear neuro-fuzzy models and the local linear model tree (LOLIMOT) identification algorithm have been introduced. The linear model parameters are estimated by a local linear least squares technique, which has been shown to possess some clear advantages over a global parameter estimation. The nonlinear model parameters are determined by an incremental tree construction algorithm in a heuristic manner in order to avoid the application of nonlinear optimization techniques. Various extensions of the LOLIMOT algorithm have been examined. Different objectives for structure and parameter optimization have been proven very useful. Optimization of the approximator's smoothness and of the splitting ratios and merging of local linear models increase the flexibility of the algorithm but have been demonstrated to enhance the overall performance only insignificantly. Nevertheless, these can be powerful tools for particular applications. The advantages and drawbacks of flat and hierarchical model structures were examined. An optional principal component analysis preprocessing step has been investigated in order to overcome some disadvantages due to the axis-orthogonal input space partitioning of LOLIMOT. Finally, the extension to models with multiple outputs was discussed.

13.5 Problems

1. Explain the relationships between local linear neuro-fuzzy models, NRBF networks, Takagi-Sugeno fuzzy systems, and singleton fuzzy systems.
2. Generate two one-dimensional Gaussian membership functions with centers at $c_1 = 0.3$ and $c_2 = 0.7$ and with standard deviations of $\sigma_1 = \sigma_2 = 0.15$. Calculate and plot these membership functions $\mu_1(u)$, $\mu_2(u)$ and the normalized validity functions $\Phi_1(u)$, $\Phi_2(u)$ in the interval [0, 1]. Assign the following linear model to the first validity function: $y = u$ and this one to the second: $y = 2u - 0.5$. Calculate and plot the weighted sum of theses linear models with their validity functions in the interval [0, 1]. Repeat this step for other second linear models: $y = 2u$ and $y = 2u - 1$. Observe the interpolation behavior, in particular at $u = 0.5$, for the three different second linear models. Repeat all these steps with the half and the double values for the standard deviations.
3. Write down the corresponding fuzzy rules for one of the models from the previous Problem.
4. How do the regression matrices look like for estimating the linear model parameters locally and globally by the method of least squares? How many parameters have to be estimated? How does the computational effort grow with the number of rules?
5. What are the advantages and drawbacks of the local and global estimation approach? Explain intuitively the regularization effect that is implicitly contained in the local estimation approach. How does this regularization effect affect the practicability of the local estimation approach?
6. Give a brief overview of the major methods for determining the positions, widths, and number of validity functions for local linear model networks. Elaborate on the key features of each strategy.
7. Explain the LOLIMOT algorithm.
8. What is the difference between a flat and a hierarchical model?
9. If two or more outputs shall be modeled, the following two strategies can be pursued: (i) use one model with several outputs or (ii) use several models with one output each. Explain the properties of both approaches in general and for LOLIMOT in particular. In what cases would you choose which alternative?
10. What benefits can be expected from a PCA preprocessing step before applying LOLIMOT? Can a PCA make things worse? Why?

Chapter 14
Local Linear Neuro-Fuzzy Models: Advanced Aspects

Abstract This chapter continues the discussion of local linear neuro-fuzzy models and extends them to local model networks where the local models can be arbitrary and the fuzzy logic interpretation vanishes in the background. A couple of attractive key features of these model architectures are analyzed in this chapter. Two of them shall be highlighted here: The separation of the inputs for rule premises and for the rule consequents (local models) allows to partly overcome the curse of dimensionality because it is often possible to choose the premise input space of low dimension even if the local models are high-dimensional. This feature also will play a crucial role when dealing with dynamic models in Part C. Another nice characteristic is due to the local nature of this model architecture. It allows for extremely robust online learning without unlearning in regions where no new data arrives. Furthermore, the extension to axis-oblique tree construction is discussed and analyzed in great detail. A new algorithm called hierarchical local model tree (HILOMOT) is introduced and compared to LOLIMOT. This opens the door to solving even higher-dimensional problems with this kind of model architectures. On the basis of HILOMOT, a new design of experiments/active learning scheme is proposed that exploits the advantages of the model structure to the fullest. It has been successfully applied to multiple real-world problems that are covered in Part D.

Keywords Axis-oblique · Premise · Consequent · Online · Hierarchy · Sigmoid · DoE · HILOMOT

This chapter continues Chap. 13 and deals with the more advanced issues of local linear neuro-fuzzy models and the LOLIMOT algorithm. It is organized as follows. Section 14.1 discusses the possibility of different input spaces for rule premises and consequents, which is a unique feature of local neuro-fuzzy models. Section 14.2 introduces the use of models that are more complex than local *linear*. An extension

---

The original version of this chapter was revised: This chapter was inadvertently published with an error in equation 14.40 in page 502 which has been corrected now. The correction to this chapter is available at https://doi.org/10.1007/978-3-030-47439-3_30

O. Nelles, *Nonlinear System Identification*,
https://doi.org/10.1007/978-3-030-47439-3_14

of the LOLIMOT algorithm that allows one to optimize the structure of the rule consequents is proposed in Sect. 14.3. Section 14.4 analyzes the interpolation and extrapolation behavior of local linear neuro-fuzzy models. This investigation reveals undesirable effects in the linearization of the model, which are demonstrated and remedied in Sect. 14.5. Methods for online identification by means of recursive algorithms are discussed in Sect. 14.6. The estimation of the reliability of the model output with errorbars, their application to the design of excitation signals, and active learning are discussed in Sect. 14.9. An outlook for a further extension of the LOLIMOT algorithm and a link to ridge construction-based approaches such as MLP networks are given in Sect. 14.7, which deals with hinging hyperplanes, and Sect. 14.8 which introduces the HILOMOT algorithm. Finally, a brief summary is given, and some conclusions about this and the preceding chapter are drawn in Sect. 14.12.

14.1 Different Input Spaces for Rule Premises and Consequents

The local linear neuro-fuzzy models discussed up to now possess identical *input spaces* in the rule premises and consequents, that is, they utilize the same variables $\underline{u} = [u_1 \; u_2 \; \cdots \; u_p]^T$. In (13.3) and Fig. 13.1, the local linear models and the validity functions both depend on $\underline{u}$. One of the major strengths of local linear neuro-fuzzy models is that premises and consequents do not have to depend on identical variables. Rather (13.3) can be extended to

$$\hat{y} = \sum_{i=1}^{M} \left(w_{i0} + w_{i1}x_1 + w_{i2}x_2 + \ldots + w_{i,nx}x_{nx} \right) \Phi_i(\underline{z}) \,, \tag{14.1}$$

where the local linear models (consequents) depend on $\underline{x} = [x_1 \; x_2 \; \cdots \; x_{nx}]^T$ and the validity functions (premises) depend on $\underline{z} = [z_1 \; z_2 \; \cdots \; z_{nz}]^T$. Figure 14.1 depicts

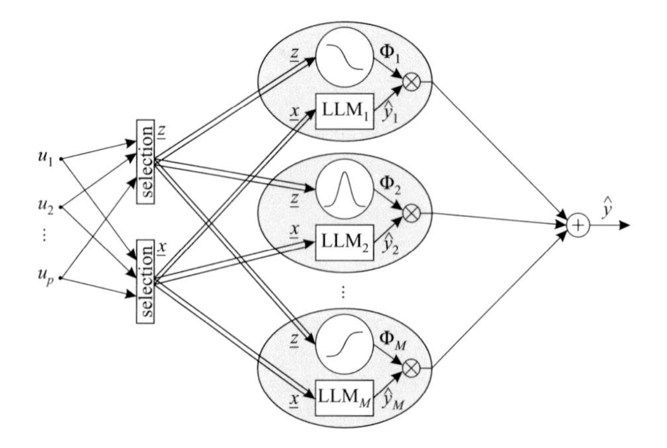

Fig. 14.1 Network structure of a local linear neuro-fuzzy model with M neurons for nx consequent inputs $\underline{x}$ and nz premise inputs $\underline{z}$

this extended network architecture; see Fig. 13.1. The following three cases can be distinguished:

1. *Identical input spaces*: This case has been discussed up to here. The consequents and premises depend on the same variables, i.e., $\underline{x} = \underline{z} = \underline{u}$. If this condition is met, the model is a universal approximator.
2. *Disjunct input spaces*: This is a *scheduling approach* in which the local linear models are interpolated by "external" variables in the premises that do not appear in the consequents. For example, the model may depend on u_1, u_2, and u_3, but the consequents may depend only on u_1 and u_3 while the premises depend only on u_2, i.e., $\underline{x} = [u_1\ u_3]^T$ and $\underline{z} = u_2$. Such a model is *no* universal approximator because it cannot generate nonlinear behavior with respect to those inputs that are not contained in $\underline{z}$.
3. *Input spaces with common variables*: The consequents and premises share some variables and possess some separate variables. For example, the model may depend on u_1, u_2, and u_3, but the consequents may depend only on u_1 and u_3, while the premises depend only on u_1 and u_2, i.e., $\underline{x} = [u_1\ u_3]^T$ and $\underline{z} = [u_1\ u_2]^T$. If the premises contain *all* variables, i.e., $\underline{z} = \underline{u}$, the model is a universal approximator independent of the choice of $\underline{x}$ since even for an empty consequent space ($\underline{x} = [\ ]$), the model deteriorates to an NRBF network that is already a universal approximator; see Sect. 13.1.4.

The distinction between the input spaces for the rule premises and consequents allows one to reduce the curse of dimensionality by the incorporation of prior knowledge about the structure of the process. In $\underline{z}$ only those variables should be gathered that are assumed to influence the process in a nonlinear way. All variables that are assumed to possess a linear effect on the process should be gathered in $\underline{x}$. In many applications (especially in dynamic systems; see Chap. 22), the number of model inputs with a significant nonlinear influence is much smaller than the overall number of inputs, i.e., $nz \ll p$. Furthermore, *qualitative knowledge* about the strengths of the nonlinear effects caused by each input is often readily available or can be easily obtained by simple nonlinearity tests; see Sect. 1.1.1 and [199]. Thus, the choice of $\underline{z}$ and $\underline{x}$ is relatively easy in practice.

Breaking down the general p-dimensional approximation problem $y = f(\underline{u})$ with $\underline{u} = [u_1\ u_2\ \cdots\ u_p]^T$ shown in Fig. 14.2a into its nonlinear (described by $\underline{z}$) and linear (described by $\underline{x}$) dependencies as shown in Fig. 14.2b can yield a significant problem complexity reduction since typically $nz < p$. With the dimensionality reduction from dim$\{\underline{z}\}$ to dim$\{\underline{u}\}$, the curse of dimensionality is reduced; see Sect. 7.6.1.

In principle, LOLIMOT is able to find the inputs with nonlinear influence by itself, so that it is not necessary to impose a reduced premise space vector $\underline{z}$; rather one can start with a complete premise space vector $\underline{z} = \underline{u}$. After LOLIMOT has converged, those inputs without any splits apparently do not contain any significant nonlinear influence. From the fuzzy logic point of view, only "don't care" membership functions are associated with these non-divided inputs. Consequently, they can be removed from the premises space vector $\underline{z}$ afterward without affecting

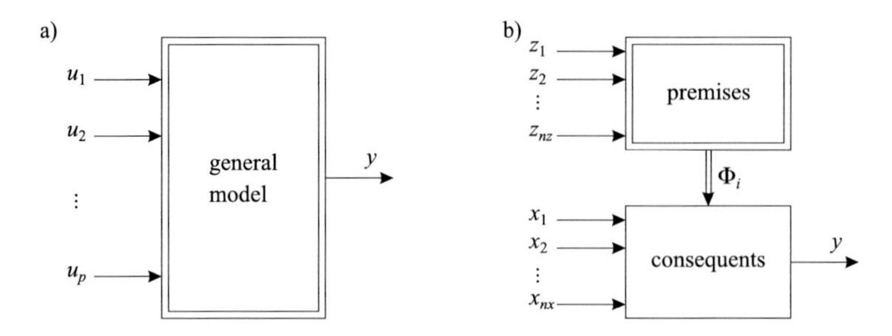

Fig. 14.2 (**a**) A general model with p inputs can typically significantly simplified when a local linear neuro-fuzzy model is utilized. (**b**) The input vector $\underline{u}$ can be separated into the nonlinearly influencing inputs $\underline{z}$ in the premises and the linearly influencing inputs $\underline{x}$ in the consequents, where $nz \leq p$ and $nx \leq p$ and often $nz \ll p$

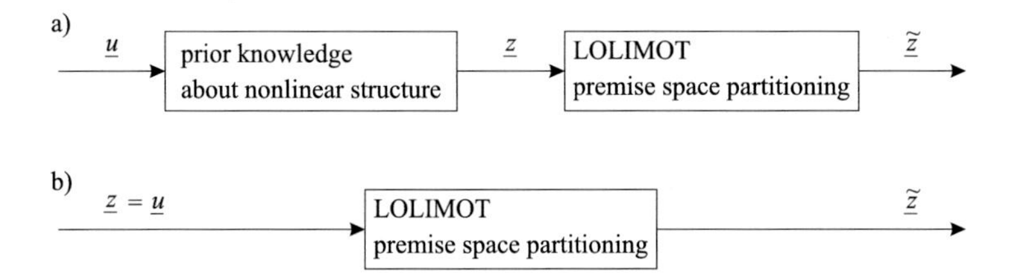

Fig. 14.3 Premise input selection by (**a**) both prior knowledge and LOLIMOT or (**b**) LOLIMOT only

the model's performance. Nevertheless, there are two important reasons why the premise space should be prestructured by the choice of $\underline{z}$ whenever possible:

- LOLIMOT operates faster the lower the dimensionality of $\underline{z}$ is. Each discarded premise space input reduces the number of alternative splits that have to be examined by LOLIMOT.
- Artifacts do not occur owing to the imperfect nature of the training data set. Because of noisy and/or insufficiently exciting data, LOLIMOT may perform splits that do not match the true process characteristics. Incorporation of prior knowledge prevents these artifacts and makes the model smaller, more easily interpretable, and more robust with respect to parameter estimation and extrapolation behavior.

Figure 14.3 compares the selection of the premise variables with and without prior knowledge, where $\underline{\tilde{z}}$ represents the final vector of premise inputs. Figure 14.4 shows how much the performance of the model is affected if the (assumed) prior knowledge about the nonlinearly influencing inputs is *wrong*. The process in Fig. 14.4a is nonlinear in both inputs u_1 and u_2. This process is approximated by LOLIMOT with 15 neurons. With the choice $\underline{x} = [u_1\ u_2]^T$ and $\underline{z} = [u_1\ u_2]^{\mathrm{T}}$, the model works fine with an RMSE of 0.006; see Fig. 14.4b. When the input u_2 is excluded from the premise space, i.e., $\underline{z} = u_1$, the approximation error increases to

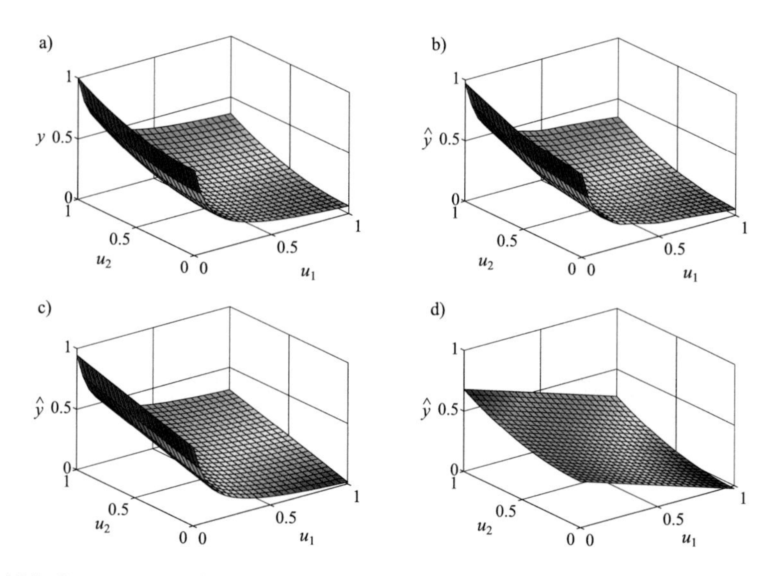

Fig. 14.4 Consequences of wrongly chosen premise inputs: (**a**) process, (**b**) approximation with $\underline{z} = [u_1\ u_2]^T$, (**c**) $\underline{z} = u_1$, (**d**) $\underline{z} = u_2$

an RMSE of 0.02 since the nonlinear behavior in u_2 cannot be modeled anymore; see Fig. 14.4c. The main characteristics of the process, however, are still covered because the process is only slightly nonlinear in u_2. In contrast, the exclusion of u_1, i.e., $\underline{z} = u_2$, deteriorates to an RMSE of 0.09; see Fig. 14.4d.

For the sake of simplicity, the premise and consequent input spaces are chosen to be equal in the remaining part of this chapter, i.e., $\underline{x} = \underline{z} = \underline{u}$, except in those cases where their distinction is important for a proper understanding.

14.1.1 *Identification of Processes with Direction-Dependent Behavior*

Many processes possess direction-dependent behavior. Often it is caused by a Coulomb type of friction, but other causes are also possible. As a simple example, the following equation of motion of an undamped mass-spring system with external force F, displacement u, mass m, spring constant c, and Coulomb friction coefficient F_0 will illustrate this effect:

$$F = m\,\ddot{u} + c(u)\,u - F_0\,\text{sign}(\dot{u})\,. \qquad (14.2)$$

If the displacement u changes slowly $m\,\ddot{u} \approx 0$, and if the nonlinear characteristics of the spring in (14.2) are progressive, the process may possess the behavior shown in Fig. 14.5a with $y \equiv F$. Note that this *hysteresis* effect depends on the direction

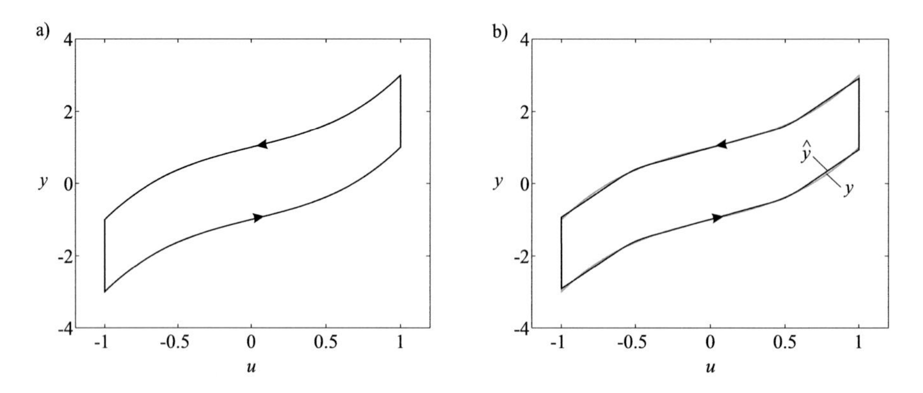

Fig. 14.5 (**a**) Hysteresis generated by (14.3). (**b**) Comparison between process output y and output $\hat{y}$ of a local linear neuro-fuzzy model trained with LOLIMOT with ten neurons and $\underline{x} = u$, $\underline{z} = [u\ \dot{u}]^T$

of u, i.e., on sign($\dot{u}$). Other physical effects such as re-magnetization can cause a hysteresis type that is much more complex than the one discussed here. In such a hysteresis, the state of the system depends on *all previous* states of the system. Modeling such hysteresis effects is a very difficult problem.

The process shown in Fig. 14.5a follows the equation

$$y = \begin{cases} u + u^3 + 1 \text{ for } \dot{u} < 0 \\ u + u^3 - 1 \text{ for } \dot{u} \geq 0 \,. \end{cases} \tag{14.3}$$

Obviously, it is not possible to find a function that describes the following relationship

$$y = f(u) \tag{14.4}$$

since the mapping from u to y is not unique. However, it is easy to find the following function:

$$y = f(u, \dot{u}) \,. \tag{14.5}$$

Indeed, LOLIMOT can easily identify the hysteresis when the premise space is chosen as $\underline{z} = [u\ \dot{u}]^T$. Alternatively, $\underline{z} = [u\ \text{sign}(\dot{u})]^T$ can be used because the absolute value of $\dot{u}$ is irrelevant. The decision whether $\dot{u} < 0$ or $\dot{u} > 0$ is made in the rule premises. Rules of the following two types are constructed by LOLIMOT (see Fig. 14.7a)

IF $u = \ldots$ AND $\dot{u} =$ negative THEN $y = w_{i0} + w_{i1}u$

IF $u = \ldots$ AND $\dot{u} =$ positive THEN $y = w_{j0} + w_{j1}u$.

The consequents do not have to contain any information about $\dot{u}$, that is, $\underline{x} = u$. In comparison with the more black box like approach $\underline{x} = \underline{z}$, fewer parameters have to be estimated, which makes LOLIMOT faster and the parameter estimates more reliable. The result of a local linear neuro-fuzzy model with ten neurons is shown in Fig. 14.5b.

In the noise-free case, there is no significant difference between the alternatives $\underline{z} = [u\ \dot{u}]^T$ and $\underline{z} = [u\ \text{sign}(\dot{u})]^T$. If, however, u is noisy, then this disturbance is amplified by the differentiation and a considerable discrepancy between both alternatives arises; see Fig. 14.6. The choice $\underline{z} = [u\ \dot{u}]^T$ is more robust against large disturbances because a wrong sign of $\dot{u}$ may cause only a limited model error when $|\dot{u}|$ is small. In contrast, the approach $\underline{z} = [u\ \text{sign}(\dot{u})]^{\text{T}}$ is extremely sensitive to large disturbances that make the model jump to the wrong hysteresis branch. However, the overall model error with the "sign" approach is smaller in this example since the "sign" operation filters out all small disturbances which keep the sign of $\dot{u}$ unchanged. In practice, the latter solution can only be used in combination with a low-pass filter that avoids undesirable sign changes. In Fig. 14.7b, c, the input data distribution of both approaches is compared for better clarity.

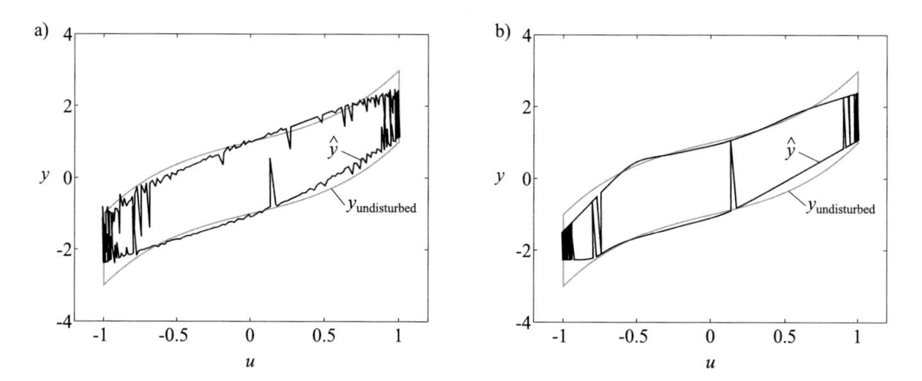

Fig. 14.6 Comparison between the two alternative premise input spaces (**a**) $\underline{z} = [u\ \dot{u}]^T$ and (**b**) $\underline{z} = [u\ \text{sign}(\dot{u})]^T$ for noisy input data u

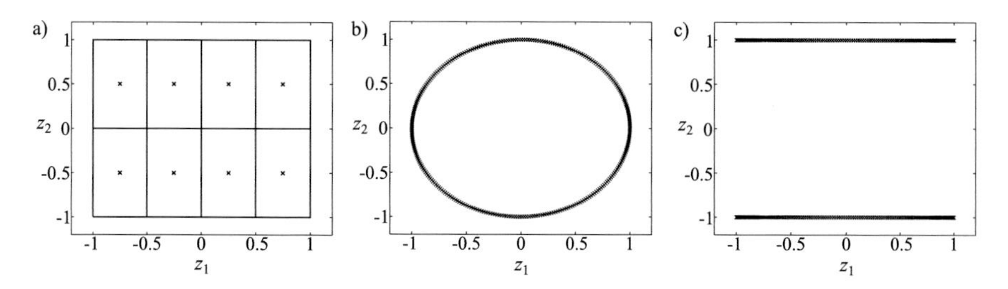

Fig. 14.7 (**a**) Input partitioning generated by LOLIMOT. Training data distribution for (**b**) $\underline{z} = [u\ \dot{u}]^T$ and (**c**) $\underline{z} = [u\ \text{sign}(\dot{u})]^T$ for the undisturbed case

14.1.2 Piecewise Affine (PWA) Models

The focus of this book is on smooth model architectures. Therefore only the case of *interpolated* (through the Φ_i's) local models is considered. For the special case of $\sigma = 0$, the interaction/overlap between the local models vanishes. These approaches are called *piecewise affine (PWA) models*. Here the terminology "affine" not "linear" is established – probably due to the more mathematical nature of the topic. These models also are called *switched models* or *hybrid models* as the partitioning of the $\underline{z}$-input space decides on the active affine model. Thus, $\underline{z}$ can be seen as switching variables, and the whole model mixes continuous variables $\underline{x}$ and discrete variables $\underline{z}$. For surveys on this topic, the reader is referred to [58, 180, 442] and in context with their extensive application toward predictive control to [8, 45].

Although PWA models are discontinuous (highly non-smooth) and therefore unattractive from many perspectives, they offer significant advantages with respect to mathematical analysis and for modeling truly hybrid systems. Important distinct features include:

- Interpolation behavior and shape of validity functions are no issues since the Φ_i's are either 0 or 1.
- The problem of estimation of the local affine models and the partitioning can be tackled simultaneously.
- Global and local estimation of the local models are identical as no overlap between the models exists. Only the data belonging to a specific local model where $\Phi_i = 1$ is used for estimation of local affine model i (no weighting necessary).

14.2 More Complex Local Models

This section discusses the choice of more complex local models than linear ones. First, the relationship between local neuro-fuzzy models to polynomials is pointed out. Section 14.2.2 demonstrates the usefulness of local quadratic models for optimization tasks. Finally, the possibility of mixing different types of local models in one overall model is discussed.

14.2.1 From Local Neuro-Fuzzy Models to Polynomials

Section 13.1.4 showed that NRBF networks can be understood as local *constant* neuro-fuzzy models, i.e., as a simplification of local *linear* neuro-fuzzy models. Clearly, in general, arbitrary local models can be utilized. Then a *local neuro-fuzzy model* is given by

$$\hat{y} = \sum_{i=1}^{M} f_i(\underline{u}) \Phi_i(\underline{u}) \tag{14.6}$$

with nonlinear functions $f_i(\cdot)$ and the input vector $\underline{u} = [u_1\ u_2\ \cdots\ u_p]^{\mathrm{T}}$. In Fig. 13.1 nonlinear functions $f_i(\cdot)$, replace the "LLM" blocks. If the $f_i(\cdot)$ are constant, an NRBF network results. For linear $f_i(\cdot)$, a local linear neuro-fuzzy model is obtained. The functions $f_i(\cdot)$ can be seen as *local approximators* of the unknown process that is to be modeled. In this context, constant and linear local models can be interpreted as zero-th and first-order *Taylor series expansions*. The local models can be made more accurate by using higher-order Taylor series expansions leading to polynomials of higher degrees. Figure 14.8 illustrates the approximation of a nonlinear process by a local neuro-fuzzy model with (a) 12 zero-th-degree, (b) 5 first-degree, (c) 2 second-degree, and (d) 1 third-degree local polynomial models. The first row in Fig. 14.8 confirms the observation already made in Sect. 13.1.4 that less powerful local models require relatively smoother validity functions to ensure a smooth overall model output.

Clearly, the more complex the local models are, the larger is the region of the input space in which the process can be described by a single local model with a given degree of accuracy. This means the fewer local models are required the more complex they are. At some point, the local models become so powerful that only a single local model can describe the whole process. Then the only remaining validity function $\Phi_1(\underline{u})$ is equal to 1 for all $\underline{u}$ owing to the normalization. In fact, the local model deteriorates to a pure high-degree polynomial. This case is illustrated in Fig. 14.8d. Thus, polynomials can be seen as one extreme case of local neuro-fuzzy models, while NRBF networks can be seen as another extreme case; see Table 14.1.

The overall model complexity depends on the number of required neurons M and the number of parameters per local model. As Table 14.1 shows, an NRBF network possesses only a single (linear) parameter for each neuron, namely, the weight or height of the basis function, but a huge number of neurons is required. In contrast, a high-degree polynomial possesses a huge number of parameters, but only a single neuron is necessary. In the experience of the author, local *linear* models are often a good compromise between these two extremes. The number of local model parameters still grows only linearly with the input space dimensionality p, but the required number of neurons is already significantly reduced compared with an NRBF network. For most applications, the next complexity step to local quadratic models does not pay off. The quadratic dependency of the number of local polynomial parameters on input space dimensionality p is usually not compensated by the smaller number of required neurons, especially for high-dimensional mappings. Of course, the choice of the best model architecture is highly problem dependent, but local linear models seem to be a well-suited candidate for a good *general* model architecture.

Although the extension of local constant and local linear models to higher-degree polynomials is straightforward, in principle, the nonlinear local functions $f_i(\cdot)$ can be of any type. It is convenient to choose them linearly parameterized in order to

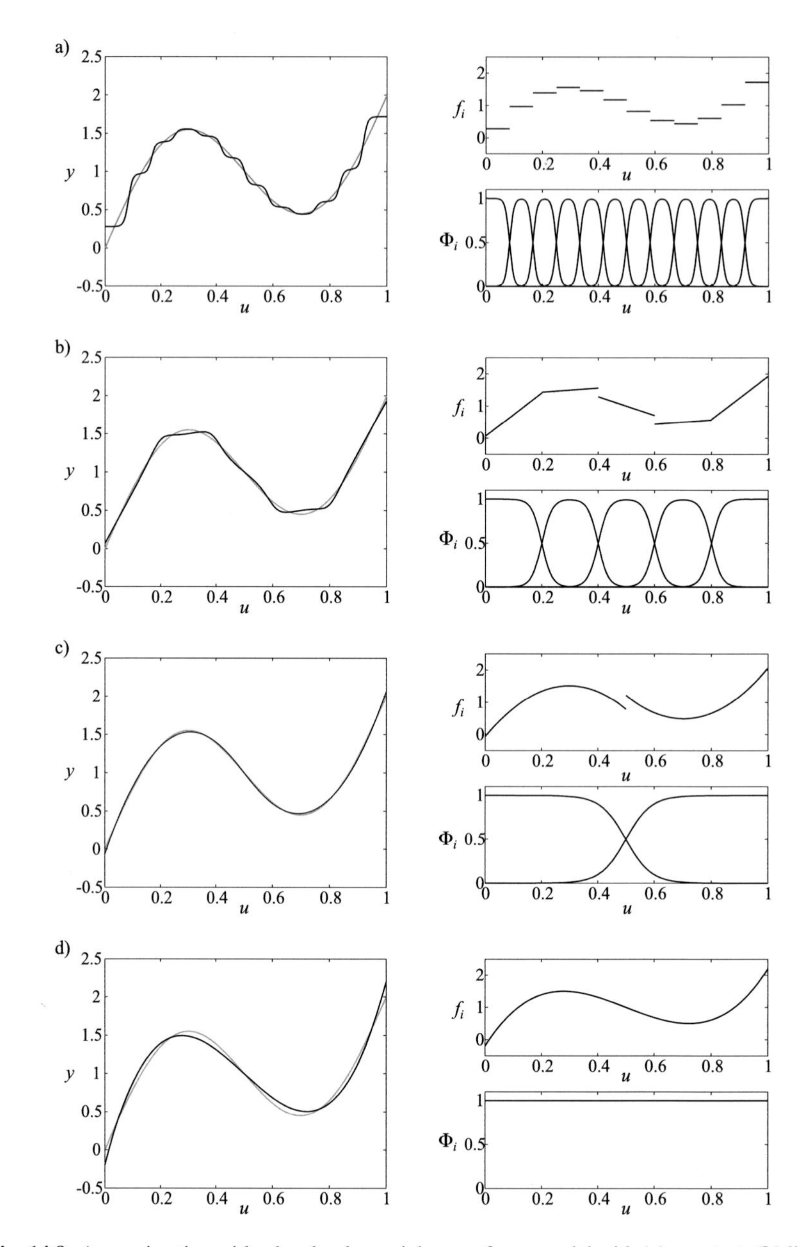

Fig. 14.8 Approximation with a local polynomial neuro-fuzzy model with (**a**) constant, (**b**) linear, (**c**) quadratic, (**d**) cubic local models. The left side shows the process and its approximation. The right side shows the local models and their corresponding validity functions

Table 14.1 From local neuro-fuzzy models to polynomials

Order	Parameters per local model	M	Name
Constant	$1 \quad\quad\quad \rightarrow \mathcal{O}(1)$	Huge	NRBF network
Linear	$p+1 \quad\quad \rightarrow \mathcal{O}(p)$	Large	Local linear NF model
Quadratic	$(p+2)(p+1)/2 \rightarrow \mathcal{O}(p^2)$	Medium	Local quadratic NF model
$\vdots$	$\vdots$	$\vdots$	$\vdots$
lth order	$(p+l)!/(p!\,l!) \quad \rightarrow \mathcal{O}(p^l)$	1	Polynomial

p = number of inputs, M = number of neurons, rules, or local models
The complexity orders $\mathcal{O}(\cdot)$ are derived under the assumption $l \ll p$

exploit the advantages of linear optimization. Section 14.2.3 discusses other choices for $f_i(\cdot)$.

It is not necessary to choose *all* local models of the same type. Typically, different types of functions $f_i(\cdot)$ are employed if prior knowledge about the process behavior in different operating regimes is available. This issue is also discussed further in Sect. 14.2.3. Additionally, in Sect. 14.3, a method for data-driven selection of the local models' structures is proposed. In [490] local polynomial neuro-fuzzy models are pursued further.

14.2.2 Local Quadratic Models for Input Optimization

An example where local quadratic models are very advantageous is presented in the following. The task may be to find the input u^* that minimizes the (unknown) function $y = f(u)$, which is approximated by estimating a local neuro-fuzzy model from data. In a more general setting, the problem can be formulated as follows:

$$\hat{y} = f(\underline{u}) \underset{\underline{u}^{(\text{var})}}{\longrightarrow \min} \qquad \text{with } \underline{u}^{(\text{fix})} = \text{constant} \tag{14.7}$$

where $\underline{u}^{(\text{var})}$ and $\underline{u}^{(\text{fix})}$ are the variable and fixed inputs taken from the input vector $\underline{u} = [\underline{u}^{(\text{var})}\ \underline{u}^{(\text{fix})}]^T$. For example, $\underline{u} = [u_1\ u_2\ u_3\ u_4]^T$ with the constant inputs defining the operating point $\underline{u}^{(\text{fix})} = [u_1\ u_2]^T$, and the variable input to be optimized $\underline{u}^{(\text{var})} = [u_3\ u_4]^T$. Such types of optimization problems occur frequently in practice. For example, in automotive control of combustion engines, a four-dimensional mapping may describe the functional relationship between the inputs engine speed u_1, injection mass u_2, injection angle u_3, exhaust gas recirculation u_4, and the output engine torque y. Then for any operating point defined by the first two inputs, the last two inputs need to be optimized in order to achieve the maximum engine torque or a minimal exhaust gas, e.g., NO_x, or some tradeoff between both goals. For these types of problems, it is advantageous to utilize local models whose outputs are quadratic in the variable inputs. In the above example, the local models thus should be chosen as

$$\hat{y}_i = w_{i0}+w_{i1}u_1+w_{i2}u_2+w_{i3}u_3+w_{i4}u_4+\underbrace{w_{i5}u_3^2 + w_{i6}u_4^2 + w_{i7}u_3u_4}_{\text{quadratic terms}} . \quad (14.8)$$

The benefits obtained by the quadratic shape with respect to the variable inputs are illustrated in Fig. 14.9 with a simple one-dimensional example. In many situations, the function to be minimized possesses a unique optimum and thus has a U-shaped character with respect to the variable inputs, similar to that depicted in Fig. 14.9a. In some cases, the optimum may be attained at the upper or lower bounds, and thus the function may be not U-shaped. Then the use of local quadratic models described below is not necessarily beneficial.

The *approximation* performance of local linear and local quadratic neuro-fuzzy models is *comparable* if the number of parameters of both models (local linear: two parameters per local model, local quadratic: three parameters per local model) is taken into account; see Fig. 14.9c.

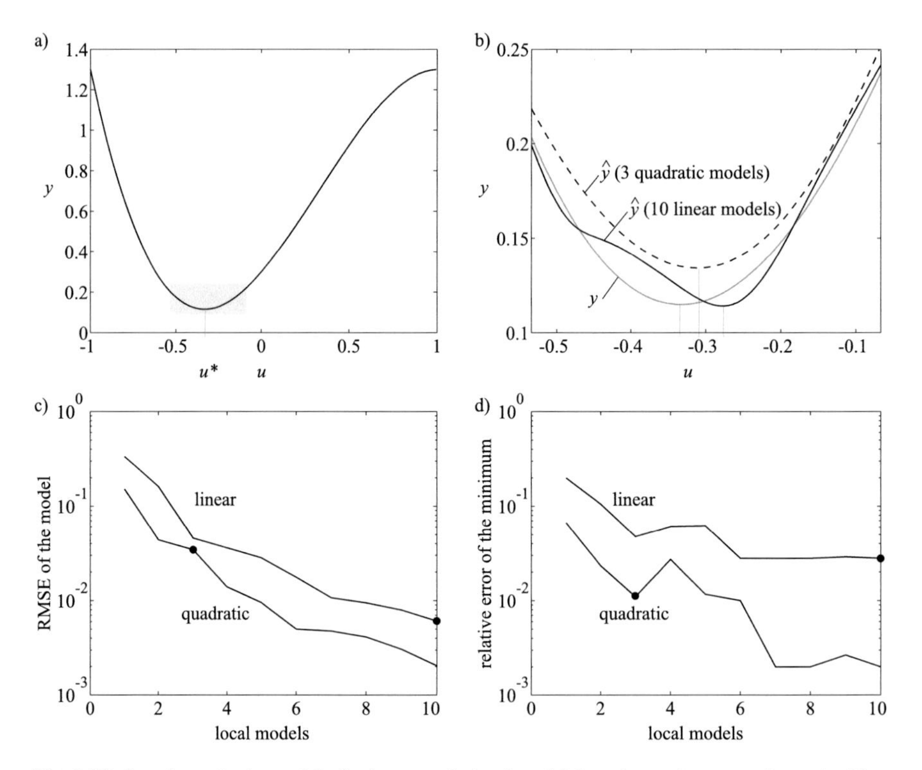

Fig. 14.9 Local quadratic models for input optimization: (**a**) function to be approximated with a minimum at $u^* = -1/3$; (**b**) zoom of the white area in (**a**) comparing the process output with a local linear and a local quadratic neuro-fuzzy model with ten and three neurons, respectively; (**c**) convergence of the approximation error; (**d**) convergence of the error of the model minimum

Contrary to the approximation performance, the local quadratic models are *superior* in the accuracy of the *minimum*, i.e., the minimum of the model is closer to the true minimum of the process; see Fig. 14.9d. The reason for this superiority is illustrated in Fig. 14.9b, which compares the process output with the local linear and local quadratic models. Although the neuro-fuzzy model with ten local linear models is much more accurate than the one with only three local quadratic models (compare also Fig. 14.9c), the minimum estimate of the local quadratic approach is much better than that of the local linear approach. Around the optimum, a quadratic shape is a very good description of any smooth function. The derivative is equal to zero at the minimum, which leads to poor performance of local linear models.

14.2.2.1 Local Sparse Quadratic Models

Local quadratic models are limited to low- and medium-dimensional problems since the number of parameters for the quadratic terms grows quadratically with the input dimension p:

$$n_{\text{fullquad}} = (p+1)(p+2)/2\,. \tag{14.9}$$

In many cases it may already yield significant benefits to model the pure quadratic behavior u_i^2, $i = 1, \ldots, p$, which describes the individual curvature for each input i but to omit the cross/interaction terms $u_i u_j$ for $i \neq j$. Such models shall be called *sparse quadratic*. Their contours are axis-orthogonal ellipsoids. The number of quadratic terms in a sparse quadratic model grows only linearly with the input dimension:

$$n_{\text{sparsequad}} = p\,. \tag{14.10}$$

The sparse quadratic version of (14.8) is

$$\hat{y}_i = w_{i0} + w_{i1}u_1 + w_{i2}u_2 + w_{i3}u_3 + w_{i4}u_4 + \underbrace{w_{i5}u_3^2 + w_{i6}u_4^2}_{\text{pure quadratic terms}}\,. \tag{14.11}$$

The difference between full and sparse quadratic models becomes increasingly pronounced with growing input dimension p.

In MATLAB the terminology, "pure" instead of "sparse" is used. In [121] "sparse quadratic" is called "quadratic," and "quadratic" is called "full quadratic."

One convincing argument for (sparse) quadratic models in the context of optimization also is the following: In order to describe an optimum in p dimensions, at least $p + 1$ local linear models are required, and the optimum lies in the (least accurate) interpolation region between them. The same task could be performed by a single local (sparse) quadratic model. This indicates how much more effective local quadratic approaches are in describing minima and maxima.

The superiority of the local quadratic approach for optimum determination is similar to the fast convergence of Newton's method in nonlinear local optimization. Newton's method is also based on a local (with first and second derivative matching the function) quadratic model of the function to be optimized; see Sect. 4.4. Sparse quadratic models correspond to considering only the diagonal of the Hessian.

Note that the additional number of parameters in each LM due to the quadratic terms can be kept relatively small because the LMs have to be extended by quadratic terms only in the optimization variable inputs. For example, in (14.8) only eight regressors are required compared with 15 regressors for a complete quadratic local model.

14.2.3 Different Types of Local Models

Up to here the use of polynomial type local models $f_i(\cdot)$ has been addressed. From the discussion in Sect. 14.2.1, first-degree polynomials, that is, local *linear* models, generally seem to offer some favorable properties and thus are most widely applied. Therefore, for all that follows, the focus is on local linear neuro-fuzzy models. However, local neuro-fuzzy models in principle allow the incorporation of arbitrary local model types.

So, one neuro-fuzzy model may include different local model structures and even architectures such as:

- Linear models of different inputs
- Polynomial models of different degrees and inputs
- Neural networks of different architectures, structures, and inputs
- Rule-based systems realized by fuzzy logic
- Algebraic or differential equations obtained by first principles

The idea is to represent each operating regime by the model that is most appropriate. For example, theoretical modeling may be possible for some operating conditions, while the process behavior in other regimes might be too complex for a thorough understanding, and thus a black box neural network approach may be the best choice. Furthermore, it may be easily possible to formulate the desired extrapolation behavior in the form of fuzzy rules because for extrapolation qualitative aspects are more important than numerical precision. Since local neuro-fuzzy models allow such integration of various modeling approaches in a straightforward manner, they are a very powerful tool for gray box modeling [572]. In particular, the combination of data-driven modeling with the design of local models by prior knowledge in regimes where data cannot be measured owing to safety restrictions or for productivity reasons is extremely important in practice.

As demonstrated in Sects. 13.2.2 and 13.2.3, local models can be optimized individually. This feature allows one to break down difficult modeling problems into a number of simpler tasks for each operating regime; this is called a *divide-and-conquer strategy*. If, for example, a local neuro-fuzzy model consists of three local models – an MLP network, a theoretical model with unknown nonlinear parameters, and a polynomial model – then the MLP network may be trained with backpropagation, the nonlinear parameters of the theoretical model may be optimized by a Gauss-Newton search, and the coefficients of the polynomial can be estimated by a linear least squares technique.

An interesting situation occurs when the local models themselves are again chosen as a local neuro-fuzzy model. This is a quite natural idea because if a certain model architecture is favorable (for whatever reasons), it is logical to employ it for the submodels as well. It yields a self-similar or fractal model structure since each model consists of submodels of the same type, which again consist of sub-submodels of the same type, etc. until the maximum depth is reached. Such hierarchical model structures are analyzed in Sect. 14.8.8.

Next, data-driven methods for determination of the rule consequents (Sect. 13.2) and premises (Sect. 13.3) are presented and analyzed. Figure 14.10 illustrates various ways in which the information from prior knowledge and measurement data can be integrated into local linear neuro-fuzzy models. One common approach is to determine the rule premises by prior knowledge since this requires information about the nonlinear structure of the process, which is often available, at least in a qualitative manner. The rule consequents are then estimated from data; see Fig. 14.10a. An alternative is that some local models are fully chosen by prior knowledge and some others are fully optimized with regard to the training data; see Fig. 14.10b. This is a particularly appropriate approach when first principles models can be built for some operating regimes, while prior knowledge does not exist for the process behavior in other regimes. Finally, Fig. 14.10c shows a two-stage strategy in which first rule premises and consequents are chosen by prior

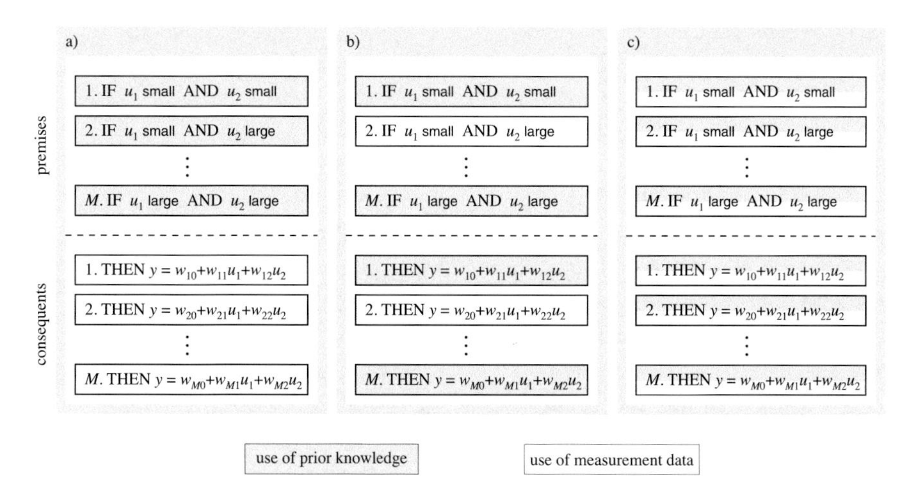

Fig. 14.10 Integration of knowledge-based and data-driven modeling: (**a**) parameter dependent, (**b**) operating point dependent, (**c**) knowledge-based initialization with subsequent data-driven fine-tuning

knowledge and subsequently a fine-tuning stage improves this initialization by an optimization method that utilizes data. Care must be taken in order to preserve the prior knowledge in the second stage; for more details, refer to Sect. 12.3.5.

14.3 Structure Optimization of the Rule Consequents

The LOLIMOT algorithm allows one to extract those variables from the premise input vector $\underline{z}$ that have a significant nonlinear influence on the process output by analyzing the generated premise input space partitioning. In analogy to Fig. 14.3, a method for structure selection of the local linear models in the rule consequents would be desirable. As proposed in [428] and extended in [414, 427, 429], a linear subset selection technique such as the *orthogonal least squares (OLS)* algorithm can be applied instead of the least squares estimation of the consequent parameters described in Sect. 13.2. For the same reasons as discussed in Sect. 13.2, it is usually advantageous to apply this structure optimization technique locally. Additionally, the benefit of lower computational demand becomes even more pronounced since the training time of the OLS grows much faster with the number of potential regressors than in the case of the LS. Similar to the premise structure selection in LOLIMOT, the OLS can be applied either directly to all potential regressors $\underline{x} = \underline{u}$ or after a preselection based on prior knowledge; see Fig. 14.11. A detailed description of the OLS and other linear subset selection schemes can be found in Sect. 3.4. These standard OLS algorithms cannot be applied directly for local estimation because the weighting of the data has to be taken into account. This can be done by carrying out the following transformations on the local regression matrices $\underline{X}_i$ and the output vector $\underline{y}$:

$$\underline{\widetilde{X}}_i = \sqrt{\underline{Q}_i}\,\underline{X}_i\,, \quad \underline{\widetilde{y}} = \sqrt{\underline{Q}_i}\,\underline{y} \tag{14.12}$$

with the diagonal weighting matrices $\underline{Q}_i$; see Sect. 13.2.2. When $\underline{\widetilde{X}}_i$ and $\underline{\widetilde{y}}$ are used in the OLS instead of $\underline{X}_i$ and $\underline{y}$, a local weighted orthogonal least squares approach results.

If all LS estimations in the LOLIMOT algorithm are replaced with an OLS structure optimization, the obtained LOLIMOT+OLS algorithm can optimize the

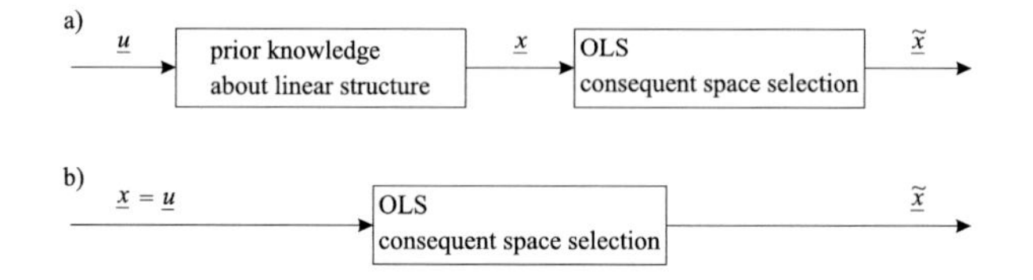

Fig. 14.11 Consequent input selection by (**a**) both prior knowledge and OLS or (**b**) OLS only

structure of the rule consequents locally. This means that different consequent structures and even consequents of different complexity can be automatically constructed. This feature is especially attractive for the identification of *dynamic* systems; see Chap. 22. Consequently, models that are easier to interpret, faster to estimate, and more parsimonious can be generated by the LOLIMOT+OLS algorithm than with standard LOLIMOT.

The operation of the LOLIMOT+OLS algorithm will be illustrated with the following example. The function

$$y = \frac{1}{u_1 + 0.1} + u_2^2 \tag{14.13}$$

will be approximated (Fig. 14.13a) utilizing five inputs $u_1, \ldots, u_5$, assuming that the relevance of these inputs is not known a priori. 625 training data samples are available in which the first two inputs u_1 and u_2 are equally distributed in the interval $[0,\ 1]$, the third input $u_3 = u_1 + 2u_2$ depends linearly on the first two inputs, u_4 is a nonlinear function of the first two inputs that is unrelated to (14.13), and u_5 is a normally distributed random variable. If no prior knowledge about the relevance of these inputs is available LOLIMOT+OLS starts with the following premise and consequent input spaces:

$$\underline{z} = [u_1\ u_2\ \cdots\ u_5]^T\,, \quad \underline{x} = [u_1\ u_2\ \cdots\ u_5]^T\,. \tag{14.14}$$

The task of the OLS is to select the significant regressors from the set of six potential regressors, i.e., all inputs gathered in $\underline{x}$ plus a constant for modeling the offset. In the first LOLIMOT iteration, the OLS estimates a global linear model. Its convergence behavior is depicted in Fig. 14.12a. This plot shows how the amount of unexplained output variance of the process varies over the selected regressors; for details refer to Sect. 3.4.2. The OLS selected the regressors in the order $u_2, u_3, 1, u_4, u_5, u_1$. While for the first three iterations of the OLS each additionally selected regressor improves the model quality, no further significant improvement is achieved for more than three regressors. Similar to the convergence behavior of LOLIMOT, the OLS algorithm can be terminated if the improvement due to an additional regressor is below a user-determined threshold. Note, however, that Fig. 14.12a illustrates an artificial example, and in practice, the convergence curve typically is not that clear. In particular, a situation may occur where a selected regressor improves the model quality only insignificantly, but the next selected regressor yields considerable improvement. Then the termination strategy described above is not optimal. In the experience of the author, it is very difficult for the user to find a suitable threshold that is appropriate for the whole training procedure. Therefore, it is recommended to fix the *number* $\widetilde{nx}$ of regressors to be selected instead of a variance threshold. In the context of *dynamic* systems, additional difficulties arise for finding a suitable termination criterion; see Chap. 22.

As Fig. 14.12a shows, the regressor u_3 is selected instead of u_1. Since u_1, u_2, and u_3 are linear dependent, it is equivalent to select any two out of these

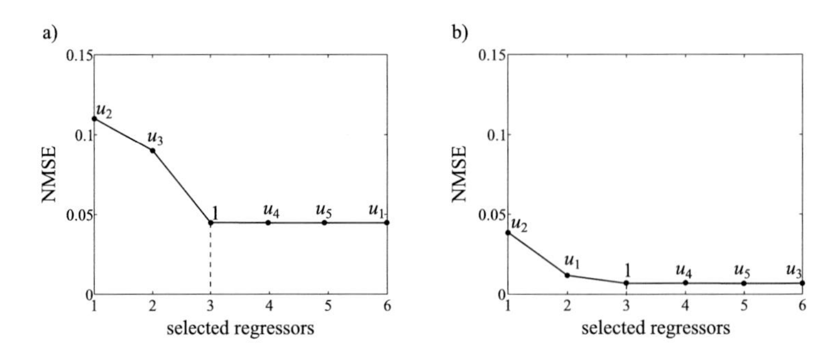

Fig. 14.12 Convergence behavior of the local OLS algorithm for regressor selection in (**a**) the first and (**b**) the tenth LOLIMOT iteration

three regressors. For this reason (when u_2 and u_3 are already chosen), u_1 is redundant and thus would be selected at last. After ten iterations, the training with LOLIMOT is terminated. The OLS optimization of one local linear model in this final LOLIMOT iteration is depicted in Fig. 14.12b. Compared with Fig. 14.12a, the level of unexplained output variance is much smaller because the linear model is valid only for a small region of the input space according to the partitioning constructed by LOLIMOT. In a smaller region, the process can be approximated more accurately by a linear model, and thus the curve in Fig. 14.12b is closer to zero. Note that the unexplained output variance depends on both the mismatch due to the nonlinear process characteristics and the local disturbance level (which is equal to zero in this example).

It is interesting to note that during the whole training procedure of LOLIMOT, u_2 is the most significant regressor for all local linear models. The reason for this lies in the input space partitioning constructed by LOLIMOT. Many more splits are carried out in the u_1-dimension, which decreases the importance of u_1 in the rule consequents since the LLMs possess a larger extension in the u_2- in the u_1-dimension.

In the final model generated by LOLIMOT+OLS, all rules have one of the following two forms:

$$R_i\text{:}\quad \text{IF } u_1 = \ldots \text{ AND } u_2 = \ldots \text{ THEN } y = w_{i0} + w_{i1}u_1 + w_{i2}u_2$$

or

$$R_j\text{:}\quad \text{IF } u_1 = \ldots \text{ AND } u_2 = \ldots \text{ THEN } y = w_{j0} + w_{j1}u_2 + w_{j2}u_3\,.$$

The irrelevant inputs u_4 and u_5 and the redundant input (either u_1 or u_3) are successfully *not* selected for any rule. LOLIMOT discovered that the nonlinear characteristics are determined only by u_1 and u_2 and does not perform a split in any other input. Thus, in subsequent investigations with the same process, this

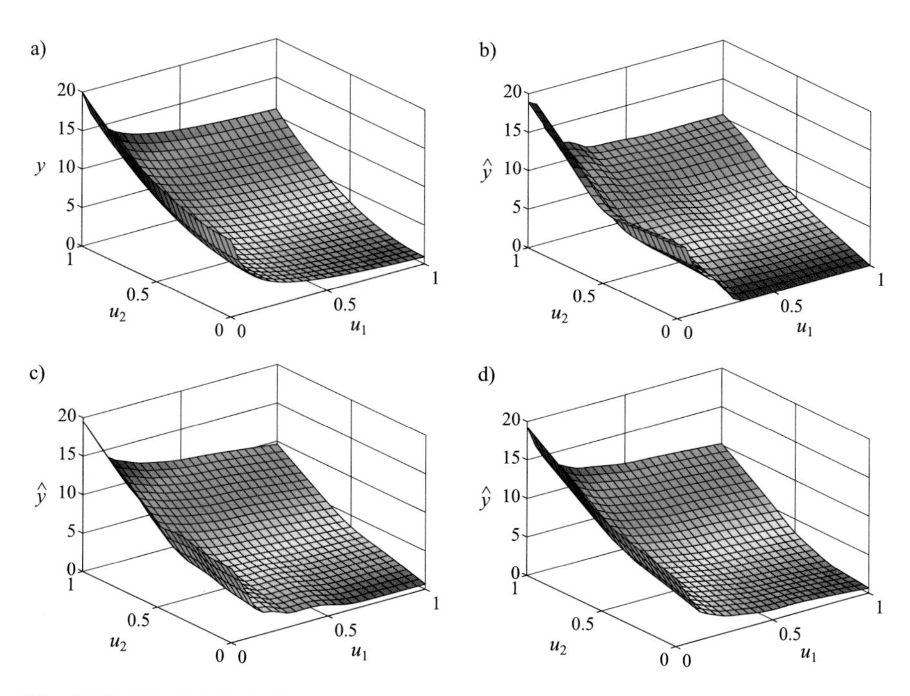

Fig. 14.13 (**a**) Original function (14.13). Local neuro-fuzzy models constructed by LOLIMOT+OLS with ten rules and (**b**) one, (**c**) two, and (**d**) three selected regressors

knowledge can be utilized by discarding the obviously irrelevant variables from the premise and consequent input spaces.

Figure 14.13 demonstrates how the model quality improves with the number of selected regressors $\widetilde{nx}$ per local linear model. The original function in Fig. 14.13a is approximated with a neuro-fuzzy model with ten rules. The RMSE decreases from 0.7 for $\widetilde{nx} = 1$ to 0.4 for $\widetilde{nx} = 2$ and 0.25 for $\widetilde{nx} = 3$. As illustrated in Fig. 14.12, a further increase of $\widetilde{nx}$ would incorporate irrelevant regressors into the model, which would not further improve the performance. Note that the input space partitioning created by LOLIMOT for Fig. 14.13b, d is different. Furthermore, it is interesting to realize that the model in Fig. 14.13b is similar to an NRBF network in the sense that it also utilizes only a single parameter (weight) for each validity function. However, the LOLIMOT+OLS approach is much more flexible and thus performs much better because the optimal regressor is selected for each LLM, while an NRBF network always utilizes a constant; see Sect. 13.1.4.

Since the computational demand of the OLS algorithm grows strongly with the number of potential regressors, their number should be limited by exploiting prior knowledge whenever it is available. An extension of the above described LOLIMOT+OLS approach may be advantageous for many practical problems. Often some variables are known a priori to be relevant for the rule consequents.

Then these variables should be chosen as regressors before a subsequent OLS phase can be used to select additional regressors.

In the proposed LOLIMOT+OLS approach, the OLS is nested within the LOLIMOT algorithm, that is, in *each iteration* of LOLIMOT, the OLS is utilized for structure optimization of rule consequents. An alternative, simplified strategy in order to save computation time is to run the conventional LOLIMOT without structure optimization of rule consequents and to subsequently apply the OLS only to the final model. In particular, this strategy is beneficial for models with many rules (large M) and not too many potential regressors ($\widetilde{nx} \not\ll nx$). (For a large number of potential regressors nx compared with the selected regressors, the least squares estimation can become more time-consuming than the OLS.) Note that compared with the LOLIMOT+OLS approach, such a simplified two-stage LOLIMOT and OLS strategy usually generates inferior models since the tree-construction algorithm constructs a different premise structure based on the assumption that all nx regressors are incorporated into the rule consequents.

14.4 Interpolation and Extrapolation Behavior

The interpolation and extrapolation behavior of local linear neuro-fuzzy models are determined by the type of validity functions used. Although the whole chapter focuses on normalized Gaussian validity functions, some examples in this section use a triangular type to illustrate some effects more clearly. The validity functions can be mainly classified according to the following two properties:

- *Differentiability*: The smoothness of the model depends on how many times the validity functions are differentiable. The smoothness also depends on the utilized local models. The higher the polynomial degree of the local models is, the smoother the overall model becomes.
- *Strict locality*: The normalization side effects (see Sect. 12.3.4) are strongly dependent on whether the validity functions are local or strictly local, i.e., have compact support and thus decrease exactly to zero.

Some undesirable extrapolation properties of local linear neuro-fuzzy models are caused by the normalization side effects and thus can be overcome or weakened by the utilization of a hierarchical model structure; see Sect. 14.8.8. The interpolation properties discussed in the next section, however, are of fundamental nature and apply equivalently for flat and hierarchical structures.

14.4.1 Interpolation Behavior

In order to illustrate the fundamental interpolation properties, an example with only two rules as shown in Fig. 14.14 is considered. The triangular and normalized

Gaussian validity functions are depicted in Fig. 14.14c(left) and (right), respectively. Two alternative scenarios are investigated. The local linear models $\hat{y}_1$ and $\hat{y}_2$ in Fig. 14.14a possess similar slopes, and their point of intersection lies beyond the interpolation region.[1] In contrast, the local linear models in Fig. 14.14b intersect within the interpolation region, and their slopes are highly different. In [17] these two cases are called *S-type* and *V-type* interpolation, respectively. While the interpolation behavior in the S-type is as expected, the V-type characteristics are unexpected and thus undesirable. The basic interpolation characteristics are similar for triangular and normalized Gaussian validity functions. However, Fig. 14.14(left)a and b reveals that the model outputs are non-differentiable at the edges of the triangles $u = 1/3$ and $u = 2/3$, while the behavior is very smooth in Fig. 14.14(right)a and b. The example in Fig. 14.14 can be extended to more than two triangular validity functions in a straightforward manner if they are defined such

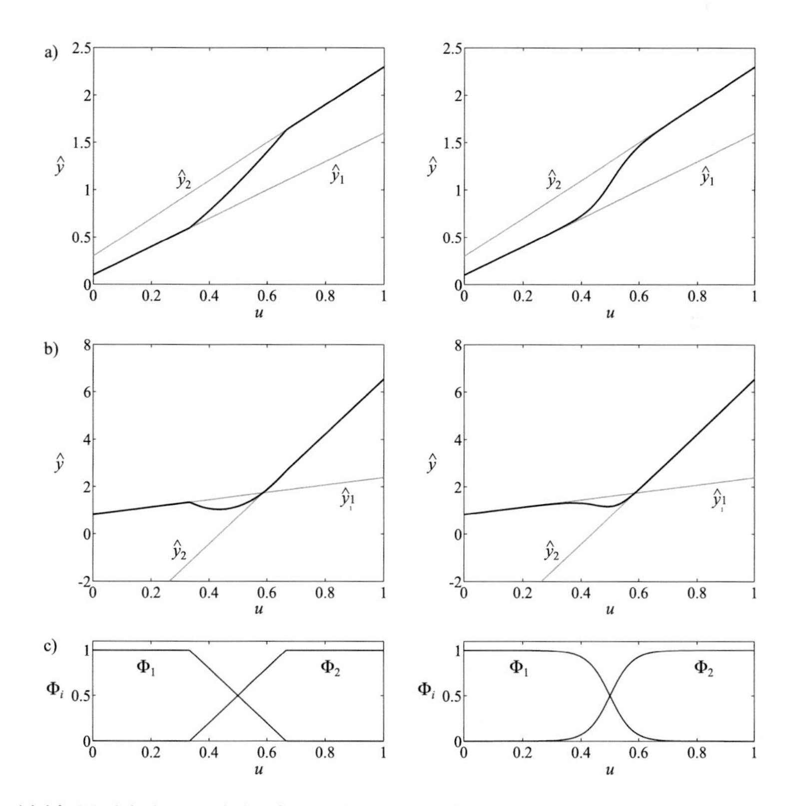

Fig. 14.14 Model characteristics for (**a**) S-type and (**b**) V-type interpolation with (**c**) triangular (left) and normalized Gaussian (right) validity functions

[1]Strictly speaking the interpolation region of the normalized Gaussians are infinitely large. However, for the degree of accuracy required in any practical consideration, it is virtually equivalent to the interpolation interval $[1/3,\ 2/3]$ of the triangular validity functions.

that only *two* adjacent validity functions overlap. In contrast, for the normalized Gaussian case, *all* validity functions always overlap. Thus, the analysis becomes more complex when the standard deviations of the Gaussians are large; otherwise it still approximately holds because only the contributions of two validity functions are significant.

Two strategies have been proposed to remedy the undesirable behavior for the V-type interpolation. Both ideas are based on the assumption of triangular validity functions and cannot be easily extended to other types such as normalized Gaussians. Note that any method that solves the interpolation problem necessarily degrades the interpretation of the model as a weighted average of local models. Either the partition of unity or the local models themselves have to be modified. Babuška et al. [16, 17] suggest modifying the inference of Takagi-Sugeno fuzzy models. The standard approach is to weight the individual rule consequent outputs $\hat{y}_i$ $(i = 1, \ldots, M)$ with their corresponding degree of rule fulfillment μ_i; see (12.12) in Sect. 12.2.3:

$$\hat{y} = \frac{\sum_{i=1}^{M} \mu_i \, \hat{y}_i}{\sum_{i=1}^{M} \mu_i} \,. \tag{14.15}$$

In [16, 17] this operation is replaced by another smooth averaging operator. The basic idea behind this new operator is that the undesirable interpolation behavior would not occur for a model output calculated as $\hat{y} = \max(\hat{y}_i)$ or $\hat{y} = \min(\hat{y}_i)$, respectively. Such a crisp switching between the rules, however, would ignore the smooth membership functions and would lead to non-continuous model output. Therefore, smoothed versions of the max or min operators are utilized that depend on the shape of the membership functions. The main drawbacks of this approach are that its complexity increases strongly with the dimensionality of the membership functions and that it is limited to cases with only two overlapping fuzzy sets.

An alternative solution to the interpolation problem is proposed by Runkler and Bezdek in [493]. It is based on a modification of the membership functions. The originally triangular membership functions are replaced by third-order polynomials. The parameters of these polynomials are determined according to the following two conditions:

- At the centers of the membership functions (cores), the output of the new model shall be identical to the output of the old model.
- The first derivative of the new model at the centers of the membership functions shall be identical to the slopes of the associated local linear models.

The main drawback of this approach is that the membership functions may become negative. This severely restricts the interpretability. Both strategies destroy the partition of unity, at least locally.

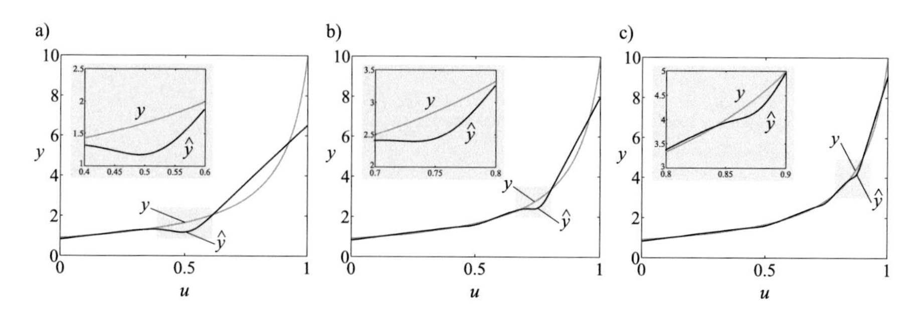

Fig. 14.15 The undesirable interpolation effects decrease with an increasing number of local linear models: model with (**a**) two, (**b**) three, and (**c**) four rules for normalized Gaussian validity functions

Another strategy is pursued by Nelles and Fischer in [423]. It is motivated by some observations that can be made from Fig. 14.15. The undesirable interpolation effects decrease as the number of rules increases because neighbored local linear models tend to become more similar. Thus, undesirable interpolation effects may be tolerably small for most applications. However, major difficulties arise when the derivative of the model is required since differentiation magnifies the interpolation effects. Therefore, it is proposed in [423] to keep the interpolation behavior but modify the differentiation of the model output by carrying out a *local linearization*. For more details, refer to Sect. 14.5.

14.4.2 Extrapolation Behavior

Extrapolation is a difficult task, which has to be carried out with extreme caution for any model architecture. If possible, extrapolation should be completely avoided by incorporating measurements from all process condition boundaries into the training data. However, in practice, it is hard to realize such a perfect coverage of all boundaries of the input space. Consequently, the extrapolation behavior of a model becomes an important issue. As further pointed out in Chap. 22, this is especially the case for *dynamic* models. "Reasonable" extrapolation properties can ensure some robustness with regard to data that lies outside the training data range.

It is not possible to define what type of extrapolation behavior is good or bad in general. Rather, it is a matter of the specific application to decide which extrapolation properties might be suitable. Prior knowledge about the process under consideration can be exploited in order to define the desired extrapolation properties. For example, the extrapolation behavior of an additive supplementary model (see Sect. 7.6.2) should tend to zero in order to recover the original first principles model in regimes where no data is available. For a multiplicative correction model, the extrapolation behavior should tend to 1 for exactly the same reason. If no prior knowledge is available, at least some smoothness assumptions on the process can be made. Common model architectures exhibit the following extrapolation behaviors:

- *None*: Look-up table, CMAC, Delaunay network
- *Zero*: RBF network
- *Constant*: MLP, NRBF, GRNN networks, linguistic, singleton fuzzy systems
- *Linear*: Linear model, local linear neuro-fuzzy model
- *High order*: Polynomial.

The local neuro-fuzzy model extrapolation behavior depends on the kind of local models employed. The focus here is on local linear models that extrapolate with a linear function. However, some unexpected effects may occur. Their explanation and a remedy are discussed in the subsequent paragraph. Finally, a strategy for enforcing an arbitrary user-defined extrapolation characteristic in local linear neuro-fuzzy models is proposed.

14.4.2.1 Ensuring Interpretable Extrapolation Behavior

Owing to normalization side effects (see Sect. 12.3.4), the normalized Gaussian validity function with the largest standard deviation reactivates for $u \to -\infty$ and $u \to \infty$. Only for two special cases is it guaranteed that the outermost validity functions maintain their activity for extrapolation:

- The standard deviations of all validity functions are equivalent in each dimension.
- The standard deviations of the outermost validity functions are equivalent and larger than the standard deviations of all inner validity functions.

Otherwise, the normalization side effects lead to undesirable extrapolation behavior because the local model that is closest to the boundary is always expected to determine the extrapolation characteristics. Figure 14.16(left) illustrates this effect for a model with three rules. The local linear model associated with the validity function Φ_3 determines the extrapolation behavior for $u > 1$, which is expected, and for $u < -0.5$, which is totally unexpected. The Gaussian membership functions in Fig. 14.16a(left) are shown in logarithmic scale in Fig. 14.17a, where it can be seen why Φ_3 reactivates for $u < -0.5$.

A solution to this problem is proposed in [411, 432] and is illustrated in Fig. 14.16(right). The degree of membership of all MSFs is frozen at the interpolation/extrapolation boundary, i.e., here at $u = 0$ and $u = 1$. Then no reactivation can occur. This remedy becomes even more important as the number of rules increases. Figure 14.17b demonstrates that otherwise, the reactivation can occur very close to the interpolation/extrapolation boundary if the width of the validity function at the boundary is very small.

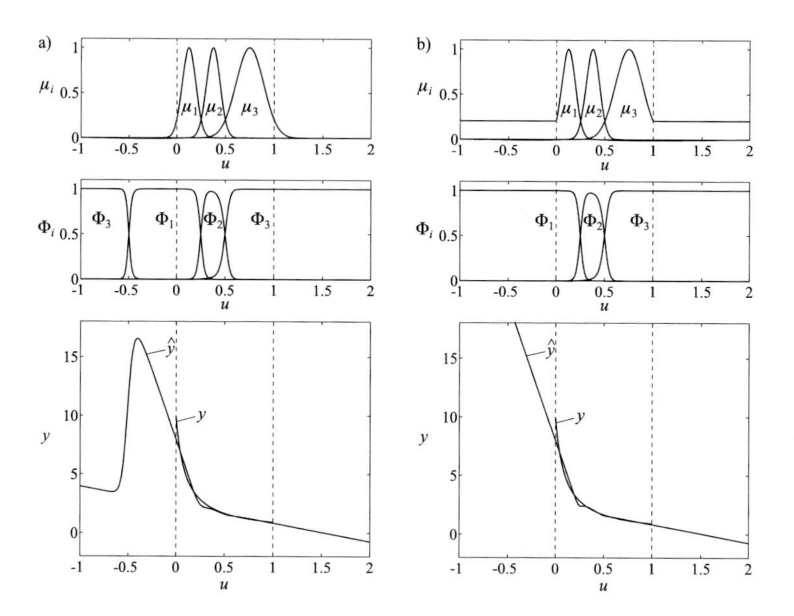

Fig. 14.16 Undesirable extrapolation effects: (**a**) original local linear neuro-fuzzy model with reactivation of Φ_3 for $u < -0.5$; (**b**) remedy of the reactivation by freezing the membership function values at the interpolation/extrapolation boundaries

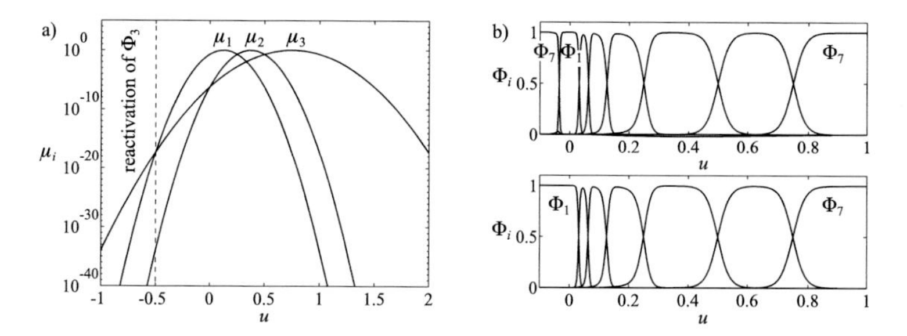

Fig. 14.17 (**a**) Logarithmically scaled membership functions from Fig. 14.16a(left). For $u < -0.5$, μ_3 dominates. (**b**) The reactivation (here of Φ_7) is close to the boundary $u = 0$ since the width of Φ_1 is small (top), no reactivation takes place due to frozen membership functions (bottom)

14.4.2.2 Incorporation of Prior Knowledge into the Extrapolation Behavior

With the strategy proposed in the previous paragraph, it can be guaranteed that the extrapolation behavior is determined by the local model that is closest to the boundary. For various applications, prior knowledge about the process output within the extrapolation regions is available in the form of lower or upper bounds or slopes. This knowledge should be exploited by incorporating it into the model. A straightforward way to do this is to define additional hyperrectangles in the input space and to construct the corresponding validity functions that describe the extrapolation regimes. The local linear models in these extrapolation regimes can be defined by the user according to the available prior knowledge. Thus, the local linear neuro-fuzzy model is extended to

$$\hat{y} = \sum_{i=1}^{M} L_i \Phi_i(\underline{u}) + \sum_{j=1}^{M_{\text{ex}}} L_j^{(\text{ex})} \Phi_j^{(\text{ex})}(\underline{u}), \tag{14.16}$$

where the first sum represents the conventional neuro-fuzzy model with the local linear models L_i and the second term summarizes M_{ex} extrapolation regimes. The extrapolation validity functions $\Phi_j^{(\text{ex})}$ are defined such that the partition of unity holds for the overall model:

$$\sum_{i=1}^{M} \Phi_i(\underline{u}) + \sum_{j=1}^{M_{\text{ex}}} \Phi_j^{(\text{ex})}(\underline{u}) = 1. \tag{14.17}$$

While the determination of the local extrapolation models $L_j^{(\text{ex})}$ is directly dependent on the desired extrapolation behavior, the choice of the corresponding validity functions $\Phi_j^{(\text{ex})}$ is not straightforward. The following difficulties arise for the choice of the extrapolation regime sizes:

- *Large extrapolation regimes* implying large widths of the extrapolation validity functions can prevent the reactivation of all interpolation validity functions and thus allow one to approach $\Phi_j^{(\text{ex})} \to 1$ as $u \to -\infty$ or $u \to \infty$, respectively. However, wide extrapolation validity functions may strongly influence the interpolation behavior of the model and thus degrade its interpolation accuracy.
- *Small extrapolation regimes* implying small widths of the extrapolation validity functions avoid this problem. However, the extrapolation validity function values must be frozen according to the strategy in the previous paragraph in order to avoid reactivation of an interpolation validity function. The drawback of the freezing procedure is that the $\Phi_j^{(\text{ex})}$ do not approach 1 arbitrarily closely. If, for example, a $\Phi_j^{(\text{ex})}$ is frozen at 0.9, then the extrapolation behavior is not solely determined by the user-defined local extrapolation model $L_j^{(\text{ex})}$ since 10% is influenced by the other local interpolation models.

Figure 14.18 illustrates the dilemma discussed above. It is assumed that the following prior knowledge about the desired extrapolation behavior is available: $y = 10$ for $u < 0$ and $y = 0$ for $u > 1$. Then the two additional extrapolation validity functions $\Phi_1^{(\text{ex})}$ and $\Phi_2^{(\text{ex})}$ shown in Fig. 14.18 are introduced with the local linear extrapolation models $L_1^{(\text{ex})} = 10$ and $L_2^{(\text{ex})} = 0$. In Fig. 14.18a, the widths of the extrapolation validity functions are chosen to be large; in Fig. 14.18b they are chosen to be small. Obviously, the desired extrapolation behavior is realized in Fig. 14.18a, but the interpolation behavior is degraded. In contrast, the interpolation properties in Fig. 14.18b are not affected. However, the desired extrapolation behavior can be

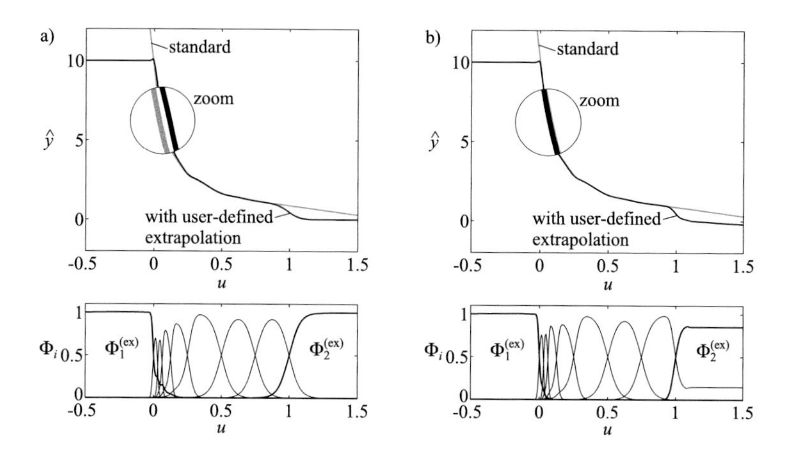

Fig. 14.18 Incorporating prior knowledge into the extrapolation behavior by additional validity functions with (**a**) large width, (**b**) small width

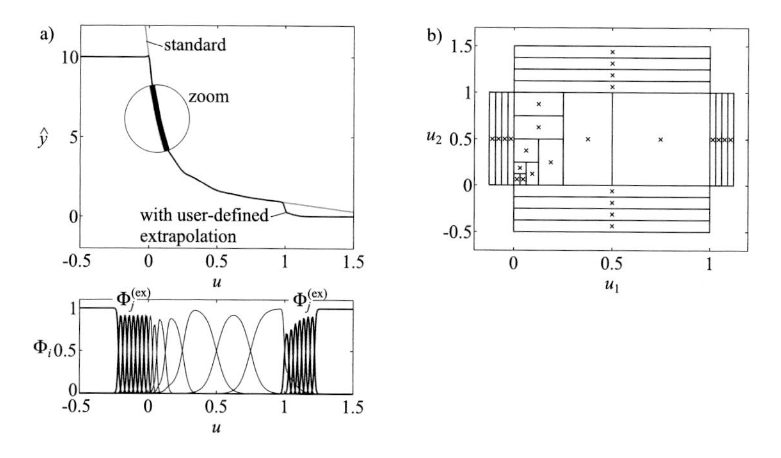

Fig. 14.19 New strategy for incorporation of prior knowledge into the extrapolation behavior: (**a**) one-dimensional example of Fig. 14.18 with eight extrapolation validity functions to the left and right; (**b**) two-dimensional example with four extrapolation regimes to the left, right, bottom, and top

achieved only for $u < 0$, while in the regime $u > 1$, the model still extrapolates with a negative slope since $\max(\Phi_2^{(ex)})$ is significantly smaller than 1.

A possible solution to the dilemma described above is to add more than one extrapolation regime at each boundary to the model. With this strategy, the advantages of both the approaches discussed above can be combined, while their drawbacks are overcome. Figure 14.19a demonstrates that the introduction of multiple extrapolation regimes with small widths successfully prevents a nearby reactivation of an interpolation validity function and furthermore ensures that the extrapolation validity functions approach 1. The price to be paid is a higher model complexity owing to the larger number of local models. The number of extrapolation regimes that are necessary depends on the width ratio of the neighboring validity functions, which can be easily assessed by the user.

The straightforward extension of this strategy to higher-dimensional premise input spaces is illustrated in Fig. 14.19b. The extrapolation (hyper)rectangles in all but one dimension extend over the whole input space. In the extrapolation dimension, their extension is chosen equal to the width of the interpolation regime with the smallest width in this dimension.

14.5 Global and Local Linearization

As analyzed in Sect. 14.4.1, the interpolation behavior of local linear neuro-fuzzy models may reveal undesirable effects. These effects can either be explicitly compensated or neglected when "enough" local models are estimated. If, however, the derivative of the model output is calculated, these interpolation effects are magnified and usually cannot be neglected anymore. Since the model's derivative is required in many applications, a remedy to this problem is of fundamental importance. Derivatives are, for example, needed for calculating the gradients in an optimization problem; see the last paragraph in Sect. 14.2.1. In the context of *dynamic* models (see Chap. 22), a wide range of linear design strategies for optimization, control, fault detection, etc. can be extended to nonlinear dynamic models in a straightforward manner by utilizing *linearization*. All these applications require a reliable calculation of the model's derivatives.

Figure 14.20a shows the derivatives of the models from the example in Fig. 14.14. While the local linear neuro-fuzzy model with triangular validity functions (left) possesses a non-continuous derivative, the derivative is smooth for the normalized Gaussian case (right). Independent of the type of validity function, however, the undesirable interpolation effects cause a partly *negative* derivative in the interpolation region although the slopes of both local linear models are positive. This behavior can have dramatic consequences since it means that the gain of a linearized model may have the wrong sign. Hence, unstable closed-loop behavior may result from a controller design that is based on such a linearized model. This effect can be overcome by calculating the derivative *locally*. Instead of differentiating the complete model equation

$$\hat{y} = \sum_{i=1}^{M} \left(w_{i0} + w_{i1}x_1 + \ldots + w_{i,nx}x_{nx} \right) \Phi_i(\underline{z}) = \sum_{i=1}^{M} L_i(\underline{x})\Phi_i(\underline{z}) \qquad (14.18)$$

analytically, i.e., *globally*, with respect to some input u_j

$$\frac{\partial \hat{y}}{\partial u_j} = \sum_{i=1}^{M} \frac{\partial}{\partial u_j} \left\{ L_i(\underline{x})\Phi_i(\underline{z}) \right\} \qquad (14.19)$$

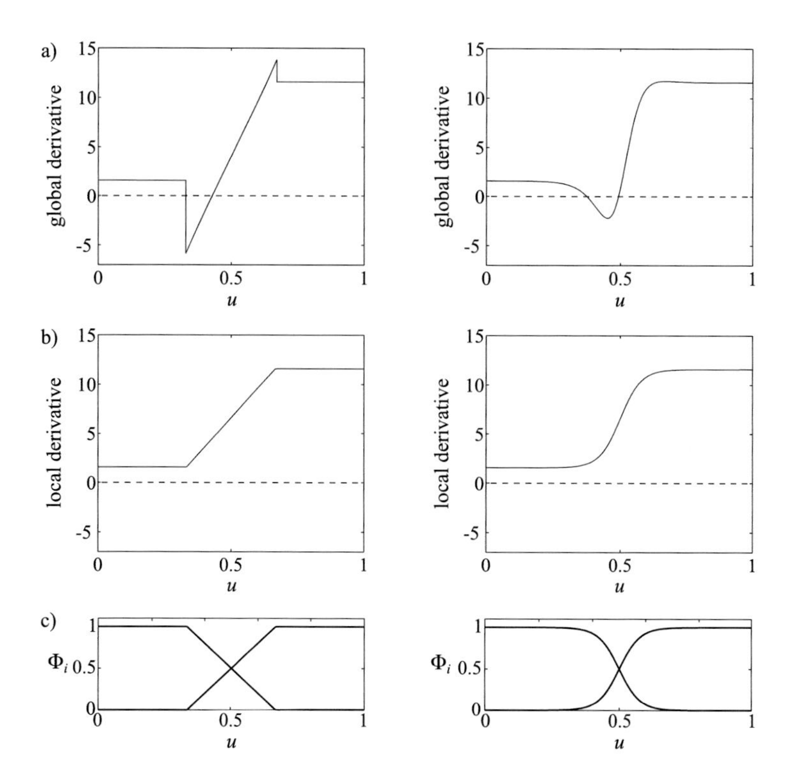

Fig. 14.20 (**a**) Global and (**b**) local derivatives of a local linear neuro-fuzzy model with (**c**) triangular (left) and normalized Gaussian (right) validity functions; with compare Fig. 14.14

only the local models are differentiated individually, i.e., *locally*,

$$\left.\frac{\partial \hat{y}}{\partial u_j}\right|_{\text{local}} = \sum_{i=1}^{M} \frac{\partial}{\partial u_j} \left\{L_i(\underline{x})\right\} \Phi_i(\underline{z}) \,. \tag{14.20}$$

This can be interpreted as an interpolation of the local model derivatives by the validity functions $\Phi_i(\underline{z})$. Similar to the relationship between local and global parameter estimation (Sect. 13.2.3), the local derivative in (14.20) offers some important advantages over the global one in (14.19) although the global derivative is the mathematically correct one. The local derivative retains the monotony of the local linear models' slopes. As shown in Fig. 14.20b, it ensures that the derivative is monotonically increasing in the interpolation region. In particular, this property implies that undesirable interpolation effects are overcome. In comparison to the approaches of Babuška et al. [16, 17] and Runkler and Bezdek in [493] discussed in Sect. 14.4.1, the local derivative approach by Nelles and Fischer [423] offers the following important advantages. It is

- Simple
- Interpretable

- Easy to compute
- Can be applied to all types of membership or validity functions and all types of local (not only linear) models

However, it solves the difficulties in the interpolation behavior only for the model derivatives, not for the model output itself. Figure 14.21 demonstrates how the local differentiation compares with the global analytic one for the example introduced in Fig. 14.15. This comparison clearly confirms that the characteristics of the local derivative are not only more intuitive but also possess a higher accuracy.

It is interesting to investigate the mathematical differences between an analytic and a local derivative. According to (14.19), the calculation of the analytic derivative with respect to an input u_j depends on whether the premise input vector $\underline{z}$ and/or the consequent input vector $\underline{x}$ contain u_j; see Sect. 14.1. The following three cases can be distinguished:

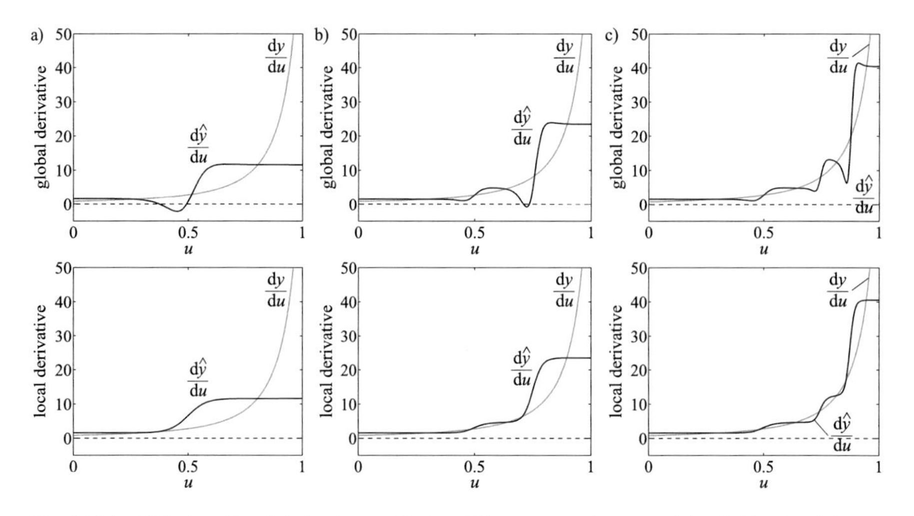

Fig. 14.21 Global and local derivatives of the local linear neuro-fuzzy models $d\hat{y}/du$ with (**a**) two, (**b**) three, and (**c**) four rules as introduced in Fig. 14.15 and derivatives of the original function dy/du

$$\frac{\partial \hat{y}}{\partial u_j} = \begin{cases} \sum_{i=1}^{M} w_{i,k} \Phi_i(\underline{z}) + L_i(\underline{x}) \dfrac{\partial \Phi_i(\underline{z})}{\partial z_l} & \text{for } x_k = u_j,\ z_l = u_j \\ \sum_{i=1}^{M} w_{i,k} \Phi_i(\underline{z}) & \text{for } x_k = u_j,\ u_j \notin \underline{z} \\ \sum_{i=1}^{M} L_i(\underline{x}) \dfrac{\partial \Phi_i(\underline{z})}{\partial z_l} & \text{for } z_l = u_j,\ u_j \notin \underline{x} \end{cases} . \tag{14.21}$$

With (14.20) the local derivative becomes

$$\left.\frac{\partial \hat{y}}{\partial u_j}\right|_{\text{local}} = \begin{cases} \sum_{i=1}^{M} w_{i,k} \Phi_i(\underline{z}) & \text{for } x_k = u_j,\ z_l = u_j \\ \sum_{i=1}^{M} w_{i,k} \Phi_i(\underline{z}) & \text{for } x_k = u_j,\ u_j \notin \underline{z} \\ 0 & \text{for } z_l = u_j,\ u_j \notin \underline{x} \end{cases} . \tag{14.22}$$

Obviously, the local and the analytic derivative are almost equivalent for

$$\frac{\partial \Phi_i(\underline{z})}{\partial z_l} \approx 0 \,. \tag{14.23}$$

This condition is met for inputs close to the centers of the validity functions and close to the extrapolation boundary. Between two validity functions where the undesirable interpolation effects occur (14.23) does not hold since the slope of the validity functions is considerable. If the standard deviations of the normalized Gaussians are decreased, the regions for which (14.23) is met are extended, i.e., analytic and local derivatives coincide for larger regions in the input space. Note, however, that then $\partial \Phi_i(\underline{z})/\partial z_l$ can become huge in the remaining small interpolation regions. This relationship is illustrated in Fig. 14.22. With the analytic (global) derivative, either large deteriorations occur for small regions of the input space (Fig. 14.22a) or small deteriorations occur for large regions of the input space (Fig. 14.22c).

If the local models $L_i(\underline{x})$ are linear, another derivation of the local differentiation can be carried out that underlines its fundamental importance. The output of a local linear neuro-fuzzy model in (14.18) can be reformulated as

$$\hat{y} = \underbrace{\sum_{i=1}^{M} w_{i0} \Phi_i(\underline{z})}_{w_0(\underline{z})} + \underbrace{\sum_{i=1}^{M} w_{i1} \Phi_i(\underline{z})}_{w_1(\underline{z})} \cdot x_1 + \ldots + \underbrace{\sum_{i=1}^{M} w_{i,nx} \Phi_i(\underline{z})}_{w_{nx}(\underline{z})} \cdot x_{nx} \,. \tag{14.24}$$

This is a pseudo-linear relationship between the rule consequent inputs x_i and the model output $\hat{y}$ with the parameters $w_i(\underline{z})$ that depend on the operating point $\underline{z}$, i.e., the premise inputs. Equation (14.24) is called a *local linearization* of the model. The operating point dependent parameters are equivalent to the local derivatives in (14.22) with respect to the inputs x_i. The local linearization can be efficiently utilized to exploit mature linear design methods for nonlinear models. The system in (14.24) is called *linear parameter varying (LPV)*; see the parameter scheduling approach discussed in Sect. 19.4.

14.6 Online Learning

The following major motivations for the application of online adaptation or learning[2] can be distinguished:

1. The process possesses time-variant behavior that would make a time-invariant model too inaccurate.
2. The model structure is too simplistic in order to be capable of describing the process in all relevant operating regimes with the desired accuracy.
3. The amount, distribution, and/or quality of measurement data that is available before the model is put to operation is not sufficient to build a model that would meet the specifications.

The first point is the classical reason for the application of an adaptive model. Truly time-variant behavior is often caused by aging or wearing of components.

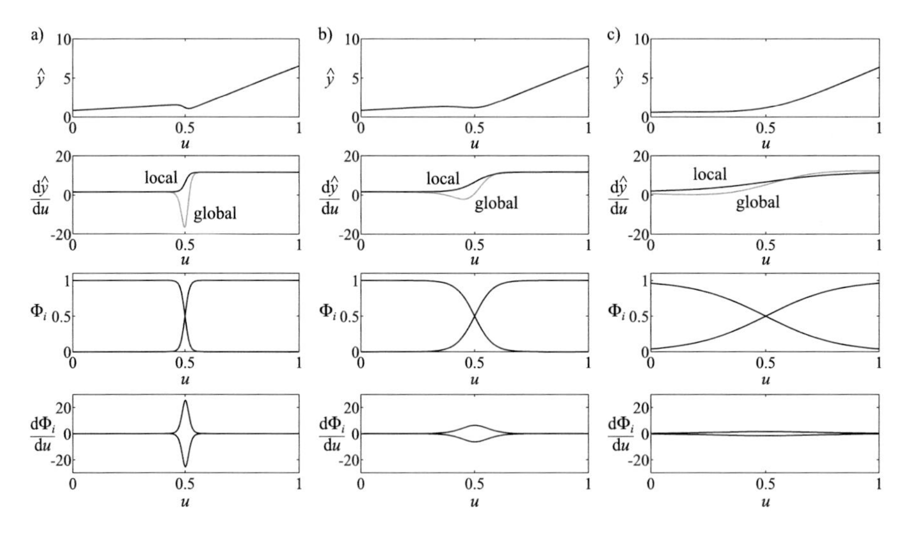

Fig. 14.22 Global and local derivatives in dependency on the model smoothness: (**a**) small, (**b**) medium, (**c**) large smoothness

[2]The term "learning" is used if the model possesses a memory in the sense that it does not forget previously learned relationships when the operating conditions change. Thus, here "learning" implies "adaptive nonlinear" plus a mechanism against arbitrary forgetting, e.g., locality.

Since the size of these effects is sometimes difficult to assess a priori, an adaptive model may be necessary in order to track the time-variant process behavior. In signal processing or control systems often, disturbances occur that can be neither modeled nor directly measured, and thus the process behavior appears to be time-variant as well. Therefore, adaptive models are widely applied in these areas. Local linear neuro-fuzzy models are particularly well suited for online learning since they are capable of solving the so-called *stability/plasticity dilemma* [91]; see Sect. 14.6.1.

The second issue also addresses an important motivation for online adaptation or learning. If, for example, a linear model is utilized for modeling a strongly nonlinear process, it has to be adapted online to the current operating point. The need for online adaptation fades as more suitable nonlinear models are employed. In fact, a good nonlinear model should make this motivation superfluous.

The third point covers a large number of realistic situations. If the amount of data that can be measured before the model goes into operation is small or the data is very noisy, only a rough model can be trained offline. (In the most extreme case, no offline measured data is available at all, which implies that no offline trained model exists.) Owing to the bias/variance tradeoff (see Sect. 7.2), both the bias and the variance error might be decreased with a subsequent online learning phase utilizing new data. A decrease in the variance error can be achieved in a relatively easy way by parameter adaptation. This approach is pursued in Sect. 14.6.1. A reduction in the bias error, however, requires an online increase of the model's flexibility, i.e., the incorporation of additional neurons into the model. This is clearly a very complex topic for future research. Some first ideas are introduced in Sect. 14.6.2.

In many applications, the distribution of the offline collected data is not perfect. Often the data might not cover all operating conditions of interest because the time for experiments is limited and the process characteristics may not be well understood a priori. Furthermore, some inputs, e.g., an environment temperature, may be measurable but cannot be actively influenced and thus excited. Therefore, the training data often cannot cover all operating conditions. Then it is important to distinguish between two cases. If the weakly excited model inputs are assumed to influence the process behavior in a mainly linear way, the approach in Sect. 14.6.1 can be pursued. If the process is assumed to depend on these inputs in a strongly nonlinear way, a complex strategy as discussed in Sect. 14.6.2 should be applied. Note that the advantageous distinction between linear and nonlinear influence of the model inputs is a feature of local neuro-fuzzy models that is *not* shared by most other model architectures; see Sect. 14.1.

For online learning in real time, the LOLIMOT algorithm is difficult to utilize directly since its computational demand grows linearly with the number of training data samples. Thus, a recursive algorithm is required that possesses constant computation time in order to guarantee execution within one (or a fixed number of) sampling interval(s). Such approaches are discussed in the following sections.

14.6.1 Online Adaptation of the Rule Consequents

In [409] a new strategy for online learning with local linear neuro-fuzzy models is proposed. It is based on the following assumptions:

1. A local linear neuro-fuzzy model has been trained offline a priori, e.g., with the LOLIMOT algorithm, and the model structure represents the nonlinear characteristics of the process sufficiently well.
2. The process is not or only negligibly time-variant in its nonlinear *structure*, i.e., the operating regions for which the process possesses strongly nonlinear behavior do not change significantly.

Both assumptions are necessary because the online learning strategy keeps the nonlinear structure of the model, which is represented by the rule premises, fixed, and adapts only the linear parameters in the rule consequents. The reason for this restriction is that nonlinear parameters cannot easily be adapted online in a reliable manner. Difficulties with local optima, extremely slow convergence, choice of step sizes, etc. usually rule out an online adaptation of nonlinear parameters in real applications. In contrast, an online adaptation of linear parameters in linear models is state of the art in adaptive signal processing and adaptive control applications. A number of robust recursive algorithms and supervision concepts are well developed; see Sects. 3.2 and 18.8. As demonstrated in the following, the step from the adaptation of linear parameters in linear models to local linear neuro-fuzzy models is still realistic even for industrial applications.

The second assumption is not crucial; it is just relevant with respect to the model accuracy that can be achieved. The more a process changes its nonlinear structure, the less optimal the offline obtained partitioning of the input space will be. Nevertheless, even in such cases, an adaptive local linear neuro-fuzzy model can be expected to outperform a simple adaptive linear model. In contrast to an adaptive linear model, an adaptive local linear neuro-fuzzy model "memorizes" the process behavior in different operating regimes. A linear model has to be newly adapted after any operating point change and thus is very inaccurate during and right after the change even if the process is time-invariant or only slowly time-variant. A local linear neuro-fuzzy model has only to adapt to true time-variance of the process and can accurately represent any operating point change without any adaptation period in which the model might be unreliable. This feature solves the so-called *stability/plasticity dilemma* coined by Carpenter and Grossberg [91]. This dilemma states that there exists a tradeoff between the speed of learning new relationships that require fast adaptation to new data (called "plasticity") and good noise attenuation that requires slow adaptation to new data (called "stability"). In other words, the dilemma expresses the fact that adaptation to new data involves the danger of *destructive learning effects*, also known as *unlearning* or *data interference* [12] for already learned relationships. By adapting locally, new relationships can be learned in one operating regime while the old information is conserved in all others. As will be shown, these benefits can be realized with very little additional computational effort in comparison with an adaptive linear model.

14.6.1.1 Local Recursive Weighted Least Squares Algorithm

For an online adaptation of the rule consequent parameters, the local estimation approach is chosen because the global version possesses additional drawbacks to those already discussed in the context of offline use in Sect. 13.2.3. So the numerical robustness becomes a critical issue since the number of parameters in global estimation can be very large. In fact, for a large number of local linear models, global estimation cannot usually be carried out with a standard recursive least squares (RLS) algorithm [278, 356]; even numerically sophisticated algorithms can run into trouble because of the poorly conditioned problem. Assuming that the rule premises and thus the validity functions Φ_i are known, the following *local recursive weighted least squares* algorithm with exponential forgetting can be applied separately for each rule consequent $i = 1, \ldots, M$:

$$\underline{\hat{w}}_i(k) = \underline{\hat{w}}_i(k-1) + \underline{\gamma}_i(k) e_i(k), \tag{14.25a}$$

$$e_i(k) = y(k) - \underline{\tilde{x}}^T(k)\underline{\hat{w}}_i(k-1),$$

$$\underline{\gamma}_i(k) = \frac{1}{\underline{\tilde{x}}^T(k)\underline{P}_i(k-1)\underline{\tilde{x}}(k) + \dfrac{\lambda_i}{\Phi_i(\underline{z}(k))}} \underline{P}_i(k-1)\underline{\tilde{x}}(k), \tag{14.25b}$$

$$\underline{P}_i(k) = \frac{1}{\lambda_i}\left(\underline{I} - \underline{\gamma}_i(k)\underline{\tilde{x}}^T(k)\right)\underline{P}_i(k-1). \tag{14.25c}$$

Compared with $\underline{x}$, the augmented consequent input vector $\underline{\tilde{x}} = [1\ x_1\ \cdots\ x_{nx}]^T$ additionally contains the regressor "1" for adaptation of the offsets w_{i0}. If prior knowledge is available, different forgetting factors λ_i and initial covariance matrices $\underline{P}_i(0)$ can be implemented for each LLM i.

14.6.1.2 How Many Local Models to Adapt

For the application of (14.25a–14.25c) to online learning of local linear neuro-fuzzy models, two strategies can be distinguished:

1. Adapt *all* local linear models at each sampling instant.
2. Adapt only those local linear models for which $\Phi_i > \Phi_{\text{thr}}$ holds, e.g., with $\Phi_{\text{thr}} = 0.1$.

The first strategy is the straightforward counterpart to the local offline estimation. All rule consequent parameters are adapted with each incoming data sample. The

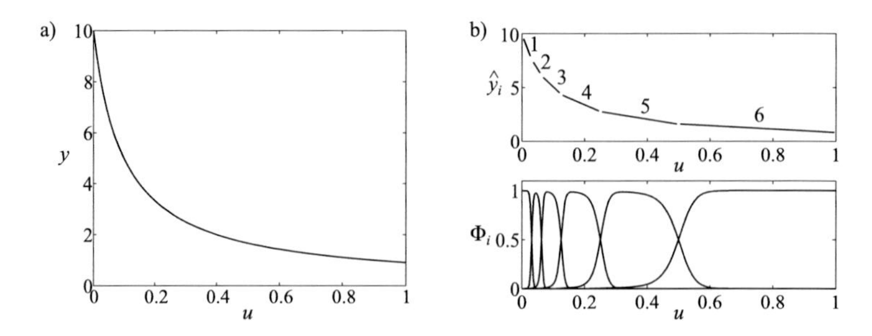

Fig. 14.23 The function in (**a**) is approximated with (**b**) six local linear models

degree of adaptation is controlled by the value of the validity function $\Phi_i(\underline{z}(k))$ in (14.25b). So nearly inactive local models are scarcely adapted.

The second strategy is a simplified version of strategy 1. For a large activity threshold Φ_{thr}, a few or even only the most active of the local linear models are adapted at each sampling instant. For a small threshold, the second strategy approaches strategy 1. As discussed in greater detail below, strategy 2 possesses some important advantages besides the obviously lower computational demand. However, a reduction in convergence speed is the price to be paid for these benefits.

As an example, the function in Fig. 14.23a is approximated by a neuro-fuzzy model with six rules; see Fig. 14.23b. Next, the parameters of the rule consequents are set to zero ($\underline{w}_i(0) = [0\ 0]^T$) in order to be able to assess the convergence behavior. In a subsequent online learning phase, 1000 data samples equally distributed in [0, 1] are generated. All covariance matrices $\underline{P}_i(0)$ are initialized with $1000\underline{I}$, and the forgetting factors are chosen as $\lambda_i = 0.98$.

14.6.1.3 Convergence Behavior

Figure 14.24 illustrates the way in which the model adapts to the process behavior. For strategy 1, Fig. 14.25a shows the convergence of the parameters of the rules representing the smallest input values (LLM 1) and the largest input values (LLM 6). All rule consequent parameters converge within the 1000 data samples to their optimal values (gray lines). Obviously, the convergence is much slower for the parameters of the first LLM than for the sixth LLM. This can be easily explained with the local model's activities shown in Fig. 14.25c. Since LLM 6 covers half of the input space it is active for about half of the data samples. In contrast, LLM 1 covers a regime that is 16 times smaller and thus is correspondingly less excited. This demonstrates that the theoretically optimal data distribution depends on the partitioning of the input space and therefore on the nonlinear characteristics of the process; see Sect. 14.9.3 for more details.

Strategy 2 is pursued in [409] and is also mentioned in [300]. As expected, it yields slower convergence behavior. The threshold in Fig. 14.25b is chosen to be $\Phi_{\text{thr}} = 0.5$, which means that only the most active local linear model is adapted.

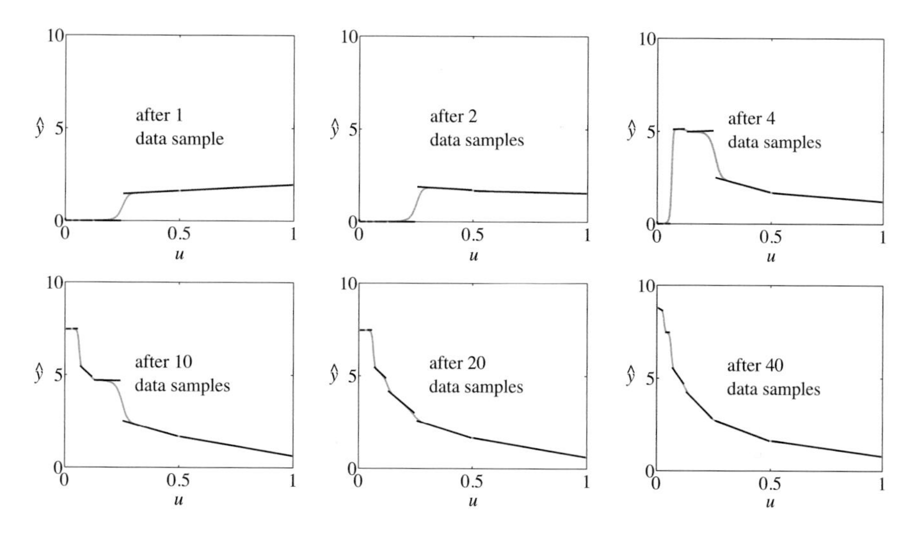

Fig. 14.24 Online adaptation with strategy 1 of the model initialized with the correct structure but wrong consequent parameters $\underline{w}_i = [0\ 0]^T$

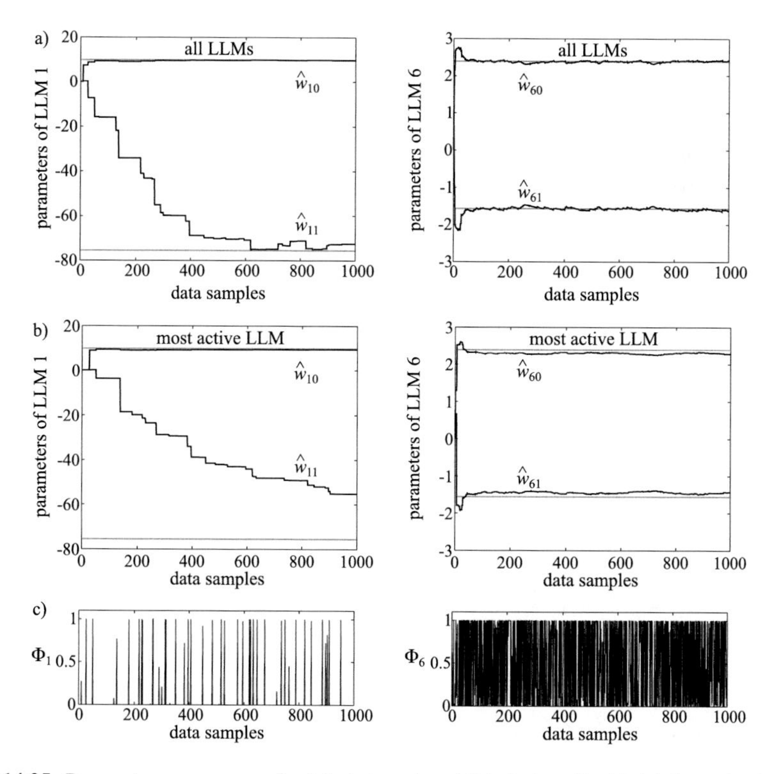

Fig. 14.25 Parameter convergence for (**a**) strategy 1 and (**b**) strategy 2 with (**c**) the activations of the corresponding local linear models

14.6.1.4 Robustness Against Insufficient Excitation

The benefits of strategy 2 are illustrated in Fig. 14.26. A model with optimal consequent parameters is adapted online with 1000 data samples. In contrast to the above example, the data samples are not "well" distributed over the whole input space. Rather all data samples lie at $u = 0.016$, which is the center of validity function Φ_1. This example imitates the very realistic situation where the process constantly stays at one operating condition for a long time, e.g., a car that is driven at a constant speed on a flat highway.

Figure 14.26a shows the model after adaptation (before adaptation the approximation was very good). Obviously, the parameters of all local linear models have been adapted such that they decrease their error at $u = 0.016$. So all lines (LLMs) go through the point (0.016, 8.5). This demonstrates that even local models with very small activations $\Phi_i \ll 1$ are considerably adapted when the process is operated long enough in one operating regime. This effect is, of course, highly undesirable because it degrades all non-active local models. It is a direct consequence of the fact that the normalized Gaussian validity functions are not strictly local, i.e., do not have compact support.

Figure 14.26b shows how fast the parameters of LLM 6 converge from their initially optimal to "wrong" values. The parameters of the other LLMs adapt even faster since their corresponding validity function values are larger than $\Phi_6(u = 0.016)$. For strategy 2 only the most active LLMs, i.e., LLM 1 and perhaps LLM 2 depending on the choice of Φ_{thr}, are adapted while all others are kept fixed. Thus, no such *destructive learning effect* as in Fig. 14.26a can occur. Therefore, in practice, strategy 2 is much more robust and should be preferred even if the convergence speed is slightly lower. The threshold Φ_{thr} can be chosen so that the most active and its neighboring LLMs are adapted. Then virtually no differences in convergence speed to strategy 1 can be observed.

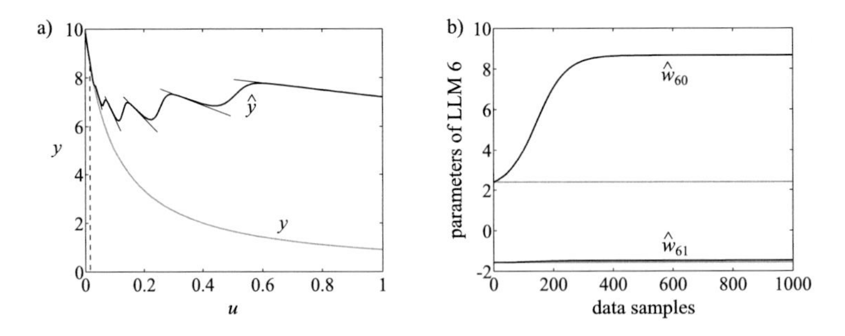

Fig. 14.26 Destructive learning effect for insufficient excitation: (**a**) process and model output (with six LLMs) after adaptation with strategy 1 with non-exciting data; (**b**) convergence of the parameters of LLM 6 from their optimal values to "wrong" values

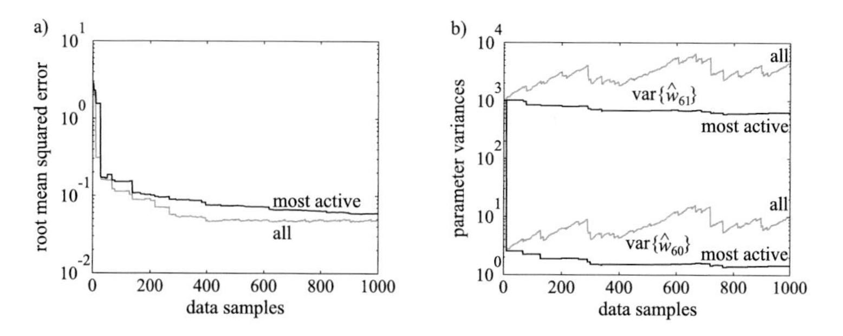

Fig. 14.27 (**a**) Convergence of the model error with strategy 1 (all) and 2 (most active). (**b**) Diagonal entries of the covariance matrix $\underline{P}_6$ for both strategies

14.6.1.5 Parameter Variances and Blow-Up Effect

Figure 14.27a compares the convergence behavior of both strategies for the original example with equally distributed online data. Obviously, the loss in convergence speed of strategy 2 (adaptation of the most active LLM) compared with strategy 1 (adaptation of all LLMs) is not very significant. In Fig. 14.27b the diagonal entries of the covariance matrix $\underline{P}_6$ are shown for both strategies. With strategy 2, the parameter variances decrease, while with strategy 1 a *blow-up effect* can be observed; see Sect. 3.2.2 for details. This effect can be prevented by controlling the forgetting factor dependent on the current excitation of the process. This issue is particularly important for dynamic models; see Sect. 28.2 and [145, 153, 165, 324].

It is interesting to note where even in the case that only a single parameter is adapted for each LLM, e.g., only the offset, the parameter variances can become very large although *no* blow-up effect can occur. This can be understood by investigating the simple case $\underline{\tilde{x}} = 1$. In steady state $\underline{\gamma}_i(k) = \underline{\gamma}_i(k-1)$ and $\underline{P}_i(k) = \underline{P}_i(k-1)$. With (14.25b) and (14.25c), this yields $\underline{P}_i(\infty) = (1-\lambda_i)/\Phi_i$, where Φ_i is the activation of LLM i. Thus, as the activation approaches zero, $\underline{P}_i(\infty) \to \infty$ without any exponential blow-up due to the forgetting factor. This is a further reason for the use of strategy 2, which prevents this effect because the less active local models are not updated.

14.6.1.6 Computational Effort

The computational demand of online learning is small when the local adaptation scheme is chosen. Already for strategy 1, the computational complexity grows only linearly with the number of local linear models. This means that the adaptation of a local linear neuro-fuzzy model requires only M times the operations needed for an adaptive linear model plus the evaluation time for the neuro-fuzzy model. When the more robust and thus recommended strategy 2 is followed, the complexity of the parameter update, in fact, becomes comparable with the linear model case.

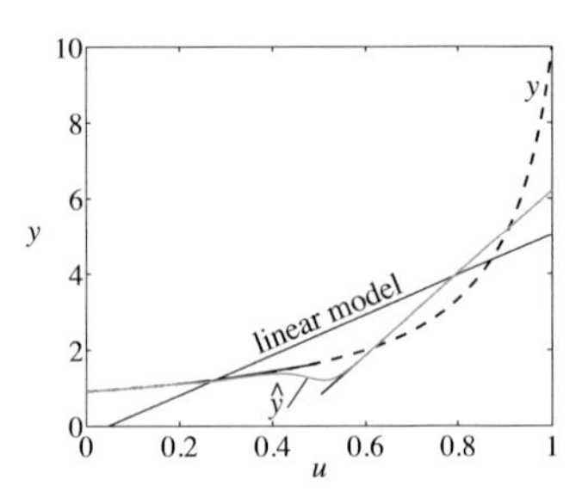

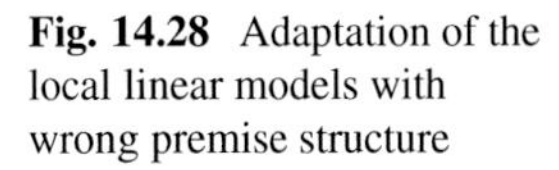
Fig. 14.28 Adaptation of the local linear models with wrong premise structure

Consequently, an online learning local linear neuro-fuzzy model can be employed in almost any application where computer technology would allow one to run an online adaptive linear model.

14.6.1.7 Structure Mismatch

The online learning strategies discussed in this section are based on a fixed rule premise structure. What model accuracy can be expected if the nonlinear structure of the process changes, that is, assumption 2 does not hold? In order to assess this effect, a worst-case scenario is considered. The model structure is optimized for the function shown in Fig. 14.23a. Then this model will be adapted online to the function shown in Fig. 14.28 (dashed line). After 1000 data samples, the model converged to its optimal consequent parameters (gray line in Fig. 14.28). Although the partitioning of the input space is not suitable for the new nonlinear characteristics of the process, the local linear neuro-fuzzy model is still significantly better than the adaptive linear model. Thus, an online learning philosophy with fixed premise structure and adaptive consequent parameters is even justified for structurally time-variant processes.

14.6.2 Online Construction of the Rule Premise Structure

If an initial model cannot be trained a priori or the available amount and/or quality of data is so low that only a rough model with very few rules can be trained, because of the bias/variance dilemma (see Sect. 7.2) or non-representative data distribution, then an online construction of the rule premise structure would be desirable. Another motivation for online construction of the rule premise structure is a process with a strongly time-variant structure. Some ideas addressing this complex task are discussed in this section.

The LOLIMOT philosophy of dividing operating regimes into two halves can be extended to a recursive online version as follows; see Fig. 14.29. A local linear neuro-fuzzy model that has been trained offline is assumed as an initial model. If no such prior model is available, the initial model is chosen as a global linear model

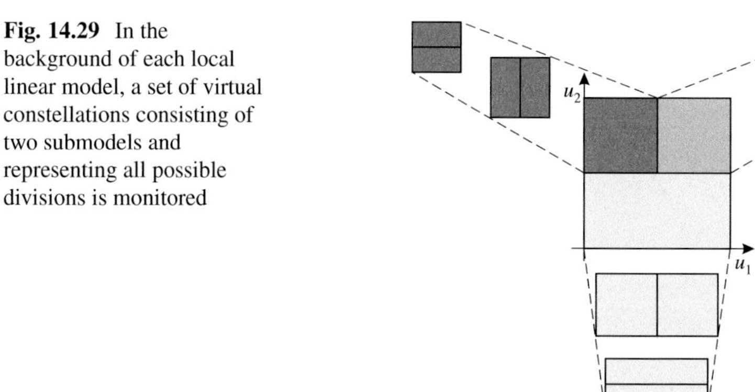

Fig. 14.29 In the background of each local linear model, a set of virtual constellations consisting of two submodels and representing all possible divisions is monitored

with the parameters $\underline{w}_1 = \underline{0}$. It is assumed that the operating range of the process is known, i.e., that lower and upper bounds are given. This information is essential for the subsequent partitioning of the input space. For a better illustration of the online construction strategy, the initial local linear neuro-fuzzy model with three rules shown in Fig. 14.29 is considered.

The existing local linear models are updated by a local recursive weighted least squares algorithm as proposed in Sect. 14.6.1. In the background, for each operating regime, two virtual local linear submodels are generated for all potential divisions. Thus, for each existing LLM, nz different constellations (each consisting of two submodels) are monitored, where nz is the number of premise inputs. In Fig. 14.29, horizontal and vertical divisions are possible, i.e., $nz = 2$. For all these constellations, local recursive weighted least squares algorithms are run in order to update the models' parameters. Note that the computational effort can be kept low if only the most active local model is considered for adaptation (strategy 2 with large Φ_{thr}, Sect. 14.6.1). Then only one actual operating LLM and $2\,nz$ virtual LLMs in the background have to be updated within each sampling period. The initial parameters of the virtual LLMs in the background can be set equal to the parameters of their "parent" LLM.

If the process behaves almost linearly within an operating regime described by one LLM, then no significant difference between this LLM and its virtual background LLMs will develop during online learning. If, however, the process is nonlinear within such a regime, then the virtual background LLMs will develop to a better process description than the single existing LLM in this regime. Depending on the nonlinear characteristics of the process, a particular virtual background constellation, i.e., division in one dimension, will outperform the others. The best virtual background constellation of two submodels replaces the existing LLM if a significant accuracy improvement can be obtained. The whole procedure can be described as follows. At each time instant, the most active local linear model is determined:

$$a = \arg\max_i \left(\Phi_i(\underline{u})\right) \quad \text{with } i = 1, \ldots, M\,. \tag{14.26}$$

Then the local loss function I_a of the most active LLM is compared with the sum of the local loss functions I_{aj}^{+} (one submodel) and I_{aj}^{-} (other submodel) of the best performing virtual background constellation with a division in dimension j, i.e.,

$$I_a > (I_{aj}^{+} + I_{aj}^{-})\, k_{\text{improve}} \tag{14.27}$$

with $k_{\text{improve}} \geq 1$ and with

$$j = \arg\min_i \left(I_{ai}^{+} + I_{ai}^{-}\right) \quad \text{with } i = 1, \ldots, nz\,. \tag{14.28}$$

If the condition in (14.27) is true, a significant improvement can be achieved by dividing LLM a into two new LLMs along dimension j where the significance level is determined by k_{improve}. Therefore, the background constellation j replaces the original LLM, that is, an online growing step is carried out. Next, new virtual background constellations are generated for the two new local linear models. Thus, the algorithm can construct an arbitrarily fine input space partitioning. In addition to this division strategy, merging of two existing LLMs can be considered with the same criteria; see Sect. 13.3.5.

The choice of the parameter k_{improve} is crucial for the behavior of this algorithm. For $k_{\text{improve}} = 1$, the algorithm tends to perform a huge number of divisions, while for large values, the algorithm stops splitting up LLMs when the local process nonlinearity is so weak that a further division cannot achieve a sufficient accuracy improvement. Another influencing factor for an appropriate choice of k_{improve} is the noise level. The noisier the data is, the larger must k_{improve} be selected to avoid a division caused by random effects instead of process/model mismatch.

Although this online structure construction algorithm works fine in theory, various difficulties may arise in practice. In the experience of the author, the choice of k_{improve} is crucial but hard to accomplish without extensive trial-and-error experiments. Thus, online learning is not very robust against variations of the noise level. Another important difficulty arises if the initial model is very rough, e.g., just linear. Then the process must run through a wide range of operating conditions in order to initiate any model division. Since the divisions are only carried out in the middle of the operating regimes, a split into two halves cannot be beneficial when a process operates only in the lower-left corner quarter (for the example with $nz = 2$) of a regime. This is an inherent drawback of the LOLIMOT decomposition philosophy; see also [310]. It can be overcome by approaches that place the centers of the validity functions on data samples, as in [399]. Such algorithms have also been proposed in the context of RBF networks [174, 175]. Grid-based approaches can be successfully applied to online learning [311, 497] as well. For offline learning, these strategies are computationally intensive or severely suffer from the curse of dimensionality, respectively. For online learning with little prior information and highly non-uniform data distributions, however, they offer some advantages.

14.7 Oblique Partitioning

Much of the remainder of this chapter deals with non-orthogonal partitioning. The focus in this book is on oblique partitioning through axis-oblique splits. Non-orthogonal partitioning also arises from clustering methods or direct numerical optimization of the centers (and possibly widths) of membership functions. In all these cases, an interpretation in terms of *one-dimensional* membership functions is not possible anymore; compare Sect. 12.3.2. From a purely mathematical point of view, such networks still represent fuzzy systems with *multi-dimensional* membership functions, but the easy interpretation is lost. Therefore the term "local model network" describes such systems better than "local linear neuro-fuzzy model" and be used for arbitrary partitioning. Also, this expression allows for local models of any type.

14.7.1 Smoothness Determination

The smoothness of the sigmoidal splitting functions shall not be optimized. The reason is exactly identical to the LOLIMOT case; compare Sect. 13.3.3. Due to the regularization character of the local estimation, the best performance (on training data) can be expected for smoothness $\to$ 0 (no overlap) because then the regularization effect vanishes. This scenario is usually undesirable in practice. Therefore the smoothness is determined heuristically and not optimized. This has the additional benefit of making the nonlinear optimization problem better conditioned and lower dimensional.

The local linear neuro-fuzzy model architecture is based on an axis-orthogonal partitioning of the input space that enables a fuzzy logic interpretation with univariate membership functions; see Sect. 13.1.3. Furthermore, efficient structure and parameter learning schemes like LOLIMOT (Sect. 13.3) can be exploited. In opposition to these major advantages, the axis-orthogonal partitioning restricts the model's flexibility, as can be observed from Fig. 14.30. The function in Fig. 14.30a is nonlinear in $u_1 + u_2$, i.e., its nonlinear characteristic stretches along the diagonal

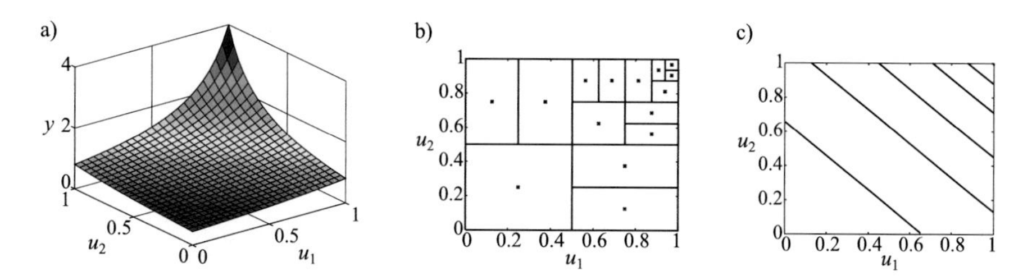

Fig. 14.30 Approximation of the function in (**a**) with (**b**) an axis-orthogonal and (**c**) an axis-oblique decomposition strategy

of the input space. As Fig. 14.30b demonstrates, this is the worst-case scenario for an axis-orthogonal decomposition technique such as LOLIMOT since a relatively large number of local linear models are required for adequate modeling. For an axis-oblique decomposition strategy, a partitioning as shown in Fig. 14.30c may result that grasps the character of the process nonlinearity and thus requires significantly fewer local linear models. In practice, hardly any process will possess an exactly axis-orthogonal or diagonal nonlinear behavior. Therefore, the possible benefit of an axis-oblique decomposition method is smaller than indicated in the above example. However, with increasing dimensionality of the problem, the limitations of an axis-orthogonal partitioning become more severe. As pointed out in the context of ridge construction in Sect. 11.1.1 and MLP networks in Sect. 11.2, the capability of following the direction of the process' nonlinearity is essential for overcoming the curse of dimensionality; see Sect. 7.6.1.

14.7.2 Hinging Hyperplanes

The development of efficient axis-oblique decomposition algorithms is an interesting topic of current research. Some new ideas are based on a recently proposed model architecture, the so-called *hinging hyperplanes* [73]. Hinging hyperplanes can be formulated in the basis functions framework, that is,

$$\hat{y} = \sum_{i=1}^{M} h_i(\underline{u}) \,. \tag{14.29}$$

The basis functions h_i take the form of an open book; see Fig. 14.32a. Each such hinge function h_i consists of two lines (one-dimensional), planes (two-dimensional), or hyperplanes (higher dimensional). The edge of a hinge function at the intersection of these two linear functions is called the *hinge*. The linear parts of a hinge function h_i can be described as

$$\hat{y}_i^+ = \underline{\tilde{u}}^T \underline{w}^+ = w_{i0}^+ + w_{i1}^+ u_1 + w_{i2}^+ u_2 + \ldots + w_{ip}^+ u_p \tag{14.30a}$$

$$\hat{y}_i^- = \underline{\tilde{u}}^T \underline{w}^- = w_{i0}^- + w_{i1}^- u_1 + w_{i2}^- u_2 + \ldots + w_{ip}^- u_p \tag{14.30b}$$

with the augmented input vector $\underline{\tilde{u}} = [1\ u_1\ u_2\ \cdots\ u_p]^T$. These two hyperplanes intersect at $\hat{y}_i^+ = \hat{y}_i^-$, which gives the following equation for the hinge:

$$\underline{\tilde{u}}_{\text{hinge}}^T \left(\underline{w}^+ - \underline{w}^-\right) = 0 \,. \tag{14.31}$$

Finally, the equations for the hinge functions are

$$h_i(\underline{u}) = \max\left(\hat{y}_i^+, \hat{y}_i^-\right) \quad \text{or} \quad h_i(\underline{u}) = \min\left(\hat{y}_i^+, \hat{y}_i^-\right) \,. \tag{14.32}$$

One characteristic feature of a hinge function is that it is nonlinear only along one direction, namely, orthogonal to the hinge. According to (14.31), the nonlinearity is in the direction $\underline{w}^+ - \underline{w}^-$.

Hinging hyperplanes have a close relationship to MLP networks and to local linear neuro-fuzzy models. With MLPs they share the ridge construction. Like the sigmoidal functions used in MLPs, the hinge functions possess one nonlinearity direction. It is determined by the weight vectors $\underline{w}^+$ and $\underline{w}^-$, which describe the slopes and offsets of both linear hyperplanes. In all directions orthogonal to $\underline{w}^+ - \underline{w}^-$, hinging hyperplanes are linear, while the sigmoidal functions used in MLPs are constant. This makes hinging hyperplanes much more powerful than MLP networks. Owing to the piecewise linear models, hinging hyperplanes are also similar to local linear neuro-fuzzy models. Figure 14.31 illustrates these relationships. Note that the model in (14.29) possesses $2M$ local linear models since each hinge function consists of two linear parts.

Hinging hyperplanes can be seen as an attempt to combine the advantages of MLP networks for high-dimensional problems and the advantages of local linear neuro-fuzzy models in the availability of fast training schemes. Indeed, when assessing hinging hyperplanes with respect to several criteria ranging from training speed to interpretability, they always lie somewhere between MLP networks and local linear neuro-fuzzy models. Originally in [73] an incremental construction algorithm similar to projection pursuit regression (Sect. 11.2.7) has been proposed for hinging hyperplanes. The parameters of the hyperplanes are estimated by a local least squares technique similar to LOLIMOT. The hinge directions are not optimized by a gradient descent type of algorithm as in MLPs since they depend on the parameters of the hyperplanes. Rather, it can be shown that an iterative application of the least squares method for estimation of the hyperplanes, with a subsequent recalculation of the new hinges, converges rapidly to a (possibly local) optimum [73].

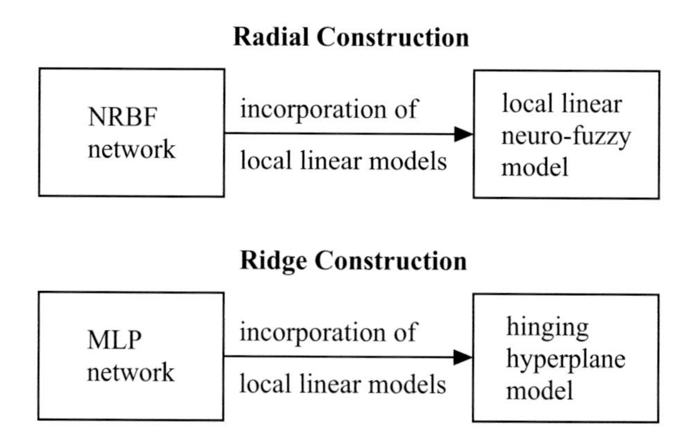

Fig. 14.31 Relationship between MLP network and hinging hyperplanes on the one hand and NRBF network and local linear neuro-fuzzy model on the other hand

14.7.3 Smooth Hinging Hyperplanes

A severe drawback of the original hinging hyperplanes is that they are not differentiable. Therefore, in [470] smooth hinge functions are proposed to overcome this problem. The min or max operators in (14.32) are replaced by a weighting with sigmoidal functions $\psi_i(\underline{u})$ known from MLP networks:

$$h_i(\underline{u}) = \underline{\tilde{u}}^T \underline{w}_i^+ \psi_i(\underline{u}) + \underline{\tilde{u}}^T \underline{w}_i^- (1 - \psi_i(\underline{u})) \,, \tag{14.33}$$

where

$$\psi_i(\underline{u}) = \frac{1}{1 + \exp\left(-\kappa \, \underline{\tilde{u}}^T (\underline{w}_i^+ - \underline{w}_i^-)\right)} \,. \tag{14.34}$$

The quantity $\underline{\tilde{u}}^T (\underline{w}_i^+ - \underline{w}_i^-)$ measures the distance of an input from the hinge. For all points on the hinge, i.e., with $\underline{\tilde{u}}^T (\underline{w}_i^+ - \underline{w}_i^-) = 0$, $\psi_i(\underline{u}) = 1/2$, which means that both hyperplanes are averaged. The $\psi_i(\cdot)$ play the role of validity functions for the hyperplanes. The parameters $\underline{w}_i^+$ and $\underline{w}_i^-$ appear in the local linear models and in the validity functions as well. Thus, they determine both the slopes and offsets of the hyperplanes and the slopes and directions of the corresponding validity functions. The parameter κ additionally allows one to adjust the degree of smoothness of the smooth hinge function (14.33) comparable to the standard deviations of the validity functions in local linear neuro-fuzzy models. For $\kappa \to \infty$ the original hinging hyperplanes are recovered. Figure 14.32 shows one- and two-dimensional examples of smooth hinge functions with their corresponding validity functions.

Similarly to the product space clustering approach for the construction of local linear neuro-fuzzy models described in Sect. 13.3, the hinging hyperplanes tie validity function parameters to the LLM parameters. The parameters $\underline{w}_i^+$ and $\underline{w}_i^-$ enter the LLMs and the validity functions. The reason for this is that the validity functions are determined by the hinges that are the intersections of the two associated LLMs. This link between the intersection of two LLMs and the validity functions is the foundation for the efficient iterative least squares training algorithm [73]. However, this link also restricts the flexibility of the model structure and causes disadvantages similar to those for the product space clustering approach; see Sect. 14.8.8.

Furthermore, the original but *not* the *extended* hinging hyperplane approach described in the next paragraph suffers from the following disadvantages:

- *Restriction to local* linear *models*: Hinging hyperplanes, like the product space clustering, approaches [16, 20, 21, 231, 335, 404, 447, 607], requires the local models to be *linear* because their construction is based on this fact. Other types of local models cannot be used, nor is exploitation of various knowledge sources by integration of different local model architectures easily possible; see Sect. 14.2.3.
- *No distinction between the inputs for the validity functions and the local linear models*: The loss of this distinction can cause severe restrictions. Note that

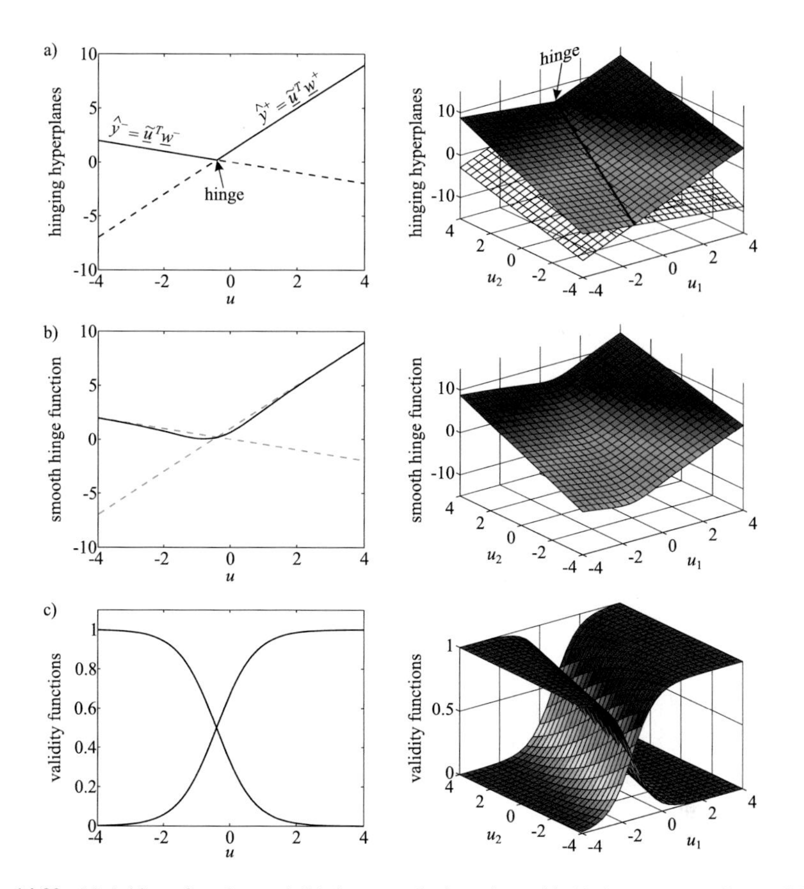

Fig. 14.32 (**a**) A hinge function and (**b**) the smoothed version with (**c**) the corresponding validity functions $\psi_i(\cdot)$ and $1 - \psi_i(\cdot)$

some construction algorithms, such as product space clustering [16, 20, 21, 231, 335, 404, 447, 607], share this drawback, although they operate on local linear neuro-fuzzy models. In hinging hyperplanes, the validity functions depend on the hinge, and the hinge depends on the parameters of the local linear models. Thus, there exists an inherent union between the inputs for the validity functions and the local linear models of the hinging hyperplanes, while they can be distinguished in rule premise inputs $\underline{z}$ and rule consequents inputs $\underline{x}$ for local linear neuro-fuzzy models. With this union, various nice features of LOLIMOT and its extensions cannot be transferred to hinging hyperplane construction algorithms. For example, no OLS structure search can be carried out for the local linear models since their parameters are coupled with the hinge. Furthermore, no structure selection like LOLIMOT can be performed for the rule premise variables for exactly the same reason. Also, the incorporation of prior knowledge in the form of different input spaces for the rule premises and consequents is not possible. This feature, however, is essential for model complexity reduction, training time reduction, and improved interpretation; see Sect. 14.1.

These difficulties can be overcome by an extension of the original hinging hyperplane approach. The validity function and LLM parameters of such an *extended hinging hyperplanes* approach are independent. This increases the number of parameters and consequently the flexibility of the model. The validity functions can be placed and oriented arbitrarily in their own $\underline{z}$-input space, independent from the local linear models that live in their $\underline{x}$-input space. For training, the validity function parameters have to be determined by nonlinear optimization, and the LLM parameters can be optimized by linear regression pursuing either the global or local estimation approach; see Sect. 13.2. The price to be paid for this extended version of the hinging hyperplanes is a higher computational training effort than for the iterative least squares algorithm. Nevertheless, it seems to be much more flexible and promising.

14.7.4 Hinging Hyperplane Trees (HHT)

A tree-construction algorithm for smooth hinging hyperplanes has been proposed by Ernst [137, 138]. It builds a hinging hyperplane tree (HHT) based on the same ideas as LOLIMOT. However, the generated model is hierarchical rather than flat; see Sect. 14.8.8. Since the hinge direction must be determined by an iterative procedure, which is computationally more expensive than the axis-orthogonal decomposition performed by LOLIMOT, the HHT algorithm is about one order of magnitude slower. This is the price to be paid for a more parsimonious model representation, i.e., the need for fewer local linear models to achieve comparable accuracy.

A hinge function can be represented as depicted in Fig. 14.33a. Then the hinging hyperplane tree can be described by a binary tree as shown in Fig. 14.33a. Each node corresponds to a split of input space into two parts. (These are "soft" splits for smooth hinging hyperplanes, i.e., they mark the area where the $\psi_i(\cdot)$ are equal.) Each leaf in the tree corresponds to a local linear model, and two leaves with the same parent belong to one hinge function. The overall model output is calculated in exactly the same manner as for hierarchical local linear neuro-fuzzy models; see Sect. 14.8.8. The overall validity functions $\Phi_i(\cdot)$ for LLM i are obtained by multiplying all $\psi_j(\cdot)$ from the root to the leaf; see Sect. 14.8.8 for more details.

For the construction of a hinging hyperplane tree, an algorithm similar to LOLIMOT can be used. Figure 14.34 illustrates three iterations of such an algorithm, which incrementally builds up the HHT shown in Fig. 14.33b. Note that in contrast to LOLIMOT, the HHT construction algorithm starts with two rather than with one LLMs, which of course is a direct consequence of the definition of a hinge function. Refer to [137, 138] for more details.

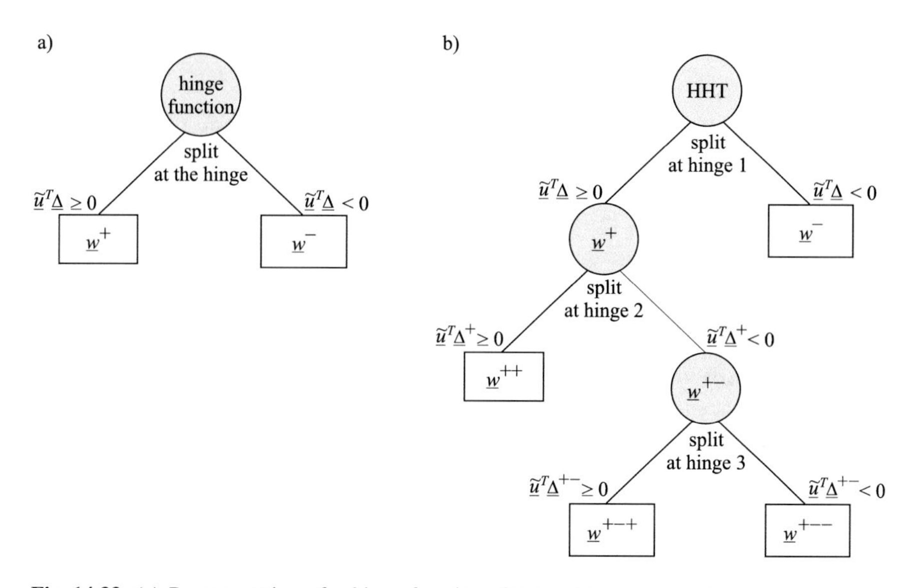

Fig. 14.33 (**a**) Representation of a hinge function. (**b**) In a hinging hyperplane tree each node represents a split of the input space into two parts, and each leaf represents a local linear model. Note that $\underline{\Delta} = \underline{w}^+ - \underline{w}^-$, $\underline{\Delta}^+ = \underline{w}^{++} - \underline{w}^{+-}$, and $\underline{\Delta}^{+-} = \underline{w}^{+-+} - \underline{w}^{+--}$

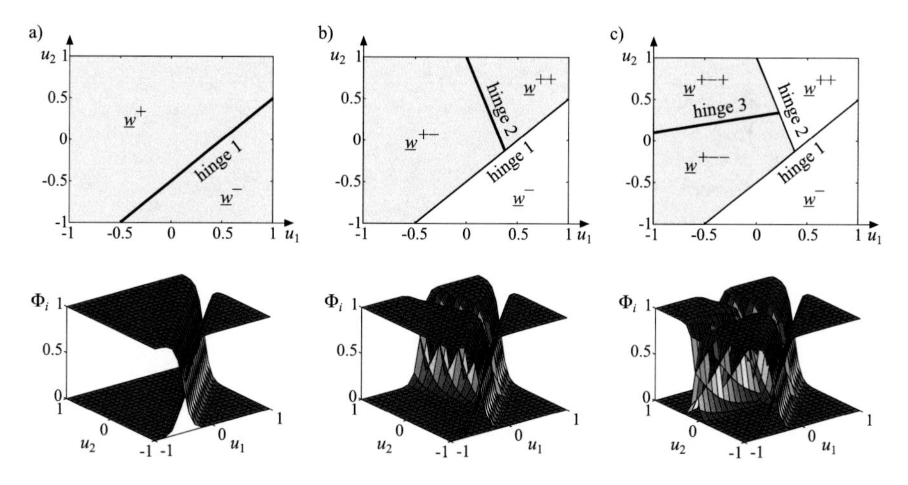

Fig. 14.34 Input space partitioning performed by the construction algorithm in three iterations as it incrementally builds the hinging hyperplane tree shown in Fig. 14.33b

14.8 Hierarchical Local Model Tree (HILOMOT) Algorithm

This section describes in detail the HILOMOT algorithm proposed in [417] that constructs an axis-oblique hinging hyperplane tree in an incremental manner very similar to LOLIMOT. It is based and expanded on the ideas discussed in [137, 138] and builds a hierarchical local model network with arbitrary $\underline{z}$-input space for the rule premises or partitioning and arbitrary $\underline{x}$-input space for the rule consequents of local models. In this chapter local *affine* models are assumed for simplicity. Note that other linear regression models like polynomials of higher degree can easily be substituted. Some implementation details are elaborated on whenever fundamental to the proper operation of the algorithm.

First, a motivating example that compares HILOMOT with LOLIMOT is discussed.

Example 14.8.1 Motivating Example for Oblique Partitioning

This example is taken from [417] and represents a worst-case scenario for LOLIMOT since the nonlinearity lies in the diagonal of the input space where axis-orthogonal partitioning performs worst. Consider the process

$$y = \frac{1}{0.1 + 0.5(1 - u_1) + 0.5(1 - u_2)} . \tag{14.35}$$

This function shall be approximated with a normalized root mean squared error of less than 5%. With LOLIMOT 14 local linear models were needed, while the axis-oblique partitioning strategy could achieve the same accuracy with 5 local linear models; see Figs. 14.35 and 14.36.

A generalization of this function shows that the advantage of the axis-oblique strategy roughly increases exponentially with the input space dimensionality. While the axis-oblique strategy for this problem is almost independent of the input space

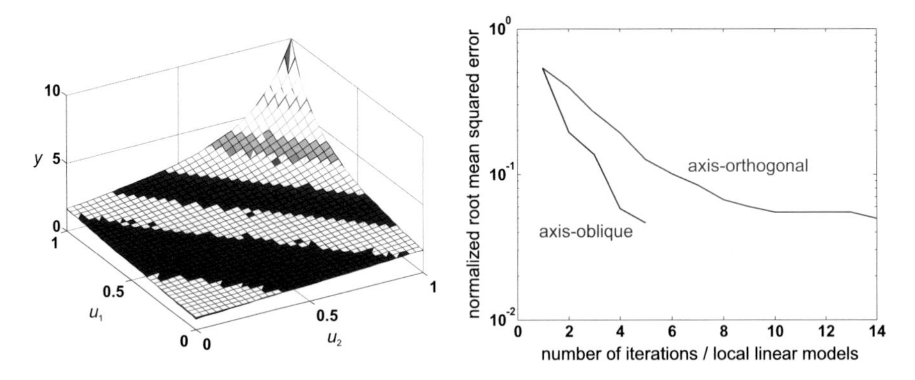

Fig. 14.35 Left: process (light) and model (solid) output with the axis-oblique partitioning strategy. Right: Convergence behavior for axis-orthogonal (LOLIMOT) and the axis-oblique (HILOMOT) partitioning strategy

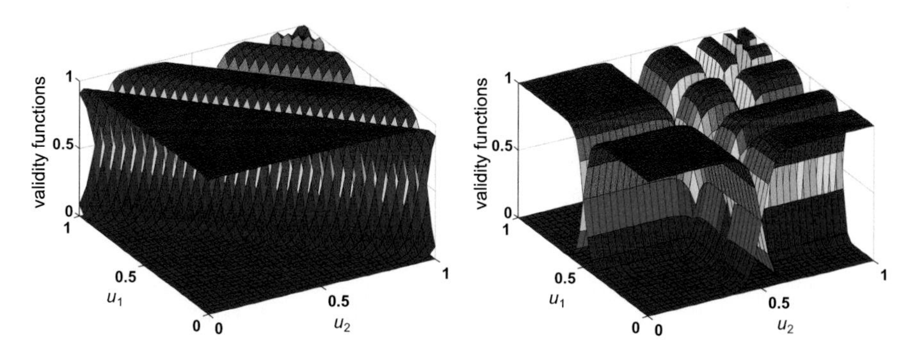

Fig. 14.36 Validity functions of the models constructed by the axis-oblique (HILOMOT) (left) and the axis-orthogonal (LOLIMOT) (right) algorithm

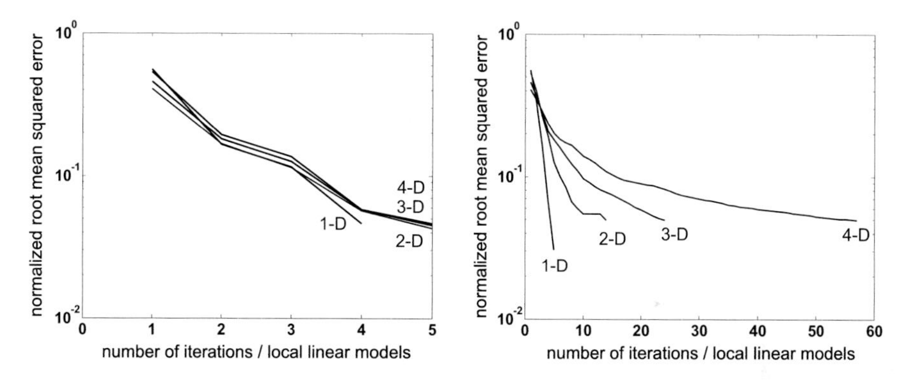

Fig. 14.37 Convergence behavior for an one-, two-, three-, and four-dimensional approximation problem with the axis-oblique (left) and axis-orthogonal (right) partitioning strategy

dimension and requires only 4 to 5 local linear models to achieve an error measure of 5%, the axis-orthogonal strategy requires 5, 14, 27, and 57 local linear models for the one-, two-, three-, and four-dimensional case, respectively. This is illustrated in Fig. 14.37. □

14.8.1 Forming the Partition of Unity

In principle, (at least) two ways of forming the partition of unity exist: (i) normalization and (ii) hierarchy. Normalization (called defuzzification in fuzzy logic) is the most widely used method. It is very simple to implement and yields a flat network structure where all partitions are on the same level. However, in spite of its popularity, normalization suffers from many severe drawbacks:

- Side effects can lead to reactivation of partitions in an obscure and often surprising manner, in opposition to the desired transparency expected from fuzzy logic; compare Sect. 12.3.4.
- Membership functions with strictly local support are difficult to handle during partition optimization if not the full tensor product is utilized to cover all membership function combinations. Care must be taken to avoid "holes" without any membership activation in the input space.
- All partitions depend on each other. If one partition changes, all others change as well due to the common normalization denominator. The coupling between the partitions can be weak (typical for LOLIMOT-trained networks due to the orthogonal partitioning) or strong (typical for clustering-based approaches due to the often densely populated covariance matrices).
- Numerical difficulties can arise if the degree of membership becomes negligibly small in order to avoid calculations close to $\frac{0}{0}$. A proper countermeasure is, e.g., to add a tiny value ε to the denominator, i.e., $\frac{0}{0+\varepsilon}$.

A beneficial alternative way of ensuring a partition of unity is to build up a hierarchy of splits where each split maintains the partition of unity by using complementary splitting functions Ψ and $1 - \Psi$. Such an approach is more complicated and requires additional computation due to the calculation of the validity functions as the product of all splitting functions in the branches of the tree from each leaf to root node. Also, it cannot efficiently be carried out in parallel. However, the following advantages strongly favor the hierarchy approach:

- No side or reactivation effects occur. Since in higher dimensions the input space partitioning cannot be visualized, it is very important to the user to reliably associate certain local models to certain regions in the input space.
- Partitions in different tree branches are strictly decoupled. Especially for any adaptive scheme (*evolving system*) or any second improvement or fine-tuning learning phase, it improves the reliability significantly if all changes in one split are limited to some small parts of the model. This is guaranteed through the hierarchical structure.

These benefits of the hierarchical approach are extremely important and often outweigh its before-mentioned drawbacks. However, for *smooth* splitting functions as discussed in this book, another (potentially severe) drawback arises in context with the degree of smoothness. Section 14.8.4 elaborates on this topic in detail.

Figure 14.38 demonstrates that normalized Gaussian membership functions always fulfill the "partition of unity" property and can look very similar to sigmoidal functions. However, strange reactivation effects can occur when the standard deviations are different in the same dimension. For the one-dimensional case, this was already discussed in Sect. 12.3.4. For *non-orthogonal* partitioning strategies, this effect is extremely likely to arise as it can occur in any dimension. Therefore normalized Gaussians cannot be recommended in the non-orthogonal partitioning case.

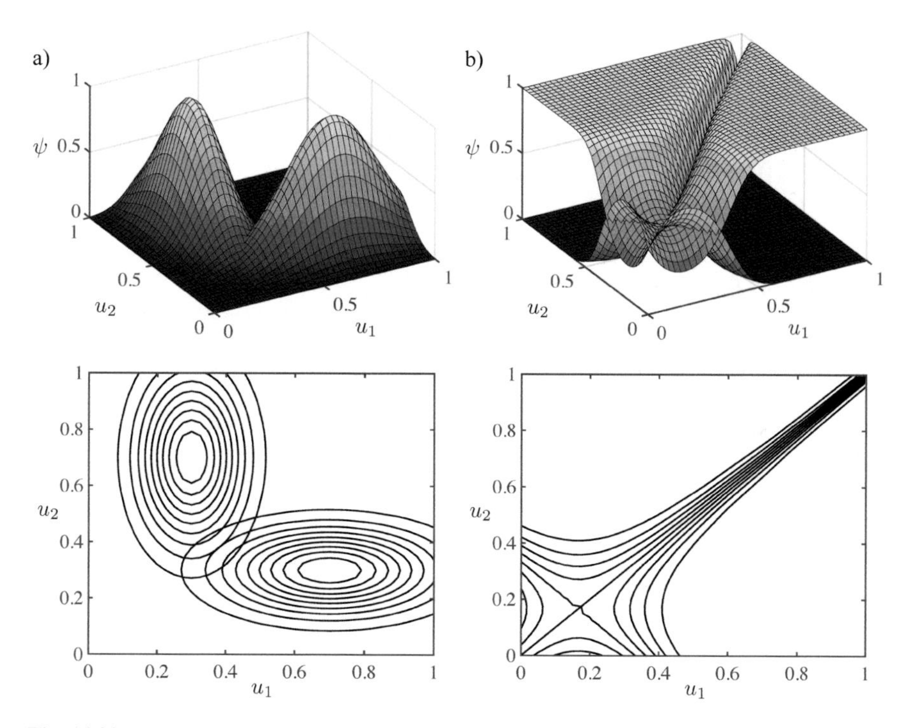

Fig. 14.38 Normalization side effects in 2D: (**a**) Two Gaussian membership functions. (**b**) Normalized Gaussian validity functions are similar to sigmoids but undesirable reactivation effects can occur (close to lower left corner in (**b**))

Finally, in principle, any axis-oblique partitioning strategy suffers from the following disadvantages:

- *Higher computational effort*: Since the nonlinear split directions must be optimized, an iterative training procedure is required that, although much more efficient than MLP network training, is about one order of magnitude slower than LOLIMOT. Furthermore, some numerical difficulties may be encountered during training, and convergence is ensured only to a local optimum. It should be stressed, however, that this has not caused any problems in practice up to now.
- *No true fuzzy logic interpretation*: Although some authors always speak of neuro-fuzzy systems when a model is evaluated according to fuzzy inference equations, no true interpretation is possible if the multivariate fuzzy sets cannot be projected to the inputs; see Sect. 12.3.4. With an axis-oblique partitioning of the input space as with hinging hyperplanes or HILOMOT, univariate membership functions cannot be constructed. Thus, interpretability is degraded for this model architecture.

Nowadays the first mentioned issue becomes more and more obsolete since very powerful computational resources are cheap. In addition, the HILOMOT algorithm described in this section is extremely efficient. The second issue remains one key weakness of any axis-oblique strategy. Thus, in applications where interpretation is in the main focus of the user, LOLIMOT may still be the better choice. If the focus tends more toward model accuracy and good handling of high-dimensional problems, HILOMOT will be a superior alternative.

14.8.2 Split Parameter Optimization

The validity functions Φ are constructed by multiplying splitting functions Ψ through the tree hierarchy. This is detailed in Sect. 14.8.3. Each individual split is performed in a smooth manner in the input space spanned by $\underline{z}$ by a sigmoidal function Ψ already known from MLP networks; compare Sect. 11.2,

$$\Psi(\underline{z}) = \frac{1}{1+\exp(-x)} \quad \text{with } x = \kappa \cdot (v_0 + v_1 z_1 + \ldots + v_{nz} z_{nz}) \tag{14.36}$$

where the splitting weights $v_1, \ldots, v_{nz}$ collected together in the vector $\underline{v}$ determine the direction of the nonlinearity and their ratio with respect to v_0 determines the distance from the origin of the coordinate system, i.e., $\underline{z} = \underline{0}$. The parameter κ can be used to adjust the steepness of the sigmoid and therefore the smoothness of the model. This parameter is redundant and therefore not strictly necessary because it can be incorporated into the $v_i, i = 0, \ldots, nz$ but it is very convenient to control the steepness with one single parameter. It plays the inverse role of the parameter k_σ for LOLIMOT; see Sect. 13.3, i.e., hard switching between the local models happens for $k_\sigma \to 0$ or $\kappa \to \infty$. Each split creates two local models where one model is weighted with Ψ and the other model with $1 - \Psi$ so that the partition of unity is always maintained. Both splitting functions cross at $\Psi = 1 - \Psi = 0.5$, i.e., the following equations hold at this point for the split:

$$\frac{1}{1+\exp(-x)} \stackrel{!}{=} \frac{1}{2} \tag{14.37}$$

$$\Rightarrow \quad x = \kappa \cdot (v_0 + v_1 z_1 + \ldots + v_{nz} z_{nz}) = 0\,. \tag{14.38}$$

For determining the direction of a split in nz dimensions, only nz degrees of freedom are necessary. Figure 14.39a illustrates this fact for $nz = 2$ dimensions. At least three different approaches for split optimization seem to be reasonable:

1. Optimization of all $nz + 1$ split parameters $v_0, v_1, \ldots v_{nz}$. This would optimize the direction and steepness of the sigmoid.
2. Optimization of all $nz + 1$ split parameters $v_0, v_1, \ldots v_{nz}$ with one constraint for fixing the steepness of the sigmoid. It seems very straightforward to enforce $||\underline{v}|| = \text{const.}$; compare Fig. 14.39a.

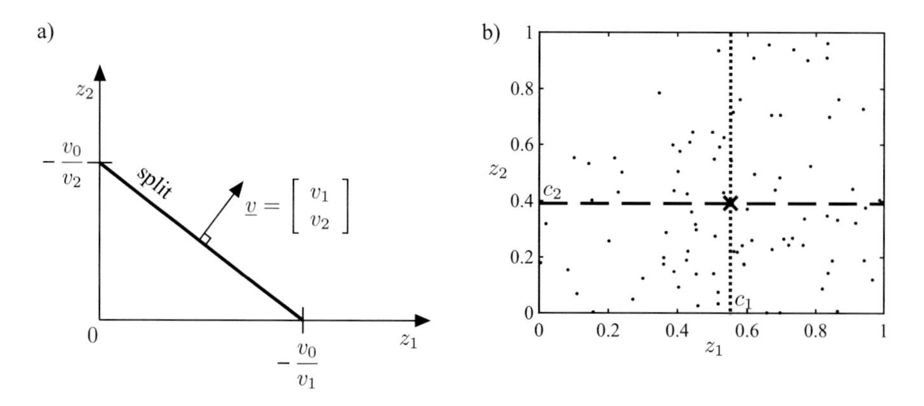

Fig. 14.39 (**a**) Split for a two-dimensional input space ($nz = 2$) with the role of the parameters v_0, v_1, v_2. (**b**) LOLIMOT splits through the center of the local model (cross) orthogonal to dimension 1 (dotted) yielding $v_0^{(\text{ini})} = -c_1, v_1^{(\text{ini})} = 1, v_2^{(\text{ini})} = 0$ and orthogonal to dimension 2 (dashed) yielding $v_0^{(\text{ini})} = -c_2, v_1^{(\text{ini})} = 0, v_2^{(\text{ini})} = 1$

3. Optimization of only nz parameters. For example, v_0 can be fixed while $v_1, \ldots v_{nz}$ are optimized.

The difficulty with approach 1 was already explained in Sect. 13.3.3 for LOLIMOT. The weights v_i would tend to infinity to minimize the regularization effect since it is disadvantageous for the training data error. Thus Approach 1 would yield hard switching networks which is undesirable. Approach 2 is the most transparent and intuitively appealing but requires the utilization of constrained optimization techniques which are less robust and mature than unconstrained ones. Approach 3 eliminates the redundant parameter which leads to coupling between the remaining ones. For example, the L_2 norm of the extended parameter vector (including v_0) can be forced to be equal to one by using normalized split parameters

$$\widetilde{v}_i = \frac{v_i}{v_0^2 + v_1^2 + \ldots + v_{nz}^2}\,, \qquad i = 0, \ldots, nz\,. \tag{14.39}$$

Note that with such a normalization the data needs to be shifted into one quadrant of the coordinate system, e.g., $\underline{z} \in [0, 1]^{nz}$ in order to rule out any divisions by 0 during optimization.

Approach 3 has the additional benefit of reducing the dimension of the optimization problem by one. Therefore HILOMOT pursues Approach 3. One degree of freedom is used for optimizing the distance of the split from the origin and the remaining $nz - 1$ degrees of freedom describe the direction of the split. All sigmoids have an identical steepness that in a second, subsequent step is scaled by a factor κ which is heuristically determined; refer to Sect. 14.8.4.

14.8.2.1 LOLIMOT Splits

Since HILOMOT optimizes the nonlinear split parameters by a local search, initial values are required. Two types of initializations suggest themselves:

- *Axes-orthogonal splits in the style of LOLIMOT*: This generates nz different initial splits.
- *Split direction of the parent node*: This is most promising if the direction of process nonlinearity does not vary significantly. This behavior can at least be expected locally. Thus, this method will be very powerful, especially in deep hierarchy levels. It cannot be applied for the very first split since the root node possesses no parent.

In HILOMOT all these $nz + 1$ initial splits are carried out and compared. The best performing one becomes the initialization for the nonlinear split optimization.

For high-dimensional $\underline{z}$-spaces, the LOLIMOT-style initializations take up significant computational resources, even dominating the whole nonlinear optimization. Thus, it seems very reasonable to limit oneself only to the parent split initialization[3] if nz is large.

14.8.2.2 Local Model Center

In LOLIMOT the splits are carried out through the center of the worst local model. The partitions have hyper-rectangular shape, and their center is geometrically well-defined. For HILOMOT the split does not necessarily go through the center of the local model since the weights v_i are optimized. However, the initial splits shall pass through the center which therefore needs to be defined.

The center of local model m shall be defined as the center of gravity of all data points where the data are weighted with their validity function value:

$$\underline{c}_m = \sum_{i=1}^{N} \underline{z}(i) \cdot \frac{\Phi_m(\underline{z}(i))}{\sum_{j=1}^{N} \Phi_m(\underline{z}(j))} . \tag{14.40}$$

In order for a split to pass through any point in $\underline{z}$, (14.38) needs to be fulfilled for this point. A split in (or orthogonal to) dimension j means that it is parallel to all other dimensions and the weights $v_i^{(\text{ini})} = 0$ except for dimension $i = j$ where

[3]Note that this is not possible for the very first split since the root possesses no parent and it significantly increases the risk of converging to a local optimum.

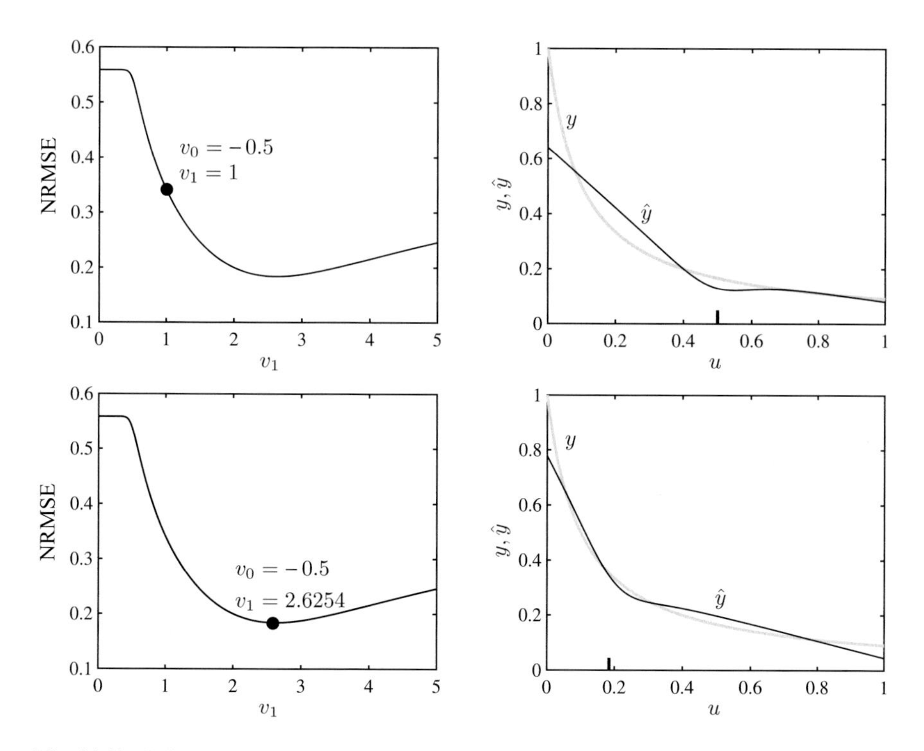

Fig. 14.40 Split optimization in 1D: First split for model of hyperbola process. Top: Initial model (LOLIMOT split). Bottom: Optimal model. Left: Loss function for optimization of v_1 (v_0 is fixed). Right: Model compared with process

$v_j^{(\text{ini})} = 1$ (or any other arbitrary[4] value $\neq 0$) and the splitting function should yield $\frac{1}{2}$ at the local model's center $\underline{z} = \underline{c}$, i.e.,

$$v_0^{(\text{ini})} + v_j^{(\text{ini})} z_j \stackrel{!}{=} 0 \;\Rightarrow\; v_0^{(\text{ini})} + 1 \cdot c_j \stackrel{!}{=} 0 \;\Rightarrow\; v_0^{(\text{ini})} = -c_j \,. \tag{14.41}$$

Example 14.8.2 Split Optimization 1D

The hyperbola process in Fig. 14.40(right) shall be approximated by a local model network with two linear models constructed by HILOMOT. Two split parameters v_0 and v_1 exist. The values of $v_1^{\text{ini}} = 1$ and $v_0^{\text{ini}} = -0.5$ are chosen according to the LOLIMOT split at $u = 0.5$, compare (14.41). The corresponding model is shown in Fig. 14.40(top, right). Its performance in terms of NRMSE can be seen on the loss function in Fig. 14.40(top, left). During the subsequent HILOMOT optimization of v_1, the value of v_0 is kept fixed. However, the sigmoid is evaluated with the normalized split parameters $\widetilde{v}_0$, $\widetilde{v}_1$; see (14.39). A nonlinear optimization algorithm

[4]Will be normalized according to (14.39) anyway.

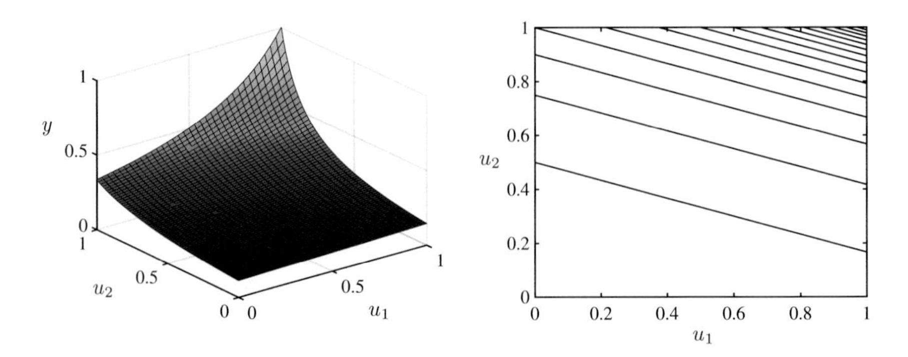

Fig. 14.41 2D hyperbola process with contours. Nonlinearity is three times more severe in u_2 than in u_1

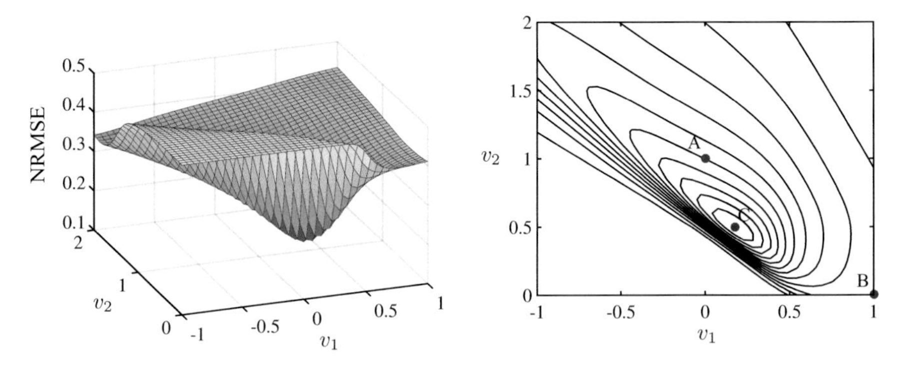

Fig. 14.42 Split optimization in 2D: First split for a model of the hyperbola process in Fig. 14.41. Loss function with contours. Point A ($v_1^{(\mathrm{ini})} = 0$, $v_2^{(\mathrm{ini})} = 1$) represents a good initial LOLIMOT split orthogonal to the more nonlinear u_2-direction with NRMSE = 0.3. Point B ($v_1^{(\mathrm{ini})} = 1$, $v_2^{(\mathrm{ini})} = 0$) represents a bad initial LOLIMOT split orthogonal to the less nonlinear u_1-direction with NRMSE = 0.41. Point C ($v_1^{(\mathrm{opt})} = 0.1667$, $v_2^{(\mathrm{opt})} = 0.502$) represents the optimal split with a loss function of NRMSE = 0.18. The ratio between $v_1^{(\mathrm{opt})}$ and $v_2^{(\mathrm{opt})}$ corresponds to the angle of the nonlinearity.

fed with the analytic gradient information (see Sect. 14.8.6) finds the optimal split parameter shown in Fig. 14.40(bottom, left). This results in the model shown in Fig. 14.40(bottom, right) with almost double performance. □

Example 14.8.3 Split Optimization 2D
The hyperbola process in Fig. 14.41 shall be approximated by a local model network with two linear models constructed by HILOMOT. As can be seen from the contours, the direction of the nonlinearity points in an angle of $\varphi \approx 72°$.

Three split parameters v_0, v_1, and v_2 exist. Figure 14.42 visualizes the loss function and its contours. The value of $v_0^{\mathrm{ini}} = -0.5$ is chosen according to the two possible LOLIMOT splits at $u_1 = 0.5$ or $u_2 = 0.5$, respectively; compare (14.41).

The LOLIMOT split in u_2-direction (Point A) is performed by $v_1^{\text{ini}} = 0$, $v_2^{\text{ini}} = 1$ and yields an NRMSE = 0.3. The LOLIMOT split in u_1-direction (Point B) is performed by $v_1^{\text{ini}} = 1$, $v_2^{\text{ini}} = 0$ and yields an NRMSE = 0.41. That split A performs better than split B was expected since the nonlinearity is more severe in u_2-direction than in u_1-direction. Thus, HILOMOT chooses split A for the initial values of the nonlinear split optimization. The optimum is marked as Point C. The NRMSE = 0.18 at the optimum is around two times better than the performance of the LOLIMOT splits.

Figure 14.42b shows a sharp increase of the loss function as either v_1 or v_2 approach $-v_0$. In fact, as $v_1 + v_2 \rightarrow -v_0$ the sigmoidal splitting function moves outside the data covered region $[0, 1]^2$, and only one model is active on the training data. Therefore the loss function goes into saturation, i.e., realizes the performance of one linear model, NRMSE = 0.48.

14.8.2.3 Convergence Behavior

Note that the loss functions are extremely smooth and possess only one optimum which ensures fast and robust convergence. Although these properties cannot be guaranteed, it is the experience of the author that the HILOMOT split optimization problems typically are well-conditioned and local optima are rare. Moreover, note that the utilization of global search methods as discussed in Chap. 5 typically requires extremely high computational resources that do not pay off in the context of HILOMOT. In a worst-case scenario, the split optimization gets stuck a local optimum, but then one of the subsequent splits will (partly) repair this suboptimality.

The plots were created with the smoothness parameter $\kappa = 30$, but the loss function hardly changes when κ is varied. In general, numerical difficulties may arise when data is scarce and the split is non-smooth (κ is large).

Figure 14.43 demonstrates that only a couple of iterations are necessary for convergence. The incremental hierarchical strategy keeps the number of nonlinear parameters very small that have to be optimized simultaneously. Often $nz \ll nx$ which also keeps the number of nonlinear parameters small while the $nx + 1$ linear parameters are estimated very efficiently by local LS. □

14.8.3 Building up the Hierarchy

This section explains how the hierarchical tree structure is constructed by HILOMOT and how the final validity functions Φ are built up from the splitting functions Ψ. Note that each leaf node in the tree corresponds to a validity function that determines the validity of the associated local model. And each node from the root up to the leaves (exclusively) corresponds to a splitting function which determines the position and direction of each split. Each validity function is just the product of all splitting functions from root to the corresponding leaf.

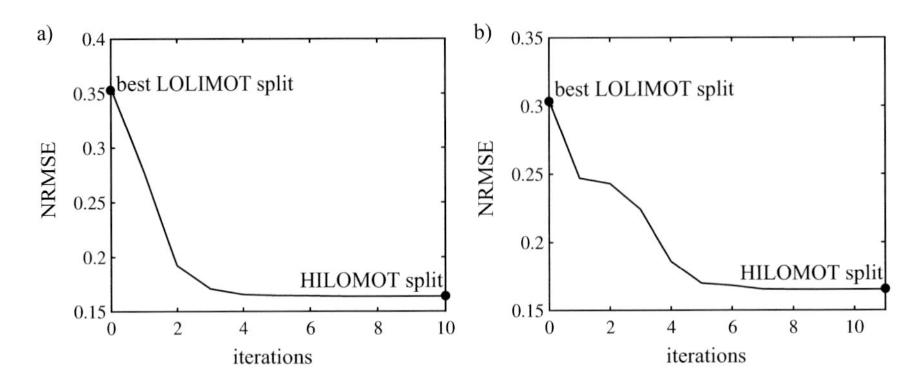

Fig. 14.43 Convergence behavior of split optimization of HILOMOT: (**a**) 1D example from Fig. 14.40 (**b**) 2D example from Fig. 14.41. One iteration may include a couple of steps within line search. Specifically, in (**a**) 13 and in (**b**) 17 loss function evaluations are computed with a Quasi-Newton algorithm

Consider the example trees shown in Fig. 14.44. They are the first five trees built up by HILOMOT operating on the process shown in Fig. 14.45a. The model output of the local model network obtained at iteration 5 which possesses 5 local linear models and corresponds to the lower right tree in Fig. 14.44 is shown in Fig. 14.45b.

The following validity functions are constructed incrementally by the first five HILOMOT iterations:

1. Initialization:

$$\Phi_1 = 1\,. \tag{14.42}$$

2. First split of LM 1 with splitting function Ψ_1:

$$\Phi_2 = \Psi_1\,, \tag{14.43}$$

$$\Phi_3 = 1 - \Psi_1\,. \tag{14.44}$$

3. Second split of LM 3 with splitting function Ψ_2:

$$\Phi_4 = (1 - \Psi_1)\Psi_2\,, \tag{14.45}$$

$$\Phi_5 = (1 - \Psi_1)(1 - \Psi_2)\,. \tag{14.46}$$

4. Third split of LM 2 with splitting function Ψ_3:

$$\Phi_6 = \Psi_1\Psi_3\,, \tag{14.47}$$

$$\Phi_7 = \Psi_1(1 - \Psi_3)\,. \tag{14.48}$$

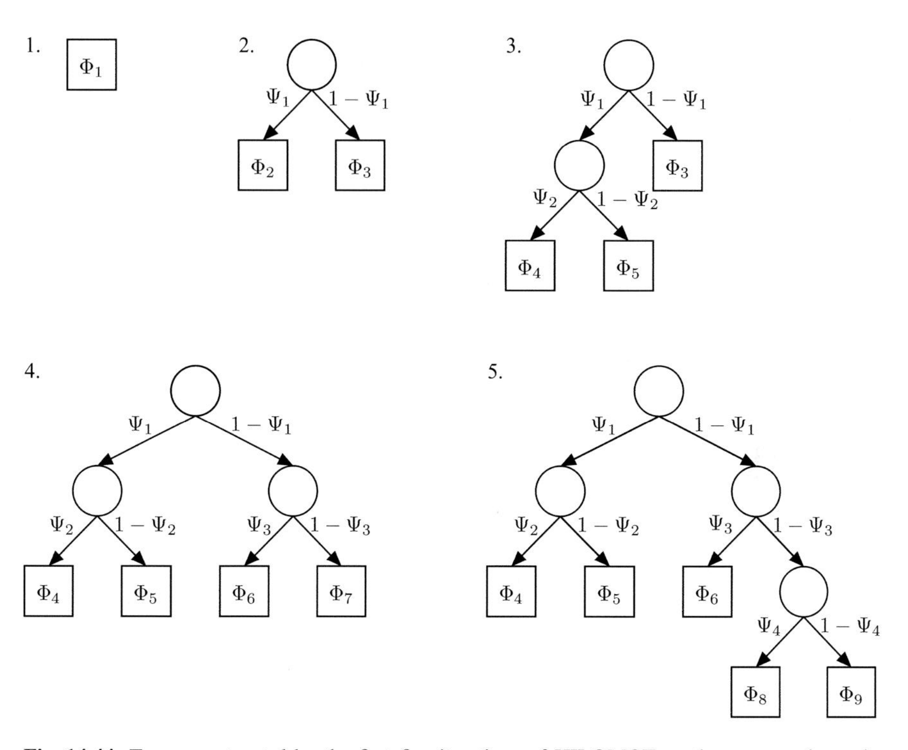

Fig. 14.44 Trees constructed by the first five iterations of HILOMOT on the process shown in Fig. 14.45a. The leaf nodes (squares) denote the validity functions Φ that the associated local model is weighted with. The split nodes (circles) divide the $\underline{z}$-input space into partitions by the corresponding splitting function Ψ. The iterations of HILOMOT are shown by the number 1., 2., …, 5.

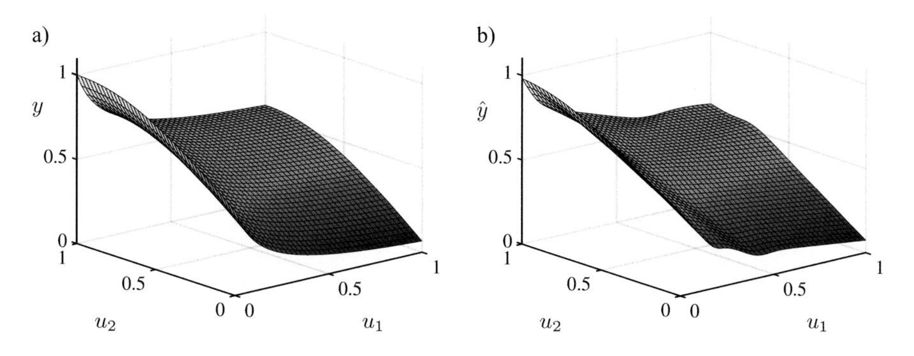

Fig. 14.45 Test function for HILOMOT demonstration. It is an additive combination of a hyperbola in u_1 and a sine in u_2. The process is shown in (**a**); the output of a local model network with five local models constructed by HILOMOT is shown in (**b**)

5. Fourth split of LM 5 with splitting function Ψ_4:

$$\Phi_8 = (1 - \Psi_1)(1 - \Psi_2)\Psi_4 \,, \tag{14.49}$$

$$\Phi_9 = (1 - \Psi_1)(1 - \Psi_2)(1 - \Psi_4) \,. \tag{14.50}$$

Note that a validity function becomes obsolete when the corresponding node is split. Then the new splitting function takes its place. Thus, the final network with 5 local models only makes use of Φ_4, Φ_5, Φ_6, Φ_8, and Φ_9.

Therefore such a local model network constructed by HILOMOT does not contain *all* local models LM_i, $i = 1, 2, \ldots, M$ weighted with *all* validity functions Φ_i, $i = 1, 2, \ldots, M$. Rather the following notation needs to be introduced.

The sets S_i contain the indices of the leaf nodes corresponding to the active local models. S_i is associated with the ith iteration and always contains i local models. For the tree in Fig. 14.48, the following sets are constructed in the first five iterations:

$$S_1 = \{1\},\ S_2 = \{2, 3\},\ S_3 = \{2, 4, 5\},\ S_4 = \{4, 5, 6, 7\},\ S_5 = \{4, 6, 7, 8, 9\}.$$

In each iteration i, the partition of unity is maintained (j runs over all members in the set S_i):

$$\sum_{j \in S_i} \Phi_j(\underline{z}) = 1 \,. \tag{14.51}$$

With this notation a network of M local models can be evaluated by

$$\hat{y} = \sum_{i \in S_M} \text{LM}_i(\underline{x}) \cdot \Phi_i(\underline{z}) \tag{14.52}$$

where LM_i denotes the ith local model.

Figure 14.46 shows the input space partitioning after five iterations together with the validity functions. Figure 14.47 illustrated how it was constructed with

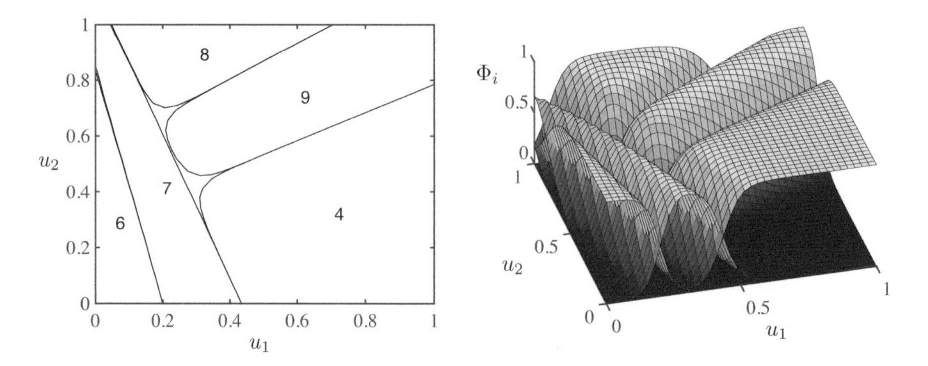

Fig. 14.46 Input space partitioning and validity functions for the model in Fig. 14.45b

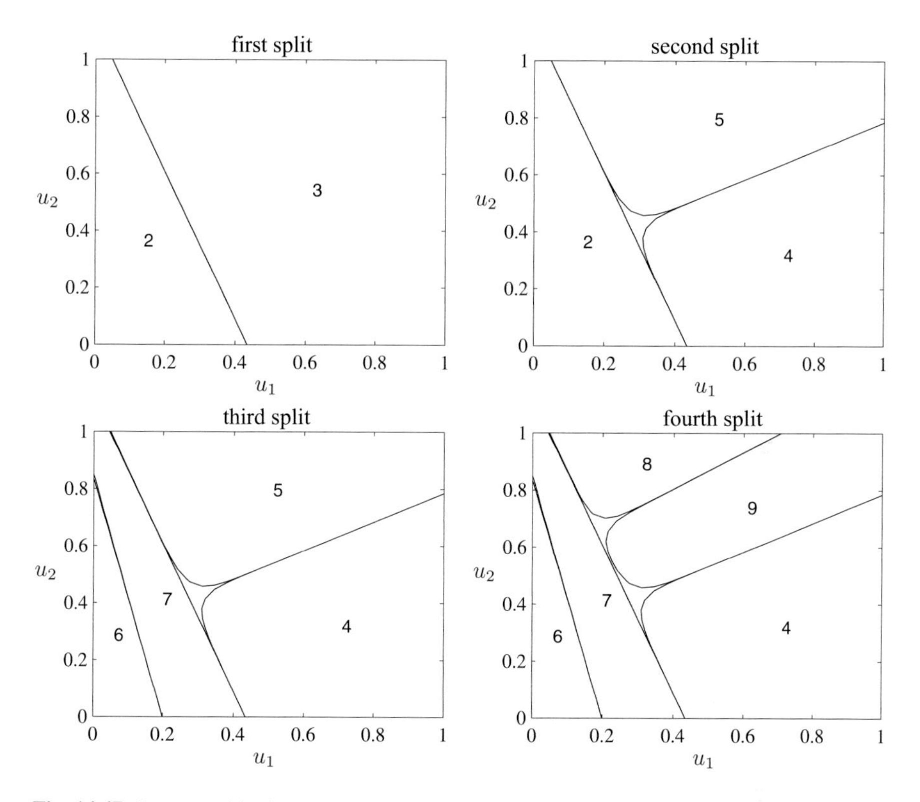

Fig. 14.47 Input partitioning constructed by HILOMOT in the first five iterations for the test function in Fig. 14.45

four splits. Figure 14.48 summarizes all validity functions that were generated during the HILOMOT run from iteration 1 to 5. These validity functions have been constructed by the product of the splitting functions shown in Fig. 14.49.

In can be seen in Fig. 14.47 that the splits of model 2 in 6, 7 and model 5 in 8, 9 are in an extremely similar direction to the parents' split. In contrast, the split of model 3 is completely different (almost orthogonal) to the parent's split. It seems reasonable to assume that – at least in the latter iterations of the algorithm – the directions of the splits stabilize as they represent the direction of the process nonlinearity in that region of the input space. Therefore it can be expected that the direction of the parent's split frequently is a superior initialization for the nonlinear split optimization to the LOLIMOT splits. It can be observed in practice that this indeed is the case.

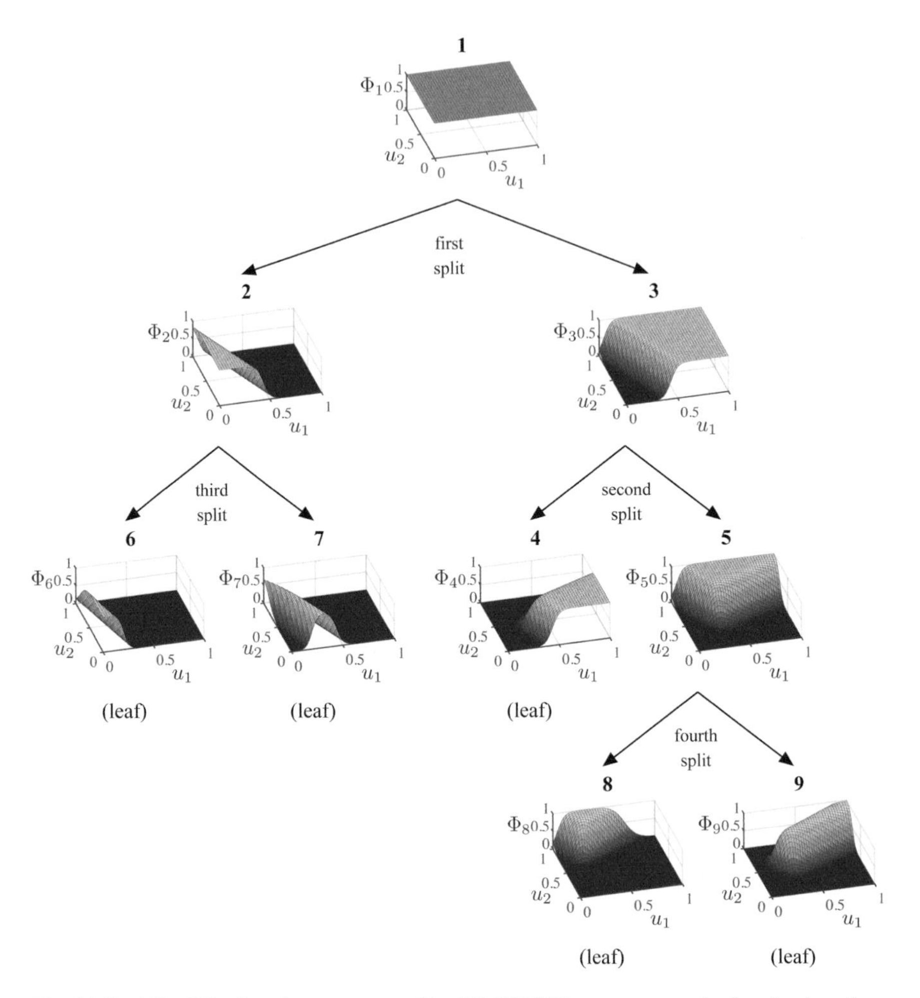

Fig. 14.48 All validity functions constructed by HILOMOT in one run over the first five iterations for the test function in Fig. 14.45

14.8.4 Smoothness Adjustment

As explained in Sect. 14.8.2, the smoothness of the sigmoids shall *not* be optimized but rather heuristically adjusted. This is a much more critical issue for hierarchical models than for flat ones. In the LOLIMOT context, the smoothness of the membership functions just determines the smoothness of the model and the amount of regularization achieved by local estimation. However, for hierarchical models as built by HILOMOT, the smoothness determines (and limits) the influence of the whole subtree below the node whose split is considered. Therefore extreme care is required in order to avoid too smooth splitting functions.

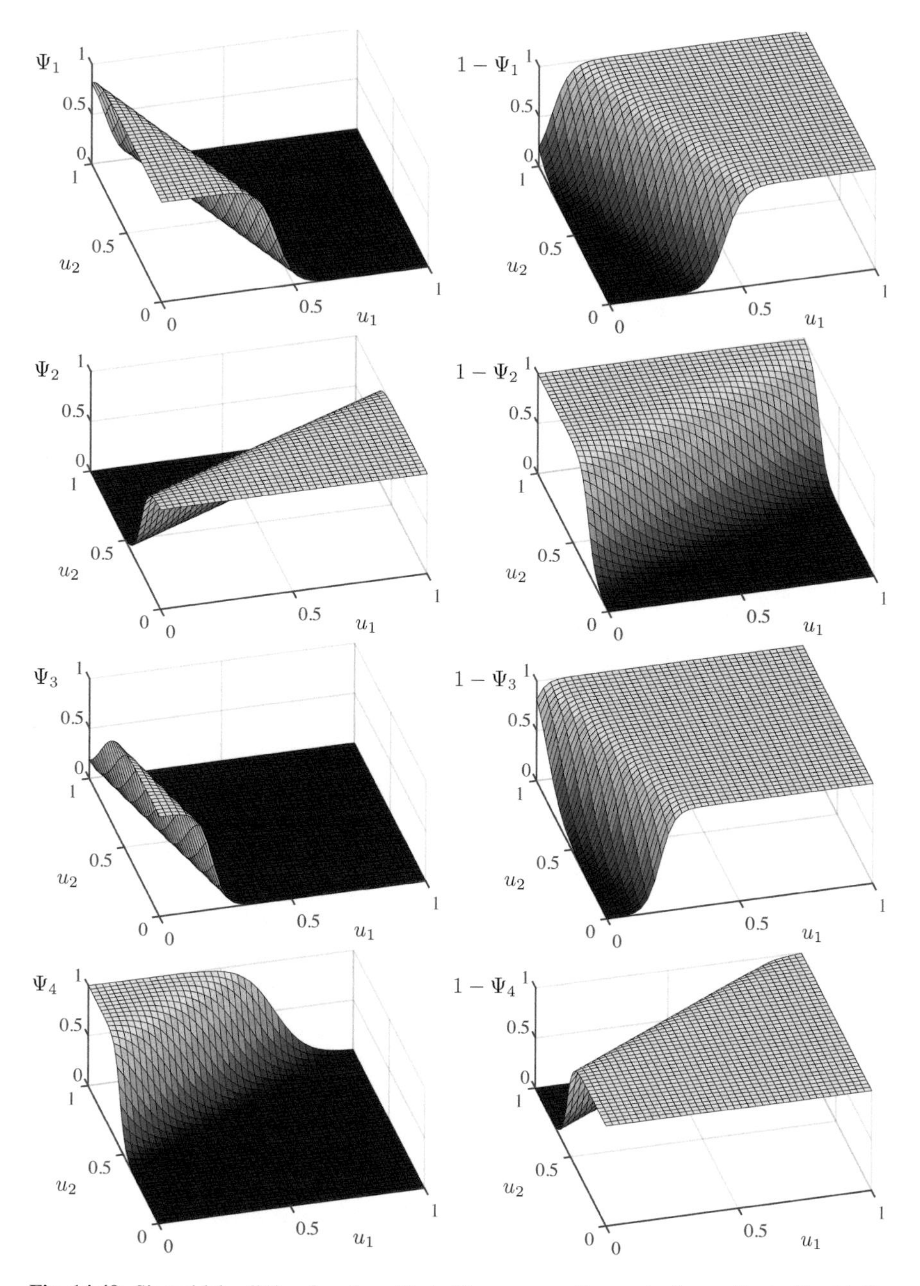

Fig. 14.49 Sigmoidal splitting functions Ψ_1 to Ψ_4 corresponding to the four splits in Fig. 14.48. Their products construct the validity functions Φ_1 to Φ_9

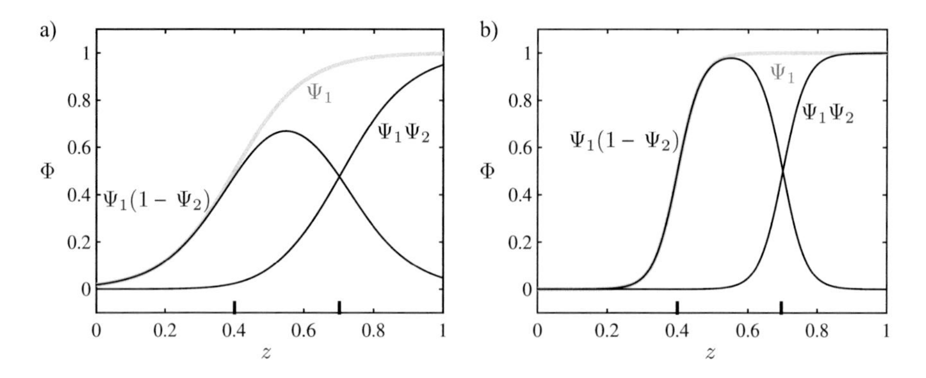

Fig. 14.50 Consequences of too smooth splitting function Ψ_1 on the validity functions Φ: First split at $z = 0.4$ (non-normalized $v_0 = -0.4$, $v_1 = 1$), second split at $z = 0.7$ (non-normalized $v_0 = -0.7$, $v_1 = 1$): (**a**) too smooth $\kappa = 10$, (**b**) reasonably smooth $\kappa = 30$

Figure 14.50 illustrates this issue. A first split is carried out at $z = 0.4$ (splitting function Ψ_1), a second split at $z = 0.7$ (splitting function Ψ_2). If the smoothness of the first split is too big as shown in Fig. 14.50a, a significant portion of the overall validity is lost for the whole subtree under Ψ_1. Therefore both validity functions $\Phi_2 = \Psi_1(1 - \Psi_2)$ and $\Phi_3 = \Psi_1\Psi_2$ cannot achieve validity values close to 1. The maximum activation of $\Psi_1(1 - \Psi_2)$ is only around 0.67. As a consequence, the associated local model describes at most 67% of the overall model output. If further splits follow, this situation becomes even more severe, and new local models basically are incapable of improving the model performance because they become almost irrelevant. Figure 14.50b shows a reasonable choice for the smoothness.

However, further splits in this region may prove this choice to be too big as well. Also, if it happens that the second split is nearby the first one, the second splitting function lies in the transition range of the first. This requires an additional tightening of this transition. Figure 14.51 illustrates the case where the second split is at $z = 0.5$ instead of $z = 0.7$. The smoothness needs to be smaller by a factor of 3 again (κ is 3 times bigger) for reasonable results. In [227] this topic is discussed in detail and an algorithm for smoothness adjustment is proposed. Nevertheless, a fully satisfactory solution to this issue that works fine for any number of dimensions and any number of local models (hierarchy levels) is still an open question.

The following observations can lead the way:

- The choice for smoothness is most critical in the upper levels of the tree (first couple of splits).
- The smoothness can be chosen individually for each split or for each hierarchy level.
- The smoothness should depend on the number of hierarchy levels of the tree.
- Too non-smooth sigmoids yield extremely badly conditioned optimization problems where the loss function becomes "staircase"-like with each step being caused by one data point.

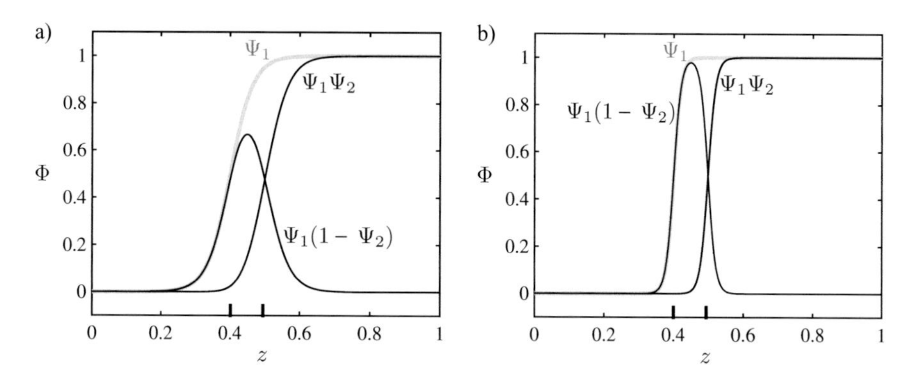

Fig. 14.51 Consequences of nearby splits on the validity functions Φ: First split at $z = 0.4$ (non-normalized $v_0 = -0.4$, $v_1 = 1$), second split in contrast to Fig. 14.50 at $z = 0.5$ (non-normalized $v_0 = -0.5$, $v_1 = 1$) (**a**) too smooth, although same $\kappa = 30$ as in Fig. 14.50b, (**b**) reasonably smooth $\kappa = 90$

- The smoothness can be altered with relatively little influence on the optimal partitioning. Thus it is possible to optimize the split with relatively smooth sigmoids in order to obtain good numerics and tighten them (and re-estimating the local models) as the optimization converges.
- It suggests itself to employ geometric features in order to determine the smoothness. One example is the distance between the centers of the two local models to be split where κ could be chosen anti-proportionally to. Note, however, that the center positions depend on the validity values of the data points which in turn depend on the split parameters. Thus, the analytical gradient, as derived in Sect. 14.8.6, would not be valid in this case. The analytical gradient derived in Sect. 14.8.6 requires κ to be independent of the split.

14.8.5 Separable Nonlinear Least Squares

During the optimization of one split discussed in Sect. 14.8.2, the regimes and the data points associated with each of both (new) local models change permanently. This effect is tremendous and needs to be taken into account.

14.8.5.1 Idea

The HILOMOT algorithm realizes an approach called ***separable nonlinear least squares*** which is illustrated in Fig. 14.52. This falls in the class of the *variable projection* (VP) methods as it projects the influence of the linear parameters (by estimating them with least squares) on the nonlinear optimization task. The

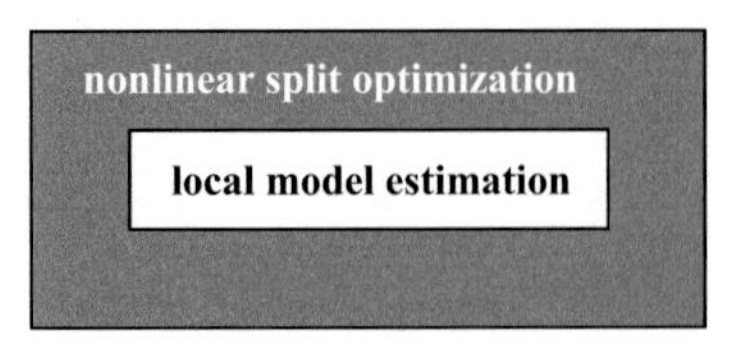

Fig. 14.52 Separable nonlinear least squares optimization for HILOMOT: LS estimation of the local models' parameters is nested into the nonlinear split optimization

algorithm nests the local LS estimations of the local models' parameters (inner loop) into the nonlinear split optimization (outer loop).

In order to perform the nonlinear optimization efficiently, the correct gradient with respect to the nonlinear split parameters is required. Since the loss function contains the LS estimation of the two local model parameters, the derivatives of the pseudoinverses have to be evaluated. This procedure was first proposed and analyzed in 1973 in [192]. It is called *separable nonlinear least squares* because two types of parameters are handled separately: (i) the linear parameters are estimated by LS (here: *two* LSs for both local models involved in the split) and (ii) the nonlinear parameters are optimized by a gradient-based scheme. Note that the number of nonlinear parameters and therefore the dimensionality of the nonlinear optimization problem is only

$$n_{\text{nonlinear}} = nz, \tag{14.53}$$

i.e., $nz + 1$ weights -1 for the normalization; compare Sect. 14.8.2. It is completely independent of the complexity of the local models. The two simple local LS estimations possess

$$n_{\text{linear}} = nx + 1 \tag{14.54}$$

parameters each which is significantly higher dimensional ($nx \gg nz \Rightarrow n_{\text{linear}} \gg n_{\text{nonlinear}}$) for many applications; compare Sect. 14.1.

The extensive, modern overview on "separable nonlinear least squares" [190] summarizes the main idea in its abstract: "These are problems for which the model function is a linear combination of nonlinear functions. Taking advantage of this special structure the method of Variable Projections [separable nonlinear least squares] eliminates the linear variables obtaining a somewhat more complicated function that involves only the nonlinear parameters. This procedure not only reduces the dimension of the parameter space but also results in a better-conditioned problem. The same optimization method applied to the original and reduced problems will always converge faster for the latter."

The following statement explains the terms *variable projections* and *separable* nonlinear least squares: "The separability aspect comes from the idea of eliminating the linear variables (i.e., either the coefficients in the linear combinations or one set of the nonlinear variables), and then minimizing the resulting Variable Projection (VP) functional that depends only on the remaining variables."

The required gradient of the pseudoinverses with respect to the split parameters is derived in Sect. 14.8.6. The pseudoinverses are needed for the LS estimation of the parameters of the two local models to be split. Note that in this context, it is important to have a fixed (i.e., independent of the split parameters) sigmoid smoothness κ.

14.8.5.2 Termination Criterion

Equivalently to the LOLIMOT algorithm (compare Sect. 13.3.1) HILOMOT may be terminated based on different criteria. The only unique aspect for HILOMOT is the assessment of the effective number of parameters as required in information criteria such as the $\mathrm{AIC_C}$. For LOLIMOT only the effective parameters of the local model have been counted which always is smaller than the nominal number dependent on the smoothing matrix. These parameters also exist in the HILOMOT case. Note, however, that typically the regularization effect is smaller because the validity functions are adjusted less smooth in order to avoid the loss of too much validity in deeper hierarchy levels; compare Fig. 14.51 in Sect. 14.8.4.

In LOLIMOT the flexibility in the partitioning was not counted as effective parameters at all. The induced error is considered negligible since the flexibility of the orthogonal LOLIMOT splits in the ratio 1:1 is relatively low. In contrast, HILOMOT partitioning is much more flexible. Fortunately, it is much easy to assess in terms of parameters. Each of the $M-1$ splits optimizes nz parameters. Thus, in the $\mathrm{AIC_C}$ or other information criteria, the effective number of parameters n_{eff} is the number of LOLIMOT parameters plus

$$(M-1)nz\,. \tag{14.55}$$

14.8.5.3 Constrained Optimization

In LOLIMOT splits that yield local models with too few data points inside are forbidden in order to guarantee a safe estimation of the local model parameters. In HILOMOT a more sophisticated strategy can be employed. It is ensured by constrained optimization that enough effective data is inside both local models, i.e., the split is optimized under the constraint

$$N_{\mathrm{eff}} = \sum_{i=1}^{N} \Phi_i(\underline{z}(i)) \geq n_{\mathrm{local}}\,. \tag{14.56}$$

For local *linear* models $n_{\mathrm{local}} = nx + 1 = p + 1$. For local polynomial models, nx can be much higher than p, thereby restricting the split optimization further.

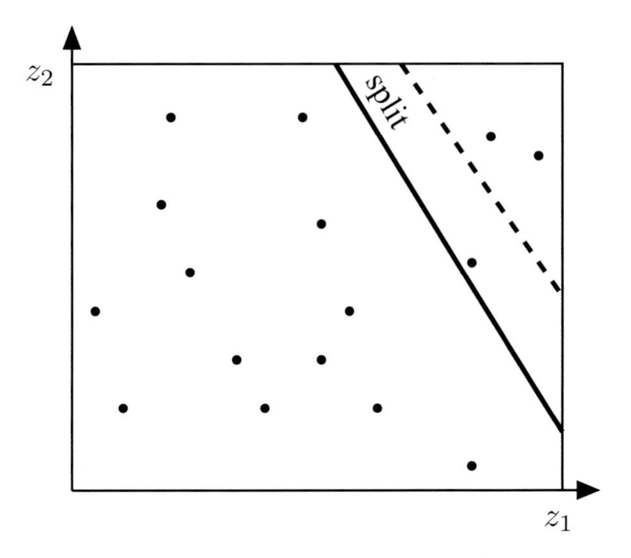

Fig. 14.53 Split with constrained optimization (solid) and unconstrained optimization (dashed) for $nz = 2$

Figure 14.53 illustrates the idea for a two-dimensional $\underline{z}$-input space and local linear models. Then constrained split optimization ensures that at least three effective data points lie inside the local model on the upper right. The unconstrained optimal split may be more extreme (dashed line). Note that it is possible to estimate the local model parameters if condition (14.56) is violated or even without any data inside a local model due to the local estimation approach. Such an approach, however, is not advisable for robustness reasons. With the proposed constrained split optimization strategy, the exiting data can be exploited much more effectively than with LOLIMOT.

14.8.5.4 Robust Estimation

Applying LS for the estimation of the local models' parameters is certainly the fastest and easiest approach. However, it is conceptionally simple to replace LS with more sophisticated approaches like robust regression; see Sect. 3.1.9. While this is extremely straightforward for LOLIMOT because the partitioning is determined by discrete optimization (selecting the best split); see Sect. 13.2.4, it is much more involved in HILOMOT. The analytic gradient calculation discussed in Sect. 14.8.6 cannot be utilized directly.

However, it is relatively easy to modify the loss function for the split optimization from the "sum of squared errors" to the "sum of absolute errors" since it is a nonlinear optimization problem anyway.

14.8.5.5 Alternatives to Separable Nonlinear Least Squares

In theory, separable nonlinear least squares can be expected to deliver the best performance in convergence speed and numerical conditioning compared to alternative approaches [190]. Two alternatives to optimization for HILOMOT have been proposed and are visualized in Fig. 14.54. The approach a) proposed in [213, 214] is based on the famous expectation maximization (EM) algorithm from statistics where first the nonlinear split parameters are initialized. Then the following two steps are alternated until convergence:

- *E-Step*: Least squares estimation of the local models' parameters while keeping the split fixed
- *M-Step*: Nonlinear optimization of the split parameters while keeping the local models fixed

This approach is very simple but also inefficient because it does not directly account for the interaction between the local models' parameters and the split. As a consequence, the convergence speed can be extremely low. On the other hand, it is very easy to implement.

More powerful is approach (b), the nonlinear optimization of all parameters (local models and split) simultaneously [215]. Unfortunately, the linear nature of the local models' parameters is not fully exploited. Also, it is very difficult (or even impossible) to incorporate any subset selection scheme for the local models' estimation.

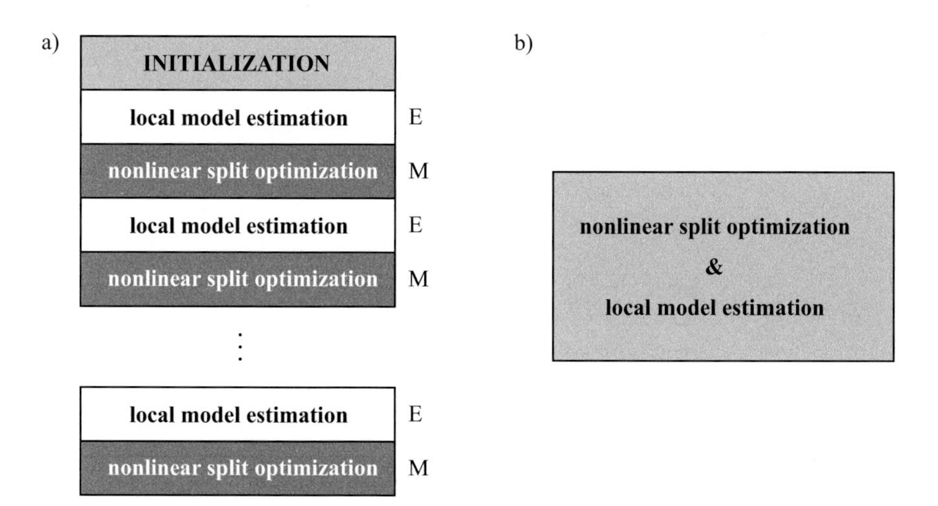

Fig. 14.54 Alternative approaches to optimization for HILOMOT according to (**a**) expectation maximization (EM) or (**b**) simultaneous nonlinear optimization

14.8.6 Analytic Gradient Derivation

In order to carry out the split optimization effectively gradient information is required. Although in principle, any gradient-free local search strategy or any global optimization scheme could be applied, this seems to be a very inefficient and unnecessary approach in the light of the typically benign shaped loss functions; compare Sect. 14.8.2. Also, note that any loss function evaluation requires two $(nx + 1)$-dimensional LS estimations and that $M - 1$ splits need to be optimized to build the whole HILOMOT tree.

CAUTION: Note that in this subsection, the splitting functions of one node of the HILOMOT tree are *not* named Ψ_k and $1 - \Psi_k$ but Ψ_k and Ψ_{k+1}. This has purely historical and software implementation reasons and hopefully does not confuse the reader.

Following [159] and the report [130], the gradient of the loss function with respect to the split parameters $\underline{v} = [v_1\ v_2\ \cdots\ v_{nz}]^T$ is derived; v_0 is fixed at its initial value. The MSE loss function is

$$I_{\text{MSE}}(\underline{v}) = \frac{1}{N}\sum_{i=1}^{N} e(i, \underline{v})^2 = \frac{1}{N}\,\underline{e}^T(\underline{v}) \cdot \underline{e}(\underline{v}) \tag{14.57}$$

with the error for all samples $e(i, \underline{v})$, $i = 1, \ldots, N$ contained in the vector $\underline{e}(\underline{v})$. The gradient with respect to $\underline{v}$ shall be calculated:

$$\underline{g}(\underline{v}) = \frac{\partial I(\underline{v})}{\partial\, \underline{v}}\,. \tag{14.58}$$

The chain rule gives

$$\underbrace{\underline{g}(\underline{v})}_{1\times nz} = \frac{1}{N}\frac{\partial}{\partial\, \underline{v}}\left(\underline{e}^T(\underline{v}) \cdot \underline{e}(\underline{v})\right) \tag{14.59}$$

$$= \frac{1}{N}\left(\underline{e}^T \cdot \frac{\partial\, \underline{e}(\underline{v})}{\partial\, \underline{v}} + \underline{e}^T \cdot \frac{\partial\, \underline{e}(\underline{v})}{\partial\, \underline{v}}\right) \tag{14.60}$$

$$= \frac{2}{N}\underbrace{\underline{e}^T}_{1\times N} \cdot \underbrace{\frac{\partial\, \underline{e}(\underline{v})}{\partial\, \underline{v}}}_{N\times nz}\,. \tag{14.61}$$

The error is

$$\underbrace{\underline{e}(\underline{v})}_{N\times 1} = \underline{y} - \underline{\hat{y}}(\underline{v})\,. \tag{14.62}$$

Since the measured output data in $\underline{y}$ are independent of $\underline{\upsilon}$

$$\frac{\partial\, \underline{e}(\underline{\upsilon})}{\partial\, \underline{\upsilon}} = \frac{\partial\, (\underline{y} - \underline{\hat{y}}(\underline{\upsilon}))}{\partial\, \underline{\upsilon}} = -\frac{\partial\, \underline{\hat{y}}(\underline{\upsilon})}{\partial\, \underline{\upsilon}} . \tag{14.63}$$

14.8.6.1 Derivative of the Local Model Network

The output of the whole local model network $\underline{\hat{y}}$ is calculated by summing up the local models $\underline{\hat{y}}_k$ ($k = 1, 2, \ldots, M$) weighted with their corresponding validity values which are the entries of the diagonal $N \times N$ weighting matrices $\underline{Q}_k$:

$$\underline{\hat{y}}(\underline{\upsilon}) = \sum_{k=1}^{M} \underbrace{\underline{Q}_k(\underline{\upsilon})}_{N\times N} \cdot \underbrace{\underline{\hat{y}}_k(\underline{\upsilon})}_{N\times 1} . \tag{14.64}$$

Both, the validity values and the local models depend on the split parameters $\underline{\upsilon}$. The derivative with respect to $\underline{\upsilon}$ is

$$\frac{\partial\, \underline{\hat{y}}(\underline{\upsilon})}{\partial\, \underline{\upsilon}} = \sum_{k=1}^{M} \frac{\partial \left(\underline{Q}_k(\underline{\upsilon}) \cdot \underline{\hat{y}}_k(\underline{\upsilon})\right)}{\partial\, \underline{\upsilon}} . \tag{14.65}$$

Only *one* split is optimized that creates the new local models $M - 1$ and M. The already existing local models $1, 2 \ldots, M - 2$ are independent of $\underline{\upsilon}$ and therefore their derivatives are zero:

$$\frac{\partial\, \underline{Q}_k(\underline{\upsilon})}{\partial\, \underline{\upsilon}} = \underline{0} \text{ and } \frac{\partial\, \underline{\hat{y}}_k(\underline{\upsilon})}{\partial\, \underline{\upsilon}} = \underline{0} \text{ with } k \notin \{M-1, M\} . \tag{14.66}$$

Therefore

$$\frac{\partial\, \underline{\hat{y}}(\underline{\upsilon})}{\partial\, \underline{\upsilon}} = \sum_{k=M-1}^{M} \frac{\partial \left(\underline{Q}_k(\underline{\upsilon}) \cdot \underline{\hat{y}}_k(\underline{\upsilon})\right)}{\partial\, \underline{\upsilon}} . \tag{14.67}$$

The derivative of the product between the validity matrix $Q_k(\underline{\upsilon})$ and the column vector of local model outputs $\underline{\hat{y}}_k(\underline{\upsilon})$ is

$$\frac{\partial \left(\underline{Q}_k(\underline{\upsilon}) \cdot \underline{\hat{y}}_k(\underline{\upsilon})\right)}{\partial\, \underline{\upsilon}} = \underbrace{\underline{Q}_k(\underline{\upsilon})}_{N\times N} \cdot \underbrace{\frac{\partial\, \underline{\hat{y}}_k(\underline{\upsilon})}{\partial\, \underline{\upsilon}}}_{N\times nz} + \underbrace{\left[\underbrace{\frac{\partial\, \underline{Q}_k(\underline{\upsilon})}{\partial\, \upsilon_1}}_{N\times N} \cdot \underbrace{\underline{\hat{y}}_k(\underline{\upsilon})}_{N\times 1} \;\cdots\; \frac{\partial\, \underline{Q}_k(\underline{\upsilon})}{\partial\, \upsilon_{nz}} \cdot \underline{\hat{y}}_k(\underline{\upsilon})\right]}_{N\times nz} \tag{14.68}$$

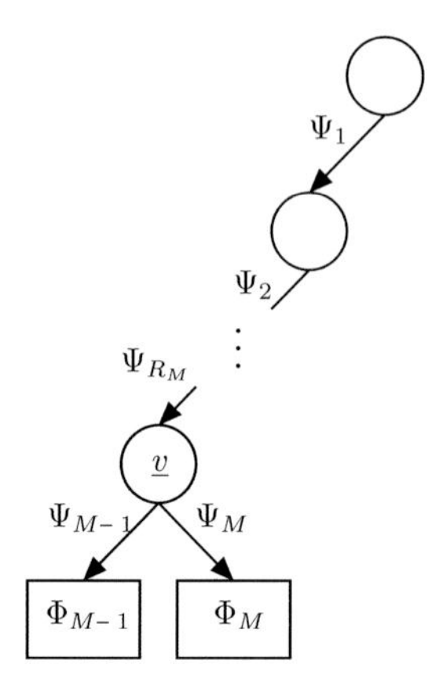

Fig. 14.55 The optimization of the recent split creating the two new local models $M-1$ and M. All splitting functions up to the root node $\Psi_1 \ldots \Psi_{R_M}$ are fixed.
The number of nodes between node $M-1$ and the root is R_{M-1} which is identical to the number of nodes between node M and the root R_M.
Only the new split performed by the splitting functions Ψ_M and $\Psi_{M-1} = 1 - \Psi_M$ is influenced by the split parameters $\underline{v}$

The validity function the kth local model is the product of all R_k splitting functions up to the root node of the tree; compare Fig. 14.55. The hierarchy level of the node to be split is denoted by R_k. Only the new split performed by Ψ_k depends on the split parameters $\underline{v}$. This relationship is expressed by

$$\underline{Q}_k(\underline{v}) = \prod_{j=1}^{R_k} \underline{P}_j \cdot \underline{P}_k(\underline{v}) = \widetilde{\underline{P}}_k \cdot \underline{P}_k(\underline{v}) \tag{14.69}$$

with

$$\widetilde{\underline{P}}_k = \prod_{j=1}^{R_k} \underline{P}_j \,. \tag{14.70}$$

The matrices $\underline{Q}_k(\underline{v})$, $\underline{P}_k(\underline{v})$, and $\widetilde{\underline{P}}_k$ are all diagonal and of dimension $N \times N$. Their diagonal entries are the validity functions, splitting functions, or products of splitting functions evaluated at the N data points.

With (14.69)

$$\frac{\partial\, \underline{\hat{y}}(\underline{v})}{\partial\, \underline{v}} = \sum_{k=M-1}^{M} \widetilde{\underline{P}}_k \cdot \underline{P}_k(\underline{v}) \cdot \underbrace{\frac{\partial\, \underline{\hat{y}}_k(\underline{v})}{\partial\, \underline{v}}}_{N \times nz} +$$

$$\underbrace{\left[\underbrace{\underline{\widetilde{P}}_k \cdot \frac{\partial \underline{P}_k(\underline{v})}{\partial v_1}}_{N\times N} \cdot \underbrace{\hat{\underline{y}}_k}_{N\times 1} \quad \cdots \quad \underline{\widetilde{P}}_k \cdot \frac{\partial \underline{P}_k(\underline{v})}{\partial v_{nz}} \cdot \hat{\underline{y}}_k\right]}_{N\times nz}$$

$$= \sum_{k=M-1}^{M} \underline{\widetilde{P}}_k \left(\underline{P}_k(\underline{v}) \cdot \frac{\partial \hat{\underline{y}}_k(\underline{v})}{\partial \underline{v}} + \left[\frac{\partial \underline{P}_k(\underline{v})}{\partial v_1} \cdot \hat{\underline{y}}_k \quad \cdots \quad \frac{\partial \underline{P}_k(\underline{v})}{\partial v_{nz}} \cdot \hat{\underline{y}}_k\right]\right) \tag{14.71}$$

Note that the splitting functions up to the root which do not depend on $\underline{v}$ are identical for model M and model $M-1$ because they possess identical parents

$$\underline{\widetilde{P}}_M = \prod_{j=1}^{R_M} \underline{P}_j = \prod_{j=1}^{R_{M-1}} \underline{P}_j = \underline{\widetilde{P}}_{M-1} \tag{14.72}$$

and therefore the validity matrices differ only in the new splitting function Ψ_M and Ψ_{M-1} or the new splitting matrices $P_M(\underline{v})$ and $\underline{P}_{M-1}(\underline{v})$:

$$\underline{Q}_M(\underline{v}) = \underline{P}_M(\underline{v}) \cdot \prod_{j=1}^{R_M} \underline{P}_j = \underline{\widetilde{P}}_M \cdot \underline{P}_M(\underline{v}) \tag{14.73}$$

$$\underline{Q}_{M-1}(\underline{v}) = \underline{P}_{M-1}(\underline{v}) \cdot \prod_{j=1}^{R_{M-1}} \underline{P}_j = \underline{\widetilde{P}}_M \cdot \underline{P}_{M-1}(\underline{v}) . \tag{14.74}$$

Since $\Psi_{M-1} = 1 - \Psi_M$ for the splitting matrices, the following relationship holds:

$$\underline{P}_{M-1}(\underline{v}) = \underline{I} - \underline{P}_M(\underline{v}) . \tag{14.75}$$

Therefore the derivatives are related by

$$\frac{\partial \underline{P}_{M-1}(\underline{v})}{\partial \underline{v}} = \frac{\partial}{\partial \underline{v}} \left(\underline{I} - \underline{P}_M(\underline{v})\right) = -\frac{\partial \underline{P}_M(\underline{v})}{\partial \underline{v}} . \tag{14.76}$$

Thus, the derivative of the validity matrix $\underline{Q}_M(\underline{v})$ with respect to $\underline{v}$ according to (14.73) yields

$$\frac{\partial \underline{Q}_M(\underline{v})}{\partial \underline{v}} = \underline{\widetilde{P}}_M \cdot \frac{\partial \underline{P}_M(\underline{v})}{\partial \underline{v}} \tag{14.77}$$

and with (14.74) and (14.76)

$$\frac{\partial \underline{Q}_{M-1}(\underline{\upsilon})}{\partial \underline{\upsilon}} = \widetilde{\underline{P}}_M \cdot \frac{\partial \underline{P}_{M-1}(\underline{\upsilon})}{\partial \underline{\upsilon}} = -\widetilde{\underline{P}}_M \cdot \frac{\partial \underline{P}_M(\underline{\upsilon})}{\partial \underline{\upsilon}} . \tag{14.78}$$

Summarizing the formulas, the derivative of the whole local model network output $\hat{\underline{y}}(\underline{\upsilon})$ with respect to the split parameters $\underline{\upsilon}$ is given by

$$\begin{aligned} \frac{\partial \hat{\underline{y}}(\underline{\upsilon})}{\partial \underline{\upsilon}} &= \widetilde{\underline{P}}_M \cdot \Bigg(\underline{P}_M(\underline{\upsilon}) \cdot \frac{\partial \hat{\underline{y}}_M(\underline{\upsilon})}{\partial \underline{\upsilon}} + \left[\frac{\partial \underline{P}_M(\underline{\upsilon})}{\partial \upsilon_1} \cdot \hat{\underline{y}}_M \cdots \frac{\partial \underline{P}_M(\underline{\upsilon})}{\partial \upsilon_{nz}} \cdot \hat{\underline{y}}_M \right] + \\ & \quad \underline{P}_{M-1}(\underline{\upsilon}) \cdot \frac{\partial \hat{\underline{y}}_{M-1}(\underline{\upsilon})}{\partial \underline{\upsilon}} - \left[\frac{\partial \underline{P}_M(\underline{\upsilon})}{\partial \upsilon_1} \cdot \hat{\underline{y}}_{M-1} \cdots \frac{\partial \underline{P}_M(\underline{\upsilon})}{\partial \upsilon_{nz}} \cdot \hat{\underline{y}}_{M-1} \right] \Bigg) \end{aligned} \tag{14.79}$$

$$\begin{aligned} &= \widetilde{\underline{P}}_M \cdot \Bigg(\underline{P}_M(\underline{\upsilon}) \cdot \frac{\partial \hat{\underline{y}}_M(\underline{\upsilon})}{\partial \underline{\upsilon}} + \underline{P}_{M-1} \frac{\partial \hat{\underline{y}}_{M-1}(\underline{\upsilon})}{\partial \underline{\upsilon}} + \\ & \left[\frac{\partial \underline{P}_M(\underline{\upsilon})}{\partial \upsilon_1} \cdot \left(\hat{\underline{y}}_M - \hat{\underline{y}}_{M-1} \right) \cdots \frac{\partial \underline{P}_M(\underline{\upsilon})}{\partial \upsilon_{nz}} \cdot \left(\hat{\underline{y}}_M - \hat{\underline{y}}_{M-1} \right) \right] \Bigg) . \end{aligned} \tag{14.80}$$

The derivatives of the splitting function with respect to $\underline{\upsilon}$ (lightly shaded in (14.80)) are detailed in the next paragraph. Subsequently, the darkly shaded derivatives of the local models with respect to $\underline{\upsilon}$ are derived.

14.8.6.2 Derivative of the Sigmoidal Splitting Function

The diagonal matrix $\underline{P}_k$ evaluates the sigmoidal splitting function Ψ_k at all data points yielding

$$\underline{P}_k = \text{diag} \left\{ \frac{1}{1 + \exp\left(-\kappa \cdot \frac{1}{\|\underline{\upsilon}^{\circ}\|} \cdot (\upsilon_0 + \underline{Z}\, \underline{\upsilon}) \right)} \right\} \tag{14.81}$$

with the data in the $\underline{z}$-input space (premise space) collected in

$$\underline{Z} = [\underline{z}_1\ \underline{z}_2\ \cdots\ \underline{z}_{nz}] \tag{14.82}$$

and the *extended* split vector $\underline{v}^\circ$ which contains the fixed (not optimized) offset v_0 and the vector of the split direction to be optimized $\underline{v}$:

$$\|\underline{v}^\circ\| = \begin{bmatrix} v_0 \\ \underline{v} \end{bmatrix} . \tag{14.83}$$

If the smoothness κ does not depend on $\underline{v}$, the derivative of the sigmoidal entries of the splitting matrix with respect to directional weights v_j yields

$$\begin{aligned} &\frac{\partial \underline{P}_k}{\partial v_j} \\ &= \text{diag}\left\{(1-\underline{P}_k)\underline{P}_k\kappa \frac{\partial \left(\frac{1}{\|\underline{v}^\circ\|}\left(v_0 + \underline{Z}\,\underline{v}\right)\right)}{\partial v_j}\right\} \\ &= \text{diag}\left\{(1-\underline{P}_k)\underline{P}_k\kappa \left(\frac{\partial\left(\frac{1}{\|\underline{v}^\circ\|}\right)}{\partial v_j}\left(v_0 + \underline{Z}\,\underline{v}\right) + \frac{1}{\|\underline{v}^\circ\|}\frac{\partial\left(v_0 + \underline{Z}\,\underline{v}\right)}{\partial v_j}\right)\right\} . \end{aligned} \tag{14.84}$$

The derivatives in the sigmoidal arguments are

$$\frac{\partial\left(v_0 + \underline{Z}\,\underline{v}\right)}{\partial v_j} = \underline{z}_j \tag{14.85}$$

and

$$\begin{aligned} \frac{\partial}{\partial v_j}\left(\frac{1}{\|\underline{v}^\circ\|}\right) &= \frac{\partial}{\partial v_j}\left(\frac{1}{\sqrt{\underline{v}^{\circ T}\cdot\underline{v}^\circ}}\right) = -\frac{1}{2}\cdot\frac{1}{\sqrt{\underline{v}^{\circ T}\cdot\underline{v}^\circ}^3}\cdot\frac{\partial}{\partial v_j}\left(\underline{v}^{\circ T}\cdot\underline{v}^\circ\right) \\ &= -\frac{v_j}{\|\underline{v}^\circ\|^3} . \end{aligned} \tag{14.86}$$

In summary

$$\frac{\partial \underline{P}_k}{\partial v_j} = \text{diag}\left\{(1-\underline{P}_k)\underline{P}_k\kappa\left(-\frac{v_j}{\|\underline{v}^\circ\|^3}\left(v_0 + \underline{Z}\,\underline{v}\right) + \frac{\underline{z}_j}{\|\underline{v}^\circ\|}\right)\right\} . \tag{14.87}$$

14.8.6.3 Derivative of the Local Model

Generally, the derivative of local model k is given by

$$\underbrace{\frac{\partial\, \hat{\underline{y}}_k(\underline{v})}{\partial\, \underline{v}}}_{N\times nz} = \underbrace{\underline{X}}_{N\times nx+1} \cdot \underbrace{\frac{\partial \underline{\theta}_k(\underline{v})}{\partial\, \underline{v}}}_{nx+1\times nz} = \underline{X} \cdot \left[\frac{\partial\, \underline{\theta}_k(\underline{v})}{\partial\, v_1} \cdots \frac{\partial\, \underline{\theta}_k(\underline{v})}{\partial\, v_{nz}} \right]. \tag{14.88}$$

Remember that nz denotes the dimensionality of the $\underline{z}$-input or premise space, while nx denotes the number of inputs of the $\underline{x}$-input or consequents space. For local affine models, thus the number of parameters $\underline{\theta}$ in the local models is $nx + 1$. This situation is assumed here for the given matrix and vector dimensions. However, nx can take any value without changing the equations. For example, for local polynomials of a higher degree $\dim\{\underline{\theta}\}$ and the number of columns in $\underline{X}$ will be much larger than $nx + 1$.

The estimation of the parameters $\underline{\theta}_k$ is carried out by local weighted least squares

$$\underline{\theta}_k(\underline{v}) = \left(\underline{X}^T \underline{Q}_k(\underline{v})\, \underline{X}\right)^{-1} \underline{X}^T \underline{Q}_k(\underline{v})\, \underline{y} \tag{14.89}$$

with the validity matrix $\underline{Q}(\underline{v})$ as weighting matrix

$$\underline{Q}_k(\underline{v}) = \begin{bmatrix} \Phi_k(1) & 0 & \cdots & 0 \\ 0 & \Phi_k(2) & \cdots & 0 \\ \vdots & \vdots & \ddots & \vdots \\ 0 & 0 & \cdots & \Phi_k(N) \end{bmatrix} = \widetilde{\underline{P}}_k \cdot \underline{P}_k(\underline{v}). \tag{14.90}$$

Evaluating (14.88) with the product law leads to

$$\frac{\partial\, \underline{\theta}_k(\underline{v})}{\partial\, v_j} = \left(\underline{X}^T \underline{Q}_k(\underline{v})\underline{X}\right)^{-1} \underline{X}^T \frac{\partial\, \underline{Q}_k(\underline{v})}{\partial\, v_j} \underline{y} + \frac{\partial \left(\underline{X}^T \underline{Q}_k(\underline{v})\underline{X}\right)^{-1}}{\partial\, v_j} \underline{X}^T \underline{Q}_k(\underline{v})\underline{y} \tag{14.91}$$

With the rules for deriving inverse matrices (14.91) can be written as

$$\frac{\partial\, \underline{\theta}_k(\underline{v})}{\partial\, v_j} = -\left(\underline{X}^T \underline{Q}_k(\underline{v})\underline{X}\right)^{-1} \underline{X}^T \frac{\partial\, \underline{Q}_k(\underline{v})}{\partial\, v_j} \underline{X} \underbrace{\left(\underline{X}^T \underline{Q}_k(\underline{v})\underline{X}\right)^{-1} \underline{X}^T \underline{Q}_k(\underline{v})\, \underline{y}}_{=\underline{\theta}_k(\underline{v})}$$

$$+ \left(\underline{X}^T \underline{Q}_k(\underline{v})\underline{X}\right)^{-1} \underline{X}^T \frac{\partial\, \underline{Q}_k(\underline{v})}{\partial\, v_j} \underline{y}. \tag{14.92}$$

It follows with (14.89) and $\underline{X}\,\underline{\theta}_k(\underline{v}) = \underline{\hat{y}}_k$:

$$\frac{\partial\,\underline{\theta}_k(\underline{v})}{\partial\,v_j} = \left(\underline{X}^T\,\underline{Q}_k(\underline{v})\underline{X}\right)^{-1}\underline{X}^T\,\frac{\partial\,\underline{Q}_k(\underline{v})}{\partial\,v_j}\left(-\underline{X}\,\underline{\theta}_k(\underline{v}) + \underline{y}\right)$$
$$= \left(\underline{X}^T\,\underline{Q}_k\,(\underline{v})\,\underline{X}\right)^{-1}\underline{X}^T\,\frac{\partial\,\underline{Q}_k(\underline{v})}{\partial\,v_j}\left(\underline{y} - \underline{\hat{y}}_k\right). \qquad (14.93)$$

With the derivative of the validity matrix

$$\frac{\partial\,\underline{Q}_k(\underline{v})}{\partial\,v_j} = \underline{\widetilde{P}}_k\cdot\frac{\partial\,\underline{P}_k(\underline{v})}{\partial\,v_j} \qquad (14.94)$$

the following local model derivative results

$$\underbrace{\frac{\partial\,\underline{\hat{y}}_k(\underline{v})}{\partial\,\underline{v}}}_{N\times nz} = \underline{X}\left(\underline{X}^T\,\underline{\widetilde{P}}_k\underline{P}_k(\underline{v})\underline{X}\right)^{-1}\underline{X}^T\,\underline{\widetilde{P}}_k\cdot$$
$$\underbrace{\left[\frac{\partial\,\underline{P}_k(\underline{v})}{\partial\,v_1}\left(\underline{y} - \underline{\hat{y}}_k\right)\ \cdots\ \frac{\partial\,\underline{P}_k\,(\underline{v})}{\partial\,v_{nz}}\left(\underline{y} - \underline{\hat{y}}_k\right)\right]}_{N\times nz}. \qquad (14.95)$$

For the derivative of the two new local models M and $M-1$, this evaluates to

$$\frac{\partial\,\underline{\hat{y}}_M(\underline{v})}{\partial\,\underline{v}} = \underline{X}\left(\underline{X}^T\,\underline{\widetilde{P}}_M\underline{P}_M\underline{X}\right)^{-1}\underline{X}^T\,\underline{\widetilde{P}}_M\cdot$$
$$\left[\frac{\partial\underline{P}_M\,(\underline{v})}{\partial\,v_1}\left(\underline{y} - \underline{\hat{y}}_M\right)\ \cdots\ \frac{\partial\underline{P}_M\,(\underline{v})}{\partial\,v_{nz}}\left(\underline{y} - \underline{\hat{y}}_M\right)\right]. \qquad (14.96)$$

$$\frac{\partial\,\underline{\hat{y}}_{M-1}(\underline{v})}{\partial\,\underline{v}} = \underline{X}\left(\underline{X}^T\,\underline{\widetilde{P}}_M\underline{P}_{M-1}\underline{X}\right)^{-1}\underline{X}^T\,\underline{\widetilde{P}}_M\cdot$$
$$\left[\frac{\partial\underline{P}_{M-1}\,(\underline{v})}{\partial\,v_1}\left(\underline{y} - \underline{\hat{y}}_{M-1}\right)\ \cdots\ \frac{\partial\,\underline{P}_{M-1}(\underline{v})}{\partial\,v_{nz}}\left(\underline{y} - \underline{\hat{y}}_{M-1}\right)\right]$$
$$= -\,\underline{X}\left(\underline{X}^T\,\underline{\widetilde{P}}_M\underline{P}_{M-1}\underline{X}\right)^{-1}\underline{X}^T\,\underline{\widetilde{P}}_M\cdot$$
$$\left[\frac{\partial\underline{P}_M(\underline{v})}{\partial\,v_1}\left(\underline{y} - \underline{\hat{y}}_{M-1}\right)\ \cdots\ \frac{\partial\underline{P}_M(\underline{v})}{\partial\,v_{nz}}\left(\underline{y} - \underline{\hat{y}}_{M-1}\right)\right]. \qquad (14.97)$$

14.8.6.4 Summary

The gradient of the local model network output with respect to the split parameters $\underline{v}$ of the split that generates local models $M - 1$ and M is given by

$$\frac{\partial \,\underline{\hat{y}}(\underline{v})}{\partial \,\underline{v}} = \sum_{k=M-1}^{M} \widetilde{\underline{P}}_k \left(\underline{P}_k(\underline{v}) \frac{\partial \,\underline{\hat{y}}_k(\underline{v})}{\partial \,\underline{v}} + \left[\frac{\partial \underline{P}_k(\underline{v})}{\partial \,v_1} \underline{\hat{y}}_k \;\; \cdots \;\; \frac{\partial \underline{P}_k(\underline{v})}{\partial \,v_{nz}} \underline{\hat{y}}_k \right] \right)$$

with

$$\frac{\partial \underline{P}_k}{\partial \,v_j} = \text{diag} \left\{ (1 - \underline{P}_k) \underline{P}_k \kappa \left(-\frac{v_j}{\|\underline{v}^{\circ}\|^3} \left(v_0 + \underline{Z}\,\underline{v} \right) + \frac{\underline{z}_j}{\|\underline{v}^{\circ}\|} \right) \right\}$$

and with

$$\frac{\partial \,\underline{\hat{y}}_k(\underline{v})}{\partial \,\underline{v}} = \underline{X} \left(\underline{X}^T \widetilde{\underline{P}}_k \underline{P}_k(\underline{v}) \underline{X} \right)^{-1} \underline{X}^T \cdot$$

$$\widetilde{\underline{P}}_k \left[\frac{\partial \underline{P}_k(\underline{v})}{\partial \,v_1} \left(\underline{y} - \underline{\hat{y}}_k \right) \;\; \cdots \;\; \frac{\partial \underline{P}_k(\underline{v})}{\partial \,v_{nz}} \left(\underline{y} - \underline{\hat{y}}_k \right) \right] .$$

14.8.7 Analyzing Input Relevance from Partitioning

This section describes an embedded method that can be used for assessing the relevance of the inputs and possibly for selecting the most significant ones. It is an *embedded* method because it relies on the special partitioning structure generated by HILOMOT and cannot directly be generalized to alternative modeling approaches. More universal methods that offer particular benefits for local model networks but are not intertwined with a specific algorithm are discussed in Chap. 15. For important issues and recent developments of model-independent input selection strategies, refer, e.g., to [263] and the references therein.

The idea behind this embedded relevance determination method proposed and analyzed in [38] comes from the fact that the directions of the splits represent the strength and direction of the process nonlinearity. If a process is linear with respect to an input, all splits performed by HILOMOT will be approximately parallel to this dimension. Therefore an analysis of the split parameters can reveal insights about the relationship between each model input and the output. However, note that even if the output depends strongly on one input but this relationship is linear, it will be considered irrelevant in this section. More precisely speaking, this section analyzes the relevance for the *nonlinear behavior* and thus for the $\underline{z}$-input space. Even if an input should not be included in $\underline{z}$, it still can be important to be included in the local models and thus in the $\underline{x}$-input space.

14.8.7.1 Relevance for One Split

As Fig. 14.56 illustrates, the split weights $v_i, i = 1, \dots, nz$ can be interpreted as measure for the strength of nonlinearity of the model output with respect to input z_i. The split weight v_0 does not influence the split direction and therefore plays no role in the subsequent discussion. It is useful to normalize the sum of all input relevances to one and thus consider their relative value. The input relevance factor ρ_i for *one* split then becomes

$$\rho_i^{(\text{split})} = \frac{|v_i|}{\sum_{j=1}^{nz} |v_j|} \tag{14.98}$$

for input $z_i, i = 1, \dots, nz$.

14.8.7.2 Relevance for the Whole Network

Since the whole model contains many splits, their overall effect has to be aggregated. The relevance for input z_i can be assessed by the accumulated weighted split parameters ζ_i. Split j possesses the split parameters $v_{1j}, \dots, v_{nz,j}$. The importance of split j for the overall model can be judged by the following two criteria as proposed in [38]:

- Number of data points within the node that is split which can be measured by $\sum_{k=1}^{N} \Phi_j(\underline{z}(k))$ due to the fuzzy nature of the splits.

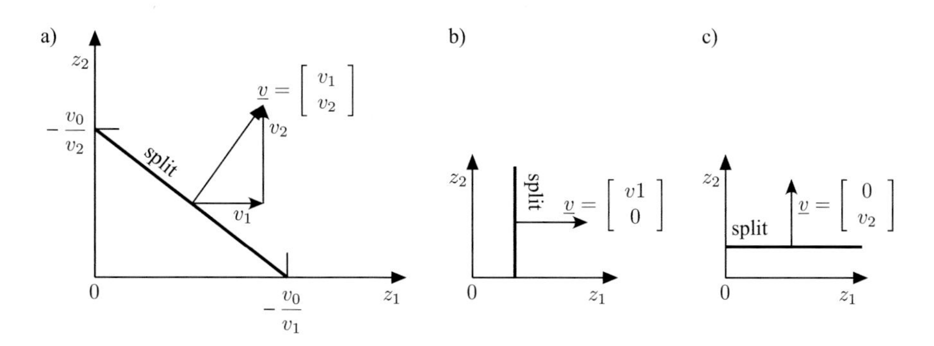

Fig. 14.56 (**a**) Split direction interpretation for relevance. (**b**) The split in z_1 yields the relevances $\rho_1 = 1$ and $\rho_2 = 0$. (**c**) The split in z_2 yields the relevances $\rho_1 = 0$ and $\rho_2 = 1$

- Performance improvement through the jth split PI_j. It can be measured, e.g., by the difference or ratio of the NRMSEs before and after the split is carried out.

These two criteria establish the weighting of each split j. As a network with M local models possesses $M-1$ splits, the (non-normalized) relevance for input z_i becomes

$$\zeta_i = \sum_{j=1}^{M-1} |v_{ij}| \cdot \mathrm{PI}_j \cdot \sum_{k=1}^{N} \Phi_j(\underline{z}(k)) \tag{14.99}$$

for $i = 1, \ldots, nz$. Then the input relevance factor for the *whole* local model network is its normalized value:

$$\rho_i = \frac{\zeta_i}{\sum_{j=1}^{nz} |\zeta_j|} \,. \tag{14.100}$$

Example 14.8.4 Relevance Evaluation
This relevance formula (14.99) together with (14.100) shall be illustrated with a simple example with two orthogonal splits as shown in Fig. 14.57. The first split cuts an area with 16 data points in two halves and therefore is weighted with $\sum_{k=1}^{16} \Phi_j(\underline{z}(k)) = 16$, while the second split cuts an area of 8 data points in two halves and therefore is weighted with $\sum_{k=1}^{8} \Phi_j(\underline{z}(k)) = 8$. If the performance improvement of both splits would be identical, i.e., $\mathrm{PI}_1 = \mathrm{PI}_2$, this yields input relevance factors of $\underline{\rho} = [\rho_1\ \rho_2] = [\frac{2}{3}\ \frac{1}{3}]$.

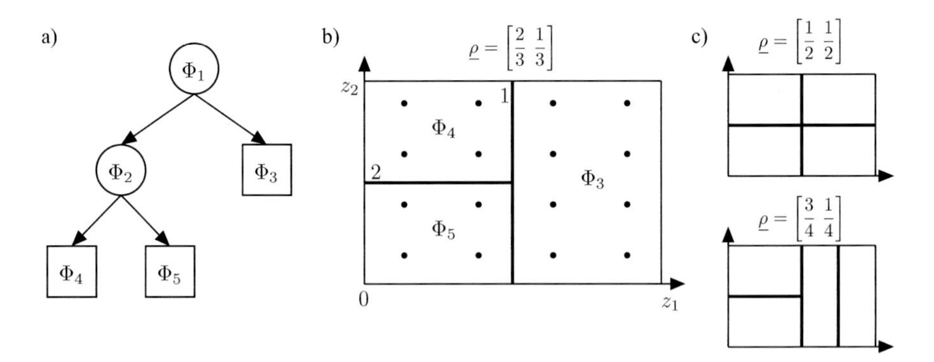

Fig. 14.57 Example for relevance formula: (**a**) HILOMOT tree.
(**b**) Partitioning of the input space with 16 data points as dots. Split 1 divides z_1 in half and subsequently split 2 divides model 2 (the left one) in z_2 in half. Since model 1 with validity function Φ_1 (whole input space) contains twice as many points as Φ_2 (half input space), split 1 is weighted twice as high as split 2. Thus, the relevance factors are $\underline{\rho} = [\rho_1\ \rho_2] = [\frac{2}{3}\ \frac{1}{3}]$.
(**c**) Two alternative third splits with corresponding relevance factors

Note that in most applications, it is realistic to have a better performance in iteration $j+1$ than in iteration j since the convergence curve is decreasing. However, although the convergence curve normally is monotonously decreasing, its slope typically is not. Therefore it is normal to obtain $\text{PI}_{j+1} < \text{PI}_j$.

Consider a potential third split as shown in Fig. 14.57c. If the third split would cut model 3 into two halves in the z_2-dimension, then split 1 would count equally (16) as split 2 (8) and 3 (8) combined which accumulates to input relevance factors of $\underline{\rho} = [\rho_1\ \rho_2] = [\frac{1}{2}\ \frac{1}{2}]$. If the third split would cut the z_1-dimension, then split 1 (16) and split 3 (8) would count in favor of z_1 opposed to (8) for split 2 in z_2 resulting in input relevance factors of $\underline{\rho} = [\rho_1\ \rho_2] = [\frac{3}{4}\ \frac{1}{4}]$. □

Example 14.8.5 Hyperbola
Two two-dimensional hyperbolic functions shall be considered. Both inputs shall enter the $\underline{z}$- and $\underline{x}$-input spaces, $\underline{z} = \underline{x} = [u_1\ u_2]^T$. Input data is in $[0,\ 1]^2$. The first function follows:

$$y = \frac{0.1}{0.1 + 0.5u_1 + 0.5u_2} \tag{14.101}$$

whose nonlinearity points in 45°-direction (both inputs are equally important). If the data is distributed uniformly on a grid, indeed relevance factors of $\underline{\rho} = [\rho_1\ \rho_2] = [0.5\ 0.5]$ are the consequence. For the hyperbola

$$y = \frac{0.1}{0.1 + 0.73u_1 + 0.27u_2} \tag{14.102}$$

which points in 20°-direction. The relevance factors of $\underline{\rho} = [\rho_1\ \rho_2] = [0.73\ 0.27]$ result. It can be observed that the relevance factor evaluation is very stable and robust with respect to noise in the data. This is due to the quite strong and consistent direction of the process nonlinearity. □

Example 14.8.6 Gaussian
In contrast, if the process is an axis-orthogonal two-dimensional Gaussian function

$$y = \exp\left(-\frac{u_1^2}{2\sigma_1^2}\right)\exp\left(-\frac{u_2^2}{2\sigma_2^2}\right) \tag{14.103}$$

the nonlinear behavior exists in all directions. Nevertheless, for $\sigma_1 = \sigma_2$, the relevance factors are around $\underline{\rho} = [\rho_1\ \rho_2] = [0.5\ 0.5]$. For $\sigma_1 = 2\sigma_2$, the nonlinearity is much weaker in z_1 than in z_2, and the relevance factors are around $\underline{\rho} = [\rho_1\ \rho_2] = [0.15\ 0.85]$. However, the results are far less stable and much easier disturbed by measurement noise on the data compared to the hyperbola example.

□

14.8.8 HILOMOT Versus LOLIMOT

The test process in Fig. 14.45 shall be used for contrasting the characteristics of the LOLIMOT and HILOMOT algorithms for construction of local model networks. The following setting is shared by the subsequent investigations:

- Process:

$$y = \frac{1}{2}\frac{0.15}{0.15 + u_1} + \frac{1}{2}\sin\left(\frac{\pi}{2}u_2\right) . \tag{14.104}$$

- Training data: Input: Maximin Latin hypercube in $[0, 1]^2$
  Output: Disturbed by additive white Gaussian noise
- Test data: Input: 10,000 points drawn from a Sobol sequence
  Output: Noise-free
- Termination criterion: Best AIC_C value
- Three different training data sets: 30, 100, and 300 points
- Three different output noise levels: $\sigma_n = 0.003$, $\sigma_n = 0.01$, and $\sigma_n = 0.03$
- Three different smoothnesses: low, medium, and high (factor 2 in between)

This section is just meant for demonstration purposes and for obtaining some "feeling" on the algorithms. One key factor that is not varied here is the dimensionality of the problem. The process considered is only 2D for visualization purposes. But the following issues will be discussed: partitioning, validity functions, training, and test errors, network complexity, the influence of network smoothness, and amount and noise level of the data.

Figure 14.58 shows the training input data distribution for the three data sets. Training is carried out according to the following procedure:

- Local linear model networks are trained with HILOMOT and LOLIMOT.
- Termination criterion is a two times successively increasing AIC_C value.
- The network with the best AIC_C value is selected.
- Test data is only used for information purposes.

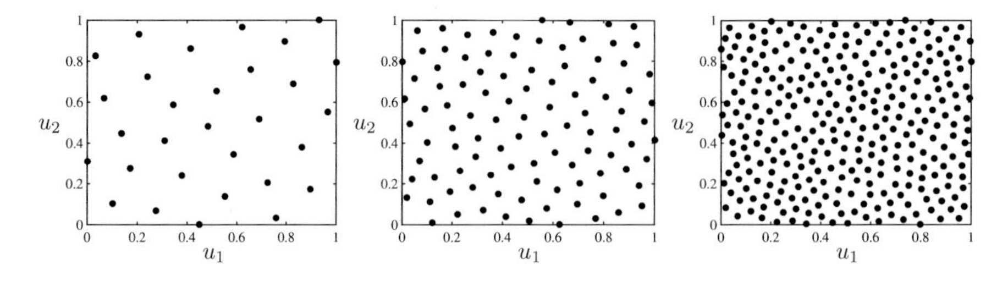

Fig. 14.58 Training data input: 2D maximin Latin hypercubes with 30, 100, and 300 points optimized according to the extended two-phase method (EDLS) in [131]

- Model error, partitioning, and validity functions are visualized on a 40×40 grid in the input space.
- The smoothness for LOLIMOT (normalized Gaussian validity functions) and HILOMOT (products of sigmoidal validity functions) is adjusted for comparable overlap.
- The results are obtained without any tedious fine-tuning or trail-and-error procedure.

The optimal model complexities for data sets of different sizes and noise levels are listed in Tables 14.2 and 14.3. Training and test errors are not listed in these tables because they are extremely close to each other and thus offer no additional information. Note that the discrepancy between 18 local models for HILOMOT (last row) in Table 14.2 and 14 local models for HILOMOT (medium row) in Table 14.3 is due to different noise realizations.

When comparing the number of local models, it is important to keep in mind that the number of *parameters* per local model is larger for HILOMOT because the splitting parameters are optimized as well. In the AIC_C this issue is taken into account by adding $2(M-1)$ parameters for a network with M local models because for two input dimensions two splitting parameters for $M-1$ splits need to be considered.

In all cases, LOLIMOT constructed more local models which is perfectly logical since it is less flexible than HILOMOT. However, the discrepancy is at most a factor of 2. This can be expected to become more pronounced when the dimensionality of the problem increases; compare Example 14.8.1.

The network complexities behave exactly as expected: More complex models for higher amounts of data, less complex models for higher noise levels.

The case $N = 100$ and $\sigma_n = 0.01$ shall be investigated in more detail for medium network smoothness. Figure 14.59 shows some results for LOLIMOT with 15 local models on the left and HILOMOT with 8 local models on the right. The top row illustrates the model error. It is remarkably similar for both networks. The NRMSE and worst-case error are very comparable. However, the patterns that can be seen in the surface plots allow concluding about the partitioning. Roughly the splits or transitions between the local models can be recognized. Indeed this can be verified by looking at the partitioning in the middle row and the validity functions in the

Table 14.2 Optimal network complexity = number of local models with best AIC_C ($\sigma_n = 0.01$)

Amount of training Data N	LOLIMOT	HILOMOT
30	6	4
100	15	8
300	25	18

Table 14.3 Optimal network complexity = number of local models with best AIC_C ($N = 300$)

Noise level σ_n	LOLIMOT	HILOMOT
0.003	33	18
0.01	25	14
0.03	9	8

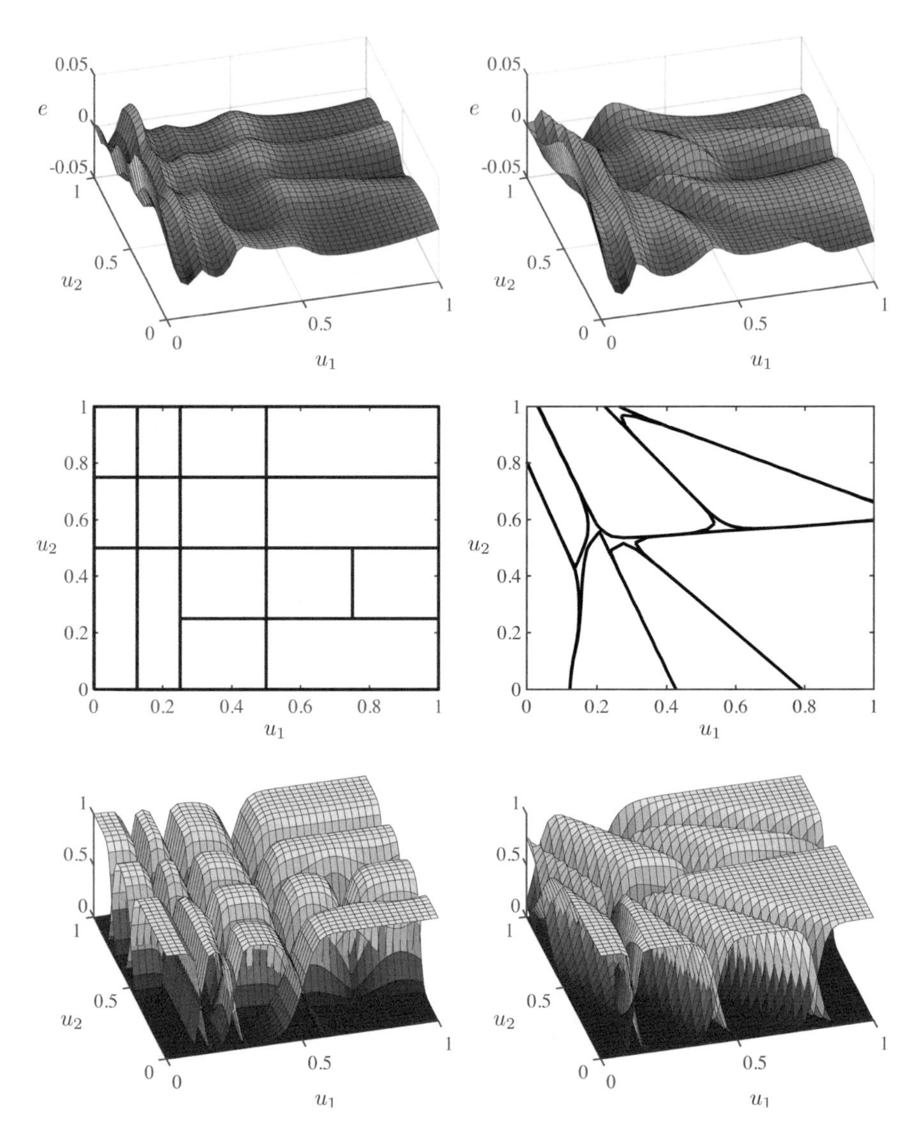

Fig. 14.59 Model error and partitioning for $N = 100$ training data points ($\sigma_n = 0.01$): LOLIMOT (left), HILOMOT (right).
Top: Model error. Middle: Partitioning. Bottom: Validity functions

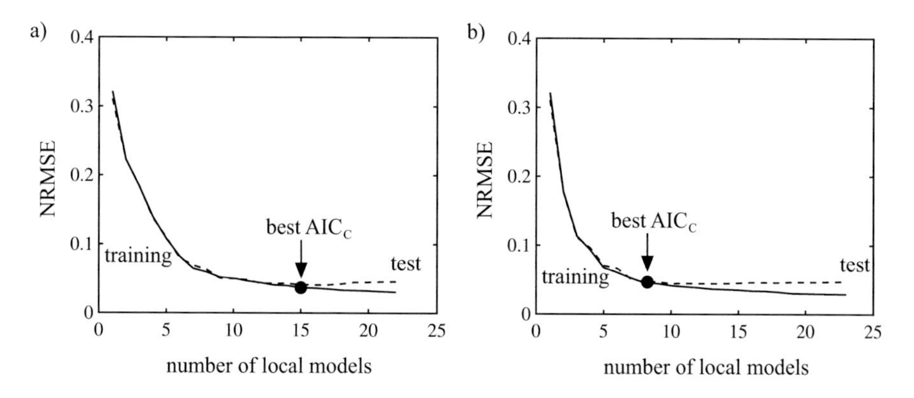

Fig. 14.60 Convergence curve for (**a**) LOLIMOT ($M_{\text{opt}} = 15$) and (**b**) HILOMOT ($M_{\text{opt}} = 8$) with selected model according to the best AIC$_C$

bottom row. The validity functions show a similar overlap. This is very important for a fair comparison because it influences two important factors:

- *Smoothness of the model output*: This can be an application-dependent important issue that does not reveal itself in the normal error measures. Rather the error measures tend to worsen for an increased smoothness (see the second factor). However, smoothness may be an important "soft" criterion. It is automatically addressed in the kernel methods discussed in Chap. 16, but in the local modeling context, it needs to be addressed explicitly by the user.
- *Amount of regularization*: The more the validity functions overlap (higher smoothness), the less flexible the model becomes if the local estimation is applied (as it is done here); compare Sect. 13.2.2. Therefore the smoothness directly influences the effective number of parameters and thus the variance error or overfitting tendency.

Figure 14.60 compares the convergence curves for the training and the test errors for LOLIMOT on the left and HILOMOT on the right. The point of the best AIC$_C$ is marked. These networks were used in Fig. 14.59. It is extremely remarkable that the test errors start to become significantly worse than the training errors exactly at this point. Of course, a similar behavior should be expected from the AIC$_C$, and it works perfectly.

It is very interesting to observe that the convergence curve of the test error departs earlier for HILOMOT than for LOLIMOT and the discrepancy between both grows bigger. This also can be expected since the difference between test and training error is very roughly[5] a measure for the effective number of parameters (or degrees of freedom) and therefore for the variance error (or overfitting tendency); see the discussion about *optimism* in Sect. 7.2.3.

[5]Not exactly, see the paragraph about optimism in Sect. 7.2.3.

Overall, Fig. 14.60 shows an impressive agreement between training and test data error up to the selected network complexity. It not very dangerous to mistakingly assess the model quality from its training error. This robustness property certainly can be attributed to the local estimation approach and the incremental split optimization that only optimizes one split at a time. Only nz split parameters and $nx + 1$ local model parameters are estimated simultaneously (here $nx = nz = 2$). Both enormously limit the overfitting tendency in practice.

Figure 14.61 illustrates the effect of the smoothness on the model trained with HILOMOT. The behavior for LOLIMOT is similar. The first thing to note is the identical optimal number of local models $M_{\text{opt}} = 8$ for the original smoothness in Fig. 14.59 and its half and double value in Fig. 14.61(left) and (right). The partitioning is different but similar in all three cases. The model error is significantly sharper (left) or smoother (right) compared to the medium case as expected.

Up to the best AIC_C value, the training and test errors hardly differ for all three smoothness cases. Obviously, the lower smoothness case exhibits higher errors close to the splits (ridges in the surface plot along the splits). Since the overall test error is comparable, this seems to be compensated by a better local model fit inside where $\Phi_i \approx 1$ (less regularization effect).

The strength of the regularization effect can be observed by the convergence curve of the test error. It deviates the earlier and the stronger from the training error convergence curve, the lower the smoothness is. The bigger gap between training and test errors in the least-smooth case is also due to a better training error. Both are manifestations of overfitting.

Note that the AIC_C contains the effective number of parameters in its penalty term. Thus the penalty is bigger for low smoothness values. Nevertheless, the best AIC_C is realized also at $M_{\text{opt}} = 8$ because the training error decrease is stronger for low smoothness values.

A final note about the wiggly contour lines in Fig. 14.61(left): They are just a consequence of the limited resolution of the 40×40 grid to generate the plot and the very high slopes of the sharp sigmoidal transitions. It is only a resolution/visualization problem and does not occur in reality.

Figure 14.62 compares the convergence curves for LOLIMOT and HILOMOT on training data (left) and test data (right). Both plots look almost identical. As expected, HILOMOT converges significantly faster than LOLIMOT. Both achieve identical performance.

Finally, Fig. 14.63 compares the test error convergence curves for LOLIMOT (left) and HILOMOT (right) on three different noise levels for $N = 300$ training data points. The convergence curves stop at the selected network complexity. Note that during training two more iterations were carried out to ensure with some certainty that the AIC_C value has reached its minimum. Again the finally achieved performances are very comparable for both algorithms.

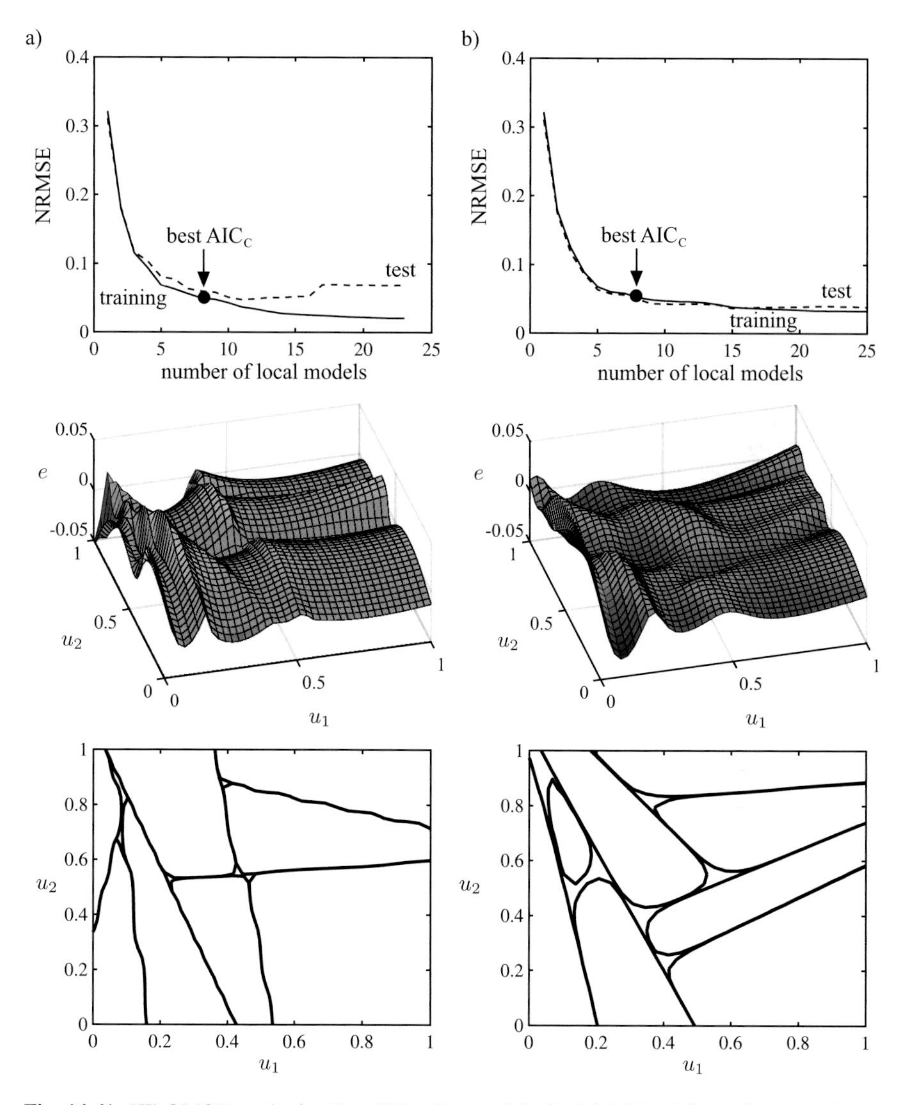

Fig. 14.61 HILOMOT results for $N = 100$ and $\sigma_n = 0.01$ for (**a**) half and (**b**) double smoothness of Fig. 14.59.
Top: Convergence curves for training and test data with best AIC$_C$ which in all cases selects $M_{opt} = 8$. Middle: Model error. Bottom: Partitioning

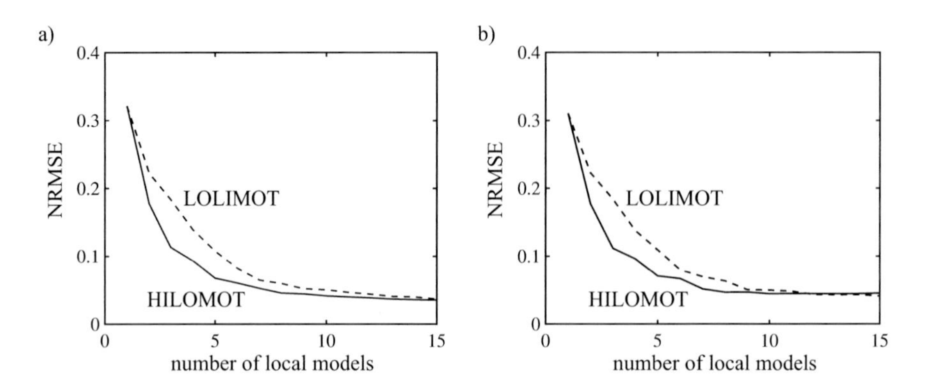

Fig. 14.62 Convergence curves on (**a**) training data and (**b**) test data

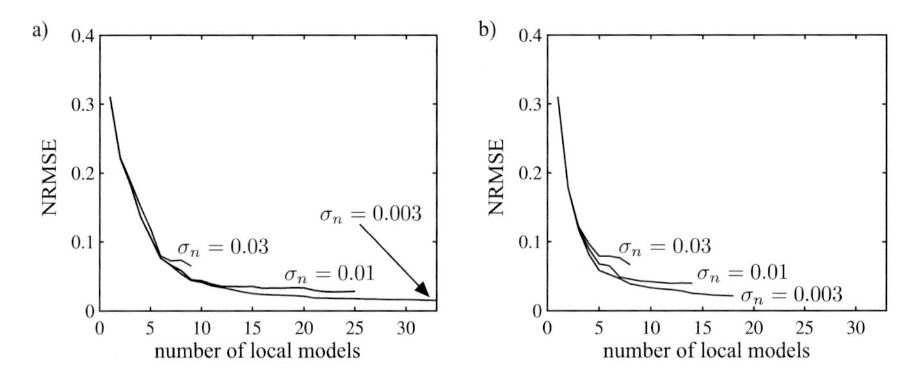

Fig. 14.63 Test error convergence curves for $N = 300$ training data points for low, medium, and high noise levels for (**a**) LOLIMOT, (**b**) HILOMOT

14.9 Errorbars, Design of Excitation Signals, and Active Learning

Any model is almost worthless without information about its accuracy. Only with sufficient confidence in the accuracy of a model can it be utilized in an application. It is possible to assess the achieved model accuracy by an examination of the model error on a test data set; see Sect. 7.3.1. However, the average error does not reveal any information about the model accuracy that can be expected for a given model input $\underline{u}$. A model may be very accurate in some operating regions and inaccurate in others; thus an estimate of the average error is not sufficient. Ideally, a model should provide its user not only with a model output y but also with an interval $[y_{\min}\ y_{\max}]$ within which the process output is guaranteed to lie with a certain probability. Close intervals then would indicate a high confidence in the model output, while large intervals would alarm the user that the model output might be uncertain. Such an information can be exploited in many applications. For simulation a number of different models may be utilized, e.g., a first principle model, a qualitative fuzzy model, and a black box neural network or several models with identical architecture but different complexity (number of neurons, rules, etc.). For each input, a decision level can assess the expected model accuracies and can use the model with the most confident output or weight between all models with confidence-dependent weights. In model-based predictive control, information about the model accuracy can be exploited to determine the prediction and control horizons or to constrain the nonlinear optimizer to operating regions in which the model can represent the process with sufficient accuracy. (A well-known problem with nonlinear predictive control is that the optimizer might excite the model in untrained operating regions and thus might lead to inferior performance since the optimal control with the model is not realized with the process [399].)

Unfortunately, it is a very difficult problem to estimate the expected model accuracy. The major reason for the difficulties is that the model error consists of a bias and a variance part; see Sect. 7.2. The bias part represents the error due to the structural mismatch between the process and the model. Since in black box or gray box modeling the true structure of the process is unknown,[6] the bias error can generally not be assessed. Usually, the best that can be done is to estimate the variance error of the model. At least this gives some information about the model accuracy. In some cases, good arguments can be found that the bias error can be neglected and thus the total model error can be assessed by the variance error. In Sect. 14.9.1, an expression is derived for the variance error of a local linear neuro-fuzzy model. It can be utilized to detect extrapolation (Sect. 14.9.2) and can serve as a very helpful tool for the design of good excitation signals (Sect. 14.9.3). Finally, Sect. 14.10.4 gives a brief overview of the field of active learning, which can be seen as an online design of excitation signals.

14.9.1 Errorbars

An estimate of a model's variance error allows one to draw so-called *errorbars*. These errorbars represent a confidence interval in which the process output lies with a certain probability (say 95%) under the assumption that the bias error can be neglected. Because the variance error is proportional to the noise variance σ_n^2 (assuming uncorrelated Gaussian noise), the calculation of absolute values for the errorbars requires either knowledge about the noise variance or an estimate. If indeed the bias error were equal to zero, the noise variance could be estimated as $I/(N - n_{\mathrm{eff}})$, where I is the sum of squared errors loss function value, N is the number of data samples, and n_{eff} is the effective number of parameters; see (3.35) in Sect. 3.1.1. However, for nonlinear models, the bias error typically cannot be neglected because of a structural mismatch between the process and

[6] Otherwise a superior white box model could be used.

the model. Thus, the noise variance cannot be estimated with sufficient accuracy. Furthermore, for nonlinear models, it is not sufficient to estimate a global value of the noise variance. Rather, the noise level may depend on the operating condition of the process. Because of all these difficulties, the errorbars are usually utilized only as a *qualitative* or *relative* measure. This means that the model accuracy for different model inputs and of different models can be compared, but no probability is assigned to the interval of absolute errorbar values, i.e., they do not represent a certain confidence interval. The calculation of a confidence interval would require assumptions about the noise probability density function, e.g., the normal distribution assumption. Previous work in this field in context with neural networks can be found in, e.g., [333, 345].

When can the bias error be neglected? In first principle models, the bias error can often be neglected since the model structure is assumed to *match* the true process structure. In contrast, black box models are based on the idea of providing a flexible model structure that *approximates* the unknown process structure. If the model structure is indeed rich enough, its bias error can be neglected. However, a good bias/variance tradeoff (Sect. 7.2.1) typically leads to a significant bias error for nonlinear models. Either the model complexity is optimized (Sect. 7.4) or a regularization technique is applied (Sect. 7.5). Thus, it is important to keep in mind that the following errorbar results do not include the bias error, which may be significant.

The variance error of a model can be measured by the variances of the model output. Therefore, the errorbars can be written as

$$\pm\sqrt{\mathrm{diag}\left(\mathrm{cov}\{\underline{\hat{y}}\}\right)}\,. \tag{14.105}$$

The covariance matrix of the model output can be derived from the parameter and noise covariance matrices. If uncorrelated noise is assumed, its covariance matrix is simply $\mathrm{cov}\{\underline{n}\} = \sigma_n^2 \underline{I}$. In order to evaluate the covariance matrix of the parameters, linear and nonlinear parameters have to be distinguished. While for linear parameters an analytic expression can be derived, numerical approximations have to be used for the nonlinear parameters. Therefore, in all that follows, it is assumed that no nonlinear parameter optimization has been utilized for building the model. Strictly speaking, the rule premises have to be fixed, and only the linear rule consequent parameters are allowed to be estimated from data. Nevertheless, the results may also be used for models trained with LOLIMOT because the variance of the premise parameters is negligibly small owing to their limited degrees of freedom. In contrast, the results do not apply, e.g., for the training method ANFIS proposed by Jang [289, 290, 292] which is based on nonlinear optimization of the membership functions, because then the variance contribution of the nonlinear parameters becomes significant.

14.9.1.1 Errorbars with Global Estimation

The variances and covariances of the globally estimated parameters are given by (see (3.34) in Sect. 3.1.1)

$$\text{cov}\{\underline{\hat{w}}\} = \sigma_n^2 \left(\underline{X}^T \underline{X}\right)^{-1}, \tag{14.106}$$

where σ_n^2 is the noise variance. The covariance matrix of the model outputs is thus given by (see Sect. 3.1.2)

$$\text{cov}\{\underline{\hat{y}}\} = \underline{X}\,\text{cov}\{\underline{\hat{w}}\}\underline{X}^T. \tag{14.107}$$

14.9.1.2 Errorbars with Local Estimation

The variances and covariances of the parameters obtained by local estimation are given by (see (3.71) in Sect. 3.1.8)

$$\text{cov}\{\underline{\hat{w}}_i\} = \sigma_n^2 \left(\underline{X}_i^T \underline{Q}_i \underline{X}_i\right)^{-1} \underline{X}^T \underline{Q}_i \underline{Q}_i \underline{X} \left(\underline{X}_i^T \underline{Q}_i \underline{X}_i\right)^{-1} \tag{14.108}$$

with $i = 1, \ldots, M$. The covariance matrices of the local linear model outputs are thus given by (see Sect. 3.1.2)

$$\text{cov}\{\underline{\hat{y}}_i\} = \underline{X}\,\text{cov}\{\underline{\hat{w}}_i\}\underline{X}^T. \tag{14.109}$$

These local covariances can be superimposed to the overall model output covariances:

$$\text{cov}\{\underline{\hat{y}}\} = \sum_{i=1}^{M} \underline{Q}_i \text{cov}\{\underline{\hat{y}}_i\}. \tag{14.110}$$

As an example, the function shown in Fig. 14.23a is approximated by a local linear neuro-fuzzy model with six rules by the LOLIMOT algorithm, which yields the local linear models and the validity functions depicted in Fig. 14.23b. The training data consists of 300 equally distributed samples in [0, 1]. The errorbars of the local models according to (14.109) are shown in Fig. 14.64a. They are combined by (14.110) to the overall model errorbar in Fig. 14.64b. The errorbars of the individual local linear models possess their minimum close to the centers of their validity functions because the data samples in these regions are highly relevant for parameter estimation. The amount of data that effectively is utilized decreases with increasing distance from the centers because the weighting factors in the weighted LS estimation tend to zero. The only exception is the two local models next to the boundaries, but they suffer from the fact that no training data is available

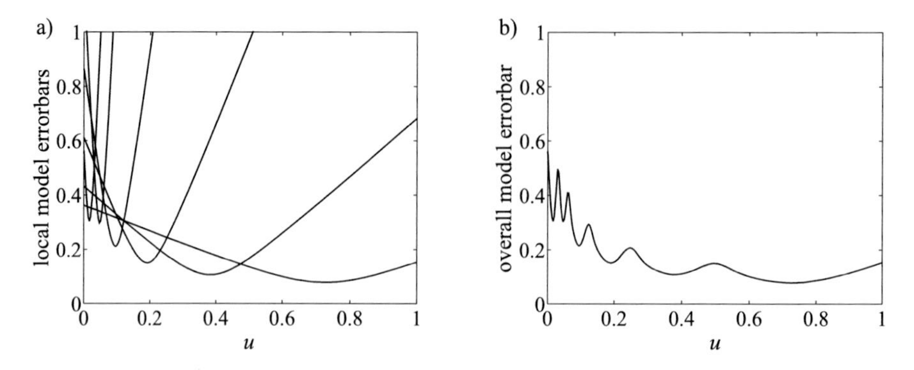

Fig. 14.64 Errorbars of (**a**) the local linear models and (**b**) the overall neuro-fuzzy model for 300 equally distributed training data samples

for $u < 0$ and $u > 1$. The overall model errorbar also contains these properties, and the interpolation regions between two local linear models are clearly visible as errorbar maxima. Furthermore, Fig. 14.64b reveals increasing variance errors for smaller inputs u. This effect is caused by the finer partitioning of the input space for small u. Thus, less data is available for the estimation of the local linear models for small u than for large u. This makes the model output less accurate for small inputs.

14.9.2 Detecting Extrapolation

For all models without built-in prior knowledge, extrapolation is dangerous; see Sect. 14.4.2. Thus, it is important to warn the user when the model is extrapolating. If the input data is just a little bit outside the region covered by the training data, no serious difficulties may arise; however, the model might not be capable of performing reasonably well if the input is far away. Therefore, it is not sufficient to detect whether the model is extrapolating or not; a confidence measure is required. As Fig. 14.65 demonstrates, errorbars are an appropriate tool for that task. The errorbar is monotonically increasing in the extrapolation regions. For an extrapolation to the right ($u > 1$) a large number of training data samples support the rightmost local linear model since its corresponding validity function stretches from 0.5 to 1. Based on this information, the model may extrapolate reasonably well for $u > 1$. However, the leftmost local linear model stretches only from 0.03125 to 0 and thus is estimated with only a few training data samples. Consequently, the errorbar tends sharply to infinity for $u < 0$. Note that a second reason for the sharp increase in the errorbar for extrapolation to the left is the large slope of the active local linear model, which makes the model output highly sensitive to parameter variations.

Figure 14.66 illustrates a two-dimensional errorbar example. The function in Fig. 14.66a is approximated with a local linear neuro-fuzzy model with six rules

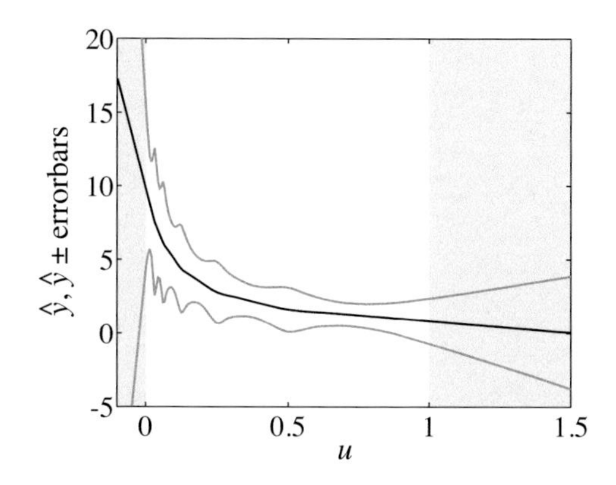

Fig. 14.65 Errorbars indicate extrapolation; see example from Fig. 14.64

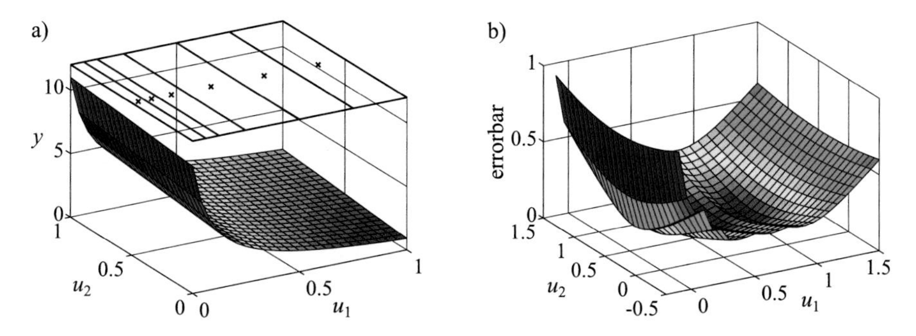

Fig. 14.66 The errorbar in (**b**) of the model of the function in (**a**) indicates a more reliable extrapolation in the u_2- than in the u_1-dimension

trained by LOLIMOT. Since the function is nonlinear only in u_1, the constructed input space partitioning does not contain any splits in the u_2-dimension. This fact leads to more accurate estimates of the slopes w_{i2} in the u_2-dimension than of the slopes w_{i1} in the u_1-dimension. As a consequence, the errorbar increases less in the u_2- than in the u_1-dimension; see Fig. 14.66b.

14.9.3 Design of Excitation Signals

The errorbars depend on the training data distribution and on the model and in particular on the partitioning of the input space. Therefore, they can serve as a valuable tool for the design of excitation signals. Errorbars allow one to detect data holes and regions that are covered too densely with data so that they might dominate the loss function. Figure 14.67 illustrates the relationship between data distribution and local model complexity for the example discussed above. The errorbar in Fig. 14.67a is calculated for a training data set consisting of 300 equally distributed

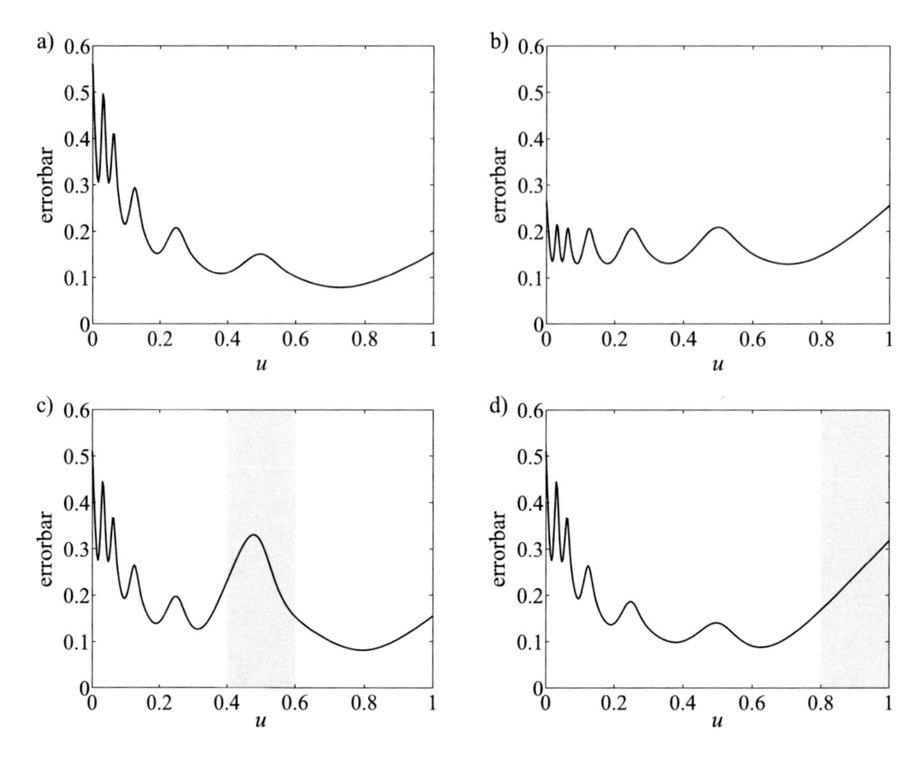

Fig. 14.67 Relationship between data distribution, model complexity, and errorbars: errorbar for (**a**) equal training data distribution (300 samples), (**b**) higher training data density in regions of small u (300 samples), (**c**) missing data in [0.4, 0.6] (240 samples), (**d**) missing data in [0.8, 1.0] (240 samples)

data samples. In regions of higher model complexity (small u), the errorbar is larger since the same amount of data has to be shared by several local linear models. Knowing this relationship, a better excitation signal can be designed that contains the same number of data samples for each local linear model, i.e., the data density decreases with increasing u. With training data generated correspondingly, the errorbar in Fig. 14.67b can be achieved with an identical number of data samples. Consequently, the variance error in Fig. 14.67b is almost independent of u in the interpolation range. The significant variance error decrease for small inputs has to be paid for only with a minor errorbar increase for large inputs.

Data holes can be easily discovered by errorbars, as demonstrated in Fig. 14.67c, where the training data in the interval [0.4 0.6] has been removed from the original set used in Fig. 14.67a. Missing data is especially harmful to the boundary of the input space (Fig. 14.67d) because it extends the extrapolation regions. Therefore, a good excitation signal should always cover all types of extreme process situations.

Note that the importance of these tools grows with increasing input space dimensionality. For models with one or two inputs, the training data distribution can be visually inspected by the user. Such a powerful visualization and the imagination capabilities of humans fade for higher-dimensional problems.

The design of excitation signals for nonlinear *dynamic* processes is a much more complex topic. Nevertheless, the tools proposed here can be applied as well; see [156] for more details.

14.10 Design of Experiments

This section gives a brief overview of common methods for the design of experiments (DoE). These are methods that establish where in the inputs space the data shall be measured, i.e., for which values $\underline{u}(i) = [u_1(i)\ u_2(i)\ \cdots\ u_p(i)]^T$, $i = 1, \ldots, N$. It is obvious that all data-driven modeling approaches crucially depend on the quality of the data. Quality, on the one hand, refers to the noise level and systematic measurement error which can be kept as small as possible by taking care, choosing the appropriate sensors and environment. One the other hand, the data distribution in the input space decides on how much information is gathered on the process by the measurements. This second issue is (almost) independent of physics and hardware and can be carried out in more or less appropriate and clever ways. This section deals with these methods and advocates an active learning approach based on local model networks, specifically on the HILOMOT algorithm introduced in Sect. 14.8.

14.10.1 Unsupervised Methods

The common feature of all unsupervised methods is that they purely focus on geometric properties in the input space. They can be generated *in advance* with an arbitrary number of data points N, and during the measurement, just a list of points needs to be processed. The unsupervised DoE methods are usually the simplest and easiest to handle. They are also the most robust with respect to the process and model behavior since they are independent of both. Thus, they are universally applicable without any other prior knowledge than the limits (min and max values) of each input variable.

14.10.1.1 Random

A uniform random distribution yields a very bad distribution of data points because the probability for the generation of points very close to each other is too big. Figure 14.68a shows that many holes and small point clusters can occur. Both are undesirable for a space-filling point distribution.

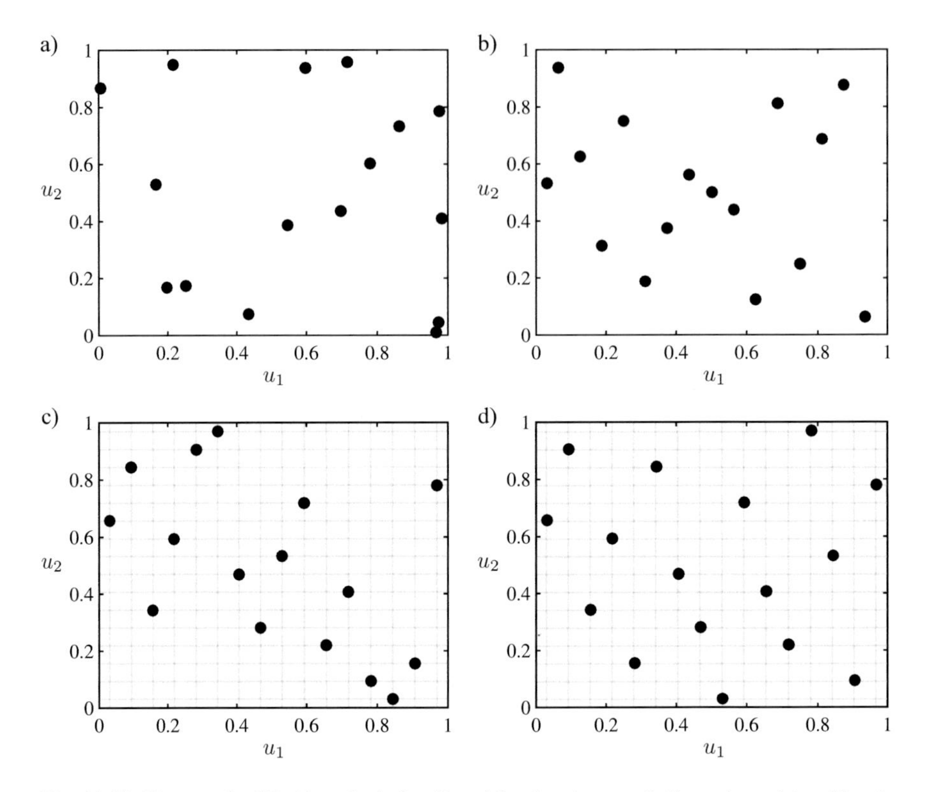

Fig. 14.68 Unsupervised DoE methods for $N = 16$ points in $p = 2$ dimensions: (**a**) uniformly random, (**b**) Sobol set, (**c**) Latin hypercube, (**d**) maximin Latin hypercube

14.10.1.2 Sobol Sequence

A number of different pseudo-random low-discrepancy sequences exist which avoid these issues. Besides the Sobol sequence shown in Fig. 14.68b, other common sequences are the van der Corput or Halton sequence or numbers drawn from a Hammersley set [71]. As can be observed in Fig. 14.68b for few data points, the input space coverage is not excellent. However, this improves significantly as N increases. Note, however, that it is rather typical to deal with sparse data similar to or even worse than shown in the figure as the dimensionality of the problem becomes larger.

14.10.1.3 Latin Hypercube (LH)

Latin hypercubes guarantee no space-filling properties. Their primary goal is to uniformly cover all inputs when the data is projected to them. In order to achieve this goal, each input is discretized into N values. Thus, a grid with N^p grid points is constructed in the whole input space where p is the input dimension, i.e., an $N \times N \times \ \ldots \ \times N$-grid. Only N points of this grid are occupied by data in a way that all discrete values for each input are taken once. The simplest Latin hypercube would be the diagonal through the input space. Obviously, a huge amount of other Latin hypercubes exists. And only those with good space-filling properties should be utilized.

The main benefit of Latin hypercubes is the *non-collapsing design*. This means that a Latin hypercube in p dimensions projected to a subset of $q < p$ dimensions is still a Latin hypercube. And for $q = 1$, the axis is covered perfectly uniformly.

The Latin hypercube shown in Fig. 14.68c was constructed by the MATLAB function `lhsdesign` which randomly creates a couple of Latin hypercubes and selects the one with best space-filling properties. One common criterion for assessing this space-filling property is the worst (i.e., closest) nearest neighbor distance or alternatively formulated the distance of the closest point pair.

In Fig. 14.68c big holes can be observed close to the lower left and upper right corners. In order to ensure good space-filling properties, the abovementioned criterion needs to be optimized. This means that during optimization, the points are driven away from each other. The nearest neighbor distance criterion makes only sense in the context of optimization. It is not useful as an objective evaluation of the quality of a Latin hypercube. To see this: imagine a perfectly space-filling Latin hypercube with just one bad point being close to another one. This single point would ruin the "quality" measure completely although the rest of the Latin hypercube is perfect.

14.10.1.4 Optimized Latin Hypercube

In [131] a relatively simple and efficient method for optimizing Latin hypercubes is proposed. The references in this paper also give an overview on alternative strategies and the state of the art of local and global Latin hypercube optimization. Latin hypercubes that maximize the minimum (worst) nearest neighbor distance are called *maximin Latin hypercubes*. Starting with the Latin hypercube from Fig. 14.68c, the maximin optimum according to the algorithm given in [131] can be seen in Fig. 14.68d. It still offers all the advantages of Latin hypercubes but guarantees good space-filling properties.

The drawbacks of maximin Latin hypercubes are as follows: (i) the optimization is time-consuming for large N and (ii) it is hard for any Latin hypercube design to extend the number of measurement points afterward. Supplementing data later is much easier to handle for Sobol or similar sequences.

14.10.2 Model Variance-Oriented Methods

The most common sophisticated methods for the design of experiments are motivated by the endeavor of making the model variance small. If the model structure adequately can describe the process behavior, then the dominant model error comes from the limitations of finite, noisy data. Therefore the idea is to place the data points in the input space in such a manner that the estimated model will be maximally robust with respect to measurement noise. For linear regression models, these "optimal" designs are particularly simple.

14.10.2.1 Optimal Design

The so-called "optimal" designs optimize a criterion based on the information matrix $\underline{X}^T \underline{X}$ because its inverse is proportional to the variances and covariances of the estimated parameters; compare Sect. 3.1.1. Therefore it reflects the sensitivity of the model with respect to noise. The most common optimal design criteria are[7]:

- *D-optimality*: seeks to minimize $\det((\underline{X}^T \underline{X})^{-1})$ or equivalently maximize the determinant of the information matrix $\underline{X}^T \underline{X}$ of the design. This criterion results in maximizing the differential Shannon information content of the parameter estimates. The quantity $\det((\underline{X}^T \underline{X})^{-1})$ also measures the volume of the confidence ellipsoid of the parameter vector. Thus it is perfectly natural to minimize this volume.
- *A-optimality*: Seeks to minimize the trace of the inverse of the information matrix $\underline{X}^T \underline{X}$. This criterion results in minimizing the average variance of the estimated parameters.
- *G-optimality*: Seeks to minimize the maximum entry in the diagonal of the smoothing matrix $\underline{S} = \underline{X}(\underline{X}^T \underline{X})^{-1} \underline{X}^T$. This has the effect of minimizing the maximum variance of the predicted values.

Further optimal design approaches exist. In practice, they all yield very comparable results. Note that the regression matrix $\underline{X}$ only contains information about the input data. The output is irrelevant. In this respect, optimal designs are also unsupervised. However, they assume a specific model structure in order to build $\underline{X}$ where both

- The model architecture, e.g., polynomial, basis function network, etc.
- The model complexity, e.g., degree, number of neurons, etc.

need to be known a priori. It is rather obscure where this knowledge should come from. In some cases, it might come from physical insights. In some cases, it might be just a guess oriented along some rough idea about the process complexity and the amount of data that shall be gathered.

[7]https://en.wikipedia.org/wiki/Optimal_design

14.10.2.2 Polynomials

Most commonly optimal designs are utilized in the context of polynomial models; typically of degrees 1, 2, or 3. Figure 14.69 shows how a D-optimal design looks like in these cases. In cases where more data points are available than necessary to cover the optimal pattern (here a grid), multiple measurements are taken at some points. Obviously, such a strategy can only be recommendable if the data is extremely noisy and/or the model structure perfectly describes the process structure.

Example 14.10.1 D-Optimal Design for Polynomials
Figure 14.70 demonstrates an example of how bad D-optimal designs perform if the model structure does not match the process. The process is a hyperbola and polynomials of degree 2 (left column) and degree 3 (right column) shall be fitted. Training data are $N = 10$ noise-free points; generalization data used for error calculation are 100 equidistantly distributed points. The top row of Fig. 14.70 shows the models fitted with equidistantly distributed training data. The bottom row shows the models fitted with D-optimal training data. For the second-degree polynomial, the D-optimal data was at $u(i) \in \{0, \frac{1}{2}, 1\}$, $i = 1, \ldots, N$ and for the third-degree polynomial at $u(i) \in \{0, \frac{1}{3}, \frac{2}{3}, 1\}$, $i = 1, \ldots, N$. All these points were sampled multiple times achieving the best D-optimal design for $N = 10$ training data points. Note that in absence of noise, this delivers just redundant data.

As can be seen from the NRMSE values, the D-optimal design is significantly worse to the equidistant design. The two reasons for that are as follows:

1. The model structures are inadequate for describing the process. Even with best parameters, the models exhibit a large bias error. Thus, minimizing the variance error is beside the point.
2. The data is noise-free. Therefore repeated measurements at the same positions offer no advantage in terms of averaging.

Issue 1 will always be encountered more or less in real-world applications. It will be severe in the case of black box approaches. It will be weak in the case of first principles modeling. Issue 2 may be relevant depending on the amount and noise level of the measurement data. The variance error tends to play a significant role if

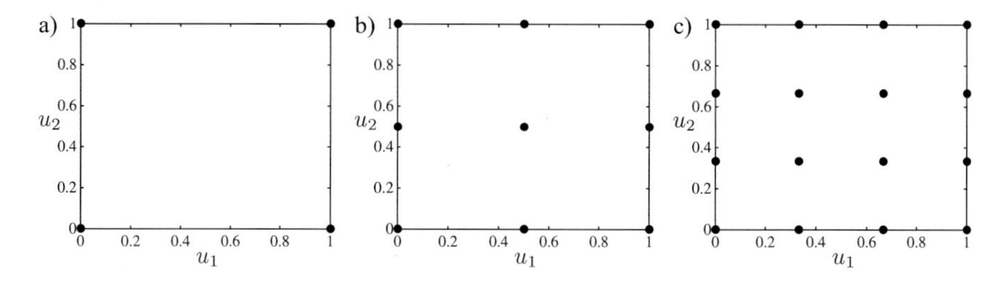

Fig. 14.69 D-optimal designs for $N = 16$ points in $p = 2$ dimensions for polynomial models of (**a**) first, (**b**) second, (**c**) third degree. Some points are chosen multiple times

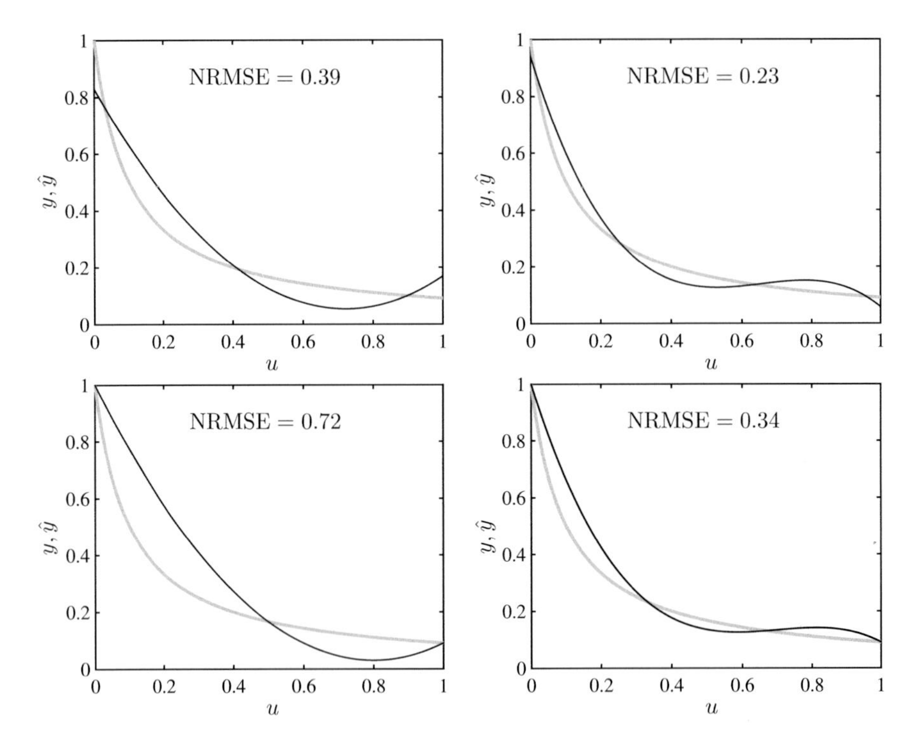

Fig. 14.70 D-optimal design for polynomials. A hyperbola is approximated by a polynomial of degree 2 (left) and degree 3 (right) with $N = 10$ noise-free data points. Top row: Equidistant data. Bottom row: D-optimal data

the number of data points is close to the effective number of parameters $N \approx n_{\text{eff}}$ and/or the noise level is high.

In summary, "optimal" designs can only be recommended if the variance error can be expected to be dominant. Whenever the bias error is the key component, model variance-based methods make no sense. □

14.10.2.3 Basis Function Network

In order to build the regressor matrix $\underline{X}$, all hidden layer parameters (centers, widths) have to be determined before. Also, the number of basis functions is fixed. It is hard to see how all these parameters can be fixed a priori. Realistically this can only be carried out by a preliminary modeling step which again requires a DoE. This would lead to an iterative DoE and modeling procedure which seems very unrealistic and time-consuming.

14.10.2.4 Multilayer Perceptron, Local Model Network, etc.

For models with complicated but performance-crucial hidden layer structure given through the hidden layer weights of an MLP or the partitioning of local model networks, those degrees of freedom need to be considered in the model variance assessment as well. They are responsible for significant parts of the model variance. As, e.g., proposed in [216], this variance part can be approximated by the Fisher information matrix. For linear regression, this corresponds to $\underline{X}^T \underline{X}$; for nonlinear parameters, it is a linearization. This linearization can cause trouble in practice. But the main drawback is that the whole network structure with the number of neurons has to be fixed a priori.

14.10.2.5 Gaussian Process Regression

Gaussian processes are discussed in Chap. 16. One of their main strengths is that the model not only yields the model output (mean value of the Gaussian distribution) but also the model variance. Thus, it naturally lends itself to carry out an optimal design. Gaussian process models are much more powerful than, e.g., polynomials. Thus it is more reasonable to apply a variance minimizing DoE because the bias error can be expected to be much smaller than for polynomials.

However, similar to the hidden layer parameters in basis function networks, the hyperparameters (typically width(s) and amount of regularization) have to choose a priori. The centers and the number of basis functions are usually determined by the training data since one kernel is placed on each data point. Therefore the nominal model complexity just grows linearly with the amount of data. However, the effective model complexity is determined by the hyperparameters which typically requires a marginal likelihood or leave-one-out-error optimization of the model. Thus, a Gaussian process also requires some iterative algorithm. However, this is more straightforward than for basis function networks due to the 1:1 mapping between data and kernels. For a discussion about such algorithms, refer to Sect. 14.10.4.

Note that the hyperparameter optimization makes the variance-oriented Gaussian process DoE partly supervised. Although the variance of the model is first and foremost dependent on the data distribution in the input space, the hyperparameter optimization is based on the model error and thus considers the model output. In practice, usually identical widths[8] are chosen for all kernels. As a consequence, the model variance will be low in areas with dense training data and will be high in areas with sparse training data. The local complexity of the process is irrelevant; it just enters in average into the hyperparameter tuning. In the end, the model variance is kept minimal through a uniform training data distribution, and the model output is not important.

[8]The widths can be chosen individually for each input dimension but are identical for all kernels.

14.10.3 Model Bias-Oriented Methods

Unsupervised DoE methods ignore the behavior of the process and model output completely. Variance-oriented DoE methods consider the model output partly because the regressors need to be fixed which typically is done in a way that promises good model performance. Variance-oriented DoE is only reasonable in high noise, few data cases where the model structure closely matches the process. This might be the case for first principles models and extremely flexible models that can realize an extremely small bias error.

This section deals with model bias-oriented methods for DoE that address the bias error of the model. These methods are not so widespread because it is difficult to assess the model bias in practice. Two approaches shall be introduced. The first one is a general-purpose approach that uses model committees. The second one is an embedded method that utilizes features of local model networks.

14.10.3.1 Model Committee

In order to calculate the bias error of a model, the process behavior needs to be known. Since it is unknown in practice, some other ways for assessing the systematic model error must be found. One idea is based on model committees; see Fig. 14.71a. The process is fit with C different models. They all yield their individual approximations of the true process. The models can represent different bias/variance tradeoffs by varying the model complexity explicitly or implicitly due to the regularization strength. Preferably different model architectures can be applied which have different types of interpolation and extrapolation behavior and individual strengths and weaknesses. If for one input data point $\underline{u}(i)$, all model outputs approximately agree, i.e., $\hat{y}^{(1)}(i) \approx \hat{y}^{(2)}(i) \approx \ldots \approx \hat{y}^{(C)}(i)$ with high likelihood, this model output will be close to the process. However, if the outputs disagree strongly obviously, the available data allows for different interpretations depending on the individual model assumptions. Thus, it would be very informative to place a measurement at exactly this point to clarify the situation. This is the strategy of learning with model committees. A measure for the disagreement between the model outputs could be their variance in the regression case or their entropy in the classification case.

14.10.3.2 Model Ensemble

The more diverse the committee model architectures are, the more universal this approach for bias assessment is. However, all models should deliver some minimal performance since bad performing models dilute the committee decision. If such a committee is utilized for DoE, it makes sense to include them in a model ensemble as shown in Fig. 14.71b. By averaging many models of different architectures, their

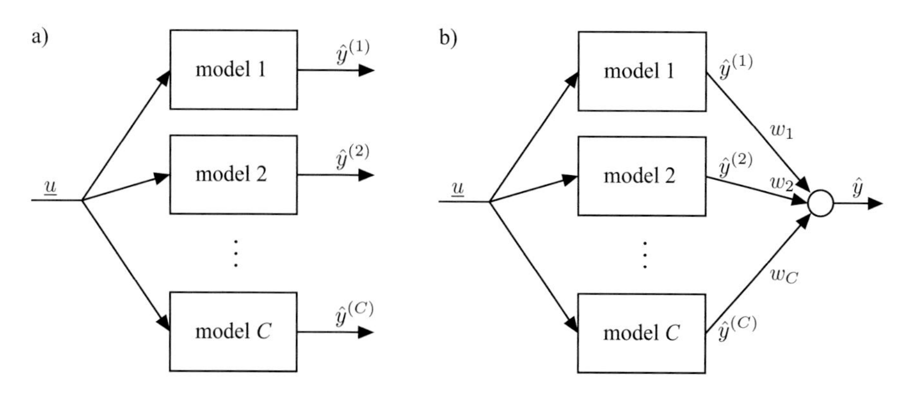

Fig. 14.71 (**a**) Model committee: A set of models typically of different architectures for the same process. (**b**) Model ensemble: The output is a (possibly weighted) average of the models in the ensemble

variance errors can be reduced [122], and their systematic weakness can compensate each other yielding a more robust overall performance. For one model architecture alone, this ensemble idea is illustrated in Sect. 14.11.

The output of a model ensemble is the weighted average of all C models within the committee:

$$\hat{y} = w_1 \hat{y}^{(1)} + w_2 \hat{y}^{(2)} + \ldots + w_C \hat{y}^{(C)} . \tag{14.111}$$

In the simplest case, the weights are chosen as

$$w_i = \frac{1}{C} \tag{14.112}$$

which simply represents an average. A more sophisticated choice would be an *inverse variance weighting* which is optimal in the statistical sense; compare Sect. 14.11.4. The weights are normalized

$$w_i = \frac{\widetilde{w}_i}{\sum_{j=1}^{C} \widetilde{w}_j} \tag{14.113}$$

and chosen proportional to the mean squared error (or error variance) of each model

$$\widetilde{w}_i = \frac{1}{\text{MSE}_i} . \tag{14.114}$$

Note that it is very important to *not* assess the MSE in (14.114) with the training data because this will be over-optimistic due to overfitting effects. Rather (14.114) has to be evaluated on validation data or with the help of cross-validation, compare Sect. 7.3.2.

14.10.3.3 HILOMOT DoE

Local model networks offer a new possibility to address the problem of model bias because information about the local distribution of the model performance (local error) is accessible. In [226] an algorithm based on the HILOMOT training called HILOMOT DoE is proposed. Its main idea is to gather data in regions where modeling is hard and leave data sparse in regions where modeling is easy. If the local models are of *linear* type, a data density will result that reflects the severeness of the process nonlinearity. It is perfectly intuitive to gather more data where the process is complex because a more complex model needs to be created in these regions with more parameters to fit which in turn requires more data. Note, however, that HILOMOT DoE discourages a self-reinforcing loop of repeated *local* data gathering and splitting by always analyzing the overall model.

For the original references of HILOMOT DoE, refer to [224] for the methodology, to [226, 384] for applications in the area of structural health monitoring and to [228, 322] for applications in the area of combustion engine measurement.

HILOMOT DoE operates in two stages:

1. The worst local model j of a local model network is found by calculating the local error measures $J_j = \sum_{i=1}^{N} e^2(i)\Phi_j$, $j = 1, \ldots, M$. Alternatively, also other norms like the ∞- or 1-norm can be chosen in order to find the local model j with the maximum error or worst average absolute error.

   Note that for calculation of the local model output values the data of the whole input vector $\{\underline{u}(1), \underline{u}(2), \ldots, \underline{u}(N)\}$ is relevant. Those input variables that enter $\underline{x}$ are responsible for the outputs of the local models. Those input variables that enter $\underline{z}$ are responsible for distinguishing which data falls in which local model.
2. A new measurement point is placed in the biggest data hole in the worst local model found under Stage 1. That is, a space-filling approach is taken inside the local model. All data inside the local model is considered for that evaluation; see the discussion below.

   Note that the biggest hole has to be found in the $\underline{z}$-input space, i.e., only those input variables that enter $\underline{z}$ are relevant for these considerations. In practice, usually $nz < nx$ and thus the dimensionality of this space often is rather low.

Figure 14.72 visualizes Stage 1 in (a) and Stage 2 in (b). Stage 2 can be realized by simply evaluating many candidate points and calculating the nearest neighbor distances. This candidate point generation becomes problematic in high dimensions and may be replaced by some random walk approach. Or more sophisticated optimization techniques may be applied.

In order to operate satisfactorily, the committee methods or HILOMOT DoE have to be utilized in an active learning framework which is described in Sect. 14.10.4.

HILOMOT DoE offers the following advantages:

1. In early measurement phases where little data is available, the local model network will be simple, perhaps consisting of only one local = global model. Then a purely space-filling approach results.

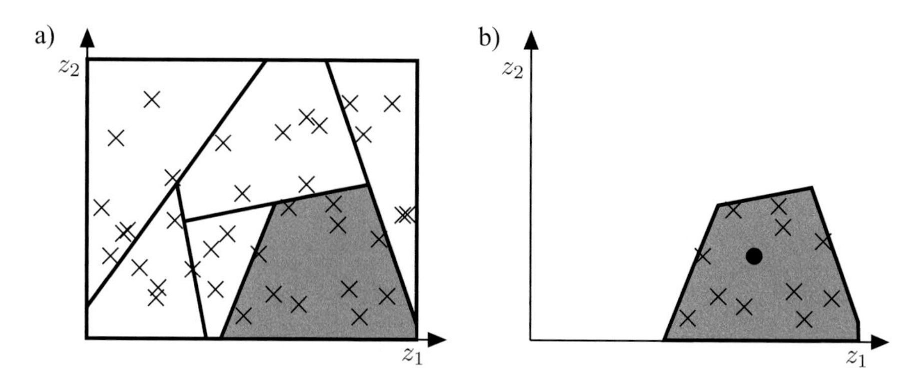

Fig. 14.72 HILOMOT DoE: (**a**) The worst local model (gray) is selected by comparing local error measures. (**b**) Inside the worst local model the new measurement (filled circle) is placed in the biggest hole (space-filling)

2. As more data is gathered, more complex networks result, and the DoE becomes more and more local, precise, and goal-oriented.
3. It encourages to place data points where they are needed for proper modeling. If a local model describes the process well in some region, there is no need to gather more data inside.
4. Future splits become more likely in local models with many data points inside as the local error is a validity weighted sum (SSE), *not* an average (MSE).
5. It is rather robust with respect to measurement noise since it never considers the performance at one (or very few) data samples. Only the errors at all points $e(i) = y(i) - \hat{y}(i)$ within each local model are evaluated. The positions $\underline{u}(i)$ of these points in the input space are required for distance calculations in search for the biggest hole.

As illustrated in Fig. 14.73, Point 2 represents an elegant solution to the *exploration vs. exploitation dilemma* [13, 46]. On the one hand, new data should be gathered in the most promising/informative areas (exploitation). On the other hand, if other areas are not yet investigated, how should the algorithm ever know how promising the associated regions are? Therefore all areas need to be explored. However, the resources for each data point can be spent only once.

Point 5 addresses a key weakness of many algorithms around, see, e.g., [288]. Oftentimes a large error at a single data sample severely cripples the performance of an algorithm. This is not the case with HILOMOT DoE. Note that it can further be robustified by choosing an appropriate error measure like the sum of *absolute* instead of squared errors.

Above the terminology "data inside a local model" was frequently used. More precisely speaking, this means all data points whose validity value Φ_j is largest for a specific local model j. However, this implies that data slightly outside a split "belongs" to a neighboring local model. It might be reasonable to define "data inside a local model" less strict as all data with $\Phi_j >$ const with const $= 0.3$ or similar.

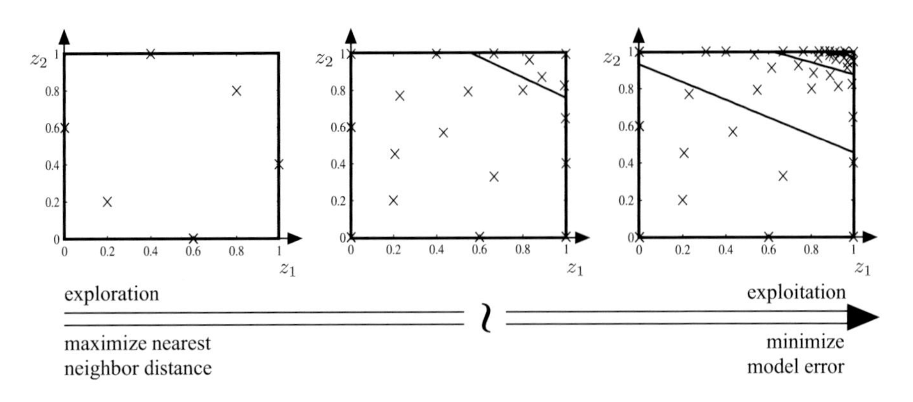

Fig. 14.73 HILOMOT DoE: Automatic tradeoff between exploration and exploitation. In early phases where local models are big (have wide validity regions), new data is placed in the biggest data hole (space-filling). The model becomes more complex as the amount of data increases, and the data placement becomes more and more local and the criterion focuses on the model error The input spaces show: Left: initial model with 6 maximin Latin hypercube data points. Center: 2 local models with 20 data points. Right: 4 local models with 40 data points (the third split can hardly be seen since it is close to the upper right corner)

This would associate points close to a split to more than one local model preventing that they are ignored in the calculation of the biggest hole.

14.10.4 Active Learning with HILOMOT DoE

In this section active learning is explained together with the HILOMOT DoE strategy. This can be seen as a place holder for any DoE strategy that can benefit from an active learning approach. A prerequisite is that new data points are chosen differently depending on the previously gathered (old) data. This rules out all simple unsupervised methods. However, results of more advanced unsupervised methods that incorporate some optimization schemes such as maximin Latin hypercubes depend on old measurement data – so they fulfill this property. Also, the model variance-oriented methods fulfill this property since the model variance decreases in regions where the data becomes denser.

But most definitions of *active learning*[9] ask for much more: The *output* needs to be processed and utilized for learning, typically within a model. This restricts the active learning DoE to model bias-oriented methods and model variance-oriented methods with adaptive basis functions such as Gaussian process models. If the model structure is fixed like for D-optimal polynomial DoE, the output does not

[9]https://en.wikipedia.org/wiki/Active_learning_(machine_learning)

influence the DoE process at all. And even Gaussian process models only exhibit a slight influence via the hyperparameter optimization.

For a general literature survey on active learning, refer to [529].

14.10.4.1 Active Learning in General

If prior knowledge is available, it should be exploited for the design of an appropriate training signal. Without any information, a uniform data distribution is optimal. Next, a model can be trained with the gathered data, and the bias error and/or variance error of the model can be approximated. They give hints for an improved redesign of the training signal. Thus, the design of the training signal is an iterative procedure, which may converge to an optimal solution. The iterative nature of this problem arises naturally from the interdependency of the two tasks: Modeling and identification on the one hand and design of the training signal on the other hand. This is closely related to the iterative identification for control approaches described in Sect. 18.11.3.

The difficulty with an iterative design of the training signal is that the process must be available for measurements during a long-time period which can be quite unrealistic in a concrete industrial situation. Active learning strategies try to overcome this dilemma by actively gathering new information about the process, while it is in operation. Instead of collecting the whole data set at once and subsequently utilizing it for modeling and identification, decisions about which data to measure are already made within the measurement phase. Active learning strategies typically try to optimize the amount of new information that can be obtained through the next measurement.

The most informative data sample which shall be measured next is usually called the *query*. A danger of such active learning methods is that they can get stuck in one operating region since locally more data allows one to construct a locally more complex model, which in turn asks for more data [399]. More details on active learning and all its variations can be found in [110–112, 462, 559] and the references therein.

Active learning can also be carried out during the normal operation phase of the process in order to further improve the model. During normal operation, however, two goals are in conflict: to operate the process optimally and to gather the maximum amount of new information. In adaptive control, this problem has led to a dual strategy that takes both components into account in a single loss function. Thus, the "curiosity" component that tries to acquire new information must be constrained. Active learning clearly is a promising topic for further future research.

More concretely, Fig. 14.74 outlines the principal idea of active learning. In the first step, an initial DoE required for building an initial model is planned and measured. This should be a space-filling design if no other prior knowledge is available. In a loop, after the measurement and modeling steps are carried out, it is checked whether the loop shall be terminated. Possible termination criteria in particular are as follows:

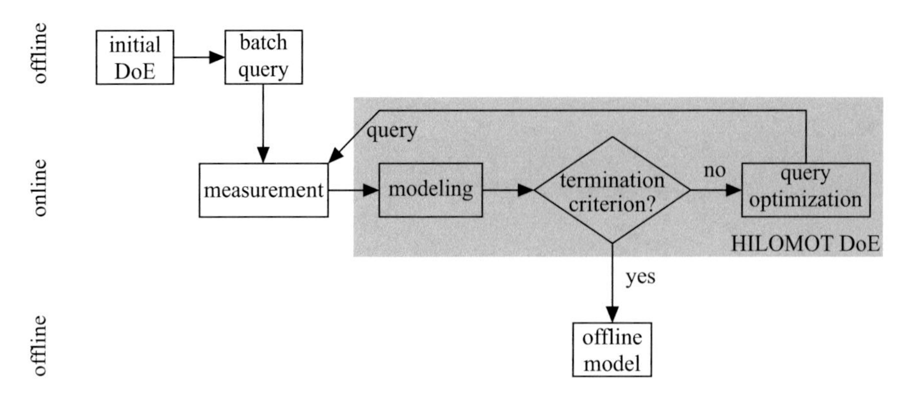

Fig. 14.74 Active learning

- Maximum number of measurement points gathered achieved or overall measurement time period elapsed.
- Model accuracy achieved on validation data or assessed by cross-validation. Many error measures are possible; particularly interesting is the ∞-norm which gives the maximum absolute error. This is interesting in order to guarantee a tolerance band.
- Model complexity which will grow as the size of the data set increases. If there is a bound on the evaluation time for the model, this might be a reasonable criterion.
- Some model interpretation issues. For example, the local models' dynamics might be unreasonable as long as the local linear models are valid in too large areas.

If no termination criterion is fulfilled, a new query is generated. This should be the point of promising the highest information gain when measured. In HILOMOT DoE, this will be the point that fills the biggest data hole in the worst-performing local model. In a model committee approach, this is the point of biggest disagreement between the models.

The modeling step usually is crucial for the computational demand for active learning and highly depends on the kind of model employed and the kind of training algorithm utilized [215]. In principle, modeling can be carried out in two ways:

- *Batch Mode*: All measurement data is stored. The model is built from scratch. This has the advantages be being (i) very robust and (ii) offering a maximum amount of flexibility as the model structure is built up completely new. The main drawbacks are (i) the memory requirements as all data has to be stored and (ii) computational demand for building the model.
- *Online Mode*: Only the new data is used to adapt the model in some recursive or iterative manner. Such models are also known under the term *evolving systems* [215, 359]. The advantages and drawbacks are the complete opposite of the batch mode. Such an approach often is extremely sensitive with respect to the ordering of the data. Moreover, for real-world applications certainly, a supervisory level

is necessary that takes actions when "things go wrong" as known from adaptive control.

In the following, only the first approach is taken. Active learning is discussed in the context of HILOMOT DoE. A model committee strategy could play the same role as well.

14.10.4.2 Active Learning with HILOMOT DoE

With HILOMOT DoE, the initial model typically will be affine, i.e., one local = global model. Thus, the initial batch query must contain at least $N = nx + 1$ data points. Since often $nx \geq nz$, this is more than enough for split optimization. However, to account for the rare case where $nx < nz$, the initial batch should possess at least $N = \max(nx + 1, nz)$ data points to have enough information for both: local model estimation and split optimization. It is advisable to start with more points than the requested minimum to attenuate the effect of measurement noise and maybe allow for a first split. These initial data points should be distributed in a space-filling manner, not optimally with respect to an affine model because as HILOMOT DoE progresses, a complete nonlinear model shall be established.

Figure 14.75a schematically displays the standard HILOMOT DoE procedure. It alternates between a measurement step and a HILOMOT DoE step until the termination criterion is fulfilled; compare Fig. 14.74. At time step i the measurement is carried out at $\underline{u}(i)$ returning the associated process output $y(i)$. Then the HILOMOT DoE step is performed. It mainly consists of (i) modeling and (ii) query optimization. At time step i, the modeling is carried out with the data available at that moment, i.e., $\{(\underline{u}(1), y(1)),\ (\underline{u}(2), y(2)),\ \ldots,\ (\underline{u}(i), y(i))\}$.

14.10.4.3 Query Optimization

The model built with this data is used to calculate the next query $\underline{u}(i + 1)$. In HILOMOT DoE, this is done by searching the biggest data hole in the worst-performing local model. The implementation of that query optimization is tricky and can alternatively be realized by evaluating a number of candidate points or by performing a numerical optimization. The candidate alternative is much simpler and approximately finds the global optimum. The numerical optimization alternative is computationally more demanding and finds only a local optimum. However, the candidate alternative fails in high dimensions due to the curse of dimensionality. Fortunately, often $nz < nx$ and consequently the $\underline{z}$-input space where the biggest hole must be found is low dimensional.

14.10.4.4 Sequential Strategy

In applications where the measurement time dominates and the HILOMOT DoE time is negligible, Fig. 14.75a seems to be the most straightforward choice. If it is the other way round, it makes sense to generate more than one query. Otherwise, the time fraction spent for the HILOMOT DoE step which can be considered as overhead becomes too big. A strategy suitable for this case as shown in Fig. 14.75b is explained further below.

Example 14.10.2 HILOMOT DoE for Learning a Hyperbola
This example illustrates the operation of HILOMOT DoE for learning the two-dimensional hyperbola:

$$y = \frac{0.3}{0.3 + (1 - u_1) + 3(1 - u_2)} . \tag{14.115}$$

The measurements are noise-free for simplicity. The process exhibits a strong nonlinear behavior which is more severe in u_2 than in u_1. The data shall lie in $[0, 1]^2$, and thus it can be expected to concentrate toward $(1, 1)$ with a stronger concentration in u_2 than in u_1.

HILOMOT DoE is run for generating optimal data for modeling the function in (14.115). The initial DoE consists of 6 data points generated by a maximin Latin hypercube. Note that $6 = 2(nx + 1)$ is the minimum number of data points where a robust network with 2 local models is possible since the constraint split optimization requires at least $nx + 1$ points inside each local model; compare Sect. 14.8.5. The left plot in Fig. 14.73 shows the distribution of these initial points. Note that due to the Latin hypercube, construction 4 of the 6 points lie on the edges of the input space. The percentage of points "inside" will be forced (by the Latin hypercube construction) to grow if the initial design is larger and the Latin hypercube grows.

HILOMOT DoE is run up to a network of 8 local models which requires 61 data points. Figure 14.76a shows how many local models were identified by HILOMOT starting from the initial DoE with 6 points to the final network for 63

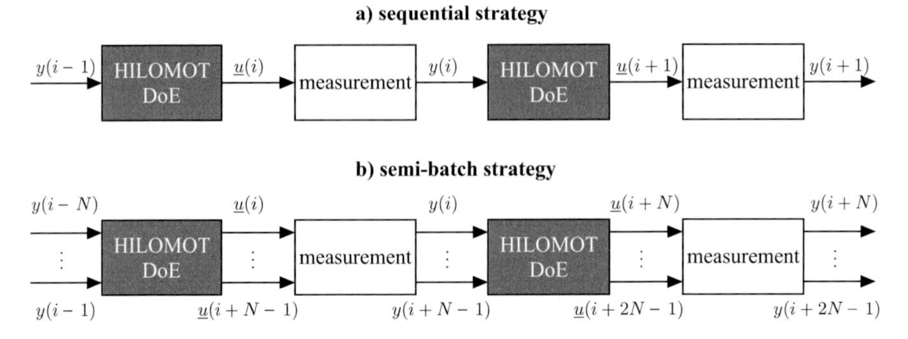

Fig. 14.75 Active learning: (**a**) Sequential strategy. (**b**) Semi-batch strategy

points. HILOMOT DoE here was terminated for the pure sake of simplicity for that demonstration example. In practice, any termination criterion from the list in Sect. 14.10.4 or others may be used. Typically the splitting of HILOMOT saturates at some point, and the number of local models hardly increases when the NRMSE has reached the noise level.

The number of local models in the network is almost a monotonous function of the number of data points. This is by no means self-evident and does not always happen in such a nice way. Although, in principle, the network complexity is expected to grow with the amount of data, there are many suboptimalities incorporated in the HILOMOT algorithm such as local estimation, local split optimization, incremental nature that only optimizes one split at a time, etc. In other cases, it can occur that the number of local models alternates. However, looking at a filtered version of the plot in Fig. 14.76a always gives a monotonously increasing curve.

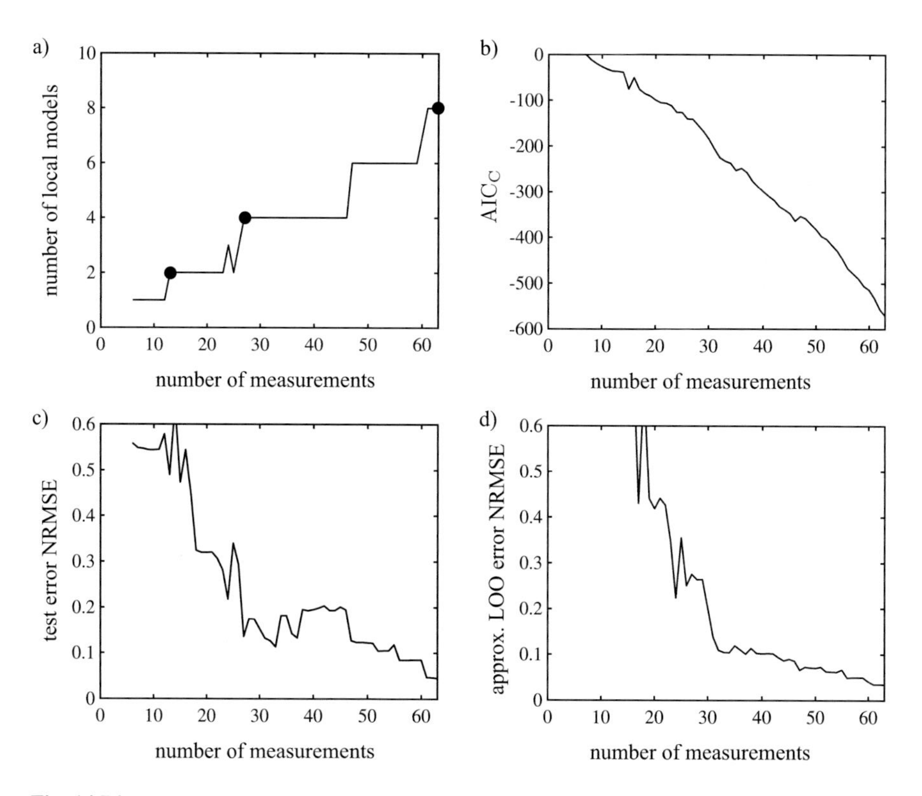

Fig. 14.76 Operation of HILOMOT DoE. All plots show different quantities over the number of measurements starting with initially 6 up to 63 data points. (**a**) Number of local models. The three dots mark networks whose behavior is shown in Figs. 14.77 and 14.78. (**b**) Best AIC_C of all networks constructed by HILOMOT for that data. (**c**) Test error (NRMSE). (**d**) Approximated leave-one-out error (NRMSE) for the local linear regression models – not the partitioning

Figure 14.76b shows the AIC_C of the local model network with suggested complexity (i.e., best AIC_C value) over the number of measurements. Except for some small "roughnesses," this is a nicely monotonously decreasing curve. Figure 14.76c shows the test error on 10,000 test data points generated by a Sobol sequence. In practice, this information is not available. However, it can be approximated by the leave-one-out (LOO) cross-validation error shown in Fig. 14.76d. Note that this is not the *true* LOO error as it would be very resource-intensive to calculate. Rather it is only the LOO error of the local linear regression models which can be calculated in a closed analytic form via the smoothness matrix, compare (7.10) in Sect. 7.3.2. It is important to understand that this quantity does not account for the variance error in the partitioning. In practice, together with the AIC_C, this gives valuable information about when to terminate HILOMOT DoE. Usually both curves saturate for $N \to \infty$. This effect cannot be observed here because the number of data points is still low and *no* measurement noise is simulated which clearly is a quite unrealistic scenario.

Figure 14.77 illustrates two networks. On the left-hand side, it is the first network that suggested two local models according to the AIC_C. On the right-hand side, it is the first network with four local models. As expected it clearly can be observed that the partitioning reflects the strength and direction of the process nonlinearity. As a consequence, HILOMOT DoE often placed new data close to the upper right corner of the input space. The concentration of points is indeed more dense in u_2 than in u_1. The algorithm operates as expected before. Note that placing many data points in the strongly nonlinear regions is a prerequisite of being able to construct and estimate many local models therein. With a space-filling design or similar, more uniform data coverage, it would be totally impossible to capture the strongly nonlinear behavior in the upper right corner.

Figure 14.78 summarizes two key issues. On the left-hand side, it compares the test error with the approximated LOO error which are in good agreement with each other from measurement 16 on or so. Since the approximated LOO can be calculated very efficiently, it should always be considered as a termination criterion together with the AIC_C.

On the right-hand side, the final model built with 63 data points generated by HILOMOT DoE is shown. It is in extremely good agreement with the process with a training error of NRMSE $\approx 1\%$ and a test error of NRMSE $\approx 4.5\%$. The wide gap between training and test error is very remarkable. It can be interpreted as a strong tendency to overfit [561]. The reason for this effect is that the test data is uniformly distributed (Sobol set) in the input space, while the training data concentrate in areas where the function is complicated. Only a small percentage of the test data lies in the upper right corner. Therefore the errors in this region influence the test error only marginally. Thus, the test NRMSE may be misleading.

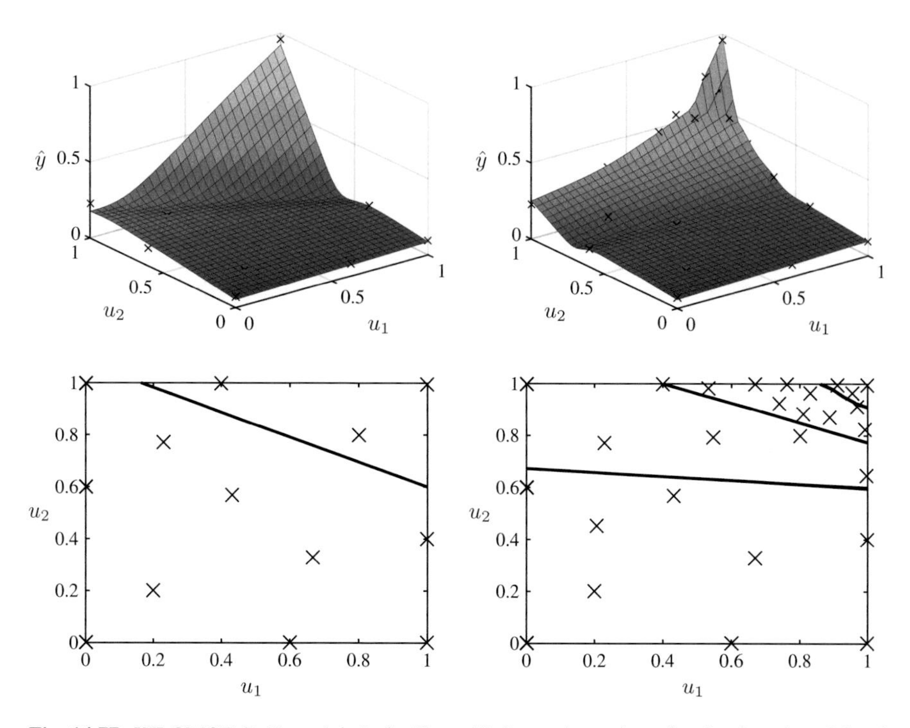

Fig. 14.77 HILOMOT DoE model: Left: $N = 13$ data points where for the first time 2 local models are constructed (first dot in Fig. 14.76a). Right: $N = 27$ data points where for the first time 4 local models are constructed (second dot in Fig. 14.76a). The plots on the top show the model behavior. The plots at the bottom show the partitioning with the data points

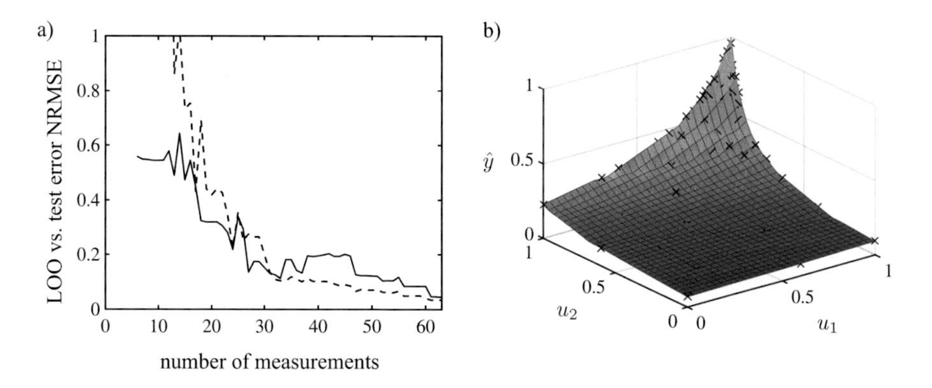

Fig. 14.78 Left: Comparison between the approximate leave-one-out error (dashed) and the test error (solid) for the HILOMOT DoE run. Right: Final model for $N = 63$ data points where 8 local models are constructed (third dot in Fig. 14.76a)

14.10.4.5 Comparison of HILOMOT DoE with Unsupervised Design

Frequently in real-world applications, the worst-case scenario is most interesting which can be described by the *maximum absolute error*. For comparison sake, three networks with eight local models are trained with different data sets: (i) a Sobol sequence, (b) a maximin Latin hypercube, and (c) a data set generated with HILOMOT DoE. The data sets contain $N = 63$ points each. The model performance on the HILOMOT DoE data is shown in Fig. 14.78b. Test data again are 10,000 Sobol set points.

Table 14.4 summarizes the training and test errors as NRMSE and additionally the maximum absolute test error. The clear ordering in model error on all three criteria is:

$$\text{Sobol} \; > \; \text{Maximin LH} \; > \; \text{HILOMOT DoE} \, .$$

But one value clearly stands out. The maximum absolute test error for HILOMOT DoE is more than one order of magnitude better than for both unsupervised designs. The reason for this extreme result can be observed in Fig. 14.79a–c. The unsupervised training data does not cover the most complicated region (upper right corner) of the process sufficiently well. Therefore the test error in this region is immense.

In return, the unsupervised designs cover the more benign process regions more densely which lead to more splits in the lower right regions. In can be observed in the Fig. 14.79a–c that the error in these benign regions is larger for HILOMOT DoE data.

Figure 14.79d shows the convergence curves for the three DoEs over the network complexity. It is clear from the plots in Fig. 14.79a–c that HILOMOT DoE outperforms both unsupervised approaches significantly. The new information in Fig. 14.79d is that with HILOMOT DoE data, HILOMOT continues to improve the maximum absolute error performance with each iteration, i.e., with each additional local model. In contrast, this is not the case for both unsupervised designs. In this example, they are incapable to improve the worst-case error beyond $M = 3$ local models because most data lies in the benign process regions and HILOMOT splits in these benign regions. The reason is twofold: (i) More data in these regions increases the local model errors and thus the worst local model is more likely found in these regions and (ii) even if the worst local model still is found in the upper right corner,

Table 14.4 Performance comparison of different DoEs

Data	Train NRMSE	Test NRMSE	Max. Abs. Test Error
Sobol set	0.028	0.13	0.25
Maximin LH	0.014	0.077	0.20
HILOMOT DoE	0.013	0.045	**0.014**

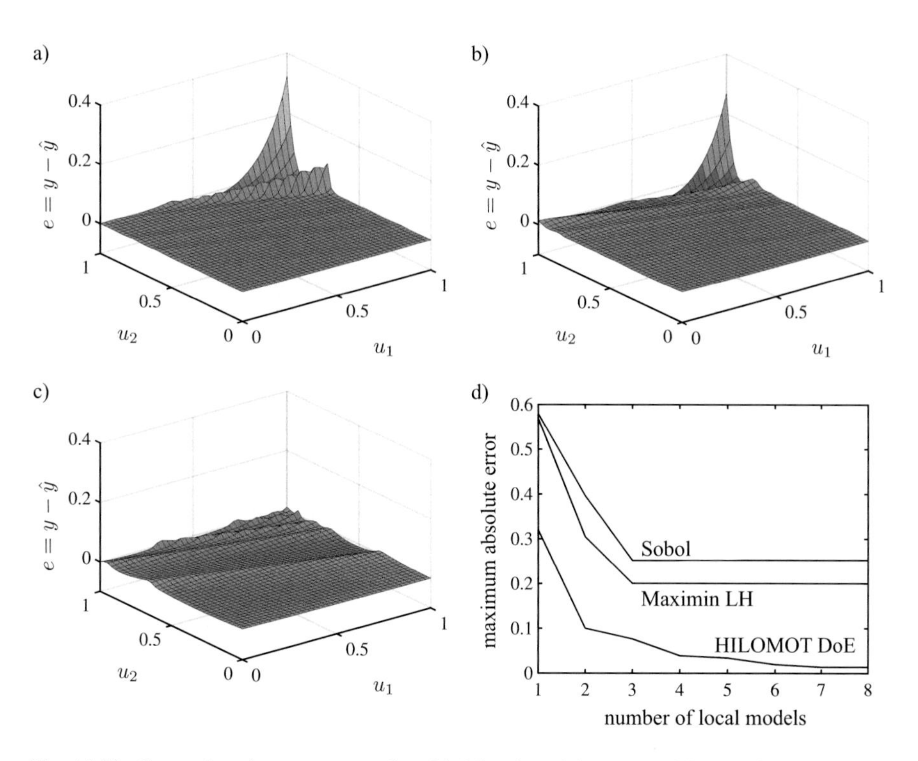

Fig. 14.79 Comparison between networks with 8 local models generated by HILOMOT with 63 data samples created with (**a**) a Sobol set, (**b**) a maximin Latin hypercube, (**c**) HILOMOT DoE. (**d**) Convergence curve comparison of the maximum absolute error for all three DoEs

most likely it does not contain enough data to be split; compare Sect. 14.8.5 on constrained split optimization. □

Finally, the following conclusions can be drawn from this example and for a generalization to other processes and higher dimensionalities:

1. HILOMOT DoE automatically solves the exploration vs. exploitation dilemma. No tuning factor or fiddle parameter (like the amount of curiosity) has to be specified.
2. The approximated leave-one-out error offers an orientation additional to the AIC_C (or other information criteria) when to terminate the data acquisition process.
3. HILOMOT DoE tries to place new data in regions where they are most useful in modeling the process with local model networks. It has a tendency to deliver excellent worst-case performance.
4. HILOMOT DoE mainly operates in the $\underline{z}$-input space. Therefore it returns only $\underline{z}$ as the next query point. In the above example where $\underline{z} = \underline{x} = \underline{u}$, this causes no difficulties. However, if only some inputs enter $\underline{z}$, the remaining ones have to be determined somehow by other means. This is an open topic for future

research. An obvious approach could be to select the remaining dimensions in a space-filling manner. Note that this is a potentially powerful feature of HILOMOT DoE since it can keep the problem dimension low, even if the process is high dimensional.

5. The example represents an ideal scenario: No noise, very nonlinear process, simple unidirectional nonlinear behavior. The performance gain can be expected to be less dramatic in a real-world application.
6. As the dimensionality of the problem increases, the benefits due to HILOMOT DoE can be expected to grow. It is well known that all unsupervised methods approach each other in high-dimensional problems as all points are far away from each other [2]. Basically, it is not possible anymore to cover the relevant portion of the input space with data points in some meaningful density. Only model bias- or output-oriented strategies may offer a way around this dilemma.

14.10.4.6 Exploiting the Separation Between Premise and Consequent Input Spaces in Local Model Networks for DoE

The issue discussed in this paragraph is analyzed in more detail in [237]. However, Point 4 from the list above deserves a more elaborate discussion. Consider the following very realistic situation: Assume all inputs u_i, $i = 1, \ldots, p$ enter the local models, i.e., for 2D this is $\underline{x} = [u_1\ u_2]^T$ and for 3D this is $\underline{x} = [u_1\ u_2\ u_3]^T$. However, the last input does not enter the $\underline{z}$-input space because the process exhibits roughly linear behavior in it, i.e., for 2D $\underline{z} = [u_1]^T$ and for 3D $\underline{z} = [u_1\ u_2]^T$.

HILOMOT DoE only can determine those dimensions of the query that enter $\underline{z}$ since that is the space where the partitioning lives. The inputs that only enter $\underline{x}$ need to fixed with some other strategy. Probably the most straightforward idea is to choose some space-filling approach for these dimensions. Note that only the dimensions in $\underline{x}$ that do not enter $\underline{z}$ need to be fixed or optimized but the calculation of the distances for evaluating the neighborship relationship can be carried out in the whole $\underline{x}$-input space.

An alternative to the space-filling strategy would be a D-optimal design based on the fact that the local models are of known type (typically linear) and thus a D-optimal design truly offers the variance optimal approach. For local *linear* models, this means that only minimum and/or maximum values are measured for the dimensions in $\underline{x}$ but not in $\underline{z}$. Although this seems to be a very compelling alternative, it is only advisable if the decision which variables enter $\underline{x}$ and which enter $\underline{z}$ is certain and final. There is no way to discover any nonlinear behavior in dimensions where only min and max values have been measured. Thus, it seems to be a risky strategy in practice where hardly anything is certain and final.

Figure 14.80 compares a space-filling design for u_2 in (a) with a D-optimal design for u_2 in (b) for two local models, four data points per local model, in the 2D case. Note that the axis-orthogonal split is not a consequence of axis-orthogonal

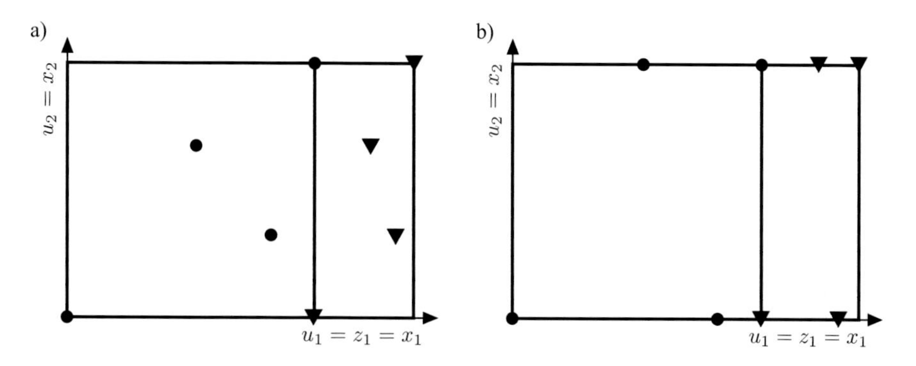

Fig. 14.80 HILOMOT DoE for different input spaces in 2D. The dots represent four points in the left local model, the triangles in the right local model. The points in u_1-direction are distributed according to HILOMOT DoE or some other model bias-oriented method. In u_2-direction which does not enter $\underline{z}$ the points can be distributed in a (**a**) space-filling or (**b**) D-optimal manner

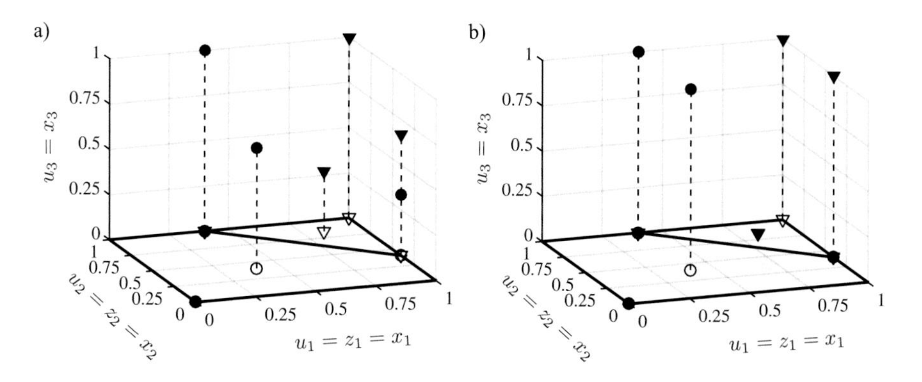

Fig. 14.81 HILOMOT DoE for different input spaces in 3D. The dots represent four points in the lower left local model, the triangles in the upper right local model. Light circles/triangles represent just the projections of the data on the plane $u_3 = 0$. Same scenario as in Fig. 14.80 with two inputs entering $\underline{z}$ and all three inputs entering $\underline{x}$: (**a**) space-filling or (**b**) D-optimal manner

splitting in LOLIMOT-style but is an implication of the fact that $\underline{z}$ is only one-dimensional.

In order to illustrate this, Fig. 14.81 shows the corresponding scenario in 3D. Here the split can be oblique as it lives in the two-dimensional $\underline{z}$.

These just are some preliminary ideas. Concrete algorithms have to be developed and applied to real-world problems in the future. However, it seems that HILOMOT DoE together with an exploitation of the separation into $\underline{x}$- and $\underline{z}$-input spaces offers excellent chances for extremely powerful DoE strategies for high-dimensional problems. These ideas are more thoroughly investigated and compared in [237]. Also in [237] a significantly superior alternative to D-optimal designs in the context of a mixed $\underline{x}$-$\underline{z}$-design is proposed. This alternative is based on a *maximin sparse level* design that essentially follows the same ideas as in maximin Latin hypercube optimization with only very few amplitude levels (only 2 levels in the limit for linear models). This is in contrast to standard Latin hypercubes where the number of levels is equal to the number of data points.

14.10.4.7 Semi-Batch Strategy

Figure 14.75b shows a semi-batch strategy where N queries are generated, and consequently the measurement time period will be N-fold. Of course, creating N queries with one model based on past data only can solve this resource problem but is inferior to the sequential strategy.

How these N queries should be selected is an open question for future research. In the context of HILOMOT DoE, two strategies are obvious: (i) Selection of the second biggest, third biggest, etc. hole in the worst-performing local model and (ii) selection of the biggest hole in the second, third, etc. worst-performing local models – and of course combinations thereof. A clever combination of both ideas that takes into account the performance differences of the local models and the distance difference of the holes (and the distances of the holes to each other) seems to be advisable.

14.10.4.8 Active Learning for Slow Modeling Approaches

The sequential strategy in Fig. 14.75a and the semi-batch strategy in Fig. 14.75b both possess the same disadvantage that no data is measured during the HILOMOT DoE calculations. Since the measurement time typically is the main bottleneck, such an approach wastes valuable time at the test stand or in the plant.

If the measurement time is about equal to the calculation time of HILOMOT DoE, the scheme in Fig. 14.82c explains how the measurement and the HILOMOT DoE calculations can be carried out in parallel. The measurements lag behind by one data point which causes suboptimal behavior, of course. The modeling and query optimization step at time step i relies on information which is from time step $i - 1$, i.e., 1 measurement back in time because the current measurement is still about to be taken. This procedure allows eliminating all overhead if the measurement time $\geq$ calculation time. Lagging one step behind seems to be a small price to pay for such an efficient setup.

If the measurement time $<$ calculation time the scheme in Fig. 14.82d can be applied. This typically occurs for non-stationary measurements and/or very complex modeling schemes like Gaussian process regression with hyperparameter optimization. Figure 14.82d shows that two measurements are carried out sequentially but in parallel to the calculation for the price of lagging two measurements behind. Of course, this idea can be generalized to $3, 4, \ldots$ measurements. At some point, the information lag becomes so big that the active learning strategy deteriorates to guesswork.

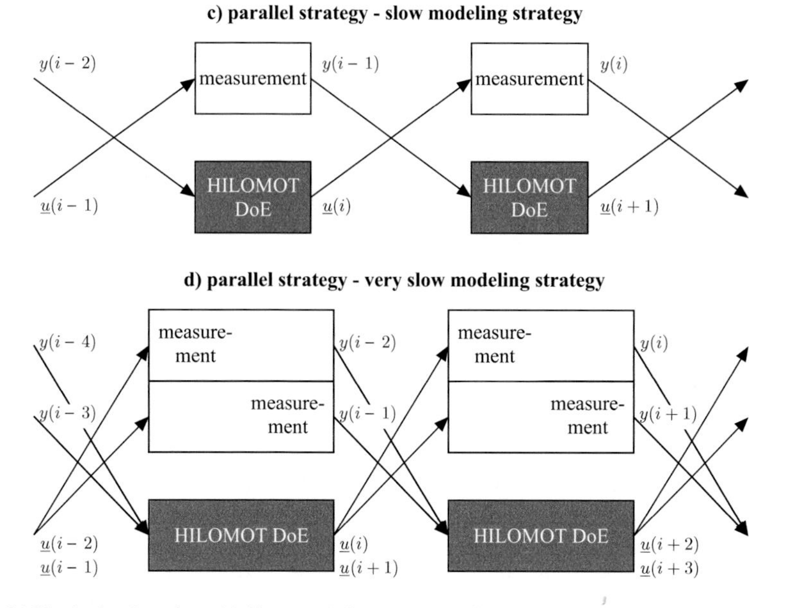

Fig. 14.82 Active learning: (**c**) Slow modeling strategy. (**d**) Very slow modeling strategy

14.10.4.9 Applications of HILOMOT DoE

HILOMOT DoE has proven good robustness and performance in some applications. For a structural health monitoring example, see Sect. 26.2 [384]. For combustion engine measurement, it reduced the measurement time by more than a factor of 2 for building a comparably performing model; see Sect. 26.3 [228].

14.11 Bagging Local Model Trees

Bagging stands for *bootstrap aggregating* and was proposed by Breiman in 1994 [76]. It is an *ensemble method* that can be utilized with any model architecture like boosting (which is explained in Sect. 3.1.13). It combines multiple models of the same architecture (but possibly different complexity) in order to improve the model accuracy. It trains N_{bag} models on bootstrap samples generated from the original training data set. Bootstrapping is explained in Sect. 7.3.2.

The bagged model output is calculated as the average of the models $\hat{y}^{(j)}$ trained on the bootstrap samples $j = 1, \ldots, N_{\text{bag}}$:

$$\hat{y}^{(\text{bag})} = \frac{1}{N_{\text{bag}}} \sum_{j=1}^{N_{\text{bag}}} \hat{y}^{(j)} . \tag{14.116}$$

The bootstrap introduces variations which are then is averaged out again in (14.116). Advantages of bagging are:

- *Can be utilized with any model architecture*, in principle, although it promises benefits only with unstable ones (see discussion in Sect. 14.11.1).
- *Conceptionally simple* and *easy-to-use* approach.
- *Model variance reduction*: By averaging models, the effect of noise will be reduced. Furthermore, the different bootstrap samples excite the process at different input locations; thereby they "simulate" different data acquisition scenarios. Both effects can reduce the variance of the model significantly.

  Since a too high variance error is a potential shortcoming of most flexible black box architectures, bagging addresses a very important issue.
- *Smoother model*: Since the final model is the average of N_{bag} models, the model output will be the smoother the higher N_{bag} is. Note, however, that if the base models are not continuous or differentiable, this situation does not change through bagging. Nevertheless, the steps in absolute value or slope usually will be smaller.
- *Inherent parallel*: No communication between different bagged models is necessary. Thus, implementation on a parallel computer architecture is straightforward and easy.
- *Out-of-bag data*: This is the data which was not drawn into the bootstrap sample, therefore is not used for training, and thus can be utilized for validation purposes [77].
- *Errorbars* can be extracted in a very simple fashion by just calculating the standard deviation of the model ensemble for a given input location $\underline{u}$. The procedure of calculating the errorbars works correctly independently of model type, training algorithm, tuning factors, selection procedures, etc. Thus, it yields the model variance including all factors that often are unfeasible to account for or simply forgotten.

  A simple way for generation of 95% errorbars is to set, e.g., $N_{\text{bag}} = 200$, and for any input value $\underline{u}$ skip the 5 lowest and 5 highest prediction values – the remaining 190 ones represent the 95% confidence interval.

Drawbacks of bagging are as follows:

- *Higher computational demand* for training and evaluation: On serial computers the needed resources (time, storage) scale linearly with the number of bagged models N_{bag}. However, on a parallel computers with N_{core} cores, the resources only scale with $N_{\text{bag}}/N_{\text{core}}$ because bagging parallelizes perfectly. Since future computers are very likely to be *extremely* parallel, bagging may come at (almost) no additional computational cost.
- *Less interpretable* models: Even if a single model offers a good interpretability, this property is almost lost with bagging, e.g., *one* fuzzy rule set may be understandable, but N_{bag} ones will be overwhelming for a human. However, new ideas for interpretation are generated through bagging by investigating the statistics of the whole model population. This is a popular approach in the context

of bagged trees or random forests [230] where, e.g., the average number of splits in each input (possibly weighted with the distance to the root node or fraction of the data samples in the subtree) can be interpreted as variable importance.

14.11.1 Unstable Models

The bagging procedure only works well on so-called *unstable* models [230]. Here "unstable" does not refer to *dynamic instability* but rather means that the model structure is very *sensitive* with respect to perturbations in the training data. If, e.g., in the training data one data point is changed or omitted, then an unstable model can possess a very different structure. Examples are the following:

- Hierarchical or tree-based approaches like CART, LOLIMOT, HILOMOT, etc.: Because in the partitioning process, a decision is made in each iteration about which local model to split further. A second source of instability is the decision about in which dimension to split. Note that this second source of instability only exists for axis-orthogonal partitioning (CART, LOLIMOT) but not for axis-oblique partitioning (HILOMOT) where instead a weight vector (direction) is optimized. Decisions can be seen as discrete optimization and usually yield unstable behavior (possibly different splits if one data sample is perturbed), while continuous optimization usually yields stable behavior (the split just slightly differs in direction if one data sample is perturbed).
- Subset selection or input selection of any kind also tend to induce instability. However, this instability is less severe than for trees. Usually the decision process happens on a global level and thus depends on all or most data, while for trees it happens on a local level which increases the sensitivity.

A sequence of decisions typically emphasizes the described effects and thus leads to very unstable models.

14.11.2 Bagging with HILOMOT

Bagging is known to be very effective on trees. It was applied to the LOLIMOT algorithm in [7] and shall be discussed in the context with HILOMOT and some comparison to LOLIMOT in the following. Since axis-oblique partitioning is far more flexible than axis-orthogonal ones, the variation within the model ensemble can be expected to be bigger. However, as mentioned above, oblique splitting yields less unstable modes. Anyway, the main differences will be observed on higher-dimensional problems. For comparison with standard CART, note that in one dimension, CART is similar to HILOMOT because both approaches optimize the location of each split. In higher dimensions, however, CART is more similar to LOLIMOT because both approaches carry our axis-orthogonal splits.

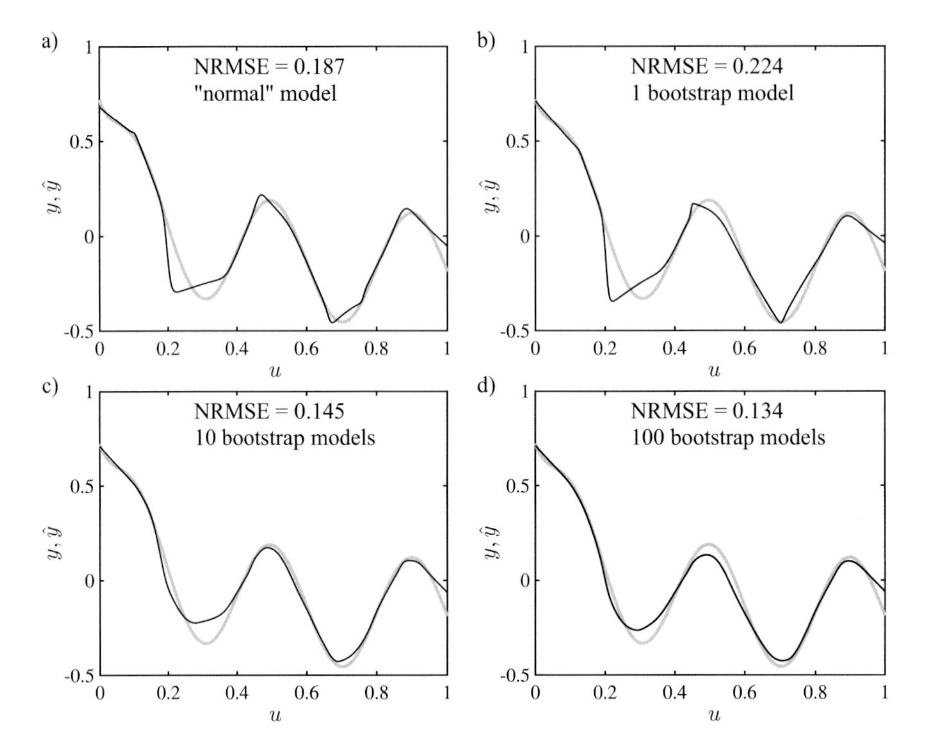

Fig. 14.83 Comparison (process in gray, model in black) between an (**a**) "normal" model with bagged models consisting of (**b**) 1 model, (**c**) 10 models, (**d**) 100 models trained on bootstrap samples. All models are local model networks trained with HILOMOT terminated at the best AIC_C value

Example 14.11.1 Bagging Local Model Networks with HILOMOT
The function shown in Fig. 14.83a (gray) shall be approximated with $N = 50$ training data points uniformly randomly distributed in the interval [0, 1]. The output is disturbed by Gaussian noise with standard deviation $\sigma_n = 0.01$. The test data contains 300 points uniformly distributed over the input range [0, 1]. A local model network trained with HILOMOT achieves its best AIC_C value for $M = 10$ local models, and its output and error is shown in Fig. 14.83a (black) as well.

The bagging procedure yields the model outputs and errors shown in Fig. 14.83b–d. Obviously, it is extremely likely that the bag with 1 model performs poorly since only about 63% of the original training data are contained in the first bootstrap sample. It can been seen that already for a couple of bootstrap samples, the bagged model performs better and is smoother than the "normal" model. Both effects are highly desirable for most applications, of course.

Figure 14.84a demonstrates the bagging performance as N_bag grows. Notice that this is *no* plot over a "number of iteration" since all points, in principle, are independent of each other. Note also that this is no monotonously decreasing curve due to the stochastic nature of bootstrapping.

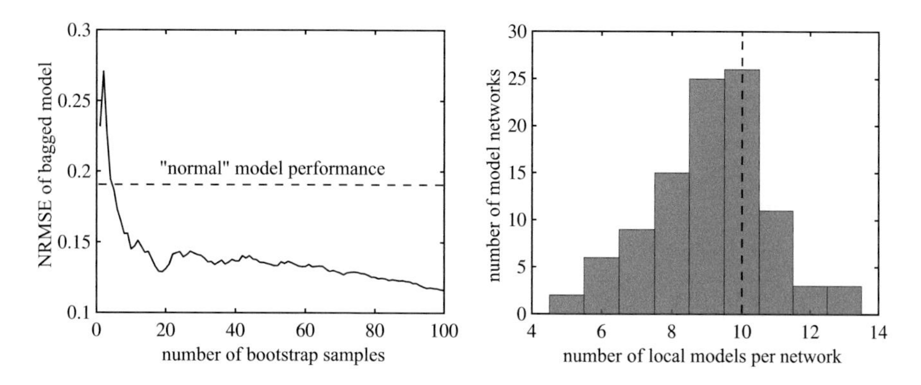

Fig. 14.84 Left: Convergence of bagged models' test performance dependent on the number of bootstrap samples. The dashed line shows the test performance of the model trained on the full training data (not the bootstrap sample).
Right: Distribution of the complexity of the 100 local model networks. The dashed line shows the complexity of the model trained on the full training data

How many bootstrap samples should be averaged? There is no general answer to this question. But it is intuitively clear that this number should be the higher the more "randomness" exist in the model ensemble. Popular approaches like *random forests* are based on the introduction of randomness *on purpose* in order to make the bagging procedure more effective by reducing the correlation between the bagged models [230].

Figure 14.84b analyzes the model with 100 bootstrap samples. The number of local models in the network ensemble varies between 5 and 13. The complexity of the "normal" model is higher than average because the original training data set is more informative than a typical bootstrap sample (where some points are contained multiple times) – although they possess an identical amount of data points, of course. From the distribution in Fig. 14.84b, widely varying model performances can be expected within the ensemble. This assumption is indeed verified by Fig. 14.85(top) which displays the model outputs for 10 and 100 bootstrap samples. □

14.11.3 Bootstrapping for Confidence Assessment

The effect of *explicit* structural decisions on the number of neurons, rules, or polynomial terms, etc. is very obvious because they change the number of parameters of the model. Therefore their effect on the model flexibility can be observed directly. In contrast, the effect of *implicit* structural decisions is more difficult to see and often hard or impossible to analyze quantitatively in an explicit manner. However, their effect can be analyzed with bootstrapping. Bootstrapping is a "model-free" approach in that it operates purely on data without any further model assumptions [230]. Note that the model flexibility is decisive for the model variance and therefore for its

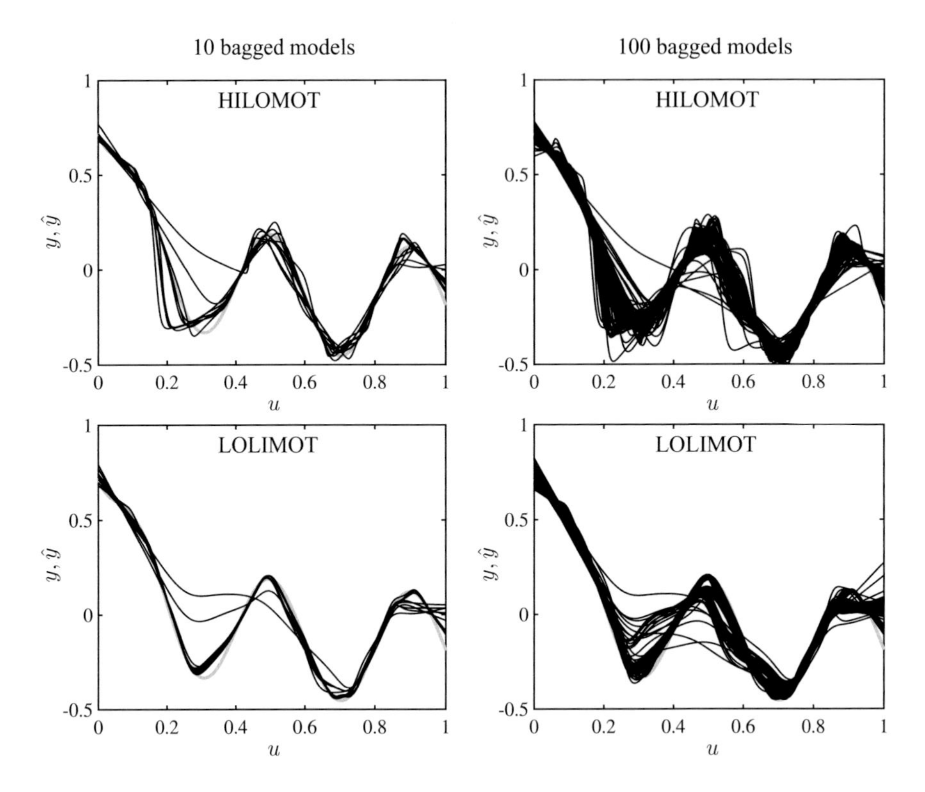

Fig. 14.85 Comparison of 10 (left) and 100 (right) local model networks trained on bootstrap samples with HILOMOT terminated at the best AIC_C value

overfitting tendency. And many structural aspects are responsible for the model flexibility.

Thus, confidence intervals need to account for the model variance due to structural aspects like

- *Partitioning* can be carried out in many different ways like axis-orthogonal splits (LOLIMOT) or axis-oblique splits (HILOMOT) or clustering in input space (k-means) or product space (Gustafson-Kessel). Sometimes explicitly parameters are optimized like the sigmoidal weights for the splitting direction in HILOMOT; sometimes just structural decisions are made like which axis to split in LOLIMOT. Anyway, this flexibility in partitioning is responsible for a (sometimes significant) part of the model variance error.
- *Hyperparameters* also influence the model flexibility, e.g., the kernel widths in GPMs: Wide kernels make the model more global and more flexible than smaller ones.
- *Regularization* of any form directly influences the model flexibility or the effective number of parameters, e.g., λ in ridge regression.

- *Subset selection* has an explicit effect due to the number of selected terms or inputs. However, the pure power and flexibility of the selection procedure is also responsible for parts of the model variance. Afterward it may be unknown whether a model with a certain term selection was chosen out of insights, luck, or subset selection. If the data set was exploited for this selection, it causes additional model variance. Often this variance contribution is ignored although it is very significant.

From the above discussion, it is clear that local model networks trained with HILOMOT should have a significantly higher variance error compared to LOLIMOT. This can indeed be observed from Fig. 14.85 for Example 14.11.1.

Example 14.11.2 Bagging: HILOMOT versus LOLIMOT
This continues Example 14.11.1. Bagging with LOLIMOT delivers much worse results than the "normal" model trained with LOLIMOT; see Fig. 14.86(top). Comparing with Fig. 14.84 where bagging with HILOMOT outperforms the "normal" model for relatively small ensembles, bagging with LOLIMOT is worse than the "normal" model.

The problem seems to be caused by too few training data and the "lucky" coincidence that LOLIMOT for the "normal" model actually performs better than HILOMOT.[10] Since LOLIMOT partitions the input space in a quite suboptimal manner, it needs to split more often than HILOMOT yielding networks with more local models. However, with only $N = 50$ training data points, the splitting procedure terminates due to lack of data (not enough data points in the local models to effectively split further). For a higher amount of training data (e.g., $N = 150$), the bagged LOLIMOT with 100 bootstrap samples outperforms standard LOLIMOT by 30%; see Fig. 14.86(bottom, left).

The larger amount of training data allows LOLIMOT to partition finer yielding more complex models; see Fig. 14.86(bottom, right). The achieved model error is significantly reduced as well. □

14.11.4 Model Weighting

By averaging the models in the bag all bootstrap samples, enter the final model with identical weight. However, some bootstrap models have a very low quality due to "unlucky" (non-representative) sampling. They deteriorate the bagged model significantly. It has been proposed to omit the bad models in the ensemble to improve the bagged model's accuracy; refer to "Ensembling neural networks: Many could be better than all" [614].

[10]This is quite unlikely to happen but for a 1D example due to the suboptimal nature of incrementally building, the tree can occur sometimes.

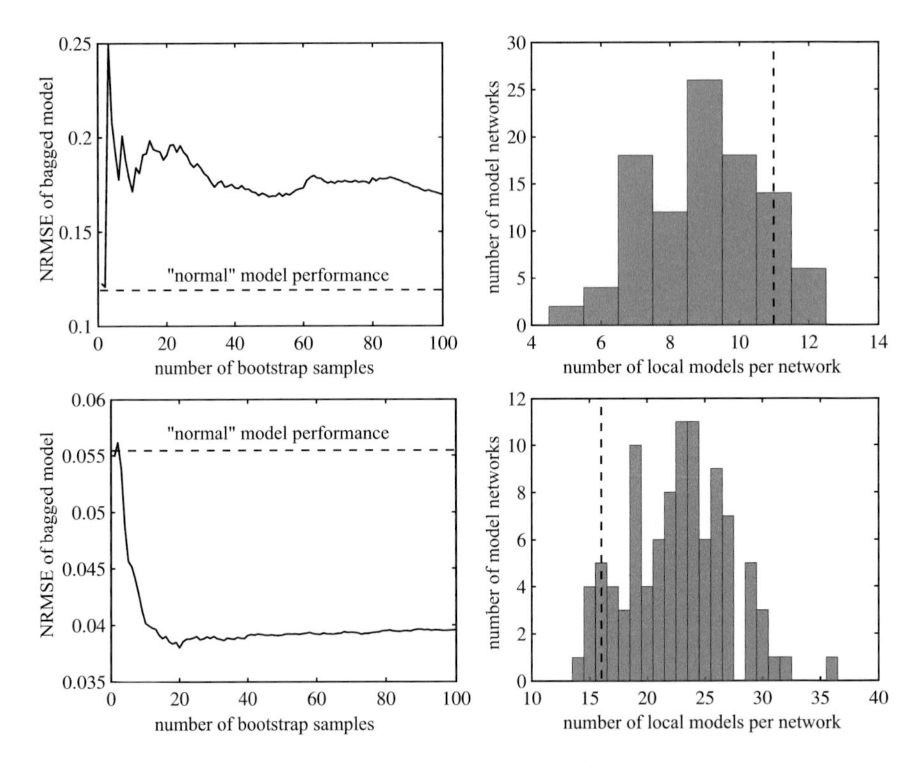

Fig. 14.86 Top: $N = 50$ training data points. Bottom: $N = 150$.
Left: Convergence of bagged models' test performance dependent on the number of bootstrap samples. The dashed line shows the test performance of the model trained on the full training data (not the bootstrap sample).
Right: Distribution of the complexity of the 100 local model networks. The dashed line shows the complexity of the model trained on the full training data.

An alternative and probably improved approach is to weight the different models according to their quality. A weight proportional to the inverse error variance would be statistically optimal[11] if the models were independent (which is not really the case here, of course). The error can be calculated on validation data, by cross-validation or simply via the out-of-bag data. Then the bagged model becomes

$$\hat{y}^{(\mathrm{bag})} = \frac{1}{\sum_{j=1}^{N_{\mathrm{bag}}} w_j} \sum_{j=1}^{N_{\mathrm{bag}}} w_j \hat{y}^{(j)} \tag{14.117}$$

with

[11] https://en.wikipedia.org/wiki/Inverse-variance_weighting

$$w_j = \frac{1}{\sigma_j^2} \quad \text{with } \sigma_j^2 = \frac{1}{N_{\text{out-of-bag}}^{(j)}} \sum_{i=1}^{N_{\text{out-of-bag}}^{(j)}} \left(y^{(j)}(i) - \hat{y}^{(j)}(i) \right)^2 . \tag{14.118}$$

Here $N_{\text{out-of-bag}}^{(j)}$ is the number of out-of-bag data points for bootstrap sample j which roughly corresponds to $0.37N$ data points with N being the available training data. The process output $y^{(j)}(i)$ and the model output $\hat{y}^{(j)}(i)$ depend on the data point i and bootstrap sample j. Note that $y^{(j)}(i)$ depends on j because each bootstrap sample contains different data points; thus the out-of-bag error has to be calculated on different data for each bootstrap sample.

It is important to realize that the training error cannot be utilized for calculating the weights w_j because they are over-optimistic due to overfitting effects. However, the out-of-bag error may be over-pessimistic because it possibly samples the process at locations far away from the training data, particularly if the data was distributed approximately space-filling. Therefore sometimes the whole original training data is used to evaluate the model quality consisting of approximately 63% used for training and 37% saved for validation due to the bootstrap sample construction.

Example 14.11.3 Weighted Bagging

This continues Example 14.11.1. Figure 14.87(left) compares the error of a standard bagged local model network with HILOMOT and a weighted one. Figure 14.87(right) displays the weights associated with the models in the ensemble. Clearly, the wide range of model qualities assumed from the distribution of model complexities and Fig. 14.85 can be verified. □

14.12 Summary and Conclusions

Advanced aspects of static local linear neuro-fuzzy models and of the LOLIMOT training algorithm have been discussed in this chapter. Different input spaces can be utilized for the rule premises and consequents. This enables the user to incorporate prior knowledge into the model. More complex local models can be advantageous for specific applications compared with local linear ones. The structure of the rule consequents can be optimized efficiently with a local OLS subset selection algorithm. The interpolation and extrapolation behavior has been studied, and possibilities for an incorporation of prior knowledge were pointed out. Different strategies for linearization of the local neuro-fuzzy models have been introduced. An efficient and robust online learning scheme that solves the so-called stability/plasticity dilemma has been proposed. The evaluation of errorbars allows one to assess the model accuracy and thus serves as a valuable tool for the design of excitation signals and the detection of extrapolation. Finally, an extensive section introduced the axis-oblique partitioning algorithm HILOMOT. The constructed models are called local model networks in order to emphasize the loss of the easy

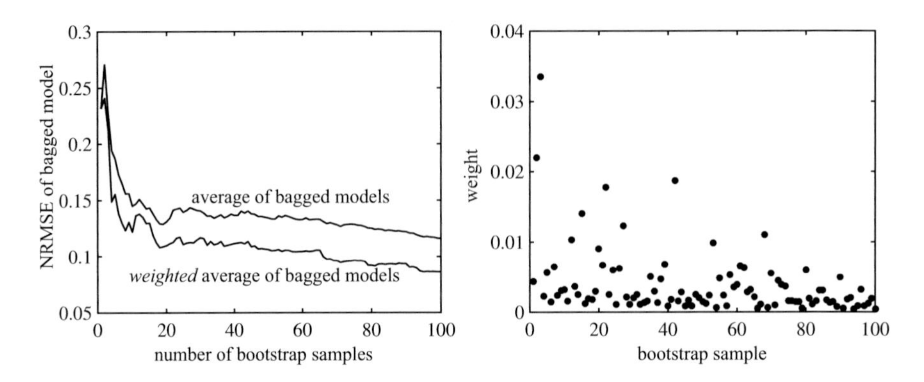

Fig. 14.87 Left: Convergence of bagged models' test performance dependent on the number of bootstrap samples. Weighted averaging can improve the overall model performance significantly. Right: Weights of the models differ enormously. Weights are chosen as the inverse of the error variance on out-of-bag data; see (14.118)

fuzzy logic interpretation in terms of one-dimensional membership functions. The similarities to local linear neuro-fuzzy models were pointed out, and the properties of both approaches were compared.

The most important features of local linear neuro-fuzzy networks can be summarized as follows:

- *Interpolation behavior* is as expected for the S-type situations where neighbored local linear models have a similar characteristics. For V-type situations, however, the interpolation behavior is not as intuitively expected; see Sect. 14.4.1. In some cases, this can cause strange model behavior and must be seen as a serious drawback of local linear neuro-fuzzy models. The smoothness of the interpolation behavior can be easily controlled by adjusting the overlap between the validity functions. The simpler the local models are chosen, the smoother the validity function should be chosen.
- *Extrapolation behavior* is linear, i.e., the slope of the model is kept constant. In many real-world problems, such a behavior is desirable. Furthermore, the user can specify a desired extrapolation behavior that is different from the standard characteristics; see Sect. 14.4.2.
- *Locality* is guaranteed if the validity functions $\Phi_i(\cdot)$ are local. This is ensured if the membership functions $\mu_i(\cdot)$ are chosen local and reactivation due to normalization effects is prevented. A local choice (e.g., Gaussians) for the membership functions is natural in order to obtain a reasonable interpretation, and thus this requirement typically is fulfilled. Reactivation can occur in the normalization procedure; see Sect. 12.3.4. It can be mostly prevented if the LOLIMOT algorithm is utilized. Typically, algorithms that do not carry out an axis-orthogonal input space partitioning and/or generate very differently sized neighbored operating regimes are much more sensitive to reactivation effects. For axis-oblique partitioning, locality can be ensured by the hierarchical

structure when the HILOMOT algorithm is applied. This is not the case for the standard hinging hyperplanes training methods; see Sect. 14.7.2. For the very flexible construction algorithm proposed in [399] or for product space clustering approaches, locality generally cannot be ensured.

- *Accuracy* is typically high. Although the validity function parameters are usually not truly optimized but rather roughly determined heuristically, only a few neurons are required to achieve high accuracy. The reason for this is that, in contrast to standard neural networks, each neuron is already a relatively good representation in quite a large operating region.
- *Smoothness* can be defined by the user. In contrast to other architectures (e.g., RBF networks), the smoothness has no decisive influence on the model performance or the approximation characteristics.
- *Sensitivity to noise* depends on the specific training algorithm. Local linear neuro-fuzzy models tend to require more parameters than other model architectures because the validity functions are typically chosen according to a heuristic and thus are suboptimal. With global estimation of the LLM parameters, the model often tends to be over-parameterized and thus is sensitive to noise; with local estimation, the opposite is true owing to its considerable regularization effect.
- *Parameter optimization* is fast if global estimation is used, and it is extremely fast if the local estimation approach is followed. By avoiding nonlinear optimization and exploiting local relationships, local linear model architectures are very efficient to train.
- *Structure optimization* is very fast for the LOLIMOT algorithm and fast for the HILOMOT algorithm and product space clustering algorithm and of medium speed for the flexible search proposed in [399]. The architecture allows one to construct several efficient structure optimization algorithms.
- *Online adaptation* is very robust and very efficient if only the LLM parameters are adapted. This can be done by a local linear recursive least squares (RLS) algorithm. Owing to the locality of the validity functions, an appropriate online adaptation scheme can adapt the local model in one operating regime without the risk of degrading the performance in the other regimes.
- *Training speed* is very fast for LOLIMOT and fast for HILOMOT and product space clustering, respectively.
- *Evaluation speed* is medium. In general, the number of neurons is relatively small since each neuron is very powerful, but the normalized validity functions can be quite tedious to evaluate.
- *Curse of dimensionality* is low for LOLIMOT and very low for HILOMOT. Product space clustering suffers from quadratic complexity of the full covariance matrix.
- *Interpretation* is especially easy when modeling dynamic systems; refer to Chap. 22. However, for static models, the rule premise structure and the parameters of the local linear models can reveal details about the process in a straightforward manner as well.
- *Incorporation of constraints* is possible because the parameters can be interpreted. In particular, LOLIMOT, which allows different objectives for parameter

Table 14.5 Comparison between different algorithms for construction of local linear neuro-fuzzy models

Properties	LOLIMOT	HILOMOT	Product space clustering
Interpolation behavior	0	+	0
Extrapolation behavior	++	+	0
Locality	++	++	+
Accuracy	+	++	++
Smoothness	+	+	+
Sensitivity to noise	++	+	+
Parameter optimization	++	++	+
Structure optimization	++	++	++
Online adaptation	++	++	++
Training speed	++	+	+
Evaluation speed	+	0	0
Curse of dimensionality	0/+	+/++	−
Interpretation	++	+	0
Incorporation of constraints	++	+	+
Incorporation of prior knowledge	++	+	+
Usage	+	0	+

* = linear optimization, ** = nonlinear optimization,
+ + / − − = model properties are very favorable/undesirable

and structure optimization, is well suited for the incorporation of constraints; see Sect. 13.3.2.

- *Incorporation of prior knowledge* is easily possible owing to the local character of the architecture. As pointed out in Sect. 14.2.3, various alternatives to integrate different kinds of knowledge can be realized.
- *Usage* generally is high, particularly for dynamic models, but HILOMOT is less popular up to now. Historically, local linear model architectures for modeling and control of dynamic systems boomed in the late 1990s. This was caused by the favorable features that these architectures offer in particular for dynamic systems; see Chap. 22.

These advantages and drawbacks of local model architectures trained by LOLIMOT, HILOMOT, and product space clustering are summarized in Table 14.5. Note that these algorithms are continuously improved and many different versions exist. Thus, this list can only give a tendency and may be inaccurate in particular cases.

Compared with many other model architectures, local linear neuro-fuzzy models offer the following features:

1. Very efficient and highly autonomous, incremental training algorithm (Sects. 13.2 and 13.3)
2. Interpretation as Takagi-Sugeno fuzzy model (Sect. 13.1.3)

3. Efficient structure selection of the rule premises (Sect. 13.3)
4. Efficient structure selection of the rule consequents (Sect. 14.3)
5. Distinction between premise and consequent inputs (Sect. 14.1)
6. Capability of integration of various knowledge sources (Sect. 14.2.3)
7. Ensuring certain extrapolation behavior (Sect. 14.4.2)
8. Efficient and robust online learning scheme (Sect. 14.6.1)
9. Easy to calculate expressions for the model accuracy (variance error) (Sect. 14.9)
10. Various additional advantages for modeling and identification of dynamic processes (Chap. 22)

Compared with alternative algorithms such as product space clustering, LOLIMOT and its extensions offer the following advantages:

1. The algorithm is extremely fast because of the exploitation of linear least squares optimization techniques, the local parameter estimation approach, and the axis-orthogonal decomposition algorithm, which allows one to utilize information from the previous iteration. The computational complexity of LOLIMOT grows very moderately with the input dimensionality and the number of rules (model complexity). LOLIMOT is the fastest construction algorithm for local linear neuro-fuzzy models available to date.
2. Since LOLIMOT is an incremental growing algorithm, it is easy to determine the optimal model complexity without additional effort.
3. The distinction between rule premise and consequent inputs can be favorably exploited for the incorporation of prior knowledge and automatic structure selection techniques.
4. Different objectives for structure and parameter optimization can be exploited in order to incorporate constraints indirectly into the modeling tasks without sacrificing convergence speed.
5. LOLIMOT+OLS allow a computationally cheap structure optimization of the rule consequents.
6. The consequent structure selection is *local*. This permits a separate structure selection for each local model. Therefore, the consequents must not contain all variables in $\underline{x}$ (while already $\dim\{\underline{x}\} \leq \dim\{\underline{u}\}$). This allows one to further reduce the number of estimated parameters. Hence, the variance error or the necessary amount of data decrease, and the interpretability of the model is improved.
7. The LOLIMOT input space decomposition strategy avoids strongly different widths of neighbored validity functions. This fact and the axis-orthogonal partitioning reduce (and in most cases totally avoid) undesirable normalization side effects.
8. The idea of freezing the validity function values at the boundaries of the input space avoids undesirable extrapolation behavior. Furthermore, a strategy to incorporate prior knowledge in the extrapolation is proposed.
9. Various additional advantages exist for modeling and identification of dynamic processes; refer to Chap. 22.

The drawbacks of the LOLIMOT algorithm are mainly as follows:

1. The axis-orthogonal input space decomposition is suboptimal and can become very inefficient for high-dimensional problems. This is particularly true if no or little knowledge is available that can be exploited for reducing the premise input space dimensionality or if the problem dimensionality cannot be reduced in principle.
2. The incremental input decomposition strategy is inferior in accuracy even when compared with other axis-orthogonal approaches such as ANFIS [289, 290, 292], because the centers (and less crucially the standard deviations) are not optimized. Therefore, LOLIMOT generally constructs models with too many rules.

These drawbacks are the price to be paid for obtaining the advantages above. Although the list of advantages is much longer than the list of the drawbacks, there are many applications where these drawbacks can outweigh the advantages. This is especially the case for high-dimensional problems with little available prior knowledge and whenever the evaluation speed of the model (basically determined by the number of rules) is more important than the training time.

LOLIMOT has been successfully used for modeling and identification of static processes in the following applications:

- Nonlinear adaptive filtering of automotive wheel speed signals for improvement of the wheel speed signal quality and its derivative [521–524]
- Nonlinear adaptive filtering of a driving cycle for automatic driving control with a guaranteed meeting of tolerances [515] (Sect. 24.1)
- Modeling and multi-criteria optimization of exhaust gases and fuel consumption of Diesel engines [203–205]
- Modeling and system identification with physical parameter extraction of a vehicle suspension characteristics [209]
- Online learning of correction mappings for an optimal cylinder individual fuel injection of a combustion engine [390]
- Modeling of the static characteristics of the torque of a truck Diesel engine in dependency on the injection mass, engine speed, and air/fuel ratio λ
- Modeling of a static correction mapping to compensate errors in a dynamic first principles roboter model [1]
- Modeling of the charging temperature generated by an intake air cooler that follows a turbocharger

14.13 Problems

1. Why is the concept of different input spaces for the rule premises and the rule consequents so powerful? Write down the equation for a local model network with different input spaces and give some examples where it might be useful.

2. Explain how local linear model networks can be extended to local polynomial model networks. Give examples where this extension might yield better results. Comment on the tradeoff between the complexity of the local models and the required number of local models. How does this tradeoff depend on the dimensionality of the input space?
3. Explain how structure selection techniques can be applied individually for each local model.
4. Illustrate the undesirable interpolation and extrapolation effects encountered in local linear model networks. How can they be avoided?
5. How do the undesirable interpolation effects affect the derivative of the model output? Explain possibilities to overcome this problem.
6. What are the key advantages of local model networks compared to other architectures when online learning shall be applied? Why is it much more robust to adapt the local models' parameters only and keep the validity functions' parameters fixed? What are the restrictions for online learning if the rule premises are kept fixed?
7. If forgetting is used in the RLS or similar algorithms, what can happen in case of insufficient excitation, e.g., if the process stays at a certain operating point for a long time? What strategy can overcome this problem? What price has to be paid for that?
8. What are errorbars used for? Why are they important? Under which assumptions is it possible to estimate the errorbars by the relatively simple equations given in this chapter?
9. What are the main drawbacks of the LOLIMOT algorithm? How does it perform for high-dimensional input spaces? Which neural network architecture does not possess these drawbacks?
10. Explain the key ideas of the HILOMOT algorithm. How is the linear estimation of the local model parameters combined with the nonlinear optimization of the splits constituting the partitioning?

Chapter 15
Input Selection for Local Model Approaches

Abstract This chapter discusses the topic of input selection in the context of local model networks. It utilizes a key advantage of this model architecture analyzed in Chap. 14 for the issue of selecting the relevant inputs to the model. Due to the curse of dimensionality, it is obvious that it can be beneficial to discard inputs with no or little relevance for the overall model performance and robustness. For local model networks, this can be done individually for the premises and the consequents. This is a major advantage compared to other model architectures and will be detailed in this chapter. Various strategies to exploit it are proposed and applied to toy and real-world examples. In addition, the visualization method "partial dependence plot" is introduced, which represents an excellent tool helping to visualize the main characteristics of high-dimensional mappings and should become more familiar in general and particularly in engineering. Also, the general issue of input relevance is addressed, which also can help in a data mining context.

Keywords Input selection · L1 regularization · Embedded · Visualization · Partial dependence plot

In this chapter input selection methods in combination with local model networks (LMNs) are utilized to weaken the effects of the curse of dimensionality. Input selection methods try to find subsets of input variables that lead to the best possible bias/variance tradeoff. Advantages due to omitting input variables are:

- Fewer inputs (might) lead to a better bias/variance tradeoff.
- Fewer inputs lead to better interpretability.
- The curse of dimensionality is weakened. Comparable model accuracies can be achieved with significantly fewer measurements.

The substantial contribution by Dr. Julian Belz for writing this chapter is gratefully acknowledged and appreciated.

O. Nelles, *Nonlinear System Identification*,
https://doi.org/10.1007/978-3-030-47439-3_15

When combined with LMNs, additional advantages arise due to the possibility to automatically separate linear from nonlinear effects (as will be discussed in more detail in the next paragraph), such as:

- The number of variables contained in the $\underline{x}$- and $\underline{z}$-input space can be limited to the necessary minimum *individually*, leading to a possibly even better bias/variance tradeoff compared to a general nonlinear model.
- Possibility to separate linear and nonlinear effects when using local affine models.
- The design of experiments can be adjusted to exploit the knowledge about the linear and the nonlinear effects.

The main disadvantage caused by the input space separation regards the increased complexity due to the fact of having two input spaces. Input selection methods have to find good input subsets for both the $\underline{x}$- and $\underline{z}$-input spaces. In fact, the number of potential inputs to choose from is virtually doubled because each process input can be assigned to each input space individually.

The possibility to separate linear from nonlinear effects is a direct consequence of the ability of LMNs to separate the input space into a $\underline{x}$- and a $\underline{z}$-input space as explained in Sect. 14.1. Variables contained in the $\underline{x}$-input space are used for the local models (rule consequents), whereas variables contained in the $\underline{z}$-input space belong to the validity functions (rule premises). If the local models are of affine type, the LMN is only able to follow a change in the slope of a process by switching to another local model. If some variables affect the process output only in an affine way, these variables are not needed in the $\underline{z}$-input space because the local models are able to capture their effects. To illustrate this, an artificial process following the equation

$$y(u_1, u_2) = \frac{0.2}{1.2 - u_1} + 0.8u_2 \,, \tag{15.1}$$

is shown in Fig. 15.1a together with three affine local models. From (15.1) it is already clear that input u_2 only has a linear effect on the process output whereas input u_1 contributes to the nonlinear process behavior. This can also be seen in Fig. 15.1. The affine local models are able to follow the process exactly in the u_2-direction. In order to follow the slope changes along the u_1-direction, a partitioning along the u_1-axis is necessary. The validity functions Φ_i belonging to the local models from Fig. 15.1a are shown in Fig. 15.1b. It is easy to see that input u_2 has no influence on the validity functions and can, therefore, be omitted for the description of the partitioning. In this example necessary inputs for the $\underline{x}$- and $\underline{z}$-input space are $\underline{x} = \begin{bmatrix} u_1 & u_2 \end{bmatrix}$ and $\underline{z} = u_1$, respectively.

Besides the rather academic process from (15.1) a combustion engine with a variable-length intake manifold can also serve as an example for which the separation into $\underline{x}$- and $\underline{z}$-inputs is useful. The position of the swirl flap of the variable-length intake manifold changes the properties of the dynamic system. As shown in Fig. 15.2, the position of the swirl flap changes the intake path. Through the intake elongation, properties of the system obviously are subject to change, like

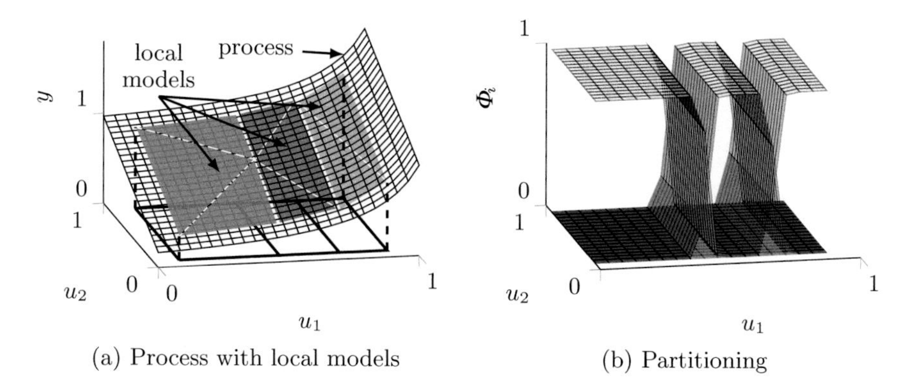

Fig. 15.1 Artificial process with one nonlinearly influencing input (u_1) and one linearly influencing input (u_2) together with three local models of an LMN (**a**) and its partitioning (**b**)

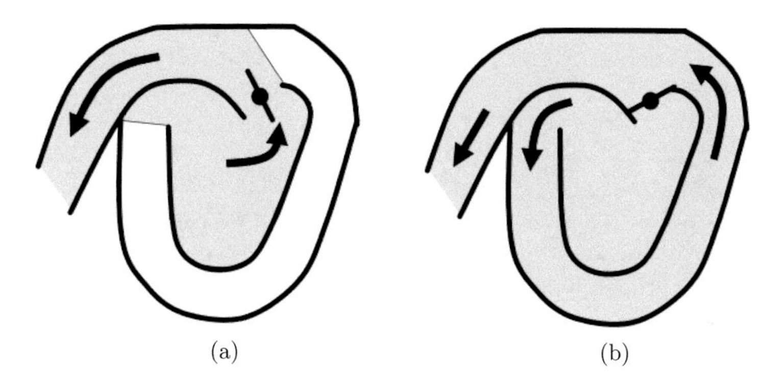

Fig. 15.2 The position of a swirl flap in a variable-length intake manifold determines the intake path (gray-shaded regions). (**a**) Swirl flap open. (**b**) Swirl flap closed

time constants, gains, dead times, etc. However, dynamic models describing the system for either an opened or closed swirl flap might not explicitly depend on the swirl flap position. As a result of this, an LMN describing the variable-length intake manifold system needs the swirl flap position only in the $\underline{z}$-input space, not in the $\underline{x}$-input space.

Another point of view is the following. The variables contained in the $\underline{z}$-input space define an operating point for which a specific and possibly affine local model is valid. The variables that define an operating point might not be needed to calculate the output of the corresponding local model. It is assumed that especially for dynamic models, the number of $\underline{z}$-inputs typically can be kept small. In contrast to that, the number of $\underline{x}$-inputs might be relatively large in order to describe the dynamics of the local model adequately.

Mainly three different methods for input selection using LMNs are developed and investigated in more detail throughout the rest of this chapter. For all methods specifically designed, test processes serve as benchmarks which are introduced in Sect. 15.1. Section 15.2 deals with a mixed wrapper-embedded input selection method. It utilizes existing training algorithms and wraps the input selection around it but simultaneously exploits the LMN structure to separate the linear from the nonlinear effects. Section 15.3 elaborates a regularization-based input selection method through an extension of the HIerarchical LOcal MOdel Tree (HILOMOT) algorithm. Section 15.4 presents an embedded input selection method that extracts information about the nonlinearity of a process directly from the partitioning of the $\underline{z}$-input space. Eventually, Sect. 15.5 deals with a visualization technique that allows inspecting dependencies between input variables and the output even for high-dimensional input spaces. This visualization technique is applicable to any type of model and is therefore not restricted to the class of LMNs.

15.1 Test Processes

Test processes are used to demonstrate the abilities of presented input selection approaches. The focus lies on the possibility to distinguish between the $\underline{x}$- and $\underline{z}$-input space and the resulting input selection schemes that are depicted in Fig. 15.9. Throughout this chapter only local affine models are used, such that a distinction between linearly and nonlinearly influencing inputs is possible. As discussed in Sect. 14.1, linearly influencing variables should be included in the $\underline{x}$-input space, and nonlinearly influencing variables should be included in the $\underline{z}$-input space of an LMN in order to describe the corresponding process adequately. The test processes are designed such that all possible scenarios for the assignment of the physical inputs occur:

1. Physical inputs that are important only for the $\underline{x}$-input space
2. Physical inputs that are important only for the $\underline{z}$-input space
3. Physical inputs that are important for both the $\underline{x}$- and $\underline{z}$-input spaces
4. Physical inputs that are not important for any of the two input spaces

While being able to cover all of the above-listed scenarios, the test processes are kept as simple as possible to enable an easy understanding. The presented input selection approaches have to find the correct assignments of the physical inputs to the two existing input spaces. Specific values in test processes are chosen to illustrate, e.g., variable importance or constraining the gain, etc.

15.1.1 *Test Process One (TP1)*

The first test process depends on four inputs, consists of three single functions that are summed up and follows the equation:

$$\begin{aligned} y(u_1, u_2, u_3, u_4) &= f_1(u_1, u_2) + f_2(u_3) + f_3(u_4) \\ &= \frac{0.1}{0.08 + 0.5(1-u_1) + 0.5(1-u_2)} + 0.8u_3 + u_4 \end{aligned} \tag{15.2}$$

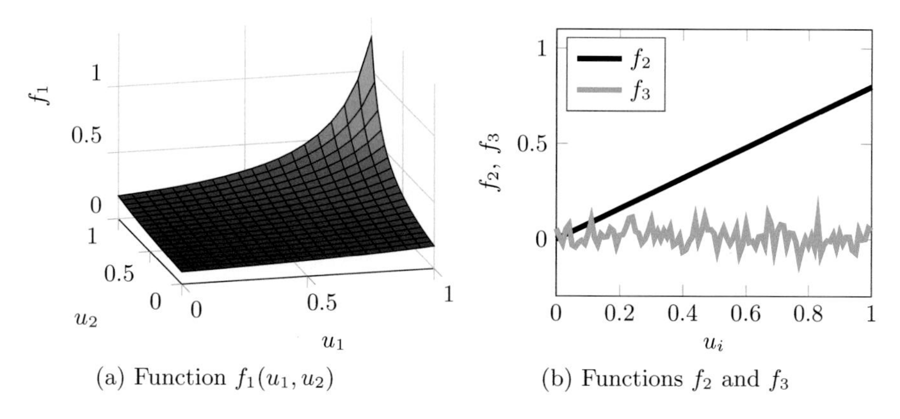

Fig. 15.3 Function f_1 (**a**) and the single functions f_2 as well as f_3 (**b**) of TP1

Input u_4 corresponds to noise which is normally distributed $\mathcal{N}(0, 0.025)$ and therefore is important for neither the $\underline{x}$- nor the $\underline{z}$-input space. The values of all other inputs are assumed to lie in the interval [0, 1]. From (15.2) it is easy to see that inputs u_1 and u_2 are interacting and influence the output in a nonlinear way. Both inputs are therefore important for both the $\underline{x}$- and $\underline{z}$-input spaces. Because input u_3 has just a linear influence, it is only needed in the $\underline{x}$-input space. Functions f_1 and f_2 as well as one realization of the output of function f_3 are shown in Fig. 15.3. The only scenario that is missing in test process one (TP1) is scenario 2 – a variable important only for the $\underline{z}$-input space – but this will be covered in the second test process.

15.1.2 Test Process Two (TP2)

The second test process consists of two different planes which are shown in Fig. 15.4a, b. Which of the two planes is currently valid is defined by the values of two operating point variables as shown in Fig. 15.4c. Note that the operating point variables and the variables on which functions f_1 and f_2 depend are different ones. The operating point variables are discrete and are either equal to 0 or 1. Variables u_1 and u_2 are continuous and lie in the interval [0, 1]. Mathematically the output of test process two (TP2) can be calculated according to the following equation:

$$y = \begin{cases} f_1(u_1, u_2) & \text{for } u_3 = 1 \text{ and } u_4 = 1 \\ f_2(u_1, u_2) & \text{else} \end{cases} \tag{15.3}$$

with

$$f_1(u_1, u_2) = -3u_1 + 2u_2 + 1 \text{ and}$$

$$f_2(u_1, u_2) = 3u_1 + 2u_2 + 0.5\,.$$

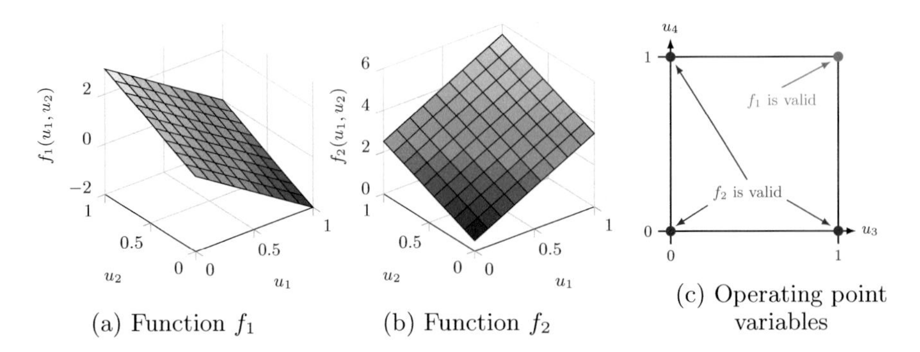

Fig. 15.4 Operating points (**a**) that define when function f_1 (**b**) and function f_2 (**c**) are valid for TP2

For TP2 the variables u_1 and u_2 are important to be included only in the $\underline{x}$-input space of an LMN, whereas variables u_3 and u_4 are important to be included only in the $\underline{z}$-input space.

15.1.3 Test Process Three (TP3)

The third process is specifically designed to test the abilities of the embedded input selection approach presented in Sect. 15.4. Test process three (TP3) follows the equation

$$y(u_1, u_2) = \frac{0.1}{0.08 + 0.73(1 - u_1) + 0.27(1 - u_2)} \tag{15.4}$$

and is shown in Fig. 15.5. All inputs are important for both the $\underline{x}$- and $\underline{z}$-input spaces of an LMN but can be qualitatively rated regarding their nonlinear influence. The main direction of the nonlinearity is visualized in Fig. 15.5b and is orthogonal to the contour lines. From this direction, it is obvious that input u_1 is more important to capture the nonlinearity than input u_2, because the angle between the u_1-axis and the nonlinearity is smaller than the angle between the u_2-axis and the nonlinearity. If the angle between the u_1-axis and the nonlinearity would be zero degrees, there would be no change in the process output in the u_2-direction, meaning that u_2 has no nonlinear influence on the output values at all.

15.1.4 Test Process Four (TP4)

The fourth test process is a dynamic one and follows the equation:

$$y(k) = 0.1867 \arctan[u(k-1)] + 0.8187 y(k-1) \,. \tag{15.5}$$

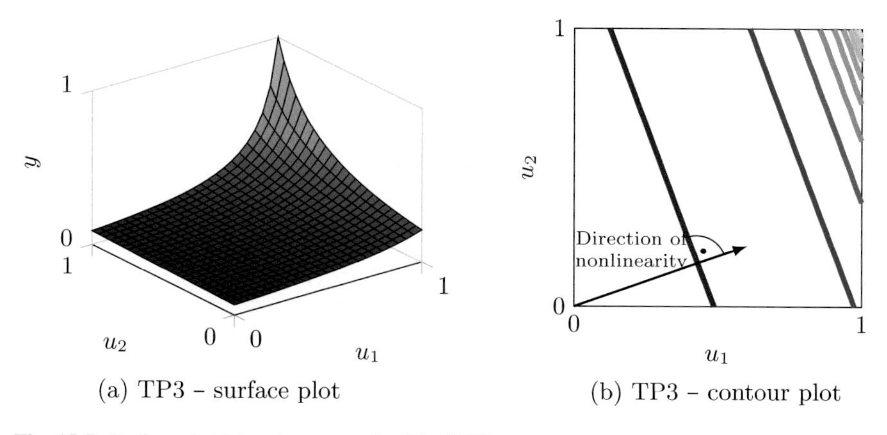

Fig. 15.5 Surface plot (**a**) and contour plot (**b**) of TP3

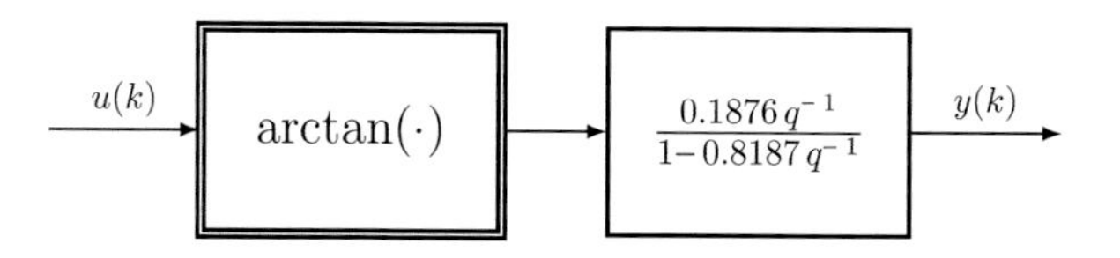

Fig. 15.6 Block diagram of the Hammerstein system used as TP4

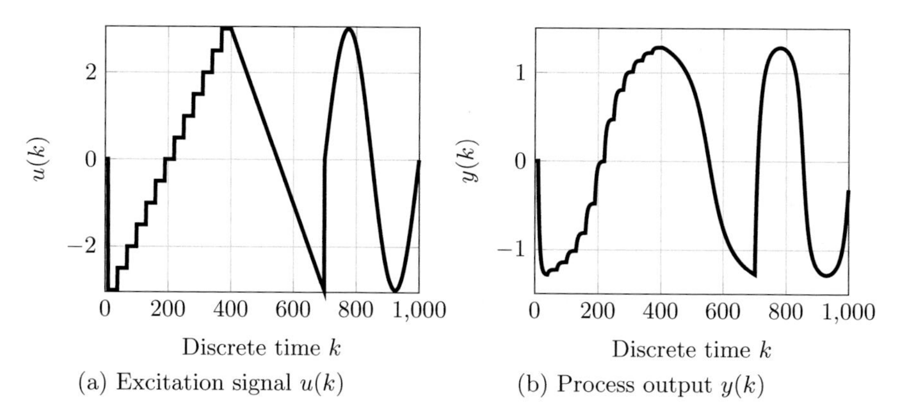

Fig. 15.7 Excitation signal (**a**) and the response of TP4 (**b**)

The input values should be in the interval $[-3, 3]$ in order to create sufficiently strong saturation effects and therefore lead to fairly nonlinear behavior. TP4 is a so-called Hammerstein process, which is a static nonlinearity followed by a linear dynamic system, as visualized in Fig. 15.6.

Here, the nonlinearity is the arctangent function succeeded by a first-order time-lag system. As can be seen, from (15.5) the nonlinearity only affects the delayed input. Therefore, only this input should be needed in the $\underline{z}$-input space, while the delayed output should only be needed in the $\underline{x}$-input space. Figure 15.7a shows an excitation signal for TP4 and Fig. 15.7b the corresponding process output.

All test processes are used throughout the remaining sections of this chapter to show the strengths and weaknesses of the presented input selection approaches. Whenever they are used, the number of data samples, possibly added noise, the distribution of the data, and other test conditions may vary and are mentioned explicitly. Note that not each test process is investigated for each input selection approach.

15.2 Mixed Wrapper-Embedded Input Selection Approach: Authored by Julian Belz

In order to weaken the effects of the curse of dimensionality, a mixed wrapper-embedded input selection approach is presented that fully exploits a unique property of LMNs, namely, the input space separation as explained in Sect. 14.1. The task of input selection can be formulated as a combinatorial, nonlinear, and discrete optimization problem. Due to the two arising input spaces for LMNs, the problem can be formulated as follows:

$$\begin{aligned} \underset{\underline{x},\underline{z}}{\text{minimize}} \quad & J(\underline{x}, \underline{z}) \\ \text{subject to} \quad & \underline{x} \subseteq \mathbb{P} \\ & \underline{z} \subseteq \mathbb{P}\,. \end{aligned} \tag{15.6}$$

J denotes an error measure of an LMN and therefore has to be minimized by choosing good $\underline{x}$- and $\underline{z}$-input subsets from all potential inputs $\mathbb{P}$. For the sake of simplicity, it should be assumed in the following that all potential inputs $\mathbb{P}$ correspond to the physical inputs $\underline{u}$ unless otherwise mentioned. Note however that in the case of dynamic models, the potential inputs $\mathbb{P}$ actually consist of filtered versions of the physical inputs $\underline{u}$ and outputs y combined in variable $\underline{\varphi}$. Theoretically (15.6) can be solved by just trying out all possible input subsets for the $\underline{x}$- and $\underline{z}$-input space and using the input combination that leads to the best model quality. Unfortunately, this is not feasible even for a relatively moderate number of candidate inputs. Assuming that this number, equivalent to the cardinality of $\mathbb{P}$, is n_p, there are 2^{2n_p} possible input subsets, because each element in $\mathbb{P}$ can be assigned to the $\underline{x}$- and $\underline{z}$-input space individually. In order to find a solution in an acceptable time, it is quite common to use heuristic search strategies that explore only parts of all possible input subsets, such as simple forward selection or backward elimination as described in more detail in Sect. 15.2.3.2.

The mixed wrapper-embedded approach tackles the optimization problem (15.6) by utilizing either LOcal LInear MOdel Tree (LOLIMOT) or HILOMOT as training algorithm and wraps the input selection around it. This is not just a wrapper approach because the possibility of both training algorithms to cope with the input space separation of LMNs is exploited. As illustrated in Fig. 15.8, a search strategy

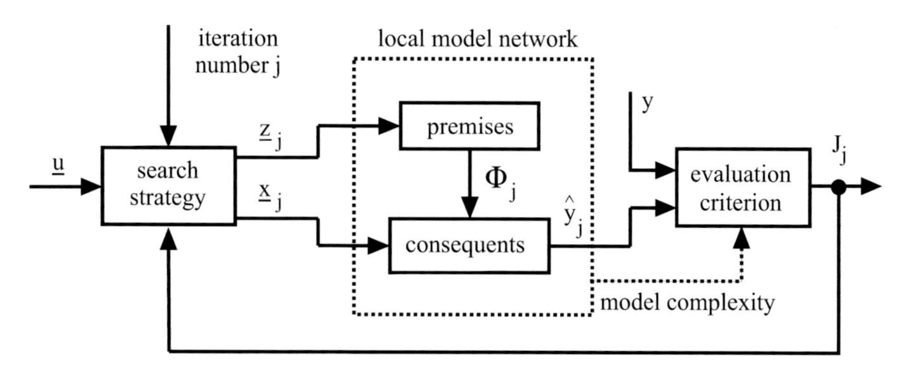

Fig. 15.8 Block diagram of the mixed wrapper-embedded input selection approach. Because of the utilization of an LMN, the search strategy has the possibility to assign the physical inputs $\underline{u}$ to the $\underline{x}$- and $\underline{z}$-input space individually

assigns physical inputs $\underline{u}$ to the $\underline{x}$- and $\underline{z}$-input space individually. With the chosen subsets $\underline{x} \subseteq \underline{u}$ and $\underline{z} \subseteq \underline{u}$, an LMN is trained. An evaluation criterion J_j is used to assess the model performance that is achieved with the subsets $\underline{x}_j$ and $\underline{z}_j$, belonging to model j. The evaluation criterion using measured (y) as well as predicted ($\hat{y}$) outputs might incorporate information about the model complexity and can be used to guide the search for good $\underline{x}$- and $\underline{z}$-input subsets. Possible choices for the evaluation criterion that assess the model accuracy are, e.g., the S-fold cross-validation error or the validation error achieved on a separate data set not used for training.

As a result of the two input spaces, four different selection schemes are possible, which are visualized in Fig. 15.9 and explained in the following.

Linked $\underline{x}$-$\underline{z}$-input selection: The $\underline{x}$- and $\underline{z}$-input spaces contain the exact same physical inputs ($\underline{x} = \underline{z}$), which are a subset of all physical inputs as shown in Fig. 15.9a. This selection scheme does not exploit the input space separability and can therefore be done with any model type.

Separated $\underline{x}$-$\underline{z}$-input selection: The $\underline{x}$- and $\underline{z}$-input spaces are completely disjunct, meaning that both might contain arbitrary subsets of the physical inputs ($\underline{x} \subseteq \underline{u}$; $\underline{z} \subseteq \underline{u}$); see Fig. 15.9b. In this case the number of inputs to choose from is virtually doubled, because each physical input can be assigned to the $\underline{x}$- and $\underline{z}$-input space individually.

$\underline{x}$-input selection: Subsets of all physical inputs are only sought for the $\underline{x}$-input space, while the inputs contained in the $\underline{z}$-input space are kept fixed. Either the inputs for the $\underline{z}$-input space have to be chosen by expert knowledge, a former $\underline{z}$-input selection, or simply all physical inputs are chosen for the $\underline{z}$-input space as shown in Fig. 15.9c.

$\underline{z}$-input selection: Subsets of all physical inputs are only sought for the $\underline{z}$-input space, while the inputs contained in the $\underline{x}$-input space are kept fixed. Either the inputs for the $\underline{x}$-input space have to be chosen by expert knowledge, a former

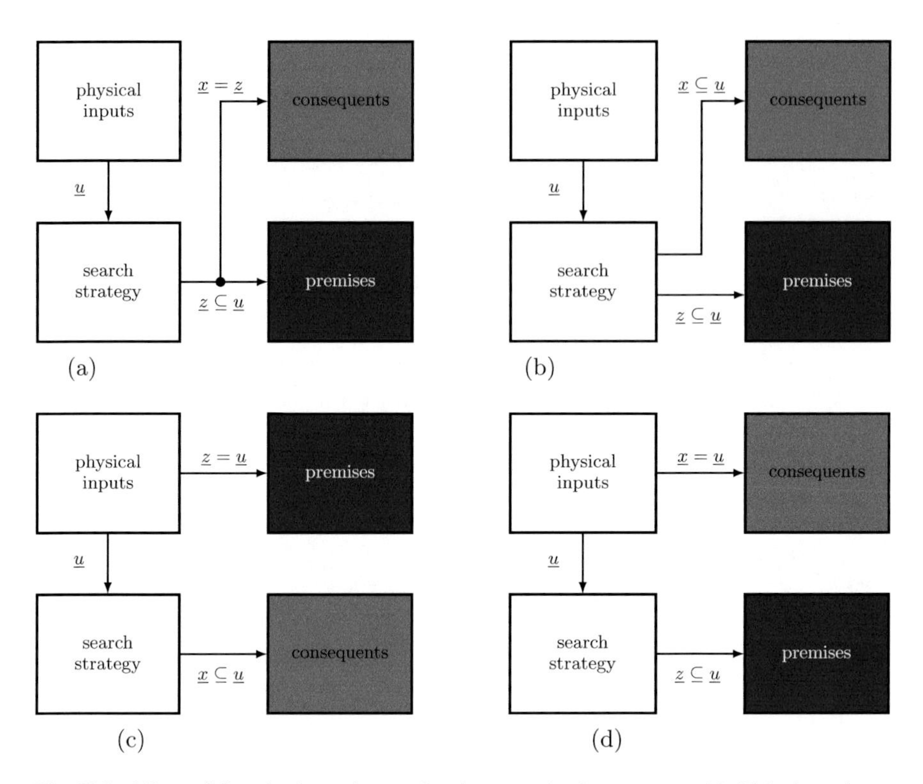

Fig. 15.9 All possible selection schemes for the $\underline{x}$- and $\underline{z}$-input space. (**a**) Linked $\underline{x}$-$\underline{z}$-input selection. (**b**) Separated $\underline{x}$-$\underline{z}$-input selection. (**c**) $\underline{x}$-input selection. (**d**) $\underline{z}$-input selection

$\underline{x}$-input selection, or simply all physical inputs are chosen for the $\underline{x}$-input space as shown in Fig. 15.9d.

Which of the presented selection schemes should be used depends on the problem at hand. The most flexible one is the separated $\underline{x}$-$\underline{z}$-input selection for which the fewest prior knowledge is necessary. For dynamic systems, the $\underline{z}$-input selection might be of special interest if the structure of the dynamic (local) models is known from first principles. In this case, the $\underline{z}$-input selection is able to reveal important inputs for the definition of the operating points. The $\underline{x}$-input selection as well as the linked $\underline{x}$-$\underline{z}$-input selection is of rather low interest. In the case of the $\underline{x}$-input selection, the influence of harmful inputs—in terms of the bias/variance tradeoff—can be weakened by regularization techniques easily as a promising alternative to selection. The linked $\underline{x}$-$\underline{z}$-input selection does not bring anything new scientifically, because it is just a wrapper method that could be pursued with any model type.

Section 15.2.1 uses the test processes described in Sect. 15.1 and carries out first investigations with the separated $\underline{x}$-$\underline{z}$-input selection. Through the artificial demonstration examples, the abilities of the mixed embedded-wrapper input selection approach should be emphasized. In Sect. 15.2.3 extensive simulation studies based on the function generator presented in Sect. 26.1.1 are performed to compare several search strategies and evaluation criteria.

15.2.1 Investigation with Test Processes

The mixed wrapper-embedded input selection approach is tested with TP1 and TP2 in this section. A separated $\underline{x}$-$\underline{z}$-input selection is carried out for both test processes with backward elimination as the search strategy and Akaike's information criterion (AIC_C) as an evaluation criterion. More details about backward elimination and the AIC_C can be found in Sect. 15.2.3. HILOMOT serves as training algorithm for the LMNs, which is described in Sect. 14.1.

15.2.1.1 Test Process One

As already described in Sect. 15.1 it is known a priori which inputs are important for the $\underline{x}$- and $\underline{z}$-input space for TP1. In particular, inputs u_1 and u_2 are important for both the $\underline{x}$- and $\underline{z}$-input spaces. Input u_3 is only important for the $\underline{x}$-input space, while input u_4 is only noise and is therefore not needed in any of the two input spaces. For this investigation a maximin optimized Latin Hypercube (LH) design with $N = 100$ samples is generated. Since input u_4 consists only of noise, no additional noise is added to the outputs. This data set is used to train the LMNs and to calculate the AIC_C values.

Figure 15.10 shows the result of the mixed wrapper-embedded input selection approach for TP1. The evaluation criterion is plotted against the number of inputs, which is simply the added number of variables contained in the $\underline{x}$- and $\underline{z}$-input space.

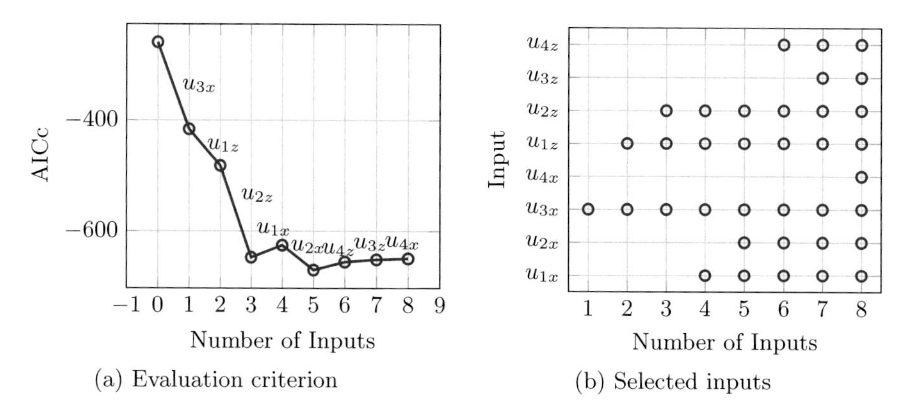

Fig. 15.10 Evaluation criterion (**a**) and selected inputs (**b**) versus the number of inputs for test process one

Because the search strategy is a backward elimination, the graph can be read from right to left. In between two subsequent AIC_C values, the input that is discarded in the corresponding backward elimination step is denoted. The index of the input u indicates the removal from the $\underline{x}$- or $\underline{z}$-input space, respectively. In Fig. 15.10b the subsets for a specific number of inputs are illustrated explicitly. For example, if the number of inputs is four, the inputs contained in the $\underline{x}$-input space are u_1 and u_3. The inputs contained in the $\underline{z}$-input space are u_1 and u_2. This illustration is not necessary for the backward elimination search strategy since the contained information is already included in Fig. 15.10a. However, for other search strategies, such as exhaustive search, this figure is necessary to represent the subsets belonging to a specific number of inputs. Since lower AIC_C values correspond to higher model qualities, the best LMN is obtained if five inputs are chosen. Three of these inputs are contained in the $\underline{x}$-input space (u_1, u_2, u_3), and two are contained in the $\underline{z}$-input space (u_1, u_2); see Fig. 15.10b. Therefore, the optimal result according to prior knowledge is obtained.

15.2.2 Test Process Two

TP2 has also four inputs which are virtually doubled due to the separated $\underline{x}$-$\underline{z}$-input selection. For this process u_1 and u_2 are important to be included only in the $\underline{x}$-input space, while u_3 and u_4 are only necessary in the $\underline{z}$-input space. As already described in Sect. 15.1, u_3 and u_4 are discrete operating point variables that are either equal to 0 or 1. The design of experiments (DoE) consists of four two-dimensional maximin optimized LH designs, one for each possible combination of the two discrete inputs. Each LH design contains 25 samples such that the total number of data samples is $N = 100$. Zero mean white Gaussian noise is added to the process output such that the signal-to-noise ratio (SNR) equals 20 dB.

Figure 15.11 shows the results for TP2. Again the expected result is obtained. The best model quality is achieved with four inputs, yielding exactly the optimal solution. Remarkable in Fig. 15.11a is the continuous slight model improvement while discarding superfluous inputs and the sharp drop in performance as soon as important inputs are thrown out.

In summary, it is shown that the mixed wrapper-embedded input selection approach is able to assign variables to the $\underline{x}$- and $\underline{z}$-input space correctly for TP1 and TP2. All types of variables occur in these two test processes. These are variables that are important in both the $\underline{x}$- and $\underline{z}$-input spaces, variables that are only important in one of the two input spaces, and variables that are unimportant for both input spaces. The next section investigates the influence of the chosen search strategy and evaluation criterion on the input selection result.

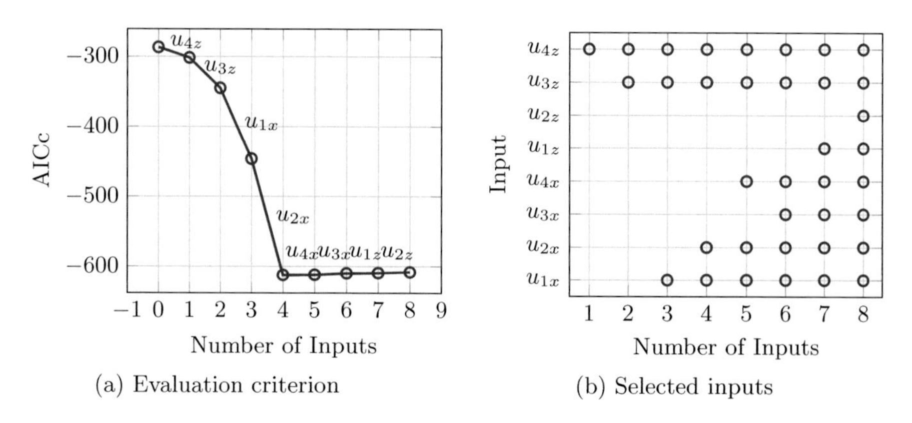

Fig. 15.11 Evaluation criterion (**a**) and selected inputs (**b**) versus the number of inputs for test process two

15.2.3 Extensive Simulation Studies

The goal of this section is to provide recommendations for the choice of a search strategy in combination with an evaluation criterion. Investigated criteria are described in more detail in Sect. 15.2.3.1 and incorporate the error on a distinct validation data set, cross-validation, and Akaike's corrected information criterion (AIC_C). Investigated search strategies are forward selection (FS), backward elimination (BE), an exhaustive search (ES), and a genetic algorithm (GA). All search strategies are described in more detail in Sect. 15.2.3.2. Combinations of these search strategies with evaluation criteria are compared in terms of the achieved model performance and the computation time needed to find the presumably best subset of inputs. The function generator described in Sect. 26.1.1 is used to randomly generate several static test processes. With the help of the test processes, training data sets and test data sets are created. Input selection is performed only with the training data sets. As a result, a presumably best subset of inputs is found for each search strategy in combination with a specific evaluation criterion. With the determined inputs, a model is trained, and for reference, its model performance is assessed with the help of the test data set. The assessed model performances can eventually be compared.

The function generator is used to randomly generate 100 static test processes for input dimensionalities $p = 2, 3, \ldots, 10$ and training sample sizes of $N = 100$, $N = 500$, and $N = 1000$. The used training data sets are created by maximin LH designs generated with the extended deterministic local search (EDLS) algorithm proposed in [131]. Zero mean white Gaussian noise is added to the outputs of the training data sets such that the SNR equals 30 dB. For each input dimensionality, one Sobol sequence is used for creating the test data sets. The number of test data samples is $N_t = 10^5$, independent of the input dimensionality. No noise is added to the outputs in the case of the test data sets.

For the investigations herein, the most flexible selection scheme is used, which is the separated $\underline{x}$-$\underline{z}$-input selection. Since all inputs can be assigned to both the $\underline{x}$- and $\underline{z}$-input spaces, the number of inputs to choose from is virtually doubled as already discussed in Sect. 15.2. HILOMOT is used as a training algorithm to build the LMNs with the subset of inputs selected by the particular search strategy; see Fig. 15.8 for a block diagram of the whole input selection procedure. Local affine models are used, and the model complexities (number of local models) of the LMNs are determined by Akaike's corrected information criterion.

15.2.3.1 Evaluation Criteria

As already mentioned, the investigated evaluation criteria that are used to find good subsets of inputs incorporate the error on validation data, cross-validation (CV), and the AIC_C. Each evaluation criterion is shortly explained in the following.

Validation Data

A separate validation data set is used to assess model quality. Here, the error on the validation data set is only used as an evaluation criterion to guide the input selection. The amount of validation data is chosen to be 20% of the overall available training data. This means that in the CV and AIC_C cases, 100% training data is available while for the validation-error-based criterion only 80% training data remains.

S-Fold Cross-Validation

This type of validation is described in Sect. 7.3.2 in detail. The number of folds for the S-fold CV is chosen to be $S = 10$ for this investigation following the recommendations in the publications [75] and [325]. Besides the number of folds, S represents also the number of runs that are necessary to assess the quality of a model. In each run one model is trained with $N - N/k$ data points that are randomly selected. Errors of the trained model are calculated on the left-out N/k data samples as visualized in Fig. 7.14. It is assumed that each data sample is selected only once for the model quality assessment throughout all runs. When all runs are finished, all calculated errors on the left-out data points can be used to form a model quality measure, like the normalized root mean squared error (NRMSE). Since all data samples have been used for the model training as well as for the model quality assessment, S-fold CV uses the available data very efficiently. The price to be paid is roughly $S+1$ times the higher computational cost. Note that after the best subset of inputs is determined according to the 10-fold CV error measure, a model is trained to incorporate all available data.

Akaikes's Information Criterion

Akaike's corrected information criterion (AIC_C) is a combination of the error on training data and an added complexity penalty as already described in Sect. 7.3.3. Note that the complexity penalty is proportional to both the number of local models and the number of inputs. Therefore, the AIC_C should be suited to compare models with different input subsets and should be able to find a good bias/variance tradeoff.

The advantage of the AIC_C and the CV as evaluation criterion compared to the usage of a separate validation data set is that they use all available data for the training. In contrast to the 10-fold CV, the AIC_C is computationally much less demanding.

15.2.3.2 Search Strategies

Four different search strategies are compared, which are forward selection (FS), backward elimination (BE), exhaustive search (ES), and a genetic algorithm (GA). FS starts with an empty subset of inputs and adds one input in each iteration of the search algorithm. Once one input is added, it cannot be removed afterward. In each iteration all not yet selected inputs are tentatively tested to be added, a model is trained, and the input leading to the best evaluation criterion is eventually selected. This is done until all inputs are selected for the presented investigations in this section.

In contrast to that, BE starts with all inputs, and in each iteration of the search algorithm, one of the inputs is discarded from the input subset. Once an input is removed from the subset, it cannot be added afterwards. To decide which input variable should be removed at the current iteration, each of the remaining input variables is tentatively discarded from the input subset, and a model is trained. The evaluation criterion for each removed input is calculated and the input that yields the best evaluation criterion is picked for removal at the current iteration. For the investigations in this section, the search algorithm stops after all inputs are removed.

The ES simply tries out all possible combinations of inputs. For each possible combination, a model is trained and the evaluation criterion is calculated. Because the number of models that have to be trained grows exponentially with the number of potential inputs, this search strategy is only used up to $p = 5$ inputs. Because the separated $\underline{x}$-$\underline{z}$-input selection is used, the number of possible input combinations and therefore the number of necessary models to be trained reach already $2^{2 \cdot 5} = 1024$.

The GA is a global optimization algorithm, meaning that it is theoretically capable of finding the globally best solution of a nonlinear optimization problem. More information about GAs in can be found in Sect. 5.2.2 and [374] or [337].

Within this section, each individual represents one subset of inputs. Therefore, each individual is a bit-string, containing just ones and zeros indicating if an input is part of the subset (1) or not (0). This GA-typical binary coding is particularly straightforward for combinatorial optimization tasks like input selection. For the calculation of the fitness scores, the evaluation criteria from Sect. 15.2.3.1 are used.

Note that for all evaluation criteria, lower values are better and therefore get higher fitness scores. The mutation rate is set to 0.1, while the recombination rate is set to 0.8. The population size is chosen to be five times the number of potential inputs. The GA stops either after three generations without any improvement in the best fitness score of all individuals or after a maximum calculation time of 24 h.

FS and BE both follow a rather simple heuristic, and finding the best combination of inputs is by no means guaranteed. However, these search strategies are often used because of their tractability and the reasonably good results that are achieved with them. In the case of ES, the best combination of inputs is guaranteed to be included in all subsets that are tried out. Nonetheless, the optimal subset might not be recognized as the best combination of inputs if the used evaluation criterion fails to do so. As mentioned in [482], one possible reason is overfitting. This does not mean that each individual model overfits. It means that the selection itself induces overfitting. If the number of different input subsets that are compared based on a specific evaluation criterion becomes huge, chances increase to achieve the best score for an objectively not optimal subset just by chance. Being a nonlinear global optimizer, the GA has also the potential to find the best possible combination of inputs for the problem at hand and should, therefore, be capable to produce the same results as the ES. Anyhow, due to time limitations or other user-specified restrictions, the results of the ES and GA do not need to be the same.

15.2.3.3 A Priori Considerations

Before the comparison starts, some thoughts are provided about what can be expected. Deviations from these expectations are considered to be interesting results and worth to be discussed in some more detail. At first expectations about the evaluation criteria are summarized.

Using validation data as an evaluation criterion is considered to perform worst in choosing good subsets of inputs. The number of samples used for the actual training is the smallest in this case. Additionally, the validation data set is used very often which increases the danger in the selection procedure of overfitting. The more models are tested on the validation data set, the more likely it becomes to assess one of these models as good just by chance and not because it has a really good generalization performance. This danger increases with an increasing number of tested models. Therefore the risk of observing this kind of overfitting is the highest for the ES and the GA search strategy.

The 10-fold cross-validation is considered to perform best in choosing good subsets of inputs. The available data is utilized in the most efficient way. Because not only one hold-out data set is used, the danger of overfitting should be far smaller compared to the validation data set. Additionally, the generalization performance is measured based on data not used for the training opposed to just adding some complexity penalty as it is done in case of the AIC_C.

For the $\mathrm{AIC_C}$ as evaluation criterion, it is hard to make an educated guess about the expected performance in choosing subsets of inputs. On the one hand, the $\mathrm{AIC_C}$ appears to be rather simple by just augmenting the error on training data with a complexity penalty. On the other hand, this criterion shows very good results in choosing an appropriate model complexity for LMNs as shown in [221].

In the case of the used search strategies, it is expected that the BE finds on average better input subsets than the FS, because all interactions between inputs of the test functions are observable from the start. Especially in the early stages of the FS, these interactions might be hidden, and more inputs are presumably needed to cover them. It is not straight away predictable if the rather simple greedy search strategies, which are FS and BE, perform better than the more extensive ones, which are ES and GA. On the one hand, FS and BE are known to be less prone to overfitting [355]; on the other hand, a lot of possibly good input subsets are never explored. Here, a strong interdependency with the used evaluation criterion is likely to be present. Therefore, no a priori guess exists about the ranking of the used search strategies.

15.2.3.4 Comparison Results

The results of the extensive simulation studies are used to compare all possible combinations of search strategies and evaluation criteria regarding the obtained model accuracies and required computation times. Additionally, the benefit of the mixed wrapper-embedded input selection (MWEIS) approaches compared to using all available inputs is pointed out. Another result of these studies is the *effective* dimensionality of the used test functions created with the function generator. Therefore the number of inputs that is actually chosen by the MWEIS approaches is reviewed.

The relative improvement RI defined as

$$\mathrm{RI} = \frac{\mathrm{NRMSE_{all}} - \mathrm{NRMSE_{subset}}}{\mathrm{NRMSE_{all}}}\,, \tag{15.7}$$

is used to demonstrate the benefit of the MWEIS approach compared to using all available inputs. The NRMSE values are calculated on the test data, either with all available inputs (index *all*) or with the subset suggested by one of the MWEIS approaches (index *subset*). Because lower NRMSE values indicate better model performances, positive RI values indicate improvements gained through the MWEIS, while deteriorations are indicated by negative RI values. The relative improvements gained through the input selection are also used to compare all possible combinations of search strategies and evaluation criteria with each other. The reference point is always the accuracy of a model using all inputs and is, therefore, the same for all MWEIS approaches. Thus, the biggest improvement corresponds to the best model accuracy in absolute terms.

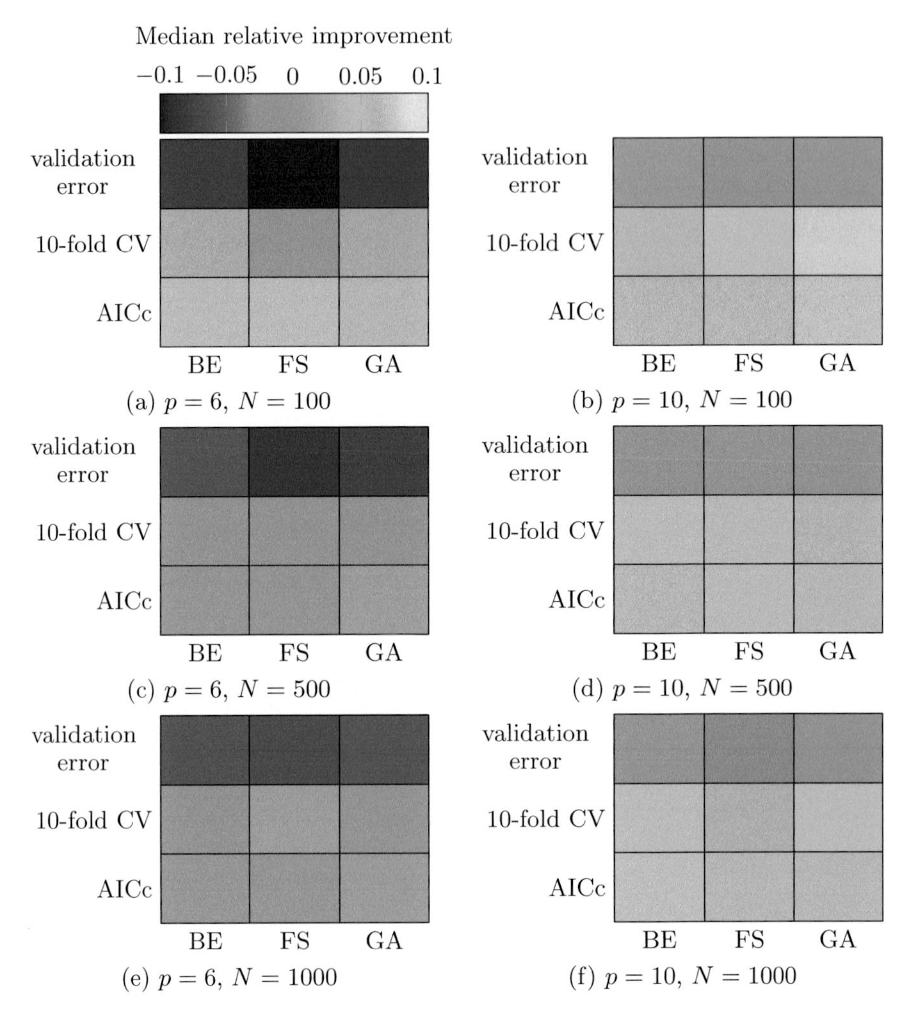

Fig. 15.12 Median relative improvements for all combinations of search strategies and evaluation criteria for input dimensionalities $p = 6$ and $p = 10$ with training data set sizes $N = 100, 500$, and 1000. (**a**) $p = 6$, $N = 100$. (**b**) $p = 10$, $N = 100$ (**c**) $p = 6$, $N = 500$. (**d**) $p = 10$, $N = 500$. (**e**) $p = 6$, $N = 1000$. (**f**) $p = 10$, $N = 1000$

Figure 15.12 shows the median relative improvements of all MWEIS approaches for all training data sizes and two input dimensionalities, which are $p = 6$ and $p = 10$. The median is determined over the 100 different test functions. The shown input dimensionalities are chosen because the advantages of the MWEIS approaches become clear from $p = 6$ on and increase with an increasing p until the highest improvements are obtained with $p = 10$. For input dimensionalities lower than 6 ($p < 6$), the median relative results look like Fig. 15.12c. No significant improvements are observable for the 10-fold cross-validation and the AIC_C, while the model accuracies with subsets obtained with the validation error as evaluation criterion are deteriorated.

In fact, subsets chosen by the validation error as evaluation criterion have almost never a substantial advantage when compared to models using all inputs. The AIC_C for the MWEIS turns out to be the best choice as an evaluation criterion among the tested ones. In most cases, the model accuracies obtained with input subsets chosen by the AIC_C and the 10-fold cross-validation are on the same level. The advantage of the AIC_C increases with a decreasing input dimensionality and decreasing training data sizes.

The BE and the GA are the best search strategies, even if the advantage compared to the FS is quite small in most cases. No results of the ES are presented in Fig. 15.12, since ES is only used for smaller input dimensionalities (up to $p = 5$). The results obtained by the ES are comparable to the outcomes of the GA for input dimensionalities $p = 2, \ldots, 5$.

Unfortunately, it is not possible to make meaningful statements about the variations from the median test errors for different combinations of search strategies and evaluation criteria. There are two sources for the variations in the achieved test errors:

1. The 100 different test functions. Even if all MWEIS approaches would find the optimal subset of inputs for each of the 100 test functions, there would still be a certain amount of variation due to various complexities in the test functions. The exact amount of variation cannot be quantified, and thus no reference exists.
2. Possibly suboptimal subsets increase the range of achieved test errors and thus the degree of variation.

For the comparison of different search strategies with different evaluation criteria, only the second source of variation is of interest. Since there is only one noise realization for each test process, it is not possible to determine the variation that comes from a specific combination of a search strategy with an evaluation criterion.

The required computation times of all MWEIS approaches are compared in Fig. 15.13 for input dimensionalities $p = 2$ and $p = 10$ with a training data size of $N = 1000$. Note that Fig. 15.13a, b are both scaled logarithmic but differently. Again, the median is taken over the number of different test functions. For the shown data amount, the highest computation times occur. In general, the required computation time increases with an increasing input dimensionality as well as with an increasing amount of training samples. As expected, the 10-fold CV requires the highest computation times among all evaluation criteria. In general, using the validation error as an evaluation criterion requires slightly less computation time than the AIC_C. Note that even though the maximum calculation time is limited to 24 h, the highest median computation time is only 210 min ($p = 10$, $N = 1000$, 10-fold CV, GA). Since the maximum required computation time at all is 9.5 h, it can be concluded that the stopping criterion for all GA runs is always the number of generations without any improvement in the best fitness score (chosen to be 3).

Typically the following ordering occurs for the search strategies regarding the required computation time in ascending order: FS, BE, ES, and GA. A remarkable

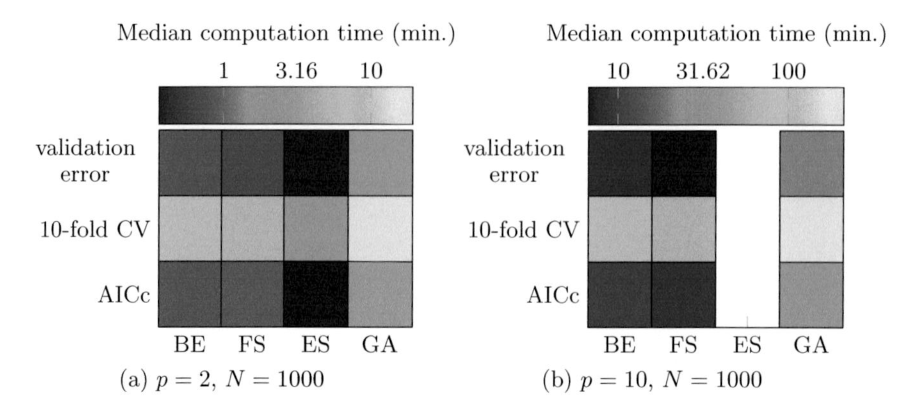

Fig. 15.13 Required median computation times for all combinations of search strategies and evaluation criteria for input dimensionalities $p = 2$ (**a**) and $p = 10$ (**b**) with $N = 1000$

exception that can be seen in Fig. 15.13a occurs for $p < 4$ where the ES search strategy requires the least computation time among all search strategies. This is due to the fact that LMNs for all possible subsets of inputs can be trained simultaneously if enough CPU cores are available. In contrast to that, the FS and BE search strategies add/eliminate one input in each iteration. Until it is known which input is added/discarded in the current iteration, no further results are calculated in parallel. Waiting on these intermediate results leads to higher computation times even though the computational effort is lower compared to the ES search strategy.

Figure 15.14 shows the median input subset size for all MWEIS approaches for selected input dimensionalities and a training data set size of $N = 100$.

Input dimensionality $p = 4$ is selected to be shown because it is the lowest input dimensionality for which differences in the chosen input subset sizes become clear. Then, the number of inputs that are chosen increases with an increasing input dimensionality until input dimensionality $p = 7$. It is interesting to see that there is no significant increase in the number of selected inputs if the input dimensionality is further increased; compare Fig. 15.14c with Fig. 15.14d. It seems as if the *effective* dimensionality of the test functions generated with the function generator saturates. Note that due to the used separated $\underline{x}$-$\underline{z}$-input selection, there are potentially 20 inputs that can be selected if the input dimensionality is $p = 10$.

Qualitatively similar results are obtained for the other two investigated training data set sizes $N = 500$ and $N = 1000$, but with generally bigger input subset sizes. Figure 15.15 shows how the median subset size increases with the number of samples for input dimensionalities $p = [2, 3, 5, 8, 9, 10]$. Median values are taken over the 100 different test functions with BE as search strategy and the AIC_C as an evaluation criterion. For rather low-dimensional input spaces, the subset size does not further increase with more available training data. For example, with a two-dimensional input space ($p = 2$), a training data set size of $N = 100$ already covers the input space well and therefore carries sufficient information such that in

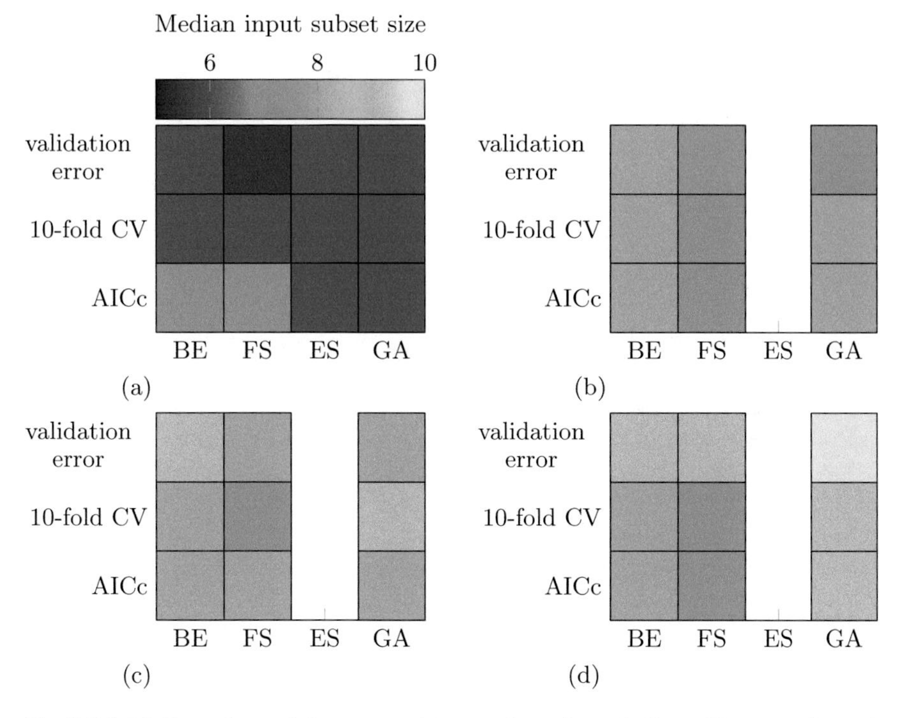

Fig. 15.14 Median values of the suggested input subset sizes of all combinations of search strategies and evaluation criteria for input dimensionalities $p = 4, 6, 7$, and 10 with $N = 100$. (**a**) $p = 4$, $N = 100$. (**b**) $p = 6$, $N = 100$. (**c**) $p = 7$, $N = 100$. (**d**) $p = 10$, $N = 100$

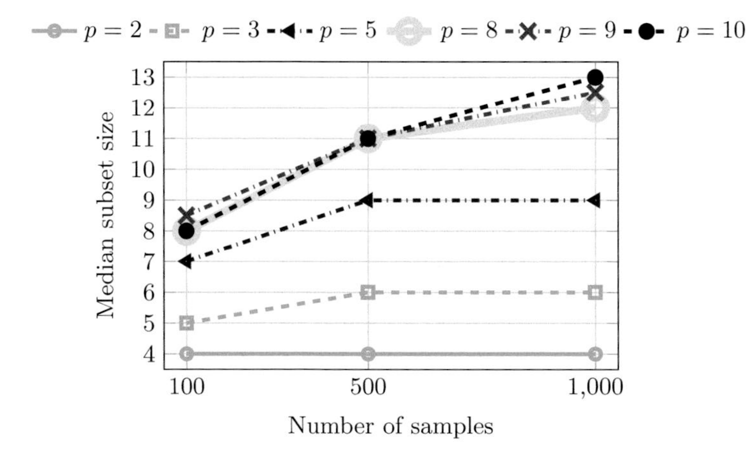

Fig. 15.15 Median subset sizes for different input dimensionalities over all 100 test functions for BE as search strategy and AIC_C as evaluation criterion

terms of a good bias/variance tradeoff, all beneficial LMN inputs could be selected. Therefore, no more increase in the subset size can be expected if the number of samples is increased. In contrast to that, the input space for $p = 10$ is almost empty in case of $N = 100$ samples. As a result, the input subset size is kept rather small in order to achieve a good bias/variance tradeoff. As more information becomes available, i.e., N increases, the best bias/variance tradeoff is shifted to more complex models leading to bigger input subset sizes.

Typically the lowest number of inputs is chosen by the FS and BE search strategies, while GA chooses more inputs to be included in the suggested subset. The ES search strategy is typically on the same level as FS and BE. Comparing the investigated evaluation criteria, the ascending ordering regarding the number of typically chosen inputs is AIC_C, 10-fold cross-validation, and the validation error.

In summary, the presented results lead to the recommendation of using the BE search strategy combined with the AIC_C as an evaluation criterion. The required computation time is only lower if FS is used, while the achieved model accuracies are typically among the best of all search strategies when BE is used. In some cases, the GA leads to slightly better model accuracies compared to the BE but therefore typically chooses more inputs and requires more computation time. The AIC_C is recommended as an evaluation criterion because it leads to model accuracies that are only seldom outperformed by input subsets selected by 10-fold cross-validation. The number of chosen inputs is typically low compared to the other evaluation criteria, and the required computation time is also among the lowest.

15.3 Regularization-Based Input Selection Approach: Authored by Julian Belz

A normalized L1 regularization for axis-oblique partitioning strategies is described, which has been previously proposed by the author [44]. In contrast to commonly used L1 regularization techniques, only the amount of obliqueness incorporated in the input space partitioning is penalized. In principle, this approach can be applied to any axis-oblique partitioning strategy based on nonlinear optimization techniques. In this section, the focus lies on an implementation for HILOMOT.

Axis-oblique partitioning strategies weaken the effects of the curse of dimensionality significantly compared to axis-orthogonal partitioning strategies. However, the additional flexibility comes along with an increased variance error. The normalized L1 regularization should account for the increased variance error and might be able to determine the most important input variables. Since only the partitioning is affected by the proposed regularization, only input variables in the $\underline{z}$-input space are selected. Additional effort is necessary to find the best subset of inputs for the $\underline{x}$-input space, e.g., a subsequent $\underline{x}$-input selection.

The main idea of regularization-based input selection is to introduce a penalization of high absolute values of the model parameters $\underline{\theta}$ such that these are pushed

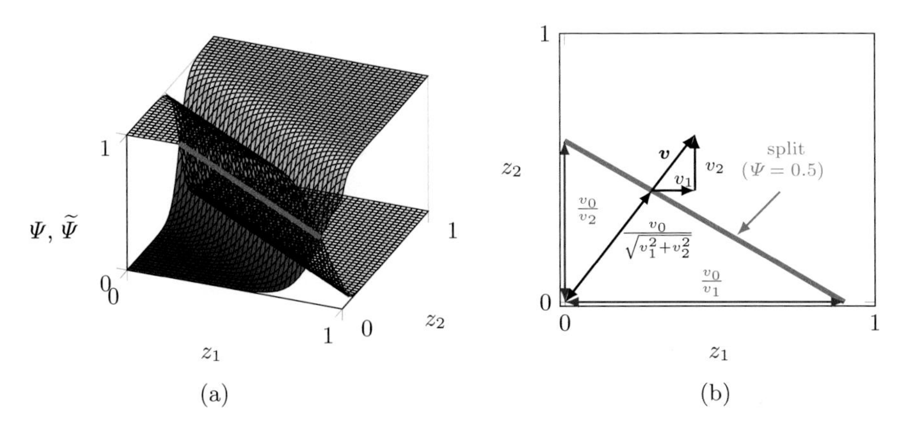

Fig. 15.16 Explanation of splitting parameters. (**a**) Sigmoid and complementary function. (**b**) Topview on z_1-z_2 plane

toward zero [230, 560]. Once a model parameter reaches zero, the input associated with it can be discarded. The whole loss function that has to be minimized is compounded of a part related to the model error $\underline{e} = \left[e(1)\ e(2)\ \cdots\ e(N)\right]^T$ and a part related to the sum of absolute values of the model parameters $\underline{\theta}$:

$$J_{L1} = \frac{1}{N}\underline{e}^T(\underline{\theta})\underline{e}(\underline{\theta}) + \lambda||\underline{\theta}||_1 . \tag{15.8}$$

The regularization strength is determined by λ, with $\lambda \geq 0$. By constantly increasing the regularization parameter λ, more and more inputs are discarded, and the model becomes less flexible. Typically, the regularization parameter λ is tuned such that the generalization performance of the model is maximized. This corresponds to the best bias/variance tradeoff.

The meaning of the parameters contained in $\underline{v}$ is illustrated for a two-dimensional $\underline{z}$-input space in Figs. 15.16 and 15.17. A sigmoid and its complementary counterpart are visualized in Fig. 15.16a, while Fig. 15.16b shows the top view on the z_1-z_2 plane together with the split and the interpretation of the splitting parameters. The orientation of the split is determined by the ratio of the splitting parameters $\underline{v}$ with respect to each other. The split orientation is important for the LMN since it determines the direction in which its slope is able to change if local affine models are used. In this case changes in the slope of the LMN can only be realized by switching to another local model and vector $\underline{v}$ points directly to the adjacent model. If one variable contained in the $\underline{z}$-input space has no nonlinear influence, its corresponding splitting parameter is zero.

If the obliqueness of a split should be penalized—as it is the goal of the normalized L1 split regularization described in the following section—the sigmoid splitting parameters $\underline{v}$ should be pushed toward zero. The offset parameter v_0 is

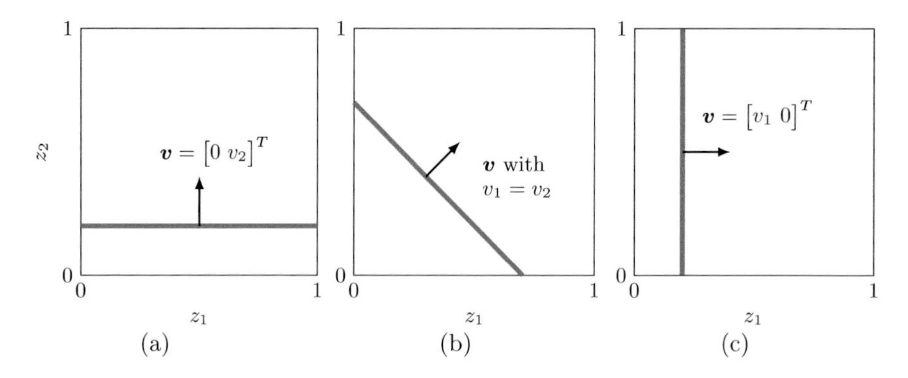

Fig. 15.17 Three special cases of the sigmoid splitting parameters $\underline{v}$ without the offset parameter v_0. (**a**) Split orientation: 0°. (**b**) Split orientation: 45°. (**c**) Split orientation: 90°

not regularized by this method, which means that the location of the split does not influence the penalty.

Before details about the normalized L1 regularization are presented in Sect. 15.3.1, the reader is referred to Sect. 14.8 where the HILOMOT algorithm is discussed in great depth. In particular, Sect. 14.8.2 which deals with the split optimization is of fundamental importance for the following discussion. Section 15.3.2 demonstrates the abilities of the normalized L1 regularization implemented for HILOMOT.

15.3.1 Normalized L1 Split Regularization

The main goal of the split regularization is to reduce the model variance introduced by the split optimization and to get a low-dimensional $\underline{z}$-input space. Therefore the regularization should favor axis-orthogonal splits, while an oblique partitioning should only be pursued if there is a significant performance gain. In order to achieve these goals, the following penalty term is proposed:

$$J_p(\underline{v}) = \frac{\lambda}{\sqrt{p_z} - 1} \left(\left\| \frac{\underline{v}}{\|\underline{v}\|_2} \right\|_1 - 1 \right) . \tag{15.9}$$

It is a 1-norm, where the splitting parameters $\underline{v}$ are normalized by their 2-norm. The regularization parameter is λ. The remaining adjustments, i.e., the subtraction of one and the division by $\sqrt{p_z} - 1$, lead to the scaling of the penalty term to the interval $[0, 1]$, if $\lambda = 1$. p_z is the number of variables contained in the $\underline{z}$-input space. The maximum of the penalty term is reached if all entries in $\underline{v}$ have the same absolute value. Its minimum is reached if all but one entries are zero, which corresponds to an

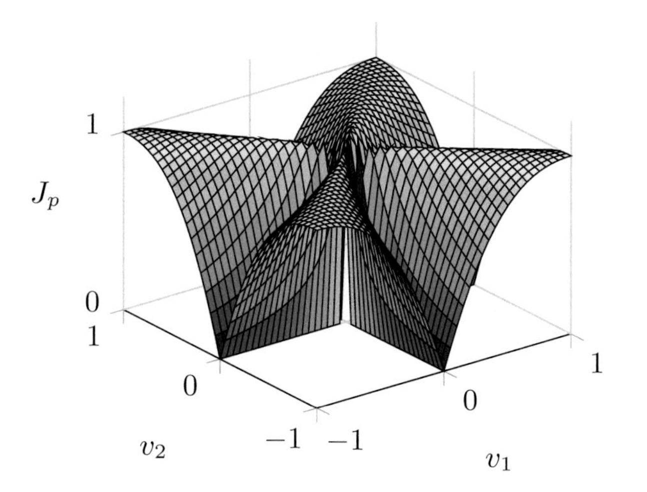

Fig. 15.18 Penalty term for a two-dimensional $\underline{z}$-input space

axis-orthogonal split. The penalty term for $\lambda = 1$ is visualized for a two-dimensional $\underline{z}$-input space in Fig. 15.18.

Note that the splitting parameter vector $\underline{v}$ does only include elements determining the orientation of the split, see Sect. 14.8.2, and therefore only the amount of obliqueness is penalized. The position in the $\underline{z}$-input space does not affect the penalty term (15.9).

In order to improve the performance of the nonlinear optimization, the analytical gradient of the whole loss function is required. Assuming the NRMSE as unregularized loss function, the whole objective including the regularization term becomes

$$J(\underline{\theta}_{LM}, \underline{v}) = \sqrt{\frac{\sum_{i=1}^{N}(y(i) - \hat{y}(i, \underline{\theta}_{LM}, \underline{v}))^2}{\sum_{i=1}^{N}(y(i) - \bar{y})^2}} + J_p(\underline{v}), \qquad (15.10)$$

with the measured outputs $y(i)$, the mean of all measured outputs $\bar{y}$, and the LMN output $\hat{y}(i)$ depending on the affine local model parameters $\underline{\theta}_{LM}$ and the splitting parameters $\underline{v}$. In [159] the analytical gradient for the unregularized case has already been derived. Therefore only the derivative of J_p with respect to the splitting parameters is derived here. A differentiable approximation for the absolute value of a given number has to be used for the calculation of the 1-norm. We use the smooth approximation proposed in [514]. The absolute value of a number x is approximately:

$$|x| \approx \frac{1}{\alpha}\left(\log\left(1 + e^{-\alpha x}\right) + \log\left(1 + e^{\alpha x}\right)\right) . \qquad (15.11)$$

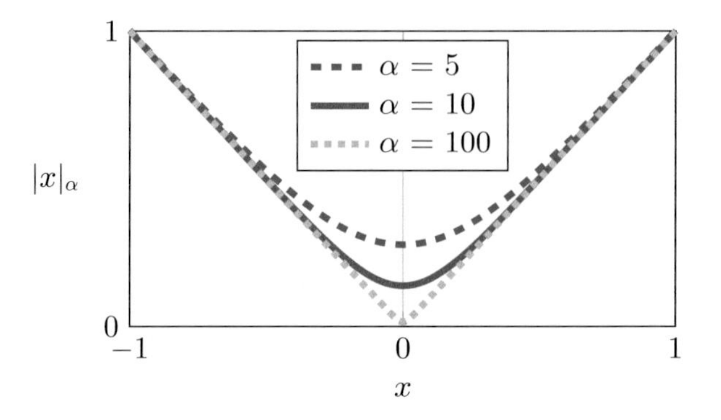

Fig. 15.19 Approximation $|x|_\alpha$ of the absolute value $|x|$ for three different parameters of α

The approximation will be denoted as $|x|_\alpha$. Here, α is a parameter that determines how close the approximation comes to the true absolute value. As stated in [514] and demonstrated in Fig. 15.19, $|x|_\alpha$ converges to $|x|$ as α approaches infinity.

With this approximation the penalty term becomes

$$\begin{aligned} J_p(\underline{\upsilon}) =& \frac{\lambda}{\sqrt{p_z}-1}\left(\left\|\frac{\underline{\upsilon}}{\|\underline{\upsilon}\|_2}\right\|_{1\alpha} - 1\right) \\ =& \frac{\lambda}{\sqrt{p_z}-1}\left[\sum_{i=1}^{p_z}\frac{1}{\alpha}\left(\log\left(1+\mathrm{e}^{-\alpha\frac{\upsilon_i}{\|\underline{\upsilon}\|_2}}\right)+\log\left(1+\mathrm{e}^{\alpha\frac{\upsilon_i}{\|\underline{\upsilon}\|_2}}\right)\right)-1\right], \end{aligned} \tag{15.12}$$

where $\|\cdot\|_{1\alpha}$ is the sum of approximated absolute values of the corresponding vector elements. In [514] this is called *Smooth L1 Approximation Method*. Variable p_z denotes the number of inputs in the $\underline{z}$-input space. It follows the detailed derivation of the analytical gradient for splitting parameter υ_j. At first, the derivative of the Smooth L1 approximation with respect to an arbitrary variable x is derived:

$$\begin{aligned} \frac{\mathrm{d}|x|_\alpha}{\mathrm{d}x} &= \frac{1}{\not\alpha}\left(\frac{-\not\alpha\cdot\mathrm{e}^{-\alpha x}}{1+\mathrm{e}^{-\alpha x}}\cdot\frac{\mathrm{e}^{\alpha x}}{\mathrm{e}^{\alpha x}}+\frac{\not\alpha\cdot\mathrm{e}^{\alpha x}}{1+\mathrm{e}^{\alpha x}}\right) \\ &= \frac{\mathrm{e}^{\alpha x}-1}{\mathrm{e}^{\alpha x}+1} = \frac{\mathrm{e}^{\frac{2\alpha x}{2}}-1}{\mathrm{e}^{\frac{2\alpha x}{2}}+1} = \tanh\left(\frac{\alpha x}{2}\right). \end{aligned} \tag{15.13}$$

Another necessary inner derivative concerns the multiplicative inverse of the 2-norm with respect to one splitting parameter υ_j:

$$\frac{\mathrm{d}\left(\|\underline{v}\|_2\right)^{-1}}{\mathrm{d}v_j} = \frac{\mathrm{d}\left(\sum\limits_{i=1}^{p_z} v_i^2\right)^{-1/2}}{\mathrm{d}v_j} = -\frac{1}{\not{2}}\left(\sum_{i=1}^{p_z} v_i^2\right)^{-3/2} \cdot \not{2} \cdot v_j$$
$$= \frac{-v_j}{\left(\|\underline{v}\|_2\right)^3} \,. \tag{15.14}$$

In the following, it is assumed that v_j is the last entry in the vector $\underline{v}$, i.e., $j = p_z$. This assumption simplifies the handling of summation indices without the loss of generality. With the help of (15.13), (15.14), and the chain rule, the partial derivative with respect to splitting parameter v_j equals

$$\frac{\partial J_p}{\partial v_j} = \frac{\lambda}{\sqrt{p_z} - 1}\left[\left(\sum_{i=1}^{p_z-1} \tanh\left(\frac{\alpha v_i}{2\|\underline{v}\|_2}\right)\frac{v_i \cdot (-v_j)}{\|\underline{v}\|_2^3}\right) + \ldots\right.$$
$$\left.\ldots + \tanh\left(\frac{\alpha v_j}{2\|\underline{v}\|_2}\right)\left(\frac{1}{\|\underline{v}\|_2} + \frac{v_j \cdot (-v_j)}{\|\underline{v}\|_2^3}\right)\right]$$
$$= \frac{\lambda}{\sqrt{p_z} - 1}\left[-v_j\left(\sum_{i=1}^{p_z} \tanh\left(\frac{\alpha v_i}{2\|\underline{v}\|_2}\right)\frac{v_i}{\|\underline{v}\|_2^3}\right) + \tanh\left(\frac{\alpha v_j}{2\|\underline{v}\|_2}\right)\frac{1}{\|\underline{v}\|_2}\right] . \tag{15.15}$$

The proposed penalization might get into trouble if all parameters get close to zero. In practice this is of minor concern since it corresponds to a split which would have an infinite distance to the origin; see Fig. 15.16b. If such a split would result from the optimization, it simply could be omitted.

As already mentioned, the main goal is to get a low-dimensional $\underline{z}$-input space and eliminate all unnecessary variables therein. A shrinkage of the remaining $\underline{z}$-inputs is not desired. Therefore the penalization is only used to determine the unnecessary $\underline{z}$-inputs. Afterward an unregularized split optimization is done with all left-over $\underline{z}$-inputs. How many splitting variables are kept depends on the size of the regularization parameter λ. Therefore λ can be used to smoothly transform the axis-oblique ($\lambda \to 0$) to an axis-orthogonal ($\lambda \to \infty$) partitioning strategy.

An important task is to find a regularization parameter λ that leads to a good bias/variance tradeoff. Throughout the training of an LMN, several values for λ are needed since in each iteration of HILOMOT a split optimization is carried out. For each of these optimizations, a different value for λ might be optimal in terms of the bias/variance tradeoff. Two different approaches for the determination of λ values are investigated and compared in more detail.

The first approach follows a simple heuristic. Typically, the model performance saturates with an increasing amount of local models because each additional split acts on a smaller subarea of the $\underline{z}$-input space. To account for this saturation, an initial regularization parameter λ is weighted with the normalized *effective* number of data samples for the current split:

$$\lambda_{split} = \frac{\lambda}{N} \cdot \sum_{i=1}^{N} \Phi_{worstLM}(\underline{z}(i)) \,. \tag{15.16}$$

The effective number of data samples is the number of points contained in the local model that is currently subdivided by the split. This local model is denoted as the worst local model (index *worstLM*), since its local error measure is worst compared to all other local models. The number of points in the worst local model can be calculated by the summation of the validity values $\Phi_{worstLM}$ of all N training data points. Through the normalization with N (15.16) measures the fraction of all data that is inside the local model to be split, and it is guaranteed that $\lambda_{split} = \lambda$ in the first split. Then, for each following split, the regularization of the current split is decreased according to the number of points that are affected by this split. For the initial regularization parameter λ, several values have to be tested. For each of these regularization parameters, an LMN is trained with activated split regularization using the heuristic adaption of λ (15.16) for each split. The resulting LMNs can then be compared in terms of the achieved model quality revealing the best of the tested regularization parameters.

The second approach for the determination of the regularization parameter optimizes λ for each split. This is computationally more expensive than the first approach but has obviously the potential to outperform it. The objective of the optimization is an approximation of the leave-one-out (LOO) error that has to be minimized. For the approximative LOO error, the partitioning is assumed to be known and fixed. This simplification helps to decrease the computational load tremendously since the problem is a linear estimation problem once the partitioning is known. In fact, the LOO error can be calculated directly based on one least squares solution with the help of the smoothing matrix $\underline{S}$ as shown in [486]. This matrix describes the direct relationship between the measured process outputs $\underline{y}$ and the model outputs $\underline{\hat{y}}$:

$$\underline{\hat{y}} = \underbrace{\underline{X} \overbrace{\left(\underline{X}^T \underline{X}\right)^{-1} \underline{X}^T \underline{y}}^{\underline{\theta}_{LS}}}_{\underline{S}} \,, \tag{15.17}$$

with the regression matrix $\underline{X}$ and the model parameters $\underline{\theta}_{LS}$ obtained by least squares. Hartmann [221] derived an equation to directly calculate the LOO error for LMNs under the assumption of a fixed partitioning if the parameters of the local models are estimated by a locally weighted least squares estimation scheme:

$$\text{LOOE} = \sum_{j=1}^{N} \frac{y(j) - \hat{y}(j)}{1 - \sum_{k=1}^{M} s_{jj,k}} . \tag{15.18}$$

N denotes the number of training data samples, and M is the number of local models. $s_{jj,k}$ is the entry in row and column j (diagonal entries) of the smoothing matrix $\underline{S}$ belonging to the parameter estimation of local model k. For more details about the derivation, the reader is referred to [221]. The approximative LOO error is chosen as an objective function for the optimization of the regularization parameter λ because it measures the generalization performance of the LMN while the computational costs are tolerable.

15.3.2 Investigation with Test Processes

Test processes one and four are used to demonstrate the abilities and shortcomings of the regularization-based input selection (RBIS) approach presented in the previous section. Both ways to determine the regularization parameter λ are tested, i.e., the heuristic from (15.16) and the optimization with respect to the approximation of the LOO error. Since the RBIS approach affects only the $\underline{z}$-input space, the results ideally reveal which inputs influence the output in a nonlinear way.

For the investigation of both test processes, the settings of the identification algorithm are identical. HILOMOT is used, but with the added penalty term for the split optimization; see (15.10). The smooth L1 approximation method with $\alpha = 650$ is used, such that an analytic gradient of the objective function can be provided, see (15.15). For the determination of the model complexity, Akaike's corrected information criterion is used. Only affine local models are used. The usage of higher-order, locally valid polynomials would be counterproductive to the intended goal of the investigation. It should reveal nonlinearly influencing inputs. Having more complex local models could make splits along specific axes obsolete even if there is a nonlinear influence. For the heuristic determination of the split regularization parameter λ, four different initial values are tested which are 10^{-4}, 10^{-3}, 10^{-2}, and 10^{-1}. In addition to the RBIS approaches, LMNs are trained with the standard HILOMOT algorithm (no split regularization or $\lambda = 0$) and HILOMOT with deactivated split optimization (LOLIMOT-type of splitting). The deactivation of the split optimization results in an input space partitioning with only axis-orthogonal splits. Note that this is *not* equivalent to the RBIS approach with $\lambda = \infty$. In case the split regularization parameter is set to infinity, the split position can still be optimized. In contrast to this, HILOMOT with deactivated split optimization does not optimize the split position and simply splits the worst local model into two halves that both contain approximately the same amount of data samples.

15.3.2.1 Test Process One

TP1 is static and therefore no dynamics have to be represented. The first and second input (u_1 and u_2) contributes nonlinearly to the process output, while input u_3 influences it only linearly. Input u_4 consists only of noise. Results obtained by the RBIS approach should indicate inputs u_1 and u_2 as nonlinear. Note that the results of the RBIS approaches contain no information about the input's usefulness for the $\underline{x}$-input space.

For the investigation of the RBIS approaches, 100 different training data sets are generated that only differ in the noise realization of the output values. The design of experiments is a maximin optimized LH design with $N = 100$ samples. Since input u_4 consists only of noise, no additional noise is added to the outputs. One Sobol sequence serves as a test data set with $N_t = 10^5$ samples, which is used to assess the quality of the resulting models. Note that for the generation of the test data set, input u_4 is neglected in order to create noise-free samples.

Because of the way the HILOMOT algorithm works, the best initial split does not necessarily lead to the best overall model performance when the training of the LMN is finished. In each iteration, the worst local model is further split, and only this new split is optimized; see Sect. 14.8 for a detailed description of HILOMOT. As a result, the current split is optimal with respect to the existing partitioning, not incorporating any further changes in it. This is the reason why the value of already existing splits to the overall model performance can change during the training procedure. Having this in mind, it makes sense to look at the generalization performances with LMNs having just two local models in order to compare all RBIS approaches directly. In addition, the benefit of all distinct RBIS approaches for the generalization performance is of interest after the optimal model complexities are chosen (most likely leading to LMNs having more than just two local models).

Figure 15.20a shows boxplots of the achieved generalization performances of all investigated scenarios for LMNs with just two local models. Each boxplot contains the achieved generalization performances of the 100 different noise realizations. The red lines mark the median values for each method. The best of these median NRMSE values on test data is achieved with the heuristic and an initial value of the split regularization parameter of $\lambda = 10^{-4}$. The generalization performance achieved with the standard HILOMOT algorithm is only outperformed by the heuristic with initial split regularization parameters $\lambda = 10^{-4}$ and $\lambda = 10^{-3}$. From all RBIS approaches, the optimization of the split regularization parameter turns out to be the second worst, performing only better than the heuristic with an initial split regularization parameter of $\lambda = 0.1$. The median of all optimized split regularization parameters is $\lambda_{split} = 0.09$ (not shown in the figure). The worst median NRMSE value on test data is achieved by HILOMOT with deactivated split optimization, abbreviated with *orth.* HILOMOT since all splits are axis-orthogonal. The lowest variations in the achieved generalization performances occur for the orthogonal HILOMOT algorithm. The variations in the model performance are also very low for the heuristic with $\lambda = 0.1$. All other investigated scenarios are comparable in terms of their variations in the generalization performance.

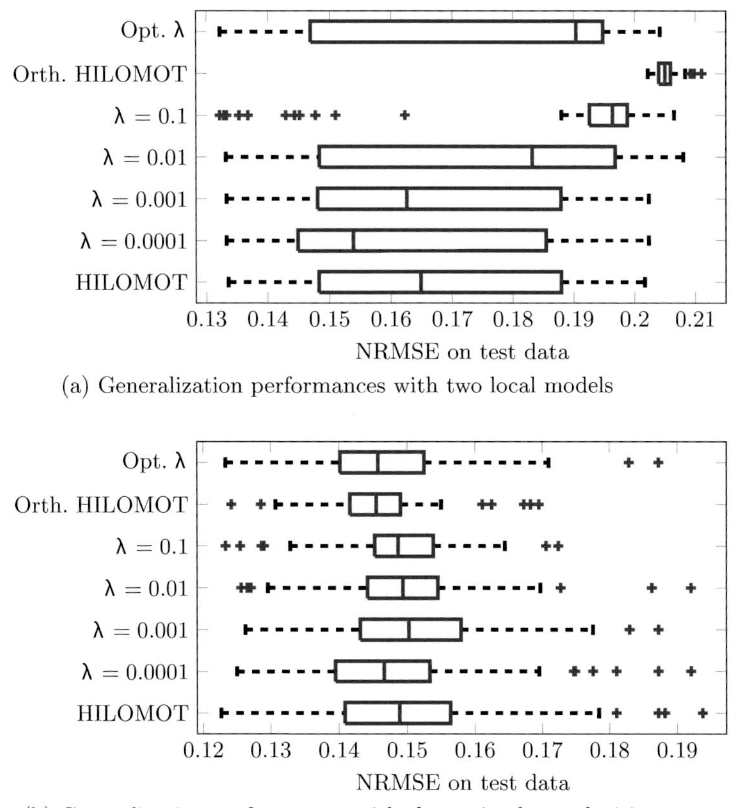

(a) Generalization performances with two local models

(b) Generalization performances with the optimal complexities

Fig. 15.20 Histograms of the generalization performances of all investigated scenarios for two local models (**a**) and the optimal model complexity according to the AIC_C (**b**)

Figure 15.20b shows boxplots of the achieved generalization performances where the model complexity of each LMN is set to its optimal value according to the AIC_C.

All median NRMSE values on test data are very close to each other. The best values are achieved with the optimized split regularization parameter, the heuristic with $\lambda = 10^{-4}$, and the orthogonal HILOMOT algorithm. The variations in the generalization performances are the lowest for the orthogonal HILOMOT algorithm, followed by the heuristic with $\lambda = 0.1$ before the heuristic with $\lambda = 0.01$. Variations of the remaining scenarios are on a similar level.

Figure 15.21 shows the convergence of the median test errors over the number of local models for all investigated scenarios. The heuristic with $\lambda = 10^{-4}$ converges the fastest, whereas the orthogonal HILOMOT algorithm has the slowest convergence rate.

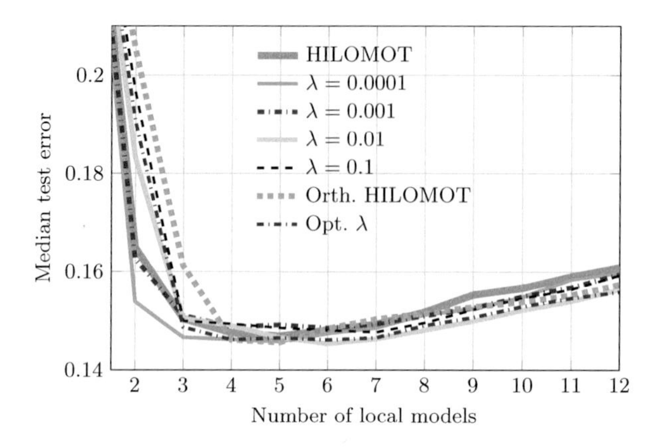

Fig. 15.21 Median NRMSE values on test data over the number of local models

Besides the generalization performance and the corresponding variations, it is interesting to see the effect of the RBIS approaches on the splitting parameters. Therefore the median splitting parameter values are shown for the first three splits in Fig. 15.22 for HILOMOT, the heuristic with $\lambda = 10^{-4}$ as well as $\lambda = 10^{-2}$, and the optimized λ values. Here, the median of the *absolute* splitting parameter values $\widetilde{v}_i$ is shown, such that a value of zero implies that for at least 50 of the 100 noise realizations, the corresponding splitting parameter is zero. The information about how many splitting parameters are actually zero is important since this corresponds to orthogonal splits. Without taking the absolute values, a zero median value would also result if the corresponding splitting parameter values would be normally distributed around zero. As a result of this, no conclusion about how many of the 100 noise realizations lead to orthogonal splits would be possible. The desired effect of all RBIS approaches is to push the splitting parameters corresponding to the third and fourth $\underline{z}$-input toward zero. These inputs have no nonlinear influence on the process output, and splits in these directions are considered to be harmful to the generalization performance. Note that the binary trees shown in Fig. 15.22 are just for visualization purposes. Not all of the binary trees resulting from the 100 noise realizations have the exact same structure, e.g., split three might as well split node four or five further instead of node two.

Only small differences are observable when comparing the median absolute splitting parameter values of HILOMOT and the heuristic with $\lambda = 10^{-4}$ shown in Fig. 15.22a, b, respectively. The desired effect of pushing v_3 and v_4 toward zero is only achieved if the initial split regularization parameter of the heuristic is raised to $\lambda = 10^{-2}$ and above. This desired effect is also achieved if the split regularization parameter λ is optimized; see Fig. 15.22d. The question arises why achieving the desired effect does not lead to higher model qualities and a significantly lower variance.

The variance in the model accuracies is likely not as significantly decreased as expected by the RBIS approaches due to the fact that there is still much variance

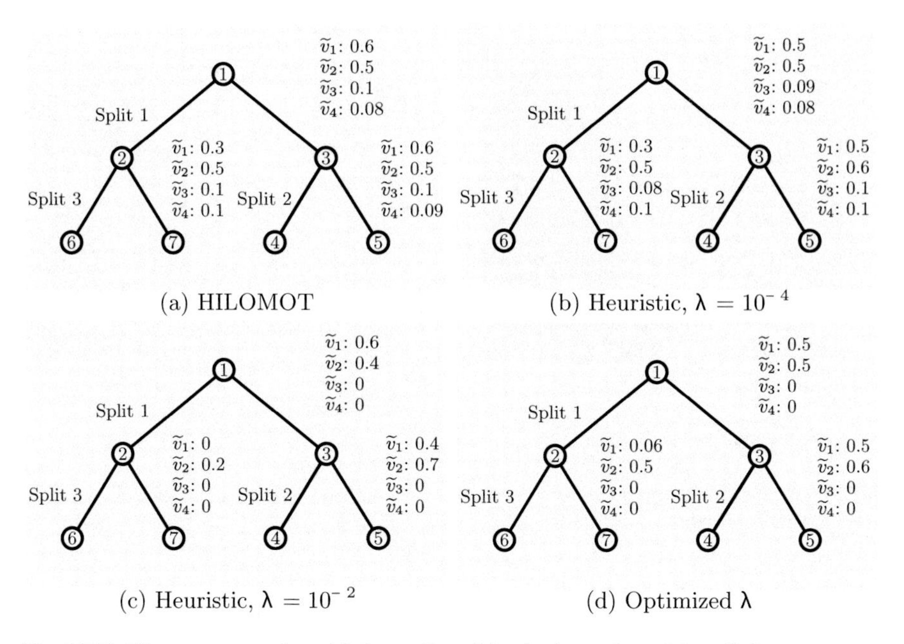

Fig. 15.22 Binary trees together with the median of the absolute values of the splitting parameters $\widetilde{v}_i$ for HILOMOT (**a**), the heuristic with $\lambda = 10^{-4}$ (**b**) as well as $\lambda = 10^{-2}$ (**c**), and the optimized λ values (**d**) for splits one to three

incorporated in the split position. Figure 15.23a shows the first split for all 100 noise realizations in the u_2–u_3 subspace for the standard HILOMOT algorithm opposed to the ones obtained by the RBIS approach with optimized λ values depicted in Fig. 15.23b. Even though most splits are in fact orthogonal, there are a lot of different splits (varying positions) and also oblique splits.

A likely cause for the insignificantly increased generalization performances is visualized in Fig. 15.24. Histograms of the values of all four splitting parameters v_1, v_2, v_3, and v_4 are shown for the first split and all 100 noise realizations for HILOMOT and the RBIS approach with optimized λ. If a splitting parameter is zero, the split is parallel to the corresponding $\underline{z}$-input axis as explained in Sect. 14.8.2 and visualized in Fig. 15.17a. From an interpretation point of view, this means the corresponding $\underline{z}$-input is irrelevant since its value has no influence on the splitting function. As can be seen in Fig. 15.23, the number of splits that are parallel to one of the $\underline{z}$-input axes is increased in general through the RBIS approach with optimized λ compared to the standard HILOMOT algorithm. In the case of $\underline{z}$-inputs u_3 and u_4, this is desired and should improve the generalization performance of the models. However, the number of splits parallel to the $\underline{z}$-input axes u_1 and u_2 is also increased, which harms the generalization performance. Considering the interaction between u_1 and u_2 in TP1, the partitioning should ideally be oblique in the u_1–u_2 subspace.

Finally, the required training times of all investigated scenarios for TP1 are compared in Fig. 15.25. As expected, the smallest required training times are needed

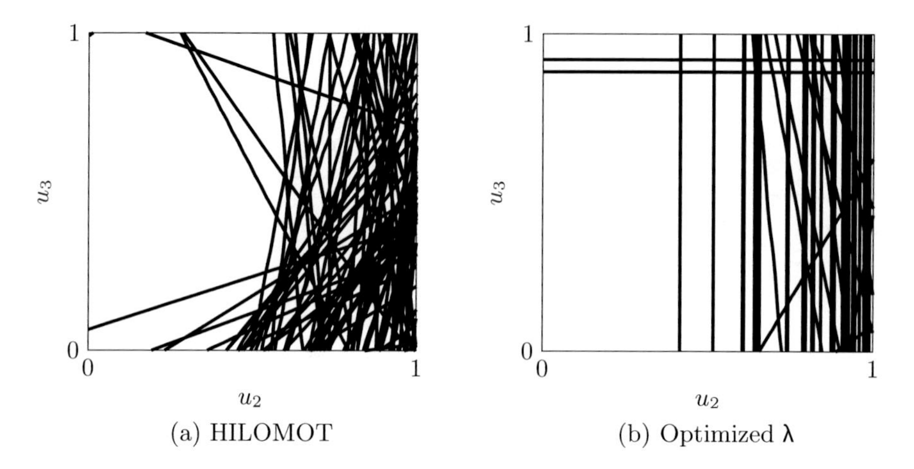

Fig. 15.23 Input space partitionings of LMNs with two local models for all noise realizations in case of HILOMOT (**a**) and the RBIS approach with optimized λ (**b**) in the u_2-u_3 subspace

by the orthogonal HILOMOT algorithm, and the largest ones are needed if the split regularization parameter is optimized. The higher the initial split regularization parameter chosen for the heuristic determination method, the smaller the required training times are.

Despite the fact that the convergence rate of the heuristic with $\lambda = 10^{-4}$ is quite impressive, a significant advantage of using any of the RBIS approaches is not observable in the case of the optimal model complexities. The median generalization performance as well as the deviations of it is only slightly improved and does not pay off the additional effort. It seems as if the introduced obliqueness penalty prevents an oblique input space partitioning in a lot of cases even if it is beneficial for the model's generalization performance.

15.3.2.2 Test Process Four

The dynamic TP4 is modeled with the external dynamics approach as explained in Sect. 19.2. Here, only simple delays q^{-1} are used as external filters, and only one filter for the physical input as well as for the physical output is used, such that the inputs for the LMNs are $u(k-1)$ and $y(k-1)$. Using only simple delays for the external dynamics approach is known as the NARX model. From the description of TP4, it is known that the nonlinearity only affects the delayed input $u(k-1)$ and not the delayed output $y(k-1)$. Therefore, the RBIS approaches should be able to detect input $u(k-1)$ as important for the $\underline{z}$-input space, while the splitting parameters corresponding to the delayed output $y(k-1)$ should be pushed to zero.

For the investigation of the RBIS approaches with TP4, 100 different training data sets are generated that only differ in the noise realization of the output values. The excitation signal used to obtain the training data sets is an amplitude modulated random binary signal (APRBS) consisting of $N = 250$ data samples. The maximum

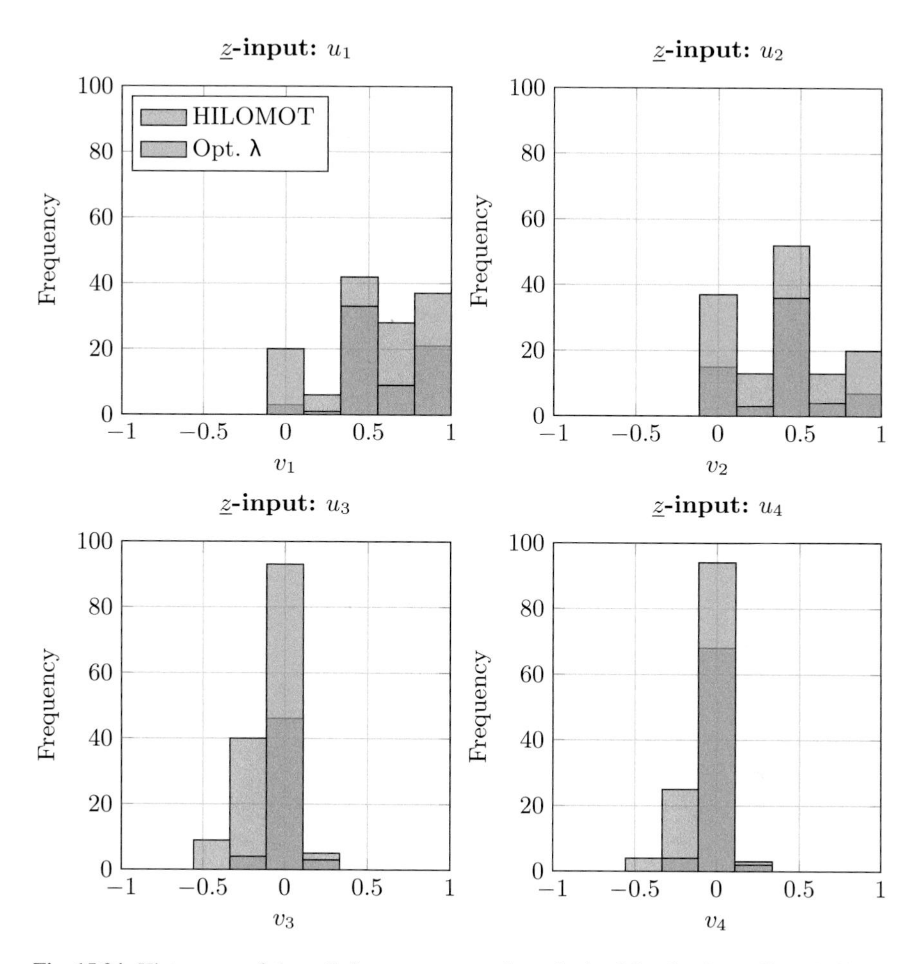

Fig. 15.24 Histograms of the splitting parameter values obtained for the first split and all 100 noise realization of TP1 in the case of HILOMOT and the RBIS approach with optimized λ

and minimum amplitude of the excitation signal is -3 and 3, respectively, in order to create sufficiently strong saturation effects. White Gaussian noise is added to the undisturbed process output to create the 100 noise realizations such that the SNR is 20 dB. The test data set consists of another APRBS input signal with length $N_t = 12{,}000$ and an undisturbed output.

Again, the achieved generalization performances are for LMNs with just two local models and for LMNs with the optimal number of local models which are shown in Fig. 15.26a, b, respectively. In the case of LMNs with two local models, the median NRMSE values are all pretty similar except for the orthogonal HILOMOT algorithm. The best values are achieved by the RBIS approaches with optimized λ as well as with the heuristic with $\lambda = 0.01$ and $\lambda = 0.1$. The worst median NRMSE value corresponds to the orthogonal HILOMOT. The variations from the

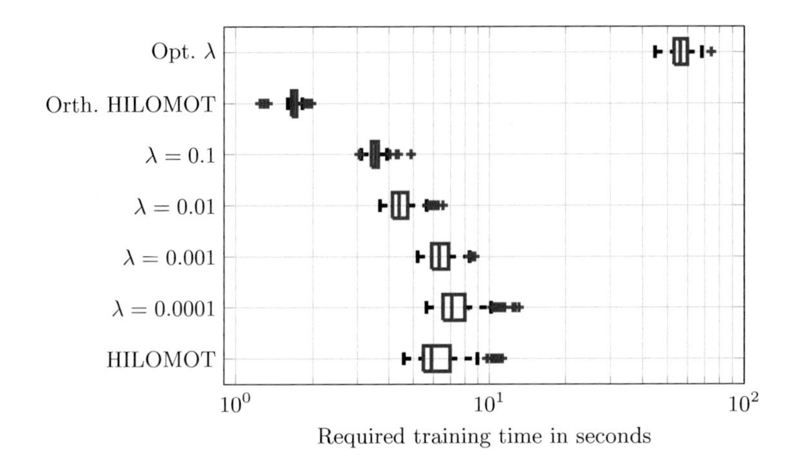

Fig. 15.25 Boxplots of the required training times for all investigated scenarios of TP1

median NRMSE values are very similar for all investigated scenarios with slightly less magnitude in thecase of the RBIS approaches with optimized λ and the heuristic with $\lambda = 0.01$ and $\lambda = 0.1$. For the LMNs with optimal model complexity, the NRMSE values are again all pretty similar except for the orthogonal HILOMOT algorithm. The best NRMSE values as well as the smallest variations from it are achieved by the RBIS approaches with optimized λ and the heuristic with $\lambda = 0.1$.

A bit surprising is the high variance of the LMNs obtained by the orthogonal HILOMOT algorithm. It results from a wider range of different optimal model complexities that are chosen by the AIC_C compared to all other training algorithms.

The inspection of the median absolute splitting parameter values reveals that the RBIS approaches with heuristic need at least an initial split regularization parameter of $\lambda = 0.01$ to push v_2 in the majority of all noise realizations to zero. The RBIS approach with optimized λ is also able to realize this desired outcome. Figure 15.27 shows exemplarily the binary trees of HILOMOT and the heuristic with $\lambda = 0.01$ together with the median absolute splitting parameter values over all noise realizations for the first three splits. Even without any regularization on the splitting parameters, the values of v_2 are very close to zero.

Plots showing the model quality over the model complexity, the input space partitioning, and histograms of the splitting parameters over all noise realizations are omitted here because they do not reveal any interesting information. Finally, the required training times are compared in Fig. 15.28. No surprises can be observed. The smallest training times are required by the orthogonal HILOMOT algorithm, while the most time-consuming variant is the RBIS approach with optimized λ. The heuristic RBIS approaches are all on a similar level independent of the used initial split regularization parameter. The required training times by HILOMOT lie in between the heuristic RBIS approaches and the orthogonal HILOMOT algorithm.

In summary, none of the RBIS approaches show convincing results. Neither the generalization performance of the LMNs nor the variance could be significantly

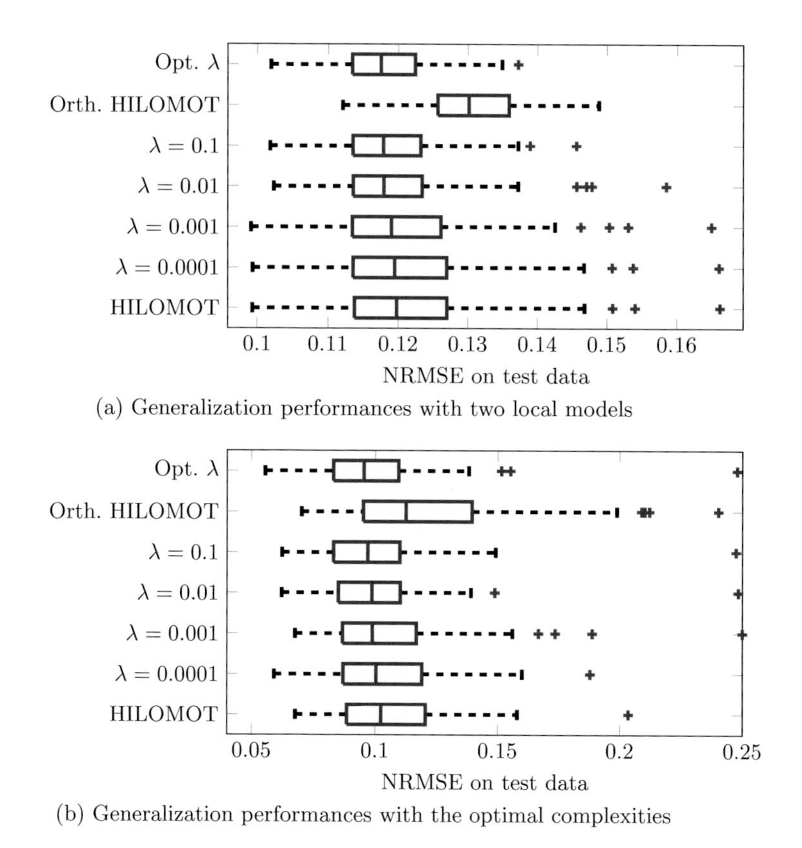

Fig. 15.26 Histograms of the generalization performances of all investigated scenarios for two local models (**a**) and the optimal model complexity according to the AIC_C

improved independently of the test process under investigation. As shown for TP1, no good compromise could be found for the split regularization parameter that is able to keep splits parallel to unimportant $\underline{z}$-inputs while being able to reliably produce oblique splits in the subspace where they are needed. In the case of TP4, no oblique splits are required at all because only input $u(k-1)$ influences the output in a nonlinear way. Even in this scenario, none of the RBIS approaches show significant advantages compared to the standard HILOMOT algorithm. Based on all obtained results for TP1 and TP4, the RBIS approach cannot be recommended to be used. However, the investigated scenarios are low-dimensional. For high-dimensional problems, as they would arise, e.g., in a nonlinear finite impulse response (FIR) modeling case, this kind of regularization may pay off.

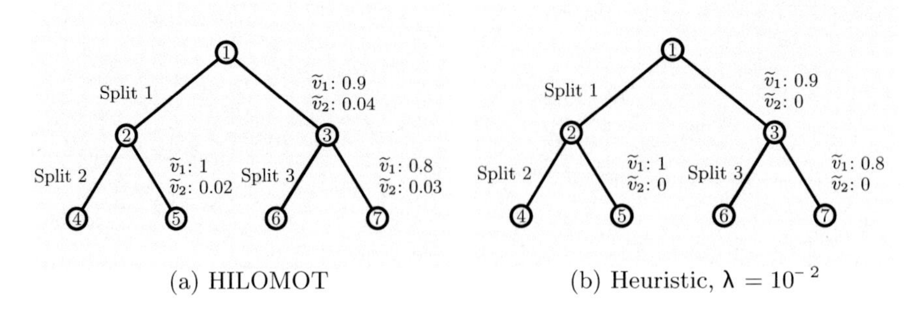

Fig. 15.27 Binary trees together with the median of the absolute values of the splitting parameters $\widetilde{v}_i$ for HILOMOT (**a**) and the heuristic with $\lambda = 10^{-4}$ (**b**) for splits one to three

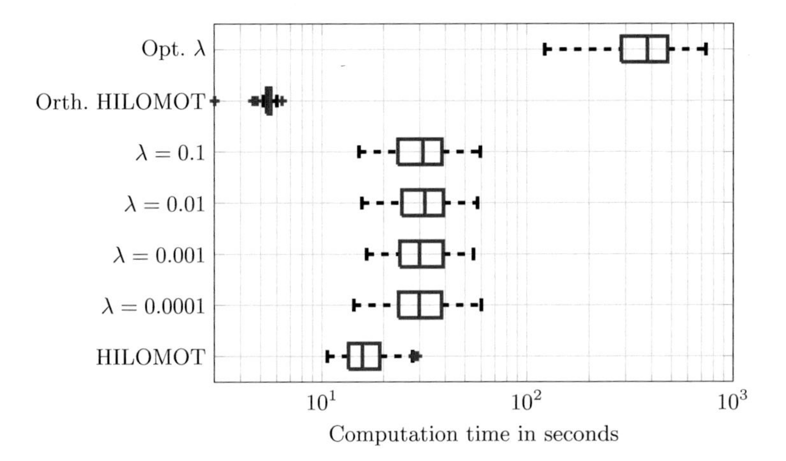

Fig. 15.28 Boxplots of the required training times for all investigated scenarios of TP4

15.4 Embedded Approach: Authored by Julian Belz

Embedded methods use model-specific or training-algorithm-specific properties to find good input subsets for the modeling task. In the following, special properties of LMNs with local affine models trained with HILOMOT are exploited. In this particular case, the directions of the splits represent the strength and main direction of the process nonlinearity. Since all local models are affine, changes in the slope of the process can only be described by the LMN by switching to another local affine model. If one input influences the process only linearly, all splits performed by HILOMOT will be approximately parallel to this dimension. The basic idea of the embedded approach is to exploit the aforementioned behavior of HILOMOT by analyzing the splitting parameters and evaluate the directions of all splits that are made. In the end, the importance of each input compared to all other dimensions can be rated. Note that importance here only refers to an input's nonlinear influence. If the output of the process depends strongly on one input, but only in a linear way, this can be described by a large slope in the local model, and it will be considered irrelevant by this embedded approach.

15.4.1 Partition Analysis

At first an LMN is trained and afterward the partitioning is analyzed. As explained in detail in Sect. 14.8.2, the orientation of each split is determined by the relative, absolute magnitude of the splitting parameters v_i, $i = 1, 2, \dots, p_z$; see Figs. 15.16 and 15.17. The offset parameter v_0 does not influence the split orientation and therefore plays no role in the subsequent discussion. First only **one single split** is considered, and the relevance factor ρ_i for the i-th $\underline{z}$-input equals

$$\rho_i = \frac{|v_i|}{\sum_{j=1}^{p_z} |v_j|} . \tag{15.19}$$

For the special cases shown in Fig. 15.17a–c, the relevance factors turn out to be $\underline{\rho}_a = \left[0\ 1\right]^T$, $\underline{\rho}_b = \left[0.5\ 0.5\right]^T$, and $\underline{\rho}_c = \left[1\ 0\right]^T$, respectively.

For the determination of the relevance factors for **the whole LMN**, the so-called importance factors are calculated. After normalizing these importance factors, the relevance factors are obtained. For the determination of the importance factors ζ_i of each input in the $\underline{z}$-input space z_i, $i = 1, 2, \dots, p_z$, the related splitting parameters of all splits are weighted and summed up:

$$\zeta_i = \sum_{j=1}^{M-1} |v_{ij}| \cdot PI_j \cdot w_{Nj} . \tag{15.20}$$

v_{ij} designates the splitting parameter of the j-th split belonging to z_i. Weighting factor PI_j is the performance improvement in the loss function achieved through split j. It is the difference between the loss function value before and after split j is carried out. Weighting factor w_{Nj} equals the number of points that are affected by split j. It is the number of points within the worst local model and can be measured by

$$w_{Nj} = \sum_{k=1}^{N} \Phi_j(\underline{z}(k)) , \tag{15.21}$$

due to the fuzzy nature of the splits. Note that an LMN with M local models possesses $M - 1$ splits. Through the two weighting factors PI_j and w_{Nj}, it is guaranteed that splitting parameter v_{ij} does only contribute to the importance factor if the corresponding split contributes to the improvement of the model performance

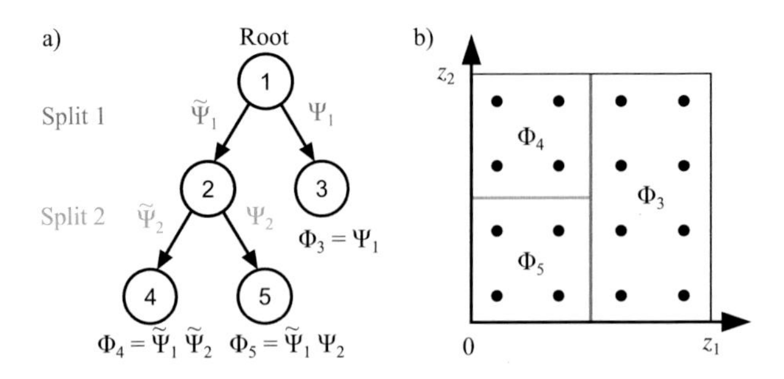

Fig. 15.29 Illustration example for the relevance factor calculation: (**a**) shows the hierarchical binary tree with highlighted splits, (**b**) shows the partitioning of the input space with 16 data points as dots

and enough data points are affected. Finally, the relevance factor for the whole LMN is obtained through the normalization of all importance factors:

$$\rho_i = \frac{\zeta_i}{\sum\limits_{j=1}^{p_z} |\zeta_j|} \,. \tag{15.22}$$

A simple example that is shown in Fig. 15.29 should further illustrate the concept of the relevance factors. The partitioning consists of two axis-orthogonal splits. For simplicity the performance improvement factors PI_j, $j = 1, 2, \ldots, M-1$, are all assumed to be one in this example. Note that usually the performance improvement is smaller than one and approaches zero with an increasing number of local models (if the loss function values saturate).

The first split possesses the splitting parameters $\underline{v}_1 = \begin{bmatrix} 1 & 0 \end{bmatrix}$ and divides 16 data points in two halves such that the weighting factor becomes $w_{N1} = 16$. The second split possesses the splitting parameters $\underline{v}_2 = \begin{bmatrix} 0 & 1 \end{bmatrix}$ and divides eight data points in two halves leading to a weighting factor of $w_{N2} = 8$. Since the performance improvement factors are assumed to be one, the importance factors are

$$\zeta_1 = v_{11} \cdot w_{N1} + v_{12} \cdot w_{N2} = 16 \text{ and}$$

$$\zeta_2 = v_{21} \cdot w_{N1} + v_{22} \cdot w_{N2} = 8 \,.$$

The final relevance factors for this example turn out to be

$$\rho_1 = \frac{\zeta_1}{\zeta_1 + \zeta_2} = \frac{2}{3} \text{ and}$$

$$\rho_2 = \frac{\zeta_2}{\zeta_1 + \zeta_2} = \frac{1}{3} \,.$$

The result of the LMN partition analysis is a ranking of all inputs according to their relevance for the $\underline{z}$-input space. In addition to the ranking, the quantitative relevance is revealed. However, this information is not further used here. One idea exists that has not been investigated so far but should be mentioned here. In tasks where distance measures are employed, e.g., to determine the largest gap in an already existing data set, the relevance factors might be used to scale the $\underline{z}$-input axes. Therefore, a weighting matrix

$$\underline{W} = \begin{bmatrix} \rho_1 & 0 & \cdots & 0 \\ 0 & \rho_2 & \ddots & \vdots \\ \vdots & \ddots & \ddots & 0 \\ 0 & \cdots & 0 & \rho_{p_z} \end{bmatrix} \quad (15.23)$$

is created with the relevance factors ρ_i, $i = 1, 2, \ldots, p_z$, as entries on its diagonal. This matrix can then be used to calculate the Mahalanobis distance between two points, $\underline{z}(i)$ and $\underline{z}(j)$:

$$d_M(\underline{z}(i), \underline{z}(j)) = \sqrt{(\underline{z}(i) - \underline{z}(j))^T \underline{W} (\underline{z}(i) - \underline{z}(j))} \,. \quad (15.24)$$

If the identity matrix $\underline{I}$ is used instead of the weighting matrix $\underline{W}$ in (15.24), the Mahalanobis distance reduces to the Euclidean distance. Using the weighting matrix $\underline{W}$ makes distances along more relevant $\underline{z}$-inputs appear larger. The larger a relevance factor of a $\underline{z}$-input is compared to the remaining $\underline{z}$-inputs, the more the focus lies on this particular $\underline{z}$-input. This might be beneficial for active learning strategies such as HILOMOT for design of experiments (HilomotDoE), which is described in more detail in Sect. 14.10.4. Through the focus on $\underline{z}$-inputs that are more relevant for the nonlinear behavior of a process, the design of experiments is potentially able to concentrate on the nonlinearity. As a result, the training data carries more information about the nonlinearity which can subsequently be exploited by experimental modeling approaches.

15.4.2 Investigation with Test Processes

TP2 and TP3 are used to demonstrate the abilities of the partition analysis explained in Sect. 15.4.1. For both test processes, three different training data set sizes, namely, $N = 50$, 100, and 200, and two different noise levels are part of the investigation. White Gaussian noise is added to the process outputs such that the SNR equals 20 dB and 25 dB. For each test process and each training data set size, 100 different noise realizations are generated. Here, neither validation nor test data is produced because the investigated LMN partition analysis does not affect the model accuracy. An LMN for each generated data set is trained, and subsequently, its partitioning is analyzed. HILOMOT is used as training algorithm, and the complexity of the LMNs is determined with the AIC_C. Only local affine models are used, such that the resulting relevance factors represent how much each $\underline{z}$-input contributes to the nonlinearity of the process. The evaluation of the results of the partition analysis is based exclusively on the knowledge about the two test processes.

15.4.2.1 Test Process Three

First, the split analysis is used for TP3, which has only two inputs. A priori it is known that both inputs contribute to the process output in a nonlinear way. However, input u_1 can be rated as more important for the nonlinear process behavior because the change in the slope in this direction is higher compared to the change in the slope in u_2-direction everywhere in the input space. In fact, the ideal values of the relevance factors can be determined exactly from prior knowledge. The derivative of (15.4) with respect to the inputs u_1 and u_2 is

$$\nabla y(u_1, u_2) = \frac{0.1}{(0.08 + 0.73(1 - u_1) + 0.27(1 - u_2))^2} \left[0.73 \; 0.27\right]^T \tag{15.25}$$

As can be seen from (15.25), the gradient of TP3 points always in the same direction defined by the ratio of the vector entries 0.73 and 0.27. Since the relevance factors are scaled such that they all sum up to one (see (15.22)), the values 0.73 and 0.27 already correspond to the ideal relevance factors. If the partitioning parameters of an LMN follow exactly this particular ratio, the local affine models should be able to adapt to the nonlinearity most efficiently. Deviations from this ratio should only occur due to noisy data and are likely to deteriorate the generalization performance of LMNs.

Figure 15.30 shows the median values of the obtained relevance factors together with the ideal ones for TP3. Additionally, the deviations around the median values are visualized as interquartile range (IQR), which is the distance between the 25th and 75th percentile. For both noise levels, the obtained results are pretty similar. The median value of the relevance factors gets closer to the ideal values, and the deviations shrink as the amount of training data increases. This is exactly the desired and expected result.

15.4.2.2 Test Process Two

TP2 is scheduled by the binary input variables u_3 and u_4 that determine which of two planes is used to calculate the process output. Both planes depend only on the input variables u_1 and u_2. Therefore, u_1 and u_2 are only relevant for the $\underline{x}$-input space of an LMN, while u_3 and u_4 are only relevant for the $\underline{z}$-input space. In the case of TP2, no ideal values for the relevance factors are available. Only qualitative relevance is known a priori. Since the partition analysis does only rate the relevance for the

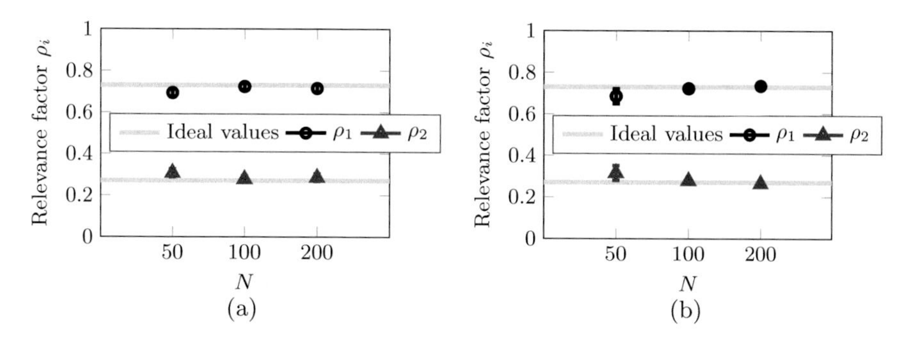

Fig. 15.30 Relevance factors ρ_i for all $\underline{z}$-inputs of TP3 for two different noise levels and three different training data set sizes $N = 50$, 100, and 200. Marker indicates the median value; vertical lines represent the IQR. (**a**) SNR = 20 dB. (**b**) SNR = 25 dB

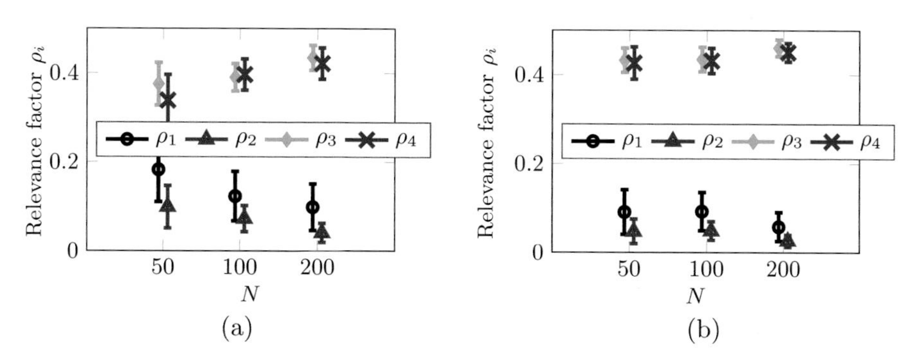

Fig. 15.31 Relevance factors ρ_i for all $\underline{z}$-inputs of TP2 for two different noise levels and three different training data set sizes $N = 50$, 100, and 200. Marker indicates the median value; vertical lines represent the IQR. (**a**) SNR = 20 dB (**b**) SNR = 25 dB

$\underline{z}$-input space, relevance factors ρ_1 and ρ_2 should have lower values compared to ρ_3 and ρ_4.

Figure 15.31 shows the results for TP2. Again, the median relevance factor, as well as a measure of the deviations from it, is shown. As expected, the deviations from the median values can be decreased either by increasing the size of the training data set or by lowering the noise level. In general, the median relevance factors for the actually irrelevant $\underline{z}$-inputs u_1 and u_2 get quite high for $N < 200$. The relative rating of the relevance of each $\underline{z}$-input is in accordance with the knowledge about TP2. However, by reviewing the obtained results, it is not clear that inputs u_1 and u_2 are irrelevant for the $\underline{z}$-input space. Furthermore, it is not clear why relevance factor ρ_1 has higher values compared to relevance factor ρ_2, since both corresponding inputs should be equally irrelevant.

In summary, it is shown that the LMN partition analysis is able to assign relevance factors that rank all $\underline{z}$-inputs according to their nonlinear influence correctly as long as there is an influence at all. The results obtained for TP3 are excellent. Even for relatively low training data set sizes, the a priori known ideal relevance factors are obtained through the LMN partition analysis with good accuracy. For TP2 it is recognized that inputs u_3 and u_4 are most relevant for the description of the nonlinearity. On the downside, it is not clear from which relevance factor value on a $\underline{z}$-input should be considered as irrelevant. As shown in Fig. 15.31a, a median relevance factor of almost 0.2 is obtained for input u_1. This is a rather high value from which one might conclude wrongly a significant nonlinear influence of u_1. The problem of having no specific threshold value from which on an input can be considered relevant for the $\underline{z}$-input space is currently unsolved.

15.5 Visualization: Partial Dependence Plots

Another idea to find relevant inputs for data-based models is the visualization of dependencies between the inputs and the output. Apparently, visualizations of data or models resulting from the data are restricted to low-dimensional views. For input space dimensions of one and two, the dependencies can easily be visualized with the help of a model. As the number of input variables increases, it gets harder to obtain this information. One popular way to deal with this problem is by examining *slices* through the model, i.e., fixing all inputs except for one or two. However, such slices can be very misleading since the effect of the fixed variables is completely concealed. Even worse, not feasible input combinations can artificially be generated. A much more powerful approach that can be applied to models of any dimension is the use of so-called partial dependence plots [230]. The idea is to view a collection of plots, where each of these plots visualizes the partial dependence of the model on a selected small subset of input variables. According to [230], such a collection can seldom provide a comprehensive depiction of the whole model, but it can lead to helpful clues. Partial dependence plots are scientifically not new but in the experience of the author seldom used and rather unknown, at least for people with an engineering background. Therefore, partial dependence plots are described in the next few paragraphs shortly following mainly the explanations from [230].

For simplicity, only the typical case is explained in the following, where the partial dependence of the model on just one single variable is considered. This input is called the partial dependence input (PDI). In order to generate a partial dependence plot, a function that estimates the partial dependence of the model on the PDI is needed and will be called the partial dependence function $\bar{f}_P$. This function represents the effect of the PDI on the model taking into account the average effects of all training data samples.

In order to determine the partial dependence function, some definitions are made in the following. If there are p different PDI variables u_j, there will be p partial dependence functions $\bar{f}_{P,j}$ with $j = 1, \cdots, p$. The vector $\underline{u}^{(-j)}(i)$ is the i-th training data sample without the PDI variable:

$$\underline{u}^{(-j)}(i) = \left[u_1(i)\ u_2(i)\ \cdots\ u_{j-1}(i)\ u_{j+1}(i)\ \cdots\ u_p(i)\right]^T . \tag{15.26}$$

Therefore $\hat{y}(u_j, \underline{u}^{(-j)}(i))$ is the model output at the i-th training data sample, but the value of the PDI can be chosen arbitrarily. With the help of these definitions, the j-th partial dependence function can be estimated by

$$\bar{f}_{P,j}(u_j) = \frac{1}{N} \sum_{i=1}^{N} \hat{y}(u_j, \underline{u}^{(-j)}(i)) \,. \tag{15.27}$$

At any specified value for the PDI, the model output $\hat{y}(\cdot)$ varies due to changes in the training data samples $\underline{u}^{(-j)}(i)$, $i = 1, 2, \ldots, N$. The mean of this variation is estimated by the partial dependence function. The estimation of the corresponding variance is straightforward:

$$\sigma^2_{P,j}(u_j) = \frac{1}{N-1} \sum_{i=1}^{N} (\hat{y}(u_j, \underline{u}^{(-j)}(i)) - \bar{f}_{P,j}(u_j))^2 \,. \tag{15.28}$$

In general there might be more than just one PDI variable; see [230] for further details. However, for visualization purposes more than two PDI variables are not very valuable. Typically, the PDI is varied equidistantly from its minimum to its maximum value occurring in the available data set. This results in a mean curve and its corresponding variances when applying (15.27) and (15.28) to each of these PDI values.

A model with two inputs is utilized to illustrate features of the partial dependence plots. As a demonstration example the following artificial process is approximated by an LMN:

$$y = u_1 + u_1 \cdot u_2^2 \,. \tag{15.29}$$

We define the first PDI variable as u_1 and the second one as u_2. So for each individual PDI variable, we obtain a partial dependence plot. The process as well as the partial dependence plots is shown in Fig. 15.32.

Both partial dependence plots contain important information about the dependencies of the process on the input variables. In Fig. 15.32b the linear influence of u_1 can be seen, and in Fig. 15.32c the quadratic characteristics of u_2 are observable. Larger standard deviations indicate stronger influences of variables not chosen as PDI. Therefore the growing standard deviation in Fig. 15.32b with increasing values of u_1 implies a growing influence of u_2. The vertical, dotted lines in Fig. 15.32b, c give hints about the data distribution along the PDI variable. In between two vertical lines lie 10 of the PDI values present in the training data set. The most left and the most right vertical lines are omitted and would occur at the beginning and the ending of the mean curves. In the example shown in Fig. 15.32, the data is distributed uniformly along both axes u_1 and u_2.

One important thing to keep in mind is that the whole visualization approach highly depends on the model of the process. Because rather the model than the

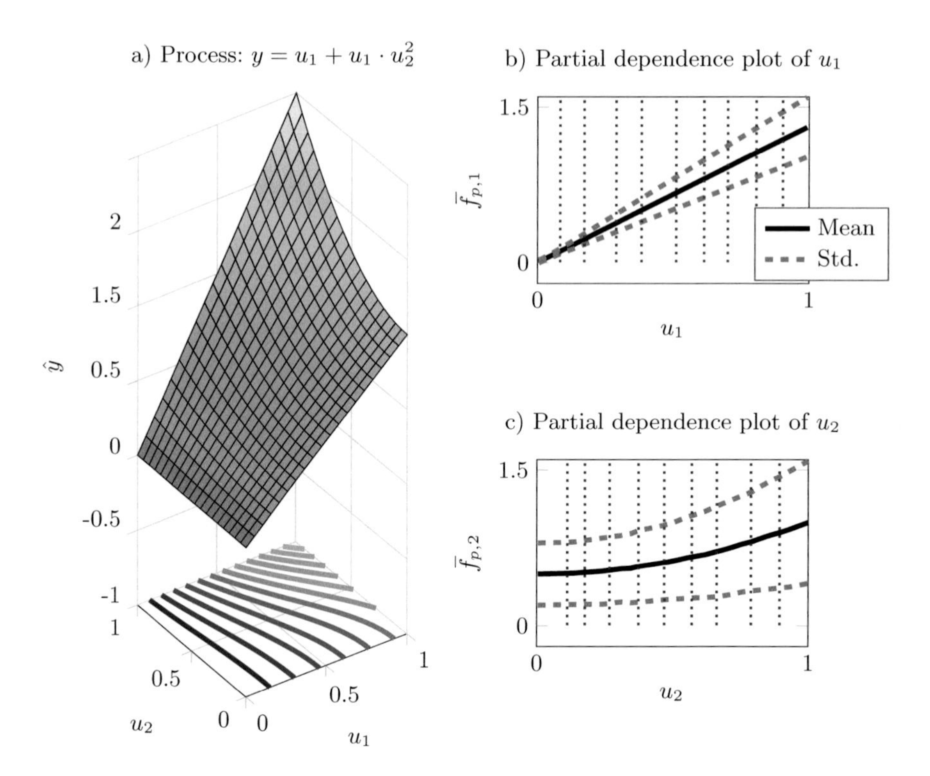

Fig. 15.32 Model with two inputs (**a**) and its partial dependence plots of u_1 (**b**) and u_2 (**c**)

process itself is visualized, the obtained results are only trustworthy if the model accuracy is high.

15.5.1 *Investigation with Test Processes*

TP1 and TP2 serve as examples to further review the capabilities of the partial dependence plots for the detection of relevant inputs. In the case of TP1, it is interesting to see if the irrelevance of input u_4 can be observed from the resulting partial dependence plots. For TP2 it is worth knowing what happens in the case of the discrete, binary inputs u_3 and u_4.

In order to create partial dependence plots, LMNs are trained with HILOMOT. Local affine models are used. The model complexity is automatically determined with the AIC_C criterion. Because both test processes have four inputs, the used training data sets share the same properties except for the added noise to the output. They consist of $N = 100$ training data samples and are obtained through a maximin LH optimization. In the case of TP1, no additional noise is added to the output because input u_4 exclusively consists of noise. For TP2 white Gaussian noise is added such that the SNR equals 20 dB. For both test processes, 100 different noise realizations are generated. The partial dependence plots shown in the following belong to just one of the 100 noise realizations. The shown plots are the typical outcome.

15.5.1.1 Test Process One

Figure 15.33 shows the partial dependence plots for all inputs of TP1. The nonlinear dependency of the output from inputs u_1 and u_2 is observable as well as the linear relationship between u_3 and the output. Furthermore, it can be recognized that inputs u_1 and u_2 influence the output in a quite similar way. The partial dependence plot of input u_4 indicates that it does not influence the output at all. The slope of the mean curve is zero independent of the value of u_4. Or in other words, the value of u_4 does not have a significant impact on the output value. These observations are in perfect accordance to the knowledge about TP1; see (15.2). Looking at the standard deviations, it is noticeable that these are almost constant for each partial dependence plot independent of the value of the PDI. This indicates that the influence of a PDI does not depend on its value.

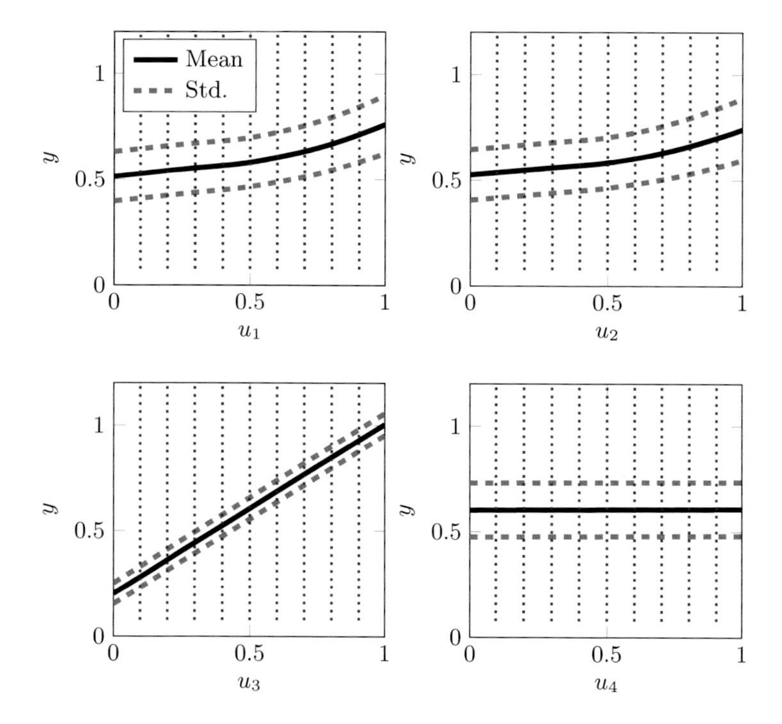

Fig. 15.33 Partial dependence plots for all inputs of TP1

15.5.1.2 Test Process Two

Figure 15.34 shows the partial dependence plots for all inputs of TP2. The mean curve for input u_1 shows a nonlinear behavior, i.e., a decrease in the slope with increasing values of u_1. This is likely due to the fact that the model output gets pretty small for high u_1 values if both discrete inputs u_3 and u_4 equal one; see Fig. 15.4 and (15.3). However, this interpretation is not based on the partial dependence plot itself but on the knowledge available about TP2. Without this knowledge input u_1 appears to have a nonlinear influence. The origin of this nonlinear behavior, which is the interaction of u_1 with inputs u_3 and u_4, is unclear. In contrast to that, input u_2 seems to influence the input in a linear way. In fact, for the influence of u_2 on the output, it is irrelevant how the discrete inputs u_3 and u_4 are chosen because the coefficients for input u_2 are the same for both of the scheduled planes; see (15.3). From the partial dependence plots of u_3 and u_4, a nonlinear relationship is observable. The tendency to obtain lower output values for high values of u_3 and u_4 is clearly visible. The lack of vertical lines indicates that the distribution of values along the axes u_3 and u_4 is by far not uniform. In fact, it can be concluded that most points lie close to the minimum and maximum values of the two axes, which of course is a result of the binary nature of these two inputs.

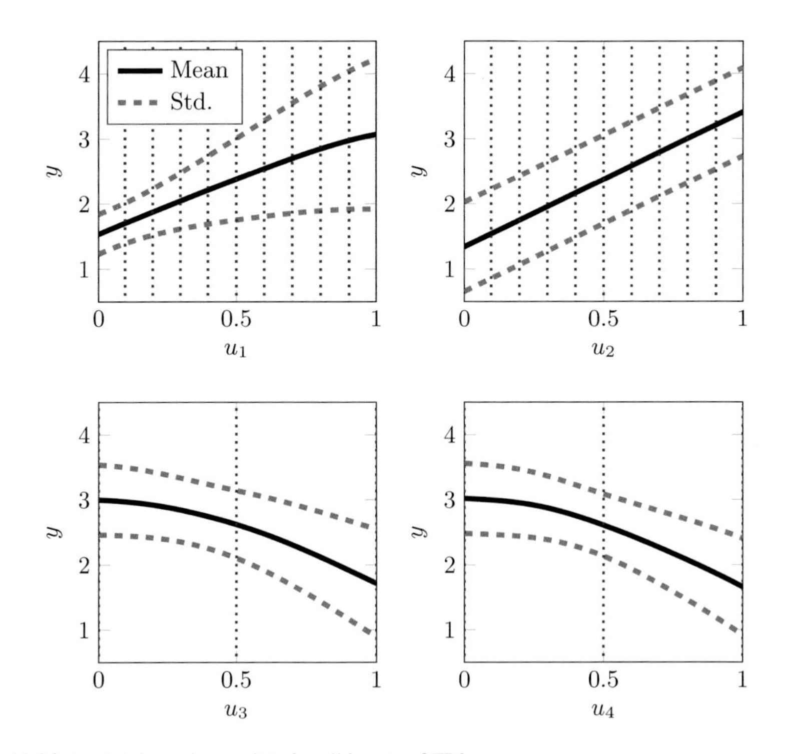

Fig. 15.34 Partial dependence plots for all inputs of TP2

The standard deviations shown in the partial dependence plots for inputs u_2, u_3, and u_4 are almost constant for all PDI values. For the partial dependence of u_1, the standard deviation increases with higher values of u_1. This indicates the growing importance of other variables. In fact, the deviations that can be expected due to switching between the two planes are higher for high values of u_1 (≈ 1) compared to low values of u_1 (≈ 0); see Fig. 15.4.

In summary, the partial dependence plots are a helpful tool to quickly review the average influence of single inputs on the model output. The distinction between linearly and nonlinearly influencing inputs seems to be possible from the observation of the mean curves. The standard deviations help to rate the influence of all non-PDIs together compared to the PDI, since variations at each specific value of the PDI are only due to the variations in the non-PDIs. Mean curves with a slope being zero for all possible values of the PDI indicate that there is no influence of that PDI variable at all. Additionally, helpful information about the data distribution along all single axes can be obtained.

Despite all the advantages, partial dependence plots have to be interpreted with care. For example, the fact that the slope of the mean curve of a partial dependence plot is always positive does not necessarily mean that this is true everywhere in the input space. This can be seen in the partial dependence plot for input u_1 of TP2 shown in Fig. 15.34. From the equation of this process, it is clear that there is a negative slope in u_1-direction if both operating point variables u_3 and u_4 equal one. This behavior is not observable from the partial dependence plot because on average the process output is still increasing with higher u_1 values. Since the training data covers the input space uniformly, only one-fourth of it leads to the usage of the plane with a negative slope in the u_1-direction, in particular when $u_3 = u_4 = 1$. Care is also necessary for the interpretation of partial dependence plots of discrete inputs. Instead of plotting whole mean curves, it might be advisable to just plot values at discrete values that do exist in the training data set.

In addition, partial dependence plots are very sensitive w.r.t. the input data distribution. In the given examples, input data were evenly distributed due to the maximin LH design. Information extraction is much more difficult or even impossible if data is "strangely" distributed. Especially correlations between inputs may lead to unreliable partial dependence plots as demonstrated for the miles per gallon data set in Sect. 15.6.

15.6 Miles per Gallon Data Set

The miles per gallon (MPG) data set concerns city-cycle fuel consumption to be predicted in terms of three multi-valued discrete and four continuous input variables [473]. The input variables are:

- Number of cylinders u_1 (multi-valued discrete)
- Displacement u_2 (continuous), in cubic inches
- Horsepower u_3 (continuous)

- Weight u_4 (continuous), in pounds (lb)
- Acceleration u_5 (continuous), 0–60 MPH times in seconds
- Model year u_6 (multi-valued discrete)
- Origin u_7 (multi-valued discrete)

The fuel consumption is measured in miles per gallon (MPG); higher MPG values correspond to lower fuel consumption. The used data set is obtained from the machine learning repository of the University of California Irvine (UCI) [349] and contains $N = 392$ samples. This data set is split into a training and test data set containing $N_{train} = 333$ and $N_{test} = 59$ samples, respectively.

The auto MPG data set is used to demonstrate the abilities of the separated $\underline{x}$-$\underline{z}$-input selection in Sect. 15.6.1. In addition to that, the RBIS approach is tested on the real-world example in Sect. 15.6.2. Finally, partial dependence plots are shown and discussed in Sect. 15.6.3.

15.6.1 Mixed Wrapper-Embedded Input Selection

For the investigation with a separated $\underline{x}$-$\underline{z}$-input selection, the search strategy as well as the evaluation criterion is chosen according to the recommendations in Sect. 15.2.3, i.e., BE and AIC$_C$, respectively. Local affine models are used for the LMNs trained with the HILOMOT. The model complexity of the LMNs is determined with the help of the AIC$_C$. Due to the input selection scheme, there are 14 possible LMN inputs.

Figure 15.35a shows the AIC$_C$ values versus the number of LMN inputs.

The best AIC$_C$ value is obtained with a subset size of seven LMN inputs, since lower AIC$_C$ values indicate better model performances. In between two subsequent AIC$_C$ values, the LMN input that is removed in the corresponding BE step is denoted. To simplify the determination of the inputs contained in the best subset, Fig. 15.35b is provided and shows which input subset corresponds to a specific number of LMN inputs. As can be seen, the best input subset contains four and three inputs in the $\underline{x}$- and $\underline{z}$-input space, respectively. The chosen LMN inputs belonging to the best subset are marked in bold type in Fig. 15.35b. Table 15.1 summarizes the assignment of each physical input variable to both the $\underline{x}$- and $\underline{z}$-input spaces.

Interestingly, all physical inputs are incorporated in the best subset, but either exclusively in the $\underline{x}$-input space or the $\underline{z}$-input space. This implies that a classic wrapper input selection approach would have failed to reduce the number of inputs. Therefore, the auto MPG example benefits from the exploited possibility to distinguish between the $\underline{x}$- and $\underline{z}$-input spaces. A comparison of the model quality

Table 15.1 Assignment of all physical inputs of the auto MPG example to the $\underline{x}$- and $\underline{z}$-input space for the best subset containing seven LMN inputs

$\underline{x}$-input space	$\underline{z}$-input space
u_1, u_3, u_4, u_7	u_2, u_5, u_6

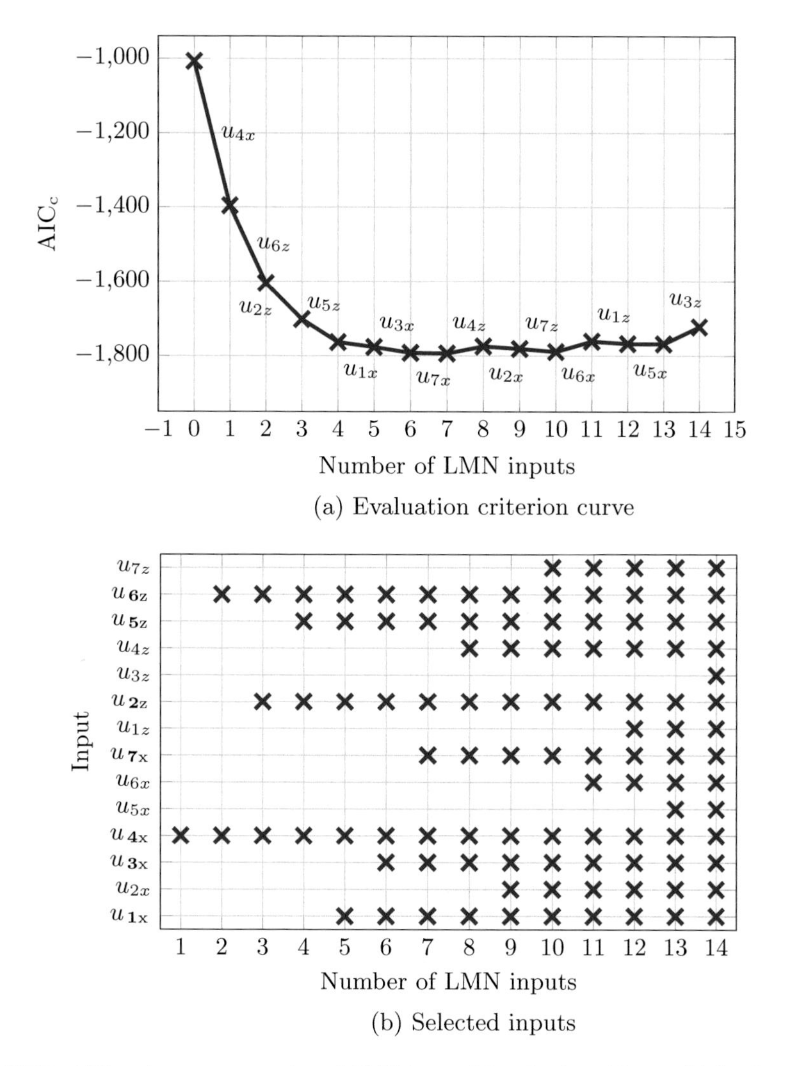

Fig. 15.35 $\mathrm{AIC_C}$ values versus number of LMN inputs (**a**) and selected inputs (**b**) for the auto MPG data set

achieved with the best input subset opposed to a model with all inputs is omitted here because it is included in the next section; see Fig. 15.36.

15.6.2 Regularization-Based Input Selection

The RBIS approaches with the heuristic as well as with the optimization-based determination of the split regularization parameters are tested on the auto MPG application. Findings from the separated $\underline{x}$-$\underline{z}$-input selection from the previous

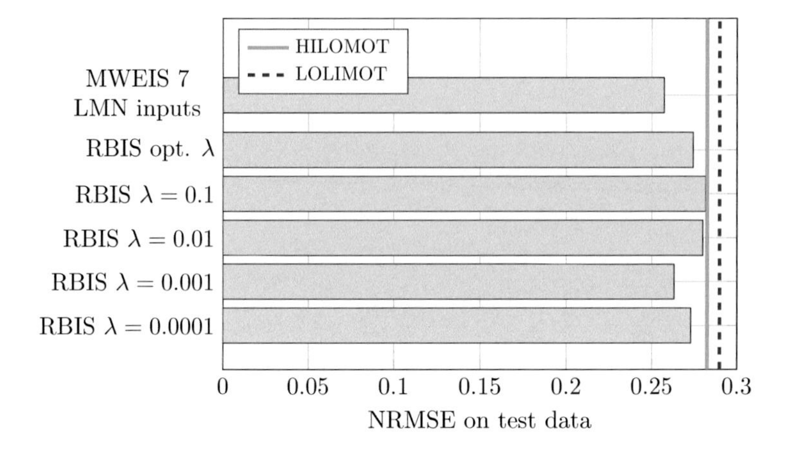

Fig. 15.36 Comparison of the achieved model qualities with all RBIS approaches, the best input subset found with an separated $\underline{x}$-$\underline{z}$-input selection, as well as LOLIMOT and HILOMOT

section are not used here. Therefore, all physical inputs u_1 up to u_7 are contained in both the $\underline{x}$- and $\underline{z}$-input spaces during the RBIS calculations. For the heuristic determination of the split regularization parameters, the same four initial values are used as previously, which are 10^{-4}, 10^{-3}, 10^{-2}, and 10^{-1}; see Sect. 15.3.2. The optimization-based determination of the split regularization parameter has the LOO error as objective.

In Fig. 15.36 the model qualities of all RBIS approaches and additionally the model quality with the best subset according to the separated $\underline{x}$-$\underline{z}$-input selection from the previous section are compared. Additionally the achieved model qualities of the LOLIMOT and HILOMOT are visualized without the use of any input selection technique, i.e., all inputs are used for both the $\underline{x}$- and $\underline{z}$-input spaces. There is no decrease in the model quality through any of the input selection techniques. The best model quality is achieved by the model with only seven LMN inputs resulting from the separated $\underline{x}$-$\underline{z}$-input selection. The best RBIS result is obtained if the heuristic is used for an initial split regularization parameter of $\lambda = 10^{-3}$.

In contrast to the investigated test processes in Sect. 15.3.2, the RBIS approach is able to increase the model accuracy more clearly. A likely cause is the higher dimensionality of the auto MPG application compared to the used test processes, where the maximum number of inputs is four instead of seven for the auto MPG application. It is assumed that through the increased input dimensionality, the benefits of regularizing the splitting parameters become more pronounced. Additional investigations of the RBIS approach with processes having higher-dimensional input spaces are suggested for future research.

15.6.3 Visualization: Partial Dependence Plot

The partial dependence plots for all inputs of the model with the best quality are shown in Fig. 15.37. The model used for the visualization is obtained through the separated $\underline{x}$-$\underline{z}$-input selection and has four $\underline{x}$-inputs and three $\underline{z}$-inputs. The nonlinear dependency on the inputs contained in the $\underline{z}$-input space, which are the displacement, the acceleration, and the model year, is clearly visible from the mean curves. The mean curves of the remaining inputs indicate an affine behavior as could be expected from the fact that these inputs are only contained in the $\underline{x}$-input space.

All mean curves show a monotonic behavior. Fuel consumption generally increases

- With fewer cylinders
- For higher values of the displacement
- For higher values of the horsepower

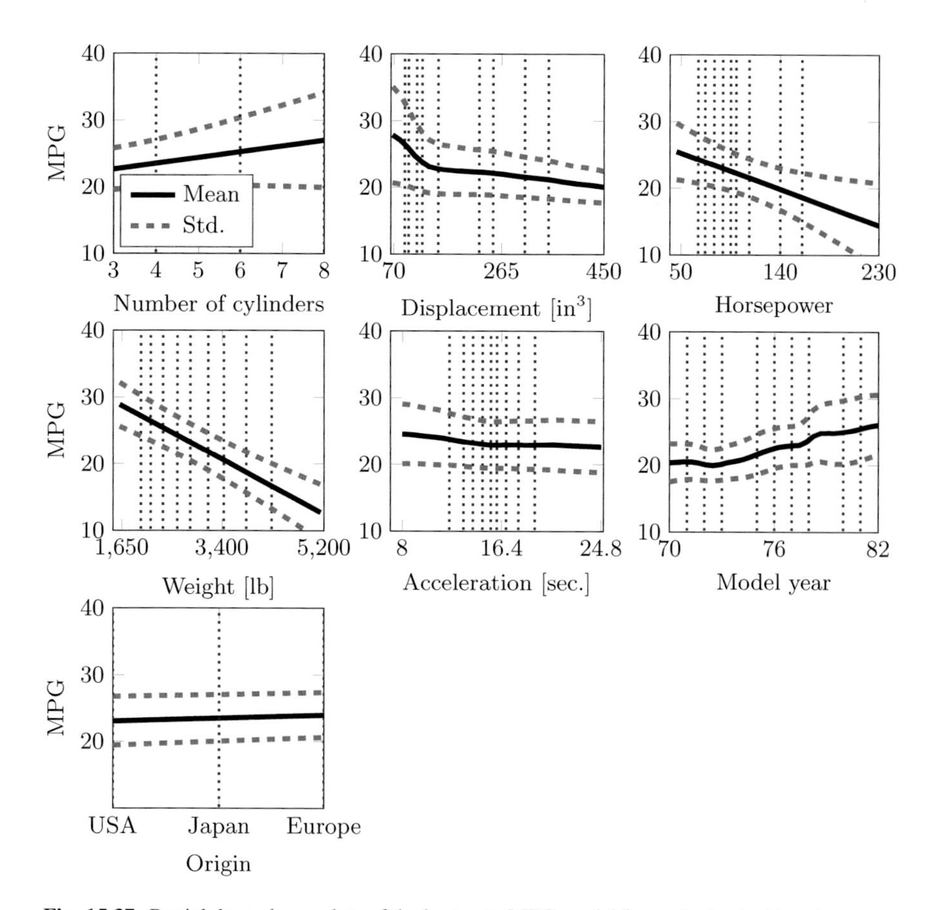

Fig. 15.37 Partial dependence plots of the best auto MPG model for each physical input

- For higher values of the weight
- For higher acceleration values
- The earlier the car is built

Additionally, the fuel consumption increases very slightly when the origin changes from Europe over Japan to the USA. Obviously, an increased fuel consumption for fewer cylinders is physically implausible. This strange effect is further elaborated in Sect. 15.6.4.

Besides the global influence of all inputs, sensitivities are observable through the slopes of the shown mean curves. The MPG values are most sensitive with respect to the weight and the horsepower. Additionally, the MPG values are sensitive to changes in the displacement but only for low displacement values.

15.6.4 Critical Assessment of Partial Dependence Plots

The partial dependence plot of the number of cylinders shown in Fig. 15.37 indicates lower MPG values, meaning an increased fuel consumption for fewer cylinders. This is obviously physically implausible, and the reason for this strange effect is further investigated in this section.

At first, the training data without the incorporation of any model is reviewed. In Fig. 15.38a, the MPG values are plotted against the number of cylinders. Although the MPG values for each number of cylinders vary widely, there is a trend toward increasing fuel consumption with more cylinders.

Since the data does not provide a reason for the partial dependence plot of the number of cylinders, it can be assumed that the underlying model delivers implausible predictions. At this point, it is not clear whether the model provides only

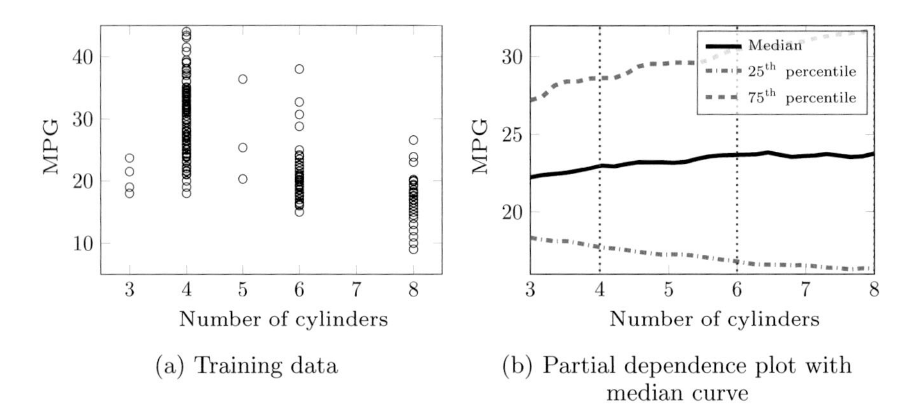

Fig. 15.38 (**a**) MPG versus number of cylinders. (**b**) Partial dependence plot for the number of cylinders with the median instead of the mean curve

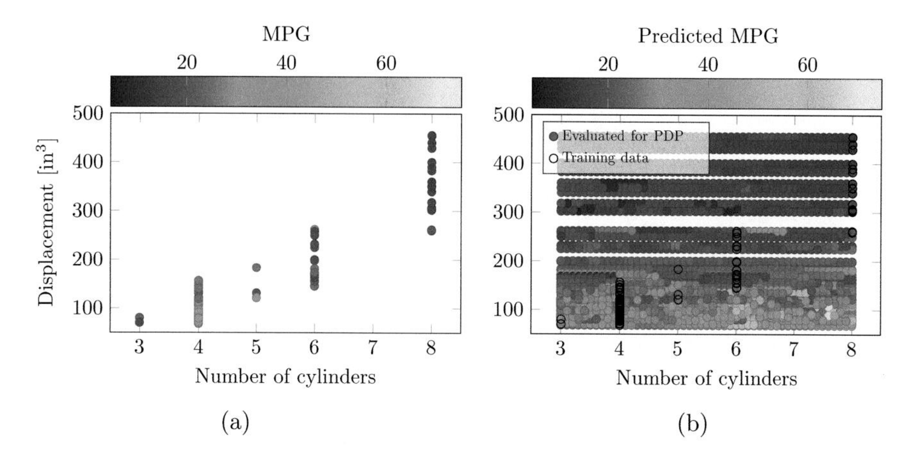

Fig. 15.39 (**a**) Displacement versus number of cylinders together with the training data and the MPG values therein. (**b**) Displacement versus number of cylinders together with points that have to be evaluated for the partial dependence plot (PDP) creation together with predicted MPG values

a few, but huge, implausible predictions that can be regarded as a kind of outlier. In order to verify this, the way the partial dependence plot is created is changed. As explained in Sect. 15.5, the *average* effects of all training data samples are taken into account in order to represent the effect of one input variable on the model. Instead of taking the average effects of all training data into account, the *median* effects are shown in Fig. 15.38b, since the median is known to be statistically more robust against outliers. Because the same implausible trend of the fuel consumption still maintains, there have to be more unreliable model predictions than just a few outliers.

Finally, the reason for the implausible model predictions could be found and is illustrated in Fig. 15.39. Naturally, the number of cylinders and the displacement are highly correlated as can be seen in Fig. 15.39a. This leads to regions in the input space in which no information is available for the model. In particular, no data and therefore no information are available for high displacements and few cylinders and vice versa. If the model is forced to make predictions in these regions of the input space, it can be considered as extrapolation with a high danger of unreliable predictions. This is exactly what happens. For the creation of the partial dependence plot, the model is indeed forced to extrapolate as can be seen in Fig. 15.39b. It can be observed that in the region of low displacements and many cylinders, the model predicts very low fuel consumption. The training data set does not contain such small consumption values; see Fig. 15.39a. In conclusion, extreme care is advised in the interpretation of partial dependence plots, in particular, if highly relevant inputs are strongly correlated which provokes extreme extrapolation in the partial dependence plots.

Chapter 16
Gaussian Process Models (GPMs)

Abstract This chapter is devoted to Gaussian processes. Compared to existing literature, it tries to approach this very abstract and complex topic from an intuitive perspective. The features of kernel methods are explained, and their characteristics are highlighted. The key ideas are illustrated with the help of many extremely simplified examples, typically just 1D or 2D, and very few data points. This should allow grasping the basic concepts involved. All toy examples are simultaneously carried out with two different kernel functions: Gaussians and inverse quadratic. The concepts are introduced step by step, starting with just the mean prediction in the noise-free case and adding complexity gradually. The relationship with RBF networks is discussed explicitly. Shedding light on Gaussian processes from various directions, they are hopefully easier to understand than from standard textbooks on this topic.

Keywords Kernel · RKHS · GP · Probability · Likelihood · Prior · Uncertainty

Gaussian process models can be employed for regression or classification where the former is in the focus of this book. In the last decade, Gaussian processes have gained significant attention in the learning community. The standard monograph on Gaussian processes for machine learning is [476]. In some way, it realizes the idea of neural networks to an extreme: an extremely powerful approximator/interpolator with as many parameters as data points $n = N$ is used and a proper bias/variance tradeoff to prevent overfitting is purely performed by regularization. Usually in typical neural network approaches, the model flexibility is determined by a two-stage procedure:

The original version of this chapter was revised: The abbreviation of the word (ACSMO) has been corrected to (ASCMO) on page no. 640. The correction to this chapter is available at https://doi.org/10.1007/978-3-030-47439-3_30

O. Nelles, *Nonlinear System Identification*,
https://doi.org/10.1007/978-3-030-47439-3_16

1. Choosing the number of neurons, rules, or terms
2. Applying some kind of regularization technique like early stopping (for MLPs), local estimation (for local model networks with LOLIMOT or HILOMOT), or adjusting the ridge regression regularization parameter

Skipping the first stage is one characteristic of the methods discussed in this chapter. The number of neurons is chosen equal to the number of training data points N. A direct consequence is that the evaluation of Gaussian process models requires major computational resources with respect to calculation and memory.

For example, the automotive industry heavily relies on data-driven models. Traditionally, the standard models are based on look-up tables. Due to their limitations specifically in regard to the input dimensionality discussed in Chap. 10, some key players in the market (particularly Bosch) advocate Gaussian process models as a promising new flexible modern successor. Automotive tool and engineering supplier ETAS (a Bosch subsidiary enterprise) provides a tool for advanced simulation for calibration, modeling and optimization (ASCMO) to train and utilize Gaussian process models [253]. However, the computational load for a microcontroller implementation is so big that dedicated hardware has been developed. This special-purpose hardware, a so-called advanced modeling unit (AMU[1]), is integrated in high-end engine control units (ECUs), for example. This highlights the possible industrial significance of this model type.

16.1 Overview on Kernel Methods

Gaussian processes (GPs) belong to the class of kernel methods. For regression, they are known under many different names:

- Kernel ridge regression or kernel regularized least squares [208, 595]: The derivation starts with the well-known ridge regression (see Sect. 3.1.4) which is a regularized version of least squares. The regressors are transformed into a new space before performing the ridge regression.
- Reproducing kernel Hilbert spaces (RKHS) [254]: It addresses the optimization problem of finding the best function that fits the data where *best* is expressed in terms of the error and a smoothness penalty. Without this smoothness penalty, the fitting problem can be solved in arbitrarily many ways. The smoothness penalty imposes certain properties on the function. It is formulated as a norm in Hilbert space. Dependent on its exact formulation, a specific functional form results.
- Regularization network [140]: This expression emphasizes the neural network or basis functions character.
- Gaussian process regression/prediction [476]: This is a probabilistic approach that starts with formulating a prior probability, i.e., a probability for the model output to take any value *before* any data is measured. This prior plays the same role in GPs as the smoothness penalty in RKHS. It describes the *best* model

[1] http://www.etas.com/download-center-files/products_ASCMO/ascmo_flyer_en.pdf

in the absence of data. Then, with the help of measurement data, the posterior probability for the model output is deduced. This represents a more abstract and more difficult-to-understand approach than a deterministic setting but offers the additional benefit of the probabilistic interpretation. Thus, not just the mean values of the posterior probability can be calculated but also its standard deviation which can directly be used for the confidence intervals.
- Kriging [366]: "Kriging was initially developed by geologists to estimate mineral concentrations over an area of interest given a set of sampled sites from the area[2]; it was also introduced about the same time in the field of statistics to include the correlations that exist in the residuals of a linear estimator."
- Wiener-Kolmogorov prediction [82]: This denotes the most general mathematical framework. In [82] it is described with "as *best* estimator of a linear functional of u, choose the linear function of observations minimizing the mean square estimation error."

All these approaches are identical, and these models shall be named Gaussian processes (GP) within this chapter. However, their derivation and interpretation differ significantly. Some of them are discussed in the subsequent sections. After training, all these models consist of a kernel function (for other networks typically called basis function) placed on each training data point like an interpolating RBF network. The associated weights are estimated by ridge regression, and the regularization parameter λ controls the flexibility of the GP. Besides λ the second decisive user choice is the type and width of the kernel function.

Due to the concept of placing a kernel on each training data point, no explicit strategy is required for partitioning the input space. Basically there exist two philosophies for controlling and distributing the model complexity in the input space:

- *Space-oriented*: Starting from the whole input space/domain, often assumed to be a p-dimensional hypercube, it is partitioned or constrained in some way in order to introduce *structure* to the model. The partitioning can be carried out by some splitting algorithm (often in a tree-based manner) or some kind of mesh in form of triangles or a grid. Alternatively or additionally constraints can be formulated as assumptions on the model such as models of additive or ANOVA type.

  The complexity of these space-oriented approaches typically is *sensitive with respect to the input space dimension p* and insensitive with respect to the number of training samples N.
- *Point-oriented*: Starting from the available data and placing a kernel on each data point ensures that the model "does something" where the data lies and therefore information is available. This principle is independent of the dimensionality of the space. Instead on *regions* in the input space, it relies on *neighborhoods* and point-to-point distances (interpreted as similarities between points).

[2]The original ideas go back to "the Master's thesis of Danie G. Krige" in 1951 [317], https://en.wikipedia.org/wiki/Kriging.

> In contrast to the space-oriented approaches, the complexity of these point-oriented approaches typically is insensitive with respect to the input space dimension p and *sensitive with respect to the number of training samples N.*

Note that in the space-oriented approaches, the modeling assumptions or prior knowledge about the process determine the model structure, i.e., how to partition the input space, how many neurons to choose, etc. In the point-oriented approach, the number of neurons typically is fixed by the number of data points, and the partitioning of the input space is determined by the data distribution. Modeling assumptions and prior knowledge enter with the decision for a kernel and the amount of regularization although the latter mostly is also optimized with the data.

Figure 16.1 gives an overview on the typical modeling philosophies. Most approaches in this book belong to the space-oriented category shown in "(c) axis-oblique partitioning" and "(d) axis-orthogonal partitioning." Category "(f) grid" fully suffers from the curse of dimensionality, while "(e) Delaunay triangulation" becomes extremely time-consuming to calculated in high dimensions (>6) so that it is restricted to relatively low-dimensional problems in practice. This chapter deals with kernel methods that belong to the point-oriented category shown in "(a) Gaussian processes." The approaches due to "(b) cluster-based approaches" are only mentioned in some sections but are not within the focus of this book.

Except for an additional offset[3] or bias[4] or intercept,[5] a well-known and widely-used machine learning technique is equivalent to GPMs and also falls under the category of kernel methods, namely, the *Least squares support vector machine* (LS-SVM) [551]: it is derived as an approximation of the classical support vector machine where one of its main features is lost, i.e., the sparse solution which means the relatively small number of support vectors. Rather *all* data points become "support vectors" and contribute to the solution due to the least squares loss function. The LS-SVM is almost identical to GPs. The only distinction is the additional offset.

The additional offset is necessary for a proper interpretation as a maximum margin classifier of support vector machines. However, it cripples the elegance and efficiency of the math involved in the offset-free variant by making the matrix-to-be-inverted non-symmetric and adding another equality constraint [484]. It is argued and empirically shown in [484] that the offset delivers no benefit in terms of performance, but many drawbacks in terms of computation and therefore should be omitted for both regression and classification tasks. The classification task which is beyond the scope of this book can be seen as regression with outputs equal to either -1 or 1 for the two classes.

In the case of a multi-class problem with M classes, two alternative strategies are common: (i) one-versus-rest or sometimes called one-versus-all strategies which are argued as being advantageous in [485] and (ii) one-versus-one or sometimes called

[3]Engineering notation

[4]Neural network notation

[5]Statistics notation

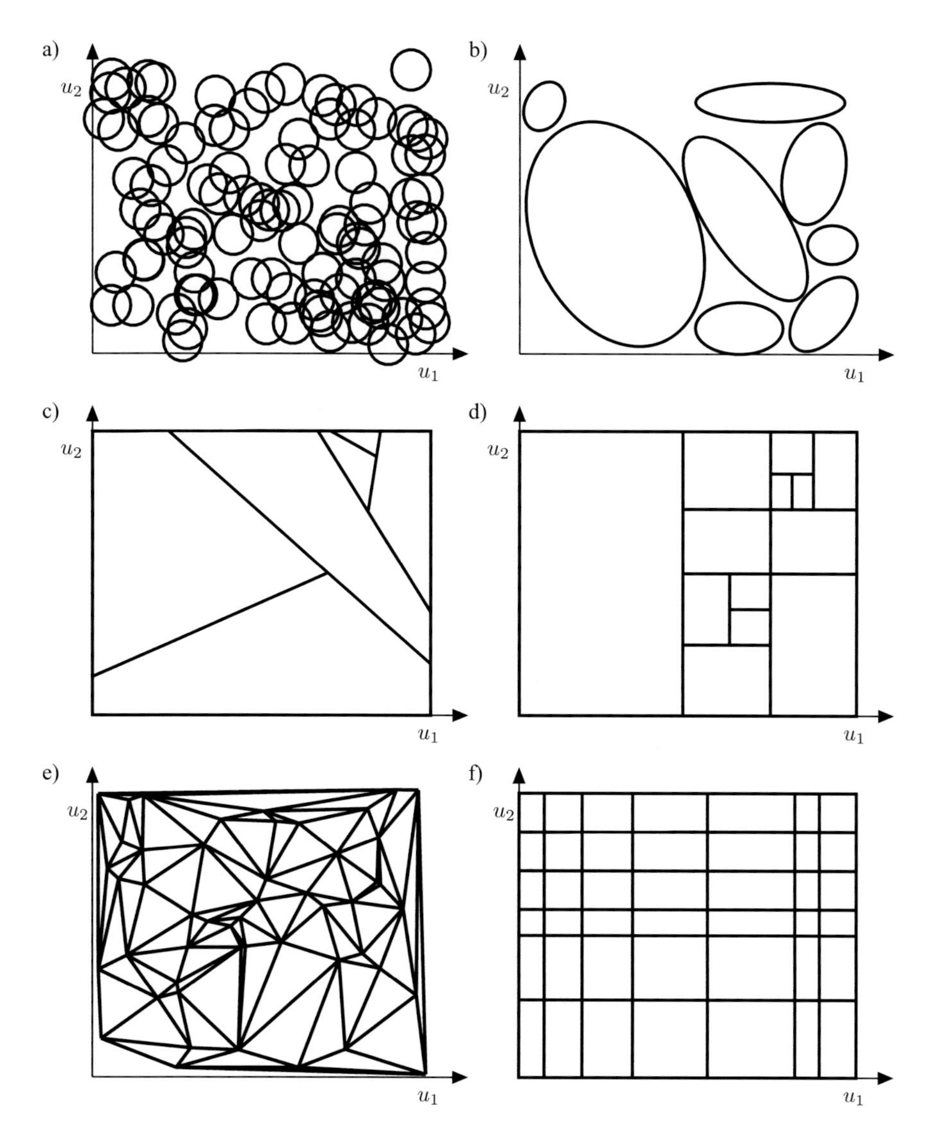

Fig. 16.1 Comparison of different modeling philosophies: Point-oriented methods like (**a**) Gaussian processes, (**b**) cluster-based approaches like k-means (unsupervised) or Gustafson-Kessel clustering (supervised). Space-oriented methods like (**c**) axis-oblique partitioning with, e.g., HILOMOT, (**d**) axis-orthogonal partitioning with, e.g., LOLIMOT, (**e**) Delaunay triangulation, (**f**) grid. The curse of dimensionality becomes more and more severe from (**a**) to (**f**)

all-versus-all strategies which [262] advocates. In one-versus-rest strategies, one classifier is trained for each class that distinguishes this class against all others; this corresponds to training M classifiers. For evaluation, the classifier with the highest score wins. In one-versus-one strategies, classifiers are trained for all pair of classes, this corresponds to training $\binom{M}{2}$ classifiers. For evaluation, the class with the largest number of votes (won pairwise competitions) wins.

16.1.1 LS Kernel Methods

The kernel methods discussed in this chapter are all least-squares-based. This offers many very important benefits, compare to the first section in Chap. 3. In the context of kernel methods, the last advantage in this list is probably the most important one: the leave-one-out error can be calculated in closed analytic form without significant additional cost.

For kernel approaches, typically at least two hyperparameters (amount of regularization and length scale) have to be tuned that control the model behavior decisively. This topic is dealt with in Sect. 16.7, compared also in Fig. 16.25. And a typical way of tuning these hyperparameters is to optimize them on validation data or by cross-validation. The first approach "loses" data that otherwise could be incorporated in the training data set. The second approach is computationally very expensive, but for all least squares kernel methods, it can be realized via the analytic leave-one-out error calculation extremely efficiently; compare Sect. 7.3.2. In the opinion of the author, this advantage outweighs all other aspects when contrasting LS with non-LS methods significantly for most applications.

16.1.2 Non-LS Kernel Methods

However, there is a wide range of non-LS kernel methods, most prominently the original support vector machine. It loses many LS benefits but still keeps the advantage of a unique global optimum. The optimization algorithms also are very mature and fast but still are outperformed by LS in terms of robustness, numerical properties, and possible problem sizes (number of parameters that can be handled).

One major benefit the original support vector machines (and similar approaches) offer is **sparsity**. This means that many/most parameters are zero and therefore their associated kernels do not enter the model. Thus, the evaluation time of such a model is significantly reduced at the price of more expensive training. The main motivation probably is that sparse models are very robust with respect to overfitting. These issues also come up in L_1 regularization; see Sect. 3.1.6.

16.2 Kernels

Kernels $K(\cdot,\cdot)$ measure the "similarity" between two points. Typically, either both points represent input data $\underline{u}(i)$ and $\underline{u}(j)$ or only one point is input data $\underline{u}(i)$, while the other can be any query point $\underline{u}$.

Throughout this chapter, two different but very similar kernels are considered in order to convey some feeling on the kernel's impact. The first kernel is by far the most frequently used Gaussian kernel:

$$K(\underline{u}, \underline{u}(i)) = \exp\left(-\frac{1}{2}||\underline{u} - \underline{u}(i)||^2/\sigma^2\right) . \tag{16.1}$$

This is no probability density function (with area 1 under the curve); it is used to describe correlations. Thus, the exponential function possesses no normalization factor. The correlation between a point and itself is 1, the correlation between a point and another point declines as their distance increases, and if both points are very far away from each other, the correlation $\to 0$. Usually, the 2-norm is used in (16.1) and all other kernels.

Note that in many publications, the standard deviation σ is called *length scale* which is a notion that can be transferred to other kernels as well. It characterizes or is proportional to the region of influence of a kernel. If the length scale is large, the model will be smooth; points nearby are highly correlated. If the length scale is small, the model will be wiggly; points nearby are only weakly correlated.

The second kernel considered here is the inverse quadratic kernel defined by

$$K(\underline{u}, \underline{u}(i)) = \frac{1}{1 + \left(||\underline{u} - \underline{u}(i)||/\alpha\right)^2} . \tag{16.2}$$

Here the length scale is α. Figure 16.2 compares the two kernels for identical σ and α. The inverse quadratic kernel obviously has a fatter tail and therefore a little bit more global influence. The inverse quadratic kernel decreases inverse proportionally with the squared distance for wide distances like many physical laws do (gravitational or electrostatic force).

As remarked in [129], the inverse quadratic "kernel is equivalent to adding together many Gaussian kernels with different standard deviations (length scales). So, GP priors with this kernel expect to see functions that vary smoothly across

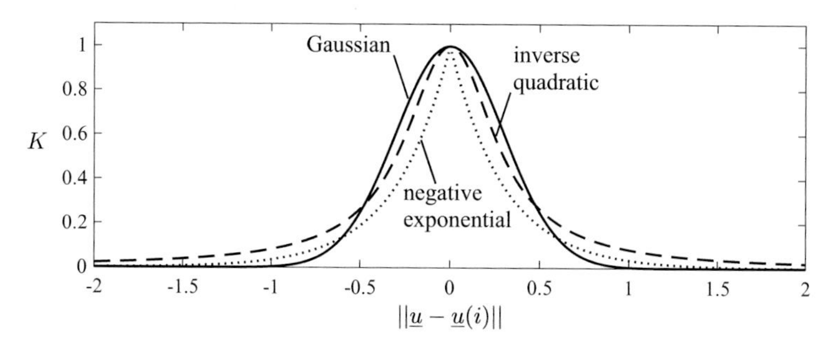

Fig. 16.2 Gaussian kernel (solid), inverse quadratic kernel (dashed), negative exponential or Laplacian kernel (dotted) for $\sigma = \alpha = \beta = 0.3$

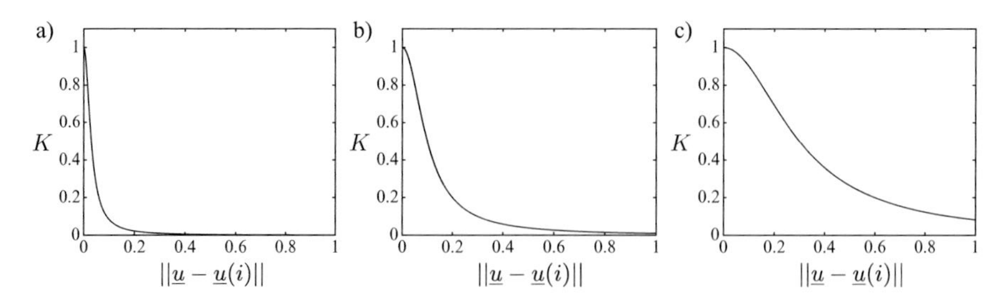

Fig. 16.3 Inverse quadratic kernels for different hyperparameters (length scales) (**a**) $\alpha = 0.03$, (**b**) $\alpha = 0.1$, (**c**) $\alpha = 0.3$. For $||\underline{u} - \underline{u}(i)|| = \alpha$, the kernel has dropped off to 0.5

many length scales." Figure 16.3 compares this less popular kernel for different values of α.

Finally, the negative exponential or Laplacian kernel shall be mentioned and is depicted in Fig. 16.2. It follows

$$K(\underline{u}, \underline{u}(i)) = \exp\left(-\left|\left|\underline{u} - \underline{u}(i)\right|\right|/\beta\right) \tag{16.3}$$

with the length scale β. Its tail is fatter than the Gaussian's but thinner than the inverse quadratic's. Note that due to the non-differentiability at 0, this kernel seems to be unattractive for function approximation. It is not considered further in this chapter.

16.3 Kernel Ridge Regression

In case of a linear model $\underline{y} = \underline{X}\,\underline{\theta}$ without offset, the regressors (columns in $\underline{X}$) are chosen just as the process inputs. If the inputs are transformed by some nonlinear mapping $\underline{\phi}(i) = \phi(\underline{u}(i))$ for each data point i, a nonlinear (but linear regression) model results $\underline{y} = \underline{\Phi}\,\underline{\theta}$. The nonlinear mapping $\phi(\cdot)$ maps the p-dimensional input space to an n-dimensional so-called *feature space*; compare Fig. 16.4. Here the transformed regression matrix and the new parameter vector look as follows (compare to (10.3) in Sect. 10.1)

$$\underline{\Phi} = \begin{bmatrix} \phi_1(\underline{u}(1)) & \phi_2(\underline{u}(1)) & \cdots & \phi_n(\underline{u}(1)) \\ \phi_1(\underline{u}(2)) & \phi_2(\underline{u}(2)) & \cdots & \phi_n(\underline{u}(2)) \\ \vdots & \vdots & & \vdots \\ \phi_1(\underline{u}(N)) & \phi_2(\underline{u}(N)) & \cdots & \phi_n(\underline{u}(N)) \end{bmatrix} \qquad \underline{\theta} = \begin{bmatrix} \theta_1 \\ \theta_2 \\ \vdots \\ \theta_n \end{bmatrix}. \tag{16.4}$$

One example for such a feature mapping is $u \rightarrow [\phi_1\ \phi_2\ \phi_3]^T = [1\ \sqrt{2}u\ u^2]^T$ from $p = 1$ to $n = 3$ dimensions. Such a mapping allows to tackle nonlinear problems but to stay in the class of linear regression similar to RBF networks.

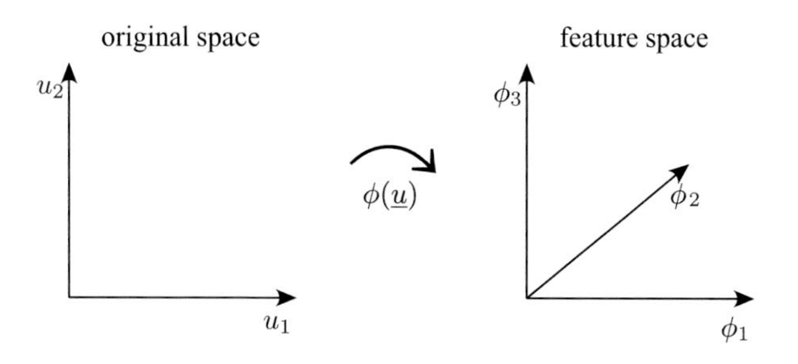

Fig. 16.4 Mapping from a $p = 2$ dimensional original space to an $n = 3$ dimensional feature space

The optimal parameter vector in the alternative form of ridge regression given in (3.1.5) in Sect. 3.1.5 of this transformed problem is

$$\underline{\hat{\theta}} = \underline{\Phi}^T \left(\underline{\Phi}\, \underline{\Phi}^T + \lambda \underline{I} \right)^{-1} \underline{y}\,. \tag{16.5}$$

The reason for writing this in the alternative version is that this requires the inversion of an $N \times N$ matrix (N is the number of data points) while the original version requires the inversion of the $n \times n$ matrix $\underline{\Phi}^T \underline{\Phi} + \lambda \underline{I}$ (n is the number of parameters). The intention is to keep the calculation as independent of n as possible. Eventually, even $n = \infty$ will usually occur which works only in the dual form.

The model output value for data point i can be calculated by taking one row of $\underline{\Phi}$ denoted as $\underline{\phi}(i)^T = [\phi_1(\underline{u}(i))\ \phi_2(\underline{u}(i))\ \cdots\ \phi_n(\underline{u}(i))]$ that corresponds to data point i multiplied with the weight vector:

$$\hat{y}(i) = \underline{\phi}(i)^T \underline{\theta} = \underline{\phi}(i)^T \underline{\Phi}^T \left(\underline{\Phi}\, \underline{\Phi}^T + \lambda \underline{I} \right)^{-1} \underline{y}\,. \tag{16.6}$$

The vector-matrix multiplication $\underline{\phi}(i)^T \underline{\Phi}^T$ and the matrix-matrix multiplication $\underline{\Phi}\, \underline{\Phi}^T$ both take place in feature space and thus solely calculate n-dimensional inner (scalar) products.

For an overview, also compare the box at the end of Sect. 3.1.5.

16.3.1 Transition to Kernels

If the nonlinear mapping to feature space $\phi(\cdot)$ is of infinite dimension ($n = \infty$), these inner products cannot be calculated explicitly. Rather a so-called kernel function $K(\underline{u}(i), \underline{u}(j))$ (which associates a kernel to the mapping $\phi(\cdot)$ and vice versa) measures the "similarity"[6] of two vectors (points) $\underline{u}(i)$ and $\underline{u}(j)$ in n-dimensional feature space.

[6] ≈ 1 for points nearby, ≈ 0 for points far away from each other

On the basis of the N data points, the kernel vector $\underline{k}(i)$ measures the similarity between point $\underline{u}(i)$ and all points in the data set $\underline{u}(j)$, $j = 1, \ldots, N$:

$$\underline{k}(i)^T = \underline{\phi}(i)^T \underline{\Phi}^T \tag{16.7}$$

and the kernel matrix $\underline{K}$ measures the similarities between all data points $\underline{u}(i)$, $i = 1, \ldots, N$ and all data points $\underline{u}(j)$, $j = 1, \ldots, N$:

$$\underline{K} = \underline{\Phi}\,\underline{\Phi}^T \tag{16.8}$$

which possesses diagonal elements $= 1$ as a point is perfectly similar to itself.

Thus, (16.6) can be written as

$$\hat{y}(i) = \underline{k}(i)^T \left(\underline{K} + \lambda \underline{I}\right)^{-1} \underline{y}\,. \tag{16.9}$$

Thereby, the evaluation of an n-dimensional problem is transformed into an N-dimensional one. Thus, it can be evaluated even if $n = \infty$ as the number of data points N is always finite. This is known as the *kernel trick*. With this trick, the size of the problem is limited to N, independent of the dimension n. Each kernel $K(\cdot, \cdot)$ corresponds to a feature mapping $\phi(\cdot)$ which does not has to be known or evaluated explicitly.

As a consequence, the n-dimensional parameter vector $\underline{\theta}$ is gone. In the following sections, it will be replaced by a new[7] N-dimensional weight vector which corresponds to the second factor in (16.9): $\underline{w} = \left(\underline{K} + \lambda \underline{I}\right)^{-1} \underline{y}$.

For the sake of clarity, please note the distinctions between:

- $K(\cdot, \cdot)$: Kernel function, takes two arguments and measures a degree of similarity between them.
  Often the second argument is set to the training data points yielding the one-dimensional functions $K(\cdot, \underline{u}(i))$.
- $\underline{k}(\cdot)$: Kernel vector function, dimension is $N \times 1$, takes one argument and measures the similarity between the argument and all data points:

$$\underline{k}(\underline{u}(i)) = \begin{bmatrix} K(\underline{u}(i), \underline{u}(1)) \\ K(\underline{u}(i), \underline{u}(2)) \\ \vdots \\ K(\underline{u}(i), \underline{u}(N)) \end{bmatrix}. \tag{16.10}$$

[7] for the dual variables

- $\underline{K}$: Kernel matrix, symmetric, positive definite, dimension is $N \times N$ and evaluates the kernel for all data point combinations:

$$\underline{K} = \begin{bmatrix} K(\underline{u}(1), \underline{u}(1)) & K(\underline{u}(1), \underline{u}(2)) & \cdots & K(\underline{u}(1), \underline{u}(N)) \\ K(\underline{u}(2), \underline{u}(1)) & K(\underline{u}(2), \underline{u}(2)) & \cdots & K(\underline{u}(2), \underline{u}(N)) \\ \vdots & \vdots & & \vdots \\ K(\underline{u}(N), \underline{u}(1)) & K(\underline{u}(N), \underline{u}(2)) & \cdots & K(\underline{u}(N), \underline{u}(N)) \end{bmatrix} . \quad (16.11)$$

Remark: It can be shown that the kernel relates to the abovementioned feature mapping $\phi(\cdot)$ in the following manner [140]:

$$K(\underline{u}(i), \underline{u}(j)) = \sum_{k=0}^{\infty} \lambda_k \phi_k(\underline{u}(i))\, \phi_k(\underline{u}(j)) \,. \quad (16.12)$$

This is the (possibly dimension-weighted with λ_k) inner product of the two vectors

$$[\phi_1(\underline{u}(i))\ \phi_2(\underline{u}(i))\ \cdots\ \phi_\infty(\underline{u}(i))] \cdot [\phi_1(\underline{u}(j))\ \phi_2(\underline{u}(j))\ \cdots\ \phi_\infty(\underline{u}(j))]^T \,.$$

As stated in [140], it depends on the type of kernel whether it can be decomposed into a finite or infinite sum. For Gaussian kernels, the feature space indeed is infinite-dimensional. For polynomial kernels of degree m $K(\underline{u}(i), \underline{u}(j)) = (1+\underline{u}(i)\cdot\underline{u}(j))^m$, the associated feature map is finite-dimensional, and the sum in (16.12) contains only a finite number of terms. The mapping $\phi(\cdot)$ from 1D to 3D mentioned at the beginning of this section $u \to [\phi_1\ \phi_2\ \phi_3]^T = [1\ \sqrt{2}u\ u^2]^T$ according to (16.12) corresponds to the polynomial kernel of degree 2: $K(u(i), u(j)) = (1 + u(i) \cdot u(j))^2 = 1 + 2u(i)u(j) + u^2(i)u^2(j)$ as

$$[1\ \sqrt{2}u(i)\ u^2(i)]^T \cdot [1\ \sqrt{2}u(j)\ u^2(j)] = 1 + 2u(i)u(j) + u^2(i)u^2(j) \,.$$

16.4 Regularizing Parameters and Functions

This section anticipates some issues from the remainder of this chapter and from Sect. 18.6.2 in order to convey its main message. For the following discussion, it is important to understand the distinction between regularizing n parameters for a *fixed* number of n basis functions and regularizing N parameters which can grow arbitrarily with the size N of the training data set. This corresponds to the distinction between a parameter optimization problem which is of finite dimension and a function optimization problem that is of infinite dimension. However, in the "learning from data" context, the function optimization problem is also limited. Its complexity is limited to N dimensions, i.e., the information contained in the data. Consequently, in practice, it can become arbitrarily large but never infinite.

Two types of problems shall be discussed in the following. If the goal is to optimize parameters, then the loss function can be augmented by an additional penalty term $\lambda\, \underline{\theta}^T \underline{C}\, \underline{\theta}$ yielding a regularization effect. This procedure is well known as ridge regression if $\underline{C} = \underline{I}$; compare Sect. 3.1.4. It is also used for regularized FIR model estimation where $\underline{C} = \underline{\Sigma}^{-1} = \underline{K}^{-1}$; see Sect. 18.6.2. Here $\underline{\Sigma}^{-1}$ denotes the *inverse* covariance matrix of the parameters corresponding to the prior knowledge. By using a full matrix $\underline{C}$, the off-diagonal terms penalize correlations between different parameters. The following equation shows that the product between θ_i and θ_{i+1} is penalized by the amount $(c_{i+1,i} + c_{i,i+1})$. Since $\underline{C}$ usually is symmetric $c_{i+1,i} = c_{i,i+1}$.

$$\underline{\theta}^T \underline{C}\, \underline{\theta} = [\cdots\ \theta_i\ \theta_{i+1}\ \cdots] \begin{bmatrix} \ddots & & & \\ \cdots & c_{i,i} & c_{i,i+1} & \cdots \\ \cdots & c_{i+1,i} & c_{i+1,i+1} & \cdots \\ & & & \ddots \end{bmatrix} \begin{bmatrix} \vdots \\ \theta_i \\ \theta_{i+1} \\ \vdots \end{bmatrix} \tag{16.13}$$

$$= \ldots + c_{i,i}\theta_i^2 + (c_{i+1,i} + c_{i,i+1})\theta_i\theta_{i+1} + c_{i+1,i+1}\theta_{i+1}^2 + \ldots$$

As an example, Fig. 16.5a illustrates the problem of estimating an impulse response in discrete time (FIR model). It is treated in Sect. 18.6.2 more thoroughly. After fixing the model order m, the basis functions are $u(k-1), \ldots, u(k-m)$, and the associated m parameters have to be estimated. The matrix $\underline{C}$ can be seen as an inverse kernel matrix $\underline{K}^{-1}$ or as an inverse covariance matrix $\underline{\Sigma}^{-1}$ in a probabilistic interpretation as Gaussian process; compare Sect. 16.6.

Assume in the above example the simple case of only two parameters which possess a large covariance of 0.9 between θ_i and θ_{i+1} (and variances $= 1$) then it follows from

$$\underline{\Sigma} = \begin{bmatrix} 1 & 0.9 \\ 0.9 & 1 \end{bmatrix} \quad \Rightarrow \underline{C} = \underline{\Sigma}^{-1} = \begin{bmatrix} 5.26 & -4.74 \\ -4.74 & 5.26 \end{bmatrix} \tag{16.14}$$

that $c_{i,i} = c_{i+1,i+1} \approx -c_{i,i+1} = -c_{i+1,i}$. This implies that the penalty terms almost cancel each other out if $\theta_i \approx \theta_{i+1}$ which can be expected if they are

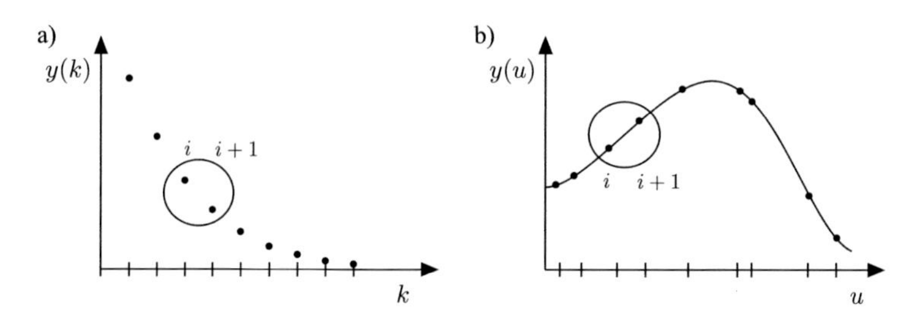

Fig. 16.5 Regularization of (**a**) parameters and (**b**) functions

highly correlated, i.e., covariance 0.9. Thus, the higher the covariance between two parameters θ_1 and θ_2 is, the lower is the penalty term if both parameters agree: covariance $\to 1 \Rightarrow$ penalty $\to 0$, if $\theta_1 \approx \theta_2$.

Figure 16.5b illustrates a function approximation problem where u is a continuous variable. Here the whole function (can be seen as infinitely many points) shall be regularized in order to make it smooth. It is known from the representer theorem (see Sect. 16.5) that this infinite-dimensional function optimization problem can be cast into an N-dimensional parameter optimization problem. This is done by placing a kernel on each training data point and estimating the associated weights. While the number of parameters in Fig. 16.5a is fixed with the model structure, their number N grows with the training data in Fig. 16.5b.

In the GP context (Sect. 16.6), the prior describes the probability of a function generated by a model *before* any data is measured. This function can be calculated by placing a kernel on every possible value for $\underline{u}$. This can be visualized, e.g., by equidistantly distributing N data points in $\underline{u}$, calculating the $N \times N$ kernel matrix $\underline{K}$ and drawing one sample from the N-dimensional normal distribution with mean $\underline{\mu} = \underline{0}$ and covariance matrix $\underline{\Sigma} = \underline{K}$

$$\mathcal{N}\left(\underline{0}, \underline{K}\right) = \frac{1}{\sqrt{(2\pi)^N |\underline{K}|}} \exp\left(-\frac{1}{2}\underline{u}^T \underline{K}^{-1} \underline{u}\right) . \tag{16.15}$$

Each sample drawn from this distribution corresponds to a vector of N output values based on the prior knowledge on the smoothness of the function to be approximated.

The property "smoothness of functions" can be interpreted as "small" higher-order derivatives. Each kernel implies a mathematical formulation of what this exactly means. For example, it can be shown that the Gaussian kernel with standard deviation σ penalizes the quadratic norm of the mth derivative of the function with a_m [463]:

$$a_m = \frac{\sigma^{2m}}{m!\, 2^m} . \tag{16.16}$$

Other kernels lead to other weightings. For example, cubic splines arise if only the quadratic norm of the 2nd derivative (curvature) is penalized.

16.4.1 Discrepancy in Penalty Terms

In the parameter estimation problem discussed above and treated in detail in Sect. 18.6.2, the loss function (18.138) is

$$I(\underline{\theta}) = \underline{e}^T \underline{e} + \lambda\, \underline{\theta}^T \underline{K}_1^{-1} \underline{\theta} \tag{16.17}$$

with the error vector $\underline{e}$, the parameter vector to be estimated $\underline{\theta}$, and the kernel matrix $\underline{K}_1$ which also can be interpreted as covariance matrix $\underline{\Sigma}_1$. In the function approximation context via kernel-based methods discussed in this chapter, the loss function is, compare (16.34),

$$I(\underline{w}) = \underline{e}^T \underline{e} + \lambda\, \underline{w}^T \underline{K}_2\, \underline{w} \tag{16.18}$$

where the vector $\underline{w}$ contains the kernel weights to be estimated and the kernel matrix $\underline{K}_2$ which also can be interpreted as covariance matrix $\underline{\Sigma}_2$.

The penalty term in (16.17) is $\lambda\, \underline{\theta}^T \underline{K}_1^{-1} \underline{\theta}$; in (16.18) it is $\lambda\, \underline{w}^T \underline{K}_2\, \underline{w}$. Where does this discrepancy come from? The explanation is that the kernel matrices $\underline{K}_1$ in (16.17) and $\underline{K}_2$ in (16.18) are different; they describe prior knowledge on different quantities.

The kernel matrix in (16.17) describes the covariances between the *parameters*:

$$K_1(i, j) = \mathrm{cov}\{\theta_i, \theta_j\}\,. \tag{16.19}$$

Therefore with (16.17) the following prior distribution of the parameters is assumed:

$$\underline{\theta} \sim \mathcal{N}(0, \underline{K}_1)\,. \tag{16.20}$$

In contrast, the kernel matrix in (16.18) describes the covariances between the *outputs*:

$$K_2(i, j) = \mathrm{cov}\{y(i), y(j)\}\,. \tag{16.21}$$

Therefore with (16.18) the following prior distribution of the outputs is assumed:

$$\underline{y} \sim \mathcal{N}(0, \underline{K}_2)\,. \tag{16.22}$$

The relationship between the distribution of outputs and weights also becomes clear by observing how $\underline{y}$ and $\underline{w}$ are related (without any regularization, i.e., for $\lambda = 0$)

$$\underline{y} = \underline{K}\, \underline{w} \;\Rightarrow\; \underline{w} = \underline{K}^{-1} \underline{y} \tag{16.23}$$

and taking into account that covariance scale quadratically. Thus, if $\underline{y} \sim \mathcal{N}(0, \underline{K})$ then $\underline{w} \sim \mathcal{N}(0, \underline{K}^{-1})$. This explain the "$\underline{K}^{-1}$" in (16.17).

16.5 Reproducing Kernel Hilbert Spaces (RKHS)

This section is mostly based on the excellent thesis [484] and the paper [140]. The conventional procedure with neural networks is as follows: (i) a model architecture (MLP, RBF, ...) is selected and (ii) the model is fit to the data. Step (ii) can be as simple as parameter optimization for a fixed model structure (number of neurons), or it can be more advanced combined structure and parameter optimization, e.g., with a growing or pruning strategy. Step (i) is determined by many factors like computational demand for training and evaluation, interpolation and extrapolation behavior, ease-of-use, available software, or just personal experience and preference. Oftentimes many architectures are tried out and compared.

RKHS tries to solve the more fundamental problem of fitting the "best" function through the data, not just fitting a given function type (architecture) best. Of course, the key issue here is how "best" is defined. Intuitively the function should be as smooth as possible but still describe the data. In RKHS the roughness (opposite of smoothness or quality) of a function is measured with the norm of this function in Hilbert space induced by a kernel K which is written as $||f||_{\mathcal{H}_K}$. At this point, it should be clear that RKHS does not really "solve" the problem in a universal manner. Rather it converts the problem of finding the best function $f(\cdot)$ into the problem of finding the best kernel K. Nevertheless, this is extremely beneficial as it moves the problem in a higher (potential very powerful) abstraction level.

16.5.1 *Norms*

For a space, the norm $||\cdot||$ induces a metric (a notion of distance, length, or size). For example, in an n-dimensional Euclidean space $\mathbb{R}^n$, the squared 2-norm is given by

$$||\underline{x}||^2 = x_1^2 + x_2^2 + \ldots + x_n^2 \,. \tag{16.24}$$

It allows to measure the length or size of a vector $\underline{x}$ by $||\underline{x}||$ and the distance between two vectors $\underline{x}_1$ and $\underline{x}_2$ by $||\underline{x}_1 - \underline{x}_2||$. In general, a norm can be defined via an inner or scalar product:

$$||\underline{x}||^2 = \langle \underline{x}, \underline{x} \rangle = \underline{x}^T \underline{x} \,. \tag{16.25}$$

In the same manner, the norm of a *function* $f(\cdot)$ can be expressed via the inner product

$$||f(\cdot)||^2 = \langle f(\cdot), f(\cdot) \rangle = \int_{-\infty}^{\infty} f(x)^2 dx \,. \tag{16.26}$$

The inner product in an RKHS between a function $f(\cdot)$ and the kernels placed on the data points $\underline{u}(i)$ is equal to the function evaluated at the data point

$$\langle K(\cdot, \underline{u}(i)), f(\cdot)\rangle_{\mathcal{H}_K} = f(\underline{u}(i)) \tag{16.27}$$

and therefore also

$$\langle K(\cdot, \underline{u}(i)), K(\cdot, \underline{u}(j))\rangle_{\mathcal{H}_K} = K(\underline{u}(i), \underline{u}(j))\,. \tag{16.28}$$

The inner product in an RKHS can be evaluated via the kernel! Therefore, it is possible to avoid calculating the inner product directly. It allows to evaluate inner products even in infinite-dimensional spaces (integrals instead of sums) by just evaluating the kernel for all point-pairs of the finite data; see below.

Equation (16.27) is called the reproducing property and explains the name "RKHS." The left side of (16.27) and (16.28) measures the norm of or "similarity" between two functions which is something very complex. And the reproducing property states that this is equivalent to something very simple. As (16.28) states, the norm of $K(\cdot, \underline{u}(i))$ and $K(\cdot, \underline{u}(j))$ is equivalent to the evaluation of the kernel. Thus, the kernel can be interpreted as a *similarity measure*.

In the literature, this property usually is described by transferring something of infinite dimension (comparing functions) to something of finite dimension N (evaluating the kernel at all data point-pairs).

16.5.2 RKHS Objective and Solution

In general, RKHS start with the following objective

$$\min_{f \in \mathcal{H}} \sum_{i=1}^{N} V\left[y(i), f(\underline{u}(i))\right] + \lambda ||f||^2_{\mathcal{H}_K}\,. \tag{16.29}$$

Here $V(\cdot, \cdot)$ can represent any kind of loss function that compares the measured outputs $y(i)$ with the model outputs $\hat{y}(i) = f(\underline{u}(i))$. Later it will be chosen as squared error loss. It may be convenient to scale V by $1/N$ to make λ independent of the amount of data.

The optimization problem in (16.29) searches for the function in Hilbert space that minimizes two conflicting objectives that are traded off by controlling the regularization parameter λ:

- Loss function $V\left[y(i), f(\underline{u}(i))\right]$ measures the *quality of the fit* of the training data. Interpolation would be optimal as it yields a zero error.
- Squared norm of the function in Hilbert space $||f||^2_{\mathcal{H}_K}$ measures the *smoothness of the function*. A constant of zero would be optimal.

$\lambda = 0$ leads to the interpolation solution, $\lambda = \infty$ yields the zero solution. In practice, adjusting λ well is decisive. Also the choice of K is extremely important. These issues are discussed in Sect. 16.7.

It is very remarkable that problem (16.29) has an extremely simple solution whose structure is independent of $V[\cdot, \cdot]$. The root cause for this is the so-called *representer theorem* from 1950. It proves that the optimal solution to the function approximation problem in (16.29) has the following form [11]:

$$\hat{y}(\underline{u}) = \sum_{i=1}^{N} w_i K(\underline{u}, \underline{u}(i)) \, . \tag{16.30}$$

This can be interpreted as a neural network with N hidden layer nodes, and the kernels can be seen as basis functions. It is basically an RBF network with the centers of the basis functions placed on the training data points. The only conceptional distinction is that usually for neural networks, the number of neurons M is not identical to the number of data points N.

Evaluated at the training data this becomes

$$\underline{\hat{y}} = \underline{K} \, \underline{w} \, . \tag{16.31}$$

Thus, the problem of finding the best function $f(\cdot)$ has been extremely simplified to the problem of finding the best parameters/weights w_i, thereby converting an infinite-dimensional problem to an N-dimensional one. The generality of this theorem is very remarkable; it does not even require a special error measure.

However, if the loss function $V(\cdot, \cdot)$ is chosen as the sum of squared errors, then the weights w_i can be obtained very easily by least squares. The objective is

$$\min_{f \in \mathcal{H}} \sum_{i=1}^{N} [y(i) - f(\underline{u}(i))]^2 + \lambda ||f||^2_{\mathcal{H}_K} \tag{16.32}$$

and the error in vector form becomes

$$\underline{e} = \underline{y} - \underline{\hat{y}} = \underline{y} - \underline{K} \, \underline{w} \tag{16.33}$$

where $\underline{K}$ is the kernel matrix in (16.11). Then, (16.32) can be written as

$$\min_{\underline{w}} \underline{e}^T \underline{e} + \lambda \, \underline{w}^T \underline{K} \, \underline{w} \tag{16.34}$$

since $\underline{w}^T \underline{K}\, \underline{w}$ measures the function norm in the RKHS with kernel $K(\cdot, \cdot)$, compare (16.28). The penalty term can alternatively be formulated as

$$||f||^2_{\mathcal{H}_K} = \sum_{i=1}^{N} \sum_{j=1}^{N} w_i w_j K(u(i), u(j)) = \underline{w}^T \underline{K}\, \underline{w}\,. \tag{16.35}$$

It is stated in [208] about (16.34): "Notice that this minimization is equivalent to a ridge regression in a new set of features, one that measures the similarity of an exemplar to each of the other exemplars." Pure ridge regression results if the kernel matrix $\underline{K}$ approaches the identity matrix $\underline{I}$ as it would happen if all data points are very far apart of each other (not similar at all). Then the penalty term degenerates to $\lambda\, \underline{w}^T \underline{I}\, \underline{w} = \lambda\, \underline{w}^T \underline{w}$. In (16.35), the parameters in $\underline{w}$ are weighted with the similarity of their associated data points. For similar points (close to each other), the penalty is big; for less similar points (far away from each other), the penalty is small. This enforces smoothness on the model.

Finally, the optimal parameters minimizing (16.34) can be calculated by least squares:

$$\underline{\hat{w}} = \left(\underline{K} + \lambda \underline{I}\right)^{-1} \underline{y}\,. \tag{16.36}$$

Note that $\underline{K}$ is a symmetric, positive definite matrix, and if this were not the case, the additional regularization term $\lambda\, \underline{I}$ would make it always invertible and well conditioned if λ is big enough. The equations (16.30) and (16.36) are identical to (16.9) but more in the style of this book.

16.5.3 Equivalent Kernels and Locality

It is explained in [140] that the representation given in (16.30), (16.31), and (16.36) has a different interpretation than the *equivalent representation* which is given by basis functions weighted with the output measurements

$$\underline{\hat{y}} = \underline{K}\, \underline{w} = \underline{K} \left(\underline{K} + \lambda \underline{I}\right)^{-1} \underline{y} = \underline{S}\, \underline{y} \tag{16.37}$$

with

$$\underline{S} = \underline{K} \left(\underline{K} + \lambda \underline{I}\right)^{-1} \tag{16.38}$$

being the smoothing or hat matrix known from Sect. 3.1.11. Note that for $\lambda = 0$ (no regularization) $\underline{S} = \underline{I}$, i.e., $\underline{\hat{y}} = \underline{y}$, because this corresponds to the *interpolation* case.

In the language of kernel methods, the columns of $\underline{S}$ are called *equivalent kernels*. Note that the roles of weights and kernels interchange going from the kernel representation to the equivalent kernel representation. In the kernel representation, the kernels are simply evaluated by plugging in the input data, while the weights are estimated by LS. The weights also depend on the strength of regularization λ. In the equivalent kernel representation, however, the "weights" simply are the measured outputs $\underline{y}$, while the equivalent kernels are estimated by LS and depend on the strength of regularization λ. Therefore the equivalent kernels reveal the overall relationship from measured output y to model output $\hat{y}$ including the influence of the estimation and regularization.

In [140] and more extensively in [186] also, the meaning of *locality* is discussed in the context of kernels. Traditionally, a model is considered local if its basis functions or kernels have local characteristics (like, e.g., Gaussians). However, good arguments can be made that it is the characteristics of the *equivalent* kernels that really count, because it determines how neighboring output measurements influence the model output. Examples can be given that global kernels can lead to local equivalent kernels [140]. Therefore locality is a more complex issue than it appears at first sight. In [543] more advanced aspects of understanding equivalent kernels are discussed.

Figure 16.6 shows the equivalent kernels for a Gaussian and inverse quadratic kernel of identical length scale $\sigma = \alpha = 0.2$ located at $y = 0.5$. This equivalent kernel illustrates the influence of the median output value on the model output; it represents the center column of the smoothing matrix $\underline{S}$ assuming equidistantly distributed inputs and outputs. Taking into account the identical length scales, it is surprising how much more global (a) is compared to (b). This clearly underlines that local kernels may act quite globally via the effect of the weight estimation.

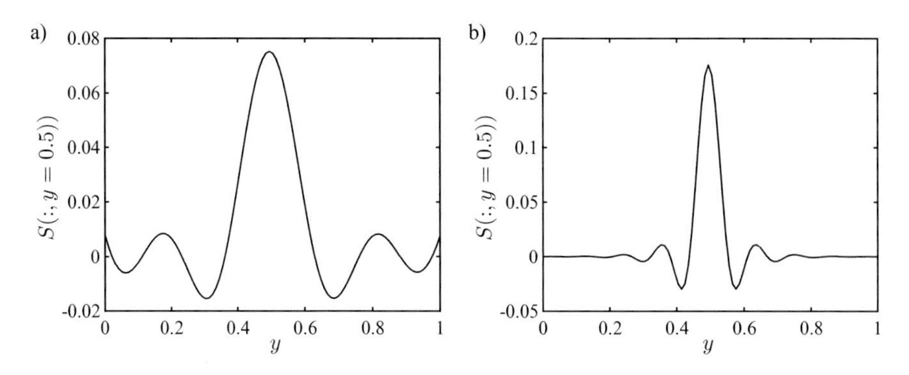

Fig. 16.6 Equivalent kernels for equidistantly spaced inputs and outputs in [0, 1] for (**a**) Gaussian and (**b**) inverse quadratic kernels for output value $y = 0.5$ with identical length scales $\sigma = \alpha = 0.2$

16.5.4 Two Points of View

The article [208] advocates the use of least squares kernel methods as *the* modern standard tool. It should replace generalized linear models that remain the workhorse in the social sciences for regression and classification. This article describes two different points of view on (16.30) which give different interpretations.

16.5.4.1 Similarity-Based View

This view starts from the data points and their relationship to each other like in the nearest neighbor approaches.

The kernel function $K(\underline{u}, \underline{u}(i))$ measures the similarity between an input vector which is $\underline{u}$ and the input vector of data point i which is $\underline{u}(i)$. Most kernels are local (compare Fig. 16.2), i.e., if both points coincide, $K(\underline{u}(i), \underline{u}(i)) = 1$, while if $\underline{u}$ and $\underline{u}(i)$ are far apart from each other $K(\underline{u}, \underline{u}(i)) \rightarrow 0$. The weight w_i is associated with the input vector $\underline{u}(i)$. It corresponds to the part of the model output which is determined by data sample i. If the model is evaluated at $\underline{u}$, the most significant portion of the model output comes from the data points similar to $\underline{u}$. What "similar" means is defined by the kernel. Typically, "similar" implies "close" in input space. However, in some applications, it might be reasonable to choose periodic kernels or locally periodic kernels in order to reflect repeating phenomena.

For estimation of the weights, w_i the kernel matrix (plus regularization term) needs to be inverted in (16.36). The kernel matrix contains the similarity measures of all training data points to each other. The diagonal entries contain the similarities of the points to themselves (typically $= 1$), while the other entries contain the similarities of two different training data points (typically < 1).

16.5.4.2 Superposition of Kernels View

This view starts with basis functions like RBF networks.

The kernel functions $K(\underline{u}, \underline{u}(i))$ in (16.30) can be seen as basis functions. The model output is just a weighted sum of these bases. In this view, a basis function is placed on each training data point resulting in an interpolating RBF network. Due to the regularization term in (16.36), the network does not really *interpolate* although its nominal number of parameters is identical to the number of data points N. For $\lambda > 0$ the effective number of parameters is smaller than N, and the training data is not fit exactly. As the number of basis functions always[8] is N, the flexibility of the network is only controlled by the choice of λ.

---

[8]The size of the training data set N is not selected to control the network's flexibility like it is usually done with the number of neurons M.

Looking at the kernel matrix $\underline{K}$ in (16.11), the first view focuses on the *rows* of the matrix (data points), while the second view focuses on the *columns* of the matrix (basis functions). Since a basis function is placed on each data point and therefore $\underline{K}$ is symmetric, both views are equivalent.

They deviate if not all basis functions shall be considered. It might be reasonable or even necessary due to limited computational resources to *select* only the significant basis functions by some subset selection technique; see Sect. 3.4. Alternatively, a couple of basis functions or data points can be fused or clustered to reduce the computational effort for training and model evaluation [538]. Refer to [474] for a unified probabilistic view on these sparse approaches.

16.6 Gaussian Processes/Kriging

Gaussian processes or Kriging is an alternative to RKHS treating the approximation problem from a probabilistic point of view. Bayesians introduce assumptions via a prior over functions instead of imposing smoothness assumptions on $f(\cdot)$ as in the RKHS approach. This leads to exactly identical equations. However, it is more difficult to understand (in the opinion of the author) but offers an additional probabilistic interpretation. Direct benefits are (i) a direct derivation of confidence intervals and (ii) an elegant way to determine optimal hyperparameters via maximizing the marginal likelihood function; see Sect. 16.7.

16.6.1 Key Idea

For an extensive and thorough treatment of Gaussian processes, refer to the standard monograph [476]. This section goes easy on the math and tries to discuss the topic in an intuitive manner. Figure 16.7 illustrates the key idea of Gaussian processes. The model output is not just described as the deterministic functional relationship between inputs and output $\hat{y} = f(\underline{u})$. Rather the output is described as a probability density function, namely, a Gaussian. The mean μ of this Gaussian then gives the most likely model output used for prediction. The standard deviation σ of this Gaussian can be used for constructing the confidence interval (errorbars) $\mu \pm k\,\sigma$ where k corresponds to the user-specified confidence level, e.g., $k = 1.96$ for a probability of 95%.

Of course, a certain Gaussian can describe the model output $\hat{y}$ only for a specific input value $\underline{u} = \underline{u}_*$. In order to describe $\hat{y}(\underline{u})$ for any input $\underline{u}$, the mean value and the standard deviation of the Gaussian need to be able to vary with $\underline{u}$. This is realized by expressing the covariance/similarity between two output values $y(i)$ and $y(j)$ by a kernel comparing their associated inputs $K(\underline{u}(i), \underline{u}(j))$.

These ideas are depicted in Fig. 16.7. If the kernels are wide (large length scale), neighboring points are highly correlated, and the change of the Gaussian over $\underline{u}$

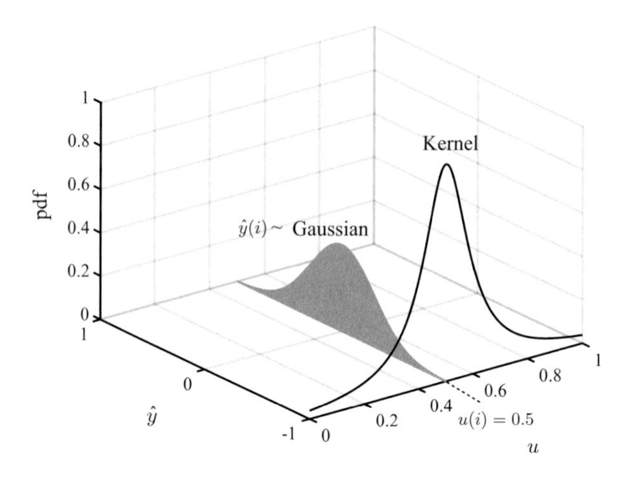

Fig. 16.7 Visualization of a Gaussian process model for data point i at location $u(i) = 0.5$ and the model output $\hat{y}(i)$ is described as a Gaussian probability density function. The kernel determines the correlations between neighboring data points in u. Typical kernels (here the inverse quadratic kernel) ≈ 1 for points nearby $u(i)$ and ≈ 0 for points far away

will be small. If the kernels are narrow (small length scale), neighboring points are weakly correlated, and the change of the Gaussian over $\underline{u}$ *can* be large. Thus, the kernel expresses the prior knowledge about the process, i.e., before any data is known. This establishes a prior pdf about $\hat{y}$. In the next step, when the measurement data comes in, this prior pdf has to be adjusted via Bayes' theorem in order to calculate the posterior pdf. The adjustment step involves a least squares estimation in order to fit the model to the data. It can be seen as the "training" of a Gaussian process model.

16.6.2 Some Basics

If an n-dimensional random vector $\underline{x}$ is normally distributed with mean $\underline{\mu}$ and covariance matrix $\underline{\Sigma}$, this is denoted by

$$\underline{x} \sim \mathcal{N}\left(\underline{\mu}, \underline{\Sigma}\right) . \tag{16.39}$$

This means, it follows the normal distribution

$$p(\underline{x}) = \frac{1}{\sqrt{(2\pi)^n \det(\underline{\Sigma})}} \exp\left(-\frac{1}{2}(\underline{x}-\underline{\mu})^T \underline{\Sigma}^{-1} (\underline{x}-\underline{\mu})\right) . \tag{16.40}$$

Note that drawing *one* sample from this distribution yields an n-dimensional vector corresponding to one realization of $\underline{x}$. In the following sections, $\underline{x}$ stands for the model or process output, while the covariance matrix $\underline{\Sigma}$ depends on the model/process inputs. Thus, the distribution is over output variables, but its shape and the geometry of its contours depend on the input variables.

The log likelihood function of such a multivariate Gaussian is the natural logarithm of this pdf which evaluates to

$$\ln p(\underline{x}) = -\frac{1}{2}(\underline{x} - \underline{\mu})^T \underline{\Sigma}^{-1}(\underline{x} - \underline{\mu}) - \frac{1}{2}\ln\det(\underline{\Sigma}) - \frac{n}{2}\ln 2\pi \, . \qquad (16.41)$$

Note that the "$-\frac{1}{2}$" in the first term comes from the exponent of the exp function; in the second and third term, it comes from the inverse square root. The likelihood is generally computed in logarithmic scale which is a monotonically increasing function – so the position of the optimum does not change. "Taking the logarithm not only simplifies the subsequent mathematical analysis, but it also helps numerically because the product of a large number of small probabilities can easily underflow the numerical precision of the computer, and this is resolved by computing instead the sum of the log probabilities."[9] The likelihood or log-likelihood is maximized or alternatively the negative one is minimized.

16.6.3 Prior

The idea in Gaussian processes is to specify the covariance between the *outputs* of two data points $(\underline{u}(i), y(i))$ and $(\underline{u}(j), y(j))$ in terms of a kernel comparing their *inputs*:

$$\text{cov}\{y(i), y(j)\} = K(\underline{u}(i), \underline{u}(j)) \, . \qquad (16.42)$$

The reasoning here is if the process is smooth, then similar inputs should yield similar outputs. The covariance is maximal if $u(i) = u(j)$ and consequently in the noise-free case $y(i) = y(j)$. Typically, the covariance decreases with increasing distance of the inputs which is fulfilled for local kernels. One noticeable exception is periodic kernels which imply periodic covariances.

The prior pdf is generated by choosing a mean of zero (best guess if nothing else is known) and a covariance matrix equal to the kernel matrix reflecting the prior

[9] http://math.stackexchange.com/questions/892832/why-we-consider-log-likelihood-instead-of-likelihood-in-gaussian-distribution

knowledge about the similarity of different points. In order to be able to visualize the prior, some test input data $\underline{u}_*(i)$, $i = 1, \ldots, N$ must be generated. In this chapter test data is marked by the subscript "$_*$" as in [476]:

$$\underline{U}_* = \begin{bmatrix} \underline{u}_*^T(1) \\ \underline{u}_*^T(2) \\ \vdots \\ \underline{u}_*^T(N) \end{bmatrix} = \begin{bmatrix} u_{*1}(1) & u_{*2}(1) & \cdots & u_{*p}(1) \\ u_{*1}(2) & u_{*2}(2) & \cdots & u_{*p}(2) \\ \vdots & \vdots & & \vdots \\ u_{*1}(N) & u_{*2}(N) & \cdots & u_{*p}(N) \end{bmatrix}, \tag{16.43}$$

Then the kernel matrix $\underline{K}(\underline{U}_*, \underline{U}_*)$ equivalent to (16.11) can be constructed that contains all (test point)-to-(test point) similarities. It is symmetric and positive-definite. So it perfectly can be utilized as covariance matrix for a normal distribution $\underline{\Sigma} = \underline{K}$. Then the prior pdf of the outputs becomes

$$\underline{\hat{y}}_* = \begin{bmatrix} \hat{y}_*(1) \\ \hat{y}_*(2) \\ \vdots \\ \hat{y}_*(N) \end{bmatrix} \sim \mathcal{N}\left(\underline{0}, \underline{K}(\underline{U}_*, \underline{U}_*)\right) . \tag{16.44}$$

Examples for such priors can be seen in Fig. 16.8. $N = 21$ test inputs were distributed equidistantly, and five samples were drawn from (16.44) for each figure. A larger length scale leads to smoother, which is less wiggly functions.

Example 16.6.1 (Gaussian Process for Two Training Data Points) This example shall illustrate the role of the kernel, the implied Gaussian process, and the resulting model for two training data points $(u(1), y(1))$ and $(u(2), y(2))$. For simplicity, the noise-free and unregularized case (i.e., interpolation $\hat{y}(i) = y(i)$) is assumed. First, let's consider only knowledge about the input values $u(1)$ and $u(2)$. Figure 16.9

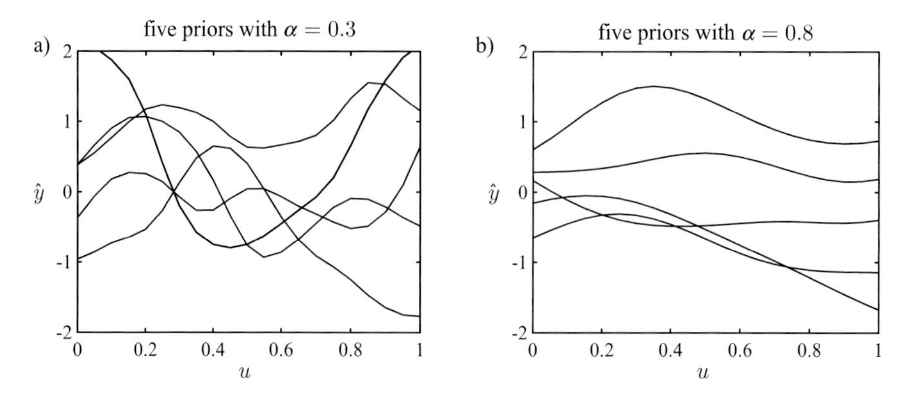

Fig. 16.8 Five prior distributions based on 21 equidistant test points in u with inverse quadratic kernel for length scale (**a**) $\alpha = 0.3$, (**b**) $\alpha = 0.8$

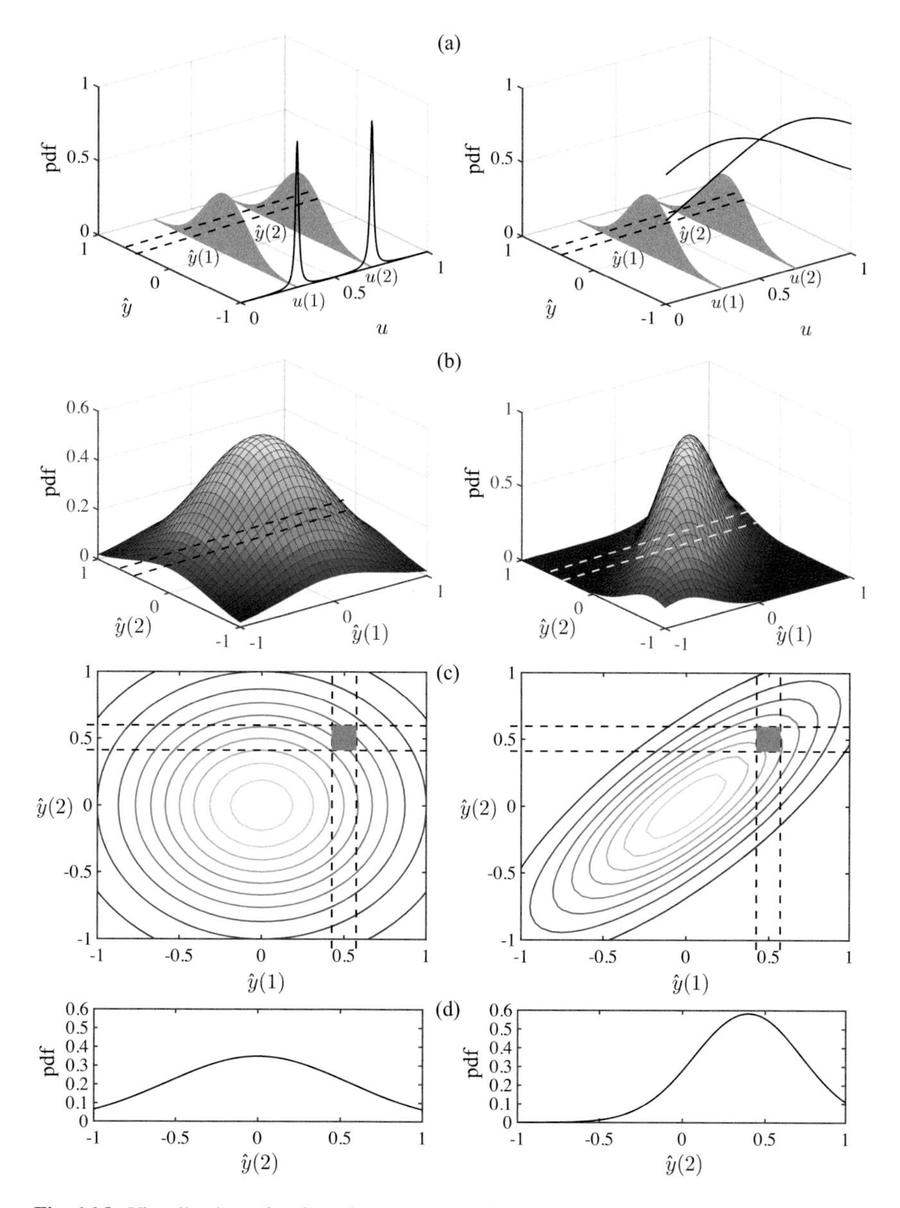

Fig. 16.9 Visualization of a Gaussian process model (**a**) and prior (**b**, **c**) for two training data point locations $u(1) = 0.3$ and $u(2) = 0.7$ with almost zero covariance (kernels are zero at the other point, respectively) (left) and covariance 0.8 (kernels have values of 0.8 at the other point, respectively) (right). The dotted lines indicate the intervals $\hat{y}(1) = 0.5 \pm 0.1$ and $\hat{y}(2) = 0.5 \pm 0.1$. Due to the high covariance on the right, the probability (pdf volume under the gray square) for almost identical values for $\hat{y}(1)$ and $\hat{y}(2)$ is much bigger than for a very low covariance on the left. (**d**) Pdf of $\hat{y}(2)$ conditioned on $\hat{y}(1) = 0.5$ which corresponds to a slice through the plot above at $\hat{y}(1) = 0.5$

shows a scenario for very low covariance (left) and very high covariance (right) for an inverse quadratic kernel. The kernel matrix $\underline{K}$ is identical with the covariance matrix $\underline{\Sigma}$ of the Gaussian distribution of the model outputs:

$$\underline{K} = \begin{bmatrix} K(u(1), u(1)) & K(u(1), u(2)) \\ K(u(2), u(1)) & K(u(2), u(2)) \end{bmatrix} = \underline{\Sigma} \tag{16.45}$$

with $u(1) = 0.3$ and $u(2) = 0.7$ for this example. The variances on the diagonal entries are $K(u(1), u(1)) = K(u(2), u(2)) = 1$ independent of the length scale. The covariances for length scale $\alpha \to 0$ (left) are roughly 0, while for length scale $\alpha = 0.8$ (right), they are

$$K(0.3, 0.7) = K(0.7, 0.3) = \frac{1}{1 + (0.7 - 0.3)/0.8)^2} = 0.8\,, \tag{16.46}$$

compare (16.2).

In the Gaussian process model, a kernel is placed on each training data point. The kernel matrix obtained by evaluating all kernels for all training data points is used as a covariance matrix for a Gaussian distribution describing the model outputs. In this example, the kernel matrix is 2×2, and the only difference between both scenarios is the covariance between both training points.

As a consequence, if the covariance is very low, then the model outputs $\hat{y}(1)$ and $\hat{y}(2)$ are almost independent of each other (covariance ≈ 0). This means they easily can have very different values. For example, if the probability of $\hat{y}(1)$ and $\hat{y}(2)$ is in the interval $[0.4, 0.6]$ (marked as a gray square in the contour plots in Fig. 16.9c) is p, then the probability of both values being in this interval concurrently is p^2 which is much lower than p. Thus, kernels with small length scales easily can yield almost independent model output values.

In contrast, if the covariance is high, then the probability of both values being in this interval concurrently is higher than p^2. In the extreme case, where both values are perfectly correlated (covariance ≈ 1), their values must be identical. The probability of both values being in this interval concurrently then is p as well.

It can be observed from the pdfs in Fig. 16.9 that the volume under the gray-shaded area is much higher on the right. On the other hand, it would be much lower in the interval $0.4 < \hat{y}(1) < 0.6 \wedge -0.6 < \hat{y}(2) < -0.4$. Similar values are very likely, dissimilar ones are very unlikely for a large positive covariance.

Figure 16.10 illustrates a GP for two data points at (0.3, 0.8) and (0.7, 0.6) for four different length scales for an inverse quadratic kernel. The kernel (and therefore covariance) matrices for the length scales (a) $\alpha = 0.1$, (b) $\alpha = 0.3$, (c) $\alpha = 0.8$, and (d) $\alpha = 2$ are

$$\text{(a) } \underline{K}_{\alpha=0.1} = \begin{bmatrix} 1 & 0.059 \\ 0.059 & 1 \end{bmatrix}, \quad \text{(b) } \underline{K}_{\alpha=0.3} = \begin{bmatrix} 1 & 0.36 \\ 0.36 & 1 \end{bmatrix}, \tag{16.47}$$

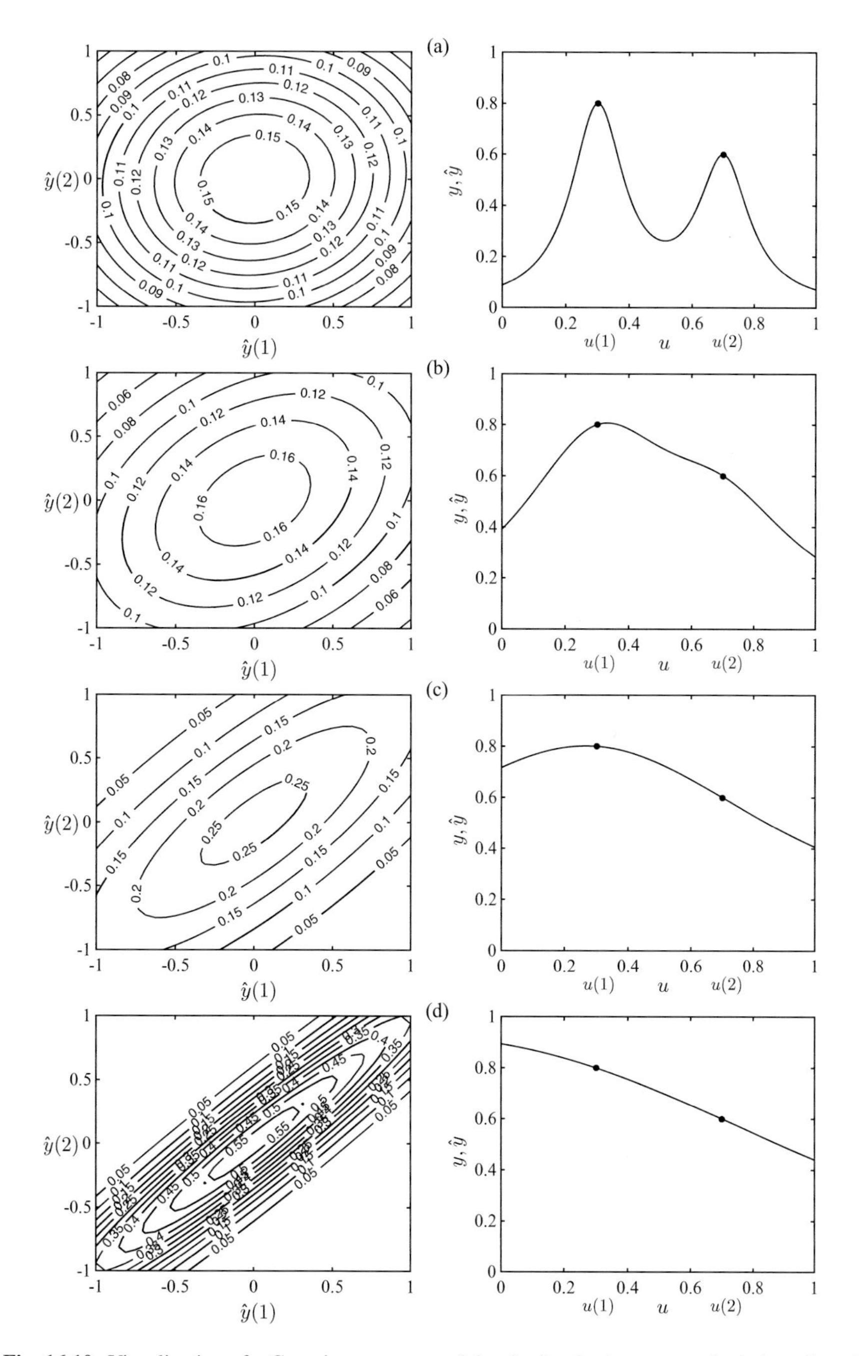

Fig. 16.10 Visualization of a Gaussian process model and prior for inverse quadratic kernels and different length scales (**a**) $\alpha = 0.1$, (**b**) $\alpha = 0.3$, (**c**) $\alpha = 0.8$, (**d**) $\alpha = 2$: Contours of GPM prior distributions (left). Two data points (0.3, 0.8) and (0.7, 0.6), and the GPM model output (right)

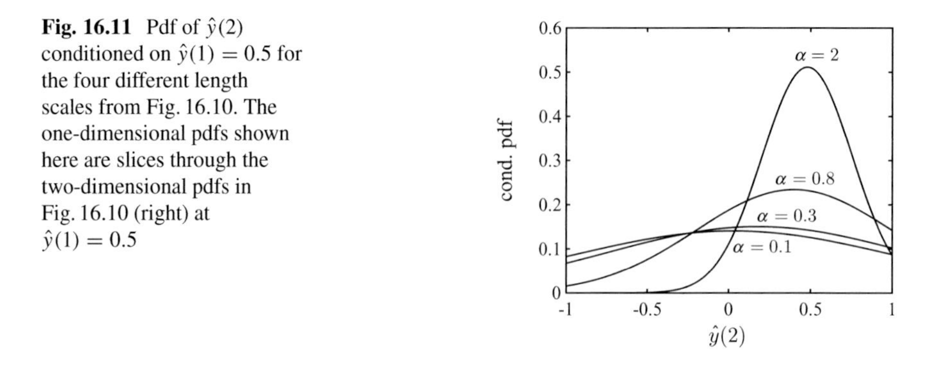

Fig. 16.11 Pdf of $\hat{y}(2)$ conditioned on $\hat{y}(1) = 0.5$ for the four different length scales from Fig. 16.10. The one-dimensional pdfs shown here are slices through the two-dimensional pdfs in Fig. 16.10 (right) at $\hat{y}(1) = 0.5$

$$\text{(c) } \underline{K}_{\alpha=0.8} = \begin{bmatrix} 1 & 0.80 \\ 0.80 & 1 \end{bmatrix}, \quad \text{(d) } \underline{K}_{\alpha=2} = \begin{bmatrix} 1 & 0.96 \\ 0.96 & 1 \end{bmatrix}. \tag{16.48}$$

First, only the measured input values $u(1)$ and $u(2)$ are necessary to construct the prior distributions. It clearly can be seen in Fig. 16.10 (left) that wider kernels lead to rotated and narrowed Gaussians and smoother models. Once the covariance has reached a critical value (around 0.8), the orientation of the Gaussian is roughly 45°; the main effect of even higher covariances then is caused by squeezing the Gaussian together.

Next, let's consider also the knowledge about the measured output values $y(1)$ and $y(2)$. The plots in Fig. 16.10 (right) were generated by first estimating the two weights associated with the two kernel functions with LS by (16.36) without regularization ($\lambda = 0$). This step is explained after this example. Then the model with two kernels (basis functions) can be evaluated at arbitrary input values by (16.30).

Consider Fig. 16.11: if the output is known only for $u(1)$ to be $\hat{y}(1) = y(1) = 0.5$, then the pdfs in Fig. 16.10 (left) will collapse to one-dimensional pdfs for $\hat{y}(2)$. For the four length scales α considered in Fig. 16.10, these pdfs are compared. The higher the covariance between both points is, the further the pdfs' centers approach $\hat{y}(2) = 0.5$, and the narrower they become.

Figure 16.12 illustrates the relationship between the prior pdf and five possible realization for two input locations $u(1) = 0.3$ and $u(2) = 0.7$ for low ($\alpha = 0.3$) and high ($\alpha = 0.8$) covariance. Note that all this can be established before the measurement of the outputs $y(1)$ and $y(2)$ is known. □

16.6.4 Posterior

Figure 16.10 (right) already shows the GPMs output $\hat{y}$ over the input u. Since this is the noise-free and therefore interpolation case (no regularization, $\lambda = 0$), the data is fit *exactly*. In order to accomplish that, the optimal weights have to be

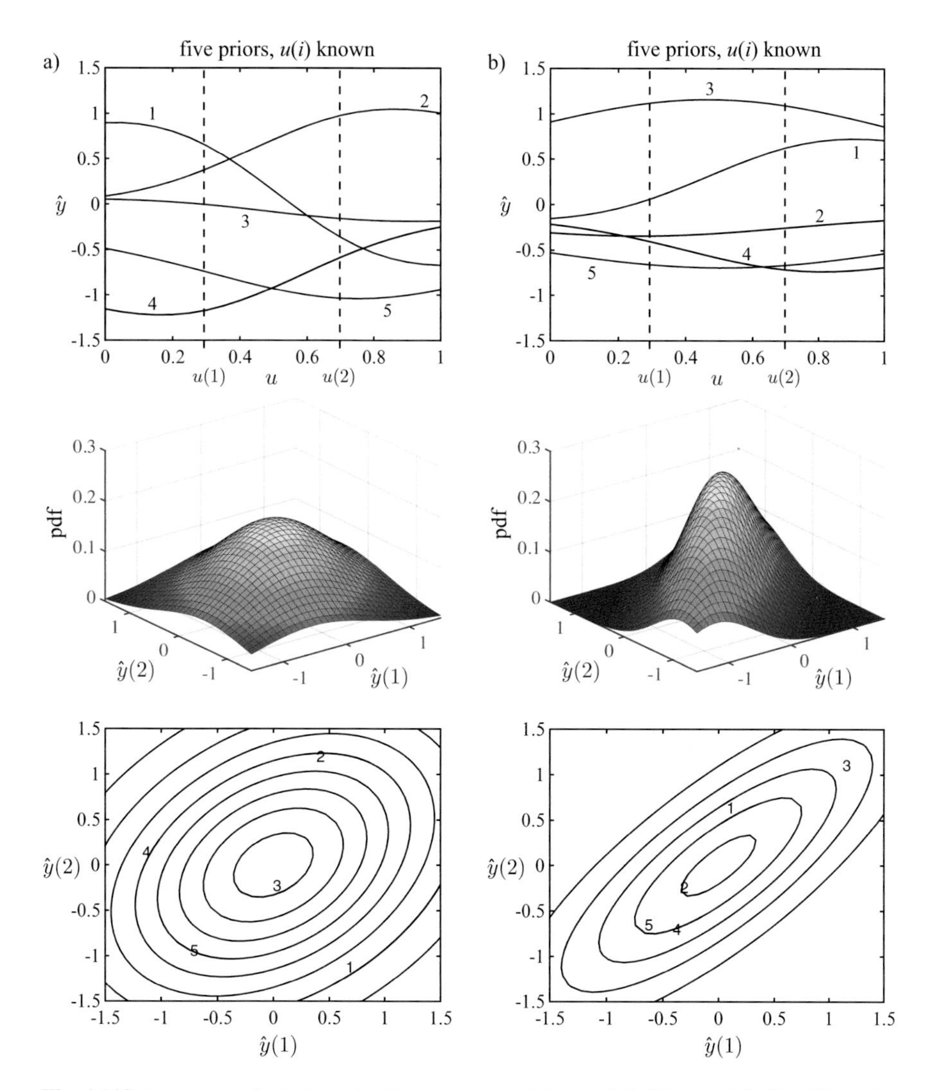

Fig. 16.12 Inverse quadratic kernel with length scale (**a**) $\alpha = 0.3$, (**b**) $\alpha = 0.8$. Top: Five prior distributions with samples for $[\hat{y}(1)\ \hat{y}(2)]^T$ drawn from the pdf below. Middle: Pdf describing the probabilistic relationship between $\hat{y}(1)$ and $\hat{y}(2)$. Bottom: Contour plot of the pdf with the locations (numbers) of the five realizations

estimated. Assume N data points $\{\underline{u}(i), y(i)\}, i = 1, \ldots, N$ have been collected, then the following steps need to be carried out: (i) building the kernel matrix $\underline{K}$, (ii) estimating the optimal weights by least squares with (16.36), and (iii) using the model in (16.30). As will be shown in the subsequent section, the additional consideration of output noise leads to exactly identical expressions as in RKHS also for $\lambda > 0$.

The equivalence between RKHS and Gaussian process models is nicely explained in [140]: "Assume that the problem is to estimate $f(\cdot)$ from sparse data $y(i)$ at location $\underline{u}(i)$. From the previous description, it is clear that choosing a kernel [matrix] $\underline{K}$ is equivalent to assuming a Gaussian prior on $f(\cdot)$ with covariance [matrix] equal to $\underline{K}$. Thus choosing a prior through $\underline{K}$ is equivalent (a) to assume a Gaussian prior and (b) to assume a correlation function associated with the family of functions $f(\cdot)$. The relation between positive definite kernels and correlation functions $\underline{K}$ of Gaussian random processes is characterized in detail in [582], theorem 5.2."

However, the statements above just give the deterministic model output which corresponds to the mean of the Gaussian distribution for $\hat{y}$. In Gaussian processes, one is interested in the *whole* distribution, i.e., also in the variance. In order to obtain that, one must go through the following steps:

1. Gather the training data: $\{\underline{u}(i), y(i)\}, i = 1, \ldots, N$.
2. Build the $N \times N$-dimensional kernel matrix $\underline{K}$ in (16.11) from the input data.
3. Build the N-dimensional joint normal distribution (Gaussian) of the outputs. **Prior:**

$$\underline{\hat{y}} = \begin{bmatrix} \hat{y}(1) \\ \hat{y}(2) \\ \vdots \\ \hat{y}(N) \end{bmatrix} \sim \mathcal{N}\left(\underline{0}, \underline{K}\right) . \tag{16.49}$$

4. Generate the test or query input $\underline{u}_*(N+1)$ at which the model shall be evaluated.
5. Build the $N+1$-dimensional joint normal distribution (Gaussian) for the combined training and test outputs

$$\begin{bmatrix} \underline{\hat{y}} \\ \hat{y}_*(N+1) \end{bmatrix} = \begin{bmatrix} \hat{y}(1) \\ \hat{y}(2) \\ \vdots \\ \hat{y}(N) \\ \hat{y}_*(N+1) \end{bmatrix} \sim \mathcal{N}\left(\underline{0}, \underline{\widetilde{K}}\right) \tag{16.50}$$

with the $(N+1) \times (N+1)$-dimensional kernel matrix

$$\underline{\widetilde{K}} = \begin{bmatrix} \underline{K} & \underline{k}(\underline{u}_*) \\ \underline{k}^T(\underline{u}_*) & K(\underline{u}_*, \underline{u}_*) \end{bmatrix} . \tag{16.51}$$

It measures the similarities between the training data inputs to each other in $\underline{K}$, between the training data inputs and the test data input in

$$\underline{k}(u_*) = \begin{bmatrix} K(u_*, u(1)) \\ K(u_*, u(2)) \\ \vdots \\ K(u_*, u(N)) \end{bmatrix}, \tag{16.52}$$

and the test data input to itself $K(\underline{u}_*, \underline{u}_*) = 1$ for standard kernels. This formulation is open to an extension for multiple test points where u_* and $\hat{y}_*$ become vectors which will not be discussed here.

6. In the final decisive step, this $N + 1$-dimensional joint pdf in (16.50) for $[\underline{\hat{y}} \;\; \hat{y}_*]^T$ must be reduced to a one-dimensional pdf for the test output $\hat{y}_*$. This is done by *conditioning* the joint Gaussian prior distribution on the measured outputs $\underline{y}$ which in the interpolation case are identical to the model outputs $\underline{\hat{y}}$: $p\left(y_* | \underline{\hat{y}}(i) = \underline{y}(i)\right)$. This conditioning means that the effects of the first N dimensions of the distribution are taken into account. The $N + 1$-dimensional Gaussian is sliced at the N output values of the data points $\hat{y}(i) = y(i), i = 1, \ldots, N$ resulting in a one-dimensional Gaussian. Figure 16.11 illustrates this procedure for an example with $N = 1$ where a 2D Gaussian is reduced to a 1D Gaussian by fixing the value in one dimension.

   This operation again yields a Gaussian.[10] According to [476], this conditioning yields the following one-dimensional pdf for the test output which is called the **posterior:**

$$\hat{y}_* \sim \mathcal{N}\left(\underline{k}^T(\underline{u}_*)\underline{K}^{-1}\underline{y},\; K(\underline{u}_*, \underline{u}_*) - \underline{k}^T(\underline{u}_*)\underline{K}^{-1}\underline{k}(\underline{u}_*)\right) \tag{16.53}$$

   where the kernel matrix $\underline{K} = \underline{K}\left(\underline{U}, \underline{U}\right)$ only depends on the training data inputs. Note that for computation of the variance, just the inputs are required, i.e., the training inputs $\underline{u}(i)$ for calculation of $\underline{K}$ and additionally the test input $\underline{u}_*$ for calculation of $\underline{k}(\underline{u}_*)$, while for the computation of the mean the training outputs $y(i)$ are required as well, $i = 1, \ldots, N$.

   The mean of the normal distribution in (16.53) exactly corresponds to the scalar product $\underline{k}^T(\underline{u}_*)\,\underline{\hat{w}}$ which is the sum of kernel basis functions evaluated at the test point $\underline{k}^T(\underline{u}_*)$ multiplied with the associated optimal weights $\underline{\hat{w}} = \underline{K}^{-1}\underline{y}$. The variance expression in (16.53) is further discussed in Sect. 16.6.6.

[10]Slices through Gaussians and marginal distributions of Gaussians always are Gaussians themselves. This is the reason why Gaussian process models are working so nicely and efficiently.

Example 16.6.2 (Gaussian Process for Two Training and One Test Point) For this example, consider again the scenario with two training data points illustrated in Fig. 16.10. In order to generate the plots on the right hand side, a test point must sweep the interval $u_* = 0 \dots 1$. The prior for the training data alone is given by

$$\begin{bmatrix} \hat{y}(1) \\ \hat{y}(2) \end{bmatrix} \sim \mathcal{N}\left(\begin{bmatrix} 0 \\ 0 \end{bmatrix}, \begin{bmatrix} K(u(1), u(1)) & K(u(1), u(2)) \\ K(u(2), u(1)) & K(u(2), u(2)) \end{bmatrix}\right). \tag{16.54}$$

With a test point at $u_*(3)$, the extended kernel matrix becomes

$$\underline{\widetilde{K}} = \begin{bmatrix} \underline{K} & \begin{bmatrix} K(u(1), u_*(3)) \\ K(u(2), u_*(3)) \end{bmatrix} \\ [K(u_*(3), u(1)) \;\; K(u_*(3), u(2))] & K(u_*(3), u_*(3)) \end{bmatrix}. \tag{16.55}$$

Starting from the kernel matrices for $u(1) = 0.3$ and $u(2) = 0.7$ given in (16.47) and (16.48), for the test input $u_*(3) = 0.4$, this extends to (for the length scales $\alpha = 0.1, 0.3, 0.8, 2$):

$$\underline{\widetilde{K}}_{\alpha=0.1} = \begin{bmatrix} 1.0 & 0.059 & 0.5 \\ 0.059 & 1.0 & 0.1 \\ 0.5 & 0.1 & 1.0 \end{bmatrix}, \quad \underline{\widetilde{K}}_{\alpha=0.3} = \begin{bmatrix} 1.0 & 0.36 & 0.9 \\ 0.36 & 1.0 & 0.5 \\ 0.9 & 0.5 & 1.0 \end{bmatrix}, \tag{16.56}$$

$$\underline{\widetilde{K}}_{\alpha=0.8} = \begin{bmatrix} 1.0 & 0.8 & 0.98 \\ 0.8 & 1.0 & 0.88 \\ 0.98 & 0.88 & 1.0 \end{bmatrix}, \quad \underline{\widetilde{K}}_{\alpha=2} = \begin{bmatrix} 1.0 & 0.96 & 1.0 \\ 0.96 & 1.0 & 0.98 \\ 1.0 & 0.98 & 1.0 \end{bmatrix}. \tag{16.57}$$

As can be observed with $\underline{\widetilde{K}}_{\alpha=2}$, the kernel matrix approaches singularity as the length scale $\alpha \to \infty$.

After measuring $y(1) = 0.8$ and $y(2) = 0.6$ the model is forced to fit these data points. As stated in [476]: “Graphically ... you may think of generating functions from the prior, and rejecting the ones that disagree with the observations, although this strategy would not be computationally very efficient.”

In terms of probability calculations, the three-dimensional pdf with kernel matrix $\underline{\widetilde{K}}$ is conditioned on the measured output values $y(1)$ and $y(2)$. Thereby, the three-dimensional pdf for $[\hat{y}(1)\ \hat{y}(2)\ \hat{y}_*(3)]^T$ reduces to a one-dimensional pdf for $\hat{y}_*(3)$ with the following mean and variance:

$$\hat{y}_*(3) \sim \mathcal{N}\left(\underline{k}^T(u_*(3))\, \underline{w},\ 1 - \underline{k}^T(u_*(3))\underline{K}^{-1}\underline{k}(u_*(3))\right) \tag{16.58}$$

with the two parameters $\underline{w} = [w_1\ w_2]^T$ estimated by least squares:

$$\underline{w} = \underline{K}^{-1}\underline{y}\,. \tag{16.59}$$

The model output is just the mean of the Gaussian distribution in (16.58). Assuming test input $u_*(3) = 0.4$ and the kernel matrices from (16.47) and (16.48), this computes for the considered length scales to $\hat{y}_{*\alpha=0.1}(3) = 0.42$, $\hat{y}_{*\alpha=0.3}(3) = 0.74$, $\hat{y}_{*\alpha=0.8}(3) = 0.77$, and $\hat{y}_{*\alpha=2}(3) = 0.79$, respectively. If u_* sweeps from 0 to 1, the model output can be seen in Fig. 16.10 (right) for all considered length scales. □

16.6.5 Incorporating Output Noise

Up to this point, only the noise-free case was considered in order to keep things simple. Figure 16.13 explains the notation where the true functional output f is disturbed by noise n which yields the measured process output y; this notation is adopted from [476]. In the noise-free case, $y = f$ and $n = 0$. The section shall explain how output noise is taken into account and therefore $y \neq f$. In statistics y is called an *observable variable* (directly measured), while f is called a *latent variable* (inferred through a mathematical model). In other parts of this book, f is denoted by y_u to emphasize that it purely is caused by the input $\underline{u}$.

If n is additive *independent* identically distributed Gaussian noise with variance σ_n^2, then the variances of each output var$\{y(i)\}$ increases by this amount:

$$\mathrm{cov}\{y(i), y(j)\} = \begin{cases} K(\underline{u}(i), \underline{u}(j)) + \sigma_n^2 & \text{for } i = j \\ K(\underline{u}(i), \underline{u}(j)) & \text{for } i \neq j \end{cases} . \tag{16.60}$$

Thus, the kernel matrix changes:

$$\underline{K} \rightarrow \underline{K} + \sigma_n^2 \underline{I} = \underline{K} + \lambda \underline{I} \tag{16.61}$$

with $\lambda = \sigma_n^2$ to see the correspondence to RKHS. In all equations simply $\underline{K}$ needs to be replaced by $\underline{K} + \sigma_n^2 \underline{I}$. This introduces a new hyperparameter $\lambda = \sigma_n^2$ which needs to be determined somehow besides the length scale σ or α. It controls the amount of regularization. Section 16.7 is devoted to these hyperparameters.

The relationship between f and y in probabilistic terms is discussed in some depth in Sect. 16.7.3.

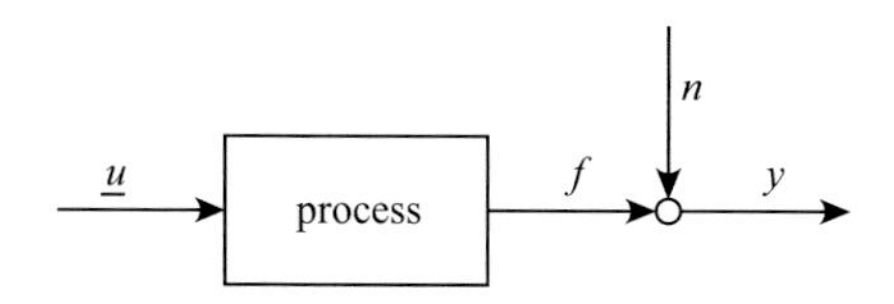

Fig. 16.13 If the process output is disturbed by noise n, it is important to distinguish between the outputs f (undisturbed) and y (disturbed)

16.6.6 Model Variance

A fundamental property of GPs is that the output is described as a random variable with a Gaussian distribution. Besides its mean which was in the focus of the discussion in the previous sections, it possesses a variance. According to (16.53), the variance of the distribution σ_*^2 (= one-dimensional covariance matrix Σ_* of the test output, i.e., a scalar) can be calculated by

$$\sigma_*^2 = \Sigma_* = K(u_*, u_*) - \underline{k}^T(u_*) \left(\underline{K} + \sigma_n^2 \underline{I}\right)^{-1} \underline{k}(u_*) \,. \tag{16.62}$$

With this knowledge the confidence interval can be computed as $\mu \pm k\sigma_*$ where, e.g., $k = 1.96$ corresponds to 95% confidence and $\mu = \underline{k}^T(\underline{u}_*) \left(\underline{K} + \sigma_n^2 \underline{I}\right)^{-1} \underline{y}$ is the mean of the distribution according to (16.53). As emphasized in [476], (16.62) gives one-dimensional variances for single test points. However, if multiple test points are used, then u_* becomes an nu_*-dimensional vector, and $\underline{k}(\underline{u}_*)$ becomes an $N \times nu_*$-dimensional matrix where nu_* is the number of test points. And then the resulting matrix becomes an $nu_* \times nu_*$-dimensional covariance matrix. Its diagonal entries are the variances of the test points $\underline{u}_*$, and the off-diagonal entries represent the covariance between different test points.

Analyzing (16.62), it can be seen that with typically $K(u_*, u_*) = 1$ being the prior variance of the test point, the output variance decreases as more training data is gathered. Thus, $\sigma_*^2 \leq 1$. In order to allow for a more realistic variance assessment, the prior variance needs to be modeled by another hyperparameter σ_f and in consequence $\sigma_*^2 \leq \sigma_f^2$ which is explained in Sect. 16.7.4.

For simplicity assume the non-regularized (interpolation) case ($\sigma_n = 0$): $\underline{K} \uparrow,\ \underline{K}^{-1} \downarrow,\ -\underline{K}^{-1} \uparrow \ \Rightarrow\ \sigma_* \downarrow$ where $\uparrow$ means increasing and $\downarrow$ decreasing.[11] If the covariance of u_* with the training data inputs in $\underline{U}$ is small as, e.g., for extrapolation, then the output variance increases and vice versa: $\underline{k}(\underline{u}_*) \downarrow,\ -\underline{k}(\underline{u}_*) \uparrow\ \Rightarrow \sigma_* \downarrow$.

Note that for all these variance aspects in GPMs, only the *inputs* of training and test data matter. *The output is irrelevant!* It may only play some indirect role via the hyperparameter optimization discussed in Sect. 16.7. This independence is due to the key assumptions of GPs expressed in (16.42) that similarities at the output level can be described by similarities at the input level.

Figure 16.14 illustrates the dependency of the variance and therefore also of the confidence intervals on the widths or length scales of the kernel. Obviously, with increasing kernel widths, the correlation between neighboring model output values increases, and in turn the uncertainty reduces. At the measurement points, the variance and the uncertainty are equal to zero (in the noise-free, unregularized case).

[11] For matrices and vectors, it can be interpreted as some increasing or decreasing norm.

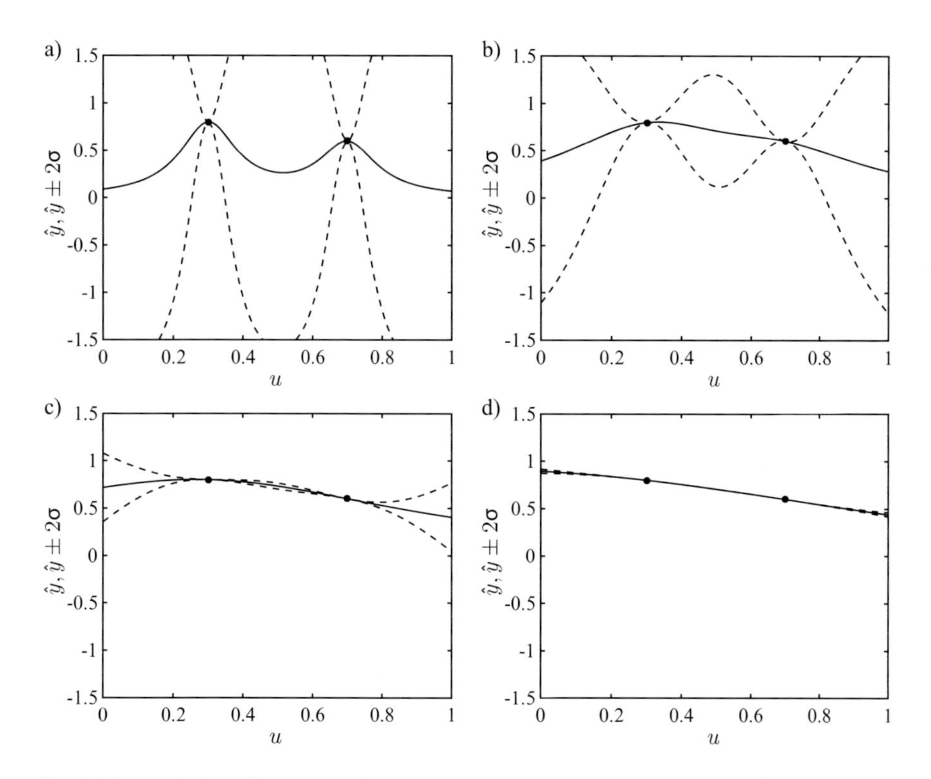

Fig. 16.14 A GP fit (solid) through the two data points from Fig. 16.10 for kernels with increasing widths from **a**) to **d**). The dashed curves represent the 95% confidence intervals

In some of the next figures, the model $\hat{y}$ is shown as a solid curve and the 95% confidence interval $\hat{y} \pm 2\sigma$ as dashed curves. Here σ denotes the standard deviation of the Gaussian process which is totally unrelated to σ used for the length scale for Gaussian kernels.

It is a very popular strategy in the design of experiments to place new measurement data points where the uncertainty measured by the confidence or some related criterion of the model is maximal [494]. For a methodical comparison with alternative approaches, refer to Sect. 14.10.

16.6.7 Incorporating a Base Model

A very simple and powerful idea to improve GP models is the incorporation of a base model, the so-called *bias space*. This is typically realized by adding a weighted sum of basis functions Φ_j, $j = 1, \ldots, M$ to the GP model

$$\hat{y}(\underline{u}) = \sum_{i=1}^{N} w_i K(\underline{u}, \underline{u}(i)) + \sum_{j=1}^{M} \theta_j \Phi_j(\underline{u}) . \tag{16.63}$$

Often the base model is chosen as a linear trend, i.e., as a linear model

$$\hat{y}(\underline{u}) = \sum_{i=1}^{N} w_i K(\underline{u}, \underline{u}(i)) + \sum_{j=0}^{p} \theta_j u_j \tag{16.64}$$

with $u_0 = 1$ for modeling the offset with θ_0 and p being the number of inputs.

Of course, the base model can be utilized to incorporate any prior knowledge in the form of a linearly parameterized model. Since the GP part declines toward zero far away from the training data for local kernels, the overall model then approaches the base model. This is very useful to ensure a desired type of *extrapolation* behavior.

Two alternative approaches can be taken to estimate the base models' parameters: subsequent or simultaneous optimization.

16.6.7.1 Subsequent Optimization

This is the simplest approach. First, the base model is estimated from data. Then the error of this base model is approximated by the GP. Of course, this approach does not yield the overall best parameters since the joint influence of the w_i and θ_j is not considered. How the base model is estimated incorporating the knowledge of the GP approximation comes next.

16.6.7.2 Simultaneous Optimization

In [458] a theorem from [582] is cited which shows that joint estimation of the w_i and θ_j can be performed by computing

$$\underline{\hat{\theta}} = \left(\underline{\Phi}^T \underline{A}^{-1} \underline{\Phi}\right)^{-1} \underline{\Phi}^T \underline{A}^{-1} \underline{y} \tag{16.65}$$

and

$$\underline{\hat{w}} = \underline{A}^{-1} \left(\underline{y} - \underline{\Phi}\,\underline{\hat{\theta}}\right) \tag{16.66}$$

with

$$\underline{A} = \underline{K} + \lambda \underline{I} \tag{16.67}$$

and $\underline{\Phi}$ being the regression matrix for the base model containing the basis functions evaluated for all data points in its columns. The weights for the kernels are estimated by (16.66) which uses the error of the base model as desired "output" $\underline{y} - \underline{\Phi}\,\underline{\hat{\theta}}$. This is perfectly logical since the kernels should compensate for the errors of the base model. In (16.65) a Markov (weighted least squares) estimation with covariance matrix $\underline{A}$ or weighting matrix $\underline{A}^{-1}$ takes place; compare Sect. 3.1.7.

16.6.8 Relationship to RBF Networks

There is a fundamental difference between the modeling approach discussed in this chapter (GPM) and the rest of the book. In (16.36) the number of parameters is equal to the number of data points. A kernel function is placed on each data point. The model flexibility is solely controlled by regularization. In all the other approaches (local model networks, RBF, MLP, fuzzy, polynomial,…), the flexibility mainly is controlled by selecting the structure (number of local models, neuron, rules, terms) and possibly fine-tuned by some regularization (early stopping, ridge regression).

Nevertheless, the approaches are closer connected than one might think. Something very similar to kernel-based methods can be realized by an RBF network by using the basis functions instead of the kernel. The equations would be almost identical: the weighted sum of kernel/basis functions

$$\underline{y} = \sum_{i=1}^{n} \theta_i K(\underline{u}, \underline{c}(i)), \tag{16.68}$$

with $n = N$ (interpolation case) and the weights estimated by ridge regression

$$\underline{\hat{\theta}} = \left(\underline{X}^T \underline{X} + \lambda \underline{I}\right)^{-1} \underline{X}^T \underline{y}. \tag{16.69}$$

The $n \times n$-dimensional regression matrix $\underline{X}$ contains the kernel/basis functions in its columns and the data points in its rows.

Without regularization ($\lambda = 0$) and a kernel/basis function placed on each data point ($\underline{c}(i) = \underline{u}(i)$), both approaches are exactly equivalent $\left(\underline{X}^T \underline{X}\right)^{-1} \underline{X}^T = \underline{X}^{-1} = \underline{K}^{-1}$.

But even with regularization ($\lambda > 0$) both approaches are almost identical. The slight difference is that for RBF networks, the tradeoff for the regularization term happens with the *squared* matrix $\underline{X}^T \underline{X}$, while for GPMs it happens with the (non-squared) kernel matrix $\underline{K}$.

For more background on this topic, refer to [10]. One main conclusion from [10] is: "With this equivalence in place, the main operational difference between the RBF and GPR techniques is the assumption on 'what is known when.' In the RBF technique the data values $y(i)$ are assumed known in advance, whereas the locations to be interpolated (or extrapolated) are only known when the interpolation is evaluated. GPR has the opposite assumption, that the data covariance is known in advance (at arbitrary locations), but the surrounding data values $y(i)$ are not known until run-time.

On the other hand, RBF and GPR models are not always equivalent. There are many choices of covariance that are not radially symmetric, and many choices of radial basis that cannot be covariances. A common instance of the latter was discussed in section ..., i.e., the radial basis functions generated as Green's functions of a regularizer.... These functions are zero at the origin, which precludes considering them as covariances."

16.6.9 High-Dimensional Kernels

Kernel-based approaches are nicely suited for high-dimensional problems because their computational complexity primarily depends on the number of data points, not on the dimension of the input space. Nevertheless, the kernels measure the *similarity* or the *neighborhood* relation between two points, and this measure deteriorates with the dimensionality of the input space [2]. Additionally, in most cases, kernels are required to be local or at least *locality* is a desirable property, and this property is also sensitive with respect to the input space dimensionality [166]. For overcoming these issues which are a manifestation of the curse of dimensionality, two strategies have been proposed. They are explained by the means of the Gaussian kernel.

Gaussians are the most widely used kernels. A "normal" Gaussian utilizes a Euclidean distance metric (2-norm) and a power of 2

$$\exp\left(-0.5\frac{||\underline{d}||_2^2}{\sigma^2}\right) \tag{16.70}$$

where $\underline{d}$ is the distance vector (in Euclidean space) between the evaluation point and the Gaussian's center. A generalized Gaussian can be written as

$$\exp\left(-0.5\frac{||\underline{d}||_k^r}{\sigma^r}\right) \tag{16.71}$$

and is based on a power of r and the k-norm

$$||\underline{d}||_k = \sqrt[k]{d_1^k + d_2^k + \ldots + d_p^k} \tag{16.72}$$

in p-dimensional space.[12]

Gaussians with powers $r > 2$ are usually called *super-Gaussians* [385] or r-Gaussians [166]. They can preserve the local character for higher dimensions. As the dimension p grows, the normal Gaussians accumulate most of their mass in outer

[12]Note that in most of the standard literature, both r and k are typically denoted by p which is reserved for the number of inputs throughout this book.

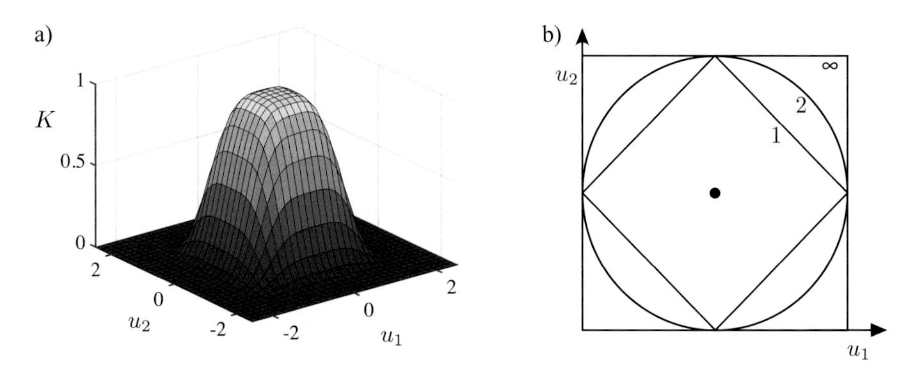

Fig. 16.15 (**a**) A two-dimensional super-Gaussian with $r = 4$, (**b**) a contour line of a "normal" Gaussian for different distance norms for $k = 1, 2, \infty$

regions far away from the center. For super-Gaussians, however, the slope becomes steeper as r increases, enforcing locality; see Fig. 16.15a. Utilizing powers with $r > 2$ in kernel machines is not investigated thoroughly. It might be a promising idea in order to preserve locality in high-dimensional spaces where meaningful neighborhood relationships tend to vanish.

The same is true for norms with $k < 2$. In [2] this is convincingly advocated for high-dimensional problems. Even fractional norms with $k < 1$ have been investigated and yielded superior results for very high-dimensional problems in preserving neighborhood relations as utilized in methods based on nearest neighbors or clustering. Contour lines of normal Gaussians with the 1-, 2-, and ∞-norm are shown in Fig. 16.15b. The distinction between them is significantly emphasized as the dimension grows.

16.7 Hyperparameters

The expression "hyperparameters" denotes all parameters besides the kernel weights $\underline{w}$. They typically control the amount of regularization by one parameter λ and the exact form of the kernel by one or more length scales (e.g., for Gaussian kernels identical σ for all dimensions or individual σ_i for each dimension i or even entries of a full covariance matrix) and possibly other shape parameters for more advanced kernels.

In the simplistic examples above, the length scale σ or α was fixed, and the amount of regularization was set to zero, i.e., $\lambda = 0$. In order to obtain good performance for any real application and avoid overfitting, the hyperparameters have to be tuned somehow. It is intuitively obvious that a good length scale will reflect the spatial frequency of the process, while a good amount of regularization is closely coupled with the noise level as the derivation suggests $\lambda = \sigma_n^2$. The tuning of these hyperparameters is much more important than for standard RBF networks because of the enormous flexibility resulting from placing one kernel function on each training data point and a bad choice for λ can yield serious overfitting.

16.7.1 Influence of the Hyperparameters

This section shall convey an impression of the influence that the length scale and regularization hyperparameters have individually and jointly and how decisive they are for the model behavior.

Example 16.7.1 (Influence of Length Scale and Regularization) This example shall illustrate the individual effect of the length scale and the amount of regularization on the model behavior.

In the *first example*, the interpolation of four training data points is considered. The points lie at $\{u(i), y(i)\}_{i=1}^4 = \{(0.2, 1); (0.4, 2); (0.6, -1); (1, 1)\}$. No regularization is applied ($\lambda = 0$). Thus, all training data is met exactly. The length scale is varied from small to large. Figure 16.16 shows the results for a Gaussian kernel on the left and a multiquadratic kernel on the right. Because both kernels are only locally active, the model always approaches 0 in extrapolation.

The interpolation behavior is problematic for very small length scales because the model drops in between data points. For the Gaussian kernel, this drop happens more pronounced since the Gaussian decays faster than the inverse quadratic.

For small length scales, the estimated weights are close to the output values of the data points; for large length scales, compensation effects become dominant and the weights grow beyond the physical range of y.

Overall, the model quality for both kernels can be considered comparable.

For very small length scales (a), the weights are almost identical to the $y(i)$-values of the data points.

For the double length scale (b), the drops in interpolation behavior are basically gone, and the model quality seems to be much higher (more appropriate). It is obvious that too small length scales severely cripple the model. For (b) the weights show first compensation "effects," i.e., neighboring kernels work together. So, w_1 and w_3 are lower than the function value, while w_2 is higher. This leads to the undesirable effect that the model overshoots around point 2 and undershoots around point 3. This effect is more pronounced in the Gaussian kernel case. However, the interpolation behavior between point 2 and 3 seems to be "strange" for the inverse quadratic kernel. Overall, the model quality to the human eye looks slightly worse for the inverse quadratic kernel.

Doubling the length scale again (c) enhances the above-explained effects. In particular, the negative undershoots around $u \approx 0$ seems undesirable. The estimated weights for the Gaussian kernel are significantly bigger. The model clearly leaves the range of function values $[-1, 2]$ which in many applications is very undesirable.

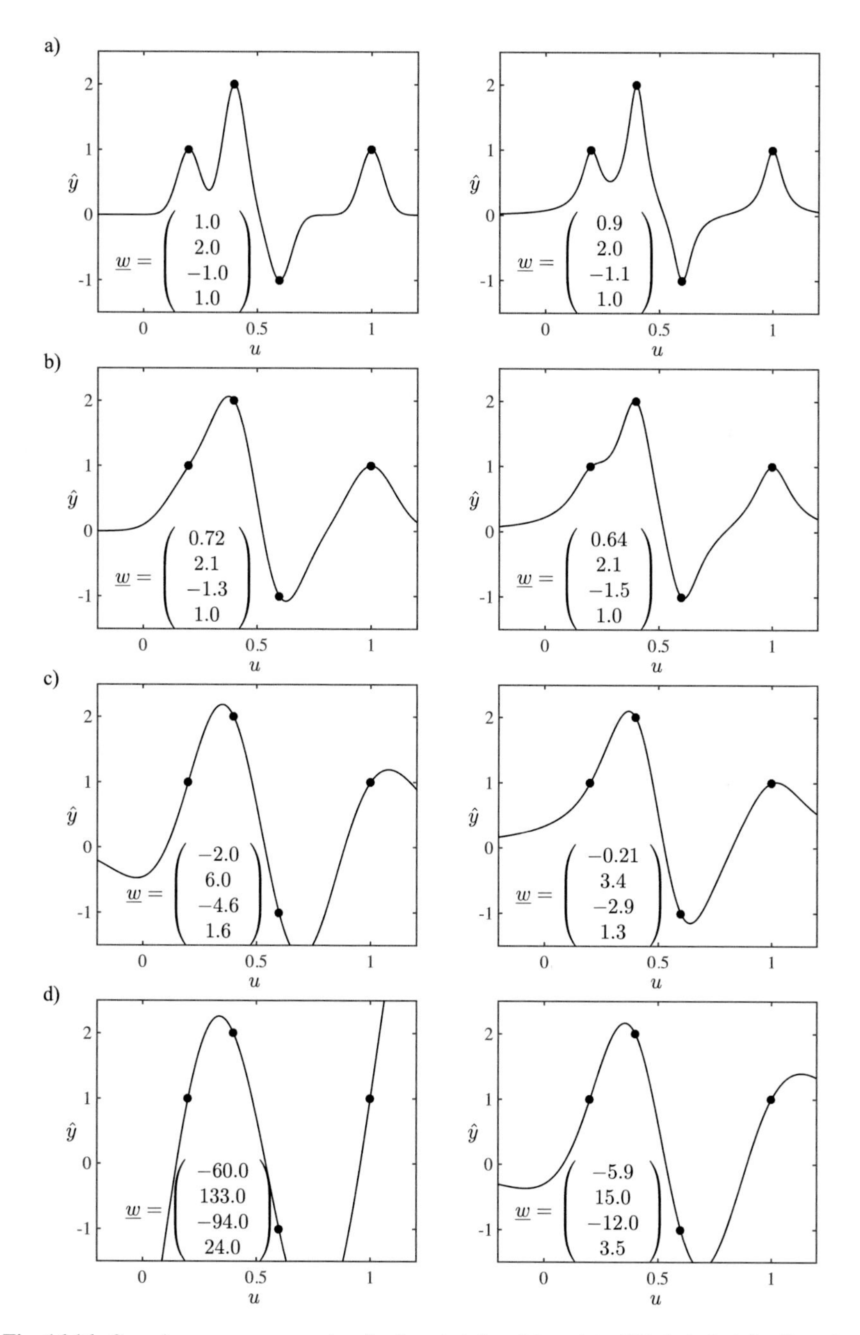

Fig. 16.16 Gaussian process regression for four training data points (filled circles) for Gaussian kernels (left) and inverse quadratic kernels (right) without regularization ($\lambda = 0$) and different length scales: (**a**) $\sigma = \alpha = 0.05$, (**b**) $\sigma = \alpha = 0.1$, (**c**) $\sigma = \alpha = 0.2$, (**d**) $\sigma = \alpha = 0.4$. The given weights are ordered from left to right

Doubling the length scale again yields very extreme weights which are orders of magnitude off. Inter- and extrapolation behavior is completely unreliable. Again it is significantly worse for the Gaussian kernel. Note that these extreme weights obviously would be damped by regularization. For example, $w_2 = 133$ ($\lambda = 0$) becomes $w_2 = 66$ ($\lambda = 0.01$) and $w_2 = 12$ ($\lambda = 0.1$). Nevertheless, too large length scales make the model completely global. All kernels balance each other which can yield disastrous interpolation behavior, particularly in data holes and undesirable extrapolation cases.

Finally, it can be observed that the length scale needs careful tuning, at least if λ is fixed. Thus, a method for automatic optimization of the length scale is required; compare Sect. 16.7.2.

The *second example* shall illustrate the influence of the amount of regularization controlled by λ. Figure 16.17 shows the models with identical length scale for increasing λ. The chosen length scale is a medium $\sigma = \alpha = 0.15$ between cases (b) and (c) in Fig. 16.16.

In (a) for $\lambda = 0.001$, no difference to the interpolation case ($\lambda = 0$) can be perceived. For one order of magnitude bigger regularization ($\lambda = 0.01$) in (b), only minor changes in the weights and in the model output are visible. For $\lambda = 0.1$ in (c), changes are still not very significant although clearly visible. However, $\lambda = 1$ forces the weights to become much smaller and the whole model is squashed toward 0. For all plots in Fig. 16.17, the difference between both kernels is negligible. This behavior is typical for well-adjusted length scales.

The main motivation of the regularization can hardly be understood from this simple four-point example. The big advantage to counter overfitting is better demonstrated in an example with more training data and different levels of process noise. This is investigated next. □

Example 16.7.2 (Process Noise, Regularization, and Overfitting) This example shall illustrate the benefits of regularization for Gaussian process models for a more realistic setting. The process is the hyperbola shown in Fig. 16.18. $N = 30$ training data points are acquired under (a) low and (b) high output noise (white Gaussian) at uniformly randomly distributed locations, identically for both scenarios. The two data sets are also shown in Fig. 16.18.

In this example, the length scale is fixed at $\sigma = \alpha = 0.2$, and the regularization strength λ is varied. All figures show the Gaussian kernel on the left, the inverse quadratic kernel on the right and the low noise case at the top, the high noise case at the bottom. Since the factor between high and low noise variance is $0.1^2/0.01^2 = 100$, this factor can roughly also be expected between the optimal regularization parameters.

First, the unregularized case ($\lambda = 0$) shall be considered; see Fig. 16.19. While the Gaussian kernel performs quite well in the low noise case, the other results are devastating. Obviously, there is a strong need for regularization to reduce overfitting. This is a typical finding for GPs because in the unregularized case, their effective number of parameters is close to the nominal one, i.e., N. Therefore in order to avoid the extremely flexible interpolation case, regularization is required. Note that

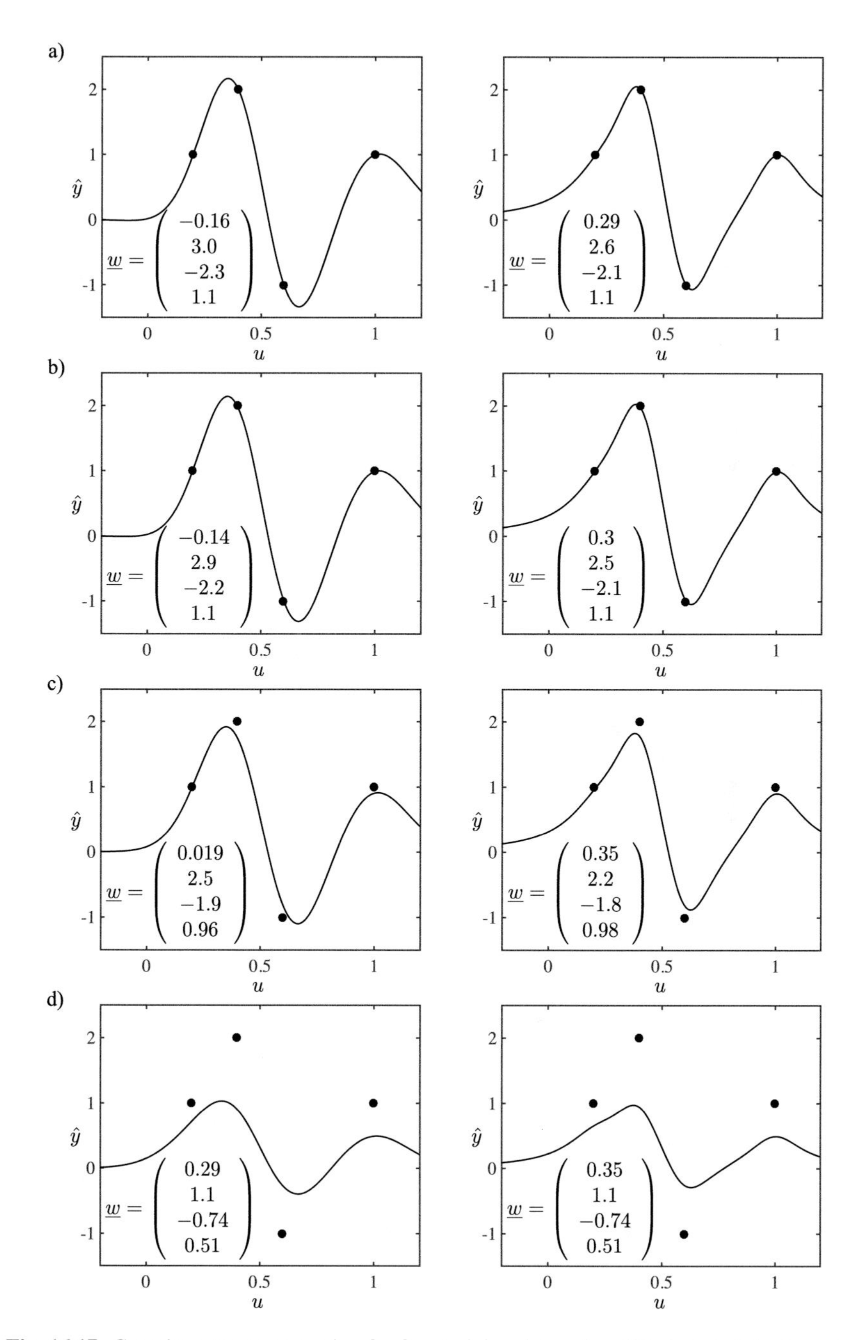

Fig. 16.17 Gaussian process regression for four training data points (filled circles) for Gaussian kernels (left) and inverse quadratic kernels (right) with identical length scale ($\sigma = \alpha = 0.3$) and different regularization strengths: (**a**) $\lambda = 0.001$, (**b**) $\lambda = 0.01$, (**c**) $\lambda = 0.1$, (**d**) $\lambda = 1$. The given weights are ordered from left to right

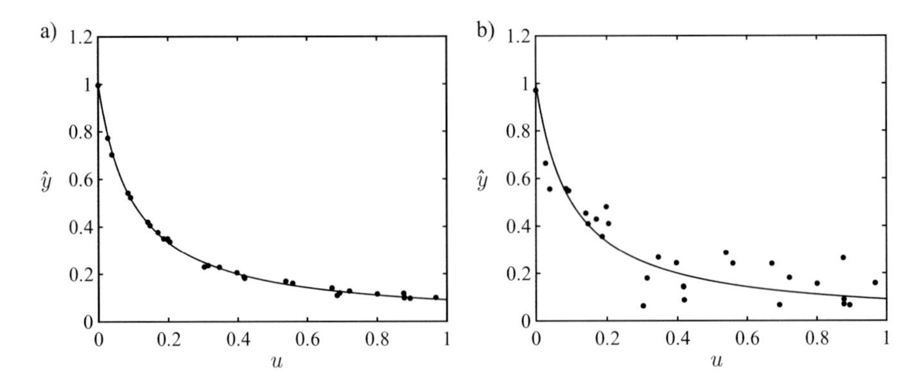

Fig. 16.18 Hyperbola process measured (dots) with (**a**) low noise ($\sigma_n = 0.01$), (**b**) high noise ($\sigma_n = 0.1$)

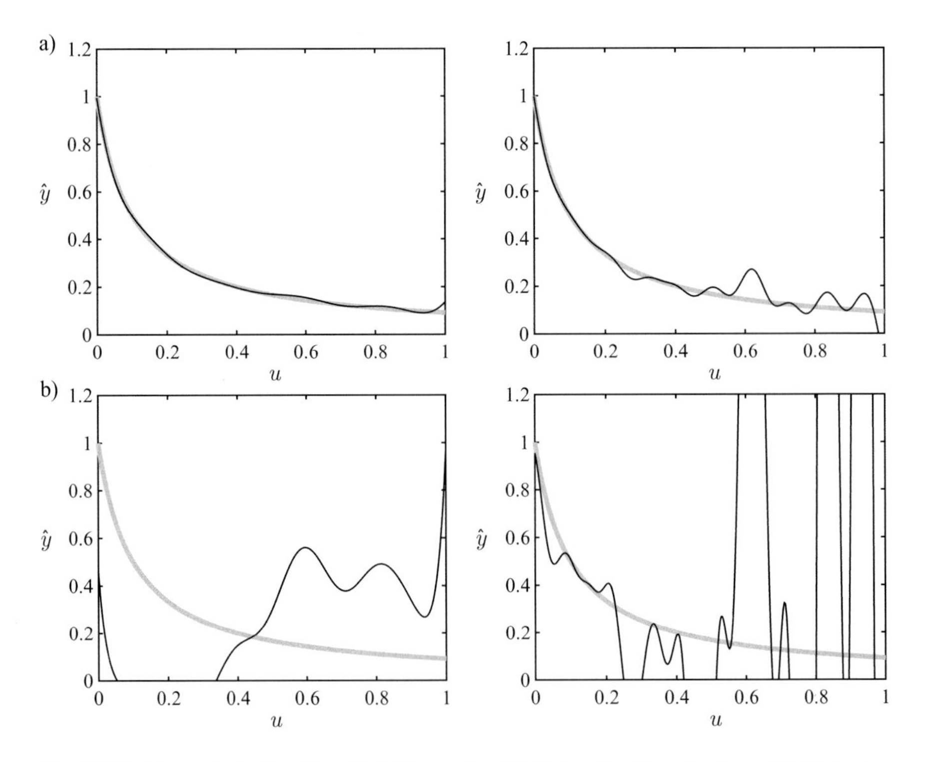

Fig. 16.19 No regularization ($\lambda = 0$): (**a**) low noise, (**b**) high noise. Gaussian kernel (left), inverse quadratic kernel (right)

the inverse quadratic kernel is much more sensitive in this respect since it has a fatter tail and therefore more global properties. Nevertheless, decreasing the length scale is counterproductive as will be seen at the end of this example in Fig. 16.24.

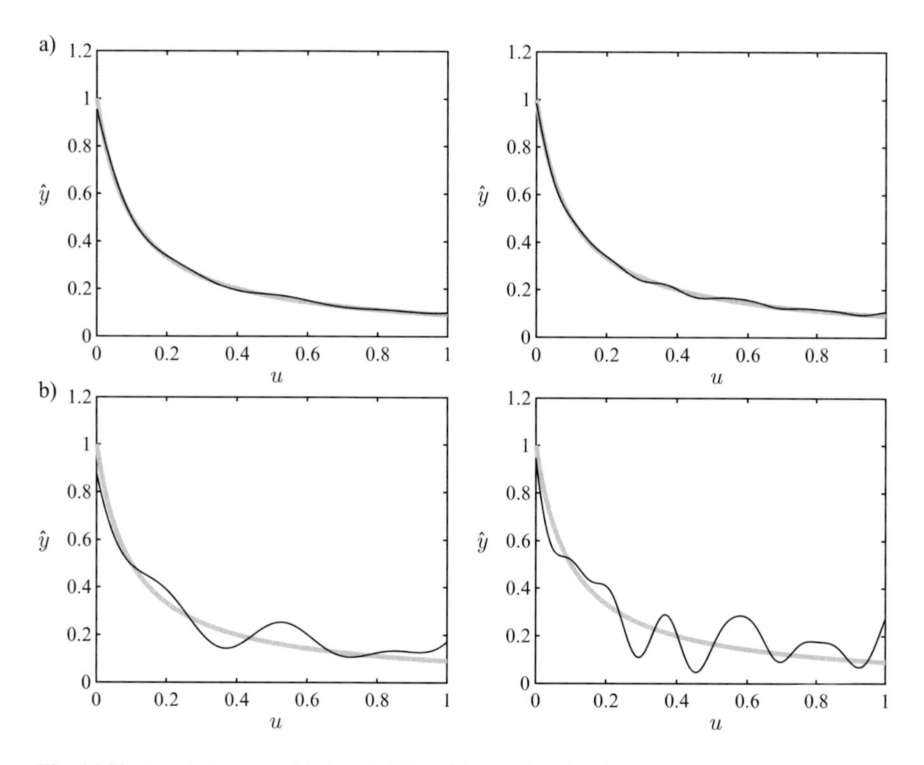

Fig. 16.20 Regularization with $\lambda = 0.001$: (**a**) low noise, (**b**) high noise. Gaussian kernel (left), inverse quadratic kernel (right)

Figure 16.20 demonstrates that even with very low regularization, the overfitting effect reduces extremely. However, the performance for the high noise case is still unsatisfactory.

In Fig. 16.21 the high noise case improves further, but the low noise case starts to deteriorate. Particularly, in the region $u \approx 0$, it can be observed that the model is biased toward zero due to the penalization term. However, for the high noise case, this bias is overcompensated by the positive effect on the model variance manifesting itself by the weaker oscillatory behavior. The model clearly loses flexibility.

Figure 16.22 illustrates this loss in flexibility further. At this point, the bias becomes too large even for the Gaussian kernel.

In order to assess the overall effect of the choice for λ, the model performance can be assessed on validation data. Here the validation data are 300 points equally distributed in [0, 1]. If no validation data is available in practice, similar plots can be generated with cross validation; see Sect. 7.3.2.

Figure 16.23 shows how the validation error performance depends on the regularization parameter. Note that the plot on the top left has a different scale for λ because its optimal value is smaller than 10^{-5}. The wiggles for very small

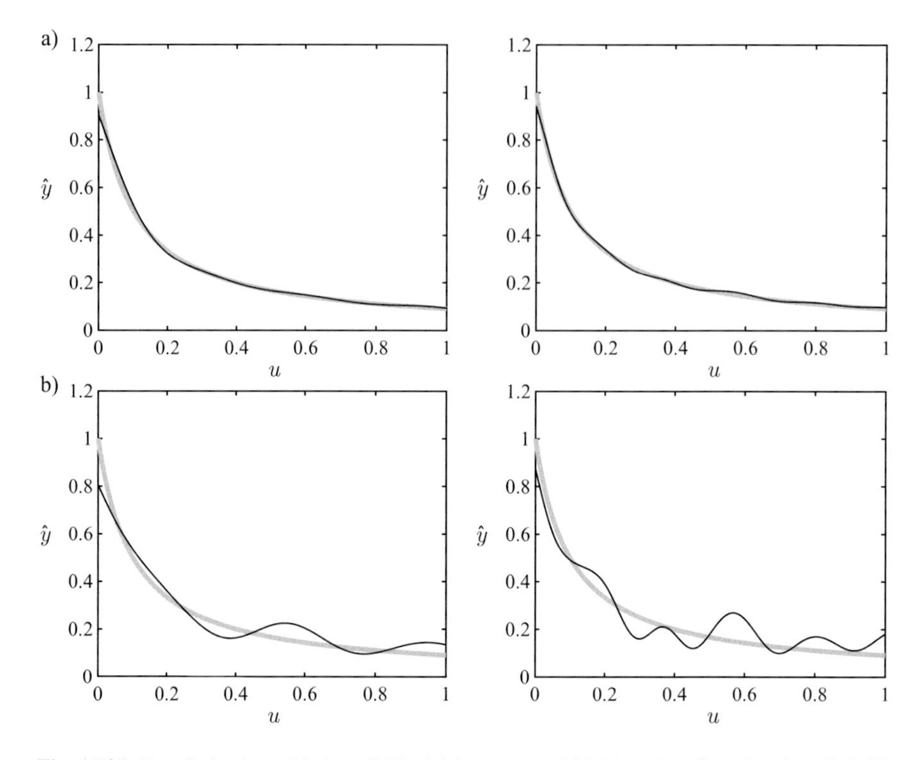

Fig. 16.21 Regularization with $\lambda = 0.01$: (**a**) low noise, (**b**) high noise. Gaussian kernel (left), inverse quadratic kernel (right)

λ values are due to numerical problems in the matrix inversion as a consequence of bad conditioning. More sophisticated matrix inversion algorithms would resolve that. State of the art is using the Cholesky decomposition [476].

It can clearly be observed, as expected, that for a low noise level (a), a weaker regularization is required than for a high noise level (b). Generally, optimal λ values are significantly smaller for the Gaussian kernel (left) than for the inverse quadratic kernel (right). This is probably a direct consequence of the more global (fatter tail) behavior of the latter. Therefore it is more sensitive with respect to overfitting.

For the inverse quadratic kernel, the factor between the optimal values for λ for the low and high noise case is indeed almost 100. For the Gaussian kernel, the factor is 10^4, but the sensitivity of the validation performance with respect to λ is extremely low making these considerations very insecure.

Although the model performance for the optimal amount of regularization is comparable for both kernels, the inverse quadratic kernel is much more sensitive to the correct choice of λ. It is necessary to optimize λ in order to obtain sufficient results. The Gaussian kernel makes a rough estimate of λ easier.

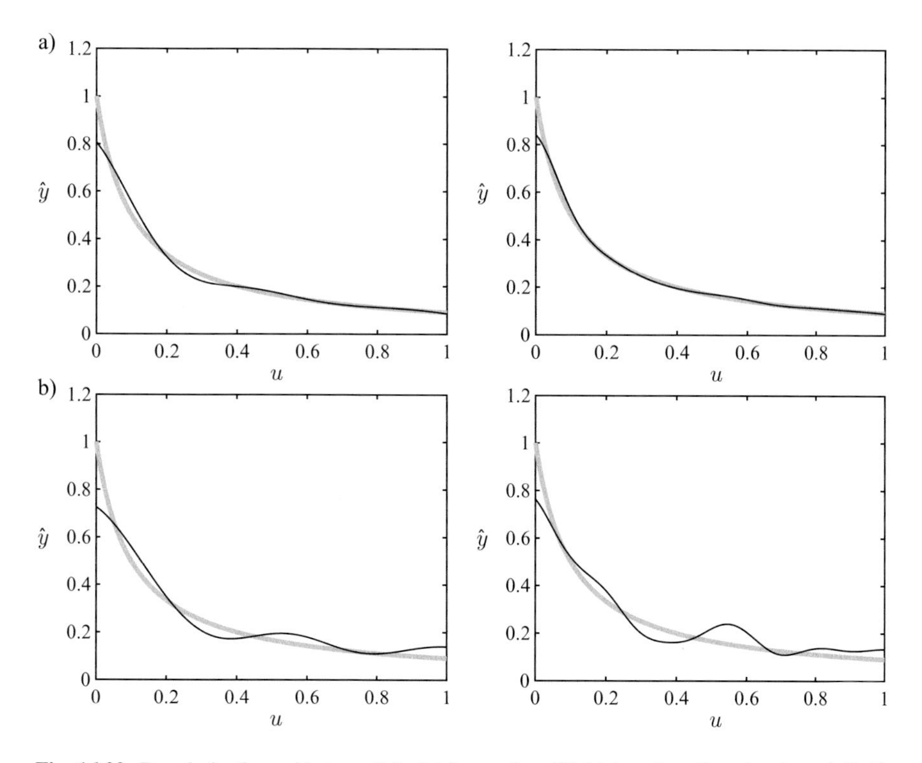

Fig. 16.22 Regularization with $\lambda = 0.1$: (**a**) low noise, (**b**) high noise. Gaussian kernel (left), inverse quadratic kernel (right)

It shall be emphasized that there is a strong interaction between the length scale σ or α and λ. For the high noise case (b), this is illustrated by Fig. 16.24. Thus, both parameters need to be tuned simultaneously.

It can be clearly observed that σ or α is highly correlated with λ. Obviously, in this setting, larger length scales call for smaller amounts of regularization. This negative correlation can be expected since an increase of each hyperparameter makes the model smoother, so they can (partly) compensate each other.

Especially for non-monotonous or even almost periodic processes, the optimum with respect to σ or α is typically very pronounced. In contrast, for a monotonous process as in this example, it is relatively flat.

For Gaussian kernels, the surface in Fig. 16.24 is more rugged, and therefore σ and λ are more difficult to tune. Notice that this observation is extremely process-dependent and it is not possible to draw universal conclusions. □

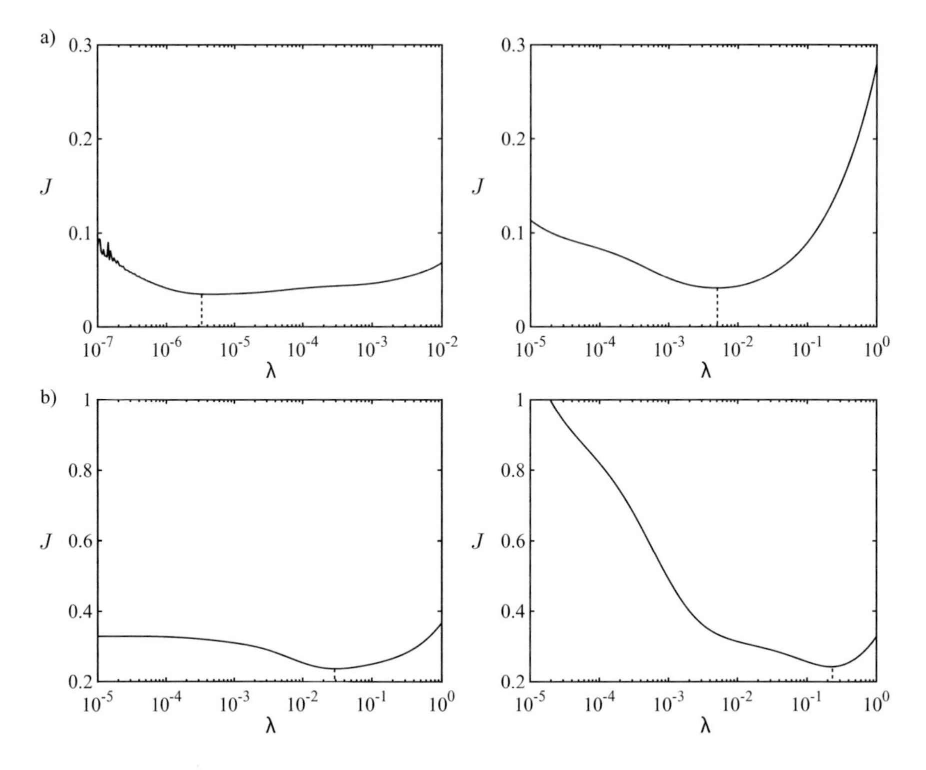

Fig. 16.23 Normalized root mean squared error over λ for $\sigma = \alpha = 0.2$ on validation data distributed equidistantly: (**a**) low noise, (**b**) high noise. Gaussian kernel (left), inverse quadratic kernel (right)

16.7.2 Optimization of the Hyperparameters

Tuning hyperparameters can be seen as a nested loop around estimating the weights; compare Fig. 16.25. It can be compared to structure *optimization* in other neural networks. Increasing λ is a kind of continuous pruning. Instead of eliminating neurons, the GP's flexibility is sucked out of the model.

In the previous section, only one-dimensional examples with two hyperparameters were considered: (i) the length scale σ or α and (ii) the amount of regularization λ. For multivariate problems, a straightforward extension is to allow individual length scales for each dimension. This is illustrated in the next example.

Example 16.7.3 (Different Length Scales for Each Dimension) For this example, 100 training data points originating from a two-dimensional function are used. GP models with Gaussian kernels are employed. The training data and thus the location of the kernels are uniformly randomly distributed in the input space. For validation, the data is distributed on a regular grid. Since the focus of this example is on

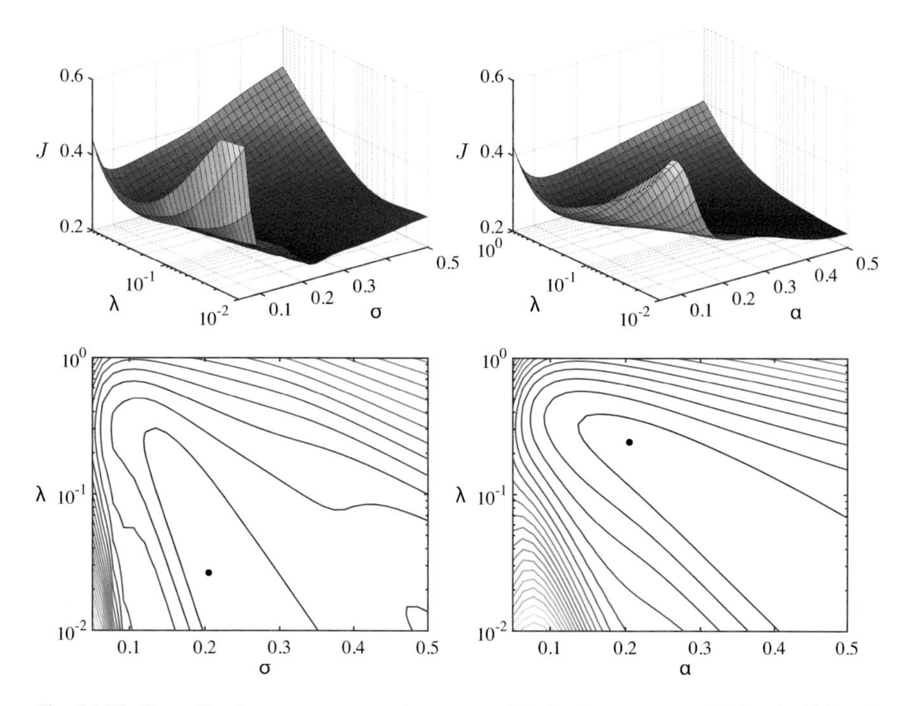

Fig. 16.24 Normalized root mean squared error on validation data over σ and λ for the high noise case $\sigma_n = 0.1$. The dots in the contour plots indicate the best λ value as already shown in Fig. 16.23 for $\sigma = \alpha = 0.2$, i.e., $\lambda_{\text{opt}} = 0.027$ (left) and $\lambda_{\text{opt}} = 0.24$ (right). Gaussian kernel (left), inverse quadratic kernel (right)

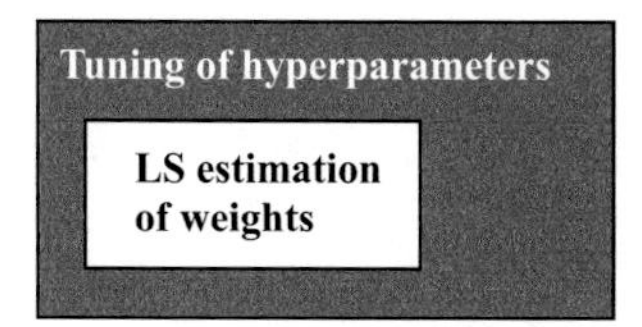

Fig. 16.25 Nested optimization: in an outer optimization loop, the hyperparameter is tuned. Within the loss function, the weights are estimated by least squares

different length scales, the regularization parameter is fixed to $\lambda = 0.01$ for all models.

On purpose, the function is chosen to be smooth in u_1 direction and oscillating in the u_2 direction. Therefore, it is expected that the choice of the length scale in dimension 2 is much more critical than in dimension 1. Indeed, as can be seen from Fig. 16.26, this is the case. Too large length scales σ_2 are not able to describe the sine wave appropriately, and as a consequence, the error on the validation data becomes huge for $\sigma_2 > 0.3$. In contrast, the influence of σ_1 is of minor importance if it does not become too small.

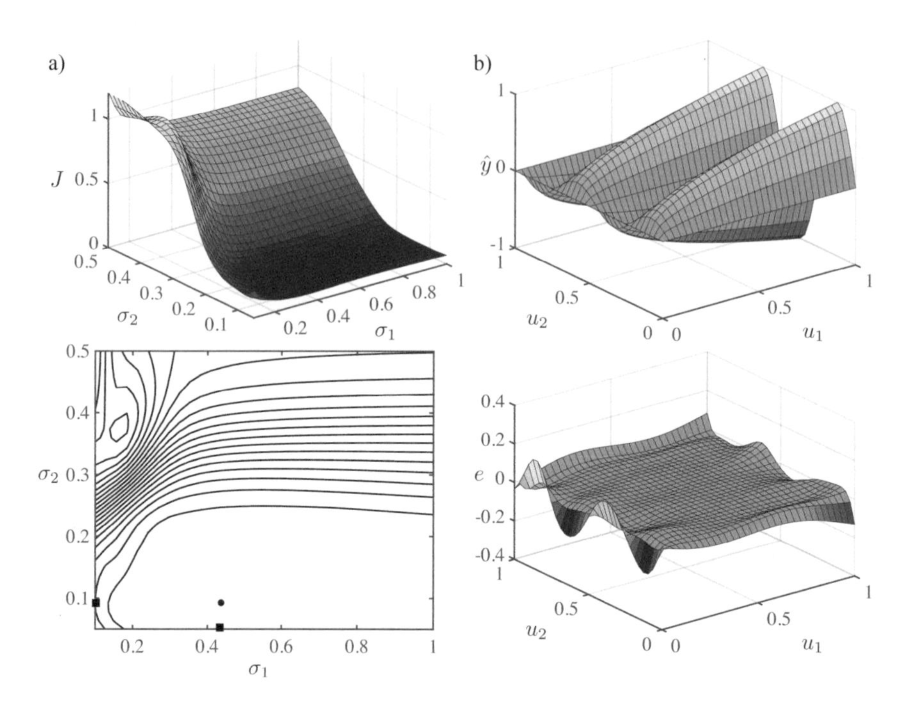

Fig. 16.26 Approximation of the two-dimensional function $y = \sqrt{u_1} \cdot \sin(4\pi u_2)$ with a GPM regularized with $\lambda = 0.01$. The Gaussian kernels are placed on 100 uniformly distributed training data points: (**a**) NRMS error and its contours over the length scales for both dimensions. The round dot in the contour plot marks the best GP model; the square dots mark the GP models in Fig. 16.27. (**b**) Best GP model and its error for $\sigma_1 = 0.44$, $\sigma_2 = 0.09$

The optimum in Fig. 16.26 is very wide/robust which means that the length scale optimization can be carried out very roughly and the tuning problem is benign. For example, a coarse grid search would be sufficient. However, grid searches (no matter how coarse) fully run in the curse of dimensionality and therefore can only be performed for low-dimensional problems. On the other hand, note that any local search method requires realistic initial values because the optimum is not unique. An extremely bad local optimum at $\sigma_1 = 0.18$, $\sigma_2 = 0.37$ yields a completely unsatisfactory model (NRMS error $\approx$ 1). It is mentioned in [476] that from practical experience, local optima generally are no severe issue for GP hyperparameter optimization. □

The plots in Figs. 16.23, 16.24, and 16.26 have been generated by evaluating the model error on validation data. Alternatively and preferably, the hyperparameters can be optimized on some quantity that also incorporates the generalization performance. This is preferable since the validation data set then is not required and can be included in the training data set. Two main quality measures for generalization performance are commonly used:

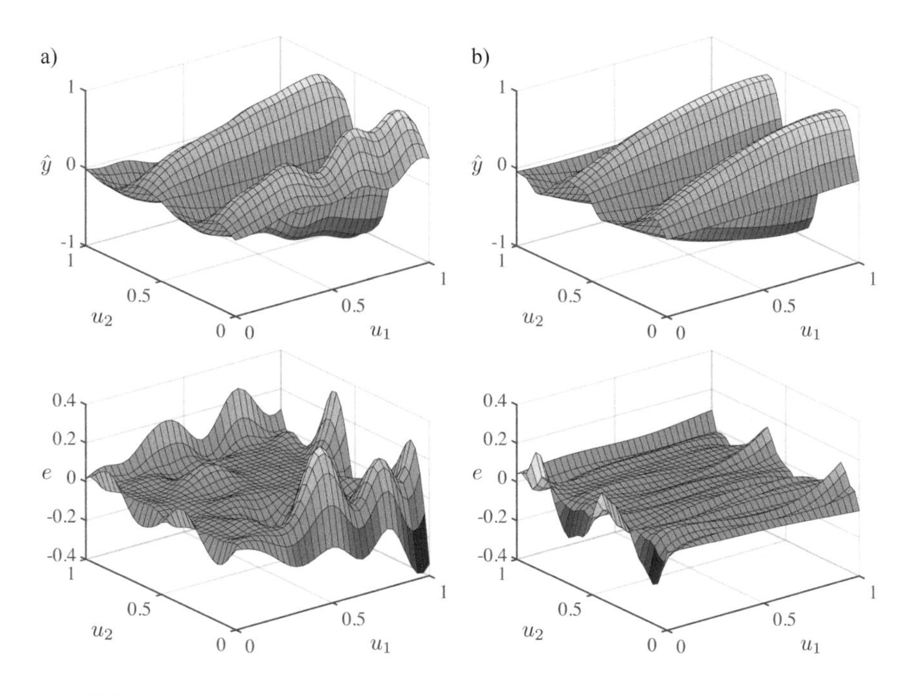

Fig. 16.27 Two subpotimal GP models (top) and their errors (bottom): (**a**) Length scale in u_1 too small: $\sigma_1 = 0.10$, $\sigma_2 = 0.09$. (**b**) Length scale in u_2 too small: $\sigma_1 = 0.44$, $\sigma_2 = 0.05$

- Marginal likelihood: This comes from the Bayesian perspective and is the standard choice for Gaussian processes.
- Leave-one-out (LOO) error or generalized cross validation (GCV) error: This comes from a deterministic perspective and is the standard choice in RKHS or support vector machine contexts.

They are discussed in some depth in Sect. 16.7.3.

In [127] the usefulness of several simple performance measures is empirically studied that are inexpensive to compute (in the sense that they do not require expensive matrix operations involving the kernel matrix).

16.7.2.1 Number of Hyperparameters

With dimension-individual length scales, the number of hyperparameters grows from 2 to $p + 1$ where p is the input dimension. Even more flexible would be to allow a full covariance matrix with an arbitrary orientation which increases the number of hyperparameters to $(N+1)N/2+1$. Also, other kernel shape-influencing hyperparameters could be introduced, e.g., allowing an arbitrary power *pow* in the exp-function. This would contain negative exponentials ($pow = 1$), Gaussians ($pow = 2$), and super-Gaussians ($pow > 2$) as special cases. In practice, the strategies with one length scale and one length scale per dimension are by far the most common choices.

It is important to understand that all data-based validation measures (the ones noted above or simply a validation data set) will give optimistic performance measures. That is the reason why an independent and totally unused test data set for objective performance evaluation is so crucial; compare Sect. 7.3.1. The severity of the overfitting effect with respect to the validation performance depends on the size of the validation model set, i.e., the number of different models compared during validation. The probability to hit a very good model validation result just by chance simply grows with the number of tries. For a thorough discussion and analysis of this topic, refer to [93].

Therefore the number of hyperparameters should be kept as low as possible. And for this reason, the approach with a full covariance matrix is hardly pursued. In literature typically, the following three hyperparameter strategies are preferred:

1. Only the amount of regularization is optimized. The length scales are simply adjusted by a rule of thumb [484]. Several such rules have been proposed, and they typically take into account the dimensionality and the number and distribution of the data, e.g., by investigating the k-nearest-neighbor distances or something similar. This follows the thought that the kernels should (on average) overlap by some desired amount
2. One length scale and the amount of regularization is optimized. Sometimes this is called the RBF kernel. A nice feature of this strategy is that the number of hyperparameters stays constant independent of the dimensionality of the problem.
3. One length scale for each dimension and the amount of regularization is optimized. In literature this strategy often is called *automatic relevance determination* (ARD) since almost irrelevant inputs imply the length scale $\to \infty$. Note the similarities to excluding an input from $\underline{z}$ in local model networks; see Sect. 14.1.

From a pure computational resources point of view, the strategies become more complex from 1 to 3. The flexibility of the modeling approach increases. The discrepancy between these strategies grows with the dimensionality of the problem.

16.7.2.2 One Versus Multiple Length Scales

From Example 16.7.3, it seems obvious that individual length scales are of paramount importance and thus Strategy 3 is preferable. This might be true for low-dimensional problems. However, particularly for high-dimensional problems, as they frequently occur in a classification context, the validation overfitting may be so severe that Strategy 2 is preferable. At least there is some research leaning in that direction [93]: "For the majority of the benchmarks, a significantly lower test error is achieved (according to the Wilcoxon signed ranks test) using the basic RBF kernel; the ARD kernel only achieves statistical superiority on one of the thirteen (image). This is perhaps a surprising result as the models are nested, the RBF kernel being a special case of the ARD kernel, so the optimal performance that can be achieved with the ARD kernel is guaranteed to be at least equal to the performance achievable using the RBF kernel. The reason for the poor performance of the ARD kernel in practice is because there are many more kernel parameters to be tuned in model selection and so many degrees of freedom available in optimising the model selection criterion... the fact that the majority of ARD models display a lower value for the PRESS statistic than the corresponding RBF model, while exhibiting a higher test error rate, is a strong indication of over-fitting the model selection criterion. This is a clear demonstration that over-fitting in model selection can be a significant problem in practical applications, especially where there are many hyper-parameters or where only a limited supply of data is available."

On the other hand, [94] finds the exact opposite. As discussed extensively in [93], the whole area of overfitting in validation or "model selection is a genuine problem in machine learning, and hence is likely to be an area that could greatly benefit from further theoretical analysis."

16.7.2.3 Hyperparameter Optimization Methods

When the criterion for hyperparameter optimization is fixed, the user needs to specify a search strategy in order to find the minimum. It is not important, maybe not even advisable to search for the minimum with high accuracy since this emphasizes the validation overfitting effect. However, multiple local optima can arise which may make a local search problematic; at least it introduces the issue of a proper initialization.

A very simple, global strategy that can easily implemented in a parallel manner is to search a *grid* of hyperparameter combinations. For the length scale(s), reasonable minimal and maximal values can be determined considering the physical limits of each axis, e.g., $\sigma = k(\max - \min)$, $k = 1/10, 2/10, \ldots$.

For the regularization parameter, it is more difficult to specify a reasonable range. An appropriate normalization makes this issue less problem-dependent, i.e., it is advisable trading off $\lambda \underline{\theta}^T \underline{\theta}/n$ against $\underline{e}^T \underline{e}/N$ after normalizing (or standardizing) the data. Nevertheless, the optimal value for λ typically can vary over orders of magnitude on different problems. Therefore it is advisable to space it logarithmically, e.g., $10^{-1}, 10^{-2}, \ldots$.

Finally, the resolution depends on the computational resources, e.g., $3 \times 3 \ldots 6 \times 6$ grids are common choices. It is obvious that the grid-based search strategies fail for significantly more hyperparameters due to the curse of dimensionality.

In [47] grid search is compared with *random search*. "This paper shows empirically and theoretically that randomly chosen trials are more efficient for hyperparameter optimization than trials on a grid."

Of course more sophisticated global strategies can be utilized for hyperparameter search like *genetic algorithms* [96], *simulated annealing* [351], or *particle swarm optimization* [197].

In [293] the problem is tackled in a totally different way. The following approach is proposed: a *clustering* algorithm run on the input data is used to find a nearly "optimal" number of clusters and their centers. The proposed clustering algorithm is applied to select the Gaussian kernel parameters. This basically combines data reduction with hyperparameter determination.

In [439] it is proposed to not determine the prior or kernel directly but elevating the problem one abstraction level higher by setting up *hyperpriors* which describe the distribution of the priors.

In spite of all these approaches for hyperparameter tuning, a direct local search of the minimum of the *marginal likelihood* or leave-one-out error is probably the computationally most efficient alternative. It is argued in [476] that local optima are hardly a severe problem in practice. Experience shows that they do not abound and if existent often represent different interpretations of the data that are reasonable.

16.7.3 Marginal Likelihood

The *marginal likelihood* or *evidence* is the probability to match the measured process output $y(i)$ with the model at all training data locations $\underline{u}(i)$ to a certain degree.[13] This marginal likelihood depends on the hyperparameters, typically regularization strength λ and kernel width(s) σ. It is typically optimized with respect to these hyperparameters in order to obtain a well-performing model.

Figure 16.28 taken from a similar figure in [476] visualizes the relationship between model complexity and marginal likelihood. The horizontal axis symbolically describes all possible data sets. The training data set takes one specific value on this axis. Three models are compared:

- Simple model (λ and/or σ large): If the model is very simple, it can describe only a small fraction of all possible processes and therefore data sets. But it is able to do that extremely well. Thus, for a small number of data sets, the likelihood of fitting the output well at the training data locations $\underline{u}(i)$ is very high. For other data sets, the model is inappropriate and the likelihood is very small.
- Complex model (λ and/or σ small): If the model is very complex, it will be capable to describe a wide range of possible scenarios and therefore data sets. Because it does not specialize in one kind of process, it is able to adjust to almost any kind of training data. On the one hand, this flexibility makes the model very powerful. On the other hand, the likelihood of fitting an individual process

[13]The probability to fit any real number *exactly* is zero, of course.

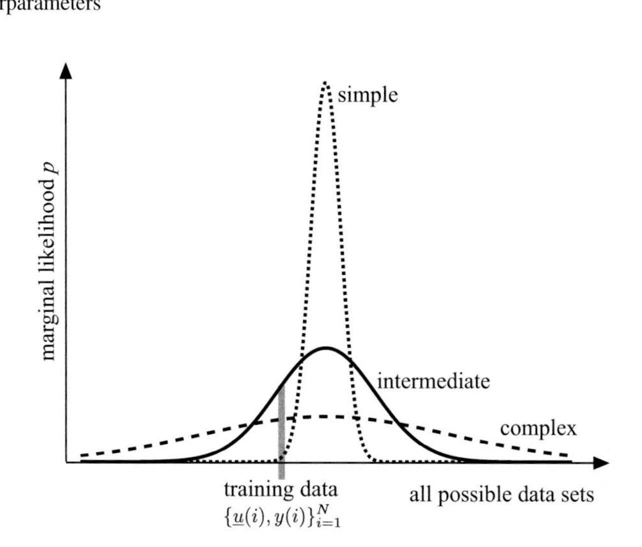

Fig. 16.28 Visualization of the marginal likelihood. A model of intermediate complexity describes the training data better than too simple or complex ones: Thick gray line for one concrete training data set achieves the highest marginal likelihood value for an intermediate model complexity. Note that since the marginal likelihood is a probability density function, it must normalize to unity [476]

becomes relatively small. The pdf in Fig. 16.28 spreads over a wide range of data sets and needs to integrate to unity.

- Intermediate model (λ and/or σ intermediate): A tradeoff between specialization and generality gives the highest marginal likelihood value for a specific training data set.

By optimizing the hyperparameters, the likelihood function with the highest value at the training data (vertical line) is found. Finding the maximal (log) marginal likelihood value automatically determines the best bias/variance tradeoff.

For a specific training data set, the optimal hyperparameters λ and σ need to be found that gives the best tradeoff model.

16.7.3.1 Likelihood for the Noise-Free Case

In the noise-free case ($n = 0, y = f$; see Fig. 16.13), the "normal" likelihood function (compare also Sect. 2.3.1) can be used to assess the probability to match the data with the model. It is given by the N-dimensional distribution

$$p(\underline{y}) = \frac{1}{\sqrt{(2\pi)^N \det(\underline{K})}} \exp\left(-\frac{1}{2}\underline{y}^T \underline{K}^{-1} \underline{y}\right) \tag{16.73}$$

because $\underline{y} \sim \mathcal{N}\left(\underline{0}, \underline{K}\right)$. For kernels with zero covariance (e.g., $\underline{K} = \sigma^2 \underline{I}$), this is equivalent to the product (because of independence) of one-dimensional distributions for each data point $p(\underline{y}) = p(y(1)) \cdot p(y(2)) \cdot \ldots \cdot p(y(N))$. Of course, the whole idea of Gaussian process models is that there *are* correlations between the data points.

Note that the kernel $\underline{K}$ is a function of the width(s)/length scale(s). The likelihood is simply the evaluation of the normal prior distribution at the N-dimensional output $\underline{y}$. When the kernel width σ or the input dimension specific kernel widths σ_i, $i = 1, \ldots, p$, are optimized by $\ln p(\underline{y}) \to \max_\sigma$, the shape of the distribution changes such that it maximizes its height at point $\underline{y}$. For a two-dimensional visualization, see, e.g., Fig. 16.10. If the covariances become huge, this is expressed by $\sigma \to \infty$, and the distribution is squashed onto the diagonal as in Fig. 16.10d and realizes extremely large values on this diagonal, i.e., $y(1) \approx y(2) \approx \ldots \approx y(N)$ (= very smooth behavior). If the data is scattered, i.e., $y(1), y(2), \ldots, y(N)$ are almost independent, then the optimal kernel width is approaching 0 ($\sigma \to 0$, covariances are tiny) because this gives the maximum height at a random point $\underline{y}$ (= very wiggly behavior).

Another line of reasoning goes as follows: if the length scale is very small (σ or $\alpha \to 0$, covariances $\to 0$), abrupt changes between neighboring outputs are easily possible (extreme flexibility). However, it is very unlikely to hit all points with the prior. If the probability hitting one output point within a certain accuracy is p, then the probability of hitting N output points is almost p^N, i.e., extremely small.

On the other hand, if the length scale is very large (σ or $\alpha \to \infty$, covariances $\to 1$), the model is forced to be extremely smooth. Therefore it is very unlikely to obtain abrupt changes. However, it is very likely to hit a certain level within a certain accuracy. The probability will be close to p, because the outputs are almost perfectly correlated.

Therefore overall, neither very small nor very large length scales will model the output points with a high likelihood. Chances are optimal for medium length scales.

16.7.3.2 Marginal Likelihood for the Noisy Case

Things get mathematically more involved if the process output is disturbed by noise. Then the above given likelihood in (16.73) is valid only for the noise-free output f and the disturbed output y is additionally corrupted by noise $y = f + n$. A simple way of argumentation is that two random variables f and n are added; thus their variances add up as well. So the covariance matrix of the noise given by $\sigma_n^2 \underline{I}$ (i.i.d. noise) has to be added to the covariance matrix for the noise-free case given by $\underline{K}$. Therefore, the disturbed output $\underline{y}$ is distributed with $\underline{y} \sim \mathcal{N}\left(\underline{0}, \underline{K} + \sigma_n^2 \underline{I}\right)$ or

$$p(\underline{y}) = \frac{1}{\sqrt{(2\pi)^N \det\left(\underline{K} + \sigma_n^2 \underline{I}\right)}} \exp\left(-\frac{1}{2}\underline{y}^T \left(\underline{K} + \sigma_n^2 \underline{I}\right)^{-1} \underline{y}\right) . \tag{16.74}$$

This is called the *marginal likelihood* function; see explanation below. Taking the natural logarithm gives, compare Sect. 16.6.2

$$\ln p(\underline{y}) = \underbrace{-\frac{1}{2}\underline{y}^T(\underline{K}+\sigma_n^2\underline{I})^{-1}\underline{y}}_{\text{data fit}} \underbrace{-\frac{1}{2}\ln\det(\underline{K}+\sigma_n^2\underline{I})}_{\text{complexity penalty}} \underbrace{-\frac{N}{2}\ln 2\pi}_{\text{constant}} . \tag{16.75}$$

The first term in the log marginal likelihood is the only one containing the measured outputs $\underline{y}$ and evaluates how good the model fits the data. $-\underline{y}^T(\underline{K}+\sigma_n^2\underline{I})^{-1}\underline{y}/2$ *decreases* with growing length scales as then the model becomes more flexible and powerful. The second term evaluates only the kernel. $-\ln\det(\underline{K}+\sigma_n^2\underline{I})/2$ is a complexity penalty that *increases* with growing length scales as then the model is more likely to overfit. The third term is just a constant. The interpretation of (16.75) is comparable to information criteria like the AIC where model fit and model complexity are balanced in order to obtain the best bias/variance tradeoff.

Optimizing (16.75) yields the optimal kernel width σ or input dimension specific kernel widths σ_i, $i = 1, \ldots, p$ and the regularization strength $\lambda = \sigma_n^2$. The geometric influence on the shape of the likelihood function is known from ridge regression (compare Sect. 3.1.4), i.e., the contours become more "circle-like" as the noise variance gets more dominant; the output variance increases in all dimensions by the same amount. This makes the pdf wider and flatter the bigger $\lambda = \sigma_n^2$ gets.

The above argumentation does not explain the expression *marginal* likelihood. A *marginal* distribution is a distribution of variables (here $\underline{f}$ and $\underline{y}$) where some variables (here $\underline{f}$) are unknown (latent variables) and therefore are integrated out to account for their average influences, compare also Appendix B.2.

The relationship between the pdfs is visualized for a one-dimensional example in Figs. 16.29. Figure 16.30 explains how the marginal likelihood (or evidence) can be understood. In equation form, it is then given by [476] as

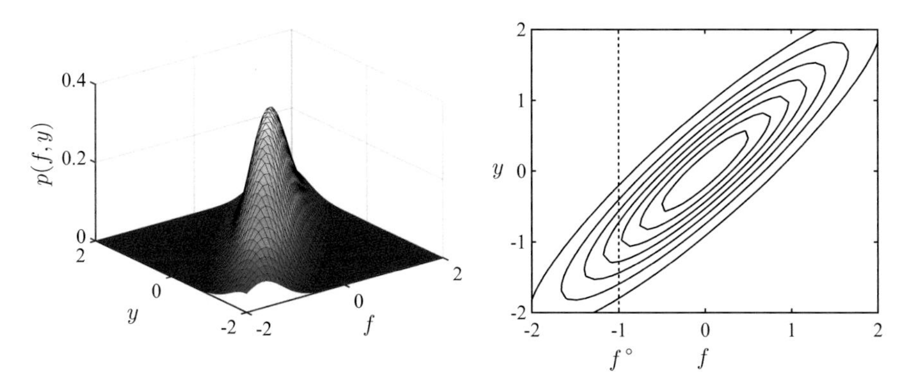

Fig. 16.29 Visualization for one data point of the joint pdf $p(f, y)$ (left) and its contours (right) that couples the noise-free process output f and the disturbed process output $y = f + n$ for a kernel width of $\sigma = 1$ and a noise variance of $\sigma_n^2 = \lambda = 0.2$. The pdf is also conditioned on the locations of the input data $\underline{U}$ which is omitted here for brevity

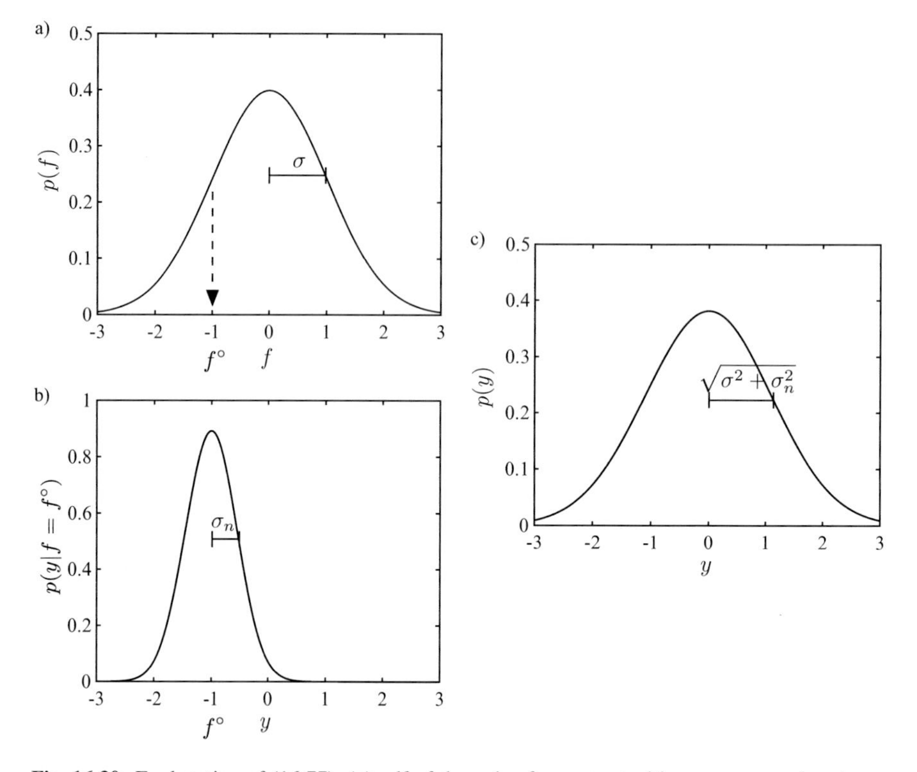

Fig. 16.30 Explanation of (16.77): (**a**) pdf of the noise-free output with zero mean and variance σ^2, (**b**) pdf of the disturbed output y conditioned on the knowledge about f – in this example the sample f° (also shown in Fig. 16.29b) is drawn from (**a**). This means that y-values around f° are most probable, (**c**) pdf of the disturbed output y with *any* mean f drawn from (**a**) and variance $\sigma^2 + \sigma_n^2$. All pdfs are also conditioned on the locations of the input data $\underline{U}$ which is omitted here for brevity

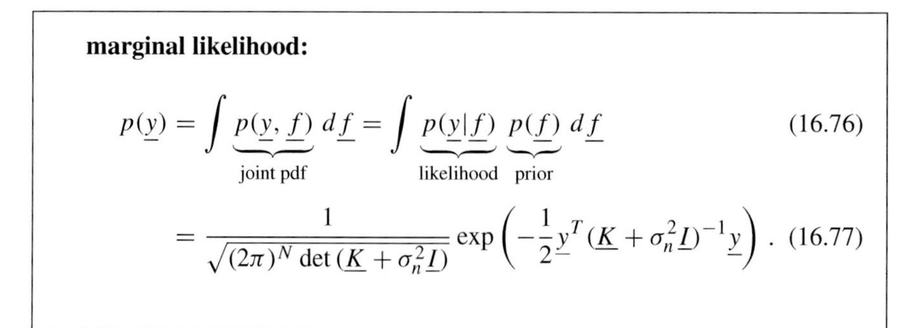

marginal likelihood:

$$p(\underline{y}) = \int \underbrace{p(\underline{y}, \underline{f})}_{\text{joint pdf}} \, d\underline{f} = \int \underbrace{p(\underline{y}|\underline{f})}_{\text{likelihood}} \, \underbrace{p(\underline{f})}_{\text{prior}} \, d\underline{f} \tag{16.76}$$

$$= \frac{1}{\sqrt{(2\pi)^N \det(\underline{K} + \sigma_n^2 \underline{I})}} \exp\left(-\frac{1}{2}\underline{y}^T (\underline{K} + \sigma_n^2 \underline{I})^{-1} \underline{y}\right) . \tag{16.77}$$

$p(\underline{y})$ is obtained by integrating $p(\underline{y}|\underline{f})$ over all possible realizations of $\underline{f}$ (all possible outcomes f° in Fig. 16.30a) weighted with their probability $p(\underline{f})$.

The distribution of $\underline{f}$ before any data is known is the

prior:

$$p(\underline{f}) = \mathcal{N}\left(\underline{0}, \underline{K}\right) = \frac{1}{\sqrt{(2\pi)^N \det(\underline{K})}} \exp\left(-\frac{1}{2}\underline{f}^T \underline{K}^{-1} \underline{f}\right) . \tag{16.78}$$

In a non-Bayesian setting, the negative log prior is thought of as a non-smoothness penalty term. The distribution of $\underline{y}$ conditioned on the knowledge of $\underline{f}$ is called the

likelihood:

$$p(\underline{y}|\underline{f}) = \mathcal{N}\left(\underline{f}, \sigma_n^2 \underline{I}\right) = \frac{1}{\sqrt{(2\pi\sigma_n^2)^N}} \exp\left(-\frac{1}{2}\frac{[\underline{y}-\underline{f}]^T[\underline{y}-\underline{f}]}{\sigma_n^2}\right) . \tag{16.79}$$

All these distributions are also conditioned on the input data $\underline{U}$ which is omitted for the sake of simplicity.

The fundamental goal of Bayesian inference is to go from the prior pdf of the quantity of interest (here $\underline{f}$) before any measurements are taken to the posterior pdf which contains all the information about the measurements $\underline{y}$:

$$\underbrace{p(\underline{f})}_{\text{prior}} \rightarrow \underbrace{p(\underline{f}|\underline{y})}_{\text{posterior}} \tag{16.80}$$

This can be accomplished when the relationship between the pdf of the noise-free f and the pdf of the noisy y is known:

$$p(\underline{f}) \longleftrightarrow p(\underline{y}) . \tag{16.81}$$

This relationship is given by the *Bayes theorem*

$$p(\underline{f}|\underline{y})\, p(\underline{y}) = p(\underline{y}|\underline{f})\, p(\underline{f}) . \tag{16.82}$$

Isolating the factor of interest: The distribution of the noise-free output $\underline{f}$ when the noisy process output $\underline{y}$ is measured gives the *posterior:*

$$p(\underline{f}|\underline{y}) = \frac{p(\underline{y}|\underline{f})\, p(\underline{f})}{p(\underline{y})} . \tag{16.83}$$

Example 16.7.4 (Amount of Data and Kernel Width) This example illustrates how the log marginal likelihood develops as the number of data points increases. The data locations are drawn from a uniform distribution. Two noise-free processes are considered: one-dimensional (i) sine and (ii) hyperbola nonlinearities shown in Figs. 16.31 and 16.32, respectively. The GPMs are estimated with Gaussian kernels without regularization ($\lambda = 0$) and for a length scale of $\sigma = 0.2$ for the sine and $\sigma = 0.4$ for the hyperbola.

Note that the optimum of the log marginal likelihood function is just marked for information purposes. The models are always built with identical length scales of 0.2 or 0.4, respectively. The following observations can be summarized:

- The confidence intervals decrease quickly with a growing number of data points.
- The log marginal likelihood forms a sharper optimum as the number of data points grows.
- For the hyperbola process, the optimal length scale decreases with increasing point density. The data locations get closer together, and the support of the kernels can become more local.
- For the sinusoidal process, something different happens: The optimal length scale converges to the best value to describe the spatial process frequency. It is less dependent on the number of data points.

It is astonishing that the likelihood function even gives reasonable optimal hyperparameters for just 2 or 3 data points! It seems perfectly reasonable that the optimal length scale is significantly larger for the hyperbola process since it is much smoother. □

Example 16.7.5 (Noise Level and Regularization Strength) This example illustrates the effect the noise level has on the model, the confidence interval, and the likelihood function. The number of data points is fixed at 5, the noise variances $\sigma_n^2 = 0.1, 0.01, 0.001$ are investigated. The same processes are considered as in the previous example: (i) sine in Fig. 16.33 (length scale $\sigma = 0.2$) and (ii) hyperbola in Fig. 16.34 (length scale $\sigma = 0.4$).

In contrast to the previous example, the optimal hyperparameters shown on the right in the figures were used for the models shown on the left. The following observations can be summarized:

- The optimal amount of regularization fits relatively closely to the noise variance $\lambda \approx \sigma_n^2$.
- The confidence interval becomes bigger with increasing regularization.
- The data points are fit less accurately with increasing regularization.

Figure 16.35 visualizes the dependence of the marginal likelihood functions on both hyperparameters for five data points for a noise variance of $\sigma_n^2 = 0.01$. The following observations can be made:

- The relationships are smooth.
- Although this is a non-convex optimization problem, no local optima arise in this case.

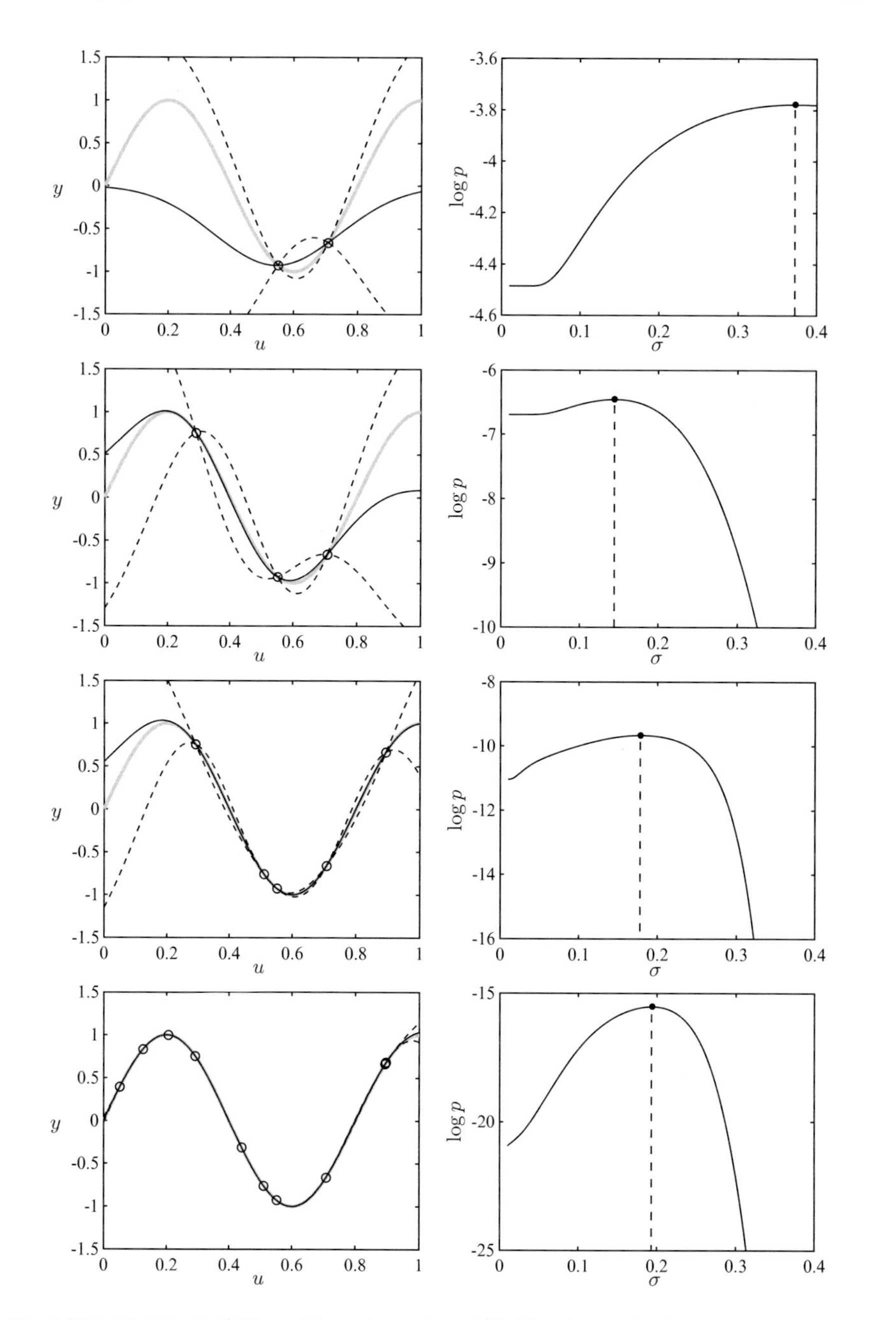

Fig. 16.31 Model with 95% confidence intervals and likelihood for noise-free sine process ($\sigma_n = 0$) for 2, 3, 5, 10 data points. Left: Process (gray) and model (black) with confidence intervals (dashed) for a length scale of $\sigma = 0.2$. Right: Log marginal likelihood function with vertical dashed line at the optimum [only for information!]

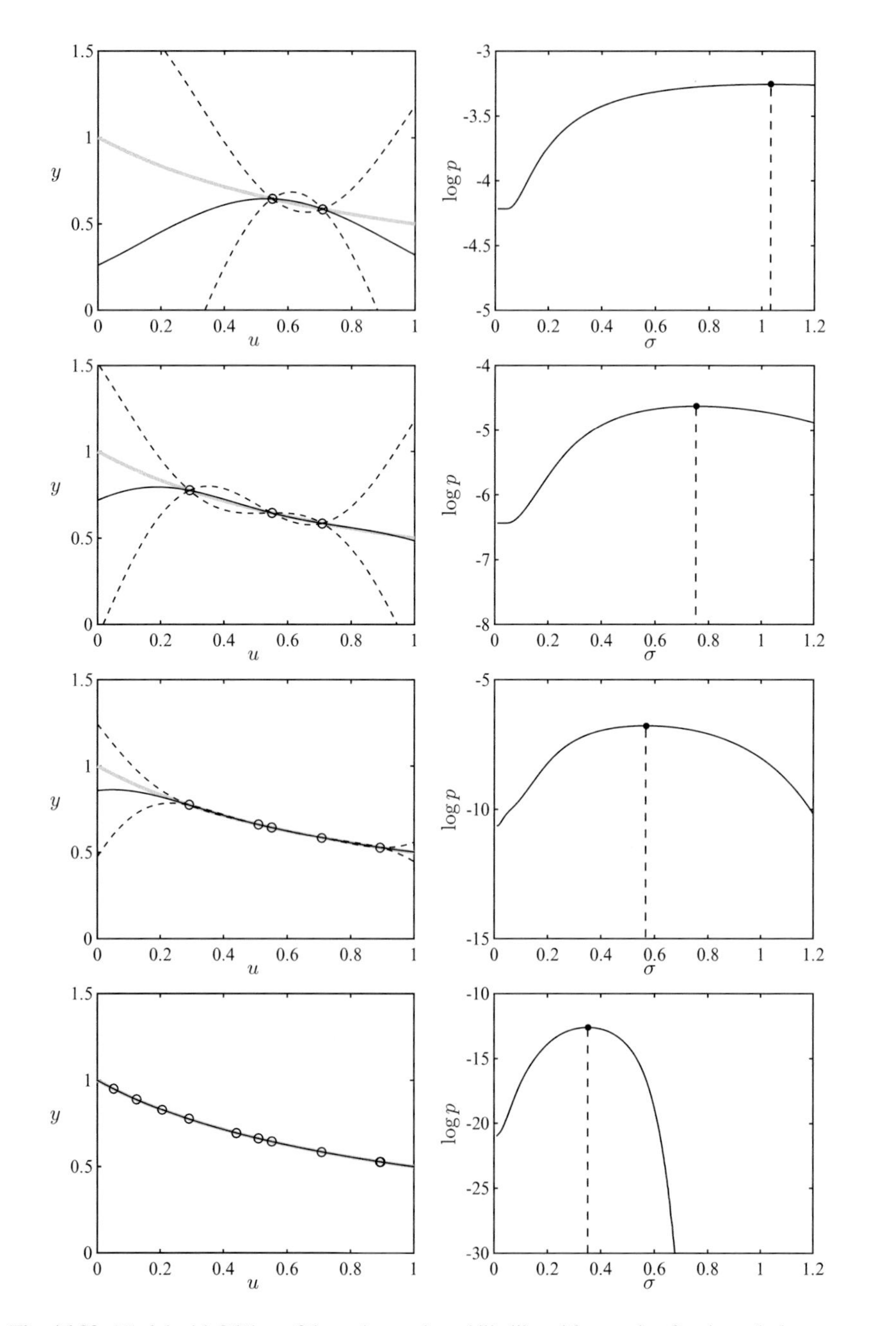

Fig. 16.32 Model with 95% confidence intervals and likelihood for a noise-free hyperbola process ($\sigma_n = 0$) for 2, 3, 5, 10 data points. Left: Process (gray) and model (black) with confidence intervals (dashed) for a length scale of $\sigma = 0.4$. Right: Log marginal likelihood function with vertical dashed line at the optimum [only for information!]

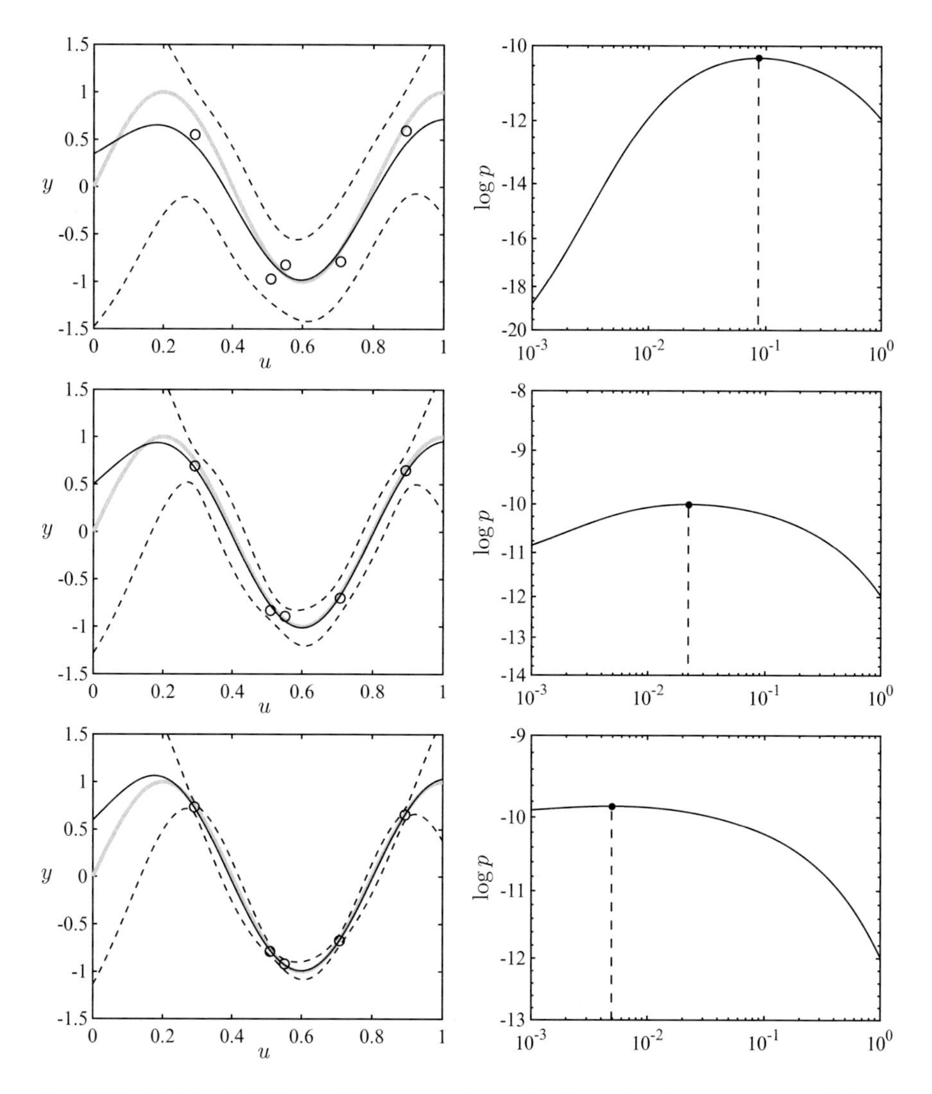

Fig. 16.33 Model with 95% confidence intervals and likelihood for sine process with different noise levels for five data points. Left: Process (gray) and model (black) with confidence intervals (dashed) for a length scale of $\sigma = 0.2$ with the optimal amount of regularization λ_{opt} shown on the right. Right: Log marginal likelihood function with vertical dashed line at the optimum used on the left Noise variances: $\sigma_n^2 = 0.1$ (top), $\sigma_n^2 = 0.01$ (middle), $\sigma_n^2 = 0.001$ (bottom)

- The theoretically optimal value $\lambda_{\text{opt}} = \sigma_n^2$ is matched excellently, remarkable for just five data points.
- The sine function (a) can be hardly modeled with wide kernels (large σ).
- The hyperbola function (b) can be hardly modeled with narrow kernels (small σ).

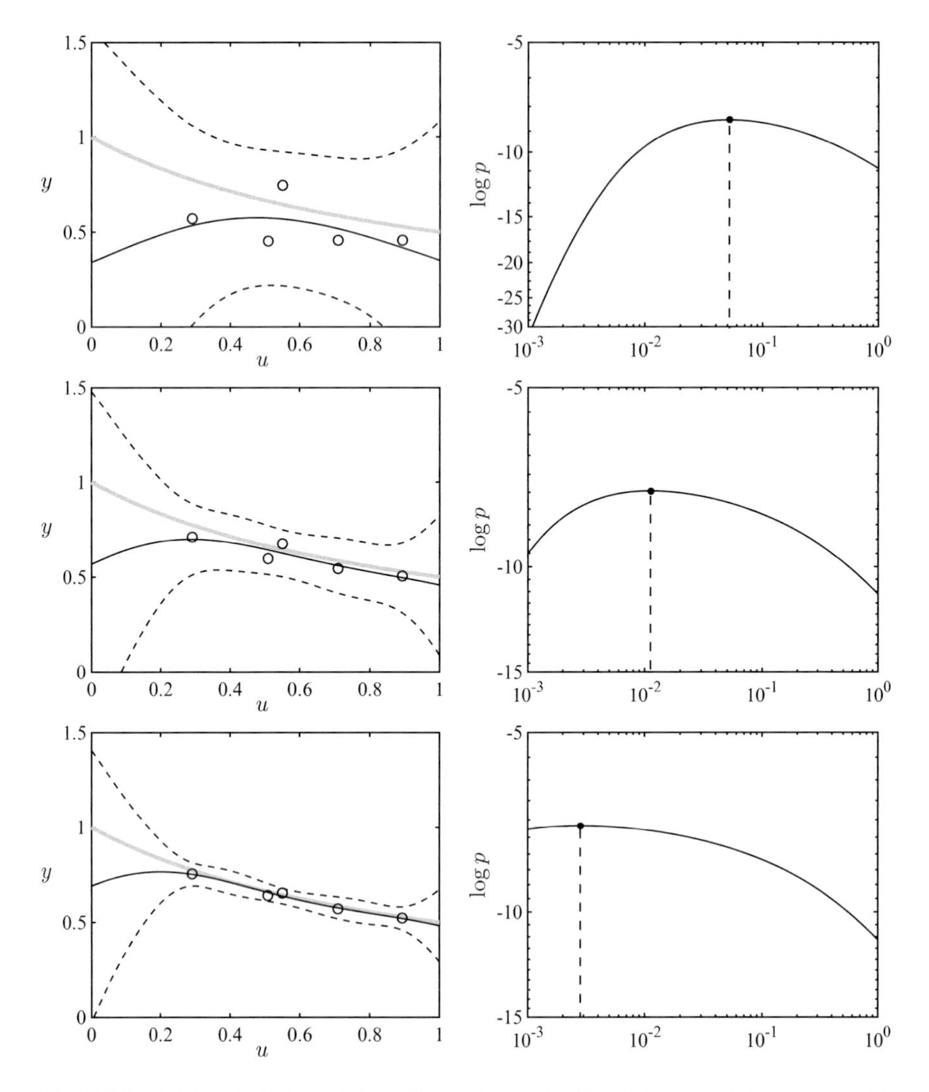

Fig. 16.34 Model with 95% confidence intervals and likelihood for hyperbola process with different noise levels for five data points. Left: Process (gray) and model (black) with confidence intervals (dashed) for a length scale of $\sigma = 0.4$ with the optimal amount of regularization λ_{opt} shown on the right. Right: Log marginal likelihood function with vertical dashed line at the optimum used on the left Noise variances: $\sigma_n^2 = 0.1$ (top), $\sigma_n^2 = 0.01$ (middle), $\sigma_n^2 = 0.001$ (bottom)

- The sine function is more sensitive (more pronounced optimum) with respect to the choice of σ than the hyperbola.
- The small dip in the surface in (b) for low λ, σ values is probably due to numerical inaccuracies.

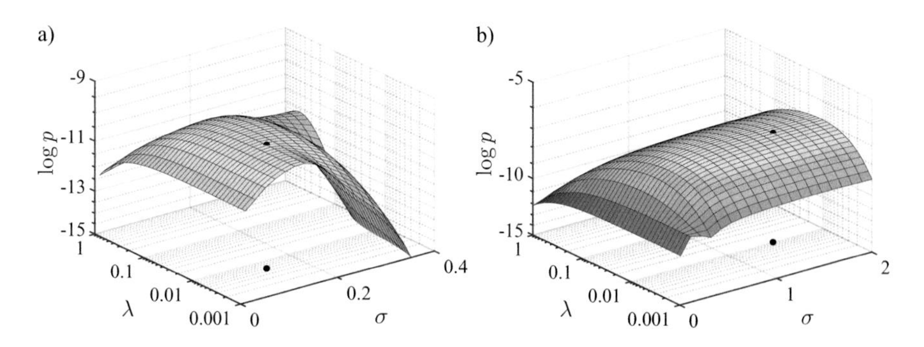

Fig. 16.35 Marginal likelihood in dependence on regularization strength λ and kernel widh σ for (**a**) sine (**b**) hyperbola. Noise variance $\sigma_n^2 = 0.01$. The filled dots mark the optimal λ, σ combinations

A general observation about the log marginal likelihood functions is that they typically are quite smooth and the optimum is "well conditioned." As a consequence, the optimization is unproblematic in terms of convergence speed and required accuracy. □

16.7.3.3 Marginal Likelihood Versus Leave-One-Out Cross Validation

For linear regression problems, applying leave-one-out cross validation is particularly attractive as it can be computed in a very efficient manner, compare Sect. 7.3.2. Thus, it is an alternative to the marginal likelihood optimization for the determination of the hyperparameters. Both approaches offer almost identical computational complexity for GPs. Experts disagree about the practical merits of both approaches. There seems to be a consensus about the fact that cross-validation is more robust against misspecification (i.e., badly chosen model structure or priors) in the sense that it still gives a useful indicator of generalization performance. "Both the marginal likelihood and cross validation performance estimates are evaluated over a finite sample of data, and hence there is always a possibility of overfitting if a model is tuned by optimizing either criterion. For small sample sizes, the difference in the variance of the two criteria may decide which works best"[14]; see [93].

In Chapter 5 of [476] Rasmussen and Williams argue: "However, this loss function is only a function of the predicted mean and ignores the validation set variances. Further, since the mean prediction eq. (2.23) is independent of the scale of the covariances (i.e. you can multiply the covariance of the signal and noise by an arbitrary positive constant without changing the mean predictions), one degree of freedom is left undetermined by a LOO-CV procedure based on squared error

[14]http://stats.stackexchange.com/questions/24799/cross-validation-vs-empirical-bayes-for-estimating-hyperparameters/24818

loss (or any other loss function which depends only on the mean predictions). But, of course, the full predictive distribution does depend on the scale of the covariance function. Also, the computation of the derivatives based on the squared error loss has similar computational complexity as the negative log validation density loss. In conclusion, **it seems unattractive to use LOO-CV based on squared error loss for hyperparameter selection.**"

In contrast, Wahba [582] advocates generalized cross-validation (GCV); see Sect. 7.3.2. In [581] Wahba demonstrated for generalized spline smoothing models that "if the function to be estimated is smooth ... then the GML [marginal likelihood] estimate undersmooths relative to the GCV estimate and the predictive mean square error using the GML estimate goes to zero at a slower rate than the mean square error using the GCV estimate."

And indeed Rasmussen and Williams relativize [476]: "Comparing the pseudo-likelihood for the LOO-CV methodology with the marginal likelihood from the previous section, it is interesting to ask under which circumstances each method might be preferable. Their computational demands are roughly identical. This issue has not been studied much empirically. However, it is interesting to note that the marginal likelihood tells us the probability of the observations given the assumptions of the model. This contrasts with the frequentist LOO-CV value, which gives an estimate for the (log) predictive probability, whether or not the assumptions of the model may be fulfilled. Thus Wahba [1990, sec. 4.8] has argued that **CV procedures should be more robust against model mis-specification.**"

16.7.4 A Note on the Prior Variance

Coming from RKHS, it seems quite natural to assume a kernel with entries of "1"s on its diagonal because $K(\underline{u}(i)\underline{u}(j)) = 1$ if $i = j$ defined the largest similarity between $\underline{u}(i)$ and $\underline{u}(j)$. However, in the GP interpretation according to (16.42) $K(\underline{u}(i), \underline{u}(j)) = \text{cov}\{y(i), y(j)\}$ means the covariance between $y(i)$ and $y(j)$ which for $y(i) = y(j)$ is just the *variance*.

The prior of the undisturbed output $\underline{f}$ is given by

$$p(\underline{f}) = \mathcal{N}\left(\underline{0}, \underline{K}\right) = \frac{1}{\sqrt{(2\pi)^N \det(\underline{K})}} \exp\left(-\frac{1}{2}\underline{f}^T \underline{K}^{-1} \underline{f}\right). \tag{16.84}$$

If the variances of the output signal $\underline{f}$ which are on the diagonal of $\underline{K}$ are not 1 but σ_f^2, then $\underline{K} = \sigma_f^2 \underline{\bar{K}}$ where $\underline{\bar{K}}$ corresponds to the matrix with "1"s on its diagonal and the prior becomes

$$p(\underline{f}) = \mathcal{N}\left(\underline{0}, \underline{K}\right) = \mathcal{N}\left(\underline{0}, \sigma_f^2 \underline{\bar{K}}\right) = \frac{1}{\sqrt{(2\pi)^N \det(\sigma_f^2 \underline{\bar{K}})}} \exp\left(-\frac{1}{2}\underline{f}^T \frac{\underline{\bar{K}}^{-1}}{\sigma_f^2} \underline{f}\right). \tag{16.85}$$

Note that this means that the output prior does not scatter between −1.96 and 1.96 (in 95% of the cases) as shown, e.g., in Fig. 16.8 but around larger or smaller values if $\sigma_f^2 > 1$ or $\sigma_f^2 < 1$. However, this does *not* change the model output's *mean*! The reason is that the estimation of the weights changes to

$$\underline{\hat{w}} = \left(\sigma_f^2 \underline{\bar{K}} + \lambda \underline{I}\right)^{-1} \underline{y}\,. \tag{16.86}$$

If also λ takes a value which is bigger by a factor σ_f^2, then all weights become smaller by a factor $1/\sigma_f^2$. Consequently, in the model output calculation

$$\underline{\hat{y}} = \underline{K}\,\underline{\hat{w}} = \sigma_f^2 \underline{\bar{K}}\,\underline{\hat{w}} = \sigma_f^2 \underline{\bar{K}} \left(\sigma_f^2 \underline{\bar{K}} + \lambda \underline{I}\right)^{-1} \underline{y} = \underline{\bar{K}} \left(\underline{\bar{K}} + \frac{\lambda}{\sigma_f^2} \underline{I}\right)^{-1} \underline{y} \tag{16.87}$$

this factor cancels out again. Just the kernel is scaled by σ_f^2 and the regularization parameter and weights by $1/\sigma_f^2$. The overall amount of regularization always is equal to $\lambda/\sigma_f^2 = \sigma_n^2/\sigma_f^2$.

Thus, from a pure "model output" perspective, σ_f^2 is irrelevant as far as only the mean of the GP is considered. For this reason, it has been assumed (without explicitly mentioning this) that $\sigma_f^2 = 1$ up to this point of this chapter. For normalized (output) data, this is a reasonable choice anyway.

For all probabilistic or hyperparameter interpretation issues, i.e., for the model variance and for the likelihood function, σ_f^2 can and should be introduced as an additional degree of freedom. The model variance for $\sigma_f^2 = 1$ is given in (16.62). When introducing an arbitrary prior variance with $\underline{K} = \sigma_f^2 \underline{\bar{K}}$, this expression becomes

$$\begin{aligned}\sigma_*^2 &= \sigma_f^2 \bar{K}(u_*, u_*) - \sigma_f^2 \underline{\bar{k}}^T(u_*) \left(\sigma_f^2 \underline{\bar{K}} + \sigma_n^2 \underline{I}\right)^{-1} \sigma_f^2 \underline{\bar{k}}(u_*) \\ &= \sigma_f^2 \left[\bar{K}(u_*, u_*) - \underline{\bar{k}}(u_*)^T \left(\underline{\bar{K}} + \frac{\sigma_n^2}{\sigma_f^2} I\right)^{-1} \underline{\bar{k}}(u_*) \right].\end{aligned} \tag{16.88}$$

Thus, the model output variance is proportional to the prior variance σ_f^2. The more data is available, the smaller the model output variance becomes. Note that the expression in brackets usually is equal to one without any data (second term then is zero) as $\bar{K}(u_*, u_*) = 1$. Therefore an appropriate choice or tuning of σ_f is essential for obtaining quantitatively interpretable confidence intervals/errorbars. The primitive formula (16.62) only allows a qualitative interpretation.

In [476] σ_f is always chosen as an individual hyperparameter, although, e.g., the provided MATLAB toolbox[15] also offers some kernels where it is fixed to $\sigma_f^2 = 1$. Of course, if σ_f^2 is optimized as an additional hyperparameter, this increases the dimensionality of the marginal likelihood optimization by one.

Finally, for alternative hyperparameter optimization schemes based on the leave-one-out error or generalized cross-validation error (Sect. 7.3.2) the hyperparameter σ_f simply does not exist; it only is meaningful in a probabilistic framework. Thus, also for GPs it seems reasonable to fix $\sigma_f = 1$, normalize or standardize the output data, and just optimize λ and the length scales/widths.

16.8 Summary

The benefits and drawbacks of kernel-based function approximation approaches are basically the opposite of input-space partitioning approaches. In black box modeling, all information comes from the data. Therefore it is very straightforward to rather focus on the data points than the input space. Without any regularization, placing a kernel on each data point allows for maximum flexibility and an interpolation of the data. Adjusting the length scale(s) determines the smoothness of the interpolator. Introducing regularization (as in ridge regression) allows for a proper bias/variance tradeoff. Thus the flexibility is not controlled explicitly by adjusting the number of neurons or rules but implicitly by tuning the regularization strength. One price to be paid for this powerful strategy is that the model evaluation effort is proportional to the number of training data points.

The most important features of Gaussian process models can be summarized as follows; compare Table 16.1:

- *Interpolation behavior* is usually very preferable since a huge number of smooth kernels superimpose.
- *Extrapolation behavior* strongly depends on the type of kernel and the (underlying) bias model if any. Typically GPs extrapolate toward 0.
- *Locality* typically is not very strict, even for local kernels, since the hyperparameter optimization balances robustness with respect to output noise (global behavior is preferable) and overfitting effects (local behavior is preferable).
- *Accuracy* is usually very high, in particular, if continuous hyperparameter optimization (not just grid search) is carried out.
- *Smoothness* properties are extremely well. The starting point of RKHS is balancing approximation error with smoothness.
- *Sensitivity to noise* is quite low if hyperparameter optimization is carried out. Otherwise, bad regularization adjustment can yield noise-sensitive behavior.

[15] http://www.gaussianprocess.org/gpml/code/matlab/doc/index.html

Table 16.1 Assessment of the pros and cons

Properties	Gaussian Process Models
Interpolation behavior	++
Extrapolation behavior	−
Locality	+
Accuracy	++
Smoothness	++
Sensitivity to noise	++
Parameter optimization	+
Structure optimization	++
Online adaptation	0
Training speed	−
Evaluation speed	−−
Curse of dimensionality	+
Interpretation	0
Incorporation of constraints	−
Incorporation of prior knowledge	−
Usage	+

* = linear optimization, ** = nonlinear optimization,
++/−− = model properties are very favorable/undesirable

- *Parameter optimization* is very efficiently done with least squares (ridge regression). However, the huge number of parameters corresponding to the number of data points can be a big handicap.
- *Structure optimization* is automatically realized by hyperparameter optimization which controls flexibility (via regularization strength) and input relevance (via individual length scales per input).
- *Online adaptation* can be realized easily by an RLS if only the weights are adapted. Online hyperparameter adaptation is a much more involved issue.
- *Training speed* Hyperparameter optimization with a nested huge least squares estimation for the weights requires significant resources.
- *Evaluation speed* suffers strongly under the huge number of kernel to be evaluated. Hardware acceleration is relatively simple, however, as a consequence of the simple parallel structure.
- *Curse of dimensionality* is weak since GPs are data point-based approaches that do not care about input space partitioning but rather data point distribution. Therefore they are not sensitive with respect to dimensionality but rather to the amount of data.
- *Interpretation* is difficult due to the huge number of kernels. But the optimal hyperparameters offer important information about the input relevances and smoothnesses.
- *Incorporation of constraints* is difficult since the weights hardly can be interpreted (due to their huge number).

- *Incorporation of prior knowledge* is possible in an indirect manner via the shape and properties of the kernel. Explicit incorporation of analytic knowledge seems difficult.
- *Usage* has skyrocketed in recent years due to their extremely favorable properties. GPs are particularly popular in the field of engine control where the company Bosch has invested significantly in technologies based on GPs.

16.9 Problems

1. What are the benefits and drawbacks of a data point-based approach compared to an input-space partitioning approach?
2. What is the idea behind function smoothness in RKHS?
3. How does the model structure of the optimal function approximator in RKHS look like?
4. What are the key assumptions that Gaussian process models are based on?
5. Where does prior knowledge enter a Gaussian process model?
6. What is the computational complexity of a Gaussian process model for fixed hyperparameters?
7. Plot a couple of potential one-dimensional Gaussian process model outputs for Gaussian priors with small and large standard deviations.
8. What is the relationship between a Gaussian process model and an RBF network?
9. What are the main weaknesses of Gaussian process models?
10. How could different noise levels in different (but unknown) operating regimes be dealt with in a Gaussian process model context?

Chapter 17
Summary of Part II

Abstract This chapter gives a summary of Part B.

Keywords Summary

The main properties of the most common nonlinear static model architectures discussed in this part are briefly summarized in the following. They all can be described within the basis functions framework introduced in Sect. 9.2.

- *Linear models* represent the simplest approach and should always be tried first. Only if the performance of a linear model is not satisfactory should one move on to apply nonlinear model architectures. Linear models also mark the lower bound of performance that has to be surpassed to justify the use of nonlinear models. These considerations can be quite helpful in order to assess whether an iterative training procedure has converged to a good or poor local optimum or whether it has converged at all.
- *Polynomial models* are the "classical" nonlinear models and include a linear model as a special case (polynomial of degree 1). The main advantage of polynomials is that they are linear in the parameters. The main drawbacks are their tendency toward oscillatory interpolation behavior as the polynomial degree grows and their often undesirable extrapolation behavior. Furthermore, they suffer severely from the curse of dimensionality, which basically makes high-dimensional mappings infeasible, although linear subset selection techniques can weaken this problem by choosing only the relevant regressors and allowing for higher input dimensionalities than with complete polynomials that contain all possible regressors.
- *Look-up table models* are the dominant nonlinear models in most industrial applications owing to severe restrictions on memory size and computational power. Look-up tables are implementable on low-cost electronics and thus match this profile very well. They are simple and fast to evaluate on a micro-controller without a floating-point unit. One property distinguishes look-up tables from all other model architectures discussed here: Their parameters can be chosen

O. Nelles, *Nonlinear System Identification*,
https://doi.org/10.1007/978-3-030-47439-3_17

directly as measured values without the utilization of any regression method for estimation. In fact, typically look-up tables are determined in this direct fashion. The advantages of look-up tables listed above are paid for by the following severe restrictions. They suffer severely from the curse of dimensionality, which makes them infeasible for more than three to four inputs and already very inefficient for more than two inputs. In practice, this implies that only one- or two-dimensional mappings can be implemented. This is the reason for the existence of many complicated and nebulous design procedures with inferior performance to realize higher-dimensional mappings. In addition to this restriction, the grid-based structure is already inefficient for two-dimensional problems since the model complexity cannot be appropriately adapted according to the process complexity. Finally, look-up tables realize non-differentiable mappings owing to their linear interpolation rule, which can cause difficulties in optimization and control. (When interpolation rules with higher than first order, i.e., linear, are utilized, the computational demand grows significantly, which would be against the philosophy of this model architecture.)

- *Multilayer perceptron (MLP) networks* are the most widely applied neural network architecture. They utilize nonlinearly parameterized global basis functions whose parameters are typically optimized together with the linear output weights. MLPs can function on a huge class for differently characterized problems, although they often might not be a very efficient solution. The main feature of MLPs is their automatic adjustment to the main directions of the process nonlinearity. This is a particular advantage for very high-dimensional problems. This advantage must be paid for by the need for nonlinear optimization techniques. Thus, parameter optimization is a tedious procedure, and structure optimization becomes even more complex. Usually, this drawback should only be accepted if the advantages become significant, i.e., for high-dimensional problems. Otherwise, the slow training, involving the risk of convergence to poor local optima dependent on the user-chosen parameter initialization, is not attractive.
- *Radial basis function (RBF) networks* are usually based on local basis functions whose nonlinear parameters determine their position and width. These nonlinear parameters are typically not optimized but determined through some heuristic approaches such as clustering. Therefore, RBF networks are less flexible and not so well suited as MLP networks for high input dimensionalities. Depending on the specific training method applied, RBF networks underlie the curse of dimensionality to a different extent. The most robust approaches in this respect are certainly clustering and linear (OLS) subset selection techniques. Nevertheless, RBF networks are best used for low- and medium-dimensional problems. The major advantage of RBFs is their fast and reliable training with linear optimization techniques. This is the compensation for fixing the hidden layer parameters a priori.

  Normalized RBF networks overcome the sensitivity of standard RBF networks with respect to their basis function widths. The interpolation behavior of normalized RBF networks is usually superior to that of the standard RBFs.

However, unfortunately, the advanced subset selection method cannot be directly applied for training normalized RBF networks.

- *Singleton neuro-fuzzy models* are (under some restrictions) equivalent to normalized RBF networks. They restrict the basis functions to lie on a grid in order to be able to interpret them by their one-dimensional projections on each input interpreted as the membership functions. The fuzzy rule interpretation can give some insights into the model (and hence into the process) behavior as long as the number of rules is small enough. However, the curse of dimensionality limits the usefulness of grid-based fuzzy models, as more than three or four inputs are practically infeasible. Therefore, extensions based on additive models or generalized rules have been developed together with efficient construction algorithms in order to overcome or weaken the curse of dimensionality while still retaining some rule-based interpretation.
- *Local linear (Takagi-Sugeno) neuro-fuzzy models* have become increasingly popular in recent years. This development is due to the efficient training algorithms that allow the heuristic determination of the hidden layer parameters, avoiding nonlinear optimization. Furthermore, some algorithms such as LOLIMOT are incremental, i.e., they construct a set of simple to complex models. This supports the user's decision about the number of neurons/rules without the need for any *additional* structure search. The main advantages of local linear neuro-fuzzy models are the availability of fast training algorithms, the interpretation as Takagi-Sugeno fuzzy rules, the various ways to incorporate many kinds of prior knowledge, and the relative insensitivity with respect to the curse of dimensionality compared with other fuzzy approaches. Many additional advantages are revealed in the context of dynamic systems in Part III. The weaknesses of local linear neuro-fuzzy models are the sometimes undesirable interpolation behavior and the inefficiency of very high-dimensional problems. Hinging hyperplanes are a promising remedy to the last drawback. In the opinion of the author, local linear neuro-fuzzy models should be more widely applied for static modeling problems (they are already widely used for dynamic ones), since their strengths could be exploited for many applications.

Part III
Dynamic Models

Chapter 18
Linear Dynamic System Identification

Abstract This chapter summarizes the most important issues of the state-of-the-art linear system identification. It introduces the standard terminology and explains its origin as it can be confusing at first. Dynamic models are structured into (i) time series, (ii) model with output feedback, and (iii) model without output feedback. The latter model class typically is underrepresented and underappreciated in most of the literature, in the opinion of the author. Therefore, it is treated more extensively here and also in other chapters of this book. Care is taken to explain the decisive difference between equation (or one-step prediction) error and output (or simulation) error and the consequences associated with their properties. Advanced issues like order determination, multivariate systems, and closed-loop identification are briefly discussed as well.

Keywords Dynamic model · Linear model · Output error · Equation error · Time series · FIR · ARX · OBF · Recursive

The term *linear system identification* often refers exclusively to the identification of linear *dynamic* systems. In this chapter's title, the term "dynamic" is explicitly mentioned to emphasize the clear distinction from static systems. An understanding of the basic concepts and the terminology of linear dynamic system identification is required in order to study the identification of *nonlinear dynamic* systems, which is the subject of all subsequent chapters. The purpose of this chapter is to introduce the terminology, concepts, and algorithms for linear system identification. Since this book deals extensively with local linear models as a very promising approach to nonlinear system identification, most of the methods discussed in this chapter can be transferred to this particular class of nonlinear models. It is one of the main motivations for the use of local linear model approaches that many existing and well-understood linear techniques can be successfully extended for nonlinear processes. A more detailed treatment of linear system identification can be found in [67, 141, 278, 279, 304, 356, 540]. Practical experience can be easily gathered by playing around with the MATLAB system identification toolbox [357].

O. Nelles, *Nonlinear System Identification*,
https://doi.org/10.1007/978-3-030-47439-3_18

This chapter is organized as follows. First, a brief overview of linear system identification is given to characterize the models and methods discussed here. Section 18.3 introduces the terminology used for naming the different linear model structures and explains the basic concept of the optimal predictor and prediction error methods for estimating linear models from data. After a brief discussion of time series models in Sect. 18.4, the linear models are classified into two categories: models with output feedback (Sect. 18.5) and models without output feedback (Sect. 18.6). Section 18.7 analyzes some advanced aspects that have been omitted in the preceding sections for the sake of an easier understanding. Recursive algorithms are summarized in Sect. 18.8. The extension to models with multiple inputs and outputs is presented in Sect. 18.10. Some specific aspects for identification with data measured in closed loop are introduced in Sect. 18.11. Finally, a summary gives some guidelines for the user.

18.1 Overview of Linear System Identification

Figure 18.1 gives an overview of linear system identification methods. They can be distinguished into parametric and non-parametric approaches. It is helpful to distinguish clearly the *model* and the type of *method* applied to determine the degrees of freedom of the model. The model can be parametric or non-parametric:

- *Parametric models* can (or are assumed to be able to) describe the true process behavior exactly with a *finite* number of parameters. A typical example is a differential or difference equation model. Often the parameters have a direct

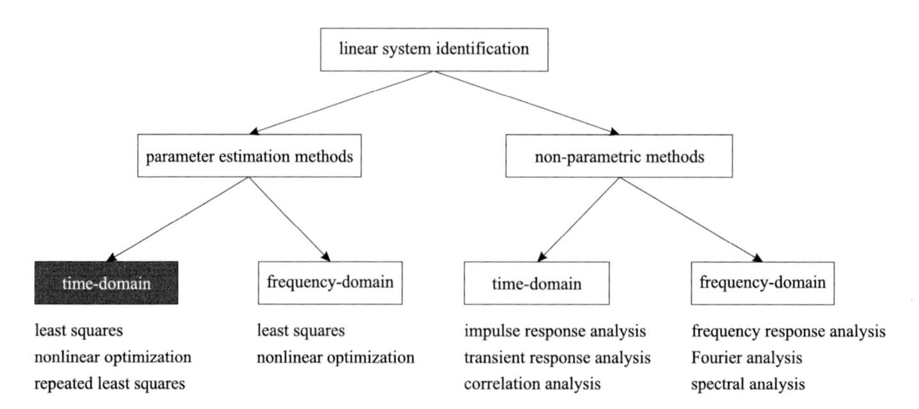

Fig. 18.1 Overview of linear system identification methods. Only the methods within the dark-shaded box are discussed in this chapter. Note that the *methods* not the models are classified into *parametric* and *non-parametric* ones. Non-parametric models, such as a finite impulse response model, may indeed be estimated with a parametric method if the infinite series is approximated by a finite number of parameters

relationship to physical quantities of the process, e.g., mass, volume, length, stiffness, and viscosity.
- *Non-parametric models* generally require an *infinite* number of parameters to describe the process exactly. A typical example is an impulse response model.

Furthermore, parametric and non-parametric methods can be distinguished:

- *Parametric methods* determine a relatively small number of parameters. Usually, these parameters are optimized according to some objective. A typical example is parameter estimation by linear regression. Parametric methods can also be used for the determination of approximate non-parametric models whose number of parameters has been reduced to a finite number. A typical example is a finite impulse response (FIR) model that approximates the infinite impulse response of a process.
- *Non-parametric methods* are more flexible than parametric methods. They are used if less structure is imposed on the model. A typical example is Fourier analysis, which yields functions of frequency and thus is not describable by a finite number of parameters. Although eventually, in their actual implementation, non-parametric methods exhibit a certain (finite) number of "parameters" (e.g., for a discrete-time Fourier analysis, the complex amplitudes for all discretized frequency intervals), this number is huge and independent of any model structure. Rather the number of "parameters" depends on factors such as the number of data samples or the quantization.

This chapter focuses on parametric models and methods for linear system identification. For a detailed discussion of non-parametric approaches, refer to [141, 278]. Furthermore, this chapter considers only time-domain approaches.

18.2 Excitation Signals

The input signal $u(k)$ of the process under consideration plays an important role in system identification. Clearly, the input signal is the only possibility to influence the process in order to gather information about its behavior. Thus, the question arises: How should the input signal be chosen?

In most real-world applications, there exist a large number of constraints and restrictions on the choice of the input signal. Certainly, for any real process, the input signal must be bounded, i.e., between a minimum u_{min} and maximum value u_{max}. Furthermore, the measurement time is always limited. Besides these basic restrictions in the ideal case, the user is free to design the input signal. This situation may arise for pilot plants or industrial processes that are not in regular operation. However, most often the situation is far from ideal. Typically safety restrictions must be obeyed, and one is not allowed to push the plant to its limits. If the plant is in normal operation, usually no or only slight changes to the standard input signal are allowed in order to meet the process goals, e.g., the specifications of the produced product. In the following, some guidelines for input signal design are given, which should be heeded whenever possible.

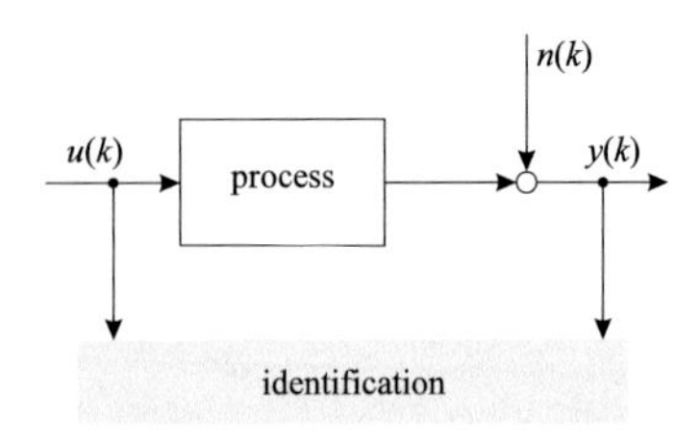

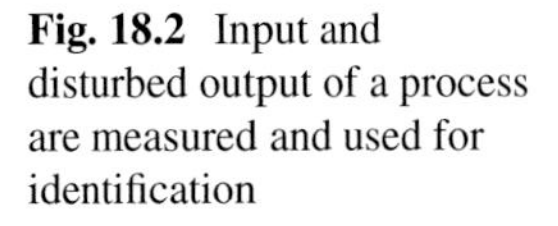
Fig. 18.2 Input and disturbed output of a process are measured and used for identification

Figure 18.2 shows a process in which all disturbances are transferred to the output in the noise $n(k)$. Disturbances that in reality affect the input or some internal process states can be transformed into the process output by a proper frequency shaping by means of a filter. Because the noise $n(k)$ cannot be influenced, the input signal is the user's only degree of freedom to determine the signal-to-noise ratio. Thus, the input *amplitudes* should exploit the full range from u_{min} to u_{max} in order to maximize the power of the input signal and consequently the signal-to-noise ratio. Therefore, it is reasonable to switch between u_{min} and u_{max}.

The spectrum of the input signal determines the *frequencies* where the power is put in. Obviously, the identified model will be of higher quality for the frequencies that are strongly excited by the input signal than for those that are not. If the input signal is a sine wave, only information about one single frequency is gathered, and the model quality at this frequency will be excellent at the cost of other frequency ranges. So, for input signal design, the purpose of the model is of fundamental importance. If the emphasis is on the static behavior, the input signal should mainly excite low frequencies. If the model is required to operate only at some specific frequencies, an additive mixture of sine waves with exactly these frequencies is the best choice for the input signal. If the model is utilized for controller design, a good match of the process around the Nyquist frequency ($-180°$ phase shift) is of particular importance. An excitation signal for model-based controller design is best generated in closed loop [184]. If very little is known about the intended use of the model and the characteristics of the process, a white input signal is the best choice since it excites all frequencies equally well. Note, however, that often very high frequencies do not play an important role, especially if the sampling time T_0 is chosen very small. Although in practice it is quite common to choose the sampling frequency as high as possible with the equipment used, it is advisable to choose the sampling time at about one-twentieth to one-tenth of the settling time of the process [277]. If sampling is performed much faster, the damping of the process typically is so large at high frequencies that it makes no sense to put too much energy in these high-frequency ranges. Furthermore, most model structures will be a simplified version of reality and thus independent of the excitation signal; structural errors will inevitably be large in the high-frequency range.

Example 18.2.1 Input Signals for Excitation

The following figures illustrate some typical input signals and the corresponding output of a first-order system with gain $K = 1$ and time constant $T = 8$ s sampled with $T_0 = 1$ s that follows the difference equation

$$y(k) = 0.1175u(k-1) + 0.8825y(k-1)\,. \tag{18.1}$$

This process is excited with each of the input signals shown in Figs. 18.3, 18.4, 18.5, 18.6, and 18.7, and 100 measurements are taken. These samples are used for identification of a first-order ARX model; see Sect. 18.3.1. The process is disturbed with filtered white noise of variance 0.01. Note that the noise filter is chosen equal to the denominator dynamics of the process $1/A$ in order to meet the assumption of

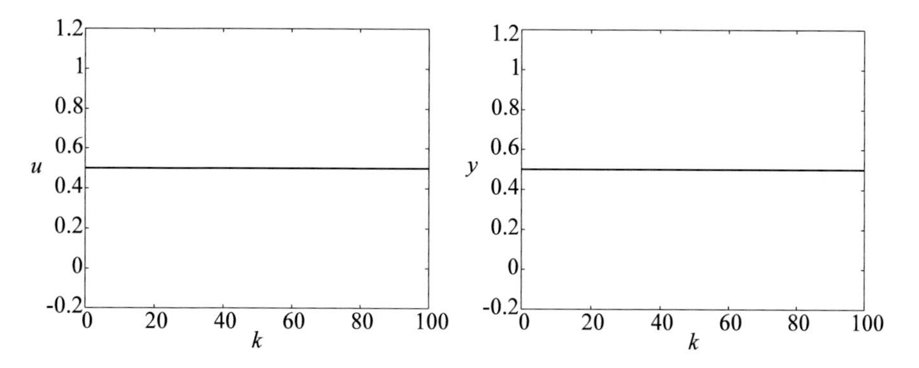

Fig. 18.3 Excitation with a constant signal (left) and the undisturbed process output (right)

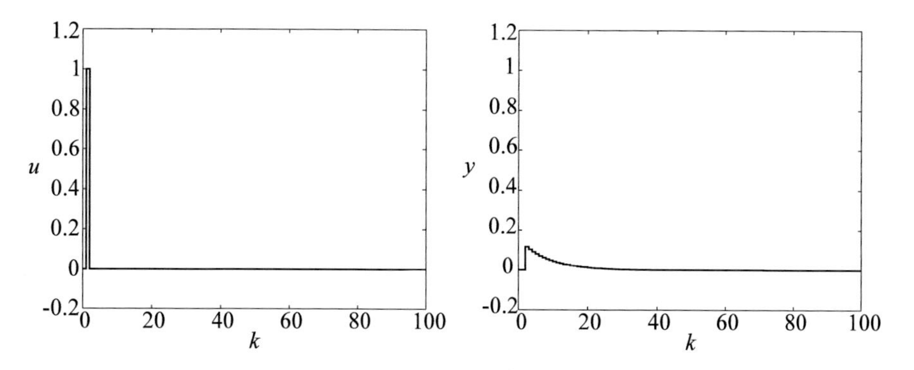

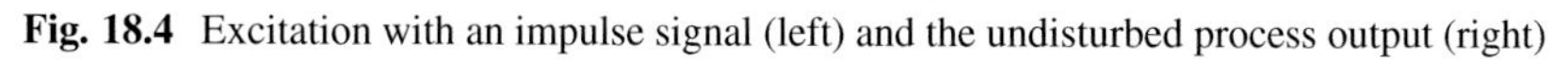

Fig. 18.4 Excitation with an impulse signal (left) and the undisturbed process output (right)

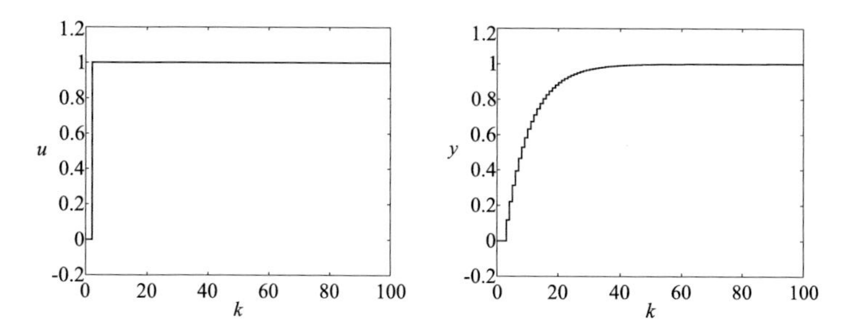

Fig. 18.5 Excitation with a step signal (left) and the undisturbed process output (right)

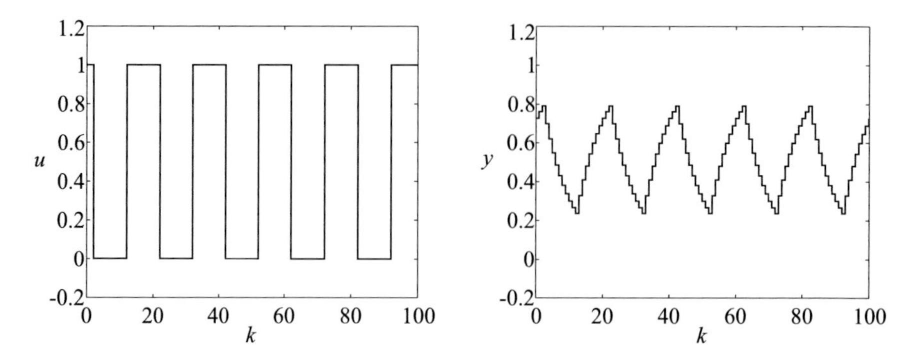

Fig. 18.6 Excitation with a rectangular signal (left) and the undisturbed process output (right)

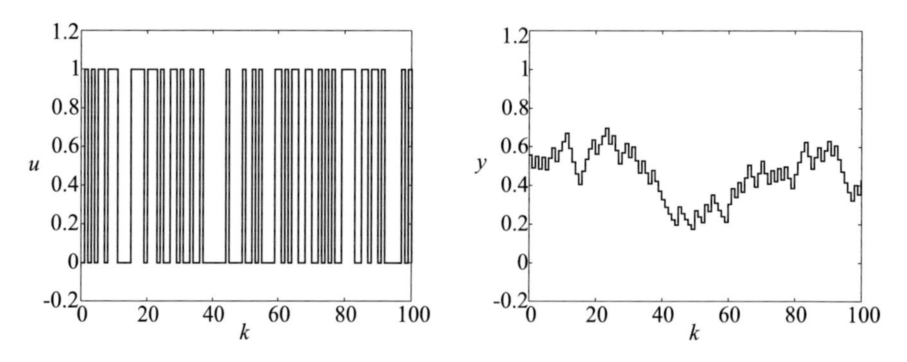

Fig. 18.7 Excitation with a PRBS (pseudo random binary signal) (left). The PRBS is an deterministic approximation of white noise in discrete time [278]. The undisturbed process output is shown at the right

the ARX model. Because the model structure and the process structure are identical, the bias error (Sect. 7.2) of the model is equal to zero, and all errors are solely due to the noise.

The results of the identification are given in each figure and summarized in Table 18.1. The comparison of the input signals demonstrates the following:

- *Constant*: Only suitable for identification of one parameter, here the static gain K, which is given by $b_1/(1-a_1)$. Not suitable for identification because no dynamics are excited. The parameters b_1 and a_1 cannot be estimated independently; only the ratio $b_1/(1-a_1)$ is correctly identified.
- *Impulse*: Not well suited for identification. In particular, the gain is estimated very inaccurately.
- *Step*: Well suited for identification. Low frequencies are emphasized. The static gain is estimated very accurately.
- *Rectangular*: Well suited for identification. Depending on the frequency of the rectangular signal, the desired frequency range can be emphasized. For the signal in Fig. 18.6, the time constant is estimated very accurately.

Table 18.1 Identification results for different excitation signals

Input signal	b_1	a_1	K	T [s]
Constant	0.2620	−0.7392	1.0048	3.3098
Impulse	0.0976	−0.8570	0.6826	6.4800
Step	0.1220	−0.8780	0.9998	7.6879
Rectangular	0.1170	−0.8834	1.0033	8.0671
PRBS	0.1201	−0.8796	0.9980	7.7964
True process	0.1175	−0.8825	1	8

- *PRBS (pseudo random binary signal)*: Well suited for identification. Imitates white noise in discrete time with a deterministic signal and thus excites all frequencies equally well.

□

18.3 General Model Structure

In this section a general linear model structure is introduced from which all linear models can be derived by simplifications. This general model is not normally used in practice; it just servefs as a unified framework. The output $y(k)$ of a deterministic linear system at time k can be computed by filtering the input $u(k)$ through a linear filter $G(q)$ (q denotes the forward shift operator, i.e., $q^{-1}x(k) = x(k-1)$, and thus it is the time domain counterpart of the $z = e^{j\omega}$-operator in the frequency domain):

$$y(k) = G(q)u(k) = \frac{\widetilde{B}(q)}{\widetilde{A}(q)}u(k) . \tag{18.2}$$

In general, the linear transfer function $G(q)$ may possess a numerator $\widetilde{B}(q)$ and a denominator $\widetilde{A}(q)$. In addition to the deterministic part, a stochastic part can be modeled. By filtering white noise $v(k)$ through a linear filter $H(q)$, any noise frequency characteristic can be modeled. Thus, an arbitrary noise signal $n(k)$ can be generated by

$$n(k) = H(q)v(k) = \frac{\widetilde{C}(q)}{\widetilde{D}(q)}v(k) . \tag{18.3}$$

A general linear model describing deterministic and stochastic influences is obtained by combining both parts (see Fig. 18.8a)

$$y(k) = G(q)u(k) + H(q)v(k) . \tag{18.4}$$

The filter $G(q)$ is called the *input transfer function*, since it relates the input $u(k)$ to the output $y(k)$, and the filter $H(q)$ is called the *noise transfer function*, since it relates the noise $v(k)$ to the output $y(k)$. These transfer functions $G(q)$ and $H(q)$ can be split into their numerator and denominator polynomials; see Fig. 18.8b. For future analysis it is helpful to separate a possibly existent common denominator

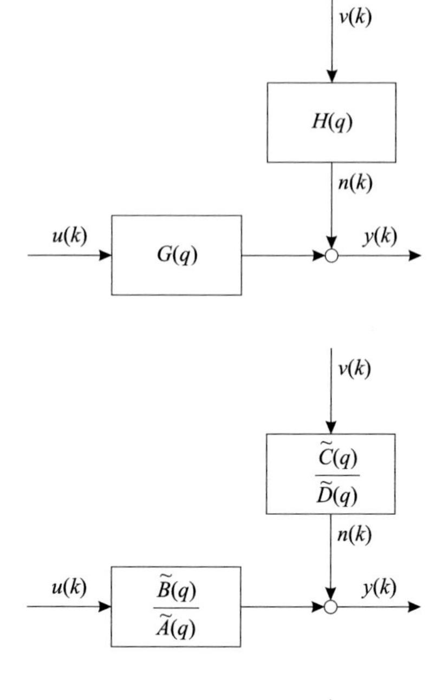

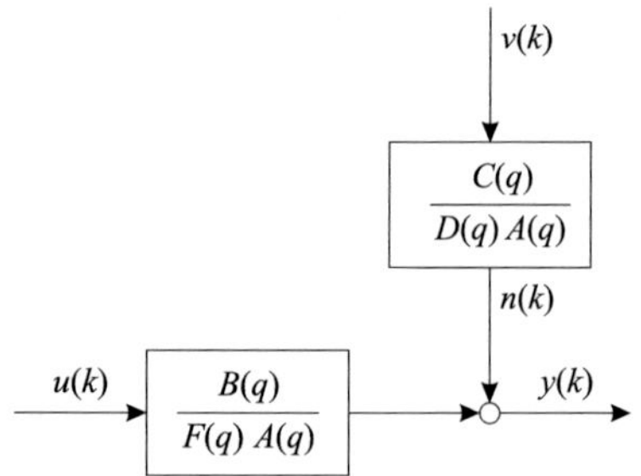

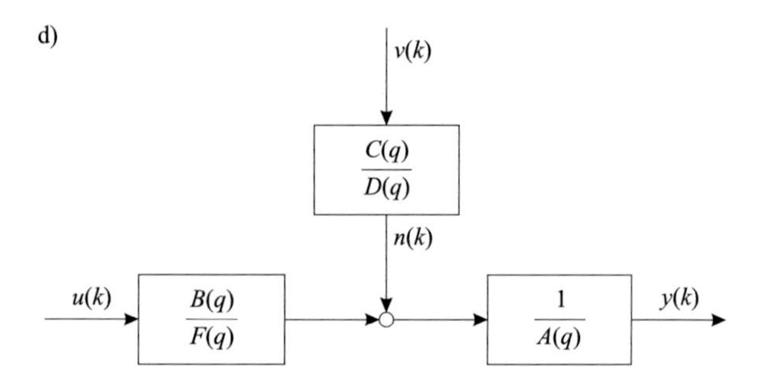

Fig. 18.8 (**a**) general transfer functions, (**b**) in numerator/denominator form, (**c**) including common numerator dynamics, (**d**) common numerator dynamics factored out

dynamics $A(q)$ from $G(q)$ and $H(q)$; see Fig. 18.8c and d. Thus, $F(q)A(q) = \widetilde{A}$ and $D(q)A(q) = \widetilde{D}$. If $\widetilde{A}(q)$ and $\widetilde{D}(q)$ do not share a common factor, then simply $A(q) = 1$. These notations of the transfer functions in Fig. 18.8a and the polynomials in Fig. 18.8d have been accepted standards since the publication of Ljung's book [356]. So the general linear model can be written as

$$y(k) = \frac{B(q)}{F(q)A(q)}u(k) + \frac{C(q)}{D(q)A(q)}v(k) \tag{18.5}$$

or equivalently as

$$A(q)y(k) = \frac{B(q)}{F(q)}u(k) + \frac{C(q)}{D(q)}v(k)\,. \tag{18.6}$$

By making special assumptions on the polynomials A, B, C, D, and F, the widely applied linear models are obtained from this general form. Before these simpler linear models are introduced, it is helpful to make a few general remarks on the terminology and to discuss some general aspects that are valid for all types of linear models.

18.3.1 Terminology and Classification

Unfortunately, the standard terminology of linear dynamic models is quite confusing. The reason for this is the historical development of these models within different disciplines. Thus, some expressions stem from time series modeling in economics. Economists typically analyze and try to predict time series such as stock prices, currency exchange rates, and unemployment rates. A common characteristic of all these applications is that the relevant input variables are hardly known and the number of possibly relevant inputs is huge. Therefore, economists started by analyzing the time series on its own without taking any input variables into account. Such models result from the general model in Fig. 18.8 and (18.6) by discarding the input, that is, $u(k) = 0$. Then the model becomes fully stochastic. Such a time series model is depicted in Fig. 18.9, opposed to the purely deterministic model shown in Fig. 18.10. From this time series point of view, the terminology used in the following is logical and straightforward. Ljung's book [356] established this as the now widely accepted standard in system identification.

A time series model with just a denominator polynomial (Fig. 18.11)

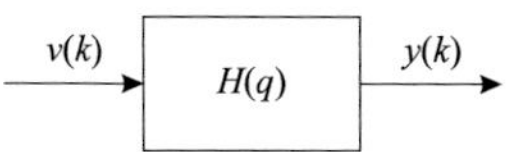

Fig. 18.9 A general linear time series model. The model input $v(k)$ is a white noise signal. There is no deterministic input $u(k)$

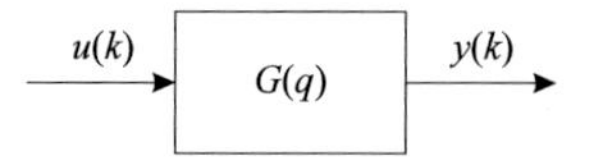

Fig. 18.10 A general linear deterministic model. The model input $u(k)$ is a deterministic signal. There is no stochastic influence such as a white noise $v(k)$

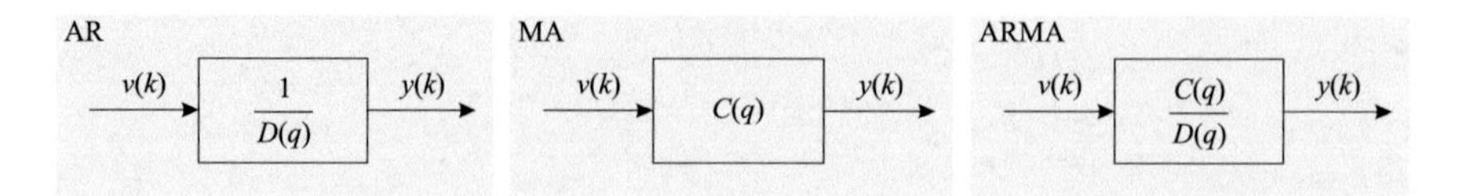

Fig. 18.11 An overview of time series models: autoregressive (AR), moving average (MA), and autoregressive moving average (ARMA) models

$$y(k) = \frac{1}{D(q)} v(k) \tag{18.7}$$

is called an *autoregressive* (AR) model.

A time series model with just a numerator polynomial (Fig. 18.11)

$$y(k) = C(q)v(k) \tag{18.8}$$

is called a *moving average* (MA) model.

A time series model with a numerator and denominator polynomial (Fig. 18.11)

$$y(k) = \frac{C(q)}{D(q)} v(k) \tag{18.9}$$

is called an *autoregressive moving average* (ARMA) model.

It is obvious that a model based on the time series only, without taking any relevant input variable into account, cannot be very accurate. Therefore, more accurate models are constructed by incorporating one (or more) input variable(s) into the model. This input $u(k)$ is called an *exogenous* input. With these considerations, the time series models in Fig. 18.11 can be extended by adding an "X" for exogenous input. To extend a moving average time series model with an exogenous input is highly uncommon. Thus, something like "MAX" is rarely used.

Figure 18.12 gives an overview of the most important linear input/output models, which are briefly discussed in the following. All models on the left-hand side of Fig. 18.12 are denoted by AR...and belong to the class of *equation error* models. Their characteristic is that the filter $1/A(q)$ is common to both the deterministic process model and the stochastic noise model. All models on the right-hand side of Fig. 18.12 belong to the class of *output error* models, which is characterized by a noise model that is independent of the deterministic process model.

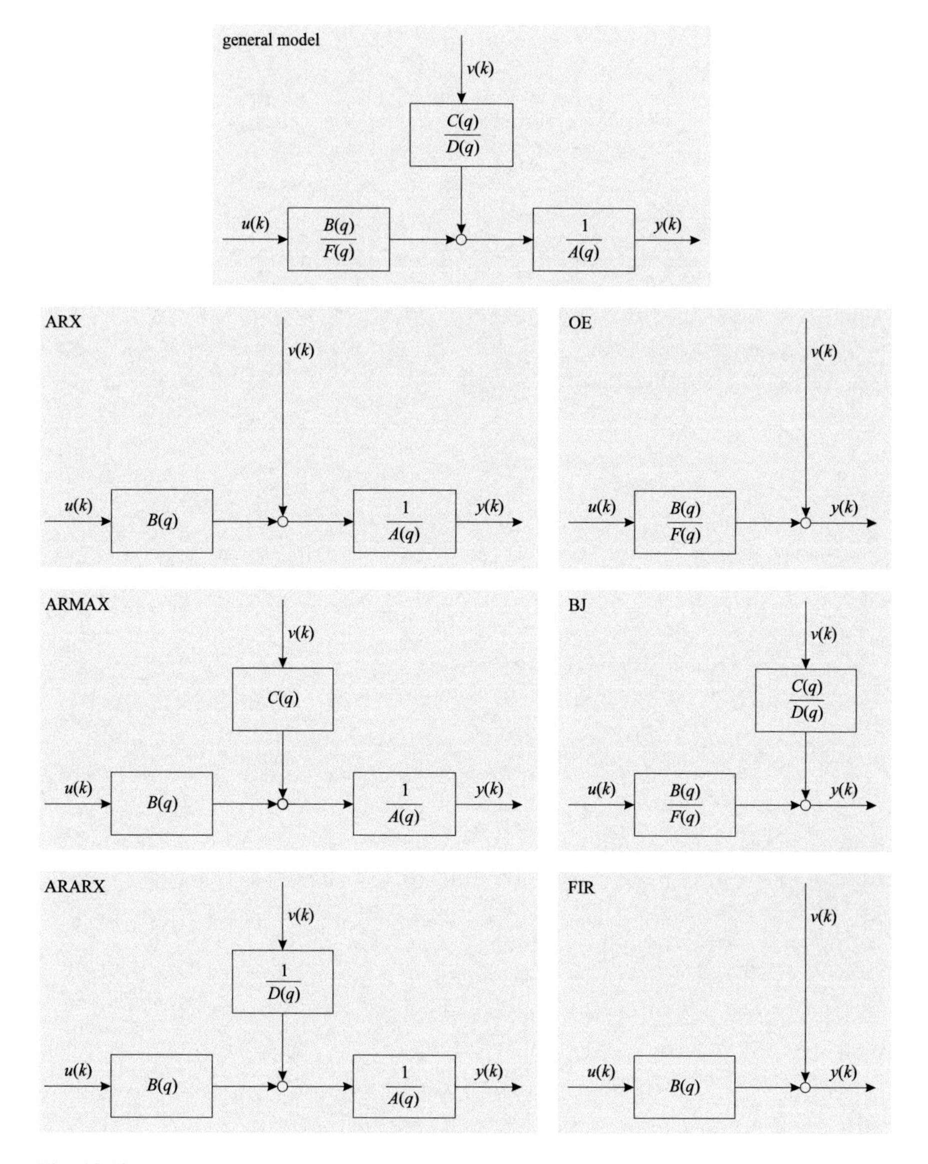

Fig. 18.12 An overview of common linear dynamic models

The *autoregressive with exogenous input* (ARX)[1] model (Fig. 18.12) is an extended AR model:

$$y(k) = \frac{B(q)}{A(q)} u(k) + \frac{1}{A(q)} v(k) . \tag{18.10}$$

Here the term "autoregressive" is related to the transfer function from the input $u(k)$ to the output $y(k)$ as well as to the noise transfer function from $v(k)$ to $y(k)$. Thus, the deterministic and the stochastic part of the ARX model possess an identical denominator dynamics. For a more detailed discussion, refer to Sect. 18.5.1.

The *autoregressive moving average with exogenous input* (ARMAX) model (Fig. 18.12) is an extended ARMA model:

$$y(k) = \frac{B(q)}{A(q)} u(k) + \frac{C(q)}{A(q)} v(k) . \tag{18.11}$$

As for the ARX model, the ARMAX model assumes identical denominator dynamics for the input and noise transfer functions. However, the noise transfer function is more flexible owing to the moving average polynomial. For a more detailed discussion, refer to Sect. 18.5.2.

The *autoregressive autoregressive with exogenous input* (ARARX) model (Fig. 18.12) is an extended AR model:

$$y(k) = \frac{B(q)}{A(q)} u(k) + \frac{1}{D(q)A(q)} v(k) . \tag{18.12}$$

This is an ARX model with additional flexibility in the denominator of the noise transfer function. Thus, instead of an additional moving average filter $C(q)$ as in the ARMAX model, the ARARX model possesses an additional autoregressive filter $1/D(q)$. For a more detailed discussion, refer to Sect. 18.5.3.

Just to complete this list, the *autoregressive autoregressive moving average with exogenous input* (ARARMAX) model can be defined as

$$y(k) = \frac{B(q)}{A(q)} u(k) + \frac{C(q)}{D(q)A(q)} v(k) . \tag{18.13}$$

Because this model type is hardly used, it is not discussed in more detail.

All these AR... models share the $A(q)$ polynomial as denominator dynamics in the input and noise transfer functions. They are also called *equation error* models. This corresponds to the fact that the noise does not directly influence the output $y(k)$

---

[1] In a considerable part of the literature and in older publications in particular, the ARX model is called an "ARMA" model to express the fact that both a numerator and a denominator polynomial exist. However, as discussed above, this book follows the current standard terminology established in [356], where ARMA stands for the time series model in (18.9).

of the model but instead enters the model before the $1/A(q)$ filter. These model assumptions are reasonable if indeed the noise enters the process early so that its frequency characteristic is shaped by the process dynamics. If, however, the noise is primarily measurement noise that typically directly disturbs the output, the so-called *output error* models are more realistic.

The output error models are characterized by noise models that do not contain the process dynamics. Thus, the noise is assumed to influence the process output directly. The terminology of these models does not follow the rules given above for an extension of the time series models. Rather, the point of view changes from time series models (where the noise model is in the focus) to input/output models (where the attention turns to the deterministic input).

The most straightforward input/output model is the *output error* (OE) model (Fig. 18.12):

$$y(k) = \frac{B(q)}{F(q)} u(k) + v(k) . \tag{18.14}$$

This OE model is one special model in the class of output error models. Unfortunately, it is difficult to distinguish between the class of output error models and this special output error model in (18.14) by the name. Therefore, it must become clear from the context whether the special model or the model class is referred to. To clarify this confusion a little bit, the abbreviation OE always refers to the special output error model in (18.14). In contrast to the ARX model, white noise enters the OE model directly without any filter. For a more detailed discussion, refer to Sect. 18.5.4.

This OE model can be enhanced in flexibility by filtering the white noise through an ARMA filter. This defines the *Box-Jenkins* (BJ) model (Fig. 18.12):

$$y(k) = \frac{B(q)}{F(q)} u(k) + \frac{C(q)}{D(q)} v(k) . \tag{18.15}$$

The BJ model relates to the ARARMAX model as the OE model relates to the ARX model. The input and noise transfer functions are separately parameterized and therefore independent. The special cases of a BJ model $C(q) = 1$ or $D(q) = 1$ do not have special names. For a more detailed discussion, refer to Sect. 18.5.5.

Finally, a quite different model belongs to the output error model class, as well. The *finite impulse response* (FIR) model is defined by (Fig. 18.12)

$$y(k) = B(q)u(k) + v(k) . \tag{18.16}$$

The FIR model is an OE or an ARX model without any feedback, that is, $F(q) = 1$ or $A(q) = 1$, respectively. As an extension of the FIR model, the *orthonormal basis functions* (OBF) model is also of significant practical interest. However, the OBF model does not fit well in the framework presented here. The FIR and OBF models are described in detail in Sect. 18.6.

At first sight, all these different model structures may be quite confusing. However, it is sufficient to remember the ARX, ARMAX, OE, FIR, and OBF models for an understanding of the rest of this book. Nevertheless, all the concepts discussed in this chapter are of fundamental importance. Figures 18.13, 18.14, 18.15, and 18.16 illustrate the described linear models from different points of view. Table 18.2 summarizes the simplifications that lead from the general model to the specific model structures.

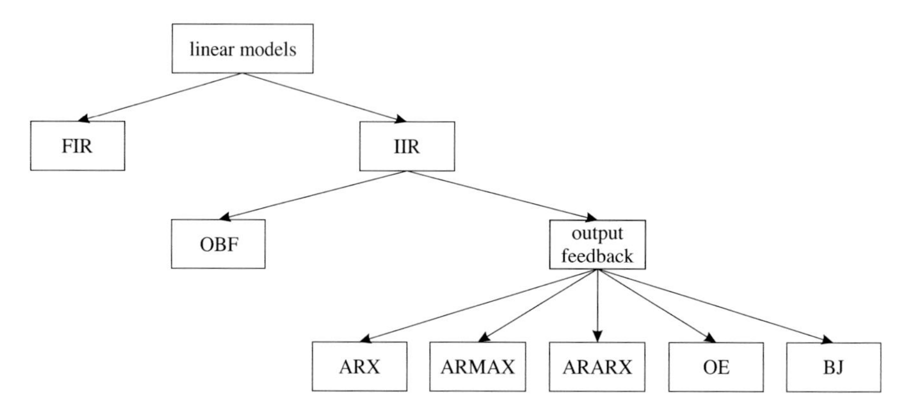

Fig. 18.13 Classification of linear models according to finite impulse response (FIR) and infinite impulse response (IIR) filters

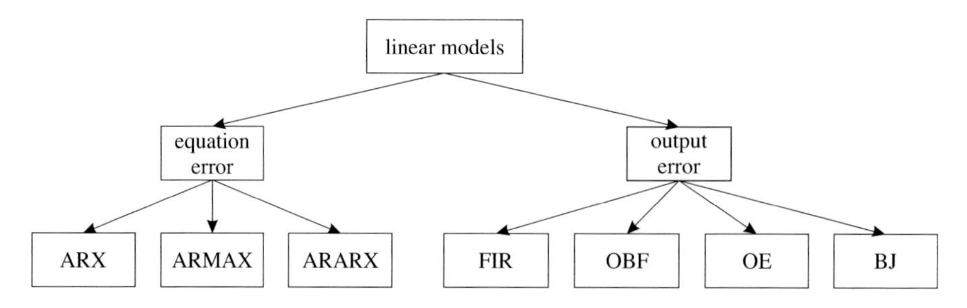

Fig. 18.14 Classification of linear models according to equation error and output error models

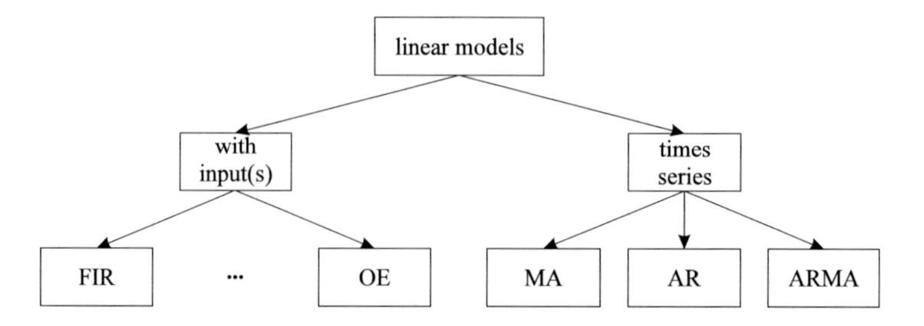

Fig. 18.15 Classification of linear models according to input/output and time series models

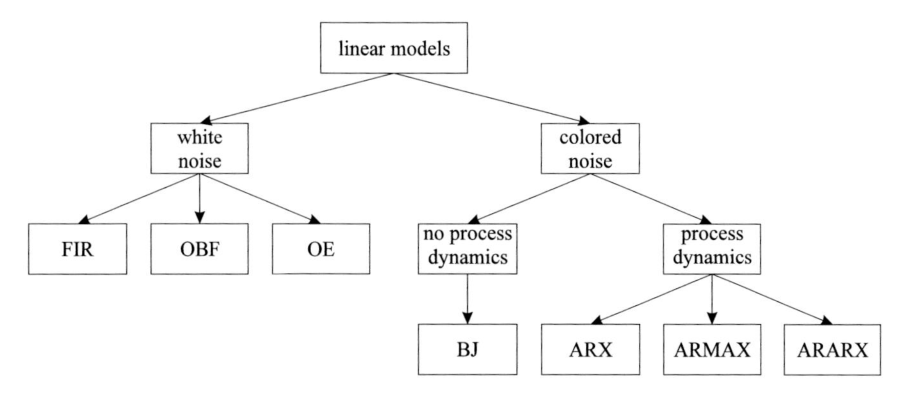

Fig. 18.16 Classification of linear models according to noise properties. The box entitled "process dynamics" refers to noise filter, which include the process denominator dynamics $1/A(q)$

Table 18.2 Common linear models

Model structures	Model equations
MA	$y(k) = C(q)\ v(k)$
AR	$y(k) = 1/D(q)\ v(k)$
ARMA	$y(k) = C(q)/D(q)\ v(k)$
ARX	$y(k) = B(q)/A(q)\ u(k) + 1/A(q)\ v(k)$
ARMAX	$y(k) = B(q)/A(q)\ u(k) + C(q)/A(q)\ v(k)$
ARARX	$y(k) = B(q)/A(q)\ u(k) + 1/D(q)A(q)\ v(k)$
ARARMAX	$y(k) = B(q)/A(q)\ u(k) + C(q)/D(q)A(q)\ v(k)$
OE	$y(k) = B(q)/F(q)\ u(k) + v(k)$
BJ	$y(k) = B(q)/F(q)\ u(k) + C(q)/D(q)\ v(k)$
FIR	$y(k) = B(q)\ u(k) + v(k)$

Note that for the sake of simplicity, the processes and models are assumed to possess no dead time. However, in any equation a dead time $d\,T_0$ can easily be introduced by replacing the input $u(k)$ with the d steps delayed input $u(k-d)$. Furthermore, it is assumed that the processes and models have no direct path from the input to the output (i.e., they are strictly proper), so that $u(k)$ does not immediately influence $y(k)$. Thus, terms like $b_0u(k)$ do not appear in the difference equations. This assumption is fulfilled for almost any real-world process.

18.3.2 Optimal Predictor

Probably the most common application of a model is forecasting the future behavior of a process. Two cases have to be distinguished: *simulation* and *prediction*. If the response of the model to an input sequence has to be calculated while the process outputs are unknown, this is called *simulation*. If, however, the process outputs are

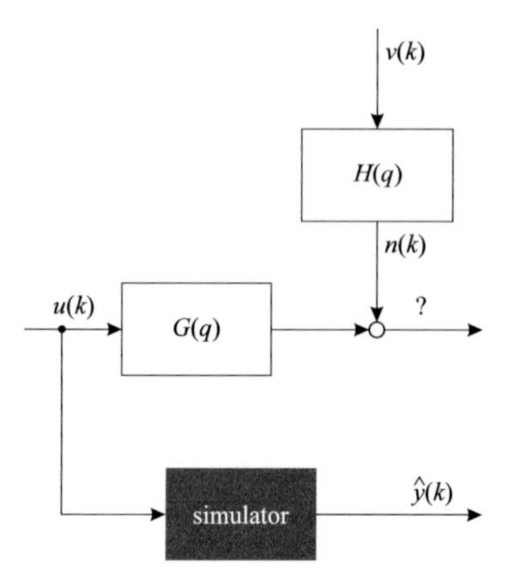

Fig. 18.17 For simulation, only the inputs are known. No information about the real process output is available

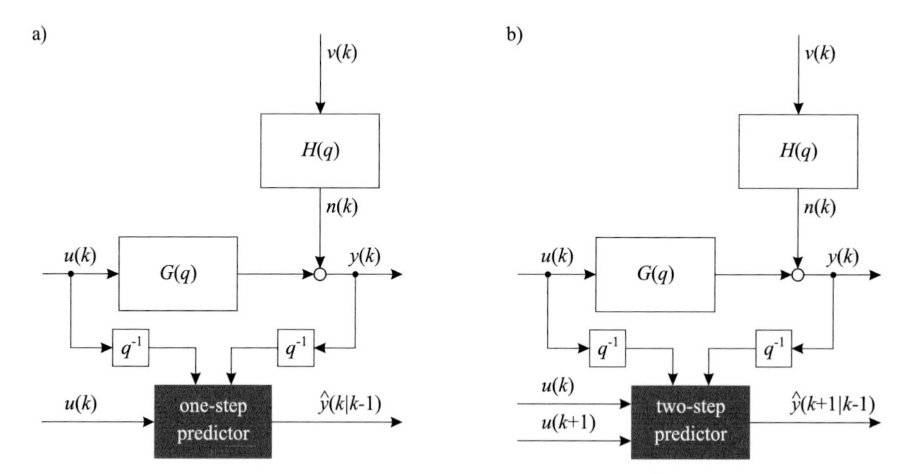

Fig. 18.18 (**a**) One-step prediction and (**b**) two-step prediction. The expression "$|k-1$" means "on the information available at time instant $k-1$." For prediction, besides the inputs, the previous process outputs are known. Note that if the prediction horizon l becomes very large, the importance of the information about the previous process outputs decreases. Thus, as $l \to \infty$, prediction approaches simulation; see Fig. 18.17

known up to some time instant, say $k-1$, and it is asked for the model output l steps in the future, this is called *prediction*. Very often one is interested in the one-step prediction, i.e., $l = 1$, and if nothing else is explicitly stated in the following, prediction will mean one-step prediction. Figures 18.17 and 18.18 illustrate the difference between simulation and prediction; see also Sects. 1.1.2 and 1.1.3.

18.3.2.1 Simulation

It is obvious from Fig. 18.17 that simulation is fully deterministic:

$$\hat{y}(k) = G(q)u(k) . \tag{18.17}$$

Thus, the noise model $H(q)$ seems irrelevant for simulation. Note, however, that the noise model $H(q)$ influences the estimation of the parameters in $G(q)$ and therefore it affects the simulation quality although $H(q)$ does not explicitly appear in (18.17).

Because the process output is unknown, no information about the disturbances is available. In order to get some "feeling" how the disturbed process output qualitatively may look, it is possible to generate a white noise signal $w(k)$ with proper variance by a computer [356], to filter this signal through the noise filter $H(q)$, and to add this filtered noise to the deterministic model output

$$\hat{y}(k) = G(q)u(k) + H(q)w(k) . \tag{18.18}$$

Note, however, that (18.18) is just a better qualitative output than (18.17). The smaller simulation error can be expected from (18.17) since $w(k)$ is a different white noise signal than the original but not measurable $v(k)$.

18.3.2.2 Prediction

In contrast to simulation, for prediction the information about the previous process output can be utilized. Thus, the *optimal predictor* should combine the inputs and previous process outputs in some way. So the optimal *linear* predictor can be defined as the linear combination of the filtered inputs and the filtered outputs

$$\begin{aligned}\hat{y}(k|k-1) = \; & s_0 u(k) + s_1 u(k-1) + \ldots + s_{ns} u(k-ns) \\ & + t_1 y(k-1) + \ldots + t_{nt} y(k-nt)\end{aligned} \tag{18.19}$$

or

$$\hat{y}(k|k-1) = S(q)u(k) + T(q)y(k) . \tag{18.20}$$

Note that the filter $T(q)$ does not contain the term t_0 since of course the value $y(k)$ is not available when predicting $\hat{y}(k|k-1)$. For most real-world processes $s_0 = 0$ as well, because the input does not instantaneously influence the output, i.e., the model is strictly proper.

The term $S(q)u(k)$ contains information about the deterministic part of the predictor, while the term $T(q)y(k)$ introduces a stochastic component into the predictor since only $y(k)$ is disturbed by noise.

The following question arises: What are the best filters $S(q)$ and $T(q)$? More exactly speaking, which filters result in the smallest squared prediction error (prediction error variance)? It can be shown that the optimal predictor is [356]

$$\hat{y}(k|k-1) = \frac{G(q)}{H(q)}u(k) + \left(1 - \frac{1}{H(q)}\right)y(k) \tag{18.21}$$

or

$$H(q)\hat{y}(k|k-1) = G(q)u(k) + (H(q)-1)\,y(k)\,. \tag{18.22}$$

Thus, $S(q) = G(q)/H(q)$ and $T(q) = 1 - 1/H(q)$.

It is very helpful to discuss some special cases in order to illustrate this optimal predictor equation.

- *ARX model*: $G(q) = B(q)/A(q)$ and $H(q) = 1/A(q)$. Therefore, the optimal predictor for an ARX model is

$$\hat{y}(k|k-1) = B(q)u(k) + (1 - A(q))\,y(k)\,. \tag{18.23}$$

  Thus, the inputs are filtered through the $B(q)$ polynomial, and the process outputs are filtered through the $1 - A(q)$ polynomial. Consequently, the predicted model output $\hat{y}(k|k-1)$ can be generated by applying simple moving average filtering. Because an ARX model implies correlated disturbances, namely, white noise filtered through $1/A(q)$, the process output contains valuable information about the disturbances. This information allows one to make a prediction on the actual disturbance at time instant k, which is implicitly performed by the term $(1 - A(q))y(k)$.
- *ARMAX model*: $G(q) = B(q)/A(q)$ and $H(q) = C(q)/A(q)$. Therefore, the optimal predictor for an ARMAX model is

$$\hat{y}(k|k-1) = \frac{B(q)}{C(q)}u(k) + \left(\frac{C(q) - A(q)}{C(q)}\right)y(k)\,. \tag{18.24}$$

  This equation is much more difficult than the ARX predictor. A characteristic is that both the input and the process output are filtered through filters with the same denominator dynamics $C(q)$. Note that the ARMAX predictor contains the ARX predictor as the special case $C(q) = 1$.
- *OE model*: $G(q) = B(q)/A(q)$ and $H(q) = 1$. Therefore, the optimal predictor for an OE model is

$$\hat{y}(k|k-1) = \frac{B(q)}{A(q)}u(k)\,. \tag{18.25}$$

  This, however, is exactly a *simulation*; see (18.17)! No information about the process output enters the predictor equation. The reason for this obviously lies in the noise filter $H(q) = 1$. Intuitively this can be explained as follows. The term $T(q)y(k)$ in (18.20) contains the information about the stochastic part of the model. If the process is disturbed by unfiltered white noise, as is assumed in the OE model, there is no correlation between disturbances $n(k)$ at different times k. Thus, knowledge about previous disturbances that is contained in $y(k)$ does not help to predict the future. Thus, the simulation of the model is the optimal prediction in the case of an OE model. At first sight, it seems strange to totally ignore the knowledge of $y(k)$. However, incorporation of the white noise corrupted $y(k)$ into the predictor would deteriorate the performance.

18.3.3 Some Remarks on the Optimal Predictor

It is important to make some remarks on the optimal predictor which have been omitted above for easier understanding.

- Equation (18.21) for the optimal predictor can be derived as follows. The starting point is the model equation

$$y(k) = G(q)u(k) + H(q)v(k)\,. \tag{18.26}$$

  The optimal predictor should be capable of extracting all information out of the signals. Thus, the prediction error, i.e., the difference between the process output $y(k)$ and the predicted output $\hat{y}(k|k-1)$, should be equal to the white noise $v(k)$, since this is the only unpredictable part in the system:

$$v(k) = y(k) - \hat{y}(k|k-1)\,. \tag{18.27}$$

  This equation can be used to eliminate $v(k)$ in (18.26). Then the following relationship results:

$$y(k) = G(q)u(k) + H(q)\left(y(k) - \hat{y}(k|k-1)\right)\,. \tag{18.28}$$

  If in this equation $\hat{y}(k|k-1)$ is isolated, the optimal predictor in (18.21) results. The optimal predictor thus leads to white residuals. Therefore, an analysis of the spectrum of the real residuals can be used to test whether the model structure is appropriate.
- It has been demonstrated above that the optimal predictor for ARX models includes previous inputs and process outputs, while the optimal predictor for OE models includes only previous inputs. This statement is correct if the transfer functions $G(q)$ and $H(q)$ are identical with the model. However, if the model only approximates the process, as is the case in all real-world applications, this statement is not necessarily valid anymore. Consider, for example, a process with integral behavior $G(q) = Kq/(q-1)$ and additive white measurement noise at

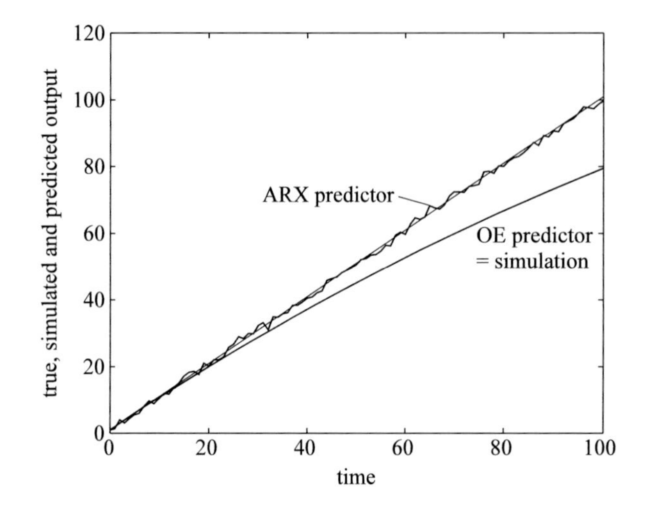

Fig. 18.19 OE and ARX predictor for a process with integral behavior. An OE model of the process $q/(q-1)$ disturbed by white output noise is assumed to be identified to $q/(q-0.995)$. The optimal OE predictor in (18.25) simulates the process

the output. Assume that an OE model is applied, which indeed is the correct structure for these noise properties, and the parameter K is not estimated exactly. Then, for this OE model, the ARX predictor may yield better results than the OE predictor. The explanation for this fact is that the error of the simulated output obtained by the OE predictor increases linearly with time proportional to the estimation error in K, while the ARX predictor is based on process outputs and thus cannot "run away" from the process. Figure 18.19 compares the behavior of an OE and an ARX predictor for this process.

Because the model parameters have not been identified exactly, the model implements no integrator but a first-order time-lag behavior (with a large time constant). Thus, the OE predictor quality becomes worse as time proceeds. In contrast, if the ARX predictor in (18.49) is utilized for the OE model, the prediction always remains close to the true process because it is also based on process outputs. The price to be paid is the introduction of the disturbances into the prediction since the process output is corrupted by noise.

Note that this example illustrates an extreme case since the investigated process is not stable. Nevertheless, in practice even for stable processes, it can be advantageous to use the ARX predictor for models belonging to the output error class. This is because non-modeled nonlinear effects can lead to significant deviations between the process and the simulated model output, while a one-step prediction with the ARX predictor will follow the operating point better. Generally, the ARX predictor can be reasonably utilized for output error models if the disturbances are small. Thus, there exists some kind of tradeoff between the wrongly assumed noise model when using the ARX predictor and the model/process mismatch when using the OE predictor.

- The optimal predictor in (18.21) is only stable if the noise filter $H(q)$ is minimum phase. The stability of the predictor is a necessary condition for the application of the prediction error methods (see next section). If $H(q)$ were non-minimum phase, $1/H(q)$ and thus the predictor would be unstable. However, then the noise filter could be replaced by its minimum phase counterpart, i.e., the conjugate complex $H^*(q) = H(q^{-1})$, because the purpose of the filter $H(q)$ is merely to shape the frequency spectrum of the disturbance $n(k)$ by filtering $v(k)$. But the spectrum of the disturbance $n(k)$ is determined only by $|H(q)|^2$, which is equal to $|H|^2 = H(q)H^*(q)$ (*spectral factorization*). Thus, both filters $H(q)$ and $H^*(q)$ result in the same frequency shaping, and therefore the filter that is stable invertible can be selected for the optimal predictor.
- Another assumption not yet mentioned is made in the optimal predictor equation (18.21). The influence of the initial conditions is neglected. Consider, for example, an OE model

$$y(k) = b_1 u(k-1) + \ldots + b_m u(k-m) - a_1 y(k-1) - \ldots - a_m y(k-m)\,. \tag{18.29}$$

For this model m initial conditions have to be assumed; at time $k = 0$ these are the values of $y(-1), \ldots, y(-m)$. Typically, these initial conditions are set to zero. This assumption is reasonable since for stable systems the initial conditions decay exponentially with time. Strictly speaking, the optimal predictor in (18.21) is only the *stationary* optimal predictor. If the initial conditions were to be taken into account, the optimal predictor would become *time-variant* and would asymptotically approach (18.21). This relationship is well known between the Kalman filter, which represents the optimal time-variant predictor considering the initial conditions and its stationary solution, the Luenberger observer, which is valid only as time $\to \infty$. Nevertheless, since the initial conditions decay rapidly, in practice the stationary counterpart, i.e., the optimal predictor in (18.21), is sufficiently accurate and much simpler to deal with.

18.3.4 Prediction Error Methods

Usually the optimal predictor is used for measuring the performance of the corresponding model. The prediction error is the difference between the desired model output (= process output) and the one-step prediction performed by the model

$$e(k) = y(k) - \hat{y}(k|k-1)\,. \tag{18.30}$$

In the following, the term *prediction error* is used as a synonym for *one-step* prediction error. With the optimal predictor in (18.21), the prediction error becomes

$$e(k) = \frac{1}{H(q)} y(k) - \frac{G(q)}{H(q)} u(k) . \tag{18.31}$$

For example, an OE model has the following prediction error: $e(k) = y(k) - G(q)u(k)$. Most identification algorithms are based on the minimization of a loss function that depends on this one-step prediction error. Although this is the most common choice, it can be reasonable to minimize another error measure. For example, *predictive control* algorithms utilize a model to predict a number of steps, say l, into the future. In this case, the performance can be improved by minimizing the error of an l-step prediction $e(k) = y(k) - \hat{y}(k|k-l)$ where l is the prediction horizon [350, 506, 507, 610]. Because the computation of such a multi-step predictor becomes more and more involved for larger prediction horizons l, even for model-based predictive control typically, the one-step prediction error in (18.30) is used for identification.

For reasons discussed in Sect. 2.3, the sum of squared prediction errors is usually used as the loss function, i.e., with N data samples

$$J = \sum_{i=1}^{N} e^2(i) . \tag{18.32}$$

It is discussed in Sect. 2.3.1 that this choice is optimal (in the maximum likelihood sense) if the noise is Gaussian distributed. Another property of the sum of squared errors loss function is that the parameters of the ARX model structure can be estimated by linear optimization techniques; see Sect. 18.5.1.

Note that a sensible minimization of the loss function in (18.32) requires the predictor to be *stable*, which in turn requires that $G(q)$ is stable and $H(q)$ is minimum phase. Otherwise, the mismatch between process and model due to different initial values would not decay exponentially (as in the stable case); rather they would influence the minimization procedure decisively.

The loss function in (18.32) can be extended by filtering the prediction errors through a linear filter $L(q)$. Since the unfiltered prediction error is (see (18.31))

$$e(k) = \frac{1}{H(q)} \left(y(k) - G(q)u(k)\right) , \tag{18.33}$$

the filtered prediction error e_F can be written as

$$e_F(k) = \frac{L(q)}{H(q)} \left(y(k) - G(q)u(k)\right) . \tag{18.34}$$

Obviously, the filter $L(q)$ has the same effect as the inverse noise model $1/H(q)$. Thus, the filter $L(q)$ can be fully incorporated into the noise model $H(q)$ or vice versa. The understanding of this relationship is important for some of the identification algorithms discussed in the following sections. For more details about this relationship, refer to Sect. 18.7.4.

18.4 Time Series Models

A time series is a signal that evolves over time,[2] such as the Dollar/Euro exchange rate, the Dow Jones index, the unemployment rate in a country, the world's population, the amount of rainfall in a particular area, or the sound of a machine received with a microphone. A characteristic of all time series is that the current value is usually dependent on previous values. Thus, a dynamic model is required for a proper description of a time series. Furthermore, typically the driving inputs, i.e., the variables that influence the time series, are not known, are not measurable, or are so huge in number that it is not feasible to include them in the model. It is no coincidence that the typical examples for time series listed above are mostly non-technical. Often in engineering applications, the relationships between different quantities are quite well understood, and at least some knowledge about the basic laws is available. Then it is more reasonable to build a model with deterministic inputs and possibly additional stochastic components. In economy and social sciences, the dependencies between different variables are typically much more complex, and thus time series modeling plays a greater role in these disciplines.

Because time series models (as defined here) do not take any deterministic input into account, the task is simply to build a model for the time series with the information about the past realization of this time series only. Because no external inputs $u(k)$ are considered, it is clear that such a model will be of relatively low quality. Nevertheless, it may be possible to identify a model that allows short-term predictions (typically one-step predictions) with sufficient accuracy or which allows gaining insights about the underlying process.

Since no inputs are available, time series models are based on the following idea. The time series is thought to be generated by a (hypothetical) white noise signal, which drives a dynamic system. This dynamic system is then identified with the time series data. The main difficulty is that the input of this system, that is, the white noise signal, is unknown. Since this chapter deals with linear system identification, the dynamic system is assumed to be linear, and the following model of the time series results (see also Fig. 18.9):

$$y(k) = H(q)v(k) = \frac{C(q)}{D(q)}v(k)\,, \tag{18.35}$$

where $y(k)$ is the time series and $v(k)$ is the artificial white noise signal. In the following three sections, two special cases of (18.35) and finally the general time series model in (18.35) are briefly discussed. The model (18.35) can be further extended by an integrator to deal with non-stationary processes. For more details, refer to [80].

[2] In some cases, the signal may not depend on time but rather on space (e.g., in geology) or some other quantity.

18.4.1 Autoregressive (AR)

The autoregressive time series model shown in Fig. 18.20 is very common since it allows one to shape the frequency characteristics of the model with a few *linear* parameters. In many technical applications of time series modeling, one is interested in resonances, i.e., weakly damped oscillations at certain frequencies which may be hidden under a high noise level. Then an AR (or ARMA) model of the time series is a powerful tool for analysis. An oscillation is represented by a weakly damped conjugate complex pole pair in $1/D(q)$. Compared with other tools for frequency analysis such as a Fourier transform, an AR (or ARMA) model does not suffer from leakage effects due to a discretization of the frequency range. Rather, AR (or ARMA) models, in principle, allow one to determine frequencies and amplitudes with arbitrary accuracy. In practice, AR (or ARMA) models are usually preferred if the number of considered resonances is small or a smoothed version of the spectrum is desired because then the order of the models can be chosen reasonably low and the parameters can be estimated accurately.

The time series is thought to be constructed by filtering white noise $v(k)$:

$$y(k) = \frac{1}{D(q)} v(k) \,. \tag{18.36}$$

The difference equation makes the linear parameterization of the AR model obvious

$$y(k) = -d_1 y(k-1) - \ldots - d_m y(k-m) + v(k) \,. \tag{18.37}$$

The prediction error simply becomes

$$e(k) = D(q) y(k) \,. \tag{18.38}$$

By taking the prediction error approach, the parameter estimation is a least squares problem, which can be easily solved. It corresponds to the estimation of an ARX model without numerator parameters, i.e., $B(q) = 1$ (see Sect. 18.5.1 for details). Another common way to estimate an AR model is to correlate (18.37) with $y(k-\kappa)$. For $\kappa > 0$ this results to

$$\text{corr}_{yy}(\kappa) = -d_1 \text{corr}_{yy}(\kappa - 1) - \ldots - d_m \text{corr}_{yy}(\kappa - m) \tag{18.39}$$

because the previous outputs $y(k-\kappa)$ do not depend on the current noise $v(k)$, i.e., $\text{E}\{y(k-\kappa)v(k)\} = 0$. For $\kappa = 0, -1, \ldots$ additional, increasingly complex terms like σ^2, $d_1\sigma^2$, etc. appear in (18.39) where σ^2 is the noise variance. Usually only

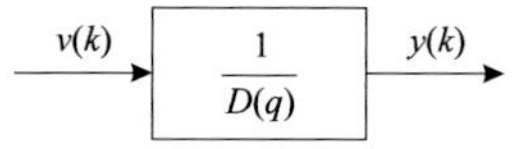

Fig. 18.20 AR model

the equations for $\kappa \geq 0$ are used to estimate the noise variance and the parameters d_i. For these different values of κ, (18.39) are called the *Yule-Walker equations*. In a second step, these Yule-Walker equations are solved by least squares; compare this with the COR-LS approach in Sect. 18.5.1.

The Yule-Walker equations are the most widely applied method for AR model estimation. Generally, most time series modeling is based on the estimation of the correlation function. This is a way to eliminate the fictitious unknown white noise signal $v(k)$ from the equations. The correlation function represents the useful information contained in the data in a compressed form.

18.4.2 Moving Average (MA)

For the sake of completeness, the moving average time series model (Fig. 18.21) will be mentioned here, too. It has less practical significance in engineering applications because a moving average filter does not allow one to model oscillations with a few parameters like an autoregressive filter does. Furthermore, in contrast to the corresponding deterministic input/output model (the FIR model), the MA model is *nonlinear* in its parameters if the prediction error approach is taken.

The MA model is given by

$$y(k) = C(q)v(k)\,. \tag{18.40}$$

The difference equation makes the nonlinear parameterization of MA model more obvious:

$$y(k) = v(k) + c_1 v(k-1) - \ldots + c_m v(k-m)\,. \tag{18.41}$$

Since $v(k-i)$ are unknown, in order to estimate the parameters c_i, the $v(k-i)$ have to be approximated by a previously built model. Thus, the approximated $\hat{v}(k-i)$, which replace the true but unknown $v(k-i)$ in (18.37), themselves depend on the parameters of a model estimated a priori. This relationship can also be understood by considering the prediction error

$$e(k) = \frac{1}{C(q)} y(k) \tag{18.42}$$

or

$$e(k) = -c_1 e(k-1) - \ldots - c_m e(k-m) + y(k)\,. \tag{18.43}$$

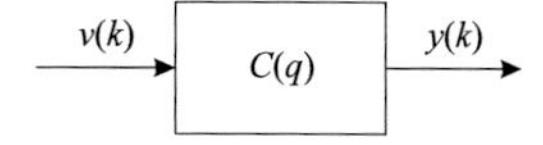

Fig. 18.21 MA model

More clever algorithms exist to estimate MA models more efficiently than via a direct minimization of the prediction errors; see [80].

18.4.3 Autoregressive Moving Average (ARMA)

The combination of an autoregressive part and a moving average part enhances the flexibility of the AR model. The resulting ARMA model shown in Fig. 18.22 is given by

$$y(k) = \frac{C(q)}{D(q)} v(k) . \tag{18.44}$$

The difference equation is

$$\begin{aligned} y(k) = &- d_1 y(k-1) - \ldots - d_m y(k-m) \\ &+ v(k) + c_1 v(k-1) - \ldots + c_m v(k-m) . \end{aligned} \tag{18.45}$$

The prediction error becomes

$$e(k) = \frac{D(q)}{C(q)} y(k) \tag{18.46}$$

or

$$\begin{aligned} e(k) = &- c_1 e(k-1) - \ldots - c_m e(k-m) \\ &+ y(k) + d_1 y(k-1) + \ldots + d_m y(k-m). \end{aligned} \tag{18.47}$$

For estimation of the nonlinear parameters in the ARMA model, the following approach can be taken; see Sect. 18.5.2. In the first step, a high-order AR model is estimated. Then the residuals $e(k)$ in (18.38) are used as an approximation of the white noise $v(k)$. With this approximation, the parameters c_i and d_i of an ARMA model are estimated by least squares. Then iteratively the following two steps are repeated: (i) approximation of $v(k)$ by (18.47) with the ARMA model obtained in the previous iteration and (ii) estimation of a new ARMA model utilizing the approximation for $v(k)$ from step (i). This two-step procedure avoids the direct nonlinear optimization of the parameters, and it is sometimes called the Hannan-Rissanen algorithm [80]. Other advanced methods are again based on the correlation

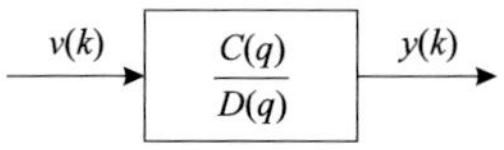

Fig. 18.22 ARMA model

idea introduced in Sect. 18.4.1, leading to the innovations algorithm or the Yule-Walker equations for ARMA models. The best (asymptotically efficient; compare Sect. B.7) estimators of AR, MA, and ARMA models are based on the maximum likelihood method [80]. However, this requires the application of a nonlinear optimization technique such as the Levenberg-Marquardt algorithm; see Chap. 4. Good initial parameter values for a local search can be obtained by any of the above strategies.

18.5 Models with Output Feedback

This section discusses linear models with output feedback. The models in this class by far are the most widely known and applied. Alternative linear models are described in the subsequent section and either do not employ any feedback or the feedback path is independent of the estimated parameter. In the following subsections on model with output feedback, the model structures are introduced together with appropriate algorithms for parameter estimation. To fully understand these algorithms, knowledge of the linear and nonlinear local optimization techniques discussed in Part I is required.

18.5.1 Autoregressive with Exogenous Input (ARX)

The ARX model is by far the most widely applied linear dynamic model. Usually, an ARX model is tried first, and only if it does not perform satisfactorily, more complex model structures are examined. The prevalence of ARX models seems puzzling at first sight since they do not match the structure of most real-world processes due to unrealistic noise assumptions. The popularity of the ARX model comes from its easy-to-compute parameters. The parameters can be estimated by a linear least squares technique since the prediction error is linear in the parameters. Consequently, a reliable recursive algorithm for online use, the RLS, exists as well; see Sect. 18.8.1.

The ARX model is depicted in Fig. 18.23 and is described by

$$A(q)y(k) = B(q)u(k) + v(k)\,. \tag{18.48}$$

The optimal ARX predictor is

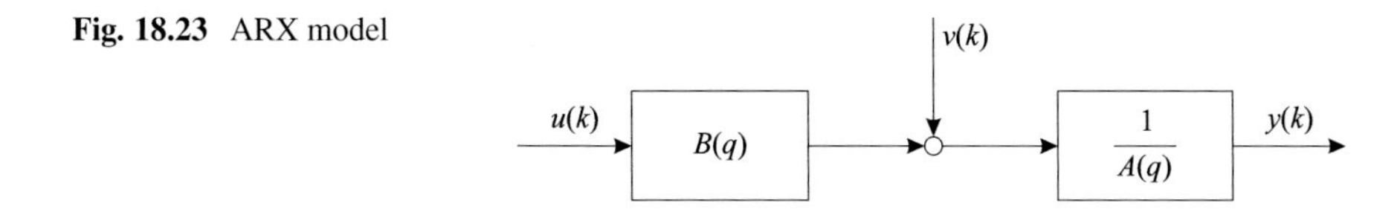

Fig. 18.23 ARX model

$$\hat{y}(k|k-1) = B(q)u(k) + (1 - A(q))y(k)\,, \tag{18.49}$$

which can be written as

$$\hat{y}(k|k-1) = b_1 u(k-1) + \ldots + b_m u(k-m)$$
$$- a_1 y(k-1) - \ldots - a_m y(k-m)\,. \tag{18.50}$$

assuming deg(A) = deg(B) = m. Note that contrary to the continuous time process description, in discrete time the numerator and denominator polynomials usually have the same order.

The ARX predictor is stable (it possesses no feedback!) even if the $A(q)$ polynomial and therefore the ARX model is unstable. This fact allows one to model unstable processes with an ARX model. However, the plant has to be stabilized in order to gather data. It is a feature of all equation error models that the $A(q)$ polynomials only appear in the numerator of their predictors, and thus the predictors are stable even if $A(q)$ is unstable.

With (18.49) the prediction error of an ARX model is

$$e(k) = A(q)y(k) - B(q)u(k)\,. \tag{18.51}$$

The term $A(q)y(k)$ acts as a whitening filter on the correlated disturbances. The measured output $y(k)$ can be split into two parts: the undisturbed process output $y_u(k)$ and the disturbance $n(k)$, where $y(k) = y_u(k) + n(k)$. Since $n(k) = 1/A(q)v(k)$ with $v(k)$ being white noise $A(q)y(k) = A(q)y_u(k) + v(k)$. Thus, the filter $A(q)$ in (18.51) makes the disturbances and consequently $e(k)$ white.

As can be seen from Fig. 18.23, one characteristic of the ARX model is that the disturbance, i.e., the white noise $v(k)$, is assumed to enter the process before the denominator dynamics $A(q)$. This fact can be expressed in another way by saying that the ARX model has a noise model of $1/A(q)$. So the noise is assumed to have denominator dynamics identical to those of the process. This assumption may be justified if the disturbance enters the process early, although even in this case the disturbance would certainly pass through some part of the numerator dynamics $B(q)$ as well. However, most often this assumption will be violated in practice. Disturbances at the process output, as assumed in an OE model in Sect. 18.5.4, are much more common.

Figure 18.24 shows three different configurations of the ARX model. Note that all three configurations represent the same ARX model, but they suggest a different interpretation. The true process polynomials are denoted as $B(q)$ and $A(q)$, while the model polynomials are denoted as $\hat{B}(q)$ and $\hat{A}(q)$.

Figure 18.24a represents the most common configuration. The prediction error $e(k)$ for an ARX model is called *equation error* because it is the difference in the equation $e(k) = \hat{A}(q)y(k) - \hat{B}(q)u(k)$; see (18.31). The term "equation error" stresses the fact that it is *not* the difference between the process output $y(k)$ and $\hat{B}(q)/\hat{A}(q)u(k)$, which is called the *output error*; see also Sect. 18.5.4. Considering

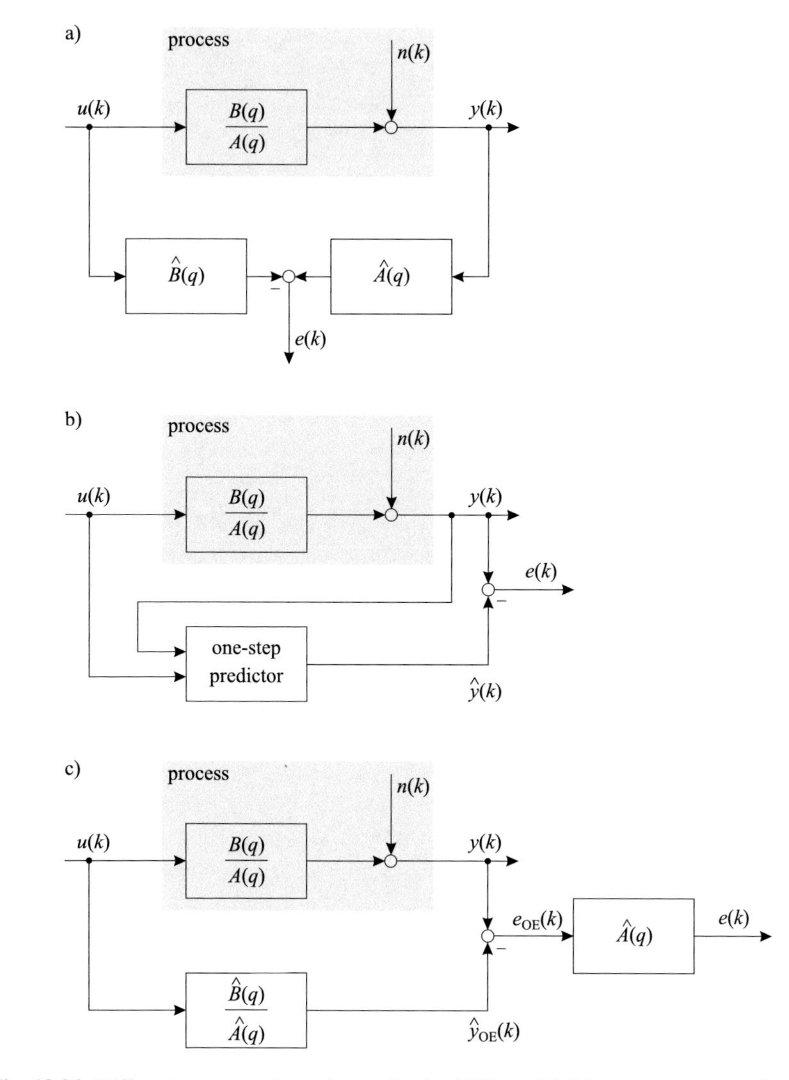

Fig. 18.24 Different representation schemes for the ARX model: (**a**) equation error configuration, (**b**) predictor configuration, and (**c**) pseudo-parallel configuration with filtering of the error signal [141]. All configurations realize the same ARX model

Fig. 18.24a, it is obvious that if the model equals the true process, i.e., $\hat{B}(q) = B(q)$ and $\hat{A}(q) = A(q)$, the equation error $e(k) = \hat{A}(q)n(k) = A(q)n(k)$. Thus, if the assumption made by the ARX model, namely, that the disturbance is white noise filtered through $1/A(q)$, is true, then the equation error $e(k)$ is white noise since $n(k) = 1/A(q)v(k)$. For each model structure, the prediction errors have to be white if all assumptions made are valid because then all information is exploited by the model.

Figure 18.24b depicts another configuration of the ARX model based on the predictor equation. With the ARX predictor in (18.49), the same equation error as in Fig. 18.24a results. Figure 18.24b can schematically represent any linear model by implementing the corresponding optimal predictor.

Figure 18.24c relates the ARX model to the OE model; see Sect. 18.5.4. This representation makes clear that the equation error $e(k)$ is a filtered version of the output error $e_{\mathrm{OE}}(k)$. Note that $e_{\mathrm{OE}}(k)$ and $\hat{y}_{\mathrm{OE}}(k)$, respectively, denote the output error and the output of an OE model; thus they are different from the prediction error $e(k)$ and the predicted output $\hat{y}(k)$ of an ARX model. Figure 18.24a, b, and c represent just different perspectives of the same ARX model and shall help us to better understand the relationships between the different model structures.

18.5.1.1 Least Squares (LS)

The reason for the popularity of the ARX model is that its parameters can be estimated by a linear least squares (LS) technique. For N available data samples, the ARX model can be written in the following matrix/vector form with $N - m$ equations for $k = m + 1, \ldots, N$ where $\underline{\hat{y}}$ is the vector of model outputs, while $\underline{y}$ is the vector of process outputs that are the desired model outputs:

$$\underline{\hat{y}} = \underline{X}\,\underline{\theta} \tag{18.52}$$

with

$$\underline{\hat{y}} = \begin{bmatrix} \hat{y}(m+1) \\ \hat{y}(m+2) \\ \vdots \\ \hat{y}(N) \end{bmatrix}, \quad \underline{y} = \begin{bmatrix} y(m+1) \\ y(m+2) \\ \vdots \\ y(N) \end{bmatrix}, \quad \underline{\theta} = \begin{bmatrix} b_1 \\ \vdots \\ b_m \\ a_1 \\ \vdots \\ a_m \end{bmatrix}, \tag{18.53}$$

$$\underline{X} = \begin{bmatrix} u(m) & \cdots & u(1) & -y(m) & \cdots & -y(1) \\ u(m+1) & \cdots & u(2) & -y(m+1) & \cdots & -y(2) \\ \vdots & & \vdots & \vdots & & \vdots \\ u(N-1) & \cdots & u(N-m) & -y(N-1) & \cdots & -y(N-m) \end{bmatrix}. \tag{18.54}$$

If the quadratic loss function in (18.32) is minimized, the optimal parameters of the ARX model can be computed by LS as (see Chap. 3)

$$\underline{\hat{\theta}} = \left(\underline{X}^T \underline{X}\right)^{-1} \underline{X}^T \underline{y}\,. \tag{18.55}$$

For the computation of the estimate in (18.55), the matrix $\underline{X}^T\underline{X}$ has to be non-singular. This is the case if the input $u(k)$ is persistently exciting. The big advantage of the ARX model is its linear-in-the-parameters structure. All features of linear optimization techniques apply, such as a fast one-shot solution that yields the global minimum of the loss function. The main drawback of the ARX model is that its noise model $1/A(q)$ is unrealistic. The additive output noise is much more common. The difficulties arising from this fact are discussed next. For more details concerning the least squares solution, refer to Chap. 3.

18.5.1.2 Consistency Problem

The ARX model and a more realistic process description are compared in Fig. 18.25. Because often the real process is not disturbed, as assumed by the ARX model, some difficulties can be expected. Indeed it can be shown that if the process does not meet the noise assumption made by the ARX model, the parameters are estimated *biased* and *non-consistent*. A bias means that the parameters systematically deviate from their optimal values, i.e., the parameters are systematically over- or underestimated. Non-consistency means that this bias does not even approach zero as the number of data samples N goes to infinity; see Sect. B.7 for more details on the bias and consistency definitions.

Even worse, the errorbars calculated from the estimate of the covariance matrix of the parameter estimate (see Chap. 3) may indicate that the estimate is quite accurate even if the bias is very large [141]. The reason for this undesirable behavior is that the derivations of many theorems about the LS in Chap. 3 assume a deterministic regression matrix $\underline{X}$. However, as can be seen in (18.54), the regression matrix $\underline{X}$ contains measured process outputs $y(k)$ that are non-deterministic owing to the disturbances. Thus, the covariance matrix cannot be calculated by (3.34), and consequently, the errorbar cannot be derived as shown in Sect. 3.1.2.

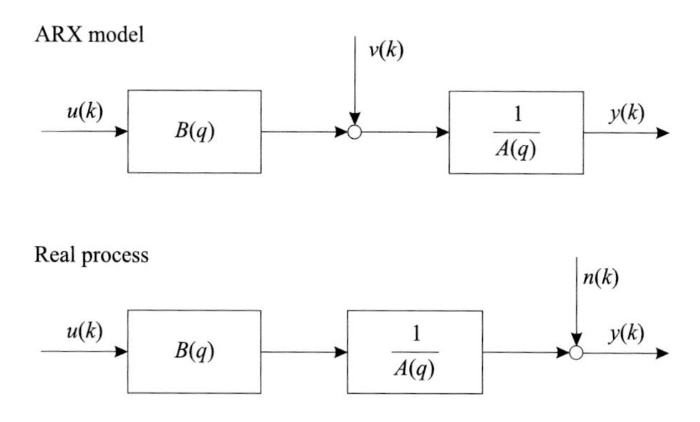

Fig. 18.25 An ARX model assumes a noise model $1/A(q)$, while more realistically a process is disturbed at the output by a noise $n(k)$, which can be white noise $v(k)$ or colored noise, e.g., $n(k) = C(q)/D(q)v(k)$

Because consistency is probably the most important property of any estimator, several strategies have been developed to avoid the non-consistent estimation for an ARX model. The idea of most of these approaches is to retain the linear-in-the-parameters property of the ARX model since this is its greatest advantage over other model structures.

Next, two such strategies are presented. The first strategy offers an alternative to the prediction error method, and the parameters are estimated with the help of instrumental variables. The idea of the second method is to work with correlation functions of the measured signals instead of the signals themselves.

Another alternative is to choose more general model structures such as ARMAX or OE that are nonlinear in their parameters and to develop algorithms that allow one to estimate the nonlinear parameters by the repeated application of a linear least squares technique. These approaches are discussed in the sections on the corresponding model structures.

18.5.1.3 Instrumental Variables (IV) Method

A very popular and simple remedy against the consistency problem of the conventional ARX model estimation is the *instrumental variables (IV)* method. It is an alternative to the prediction error methods. The starting point is the difference $\underline{e}$ of the process output $\underline{y}$ and the ARX model output $\underline{\hat{y}}$ for all data samples in matrix/vector form (see (18.52))

$$\underline{e} = \underline{y} - \underline{\hat{y}} = \underline{y} - \underline{X}\,\underline{\theta}\,. \tag{18.56}$$

The least squares estimate that results from a minimization of the sum of squared prediction errors ($\underline{e}^T \underline{e} \longrightarrow \min$) is

$$\underline{\hat{\theta}} = \left(\underline{X}^T \underline{X}\right)^{-1} \underline{X}^T \underline{y}\,. \tag{18.57}$$

The idea of the IV method is to multiply (18.56) with a matrix $\underline{Z}$ that has the same dimension as the regression matrix $\underline{X}$. The columns of $\underline{Z}$ are called *instrumental variables* and are chosen by the user to be uncorrelated with the noise, and therefore if all information is exploited, they are also uncorrelated with $\underline{e}$. This means that $\underline{Z}^T \underline{e} = \underline{0}$ because each row in $\underline{Z}^T$ is orthogonal to $\underline{e}$ (since they are uncorrelated and $\underline{e}$ has zero mean). Multiplying (18.56) with $\underline{Z}^T$ from the left yields

$$\underline{0} = \underline{Z}^T \underline{y} - \underline{Z}^T \underline{X}\,\underline{\theta} \tag{18.58}$$

and consequently

$$\underline{Z}^T \underline{y} = \underline{Z}^T \underline{X}\,\underline{\theta}\,. \tag{18.59}$$

If $\underline{Z}^T \underline{X}$ is non-singular, which is the case for persistent excitation and a proper choice of $\underline{Z}$, the IV estimate becomes

$$\hat{\underline{\theta}} = \left(\underline{Z}^T \underline{X}\right)^{-1} \underline{Z}^T \underline{y} . \tag{18.60}$$

Obviously, the IV estimate is equivalent to the LS estimate if $\underline{Z}^T = \underline{X}^T$. Note, however, that the columns in $\underline{X}$ cannot be used as instrumental variables since the columns containing $y(k-i)$ regressors are disturbed by noise. Thus, $\underline{X}$ is correlated with $\underline{e}$, i.e., $\underline{X}^T \underline{e} \neq \underline{0}$.

If the instrumental variables in $\underline{Z}$ are uncorrelated with the noise, the IV estimate is *consistent*. Although all choices of $\underline{Z}$ that fulfill this requirement lead to a consistent estimate, the variance of the estimate depends strongly on $\underline{Z}$. Recall that the parameter variance is proportional to $(\underline{Z}^T \underline{X})^{-1}$; see Sect. 3.1.1. Thus, the higher the correlation between the instrumental variables in $\underline{Z}$ and the regressors in $\underline{X}$, the smaller the variance error.

Now, the question arises, how to choose $\underline{Z}$? The answer is that the instrumental variables should be highly correlated with the regressors (columns in $\underline{X}$) in order to make the variance error small. For an easier understanding of a suitable choice of $\underline{Z}$, it is convenient to reconsider the ARX regression matrix in (18.54):

$$\underline{X} = \begin{bmatrix} u(m) & \cdots & u(1) & -y(m) & \cdots & -y(1) \\ u(m+1) & \cdots & u(2) & -y(m+1) & \cdots & -y(2) \\ \vdots & & \vdots & \vdots & & \vdots \\ u(N-1) & \cdots & u(N-m) & -y(N-1) & \cdots & -y(N-m) \end{bmatrix} . \tag{18.61}$$

The second half of $\underline{X}$ consists of delayed input signals, which are undisturbed. Consequently, the best instrumental variables for these regressors are the regressors themselves. However, for the first half of $\underline{X}$, the $y(k-i)$ regressors cannot be used in $\underline{Z}$ because the uncorrelation conditions have to be met. Good instrumental variables for the $y(k-i)$ would be an undisturbed version of these regressors. They can be approximated by filtering $u(k)$ through a process model. Thus, the following four-step algorithm can be proposed:

1. Estimate an ARX model from the data $\{\underline{u}(k), \underline{y}(k)\}$ by

$$\hat{\underline{\theta}}_{\text{ARX}} = \left(\underline{X}^T \underline{X}\right)^{-1} \underline{X}^T \underline{y} . \tag{18.62}$$

2. Simulate this model:

$$y_u(k) = \frac{\hat{B}(q)}{\hat{A}(q)} u(k) \tag{18.63}$$

where $\hat{B}(q)$ and $\hat{A}(q)$ are determined by $\hat{\underline{\theta}}_{\text{ARX}}$.

3. Construct the following instrumental variables:

$$\underline{Z} = \begin{bmatrix} u(m) & \cdots & u(1) & -y_u(m) & \cdots & -y_u(1) \\ u(m+1) & \cdots & u(2) & -y_u(m+1) & \cdots & -y_u(2) \\ \vdots & & \vdots & \vdots & & \vdots \\ u(N-1) & \cdots & u(N-m) & -y_u(N-1) & \cdots & -y_u(N-m) \end{bmatrix}.$$

4. Estimate the parameters with the IV method by

$$\underline{\hat{\theta}}_{\text{IV}} = \left(\underline{Z}^T \underline{X}\right)^{-1} \underline{Z}^T \underline{y}. \tag{18.64}$$

Because the ARX model parameters estimated in the first step are biased, the ARX model may not be a good model of the process. Nevertheless, the simulated process output $y_u(k)$ can be expected to be reasonably close to the measured process output $y(k)$ so that the correlation between $\underline{Z}$ and $\underline{X}$ is high. The IV method can be further improved by repeating Steps 2–4. Then in each iteration for the simulation in Step 2, the IV estimated model from the previous Step 4 can be applied. This procedure converges very fast, and experience teaches that more than two or three iterations are not worth any effort.

Ljung proposes performing the following additional five steps after going through Steps 1–4 [356].

5. Compute the residuals:

$$e_{\text{IV}}(k) = \hat{A}(q)y(k) - \hat{B}(q)u(k) \tag{18.65}$$

where $\hat{B}(q)$ and $\hat{A}(q)$ are determined by $\underline{\hat{\theta}}_{\text{IV}}$.

6. Estimate an AR time series model for the residuals to extract the remaining information from $e_{\text{IV}}(k)$. The AR filter acts as a whitening filter, i.e., it is supposed to decorrelate the residuals. Remember that the residuals should be as close to white noise as possible since then the process output is fully explained by the model besides the unpredictable part of the noise. The dynamic order of the AR time series model is chosen as $2m$ (or $n_a + n_b$ if $n_a = \deg(A)$ and $n_b = \deg(B)$ are not identical). Thus, the following relationship is postulated:

$$e_{\text{IV}}(k) = \frac{1}{L(q)} v(k) \qquad \text{or} \quad L(q)e_{\text{IV}}(k) = v(k) \tag{18.66}$$

with the white noise $v(k)$. Refer to Sect. 18.4.1 for a more detailed description of the estimation of AR time series models.

7. Filter the instruments calculated in Step 3 with the filter $\hat{L}(q)$ estimated in Step 6:

$$y_M^L(k) = L(q)y_u(k) \qquad \text{and} \quad u^L(k) = L(q)u(k). \tag{18.67}$$

Filter the process output $y^L(k) = L(q)y(k)$ and the regressors (columns in $\underline{X}$) denoted as $\underline{X}^L$.

8. Construct the following instrumental variables: $\underline{Z}^L =$

$$\begin{bmatrix} u^L(m) & \cdots & u^L(1) & -y_M^L(m) & \cdots & -y_M^L(1) \\ u^L(m+1) & \cdots & u^L(2) & -y_M^L(m+1) & \cdots & -y_M^L(2) \\ \vdots & & \vdots & \vdots & & \vdots \\ u^L(N-1) & \cdots & u^L(N-m) & -y_M^L(N-1) & \cdots & -y_M^L(N-m) \end{bmatrix}. \tag{18.68}$$

9. Estimate the parameters with the IV method by

$$\underline{\hat{\theta}}_{\mathrm{IV}}^L = \left((\underline{Z}^L)^T \underline{X}^L\right)^{-1} (\underline{Z}^L)^T \underline{y}^L . \tag{18.69}$$

Note that the instrumental variables introduced above are *model dependent*, i.e., they are calculated on the basis of the actual model; see (18.63). A simpler (but less effective) approach is to use *model-independent* instruments. This avoids the first LS estimation step, which computes a first model to generate the instruments. A typical choice for model-independent instrumental variables is

$$\underline{z} = [u(k-1) \ \cdots \ u(k-2m)]^T . \tag{18.70}$$

For more details about the IV method, the optimal choice of the instrumental variables, and its mathematical relationship to the prediction error method, refer to [356, 540].

18.5.1.4 Correlation Functions Least Squares (COR-LS)

The correlation function least squares (COR-LS) method proposed in [279] avoids the consistency problem by the following idea. Instead of computing the LS estimate directly from the signals $u(k)$ and $y(k)$ as is done in (18.52), the COR-LS method calculates correlation functions first. The starting point is the linear difference equation

$$y(k) = b_1 u(k-1) + \ldots + b_m u(k-m) - a_1 y(k-1) - \ldots - a_m y(k-m) . \tag{18.71}$$

This equation is multiplied by the term $u(k-\kappa)$:

$$\begin{aligned} u(k-\kappa)y(k) &= b_1 u(k-\kappa)u(k-1) + \ldots + b_m u(k-\kappa)u(k-m) \\ &\quad - a_1 u(k-\kappa)y(k-1) - \ldots - a_m u(k-\kappa)y(k-m). \end{aligned} \tag{18.72}$$

Now the sum over $N - \kappa$ data samples, e.g., $k = \kappa + 1, \ldots, N$, can be calculated in order to generate estimates of correlation functions (see Sect. B.6)

$$\sum_{k=\kappa+1}^{N} u(k-\kappa)y(k) = \tag{18.73}$$

$$b_1 \sum_{k=\kappa+1}^{N} u(k-\kappa)u(k-1) + \ldots + b_m \sum_{k=\kappa+1}^{N} u(k-\kappa)u(k-m)$$

$$-a_1 \sum_{k=\kappa+1}^{N} u(k-\kappa)y(k-1) - \ldots - a_m \sum_{k=\kappa+1}^{N} u(k-\kappa)y(k-m).$$

Thus, this equation can be written as

$$\begin{aligned} \mathrm{corr}_{uy}(\kappa) = {} & b_1 \ \mathrm{corr}_{uu}(\kappa-1) + \ldots + b_m \, \mathrm{corr}_{uu}(\kappa-m) \\ & - a_1 \ \mathrm{corr}_{uy}(\kappa-1) - \ldots - a_m \, \mathrm{corr}_{uy}(\kappa-m). \end{aligned} \tag{18.74}$$

Obviously, (18.74) possesses the same form as (18.71); only the signals $u(k)$ and $y(k)$ are replaced by the autocorrelation functions $\mathrm{corr}_{uu}(\kappa)$ and the cross-correlation functions $\mathrm{corr}_{uy}(\kappa)$. Thus, the least squares estimation in (18.55) can be applied on the level of correlation functions as well by changing the the regression matrix and the output vector to $\underline{X}_{\mathrm{corr}} =$

$$\begin{bmatrix} \mathrm{corr}_{uu}(0) & \cdots \ \mathrm{corr}_{uu}(1-m) & -\mathrm{corr}_{uy}(0) & \cdots \ -\mathrm{corr}_{uy}(1-m) \\ \mathrm{corr}_{uu}(1) & \cdots \ \mathrm{corr}_{uu}(2-m) & -\mathrm{corr}_{uy}(1) & \cdots \ -\mathrm{corr}_{uy}(2-m) \\ \vdots & \vdots & \vdots & \vdots \\ \mathrm{corr}_{uu}(l-1) & \cdots \ \mathrm{corr}_{uu}(l-m) & -\mathrm{corr}_{uy}(l-1) & \cdots \ -\mathrm{corr}_{uy}(l-m) \end{bmatrix} \tag{18.75}$$

$$\underline{y}_{\mathrm{corr}} = \begin{bmatrix} \mathrm{corr}_{uy}(1) \\ \mathrm{corr}_{uy}(2) \\ \vdots \\ \mathrm{corr}_{uy}(l) \end{bmatrix}, \tag{18.76}$$

where it is assumed that the correlation functions are used from $\kappa = 1-m$ to $\kappa = l$. Note that the number of terms in the sum that approximates the correlation functions decreases as the time shift $|\kappa|$ increases. Therefore, as l in (18.75) increases, the effect of the correlation decreases; in the extreme case, the sum contains only one term. Nevertheless, the full range of possible correlation functions can be utilized, and then the number of rows in $\underline{X}_{\mathrm{corr}}$ becomes $N - 1$.

Figures 18.26 and 18.27 show examples for the auto- and cross-correlation functions. The simulated process follows the first-order difference equation $y(k) =$

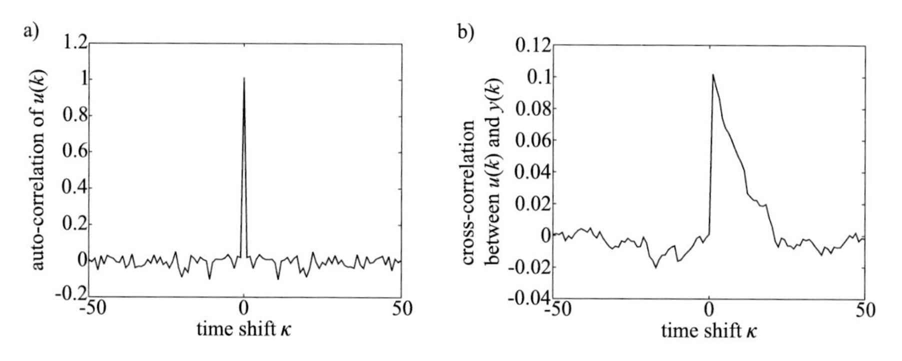

Fig. 18.26 (**a**) Autocorrelation and (**b**) cross-correlation functions for a white input sequence $\{u(k)\}$ of 1000 data samples and time shifts κ between -50 and 50. The process used is $y(k) = 0.1u(k-1) + 0.9y(k-1)$

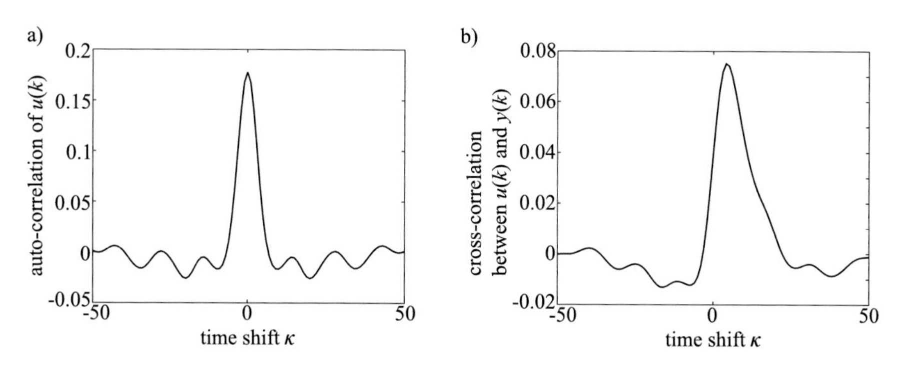

Fig. 18.27 (**a**) Autocorrelation and (**b**) cross-correlation functions for a low-frequency input sequence $\{u(k)\}$ of 1000 data samples and time shifts κ between -50 and 50. The process used is $y(k) = 0.1u(k-1) + 0.9y(k-1)$. Compare with Fig. 18.26

$0.1u(k-1) + 0.9y(k-1)$. In Fig. 18.26 the input signal is white. Therefore the autocorrelation function $\text{corr}_{uu}(\kappa)$ is a Dirac impulse. In Fig. 18.27 the input signal is low-pass filtered, and therefore the autocorrelation function is wider. The cross-correlation functions of both figures look similar; the one in Fig. 18.27 is smoother owing to the lower-frequency input signal. For non-positive time shifts, the cross-correlations are about zero since $y(k)$ (for a causal process) does not depend on the future inputs $u(k-\kappa)$, $\kappa \leq 0$. Thus, the cross-correlation function in Fig. 18.26 jumps at $\kappa = 1$ on its maximum value and decays as the correlation between $y(k)$ and inputs κ time steps in the past decreases. As the number of samples increases, the random fluctuations in the correlation functions decrease. For an infinite number of data samples, the cross-correlation function in Fig. 18.26 would be identical to the impulse response of the process. This makes the relationship between the signals and the correlation functions obvious.

The drawback of the COR-LS method is the higher computational effort. However, the correlation functions can possibly be exploited for the estimation of the dynamic process order as well; see Sects. 18.9 and B.6. So the additional effort may be justified. The advantage of this COR-LS compared with the conventional ARX method is that the regression matrix X consists of virtually deterministic values since the correlation with $u(k-\kappa)$ eliminates the noise in $y(k)$ because $u(k)$ is uncorrelated with $n(k)$. Consequently, the COR-LS method yields *consistent* estimates. Experience shows that the COR-LS method is well capable of attenuating noise, and it is especially powerful if the noise spectrum lies in the same frequency range as the process dynamics and thus filtering cannot be applied to separate the disturbance for the signal [279].

18.5.2 Autoregressive Moving Average with Exogenous Input (ARMAX)

The ARMAX model is probably the second most popular linear model after the ARX model. Some controller designs such as minimum variance control are based on an ARMAX model and exploit the information in the noise model [282]. Compared with the ARX, the ARMAX model is more flexible because it possesses an extended noise model. Although with this extension the ARMAX model becomes nonlinear in its parameters, quite efficient multistage linear least squares algorithms are available for parameter estimation, circumventing nonlinear optimization techniques. Furthermore, a straightforward recursive algorithm (RELS) exists; see Sect. 18.8.1.

The ARMAX model is depicted in Fig. 18.28 and is described by

$$A(q)y(k) = B(q)u(k) + C(q)v(k)\,. \tag{18.77}$$

The optimal ARMAX predictor is

$$\hat{y}(k|k-1) = \frac{B(q)}{C(q)}u(k) + \left(1 - \frac{A(q)}{C(q)}\right)y(k)\,. \tag{18.78}$$

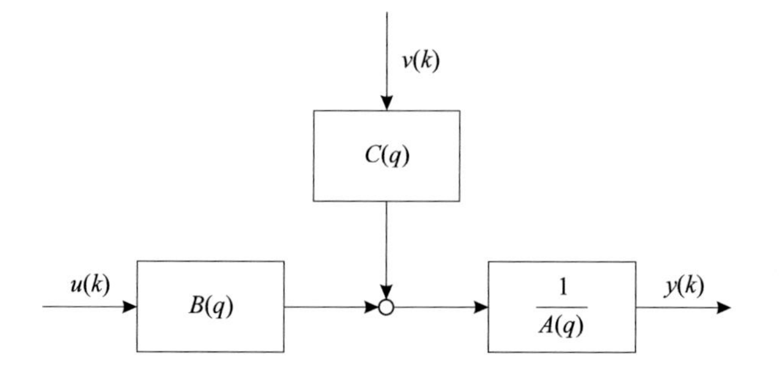

Fig. 18.28 ARMAX model

The ARMAX predictor is stable even if the $A(q)$ polynomial and therefore the ARMAX model is unstable. However, the polynomial $C(q)$ is required to be stable.

With (18.78) the prediction error of an ARMAX model is

$$e(k) = \frac{A(q)}{C(q)}y(k) - \frac{B(q)}{C(q)}u(k)\,. \tag{18.79}$$

Studying the above equations reveals that the ARMAX model is an extended ARX model owing to the introduction of the filter $C(q)$. If $C(q) = 1$, the ARMAX simplifies to the ARX model. Owing to the additional filter $C(q)$, the ARMAX model is very flexible. For example, with $C(q) = A(q)$, the ARMAX model can imitate an OE model; see Sect. 18.5.4.

Because the noise filter $C(q)/A(q)$ contains the model denominator dynamics, the ARMAX model belongs to the class of equation error models. This is also obvious from the ARMAX configuration depicted in Fig. 18.29; see Fig. 18.24. If $\hat{A}(q) = A(q)$, $\hat{B}(q) = B(q)$, and $\hat{C}(q) = C(q)$, the residuals $e(k)$ are white. Thus $\hat{A}(q)/\hat{C}(q)$ acts as a whitening filter.

18.5.2.1 Estimation of ARMAX Models

The prediction error (18.79) of an ARMAX model is nonlinear in its parameters owing to the filtering with $1/C(q)$. However, the prediction error can be expressed in the following *pseudo-linear* form:

$$C(q)e(k) = A(q)y(k) - B(q)u(k)\,, \tag{18.80}$$

which can be written as

$$e(k) = A(q)y(k) - B(q)u(k) + (1 - C(q))\,e(k)\,. \tag{18.81}$$

This results in the following difference equation:

$$\begin{aligned} e(k) &= a_1 y(k-1) + \ldots + a_m y(k-m) \\ &- b_1 u(k-1) - \ldots - b_m u(k-m) \\ &- c_1 e(k-1) - \ldots - c_m e(k-m)\,. \end{aligned} \tag{18.82}$$

The above equation formally represents a linear regression. However, because the $e(k-i)$ that estimate the unknown $v(k-i)$ (compare the first point in Sect. 18.3.3) are not measured but have to be calculated from previous residuals, the corresponding parameters are called to be pseudo-linear. Therefore, (18.81) and (18.82) allow two approaches for parameter estimation. The most straightforward approach is based on nonlinear optimization, while the second strategy exploits the pseudo-linear form of the prediction error.

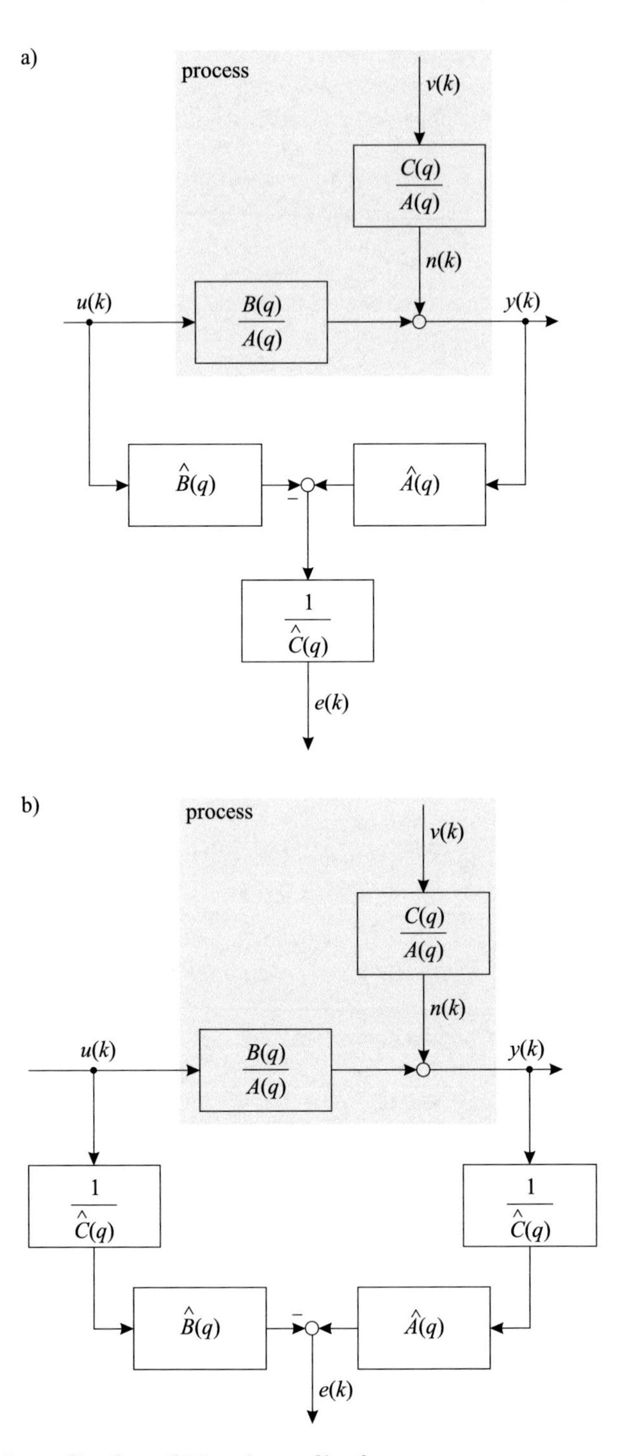

Fig. 18.29 (**a**) error filter form, (**b**) input/output filter form

Nonlinear Optimization of the ARMAX Model Parameters

1. Estimate an ARX model $A(q)y(k) = B(q)u(k)+v(k)$ from the data $\{\underline{u}(k), \underline{y}(k)\}$ by

$$\underline{\hat{\theta}}_{\text{ARX}} = \left(\underline{X}^T \underline{X}\right)^{-1} \underline{X}^T \underline{y}. \tag{18.83}$$

2. Optimize the ARMAX model parameters with a nonlinear optimization technique, e.g., with a nonlinear least squares method such as the Levenberg-Marquardt algorithm; see Chap. 3. The ARX model parameters obtained in Step 1 can be used as initial values for the a_i and b_i parameters.

   An efficient nonlinear optimization requires the computation of the gradients. The gradient of the squared prediction error $e^2(k) = (y(k) - \hat{y}(k))^2$ is $-2\, e(k)\, \partial \hat{y}(k)/\partial\theta$. Thus, the gradients of the predicted model output have to be calculated. It is convenient to multiply (18.78) by $C(q)$ in order to get rid of the denominators:

$$C(q)\hat{y}(k|k-1) = B(q)u(k) + (C(q) - A(q))y(k). \tag{18.84}$$

   Differentiation of (18.84) with respect to a_i yields [356]

$$C(q)\frac{\partial \hat{y}(k|k-1)}{\partial a_i} = -y(k-i), \tag{18.85}$$

   which leads to

$$\frac{\partial \hat{y}(k|k-1)}{\partial a_i} = -\frac{1}{C(q)}y(k-i). \tag{18.86}$$

   Differentiation of (18.84) with respect to b_i yields [356]

$$C(q)\frac{\partial \hat{y}(k|k-1)}{\partial b_i} = u(k-i), \tag{18.87}$$

   which leads to

$$\frac{\partial \hat{y}(k|k-1)}{\partial b_i} = \frac{1}{C(q)}u(k-i). \tag{18.88}$$

   Differentiation of (18.84) with respect to c_i yields [356]

$$\hat{y}(k-i|k-i-1) + C(q)\frac{\partial \hat{y}(k|k-1)}{\partial c_i} = y(k-i), \tag{18.89}$$

which leads to

$$\frac{\partial \hat{y}(k|k-1)}{\partial c_i} = \frac{1}{C(q)} \left(y(k-i) - \hat{y}(k-i|k-i-1) \right) . \tag{18.90}$$

Thus, the gradient can be easily computed by filtering the regressors $-y(k-i)$, $u(k-i)$, and $e(k-i) = y(k-i) - \hat{y}(k-i|k-i-1)$ through the filter $1/C(q)$. The residuals $e(k)$ approach the white noise $v(k)$ as the algorithm converges.

The drawbacks of the nonlinear optimization approach are the high computational demand and the existence of local optima. The danger of convergence to a local optimum is reduced, however, if the initial parameter values are close to the optimal ones. In [356] experiences are reported that the globally optimal parameters of ARMAX models are "usually found without too much problem," while for OE and BJ models "convergence to false local minima is not uncommon."

Multistage Least Squares for ARMAX Model Estimation

This algorithm is sometimes called extended least squares (ELS).[3]

1. Estimate an ARX model $A(q)y(k) = B(q)u(k)+v(k)$ from the data $\{\underline{u}(k), \underline{y}(k)\}$ by

$$\underline{\hat{\theta}}_{\text{ARX}} = \left(\underline{X}^T \underline{X}\right)^{-1} \underline{X}^T \underline{y} . \tag{18.91}$$

2. Calculate the prediction errors of this ARX model:

$$e_{\text{ARX}}(k) = \hat{A}(q)y(k) - \hat{B}(q)u(k) , \tag{18.92}$$

   where $\hat{B}(q)$ and $\hat{A}(q)$ are determined by $\underline{\hat{\theta}}_{\text{ARX}}$.
3. Estimate the ARMAX model parameters a_i, b_i, and c_i from (18.82) with LS by approximating the ARMAX residuals as $e(k-i) \approx e_{\text{ARX}}(k-i)$.

Steps 2–3 of the ELS algorithm can be iterated until convergence is reached. Then, of course, in Step 2 the residuals from the previously (in Step 3) estimated ARMAX model are used, and in Step 3 the ARMAX residuals are approximated by the residuals of the ARMAX model from Step 2. The ARMAX prediction error should approach white noise as all information is going to be exploited by the model and then $e(k)$ approaches the white noise $v(k)$. Note that the prediction error of the ARMAX model can be obtained by filtering either the ARX model error or $u(k)$ and $y(k)$ in (18.92) with $1/C(q)$ as shown in Fig. 18.29. The speed of convergence

[3]Often ELS denotes the *recursive* version of this algorithm. Here, for the sake of clarity, the recursive algorithm is named RELS; see Sect. 18.8.3.

with the ELS algorithm may be somewhat faster than with nonlinear optimization. However, the (mild) local optima problem can, of course, not be solved.

In [356] an ARX model of higher order than m is proposed for Step 1 to obtain a better approximation of the white noise $v(k)$. Ideally, $e(k)$ converges to $v(k)$.

The ARMAX model can be extended to the ARIMAX model, where “I” stands for integration. The noise model is extended by an integrator to $C(q)/(1 - q^{-1})A(q)$. This allows for drifts in the output signal. Alternatively, the data can be filtered with the inverse integrator $1 - q^{-1}$ (see Sects. 18.3.4 and 18.7.5), or the noise model can be made flexible enough that the integrator is found automatically [356].

18.5.3 Autoregressive Autoregressive with Exogenous Input (ARARX)

The ARARX model can be seen as the counterpart of the ARMAX model. While the disturbance is filtered through an MA filter $C(q)v(k)$ for the ARMAX model, it goes through an AR filter $1/D(q)v(k)$ for the ARARX model. The ARARX model is not as common as the ARX or ARMAX model since the additional model complexity often does not pay off.

The ARARX model is depicted in Fig. 18.30 and is described by

$$A(q)y(k) = B(q)u(k) + \frac{1}{D(q)}v(k)\,. \tag{18.93}$$

The optimal ARARX predictor is

$$\hat{y}(k|k-1) = D(q)B(q)u(k) + (1 - D(q)A(q))\,y(k)\,. \tag{18.94}$$

The ARARX predictor is stable even if the $A(q)$ or $D(q)$ polynomials and therefore the ARARX model itself are unstable.

With (18.94) the prediction error of an ARARX model is

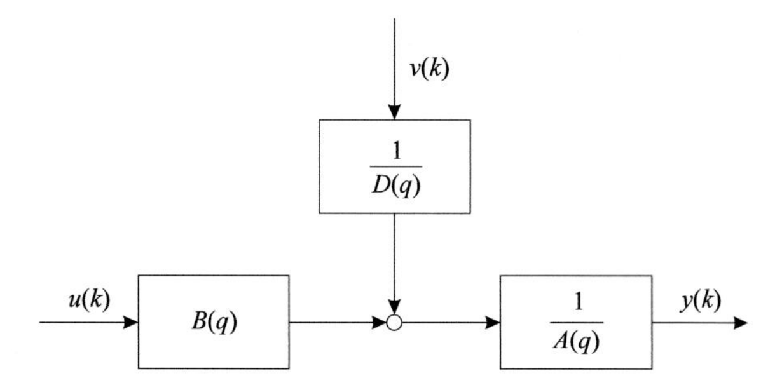

Fig. 18.30 ARARX model

$$e(k) = D(q)A(q)y(k) - D(q)B(q)u(k)\,. \qquad (18.95)$$

Studying the above equations reveals that the ARMAX model, like the ARMAX model, is an extended ARX model owing to the introduction of the filter $D(q)$. If $D(q) = 1$ the ARARX simplifies to the ARX model. Owing to the additional filter $D(q)$, the ARARX model is more flexible than the ARX model. However, because $D(q)$ extends the denominator dynamics compared with the extension of the numerator dynamics in the ARMAX model, the denominator dynamics $A(q)$ cannot be (partly) canceled in the noise model.

Because the noise filter $1/D(q)A(q)$ contains the model denominator dynamics, the ARARX model belongs to the class of equation error models. This is also obvious from the ARARX configuration depicted in Fig. 18.31; see Fig. 18.24. If $\hat{A}(q) = A(q)$, $\hat{B}(q) = B(q)$, and $\hat{D}(q) = D(q)$, the residuals $e(k)$ are white. Thus $\hat{D}(q)\hat{A}(q)$ acts as a whitening filter.

The parameters of the ARARX model can be estimated either by a nonlinear optimization technique or by a repeated least squares and filtering approach [279].

Nonlinear Optimization of the ARARX Model Parameters

1. Estimate an ARX model $A(q)y(k) = B(q)u(k)+v(k)$ from the data $\{\underline{u}(k), \underline{y}(k)\}$ by

$$\underline{\hat{\theta}}_{\text{ARX}} = \left(\underline{X}^T\underline{X}\right)^{-1}\underline{X}^T\underline{y}\,. \qquad (18.96)$$

2. Optimize the ARARX model parameters with a nonlinear optimization technique. The ARX model parameters obtained in Step 1 can be used as initial values for the a_i and b_i parameters. The gradients of the model's prediction (18.94) can be computed as follows.

   Differentiation of (18.94) with respect to a_i yields [356]

$$\frac{\partial \hat{y}(k|k-1)}{\partial a_i} = -D(q)y(k-i)\,. \qquad (18.97)$$

   Differentiation of (18.94) with respect to b_i yields [356]

$$\frac{\partial \hat{y}(k|k-1)}{\partial b_i} = D(q)u(k-i)\,. \qquad (18.98)$$

   Differentiation of (18.94) with respect to d_i yields [356]

$$\frac{\partial \hat{y}(k|k-1)}{\partial d_i} = B(q)u(k-i) - A(q)y(k-i) = -e_{\text{ARX}}(k-i)\,. \qquad (18.99)$$

Fig. 18.31 (**a**) error filter form, (**b**) input/output filter form

a)

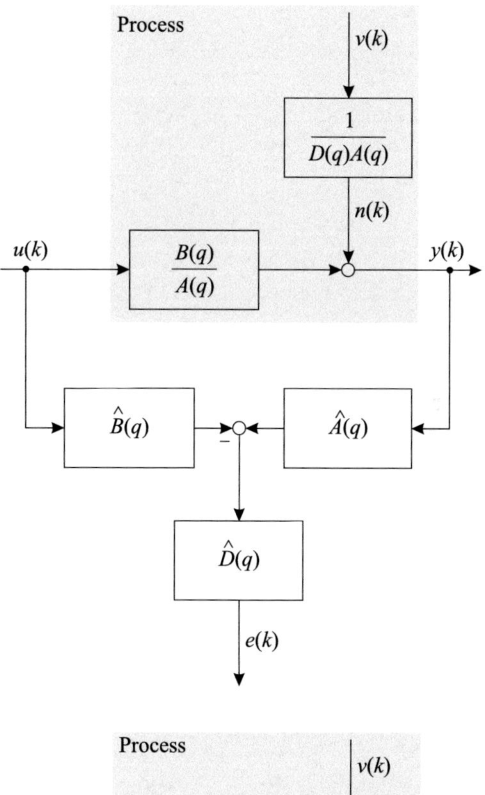

b)

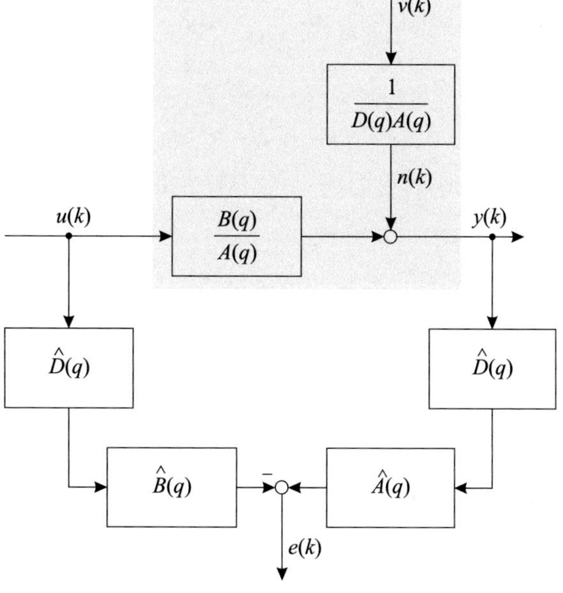

Repeated Least Squares and Filtering for ARARX Model Estimation
(Generalized Least Squares (GLS))

1. Estimate an ARX model $A(q)y(k) = B(q)u(k)+v(k)$ from the data $\{\underline{u}(k), \underline{y}(k)\}$ by

$$\underline{\hat{\theta}}_{\text{ARX}} = \left(\underline{X}^T \underline{X}\right)^{-1} \underline{X}^T \underline{y}\,. \tag{18.100}$$

2. Calculate the prediction errors of this ARX model:

$$e_{\text{ARX}}(k) \;=\; \hat{A}(q)y(k) - \hat{B}(q)u(k)\,, \tag{18.101}$$

   where $\hat{B}(q)$ and $\hat{A}(q)$ are determined by $\underline{\hat{\theta}}_{\text{ARX}}$.
3. Estimate the d_i parameters of the following AR model by least squares (see Sect. 18.4.1)

$$e_{\text{ARX}}(k) = \frac{1}{D(q)} v(k)\,. \tag{18.102}$$

   Compare (18.93) and Fig. 18.31a for a motivation of this AR model. The prediction error $e(k)$ in Fig. 18.31a becomes white, i.e., equal to $v(k)$, if $e_{\text{ARX}}(k)$ in (18.101) is filtered through $D(q)$.
4. Filter the input $u(k)$ and process output $y(k)$ through the estimated filter: $\hat{D}(q)$

$$u^D(k) = \hat{D}(q)u(k) \qquad \text{and} \qquad y^D(k) = \hat{D}(q)y(k)\,. \tag{18.103}$$

5. Estimate the ARARX model parameters a_i and b_i by an ARX model estimation with the filtered input $u^D(k)$ and output $y^D(k)$; see Fig. 18.31b.

   Steps 3–5 of the GLS algorithm can be iterated until convergence is reached.

18.5.4 Output Error (OE)

Together with the ARX and ARMAX model, the OE model is the most widely used structure. It is the simplest representative of the output error model class. The noise is assumed to disturb the process additively at the output, not somewhere inside the process as is assumed for the equation error models. Output error models are often more realistic models of reality, and thus they often perform better than equation error models. However, because the noise models do not include the process denominator dynamics $1/A(q)$, all output error models are nonlinear in their parameters, and consequently, they are harder to estimate.

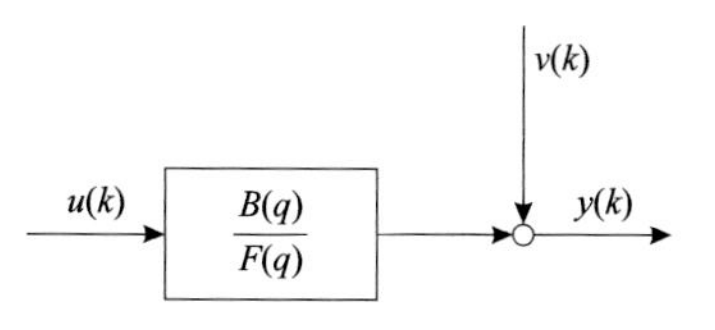

Fig. 18.32 OE model

The OE model is depicted in Fig. 18.32 and is described by

$$y(k) = \frac{B(q)}{F(q)} u(k) + v(k). \tag{18.104}$$

It is standard in linear system identification literature to denote the denominator of process models belonging to the output error class as $F(q)$, while the denominators of equation error models such as ARX, ARMAX, and ARARX are denoted as $A(q)$ [356]. Of course, these are just notational conventions to emphasize the different noise assumptions; a model denoted as $B(q)/A(q)$ can be exactly identical to a model denoted as $B(q)/F(q)$.

The optimal OE predictor is in fact a simulator because it does not make any use of the measurable process output $y(k)$:

$$\hat{y}(k|k-1) = \hat{y}(k) = \frac{B(q)}{F(q)} u(k). \tag{18.105}$$

Note that the notation "$|k-1$" can be discarded for the OE model because the optimal prediction is not based on previous process outputs.

Furthermore, note that the OE predictor is unstable if the $F(q)$ polynomial is unstable. Therefore the OE model cannot be used for modeling unstable processes. The same holds for all other models belonging to the class of output error models.

With (18.105) the prediction error of an OE model is

$$e(k) = y(k) - \frac{B(q)}{F(q)} u(k). \tag{18.106}$$

Figure 18.33 depicts the OE model in parallel to the process. The prediction error of the OE model is the difference between the process output and the simulated model output. The disturbance $n(k)$ is assumed to be white.

Figure 18.34 relates the residuals of an OE model to the residuals of an ARX model. Owing to the equation error configuration of the ARX model (see Fig. 18.24c), the ARX model residuals can be interpreted as filtered OE residuals:

$$e_{\mathrm{ARX}}(k) = F(q)\, e_{\mathrm{OE}}(k). \tag{18.107}$$

Assume that $\hat{F}(q) = F(q)$ and $\hat{B}(q) = B(q)$. If the process noise is white ($n(k) = v(k)$), then $e_{\mathrm{OE}}(k) = v(k)$ is white as well, while $e_{\mathrm{ARX}}(k) = F(q)v(k)$ is correlated. If, however, the process noise is correlated such that $n(k) = 1/F(q)v(k)$,

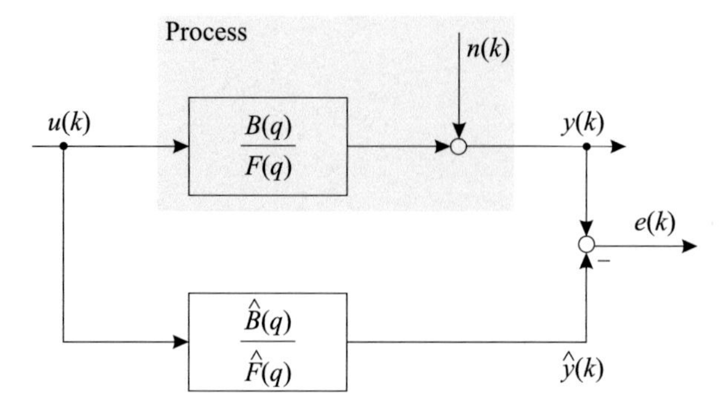

Fig. 18.33 OE model in parallel to the process

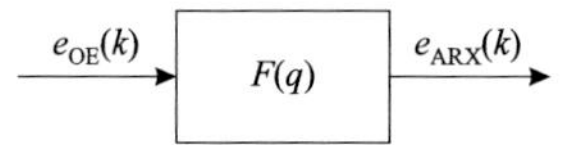

Fig. 18.34 Relationship between ARX model residuals and OE model residuals. The ARX model residuals can be obtained by filtering the OE model residuals through $F(q)$

then $e_{\mathrm{OE}}(k) = 1/F(q)v(k)$ is correlated, while $e_{\mathrm{ARX}}(k) = v(k)$ is white. This relationship allows an output error parameter estimation based on repeated linear least squares and filtering, although the parameters are nonlinear. In the above discussion, $F(q)$ and $\hat{F}(q)$ can be replaced by $A(q)$ and $\hat{A}(q)$ if the argumentation is starting from the ARX model point of view.

It is helpful to illuminate why the predicted output of an OE model is nonlinear in its parameters (see (18.105))

$$\begin{aligned} \hat{y}(k) = {} & b_1 u(k-1) + \ldots + b_m u(k-m) \\ & - f_1 \hat{y}(k-1) - \ldots - f_m \hat{y}(k-m) \,. \end{aligned} \tag{18.108}$$

Compared with the ARX model, the measured output in (18.50) is replaced with the predicted (or the simulated, which is the same for OE) output in (18.108). Here lies the reason for the nonlinearity of the parameters in (18.108). The predicted model outputs $\hat{y}(k-i)$ depend themselves on the model parameters. So in the terms $f_i\, \hat{y}(k-i)$, both factors depend on model parameters, which results in a nonlinear dependency. To overcome these difficulties, one may be tempted to approximate in (18.108) the model outputs $\hat{y}(k-i)$ by the measured process outputs $y(k-i)$. Then the OE model simplifies to the ARX model, which is indeed linear in its parameters.

The parameters of the OE model can be estimated either by a nonlinear optimization technique or by a repeated least squares and filtering approach exploiting the relationship to the ARX model [304].

18.5.4.1 Nonlinear Optimization of the OE Model Parameters

1. Estimate an ARX model $F(q)y(k) = B(q)u(k)+v(k)$ from the data $\{\underline{u}(k), \underline{y}(k)\}$ by

$$\hat{\underline{\theta}}_{\text{ARX}} = \left(\underline{X}^T \underline{X}\right)^{-1} \underline{X}^T \underline{y}\,, \tag{18.109}$$

where the parameters in $\hat{\underline{\theta}}$ are now denoted as f_i and b_i instead of a_i and b_i.

2. Optimize the ARARX model parameters with a nonlinear optimization technique. The ARX model parameters obtained in Step 1 can be used as initial values for the f_i and b_i parameters. The gradients of the model's prediction (18.105) can be computed as follows. First, (18.105) is written in the following form:

$$F(q)\hat{y}(k) = B(q)u(k)\,. \tag{18.110}$$

Differentiation of (18.110) with respect to b_i yields

$$F(q)\frac{\partial \hat{y}(k)}{\partial b_i} = u(k-i)\,, \tag{18.111}$$

which leads to

$$\frac{\partial \hat{y}(k)}{\partial b_i} = \frac{1}{F(q)}u(k-i)\,. \tag{18.112}$$

Differentiation of (18.110) with respect to f_i yields

$$\hat{y}(k-i) + F(q)\frac{\partial \hat{y}(k)}{\partial f_i} = 0\,, \tag{18.113}$$

which leads to

$$\frac{\partial \hat{y}(k)}{\partial f_i} = -\frac{1}{F(q)}\hat{y}(k-i)\,. \tag{18.114}$$

18.5.4.2 Repeated Least Squares and Filtering for OE Model Estimation

1. Estimate an ARX model $F(q)y(k) = B(q)u(k)+v(k)$ from the data $\{\underline{u}(k), \underline{y}(k)\}$ by

$$\hat{\underline{\theta}}_{\text{ARX}} = \left(\underline{X}^T \underline{X}\right)^{-1} \underline{X}^T \underline{y}\,, \tag{18.115}$$

where the parameters in $\hat{\underline{\theta}}$ are now denoted as f_i and b_i instead of a_i and b_i.

2. Filter the input $u(k)$ and process output $y(k)$ through the estimated filter $\hat{F}(q)$:

$$u^F(k) = \frac{1}{\hat{F}(q)} u(k) \qquad \text{and} \quad y^F(k) = \frac{1}{\hat{F}(q)} y(k) \, . \tag{18.116}$$

3. Estimate the OE model parameters f_i and b_i by an ARX model estimation with the filtered input $u^F(k)$ and output $y^F(k)$; see Fig. 18.34.

Steps 2–3 of this algorithm can be iterated until convergence is reached. This algorithm exploits the relationship between the ARX and OE model prediction errors. It becomes intuitively clear from another point of view as well. In Sects. 18.3.4 and 18.7.4, it is shown that a noise model and filtering with the inverse noise model are equivalent. Thus, the ARX noise model $1/A(q)$ has the same effect as filtering of the data through $A(q)$. The filtering with $1/F(q)$ in (18.116) tries to compensate this effect, leading to the noise model 1, which corresponds to an OE model.

18.5.5 Box-Jenkins (BJ)

The Box-Jenkins (BJ) model belongs to the class of output error models. It is an OE model with additional degrees of freedom for the noise model. While the OE model assumes an additive white disturbance at the process output, the BJ allows any colored disturbance. It may be generated by filtering white noise through a linear filter with arbitrary numerator and denominator.

The BJ model is depicted in Fig. 18.35 and is described by

$$y(k) = \frac{B(q)}{F(q)} u(k) + \frac{C(q)}{D(q)} v(k) \, . \tag{18.117}$$

Thus, the BJ model can be seen as the output error class counterpart of the ARARMAX model, which belongs to the equation error class. For the equation error models, the special case $D(q) = 1$ corresponds to the ARMAX model, and the special case $C(q) = 1$ corresponds to the ARARX model. These special cases

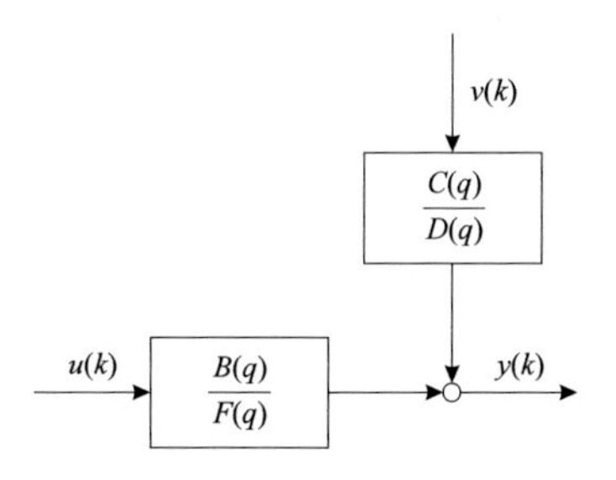

Fig. 18.35 Box-Jenkins model

for the BJ model do not have specific names. For $C(q) = D(q)$, the BJ simplifies to the OE model. Note that the BJ model can imitate all equation error models if the order of the noise model is high enough. Then the denominator of the noise model $D(q)$ may (but of course does not have to) include the process denominator dynamics $F(q)$.

Of all linear models discussed so far, the BJ model is the most general and flexible. It allows one to estimate separate transfer functions with arbitrary numerators and denominators from the input to the output and from the disturbance to the output. However, on the other hand, the flexibility of the BJ model requires one to estimate a large number of parameters. For most applications, this is either not worth the price or not possible owing to data sets that are too small and noisy. Consequently, the BJ model is seldom applied in practice.

The optimal BJ predictor is

$$\hat{y}(k|k-1) = \frac{B(q)D(q)}{F(q)C(q)}u(k) + \frac{C(q)-D(q)}{C(q)}y(k)\,. \tag{18.118}$$

Note that the notation "$|k - 1$" cannot be discarded as for the OE model because the optimal prediction of a BJ model utilizes previous process outputs in order to extract the information contained in the correlated disturbances $n(k) = C(q)/D(q)v(k)$.

With (18.118) the prediction error of a BJ model is

$$e(k) = \frac{D(q)}{C(q)}y(k) - \frac{B(q)D(q)}{F(q)C(q)}u(k)\,. \tag{18.119}$$

Typically, a BJ model is estimated by nonlinear optimization, where first an ARX model is estimated in order to determine the initial parameter values for b_i and f_i. The gradients of the model's prediction (18.118) can be computed as follows. First, (18.118) is written in the following form:

$$F(q)C(q)\hat{y}(k|k-1) = B(q)D(q)u(k) + F(q)(C(q)-D(q))y(k). \tag{18.120}$$

Differentiation of (18.120) with respect to b_i yields

$$F(q)C(q)\frac{\partial\hat{y}(k|k-1)}{\partial b_i} = D(q)u(k-i)\,, \tag{18.121}$$

which leads to

$$\frac{\partial\hat{y}(k|k-1)}{\partial b_i} = \frac{D(q)}{F(q)C(q)}u(k-i)\,. \tag{18.122}$$

Differentiation of (18.120) with respect to c_i yields

$$F(q)\left(\hat{y}(k-i|k-i-1)+C(q)\frac{\partial\hat{y}(k|k-1)}{\partial c_i}\right) = F(q)y(k-i), \qquad (18.123)$$

which leads to

$$\frac{\partial\hat{y}(k|k-1)}{\partial c_i} = \frac{1}{C(q)}\left(y(k-i)-\hat{y}(k-i|k-i-1)\right). \qquad (18.124)$$

Differentiation of (18.120) with respect to d_i yields

$$F(q)C(q)\frac{\partial\hat{y}(k|k-1)}{\partial d_i} = B(q)u(k-i)-F(q)y(k-i), \qquad (18.125)$$

which leads to

$$\frac{\partial\hat{y}(k|k-1)}{\partial d_i} = \frac{B(q)}{F(q)C(q)}u(k-i)-\frac{1}{C(q)}y(k-i). \qquad (18.126)$$

Differentiation of (18.120) with respect to f_i yields

$$C(q)\left(\hat{y}(k-i|k-i-1)+F(q)\frac{\partial\hat{y}(k|k-1)}{\partial f_i}\right) = -D(q)y(k-i), \qquad (18.127)$$

which leads to

$$\frac{\partial\hat{y}(k|k-1)}{\partial f_i} = -\frac{1}{F(q)}\left(\frac{D(q)}{C(q)}y(k-i)+\hat{y}(k-i|k-i-1)\right). \qquad (18.128)$$

18.5.6 State Space Models

Instead of input/output state space models can also be considered. A state space OE model takes the following form:

$$\underline{x}(k+1) = \underline{A}\,\underline{x}(k)+\underline{b}\,u(k) \qquad (18.129a)$$

$$y(k) = \underline{c}^T\,\underline{x}(k)+v(k). \qquad (18.129b)$$

The easiest and most straightforward way to obtain a state space model from data is to estimate an input/output model, e.g., an OE model (see Sect. 18.5.4),

$$y(k) = \frac{B(q)}{F(q)}u(k) + v(k) \qquad (18.130)$$

and use these parameters in a canonical state space form, e.g.,

$$\underline{x}(k+1) = \begin{bmatrix} 0 & 1 & \cdots & 0 \\ \vdots & \vdots & \ddots & 0 \\ 0 & 0 & \cdots & 1 \\ -f_m & -f_{m-1} & \cdots & -f_1 \end{bmatrix} \underline{x}(k) + \begin{bmatrix} 0 \\ \vdots \\ 0 \\ 1 \end{bmatrix} u(k) \qquad (18.131a)$$

$$y(k) = \begin{bmatrix} b_m \; b_{m-1} \; \cdots \; b_1 \end{bmatrix} \underline{x}(k) + v(k) \,. \qquad (18.131b)$$

The major advantage of a state space representation is that prior knowledge from first principles can be incorporated in the form equations and can be utilized to pre-structure the model [297]. Furthermore, the number of regressors is usually smaller in state space models than in input/output models. For a system of mth order, a state space model possesses $m + 1$ regressors ($x_1(k), \ldots, x_m(k)$ and $u(k)$), while an input/output model requires $2m$ regressors ($u(k-1), \ldots, u(k-m)$ and $y(k-1), \ldots, y(k-m)$). The smaller number of regressors is not very important for linear systems. However, for nonlinear models, this is a significant advantage since the number of regressors corresponds to the input dimensionality; see Sect. 19.1. Finally, for processes with multiple inputs and outputs, the state space representation is well suited. For a direct identification of state space models, the following cases can be distinguished:

- If all states are measurable, the parameters in $\underline{A}$, $\underline{b}$, and $\underline{c}$ can be estimated by a linear optimization technique. Unfortunately, the true states of the process will seldom lead to a canonical state space realization as in (18.131a) and (18.131b). Thus, without any incorporation of prior knowledge, all entries of the system matrix and vectors must be assumed to be non-zero. For an mth-order model with such a full parameterization, $m^2 + 2m$ parameters have to be estimated. Usually, this can only be done if a regularization technique is applied to reduce the variance error of the model; see Sects. 7.5 and 3.1.4.
- If the initial values $\underline{x}(0)$ for the states are known but the states cannot be measured over time, the parameter estimation problem becomes more difficult. This situation may occur for batch processes where many variables can be measured before the batch is started (initial values) but only a few variables can be measured while the process is active [508]. Since $\underline{x}(k)$ is unknown for $k > 0$, the states must be determined by a simulation of the model with fixed parameters. Thus, the model can be evaluated for a given input signal $u(k)$, the initial state $\underline{x}(0)$, and given $\underline{A}$, $\underline{b}$, and $\underline{c}$. The parameters can be iteratively optimized with a nonlinear optimization technique with regularization (for the same reasons as stated above).
- If no states are measurable, the problem can be treated similar to the case where $\underline{x}(0)$ is available. Some initial state can be assumed, say $\underline{x}(0) = \underline{0}$, and the error induced by the wrong initial values decays exponentially fast and thus can be neglected after a reasonable number of samples if the system is stable.

An alternative is the application of modern subspace identification methods; see Sect. 18.10.3.

18.5.7 Simulation Example

Consider the following second-order process with gain $K = 1$ and the time constants $T_1 = 10\,\text{s}$ and $T_2 = 5\,\text{s}$:

$$G(s) = \frac{1}{(10s + 1)(5s + 1)} \,. \tag{18.132}$$

It will be approximated with a first-order ARX and OE model in order to illustrate an important property of these two most commonly used model structures. The input signal is chosen as a PRBS with 255 data samples, which excites all frequency ranges equally well, and the process is sampled with $T_0 = 1\,\text{s}$. No disturbance is added in order not to obscure the effect of order reduction from the second-order process to the first-order models. The effect of disturbances on ARX and OE models is illustrated in the example in Sect. 18.6.6.

Figure 18.36 shows the loss functions for the ARX and OE models. For the ARX model according to Sect. 18.5.1, the *one-step prediction* errors are used, and thus the loss function is a parabola. This can be observed immediately from the elliptic shape of the contour lines. For the OE model according to Sect. 18.5.4, the *simulation* errors are used, making the loss function nonlinearly dependent on the parameters.

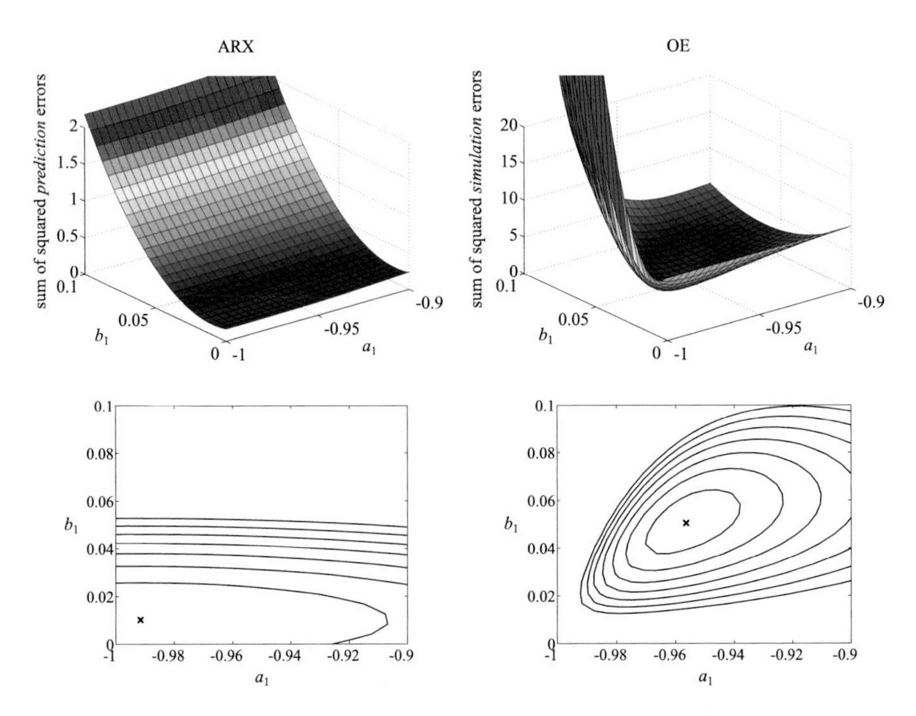

Fig. 18.36 ARX and OE model loss functions (top) and their contour plots (bottom)

The reason for the strong increase of the OE model loss function for $a_1 \to 1$ is that $a_1 = 1$ represents the stability boundary where the model becomes unstable. Therefore, the estimation of an OE model guarantees a stable model even in the case of strong disturbances, while an ARX model can well become unstable because, for one-step prediction, stable and unstable models are not fundamentally different. Besides the distinct characteristics of the ARX and OE model loss functions, their absolute values are very different because one-step prediction is a much simpler task than simulation, and consequently the errors are smaller.

The optimal parameters for the ARX and OE models are significantly different. Besides the difference in their values, they will also be estimated with different accuracy (in the case of noisy data) since the loss functions' shapes around the optimum are very different. For the OE model, both parameters can be estimated with about the same degree of accuracy because the loss function at the optimum is about equally sensitive with respect to both parameters. This can be clearly observed from the contour lines, which are roughly circles around the optimum. For the ARX model, however, the contour lines are stretched ellipses, illustrating that the loss function is very sensitive with respect to b_1 but barely sensitive with respect to a_1. Thus, for the ARX model, b_1 can be expected to be estimated with a high degree of accuracy and a_1 with a low one. For a proper understanding of the following discussion, note that in this example the optimal values are exactly reached because undisturbed data is used for identification.

The reasons for the different optimal parameter values for ARX and OE models can be best understood in the frequency domain; see Fig. 18.37(top). Obviously, the ARX model sacrifices a lot of accuracy in the low- and medium-frequency range in order to describe the process better in the high-frequency range. Indeed it can be shown that the ARX model over-emphasizes high frequencies in the model fit, while the OE model gives all frequencies equal weight; for more details, refer to Sect. 18.7.3. Because a reduced-order model is necessarily inaccurate at high frequencies, this slight improvement does not usually pay off. Thus, an OE model in most cases will be a significantly better choice.

For faster sampling rates, the ARX model's emphasis on high frequencies increasingly degrades its accuracy at low and medium frequencies. The simple reason for this fact is that faster sampling pushes the Nyquist frequency in Fig. 18.37 to the right, and thus the ARX model focuses its model fit toward even higher frequencies. In the bottom of Fig. 18.37, the sampling rate is increased by a factor of 2.5, demonstrating the further deterioration of the ARX model fit. For example, the gains of the OE models for $T_0 = 1$ s and $T_0 = 0.4$ s are 1.15 and 1.25, respectively. The gains of the ARX models are 1.2 and 1.5, respectively (the true process gain is $K = 1$).

Note that in this simple example, of course, a second-order model could have been chosen. However, the goal of this example is to illustrate the properties of ARX and OE models for order reduction. In reality (almost) any model will be of a lower order than the real process.

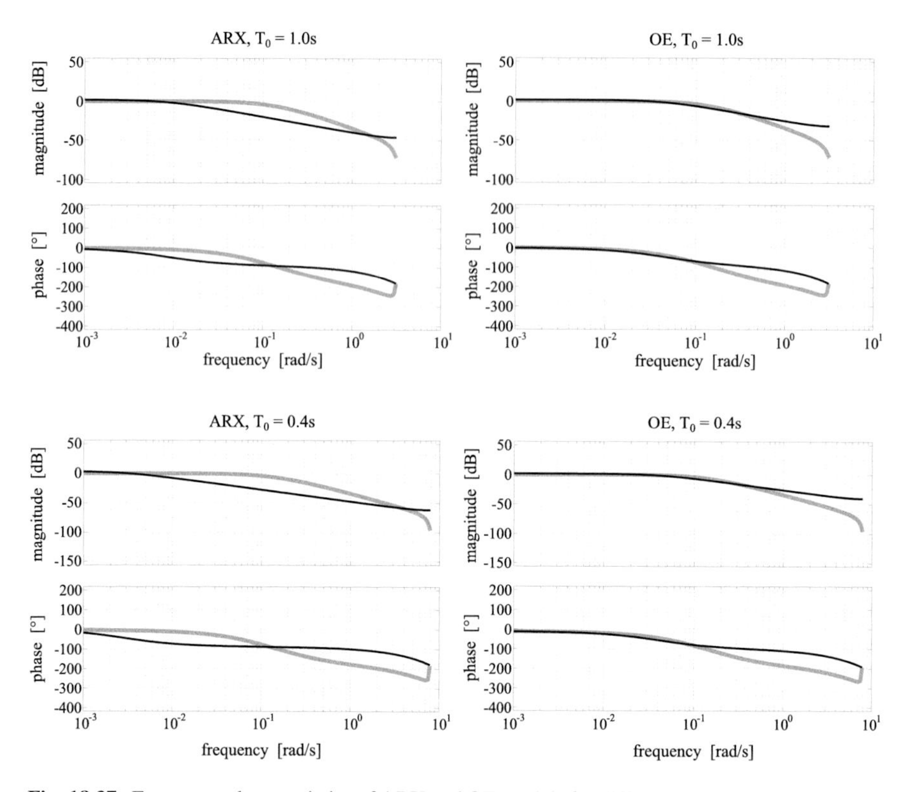

Fig. 18.37 Frequency characteristics of ARX and OE models for different sampling times T_0. The gray lines represent the process; the black lines represent the models

18.6 Models Without Output Feedback

In the previous section, linear models with output feedback have been discussed, which are by far the most common model types. They are characterized by transfer functions where the parameters of both the numerator and the denominator polynomials are estimated. By the estimation of the parameters for the $A(q)$ or $F(q)$ polynomials, respectively, the poles of the process are approximated, and therefore its dynamics can be usually described with rather low polynomial degrees and consequently few parameters.

In contrast, this section focuses on models that do not incorporate output feedback. Clearly, this restriction makes those model types less flexible and often leads to a higher number of required parameters. On the other hand, some interesting advantages can be expected from the finite impulse response (FIR) (Sect. 18.6.1) and orthonormal basis functions (OBF) (Sect. 18.6.5) models discussed in this section:

- They belong to the *output error class* and are *linear in their parameters.* In the previous section, only the ARX model was linear in the parameters, which is the main reason for its popularity. However, the assumptions about the noise model inherent in the ARX model are usually not fulfilled, and consequently, a non-consistent estimation of the parameters can be expected. All these problems are avoided with FIR and OBF models.

- *Stability* is guaranteed. Since the estimated parameters do not determine the poles of the model, stability of the model can be guaranteed independent of the estimated parameter values. In contrast, the models discussed in Sect. 18.5 may become unstable depending on the estimated parameters.

  Note that the advantage of guaranteed stability of FIR and OBF models, of course, turns into a drawback if the process under investigation itself is unstable and the model is required to be unstable as well. FIR and (with some restrictions) OBF models are not suited for identification of an unstable process. Nevertheless, in the overwhelming majority of applications, the systems[4] are stable, and a stability guarantee can be seen as an advantage.

  Since the stability of linear models can be easily checked by calculation of the poles, one may wonder about the relevance of this stability issue. However, the above discussion becomes increasingly important when these linear model types are to be adapted online or when they are generalized to nonlinear dynamic models in Chap. 19.
- They are quite *simple*. The model structure is much simpler than for models with output feedback. Consequently, these models are especially attractive if the accuracy requirements are not very tight but simplicity is an important issue. Especially for signal and image processing applications, the FIR or OBF models are extensively used. In particular, FIR models are widely utilized in adaptive filtering applications, e.g., for channel equalization in digital mobile phone communication.

In the next two sections, the FIR and OBF models are discussed.

18.6.1 Finite Impulse Response (FIR)

The finite impulse response (FIR) model is the simplest linear model. All other models possess an infinite impulse response since they incorporate some kind of feedback: output feedback in the case of all models discussed in Sect. 18.5 and a kind of internal feedback in the case of the OBF model analyzed below.

The FIR model is simply a moving average filtering of the input. Thus, the model output is a weighted sum of previous inputs (see Fig. 18.38)

$$y(k) = b_1 u(k-1) + b_2 u(k-2) + \ldots + b_m u(k-m) + v(k)\,. \qquad (18.133)$$

[4]Often for unstable processes, a simple stabilizing controller is designed, and subsequently, the resulting stable closed-loop system is identified to design a second, more advanced controller for the inner closed-loop system.

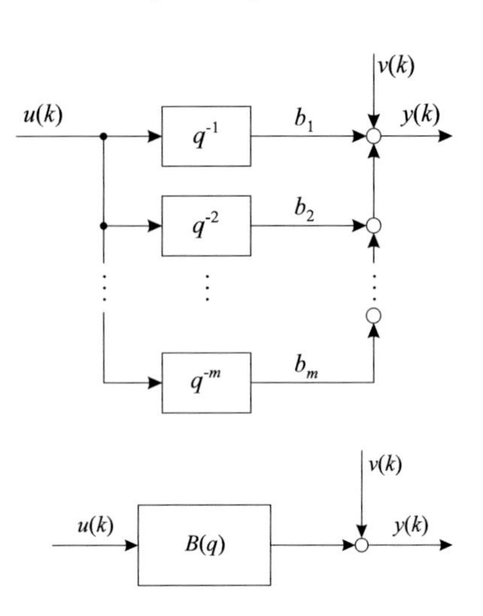

Fig. 18.38 FIR model in filter representation

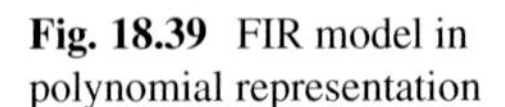

Fig. 18.39 FIR model in polynomial representation

In polynomial form, it becomes (see Fig. 18.39)

$$y(k) = B(q)u(k) + v(k)\,. \tag{18.134}$$

The terminology is very inconsequential here, since "FIR" (finite impulse response) is the counterpart to "IIR" (infinite impulse response). So, "FIR" represents another level of abstraction than "ARX," "ARMAX," etc.; see Fig. 18.14. Note that "MA" is reserved for the moving average time series model $y(k) = C(q)v(k)$; see Sect. 18.4.2.

The white noise enters the output in (18.134) additively, and consequently, an FIR model belongs to the class of output error models. The optimal predictor is simply

$$\hat{y}(k|k-1) = B(q)u(k) \tag{18.135}$$

and thus linear in the parameters. Because feedback is not involved, it is possible to have a linear parameterized model that is of output error type. Since the predictor utilizes only the input sequence $u(k)$, as for the OE model, the optimal one-step prediction is equivalent to a simulation.

The motivation for the FIR model comes from the fact that the output of each linear system can be expressed in terms of the following convolution sum:

$$y(k) = \sum_{i=1}^{\infty} g_i u(k-i) = g_1 u(k-1) + g_2 u(k-2) + \ldots\,, \tag{18.136}$$

where g_i is the impulse response. The term $g_0 u(k)$ is missing because the system is assumed to have no direct feedthrough path from the input to the output, i.e., it is strictly proper. Obviously, the FIR model is just an approximation of the convolution sum in (18.136) with the first m terms of the infinite series. Since for stable systems the coefficients g_i decay to zero as $i \to \infty$, such an approximation is possible. However, for marginally stable or unstable systems, a reasonable approximation is not possible since the g_i tends to a constant value or infinity, respectively, as $i \to \infty$. Thus, an FIR model can only be applied to model stable processes (although it can represent an unstable process for the first m sampling instants in a step or impulse response). On the other hand, the inherent stability of FIR models can be considered as an advantage in the overwhelming majority of cases where indeed the process is stable.

Like the ARX model, the FIR model is linear in the parameters. They can be estimated by least squares. The parameter vector and regression matrix are

$$\underline{\theta} = \begin{bmatrix} b_1 \\ b_2 \\ \vdots \\ b_m \end{bmatrix}, \quad \underline{X} = \begin{bmatrix} u(m) & u(m-1) & \cdots & u(1) \\ u(m+1) & u(m) & \cdots & u(2) \\ \vdots & \vdots & & \vdots \\ u(N-1) & u(N-2) & \cdots & u(N-m) \end{bmatrix}. \tag{18.137}$$

So far a number of advantages of the FIR model have been mentioned. It is linear in the parameters, it belongs to the class of output error models (thus the noise model is more realistic than for ARX), and it is simple. However, there is one big drawback of the FIR model that severely restricts its applicability. The order m has to be chosen very large. It is clear from (18.136) that m must be chosen large enough to include all g_i (modeled by the b_i) that are significantly different from zero. Otherwise, the approximation error would become too large, and the dynamic representation of the model would be poor.

In order to get an idea how large m has to be chosen, assume the following case. The sampling time is chosen reasonably, say 1/5 of the slowest time constant T of the process. Then during the approximate settling time T_{95} (the time required for the process to reach 95% of its final value), the process is sampled 15 times. This means that a reasonable choice for m is 15. If the model is utilized only for controller design, m may be chosen smaller because the accuracy requirements for the model's static behavior, which the last coefficients (b_i with large i) mainly influence, are not so important. If the purpose of the model is simulation, much smaller values than $m = 15$ would significantly degrade the model's performance. Since $m = 15$ means that 15 parameters have to be estimated, typically the degrees of freedom of an FIR model are considerably larger than for, e.g., an ARX model with a model order that yields the same accuracy. To make things worse, in practice, usually, the sampling rate is chosen much higher than 1/5 of the slowest time constant. (The reason for this lies in the fact that the drawbacks of a sampling rate that is too large are less severe than those of one that is too small. So in a quick-and-dirty approach, often the sampling rate is chosen very high.) Of course, higher sampling rates (for the same process) proportionally require larger-orders m. Thus, in practice, FIR models are often significantly over-parameterized and exhibit extremely high variance errors.

This main drawback of FIR models may be overcome or at least weakened by regularized FIR models and OBF models which both are discussed in the next sections.

18.6.1.1 Comparison ARX Versus FIR

It is interesting to contrast ARX and FIR model structures in theory and application. Here are some aspects that can be derived methodologically:

- Both are linear regression structures and thus can be estimated by least squares.
- Both are very popular, simple, and relatively easy-to-handle model structures. In the assessment of the author, the number of ARX versus FIR model applications might be around a ratio of 2:1.
- ARX optimizes the equation (one-step prediction) error, while FIR optimizes the output (simulation) error. Usually, the simulation error is the true objective, and the equation error is just required as a proxy to keep the model linear in its parameters.
- Parameter estimation for ARX is biased for white output disturbances favoring high frequencies; FIR is not.
- ARX requires much fewer parameters (lower model orders) than FIR.
- ARX typically is very sensitive with respect to the "correct" model structure, that is, order and dead time; FIR is not—the order corresponds just to the degree of accuracy. Dead time and order of the process are automatically detected by FIR by estimation of its coefficients, i.e., the coefficients $b_1, \ldots, b_d$ will be approximately 0 for dead time d (at least in principle). Also, FIR does not require critical decisions about the model order to describe a given number of resonance frequencies, say r, while ARX requires a model order $n \geq 2r$.
- ARX can model unstable or marginally stable processes; FIR cannot. But this also means: ARX models can become unstable, while FIR models cannot. This is most important in all adaptive scenarios.
- ARX becomes extremely suboptimal for multiple-output systems because a common denominator polynomial is required for the transfer functions to all outputs. This is not the case for FIR. This is one reason for the popularity of subspace methods for MIMO cases which are of FIR type [356] in state space.

In summary, ARX and FIR approaches have their own, often opposite, advantages and drawbacks. In the linear case, the overall assessment depends on the circumstances. Both approaches are realistic for applications. This assessment changes in the context of nonlinear system identification; compare Chap. 19.

18.6.2 Regularized FIR Models

As discussed in the previous subsection, FIR models have many advantages compared to models with output feedback. However, one very significant drawback is the huge number of required parameters because the model order m must be chosen to cover the significant part of the impulse response. The problem here is not so much the computational effort since it is just a linear regression. Rather the resulting high variance error of the model causes overfitting in most cases. In order to avoid this and balance the bias and variance appropriately, regularization must be employed, i.e., a penalty term can be added to the sum of squared errors:

$$I(\underline{\theta}) = \underline{e}^T \underline{e} + \lambda \underline{\theta}^T \underline{R} \underline{\theta} . \tag{18.138}$$

The optimal parameter vector then becomes

$$\underline{\hat{\theta}} = \left(\underline{X}^T \underline{X} + \lambda \underline{R}\right)^{-1} \underline{X}^T \underline{y} . \tag{18.139}$$

In a Bayesian interpretation, the penalty matrix $\underline{R}$ corresponds to the inverse covariance matrix of the parameters:

$$\underline{R} = \underline{P}^{-1} . \tag{18.140}$$

Then the estimated parameter vector in (18.139) becomes equal to [458]:

$$\underline{\hat{\theta}} = \left(\underline{X}^T \underline{X} + \lambda \underline{P}^{-1}\right)^{-1} \underline{X}^T \underline{y} = \left(\underline{P}\,\underline{X}^T \underline{X} + \lambda \underline{I}\right)^{-1} \underline{P}\,\underline{X}^T \underline{y} . \tag{18.141}$$

If the regularization parameter corresponds to the variance of the measurement noise ($\lambda = \sigma^2$), then the optimal regularization matrix becomes

$$\underline{P} = \underline{\vartheta}\,\underline{\vartheta}^T \tag{18.142}$$

where $\underline{\vartheta} = \underline{\theta}_{\text{opt}}$ is the optimal parameter vector. The regularization matrix $\underline{P}$ can be interpreted as prior covariance matrix of $\underline{\theta}$ in a stochastic Bayesian setting where $\underline{\theta}$ is considered to be a random variable. The diagonal terms of $\underline{P}$ can be interpreted as parameter variances $p_{kk} = \vartheta_k^2$ and on the secondary diagonals are the parameter covariances $p_{kl} = \vartheta_k \vartheta_l$.

Of course, the optimal FIR parameters $\underline{\vartheta}$ are unknown. Thus, $\underline{P}$ needs to be chosen by prior knowledge. A trivial choice would be $\underline{P} = \underline{0}$ which yields the unregularized FIR model. Another simple alternative would be $\underline{P} = \underline{I}$ which is known as ridge regression (see Sect. 3.1.4) and would drive the parameters toward zero. This would decrease the variance but biases the model strongly toward small values.

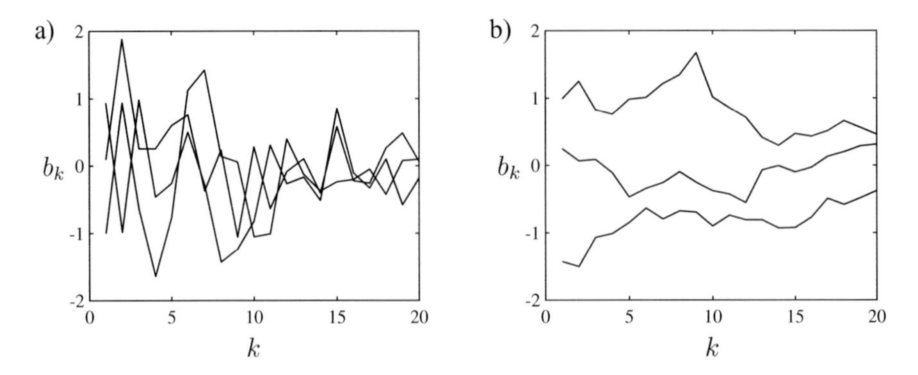

Fig. 18.40 Three possible realizations of the diagonal kernel (**a**) and TC kernel (**b**) prior with $\alpha = 0.9$ generated by a 20-dimensional normal distribution with covariance matrix $\underline{P}$ according to (**a**) (18.143) and (**b**) (18.144)

The basic idea of a suitable regularization of FIR models is to penalize the first coefficients $b_1, b_2, \ldots$ stronger than the later ones $\ldots, b_{m-1}, b_m$. It is reasonable to choose exponential weighting to prefer an exponentially decaying impulse response, e.g., α^i for $i = 1, 2, \ldots, m$, i.e.,

$$\underline{P} = \text{diag}\{\alpha^1, \alpha^2, \ldots, \alpha^m\}\,. \tag{18.143}$$

This corresponds to the first m values of an impulse response of a first-order IIR system with $y(k) = \alpha\, y(k-1) + \ldots$. Since such a regularization matrix is diagonal, all FIR parameters are penalized independent of each other.

A much more sophisticated variant is to tie the parameters together by specifying covariances between them. It seems obviously appealing to tie two parameters b_k and b_l the closer together, the smaller $|k - l|$ is. Therefore the covariance between b_k and b_l can, for example, be chosen as $\alpha^{|k-l|}$. This can be visualized as an elastic band coupling, e.g., b_1 and b_2 are coupled closer together than b_1 and b_{10}. Figure 18.40 shows how the first 20 coefficients may look like. Note that realizations of an unregularized FIR ($\lambda = 0$) or a simple ridge regression ($\underline{P} = \underline{I}$) would just look white noise, i.e., all b_k would be independent of each other and non-decreasing (before the estimation).

18.6.2.1 TC Kernel

Both effects, exponential decay and exponentially decreasing covariance, can be combined multiplicatively to

$$p_{kl} = \alpha^{|k-l|/2} \alpha^{k/2} \alpha^{l/2}\,. \tag{18.144}$$

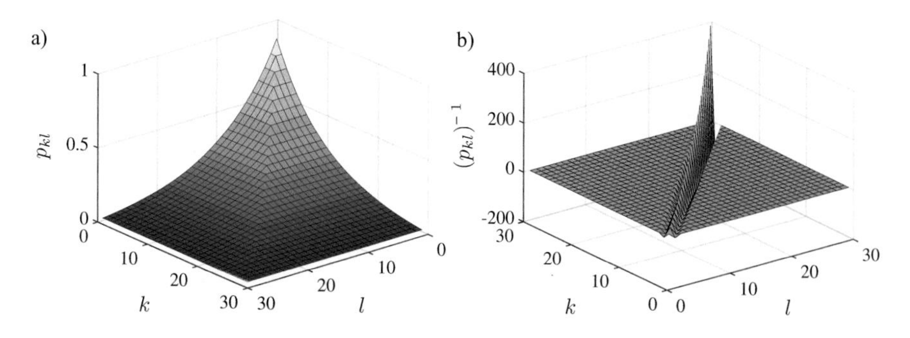

Fig. 18.41 Entries of the regularization matrix for the TC kernel (**a**) and entries of its inverse (**b**)

The square root is taken to preserve the interpretation of α; $k = l$ yields $p_{kk} = \alpha^k$. It is easy to rewrite (18.144) as

$$p_{kl} = \alpha^{\max(k,l)}\,. \tag{18.145}$$

This regularization matrix is known as the *TC kernel* which stands for Tuned/Correlated [458]. Figure 18.41a shows the entries of the regularization matrix $\underline{P}$ for $\alpha = 0.9$ and its inverse in (b) (note the reversed axes for k and l compared to a) for a better visualization.

The regularization matrix and its inverse for a TC kernel regularized FIR model of order $m = 7$ and $\alpha = 0.9$ are as follows:

$$\underline{P} = \begin{pmatrix} 0.90 & 0.81 & 0.73 & 0.66 & 0.59 & 0.53 & 0.48 \\ 0.81 & 0.81 & 0.73 & 0.66 & 0.59 & 0.53 & 0.48 \\ 0.73 & 0.73 & 0.73 & 0.66 & 0.59 & 0.53 & 0.48 \\ 0.66 & 0.66 & 0.66 & 0.66 & 0.59 & 0.53 & 0.48 \\ 0.59 & 0.59 & 0.59 & 0.59 & 0.59 & 0.53 & 0.48 \\ 0.53 & 0.53 & 0.53 & 0.53 & 0.53 & 0.53 & 0.48 \\ 0.48 & 0.48 & 0.48 & 0.48 & 0.48 & 0.48 & 0.48 \end{pmatrix} \tag{18.146}$$

and in consequence

$$\underline{P}^{-1} = \begin{pmatrix} 11.0 & -11.0 & 0 & 0 & 0 & 0 & 0 \\ -11.0 & 23.0 & -12.0 & 0 & 0 & 0 & 0 \\ 0 & -12.0 & 26.0 & -14.0 & 0 & 0 & 0 \\ 0 & 0 & -14.0 & 29.0 & -15.0 & 0 & 0 \\ 0 & 0 & 0 & -15.0 & 32.0 & -17.0 & 0 \\ 0 & 0 & 0 & 0 & -17.0 & 36.0 & -19.0 \\ 0 & 0 & 0 & 0 & 0 & -19.0 & 21.0 \end{pmatrix}. \tag{18.147}$$

Notice that the negative entries in (18.147) favor identical subsequent parameters because terms like $b_1 b_2$ get a negative penalty. Thus, a significant preference toward $b_1 \approx b_2 \approx \ldots \approx b_m$ exists, thereby stabilizing the whole impulse response.

18.6.2.2 Filter Interpretation

These ideas are further elaborated in [362] where it is proposed to decompose the penalty matrix $\underline{R}$ in (18.138) into

$$\underline{R} = \underline{P}^{-1} = \underline{F}^T \underline{F}\,. \tag{18.148}$$

Instead of choosing a specific covariance matrix $\underline{P}$ in order to incorporate the prior knowledge, it is proposed to directly select a specific filter matrix $\underline{F}$. As stated in [362]: "$\underline{F}$ can be seen as a prefiltering operator on the coefficients of the impulse response, before they enter the cost function and are penalized through the regularization term. This means that the regularization filter matrix $\underline{F}$ should be defined in such a way that it incorporates the system properties one needs to penalize in order to obtain the desired model." Each row in $\underline{F}$ is multiplied with the parameter vector $\underline{\theta}$ in the penalty term $\lambda\, \underline{\theta}^T \underline{F}^T \underline{F}\, \underline{\theta} = \lambda\, ||\underline{F}\, \underline{\theta}||^2$. A Cholesky decomposition of (18.147) delivers:

$$\underline{F} = \begin{pmatrix} 3.33 & -3.33 & 0 & 0 & 0 & 0 & 0 \\ 0 & 3.51 & -3.51 & 0 & 0 & 0 & 0 \\ 0 & 0 & 3.70 & -3.70 & 0 & 0 & 0 \\ 0 & 0 & 0 & 3.90 & -3.90 & 0 & 0 \\ 0 & 0 & 0 & 0 & 4.12 & -4.12 & 0 \\ 0 & 0 & 0 & 0 & 0 & 4.34 & -4.34 \\ 0 & 0 & 0 & 0 & 0 & 0 & 1.45 \end{pmatrix}. \tag{18.149}$$

This filter matrix reveals most obviously what is going on through the penalization. The difference of subsequent parameter values $b_i - b_{i+1}$ is penalized. Furthermore, this penalty increases exponentially with $1/\sqrt{\alpha} = 1/\sqrt{0.9}$ from row to row driving latter coefficients in the impulse response toward zero heavily.

The TC kernel possesses a single parameter α that corresponds to the time constant of the exponential decay. A second parameter is the overall strength of the penalty λ in (18.138). Since α and λ control the overall FIR model behavior, they are called *hyperparameters*. More advanced kernels exist that offer better smoothing properties but typically require more hyperparameters; refer to [458] for details.

It is possible to play with the hyperparameters manually in order to obtain a good FIR model. Often approximate prior knowledge of the process dynamics is available and can be utilized for fixing α. Then this approach becomes very similar to orthonormal basis functions discussed in Sect. 18.6.5. The amount of regularization is more difficult to assess and can be determined by (tedious) cross-validation.

However, a much more sophisticated approach is to optimize the hyperparameters on the data. One efficient way of doing this is known as maximization of the marginalized likelihood function or empirical Bayes. As discussed in [458], the following negative marginal log-likelihood function has to be minimized

$$I(\lambda, \alpha) = \underline{y}^T \underline{Z}(\lambda, \alpha)^{-1} \underline{y} + \log \det \underline{Z}(\lambda, \alpha) \longrightarrow \min_{\lambda, \alpha} \quad (18.150)$$

with the matrix

$$\underline{Z}(\lambda, \alpha) = \underline{X}\, \underline{P}(\alpha) \underline{X}^T + \lambda \underline{I} \,. \quad (18.151)$$

This is a non-convex nonlinear optimization problem which has been studied thoroughly and can be solved relatively efficiently; advanced numerical implementations exit. Nevertheless it is demanding to calculate $\underline{Z}(\lambda, \alpha)^{-1}$ and $\det \underline{Z}(\lambda, \alpha)$ (which can be computed together with the inverse) for each value of the hyperparameters. Because of the non-convexity of (18.150), local optima can arise dependent on the type of kernel and data distribution. However, this problem generally is observed as being no big issue.

The idea of optimizing the marginal log-likelihood function is discussed a little bit more elaborately in Chap. 16. In contrast to most kernel methods in machine learning, regularized FIR models possess only a few parameters (typically $n < 100$). Thus, usually, $N \gg n$ which allows implementing the marginalized likelihood optimization much more efficiently with the help of a reduced QR factorization [101]. In contrast, for Gaussian processes, support vector machines or similar kernel-based approaches typically $N \approx n$, often $N = n$, holds.

18.6.3 Bias and Variance of Regularized FIR Models

Figure 18.42 compares the typical ranges of the bias and variance error components for different dynamic structures. ARX and OE models are limited in their flexibility as they represent a (typically low-order) transfer function which usually is just a rough approximation of the more complex process behavior. Thus their bias error component is usually large. In addition, the ARX model represents an equation error model whose unrealistic noise assumption rarely holds in practice; compare Sect. 18.5.1. Thus, their parameter estimates possess a bias when estimated by least squares yielding inferior models compared to OE.

In contrast, FIR models are non-parametric and much more flexible. They are capable to describe tiny nuisances of the process behavior, in principle. However, due to their huge number of parameters, they are plagued with a high variance error component which restricts their application to large data sets with low noise levels.

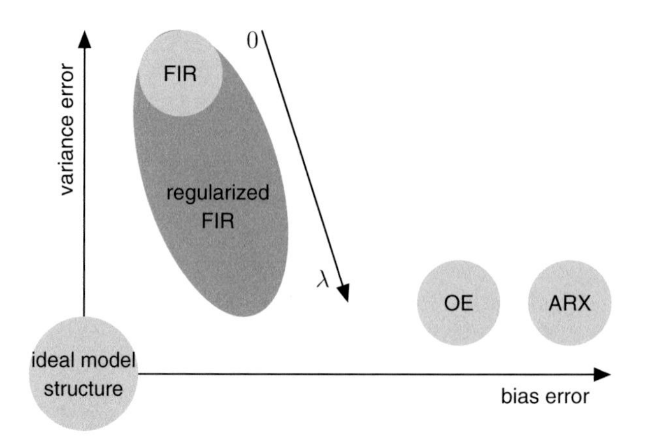

Fig. 18.42 Typical bias/variance error profile for different dynamic structures

The model structure promoted in this section, the regularized FIR model, counters this overwhelming drawback by regularization which decreases the effective number of parameters well below their nominal number, thereby reducing the model flexibility and variance error; see Sect. 3.1.12. The amount of regularization is controlled via the hyperparameter λ which typically is tuned on the data to achieve a (near-)optimal bias/variance tradeoff.

18.6.4 Impulse Response Preservation (IRP) FIR Approach

The regularization term $\lambda\, \underline{\theta}^T \underline{R}\, \underline{\theta} = \lambda\, \underline{\theta}^T \underline{P}^{-1} \underline{\theta} = \lambda\, ||\underline{F}\, \underline{\theta}||^2$ in (18.138) is traditionally chosen based on consideration about the correlations between impulse response values $g(i)$ and $g(j)$ depending on i and j [458]. Typically, the correlation is assumed to decrease with increasing distance $|i - j|$, while an exponentially decaying behavior with i and j is encouraged. This leads to the widely used TC kernel which corresponds to the following *pure* filter matrix [391]:

$$\underline{F}_{\text{pure}} = \begin{pmatrix} 1 & -1 & 0 \cdots & & 0 \\ 0 & 1 & -1 & \cdots & 0 \\ \vdots & & \ddots & & \vdots \\ 0 & \cdots & 0 & 1 & -1 \\ 0 & \cdots & 0 & 0 & 1 \end{pmatrix} \tag{18.152}$$

which is weighted with the exponentially decaying weighting matrix $\underline{W}$ yielding the filter matrix in (18.148):

$$\underline{F} = \underline{W}\, \underline{F}_{\text{pure}} \,. \tag{18.153}$$

Here the weighting matrix corresponds to

$$\underline{W} = \text{diag}(\alpha^{0/2}, \alpha^{-1/2}, \ldots, \alpha^{-L/2}) \tag{18.154}$$

with FIR order L and tuning parameter $\alpha < 1$ encouraging exponentially stabilizing impulse responses. Note that square root in the entries in (18.154) comes from the factorization in (18.148). The filter and weighting matrices are of size $(L + 1) \times (L + 1)$. The effect of the weighting with $\underline{W}$ is to scale the kth row of $\underline{F}_{\text{pure}}$ with $\alpha^{-k/2}$, thereby exponentially increasing the penalty for a deviation from the prior over the course of the impulse response from $k = 0$ to $k = L$.

In [361, 362], the filer matrix $\underline{F}$ is constructed directly via designing a filter, modeling the assumed inverse of the process under consideration.

18.6.4.1 Impulse Response Preservation (IRP)

The filter matrix is chosen based on completely different considerations, and it is shown that this approach is superior with respect to both, interpretation and performance. The approach also starts with the penalty term $\lambda \, ||\underline{F}\, \underline{\theta}||^2$. The idea is to penalize impulse responses deviating from a prior. The prior impulse response corresponds to the impulse response of linear dynamic process of order n with arbitrary numerator $B(z)$ and known denominator $A(z)$:

$$\frac{Y(z)}{U(z)} = \frac{B(z)}{A(z)} = \frac{B(z)}{z^n + a_{n-1}z^{n-1} + \ldots + a_1 z + a_0} \tag{18.155}$$

assuming $a_n = 1$ without loss of generality. The impulse response $g(k)$, for $k = 0, 1, \ldots, L$ of this system fulfills the equations

$$g(k+n) + a_{n-1}g(k+n-1) + \ldots + a_1 g(k+1) + a_0 g(k) = 0 \tag{18.156}$$

for $k = 0, 1, \ldots, L - n$ or equivalently

$$g(k) + a_{n-1}g(k-1) + \ldots + a_1 g(k-n+1) + a_0 g(k-n) = 0 \tag{18.157}$$

for $k = n, n + 1, \ldots, L$. Equations (18.156) and (18.157) shall be called impulse response preserving equations. If these equations are utilized in order to represent the penalty term, then any model matching the prior exactly is not penalized. For a first-order system ($n = 1$), the following filter matrix of size $L \times (L + 1)$ accomplishes this:

$$\underline{F}_{\text{pure}} = \begin{pmatrix} a_0 & 1 & 0 & \cdots & 0 \\ 0 & a_0 & 1 & \cdots & 0 \\ \vdots & & \ddots & \ddots & \vdots \\ 0 & \cdots & 0 & a_0 & 1 \end{pmatrix}. \tag{18.158}$$

More generally, for an nth-order system prior, the corresponding pure filter matrix of size $(L+1-n) \times (L+1)$ is

$$\underline{F}_{\text{pure}} = \begin{pmatrix} a_0 & a_1 & \cdots & a_{n-1} & 1 & 0 & \cdots & 0 \\ 0 & a_0 & a_1 & \cdots & a_{n-1} & 1 & \cdots & 0 \\ \vdots & & \ddots & \ddots & & \ddots & \ddots & \vdots \\ 0 & \cdots & 0 & a_0 & a_1 & \cdots & a_{n-1} & 1 \end{pmatrix} . \tag{18.159}$$

These matrices shall be called *impulse responses preserving (IRP)* filter matrices. The final filter matrix $\underline{F}$ is calculated by weighting with $\underline{W}$ according to (18.153) where $\underline{W} = \text{diag}(\alpha^0, \alpha^{-1}, \ldots, \alpha^{n-L})$ with n being the system order.

Of course, in most cases the exact values for the a_i coefficients are unknown. They are hyperparameters of the identification problem and have to be optimized from data. Also the weighting matrix $\underline{W}$ is required for the calculation of $\underline{F}$. The required value for the exponential decay factor α can be either (i) fixed a priori by prior knowledge which hardly will be available, (ii) tuned by optimization (then it becomes a hyperparameter), or (iii) related to the parameters of the prior system, e.g., $\alpha = -a_0$ for the first-order case similar to the TC kernel.

Figure 18.43 visualizes the mechanism of the proposed penalty. It encourages that n subsequent values of the impulse response fulfill (18.157) where n is the prior system's order. Here k moves a sliding window of width n by running from the beginning $k = n$ to the end $k = L$ of the FIR model.

Comparing the proposed approach with the TC kernel, the following observations can be made:

- The filter matrix for the TC kernel penalizes the last value of the FIR model $g(L)$ absolutely, while the IRP approach does not. In consequence, the TC kernel

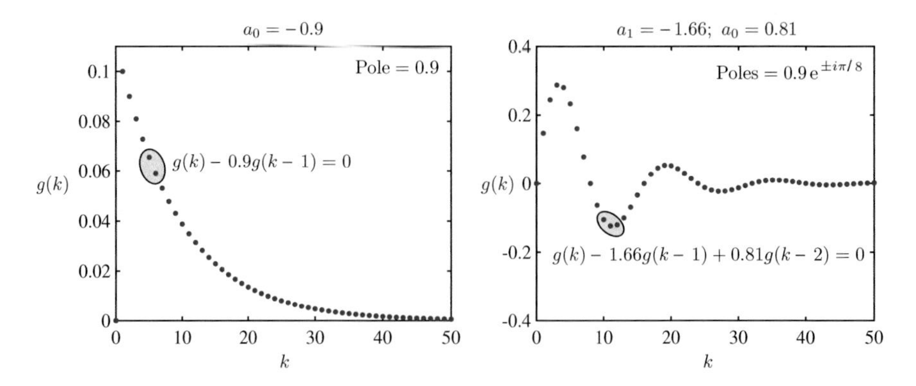

Fig. 18.43 Examples for impulse responses of a first- (top) and second-order (bottom) process with prior knowledge preserving the shape of the impulse response via incorporation in the FIR model by regularization or a penalty term, respectively. For order n, it links $n+1$ successive values of the impulse response together with $k = n, n+1, \ldots, L$

matrix has full rank $L + 1$, while the IRP penalty matrix whose inverse would correspond to a "kernel matrix" has only rank $L + 1 - n$ and is not directly invertible, and therefore no true "kernel" in the mathematical sense exists.

- The TC kernel can be seen as an approximation of the first-order IRP approach for slow prior systems with $a_0 \to -1$. Interpreted in IRP terms, the TC kernels preserve the impulse response of an integrator. The opposite sign of $\underline{F}\,\underline{\theta}$ is eliminated by forming the quadratic expression for the penalty term $\lambda\,\underline{\theta}^T\,\underline{F}^T\,\underline{F}\,\underline{\theta}$. Thus, the benefits of the IRP approach should be the more significant, the faster the process under investigation is.
- With the IRP approach, also prior knowledge about higher-order system behavior can be incorporated elegantly and efficiently. A particularly important choice is $n = 2$ for modeling oscillatory behavior of arbitrary frequency and damping by tuning the hyperparameters a_1 and a_0. For the TC kernel, this is not possible.

Remarks:

- It is important to note that the numerator $B(z)$ in (18.155) is irrelevant for the impulse response preserving equations in (18.156) and (18.157) if the system is *not* strictly proper, i.e., numerator degree equals denominator degree. In case of lower degree numerators, $B(z)$ can always be made irrelevant by shifting the sliding window toward latter times, i.e., begin $m - n + 1$ time steps latter than $k = n$ establishing (18.157) if $m < n$.
- The impulse response preserving equations utilized in $\underline{F}_{\text{pure}}$ match the order L of the FIR model. Therefore, if the model order L is selected too small and consequently significant truncation errors arise, still the penalty term will be exactly zero if the first $L + 1$ coefficients of the FIR model match the prior.
- For huge regularization strengths ($\lambda \to \infty$), the characteristics of the denominator polynomial $A(z)$ of the prior is enforced. For first-order priors, the penalty matrix $\underline{R} = \underline{F}^T\underline{F}$ has rank L and thus practically fixes L parameters of the impulse response, leaving just one degree of freedom for adjusting for the gain which represents the flexibility of a numerator polynomial $B(z)$ and can be estimated from the data. For second-order priors, the penalty matrix has rank $L-1$ and thus practically fixes $L-1$ parameters of the impulse response, leaving just two degrees of freedom for adjusting for the gain and one zero, and so on.
- The first row in $\underline{F}$ penalizes the relationship between $g(0)$ and the subsequent coefficients of the impulse response $g(k)$ for $k = 1, \ldots, n$. Whether this is correct or not depends on the properness or strict properness of the system since $u(k = 0)$ may (direct feedthrough) or may not enter the impulse response equation (18.157). Therefore, in order to keep the IRP property intact for both cases, it is recommended to remove the first row from $\underline{F}_{\text{pure}}$ and $\underline{F}$ if it is unknown whether the process is proper or strictly proper. Consequently, the ranks of the filter and penalty matrices reduce to $L - n$.

Table 18.3 compares the model errors of IRP FIR model with full filter matrix $\underline{F}$ and its first row removed, for a process with and without direct feedthrough. The example process was chosen of first-order because then the discrepancy between both cases is most severe. The data generation scenario otherwise is

Table 18.3 Error of the impulse response of an IRP FIR model for a first-order process

IR RMSE [%]	Full $\underline{F}$	$\underline{F}$ without 1st row
Process with direct feedthrough	0.041	0.053
Process without direct feedthrough	**0.32**	0.054

identical to the one explained in Sect. 18.6.4.3. It can be clearly seen in Table 18.3 that removing the first row from $\underline{F}$ costs performance, but keeping the first row risks a significant performance drop if the process has no direct feedthrough.

18.6.4.2 Hyperparameter Optimization

IRP FIR modeling follows though the steps:

1. Assume that the process can be described by an unknown transfer function $B(z)/A(z)$ of order n. This is the system prior.
2. Choose some exponential decay factor α that very roughly corresponds to the dominant pole of the process.
3. Optimize a regularized FIR model of order L with penalty matrix $\underline{P}$ which contains n hyperparameters, namely, the denominator coefficients of $A(z)$: a_i, $i = 0, 1, \ldots, n-1$. The FIR model order L should be selected "high enough." Its choice is very robust since due to the regularization, the model variance is well-controlled even if L is huge. The only drawback of increasing L is a higher computational demand. The FIR model optimization typically tunes the $n+1$ hyperparameters λ and a_i, $i = 0, 1, \ldots, n-1$ by minimizing the sum of squared generalized cross-validation (GCV) errors. In [397] an alternative, promising approach is demonstrated. The hyperparameters are derived from first principles, i.e., from knowledge about the physics of the process. The only remaining tuning parameter is the regularization strength λ. This constitutes a new type of gray-box modeling where the (optimized) size of λ represents the accuracy and validity of the first principles knowledge.

   The calculation of the loss function comprises the following steps:

   (a) Construct a penalty matrix $\underline{P} = \underline{F}^T \underline{F}$ where the filter matrix corresponds to the exponentially weighted pure filter matrix

   $$\underline{F} = \underline{W}\,\underline{F}_{\text{pure}}\,. \tag{18.160}$$

   with $\underline{W} = \text{diag}(\alpha^0, \alpha^{-1}, \ldots, \alpha^{-L})$ and the chosen exponential decay factor α. For an nth-order system prior, the corresponding pure filter matrix $\underline{F}_{\text{pure}}$ is given by (18.159).

   (b) Estimate the optimal FIR coefficients by

   $$\hat{\underline{\theta}} = \left(\underline{X}^T \underline{X} + \lambda \underline{R}\right)^{-1} \underline{X}^T \underline{y}\,. \tag{18.161}$$

(c) Calculate the model training error $\underline{e}_{\text{Train}} = \underline{y} - \underline{X}\,\hat{\underline{\theta}}$ and correct it to obtain the GCV error (compare Sect. 7.3.2):

$$\underline{e}_{\text{GCV}} = \frac{\underline{e}_{\text{Train}}}{1 - \text{tr}(\underline{S})/N} \tag{18.162}$$

with N being the number of data points and where the smoothness matrix corresponds to

$$\underline{S} = \underline{X}\left(\underline{X}^T\underline{X} + \lambda\underline{R}\right)^{-1}\underline{X}^T\,. \tag{18.163}$$

Return the sum of squares GCV errors

$$J(\lambda, a_0, \ldots, a_{n-1}) = \underline{e}_{\text{GCV}}^T\underline{e}_{\text{GCV}}\,. \tag{18.164}$$

Relationship to OE Models

This IRP FIR approach is similar to OE in that it requires the selection of a model order n and carries out a non-convex nonlinear optimization of the (hyper)parameters. However, multiple benefits of IRP FIR arise over OE:

- Only $n + 1$ parameters are optimized compared to $2n$ parameters in the OE case.
- The final model is of FIR type and therefore inherently stable, very robust w.r.t. wrong orders n and unknown dead times, and easier to adapt online due to its linear regression structure w.r.t. its coefficients $\underline{\theta}$ (but not w.r.t. its hyperparameters, of course). This makes the proposed IRP approach superior to classical OE modeling in the real world although in the perfect scenario (no plant/model mismatch) the OE model may perform slightly better.
- The properties are comparable to Gaussian processes for nonlinear regression in machine learning whose power also comes from the efficient, robust, numerically mature least squares estimation of most parameters (corresponding to the impulse response coefficients for IRP FIR or the kernel weights for GP, respectively), and only very few decisive structural parameters are nonlinearly optimized (the regularization strength λ in both cases; the denominator parameters a_i for IRP FIR or the length scales for GP, respectively).

Choice of α and Alternatives

In contrast to the TC kernel, the hyperparameter α in the weighting matrix $\underline{W}$ is not strictly required in the IRP approach to impose exponential decay. This is already realized through the pure filter matrix $\underline{F}_{\text{pure}}$. Nevertheless, in practice, it makes sense to allow more flexibility in the earlier phases of the impulse response and tightens the IRP property with increasing k. In earlier phases, bigger deviations can be expected due to possibly unknown dead times.

It is the experience of the authors that optimization of α in the IRP approach is difficult to carry out robustly because the exponential decay induced by α interacts with the exponential decay of the stable impulse response. In fact, α basically acts similar to an additional real pole. If α is not optimized but fixed by the user, then the results frequently are very sensitive w.r.t. the choice of α.

Considering the issues discussed above, it is proposed to impose a *linear* instead of an exponential decay proportional to $1 - k/(L+1)$, $k = 0, 1, \ldots, L$. This decay naturally reaches 0 at $k = L + 1$ where the impulse response vanishes anyway. This approach has the following benefits: (i) it allows for the larger deviation from the prior the earlier the coefficients are (the smaller k is) and (ii) no tuning parameter α is required—thus no user-specified fiddle parameter besides the choice of L and the prior order n exist. This linear decay approach is pursued in this section.

18.6.4.3 Order Selection

This section illustrates the sensitivity of OE and IRP FIR models w.r.t. the choice of model and prior order, respectively. Additionally, a new approach to sensitive order determination is introduced and compared the before mentioned models. For this purpose, one demonstration example shall be utilized. Very similar results have been obtained for other processes.

In the investigated setting, $N = 1000$ data points were generated by simulating a third-order process with unity gain, without zeros, and poles at $p_1 = 0.9$ and $p_{2/3} = 0.9\exp(\pm i\pi/4)$ excited by a PRBS. It is disturbed by additive white Gaussian output noise with a 30 dB signal-to-noise ratio. Three hundred noise realizations are averaged. The FIR model order (length of the impulse response) is fixed to $L = 50$. Model and prior orders range from 1 to 5.

The idea behind the proposed order selection procedure is as follows: In the IRP FIR approach, the optimal prior denominator coefficients of nth-order a_i, $i = 0, 1, \ldots, n-1$ correspond to n poles. From a performance point of view, it is not really important whether they are stable or physically plausible. However, from an interpretation point of view, this issue is decisive. For OE models, these "strange" poles do not arise since the simulation of the model during optimization would reveal the unstable or alternating characteristics.

It is proposed to penalize unstable poles p in the IRP FIR hyperparameter optimization with an additional term

$$J^{(\text{penalty})} = (1 - |p|)^2 \quad \text{if } |p| > 1 \tag{18.165}$$

and negative real poles by

$$J^{(\text{penalty})} = \text{Re}\{p\}^2 \quad \text{if } \text{Im}\{p\} = 0\,. \tag{18.166}$$

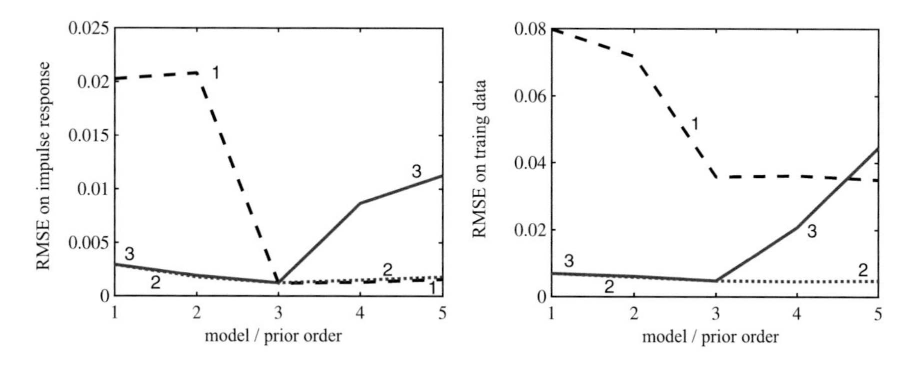

Fig. 18.44 (1) OE model, (2) regularized FIR model, (3) regularized FIR model with pole plausibility penalty

The subsequent results illustrate that such penalties have an overwhelming effect on the IRP FIR model performance if the model is over-parameterized, i.e., possesses a too large prior order. In contrast, without such penalties the IRP FIR model exhibits an extremely robust performance that is very insensitive w.r.t. the prior order.

Figure 18.44 compares the averaged results for the OE models (1) with orders 1–5 and two types of regularized FIR models with prior orders 1–5. The prior poles of one FIR model (2) were unconstrained optimized, while for the other (3) two penalties were added to the loss function for poles outside the unit circle and for negative, real poles that have no analog correspondence.

Figure 18.44(top) shows the error of the impulse responses of the models. In the correct order, all models perform best and equally well. However, the OE model (1) breaks down heavily for too low orders, while the penalized regularized FIR model (3) breaks down for too high orders. The regularized FIR model (2) exhibits a very graceful degradation w.r.t. wrong model orders. It is an extremely robust choice in practice when the model order is unknown and cannot be figured out appropriately with high reliability. For order selection, however, a very sensitive behavior w.r.t. the model order is a desirable property. Thus, (2) cannot be utilized for this task.

In a real-life scenario, the error on the impulse response is not available, of course. Usually, only the error on training data as shown in Fig. 18.44(bottom) is at the user's disposal if no separate validation data is available. Here, the OE model (1) and the standard regularized FIR model (2) keep improving with growing model orders while overfitting is increasing. This behavior is inconsistent with the true model performance in terms of impulse response quality. Only for the penalized regularized FIR model (3), the behavior for model quality and training error is consistent.

The best bias/variance tradeoff in most standard approaches shall be discovered by means of an information criterion like AIC or MDL. These try to predict the performance on test data by combining the training error with a complexity penalty. Depending on the specific information criterion applied, both models (1) and (2) could lead to various optimal model orders ≥ 3. It comes down to the very delicate question, whether improvement on training data is significant or not, i.e., larger or smaller than the complexity penalty.

A typical approach in practice in commercially available tools is to use information criteria like the AIC or MDL for order selection. They try to predict the performance of test data by combining the training error with a complexity penalty. Depending on the specific information criterion applied, the selected model orders of the OE model could be 3 or bigger and for the IRP FIR model somewhere close to 3. The decision usually is not distinct.

18.6.4.4 Consequences of Undermodeling

This section compares the model/prior orders 2 and 3 with each other in more detail for modeling the third-order process. Note that both types of FIR models with (2) and without (3) pole plausibility penalty exhibit almost identical behavior if model order $\leq$ process order but the penalized variant severely degrades if model order $>$ process order as demonstrated in Fig. 18.44.

Figure 18.45 compares for 300 noise realizations the poles of the OE models with the optimized prior poles of the IRP FIR models of orders 2 and 3, respectively. For order 3, most OE model poles almost coincide with the process poles (hard to see), and a few outlier models can be observed with almost random but stable pole configurations. The IRP FIR prior poles also concentrate close around the true process poles but scatter more significantly. Thus in most cases, the pole accuracy is inferior to the OE model. On the other hand, no outlier models exist. Therefore the reliability of the IRP FIR model is superior.

For order 2, the OE model performs terribly. In all cases, two real poles were optimized. Thus, these OE models are only capable of describing the time constants of the process. In contrast, all IRP FIR model priors possess a conjugate complex pole pair being capable of roughly describing the oscillatory behavior of the process. The LS estimation of the FIR coefficients obviously easily compensates for the deviations due to the lacking real pole.

Figure 18.46 shows typical representatives, taken from Fig. 18.45 (right), of second-order OE and regularized FIR impulse response models of the third-order demonstration process. The OE model possesses two real poles and thus is incapable of describing the oscillatory behavior. The regularized FIR model's prior even has an unstable conjugate complex pole pair. Note, however, that the model performs excellently and itself is inherently stable, of course.

18.6.4.5 Summary

Via the filter interpretation, a new class of regularized FIR models has been developed which preserve the impulse response (IRP) of a prior system of order n. With this approach, it is possible to combine the most advantages of OE models with

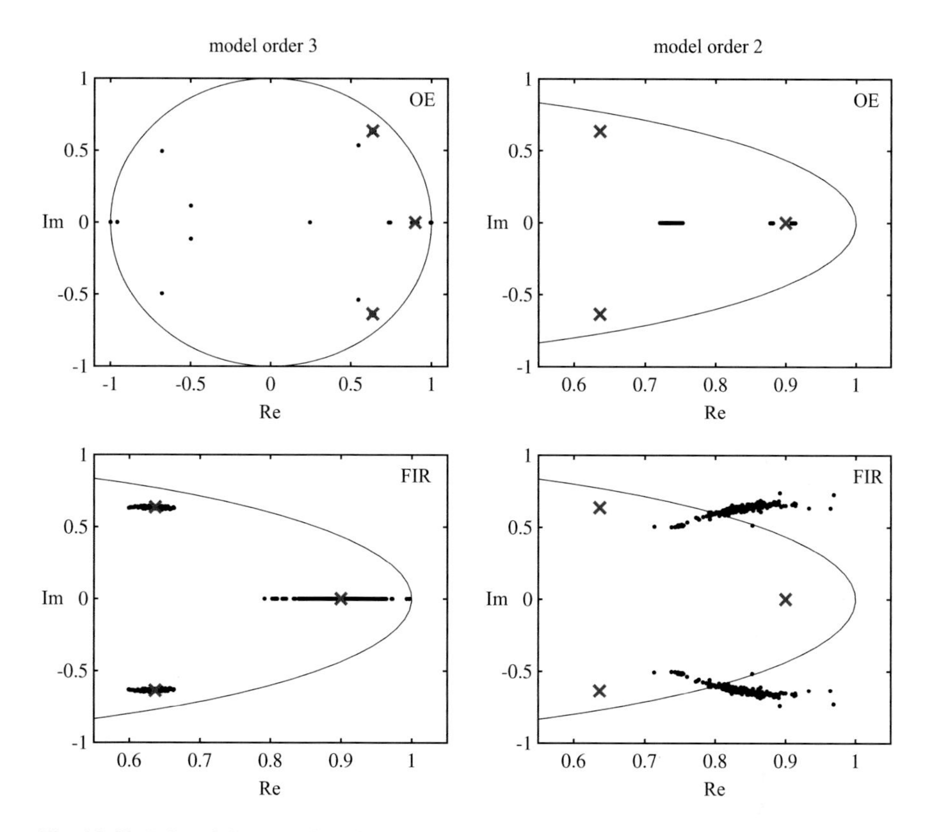

Fig. 18.45 Poles of the second- and third-order OE and unpenalized regularized FIR models for the 300 noise realizations. The process poles are shown as crosses and the model poles as dots

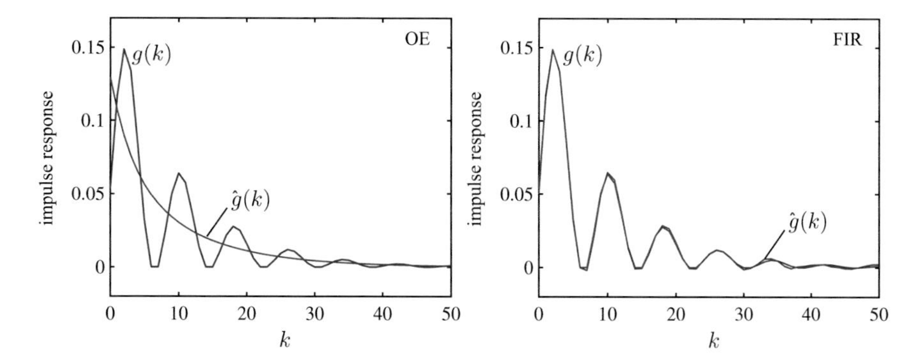

Fig. 18.46 Typical impulse responses for an OE model of order 2 (left) with poles at $p_1 = 0.91$ and $p_2 = 0.73$ and a regularized IRP FIR model with prior order 2 (right) with poles at $p_{1/2} = 0.83 \pm i0.64$, ($|p_{1/2}| = 1.04$)

the advantages of regularized FIR models. It has been shown that the IRP FIR model is very insensitive w.r.t. the choice of the prior order. This is a big plus for robust system identification in practice. Furthermore, it was shown that it is possible to extract poles with higher reliability than with OE models. This feature significantly emphasizes if the model/prior order is incorrect.

Order selection can be carried out in a very sensitive way by penalizing physically implausible poles. Then the performance of such a penalized IRP FIR model breaks down in the over-modeling case (too high orders).

In practice, it is advisable to employ regularized FIR models with the proposed IRP approach for modeling and potentially utilize the pole plausibility penalty for optimal structure determination of the prior before.

18.6.5 Orthonormal Basis Functions (OBF)

Orthonormal basis functions (OBF) models can be seen as a generalization of the FIR model. Alternatively the FIR model is a special type of OBF model. In order to illustrate this relationship, the FIR model in (18.133) can be written in the following form:

$$y(k) \;=\; b_1 q^{-1} u(k) + b_2 q^{-2} u(k) + \ldots + b_m q^{-m} u(k) \;+\; v(k)\,. \qquad (18.167)$$

Thus, the model output can be seen as a linear combination of filtered versions of the actual input $u(k)$, where the filters are $q^{-1}, q^{-2}, \ldots, q^{-m}$, respectively. These filters have all their poles at zero, that is, in the exact center of the unit disk. Because the impulse responses of these filters (see Fig. 18.47) are orthonormal, this is the simplest form of an orthonormal basis function model.

Two signals $u_1(k)$ and $u_2(k)$ in discrete time are said to be *orthogonal* if

$$\sum_{k=-\infty}^{\infty} u_1(k) u_2(k) = 0 \qquad \text{and} \qquad (18.168a)$$

$$\sum_{k=-\infty}^{\infty} u_1^2(k) = \text{constant} \qquad \text{and} \qquad \sum_{k=-\infty}^{\infty} u_2^2(k) = \text{constant}\,. \qquad (18.168b)$$

Two signal $u_1(k)$ and $u_2(k)$ in discrete time are said to be *orthonormal* if

$$\sum_{k=-\infty}^{\infty} u_1(k) u_2(k) = 0 \qquad \text{and} \qquad (18.169a)$$

$$\sum_{k=-\infty}^{\infty} u_1^2(k) = 1 \qquad \text{and} \qquad \sum_{k=-\infty}^{\infty} u_2^2(k) = 1\,. \qquad (18.169b)$$

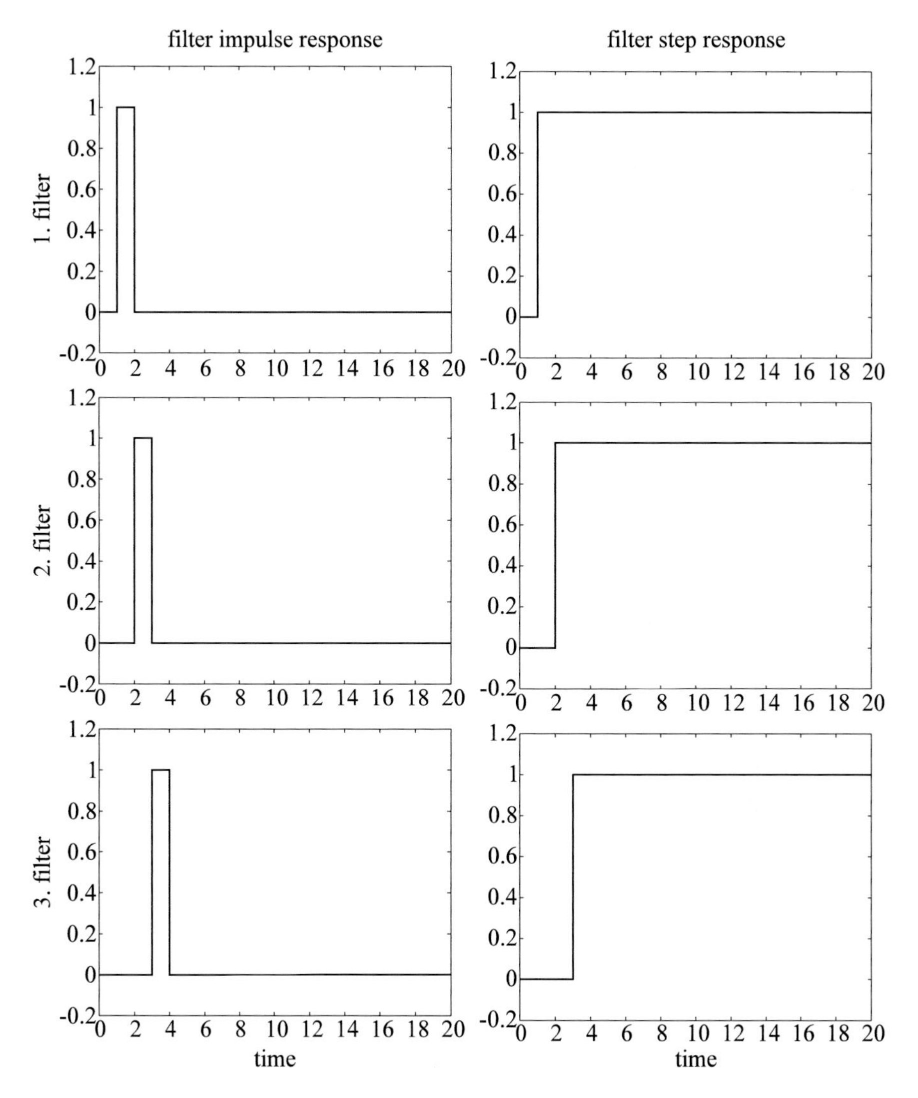

Fig. 18.47 Impulse and step responses of the first three orthonormal filters used for the FIR model

In the time domain, the output signal $y(k)$ can be represented as a weighted (with the b_i) sum of these basis functions, namely, the filters' impulse responses. In the FIR model, this weighted sum is especially simple because the basis functions are Dirac impulses and consequently do not overlap. In the previous subsection, it was discussed that the order m of FIR models must be chosen very high. In the light of OBFs, there exists a new interpretation of this fact. The basis functions of the FIR model are delayed Dirac impulses. This may be not a very realistic description of the expected output. If other, more realistic, basis functions can be chosen, it

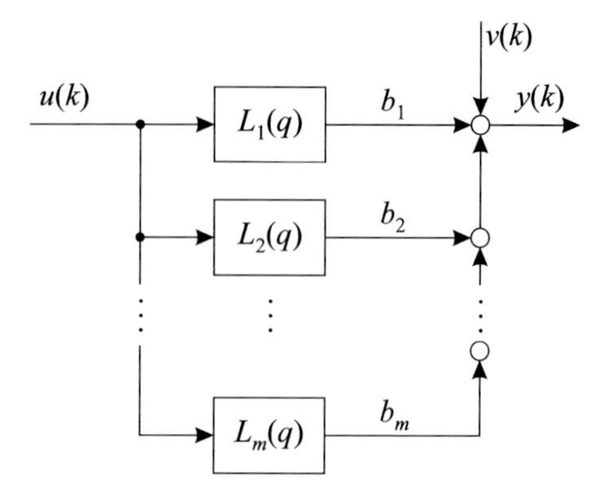

Fig. 18.48 OBF model

could be expected that fewer basis functions would be required for a satisfactory approximation. So, in the OBF model, the trivial filters q^{-i} are replaced by more general and complex orthonormal filters $L_i(q)$. The OBF model consequently becomes (see Fig. 18.48)

$$y(k) = b_1 L_1(q)u(k) + b_2 L_2(q)u(k) + \ldots + b_m L_m(q)u(k) + v(k) \qquad (18.170)$$

with the orthonormal filters $L_1(q), L_2(q), \ldots, L_m(q)$. The OBF model in (18.170) is an approximation of the following series expansion (see (18.136)):

$$y(k) = \sum_{i=1}^{\infty} g_i L_i(q)u(k) . \qquad (18.171)$$

The goal is to find filters $L_i(q)$ that yield fast converging coefficients g_i, so that this infinite series expansion can be approximated to a required degree of accuracy by (18.170) with an order m as small as possible.

The FIR model is retained for $L_1(q) = q^{-1}$, $L_2(q) = q^{-2}, \ldots, L_m(q) = q^{-m}$. The choice of the filters $L_i(q)$ is done a priori, i.e., before the b_i parameters are estimated. So the OBF model stays linear in the parameters. The choice of $L_i(q)$ can be seen as the incorporation of *prior knowledge*. For example, the choice of the FIR basis functions is optimal if nothing about the process dynamics is known but its stability. If a process is stable, its poles can lie anywhere within the unit disk. Then the FIR model filters q^{-i} with poles at zero are (in the mean) the best choice. However, often more information about the process is available. From step responses of the process, it is usually known if it exhibits oscillatory or aperiodic behavior. Additionally, some rough knowledge about the time constants of the process is typically available. (Otherwise, there would not even be enough information about how to choose the sampling rate.) This prior knowledge about the approximate process dynamics can be incorporated into the $L_i(q)$ filters. The more precise the knowledge is, the higher is the accuracy that can be expected from the approximation, and consequently the order m can be decreased. Thus, the main drawback of the FIR approach can be overcome with the help of prior knowledge about the process dynamics. It can be shown [575] that the quality of the OBF model increases rapidly as the dynamics built into the $L_i(q)$ filters approaches the true process dynamics.

The following paragraphs discuss different choices of the $L_i(q)$ filters. *Laguerre* filters allow the incorporation of one real pole for processes with aperiodic behavior, *Kautz* filters allow the incorporation of one conjugate complex pole pair for processes with oscillatory behavior, and finally the generalized OBF approach includes the Laguerre and Kautz approaches as special cases and allows the incorporation of an arbitrary number of real and conjugate complex poles. The following possible sources of prior knowledge about the approximate process dynamics exist:

- *First principles*: Fundamental analysis of laws or experience of experts usually allow one to estimate an upper and lower bound on the major time constant.
- *Step responses*: Even if the order of the process is not known, a step response gives a rough approximation of the dominant time constant and of the damping if the process has oscillatory behavior.
- *Correlation analysis*: The impulse response can be recovered in quite good quality with correlation methods [278]. This allows a good approximation of the major time constant of the process.
- *Previous identification of a model with output feedback*: In a first step, a model with output feedback, say an ARX or OE model, as discussed in Sect. 18.5 may be estimated from data. Then in a subsequent step the poles (roots of the $A(q)$ or $F(q)$ polynomials, respectively) can be estimated. Finally, these poles can be incorporated into an OBF model.

The last alternative may appear a little far fetched at first sight. What would be the advantage of transforming an ARX model into an OBF model? Clearly, this increases the effort (by a factor of about 2, depending on the order of the ARX and OBF model). One advantage is that the OBF model belongs to the output error class, and consistent parameter estimation of the OBF model may be possible, while the ARX model parameters are biased. Second, the OBF models are infinite impulse response models without output feedback. In contrast to all models described in Sect. 18.5, for simulation OBFs models feed back not the model output $\hat{y}$ but the individual filter outputs. This property becomes important when the linear model structures are generalized to nonlinear ones. Then OBF models have a significant advantage compared with output feedback models with respect to their stability properties; see Chap. 19.

18.6.5.1 Laguerre Filters

The term "Laguerre filter" stems from the fact that in continuous time the impulse responses of Laguerre filters are equal to the orthonormal Laguerre polynomials. Laguerre filters allow the incorporationof knowledge about one real pole. Consequently, they are suited for the identification of well-damped processes. The first filter $L_1(q)$ is a simple first-order time lag system with a pole at p:

$$L_1(q, p) = \frac{\sqrt{1 - p^2}}{q - p}. \tag{18.172}$$

This real pole p is the only degree of freedom of Laguerre filters. The higher-order filters are generated by cascades of the following all-pass filter with a pole at p and a zero at $1/p$:

$$\frac{1 - pq}{q - p}. \tag{18.173}$$

So, the ith Laguerre filter is computed by

$$L_i(q, p) = \frac{\sqrt{1 - p^2}}{q - p} \left(\frac{1 - pq}{q - p} \right)^{i-1}. \tag{18.174}$$

Alternatively, the ith Laguerre filter can be computed recursively by

$$L_i(q, p) = \frac{1 - pq}{q - p} L_{i-1}(q). \tag{18.175}$$

A scheme of the OBF model with Laguerre filters is depicted in Fig. 18.49. Note that for the pole $p = 0$ the Laguerre filters simplify to $L_i(q) = q^{-i}$ and the FIR

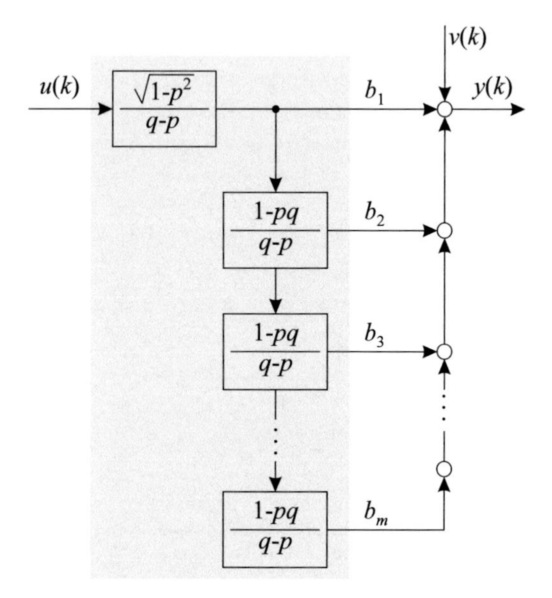

Fig. 18.49 OBF model with Laguerre filters that have a real pole at p

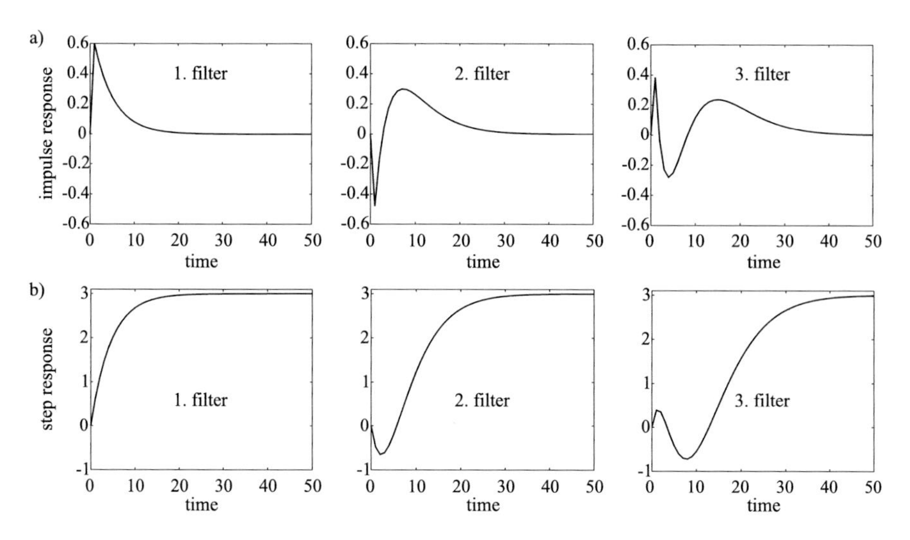

Fig. 18.50 Impulse (**a**) and step (**b**) responses of the first three Laguerre filters with the pole $p = 0.8$

model is recovered. The impulse and step responses for the first three Laguerre filters $L_1(q, p)$, $L_2(q, p)$, and $L_3(q, p)$ with a pole p at 0.8 are shown in Fig. 18.50.

The estimation of the parameters of a Laguerre OBF model works basically as for the FIR model. The regressors (columns in $\underline{X}$) in (18.137) are the input $u(k)$ filtered with q^{-i}. It is clear from Fig. 18.49 that for the estimation of a Laguerre OBF model, these regressors are simply the input $u(k)$ filtered with the more complex filters $L_i(q)$ in (18.174). Thus, the regression matrix for an OBF model is

$$\underline{X} = \begin{bmatrix} u^{L_1}(m+1) & u^{L_2}(m+1) & \cdots & u^{L_m}(m+1) \\ u^{L_1}(m+2) & u^{L_2}(m+2) & \cdots & u^{L_m}(m+2) \\ \vdots & \vdots & & \vdots \\ u^{L_1}(N) & u^{L_2}(N) & \cdots & u^{L_m}(N) \end{bmatrix}, \tag{18.176}$$

where $u^{L_i}(k) = L_i(q, p)u(k)$ are input $u(k)$ filtered with the corresponding orthonormal filters $L_i(q, p)$. *The parameter estimation according to* (18.176) *is directly applicable to all types of OBF models.* The Laguerre filters $L_i(q, p)$ simply have to be replaced by Kautz or generalized orthonormal filters. For a thorough theoretical analysis of the approximation behavior of Laguerre filters, refer to [583].

18.6.5.2 Poisson Filters

One might be irritated by the *non-minimum phase* behavior introduced by the all-pass filters in the Laguerre expansion. Notice that orthogonality or even orthonormality of the filter impulse responses is no big advantage nowadays. It offers the advantage of yielding a diagonal $\underline{X}^T \underline{X}$ for an impulse or white input signal, and in older times the parameters could be determined without matrix inversion—even by an analog circuit. These days it is no problem to compute $(\underline{X}^T \underline{X})^{-1}$ even if $\underline{X}^T \underline{X}$ is arbitrary. This reduces the benefits of orthogonality or orthonormality to better parameter variances and covariances (see (3.34) in Sect. 3.1.1) and to the numerical advantages of well-conditioned matrices. Of course, these still are desirable benefits.

In principle, it is possible to replace the Laguerre filters by just a series of first-order time lag systems:

$$L_i(q, p) = \frac{1}{q-p} \left(\frac{q}{q-p} \right)^{i-1} . \tag{18.177}$$

These are called *Poisson filters*. Note that models based on these Poisson filters have a 1:1 mapping to Laguerre filter models; just the associated parameter values are different. For example, the equivalence can be seen for a second-order filter by

$$b_1 \frac{\sqrt{1-p^2}}{q-p} + b_2 \frac{\sqrt{1-p^2}}{q-p} \frac{1-pq}{q-p} \stackrel{!}{=} \widetilde{b}_1 \frac{1}{q-p} + \widetilde{b}_2 \frac{1}{q-p} \frac{q}{q-p} \tag{18.178}$$

where $\widetilde{b}_1$ and $\widetilde{b}_2$ can be calculated from b_1 and b_2 and vice versa. Both approaches in (18.178) (Poisson and Laguerre) represent a second-order system with a double pole at p and a zero and gain that depend on the two parameters. Thus, if the parameters b_1 and b_2 or $\widetilde{b}_1$ and $\widetilde{b}_2$ are estimated anyway, only a difference in parameter variances/covariances and numerics exists.

Note that this equivalence between Laguerre and Poisson filters only holds for *linear* considerations. If nonlinear relationships come into play, like feeding a nonlinear approximator with the filter outputs (NOBF), significant differences can occur. For example, the data distribution in the input space (spanned by the filter outputs) will be very different, typically much wider and with better coverage for Laguerre filters.

18.6.5.3 Kautz Filters

If the process possesses weakly damped oscillatoric behavior, any Laguerre filter-based approach will require a large number of parameters. The required number of basis functions will be high since all real poles are far away from the weakly damped conjugate complex poles that describe the process. Kautz filters can be seen as an extension of Laguerre filters and allow the incorporation of knowledge on one conjugate complex pole pair. Consequently, they are well suited for the identification of resonant processes. The first two Kautz filters $L_1(q)$ and $L_2(q)$ can be calculated by

$$L_1(q, a, b) = \frac{\sqrt{(1 - a^2)(1 - b^2)}}{q^2 + a(b - 1)q - b} \tag{18.179a}$$

$$L_2(q, a, b) = \frac{\sqrt{(1 - b^2)}(q - a)}{q^2 + a(b - 1)q - b} \tag{18.179b}$$

with $-1 < a < 1$ and $-1 < b < 1$. The real and imaginary parts of the conjugate complex pole $p_{1/2} = p_r \pm ip_i$ are related to the Kautz filter coefficients a and b as follows:

$$a = \frac{2p_r}{1 + p_r^2 + p_i^2} \quad \text{and} \quad b = -(p_r^2 + p_i^2)\,. \tag{18.180}$$

This conjugate complex pole pair $p_{1/2} = p_r \pm ip_i$ are the only degrees of freedom of the Kautz filters. The higher-order filters are generated by cascading all-pass filter with a pole pair at $p_{1/2}$ and zeros at $1/p_{1/2}$:

$$L_{2i-1}(q, a, b) = \frac{\sqrt{(1 - a^2)(1 - b^2)}}{q^2 + a(b - 1)q - b} \left(\frac{-bq^2 + a(b - 1)q + 1}{q^2 + a(b - 1)q - b} \right)^{i-1} \tag{18.181a}$$

$$L_{2i}(q, a, b) = \frac{\sqrt{(1 - b^2)}(q - a)}{q^2 + a(b - 1)q - b} \left(\frac{-bq^2 + a(b - 1)q + 1}{q^2 + a(b - 1)q - b} \right)^{i-1} . \tag{18.181b}$$

Alternatively, the Kautz filters can be computed recursively by

$$L_{2i-1}(q, a, b) = L_{2(i-1)-1}(q, a, b) \left(\frac{-bq^2 + a(b - 1)q + 1}{q^2 + a(b - 1)q - b} \right) \tag{18.182a}$$

$$L_{2i}(q, a, b) = L_{2(i-1)}(q, a, b) \left(\frac{-bq^2 + a(b - 1)q + 1}{q^2 + a(b - 1)q - b} \right) . \tag{18.182b}$$

A scheme of the OBF model with Kautz filters is depicted in Fig. 18.51. The impulse and step responses for the first three Kautz filters $L_1(q, a, b)$, $L_2(q, a, b)$, and $L_3(q, a, b)$ with $a = 0.70$ and $b = -0.72$, which corresponds to a conjugate complex pole pair at $p_{1/2} = 0.6 \pm \mathrm{i}0.6$, are shown in Fig. 18.52.

For a thorough theoretical analysis of the approximation behavior of Kautz filters, refer to [584].

18.6.5.4 Generalized Filters

Although for many applications Laguerre and Kautz filter-based OBF models may be sufficiently accurate, high-order processes can require the consideration of more than just one pole or pole pair. However, note that in principle all OBF models (including FIR) are able to model *all* stable linear systems independent of their

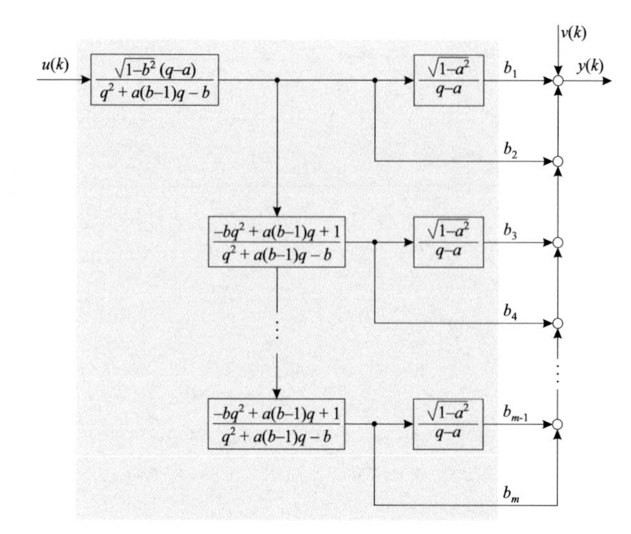

Fig. 18.51 OBF model with Kautz filters, which have a conjugate complex pole pair at $p_{1/2} = p_r \pm i p_i$

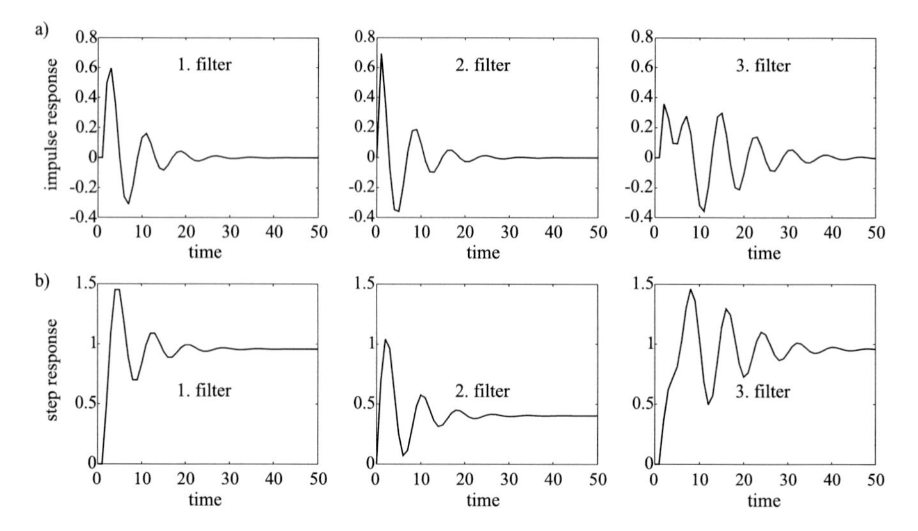

Fig. 18.52 Impulse (**a**) and step (**b**) responses of the first three Kautz filters with poles $p_{1/2} = 0.6 \pm \mathrm{i}0.6$

dynamic order. Laguerre and Kautz filter-based approaches are not only suitable for first- and second-order systems. Nevertheless, for processes with many distinct poles, the necessary model order m may become infeasible in practice for FIR, Laguerre, and Kautz OBF models. In these cases, it is useful to exploit information on more than one pole (pair) in order to build a good model.

The simplest and most straightforward strategy is to add Laguerre and Kautz models for different poles [352]:

$$y(k) = \sum_{l=1}^{l_L} \sum_{i=1}^{m_l^{(L)}} b_{l,i}^{(L)} L_{l,i}(q, p_l) u(k) + \sum_{l=1}^{l_K} \sum_{i=1}^{m_l^{(K)}} b_{l,i}^{(K)} L_{l,i}(q, a_l, b_l) u(k) , \tag{18.183}$$

where l runs over the l_L Laguerre and l_K Kautz models that represent different dynamics, $m_l^{(L)}$ and $m_l^{(K)}$ are the orders of these models, and $b_{l,i}^{(L)}$ and $b_{l,i}^{(K)}$ are the corresponding linear parameters. The drawback of this approach is that by using several OBF models in parallel, the basis functions of the overall model in (18.183) are no longer orthonormal.

An alternative approach is taken in [245, 246, 575] where the OBF models are extended. These generalized OBF models allow the incorporation of knowledge on an arbitrary number of real poles and conjugate complex pole pairs. The Laguerre and Kautz filters represent special cases of this approach.

18.6.6 Simulation Example

Consider the following third-order process with gain $K = 1$ and time constants $T_1 = 20$ s, $T_2 = 10$ s, and $T_3 = 5$ s:

$$G(s) = \frac{1}{(20\,s + 1)(10\,s + 1)(5\,s + 1)} \tag{18.184}$$

sampled with $T_0 = 1$ s. As excitation signal a PRBS 255 samples long is used, which excites the whole frequency range equally well. The goal of this example is to illustrate the functioning of Laguerre OBF models and to compare them with ARX and OE models.

The discrete-time process is identified with a Laguerre model with $m = 10$ filters whose time constant T is chosen equal to 10 s (corresponding to a pole at $p = 0.9048$). The optimal parameters b_i ($i = 1, \ldots, 10$) are shown in Fig. 18.53a. Obviously, the series expansion converges exponentially; the influence of the higher-order filters becomes insignificant.

Figure 18.53b depicts the step responses of the process and a first-, second-, and third-order Laguerre model. The second-order model already captures the main dynamics of the process although it exhibits non-minimum phase behavior. The third-order model is minimum phase and has only a slight d.c. error and a negligible error in the slow dynamics range. Figure 18.54(left) illustrates the model fit for these

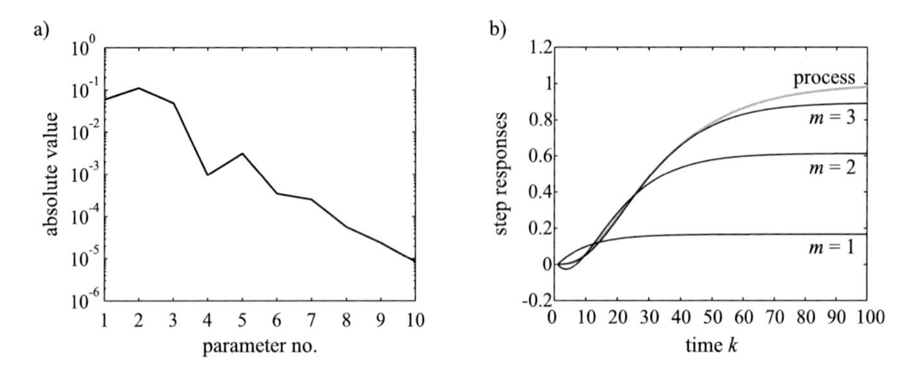

Fig. 18.53 (**a**) Optimal b_i parameter values of a tenth-order Laguerre model with time constant $T = 10$ s. (**b**) Step responses of the process and Laguerre models of first, second, and third order ($T = 10$ s)

three Laguerre models in the frequency domain. Obviously, the low frequencies are emphasized corresponding to the prior knowledge (or assumption) built into the model that the process time constant is close to $T = 10$ s. The major approximation error is in the high-frequency range, and it decreases as the order m of the Laguerre model increases.

An important issue for OBF models is how sensitive the obtained results are on the chosen filter parameters, e.g., the time constant T (or equivalently the pole p) for Laguerre models. Figure 18.54 (right) demonstrates for sixth-order Laguerre models with different time constants that the performance deterioration is moderate, and consequently, a very approximate choice of T is sufficient. No effort has been made to determine the optimal value for T. Furthermore, a poorly chosen time constant can always be compensated by selecting a higher model order. Of course, one will face severe problems due to a high variance error with model orders that are too high, which in practice restricts the required accuracy on T.

The sixth-order Laguerre model with $T = 10$ s will be compared with an ARX and an OE model of correct (i.e., third) order. For an illustration of reduced-order ARX and OE models, refer to the example in Sect. 18.5.7. Note that all models have six parameters to be estimated. While the ARX and Laguerre models are linearly parameterized, the OE model is nonlinear in its parameters, and thus it is much harder to identify since it is computationally more demanding and possibly difficulties with local optima may arise. If the data is not disturbed by noise, both ARX and OE models yield exactly the process, while the Laguerre model possesses some small approximation error; see Fig. 18.54 (right, center).

Figure 18.55 compares the ARX, OE, and Laguerre models for weakly and strongly disturbed processes. For the weak disturbance, the signal-to-noise amplitude ratio is chosen equal to 1000; for the strong disturbance, it is 10. The noise is chosen white, and it is added to the process output. Although for the weakly disturbed case all three models look very good (owing to the low resolution), a closer examination of the ARX model's frequency characteristics reveals significant

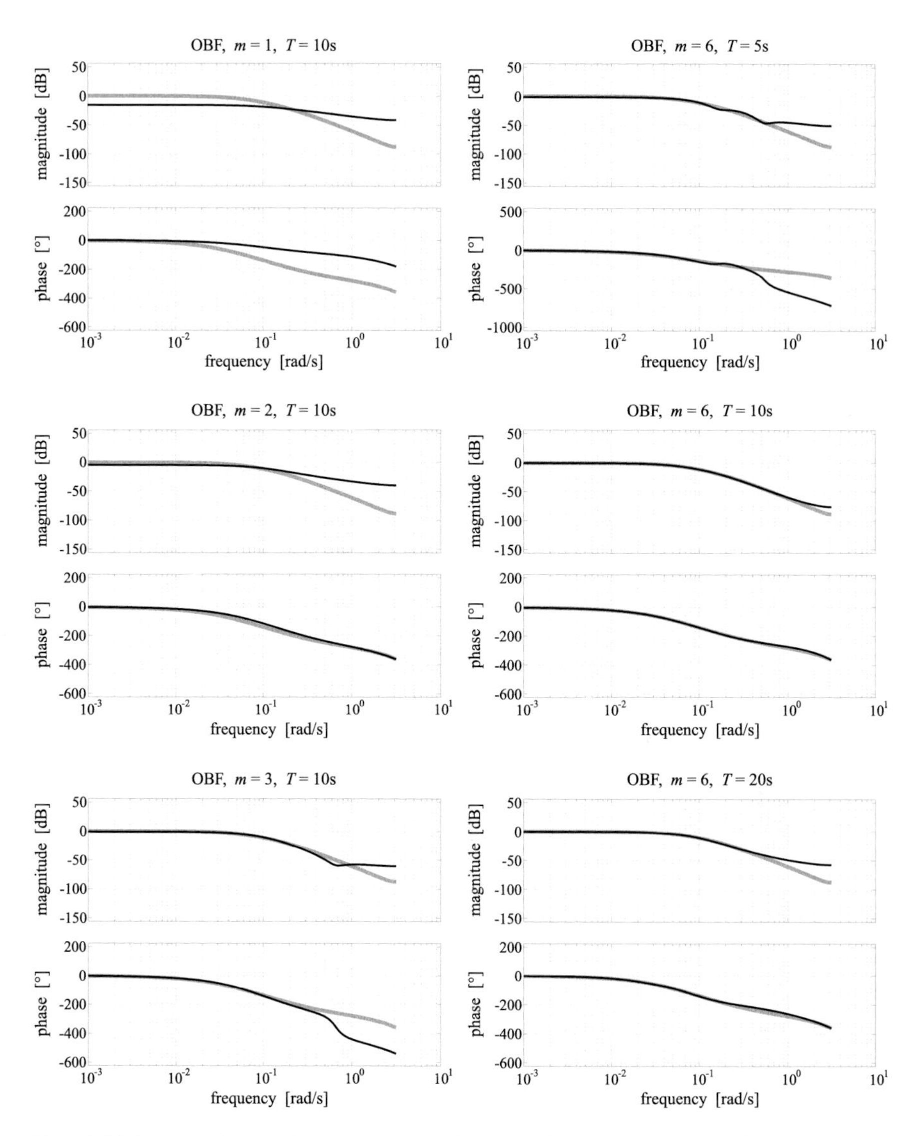

Fig. 18.54 Left: Bode plot of the process (gray) and the first-, second-, and third-order Laguerre models with time constant $T = 10$ s. Right: Bode plot of the process (gray) and sixth-order Laguerre models with time constants $T = 5, 10, 20$ s

errors in the low- and medium-frequency range. The step responses shown in Fig. 18.56(left) confirm this observation. In fact, the gain of the ARX model is 20% inaccurate, and (strongly damped) conjugate complex poles are estimated. In contrast, the OE and Laguerre models perform very well. The reason for the poor quality of the ARX model is its *biased* parameter estimates in the presence of the white disturbance; see Sect. 18.5.1.

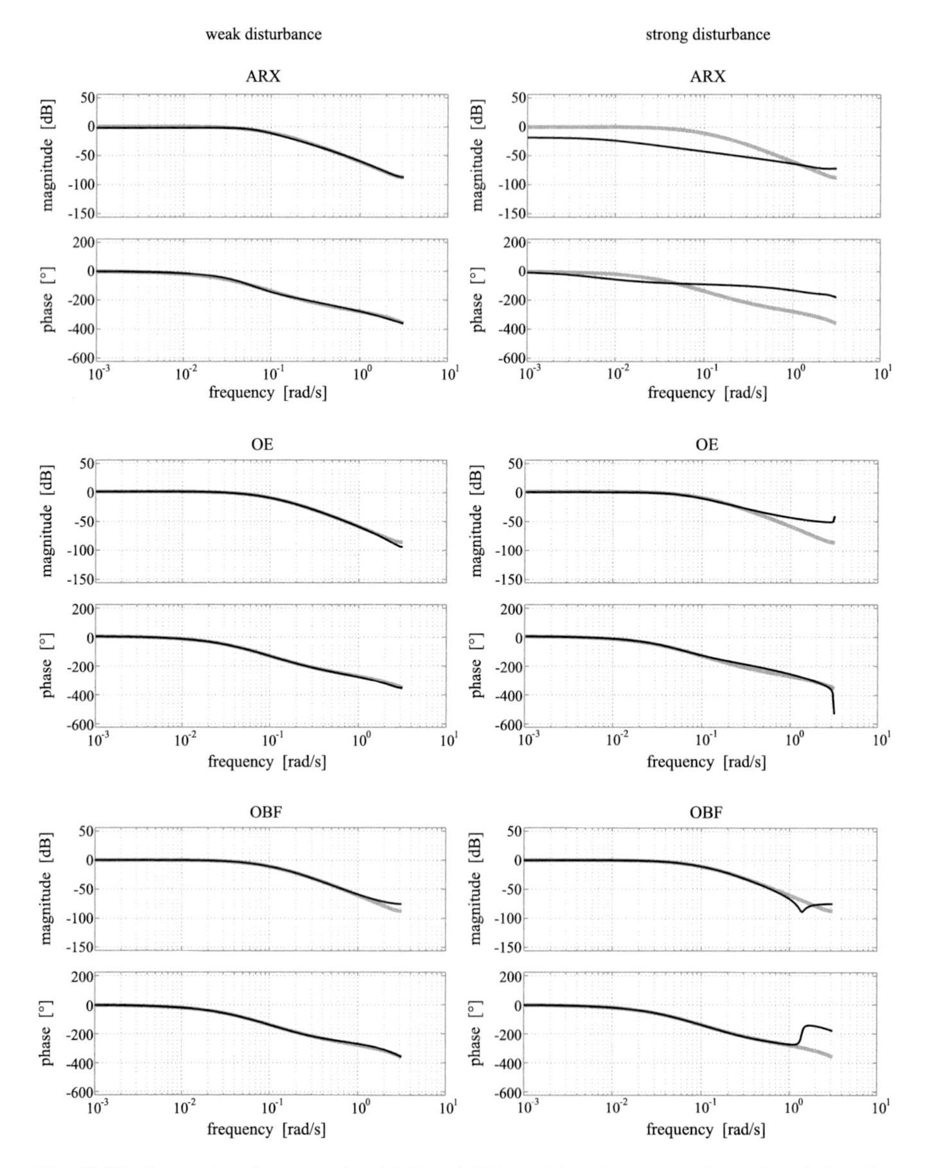

Fig. 18.55 Comparison between the ARX and OE models of correct order and a sixth-order Laguerre model with $T = 10$ s for weak (left) and strong (right) disturbances in the frequency domain. The gray lines represent the process; the black lines represent the models

For the strongly disturbed case, the ARX model yields totally unacceptable results. Interestingly, the Laguerre model performs better than the OE model. Of course, these results depend on a reasonably good choice for the Laguerre model's time constant, and in practice, the OE model may be superior if little knowledge about the process dynamics is available. Furthermore, significant improvements in

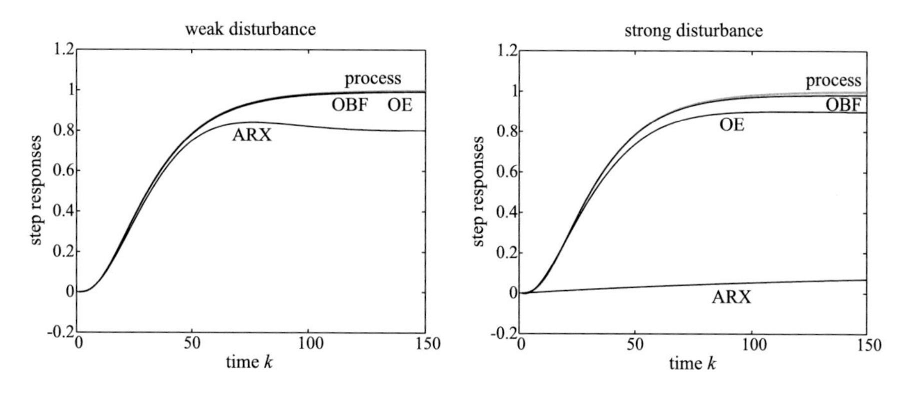

Fig. 18.56 Comparison between step responses of an ARX and OE model of correct order and a sixth-order Laguerre model with $T = 10$ s for weak (left) and strong (right) disturbances

the ARX model results can be achieved by filtering the data.[5] Nevertheless, it is impressive to observe that a linearly parameterized OBF model can perform better than an OE model with the correct model order. In practice, the OE model will be of a lower-order than the process, and this introduces an additional bias error. In this light, the approximation error introduced by a finite Laguerre series seems to be only a slight disadvantage.

18.7 Some Advanced Aspects

This section briefly addresses some more advanced aspects that have been omitted in the previous sections for the sake of simplicity.

18.7.1 Initial Conditions

In practice, only a finite amount of data is available, say for $k = 1, \ldots, N$. For a *simulation* of a model with infinite impulse response, however, all previous values back to minus infinity are (theoretically) required:

$$\hat{y}(k) = G(q)u(k) \tag{18.185}$$

[5]Filtering requires the choice of a filter bandwidth, and with this choice, certain frequency ranges are emphasized in the model fit; see Sect. 18.7.4. Note that a reasonable choice of the filter bandwidth also requires prior knowledge of the process dynamics.

or

$$\hat{y}(k) = \sum_{i=1}^{\infty} g_i u(k-i) \,. \tag{18.186}$$

The easiest solution to this problem is to assume all unknown data for $k = -\infty, \ldots, 0$ to be equal to zero:

$$\hat{y}(k) \approx \sum_{i=1}^{k-1} g_i u(k-i) \,. \tag{18.187}$$

Obviously, these unknown initial conditions degrade the simulation performance. However, if the model is stable, the initial conditions decay exponentially with time k. Thus, $\hat{y}(k)$ is reasonably accurate for $k > 3T/T_0$, where T is the dominating time constant of the model and T_0 is the sampling time.

The same difficulties with the initial conditions can occur for *prediction*. The optimal predictor is (see (18.20) and [356])

$$\hat{y}(k|k-1) = S(q)u(k) + T(q)y(k) \tag{18.188}$$

or

$$\hat{y}(k|k-1) = \sum_{i=1}^{\infty} s_i u(k-i) + \sum_{i=1}^{\infty} t_i y(k-i) \tag{18.189}$$

with $S(q) = G(q)/H(q)$ and $T(q) = 1 - 1/H(q)$. Again, by summing up to $k - 1$ instead of ∞, (18.189) can be evaluated approximately in practice. This approximation may become too inaccurate if the available data set is very short, i.e., not significantly longer than $3T/T_0$ samples. Then it might be worth performing the prediction with the optimal time-variant predictor that takes into account the uncertainty of the (assumed) initial conditions. This optimal time-variant predictor is realized by the Kalman filter and converges to the optimal time-invariant predictor (18.189) as $k \to \infty$; see Sect. 3.2.3 and [356] for more details.

Of course, the inaccuracies caused by the unknown initial values transfer to the parameter estimation utilizing the prediction error method. Special cases are the ARX and ARARX model structures. Their predictor transfer functions $S(q)$ and $T(q)$ possess only a numerator polynomial. For example, the ARX predictor is (the ARARX predictor is simply multiplied with $D(q)$; see (18.49))

$$\hat{y}(k|k-1) = B(q)u(k) + (1 - A(q))y(k) \tag{18.190}$$

or

$$\hat{y}(k|k-1) = \sum_{i=1}^{m} b_i u(k-i) + \sum_{i=1}^{m} -a_i y(k-i) . \qquad (18.191)$$

Consequently, $S(q) = B(q)$ and $T(q) = 1 - A(q)$ are finite impulse response filters, and the optimal predictor (18.189) can be calculated *exactly* since the sums run only up to the model order m. Therefore, for these model structures, the initial conditions can be described exactly by omitting the first m samples in the parameter estimation. This idea has been pursued in the LS estimation of an ARX model in (18.54); see Sect. 18.5.1. The equations start with $k = m+1$ since the first m samples $k = 1, \ldots, m$ are required to determine the initial values for $u(k)$ and $y(k)$. Note again that this approach is feasible only if the data set contains significantly more than m samples. For all other model structures, the effect of the initial conditions on the parameter estimates can be neglected by ignoring the first $3T/T_0$ data samples in the loss function.

The issue of initial conditions arises quite often in practice when different data sets are merged together. Usually, owing to several restrictions such as limited available time or memory, it is not possible to measure all required data in a single measurement. Typically, different measurements are tied together offline, and jumps can occur at the transitions. To avoid any difficulties, it is advisable to start and end each measurement with a defined constant operating condition. Otherwise, the first part of each data set has to be "wasted" for the adjustment to the new initial conditions.

18.7.2 Consistency

For consistency of the parameter estimates, the transfer function model $\hat{G}(q)$ must in principle be able to describe the true process $G(q)$. This means that the model has to be flexible "enough"; in other words $\hat{G}(q)$ has to be of a sufficiently high order. In the consistency analysis, it is assumed that this condition is fulfilled. Otherwise, the best one can hope for is a good *approximation* of the process behavior with a model that is "too simple."

In Sect. 18.5.1 the consistency problem for ARX models has been analyzed. The parameters or the ARX model is estimated consistently (see Sect. B.7) only if the true measurement noise is properly modeled by the noise model $1/A(q)$. This restriction is rarely fulfilled in practice. If the noise model structure is correct, i.e., if it is capable of describing the real measurement noise, then all models discussed here are estimated consistently. However, since reliable knowledge about the noise properties is rarely available in practice, the question arises: Which model structures allow consistent estimation of their parameters even if the noise model structure does not match the reality?

All model structures that have an independently parameterized transfer function $G(q)$ and noise model $H(q)$ allow one to estimate the parameters of the transfer function consistently even if the noise model is not appropriate [356]. "Independently parameterized" means that $G(q)$ and $H(q)$ do not contain common parameters. The class of output error models is independently parameterized, e.g., OE and BJ, while the class of equation error models, e.g., ARX and ARMAX, is *not* since $G(q)$ and $H(q)$ share the polynomial $A(q)$. This is a fundamental advantage of OE and BJ models over ARX and ARMAX models. Unfortunately, the parameters for OE and BJ models are more difficult to estimate than those for ARX and ARMAX models; see Sect. 18.5.

18.7.3 Frequency-Domain Interpretation

When a model is fitted to data by minimizing a quadratic loss function with respect to the model parameters, the following question arises: How well can the model be expected to describe the process in different frequency ranges? An answer to this question should allow the user to design a model that is better suited for the intended application. For example, a model utilized for controller design should be especially accurate around the crossover frequency, while low-frequency modeling errors can be compensated by the integrator in the controller, and high-frequency modeling errors are usually less significant since the typical low-pass characteristic damps these errors highly anyway. For other applications such as fault detection, the model error at low frequencies may be more significant.

In order to understand how the expected model quality depends on the frequency, a frequency-domain expression of the loss function has to be derived. The loss function $I(\underline{\theta})$ is related to the spectrum of the prediction error Φ_e by the inverse z-transform [356]

$$\mathrm{E}\{I(\underline{\theta})\} = \frac{1}{4\pi}\int_{-\pi}^{\pi} \Phi_e(\omega, \underline{\theta})\, d\omega\,. \tag{18.192}$$

The spectrum of the prediction error $\Phi_e(\omega, \underline{\theta})$ describes the prediction error in the frequency domain. Under ideal conditions, that is, the model (the transfer function $G(q)$ and the noise model $H(q)$) matches the process perfectly, this spectrum is white, i.e., $\Phi_e(\omega, \underline{\theta}) = \text{constant}$.

By substituting the prediction error in (18.192) with (18.31), the expectation of the loss function can be expressed as [356]

$$\begin{aligned}\mathrm{E}\{I(\underline{\theta})\} = \frac{1}{4\pi}\int_{-\pi}^{\pi} & \Big(|G(e^{j\omega}) - \hat{G}(e^{j\omega}, \underline{\theta})|^2 \Phi_u(\omega) \\ & + \Phi_n(\omega)\Big)\frac{1}{|\hat{H}(e^{j\omega}, \underline{\theta})|^2}\, d\omega\,,\end{aligned} \tag{18.193}$$

where Φ_u and Φ_n are the spectra of the input and the measurement noise, respectively, and G, $\hat{G}$, and $\hat{H}$ are the true process transfer function, the model transfer function, and the noise model. Note that, since the right side of the equation is in the frequency domain, G, $\hat{G}$, and $\hat{H}$ are written as functions of the z-transform variable $z = e^{j\omega}$ and not of q.

If the process and model transfer functions are identical, that is, the transfer function of the model has the same structure (zero bias error) and the same parameters (zero variance error) as the true process, then the expression under the integral in (18.193) simplifies to

$$\frac{\Phi_n(\omega)}{|\hat{H}(e^{j\omega}, \underline{\theta})|^2}. \tag{18.194}$$

This is exactly white noise if the noise model is identical to the true process noise, since $\Phi_n(\omega) = \sigma^2 |H(e^{j\omega})|^2$, where H describes the true process noise and σ^2 is the variance of the white noise that drives it. In this case, the loss function realizes its smallest possible value. A process/model mismatch for the noise description increases the loss function value, and the prediction error becomes colored. Nevertheless, the transfer functions G and $\hat{G}$ may be identical; see Sect. 18.7.2.

If the noise can be neglected, the expression under the integral becomes

$$\left(|G(e^{j\omega}) - \hat{G}(e^{j\omega}, \underline{\theta})|^2\right) \frac{\Phi_u(\omega)}{|\hat{H}(e^{j\omega}, \underline{\theta})|^2}. \tag{18.195}$$

$G(e^{j\omega}) - \hat{G}(e^{j\omega}, \underline{\theta})$ measures the difference between the process and the model in dependency on the frequency ω. The second factor can be interpreted as a frequency-dependent weighting factor. It shapes the accuracy of the model in the frequency domain. For example, a model can be made more accurate around the frequency ω^* if

- the input signal contains a high energy in this frequency, i.e., $\Phi_u(\omega^*)$ is large, and/or
- the amplitude of the noise model's frequency response for this frequency is small, i.e., $|\hat{H}(e^{j\omega^*}, \underline{\theta})|$ is small.

If the input signal is white, that is, $\Phi_u(\omega) = \text{constant}$, the effect of the noise model on the fit of the transfer functions becomes more obvious. For example, the OE model with $\hat{H}(e^{j\omega}) = 1$ weights all frequencies identically. In contrast, the ARX model with $\hat{H}(e^{j\omega}) = 1/A(e^{j\omega})$ weights the model error with $A(e^{j\omega})$! Since $1/A(e^{j\omega})$ has a low-pass characteristic, $A(e^{j\omega})$ is high-pass and the high frequencies are emphasized. This is the reason why ARX models usually have larger d.c. ($\omega = 0$) errors as OE models. It is important to understand that *the noise model allows one to shape the model transfer function accuracy in the frequency range.*

Noise models with low-pass characteristics result in transfer function models with good accuracy in high frequencies and vice versa.

Note that even if the weighting factor $\Phi_u(\omega)/|\hat{H}(e^{j\omega}, \underline{\theta})|^2$ is constant, as would be the case for a white input signal and an OE model, very high frequencies are less significant since G and $\hat{G}$ are typically well-damped for high frequencies (low-pass characteristic) [356].

Of course, the overall accuracy of the model also depends on the structure of G. If a high-order process G is to be approximated by a low-order transfer function model $\hat{G}$, the accuracy for high frequencies is in principle limited.

18.7.4 Relationship Between Noise Model and Filtering

The previous section showed that the noise model influences the frequency weighting of the model fit. Another possibility for frequency weighting is to filter the prediction errors. Indeed, prefiltering and incorporation of an appropriate noise model are equivalent. If the prediction error is filtered with $L(q)$, it becomes (see (18.34))

$$e_F(k) = \frac{L(q)}{H(q)} \left(y(k) - G(q)u(k)\right) . \tag{18.196}$$

Thus, instead of using the noise model $H(q)$, the prediction error can be filtered with

$$L(q) = 1/H(q) . \tag{18.197}$$

This relationship can be intuitively explained as follows. For frequencies where the noise model amplitudes are small, low noise levels are expected; so the signal-to-noise ratio is high. Consequently, this frequency range is strongly exploited for fitting the model since the data quality is good. On the other hand, for frequencies where large noise amplitudes are expected, the data is utilized more carefully, i.e., with less weight. Instead of using the noise model, a filter can be used to emphasize and deemphasize certain frequency ranges. If the noise model is low-pass, the corresponding filter is high-pass and vice versa.

To summarize, instead of using a noise model $H(q)$, a filter according to (18.197) can be employed with exactly the same effect. On the other hand, the effect of an existing noise model $H(q)$ can be *canceled* by additionally filtering the prediction error with

$$L(q) = H(q) . \tag{18.198}$$

This relationship helps us to understand algorithms such as the "repeated least squares and filtering" for OE model estimation in Sect. 18.5.4. It can be explained as follows. The ARX model prediction error is

$$e(k) = A(q)\,(y(k) - G(q)u(k))\,. \tag{18.199}$$

Filtering with $L(q) = H(q) = 1/A(q)$ cancels the effect of the ARX noise model and leads to the OE model prediction errors

$$e_F(k) = (y(k) - G(q)u(k))\,. \tag{18.200}$$

Of course, the procedure of ARX model estimation and filtering with $L(q)$ must be repeated several times until convergence. The initial ARX model denominator $A(q)$ converges to the OE model denominator $F(q)$.

18.7.5 *Offsets*

Data for linear systems is typically measured around an equilibrium point $\bar{U}$ and $\bar{Y}$. Figure 18.57 shows an equilibrium point lying on the static nonlinearity of the process. In order to obtain data that can be approximately described by a linear model, the deviations u and y from this equilibrium point must stay small enough, depending on the strength of the nonlinear behavior of the process around the equilibrium point $\bar{U}$ and $\bar{Y}$. For linear system identification, the difficulty arises that $U(k)$ and $Y(k)$ are measured but $u(k)$ and $y(k)$ are required for identification. With the deviations from the equilibrium $u(k) = U(k) - \bar{U}$ and $y(k) = Y(k) - \bar{Y}$, the following linear difference equation results [278]:

$$\begin{aligned}(Y(k) - \bar{Y}) + a_1(Y(k-1) - \bar{Y}) + \ldots + a_m(Y(k-m) - \bar{Y}) &= \\ b_1(U(k-1) - \bar{U}) + \ldots + b_m(U(k-m) - \bar{U})\,, &\end{aligned} \tag{18.201}$$

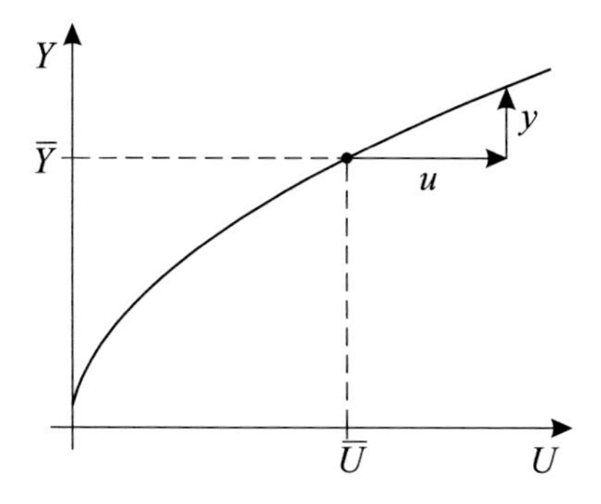

Fig. 18.57 Data for linear system identification is required around an equilibrium point $\bar{U}$, $\bar{Y}$

which can also be written as

$$\begin{aligned} Y(k) = & - a_1 Y(k-1) - \ldots - a_m Y(k-m) \\ & + b_1 U(k-1) + \ldots + b_m U(k-m) \\ & + \underbrace{(1 + a_1 + \ldots + a_m)\bar{Y} - (b_1 + \ldots + b_m)\bar{U}}_{C} \, . \end{aligned} \tag{18.202}$$

The offset C incorporates the information about the equilibrium point $\bar{U}$, $\bar{Y}$. There exist three possibilities to take this offset into account:

- removal of the offset by explicit preprocessing of the data,
- estimation of the offset,
- extension of the noise model.

From the following five approaches for dealing with offsets [356], the first two belong to category one, the third belongs to the second category, and the last two realize the third idea.

1. If the equilibrium is known, the deviations can be calculated explicitly as

$$u(k) = U(k) - \bar{U} \, , \qquad y(k) = Y(k) - \bar{Y} \, , \tag{18.203}$$

   and $u(k)$, $y(k)$ can be used for identification.
2. If the equilibrium is unknown, it can be approximated by means of $U(k)$ and $Y(k)$

$$\bar{U} = \frac{1}{N} \sum_{i=1}^{N} U(i) \, , \qquad \bar{Y} = \frac{1}{N} \sum_{i=1}^{N} Y(i) \, , \tag{18.204}$$

   and this approximated equilibrium can be used for approach 1.
3. The offset can be estimated explicitly by basing the parameter estimation on (18.202). Then for an ARX model the extended regression and parameter vectors become

$$\underline{x} = [U(k-1) \; \cdots \; U(k-m) \quad -Y(k-1) \; \cdots \; -Y(k-m) \quad 1]^T \, , \tag{18.205}$$

$$\underline{\theta} = [b_1 \; \cdots \; b_m \quad a_1 \; \cdots \; a_m \quad C]^T \, . \tag{18.206}$$

   An extension to other linear model structures is straightforward. In comparison to approach 2, an additional parameter must be estimated. This extra effort cannot usually be justified. However, approach 3 has the advantage of being less sensitive to data distribution. This can pay off if the data contains significant nonlinear behavior.

4. The offset can be eliminated by differencing the data. This can be done either by prefiltering $U(k)$ and $Y(k)$ with

$$L(q) = 1 - q^{-1}, \tag{18.207}$$

which generates $U(k) - U(k-1)$ and $Y(k) - Y(k-1)$ from $U(k)$ and $Y(k)$, respectively, or equivalently by extending the noise model with an integrator (e.g., ARIMAX model; see Sect. 18.7.4)

$$\widetilde{H}(q) = \frac{1}{1 - q^{-1}} H(q). \tag{18.208}$$

The main drawback of this simple approach is that this high-pass filter or low-pass noise model emphasizes high frequencies in the model fit; see Sects. 18.7.3 and 18.7.4.
5. The dynamic order of the noise model can be extended to allow the parameter estimation method to find the pole at $q = 1$ automatically. Compared with approach 4, this requires the estimation of additional parameters and thus is computationally more demanding.

These approaches can be extended to cope with disturbances such as drifts or oscillations [356].

18.8 Recursive Algorithms

The algorithms for linear system identification that have been discussed so far are based on least squares, repeated least squares, or nonlinear optimization methods. These methods operate on the whole data set, and their computational complexity typically increases linearly with the number of data samples. Thus, they are not well suited for an online application where a new model will be identified within each sampling instant exploiting the information contained in the new measured data sample. If windowing of the data is used, i.e., only the last (say N) data samples are taken into account for identification, it is possible to guarantee that one LS estimation is carried out within one sampling interval. Owing to the iterative character of the repeated LS and nonlinear optimization techniques, this guarantee cannot usually be given. If all data is to be used for identification, none of these methods can be applied online since at some point their computation time will exceed the sampling time.

Recursive algorithms compute the new parameters at time k in dependency on the parameters at the previous sampling instant $k-1$ and the new incoming information. Thus, their computational demand is constant, and they are well suited for online identification. In the following sections, the most common recursive methods are briefly summarized. For a more detailed analysis of the recursive least squares method, refer to Sect. 3.2 and [278, 279, 304, 356, 540]. The algorithms discussed here are simple and easy to understand. However, their numerical robustness is quite low, i.e., they are sensitive to round-off errors. For a detailed discussion of more robust and faster algorithms, refer to [232].

Section 18.8.1 briefly summarizes the recursive least squares algorithm, which can be used for online identification of ARX models. The recursive version of the instrumental variable method is presented in Sect. 18.8.2. For identification of ARMAX models, a recursive variant of the ELS (Sect. 18.5.2) algorithm is treated in Sect. 18.8.3. Finally, Sect. 18.8.4 deals with a general recursive prediction error method that can be applied to all linear model structures. In [279, 281, 498], a comparison of six recursive algorithms can be found.

18.8.1 Recursive Least Squares (RLS) Method

The recursive least squares (RLS) algorithm with exponential forgetting as discussed in Sect. 3.2 can be utilized directly for identification of ARX models since they are linear in their parameters (see (3.89a)–(3.89c)):

$$\underline{\hat{\theta}}(k) = \underline{\hat{\theta}}(k-1) + \underline{\gamma}(k)\, e(k), \quad e(k) = y(k) - \underline{x}^T(k)\underline{\hat{\theta}}(k-1), \tag{18.209a}$$

$$\underline{\gamma}(k) = \frac{1}{\underline{x}^T(k)\underline{P}(k-1)\underline{x}(k) + \lambda}\, \underline{P}(k-1)\underline{x}(k), \tag{18.209b}$$

$$\underline{P}(k) = \frac{1}{\lambda}\left(\underline{I} - \underline{\gamma}(k)\,\underline{x}^T(k)\right)\underline{P}(k-1). \tag{18.209c}$$

The regressors and parameters are

$$\underline{x}(k) = [u(k-1) \ \cdots \ u(k-m) \ \ -y(k-1) \ \cdots \ -y(k-m)]^T, \tag{18.210}$$

$$\underline{\hat{\theta}}(k) = \left[\hat{b}_1(k) \ \cdots \ \hat{b}_m(k) \ \hat{a}_1(k) \ \cdots \ \hat{a}_m(k)\right]^T. \tag{18.211}$$

18.8.2 Recursive Instrumental Variables (RIV) Method

The RLS algorithm generally yields inconsistent estimates of the true process parameters as the LS does; see Sect. 18.5.1. One common solution to this problem is the introduction of instrumental variables (IVs). With the IVs $\underline{z}(k)$ the recursive version of the instrumental variable (RIV) method becomes

$$\underline{\hat{\theta}}(k) = \underline{\hat{\theta}}(k-1) + \underline{\gamma}(k)\, e(k), \quad e(k) = y(k) - \underline{x}^T(k)\underline{\hat{\theta}}(k-1), \tag{18.212a}$$

$$\underline{\gamma}(k) = \frac{1}{\underline{x}^T(k)\underline{P}(k-1)\underline{z}(k) + \lambda} \underline{P}(k-1)\underline{z}(k), \tag{18.212b}$$

$$\underline{P}(k) = \frac{1}{\lambda}\left(\underline{I} - \underline{\gamma}(k)\underline{x}^T(k)\right)\underline{P}(k-1). \tag{18.212c}$$

A typical choice of model-independent IVs is

$$\underline{z}(k) = [u(k-1) \ \cdots \ u(k-2m)]^T . \tag{18.213}$$

It is also possible to use *model-dependent* instruments, which offer the advantage of being more highly correlated with the regressors $\underline{x}(k)$, for example,

$$\underline{z}(k) = [u(k-1) \ \cdots \ u(k-m) \quad -y_u(k-1) \ \cdots \ -y_u(k-m)]^T \tag{18.214}$$

with

$$y_u(k) = \frac{\hat{B}(q,k)}{\hat{A}(q,k)} u(k) . \tag{18.215}$$

Note that the *exact* realization of these instruments would require one to filter the input completely with the current model $\hat{B}(q,k)/\hat{A}(q,k)$, that is, not using $y_u(k-i)$ for $i > 0$ from previous recursions. This, of course, is not practicable since the computational effort increases linearly with the length of the data set. Therefore, instead of using (18.215), the IVs can be generated by calculating the following difference equation with the current parameter estimates:

$$\begin{aligned} y_u(k) = {} & \hat{b}_1(k)u(k-1) + \ldots + \hat{b}_m(k)u(k-m) \\ & - \hat{a}_1(k)y_u(k-1) - \ldots - \hat{a}_m(k)y_u(k-m) . \end{aligned} \tag{18.216}$$

In contrast to (18.215), in (18.216) the delayed filtered outputs $y_u(k-i)$ for $i > 0$ are used from the previous recursions. As long as $\hat{b}_i$ and $\hat{a}_i$ change slowly, both approaches yield similar results.

The use of model-dependent IVs can cause stability problems because they are based on the parameter estimates and thus represent a loop within a loop. Nevertheless, the application of RIV methods is quite successful in practice. During the first few iterations of the RIV, the parameter estimates $\hat{b}_i(k)$ and $\hat{a}_i(k)$ are unreliable. Therefore, the RLS algorithm is used in the starting phase. The RLS is run for the first few, say $4m$, iterations until the parameter estimates are somewhat stable. This start-up procedure is also used for the RELS and RPEM described below.

18.8.3 Recursive Extended Least Squares (RELS) Method

ARMAX models can be estimated online by the recursive extended least squares (RELS)[6] algorithm. Formally it takes the same form as the RLS. However, the regression and parameter vectors are different (see (18.82) in Sect. 18.5.2 and Fig. 18.28):

$$\underline{x}(k) = [u(k-1) \cdots u(k-m) \;\; -y(k-1) \cdots -y(k-m) \\ \widetilde{e}(k-1) \cdots \widetilde{e}(k-m)]^T , \tag{18.217}$$

$$\hat{\underline{\theta}}(k) = \left[\hat{b}_1(k) \;\cdots\; \hat{b}_m(k) \;\; \hat{a}_1(k) \;\cdots\; \hat{a}_m(k) \;\; \hat{c}_1(k) \;\cdots\; \hat{c}_m(k)\right]^T , \tag{18.218}$$

where $\widetilde{e}(k-i)$ denote the previous residuals. Note that these previous residuals in (18.217) are approximations for the unknown white noise $v(k)$ that drives the noise filter, i.e., $\widetilde{e}(k-i) \approx v(k-i)$. The residual $\widetilde{e}(k)$ is also called the *a posteriori error*, defined by

$$\widetilde{e}(k) = y(k) - \underline{x}^T(k)\hat{\underline{\theta}}(k) \tag{18.219}$$

since it utilizes the information about $\hat{\underline{\theta}}(k)$, i.e., after (a posteriori) the parameter update. In contrast, the prediction error is also called the *a priori error* because it is based on the old parameter estimate:

$$e(k) = y(k) - \underline{x}^T(k)\hat{\underline{\theta}}(k-1) . \tag{18.220}$$

The a posteriori error is known at time k because the regression vector $\underline{x}(k)$ requires knowledge of the residuals only up to time $k-1$; see (18.217). It usually speeds up the convergence of the RELS algorithm if the a posteriori rather than the a priori errors are used [540].

The RELS is also called *recursive pseudo-linear regression (RPLR)* because formally it is identical to the linear RLS algorithm although the ARMAX model is nonlinear in its parameters. The influence of this nonlinear character becomes obvious in a convergence analysis of the RELS algorithm [356, 540]. The RELS converges much more slowly than the RLS. In particular, the noise model parameters $\hat{c}_i$ converge slowly. Intuitively this can be explained by the fact that within the parameter estimation procedure, an approximation of the white noise $v(k)$ is required. In particular, during the first few iterations where the parameter estimates $\hat{a}_i$ and $\hat{b}_i$ are unreliable, the approximation of the white noise $v(k)$ is poor, which slows down the convergence of the algorithm. As for the RIV, the RELS is started with an RLS.

---

[6] Sometimes denoted only as extended least squares (ELS) [356].

18.8.4 Recursive Prediction Error Methods (RPEM)

The recursive prediction error method (RPEM) allows the online identification of all linear model structures described in this chapter. Since all model structures except ARX and FIR/OBF are nonlinearly parameterized, no exact recursive algorithm can exist; rather some approximations must be made. In fact, the RPEM can be seen as a nonlinear least squares Gauss-Newton method with sample adaptation. Refer to Sect. 4.1 for a discussion of sample and batch adaptation and to Sect. 4.5.1 for a description of the Gauss-Newton algorithm with line search for batch adaptation.

The Gauss-Newton technique is based on the approximation of the Hessian by the gradients (strictly speaking the Jacobian). Thus, the RPEM requires the calculation of the gradients of the loss function, which in turn requires the calculation of the gradients $\underline{g}(k)$ of the model output with respect to its parameters:

$$\underline{g}(k) = \frac{\partial \hat{y}(k)}{\partial \underline{\theta}(k)} = \left[\frac{\partial \hat{y}(k)}{\partial \theta_1(k)} \quad \frac{\partial \hat{y}(k)}{\partial \theta_2(k)} \quad \cdots \quad \frac{\partial \hat{y}(k)}{\partial \theta_n(k)} \right]^T . \tag{18.221}$$

With these gradients the RPEM becomes

$$\hat{\underline{\theta}}(k) = \hat{\underline{\theta}}(k-1) + \underline{\gamma}(k)\, e(k), \quad e(k) = y(k) - \underline{x}^T(k)\hat{\underline{\theta}}(k-1), \tag{18.222a}$$

$$\underline{\gamma}(k) = \frac{1}{\underline{g}^T(k)\underline{P}(k-1)\underline{g}(k) + \lambda} \underline{P}(k-1)\underline{g}(k), \tag{18.222b}$$

$$\underline{P}(k) = \frac{1}{\lambda} \left(\underline{I} - \underline{\gamma}(k)\, \underline{g}^T(k) \right) \underline{P}(k-1), \tag{18.222c}$$

which is identical to the RLS except that in (18.222b) and (18.222c) the model gradients $\underline{g}(k)$ replace the regressors $\underline{x}(k)$ in (18.209b) and (18.209c).

Example 18.8.1 RPEM for ARX models
For ARX models the gradients are

$$\begin{aligned} \underline{g}(k) &= \left[\frac{\partial \hat{y}(k)}{\partial a_1(k)} \quad \cdots \quad \frac{\partial \hat{y}(k)}{\partial a_m(k)} \quad \frac{\partial \hat{y}(k)}{\partial b_1(k)} \quad \cdots \quad \frac{\partial \hat{y}(k)}{\partial b_m(k)} \right]^T \\ &= [u(k-1) \ \cdots \ u(k-m) \quad -y(k-1) \ \cdots \ -y(k-m)]^T \end{aligned} \tag{18.223}$$

Obviously, the gradients are identical to the regressors, i.e., $\underline{g}(k) = \underline{x}(k)$. Thus, the RPEM for ARX models is equivalent to the RLS. □

Example 18.8.2 RPEM for ARMAX models
The application of the RPEM to ARMAX models is also known as the *recursive maximum likelihood (RML)* method [356]. For ARMAX models the gradients are (see (18.86), (18.88), (18.90) in Sect. 18.5.2)

$$\underline{g}(k) = \frac{1}{\hat{C}(q,k)} [u(k-1) \cdots u(k-m) \quad -y(k-1) \cdots -y(k-m)$$

$$\tilde{e}(k-1) \cdots \tilde{e}(k-m)]^T . \tag{18.224}$$

Obviously, the gradients are identical to the regressors filtered with $1/\hat{C}(q,k)$, i.e., $\underline{g}(k) = 1/\hat{C}(q,k)\underline{x}(k)$. Thus, the RPEM for ARMAX models is very similar to the RELS. The additional filtering of the regressors usually speeds up the convergence since it has a decorrelating effect. □

The application of the RPEM to other model structures is straightforward utilizing the given gradient equations in the corresponding subsections in Sect. 18.5. As for the RIV and the RELS, the RPEM is started with an RLS.

The gradients $\underline{g}(k)$ cannot be evaluated exactly in practice because the computational effort increases linearly with the length of the data set if no windowing is used. As for the model-dependent instruments in the RIV method in Sect. 18.8.2, the gradients are evaluated approximately. For example, the gradients for the ARMAX model given in (18.224) can be approximate by the following difference equation:

$$\underline{g}(k) = \underline{x}(k) - \hat{c}_1 \underline{g}(k-1) - \ldots - \hat{c}_m \underline{g}(k-m) . \tag{18.225}$$

If the parameters do not change during adaptation, (18.225) is exact. If the parameters change, (18.225) is an approximation for the following reason. $\underline{g}(k-i)$ has been evaluated with the parameters at time instant $k-i$, not with the actual ones. Thus, the approximation

$$\frac{\partial \hat{y}(k)}{\partial \underline{\theta}(k)} \approx \frac{\partial \hat{y}(k)}{\partial \underline{\theta}(k-i)} \tag{18.226}$$

is made. In the context of dynamic neural networks, this strategy is called *real-time recurrent learning*; see Sect. 19.5.2. It is approximate only if sample adaptation is applied. For batch adaptation the parameters are kept fixed during a sweep through the data set, and the adaptation is carried out only at the end of each batch. Then (18.226) is exact since $\underline{\theta}(k) = \underline{\theta}(k-i)$.

18.9 Determination of Dynamic Orders

A simple and probably the most widely applied approach to order selection is to identify several models with increasing orders and to choose the best one with respect to some model validation technique such as testing on fresh data or evaluation of information criteria; see Sect. 7.3.

The utilization of correlation functions is a powerful tool for order determination since it exploits the linear relationships assumed by choosing a linear model. Thus, the cross-correlation function between $u(k)$ and $y(k)$ gives a clear indication about the dead time d. Since the dead time is the smallest time delay with which $u(k-d)$ influences $y(k)$ directly, the cross-correlation function $\text{corr}_{uy}(\kappa)$ is expected to possess a peak at $\kappa = d$, while it is expected to be close to zero for all $\kappa < d$. Similarly, the correlation function between the input $u(k)$ and the model error $e(k) = y(k) - \hat{y}$ can reveal missing terms. Ideally, it should be close to zero when all information is exploited by the model. Peaks indicate missing input terms $u(k-i)$, while smaller but consistent deviations from zero stretching over many time lags κ indicate missing output terms $y(k-i)$.

An indication for a model order that is too high is given by approximate pole/zero cancelations, i.e., if o zeros of the estimated transfer function (almost) compensate o of its poles, then it is likely that the model order m is chosen too high and the true order of the system is merely $m - o$.

18.10 Multivariable Systems

Up to here only single input, single output (SISO) models have been discussed. In many real-world situations, the output of a process is not influenced solely by a *single* input. Rather it depends on multiple variables. The user has to decide how these variables shall be incorporated into the model. At least the following three different situations have to be distinguished:

1. A variable is measured and can be manipulated, e.g., a control signal.
2. A variable is measured but cannot be manipulated, e.g., an external disturbance such as the environment temperature.
3. A variable is not measured and cannot be manipulated, e.g., an external disturbance such as wind.

The variables of the first category should be incorporated into the model as inputs. Variables of the second category can be difficult to incorporate into a black box experimental model since they cannot be excited, and thus it sometimes may not be possible to gather data, which reflects the influence of this variable in a representative way. Alternatively, if this knowledge is not available, the variable's influence can be taken into account by the noise model. A variable of the third category can only be considered by properly structuring the noise model. Note that the distinction between variables of types 2 and 3 can be caused by fundamental reasons, e.g., in principle it may not be possible to measure a variable, or it can be caused by a benefit/cost tradeoff, e.g., the information about a signal may not be worth the cost for the sensor. The latter issue is by far the more common one. It typically arises if the influence of the variable is not very significant or the cost for a reliable sensor is extremely high (possibly owing to the environmental conditions).

A multiple input, multiple output (MIMO) model can be decomposed into several multiple input, single output (MISO) models, one for each output; see Fig. 18.58. Such a decomposition offers a number of advantages over handling the full MIMO

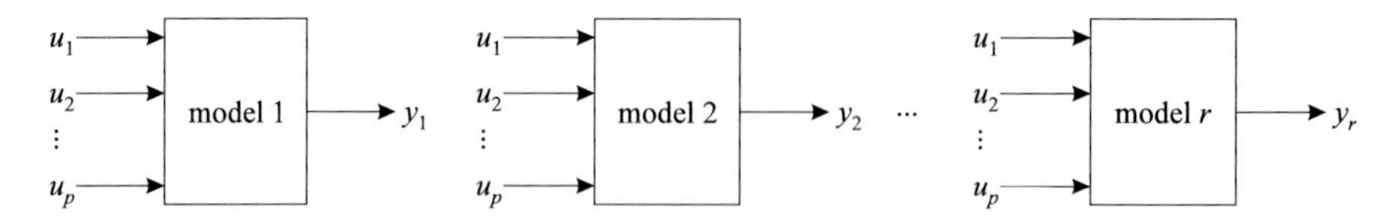

Fig. 18.58 A MIMO model can be decomposed into MISO models

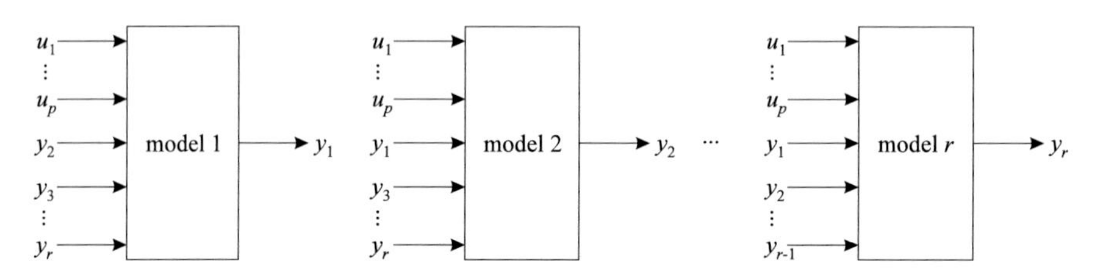

Fig. 18.59 A MIMO model can be decomposed alternatively to Fig. 9.2 into MISO models by also considering the other process outputs

model. Each output does not necessarily depend on all inputs $u_1, \ldots, u_p$. Thus, each MISO model can be simpler if it utilizes only the relevant inputs for the corresponding output. Each MISO model can be structured separately with respect to model structures, dynamic orders, dead times, etc. The design of excitation signals can be performed independently for each output. Finally, the system analysis is easier for the user when the MIMO structure is broken down into several MISO structures.

Alternatively to Fig. 18.58, the MIMO structure can be decomposed as shown in Fig. 18.59. Using other process outputs as model inputs is not necessary. Nevertheless, it can improve the model accuracy because it may allow one to apply models of lower order. Whether the approach in Fig. 18.59 is favored over that of Fig. 18.58 depends on the specific application. Clearly, it is reasonable to utilize the other process outputs (or some of them) as model inputs only when their influence is significant, e.g., if MISO model outputs are correlated. The intended use of the model also plays a crucial role in the decision about which approach is preferable. For example, a one-step prediction with the model in Fig. 18.59 can be performed with previous measurements of the other *process outputs*, while for simulation the previous *model outputs* have to be fed back. Feeding back the model outputs to the inputs of other MISO models which again feed back their model outputs as shown in Fig. 18.60 can cause stability problems and accumulation of modeling errors. No such difficulties arise with the MIMO model decomposition according to Fig. 18.58.

In the following three sections, the most important modeling and identification methods are briefly presented. The p-canonical model and the simplified matrix polynomial model represent input/output approaches, while the more sophisticated and increasingly popular subspace methods are based on a state space representation.

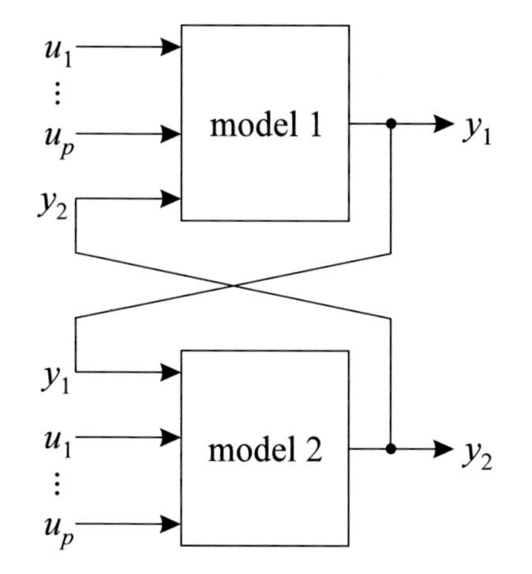

Fig. 18.60 Simulation of two coupled MISO models of the type shown in Fig. 18.59

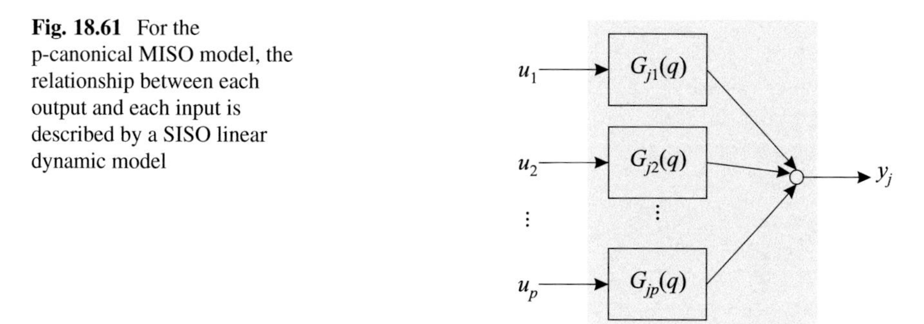

Fig. 18.61 For the p-canonical MISO model, the relationship between each output and each input is described by a SISO linear dynamic model

18.10.1 P-Canonical Model

The most straightforward approach to MISO or MIMO modeling is to describe the relationship between each input and each output by a SISO linear dynamic model, as depicted in Fig. 18.61. This is the so-called p-canonical structure into which other structures such as the v-canonical one can be transformed [282]. The models $G_{jj}(q)$ are called the *main models* and the $G_{ji}(q)$ with $i \neq j$ are called the *coupling models*. If the gains of the coupling models are very small, they may be neglected, and the MIMO model simplifies to several SISO models. The p-canonical model can be described by

$$\begin{bmatrix} y_1(k) \\ y_2(k) \\ \vdots \\ y_r(k) \end{bmatrix} = \begin{bmatrix} G_{11}(q) & G_{12}(q) & \cdots & G_{1p}(q) \\ G_{21}(q) & G_{22}(q) & \cdots & G_{2p}(q) \\ \vdots & \vdots & & \vdots \\ G_{r1}(q) & G_{r2}(q) & \cdots & G_{rp}(q) \end{bmatrix} \cdot \begin{bmatrix} u_1(k) \\ u_2(k) \\ \vdots \\ u_p(k) \end{bmatrix} . \tag{18.227}$$

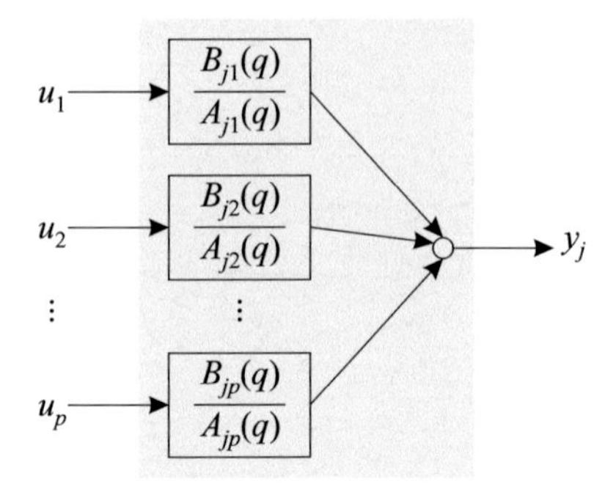

Fig. 18.62 For a p-canonical MISO model, each transfer function can be described by an output feedback model, e.g., ARX, ARMAX, OE, etc.

Thus, output $y_j(k)$ is modeled by

$$y_j(k) = G_{j1}(q)u_1(k) + G_{j2}(q)u_2(k) + \ldots + G_{jp}(q)u_p(k)\,. \qquad (18.228)$$

This model is linear in its parameters and thus easy to determine if the transfer functions $G_{ji}(q)$ are modeled by FIR or OBF structures. The regression vector is simply extended by the filtered additional inputs; see Sect. 18.5. However, the more common models with output feedback (Sect. 18.5) result in a nonlinear parameterized model (see Fig. 18.62)

$$y_j(k) = \frac{B_{j1}(q)}{A_{j1}(q)}u_1(k) + \frac{B_{j2}(q)}{A_{j2}(q)}u_2(k) + \ldots + \frac{B_{jp}(q)}{A_{jp}(q)}u_p(k)\,. \qquad (18.229)$$

This can be seen as a multivariable OE model, and its parameters have to be estimated with nonlinear optimization techniques; see Sect. 18.5.4. The generalization to a multivariable BJ model is possible via explicit incorporation of a noise model $H(q) = C(q)/D(q)$ in (18.229); see Sect. 18.5.5.

In order to be able to utilize efficient, linear parameter estimation techniques, (18.229) can be modified to the matrix polynomial model, which is discussed in the next section.

18.10.2 Matrix Polynomial Model

If it is assumed that all denominator polynomials in (18.229) are identical ($A_j(q) = A_{ji}(q)$ for all i) the matrix polynomial model is obtained (see Fig. 18.63):

$$A_j(q)y_j(k) = B_{j1}(q)u_1(k) + B_{j2}(q)u_2(k) + \ldots + B_{jp}(q)u_p(k)\,. \qquad (18.230)$$

This can be seen as a multivariable ARX model, and its parameters can be identified by linear regression techniques; see Sect. 18.5.1. The assumption of identical denominator polynomials in (18.229) is highly unrealistic. Only if the

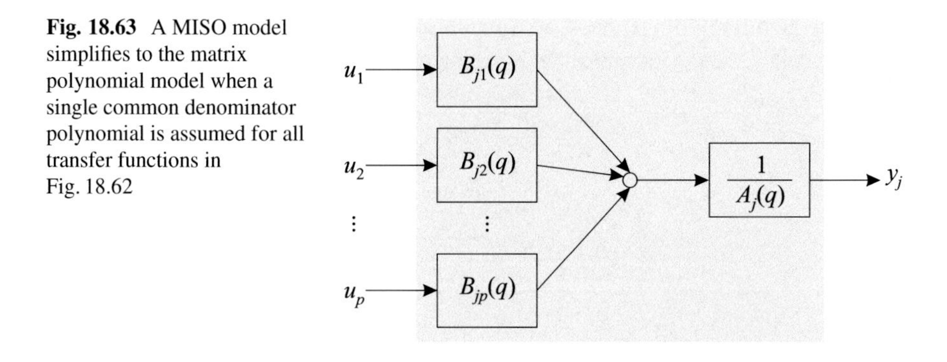

Fig. 18.63 A MISO model simplifies to the matrix polynomial model when a single common denominator polynomial is assumed for all transfer functions in Fig. 18.62

dynamics from all inputs to output j are similar, (18.230) can yield a reasonable approximation of (18.229). For many processes, this condition is not fulfilled.

However, there is an alternative way to obtain the model structure in (18.230). When (18.229) is multiplied by its common denominator, the following model results (assuming no common factors in the denominator polynomials $A_{ji}(q)$):

$$
\begin{aligned}
A_{j1}(q) \cdot \ldots \cdot A_{jp} y_j(k) = B_{j1}(q) A_{j2}(q) \cdot \ldots \cdot A_{jp}(q) u_1(k) + \\
B_{j2}(q) A_{j1}(q) \cdot A_{j3}(q) \cdot \ldots \cdot A_{jp}(q) u_2(k) + \ldots + \\
B_{jp}(q) A_{j1}(q) \cdot \ldots \cdot A_{jp-1}(q) u_p(k)
\end{aligned}
\tag{18.231a}
$$

or

$$
\widetilde{A}_j(q) y_j(k) = \widetilde{B}_{j1}(q) u_1(k) + \widetilde{B}_{j2}(q) u_2(k) + \ldots + \widetilde{B}_{jp}(q) u_p(k), \tag{18.232}
$$

where the orders of the polynomials $\widetilde{A}_j(q)$ and $\widetilde{B}_{ji}(q)$ are equal to $p \cdot m$ with m being the order of all original transfer functions in (18.229). Since (18.232) is linear in its parameters, it can be estimated easily. However, because the polynomials are of much higher order than the original ones, significantly more parameters have to be estimated. This causes a higher variance error and is the price to be paid for the linearity in the parameters.

Ideally, it should be possible to calculate the original polynomials $A_{ji}(q)$ and $B_{ji}(q)$ by pole-zero cancelations from $\widetilde{B}_{ji}(q)$ and $\widetilde{A}_j(q)$. In practice, however, this is rarely possible owing to disturbances and structural mismatch. If the process has a large number of inputs (p is large), the model (18.232) possesses a huge number of parameters and often becomes too complex for its intended use, or the high variance error makes a parameter estimation infeasible. Then one can try to estimate a model of structure (18.232) with a lower order somewhere between m and $p \cdot m$. It is important to realize that all models of type (18.229) can be transformed to (18.232) but not vice versa, because (18.232) is more flexible than (18.229).

Example 18.10.1 ARX Model for MISO Process
This example illustrates how inferior ARX models usually are for modeling

processes with multiple inputs. The situation becomes worse, the more inputs need to be modeled and the more distinct the dynamics are from each input to the output.

Consider the 2-input, 1-output process where the output is the sum of the following two transfer functions with gain 1 and poles at $p_1 = 0.9$ and $p_2 = 0.7$, respectively:

$$G_1(q) = \frac{0.1q^{-1}}{1 - 0.9q^{-1}}, \quad G_2(q) = \frac{0.3q^{-1}}{1 - 0.7q^{-1}}. \tag{18.233}$$

It is excited by two PRBS of length 630 shifted by 315 time steps to make the inputs uncorrelated; the inputs and the output is measured noise-free. Identification of a first-order ARX model yields

$$\hat{G}_1(q) = \frac{0.1348q^{-1}}{1 - 0.7953q^{-1}}, \quad \hat{G}_2(q) = \frac{0.2359q^{-1}}{1 - 0.7953q^{-1}}. \tag{18.234}$$

Thus, due to the common denominator, a compromise between the two dynamics is found. Figure 18.64a shows that even under these ideal conditions (rich, independent excitation, no noise), the process cannot be described by an ARX model of the true order. Rather a second-order ARX is able to accomplish this with the help of zero/pole cancelations. Note that such an approach would require an pth-order ARX model for modeling a p-input, 1-output process of first order. The estimated transfer functions are

$$\hat{G}_1(q) = \frac{0.1q^{-1} - 0.07q^{-2}}{1 - 1.6q^{-1} + 0.63q^{-2}}, \quad \hat{G}_2(q) = \frac{0.3q^{-1} - 0.27q^{-2}}{1 - 1.6q^{-1} + 0.63q^{-2}}. \tag{18.235}$$

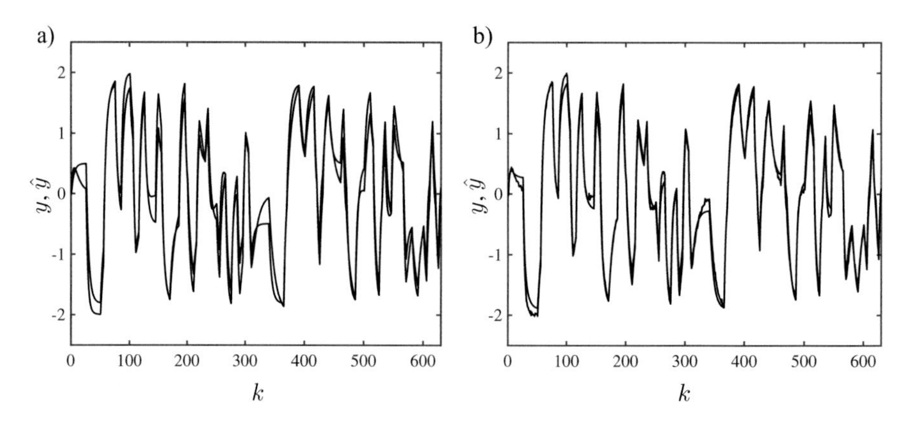

Fig. 18.64 First-order MISO process versus (**a**) first-order ARX model (noise-free case) and (**b**) second-order ARX model (low noise case)

Since the roots of the denominator polynomial are at $p_1 = 0.9$ and $p_2 = 0.7$, the model is perfect. However, if the process output is disturbed by small Gaussian noise with standard deviation 0.02, the model parameters deviate significantly although the model output still is very reasonable (see Fig. 18.64b): poles at $p_1 = 0.8451$, $p_2 = 0.6160$, gains $K_1 = 0.8265$, and $K_2 = 1.0613$. Therefore MISO ARX models are appropriate for generation of simulation models, but parameter interpretation for the individual transfer functions usually is hardly possible. □

18.10.3 Subspace Methods

The difficulties with MISO and MIMO ARX models call for a more suitable approach. OE models offer the possibility of individual poles for each transfer function avoiding the over-parameterization of multivariable ARX models. However, they require nonlinear optimization with all the drawbacks involved. State space-based approaches (see Sect. 18.5.6) are typically better suited to model MIMO systems and stay in the territory of linear regression. The main difficulty for the application of prediction error methods to state space models is to find a numerically robust canonical realization, since the alternative, a full parameterization of the state space model, would involve a huge number of parameters.

In recent years, the class of subspace identification methods has attracted much attention. The most prominent representatives of these approaches are the so-called 4SID, pronounced "foursid" (state space system identification), algorithms. They extract an extended observability matrix either directly from input/output data or via the estimation of impulse responses [579]. From this extended observability matrix, the state space description of the system can be recovered. They can be seen as an FIR approach. Thus, the poles of the systems do not cause any problems as in ARX structures.

The main advantages of subspace identification methods are (i) their low computational demand since they are based on linear algebra tools (QR and singular value decomposition), (ii) their ability to deal with systems with many inputs and outputs, and (iii) their good numerical robustness. For an overview and more details on subspace identification methods, refer to [370, 576, 578, 579].

18.11 Closed-Loop Identification

Up to now it has been implicitly assumed that the identification data is measured in open loop. There are, however, many reasons why the user may like to or has to use data for model identification that is measured in closed loop, i.e., in the presence of a feedback controller [574]:

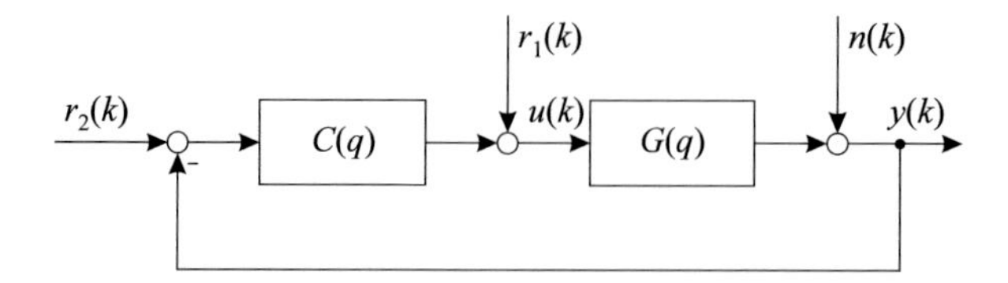

Fig. 18.65 In closed-loop identification, the process input u is correlated with the noise n [574]

1. The process is unstable. Thus, data can be collected only if a feedback controller stabilizes the process.
2. Safety, product quality, or efficiency considerations do not allow the process to be run in open loop.
3. The system may inherently contain underlying feedback loops that cannot be manipulated or removed.
4. The excited frequencies in closed-loop operation are better suited than the frequency band in open-loop operation.
5. The linearization effect of the controller is desired in order to employ linear models even for (weakly) nonlinear processes.
6. The model is to be used for the design of an improved controller.

Figure 18.65 shows the process $G(q)$ in closed-loop control with the controller $C(q)$. The typical difficulty in closed-loop identification is that the process input u is correlated with the output noise disturbance n. The excitation r_1 and the reference r_2 are external signals that may be utilized for identification.

For closed-loop identification, standard methods for open-loop identification can be applied directly, ignoring the correlation between u and n. These approaches are called *direct* or *classical methods* and are addressed in Sect. 18.11.1. An alternative is to utilize the external signals r_1 and/or r_2. These modern approaches are called *indirect methods*; see Sect. 18.11.2. This section ends with some remarks about *identification for control*, which focuses on the intended use of the model solely as a basis for controller design (Sect. 18.11.3). This section is based on the overview paper by Van den Hof [574].

18.11.1 Direct Methods

Direct approaches apply a standard identification method utilizing the process inputs and outputs u and y as if the measurement were taken in open loop. Owing to the ignored correlation between u and n, all methods, which are based on correlations such as correlation analysis, spectral estimation, and the instrumental variable method, are not suited. Typically they estimate a weighted average of the process $G(q)$ and the negative inverse controller $-1/C(q)$ [540].

However, prediction error methods can be used directly for closed-loop identification. Compared with open-loop identification, the following restrictions apply:

- For consistent estimation of $G(q)$, the external signals r_1 and r_2 must be sufficiently exciting, and the controller must be of sufficiently high order or switching between different settings during operation (the controller may be nonlinear and/or time-varying). Furthermore, the transfer function model $\hat{G}(q)$ *and* the noise model $\hat{H}(q)$ must be flexible enough to describe the true process behavior. In open-loop identification, this is not required for the noise model if different parameterizations for $\hat{G}(q)$ and $\hat{H}(q)$ are chosen as for OE and BJ model structures; see Sect. 18.7.2. Thus, in contrast to open-loop, closed-loop identification with too simple a noise model results in inconsistently estimated transfer function parameters even in the case of separate parameterization. This is the most important drawback of the direct methods.
- The frequency-domain expression for closed-loop identification becomes (see (18.192) and (18.193) in Sect. 18.7.3):

$$\Phi_e(\omega, \underline{\theta}) = \frac{|G - \hat{G}(\underline{\theta})|^2}{|H(\underline{\theta})|^2 |1 + CG|^2} \Phi_r(\omega) + \frac{|1 + C\hat{G}(\underline{\theta})|^2}{|H(\underline{\theta})|^2 |1 + CG|^2} \Phi_n(\omega) \qquad (18.236)$$

  where Φ_r is the spectrum of the collected external signals $r = r_1 + C(q) r_2$, Φ_n is the spectrum of the noise, and the argument $(e^{-i\omega})$ is suppressed for better readability. It is important to see that the minimum of $\Phi_e(\omega, \underline{\theta})$ is not necessarily obtained for a consistent estimation, i.e., for $\hat{G}(\underline{\theta}) = G$, because the $\hat{G}(\underline{\theta})$ appears also in the second term of (18.236). Consequently, some tradeoff between both terms will be performed, resulting in an inconsistent estimate of G. This also makes it difficult to adjust the frequency characteristics of $\hat{G}(\underline{\theta})$ when a low-order model is used for approximation of the process.

  Only if the noise model is flexible enough so that $\hat{H}(\underline{\theta}) = H$ does the second term become $\sigma^2 |1 + C\hat{G}(\underline{\theta})/|^2 / |1 + CG|^2$ since $\Phi_n = \sigma^2 H$ with σ^2 as the variance of the white noise v. Then a consistent estimate of G minimizes (18.236) to σ^2. This also holds when the noise variance σ^2 is small compared with the signal power [574].

  Furthermore, it is interesting to note that the transfer function model error $G - \hat{G}(\underline{\theta})$ in (18.236) is weighted not only with the inverse noise model as in the open-loop case but also with the control loop sensitivity function $1/(1 + CG)$. This typically deemphasizes the weight (and thus decreases the model accuracy) for low frequencies.
- Unstable processes can be estimated with equation error model structures such as ARX or ARMAX. Output error models cannot be utilized since unstable filters would appear in the prediction error:

$$e(k) = \frac{1}{H(q)} y(k) - \frac{G(q)}{H(q)} u(k) . \qquad (18.237)$$

For equation error models, the denominator $A(q)$ in $G(q)$ is canceled by the noise model $H(q)$, which is proportional to $1/A(q)$, and the prediction error can be calculated from $u(k)$ and $y(k)$ by stable filtering.

18.11.2 Indirect Methods

Indirect methods rely on the information about the external excitation signals r_1 and/or r_2. In the following, it is assumed that r_1 is used as external signal. However, by replacing r_1 with $C(q)r_2$, the extension to r_2 as external signal is straightforward. In [574] a lot of different indirect approaches are analyzed and compared with respect to the following criteria:

- consistent estimation of the transfer function $G(q)$ if both $\hat{G}(q)$ and $\hat{H}(q)$ are flexible enough to describe the true process;
- consistent estimation of the transfer function $G(q)$ if only $\hat{G}(q)$ is flexible enough to describe the true process;
- free choice of the model order by the user;
- easy tuning of the frequency characteristics of the model;
- identification of unstable processes;
- guarantee of closed-loop stability with the identified model $\hat{G}(q)$ with the present controller $C(q)$;
- necessity of the knowledge of the implemented controller $C(q)$;
- the accuracy (variance error) of the estimated model.

Here, as examples, only the two-stage and the coprime factor method are discussed [574].

18.11.2.1 Two-Stage Method

The two-stage method is based on the idea of replacing the process input u, which is correlated with noise n, by a simulated $u^{(r)}$ that is uncorrelated with n. This is the same idea as in the instrumental variable method; see Sect. 18.5.1. Then the standard prediction error method can be utilized with the prediction errors

$$e^{(r)}(k) = \frac{1}{H^{(r)}(q)}\left(y(k) - G(q)u^{(r)}(k)\right). \tag{18.238}$$

The estimated noise model $H^{(r)}(q)$ describes the influence of the disturbance in closed-loop $H^{(r)}(q) = H(q)/(1 + C(q)G(q))$. Thus, the noise model can be obtained by

$$\hat{H}(q) = \hat{H}^{(r)}(q)\left(1 + C(q)\hat{G}(q)\right). \tag{18.239}$$

Since $u^{(r)}$ is not correlated with n, all properties from open-loop identification hold. The simulated process input $u^{(r)}$ can be obtained as follows. First, the transfer function $S(q) = 1/(1+C(q)G(q))$ from r_1 to u is estimated. Second, the simulated process input is calculated as $u^{(r)}(k) = S(q)r_1(k)$. Finally, $u^{(r)}(k)$ is used in (18.238).

18.11.2.2 Coprime Factor Identification

The idea of coprime factor identification is to identify the transfer functions from r_1 to y and from r_1 to u:

$$y(k) = G(q)S(q)r_1(k) + S(q)n(k) = G^{(y)}r_1(k) + S\,n(k), \qquad (18.240)$$

$$u(k) = S(q)r_1(k) + C(q)S(q)n(k) = G^{(u)}r_1(k) + C\,S\,n(k) \qquad (18.241)$$

with the sensitivity function $S(q) = 1/(1 + C(q)G(q))$. First, the transfer function $G^{(y)}(q)$ from r_1 to y is estimated. Second, the transfer function $G^{(u)}(q)$ from r_1 to u is estimated. Finally, the transfer function for the process can be calculated as

$$\hat{G}(q) = \frac{\hat{G}^{(y)}(q)}{\hat{G}^{(u)}(q)}. \qquad (18.242)$$

Both $\hat{G}^{(y)}(q)$ and $\hat{G}^{(u)}(q)$ contain an estimate of $S(q)$, which ideally should cancel in (18.242). In practice, however, the two estimates will be slightly different, and the order of $\hat{G}(q)$ will be equal to twice the order of $\hat{G}^{(y)}(q)$ or $\hat{G}^{(u)}(q)$ since $S(q)$ does not cancel exactly. This drawback can be overcome by considering the more advanced approach discussed in [574].

18.11.3 Identification for Control

Identification for control deals with the question of how to identify a model that serves as a basis for controller design. The simulation or prediction accuracy of a model may not be a reliable measure of the expected performance of the controller. Simple (artificial) examples can be constructed where a controller based on a low-order model with poor simulation or prediction performance works much better than a controller based on a more accurate high-order model. Thus, in identification for control, the goal is not a very accurate model but good control performance. It can be shown that the model accuracy around the crossover frequency is most important for model-based control design [276]. This is intuitively clear since the amplitude margin (absolute value of the open-loop frequency response at the crossover frequency) determines the stability properties of the control loop.

It can be shown under some conditions that the best model for control design is identified in closed loop with an indirect identification method [574]. The best controller for doing the closed-loop measurements is the (unknown) optimal controller. This leads to the so-called *iterative identification and control* schemes, which work as follows. First, closed-loop measurements are gathered with a simple stabilizing controller whose design is possibly based on heuristics and does not require an explicit model. Second, a model is identified with this data. Third, a new controller is designed with this model. Then all steps are repeated with the new controller until the model and thus the controller has converged. For more details on identification for control and optimal experimental design, refer to [184, 251], respectively.

18.12 Summary

As a general guideline for linear system identification, the user should remember the following issues:

- If little knowledge about the process is available, then start with an ARX model, which can be easily estimated with LS.
- If the available data set is small and/or very noisy, then try filtering and/or the COR-LS or the IV method to reduce the bias of the ARX model's parameters.
- If this does not give satisfactory results, try the ARMAX or the OE model.
- Try other model structures only if necessary since their estimation is more involved.
- Make use of prefiltering or specific noise models to shape the frequency characteristics of the model.
- If knowledge about the approximate process dynamics is available, then try an OBF model. In particular, processes with many resonances can be well modeled with an OBF approach based on Kautz filters.
- Do not use correlation-based methods for closed-loop identification.
- Identify the model with closed-loop data when the only goal of modeling is controller design.

For the extension to nonlinear model structures, the ARX and OE models are particularly important. FIR and OBF models are also used for nonlinear system identification. More advanced noise models as included in ARMAX, ARARX, and BJ models are rarely applied for nonlinear system identification. This is mainly due to the fact that nonlinear models are much more complex than linear ones. The flexibility and thus the variance error tend to be a considerable restriction in the nonlinear case. Thus, simple dynamics representations are more common for nonlinear models. For models with limited flexibility, a tradeoff has to be found between approximation errors due to unmodeled nonlinear behavior and those due to unmodeled dynamics.

18.13 Problems

1. Calculate and plot the frequency spectrum of an impulse, step, and ramp signal and comment on their suitability as excitation signals.
2. What is the difference between a parametric and non-parametric model? Give some examples.
3. Explain the term "stochastic model" and give examples for common model structures belonging to this class. For what kind of application can stochastic models be used?
4. Explain the term "deterministic model" and give examples for common model structures belonging to this class. For what kind of application can deterministic models be used?
5. Compare the ARX and OE model. Write down the model equations and draw a block diagram. Derive the optimal predictor. Explain the terms "equation error" and "output error." How can the model parameters be estimated? Which model is model realistic, i.e., which model's assumption about the process structure is better suited for most applications? What happens if these assumptions are violated? Why is the ARX model so widely used?
6. What is the key shortcoming of the ARX model and what causes it? Explain the most popular methods for overcoming this drawback.
7. Calculate the first 20 time steps of the step response of the following process: $y(k) = b_1 u(k-1) - a_1 y(k-1)$ with $b_1 = 0.3$ and $a_1 = -0.8$. Write two subroutines, one for the estimation of the parameters of a first-order ARX model and one for a first-order OE model. Inputs for the subroutines are the signals $u(k)$ and $y(k)$ for $k = 0, 1, 2, \ldots, N$ (here $N = 19$); outputs are the estimated parameters $\hat{b}_1$ and $\hat{a}_1$. Estimate the parameters of the ARX and the OE model. Plot the sum of squared errors for the ARX and the OE model for the parameters in the intervals $0 < b_1 < 1$ and $-1 < a_1 < 0$.

   Hints: For the ARX subroutine, set up the regression matrix first and then solve the LS problem. For the OE subroutine, write a loss function that takes the $u(k)$ data and the current parameter values $\hat{b}_1, \hat{a}_1$ as inputs, simulates the model output $\hat{y}(k)$, and returns the sum of squared errors $\sum_{k=0}^{N}(y(k) - \hat{y}(k))^2$. Then use a standard nonlinear optimization tool to minimize this loss function. This requires an initial guess of the parameter values, e.g., $\hat{b}_1^{(0)} = 0.5$, $\hat{a}_1^{(0)} = -0.5$.
8. Disturb the output data $y(k)$ from Problem 7 with a white noise signal of small, medium, and large variance σ^2. Estimate again the parameters of the ARX and the OE model. How do they differ from the noise-free results in Problem 7?
9. What remedies exist for the poor quality of the ARX parameter estimation? Explain their key ideas.
10. FIR and OBF models belong to the output error class. Why?
11. Explain the differences between FIR and OBF models. In which cases are Laguerre models suitable and in which cases do Kautz models offer advantages? What advantages do OBF models offer over FIR models and what price has to be paid?

12. Discuss the differences of an ARX and an OE model from a frequency domain analysis of their loss function given in (18.193).
13. What methods exist to handle offsets and what are their key ideas?
14. Consider the RLS algorithm with forgetting. Which additional conditions have to be fulfilled compared to an application of the RLS with forgetting to a static problem (e.g., estimation of polynomial coefficients)?
15. Consider a p-canonical model with p inputs and one output. What assumptions have to be made in order to apply an ARX model?

Chapter 19
Nonlinear Dynamic System Identification

Abstract This chapter addresses many fundamental issues arising when transitioning from nonlinear static to nonlinear dynamic models. Many aspects are very general in nature and independent of the specific model architecture. They are analyzed here. The two competing concepts of external and internal dynamics are contrasted. It is explained how equation and output errors are traditionally treated in the neural network terminology as series-parallel and parallel model structures. The additional difficulties feedback causes are discussed and how they are dealt with in learning as well. A large section discusses the much-neglected topic of suitable excitation signals for nonlinear dynamic processes and new ideas for their analysis. Finally, a new signal generator is proposed that is capable of generating very good excitation signals, which is much more flexible and in contrast to the conventional approach of postulating a specific parameterized signal type.

Keywords Nonlinear dynamics · Stability · External dynamics · Internal dynamics · Backpropagation-through-time · Excitation signal · Signal generator

This chapter gives an overview of the concepts for identification of nonlinear dynamic systems. Basic approaches and properties are discussed that are independent of the specific choice of the model architecture. Thus, this chapter is the foundation for both classical polynomial-based and modern neural network and fuzzy-based nonlinear dynamic models.

First, in Sect. 19.1 the transition from linear to nonlinear system identification is discussed. Next, two fundamentally different approaches for nonlinear dynamic modeling are presented in Sects. 19.2 and 19.3. Then Sect. 19.4 introduces a parameter scheduling approach that is a special case of the local linear model architectures analyzed in Chap. 22. The necessary extensions to deal with multivariable systems are discussed in Sect. 19.6. In the Sect. 19.7 some guidelines for the design of excitation signals are given, and in Sect. 19.9 the topic of dynamic order determination is addressed. Finally, a brief summary is given in Sect. 19.10.

O. Nelles, *Nonlinear System Identification*,
https://doi.org/10.1007/978-3-030-47439-3_19

19.1 From Linear to Nonlinear System Identification

For the sake of simplicity, the following equations are formulated for SISO systems. The extension to multivariable systems is straightforward; see Sect. 19.6. Furthermore, in order to keep the notation simple, it is assumed that all inputs and outputs possess identical dynamic order m, that the systems have no direct path from the input to the output (so that $u(k)$ does not immediately influence $y(k)$), and that no dead times exist. All these assumptions are in no respect essential; they just simplify the notation.

The simplest linear discrete-time input/output model is the autoregressive with exogenous input (ARX) or equation error model; see Sect. 18.5.1. The optimal predictor of an mth order ARX model is

$$\hat{y}(k) = b_1 u(k-1) + \ldots + b_m u(k-m) - a_1 y(k-1) - \ldots - a_m y(k-m). \quad (19.1)$$

This can be extended to a NARX (nonlinear ARX) model in a straightforward manner by replacing the linear relationship in (19.1) with some (unknown) nonlinear function $f(\cdot)$, that is,

$$\hat{y}(k) = f\left(u(k-1), \ldots, u(k-m), y(k-1), \ldots, y(k-m)\right). \quad (19.2)$$

This is the standard approach pursued in most engineering applications. If the model is to be utilized for controller design, often the less general form

$$\hat{y}(k) = b_1 u(k-1) + \tilde{f}\left(u(k-2), \ldots, u(k-m), y(k-1), \ldots, y(k-m)\right) \quad (19.3)$$

is chosen (assumed) because it is affine (linear plus possibly an offset) in the control input $u(k-1)$. Then the control law of an inverting controller can be directly calculated as

$$u(k) = [r(k+1) - \tilde{f}(u(k-1), \ldots, u(k-m+1), y(k), \ldots, y(k-m+1))]/b_1, \quad (19.4)$$

where $r(k+1)$ denotes the desired control output (reference signal) in the next time instant.

The transition to other, more complex model structures well known from linear system identification (Chap. 18) is discussed in Sect. 19.2. A comparison between the identification of linear and nonlinear input/output models shows that the problem of *estimating the parameters* b_i and a_i extends to the problem of *approximating the function* $f(\cdot)$. This class of nonlinear input/output models is called *external dynamics* models and is discussed in Sect. 19.2.

Although the NARX model approach in (19.2) and other input/output model structures with extended noise models cover a wide class of nonlinear systems [98, 346], some restrictions apply; for details refer to Sect. 19.2.4. A more general framework is given by state space approaches. A linear state space model

$$\underline{\hat{x}}(k+1) = \underline{A}\,\underline{x}(k) + \underline{b}\,u(k) \tag{19.5a}$$

$$\hat{y}(k) = \underline{c}^T\,\underline{x}(k) \tag{19.5b}$$

can also be extended to nonlinear dynamic systems in a straightforward manner:

$$\underline{\hat{x}}(k+1) = \underline{h}\left(\underline{x}(k), u(k)\right) \tag{19.6a}$$

$$\hat{y}(k) = g\left(\underline{\hat{x}}(k)\right)\,. \tag{19.6b}$$

If all states of the process $\underline{x}$ can be measured, the identification of a nonlinear state space model is equivalent to the approximation of the functions $\underline{h}(\cdot) = [h_1(\cdot)\ h_2(\cdot)\ \cdots\ h_m(\cdot)]^T$ that describe the m state updates and the function $g(\cdot)$ that represents the output equation. In the case where the state is measurable, a nonlinear state space model is to be preferred over an input/output model since the identification of an mth order process requires only the approximation of the $m+1$-dimensional (m states plus one input) functions $h_i(\cdot)$ for each state i ($i = 1, \ldots, m$) and a one-dimensional function $g(\cdot)$, while for input/output modeling, the function $f(\cdot)$ is $2m$-dimensional. Since the model complexity, the computational demand, and the required amount of data for function approximation increase strongly with the input space dimensionality (see the curse of dimensionality in Sect. 7.6.1), state space models are typically superior in these respects for $m \geq 2$. For a more extensive analysis on nonlinear state space models, refer to [535].

Unfortunately, complete state measurements are rarely realistic in practice. Therefore, at least some states are unknown, and thus the nonlinear state space model in (19.6a)–(19.6b) is much tougher to identify in reality. The states have to be considered as unknown quantities and must be estimated as well. This leads to modeling approaches with internal states. They are subsumed under the class of so-called *internal dynamics* models; see Sect. 19.3. The high complexity involved in the simultaneous estimation of the model states and parameters is the reason for the dominance of the much simpler input/output approaches.

A more recent alternative proposal on how to deal with the identification of nonlinear state space models can be found in [363, 441]. The scheme shown in Fig. 19.1 depicts the setting. The functions $\underline{h}(\cdot,\cdot)$ and $g(\cdot)$ are parametrized, e.g., in form of polynomials. First a *linear* state space model is identified from input/output data, e.g., by a subspace identification algorithm; see Sect. 18.10.3. Utilizing this state realization $\underline{x}(k)$ as starting point, the parameters of $\underline{h}(\cdot,\cdot)$ and $g(\cdot)$ are optimized nonlinearly with real-time recurrent learning or backpropagation-through-time; compare Sect. 19.5. Note that these nonlinear state space approaches are very powerful since they are capable to model *inner states* necessary, e.g., for hysteresis modeling which are extremely though to reconstruct from inputs and/or

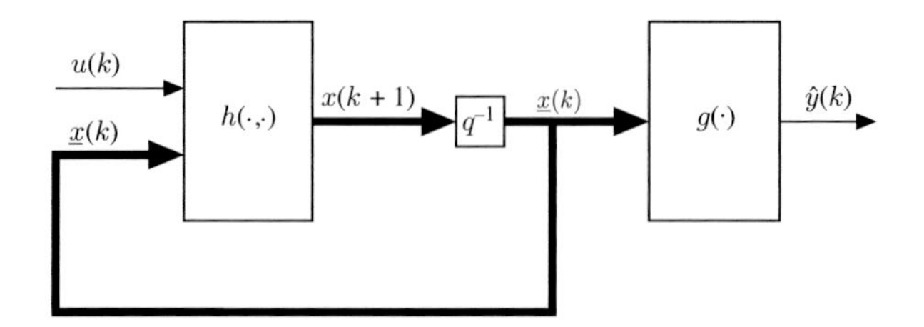

Fig. 19.1 Nonlinear state space approaches require the direct measurement or the estimation of the (internal) state vector $\underline{x}(k)$

outputs directly. Thus, the additional effort may pay off in some or even many cases. It is a topic of current research.

19.2 External Dynamics

The external dynamics strategy is by far the most frequently applied nonlinear dynamic system modeling and identification approach. It is based on the nonlinear input/output model in (19.2). The name "external dynamics" stems from the fact that the nonlinear dynamic model can be clearly separated into two parts: a nonlinear static approximator and an external dynamic filter bank; see Fig. 19.2. In principle, any model architecture can be chosen for the approximator $f(\cdot)$. However, from the large number of approximator inputs in Fig. 19.2, it is obvious that the approximator should be able to cope with relatively high-dimensional mappings, at least for high-order systems, i.e., for large m. Typically, the filters are chosen as simple time delays q^{-1}. Then they are referred to as *tapped-delay lines*, and if the approximator $f(\cdot)$ is chosen as a neural network, the whole model is usually called a *time-delay neural network (TDNN)* [347, 406]. Many properties of the external dynamics approach are independent of the specific choice of the approximator. These properties are analyzed in this section.

19.2.1 Illustration of the External Dynamics Approach

From a thorough understanding of the external dynamics approach, various important conclusions can be drawn with respect to the desirable features of a potential model architecture for the static approximator in Fig. 19.2.

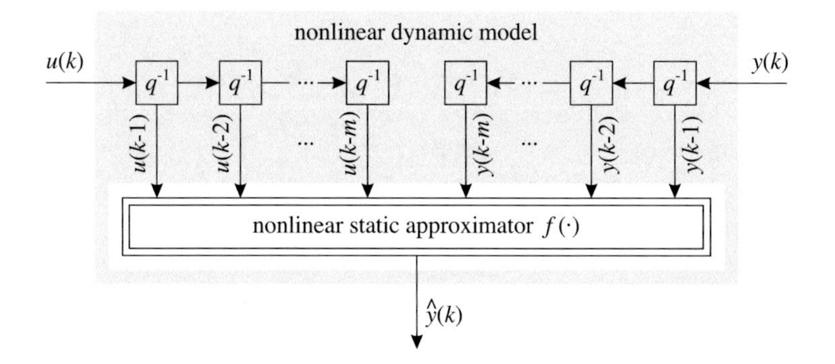

Fig. 19.2 External dynamics approach: the model can be separated into a static approximator and an external filter bank, which here is realized as a tapped-delay line

19.2.1.1 Relationship Between the Input/Output Signals and the Approximator Input Space

It is helpful to understand how a given input signal $u(k)$ and output signal $y(k)$ over time k correspond to the data distribution in the input space of the approximator $f(\cdot)$ spanned by previous values of the input and output. For a first-order system, it is possible to visualize this input space. As an example, the following Hammerstein system consisting of a static nonlinearity in series with a first-order time-lag system

$$y(k) = 0.1 \arctan(u(k-1)) + 0.9\, y(k-1) \tag{19.7}$$

will be considered; see Fig. 19.5a(left). For a step-like input signal as shown in Fig. 19.3a, the input space of the approximator is covered along lines with constant $u(k)$. For sine-like excitations of low and high frequency, the data distribution is illustrated in Fig. 19.3b and c. The following observations can be made:

- The approximator inputs cannot all be directly influenced independently. Rather, only $u(k)$ is chosen by the user, and all other delayed approximator inputs and outputs follow as a consequence.
- The lower the frequency of the input signal, the closer the data will be to the static nonlinearity (equilibrium) of the system.
- Naturally, the data distribution is denser close to the static nonlinearity than it is in off-equilibrium regions since systems with autoregressive components approach their equilibrium infinitely slowly.
- Highly dynamic input excitation is required in order to cover wide regions of the input space with data.

All these points make it clear that the excitation signal $u(k)$ has to be chosen with extreme care in order to gather as much information about the system as possible. This issue is analyzed in detail in Sect. 19.7.

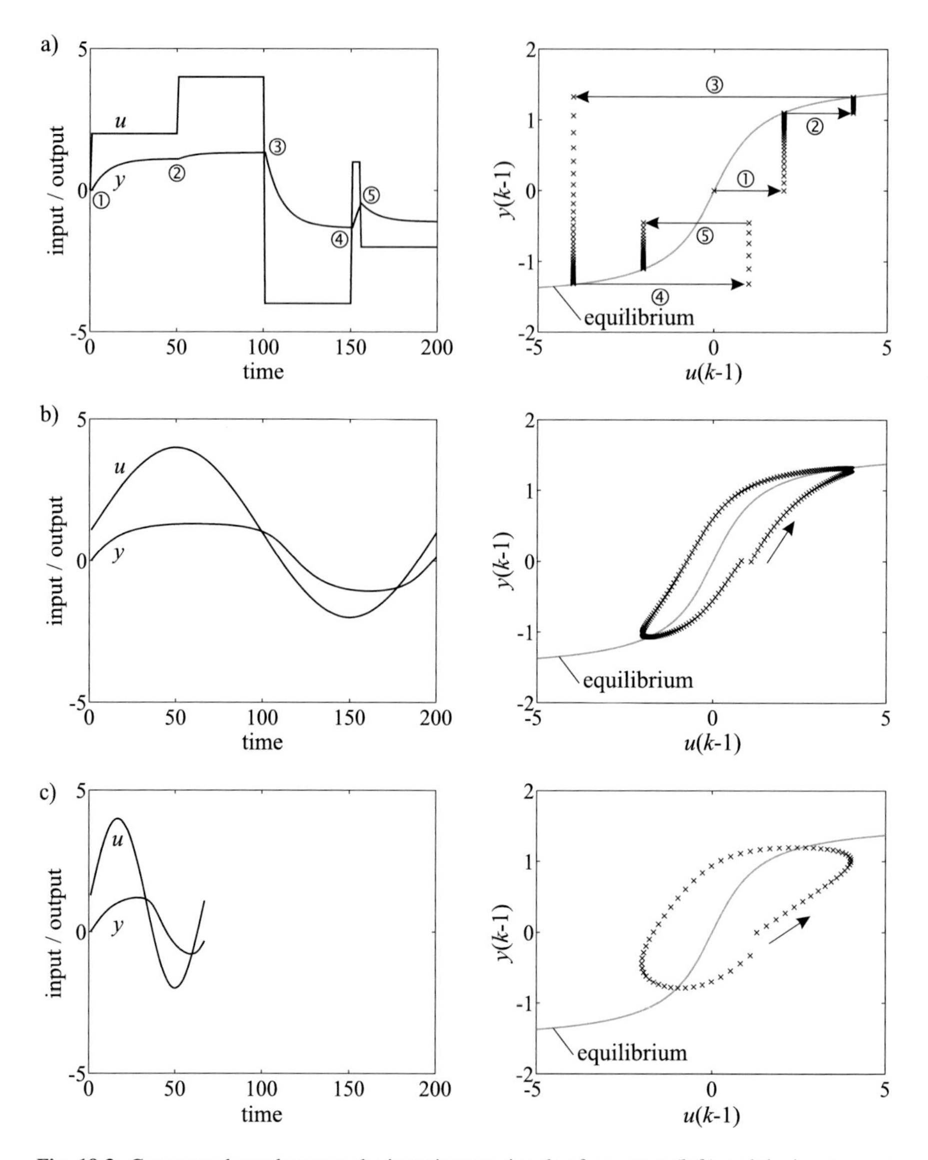

Fig. 19.3 Correspondence between the input/output signals of a system (left) and the input space of the approximator in external dynamics approaches (right): (**a**) step-like excitation, (**b**) sine excitation of low frequency, (**c**) sine excitation of high frequency. Note that the far off-equilibrium regions for small $u(k-1)$ and large $y(k-1)$ or vice versa can only be reached by large input steps. The region above the static nonlinearity represents decreasing inputs (steps ③ and ⑤), and the region below the equilibrium can be reached by increasing inputs (steps ①, ②, and ④)

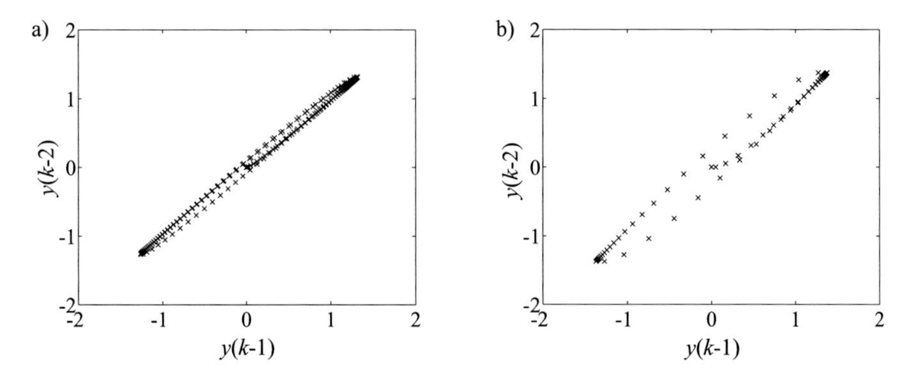

Fig. 19.4 Correlation between subsequent system outputs of a second-order system for (**a**) small and (**b**) large sampling times

For higher-order systems, the input space becomes higher dimensional, but the data distribution characteristics basically stay the same. However, it is worth mentioning that for systems of order $m \geq 2$, the previous outputs $y(k-i)$ for $i = 1, 2, \ldots, m$ are highly correlated, as Fig. 19.4a demonstrates. Although for increasing sampling times the correlation decreases, the data is always distributed along the diagonal of the $y(k-1)$-$y(k-2)$-...-$y(k-m)$-space; see Fig. 19.4b. This follows directly from the fact that the sampling time has to be chosen small enough to capture variations in the system output, and consequently $y(k-i) \approx y(k-i-1)$. A similar property holds for the inputs except at the time instants where steps take place. Therefore, in the external dynamics approach, wide regions of the input space are empty. In principle, they cannot be reached, independent of the choice of $u(k)$. This property can be exploited in order to weaken the curse of dimensionality; see Sect. 7.6.1. Obviously, the model architectures that uniformly cover the input space with basis functions (grid-based partitioning) are not well suited for modeling dynamic systems with the external dynamics approach.

19.2.1.2 Principal Component Analysis and Higher-Order Differences

The external dynamics approach leads to diagonal data distributions for higher-order ($m \geq 2$) systems, as shown in Fig. 19.4. This may motivate the application of a principal component analysis (PCA) as a preprocessing technique in order to transform the input axes; see Sect. 6.1 and 13.3.6. The dilemma with unsupervised methods such as PCA is that they generate a more uniform data distribution, which is particularly important for grid-based approaches, but there is no guarantee that the problem is easier to solve with the new axes. It rather depends on the (unknown) structure of the process nonlinearity. For dynamic systems, a PCA yields similar results as the utilization of higher-order differences which are explained in the following.

The external dynamics approach in (19.2) is based on the mapping of previous inputs and outputs to the actual output. Instead the output can also be expressed as

$$\hat{y}(k) = \tilde{f}\Big(u(k-1), \Delta u(k-1), \ldots, \Delta^{m-1}u(k-1),$$
$$y(k-1), \Delta y(k-1) \ldots, \Delta^{m-1}y(k-1)\Big), \tag{19.8}$$

where Δ is the difference operator, i.e., $\Delta = 1 - q^{-1}$, $\Delta^2 = (1 - q^{-1})^2 = 1 - 2q^{-1} + q^{-2}$, etc. Thus, a second-order system would be described as

$$\hat{y}(k) = \tilde{f}\,(u(k-1), u(k-1) - u(k-2), y(k-1), y(k-1) - y(k-2))\,. \tag{19.9}$$

In comparison to (19.2), this approach may yield two advantages. First, the input data distribution is more uniform. Second, the differences can be interpreted as changes and changes of changes, that is, as the derivatives of the signals. In particular, when a fuzzy or neuro-fuzzy model architecture is used for approximation of the function $\tilde{f}$, this interpretation can yield a higher transparency of the rule base. On the other hand, noise effects become larger with increasing order of the differences. Nevertheless, as for PCA it can well happen that the approximation problem becomes harder owing to the higher differences approach. These ideas are not yet examined completely and require future research.

19.2.1.3 One-Step Prediction Surfaces

The question arises: What does the function $f(\cdot)$ in (19.2), which has to be described by a nonlinear static approximator, look like? For first-order systems, it can be visualized as $y(k) = f(u(k-1), y(k-1))$. The function $f(\cdot)$ is a one-step predictor since it maps previous inputs and outputs to the actual model output. Figure 19.5 compares these one-step prediction surfaces of a first-order Hammerstein and Wiener system. The Hammerstein system depicted in Fig. 19.5a(left) is the same as that described in (19.7). The Wiener system is obtained by swapping the static nonlinearity and the dynamic block as shown in Fig.19.5a(right) and follows the equation

$$y(k) = \arctan\left[0.1\, u(k-1) + 0.9 \tan(y(k-1))\right]\,. \tag{19.10}$$

The one-step prediction surface of a linear system would be a (hyper)plane where the slopes are determined by its linear parameters b_i and $-a_i$. For the Hammerstein and Wiener systems, the one-step prediction surfaces are shown in Fig. 19.5b. They

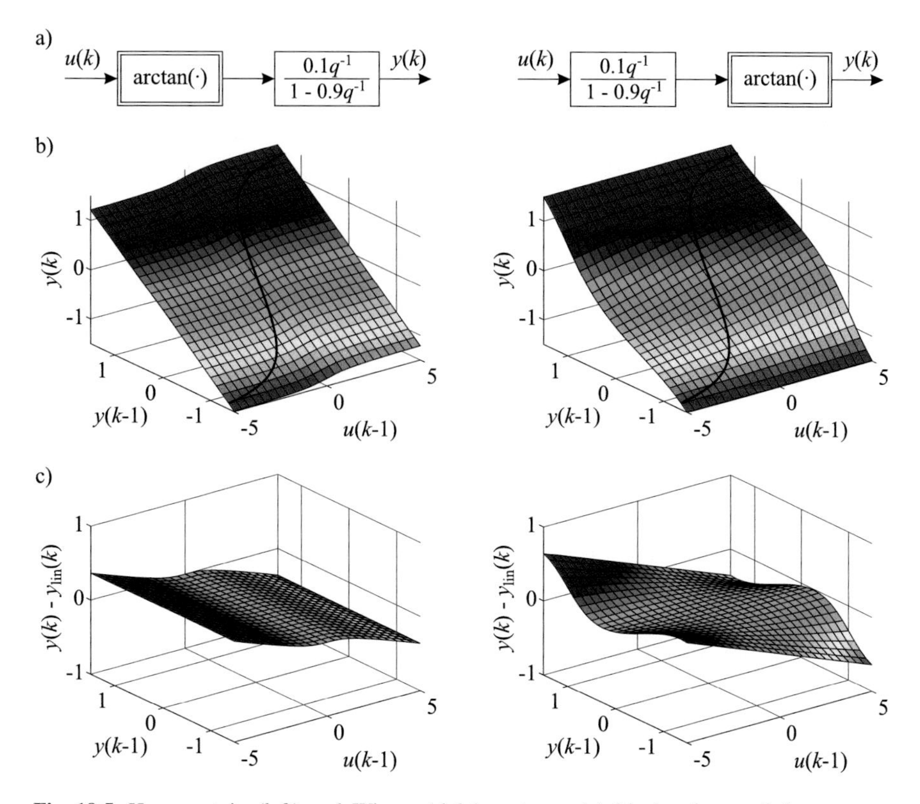

Fig. 19.5 Hammerstein (left) and Wiener (right) systems: (**a**) block scheme of the systems, (**b**) one-step prediction surfaces with static nonlinearity as bold curves, (**c**) difference surfaces representing the nonlinear system parts

represent all information about the systems. At any operating point (u_0, y_0), a linearized model

$$\Delta y(k) = b_1 \Delta u(k-1) - a_1 \Delta y(k-1) \tag{19.11}$$

with $\Delta u(k) = u(k) - u_0$ and $\Delta y(k) = y(k) - y_0$ can be obtained by evaluating the local slopes of the one-step prediction surface:

$$b_1 = \left. \frac{\partial f(u(k-1), y(k-1))}{\partial u(k-1)} \right|_{(u_0,\, y_0)} ,$$

$$-a_1 = \left. \frac{\partial f(u(k-1), y(k-1))}{\partial y(k-1)} \right|_{(u_0,\, y_0)} . \tag{19.12}$$

Although both systems possess strongly nonlinear character, with a gain varying from 1 at $u_0 = 0$ to 1/25 at $u_0 = \pm 5$, the one-step prediction surfaces are only slightly nonlinear. Note that this is no consequence of the special types of systems considered here. Rather, almost all nonlinear dynamic systems possess relatively slightly nonlinear one-step prediction surfaces. The next paragraph on the effect of the sampling time analyzes this issue in more detail. The one-step prediction surface of the Wiener system is more strongly nonlinear than that of the Hammerstein system. Both surfaces are equivalent at the static nonlinearity (bold curves in Fig. 19.5b) but differ significantly in the off-equilibrium regions.

If the one-step prediction surfaces of the linear dynamic block $y(k) = 0.1\,u(k-1) + 0.9\,y(k-1)$ of the Hammerstein and Wiener systems are subtracted from Fig. 19.5b, the difference surfaces in Fig. 19.5c result. They represent the "nonlinear part" of the systems. Obviously, for the Hammerstein system, this difference surface is nonlinear only in $u(k-1)$ but *linear* in $y(k-1)$, while for the Wiener system, it is nonlinear in both inputs $u(k-1)$ and $y(k-1)$. These observations are in agreement with (19.7) and (19.10).

From the above discussions, it is clear that Hammerstein systems generally are much easier to handle than Wiener systems, because they possess a weaker nonlinear one-step prediction surface and depend only on the previous system inputs $u(k-i)$, $i = 1, 2, \ldots, m$, in a nonlinear way, while they are linear in the previous system outputs $y(k-i)$.

Another important conclusion can be drawn from this paragraph. The one-step prediction surfaces and thus the function $f(\cdot)$ possess relatively slight nonlinear characteristics even if the underlying system is strongly nonlinear. This observation is a powerful argument for the utilization of *local linear* modeling schemes.

19.2.1.4 Effect of the Sampling Time

The question arises: Why are the one-step prediction surfaces only slightly nonlinear even if the underlying system is strongly nonlinear? The answer to this question is closely related to the equation error and output error discussion in the next sections. The function $f(\cdot)$ describes a one-step prediction. In order to predict many steps into the future, several subsequent one-step predictions have to be carried out. For example, for a first-order system, the one-step prediction is given by

$$y(k+1|k) = f\left(u(k), y(k)\right) . \tag{19.13}$$

The two-step prediction becomes

$$y(k+2|k) = f\left(u(k+1), y(k+1)\right) = f\left[u(k+1), f\left(u(k), y(k)\right)\right] . \tag{19.14}$$

This means that the functions are nested. With increasing prediction horizon, the functions are nested more and more times. For simplicity it will be assumed that the input is constant as it would be in case of a step excitation, i.e., $u(k) = u(k+1) = \ldots = u(k+h)$. Then (19.14) becomes $y(k+2|k) = f(f(u(k), y(k)))$, and the

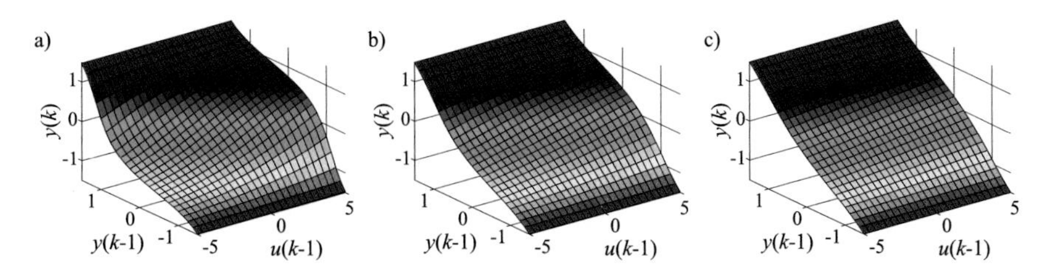

Fig. 19.6 One-step prediction surface of the Wiener system in Fig. 19.5(right) for different sampling times: $T_{95}/T_0 =$ (**a**) 15, (**b**) 30, (**c**) 60

system output can be calculated h steps into the future by nesting the function $f(\cdot)$ h times:

$$y(k+h|k) = \underbrace{f(\cdots f(f}_{h \text{ times}}(u(k), y(k))) \cdots) . \tag{19.15}$$

In order to evaluate a step response of the system approximately, $h = T_{95}/T_0$ steps have to be predicted into the future (T_{95} = settling time of the system, T_0 = sampling time). By nesting a function so many times, even slightly nonlinear functions $f(\cdot)$ yield significantly nonlinear system behavior. The smaller the sampling time T_0 chosen, the more the one-step prediction surface approaches a linear function. This relationship is illustrated in Fig. 19.6, where the one-step prediction surfaces of the Wiener system from the above example are compared for three different sampling times. The relationships are similar for other systems.

19.2.2 Series-Parallel and Parallel Models

In analogy to linear system identification (Chap. 18), a nonlinear dynamic model can be used in two configurations: for *prediction* and for *simulation*. *Prediction* means that on the basis of previous *process inputs* $u(k-i)$ *and process outputs* $y(k-i)$, the model predicts one or several steps into the future. A requirement for prediction is that the process output is measured during operation. In contrast, *simulation* means that on the basis of previous *process inputs* $u(k-i)$, *only* the model simulates future outputs. Thus, simulation does not require process output measurements during operation. Figure 19.7 compares the model configuration for prediction (a) and simulation (b). In former linear system identification literature [141] and in the context of neural networks, fuzzy systems, and other modern nonlinear models, the one-step prediction configuration is called a *series-parallel model*, while the simulation configuration is called a *parallel model*.

Typical applications for prediction are weather forecast and stock market predictions, where the current state of the system can be measured; for an extensive

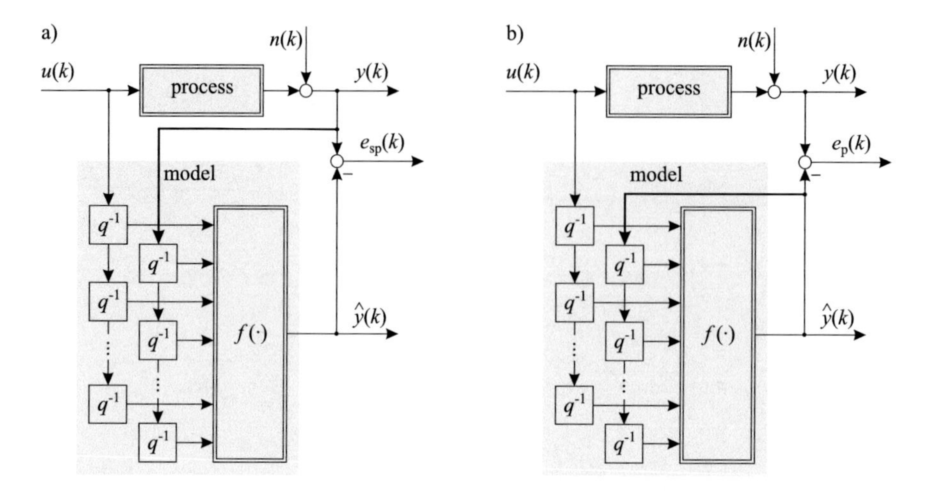

Fig. 19.7 (**a**) One-step prediction with a series-parallel model. (**b**) Simulation with a parallel model

discussion, see, e.g., [592] and Sect. 1.1.2. Also, in control engineering applications, prediction plays an important role, e.g., for the design of a minimum variance or a predictive controller.

Simulation is required whenever the process output cannot be measured during operation; see Sect. 1.1.3. This is the case when a process is to be simulated without coupling to the real system or when a sensor is to be replaced by a model. Also, for fault detection and diagnosis, the process output may be compared with the simulated model output in order to extract information from the residuals. Furthermore, as explained in the next section, the utilization of prediction or simulation is connected with special assumptions on the noise properties.

The two configurations shown in Fig. 19.7 cannot only be distinguished for the model operation phase but also during training. The model is trained by minimizing a loss function dependent on the error $e(k)$. For the series-parallel model, $e_{\text{sp}}(k)$ is called the *equation error*, and for the parallel model, $e_{\text{p}}(k)$ is called the *output error*; see also Sect. 19.2.3. This corresponds to the standard terminology in linear system identification; see Figs. 18.24 and 18.33 and Sect. 18.3.1.

For a second-order model, the one-step prediction is calculated with the previous process outputs as

$$\hat{y}(k) = f(u(k-1), u(k-2), \; y(k-1), y(k-2)) \,, \tag{19.16}$$

while the simulation is evaluated with the previous model outputs as

$$\hat{y}(k) = f(u(k-1), u(k-2), \; \hat{y}(k-1), \hat{y}(k-2)) \,. \tag{19.17}$$

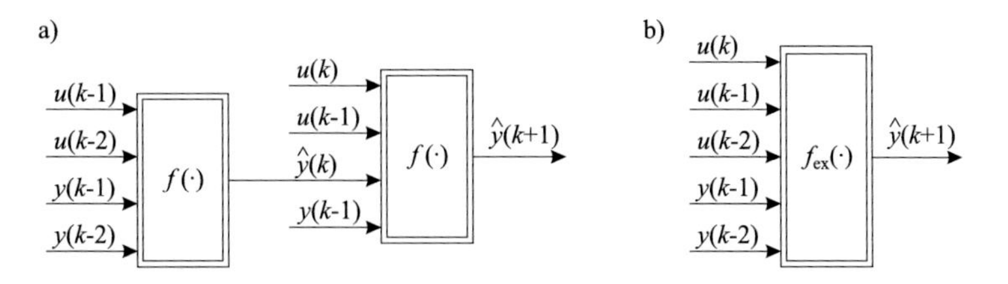

Fig. 19.8 Two-step prediction with (**a**) subsequent application of a one-step predictor and (**b**) an extended predictor with an additional input that performs the two-step prediction at once

The one-step prediction is purely *feedforward*, while the simulation is *recurrent*. Thus, (19.16) is called a feedforward model, and (19.17) is called a recurrent model. The case of multi-step prediction can be solved in two different ways; see Fig. 19.8. The standard approach is shown in Fig. 19.8a, where two one-step predictors are used in series. The second predictor is a hybrid between (19.16) and (19.17) because one process output $y(k-1)$ is utilized, but also a model output $\hat{y}(k)$ has to be used since $y(k)$ is not available. For a three-step predictor, only the model outputs can be used by the last predictor since no process outputs would be available. Generally, for an h-step predictor, the last prediction step is fully based on model outputs if $h > m$. Thus, for long prediction horizons h, the difference between prediction and simulation fades. The alternative prediction configuration shown in Fig. 19.8b predicts two steps at once. However, this requires a different approximator $f_{\rm ex}(\cdot)$ with an additional input for $u(k)$. The larger the prediction horizon is, the higher dimensional $f_{\rm ex}(\cdot)$ becomes. Furthermore, a separate predictor has to be trained for each prediction horizon. These drawbacks make approach (b) usually impractical.

In this book it is assumed that the goal for a model is to perform simulation, i.e., it will be used in a parallel configuration as shown in Fig. 19.7b. This is a much harder task than a one-step prediction since feedback is involved. Note that a model that is *used* in parallel configuration does *not* necessarily have to be *trained* in a parallel configuration as well.

19.2.3 Nonlinear Dynamic Input/Output Model Classes

In analogy to linear dynamic models (Chap. 18), nonlinear counterparts can be defined. In order to distinguish between nonlinear and linear models, it is a common notation to add an "N" for "nonlinear" in front of the linear model class name.

All nonlinear dynamic input/output models can be written in the form

$$\hat{y}(k) = f\left(\underline{\varphi}(k)\right) \tag{19.18}$$

where the regression vector $\underline{\varphi}(k)$ can contain previous and possibly current process inputs, previous process or model outputs, and previous prediction errors. It can be distinguished between models with and without output feedback. A more detailed overview can be found in [535, 536].

19.2.3.1 Models with Output Feedback

As for linear systems, models with output feedback are the most common ones. The regression vector $\underline{\varphi}(k)$ contains previous process or model outputs and possibly prediction errors. The three most common linear model structures are ARX, ARMAX, and OE models; see Sect. 18.5. Their nonlinear counterparts possess the following regression vectors (with $e(k) = y(k) - \hat{y}(k)$):

$$\begin{aligned} \text{NARX}: \ \underline{\varphi}(k) &= [u(k-1) \ \cdots \ u(k-m) \ \ y(k-1) \ \cdots \ y(k-m)]^T \\ \text{NARMAX}: \ \underline{\varphi}(k) &= [u(k-1) \ \cdots \ u(k-m) \ \ y(k-1) \ \cdots \ y(k-m) \\ &\quad\ \ e(k-1) \ \cdots \ e(k-m)]^T \\ \text{NOE}: \ \underline{\varphi}(k) &= [u(k-1) \ \cdots \ u(k-m) \ \ \hat{y}(k-1) \ \cdots \ \hat{y}(k-m)]^T \end{aligned}$$

Thus, the NARX model is trained in series-parallel configuration (Fig. 19.7a), and the NOE model is trained in parallel configuration. The NARMAX model requires both process outputs $y(k-i)$ and model outputs $\hat{y}(k-i)$ contained in $e(k-i)$. Note that explicit noise modeling as for NARMAX models implies additional inputs for the approximator $f(\cdot)$ for the $e(k-i)$. Since for nonlinear problems the complexity usually increases strongly with the input space dimensionality (curse of dimensionality), the application of lower-dimensional NARX or NOE models is more widespread. More complex noise models like NBJ or NARARX structures are uncommon because the increase in input space dimensionality does not usually pay off in terms of additional flexibility.

One drawback of models with output feedback is that the choice of the dynamic order m is crucial for the performance and no really efficient methods for its determination are available. Often the user is left with a trial-and-error approach. This becomes particularly bothersome when different orders n_u, n_y instead of m are considered for the input and output, and also a dead time d has to be taken into account. For a discussion of more sophisticated methods than trial and error, refer to Sect. 19.9.

Another disadvantage of output feedback is that, in general, stability cannot be proven for this kind of model. For special approximator architectures such as local linear modeling schemes, special tools are available that might allow one to prove the stability of the model; refer to Sect. 22.4. Generally, however, the user is left with extensive simulations in order to check stability.

A further inconvenience caused by the feedback is that the static nonlinear behavior of the model has to be computed iteratively by solving the following nonlinear equation:

$$y_0 = f(\underbrace{u_0, \ \dots, \ u_0}_{m \text{ times}}, \underbrace{y_0, \ \dots, \ y_0}_{m \text{ times}}) \tag{19.19}$$

with $u(k-1) = \dots = u(k-m) = u_0$, $y(k-1) = \dots = y(k-m) = y_0$ and where (u_0, y_0) is the static operating point.

In opposition to these drawbacks, models with output feedback compared with those without output feedback have the strong advantage of being a very compact description of the process. As a consequence, the regression vector $\underline{\varphi}(k)$ contains only a few entries, and thus the input space for the approximator $f(\cdot)$ is relatively low dimensional. Owing to the curse of dimensionality (Sect. 7.6.1), this advantage is even more important when dealing with nonlinear than with linear systems.

The advantages and drawbacks of NARX and NOE models are similar to the linear case; see Sects. 18.5.1 and 18.5.4. On the one hand, the NARX structure suffers from unrealistic noise assumptions. This leads to biased parameter estimates in the presence of disturbances, to a strong sensitivity with respect to the sampling time (too fast sampling degrades performance), and to an emphasis of the model fit on high frequencies. On the other hand, the NARX structure allows the utilization of linear optimization techniques since the equation error is linear in the parameters. This advantage, however, carries over to nonlinear models only if the approximator is linearly parameterized like polynomials, RBF networks, or local model approaches. For nonlinearly parameterized approximators such as MLP networks, nonlinear optimization techniques have to be applied anyway. In such cases, the NARX model training is still simpler because no feedback components have to be considered in the gradient calculations, but this advantage over the NOE approach is not very significant. These relationships are summarized in Table 19.1. Note that the nonlinear optimization problems arising for NOE models are more involved owing to the dynamic gradient calculations required to account for the recurrency; see Sect. 19.5 for more details.

Table 19.2 illustrates the use of NARX or NOE models from the perspective of the intended model use. Clearly, the NARX model should be used for training if the model is to be applied for one-step prediction. The NARX model minimizes exactly this one-step prediction error and is simpler to train than NOE. If the

Table 19.1 Type of optimization problem arising for NARX and NOE models when the approximator $f(\cdot)$ is linear or nonlinear parameterized

	NARX	NOE
Linear parameterized $f(\cdot)$	Linear (LS, no gradients required)	Nonlinear (dynamic gradients)
Nonlinear parameterized $f(\cdot)$	Nonlinear (static gradients)	Nonlinear (dynamic gradients)

Table 19.2 Training of NARX and NOE models, depending on the intended model use

	NARX (series-parallel)	NOE (parallel)
Model used for one-step prediction	×	Not sensible
Model used for simulation	As approximation	×

intended model is a simulation, the situation is less clear. On the one hand, the NOE model is advantageous because it yields the optimal simulation error, which is exactly the goal of modeling. On the other hand, the NARX model is simpler to train, in particular if $f(\cdot)$ is linearly parameterized; see Table 19.1. Thus, both model structures can be reasonably used. NOE matches the modeling goal better; NARX is only an approximation since it minimizes the one-step prediction error, but as compensation it offers other advantages such as simpler and faster training. A recommended strategy is to train a NARX model first and possibly utilize it as an initial model for a subsequent NOE model optimization.

As mentioned above, the training of an NOE model independently of the chosen approximator always requires nonlinear optimization schemes with a quite complex gradient calculation due to their recurrent structure; see Sect. 19.5. This disadvantage may be compensated by the more accurately estimated parameters. In contrast to NARX models, an NOE model can discover an *error accumulation* that might lead to inferior accuracy or even model instability. This can occur when small prediction errors accumulate to larger ones owing to the model output feedback (the model is fed with the already predicted and thus inaccurate signals). This effect can be observed and diminished with an NOE model during training. In contrast, a NARX model that is fed with previous process outputs during training experiences this effect due to feedback the first time when it is used for simulation.

As demonstrated in [407, 408], the error accumulation is a serious problem for NARX models. The one-step prediction error (on which the NARX model optimization is based) may decrease, while simultaneously the output error (on which the NOE model optimization is based) increases. In extreme situations, the model can become unstable but nevertheless can possess very good one-step prediction performance.

It is especially important to avoid the error accumulation effect in the extrapolation regions since there the danger of instability is large. This can be avoided by ensuring stable extrapolation behavior, which can relatively easily be realized with a *local linear* modeling approach; see Sect. 22.4.3.

19.2.3.2 Models Without Output Feedback

When no output feedback is involved, the regression vector $\underline{\varphi}(k)$ contains only previous or filtered inputs. The number of required regressors for models without output feedback is significantly higher than for models with output feedback. This drawback is known from linear models (Chap. 18), but because of the curse of dimensionality, it has more severe consequences in the nonlinear case. Therefore, only approximators that can deal well with high-dimensional input spaces can be applied.

Nonlinear Finite Impulse Response (NFIR) Models

As discussed in the previous paragraph, the output feedback leads to various drawbacks. These problems are circumvented by a nonlinear finite impulse response (NFIR) model because it employs no feedback at all. The regression vector of the NFIR model consists only of previous inputs:

$$\text{NFIR}: \ \underline{\varphi}(k) = [u(k-1)\ u(k-2)\ \cdots\ u(k-m)]^T . \tag{19.20}$$

The price to be paid for the missing feedback is that the dynamic order m has to be chosen very large to describe the process dynamics properly. Theoretically, the dynamic order must tend to infinity ($m \to \infty$). In practice, m is chosen according to the settling time of the process, typically in the range of $T_{95} = 10\,T_0 - 50\,T_0$, where T_0 is the sampling time. Since NFIR models are not recurrent but purely feedforward, their stability is ensured (although an NFIR model can represent an unstable process for the first m sampling instants in a step or impulse response). Furthermore, the static nonlinear behavior can be calculated simply by setting $u(k-i) = u_0$ for all $i = 1, \ldots, m$ in (19.20). Although these advantages are appealing, NFIR models find only few applications because of the required high dynamic order m. This leads to a high-dimensional input space for the function $f(\cdot)$, resulting in a large number of parameters for any approximator. However, for some applications an NFIR approach is useful, e.g., in signal processing applications [233] or if *inverse process models* are required, which usually possess highly differentiating character [202].

Extending the comparison between ARX and FIR in Sect. 18.6.1 to NARX versus NFIR, the following specific points can be made for *nonlinear* system identification:

- While in the linear case, FIR requires much more parameters than ARX, in the nonlinear case, it requires much more dimensions. Estimating more parameters is problematic because the complexity of a matrix inversion is $\mathcal{O}(n^3)$, and additionally FIR models often exhibit too high variance errors. However, more *dimensions* are often catastrophic due to the curse of dimensionality.
- The above point is the main reason why NARX is much more widely used than NFIR. In the assessment of the authors, the number of NARX vs. NFIR model applications might be around a ratio of at least 20:1.
- NARX models may run into stability problems. While in the linear case, model stability can at least be readily checked by calculating the poles, this is not so easy in the nonlinear case. This issue is a constant concern in industry and calls for better solutions.

- Most nonlinear IIR structures (ARX, OE, ...) approximate the one-step prediction surface. Therefore, they interpolate transfer functions. In contrast, all nonlinear orthonormal basis function approaches (NFIR, NLaguerre, ...) interpolate *signals*. Interpolating transfer functions can lead to very strange effects for models of order higher than 1. This is due to the fact that interpolating polynomials can yield unexpected roots and thus dynamics.
- The advantages of linear regression structures like ARX and FIR in the linear world are lost or at least diminish in the nonlinear world. Only for certain model architectures (like local model networks or Gaussian process models) that heavily rely on least squares, big benefits can be drawn from ARX or FIR. Generally, i.e., for nonlinearly parameterized approximators, the computational advantages compared to OE structures fade.
- The shape of the function to be approximated is highly different for NARX and NFIR. NARX requires a low-dimensional one-step prediction surface which often differs only slightly (but decisively) from a hyperplane, and the "strength" of the nonlinear behavior is generated by nesting this function as a consequence of feedback. NFIR generally requires a very high-dimensional function that is extremely though to approximate due to the curse of dimensionality.

Most of the above arguments for (N)FIR models generalize to any orthonormal basis function model (like Laguerre or Kautz). However, important new and complex issues arise with respect to the determination of the filter pole(s) that are beyond the scope of this book.

In the linear world, both approaches are heavily used in many applications. In the nonlinear world, the high dimensionality of NFIR practically ruins this approach.

Local linear model networks offer a loophole to this dilemma. The input space for the local linear models may be chosen high-dimensional, therefore yielding many parameters to be estimated for the local FIR models. But the input space for the validity functions that control which local linear model is how active can often be chosen very low-dimensional comparable (or identical) to NARX. For more details on these different input spaces, refer to Sect. 14.1 and Chap. 15. This brings the tradeoffs between NARX and NFIR back to (or at least closer to) the tradeoffs in the linear scenario.

Nonlinear Orthonormal Basis Function (NOBF) Models

The main drawback of the NFIR model is the need for a large dynamic order m. The NOBF model reduces this disadvantage by incorporating prior knowledge about the process dynamics into the linear filters. Typically, orthonormal Laguerre and Kautz filters are utilized; see Sect. 18.6.5. These filters are called orthonormal because they have orthonormal impulse responses. For processes with well-damped behavior (real pole), Laguerre filters are applied, while for processes with resonant behavior (dominant conjugate complex poles), Kautz filters are used. Information about more than one pole or pole pair, if available, can be included by generalized orthonormal basis functions introduced in [245, 246, 575]. The Laguerre and Kautz filters are completely determined by the real pole and complex pole pair, respectively. With these filters, the regression vector of the NOBF model becomes

$$\underline{\varphi}(k) = [L_1(q)u(k) \;\; L_2(q)u(k) \;\; \cdots \;\; L_m(q)u(k)]^T \,, \tag{19.21}$$

where $L_i(q)$ denote the orthonormal filters.

The choice of the number of filters m depends strongly on the quality of the prior knowledge, i.e., the assumed poles; see Sect. 18.6.5 for more details. The more accurate the assumed poles are, the lower m can be chosen. In the special case, where the assumed pole p lies in the z-plane origin ($p = 0$), the NOBF model recovers the NFIR model ($L_i(q) = q^{-i}$). Because $p = 0$ lies in the center of the unit disk, the NFIR model is optimal under all non-recurrent structures if nothing about the process is known but its stability.

For NOBF models all advantages of the NFIR structure are retained. Stability is guaranteed if the assumed poles for the Laguerre and Kautz filters design are stable. The most severe drawback of NFIR models, i.e., the high dynamic order, can be overcome with NOBF models by the incorporation of prior knowledge. This knowledge can, for instance, be obtained by physical insights about the process, or it can be extracted from measured step responses; see Sect. 18.6.5. However, if the prior knowledge is very inaccurate, a high dynamic order m, as for NFIR models, might be required in order to describe the process dynamics sufficiently well. Alternatively, the OBF poles can be optimized. In [238] a new method is proposed based on the low correlation between the poles corresponding to different physical model inputs.

While it is sufficient for linear OBF models to assume one pole or pole pair of the linear process, NOBF models may have to include various poles since the nonlinear dynamics can be dependent on the operating point. This is not the case if only the process gain varies significantly with the operating point, as for Hammerstein and Wiener systems. But for processes with significant dynamics that depend on the operating point, a number of different orthonormal filter sets may have to be incorporated in the NOBF model, or m must be increased. Then, however, the number of regressors can become huge, as in the NFIR approach. A possible solution to this dilemma is proposed in Sect. 22.8 on the basis of a *local linear* model architecture.

19.2.4 Restrictions of Nonlinear Dynamic Input/Output Models

As already mentioned in Sect. 19.1, nonlinear dynamic input/output models can describe a large class of systems [346] but are not so general as the nonlinear state space models. In particular, limitations arise for processes with non-unique nonlinearities such as hysteresis and backlash, where internal non-measurable states play a decisive role and partly for processes with non-invertible nonlinearities.

The latter restriction will be illustrated for the example depicted in Fig. 19.9. This process is of Wiener type with a non-invertible static nonlinearity $g(\cdot)$. This type of process cannot be described in terms of nonlinear dynamic input/output models when the state $x(k)$ cannot be measured. For example, consider a first-order dynamic system. Then the Wiener process follows the equation

$$x(k) = b_1 u(k-1) - a_1 x(k-1)\,, \qquad y(k) = g\left(x(k)\right)\,. \tag{19.22}$$

The Wiener process output can be written as

$$\begin{aligned} y(k) &= g\left(b_1 u(k-1) - a_1 x(k-1)\right) \\ &= g\left(b_1 u(k-1) - a_1 g^{-1}(y(k-1))\right)\,, \end{aligned} \tag{19.23}$$

where $g^{-1}(\cdot)$ is the inverse of $g(\cdot)$. If $g^{-1}(\cdot)$ exists, (19.23) can be written as a nonlinear first-order dynamic input/output model according to (19.2). Otherwise, it cannot be formulated as an input/output model. Rather internal states are required in order to describe $x(k)$. This requirement is met by an explicit state space model, possibly by NFIR and NOBF models depending on the specific optimization scheme, and by the internal dynamics approaches discussed in Chap. 23. All these strategies necessarily call for complex nonlinear optimization schemes because the input data distribution (partly) changed (the trajectory of $x(k)$) during the model parameter tuning. Therefore learning strategies like backpropagation-through-time or real-time recurrent learning are required; compare Sect. 19.5.

However, note that an input/output model can still be capable of *approximating* the system; the approximation error depends on the specific choice of $g(\cdot)$. Furthermore, no processes that contain such Wiener subsystems can be properly modeled. Demanding the invertibility of $g(\cdot)$ is a strong restriction. It necessarily implies the monotonicity of $g(\cdot)$.

The somewhat formal discussion above can be easily verified by the following intuitive argument. Suppose $g(\cdot)$ is non-invertible (e.g., $g(x) = x^2$ for both positive and negative x). It is impossible to unequivocally conclude from measurements of $y(k)$ on $x(k)$. However, a model has to be able to implicitly reconstruct $x(k)$ in order to properly identify the dynamic system between $u(k)$ and $x(k)$. If the input of the static nonlinearity $x(k)$ could be measured, those problems would vanish.

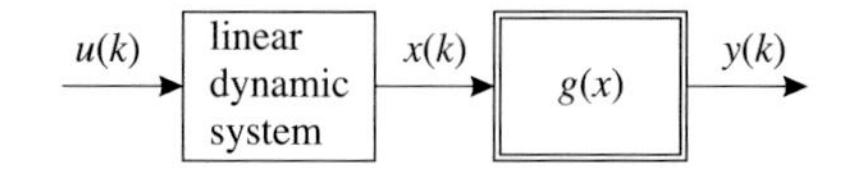

Fig. 19.9 A Wiener system cannot be modeled as input/output model if $g(\cdot)$ is not invertible

From this discussion, it is obvious that the NFIR and NOBF approaches are more general and will be successful in describing the system in Fig. 19.9 since they are based on filtered versions of $u(k)$ only which can reconstruct $x(k)$ successfully.

19.3 Internal Dynamics

Models with internal dynamics are based on the extension of static models with internal memory; see Fig. 19.10. Models with internal dynamics can be written in the following state space representation:

$$\underline{\hat{x}}(k+1) = \underline{h}\left(\underline{\hat{x}}(k), u(k)\right) \tag{19.24a}$$

$$\hat{y}(k) = g\left(\underline{\hat{x}}(k)\right), \tag{19.24b}$$

where the states $\underline{\hat{x}}$ describe the internal model states, which usually have no direct relationship to the physical states of the process. The number of the model states is typically related to the model structure and is often chosen to be much larger than the assumed dynamic order of the process. Thus, the process order determination (Sect. 19.9) is not as crucial an issue as it is for external dynamics approaches.

In contrast to models with external dynamics, the use of past inputs and past outputs at the model input is not necessary. Therefore, the application of internal dynamic models leads to a desirable reduction of the input space dimensionality. Since internal dynamic models possess no external feedback, only the parallel model approach in Fig. 19.7b can be applied. Consequently, these models are not well suited for one-step prediction tasks. The internal dynamics approach is discussed further in Chap. 23.

19.4 Parameter Scheduling Approach

Another possibility for modeling nonlinear dynamic systems is by scheduling the parameters of a "linear" model in dependency on the operating point; see Fig. 19.11.

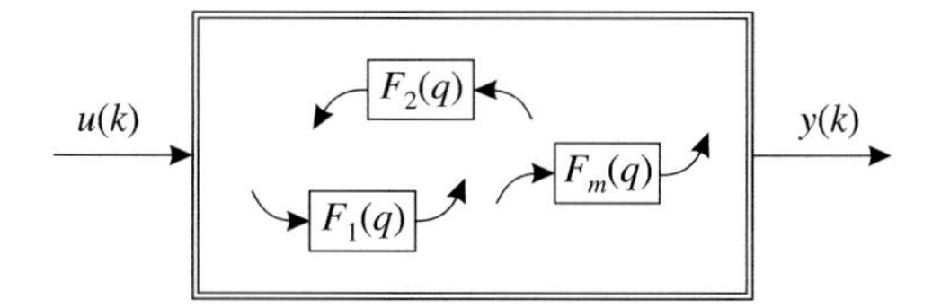

Fig. 19.10 Symbolic scheme of an internal dynamic model. Filters $F_i(q)$ (in the simplest case delays $F_i(q) = q^{-1}$) are used inside the model structure to introduce dynamics. Typically, neural networks (mostly MLPs) are utilized for the model architecture

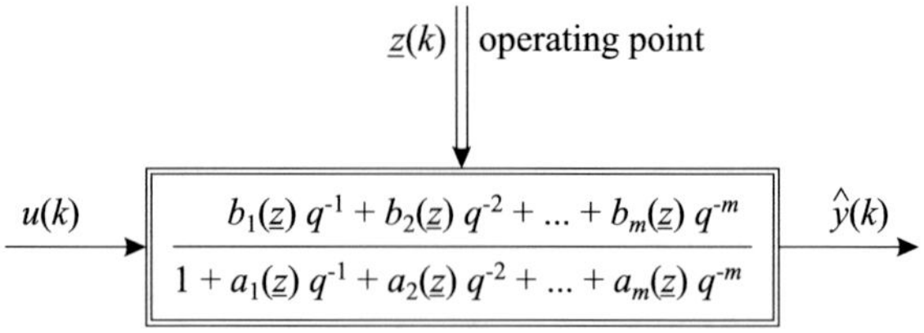

Fig. 19.11 A nonlinear dynamic model can be generated by scheduling the parameters in dependency on the operating point

This approach with operating point-dependent parameters of linear models is, for example, pursued in [200, 573]. The model output can be written as (see Fig. 19.11):

$$\hat{y}(k) = b_1(\underline{z})u(k-1) + \ldots + b_m(\underline{z})u(k-m) \\ - a_1(\underline{z})\hat{y}(k-1) - \ldots - a_m(\underline{z})\hat{y}(k-m)\,, \qquad (19.25)$$

where $\underline{z}$ denotes the operating point. The system in (19.25) is said to be *linear parameter varying (LPV)*. Interestingly, as pointed out in Sect. 22.3, (19.25) is equivalent to the *external dynamics* local linear modeling approach with ($j = 1, 2, \ldots, m$):

$$b_j(\underline{z}) = \sum_{i=1}^{M} b_{ij}\Phi_i(\underline{z})\,, \qquad a_j(\underline{z}) = \sum_{i=1}^{M} a_{ij}\Phi_i(\underline{z})\,, \qquad (19.26)$$

where b_{ij}, a_{ij} are the parameters of the local linear models and $\Phi_i(\underline{z})$ are the validity functions defining the operating regimes. In fact, the parameter scheduling approach in (19.25) is the special case of a local modeling scheme that possesses local *linear* models of *identical structure* (same order m); refer to Chap. 22.

19.5 Training Recurrent Structures

From the internal dynamics approach, recurrent structures always arise. From the external dynamics approach, NARX, NOBF, and NFIR models are feedforward during training, but NOE models are recurrent since they apply the parallel model configuration. NARMAX models and models with other more complex noise descriptions are also recurrent because a feedback path exists from the model output $\hat{y}(k)$ to the prediction errors $e(k)$ to the model inputs.

Training of feedforward structures is equivalent to the training of static models. However, the training of recurrent structures is more complicated because the feedback has to be taken into account. In particular, training recurrent models is always a nonlinear optimization problem independent of whether the utilized static model architecture is linear or nonlinear in the parameters. This severe drawback is a further reason for the popularity of feedforward structures such as the NARX model.

Basically, two strategies for training of recurrent structures can be distinguished. They are discussed in the following paragraphs.

In opposition to static models, the gradient calculation of recurrent models depends on past model states. There exist two general approaches: the backpropagation-through-time algorithm and the real-time recurrent learning algorithm [449, 450, 600]. Both algorithms calculate exact gradients in batch mode adaptation. However, they differ in the way in which they process information about past network states. The backpropagation-through-time algorithm and the real-time recurrent learning algorithm are first-order gradient algorithms and therefore are an extension of the backpropagation algorithm for static neural networks. Their principles can be generalized to the calculation of higher-order derivatives as well. They can be utilized in a straightforward manner for sophisticated gradient-based optimization techniques such as conjugate gradient or quasi-Newton methods; see Sect. 4.4.

19.5.1 Backpropagation-Through-Time (BPTT) Algorithm

The *backpropagation-through-time (BPTT)* algorithm has been developed by Rummelhardt, Hinton, and Williams [491]. It is an extension of the standard (static) backpropagation algorithm, which is discussed in Sect. 11.2.3, to recurrent models. In [491] BPTT has been introduced for MLP networks. However, it can be generalized to any kind of model architecture.

The basic idea of BPTT is to *unfold* the recurrent model structure. Figure 19.12 illustrates this idea for a first-order NOE model. The extension to higher-order models is straightforward. The nonlinear dynamic model in Fig. 19.12 at time instant k can be written as

$$\hat{y}(k) = f(\underline{\theta}(k), u(k-1), \hat{y}(k-1)), \tag{19.27}$$

where $\underline{\theta}$ denotes the parameter vector containing all model parameters. In order to train the model, these parameters have to be optimized. For an efficient training, the gradient of the loss function with respect to the parameters is required. This loss

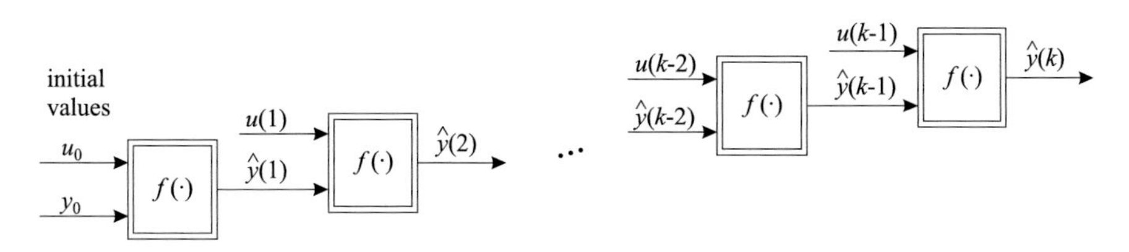

Fig. 19.12 The backpropagation-through-time algorithm unfolds a recurrent model back into the past until the initial values for the model inputs at time $k = 0$ are reached

function gradient depends on the derivative of the model output $\hat{y}(k)$ with respect to the parameters $\underline{\theta}(k)$; see (9.6) in Sect. 9.2.2. The evaluation of this derivative leads to

$$\frac{\partial \hat{y}(k)}{\partial \underline{\theta}(k)} = \underbrace{\frac{\partial f(\cdot)}{\partial \underline{\theta}(k)}}_{\text{static}} + \underbrace{\frac{\partial f(\cdot)}{\partial \hat{y}(k-1)} \cdot \frac{\partial \hat{y}(k-1)}{\partial \underline{\theta}(k)}}_{\text{dynamic}} . \tag{19.28}$$

The first term in (19.28) is the conventional model output gradient with respect to the parameters, which also appears in static backpropagation. The second term, however, arises from the feedback component. If the model input $\hat{y}(k-1)$ was an external signal, e.g., the measured process output $y(k-1)$ as is the case for the series-parallel configuration, then no dependency on the model parameters would exist, and this derivative would be equal to zero. For recurrent models, however, $\hat{y}(k-1)$ is the previous model output, which depends on the model parameters. For higher-order recurrent models and internal dynamics state space models, an expression like the second term in (19.28) appears for each model state $\hat{\underline{x}}$.

The evaluation of the second term in (19.28) requires the derivative of the previous model output with respect to the *actual* parameters. It can be calculated from

$$\hat{y}(k-1) = f(\underline{\theta}(k), u(k-2), \hat{y}(k-2)) , \tag{19.29}$$

which corresponds to (19.27) shifted one time step into the past. Note that (19.29) is written with the actual parameters $\underline{\theta}(k)$, which differs from the evaluation of the model output carried out one time instant before since this was based on the previous parameters $\underline{\theta}(k-1)$. With (19.29) the derivative in (19.28) becomes

$$\frac{\partial \hat{y}(k-1)}{\partial \underline{\theta}(k)} = \frac{\partial f(\cdot)}{\partial \underline{\theta}(k)} + \frac{\partial f(\cdot)}{\partial \hat{y}(k-2)} \cdot \frac{\partial \hat{y}(k-2)}{\partial \underline{\theta}(k)} . \tag{19.30}$$

The second term in (19.30) again requires the derivative of the model output one time instant before. This procedure can be carried out until time $k = 0$ is reached with the initial value (see Fig. 19.12):

$$\hat{y}(0) = y_0 , \tag{19.31}$$

which does not depend on the parameters

$$\frac{\partial \hat{y}(0)}{\partial \underline{\theta}(k)} = 0 . \tag{19.32}$$

Thus, the BPTT algorithm calculates the exact model derivatives by pursuing all k steps back into the past. The gradient calculation with BPTT is exact for both sample and batch mode; see Sect. 4.1. The BPTT procedure involved has to be repeated for all time instants $k = 1, 2, \ldots, N$ where N is the number of training data samples. This means that for the last training data sample, N derivatives have to be calculated for each parameter. Note that this whole procedure must be carried out many times because nonlinear gradient-based optimization techniques are iterative. Since N usually is quite large, the computational effort and the memory requirements of BPTT are usually unacceptable in practice.

In order to make BPTT feasible in practice, an approximate version can be applied. Since the contributions of the former model derivatives become smaller the more BPTT goes into the past, the algorithm can be truncated at an early stage. Thus, in approximate BPTT only the past K steps are unfolded. A tradeoff exists between the computational effort and the accuracy of the derivative calculation. In the extreme case $K = 0$, the static backpropagation algorithm is recovered since the second term in (19.28) is already neglected. This method is called *ordinary truncation*. For $0 < K < N$, the method is called *multi-step truncation*, while for $K = N$ the BPTT algorithm is recovered.

19.5.2 Real-Time Recurrent Learning

Owing to the shortcomings of BPTT, the *real-time recurrent learning* algorithm, also known as *simultaneous backpropagation*, has been proposed by Williams and Zipser [601]. It is much more efficient than BPTT since it avoids unfolding the model into the past. Rather, a recursive formulation can be derived that is well suited for a practical application. Real-time recurrent learning is based on the assumptions that the model parameters do not change during one sweep through the training data, that is, $\underline{\theta}(k) = \underline{\theta}(k-1) = \ldots = \underline{\theta}(1)$. This assumption is exactly fulfilled for batch adaptation, where the parameters are updated only after a full sweep through the training data. It is not fulfilled for sample adaptation, where the parameters are updated at each time instant. However, even for sample mode real-time recurrent learning can be applied when the step size (learning rate) is small, since then the parameter changes can be neglected, i.e., $\underline{\theta}(k) \approx \underline{\theta}(k-1) \approx \ldots \approx \underline{\theta}(1)$. With this assumption the following derivatives are (approximately) equivalent:

$$\frac{\partial \underline{x}(k-1)}{\partial \underline{\theta}(k)} = \frac{\partial \underline{x}(k-1)}{\partial \underline{\theta}(k-1)} . \tag{19.33}$$

With the identity (19.33), (19.28) can be written as

$$\frac{\partial \hat{y}(k)}{\partial \underline{\theta}(k)} = \frac{\partial f(\cdot)}{\partial \underline{\theta}(k)} + \frac{\partial f(\cdot)}{\partial \hat{y}(k-1)} \cdot \frac{\partial \hat{y}(k-1)}{\partial \underline{\theta}(k-1)} . \tag{19.34}$$

The property exploited by real-time recurrent learning is that the expression $\partial \hat{y}(k-1)/\partial \underline{\theta}(k-1)$ is the previous model gradient, which has already been evaluated at time $k - 1$ and thus is available. Consequently, (19.34) represents a dynamic system for the gradient calculation; the new gradient is a filtered version of the old gradient:

$$\underline{g}(k) = \alpha + \beta \, \underline{g}(k-1) \, . \tag{19.35}$$

Care has to be taken during learning that $|\partial f(\cdot)/\partial \hat{y}(k-1)| < 1$ because otherwise the gradient update becomes unstable. Equivalently to the BPTT algorithm, the model derivatives are equal to zero for the initial values, i.e., $\partial \hat{y}(0)/\partial \underline{\theta}(0) = 0$. For higher-order recurrent models and internal dynamics state space models, an expression like the second term in (19.34) appears for each model state $\hat{\underline{x}}$. Thus, the dynamic system of the gradient update possesses the same dynamic order as the model.

Compared with BPTT, real-time recurrent learning is much simpler and faster, it requires less memory, and most importantly its complexity does not depend on the size N of the training data set. In comparison with static backpropagation, however, it requires significantly higher effort for both implementation and computational demand.

In the case of complex internal dynamics models, the gradient calculation can become so complicated that it is not worth computing the derivatives at all. Alternatively, zeroth-order direct search techniques or global optimization schemes as discussed in Sect. 4.3 and Chap. 5, respectively, can be applied because they do not require gradients.

19.6 Multivariable Systems

The extension of nonlinear dynamic models to the multivariable case is straightforward. A process with multiple outputs is commonly described by a set of models, each with a single output; see the Figs. 18.59 and 18.60 in Sect. 9.1. Multiple inputs can be directly incorporated into the model. In internal dynamics approaches and in external dynamics approaches *without* output feedback, no restrictions apply. However, for external dynamics structures *with* output feedback, the model flexibility is limited. For example, the NOE model for p physical inputs extends to

$$\begin{aligned}\hat{y}(k) = \; & f\big(u_1(k-1), \ldots, u_1(k-m), \ldots, u_p(k-1), \ldots, u_p(k-m), \\ & \hat{y}(k-1), \ldots, \hat{y}(k-m)\big) \, . \end{aligned} \tag{19.36}$$

As demonstrated in the following, this model possesses identical denominator dynamics for the linearized transfer functions of all inputs to the output. The linearized NOE model is

$$\begin{aligned}\Delta\hat{y}(k) = {} & b_1^{(1)}\Delta u_1(k-1) + \ldots + b_m^{(1)}\Delta u_1(k-m) + \ldots \\ & + b_1^{(p)}\Delta u_p(k-1) + \ldots + b_m^{(p)}\Delta u_p(k-m) \\ & - a_1\Delta\hat{y}(k-1) - \ldots - a_m\Delta\hat{y}(k-m)\,.\end{aligned} \tag{19.37a}$$

In transfer function form this becomes

$$\Delta\hat{y}(k) = \frac{B^{(1)}(q)}{A(q)}\Delta u_1(k) + \ldots + \frac{B^{(p)}(q)}{A(q)}\Delta u_p(k) \tag{19.38}$$

with $A(q) = 1 + a_1 q^{-1} + \ldots + a_m q^{-m}$ and $B^{(i)}(q) = b_1^{(i)} q^{-1} + \ldots + b_m^{(i)} q^{-m}$.

19.6.1 Issues with Multiple Inputs

As explained in Sect. 18.10.2, (N)ARX models have severe difficulties to describe processes with multiple inputs because all transfer functions from each input to the output share a common dynamics, i.e., in the linear case the common denominator polynomial $A(q)$. Different dynamics can only be created by distinct numerator polynomials $B_i(q)$ that may (almost) cancel different poles in $A(q)$. However, this approach requires flexible, high-order models.

A first-order NARX model as shown for two inputs in Fig. 19.13 is not capable of realizing different dynamics for $u_1 \to y$ and $u_2 \to y$. The denominator polynomial of the transfer function obtained through linearization at any operating point is identical for both inputs, and the numerator is just a constant which can only change the gain but not the dynamics. In order to be capable of realizing different time constants for $u_1 \to y$ and $u_2 \to y$, a second-order (N)ARX is necessary even if both dynamics are just first order. This implies a complexity boost going from three inputs to six inputs, adding $u_1(k-2)$, $u_2(k-2)$, and $y(k-2)$.

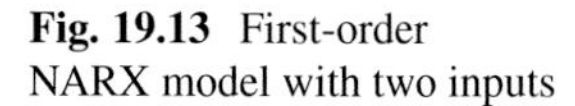
Fig. 19.13 First-order NARX model with two inputs

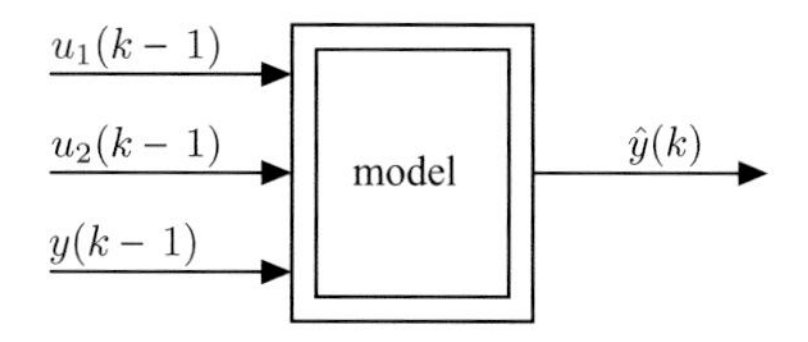

19.6.1.1 Asymmetry Going from ARX → OE to NARX → NOE

In the linear case, the problem with identical denominator dynamics for ARX models can be remedied by utilizing the OE model structure. For the two input cases, this means going from the ARX model

$$A(q)y(k) = B_1(q)u_1(k) + B_2(q)u_2(k) \tag{19.39}$$

to the OE model

$$y(k) = \frac{B_1(q)}{A_1(q)}u_1(k) + \frac{B_2(q)}{A_2(q)}u_2(k)\,. \tag{19.40}$$

If $A(q)$, $A_1(q)$, and $A_2(q)$ are of identical order, then the number of states has doubled going from ARX to OE. In the general case of p inputs, the number of states would multiply by p. Therefore it is possible to realize individual dynamics for each input to the output.

This is *not* true for the nonlinear case. As an example consider a first-order model. Going from a NARX model

$$\hat{y}(k) = f\left(u_1(k-1), u_2(k-1), y(k-1)\right) \tag{19.41}$$

to an NOE model

$$\hat{y}(k) = f\left(u_1(k-1), u_2(k-1), \hat{y}(k-1)\right) \tag{19.42}$$

keeps the number of states unchanged. Both models are utilized in simulation in exactly the same way. The model output $\hat{y}$ is delayed and fed back to the approximator input resulting in one state. Thus, the NOE model does not gain additional flexibility compared to the NARX model, as the OE does in comparison to the ARX model. In the nonlinear case, the number of states always is equal to the order of the model. As a consequence, the difficulties arising with the common denominator in MISO ARX models can be fixed with MISO OE models. This is not the case going from NARX to NOE. This fact cripples all nonlinear output feedback-based approaches.

19.6.1.2 Mixed Dynamic and Static Behavior

To make this point obvious, consider the case where there is a complex *dynamic* behavior $u_1 \to y$ while the behavior $u_2 \to y$ is (almost) static. In the context of local model network architectures, such a situation can be coped with by choosing

$$\underline{x} = [u_1(k-1) \ \cdots \ u_1(k-n) \ \ y(k-1) \ \cdots \ y(k-n)]^T \tag{19.43}$$

(no u_2) and

$$\underline{z} = [u_1(k-1) \ \cdots \ u_1(k-n) \ u_2(k) \ y(k-1) \ \cdots \ y(k-n)]^T . \qquad (19.44)$$

Then a change of u_2 just influences the scheduling variable. However, such a differentiation is not possible for other architectures. Then the model would answer a (small) step in u_2 with the linearized denominator dynamics $A(q)$.

These difficulties do not arise for internal dynamics models and for external dynamics models without output feedback such as NFIR and NOBF because they possess no common output feedback path. For the mixed static/dynamic case, NFIR and NOBF offer the simple solution of only delaying u_1 through a filter chain $L_i(q)$ or $q^{-i}, i = 1, \ldots, m$ while feeding u_2 directly (unfiltered) into the approximator. Another advantage of this approach is this keeps the dimensionality low.

19.7 Excitation Signals

One of the most crucial tasks in system identification is the design of appropriate excitation signals for gathering identification data. This step is even more decisive for nonlinear than for linear models (see Sect. 18.2) because nonlinear dynamic models are significantly more complex, and thus the data must contain considerably more information. Consequently, for the identification of nonlinear dynamic systems, the requirements on a suitable data set are very high. In many practical situations, even if extreme effort and care have been taken, the gathered data may not be informative enough to identify a black box model that is capable of describing the process in all relevant operating conditions. This fact underlines the important role that prior knowledge (and model architectures that allow its incorporation) play in nonlinear system identification.

Independent of the chosen model architecture and structure, the quality of the identification signal determines an upper bound on the accuracy that in the best case can be achieved by the model. For linear systems, guidelines for the design of excitation signals are presented in Sect. 18.2. Quite frequently, the so-called pseudo random binary signal (PRBS) is applied; see, e.g., [278, 356, 540]. The parameters of this signal, whose spectrum can be easily derived, are chosen according to the dynamics of the process. For nonlinear systems, however, besides the frequency properties of the excitation signal, the amplitudes have to be chosen properly to cover all operating conditions of interest. Therefore, the synthesis of the excitation signal cannot be carried out as mechanistically as for linear processes (although even for linear systems an individual design can yield significant improvements); each process requires an individual design. Nevertheless, the aspects discussed in the following should always be considered because they are of quite general interest.

As example, a first-order system of *Hammerstein* structure is considered whose input lies in the interval $[-4,\ 4]$ with a time constant of 16 s following the nonlinear difference equation

$$y(k) \;=\; 0.06\arctan(u(k-1)) + 0.94\,y(k-1) \qquad (19.45)$$

when sampled with $T_0 = 1$ s. For some discussions a second example of Wiener structure shall be considered. Swapping the static nonlinearity and linear dynamics in the Hammerstein process (19.45) yields the *Wiener* process

$$y(k) \;=\; \arctan\left[0.06\,u(k-1) + 0.94\,\tan(y(k-1))\right]\,. \qquad (19.46)$$

Both processes exhibit identical stationary behavior but are very different in the transient regions. Thus, the data point distribution obtained for a specific excitation signal can be different as well in the NARX input space spanned by $[u(k-1)\ y(k-1)]$.

The following sections explain why standard signals for linear system identification often cannot be directly used for nonlinear system identification and illustrate some important signal types. Also, alternative dynamic realizations are analyzed, and the multivariable case is briefly discussed. Finally, an optimal signal generator is introduced that represents a new approach toward excitation signal design.

19.7.1 From PRBS to APRBS

In this section, it will be illustrated why a PRBS is inappropriate for nonlinear dynamic systems. Figure 19.14a shows the most common standard excitation signal for linear systems in the time domain, a PRBS. The plot below depicts the resulting data points in the NARX input space, spanned by $u(k-1)$ and $y(k-1)$.[1] Clearly, this data would be well suited for estimating a plane, which is the task in linear system identification. Note that, to estimate the parameters b_1 and a_1 of a linear model $y(k) = b_1 u(k-1) - a_1 y(k-1)$, the data should stretch as widely as possible in the $u(k-1)$ and $y(k-1)$ directions. Such a distribution yields the smallest possible parameter variance in both estimates for b_1 and a_1. Exactly this property is achieved by the PRBS since it alternates between the minimum and maximum value (here -4 and 4) in $u(k-1)$ and also covers the full range for $y(k-1)$ between -1.4 and 1.4. Although a PRBS is well suited for linear system identification, i.e., if the one-step prediction function is known to be a plane, it is inappropriate for nonlinear systems. No information about the system behavior for input amplitudes other than -4 and 4 is gathered.

[1] This analysis is carried out for the external dynamics approach because it allows us to gain some important insights about the desirable properties of the excitation signals. Although with

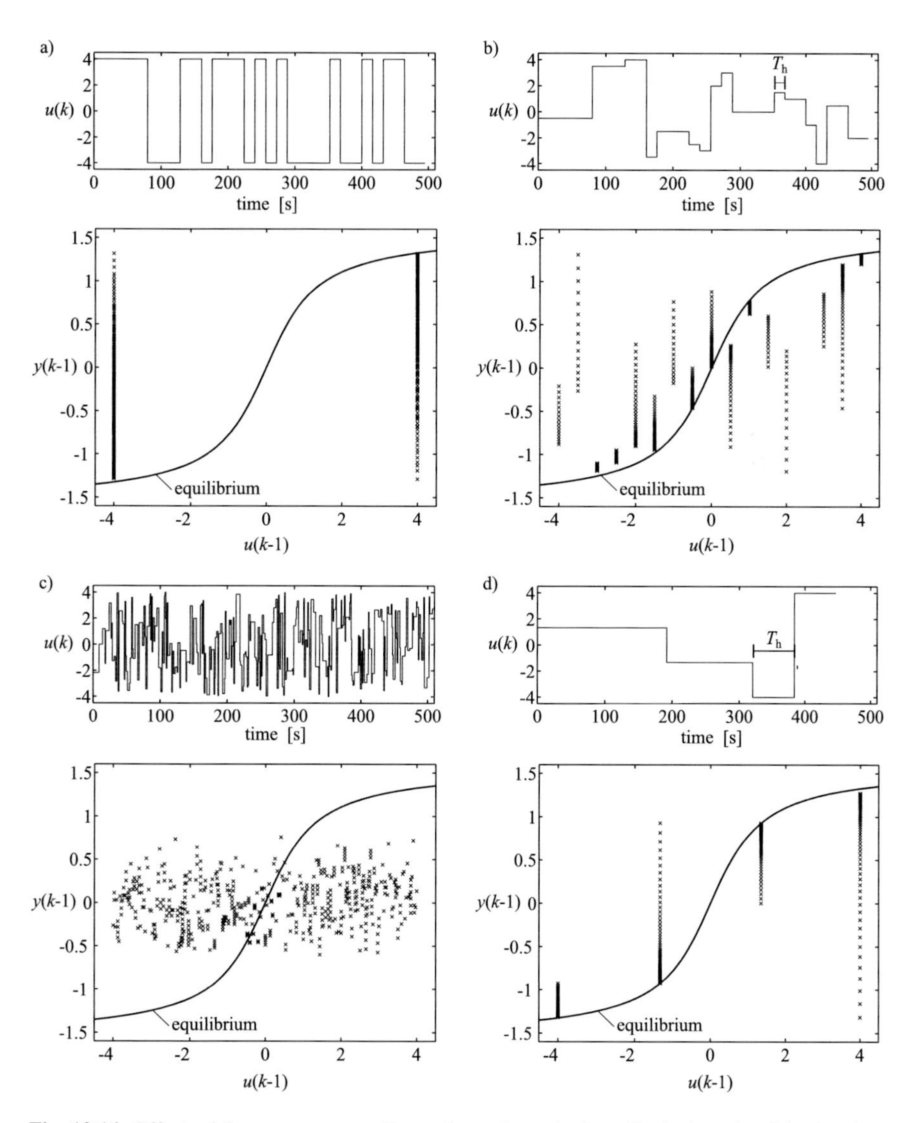

Fig. 19.14 Effect of frequency on nonlinear dynamic excitation. Excitation signal in the time domain (top). Data distribution in the NARX input space (bottom): (**a**) binary PRBS, (**b**) APRBS with appropriate minimum hold time T_h, (**c**) APRBS with too short a minimum hold time T_h, (**d**) APRBS with too large a minimum hold time T_h

19.7.1.1 APRBS Construction

An obvious solution to this problem is to extend the PRBS to different amplitudes. The arguably simplest approach proposed in [430, 431] is to give each step in the PRBS a different amplitude, leading to an amplitude modulated PRBS (APRBS). Note that this does not mean to multiply each PRBS step with some random amplitude value drawn from a uniform distribution in $[0, A]$. It is better to establish a uniform discretization of the interval $[0, A]$ in order to guarantee uniform coverage. In addition, multiplying a PRBS yields an alternating signal changing the sign in each step. This would leave empty the regions of the input space below the equilibrium for $u(k) < 0$ and above the equilibrium for $u(k) > 0$ because the signal would alternate between these regions.

Rather the following construction procedure is advised: First, a standard PRBS is generated. Then, the number of steps is counted, and the interval from the minimal to the maximum input is divided into as many levels. Finally, each step in the PRBS is given one of these levels at random.

19.7.1.2 APRBS: Smoothing the Steps

For many real-world applications, the step changes in the APRBS pose problems. Often rate constraints limit the maximal absolute value of the gradient of the input signal due to physical actuator limitations or security reasons. It is proposed in [118] to replace the steps by ramps in order to meet these constraints; see Fig. 19.15. In [239] it is proposed to replace the steps with sinusoidal-shaped transitions which additionally keeps the first derivatives smooth.

This approach can be significantly improved further by replacing the step or ramp by a sinusoidal signal as also shown in Fig. 19.15. This means using a function $\sin(\omega t)$ where the step takes place at $t = 0$ for $\omega t = -\pi/2 \ldots \pi/2$. The frequency

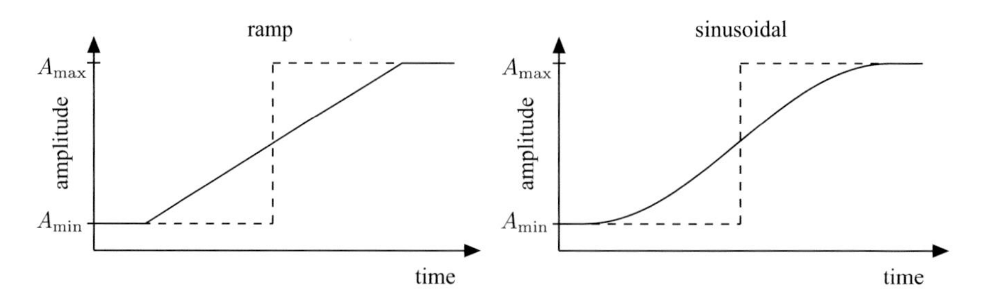

Fig. 19.15 Smoothing APRBS steps by a ramp or a sinusoidal signal

the internal dynamics approach the one-step prediction function is not explicitly approximated, this analysis based on information content considerations is also valid for this class of approaches.

ω can be adjusted in order to meet the rate constraint and thus will depend on the step size $|A_{\max} - A_{\min}|$. Advantages of such an approach are:

- The sine half-wave smoothly continues the signal (gradient = 0 at $\pm\pi/2$).
- The maximal slope occurs at $t = 0$ and can be calculated analytically.
- All higher-order derivatives can be calculated analytically as well.

All derivatives, not just the first, are smooth. Consider the second derivative of the signal. For the ramp APRBS, it just takes three values: min, 0, and max. For a sinusoidal APRBS, these values are smeared over the whole range [min, max]. The same is true for all even higher-order derivatives. Therefore a much improved data point distribution can be expected for higher-order models. Note that either a higher-order model directly utilizes higher-order derivatives as inputs (compare Sect. 19.2.1) or the higher-order derivatives are implicitly relevant through the subsequent time delays $u(k-1)$, $u(k-2)$, ... The best suited APRBS needs to be tailored to the specific requirements of each application.

An original APRBS (with steps) and the resulting input space data distribution are illustrated in Fig. 19.14b. In general, the input space is well covered with data. Some "holes" may exist, however, and their location depends on the random assignment of the amplitude level to the PRBS step. Clearly, here is some room for improvement. Refer to Sect. 19.8.2 for a proposal on how to generate an optimal APRBS. Nevertheless, the holes disappear or at least become smaller as the length of the signal increases.

Besides the minimal and maximum amplitudes and the length of the signal (controlled by the number of registers; see [278, 356, 540]), one additional design parameter exists: the minimum hold time, i.e., the shortest period of time for which the signal stays constant. Given the length of the signal, the minimum hold time determines the number of steps in the signal, and thus it influences the frequency characteristics. In linear system identification, the minimum hold time is typically chosen equal to the sampling time [278]. For nonlinear system identification, it should be chosen neither too small nor too large. On the one hand, if it is too small, the process will have no time to settle, and only operating conditions around $y_0 \approx (u_{\max} + u_{\min})/2$ will be covered; see Fig. 19.14c. A model identified from such data would not be able to describe the static process behavior well. On the other hand, if the minimum hold time is too large, only a very few operating points can be covered for a given signal length; see Fig. 19.14d. This would overemphasize low frequencies, but, much worse, it would leave large areas of the input space uncovered with data, and thus the model could not properly capture the process behavior in these regions since the data simply contains no information on them.

In the experience of the author, it is reasonable to choose the minimum hold time of the APRBS about equal to the dominant (largest) time constant of the process:

$$T_{\mathrm{h}} \approx T_{\max} . \tag{19.47}$$

A similar situation as in Fig. 19.14d occurs if multi-valued PRBS signals according to [105, 124, 279, 571] are designed. These signals retain some nice correlation properties similar to those of the binary PRBS. However, as illustrated in Fig. 19.14d, they cover only a small number of amplitude levels if the signal is required to be relatively short. Note that besides the lack of information about the process behavior between these amplitude levels, some model architectures are not robust with respect to such a data distribution. In particular, the training of local modeling approaches such as RBF networks or neuro-fuzzy models can easily become ill-conditioned for such data distributions. Global approximators such as polynomials and MLP networks are more robust in this respect, but nevertheless, they will show a strong tendency to overfitting in the $u(k-i)$-dimensions. From this discussion, it becomes clear that besides the properties of the process, the properties of the applied model architecture also play an important role in excitation signal design. In [156] some guidelines are given for local linear neuro-fuzzy models.

19.7.2 Ramp

A ramp signal is not really a *dynamic* excitation. It can only be used for fast *quasistatic* measurements. This is a fast and efficient alternative to applying input steps, waiting for the measured signal to settle, and recording one stationary measurement value per step. The faster the ramp is, the more the measured output signal will lag behind. Therefore it is important to measure both *ramp up* and *ramp down*, such that the lag equally often goes both ways. A model will then (almost) compensate for the lag error (if time constants in both directions are similar).

The ramp shown in Fig. 19.16a delivers data points in the NARX input space only close to the equilibrium as can be seen in Fig. 19.17. Of course, this allows only to model the static nonlinearity since no information about transients is gathered. In principle, ramps of different (and faster) speeds can be combined which should yield more dynamic results similar to a multisine or chirp signal.

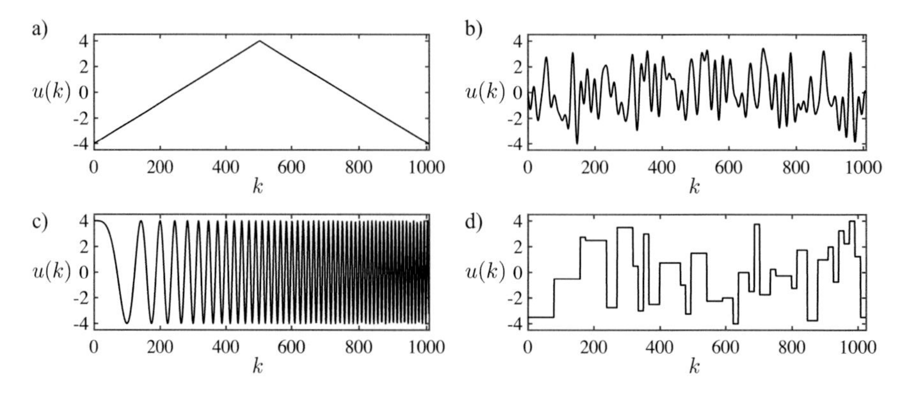

Fig. 19.16 Excitation signals for nonlinear dynamic systems: (**a**) ramp, (**b**) multisine, (**c**) chirp, (**d**) APRBS

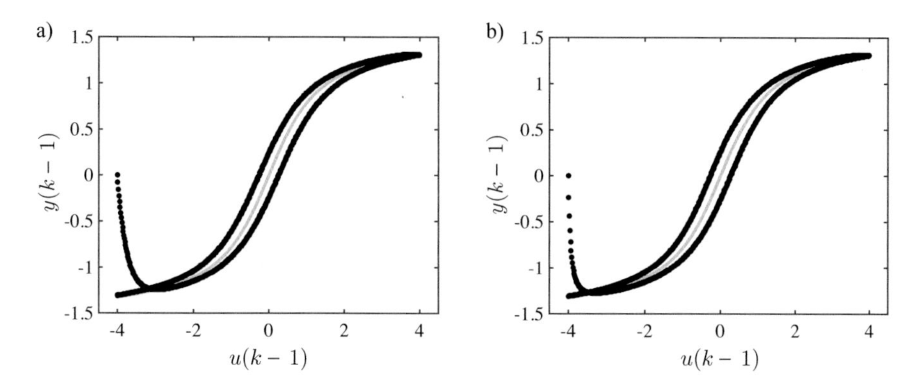

Fig. 19.17 NARX input space for ramp for (**a**) Hammerstein and (**b**) Wiener processes

19.7.3 Multisine

A multisine signal with s components has the following form:

$$u(t) = \sum_{i=1}^{s} A_i \sin(\omega_i t + \varphi_i) . \tag{19.48}$$

Typically the amplitudes A_i are selected identically, and the frequencies are chosen as multiples of a fundamental frequency $\omega_i = i \cdot \omega_0, i = 1, \ldots, s$. Multisine excitations are commonly employed to measure the frequency response of a linear or nonlinear system. In [517] it is demonstrated how powerful these types of signals are for linear frequency domain system identification in a nonlinear setting. For *linear* system identification, the main concern is to superpose the sine waves in a manner that maximizes the power of the signal with given constraints on amplitudes. This is called the crest factor maximization. This goal can be achieved by a clever combination of the phases φ_i so that the sine waves are as much out of phase as possible, e.g., so-called Schroeder phases [356].

For *nonlinear* system identification, such a crest factor maximization is contra-productive because it drives the excitation signal toward its limits similar to a PRBS and leaves the medium amplitude range only sparsely covered. Figure 19.16b shows a multisine signal generated with the MATLAB system identification toolbox which selects the maximal crest factor phases found from a few random trails [357]. Its NARX input space distribution can be seen in Fig. 19.18.

Most of the input space is covered with data points in a relatively uniform manner. The highly dynamic areas (top left and bottom right) are empty due to the maximum frequency of the signal. This property will be shared with the chirp signal discussed next. A unique weakness of the multisine is that the full range accessible within the amplitude level constraints (here $[-4, 4]$) is not completely

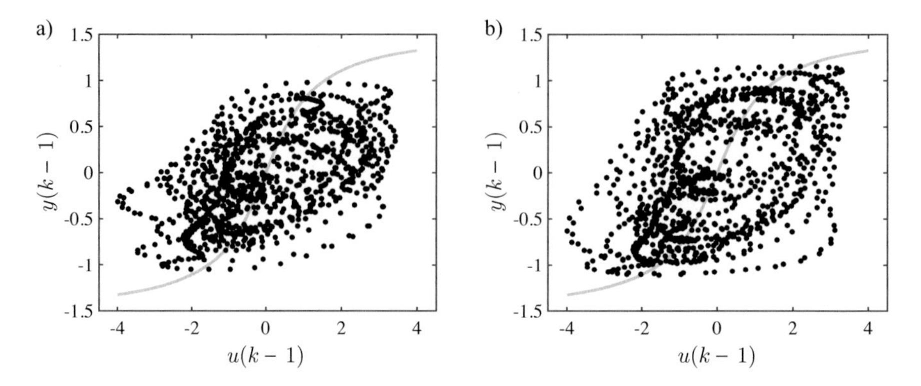

Fig. 19.18 NARX input space for multisine for (**a**) Hammerstein and (**b**) Wiener processes

explored. This is due to the fact the only very small parts of the signal stay close to the minimum and maximum amplitude levels.

19.7.4 Chirp

A chirp signal as shown in Fig. 19.16c is an oscillation with increasing (or decreasing) frequency over time. It offers many attractive features for nonlinear system excitation:

- The excited frequency range can be specified exactly.
- The coverage of the NARX input space is relatively good.
- Constraints with respect to amplitude levels can be met in a frequency-specified manner. The signal can be designed such that rate/velocity constraints are automatically satisfied since the gradient can be calculated analytically.

The property of directly complying with rate constraints is very attractive in many industrial environments. Chirps are therefore popular, e.g., for combustion engine measurement [35].

Figure 19.19 shows the coverage of the NARX input space for the chirp signal from Fig. 19.16c. Really fast dynamics in combination with large amplitude changes (top left and bottom right corners) are missing due to the nature of sine waves. The obvious "hole" in the middle is significantly bigger for the Wiener than for the Hammerstein process. It is caused by the fast dynamics around the zero crossings and can be covered by an additional multisine or various ramp signals. Other than that the data coverage is relatively even.

Nevertheless, chirp signals also suffer from some severe shortcomings:

- The point density is very non-uniform although the input space is well covered. Obviously, the gradient of the chirp signal approaches zero at the minimum

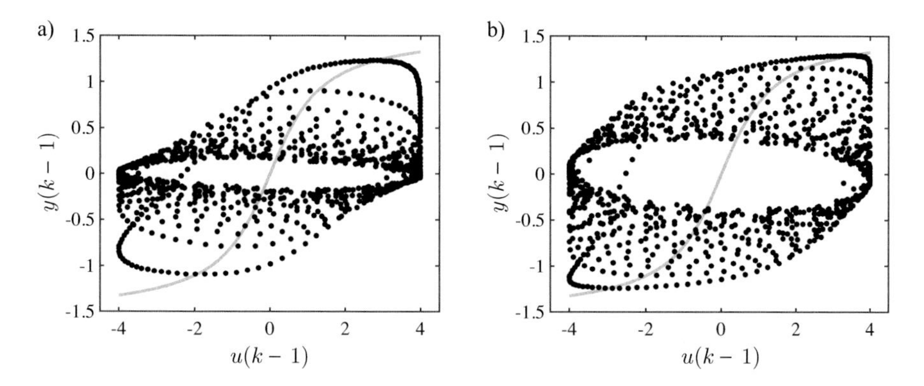

Fig. 19.19 NARX input space for chirp for (**a**) Hammerstein and (**b**) Wiener processes

and maximum amplitudes. In contrast, the signal is very fast around the zero crossings.

- The coverage of the NOBF input space is extremely sensitive with respect to the selected OBF pole(s); compare Fig. 19.23 below.
- The signal is very smooth. This can be seen as an advantage since it prevents the measured object from being damaged. However, it is not realistic for many applications and often does not drive the process to its limits implying that no information is gathered in these extreme situations.

19.7.5 APRBS

The motivation, construction, and parameterization of an APRBS were already explained in Sect. 19.7.1. The APRBS shown in Fig. 19.16d yields the NARX input space data distribution displayed in Fig. 19.20. In contrast to the other signals discussed above, the whole input space can be covered without any "holes." Particularly the regions of extreme amplitudes and high dynamics are covered as well. However, due to the probabilistic nature of the APRBS, it can happen that sparsely covered regions occur; see, e.g., $u(k-1) \approx 2$ for small $y(k-1)$ or $u(k-1) \approx 0$ for large $y(k-1)$ in Fig. 19.20a. Obviously, assigning the amplitude levels to a PRBS randomly is suboptimal. Section 19.8.2 deals with optimal signal design to overcome this drawback.

Although the data coverage for the Hammerstein and Wiener process in Fig. 19.20 is very similar (closer to each other than for the sinusoidal signals), there is a notable difference which also exists but is less visible for other signals. For the Hammerstein process, the closer the data points to each other, the more the measurement signal settles, i.e., approaches the equilibrium.

For the Wiener process, the behavior is much more complicated. The data points are not only denser close to the equilibrium; they are also denser in the extreme

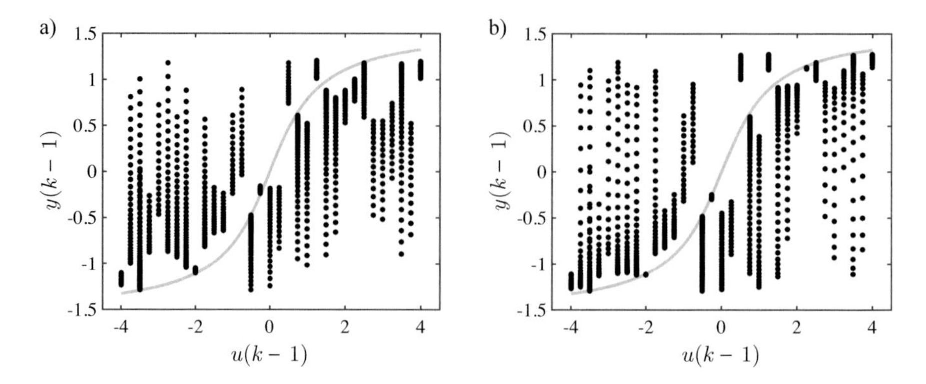

Fig. 19.20 NARX input space for APRBS for (a) Hammerstein and (b) Wiener processes

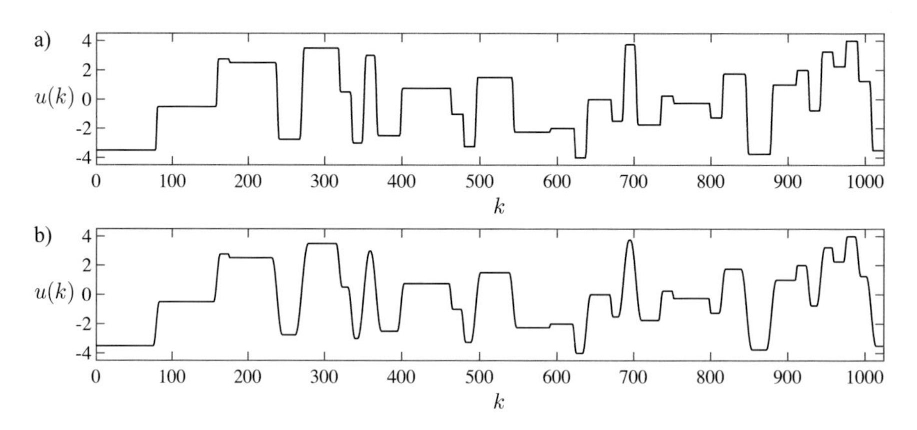

Fig. 19.21 Sinusoidal APRBS with (a) weak rate constraint (3.6 per time step) and (b) strong rate constraint (1.2 per time step)

amplitude, high dynamics regions (top left and bottom right). This is caused by the low gain of the saturation-type nonlinearity for very small and very large input values that exist in early and late phases of a step from -4 to 4, for example. In between the gain is higher (medium input values for the nonlinearity), and thus the points sparser.

19.7.5.1 Sinusoidal APRBS

As explained in Sect. 19.7.1, it is reasonable to smooth the APRBS in order to meet rate constraints and also keep higher-order derivatives small. Figure 19.21b shows such a sinusoidal APRBS with the strongest possible rate constraint of 1.2 per time step. For tighter rate constraints, the hold time of the APRBS and therefore the minimal interval between the steps are not sufficient to enforce the associated

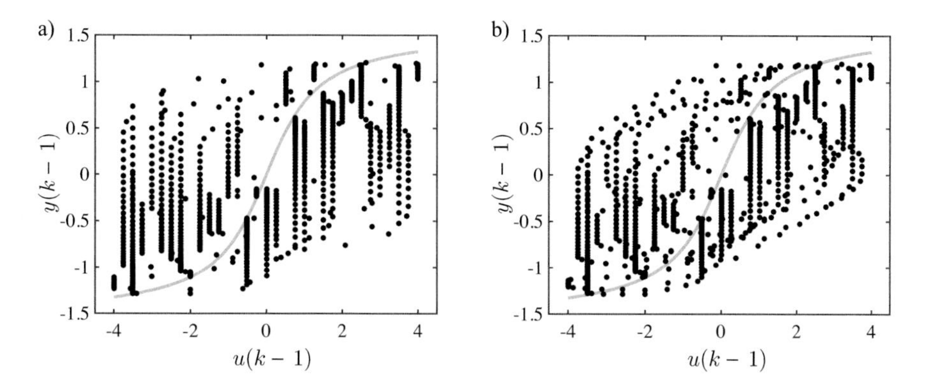

Fig. 19.22 NARX input space (Hammerstein process) for sinusoidal APRBS with (**a**) weak rate constraint (3.6 per time step) and (**b**) strong rate constraint (1.2 per time step)

smoothness in this particular example. Figure 19.21a shows a sinusoidal APRBS for three times relaxed rate constraint.

The effect of the smoothing can be observed from the corresponding NARX input space data distributions shown in Fig. 19.22. Clearly, the high dynamics regions (top left and bottom right) are not reachable anymore. The stronger the rate constraint is, the closer the data is forced toward the equilibrium. Also, the point distribution along lines with fixed discrete $u(k-1)$-values dissolves due to the sinusoidal level transitions. The data distribution gets a flavor similar to a multisine or chirp signal.

19.7.6 NARX and NOBF Input Spaces

External dynamics realizations *with* output feedback like NARX are the most common and important ones. However, realizations *without* output feedback like NFIR or NOBF offer many promising features. One unique feature is that the data distribution in the input space is completely determined by the input excitation signal alone. The process behavior is irrelevant for the model inputs; it only determines the output.

This fact makes the design of experiment procedure significantly simpler. Good properties of the excitation signal can be evaluated and optimized without taking the process behavior into account. Strictly speaking, this observation is only true for NFIRs and for NOBFs with fixed filter pole(s). If the filter pole(s) is adjusted/optimized in some scheme utilizing information about the process behavior, the data distribution varies (in some cases significantly) with this pole(s).

It is difficult to visualize the input space of NFIR models since they need to be very high-dimensional to appropriately capture the dynamics. However, a two-dimensional, i.e., second-order, NOBF model is a quite realistic choice. Consider Laguerre filters as OBFs with poles at (a) $p = 0.9$ or (b) $p = 0.95$. The first pole is faster than the processes considered above; the second one is slightly slower. The Laguerre filters are normalized to gain one which makes them only orthogonal but not orthonormal in order to keep the value range interval of the signals constant.

Figure 19.23 shows the data distribution of the chirp signal in the NOBF input space. A huge influence on the pole can be observed. Although the two poles are not that different, their effect on the data distribution is enormous. The data distribution in Fig. 19.23b can be considered mediocre at best.

For the APRBS the data distribution is not that sensitive with respect to the Laguerre filter pole; see Fig. 19.24. The character of the data distribution is completely different than for the NARX input space. For NARX the steps in the APRBS led to data points distributed along vertical lines (constant $u(k)$, settling $y(k)$). As mentioned previously this may be problematic for certain (special local) model architectures. This issue disappears for NOBF.

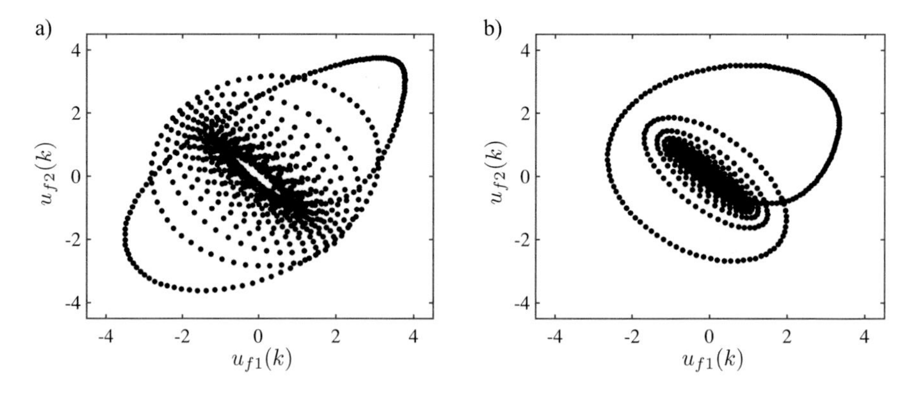

Fig. 19.23 Laguerre input space for chirp for (**a**) pole $p = 0.9$ and (**b**) pole $p = 0.95$

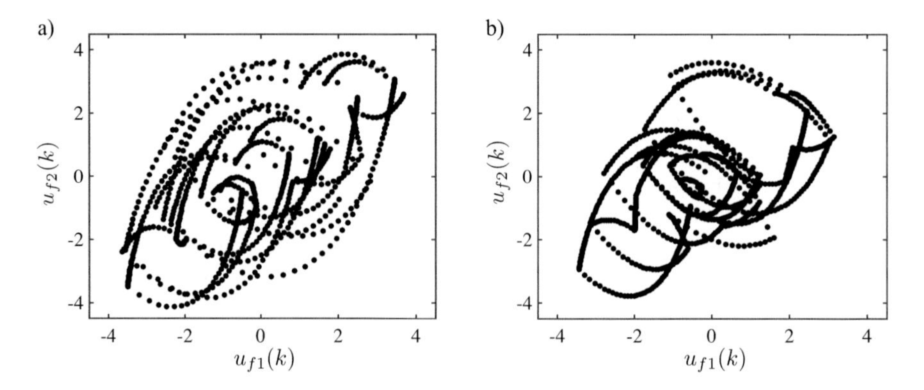

Fig. 19.24 Laguerre input space for APRBS for (**a**) pole $p = 0.9$ and (**b**) pole $p = 0.95$

19.7.7 MISO Systems

For multivariate systems, two types of input signals must be distinguished: actuation signals that the user has command over and measurable disturbances that cannot be actively influenced. Obviously, the question of excitation signal design is only relevant to the first type of inputs.

19.7.7.1 Excitation of One Input at a Time

One simple and transparent way to deal with multiple inputs is to *in turn* excite one input only and keep all other inputs constant. The value of the constant input(s) has to vary through the whole physically relevant range to minimize extrapolation of the built model during latter use. If the number of constantly held inputs is low (1 or 2), they can be varied on a grid. For higher-dimensional problems (starting with 2 dimensions), a variation according to a *Latin hypercube* [386] is advisable to escape the combinatorial explosion (at least for the design of experiments step).

One main advantage of this excitation approach is that a human can oversee the consequences of the dynamic excitation individually for each input. However, severe drawbacks make this approach not advisable in general:

- The signal will be very long in order to cover all combinations of the inputs.
- The effect of dynamic excitation in two (or more) inputs simultaneously (dynamic interaction) is never observed.

It may be possible to attenuate the first drawback by utilizing a ramp instead of constant measurements.

19.7.7.2 Excitation of All Inputs Simultaneously

The two drawbacks of the previous approach can be overcome at the price of losing its clarity by exciting all inputs dynamically at the same time. For the identification algorithms, the separation between the effects caused by different inputs is easiest (and most accurate) if the inputs are chosen in an *orthogonal* or at least *uncorrelated* manner. For PRBS or APRBS, this can be achieved by shifting them with respect to each other by at least the hold time T_h. The MATLAB system identification toolbox proposes to shift the signals by a fraction of $1/p$ for p inputs.

19.7.7.3 Hold Time

It is advisable to multiply the hold time with the number of inputs p, i.e.,

$$T_h \approx p\, T_{max}\,. \tag{19.49}$$

Then the average number of steps in a one-dimensional APRBS is about identical to the number of steps in any input in a p-dimensional APRBS. Otherwise, the output signal would have not enough time to settle in the multivariate case.

Of course, the above discussions are very general and rough. In any specific application, it is necessary to investigate how the typical signals look like during real operation. The excitation signals should reflect these characteristics. It makes no sense to gather data and carry out modeling in regimes that are not entered during the later use of the model.

19.7.8 Tradeoffs

In addition to the more general guidelines given above, the following issues influence the design of excitation signals:

1. *Purpose of modeling*: First of all, the purpose of modeling should be specified, e.g., is the model used for control, for fault diagnosis, for prediction, or for optimization. Thereby, the required model precision for the different operating conditions and frequency ranges is determined. For example, a model utilized for control should be most accurate around the crossover frequency, while errors at low frequencies may be attenuated by the integral action of the controller and errors at high frequencies are beyond the actuator and closed-loop bandwidth anyway.
2. *Maximum length of the training data set*: The more training data can be measured, the more precise the model will be if a reasonable data distribution is assumed. However, in industrial applications, the length of the signal depends on the availability of the process. Usually, the time for configuration experiments is limited. Furthermore, the maximum length of the signal might be given by memory restrictions during signal processing and/or model building.
3. *Characteristics of different input signals*: For each input of the system, it must be checked whether dynamic excitation is necessary (e.g., for the manipulated variables in control systems) or if a static signal is sufficient (e.g., slowly changing measurable disturbances in control systems).
4. *Range of input signals*: The process should be driven through all operating regimes that might occur in the real operation of the plant. Unrealistic operating conditions need not be considered. It is important that the data covers the limits of the input range because model extrapolation is much more inaccurate than interpolation.
5. *Equal data distribution*: In particular for control purposes, the data at the process output should be equally distributed in order to contain the same amount of information about each setpoint.
6. *Dynamic properties*: Dynamic signals must be designed in a way that they properly excite variant dynamics in different operating points.

From this list of general ideas, it follows that prior knowledge about the process is required for the design of an identification signal. In fact, some basic properties of the plant to be identified are usually known, namely, the static mapping from the system's inputs to the output, at least qualitatively, as well as the major time constants. If the system behavior is completely unknown, some experiments such as recording of step responses can provide the desired information.

In practice, the operator of a process restricts the period of time and the kind of measurements that can be taken. Often one will be allowed only to observe the process in normal operation without any possibility of actively gathering information. In this case, extreme care has to be taken to make the system robust against model extrapolation, which is almost unavoidable when available data is so limited. Several strategies for that purpose can be pursued: incorporation of prior knowledge into the model (e.g., a rough first-principles model or qualitative knowledge in the form of rules to describe the extrapolation behavior), detection of extrapolation, and switching to a robust backup model, etc.

Finally, it is certainly a good idea to gather more information (i.e., collect more data) in operating regimes that are assumed (i) to behave more complex and/or (ii) to be more relevant than others. The reason for (i) is that the less smooth the behavior is in some region, the more complex the model has to become there, and thus the more parameters have to be estimated requiring more data. The reason for (ii) is that more relevant operating conditions should be modeled with higher accuracy than others.

It is important to understand that a high data density in one region forces a flexible model to "spend" a great part of its complexity on describing this region. While this effect is desirable owing to reasons (i) and (ii), it can also be undesirable whenever the high data density was not generated on purpose but just accidentally exists. The latter situation almost always occurs if the data was not actively gathered by exciting the process but rather was observed during normal process operation. Then, rarely occurring operating conditions are under-represented in the data set although they are given as much importance as the standard situations. In such a case, the data can be weighted in the loss function in order to force the model to describe these regimes with equal accuracy and to prevent the model from spending almost all degrees of freedom on the regimes that are densely covered with data; see (2.2) in Sect. 2.3.

19.8 Optimal Excitation Signal Generator: Coauthored by Tim O. Heinz

Traditionally, two approaches to the design of nonlinear dynamic excitation signals have been explored:

- *Adjusting a signal of a certain type*: First a decision is made for an excitation signal of a certain type (APRBS, chirp, ...). Next, the tuning parameters (step lengths, frequency ranges, ...) are adjusted corresponding to the specific process under investigation.

- *Maximizing the Fisher information*: First a decision is made for a certain model architecture (local linear model network, MLP, ...) *and* model structure (number of neurons, polynomial terms, ...). Next, the excitation signal is designed in such a manner that it maximizes some criterion (e.g., the determinant for D-optimality) calculated from the Fisher information matrix.

The signal-oriented approaches have been discussed in the sections before. The idea of the potentially much more powerful supervised Fisher-information-based approach is outlined next. As a consequence of its drawbacks, a new idea based on a signal *generator* is proposed and discussed. It combines a signal-based approach with geometric considerations about the input space data distribution (unsupervised).

19.8.1 Approaches with Fisher Information

The Fisher information matrix measures the amount of information that the model output carries about the parameter vector $\underline{\theta}$:

$$\underline{F} = \frac{1}{\sigma^2} \underline{\Psi}^T \underline{\Psi} \tag{19.50}$$

where σ^2 represents the noise variance. In standard literature, the Fisher information matrix is named $\underline{I}$ which is avoided here to counter any confusion with the identity matrix. Note that the size of $\underline{F}$ is $n \times n$. For N data points and n parameters, $\underline{\Psi}^T$ can be written as

$$\underline{\Psi}^T = \left[\left(\frac{\partial\, \hat{y}(\underline{u}(1))}{\partial\, \underline{\theta}} \right) \left(\frac{\partial\, \hat{y}(\underline{u}(2))}{\partial\, \underline{\theta}} \right) \cdots \left(\frac{\partial\, \hat{y}(\underline{u}(N))}{\partial\, \underline{\theta}} \right) \right] \tag{19.51}$$

or

$$\underline{\Psi} = \begin{bmatrix} \dfrac{\partial\, \hat{y}(\underline{u}(1))}{\partial\, \theta_1} & \dfrac{\partial\, \hat{y}(\underline{u}(1))}{\partial\, \theta_2} & \cdots & \dfrac{\partial\, \hat{y}(\underline{u}(1))}{\partial\, \theta_n} \\ \dfrac{\partial\, \hat{y}(\underline{u}(2))}{\partial\, \theta_1} & \dfrac{\partial\, \hat{y}(\underline{u}(2))}{\partial\, \theta_2} & \cdots & \dfrac{\partial\, \hat{y}(\underline{u}(2))}{\partial\, \theta_n} \\ \vdots & \vdots & & \vdots \\ \dfrac{\partial\, \hat{y}(\underline{u}(N))}{\partial\, \theta_1} & \dfrac{\partial\, \hat{y}(\underline{u}(N))}{\partial\, \theta_2} & \cdots & \dfrac{\partial\, \hat{y}(\underline{u}(N))}{\partial\, \theta_n} \end{bmatrix} . \tag{19.52}$$

This is the gradient (or sensitivity) of the model output with respect to the model parameters for all data samples (Jacobian); compare Appendix A. In the linear regression case, $\hat{y} = \underline{X}\,\underline{\theta}$ and therefore

$$\underline{\Psi} = \underline{X}\,. \tag{19.53}$$

Then, the inverse of (19.50) is identical to the covariance matrix of the parameters (3.34); see Sect. 3.1.1.

The idea of Fisher-information-based approaches is to optimize the input sequence by maximizing some criterion, e.g., $\det(\underline{F})$, with respect to $u(1), u(2), \ldots, u(N)$. It is possible/advisable to effectively parameterize the input signal (e.g., in APRBS form) in order to reduce the dimensionality of this optimization problem.

One major advantage of this approach is that *constraints* can be easily built-in [216] which is of fundamental importance in practice. Furthermore, it exploits knowledge on the model structure in an optimal manner. The key drawback lies in the following difficulty: In order to build up the Fisher information matrix $\underline{F}$, the model structure must be fixed. Thus, an initial model needs to be built to start with. Then an iterative modeling/DoE strategy can be run similarly to "iterative identification and control" schemes (see Sect. 18.11.3); compare Fig. 19.25. With each iteration of this loop, the model and quality of excitation will improve. Besides the tedious nature of this approach, a more fundamental difficulty exists.

Due to the bias/variance tradeoff, more data and more informative data imply more complex models. Thus, the identified model on the "old" data (from the previous iteration) which the DoE typically is based on will be too simple.

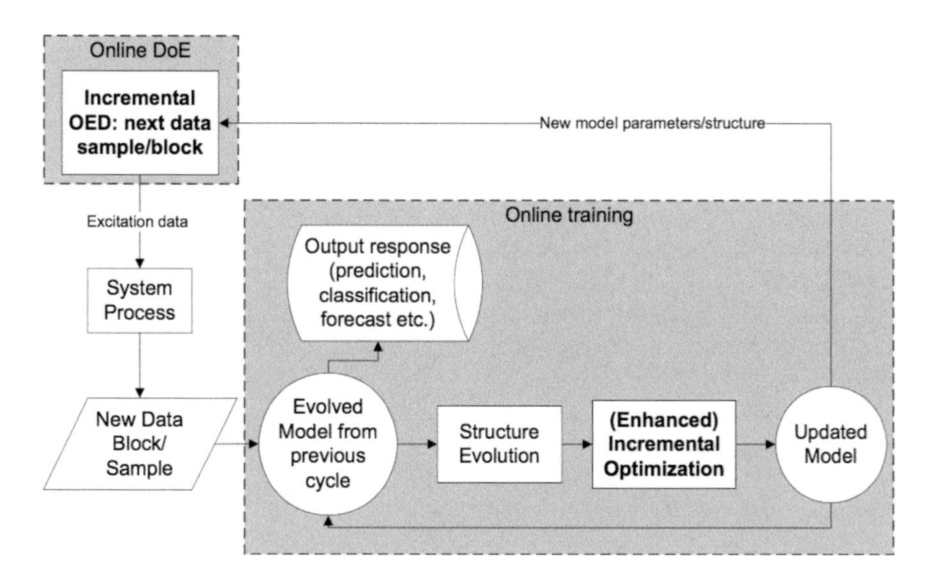

Fig. 19.25 Iterative nonlinear dynamic DoE scheme. (This figure is taken from [216])

Consequently, the data and the model structure which has been identified with it do not really match. Also, in practice, it is often unrealistic to run a couple of iterations which are required for the loop in Fig. 19.25.

For more details on Fisher-information-based algorithms in the context of local model networks (and other neural network architectures), refer to [216, 537].

19.8.2 Optimized Nonlinear Input Signal (OMNIPUS) for SISO Systems

This section presents a new approach to the generation of excitation signals for nonlinear dynamic processes. An application of the introduced algorithm to a real high-pressure fuel supply system can be found in Sect. 26.4 in Part IV. By designing a dynamic DoE, typically a signal type (chirp, multisine, APRBS, etc.) or a combination of these is used. The parameters of the signals are roughly adjusted with respect to some prior knowledge of the process (hold time of an APRBS, frequency range of a chirp or multisine, etc.). Here we propose an excitation signal generator that is not limited to any specific kind of excitation signal. The objective of the excitation signal optimization is a good space-filling property in the input space. For dynamic realizations with output feedback like NARX, the values in the input space dimensions $y(k-i)$, $i = 1, \ldots, n$ are unknown before measurement. They have to be roughly approximated by a proxy model as $\tilde{y}(k-i)$ which can be chosen simply as a first-order linear model with unit gain and a roughly estimated dominating time constant if no other prior knowledge is available. This time constant plays the same role as the hold time in an APRBS or the frequency range in a chirp or multisine signal. The space spanned by $u(k-i)$ and $\tilde{y}(k-i)$ is called *proxy* input space.

An *OptiMized Nonlinear InPUt Signal* (OMNIPUS) is the result of the signal generator which carries out an optimization. When optimizing an excitation signal with length N, all amplitude values are determined by nonlincar optimization. Global optimization methods try to yield the global optimum but are known to be computationally expensive, in particular, since the signal length N usually is large. To simplify the problem, either local optimization can be utilized, or the whole optimization can be split into subproblems. Because the problem is very rich in bad local optima, the latter simplification strategy is much more promising. The sequential optimization as shown in Fig. 19.26 splits the signal into sequences that are optimized one by one. It is the more greedy, the shorter the sequence length L is. In general, it is only suboptimal, but neglecting the interactions between sequences increases the robustness of the approach.

The idea for the optimization of one sequence with length L is to maximize the average distance between the new sequence and the already existing N data points $\underline{\tilde{X}}$ in the proxy input space. With a given measurement (cf. [240])or just an initial condition ($N = 1$), the sequences are optimized by using the following quality function inspired by the maximin criterion proposed in [305]:

$$\underset{u(N+1),\ldots,u(N+L),\,L}{\arg\max} \left(\frac{1}{L} \sum_{k=N+1}^{N+L} d_{\mathrm{NN}} \left(\underline{\tilde{X}}, \underline{\tilde{x}}\,(k) \right) \right) . \tag{19.54}$$

The function $d_{\mathrm{NN}}(\underline{\tilde{X}}, \underline{\tilde{x}}(k))$ calculates the smallest distance between each existing point in $\underline{\tilde{X}}$ and the new points $\underline{\tilde{x}}(k)$ in the proxy input space. The new points $\underline{\tilde{x}}(k)$ consist of the sequence inputs $u(k)$, $k = N+1, \ldots, N+L$ and the corresponding proxy outputs $\tilde{y}(k)$. The optimal sequence with inputs $u(N+1), \ldots, u(N+L)$ in average possesses all data points $\underline{\tilde{x}}(k)$ farthest away from the existing data points in $\underline{\tilde{X}}$. Thus, the sequence which produces data points in the sparsest area of the proxy input space is appended to the already optimized signal according to Fig. 19.26. A sequence lying in areas of the input space with no or just a few existing data points is assumed to give most "new information" about the process.

All examples discussed in the following are restricted to first-order systems. The algorithms are suitable for arbitrary system orders, and the examples can be extended to higher orders in a straightforward manner. However, it might not be necessary or even beneficial to design an OMNIPUS in a high-order input space. The difficulties arising for high dimensionalities then get significantly worse. Often it is sufficient to just choose the model order high enough and keeping excitation signal design at low order, particularly if the nonlinear behavior does not depend on higher-order derivatives.

For a SISO example, a simple proxy model with dynamic order $m = 1$ can be written as

$$\tilde{y}(k) = b_1 u(k-1) - a_1 \tilde{y}(k-1) \tag{19.55}$$

with a_1 and b_1 determining the time constant T and gain of the model. To not prefer either the input or the proxy output in the Euclidean distance calculations, a unity gain is used. This is in agreement with the typical procedure to normalize or standardize the data before system identification is carried out.

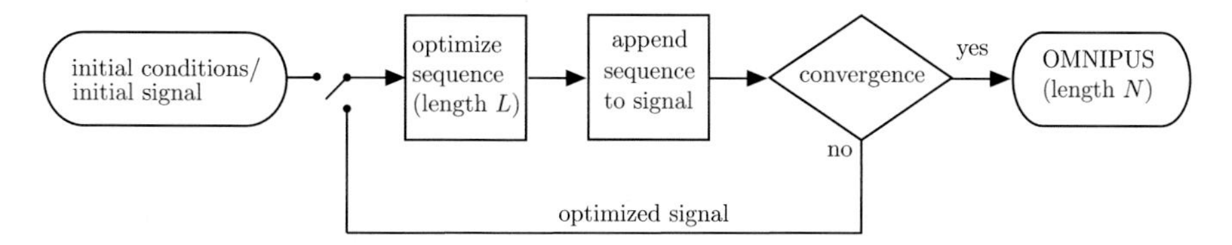

Fig. 19.26 Flow chart of the sequential optimization starting with an initial signal or initial conditions

The matrix $\underline{\tilde{X}}$ in (19.54) contains the existing data points in the proxy input space with length N

$$\underline{\tilde{X}} = \begin{bmatrix} u(1) & \tilde{y}(1) \\ u(2) & \tilde{y}(2) \\ \vdots & \vdots \\ u(N) & \tilde{y}(N) \end{bmatrix} . \tag{19.56}$$

In the case of an initial condition instead of a prior measurement, the matrix reduces to a single row.

The data points in $\underline{\tilde{x}}(k)$ depend on the to-be-optimized signal inputs $u(k)$ and the simulated proxy outputs $\tilde{y}(k)$ for $k = N+1, \ldots, N+L$

$$\underline{\tilde{x}}(k) = \begin{bmatrix} u(k) & \tilde{y}(k) \end{bmatrix} . \tag{19.57}$$

In general, the shape of each sequence is not restricted. In order to make the optimization problem in (19.54) manageable, a specific shape for the sequences has to be imposed. Here, the sequence shall be chosen as a step that yields an APRBS-like signal. In Sect. 19.7.1 it is demonstrated that an APRBS has the ability to reach every point of the input space by assuming a piecewise constant signal.

As a consequence only the amplitude level or step size and the sequence length L remain as two optimization parameters with $u(k) = u^{(\text{step})}$, $k = N+1, \ldots, N+L$. The amplitude levels are bounded by u_{min} and u_{max} given by the physics of the process. To prevent high dynamic (e.g., $L = 1$) and very low frequent (e.g., $L = N$) signals, the sequence length is restricted to be in the range

$$T/T_0 \leq l \leq 3T/T_0 \tag{19.58}$$

where T represents the dominating time constant of the process and T_0 is the sampling time.

The optimal combination of amplitude level $u^{(\text{step})}$ and sequence length L can hardly be found with a local search method. Local optima are abundant, typically (at least) between two neighboring amplitude levels in the already existing signal. Therefore a global search seems to be mandatory. A simple grid search yields good results as the problem is only two-dimensional and the accuracy requirements are only moderate.

19.8.3 *Optimized Nonlinear Input Signal (OMNIPUS) for MISO Systems*

The output of most dynamic processes may be manipulated by more than one input. Each of the inputs has an effect on the output of the process which may differ in significance and dynamics. The excitation signal design has to take these properties into account. For linear dynamic systems, some approaches to generate low-correlated, multivariate input signals exist, e.g., shifted PRBSs. For nonlinear systems, the design of multiple-input excitation signals is only marginally covered in the literature. Shifting and repeating signals are typical approaches used for real-world applications. But a complete coverage of the input space can hardly be achieved with these simple approaches. This leads to unexplored operating regions of the process; cf. [242].

With the extension to multiple inputs, the proxy input space grows from $2m$ dimensions for $u(k-i)$ and $\tilde{y}(k-i), i = 1, \ldots, m$ to $(p+1)m$ dimensions including all p inputs $u_j(k-i),\ j = 1, \ldots, p$. For example, for a process with two inputs and one output, the proxy output can be calculated by:

$$\begin{aligned} \tilde{y}_1(k) &= b_{1,1}u_1(k-1) - a_{1,1}\tilde{y}_1(k-1) \\ \tilde{y}_2(k) &= b_{1,2}u_2(k-1) - a_{1,2}\tilde{y}_2(k-1) \\ \tilde{y}(k) &= \tilde{y}_1(k) + \tilde{y}_2(k)\,. \end{aligned} \tag{19.59}$$

The additive combination of both linear proxy models is the simplest approach to construct an overall proxy output. If available, prior knowledge can be incorporated in more advanced proxy models. For example, the importance of each input can be accounted for by the gain of each SISO system, thereby constructing a *weighted* sum. To keep the range limits of the proxy output, the sum of all gains should be equal to one.

By extending the optimization approach from Sect. 19.8.2 to MISO systems, two main issues arise:

1. Computational demand: The number of optimization parameters increases by a factor of p. This increases the computational effort significantly. For example, the demand for a grid search grows exponentially with p.
2. Point distribution: The size of the (proxy) input space used for the optimization increases as well. As shown in [389], the optimization according to the maximin criterion generates data distributions covering mostly the boundaries of the high-dimensional design space. This effect intensifies with increasing input dimension and is usually undesirable.

Both issues are encountered in a naive extension of the OMNIPUS algorithm from SISO to MISO but can be avoided by two regularizing restrictions proposed in the following. Additionally, in the optimization, it is proposed to alter the limits for the sequence length in the MISO case as explained later.

19.8.3.1 Separate Optimization of Each Input

Instead of optimizing all inputs simultaneously, the optimization of each input is done sequentially inside the "optimize sequence" box in Fig. 19.26. More precisely, within the optimization loop in each iteration, the input with the shortest (already optimized) signal will be optimized next. The signals for the other inputs are

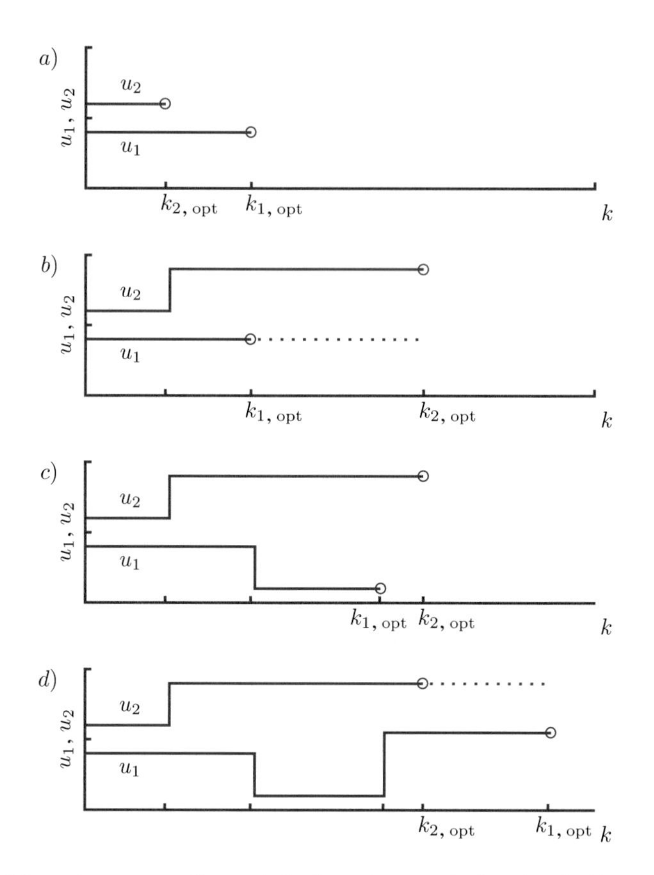

Fig. 19.27 Schematic progress of an optimized two-input excitation signal with u_1 and u_2. The ticks on the time axis mark where a new amplitude level and sequence length optimization is carried out (**a**) first, (**b**) second, (**c**) third, (**d**) forth step of OMNIPUS

kept constant on their last optimized values. This strategy is reasonable because it almost decouples the effect from each input to the output since the dynamic effects of the other inputs have settled. As a consequence, the operating points in the $u_1(k-1)$-$u_2(k-1)$-...-$u_p(k-1)$ input space typically change in an axis-parallel manner. Rarely, two or more input signals have the same length and thus change simultaneously leading to oblique transitions.

The quality function in (19.54) extends slightly when optimizing multiple inputs. Instead of optimizing w.r.t. the single input $u^{(\mathrm{step})}$, now it is done w.r.t. input $u_j^{(\mathrm{step})}$ corresponding to the shortest signal.

$$\underset{u_j^{(\mathrm{step})},\,L_j}{\arg\max}\left(\frac{1}{L_j}\sum_{k=N_j+1}^{N_j+L_j} d_{\mathrm{NN}}\left(\underline{\tilde{X}},\underline{\tilde{x}}(k)\right)\right). \tag{19.60}$$

Figure 19.27 visualizes the progress of the excitation signals for two inputs and four iterations in Fig. 19.26.

- Figure 19.27a: The signal u_1 is the longest; thus the input u_2 is chosen to be optimized.
- Figure 19.27b: The signal u_1 is elongated at its last value, and the input signal u_2 is optimized. After the optimization, signal u_1 is the shortest and subject of optimization in the next iteration.
- Figure 19.27c: The signal u_1 is optimized without the necessity to elongate u_2. The length of u_1 including the optimized sequence does not exceed the signal length of u_2. Thus, in the next iteration, signal u_1 is optimized again.
- Figure 19.27d: For the optimization of u_1, the signal u_2 has to be elongated as in Fig. 19.27b.

In the case that two input signals have the same length, both are optimized sequentially; the optimization order is chosen randomly.

Generally with this strategy, dealing with p inputs compared to one does not increase the dimensionality of the optimization problems. Instead of optimizing $2p$ parameters simultaneously, the optimization is split into p *two*-dimensional problems. The optimization of only one input at a time makes this approach feasible for a high number of input variables.

In addition to the amplitude levels $u^{(\text{step})}$, the sequence length L needs to be optimized. On average p amplitude steps occur within the sequence of length L. Thus, the frequency representation of the optimized input signal shifts to higher frequencies linearly with p. To compensate this undesirable behavior, it is reasonable to also scale the range for L with p. Furthermore, individual time constants can be associated to each input yielding distinct sequence lengths L_j in the ranges

$$pT_j/T_0 \leq L_j \leq 3pT_j/T_0 \tag{19.61}$$

for $j = 1, \ldots, p$.

19.8.3.2 Escaping the Curse of Dimensionality

For higher-dimensional input spaces, the optimization according to the maximin criterion generates designs with most data points lying at the boundaries and corners of the input space [389]. A maximin design, therefore, is beneficial if all interactions between the inputs are relevant. Since the number of corners increases exponentially with the number of input dimensions, such an approach proves unreliable in practice as the nonlinear process behavior cannot be recovered for all operating regimes. Therefore it is recommended to force more data points in the sparsely covered region around the center. This strategy works because the sparsity of effects [66] holds for most applications, i.e., nonlinear process behavior is dominated by low-order interactions.

By using latin hypercubes (LH) for static DoEs, many data points can be forced toward the center of the input space. They also represent non-collapsing designs which means that all points projected on an arbitrary input axis are distinct. Furthermore, these projected points uniformly cover this axis. With these properties an LH offers many crucial advantages. To also realize good space-filling properties, the LH must be optimized such that the points are far away from each other. This can be done w.r.t. a maximin criterion; see, e.g., [131]. In [241] the extension of OMNIPUS to multiple inputs and its application to a common rail injection system are described.

Since dynamic excitation signal design with OMNIPUS is based on the maximin criterion (19.60), the problems for many inputs discussed above occur here as well. This leads to input signals switching mostly between their extreme values. Therefore it is proposed to utilize an LH-inspired approach to overcome this weakness. By restricting the optimized signal to operate on predefined, equidistant amplitude levels, a similar restriction as for static LHs can be realized. Each such predefined amplitude level can be visited by the excitation signal only once. After an amplitude level was chosen, this level is blocked for further optimization. A signal generated by this strategy will be called an LH-based OMNIPUS. Note that it constructs no "true" LH since each amplitude level may contain several operating points due to the fact that typically only one input changes at a time.

The amplitude resolution (smallest amplitude level difference) of each input has to be fixed prior to the optimization. While the sequence lengths are unknown before the optimization, the average sequence lengths can be estimated. From (19.61) a reasonable choice for the average sequence lengths is $\bar{L}_j \approx 2pT_j/T_0$. Consequently, the resolution for each input can be calculated to (compare Fig. 19.28)

$$\Delta u_j = \frac{\bar{L}_i \left(u_{j,\max} - u_{j,\min}\right)}{N - \bar{L}_j} . \tag{19.62}$$

All available amplitude levels indexed by $l_j = 1, \ldots, M_j$ for input j shall be gathered in a vector $\underline{u}_j = [u_{j,1}\ u_{j,2}\ \cdots u_{j,M_j}]^T$ where M_j represents the number of LH levels for input j. It is advantageous to schedule the more amplitude levels M_j the more complex behavior $u_j \rightarrow y$ is expected.

The amplitude resolutions affect the sequence lengths and thus the length of the whole signal: if all sequences are short, all amplitude levels may be chosen once before reaching the user-selected overall signal length N. Hence the restriction that each level has to be chosen only once may be too strong in some cases. To weaken this restriction, the quality function in (19.54), for example, can be extended by a penalty term depending on an amplitude level counter $\#u_{j,l_j}$:

$$\underset{u_{j,l_j}^{(\text{step})},\, L_j}{\arg\max} \left(\frac{1}{L_j} \sum_{k=N_j+1}^{N_j+L_j} d_{\text{NN}} \left(\underline{\tilde{X}}, \underline{\tilde{x}}(k) \right) - \#u_{j,l_j} d_{\max} \right) . \tag{19.63}$$

Here $\#u_{j,l_j}$ represents how often input j was optimized to the level with index l_j. To penalize the already chosen amplitude levels, the counter $\#u_{j,l_j}$ is multiplied by the factor $d_{\max}$. The distance $d_{\max}$ is chosen according to the maximum distance between two reachable points in the proxy input space. Thus, each level of an input is chosen C times before any level is chosen $C + 1$ times.

19.8.3.3 Results for Two Inputs

In this paragraph OMNIPUS without restrictions and with the LH-based strategy shall be illustrated and compared. It is remarkable that the shortcomings of the pure maximin approach without restrictions that should reveal themselves for high-dimensional problems are already severe for $p = 2$ inputs. Obviously, for any $p > 1$, the LH-based OMNIPUS is significantly superior.

Figures 19.29 and 19.30 show examples for OMNIPUS excitation signals. The tendency for dense coverage of the boundaries (amplitudes close to 0 and 1) and sparse coverage of the center (amplitudes around 0.5) can already be observed in Fig. 19.29. The amplitude distribution is much more even with the LH-based OMNIPUS in Fig. 19.30.

In the two-dimensional subspace of the input space spanned by $u_1(k-1)$ and $u_2(k-1)$, this property can be seen even more expressively. Figure 19.31 shows the OMNIPUS without restrictions of the amplitude levels in this subspace. Most of the data points are lying at the boundary of the input space (73%); only a small

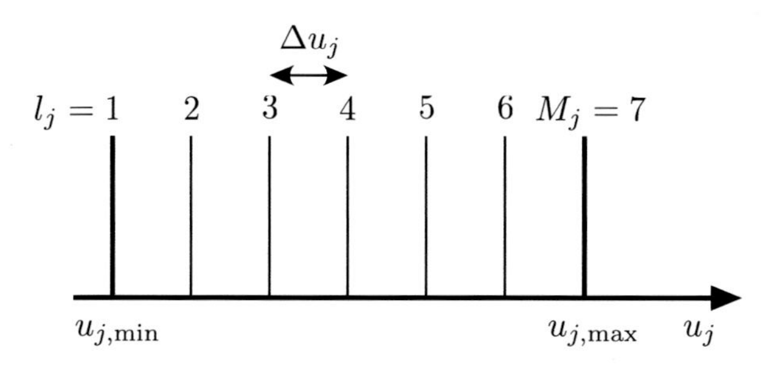

Fig. 19.28 LH resolution quantities explained for input u_j with $M_j = 7$ levels

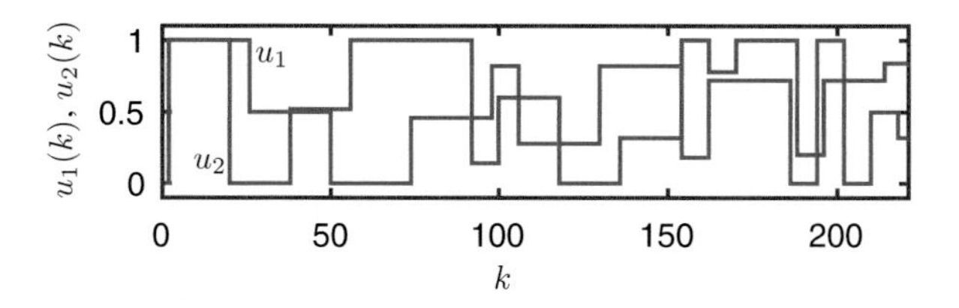

Fig. 19.29 OMNIPUS without restrictions. Extreme amplitude levels close to 0 and 1 dominate

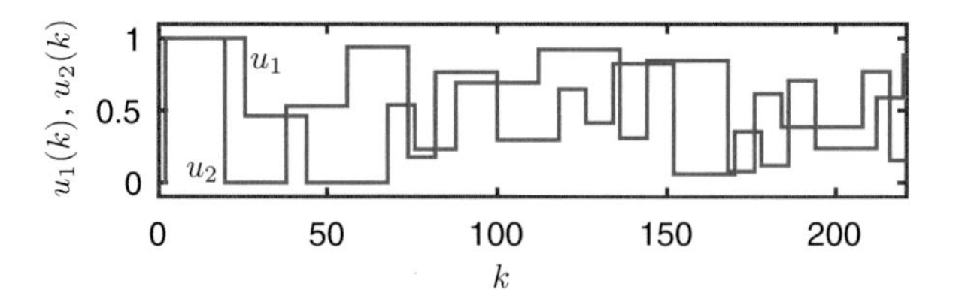

Fig. 19.30 OMNIPUS with the LH approach. At the beginning it is very similar to OMNIPUS without restrictions in Fig. 19.29; latter medium amplitude levels around 0.5 are much more densely covered because the signal is forced to still unlocked amplitude levels on the LH grid

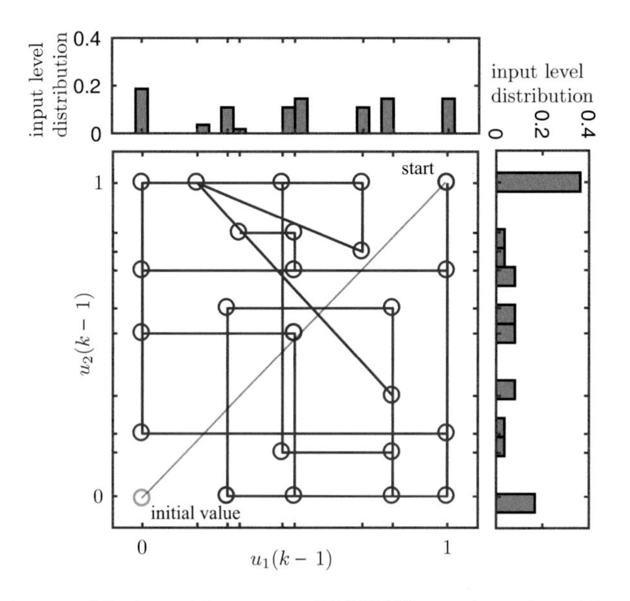

Fig. 19.31 Subspace of the (proxy) input space: OMNIPUS operating points without restrictions. If both inputs change, concurrently oblique transitions occur—this happens at $k = 154$ and $k = 162$; compare Fig. 19.29

amount of data points are inside (27%). Similar to static DoE optimization w.r.t. the maximin criterion, OMNIPUS emphasizes the boundaries of the input space.

The LH-based OMNIPUS generates more data points in the inner regions of the input space. Figure 19.32 shows this excitation signal in the $u_1(k-1)$-$u_2(k-1)$-subspace. With the LH restriction, only 23% of the data points are placed at the boundaries of the input space. The remaining 77% are points in the inner regions of the input space. An additional important advantage of the LH-based OMNIPUS is the much higher spatial resolution (finer sampling) w.r.t. all input axes due to the non-collapsing design of LHs. This feature is particularly important for accurate modeling of the nonlinear static behavior of the process.

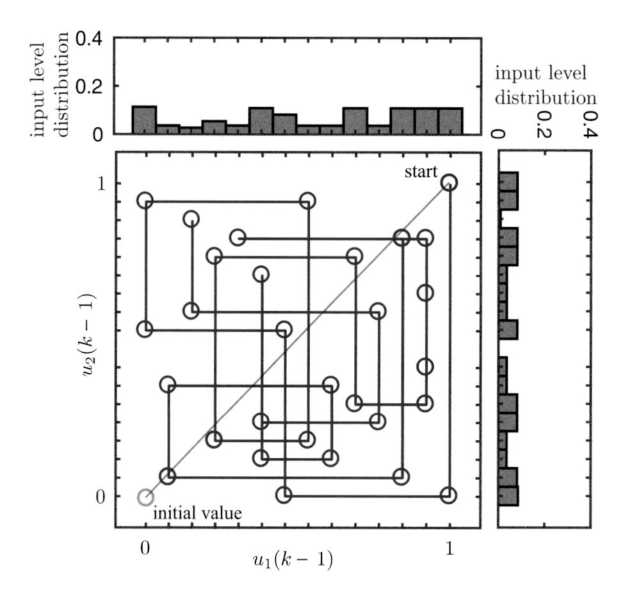

Fig. 19.32 Subspace of the (proxy) input space: OMNIPUS operating points with the LH approach

Note that the initial value at $[u_1(0)\ u_2(0)] = [0\ 0]$ is not contained in any NARX training data set since the NARX input space is spanned by the *previous* inputs and output at $k - i, i = 1, \ldots, m$.

For determining the quality of the multiple input excitation signal, the correlation and the input value distributions are discussed in the following. In this section, a brief simulation analysis of the correlation and data distribution aspects is given. The modeling quality of a NARX model with different excitation signals is analyzed in Sect. 26.4 by means of a real-world application—a high-pressure fuel supply system.

19.8.3.4 Input Signal Correlation

To distinguish the effects of each input u_j to the output y, the absolute value of the correlation between the input signals should be minimal. This property is not explicitly considered in the optimization criterion and unrelated to the space-filling objective of OMNIPUS. But Fig. 19.33 demonstrates that the absolute value of the correlation tends to zero with an increasing number of samples for an example with three inputs.

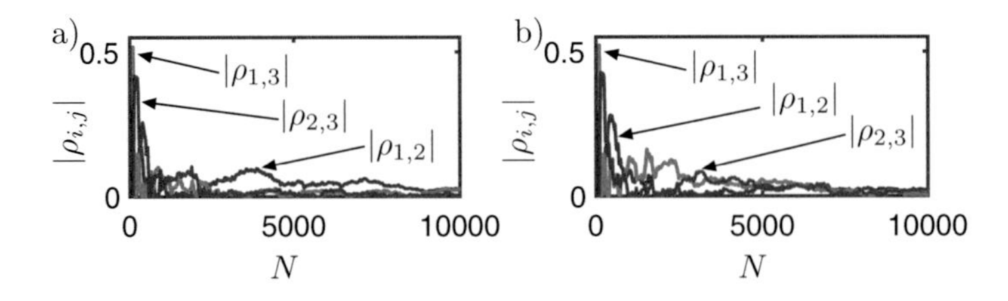

Fig. 19.33 Pairwise cross-correlations between the OMNIPUS input signals using the first N samples of the input signal for (**a**) without restrictions and (**b**) LH-based restrictions

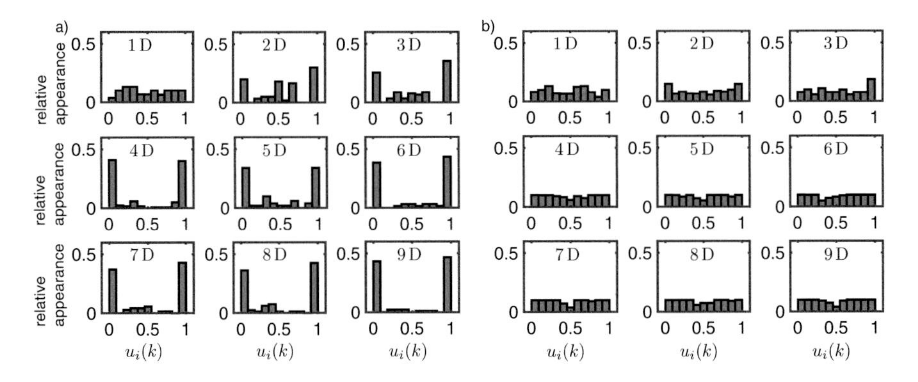

Fig. 19.34 Relative appearance of amplitude values of the input signals for 1 to 9 process inputs optimizing (**a**) a standard OMNIPUS according to (19.60) and (**b**) an LH-based OMNIPUS according to (19.63)

19.8.3.5 Input Value Distribution

An excitation signal used for the identification of nonlinear dynamic systems should cover all nonlinear effects of the underlying process. The optimization of a multiple-input excitation according to the objective in (19.60) leads to a dynamic design where most signal values are at the amplitude limits and consequently at the boundaries of the input space. In Fig. 19.34a the input value distribution is shown for different input dimensions. The signal length of each OMNIPUS was chosen w.r.t. the input dimensionality as $N_\text{n} = 1000p$. The poles used for the proxy model are chosen as $z_j = 0.8$, $j = 1, \ldots, p$. With an increasing number of inputs, the amplitude values concentrate at the limits. The resulting signal for each input is similar to a linear PRBS. Nonlinearities can hardly be recovered from such training data.

Figure 19.34b shows the amplitude values of the LH-based OMNIPUS. Compared to the classical OMNIPUS, a more homogeneous amplitude distribution is achieved independent of the input dimensionality. The superiority of the LH-based approach for dimensionalities $p > 1$ is obvious.

19.8.3.6 Extensions

In practice, often rate constraints prevent sudden input signal changes. Then OMNIPUS should be modified to perform sinusoidal transitions (as in the sinusoidal APRBS case, see Sect. 19.7.5) instead of jumping instantaneously from one amplitude level to another.

The simple linear proxy model can be replaced by a more advanced model if prior knowledge in this respect is available. Also, a more sophisticated active learning scheme (compare Sect. 14.10.4) for dynamic DoE can be established by iterating OMNIPUS with constructing data-driven models and using them as proxies.

19.9 Determination of Dynamic Orders

If the external dynamics approach is taken, the problem of order determination is basically equivalent to the determination of the relevant inputs for the function $f(\cdot)$ in (19.18). Thus, order determination is actually an input selection problem, and the algorithm given below can equally well be applied for input selection of static or dynamic systems. It is important to understand that although the previous inputs $u(k-i)$ and outputs $y(k-i)$ can formally be considered as separate inputs for the function $f(\cdot)$, they possess certain properties that make order determination, in fact, a much harder problem than the selection of physically distinct inputs. For example, $y(k-1)$ and $y(k-2)$ are typically highly correlated (indicating redundancy) but nevertheless may both be relevant. Up to now, no order determination method has been developed that fully takes into account the special properties arising from the external dynamics approach.

The problem of order determination for nonlinear dynamic systems is still not satisfactorily solved. Surprisingly, very little research seems to be devoted to this important area. It is a common practice to select the dynamic order of the model by a combination of trial and error and prior knowledge about the process (when available). Some basic observations can support this procedure. Obviously, if oscillatory behavior is observed, the process must be at least of second order. Step responses at some operating points can be investigated, and linear order determination methods can be applied; see Sect. 18.9. By these means, an approximate order determination of the nonlinear process may be possible. This is, however, a tedious procedure, and a reliable automatic data-based determination method would certainly be desirable. In [54, 56] methods based on higher-order correlations are proposed. But these approaches are merely model validation tools that require building a model with a specific order first and then indicating which information may be missing.

He and Asada [234] proposed a strategy which is based directly on measurement data and does not make any assumptions about the intended model architecture or structure. It requires only that the process behavior can be described by a smooth function, which is an assumption that has to be made anyway in black box nonlinear

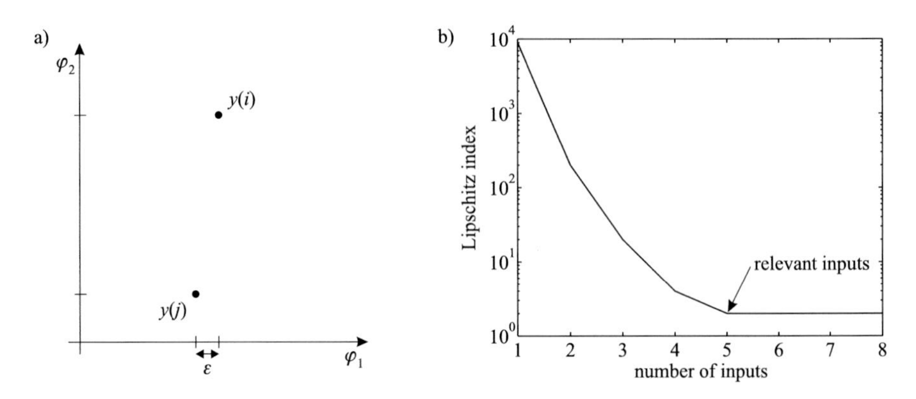

Fig. 19.35 (**a**) Two data points that are close in φ_1 but distant in φ_2 can have very different output values $y(i)$ and $y(j)$ if the function depends on both inputs while $y(i) \approx y(j)$ if the function only depends on φ_1. (**b**) The Lipschitz indices indicate the case where all relevant inputs are included

system identification. The main idea of this strategy is illustrated in Fig. 19.35a and is explained in the following. In the general case, the task is to determine the relevant inputs of the function

$$y = f(\varphi_1, \varphi_2, \ldots, \varphi_n) \tag{19.64}$$

from a set of potential inputs $\varphi_1, \varphi_2, \ldots, \varphi_o$ $(o > n)$ that is given. If the φ_i are distinct physical inputs, (19.64) describes a static function approximation problem; if they are delayed inputs and outputs, it describes an external dynamics model; see Sect. 19.2.

The idea is as follows. If the function in (19.64) is assumed to depend on only $n-1$ inputs although it actually depends on n inputs, the data set may contain two (or more) points that are very close (in the extreme case, they can be identical) in the space spanned by the $n-1$ inputs but differ significantly in the nth input. This situation is shown in Fig. 19.35a for the case $n = 1$. The two points i and j are close in the input space spanned by φ_1 alone, but they are distant in the φ_1-φ_2-input space. Because these points are very close in the space spanned by the $n-1$ inputs (φ_1), it can be expected that the associated process outputs $y(i)$ and $y(j)$ are also close (assuming that the function $f(\cdot)$ is smooth). If one (or several) relevant inputs are missing, then obviously $y(i)$ and $y(j)$ are expected to take totally different values. In this case, it is possible to conclude that the $n-1$ inputs are not sufficient. Thus, the nth input should be included and the investigation can start again.

In [234] an index is defined based on so-called Lipschitz quotients, which is large if one or several inputs are missing (the larger, the more inputs are missing) and is small otherwise. Using this Lipschitz index, a curve as shown in Fig. 19.35b may result for $n = 5$ and $o = 8$ when the following input spaces are checked: 1. φ_1, 2. φ_1-φ_2, ..., 8. φ_1-φ_2-...-φ_8. Thus, the correct inputs ($n = 5$) can be detected at the point where the Lipschitz index stops to decrease.

The Lipschitz quotients in the one-dimensional case (input φ) are defined as

$$l_{ij} = \frac{|y(i) - y(j)|}{|\varphi(i) - \varphi(j)|} \quad \text{for } i = 1, \ldots, N, \; j = 1, \ldots, N \; \text{ and } i \neq j \,, \tag{19.65}$$

where N is the number of samples in the data set. Note that $N(N-1)$ such Lipschitz quotients exist, but only $N(N-1)/2$ have to be calculated because $l_{ij} = l_{ji}$. Since (19.65) is a finite difference approximation of the absolute value of the derivative $df(\varphi)/d\varphi$, it must be bounded by the maximum slope of $f(\cdot)$ if $f(\cdot)$ is smooth. For the multidimensional case, the Lipschitz quotients can be calculated by the straightforward extension of (19.65):

$$l_{ij}^{(n)} = \frac{|y(i) - y(j)|}{\sqrt{(\varphi_1(i) - \varphi_1(j))^2 + \ldots + (\varphi_n(i) - \varphi_n(j))^2}} \tag{19.66}$$

for $i = 1, \ldots, N$, $j = 1, \ldots, N$ and $i \neq j$ and the superscript "(n)" in $l_{ij}^{(n)}$ stands for the number of inputs.

The Lipschitz index can be defined as the maximum occurring Lipschitz quotient

$$l^{(n)} = \max_{i,j,(i \neq j)} \left(l_{ij}^{(n)} \right) \tag{19.67}$$

or as proposed in [234] as the geometric average of the c (with c being a design parameter) largest Lipschitz quotients in order to make this index less sensitive to noise. As long as n is too small and thus not all relevant inputs are included, the Lipschitz index will be large because situations as shown in Fig. 19.35a will occur. As soon as all relevant inputs are included, (19.67) stays about constant.

This strategy requires the ordering of the inputs. For example, for a nonlinear dynamic system, the inputs may be chosen as $\varphi_1 = u(k-1)$, $\varphi_2 = y(k-1)$, $\varphi_3 = u(k-2)$, $\varphi_4 = y(k-2)$, etc. That would allow one to find the model order, but it would not be able to yield a possible dead time since all input regressors starting from $u(k-1)$ are automatically included. Thus, such an ordering severely restricts the flexibility of the method. It can be extended to overcome this limitation by comparing the Lipschitz indices for all input combinations. However, then the number of Lipschitz indices grows in a combinatorial way with the number of potential inputs.

Two drawbacks of this method are its sensitivity with respect to noise and the data distribution. The noise sensitivity can be reduced by choosing a $c > 1$, but this also decreases the order detection sensitivity since it averages between different data points. Nevertheless, the Lipschitz index method is a valuable tool and requires only moderate computational effort, at least for small data sets and a small number of potential inputs if the combinatorial version is used.

19.10 Summary

External and internal dynamics approaches for the generation of nonlinear dynamic models can be distinguished. The external dynamics approach represents the straightforward extension of linear dynamic input/output models and is more widely applied. For training, the NARX and NOE (and other less widespread) model structures can be distinguished. For the NARX model, the one-step prediction error is minimized, which allows one to utilize linear optimization techniques if the model architecture employed is linear in the parameters. For the NOE model, the simulation error is minimized, which requires the application of recurrent learning schemes.

The internal dynamics approach represents a nonlinear state space model. The model possesses its own internal states, which are generated by dynamic filters built into the model structure. In particular, neural networks of MLP architecture are used for this approach. Owing to their recurrent structure, these models are always trained by minimizing the simulation error.

19.11 Problems

1. A two-step prediction can be implemented in two different ways: (i) build a two-step prediction model and use it, or (ii) build a one-step prediction model and use it twice sequentially. Draw block schemes for both alternatives for the case of a second-order system. What are the advantages and drawbacks of both strategies?
2. For predictive control a 1-, 2-, …, and N_p-step prediction is required with relatively high values N_p. Which one of the two alternatives discussed in Problem 1 can be applied with a reasonable effort?
3. Explain the relationship between the sampling rate and the shape of the one-step prediction surface $f(\cdot)$.
4. What is the difference between a parallel and a series-parallel model, and which type is used in what kind of application?
5. What task is more difficult to perform: simulation or prediction? Why?
6. What are the fundamental differences between a model with and without output feedback? What does this imply for the applicability of these approaches?
7. How does the internal dynamics approach differ from the external one? What advantages and drawbacks can be derived independently from the specific model architecture?
8. What would be incorrect in training a recurrent structure with a standard (non-recurrent) training algorithm? How can the recurrency be taken into account during training? Explain the key ideas of both available solutions to this problem.

9. What do the terms “static excitation” and “dynamic excitation” mean? Explain the tradeoff involved in excitation signal design for nonlinear dynamic systems between sufficient static and dynamic excitation.
10. What is the “minimum hold time” in an APRBS? What is a reasonable choice for the minimum hold time in dependency on the time constant of the process? Why? How can the minimum hold time be chosen if the time constant of the process is strongly operating point dependent?

Chapter 20
Classical Polynomial Approaches

Abstract In this short chapter, classical polynomial approaches for modeling nonlinear dynamic processes are introduced. General approaches that fulfill universal approximation properties are discussed first. Then block-structured models of Hammerstein and Wiener type are discussed. They are traditionally realized with polynomials but, of course, could also be combined with other approximators as well. The remainder of the book focuses on universal structures that are not limited to a very specific process structure. Although these special structures can be very effective in many real-world applications, the universalness is usually the appeal of neural network-based approaches.

Keywords Dynamics · Polynomials · Volterra-series · Curse of dimensionality · Extrapolation · Hammerstein model · Wiener model

This chapter gives an overview of some common classical approaches for nonlinear system identification. A shared characteristic of all these approaches is that they are based on polynomials for the realization of the nonlinear mapping. As analyzed in Sect. 10.2, polynomials possess some usually undesirable properties with respect to their interpolation and extrapolation behavior, and they suffer severely under the curse of dimensionality. The following section evaluates the consequences arising from these properties for *dynamic* polynomial models.

The most straightforward way to utilize polynomials for nonlinear system identification is to apply a polynomial for the approximation of the one-step prediction function $f(\cdot)$ in (19.2); compare also Fig. 19.2 in Chap. 19. This NARX modeling approach is known as the *Kolmogorov-Gabor polynomial* and is treated in Sect. 20.2. When polynomial NFIR and NOBF models are identified, this leads to the *non-parametric Volterra-series*; see Sect. 20.3. These general approaches, in which the polynomials are utilized as universal approximators, can be simplified by making specific assumptions about the nonlinear structure of the process. It is important to note that these simplified models like the *parametric Volterra-series*, *NDE*, *Hammerstein*, and *Wiener* structures are suitable only for a restricted classes of processes; see Sects. 20.4, 20.5, 20.6, and 20.7 and [53]. However, even when

O. Nelles, *Nonlinear System Identification*,
https://doi.org/10.1007/978-3-030-47439-3_20

there exists some structural process/model mismatch, these simplified models may be sufficiently accurate for many applications, and thus some of them, in particular, the Hammerstein structure, are widely utilized in practice.

Strictly speaking, the assumptions about the process structure made for Hammerstein and Wiener models and the choice of a specific approximator are two separate issues. For example, it is possible to decide on a Hammerstein structure and utilize a neural network for the approximation of the static nonlinearity. Nevertheless, historically polynomials played the dominant role for these model structures, and therefore they are covered in this chapter.

In order to keep the notation in this chapter as simple as possible, only SISO systems are addressed. An extension to MIMO is straightforward. The user should be aware, however, that each additional input increases the input space dimensionality. In the following, a pragmatic point of view is taken, and the theory on functionals and Volterra kernels is omitted. For an extensive treatment including more theory about dynamic polynomial models and their historic developments, refer to [201, 282].

20.1 Properties of Dynamic Polynomial Models

The properties of static polynomial models have already been discussed in Sect. 10.2. In the context of dynamic systems, some additional considerations are necessary. One weakness of polynomials is that they fully underlie the curse of dimensionality. This makes them an inferior candidate for the external dynamics approach, which easily leads to relatively high-dimensional mappings. A second drawback of polynomials is their tendency to oscillatory interpolation behavior as the polynomial degree increases. This is especially problematic for dynamic systems, as such a spatial "oscillation" can lead to a model gain with wrong sign if it takes place in a $u(k-i)$-direction, and it can lead to totally wrong dynamics (changing the signs of the numerator coefficients of a linearized model) with a high risk of instability if it is in a $y(k-i)$-direction. Note, however, that these effects typically occur only for polynomials with a high degree and that because of the curse of dimensionality, the polynomial degree must be chosen low, e.g., at maximum 3. Thus, these difficulties may be not so important in practice. A third drawback of polynomials is their more than linearly increasing or decreasing extrapolation behavior. This means that for extrapolation the gain of the model is unlimited (i.e., its absolute value increases to ∞), and furthermore, the model certainly becomes unstable. This first property reflects the extrapolation behavior in the $u(k-i)$-directions; the second property corresponds to the extrapolation in the $y(k-i)$-directions. Therefore, extreme care has to be taken to avoid any type of extrapolation, which together with the other shortcomings makes polynomials highly unattractive for modeling dynamic systems. Note, however, that the above analysis addresses the case where the one-step prediction surface is approximated by a polynomial, as introduced in Sect. 20.2. For the less general, simplified structures, this analysis holds only partly. Although according to these fundamental considerations polynomials seem to be poorly suited for identification of dynamic systems, at least in their most general form, in some cases their application might be justified by the underlying process structure. In particular, bilinear terms (i.e., the product of two state variables or a state variable and an input) arise quite often. Then a polynomial approach with degree 2 is highly favorable over most alternatives because it matches the true process structure.

20.2 Kolmogorov-Gabor Polynomial Models

The Kolmogorov-Gabor polynomial represents a nonlinear model *with* output feedback (NARX, NOE, NARMAX, ...). The most straightforward way to use polynomials is to directly approximate the one-step prediction function, i.e.,

$$y(k) = f\left(u(k-1), \ldots, u(k-m), y(k-1), \ldots, y(k-m)\right) . \tag{20.1}$$

Therefore, all properties discussed in Sect. 20.1 apply. For a second-order model ($m = 2$) and a polynomial with degree $l = 2$, the following function results:

$$\begin{aligned} y(k) = {} & \theta_1 + \theta_2 u(k-1) + \theta_3 u(k-2) + \theta_4 y(k-1) + \theta_5 y(k-2) + \\ & \theta_6 u^2(k-1) + \theta_7 u^2(k-2) + \theta_8 y^2(k-1) + \theta_9 y^2(k-2) + \\ & \theta_{10} u(k-1)u(k-2) + \theta_{11} u(k-1)y(k-1) + \theta_{12} u(k-1)y(k-2) + \\ & \theta_{13} u(k-2)y(k-1) + \theta_{14} u(k-2)y(k-2) + \theta_{15} y(k-1)y(k-2) . \end{aligned} \tag{20.2}$$

According to (10.5) in Sect. 10.2, the number of regressors increases strongly as the dynamic order m or the polynomial degree l grows (note that the number of inputs for the approximator in (10.5) is $p = 2m$). The overwhelming number of parameters even for quite simple modeling problems is the main motivation for the simplified approaches in Sects. 20.4, 20.5, 20.6, and 20.7.

Owing to the huge model complexity of Kolmogorov-Gabor polynomials even for moderately sized problems, the utilization of linear subset selection techniques is proposed in [327]; refer also to Sect. 3.4 for an introduction to subset selection methods. They allow one to construct a reduced polynomial model that contains only the most relevant regressors. Although this extends the applicability of Kolmogorov-Gabor polynomials to more complex problems, two limitations have to be mentioned:

- The computational effort for subset selection increases strongly with the number of potential regressors, i.e., with the complexity of the full polynomial model. Thus, this approach is still feasible only for moderately sized problems.
- The selection is based on the one-step prediction error (series-parallel model error). Therefore, it is extremely sensitive with respect to noise and too fast

sampling. It is demonstrated in Sect. 22.9 that subset selection does not work satisfactorily in practice, if the set of candidate regressors contains autoregressive terms such as $y(k-i)$.

Consequently, Kolmogorov-Gabor polynomials should be applied with extreme care.

20.3 Volterra-Series Models

In contrast to the Kolmogorov-Gabor polynomial, the Volterra-series model represents a nonlinear model *without* output feedback (NFIR). Thus, the following function

$$y(k) = f\left(u(k-1), \ldots, u(k-m)\right) \tag{20.3}$$

is approximated by a polynomial. For example, for a second-order model ($m = 2$) and a polynomial with degree $l = 2$, the following function results:

$$\begin{aligned} y(k) &= \theta_1 + \theta_2 u(k-1) + \theta_3 u(k-2) \\ &\quad + \theta_6 u^2(k-1) + \theta_7 u^2(k-2) + \theta_{10} u(k-1)u(k-2)\,. \end{aligned} \tag{20.4}$$

It is important to note, however, that (20.4) is completely incapable of describing any realistic process. As discussed in Sect. 19.2.3, the dynamic order m for NFIR model structures has to be chosen in relation to the settling time of the process. Since the settling time typically is 10–50 times the sampling time ($T_{95} = 10\,T_0 - 50\,T_0$), a realistic choice for m lies between 10 and 50. Clearly, the number of regressors for a Volterra-series model of this order is exorbitant, e.g., for a polynomial of degree $l = 2$ and a dynamic order $m = 30$, already 496 regressors exist according to (10.5) in Sect. 10.2 with $p = m$. Thus, the Volterra-series model is feasible only for simple problems, even in combination with a subset selection technique; compare with the comments in the previous section. On the other hand, the advantages of the Volterra-series model are quite appealing; see Sect. 19.2.3.

- The Volterra-series model belongs to the class of output error models. This implies that both parameter estimation and subset selection are based on the simulation error, not on the one-step prediction error.
- All drawbacks of dynamic polynomial models mentioned in Sect. 20.1 associated with $y(k-i)$-directions are not valid for the Volterra-series model because the input space is spanned solely by the $u(k-i)$.
- Since no feedback is involved, the Volterra-series model is guaranteed to be stable.

Motivated by these attractive features, some effort has been made in order to overcome the Volterra-series model's most severe drawback by reducing the overwhelming number of regressors. In [338, 339, 594] it is proposed to approximate the Volterra kernels[1] by basis functions with far fewer parameters, that is, with a model within a model approach. Another idea is to extend the approach (20.3) from an NFIR to an NOBF model structure, which would allow one to reduce the order m significantly [113, 580]; compare also the NOBF paragraph in Sect. 19.2.3.

20.4 Parametric Volterra-Series Models

The parametric Volterra-series model is a simplified version of the Kolmogorov-Gabor polynomial. It realizes a linear feedback and models a nonlinearity only for the inputs:

$$y(k) = f\left(u(k-1), \ldots, u(k-m)\right) - a_1 y(k-1) - \ldots - a_m y(k-m). \qquad (20.5)$$

For example, for a second-order model ($m = 2$) and a polynomial with degree $l = 2$, the following function results:

$$\begin{aligned} y(k) = {} & \theta_1 + \theta_2 u(k-1) + \theta_3 u(k-2) + \theta_4 y(k-1) + \theta_5 y(k-2) + \\ & \theta_6 u^2(k-1) + \theta_7 u^2(k-2) + \theta_{10} u(k-1) u(k-2)\,. \end{aligned} \qquad (20.6)$$

With this simplification, the model leads to a reduced number of regressors and avoids all drawbacks discussed in Sect. 20.1 that are associated with the $y(k-i)$-directions. Also, stability can be easily proven by checking the dynamics of linear feedback. Thus, the parametric Volterra-series model avoids several disadvantages of the Kolmogorov-Gabor polynomial. The price to be paid for this is a restriction of the generality. Systems whose nonlinear behavior strongly depends on their output cannot be described by (20.5) if the order m is chosen small, i.e., comparable to the order of a Kolmogorov-Gabor polynomial. The parametric Volterra-series model, however, can also be seen as an extension of the Volterra-series model if the order m is chosen large. It can be argued that in this case, the additional linear feedback would help to reduce the dynamic order compared with the non-parametric Volterra-series model.

[1] The Volterra kernels are the coefficients associated with the regressors $u(k-i)$, $u(k-i)u(k-j)$, $u(k-i)u(k-j)u(k-h)$, etc. as 1-, 2-, 3-, etc. dimensional functions dependent on i and i, j and i, j, h, etc. For example, for a linear model, the first kernel (corresponding to $u(k-i)$) is the impulse response.

20.5 NDE Models

The nonlinear differential equation (NDE) model structure arises frequently from modeling based on first principles. It can be considered as the counterpart of the parametric Volterra-series model since it is linear in the inputs but nonlinear in the outputs:

$$y(k) = b_1 u(k-1) + \ldots + b_m u(k-m) + f\left(y(k-1), \ldots, y(k-m)\right) . \quad (20.7)$$

For example, for a second-order model ($m = 2$) and a polynomial with degree $l = 2$, the following function results:

$$\begin{aligned} y(k) = {} & \theta_1 + \theta_2 u(k-1) + \theta_3 u(k-2) + \theta_4 y(k-1) + \theta_5 y(k-2) + \\ & \theta_8 y^2(k-1) + \theta_9 y^2(k-2) + \theta_{15} y(k-1) y(k-2) . \end{aligned} \quad (20.8)$$

The NDE model possesses restrictions that are contrary to the parametric Volterra-series model. However, it does not share its benefits because all drawbacks associated with the $y(k-i)$-directions hold. Consequently, the NDE model should be applied only if its structure matches the process structure really well.

20.6 Hammerstein Models

The Hammerstein model is probably the most widely known and applied nonlinear dynamic modeling approach. It assumes a separation between the nonlinearity and the dynamics of the process. The Hammerstein structure consists of a nonlinear static block followed by a linear dynamic block (see Fig. 20.1a) and can be described by the equations

$$x(k) = g\left(u(k)\right) , \quad (20.9a)$$

$$\begin{aligned} y(k) = {} & b_1 x(k-1) + \ldots + b_m x(k-m) \\ & - a_1 y(k-1) - \ldots - a_m y(k-m) . \end{aligned} \quad (20.9b)$$

In order to avoid redundancy, the gain of the linear system may be fixed at 1, which gives the following constraint (reducing the effective number of parameters by one):

$$\frac{\sum_{i=1}^{m} b_i}{1 + \sum_{i=1}^{m} a_i} = 1 . \quad (20.10)$$

The structure describes (besides others) all systems where the actuator's nonlinearity, for example, the characteristics of a valve, the saturation of an electromagnetic motor, etc., is dominant and other nonlinear effects can be neglected. For this reason, Hammerstein models are popular in control engineering. Furthermore, it is easy to compensate the nonlinear process behavior by a controller that implements the inverse static nonlinearity $g^{-1}(\cdot)$ at its output.[2] Another advantage of the distinction between nonlinear and linear blocks is that stability is determined solely by the linear part of the model, which can be easily checked. Thus, the Hammerstein model has many appealing features. However, the structural assumptions about the processes are very restrictive, and therefore it can be applied only to a limited class of systems.

The static nonlinearity is classically approximated by a polynomial. Any other static approximator can also be utilized, which indeed is a good idea in order to avoid the inferior interpolation and extrapolation capabilities of polynomials. For an efficient identification, it is recommended that a linearly parameterized approximator is used. Note that for systems with multiple inputs, the static nonlinearity $g(\cdot)$ becomes a higher-dimensional function.

For a polynomial of degree $l = 2$, the static nonlinearity becomes

$$x(k) = c_0 + c_1 u(k) + c_2 u^2(k)\,. \tag{20.11}$$

For a second-order model ($m = 2$), the input/output relationship then is

$$\begin{aligned} y(k) &= b_1 c_0 + b_1 c_1 u(k-1) + b_2 c_0 + b_2 c_1 u(k-2) \\ &\quad + b_1 c_2 u^2(k-1) + b_2 c_2 u^2(k-2) - a_1 y(k-1) - a_2 y(k-2). \end{aligned} \tag{20.12}$$

Obviously, although both the polynomial and the linear dynamic model are linearly parameterized, the Hammerstein model is nonlinear in the parameters because products between parameters such as $b_1 c_1$ appear in (20.12). In order to avoid the application of nonlinear optimization techniques, the Hammerstein model is usually not identified directly. Rather the *generalized Hammerstein* model structure is introduced. The generalized Hammerstein model is constructed by summarizing identical terms and re-parameterizing the model in a manner that yields linear parameters. For example, in (20.12) the terms $b_1 c_0$ and $b_2 c_0$ both represent offsets and can be summarized. Then one independent parameter is assigned to each regressor. For $l = 2$ and $m = 2$, this procedure yields the generalized Hammerstein model

$$\begin{aligned} y(k) &= \theta_1 + \theta_2 u(k-1) + \theta_3 u(k-2) + \theta_4 y(k-1) + \theta_5 y(k-2) + \\ &\quad \theta_6 u^2(k-1) + \theta_7 u^2(k-2)\,, \end{aligned} \tag{20.13}$$

[2] If the inverse exists.

whose parameters can be estimated by linear optimization. This model can also be seen as a simplified parametric Volterra-series model without the cross terms between regressors that depend on different time delays.

For the original Hammerstein model, the number of parameters is $l + 2m$ (one less than the number of nominal parameters due to the constraint (20.10)), while for the generalized version, it is $m(l + 1) + 1$. For the above example, the generalized Hammerstein model has only one parameter more than the original one (7 compared with 6). However, as the polynomial degree l and the dynamic order m grow, the generalized Hammerstein model possesses significantly more parameters than the original one. Therefore, no one-to-one mapping between both model structures can exist. However, it is possible to calculate the parameters of a Hammerstein model that *approximates* a generalized Hammerstein model.

20.7 Wiener Models

The Wiener model structure is the Hammerstein structure reversed, that is, a linear dynamic block is followed by a nonlinear static block; see Fig. 20.1b. Only a few processes match these structural assumptions. A prominent example is the pH titration process. Also, control systems whose major nonlinearity is in the sensor can be described by a Wiener model. It follows the equation

$$\begin{aligned} x(k) &= b_1 u(k-1) + \ldots + b_m u(k-m) \\ &\quad - a_1 x(k-1) - \ldots - a_m x(k-m) \qquad (20.14a) \\ y(k) &= g\left(x(k)\right) \qquad (20.14b) \end{aligned}$$

with the same constraint (20.10) as for the Hammerstein model.

It is interesting to note that it is not necessarily possible to write a Wiener model structure in an input/output form. Elimination of x in (20.14a), (20.14b) yields

$$\begin{aligned} y(k) = g\big(&b_1 u(k-1) + \ldots + b_m u(k-m) \\ &-a_1 g^{-1}(y(k-1)) - \ldots - a_m g^{-1}(y(k-m))\big) , \end{aligned} \qquad (20.15)$$

where $g^{-1}(\cdot)$ is the inverse of $g(\cdot)$, i.e., $x = g^{-1}(g(x))$. The input/output relationship (20.15) does exist only if the static nonlinearity $g(\cdot)$ is invertible.

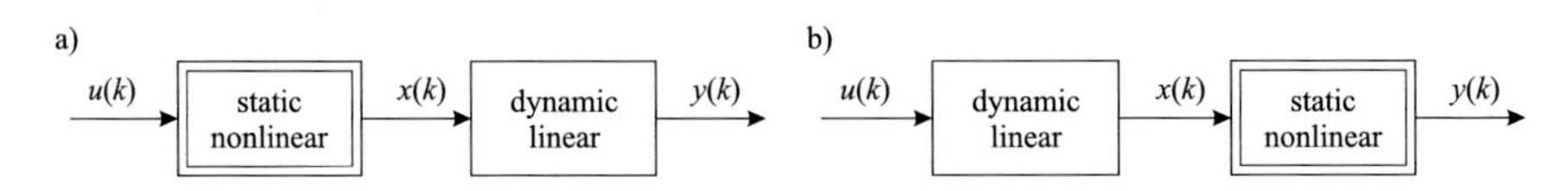

Fig. 20.1 (**a**) Hammerstein model structure. (**b**) Wiener model structure

This issue has already been addressed in Sect. 19.2.4. As can be observed from (20.15) for Wiener model structures, no straightforward linear parameterization exists. This means that nonlinear optimization methods have to be applied for parameter estimation. As for Hammerstein models, polynomials are classically used for approximation of $g(\cdot)$, but any other approximator can be applied as well.

20.8 Problems

1. Explain the extrapolation behavior of polynomials? How well is it suited for modeling dynamic systems? Why?
2. How many parameters does a Kolmogorov-Gabor polynomial of degree 2 and dynamic order 3 possess? Comment on the practicability of this model type for high-order systems.
3. What is the difference between a parametric Volterra-series and a Kolmogorov-Gabor polynomial model? Which advantages and drawbacks follow directly from this difference?
4. Write a simulation program for a linear dynamic first-order system in series with some static nonlinearity. How is such a type of nonlinear dynamic system called? Simulate the impulse, step, and ramp response of this system.
5. Swap the order of the two blocks from Problem 3, i.e., the input goes into the static nonlinearity, and its output feeds the linear dynamic system. How is such a type of nonlinear dynamic system called? Will the simulations be different for the swapped system? Why? Which responses will differ more and which less? Simulate the impulse, step, and ramp response of the swapped system for verification.
6. Consider a process that can be described by a Hammerstein model. What control strategy is advisable in order to reduce the control problem to a linear one?

Chapter 21
Dynamic Neural and Fuzzy Models

Abstract This short chapter extends the neural networks and neuro-fuzzy models from Part B to the dynamic case. The newly arising issues in that context are discussed, such as the sensitivity with respect to the curse of dimensionality. Due to the tapped delay lines in external dynamics approaches, even SISO processes require the model to deal with quite high-dimensional input spaces. The question about the tradeoff between dynamic errors and static errors arises. The former may be due to too low model orders; the latter may be due to too high-dimensional input spaces. This tradeoff is not thoroughly enough investigated up to date.

Keywords Dynamics · Curse of dimensionality · Interpolation · Extrapolation · RBF · MLP

This chapter discusses the use of static neural network and fuzzy model architectures for building nonlinear dynamic models with the external dynamics approach described in Chap. 19. The emphasis is on the following three model architectures, although many observations generalize to other neural network and fuzzy model architectures as well:

- Multilayer perceptron (MLP) networks.
- Radial basis function (RBF) networks.
- Singleton fuzzy models and normalized RBF networks, which are equivalent under some conditions; see Sect. 12.3.2.

Again, a special chapter is devoted to dynamic local linear neuro-fuzzy models (Chap. 22) because they possess various interesting features that require a more extensive study.

In this chapter those properties of the above-listed model architectures will be addressed that have important consequences for the resulting *dynamic* model. The properties of the *static* model architectures have been already discussed in Chaps. 11 and 12.

This chapter is organized according to the addressed model features rather than the model architectures in order to facilitate understanding and comparison between

O. Nelles, *Nonlinear System Identification*,
https://doi.org/10.1007/978-3-030-47439-3_21

the different approaches. The topics covered range from the curse of dimensionality in Sect. 21.1 to interpolation and extrapolation issues (Sect. 21.2) and the training algorithms in Sect. 21.3. The idea of using a linear model in parallel to the nonlinear approximator, which has already been mentioned in Sect. 1.1.1, is analyzed in Sect. 21.4. Finally, some simulation examples are presented in Sect. 21.5 to assist a better understanding of the different architectures' characteristics.

All comments concerning the one-step prediction and the simulation error assume that the intended The use of the model is *simulation* rather than prediction.

21.1 Curse of Dimensionality

As illustrated in Fig. 19.2, the external dynamics approach results in a high-dimensional mapping that has to be performed by the static approximator, in particular for high-order and multivariable systems. This basically rules out all model architectures that fully underlie the curse of dimensionality; see Sect. 7.6.1. Thus, other than for some trivial problems, model architectures such as look-up tables and other grid-based approaches cannot really be utilized. Delaunay networks suffer from shortcomings in their training algorithm generating the triangulation since its complexity and numerical ill-conditioning increase strongly with the input dimensionality. Thus, Delaunay networks are restricted by their training algorithm rather than by their model structure. The assessment of the main three model architectures addressed here with respect to the curse of dimensionality is as follows.

21.1.1 MLP Networks

Since MLP networks are able to find the main directions of the nonlinearity of a process, they avoid the curse of dimensionality. The number of parameters (input layer weights) increases only linearly with the number of inputs if the number of hidden layer neurons is fixed. Thus, MLP networks are well suited for the external dynamics approach. Furthermore, they are relatively insensitive with respect to too high a choice of dynamic orders because they can cope well with redundant inputs by driving the corresponding hidden layer weights toward zero. Also, the uneven data distribution that typically arises with the external dynamics approach (see Fig. 19.4 in Sect. 19.2.1) can easily be handled by MLP networks because the optimization of the hidden layer weights transforms the input axes in a suitable coordinate system anyway. Altogether, many of the drawbacks of MLP networks that make them quite unattractive for most static approximation problems are compensated for by their features when dealing with dynamic systems.

21.1.2 RBF Networks

The extent to which RBF networks underlie the curse of dimensionality depends on the training method chosen and can be somewhere between high and medium. Random and grid-based center placement is not suitable for the external dynamics approach; see the comments above. Clustering-based methods are better suited because they allow one to represent the uneven data distribution in the input space. The OLS structure selection probably yields the best results for training RBF networks since the center selection is done supervised, compared with the (at least partly) unsupervised clustering approaches.

21.1.3 Singleton Fuzzy and NRBF Models

As for RBF networks, the degree to which singleton fuzzy systems and NRBF networks underlie the curse of dimensionality depends on the training method. The difficulty with regard to fuzzy systems is that a grid-based approach has to be avoided (see above), but it is exactly the grid-based structure that makes fuzzy systems so easily interpretable in terms of rules formed with just one-dimensional membership functions; see Sect. 12.3.4. However, with additive, hierarchical, or other complexity-reducing strategies (compare with Sects. 7.4 and 7.5), some parts of the interpretability in terms of fuzzy rules can be recovered. Algorithms for solving that task include FUREGA and ASMOD described in Sects. 12.4.4 and 12.4.5, respectively. Unfortunately, the OLS structure selection algorithm cannot be applied in a direct and efficient way to singleton fuzzy models and NRBF networks owing to the normalization or defuzzification denominator; see Sect. 12.4.3. Although these complexity reduction algorithms allow the construction of fuzzy models with a medium number of inputs, they usually sacrifice either accuracy or interpretability in trying to find the most appropriate tradeoff between these two factors. This procedure becomes more difficult as the input space dimensionality increases. Thus, singleton fuzzy models usually offer interpretability advantages over other model architectures only for small to moderately sized problems.

21.2 Interpolation and Extrapolation Behavior

The interpolation and extrapolation behavior of the applied static model architecture has important consequences for the dynamic characteristic. It is essential to recall from Sect. 19.2 that with the external dynamics approach, the slopes of the model's one-step prediction surface in the $u(k-i)$-dimensions determine the gains and zeros of the model when linearized around an operating condition while the slopes in the $y(k-i)$-dimensions determine the poles. The most intuitive understanding of these relationships is obtained by examining a first-order system for which the static approximator has the two inputs $u(k-1)$ and $y(k-1)$. It can be seen directly that the b_1 coefficient is equal to slope in $u(k-1)$ while the $-a_1$ coefficient is equal to the slope in $y(k-1)$ assuming a linearized transfer function $G(q) = b_1 q^{-1}/(1+a_1 q^{-1})$ or equivalently the difference equation $y(k) = b_1 u(k-1) - a_1 y(k-1)$; see Sect. 19.2 for a more detailed analysis.

As a consequence of these observations, the following conclusions can be drawn:

- If the slope in a $u(k-i)$-dimension changes its sign, the model may change its gain (certainly for first order).
- If the slope in a $y(k-i)$-dimension changes its sign, the model may totally change its dynamic characteristics (it becomes oscillatory for first order).
- If the slope in a $y(k-i)$-dimension becomes large, the model may become (locally) unstable (certainly if the slope becomes larger than 1 for first order).

The first two points are particularly problematic for RBF networks. While MLP and NRBF networks and singleton fuzzy systems tend to have monotonic interpolation behavior, it is very sensitive with respect to the widths of the basis functions in RBF networks; see Sect. 11.3.5. If the widths are too small, then "dips" will occur in the interpolation behavior. If the widths are too large, then locality may be lost and numerical difficulties will emerge. In practice, it will be hardly possible to find widths that avoid both effects in all dimensions. Thus, "dips" can be expected in the interpolation behavior of RBF networks. This is a significant drawback for dynamic RBF networks. It can be weakened, however, if the RBF network is used in parallel to a first principles or linear model; see Sect. 21.4 and 7.6.2.

It is interesting to see how the extrapolation behavior influences the model dynamics. The following types of extrapolation behavior can be distinguished:

- *None*[1]: This occurs for look-up tables[2] and CMAC and Delaunay networks. For these model architectures, extrapolation must be avoided at all, or some backup system has to become active to cope with the situation. In practice, extrapolation can hardly be avoided when dealing with complex dynamic systems because it is rarely possible to cover all extreme operating conditions with the training data.
- *Zero*: This occurs for RBF networks. A model output approaching zero can hardly be considered a realistic or reasonable behavior for most applications. It is a nice feature, however, when the model is run in parallel (additively) to a first principles or linear model. Then the extrapolation behavior of this underlying model is recovered; see Sect. 21.4. It guarantees that the underlying model is

[1]It is assumed here that the upper and lower bounds are chosen corresponding to the minimal and maximal values within the training data set. If they are defined as the theoretical maximal and minimal values of the inputs (and outputs), the CMAC network extrapolates with zero, while the Delaunay network would require training data in all corners of the input space, which is an assumption that frequently cannot be fulfilled.

[2]Look-up tables can be extended so that they possess constant extrapolation behavior.

not degraded outside a certain region influenced by the RBF network. That makes RBF networks an attractive choice for *additive supplementary models* in those situations where already existing models should be improved; see also Sect. 7.6.2.

- *Constant*: This occurs for MLP, NRBF, GRNN networks and linguistic, singleton fuzzy systems. Constant extrapolation is quite reasonable behavior for many static modeling problems. For dynamic models, however, this means static extrapolation since all a_i coefficients of the transfer function of a linearized model tend to zero (constant behavior implies slope zero). This is clearly unrealistic.
- *Linear*: This occurs for linear models and local linear neuro-fuzzy models. Because the dynamic behavior is preserved, this can be considered as the most reasonable extrapolation behavior for dynamic models. Section 22.4.3 discusses this issue in greater detail.
- *High order*: This occurs for polynomials. Because the slopes of the model's one-step prediction surface are unbounded and grow severely in the extrapolation regions, extrapolation must be avoided at all costs to ensure a reasonable model behavior; see Chap. 20.

21.3 Training

For static model architectures, it is helpful to distinguish between linear and nonlinear parameters and optimization techniques. For dynamic models an additional distinction between the optimization of the simulation performance (NOE representation) and of the one-step prediction performance (NARX representation) is important. The NARX representation is the only one that keeps the parameterization of the static model. That is, with the NARX representation, linear parameters stay linear parameters. All other dynamic model representations, i.e., NOE or those with more complex noise descriptions such as NARMAX or NBJ, make all parameters nonlinear by introducing feedback. The only way to exploit the advantages of linear optimization techniques for nonlinear dynamic models is to (i) choose a linear parameterized model architecture *and* (ii) choose the NARX representation. The reason for the frequent employment of NARX models is more their computational benefits than their realistic process description. As was pointed out in Chap. 18 on linear system identification, (N)OE representations typically offer a superior process description because of their more realistic noise assumptions.

To summarize, the following cases can be distinguished (see Table 19.1):

- *NARX representation and parameters linear in the static model*: The dynamic model is linear in the parameters.
- *NARX representation and parameters nonlinear in the static model*: The dynamic model is nonlinear in the parameters, and the gradients can be calculated as for static models.

- *NOE (or other) representation and parameters linear in the static model*: The dynamic model is nonlinear in the parameters, and the gradients have to be calculated taking the feedback into account, e.g., with BPTT or real-time recurrent learning (Sect. 19.5).
- *NOE (or other) representation and parameters nonlinear in the static model*: The dynamic model is nonlinear in the parameters, and the gradients have to be calculated taking the feedback into account.

Thus, for NARX models the training is equivalent to the static case, while for NOE (or other) models, training becomes more complicated, and the distinction between (in the static case) linear and nonlinear parameterized models is lost. This means that for the NOE representation, all advantages of certain model architectures that are a consequence of their linear parameterization (in the static case) are lost. Thus, as a rule of thumb, one tends to use nonlinearly parameterized architectures such as MLP networks together with an NOE representation and linear parameterized architectures such as RBF networks together with a NARX representation. Therefore, in nonlinear system identification, the choice of the dynamic representation is highly interconnected with the choice of the model architecture.

21.3.1 MLP Networks

From the above discussion, it is clear that the MLP networks can fully play out their advantages compared with alternative linear parameterized model architectures when the NOE representation is chosen. Most of the MLP drawbacks stem from the nonlinear parameterization, which is a property shared with the alternatives when the NOE representation is used. In the NARX representation, however, the MLP network still suffers from serious drawbacks compared with the alternatives and typically becomes interesting only for high-order and/or multivariable systems that lead to very high-dimensional input spaces; compare the advantages explained in Sect. 21.1.1.

21.3.2 RBF Networks

For RBF networks, the assessment is opposite to that for MLP networks. For the NARX representation, RBF networks offer the same favorable properties as in the static case, while for the NOE representation, most of the advantages are lost and the drawbacks are retained.

When used together with the NARX representation, some important possibilities and restrictions will be pointed out with regard to the construction algorithms. In particular, the complexity-controlled clustering algorithms are promising since they advantageously allow one to incorporate other objectives into the clustering process. For dynamic models, a good strategy is to choose an objective that depends on the simulation performance (output error) rather than the one-step prediction performance.

However, when using a subset selection technique, the structure and parameter optimizations are both based on the one-step prediction performance, which does not necessarily ensure good simulation capabilities of the model [407]. Thus, the OLS algorithm (or any other subset selection algorithm) should be extended by monitoring the model's simulation performance and avoiding the construction of models that are unstable (which would not necessarily be noticed in the one-step prediction error).

21.3.3 Singleton Fuzzy and NRBF Models

For singleton fuzzy and NRBF models, basically, the same comments can be made as for RBF networks. Also, fuzzy model construction algorithms are typically organized into a structure optimization and a parameter optimization part, which often is nested within each other. Then the simulation performance can advantageously be used as an objective in the structure optimization part, while the parameter estimation is preferably based on the one-step prediction error to allow the application of linear regression schemes.

21.4 Integration of a Linear Model

As explained in Sect. 21.2, the extrapolation behavior of most model architectures is generally not well suited for modeling dynamic systems with the external dynamics approach, and the interpolation behavior of RBF networks is particularly problematic. One simple strategy to overcome or at least weaken this disadvantage is to use the nonlinear model in parallel to a linear model, as shown in Fig. 21.1. The linear model can be obtained either by identification or by first principles modeling.

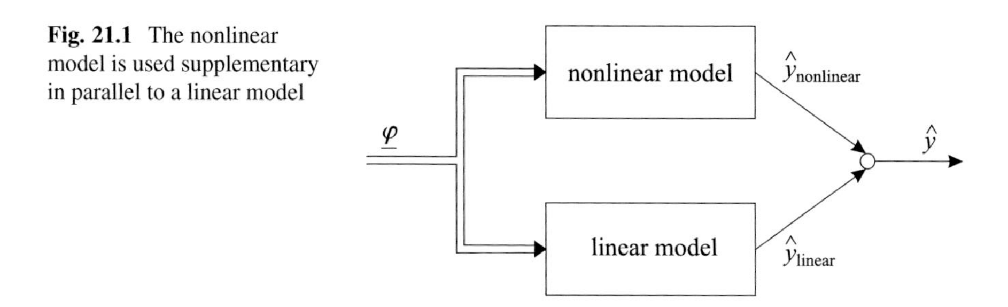

Fig. 21.1 The nonlinear model is used supplementary in parallel to a linear model

The advantages of this approach are that for extrapolation the slopes of the linear model are recovered and, for interpolation, the "dips" caused by an RBF network may be avoided in the overall model. A condition for this latter feature is that the "dips" are not too severe, i.e., the slopes of the linear model must be (in absolute value) larger than the maximal derivative of the RBF network output with the opposite sign.

Two alternative strategies for identification of a model as shown in Fig. 21.1 can be distinguished:

- First, estimate the parameter of the linear model. Second, train the supplementary model with the error of the linear model keeping the linear model fixed.
- Train the supplementary and the linear model together.

The second strategy is more flexible because the effective number of parameters equals the sum of the number of parameters of both models. Thus, it will generally lead to a higher model accuracy (at least on the training data). However, the linear model is identified only in combination with the nonlinear one. Therefore, it is not necessarily a good representation of the process on its own, and its extrapolation behavior cannot be expected to be realistic. In fact, it can easily happen that the linear model is unstable.

The first strategy is less flexible because the effective number of parameters is smaller than those of both models combined. It can be seen as the first iteration of a staggered optimization approach in which the parameters of submodels are iteratively estimated separately; see Sect. 7.5.5, where it is explained why this is a regularization technique. In contrast to the second strategy, the linear model usually captures something like the average dynamics of the nonlinear process and thus can be expected to be stable (although this is not guaranteed). Consequently, the first approach is much more reliable in producing realistic extrapolation behavior.

One limitation of the idea of putting a linear model in parallel to a nonlinear model and letting the linear model take care of the extrapolation is that the extrapolation behavior in all input space dimensions is equivalent. The straightforward extension of this idea toward different behaviors for different extrapolation regimes is given by the local linear neuro-fuzzy model architecture discussed in the next chapter.

21.5 Simulation Examples

To illustrate the functioning of the three model architectures discussed in this chapter, the following simple first-order nonlinear dynamic process of Hammerstein structure will be considered:

$$y(k) = 0.1 \arctan(u(k-1)) + 0.9y(k-1) . \qquad (21.1)$$

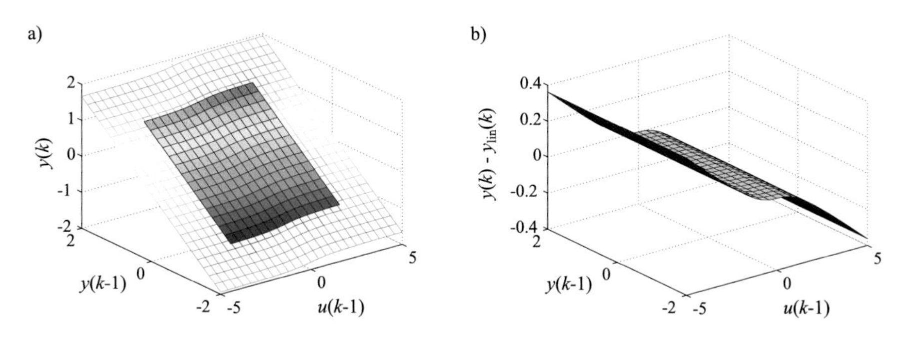

Fig. 21.2 (**a**) One-step prediction surface of the first-order Hammerstein test process. (**b**) Nonlinear component of the one-step prediction surface, $y_{\text{lin}}(k) = 0.1u(k-1) + 0.9y(k-1)$

The choice for the arctan(·)-nonlinearity is motivated by the observation that many real processes exhibit such a type of saturation behavior. The Hammerstein structure arises, for example, when the actuator of a plant introduces the dominant nonlinear effect to the overall system and possesses a saturation characteristics. The input u will vary between -3 and 3, and the process is excited by an APRBS; see Sect. 19.7. The one-step prediction surface of this process is shown in Fig. 21.2a. The training data lies in the filled part of the surface, while the outer parts represent extrapolation areas. Because the nonlinear characteristics of the one-step prediction surfaces sometimes can hardly be observed, it can be useful to examine the difference between the one-step prediction of the nonlinear model and the linear model $y_{\text{lin}}(k) = 0.1u(k-1) + 0.9y(k-1)$. Figure 21.2b shows the one-step prediction surface for the nonlinear component $y(k) - y_{\text{lin}}(k)$ of the process.

In order to illustrate the intrinsic properties of the different model architectures, the training data is not disturbed by noise. The goal for the models shall be good simulation performance. Since models with low complexity are chosen and the training data is noise-free, no overfitting can occur, and the simulation performance of the models can be assessed directly on the training data.

21.5.1 MLP Networks

MLP networks can be assumed to be well suited for the given process because their sigmoid activation functions are very close to the arctan(·) function in (21.1). Indeed, a network with just two hidden neurons and thus nine parameters yields very good approximation results. Figure 21.3 shows the simulation performance of the network on the training data and the accuracy of the static model characteristics for interpolation (gray) and extrapolation (white). For the results shown on the right, the optimization technique converged to the global optimum. For the results shown on the left, obtained with a different parameter initialization, the optimization got stuck in a local optimum. The global optimum was reached in about 30% of the trials. Note

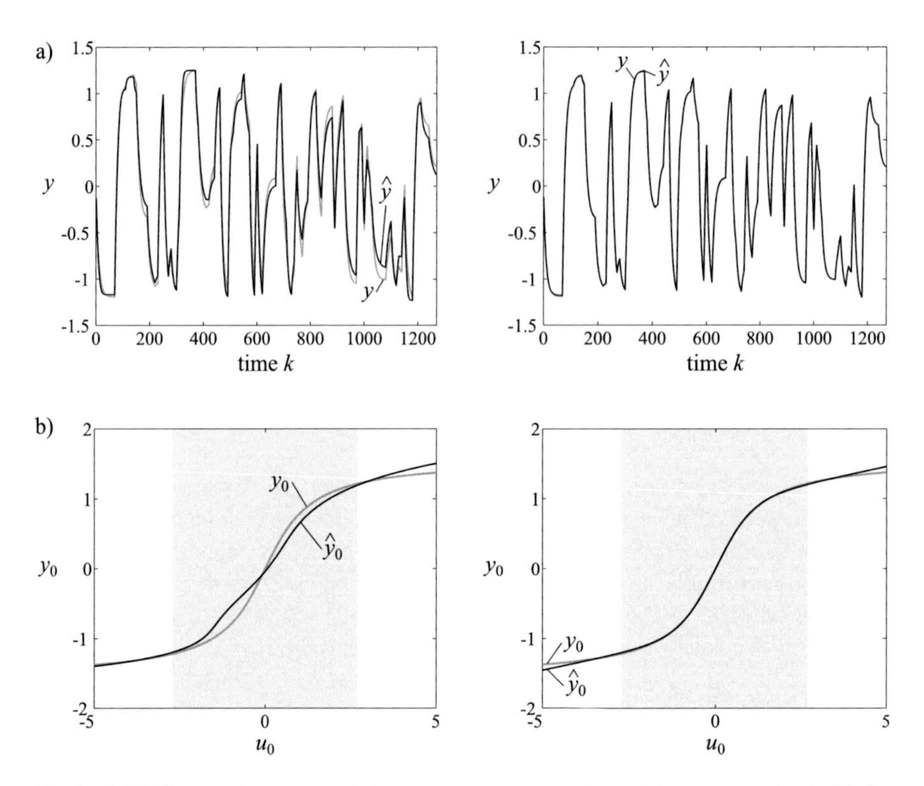

Fig. 21.3 MLP network with two hidden neurons: (**a**) comparison of the process output with the simulated MLP network output; (**b**) static behavior of the MLP network compared with the process equilibrium. The results on the left represent a local optimum; for the results on the right, the global optimum is attained

that this number can go down considerably if it is not ensured that the parameters are initialized in a manner that avoids saturation of the sigmoid functions. Nevertheless, even the local optima results are remarkably good.

Figure 21.4 shows the nonlinear part of the one-step prediction surfaces of both MLP networks and their weighted basis functions $w_i \Phi_i(\cdot)$. While the one-step prediction surface corresponding to the globally optimal solution (right) very accurately describes the true process in Fig. 21.2b, the locally optimal solution possesses a significant nonlinear behavior that does not describe the process, particularly in the extrapolation areas. Interestingly, both solutions generate one basis function that basically represents the linear part of the process y_{lin}, while the other basis function introduces the major nonlinear characteristics. This observation is one motivation for the a priori integration of a linear model as proposed in Sect. 21.4. The shape of the basis function Φ_1 ensures a very good extrapolation behavior.

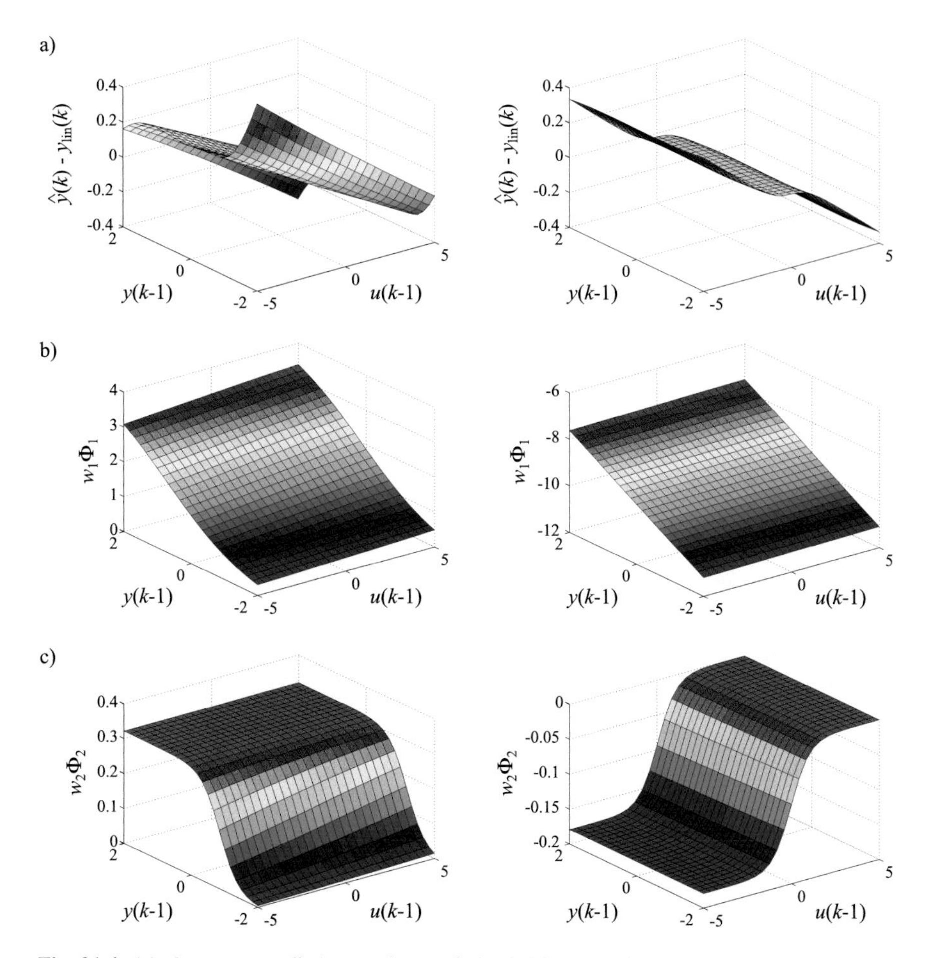

Fig. 21.4 (**a**) One-step prediction surfaces of the MLP networks with two hidden neurons. (**b**), (**c**) Weighted basis functions of the network. Note that the absolute values of the network output are adjusted by the offset parameter in the output neuron. The results on the left represent a local optimum; for the results on the right, the global optimum is attained

21.5.2 RBF Networks

RBF networks can be expected to perform worse than the MLP networks on the test process because their local basis functions do not match well its nonlinear characteristics. Indeed, it proves to be difficult to choose the standard deviations of the Gaussian RBFs. In fact, satisfactory results can be obtained only with relatively large RBF widths; otherwise "dips" in the interpolation behavior deteriorate the model performance (Sect. 11.3.4). While for one-step prediction these "dips" cause only minor accuracy deteriorations, they can be catastrophic for simulation because they can cause a model gain with a wrong sign.

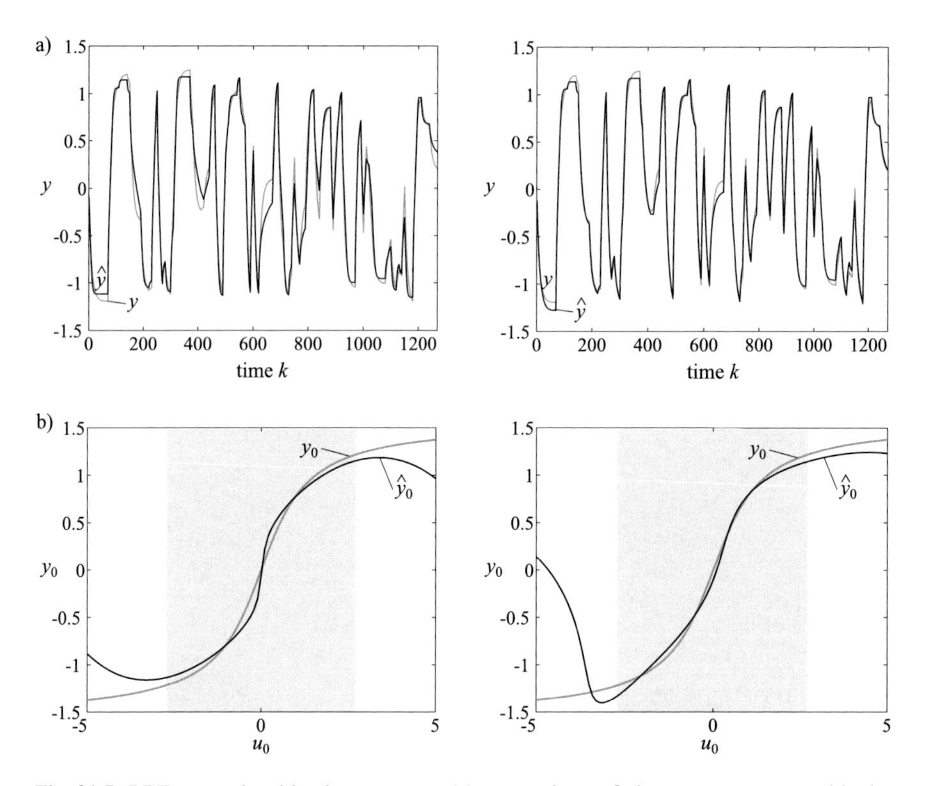

Fig. 21.5 RBF network with nine neurons: (**a**) comparison of the process output with the simulated RBF network output; (**b**) static behavior of the RBF network compared with the process equilibrium. The results on the left were obtained by placing the RBF centers on a grid; for the results on the right, the OLS algorithm was used

An RBF network with nine basis functions corresponding to the nine parameters of the MLP network was trained with the grid-based and the OLS center selection strategy. Figure 21.5 shows the simulation performance and the static model characteristics in comparison with the process. Obviously, the accuracy is significantly worse than with the MLP network. Nevertheless, the results obtained with the OLS approach (right) are satisfactory. The static model behavior reveals a severe weakness of RBF networks when applied for dynamic systems: the extrapolation behavior tends toward zero. The danger of "dips" and the unrealistic extrapolation behavior are motivations for the a priori integration of a linear model, especially with RBF networks, as proposed in Sect. 21.4. The extrapolation behavior with the OLS approach is particularly poor because the basis functions centers are more unequally distributed and optimized with respect to the interpolation behavior.

Figure 21.6 shows the one-step prediction surfaces of the RBF networks (not their nonlinear components because the model errors are large enough to be investigated on the original surface). It is obvious that the locality of the basis functions becomes a clear drawback when the function to be approximated has

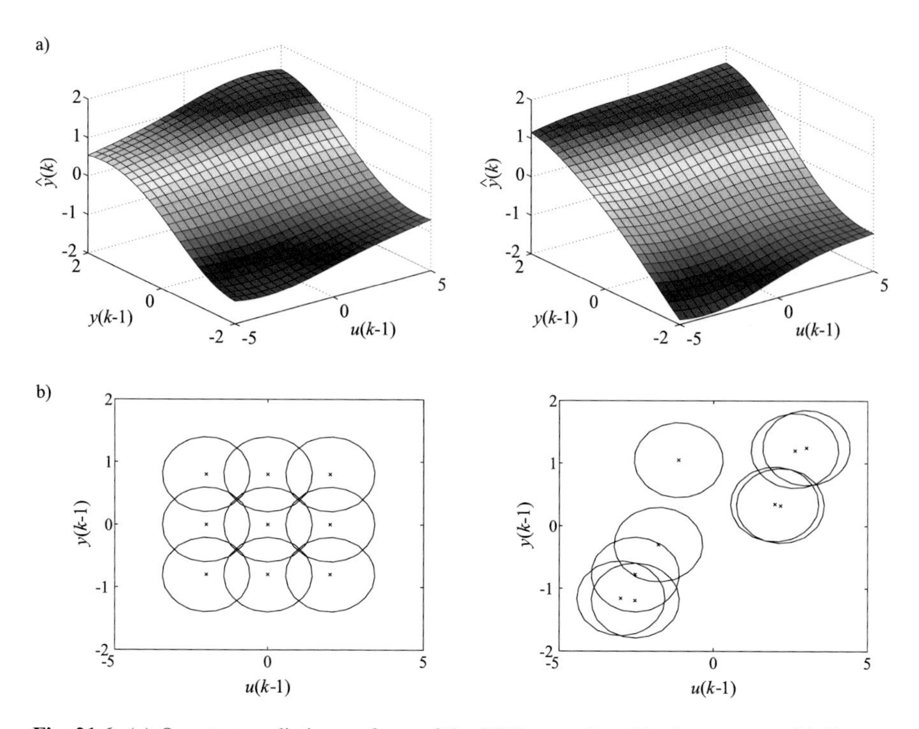

Fig. 21.6 (**a**) One-step prediction surfaces of the RBF networks with nine neurons. (**b**) Centers and contour plots of the RBFs. The results on the left were obtained by placing the RBF centers on a grid; for the results on the right, the OLS algorithm was used

such a "plane"-like characteristic. The one-step prediction surface realized with the grid-based approach is symmetric. The selection of the basis functions by the OLS algorithm emphasizes the regions close to the equilibrium because the training data distribution is inherently denser in this region.

The main problem with RBF networks is finding a good choice for the basis function widths. If the advantages of linear optimization are to be fully exploited, these widths have to be chosen by the user. This requires some trial and error. Thus, the question arises whether one advantage of RBF compared with MLP networks still exists. The advantage of not requiring several runs with different initializations (to avoid poor local optima) seems to be compensated by the trial-and-error approach required for finding good basis function widths.

21.5.3 Singleton Fuzzy and NRBF Models

The singleton neuro-fuzzy or normalized RBF model architecture possesses better-suited basis functions for the test process. Furthermore, it has already been demonstrated in Sect. 12.3.6 that the "dips" in the interpolation behavior can usually be avoided and the choice of the basis functions' widths is not as crucial

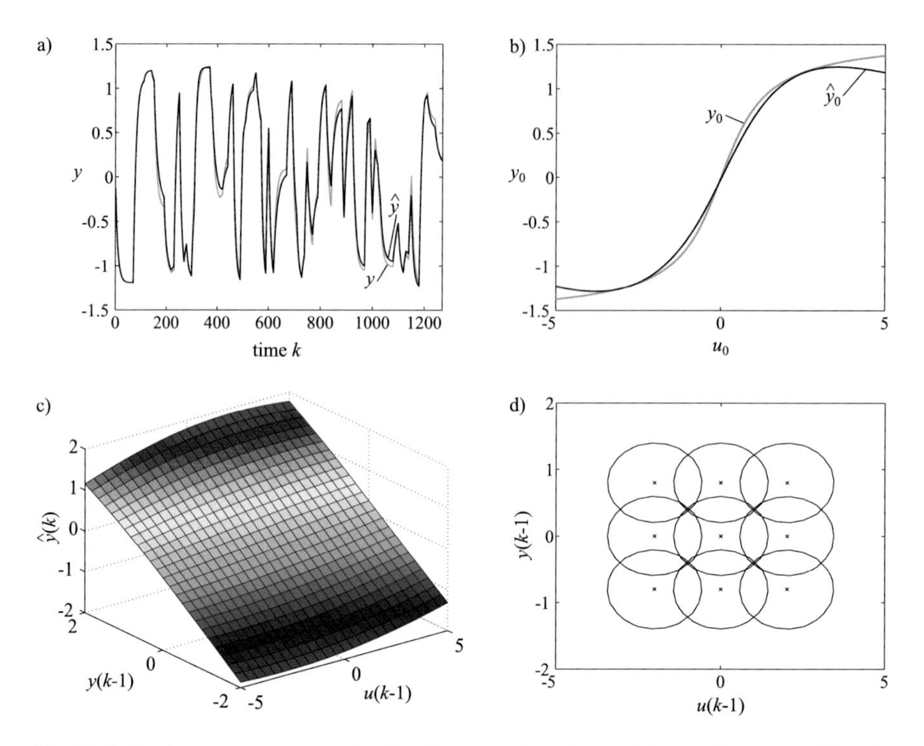

Fig. 21.7 Singleton neuro-fuzzy network with nine rules: (**a**) simulation performance; (**b**) static characteristics; (**c**) one-step prediction surface; (**d**) centers and contour plots of the *non*-normalized basis functions

for performance as it is for the standard RBF network. On the other hand, the advantageous OLS algorithm does not allow a direct center selection for this neuro-fuzzy model architecture; see Sect. 12.4.3.

Here, the grid-based approach is taken, which is required in order to enable a true fuzzy rule interpretation in terms of one-dimensional membership functions; see Sect. 12.3.4. Figure 21.7a demonstrates a significant performance gain compared with the standard RBF network with the grid-based approach but inferior performance compared with the RBF network with OLS center selection. The static model characteristics are comparable with both RBF networks for interpolation but considerably better for extrapolation. This is a clear benefit of normalization.

21.6 Summary

The curse of dimensionality is a critical problem for dynamic systems because the external dynamics approach leads to additional inputs for the static approximator. The MLP network is best suited to deal with the high-dimensional input spaces that result in multivariable and high-order systems.

The interpolation behavior of RBF networks may include "dips" which can cause undesirable effects in the model dynamics, and all main model architectures discussed in this chapter possess a static extrapolation behavior.

When a NARX representation is chosen, the features of the static model architectures are retained for the dynamic models. The NOE or other representations with more complex noise models, however, lead to nonlinear parameterized dynamic models independent of the static model architecture. Thus, for NOE (or more complex noise models) structures the advantages of linear parameterized static approximators vanish.

A linear model can be used additively in parallel to a nonlinear model architecture. Such a strategy can overcome some of the nonlinear model architecture's drawbacks with regard to interpolation and extrapolation behavior.

21.7 Problems

1. Why is the curse of dimensionality a more critical issue for dynamic systems than for static systems? Which model architectures are therefore preferable?
2. How does extrapolation behavior influence the dynamic properties of external dynamics models? What happens for constant and for linear extrapolation behavior? What type of behavior is advantageous?
3. Explain the difference between a NARX and an NOE model.

Chapter 22
Dynamic Local Linear Neuro-Fuzzy Models

Abstract This chapter covers dynamic aspects specific to local model networks. The careful interpretation of the local model parameters, which can be understood as local gain and poles, is discussed. It is demonstrated that this interpretation might be dangerous in off-equilibrium regions. Also, the potentially strange effects of interpolating denominator polynomials are analyzed. It can be concluded that extreme care must be taken when models with output feedback are involved with local model architectures. Therefore, the author argues in favor of dynamics representations without output feedback such as NFIR or NOBF. Furthermore, the issue of stability is discussed where local model networks allow for a deeper analysis than alternative architectures. And finally, various more complex approaches known from linear system identification, such as the instrumental variable method or advanced noise models, are transferred to the nonlinear world via local model networks.

Keywords Dynamics · Linearization · Stability · Equation error · Output error · ARX · OE · FIR · LOLIMOT

Chapters 13 and 14 demonstrated that local linear neuro-fuzzy models are a versatile model architecture and LOLIMOT is a powerful construction algorithm with many advantages over conventional approaches and standard neural networks or fuzzy systems. In this chapter, this model architecture and training algorithm will be applied to nonlinear *dynamic* system identification by pursuing the external dynamics approach introduced in Chap. 19. It will be shown that local linear neuro-fuzzy models offer some distinct advantages over other architectures, in particular for modeling of *dynamic* systems.

This chapter is structured as follows. After an introduction to dynamic local linear neuro-fuzzy models, Sect. 22.1 addresses the different goals of prediction and simulation. Section 22.2 shows how LOLIMOT can extract information about the process structure from data and how prior knowledge can be exploited to reduce the

O. Nelles, *Nonlinear System Identification*,
https://doi.org/10.1007/978-3-030-47439-3_22

problem complexity. Special features for the linearization of dynamic neuro-fuzzy models are treated in Sect. 22.3. Section 22.4 discusses some stability issues. The operation of the LOLIMOT algorithm for dynamic systems is illustrated with some simulation studies in Sect. 22.5. Sections 22.6 and 22.8 extend advanced methods known from linear system identification literature to local linear neuro-fuzzy models and integrate them into the LOLIMOT algorithm. The specific difficulties arising from an extension of the OLS rule consequent structure optimization to dynamic models are analyzed in Sect. 22.9. Finally, a brief summary is given and some conclusions are drawn.

First, the extension from static to dynamic local linear neuro-fuzzy model will be explained. A *static* local linear neuro-fuzzy model is defined as (see (14.1) in Sect. 14.1):

$$\hat{y} = \sum_{i=1}^{M} \left(w_{i0} + w_{i1}x_1 + w_{i2}x_2 + \ldots + w_{ip}x_{nx} \right) \Phi_i(\underline{z}) \,, \tag{22.1}$$

where in the most general case the rule consequent input vector $\underline{x} = [x_1\ x_2\ \cdots\ x_{nx}]^T$ and the rule premise input vector $\underline{z} = [z_1\ z_2\ \cdots\ z_{nz}]^T$, both are equivalent to a vector containing the p physical inputs $\underline{u} = [u_1\ u_2\ \cdots\ u_p]^T$. Pursuing the *external dynamics* approach introduced in Sect. 19.2 a dynamic local linear neuro-fuzzy model for p inputs and of order m is obtained by setting

$$\underline{x} = \underline{\varphi}(k) \,, \quad \underline{z} = \underline{\varphi}(k) \tag{22.2}$$

with

$$\underline{\varphi}(k) = [u_1(k-1)\ \cdots\ u_1(k-m)\ \cdots\ u_p(k-1)\ \cdots\ u_p(k-m) \\ y(k-1)\ \cdots\ y(k-m)]^T \,. \tag{22.3}$$

The choice of *different* input vectors for the rule consequents $\underline{x}$ and the rule premises $\underline{z}$ is of even greater practical significance for dynamic models than for static ones (Sect. 14.1). This topic is discussed in Sect. 22.2.

With the regressors in (22.2) and (22.3), a single input local linear neuro-fuzzy model in parallel configuration can be written as

$$\hat{y}(k) = \sum_{i=1}^{M} \big(b_{i1}u(k-1) + \ldots + b_{im}u(k-m) \\ -a_{i1}\hat{y}(k-1) - \ldots - a_{im}\hat{y}(k-m) + \zeta_i \big) \Phi_i(\underline{z}) \,, \tag{22.4}$$

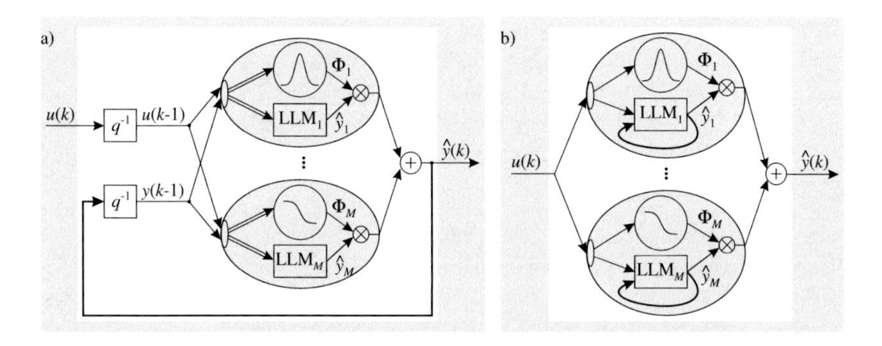

Fig. 22.1 (**a**) External dynamics approach for a local linear neuro-fuzzy model of first order with a single input and M neurons: The model output is fed back to the model input (*global state feedback*). (**b**) Internal dynamics approach for a local linear modeling scheme: The outputs of the local models (filters) are fed back individually (*local state feedback*)

where b_{ij} and a_{ij} represent the numerator and denominator coefficients and ζ_i is the offset[1] of the local linear model i. The extension to the multiple-input case is straightforward.

Figure 22.1a illustrates for a simple example how such a dynamic local linear neuro-fuzzy model operates during simulation, that is, in parallel configuration. The predicted output of all local linear models (LLMs) is calculated as

$$\hat{y}_i(k) = b_{i1}u(k-1) - a_{i1}\underbrace{\hat{y}(k-1)}_{\text{global state}} + \zeta_i\,. \tag{22.5}$$

The outputs of the LLMs $\hat{y}_i$ are weighted with their corresponding validity function values Φ_i, and these contributions are summed up to generate the overall model output $\hat{y}$. This overall model output then is fed back to external dynamics filters. In this context, the external dynamics approach is sometimes called the *global state feedback* approach to stress that the overall model output (whose delayed versions represent the global state of the model) is fed back [400].

In opposition, Fig. 22.1b depicts an internal dynamics approach also called the *local state feedback*. The fundamental difference from Fig. 22.1a is that the individual states of the local linear models (filters) are fed back locally, that is,

$$\hat{y}_i(k) = b_{i1}u(k-1) - a_{i1}\underbrace{\hat{y}_i(k-1)}_{\text{local state}} + \zeta_i\,. \tag{22.6}$$

The model in Fig. 22.1b is similar to a normalized version of an internal dynamic RBF network that is proposed in [14, 15]. Since the validity functions depend only

[1] The offsets are *not* denoted as "c_i" in order to avoid confusion with the validity function centers.

on $u(k)$ such an architecture can only model systems that are solely nonlinear in their input, i.e., Hammerstein structures. This limitation severely restricts the applicability of that approach and makes it practically ineffective. Therefore, other, more general local state feedback architectures have been proposed; see [14] and Chap. 23. Some fundamental differences between the global and local state feedback strategies should be mentioned:

- For global feedback, the dynamic order of the model is equal to the order of the LLMs. For local feedback, it is equal to the *sum* of the orders of *all* LLMs.
- If all LLMs are stable, then the overall model is stable when local feedback is applied, while this is not necessarily true for global feedback; refer to Sect. 22.4.
- If at least one LLM is unstable, then the overall model is unstable when local feedback is applied, while this is not necessarily the case for global feedback; refer to Sect. 22.4.
- Complex nonlinear dynamic phenomena such as limit cycles require locally unstable models and thus can only be realized with global feedback.
- The global feedback approach can be utilized for both one-step prediction and simulation and can be trained as a NARX or NOE structure (or with other more complex noise models). The local feedback approach can only be trained as an output error model and can only be used for simulation.
- As a consequence of the previous point, the LLM parameters are *linear* with global feedback when trained in a series-parallel (NARX) structure. With local feedback, the LLM parameters are *nonlinear* except for the special (and restricting) case where the denominator dynamics of all LLMs are identical.
- With global feedback, unstable processes can be modeled because the identification of a NARX model is possible (the one-step NARX predictor is always stable, even for unstable models; see Sect. 18.3.3). Local feedback always implies an output error model that cannot be used for the identification of unstable processes because the optimal predictor would become unstable (Sect. 18.3.3).

In the sequel only the external dynamics approach will be pursued. Internal dynamics approaches are discussed further in Chap. 23. The static global and local parameter estimation formulas in Sect. 13.2 can be extended to dynamic local linear neuro-fuzzy models in a straightforward manner. For example, the regression matrix and parameter vector for the local estimation change from (13.22) and (13.17), respectively, to

$$\underline{X}_i = \begin{bmatrix} u(m) & \cdots & u(1) & -y(m) & \cdots & -y(1) & 1 \\ u(m+1) & \cdots & u(2) & -y(m+2) & \cdots & -y(2) & 1 \\ \vdots & & \vdots & \vdots & & \vdots & \vdots \\ u(N-1) & \cdots & u(N-m) & -y(N-1) & \cdots & -y(N-m) & 1 \end{bmatrix} \tag{22.7}$$

and

$$\underline{w}_i = [b_{i1} \ \cdots \ b_{im} \ a_{i1} \ \cdots \ a_{im} \ \zeta_i]^T \, . \tag{22.8}$$

22.1 One-Step Prediction Error Versus Simulation Error

The high computational efficiency of the LOLIMOT algorithm is to a great part a consequence of the utilization of linear parameter estimation methods. For dynamic models, this requires the estimation of local linear ARX models because only the equation error is linear in the parameters. Therefore, the NARX neuro-fuzzy model is trained for optimal one-step prediction rather than simulation performance. From the discussion in the context of linear systems in Sect. 19.2.3, it is clear that this implies some drawbacks when the intended use of the model is a simulation. The drawbacks are the emphasis of high-frequency components in the model fit and the non-detection of a possible error accumulation, which is an effect of the low model accuracy in the low-frequency range. These drawbacks can be expected to carry over to the nonlinear case.

In order to weaken these drawbacks of the NARX approach and simultaneously to avoid the application of nonlinear optimization techniques that would be required for a NARMAX or NOE approach, the following strategy is proposed [411, 415, 432] (see Sect. 13.3.2):
Strategy I:

- The local linear models in the rule consequents are estimated as ARX models by minimizing the one-step prediction error in series-parallel model configuration with a local linear least squares technique.
- The criterion for structure optimization is based on the simulation performance of the model in parallel configuration. Structure optimization consists of two parts: the choice of the worst-performing LLM, which is considered for further subdivision, and the selection of the best splitting dimension.

With this combination of equation error- and output error-based optimization criteria, some advantages of both approaches can be combined. However, note that the undesirable frequency weighting of the parameter estimation still remains. Also, this combined strategy, of course, cannot be applied for the identification of unstable processes. Other choices for the structure optimization criterion may be favorable depending on the specific application of the model. If the model is used as a basis for the design of a controller or a fault diagnosis system, the control or fault diagnosis performances may be utilized directly as structure optimization criteria. Such strategies allow one to make the complete modeling procedure more goal-oriented. (It is, for example, well known that a good simulation model does not guarantee the design of a good controller; see Sect. 18.11.3.) This topic is open for future research, and in particular for the nonlinear case.

The combined strategy described above for the construction of dynamic local linear neuro-fuzzy models possesses another important advantage. The evaluation of the simulation error for structure optimization requires one to feed back the model output. Since the model output differs from the process output, this feedback generates new values at the model inputs $\hat{y}(k-i)$ that are not contained in the training data set. Thus, the evaluation of the simulation error involves a *generalization* of the model. Consequently, the combined strategy is also suitable to detect *overfitting* (which scarcely occurs owing to the regularization effect of the local parameter estimation; see Sect. 13.2.2).

In order to cope with the remaining shortcomings of the NARX model identification, the following two-stage procedure is proposed:
Strategy II:

- First, develop a model according to the combined criteria in Strategy I.
- Second, tune the obtained model by replacing the local ARX models with local OE, ARMAX, etc. models. The ARX model parameters can be utilized as initial values for the required nonlinear optimization.

This two-stage strategy offers a great reduction in computational demand compared with the complete construction of an NOE model with LOLIMOT. The tree construction is carried out with local ARX models, and only at the final stage are the local models improved by an advanced and computationally more expensive approach such as the estimation of local OE or ARMAX models. Although this strategy may lead to a slightly inferior decomposition of the input space by LOLIMOT compared with a complete NOE, NARMAX, etc. approach, this is usually overcompensated by smaller computational demand. Strategy II is utilized in Sects. 22.6 and 22.8.

22.2 Determination of the Rule Premises

In the most general case, the rule premise and the rule consequent inputs contain all regressors, i.e., $\underline{z} = \underline{\varphi}(k)$ and $\underline{x} = \underline{\varphi}(k)$ with $\underline{\varphi}(k)$ according to (22.3). The dimensionality of $\underline{\varphi}(k)$ and thus of the input spaces for the rule premises and consequents can be quite high and in particular for multivariable systems and for high dynamic order m. Therefore, external dynamics approaches require the application of model architectures that can deal with high-dimensional problems. The LOLIMOT algorithm can automatically detect those inputs that have a significant nonlinear influence on the output and so the premise input space spanned by $\underline{z}$ can be reduced. In combination with the OLS algorithm, a structure optimization of the rule consequents is possible in order to reduce the input space of the rule consequents spanned by $\underline{x}$. Besides these automatic algorithms for structure selection, the user may be able to restrict the complexity of the problem a priori by exploiting available knowledge. The distinction between rule premise and consequent inputs is an important feature of LOLIMOT that allows one to prestructure the model in various ways. For example, the following situations can be distinguished:

- *Full operating point* $\underline{z} = \underline{\varphi}(k)$: This represents a universal approximator. The operating point can represent the full dynamics of a process. If, for example, $\underline{z}$ contains the process input with several delays ($\underline{z} = [u(k-1)\ u(k-2)\ \cdots]^T$), then complex dynamic effects can be modeled. This includes direction-dependent behavior (Sect. 14.1.1) or behavior that depends on the change of the input signal because the operating point implicitly contains information about the derivative of the model input $\dot{u} \sim u(k-1) - u(k-2)$. This approach is, e.g., pursued in Sect. 25.1 for modeling a cooling blast.

- *Low dynamic order operating point*: In many applications, the nonlinear effects are simpler. Often it may be sufficient to realize an operating point of a low order, say $\underline{z} = [u(k-1)\ y(k-1)]^T$, while the local linear models in the rule consequents possess higher order.

  Another important situation where a reduced operating point is very advantageous occurs if the model is to be utilized for controller design. In this case many design methods require a model of the following form (see (19.3) in Sect. 19.1):

$$\hat{y}(k) = b_1 u(k-1) \\ + \tilde{f}\left(u(k-2), \ldots, u(k-m), y(k-1), \ldots, y(k-m)\right) . \quad (22.9)$$

  LOLIMOT can be forced to generate a model similar to type (22.9) that is affine in $u(k-1)$ by simply excluding $u(k-1)$ from the operating point, i.e., $\underline{z} = [u(k-2)\ \cdots\ u(k-m)\ y(k-1)\ \cdots\ y(k-m)]^T$. Since different b_1 parameters can be estimated for each LLM, such a local linear neuro-fuzzy model does not exactly fulfill the property (22.9), but nevertheless it is possible to solve the model symbolically for $u(k-1)$ in an operating point dependent manner by pursuing the local linearization approach. This leads to the following control law for an inverting controller (see (19.4) in Sect. 19.1):

$$u(k) = \left[r(k+1) - (b_1(\underline{z})u(k-2) + \ldots + b_m(\underline{z})u(k-m+1) \right. \\ \left. -a_1(\underline{z})y(k) - \ldots - a_m(\underline{z})y(k-m+1) + \zeta(\underline{z}))\right] / b_1(z) \quad (22.10)$$

  with $\underline{z} = [u(k-1)\ \cdots\ u(k-m+1)\ y(k)\ \cdots\ y(k-m+1)]^T$ (time is shifted by one sampling instant, i.e., $k \to k+1$), where $r(k+1)$ denotes the desired control output, that is, the reference signal. Note that it is crucial that $\underline{z}$ does *not* depend on $u(k)$ (or without time shift on $u(k-1)$), otherwise (22.10) would not solve for $u(k)$. The only structural difference between (22.10) and (19.4) is that $b_1(z)$ is not constant but operating point dependent.

- *Operating point includes only inputs*: For the Hammerstein systems discussed in Sect. 19.2.1 (see Fig. 19.5) and a much wider class of nonlinear dynamic systems, the nonlinear behavior depends only on the inputs $\underline{z} = [u(k-1)\ u(k-2)\ \cdots\ u(k-m)]$. This, e.g., is the case for the Diesel engine turbocharger discussed in Sect. 25.2.
- *Operating point includes only outputs*: For nonlinear differential equation (NDE) models that arise quite often [282, 340], the nonlinear behavior depends only on the outputs $\underline{z} = [y(k-1)\ y(k-2)\ \cdots\ y(k-m)]$.
- *Static operating point*: A further simplified approach that nevertheless is quite often successful in practice is to assume a static operating point, i.e., $\underline{z} = u(k)$

or for multivariable systems $\underline{z} = [u_1(k)\ u_2(k)\ \cdots\ u_p(k)]^T$. This approach is pursued, e.g., in [267] for modeling the longitudinal dynamics of a truck.

- *Operating point includes only external variables*: For a large class of systems, the nonlinearity depends on an external signal that does not have to be contained in the local linear models in the rule consequents. For example, the dynamics of a plane depend on its flight height, or transport processes depend on the speed of the medium; see Sects. 22.9 and 25.3.2. In these situations, the rule premise input vector $\underline{z}$ and the consequent vector $\underline{x}$ do not possess common variables. This represents a pure *scheduling* approach. The external variable(s) in $\underline{z}$ schedule the local linear models. This is known as *parameter scheduling* [282]. In the special case where all LLMs have identical dynamics and just different gains the well-known *gain scheduling*[2] approach is recovered.

22.3 Linearization

Linearization of nonlinear dynamic models allows one to exploit the huge variety of linear design techniques for the development of all kinds of model-based methods for the design of controllers, fault diagnosis systems, etc. For local linear neuro-fuzzy models, basically, two practicable ways exist to make use of linear design techniques. This is illustrated for the example of controller design. For a comparison of these two approaches, refer to [146].

- *Local model individual design*: For each local linear model, an individual design step is carried out; see Fig. 22.2a. The local controllers are merged operating

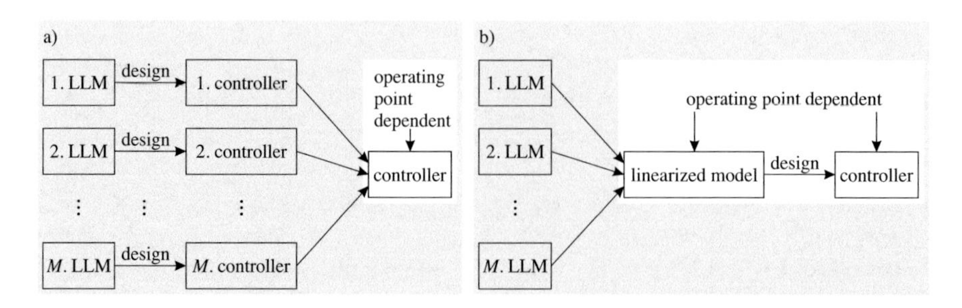

Fig. 22.2 Two strategies for the application of linear design techniques with local linear neuro-fuzzy models: (**a**) Controllers can be designed for each local linear model and subsequently merged to one operating point dependent controller. (**b**) A linearized model with operating dependent parameters can be obtained from all LLMs by linearization, and subsequently, a controller can be designed on the basis of this linearized model

[2]"Gain scheduling" is the commonly used terminology for parameter scheduling as well. The term "parameter scheduling" is seldom used although it is more exact.

point-dependently to a single controller [296, 297, 552]. This concept is often called *parallel distributed compensation*. The advantage of this strategy is that only once M controllers have to be designed, which can be done offline. During operation, only the local controllers have to be combined. Furthermore, powerful methods for proving closed-loop stability via solving linear matrix inequalities (LMIs) are available [555].

- *Local linearization of the overall model*: By linearization, an operating point dependent linearized model is generated from all local linear models [147]. This linearized model is the basis for the controller design; see Fig. 22.2b. A drawback of this strategy is that the linearization and controller design step has to be carried out online. An advantage of this strategy is its higher flexibility and slightly superior performance.

22.3.1 Static and Dynamic Linearization

For linearization of local linear neuro-fuzzy models, *local linearization* should be employed in order to avoid the magnification of undesirable interpolation effects; see Sect. 14.5. The local linearization of (22.4) becomes (see (14.24) in Sect. 14.5):

$$\begin{aligned}\hat{y}(k) = {} & b_1(\underline{z})u(k-1) + \ldots + b_m(\underline{z})u(k-m) \\ & - a_1(\underline{z})\hat{y}(k-1) - \ldots - a_m(\underline{z})\hat{y}(k-m) + \zeta(\underline{z})\end{aligned} \tag{22.11}$$

with

$$b_j(\underline{z}) = \sum_{i=1}^{M} b_{ij}\Phi_i(\underline{z}), \quad a_j(\underline{z}) = \sum_{i=1}^{M} a_{ij}\Phi_i(\underline{z}), \quad \zeta(\underline{z}) = \sum_{i=1}^{M} \zeta_i\Phi_i(\underline{z}). \tag{22.12}$$

When linearizing a model two alternatives can be distinguished.

- *Static linearization*: The operating point (u_0, y_0) is at the equilibrium and thus lies on the static nonlinearity; see small dark gray ellipses in Fig. 22.3.

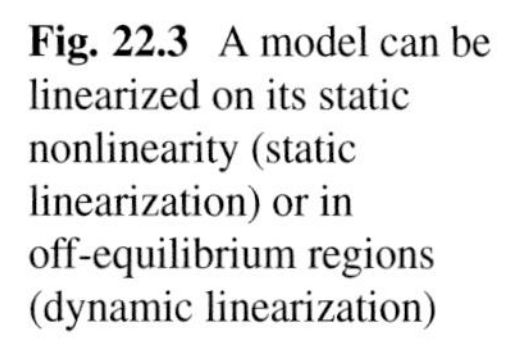

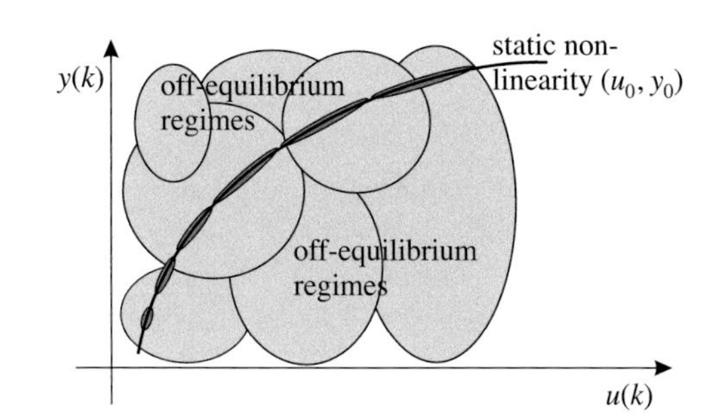

Fig. 22.3 A model can be linearized on its static nonlinearity (static linearization) or in off-equilibrium regions (dynamic linearization)

- *Dynamic linearization*: The operating point $(u(k), y(k))$ represents a transient and thus lies in the off-equilibrium regions; see large gray ellipses in Fig. 22.3.

As Fig. 22.3 illustrates, the validity functions determine whether the corresponding local linear models describe the process behavior around the static nonlinearity or for transients. When the center of a validity function is placed on the static nonlinearity, e.g., $\underline{c} = [u_0\ y_0]^T$ for a first-order system, and additionally possesses a small width, then the corresponding local linear model is activated only for low-frequency excitation with $u(k) \approx u_0$. In contrast, the local linear model that corresponds to the upper left operating regime in Fig. 22.3 is activated only in the early phase of a transient when the input steps from a very large $u(k)$ to a very small $u(k)$.

In order to explain the major importance of the transient operating regimes, the Hammerstein and Wiener systems shown in Fig. 19.5 in Sect. 19.2.1 will be considered again. The two systems behave quite differently although they possess identical static behavior. They just differ in the off-equilibrium regions. Consequently, it is fundamentally important to allow off-equilibrium regimes in modeling by choosing the rule premise input space rich enough. For example, for a first-order system $\underline{z} = [u(k-1)\ y(k-1)]$ is required in order to describe the whole space shown in Fig. 22.3.

For an extensive theoretical discussion of off-equilibrium linearization in the context of local linear model architectures, refer to [301, 302, 531].

22.3.2 Dynamics of the Linearized Model

The local linearization in (22.11) and (22.12) (and the global linearization similarly) interpolate with the validity functions between the parameters of the local linear models. What does this mean for the dynamics of the resulting linearized model? In order to keep the analysis simple, a model with only two LLMs is considered. With an extension to more than two LLMs, no qualitatively new aspects are involved.

Figure 22.4 illustrates how the interpolation between two LLMs affects the poles of the linearized model. An equivalent analysis can be performed for the zeros of the linearized model. However, the zeros can hardly be interpreted in the z-domain. Each cross represents a pole of the linearized model. For simplicity, the interpolation between the two LLMs is performed linearly, as would occur for piecewise linear validity functions. A *first-order* model is shown in Fig. 22.4a. The pole of the linearized model changes uniformly between the poles of the individual LLMs. This behavior is obvious because in first-order systems, the denominator parameter a_1 is equal to the pole since the interpolated model becomes (with $\Phi_1 + \Phi_2 = 1$)

$$\hat{y}(k) = (\Phi_1 b_{11} + (1-\Phi_1) b_{21}) u(k-1) - (\Phi_1 a_{11} + (1-\Phi_1) a_{21}) \hat{y}(k-1). \quad (22.13)$$

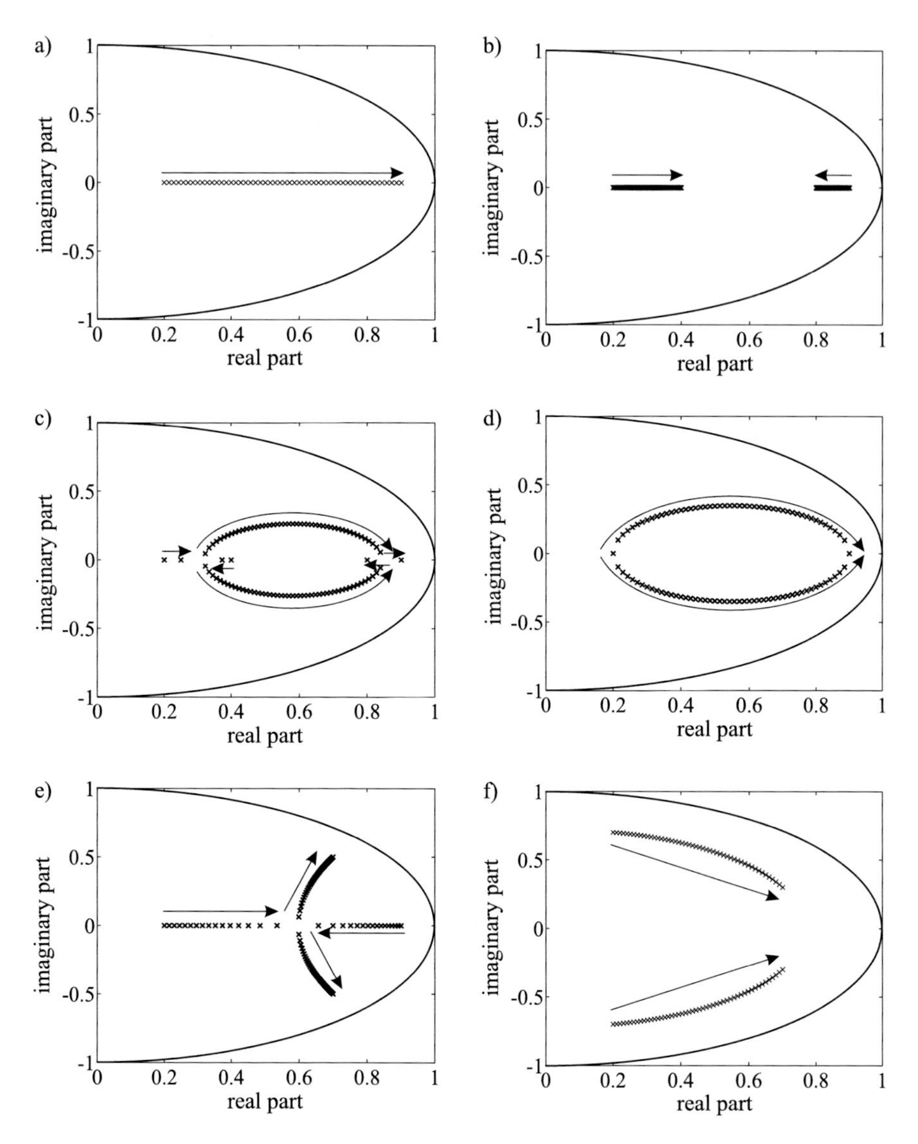

Fig. 22.4 Poles of a linearized model obtained by the linear interpolation of two local linear models: (**a**) $p = 0.2 \rightarrow p = 0.9$ (**b**) $p_{1/2} = 0.2\,/\,0.9 \rightarrow p_{1/2} = 0.4\,/\,0.8$ (**c**) $p_{1/2} = 0.2\,/\,0.4 \rightarrow p_{1/2} = 0.8\,/\,0.9$ (**d**) $p_{1/2} = 0.2\,/\,0.2 \rightarrow p_{1/2} = 0.9\,/\,0.9$ (**e**) $p_{1/2} = 0.2\,/\,0.9 \rightarrow p_{1/2} = 0.7 \pm i0.5$ (**f**) $p_{1/2} = 0.2 \pm i0.7 \rightarrow p_{1/2} = 0.7 \pm i0.3$

Thus, a linear interpolation between the parameters implies a linear interpolation between the poles. These simple characteristics change for higher-order systems.

The other plots in Fig. 22.4 represent different pole configurations for a *second-order* system, which follows

$$\hat{y}(k) = (\Phi_1 b_{11} + (1 - \Phi_1) b_{21}) u(k-1) + (\Phi_1 b_{12} + (1 - \Phi_1) b_{22}) u(k-2)$$
$$- (\Phi_1 a_{11} + (1 - \Phi_1) a_{21}) \hat{y}(k-1) - (\Phi_1 a_{12} + (1 - \Phi_1) a_{22}) \hat{y}(k-2).$$

In Fig. 22.4b each LLM possesses a fast and a slow real pole. Again the poles of the linearized system behave as expected. In Figs. 22.4c and d one LLM possesses two slow poles, while the other LLM possesses two fast poles. In Fig. 22.4c the poles of each LLM are distinct; in Fig. 22.4d they are identical. In both cases the poles of the linearized model can become complex. It is highly unexpected that the combination of two aperiodic LLMs can yield an oscillatory model behavior. The interpolation between one local linear model with two real poles and another LLM with a conjugate complex pole pair in Fig. 22.4e is as expected. Finally, the poles of the linearized model obtained by an interpolation of two LLMs with a conjugate complex pole pair each are also as anticipated.

The undesirable behavior in Fig. 22.4c, d can be overcome by directly interpolating the poles of the LLMs rather than the LLMs' parameters. This approach is suggested in [106]. However, this strategy requires the local models to be linear and of identical structure. This restriction is quite severe because the possibility of incorporating different types of local models into the overall model is one of the major strengths of local modeling schemes; see Sect. 14.2.3. Moreover, for higher-order systems, it is not quite clear between which poles the interpolation should take place. An alternative is based on the fact that the situation shown in Fig. 22.4c, d rarely occurs in practice since two neighbored LLMs represent similar dynamic behaviors if the overall model is sufficiently accurate, i.e., enough local linear models are constructed. This might allow one to neglect the imaginary parts of the poles because they decrease as the poles of the LLMs approach each other.

22.3.3 Different Rule Consequent Structures

What happens if the structures of the interpolated local linear models are not identical? Such a situation may occur for processes that possess second-order oscillatory behavior in one operating regime and first-order dynamics in another regime, such as the cooling blast investigated in Sect. 25.1. In such case, the interpolation operates on the first-order model as if it were of second order. This means that the first-order system is treated as

$$G_i(q) = \frac{b_{i1} q^{-1} + 0 q^{-2}}{1 + a_{i1} q^{-1} + 0 q^{-2}}, \tag{22.14}$$

which in practice introduces a "dummy" zero and pole at 0. This usually leads to reasonable model dynamics.

More frequently, different rule consequent structures arise from operating point dependent dead times. As an example, a first-order local linear neuro-fuzzy model with two LLMs will be considered. Each LLM describes its own operating regime with individual gains, dynamics, and dead times:

$$1.\ \text{LLM:}\quad \hat{y}_1(k) = b_{11}u(k-d_1-1) + a_{11}y(k-1) \tag{22.15a}$$

$$2.\ \text{LLM:}\quad \hat{y}_2(k) = b_{21}u(k-d_2-1) + a_{21}y(k-1) \tag{22.15b}$$

Figure 22.5a shows which behavior is expected from the model. When only LLM 1 is valid, the dead time should be equal to d_1. When only LLM 2 is valid, the dead time should be equal to d_2. When both LLMs are interpolated, the dead time should be in between, i.e., $d_1 \le d \le d_2$ (assuming $d_1 < d_2$). However, Fig. 22.5b shows what really happens when the standard interpolation method is used. An interpolation of the two LLMs (22.15a) and (22.15b) yields

$$\begin{aligned}\hat{y}(k) &= \Phi_1 b_{11}u(k-d_1-1) + \Phi_2 b_{21}u(k-d_2-1) \\ &\quad + (\Phi_1 a_{11} + \Phi_2 a_{21})\, y(k-1)\,,\end{aligned} \tag{22.16}$$

where Φ_1 and Φ_2 are the validity function values with $\Phi_2 = 1 - \Phi_1$. Thus, for $\Phi_1 \neq 0$ the dead time of the model is always equal to d_1. Although in many applications, the undesirable effect is not as drastic as shown in Fig. 22.5b, this behavior often is not acceptable. It can be overcome by altering the interpolation

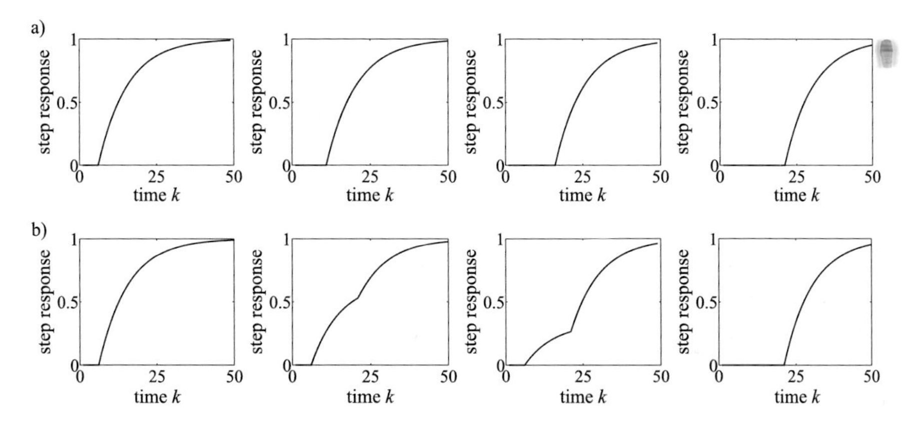

Fig. 22.5 Interpolation of local linear models with different dead times: (**a**) expected behavior, (**b**) model behavior with the standard interpolation method. From left to right, the validity function values of the two local linear models are Φ_1/Φ_2 = 0.0/1.0, 0.3/0.7, 0.7/0.3, and 1.0/0.0

method for dead times such that the dead times are directly interpolated rather than the linear parameters. This approach leads to the expected result

$$\hat{y}(k) = (\Phi_1 b_{11} + \Phi_2 b_{21})\, u(k - d - 1) + (\Phi_1 a_{11} + \Phi_2 a_{21})\, y(k - 1) \tag{22.17}$$

with

$$d = \text{int}\,(\Phi_1 d_1 + \Phi_2 d_2)\ . \tag{22.18}$$

The major benefit from this dead time adjusted interpolation method is not the improved model accuracy. More importantly, the qualitative model behavior is correct, which can be a significant advantage when the model is further exploited, e.g., for control design.

22.4 Model Stability

This section will discuss some stability issues in an informal manner. In Sect. 22.4.1, the basic properties are addressed. Section 22.4.2 briefly introduces numerical tools for proving Lyapunov stability for local linear neuro-fuzzy models. Finally, the practically very important topic of stable extrapolation behavior is addressed in Sect. 22.4.3.

As an illustrative example for the sometimes non-intuitive nature of the problem the Wiener system introduced in Fig. 19.5 in Sect. 19.2.1 will be considered again. Figure 22.6a shows the one-step prediction surface of this first order system $y(k) = f(u(k-1), y(k-1))$. The local poles of this Wiener system, or more correctly the poles of the linearized system, are shown in Fig. 22.6b. They can be calculated from the one-step prediction surface as (see (19.12) in Sect. 19.2.1):

$$p(u(k-1), y(k-1)) = \frac{\partial f(u(k-1), y(k-1))}{\partial y(k-1)}\ . \tag{22.19}$$

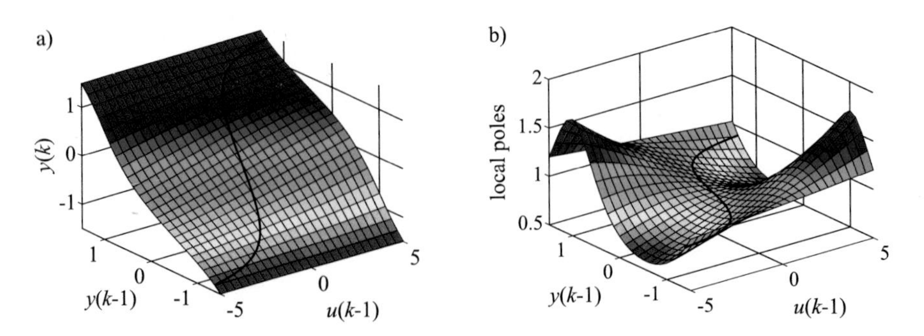

Fig. 22.6 First-order Wiener system: (**a**) one-step prediction surface, (**b**) local poles

Along the static nonlinearity shown as the thick line in Fig. 22.6, the local poles are equal to the pole of the linear block of the Wiener system in Fig. 19.5a:

$$p(u_0, y_0) = 0.9\,. \tag{22.20}$$

For the transient regions, the Wiener system possesses different local poles. In particular, it is very interesting to observe that in some off-equilibrium regions, unstable linearized models occur with $p > 1$ although the Wiener system is stable, of course. For a local linear modeling scheme, this observation means that stable overall models can result even if some unstable local linear models exist in off-equilibrium regimes. Indeed, identification of this Wiener system with LOLIMOT yields unstable off-equilibrium local models if sufficiently many neurons are generated. Note that this is a critical issue in practice since inaccurately identified operating regimes can make the model unstable and an interpretation of the LLM poles is not necessarily as straightforward as one might assume.

22.4.1 Influence of Rule Premise Inputs on Stability

The stability analysis depends on the inputs contained in the rule premises. If no model outputs are fed back to the rule premises, then a local linear neuro-fuzzy model can be seen as a simple parameter scheduling approach with the premise inputs as scheduling variables. If, however, the previous model outputs are used as premise inputs, then the stability analysis becomes more involved owing to this feedback component.

22.4.1.1 Rule Premise Inputs Without Output Feedback

It is assumed that the rule premises depend only on the process inputs and possibly on external signals, i.e., $\underline{z}$ does not contain any $\hat{y}(k-i)$. Then it is sufficient to investigate the stability of all linearized models that may occur by the interpolation of the local linear models. The following three cases can be distinguished.

- *First-order dynamics*: The parameters a_{i1} directly represent the poles. Thus, an interpolation of the parameters is equal to an interpolation of the poles; see Sect. 22.3.2. An interpolation between stable poles $-1 < p_i < 1$ yields a stable pole again. Consequently, if all local linear models are stable, then the neuro-fuzzy model is stable as well.
- *Second-order dynamics*: As Fig. 22.4 demonstrates, unexpected effects may occur when interpolating between second-order local linear models. Nevertheless, it can be guaranteed that the neuro-fuzzy model is stable when all LLMs are stable. This can be explained with Fig. 22.7a, which depicts the stability region in

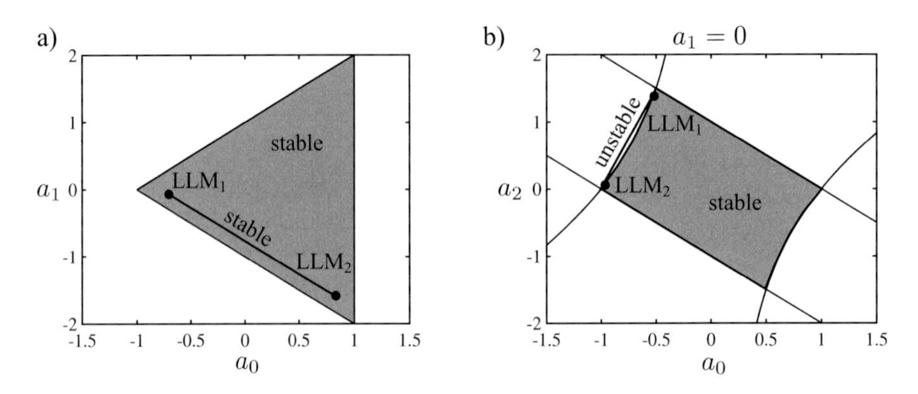

Fig. 22.7 Stability regions for (**a**) second-order systems and (**b**) third-order systems with $a_1 = 0$

the a_0-a_1-plane[3] according to the Schur-Cohn-Jury criteria. The denominator of a local linear model can be represented by a point in the a_0-a_1-plane. Figure 22.7a shows two stable LLMs, and the line represents all possible models that can arise by interpolation between these two LLMs. Obviously, as long as the two LLMs lie inside the stability region, any interpolation does so as well. This is a consequence of the convexity of the stability region.

- *Higher than second-order dynamics*: According to the Schur-Cohn-Jury criteria, the stability regions are not necessarily convex for third- and higher-order systems. Figure 22.7b depicts the stability region for a third-order system in the a_0-a_2-plane for the special case $a_1 = 0$ as an example. As illustrated in this figure, it is possible to obtain an unstable linear model by interpolating two stable ones. Although this is not very likely to happen in practice because the two LLMs must possess quite different dynamics and they must be very close to the stability boundary, the stability of all LLMs does not guarantee the stability of the neuro-fuzzy model for higher-order systems.

22.4.1.2 Rule Premise Inputs with Output Feedback

When the model output is fed back to the rule premises, i.e., $\underline{z}$ contains $\hat{y}(k-i)$ regressors, the overall model can become unstable even when it is of low order and all local linear models are stable. Intuitively, this can be understood by imagining that it is possible to generate an unstable system by switching between two stable linear systems in an appropriate manner. Since the rule premises are just "soft" switches for the rule consequents (local linear models), this example extends to neuro-fuzzy models. An in-depth analysis can be found in [555], where it is shown

[3]The denominator polynomial is assumed to be
$N(q) = q^n + a_{n-1}q^{n-1} + \ldots + a_1 q + a_0$ (by normalization $a_n = 1$).

that the smoothing realized by the non-crisp validity functions has a stabilizing effect. Although a stability proof for a local linear neuro-fuzzy model with output feedback to the rule premises is a difficult task, efficient tools are available that are capable of proving stability in many cases. These tools are briefly discussed in the next subsection.

22.4.2 Lyapunov Stability and Linear Matrix Inequalities (LMIs)

The stability of a local linear neuro-fuzzy model might be proven by Lyapunov's direct method. This general and widely applied approach to prove the stability of nonlinear dynamic systems is based on a state space formulation of the system under investigation. One tries to find a Lyapunov function, and since this is a very hard problem it is common to restrict this search to the class of quadratic Lyapunov functions

$$V(k) = \underline{x}^T(k)\underline{P}\,\underline{x}(k) > 0\,, \tag{22.21}$$

where $\underline{P}$ is a positive definite matrix and $\underline{x}(k)$ is the state of the model. For a simple linear model or equivalently for a local linear neuro-fuzzy model with a single rule, the following stability condition can be derived. For stability, the Lyapunov function must be monotonically decreasing over time k, that is,

$$\begin{aligned} V(k+1) - V(k) &= \underline{x}^T(k+1)\underline{P}\,\underline{x}(k+1) - \underline{x}^T(k)\underline{P}\,\underline{x}(k) \\ &= \left(\underline{A}\,\underline{x}(k)\right)^T \underline{P}\,\underline{A}\,\underline{x}(k) - \underline{x}^T(k)\underline{P}\,\underline{x}(k) \\ &= \underline{x}(k)^T \left(\underline{A}^T\underline{P}\,\underline{A} - \underline{P}\right)\underline{x}(k) < 0\,, \end{aligned} \tag{22.22}$$

since $\underline{x}(k+1) = \underline{A}\,\underline{x}(k)$, ignoring the input $u(k)$ that is irrelevant for stability considerations [161]. Thus, the linear model is stable if the following matrix is negative definite:

$$\underline{A}^T\underline{P}\,\underline{A} - \underline{P} \prec 0\,. \tag{22.23}$$

For *nonlinear* systems, stability is investigated for an *equilibrium point*. Typically, an autonomous system is considered where $u(k) = 0$, i.e., $\underline{x}(k+1) = \underline{f}(\underline{x}(k))$.

As shown by Tanaka and Sugeno in [555], stability of a local linear neuro-fuzzy model with M rules is guaranteed by fulfilling

$$\underline{A}_i^T \underline{P} \underline{A}_i - \underline{P} \prec 0 \qquad \text{for } i = 1, 2, \ldots, M \tag{22.24}$$

with a positive definite $\underline{P}$, where the $\underline{A}_i$ represent the system matrices of the local linear models. It is important to realize that the conditions (22.24) are valid only for local *linear* but not local *affine* models, i.e., the offsets are assumed to be zero. For affine models the equilibrium will not be at 0 and even multiple equilibria are possible, making it a much harder problem to prove stability. Therefore the usefulness of (22.24) in the context of this book is limited.

Note that in (22.24) a *common* $\underline{P}$ matrix must be found for *all* local linear models $\underline{A}_i$. This is a much stronger condition than the individual stability of the local linear models, for which only $\underline{A}_i^T \underline{P}_i \underline{A}_i - \underline{P}_i \prec 0$ with individual $\underline{P}_i$ would be required. The system matrices $\underline{A}_i$ of the LLMs required for the evaluation of (22.24) can be easily obtained from the local linear input/output models by a transformation to a canonical state space form, e.g., the observable canonical form [85, 89].

The stability condition (22.24) is remarkably simple and follows from the local linear model architecture. For other nonlinear dynamic models, no comparable results exist. This is another strong advantage of local linear modeling schemes when dealing with dynamical systems.

How can a positive definite matrix $\underline{P}$ be found that fulfills (22.24)? Fortunately, this problem can be solved automatically in a very efficient way. The stability condition (22.24) represents a so-called *linear matrix inequality (LMI)*; for a monograph on this subject, refer to [69]. These linear matrix inequalities can be solved by numerical optimization. Furthermore, it can be guaranteed to find the globally optimal solution because the optimization problem is *convex*. Efficient tools for solving LMIs are available; see e.g., [178]. Note that it may not be possible to find a $\underline{P}$ that meets (22.24), although the model may be stable. The condition (22.24) is sufficient but not necessary. As an example, consider the stable Wiener system from the beginning of Sect. 22.4, which provokes unstable LLMs in off-equilibrium regimes. The reasons for failure in finding a $\underline{P}$ matrix for a stable system can be twofold. First, only quadratic Lyapunov functions are considered. Even if no Lyapunov function of quadratic shape exists, Lyapunov functions of another type might exist. Second, (22.24) does not take the specific shape of the validity functions into account. It considers the worst-case scenario where all LLMs can be interpolated in any way. In reality, however, the validity functions are structured and constrain the way in which the LLMs can be interpolated. Current research tries to exploit this knowledge to make the stability conditions less conservative, in particular for the simple case of piecewise linear validity functions.

The same concepts that have been discussed here in the context of *model* stability can be extended to prove *closed-loop* stability when a linear controller is designed for each local linear model (parallel distributed compensation; see Sect. 22.3). Then each local linear controller in combination with each local linear model must meet a stability condition similar to (22.24) [555]. This usually leads to very conservative conditions, which do not allow one to prove stability for large classes of stable control loops. Also, for the closed-loop stability, current research focuses on finding less conservative stability conditions [367, 554]. Note again that no comparable powerful results for ensuring closed-loop stability are available for most other general nonlinear model-based controller designs.

22.4.3 Ensuring Stable Extrapolation

The extrapolation behavior of local linear neuro-fuzzy models is linear. It is determined by the local linear model at the interpolation/extrapolation boundary; see Sect. 14.4.2. This means that for extrapolation the neuro-fuzzy model behaves like a linear dynamic system that is equivalent to this "boundary LLM." This behavior is expected and usually desired by the user. Because the value of this LLM's validity function is not exactly equal to 1, the other surrounding LLMs may also have a small influence, but these effects can often be neglected.

The extrapolation behavior plays an important role in the robustness in modeling and identification of nonlinear dynamic systems. As pointed out in Sect. 19.2.3, models with output feedback underlie the risk of error accumulation, i.e., the error on the predicted output $\hat{y}(k)$ increases through the feedback. This risk is particularly grave if the prediction error drives the model into an extrapolation region where the model is usually less accurate. Therefore, dynamically stable extrapolation behavior is essential for high robustness against poorly distributed training data. For a more detailed discussion and an example on this topic, refer to [407].

For local linear neuro-fuzzy models, any desired statics and dynamics can be imposed outside the training data range by the incorporation of prior knowledge into the extrapolation behavior as proposed in Sect. 14.4.2. By pursuing this strategy, the stability of the model can be guaranteed.

The linear dynamic extrapolation behavior is a further strong advantage of local linear modeling schemes. Almost all other model architectures except polynomials extrapolate constantly; see Chaps. 18, 11, and 12. Constant extrapolation implies that the local derivatives tend to zero. Consequently, according to (22.19), the poles of a linearized model in an extrapolation region tend to zero. This means that these model architectures possess *static* extrapolation behavior which is certainly an undesirable property. For polynomial models, the situation is even worse. Their local derivative tends to ∞ or $-\infty$ when extrapolating. Thus, polynomials have *unstable* extrapolation behavior which makes nonlinear dynamic polynomial modeling approaches at least questionable for any practical application; see Chap. 20. Figure 22.8 illustrates these relationships.

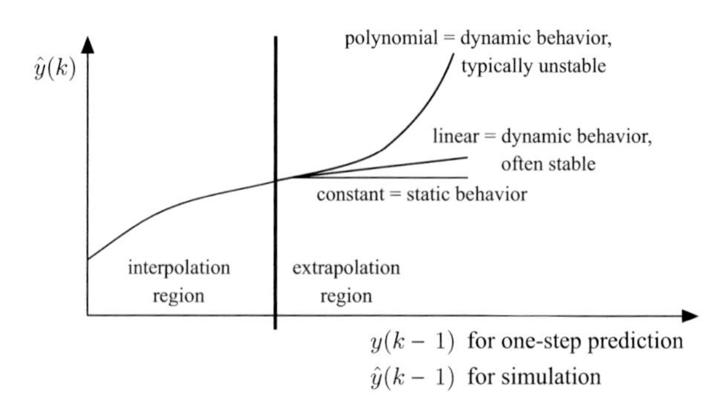

Fig. 22.8 Effect of extrapolation properties on the dynamic behavior of the model for a first-order system

It is obvious, for first-order systems only, that the slope of the one-step prediction surface in $y(k-1)$-direction needs to be smaller than 1 for stable extrapolation behavior because the a_1 coefficient is identical to the pole. However, similar, even more severe, restrictions must be respected for higher-order systems. For example, one universal condition from the Schur-Cohn-Jury stability criteria is that the sum of all denominator polynomial coefficients $a_i, i = 1, \ldots, m$ (i.e., the sum of all slopes in the $y(k-i)$-directions) must be smaller than 1. Thus, Fig. 22.8 can be qualitatively extended for higher-order systems as well.

22.5 Dynamic LOLIMOT Simulation Studies

In this section, the LOLIMOT algorithm for construction of dynamic local linear neuro-fuzzy models is demonstrated for nonlinear system identification of the four test processes introduced in Sect. 22.5.1. The purpose of this section is to illustrate interesting features of LOLIMOT and to assess how the number of required local linear models depends on the strength of nonlinear behavior and the process structure. For all simulations, the process output was disturbed by white Gaussian noise with a signal-to-noise power ratio of 200. This relatively small noise variance was chosen to make the essential effects clear and to keep the estimation bias small, which is always present with the NARX model structure. Methods for avoiding the inconsistent parameter estimates resulting from the (N)ARX model structures are discussed in the next section.

The results obtained are compared in Table 22.1 at the end of this section.

Table 22.1 Comparison of the identification results with LOLIMOT for the test processes

	Hammerstein system	Wiener system	NDE system	Dynamic nonlinearity
Number of rules	8	12	4	15
Nonlinear in	$u(k-1)$, $u(k-2)$	all	$y(k-1)$, $y(k-2)$	$u(k)$
NRMSE (training)	0.025	0.086	0.018	0.012
NRMSE (test)	0.017	0.053	0.009	0.012
Training time	12 s	21 s	4 s	40 s

22.5.1 *Nonlinear Dynamic Test Processes*

The four nonlinear dynamic test processes introduced in the sequel serve as examples for the illustration of the LOLIMOT algorithm. They cover different types of nonlinear behavior in order to demonstrate the universality of the approach.

1. A *Hammerstein* system is characterized by a static nonlinearity in series with a linear dynamic system. For the static nonlinearity, a saturation type function described by an arctan(·) is used. For the subsequent linear system, a second-order oscillatory process with gain 1, damping 0.5, and a time constant of 5 s is chosen. With a sampling time of $T_0 = 1$ s, this system follows the nonlinear difference equation

$$y(k) = 0.01867\arctan[u(k-1)] + 0.01746\arctan[u(k-2)] \\ +1.7826\,y(k-1) - 0.8187\,y(k-2)\,. \qquad (22.25)$$

   The inputs lie in the interval $[-3,\ 3]$.
2. A *Wiener* system is characterized by a linear dynamic system in series with a static nonlinearity, i.e., it is the counterpart to a Hammerstein system. The same linear dynamic and nonlinear static blocks as for the Hammerstein system are chosen. This Wiener system follows the nonlinear difference equation

$$y(k) = \arctan[0.01867\,u(k-1) + 0.01746\,u(k-2) \\ +1.7826\tan(y(k-1)) - 0.8187\tan(y(k-2))]\,. \qquad (22.26)$$

   The inputs lie in the interval $[-3,\ 3]$.
3. Another very important structure of nonlinear dynamic systems is the so-called *NDE* (nonlinear differential equation) model, which often arises directly or by approximation from physical laws [282, 340]. A second-order non-minimum phase system with gain 1, time constants 4 s and 10 s, a zero at 1/4 s, and output

feedback with a parabolic nonlinearity is chosen. With sampling time $T_0 = 1$ s, this NDE system follows the nonlinear difference equation

$$\begin{aligned} y(k) = & -0.07289 \left[u(k-1) - 0.2\, y^2(k-1) \right] \\ & +0.09394 \left[u(k-2) - 0.2\, y^2(k-2) \right] \\ & +1.68364\, y(k-1) - 0.70469\, y(k-2)\,. \end{aligned} \quad (22.27)$$

The inputs lie in the interval $[-1,\ 1]$.

4. In contrast to the previous test processes, as a fourth system, a *dynamic nonlinearity* not separable into static and dynamic blocks is considered. The eigenbehavior of the system depends on the input variable. This process is adopted from [346] and can be described by the following difference equation:

$$\begin{aligned} y(k) = & \; 0.133\, u(k-1) - 0.0667\, u(k-2) + 1.5\, y(k-1) \\ & -0.7\, y(k-2) + u(k) \left[0.1\, y(k-1) - 0.2\, y(k-2) \right]. \end{aligned} \quad (22.28)$$

The inputs lie in the interval $[-1.5,\ 0.5]$.

These test processes are excited with an amplitude modulated pseudo random binary signal (APRBS) as shown in Fig. 22.9a(left); see Sect. 19.7 and [431]. This sequence excites the processes in various operating conditions and thus is suitable for the generation of training data. The signal shown in Fig. 22.9a(right) is used for the generation of the test data. Note that for the Hammerstein and Wiener process, these signals are scaled to lie in $[-3,\ 3]$ and for the dynamic nonlinearity process they lie in $[-1.5,\ 0.5]$. This is necessary in order to create sufficiently strong nonlinear behavior to make the identification problem challenging.

22.5.2 *Hammerstein Process*

The second-order Hammerstein process with an operating point dependent gain between $K = 1$ and $K = 1/10$ can be easily identified by a local linear neuro-fuzzy model trained with LOLIMOT with the following second-order nonlinear dynamic input/output approach:

$$\hat{y}(k) = f\left(u(k-1), u(k-2), y(k-1), y(k-2)\right)\,. \quad (22.29)$$

The convergence curve shown in Fig. 22.10a reveals a rapid performance improvement in the first few iterations. Since no further significant improvement can be achieved for $M > 8$, the optimal model complexity, i.e., the number of rules, neurons, or LLMs, is chosen as $M_{\text{opt}} = 8$. Note that this choice for the optimal model complexity is a direct consequence of the noise level; larger signal-to-noise ratios would allow more complex models and vice versa (see Sect. 13.3.1). The

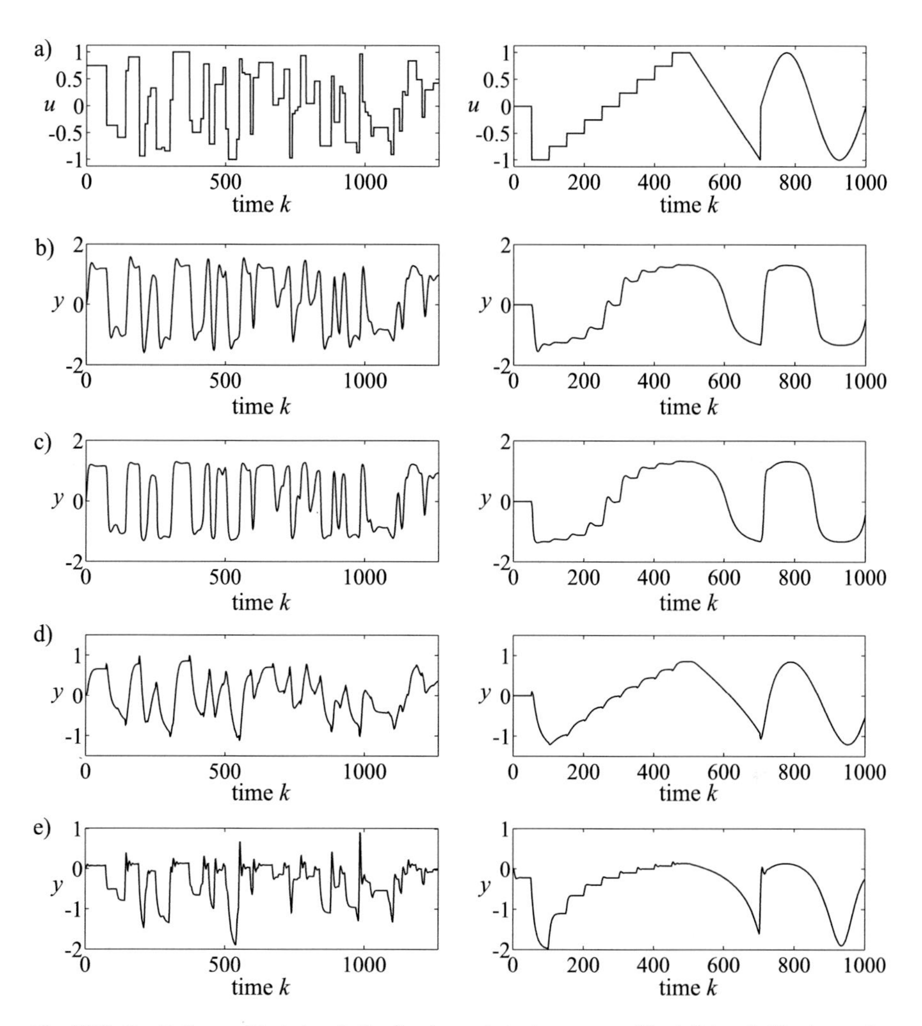

Fig. 22.9 Excitation and test signals for the dynamic test processes. The left-hand side shows the training data, the right-hand side the test data. The excitation and test input signals in (**a**) are scaled to the interval $[-3\ 3]$ for the Hammerstein process in (**b**) and the Wiener process in (**c**). They are not scaled for the NDE process in (**d**) and scaled to $[-1.5\ 0.5]$ for the dynamic nonlinearity in (**e**)

simulation performance of the model is extremely good on both training and test data; see Fig. 22.11a. The one-step prediction performance is even much better.

LOLIMOT performed only divisions in the $u(k-1)$-dimension although the process is nonlinear in both previous inputs $u(k-1)$ and $u(k-2)$; see Fig. 22.12a. The reason for this is that the input signal $u(k)$ possesses relatively few steps where $u(k-1) \neq u(k-2)$; for most training data samples, it is irrelevant whether LOLIMOT splits in the $u(k-1)$- or $u(k-2)$-dimension. But even when a rapidly varying training input signal like white noise is used, the model performance does not crucially depend on which of these two input dimensions is

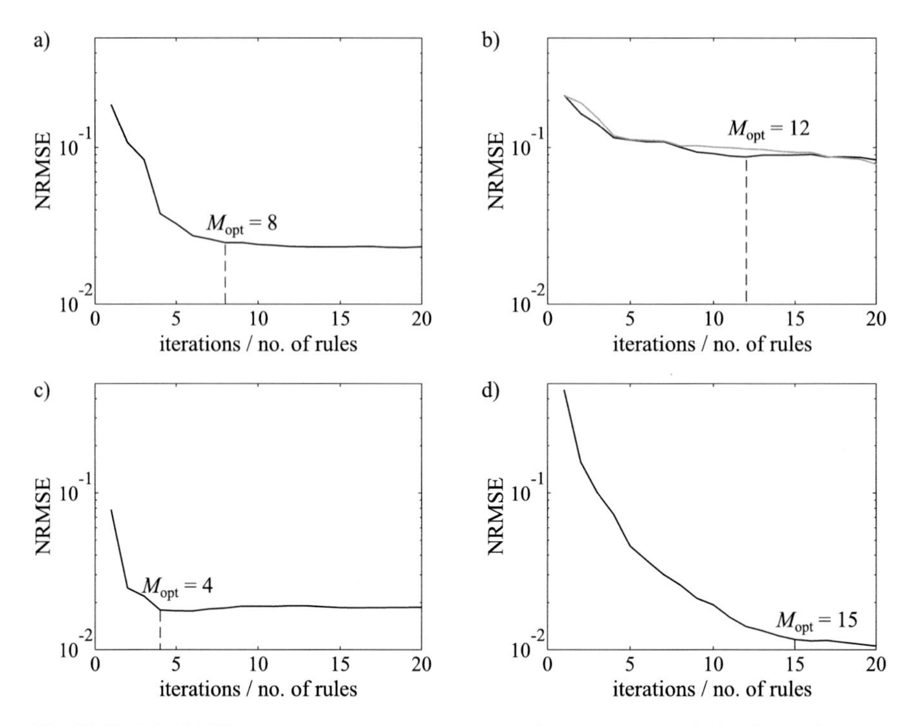

Fig. 22.10 LOLIMOT convergence curves and choice of the model complexity for the four test processes: (**a**) Hammerstein, (**b**) Wiener, (**c**) NDE, and (**d**) dynamic nonlinearity processes

split. LOLIMOT correctly recognizes that the previous outputs have no nonlinear influence on the process behavior, i.e., it does not partition the $y(k-1)$ and $y(k-2)$-dimensions. If this fact were known in advance, these model inputs could be excluded from the premise input vector, i.e., $\underline{z} = [u(k-1)\ u(k-2)]^T$ but $\underline{x} = [u(k-1)\ u(k-2)\ y(k-1)\ y(k-2)]^T$.

The static nonlinearity and the step responses depicted in Fig. 22.13a underline the high quality of the obtained model. Moderate deviations between the static behavior of the process and the model can only be observed in wide extrapolation.

22.5.3 *Wiener Process*

The Wiener process is also modeled with the approach (22.29). The convergence curve in Fig. 22.10b demonstrates that the rate of convergence and the final model quality are significantly worse than for the Hammerstein process. It is not so easy to select the optimal model complexity. Here, $M_{opt} = 12$ was chosen. The different characteristics in comparison with the Hammerstein process can be explained as follows. The one-step prediction function $f(\cdot)$ of a Wiener system is nonlinear

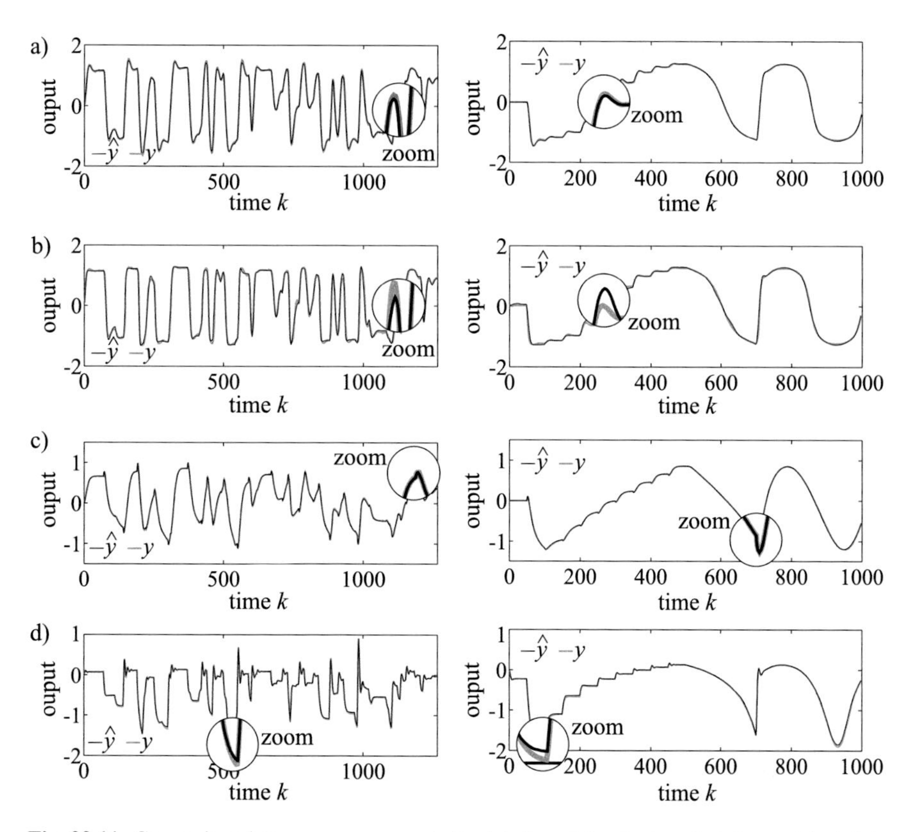

Fig. 22.11 Comparison between process and simulated model output on training and test data for the four test processes: (**a**) Hammerstein, (**b**) Wiener, (**c**) NDE, and (**d**) dynamic nonlinearity processes. (For the input signals, refer to Sect. 22.5.1)

with respect to all inputs. First, LOLIMOT cannot exploit the Wiener structure in a similar way as for the Hammerstein structure, and more decompositions are required. Second, the dynamics of the linear block are weakly observable from the process output when the static nonlinearity is in saturation. Small disturbances on the process output can thus cause significant errors in the estimation of the model dynamics for those operating regimes associated with the saturation. This effect cannot be observed for progressive static nonlinearities. Nevertheless, the model performs satisfactorily on training and validation data; see Fig. 22.11b.

As expected from the nonlinear difference equation of the process, LOLIMOT decomposes the input space in all four model inputs. However, comparable results can be achieved with the simplified reduced operating point strategy discussed in Sect. 22.2. If the premise input space is reduced to a first-order dynamics operating point $\underline{z} = [u(k-1)\ y(k-1)]^T$, similar model performance can be achieved as indicated by the second (gray) convergence curve in Fig. 22.10b. The input space partitioning for this case is shown in Fig. 22.12b. Although the one-step prediction

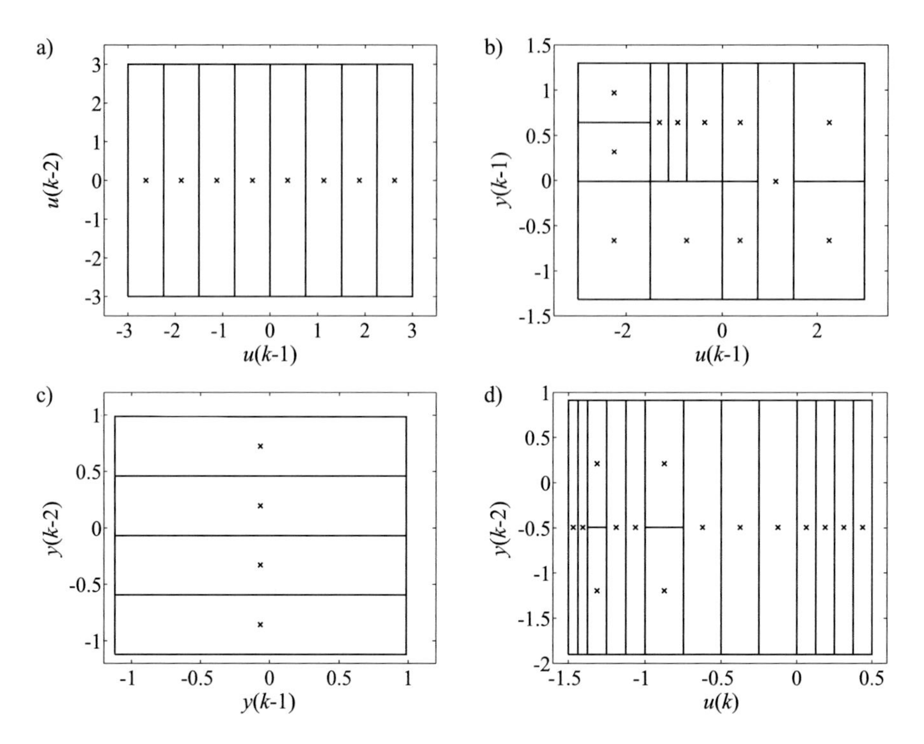

Fig. 22.12 Input space partitioning generated by LOLIMOT for the four test processes: (**a**) Hammerstein, (**b**) Wiener, (**c**) NDE, and (**d**) dynamic nonlinearity processes. Note that here the Wiener model was identified with a reduced order operating point $\underline{z} = [u(k-1)\ y(k-1)]^T$; with a complete operating point LOLIMOT also performs divisions in $u(k-2)$ and $y(k-2)$ and with negligibly better performance

surface is symmetric, the input space partitioning carried out by LOLIMOT is not, because the training data is not exactly symmetrically distributed, and it is corrupted with noise.

The static nonlinearity of the model (Fig. 22.13b(left)) shows larger deviations from the process statics than for the Hammerstein process. This is caused by the rougher decomposition of the $u(k-i)$-dimensions due to the higher-dimensional operating point, which also depends on $y(k-i)$. The step responses in Fig. 22.13b(right) clearly underline the above argument that the process dynamics cannot be estimated very accurately when the nonlinearity is in saturation.

22.5.4 NDE Process

The NDE process can be modeled according to (22.29), as well. The convergence curve in Fig. 22.10c reveals that the process is less nonlinear than the others since

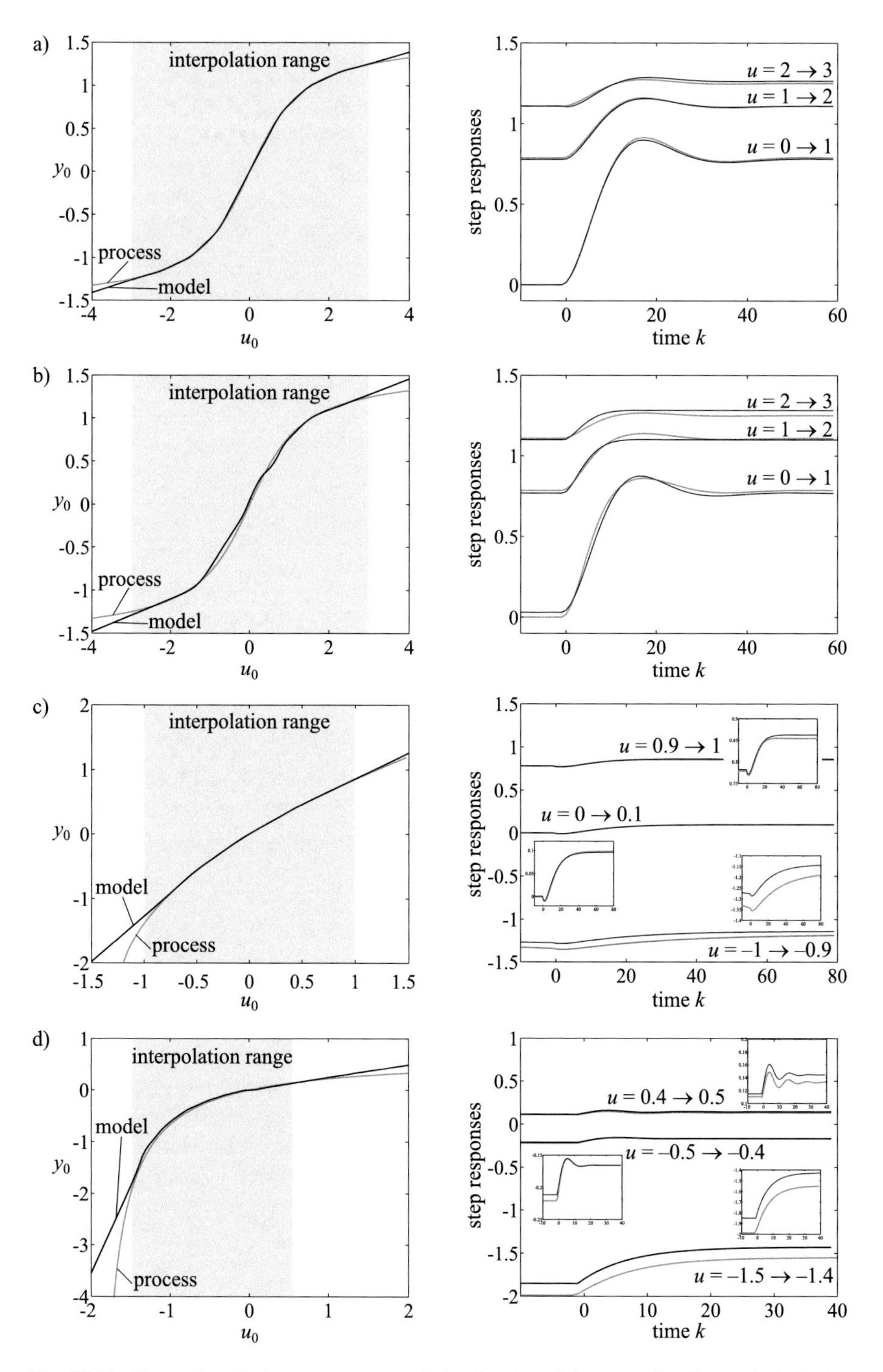

Fig. 22.13 Comparison between process and simulated model output for the static behavior (left) and some step responses (right): (**a**) Hammerstein, (**b**) Wiener, (**c**) NDE, and (**d**) dynamic nonlinearity process

the global linear model (first iteration) performs quite well, and no improvement can be achieved for more than four rules. Thus, the optimal model complexity is chosen as $M_{\text{opt}} = 4$. The model performs very well (Fig. 22.11b), and LOLIMOT decomposed the input space only in the $y(k-2)$-dimension; see Fig. 22.12c. Ad for the Hammerstein model, a decomposition in $y(k-1)$ would yield comparable results. The NDE process which is linear in $u(k-i)$ and nonlinear in $y(k-i)$ can be seen as the counterpart of the Hammerstein process, which is linear in $y(k-i)$ and nonlinear in $u(k-i)$. Therefore, LOLIMOT can exploit the NDE structure in an analogous way. The comparison between process and model statics and step responses in Fig. 22.13c shows the high model quality. However, for the regime with small output $y(k-2)$, a considerable static modeling error can be observed. It is due to the fact that the static process behavior tends to infinite slope for $u \approx -1.2$, and thus the process becomes unstable.

22.5.5 Dynamic Nonlinearity Process

The dynamic nonlinearity process possesses strongly operating point dependent dynamics. Because the input instantaneously influences the output, the following modeling approach is taken:

$$\hat{y}(k) = f\left(u(k), u(k-1), u(k-2), y(k-1), y(k-2)\right) . \tag{22.30}$$

The strongly nonlinear process characteristics can be recognized by the convergence curve in Fig. 22.10d. The linear model in the first iteration performs much worse than for the other test processes, and the convergence speed is extremely high. The model complexity $M_{\text{opt}} = 15$ is chosen. Figures 22.11d and 22.13d demonstrate that the oscillatory behavior for large inputs and the highly damped behavior for small inputs are accurately modeled. The partitioning of the input space indicated an almost solely nonlinear dependency on $u(k)$. However, in the last two iterations two splits are carried out by LOLIMOT along the $y(k-2)$-dimension; see Fig. 22.12d. These artifacts are due to the noise on the training data. The fine decomposition of the input space leaves relatively few data samples for the parameter estimation of the local linear models. Thus, the sensitivity with respect to noise grows with increasing model complexity; see Sect. 14.9.

Figure 22.14 compares the poles of the linearized process for various operating points with the poles of the 15 local linear models identified by LOLIMOT. This shows remarkably good agreement, underlining the excellent interpretability of local linear neuro-fuzzy models.

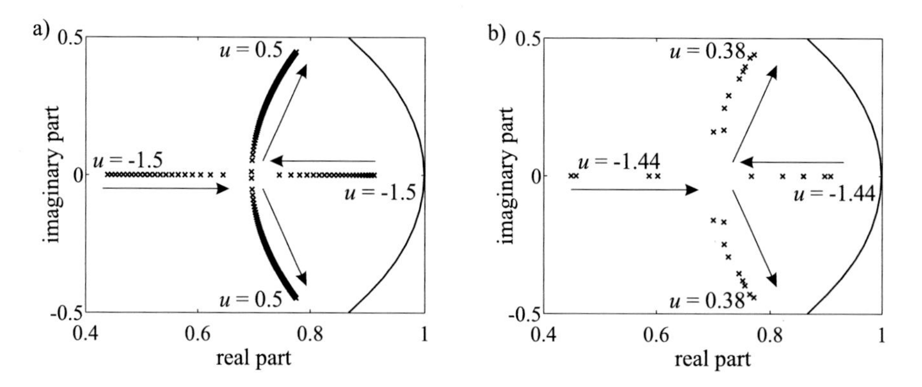

Fig. 22.14 Poles of (**a**) the linearized dynamic nonlinearity process in and of (**b**) the 15 LLMs of the identified local linear neuro-fuzzy model

22.6 Advanced Local Linear Methods and Models

Nonlinear system identification with NARX models has basically the same advantages and drawbacks as in the linear case; see Sect. 18.5.1. On the one hand, linear regression techniques can be utilized, which makes the parameter estimation computationally efficient. On the other hand, the parameters cannot be consistently estimated under realistic disturbances, and the bias increases with the noise variance.

Basically, two strategies exist to overcome this consistency problem. Either the prediction error method for parameter estimation (Sect. 18.3.4) is replaced by some correlation-based approach or a more realistic noise model is assumed. The first strategy is pursued in Sect. 22.6.1 by extending the instrumental variables (IV) method to local linear neuro-fuzzy models. The second strategy leads to NOE or NARMAX models that require nonlinear optimization techniques for parameter estimation; see Sects. 22.6.2 and 22.6.3. In all that follows, knowledge about the linear variants of the methods is presumed. For an introduction to these linear foundations, refer to Chap. 18 and [278, 356, 540].

It is a major advantage of local linear neuro-fuzzy models that these advanced concepts known from *linear* system identification literature can be extended to *nonlinear* dynamic systems in a straightforward manner. Note, however, that the application of these advanced concepts is usually only worthwhile when the measurements are significantly disturbed. Utilization of an NARX structure with a low-pass prefiltering is an alternative approach in order to increase the signal-to-noise ratio and to compensate (at least partly) for the high-frequency emphasis inherent in the NARX model. A main advantage of the advanced concepts discussed below is that they avoid the tedious tuning of the low-pass filter.

Generally, two strategies for the utilization of these advanced methods can be distinguished; see Fig. 22.15. They can be applied within the LOLIMOT algorithm: that is, after the least squares estimation of each LLM, a subsequent tuning step improves this LLM by the IV method or the optimization of local OE/ARMAX

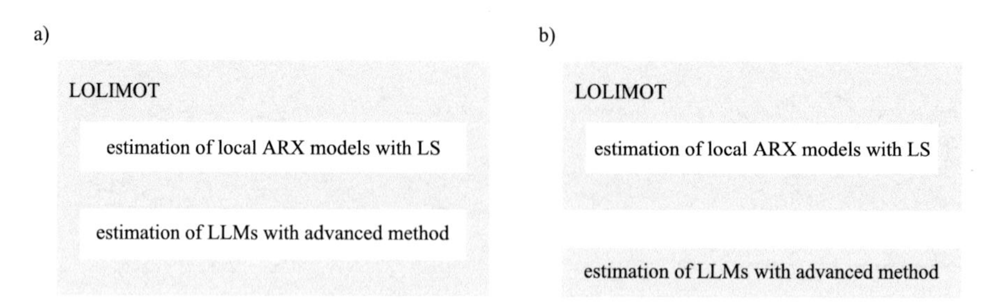

Fig. 22.15 Advanced local linear estimation methods and models can be incorporated into LOLIMOT in two ways: (**a**) they are nested into the algorithm, (**b**) the final model is tuned

models. The parameters of the local ARX models can be exploited as initial parameter values. This nested strategy is illustrated in Fig. 22.15a. It can become computationally demanding owing to the iterative nature of IV, OE, or ARMAX model estimation which is carried out in each iteration of the model construction algorithm. Alternatively, a simplified and computationally much more efficient strategy is shown in Fig. 22.15b, where the LOLIMOT algorithm is run conventionally with a local least squares ARX model estimation. Only in a subsequent phase is the final local linear neuro-fuzzy model improved by the application of an advanced method. The premise structure from the NARX model is retained. The first strategy in Fig. 22.15a can be expected to perform better because the improved LLMs are taken into account by LOLIMOT during the input space decomposition. However, experiments show that the slightly inferior input space partitioning of the strategy shown in Fig. 22.15b is usually insignificant. Thus, in practice, the second strategy delivers similar results with much lower computational demand.

22.6.1 *Local Linear Instrumental Variables (IV) Method*

The idea of the instrumental variables (IV) method is to correlate the residuals with so-called instrument variables that are uncorrelated with the disturbance in the process output. This idea changes the local parameter estimation in (13.22) to (see Sect. 18.5.1 and [278, 356, 540]):

$$\underline{\hat{w}}_i = \left(\underline{Z}_i^T \underline{Q}_i \underline{X}_i\right)^{-1} \underline{Z}_i^T \underline{Q}_i \underline{y}\,, \tag{22.31}$$

where the columns in $\underline{Z}_i$ are the instrumental variables. The IV estimate (22.31) replaces the standard LS estimate in (13.22).

The local regression matrix of a single input mth order model is

$$\underline{X}_i = \begin{bmatrix} u(m) & \cdots & u(1) & -y(m) & \cdots & -y(1) & 1 \\ u(m+1) & \cdots & u(2) & -y(m+2) & \cdots & -y(2) & 1 \\ \vdots & & \vdots & \vdots & & \vdots & \vdots \\ u(N-1) & \cdots & u(N-m) & -y(N-1) & \cdots & -y(N-m) & 1 \end{bmatrix} . \quad (22.32)$$

The instrumental variables should be chosen highly correlated with the regressors (columns in $\underline{X}_i$) but uncorrelated with the noise contained in $y(k-i)$. Thus, the instrumental variables (columns in $\underline{Z}_i$) are usually chosen as

$$\underline{Z}_i = \begin{bmatrix} u(m) & \cdots & u(1) & -\hat{y}(m) & \cdots & -\hat{y}(1) & 1 \\ u(m+1) & \cdots & u(2) & -\hat{y}(m+2) & \cdots & -\hat{y}(2) & 1 \\ \vdots & & \vdots & \vdots & & \vdots & \vdots \\ u(N-1) & \cdots & u(N-m) & -\hat{y}(N-1) & \cdots & -\hat{y}(N-m) & 1 \end{bmatrix} , \quad (22.33)$$

which is almost equivalent to $\underline{X}$, but the measured process outputs $y(k-i)$ are replaced by the *simulated* model outputs $\hat{y}(k-i)$. Therefore, before the IV estimate can be evaluated, a NARX model has to be estimated to provide a basis model for the simulation of $\hat{y}$. The application of the non-recursive IV method in this simple form requires about twice the computation time as the standard LS estimation of local ARX models because two least squares estimate have to be calculated (one for the original ARX model and the one in (22.31)) [36].

It is important to note the following discrepancy between the applications of the IV method to linear and nonlinear models. During the first LOLIMOT iterations, the simulated output $\hat{y}$ of the neuro-fuzzy model will hardly be close to the process output y because the nonlinear process behavior will not be fully described by the model. Consequently, the correlation between the simulated model output IVs and the process outputs may be smaller than expected, which leads to a higher variance error. Thus, after LOLIMOT has converged, the parameters of all LLMs should be re-estimated with the IV method based on the actual simulated model output with high accuracy.

From the four nonlinear dynamic test processes compared above, the dynamic nonlinearity introduced in Sect. 22.5.5 will be used as an example. The process output is disturbed by white Gaussian noise with signal-to-noise amplitude ratios of 100, 50, and 25. The performance of a NARX approach is summarized in the first row of Table 22.2. The parameter estimates of the local ARX models can be expected to be biased [278, 356, 540], and this bias increases with a decreasing signal-to-noise ratio. Thus, the benefits of the instrumental variables method can be expected to grow. Indeed, the LOLIMOT convergence curves for a nested IV parameter estimation in Fig. 22.16 underline the fact that virtually no improvement can be achieved for small noise levels (a) but significant benefits are obtained for high noise levels (b).

Table 22.2 Comparison of LOLIMOT on the dynamic nonlinearity test process with local equation error models estimated with least squares (NARX), estimated with the instrumental variables method (NIV), and local output error models (NOE)

$100 \cdot$ NRMSE training/test	$\sigma_n = 0.01\,\sigma_y$	$\sigma_n = 0.02\,\sigma_y$	$\sigma_n = 0.04\,\sigma_y$
NARX	2.1/1.8	4.3/2.5	9.0/5.2
NIV	1.8/2.2	2.6/2.9	4.7/5.1
NOE	1.4/1.6	2.4/2.6	5.0/3.8

σ_n = noise standard deviation, σ_y = process output standard deviation

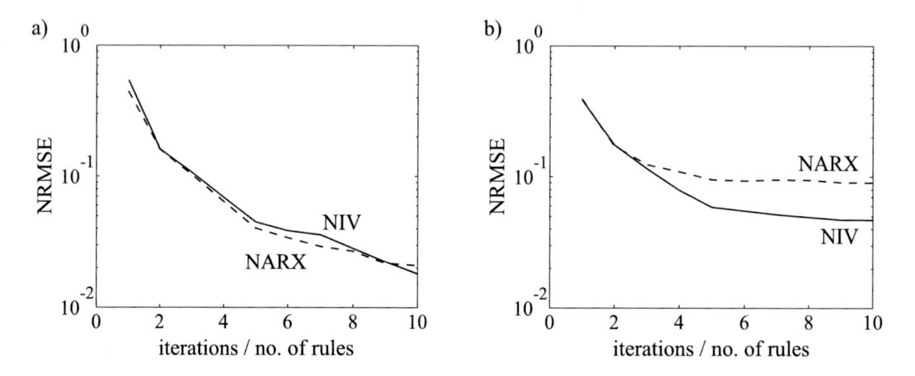

Fig. 22.16 Convergence curves for the conventional dynamic LOLIMOT algorithm (NARX) and the nested application of the instrumental variables method (NIV): (**a**) $\sigma_n = 0.01\,\sigma_y$, (**b**) $\sigma_n = 0.04\,\sigma_y$, where σ_n = noise standard deviation, σ_y = process output standard deviation

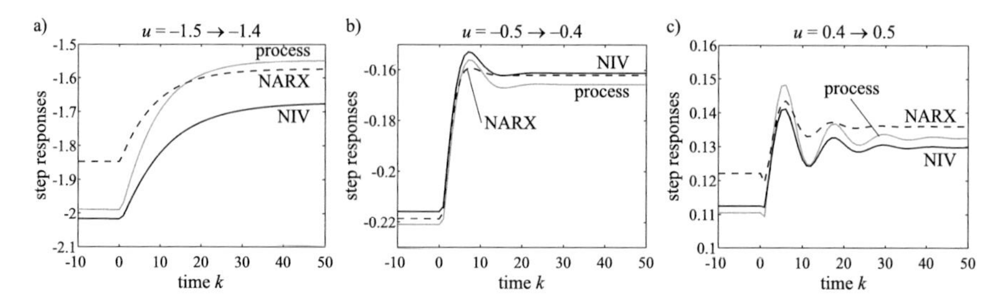

Fig. 22.17 Step responses of models with ten rules identified with the conventional dynamic LOLIMOT algorithm (NARX) and with the nested instrumental variables method (NIV). Step responses for (**a**) low, (**b**) medium, (**c**) high input values

The step responses of the models (Fig. 22.17) obtained for the highest noise level with the NARX and NIV approaches also show that the nonlinear instrumental variable method improves the model performance significantly, especially in the dynamics. Owing to unmodeled nonlinear effects, however, this improvement cannot be guaranteed for all operating conditions. Note that Fig. 22.17a, c represent the boundaries of the interpolation range, and thus the model quality is worse than in Fig. 22.17b, where the operating point is close to the center of a validity function.

22.6.2 Local Linear Output Error (OE) Models

The consistency problem of NARX models can also be overcome by the optimization of NOE models. Since the output error is nonlinear in the parameters, a nonlinear optimization technique has to be utilized. In order to exploit the quadratic form of the loss function, a nonlinear least squares optimization technique, e.g., the Levenberg-Marquardt algorithm, can favorably be applied; see Sect. 4.5.2. Nevertheless, this implies an increase of the computational demand compared with the NARX approach of at least one or two orders of magnitude.

Similar to the estimation of local linear ARX models, the optimization of local linear OE models can be carried out globally or locally. In contrast to the ARX model case, however, the local optimization of the individual LLMs cannot be performed independently of each other. Rather the output error contains a contribution of *all* LLMs because the overall model output is fed back. Thus, local parameter optimization of local OE models depends on the order in which the LLMs are optimized. In fact, it can be seen as a staggered optimization approach; see Sect. 7.5.5.

The results of the NOE approach are shown in Table 22.2, and the step responses are close to those obtained with the NIV method in Fig. 22.17. The much higher computational demand required for the optimization of an NOE model compared with the NIV method is not justified for this example; see Table 22.2. However, in some cases, it can yield significantly better results. Furthermore, the local nonlinear parameter optimization is useful in situations where the desired model output is not directly available. If, for example, an *inverse process model* is to be trained, then the output of the model is used as the input for the process, and the loss function is calculated at the process output. Thus, the error must be propagated back through the nonlinear process behavior, which requires a nonlinear optimization technique for the estimation of the model parameters. For more details on this topic, refer to [147, 268, 377].

22.6.3 Local Linear ARMAX Models

For the estimation of local ARMAX models basically, all remarks made in the previous subsection hold as well. An additional difficulty arises with the approximation of the unknown white disturbance that drives the noise filter; see Sect. 18.5.2. This disturbance is approximated by the residuals $e(k) = y(k) - \hat{y}(k)$, that is,

$$e(k) = y(k) - f(u(k-1), \ldots, u(k-m), y(k-1), \ldots, y(k-m), e(k-1), \ldots, e(k-m)) \, . \quad (22.34)$$

In contrast to linear ARMAX models, a significant part of the prediction error (residual) of NARMAX models can be expected to be caused by unmodeled nonlinear effects. This means that $e(k)$ is typically not dominated by the stochastic effects but rather the systematic model error (bias error) plays at least an equally important role. As a consequence, the unknown disturbance can hardly be well approximated by (22.34), and the benefit achieved by the application of NARMAX instead of NARX models can deteriorate. In fact, it is the experience of the author that NARMAX models do not reliably perform better than NARX models. Therefore, they are not pursued further.

22.7 Local Regularized Finite Impulse Response Models: Coauthored by Tobias Münker

As discussed at the beginning of Sect. 22.8, until recently, NFIR models seemed to be infeasible in practice due to their huge input space dimensionality. Two developments changed this assessment significantly making NFIR models a feasible and very promising option in the context of local linear neuro-fuzzy networks:

- The structural decoupling between the inputs $\underline{x}$ for the rule consequents (typically high-dimensional) and the inputs $\underline{z}$ for the rule premises (typically low-dimensional) allows confining the dimensionality issue to the huge parameter estimation problem for the local linear models. The premise input space can be spanned low-dimensionally by NARX or NOBF inputs, compare also Sects. 14.1 and 22.9.
- The high variance typically encountered when estimating a high-order FIR model is avoided by new clever regularization approaches that encourage (i) stable impulse responses decaying with a given time constant and (ii) close parameter values for neighboring impulse response coefficients, compare Sect. 18.6.2.

The combination of these two ideas has been investigated for SISO systems in [394] and for MISO systems in [393, 395] and compared for a couple of benchmark problems in [41, 392]. It is detailed in this section.

22.7.1 Structure

Local model networks can be used to transfer ideas employed for linear system identification to the nonlinear setup. This is due to the fact that for local model networks, identification problems which are linear in their parameters in a linear system identification framework, are transferred to a weighted least squares problem in the nonlinear case. For an NFIR identification problem the regression matrix for the estimation of a LLM becomes

$$\underline{X} = \begin{bmatrix} 1 & u(m) & u(m-1) & \ldots & u(1) \\ 1 & u(m+1) & u(m) & \ldots & u(2) \\ \vdots & \vdots & & \vdots & \\ 1 & u(N) & u(N-1) & \vdots & u(N-m+1) \end{bmatrix}. \qquad (22.35)$$

These regressors are weighted locally and then used for the estimation of the parameters of the FIR system. This has the drawback that a high number of parameters has to be estimated, and the variance error of the estimate is significantly increased. For the control of the variance error for linear FIR models, regularized identification techniques described in Sect. 18.6.2 can be utilized. This regularization technique is transferred to the nonlinear world as depicted in Fig. 22.18. The structure of the local model network simply employs local models of regularized

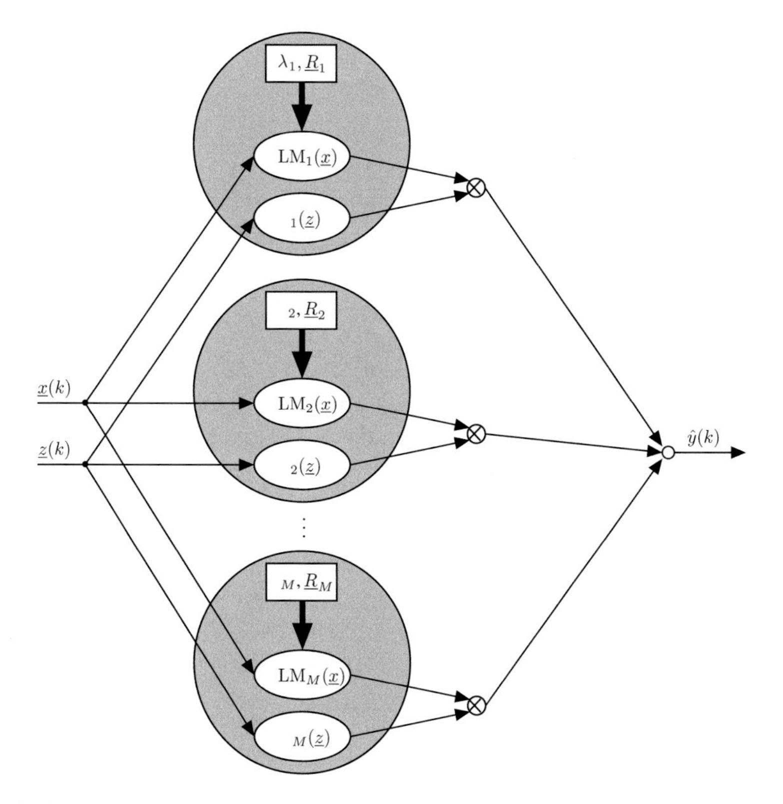

Fig. 22.18 Structure of a local model network with local regularized FIR models. Different regularization strengths $\lambda_1, \lambda_2, \ldots, \lambda_M$ and penalty matrices $\underline{R}_1, \underline{R}_2, \ldots, \underline{R}_M$ can used for the regularization of the local models (LM)

FIR type. The main difference is that for the estimation of the local models' penalty matrices $\underline{R}_1, \underline{R}_2, \ldots, \underline{R}_M$ with regularization strengths $\lambda_1, \lambda_2, \ldots, \lambda_M$ are applied during the estimation procedure for regularization. Note that furthermore, the regularization matrices may depend on additional hyperparameters like the rate of impulse response decay or poles describing a potential denominator dynamics which may be fixed by the user or optimized on data by hyperparameter tuning.

22.7.2 Local Estimation

The regularized local estimation is the solution of the optimization problem

$$(\underline{y} - \underline{X}\,\underline{w}_i)^T \underline{Q}_i (\underline{y} - \underline{X}\,\underline{w}_i) + \lambda\,\underline{w}_i^T \underline{R}\,\underline{w}_i \rightarrow \min_{\underline{w}_i} \tag{22.36}$$

with the parameters of the i-th local model denoted as $\underline{w}_i$ and the corresponding validity function values of the data is in the diagonal weighting matrix $\underline{Q}_i$. To simplify the notation the weighted regression matrix and the weighted output vector are introduced as

$$\underline{\tilde{X}}_i = \underline{Q}_i^{\frac{1}{2}} \underline{X} \qquad \underline{\tilde{y}}_i = \underline{Q}_i^{\frac{1}{2}} \underline{y}. \tag{22.37}$$

The optimal solution can then be found simmilar to the solution of an LS problem as

$$\underline{w}_i = \left(\underline{\tilde{X}_i}^T \underline{\tilde{X}_i} + \lambda\,\underline{R}\right)^{-1} \underline{\tilde{X}_i}^T \underline{\tilde{y}_i}\,. \tag{22.38}$$

For the penalty matrix $\underline{R}$ in principle every parameterization described in Sect. 18.6.2 can be used. Often, the TC kernel of first order [458] works well. It is important to notice that it is still possible to choose input space of the local models $\underline{x}(k)$ and the input space of the validity functions $\underline{z}(k)$ differently. It is thus possible and often advantageous to use delayed values of the input $u(k-i)$ and output $y(k-i)$ with $i = 1, 2, \ldots, n$ for $\underline{z}(k)$ while $\underline{x}(k)$ represents the input only for the local FIR models identified with regularization.

22.7.3 Hyperparamter Tuning

To perform the local estimation, the hyperparameters λ_i for the penalty matrices $\underline{R}_i, i = 1, 2, \ldots, M$ have to be set. Furthermore, the offsets of the local models play a special role. Usually, there is no prior knowledge for the offsets available, and thus they are excluded from the penalty terms in the optimization problem by introducing zeros for the respective entries in the $\underline{R}_i$.

In general, several strategies for the choice of the hyperparameters are possible. One strategy is to use a priori defined hyperparameters for all local models. This can be useful if the dominant time constant of the process is known beforehand. However, the flexibility of the model is limited significantly by this approach. An approach that allows for more flexibility is to obtain individual $\underline{R}_i$ for each local model. In general, both generalized cross validation and marginal likelihood optimization can be used to calculate the hyperparameters. For the marginal likelihood case, the estimation of the offset cannot be done straightforwardly. In this case, a two-step procedure is favorable. In the first step, a high order unregularized local FIR model is used to estimate the noise and offset of the process. Then in a second step, the remaining hyperparameters (for the TC kernel this is regularization strength and decay rate) are optimized within the marginal likelihood optimization.

To describe the marginal likelihood optimization problem, the output for the i-th local model is computed as $\underline{\bar{y}}_i = \underline{Q}_i^{\frac{1}{2}} \left(\underline{y} - w_{0i} \right)$ with the previously identified offset w_{0i}. Furthermore, the regression matrix $\underline{\bar{X}}_i$ is used to denote the weighted version of the regressor given by (22.35) without the first column for the offset (1's before the weighting). The negative marginal likelihood can then be written as

$$L_i = \underline{\bar{y}}_i^T \underline{Z}_i^{-1} \underline{\bar{y}}_i + \log\det(\underline{Z}_i) \tag{22.39}$$

with

$$\underline{Z}_i = \frac{1}{\lambda_i} \underline{\bar{X}}_i^T \underline{R}_i^{-1} \underline{\bar{X}}_i + \sigma^2 \underline{I} \,. \tag{22.40}$$

A significant issue of the maximum likelihood optimization is the inverse of $\underline{Z}_i$, which is a $N \times N$ matrix. This inversion can be avoided by the usage of a matrix inversion lemma and the QR factorization. This is described in detail in [101].

22.7.4 *Evaluation of Performance*

The performance of the regularized local FIR network is evaluated on a Wiener example system. Therefore, N = 1000 data points are generated with an APRBS with a holding time of $8T_0$ as the input. The performance of the model is evaluated at 10,000 time steps on a noise-free test data set. For comparison, the following models are evaluated

- **ARX2**: A second order ARX local model network with two delayed values of input and output for the $\underline{z}$ input space.
- **ARX4**: A fourth order ARX local model network with four delayed values of input and output for the $\underline{z}$ input space.
- **FIR**: A local model network with unregularized FIR models as local models and two delayed values of input and output for the $\underline{z}$ input space.

- **RFIR**: A local model network with TC kernel regularized local FIR models with two delayed values of input and output for the $\underline{z}$ input space.

To study the behavior, the output of the Wiener system is corrupted by white noise. The amplitude of the noise is chosen such that the signal has a defined signal to noise ratio (SNR), which is calculated as

$$SNR = \frac{\sum_{k=1}^{N} n(k)^2}{\sum_{k=1}^{N} y(k)^2}. \tag{22.41}$$

This is the ratio between signal and noise power. As a termination criterion for the LOLIMOT algorithm, the corrected AIC is used.

An excerpt from the obtained output data for the four models is shown in Fig. 22.19 for a low SNR of 100 and a high SNR of 10^4. The local models of ARX type have difficulties in identifying the system. For the local FIR models, especially in the low SNR case, the variance error of the parameter dominates, which results in a spiky response (B in the figure). In the low SNR case, the biased estimate of the model's parameters results in high errors of the output (C in the figure). Also, in the high SNR case, local ARX models perform poorly and in the second order case become unstable (A in the figure). In contrast, the regularized local FIR models circumvent both of the problems. The stability of the local FIR models is inherently guaranteed. An appropriate regularization circumvents the variance error due to many coefficients and high noise.

The positive effect of regularization can also be well observed in the impulse responses of the local models shown in Fig. 22.20. Here, the impulse responses

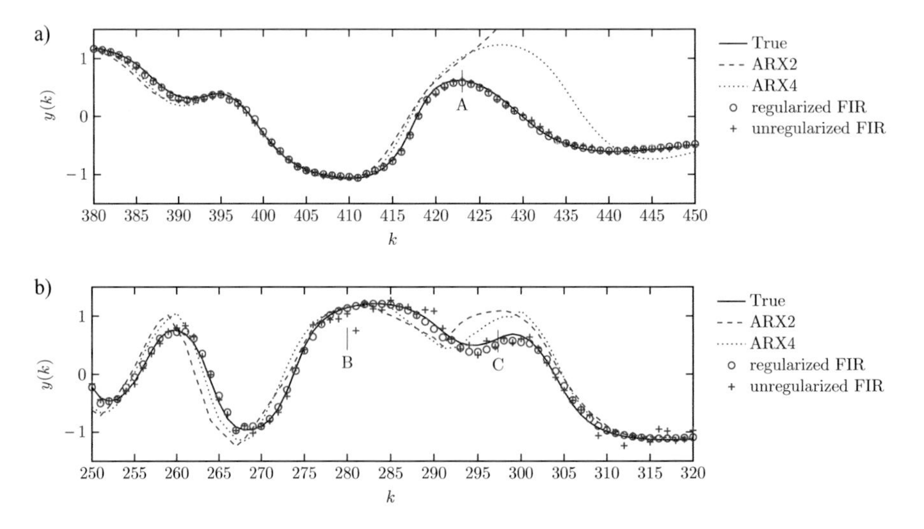

Fig. 22.19 Comparison of results for LOLIMOT with different local models. In (**a**) the case with a low SNR of 100 is shown, while in (**b**) the SNR high at a value of 10^4

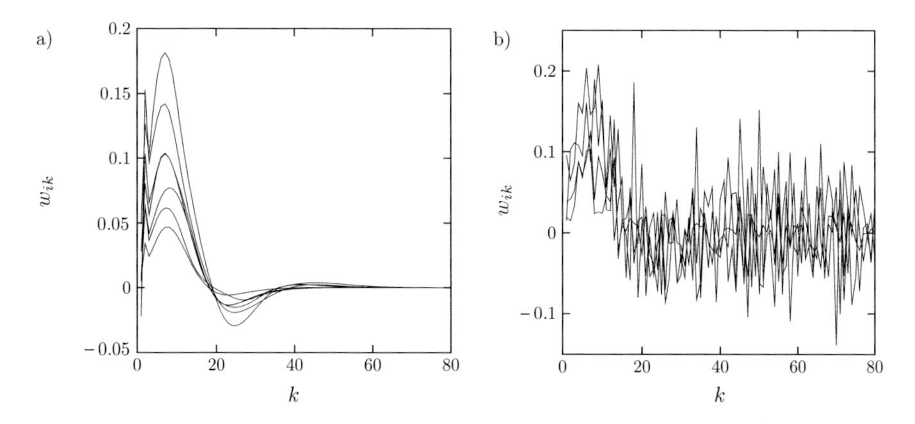

Fig. 22.20 Impulse response coefficients of the local models for a SNR of 100. In (**a**) the coefficients in the regularized case are shown and in (**b**) the coefficients of the unregularized local models are depicted

of the local models of the final models determined by AIC_C are shown. For the unregularized local FIR models, the effect of the variance error is visible. The impulse responses are very non-smooth. For the regularized case, this does not hold. Here, the local impulse responses are smoothed by the penalty term in the regularization. Thus, regularized local FIR models provide significant advantages, especially in cases of high noise at the output.

For more details on local regularized FIR model networks, refer to the references [391, 393–395]. Benchmark problems are treated in [392]. For a comparison with ARX and OE models and a discussion on the order reduction potential, refer to [396]. First, gray-box modeling ideas in this context are pursued in [397, 452].

22.8 Local Linear Orthonormal Basis Functions Models

A more radical strategy to overcome the consistency problem of NARX models is to discard the output feedback structure. Since NFIR models suffer from the problem of requiring huge input space dimensionalities (Sect. 19.2.3), nonlinear orthonormal basis function (NOBF) models have recently gained more interest [413, 502, 503, 518]. Because linear OBF models are an active topic of current research, the status for NOBF models is still premature. An overview of linear OBF models can be found in Sect. 18.6.5 and [575]. In particular, the combination of local linear modeling schemes and OBFs promises the following important advantages:

1. Low sensitivity with respect to the (typically unknown) dynamic order of the process
2. Linear parameterized nonlinear output error model

3. Inherent stability of the nonlinear dynamic model
4. Incorporation of prior knowledge about the process dynamics

The first three points are clearly positive. In particular, the second advantage solves the dilemma that NARX models are linear parameterized but unfortunately rely on the one-step prediction error, and NOE models are based on the simulation error, which is the actual modeling goal, but they are nonlinearly parameterized. With OBFs both advantages can be combined. The stability of an NOBF model is ensured because the orthonormal filters are designed a priori in a stable manner and thus during training no feedback component is adapted. This feature in combination with the linear parameterization is particularly attractive for online learning with a recursive least squares algorithm. The fourth advantage, however, turns into a drawback when no such prior knowledge is available. In addition to linear OBF models, NOBF models require knowledge about the approximate process dynamics for different operating conditions, at least when the process dynamics vary strongly with the operating point. One major drawback of NOBF models is that as for NFIR models, an infinite number of filters is theoretically required. In practice, of course, it is approximated with a finite and relatively small number of filters in order to keep the number of estimated parameters small. This introduces an approximation error.

Comparing the NOBF model in (19.21) with the NARX model in (19.2), the following observations can be made, assuming that the models will be used for simulation not for one-step prediction:

1. Both model structures have an infinite impulse response (IIR). The NARX model possesses external feedback, which is determined by the approximation of $f(\cdot)$, while the NOBF model has internal feedback, which is fixed by the user a priori by choosing the characteristics of the filter $L_i(q)$.
2. As a consequence, depending on $f(\cdot)$ the NARX model can become unstable, while the NOBF model is guaranteed to be stable for all possible $f(\cdot)$.
3. NARX models can identify unstable systems; NOBF models cannot because their optimal predictor would be unstable.
4. The NOBF model can only approximate the process dynamics (the higher the order m is chosen the better the approximation), whereas the NARX model can exactly capture the process dynamics if the model order is chosen equivalent to the process order. The importance of this issue fades if one realizes that, in practice, models will usually be low-order approximations of the process and that further presumably more significant errors are introduced by the approximation of $f(\cdot)$.

5. The NOBF model requires rough prior knowledge about the process dynamics.[4] As the quality of the knowledge decreases, the number of filters required for the same accuracy increases, and as a consequence, the input space dimensionality of $f(\cdot)$ increases. Owing to the curse of dimensionality, overly inaccurate prior information about the process dynamics may let the NOBF approach fail. Furthermore, processes with weakly damped or dispersed poles require Kautz functions or generalized OBFs respectively, with an increasing demand on the prior knowledge.
6. Both model structures are linear in their parameters if a linearly parameterized function approximator for $f(\cdot)$ is selected.
7. The NOBF model structure represents an output error model; the NARX model structure represents an equation error model. Therefore, the estimates of the NARX model's parameters are biased, while the NOBF model's parameters are estimated unbiasedly in the presence of uncorrelated additive output noise. Furthermore, an exact expression for the variance error of NOBF models can be derived if the approximator of $f(\cdot)$ is linearly parameterized because the regression matrix is deterministic. This also allows the calculation of the NOBF model's confidence intervals (errorbars), whereas this is not (as easily) possible for NARX models.
8. The data distribution in the input space of $f(\cdot)$ is different for NOBF and NARX approaches. This can have a significant influence on the achievable approximation quality of $f(\cdot)$, in particular, if axis-orthogonal partitioning strategies are applied.

The standard NOBF approach shown in Fig. 22.21a that also is pursued in [518] is based on the following model:

$$\hat{y}(k) = f\left(L_1(q)u(k), L_2(q)u(k), \ldots, L_m(q)u(k)\right) \tag{22.42}$$

where $L_i(q)$ are the orthonormal filters and $f(\cdot)$ can be any type of static approximator. The parameter(s) of the linear filters $L_i(q)$ are not the subject of the training; rather they are chosen a priori by the user. The following cases can be distinguished:

- *No knowledge about the process dynamics available*: The filters are chosen to $L_i(q) = q^{-i}$, which is equivalent to the nonlinear finite impulse response (NFIR) model. Then the order m has to be chosen huge to describe the process dynamics appropriately. As a guideline the model order should be chosen as

[4] Approaches that are less sensitive to prior knowledge have been proposed. For example, the Laguerre pole can be estimated by nonlinear optimization, or from an initial NOBF model based on prior knowledge, a better choice for the pole can be found by model reduction techniques (this idea can be applied in an iterative manner). These ideas clearly require further investigation and are not pursued here because they are computationally more demanding than the calculation of a least squares solution.

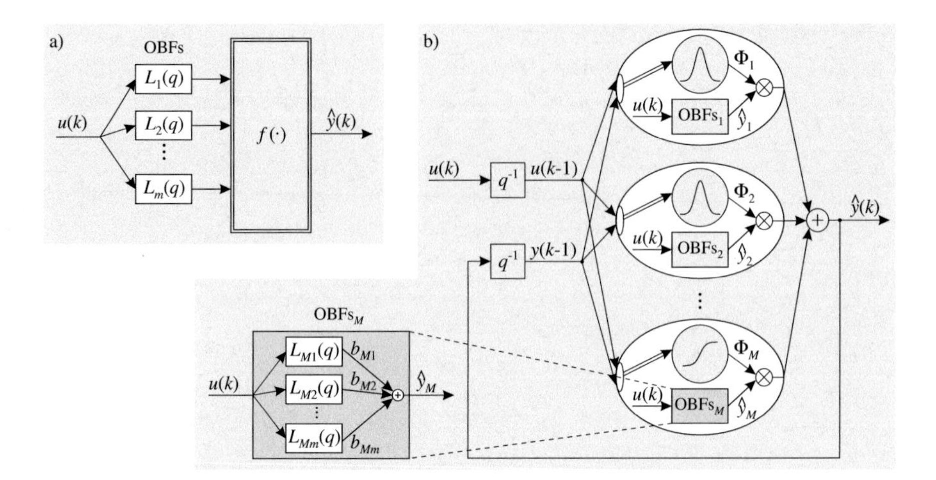

Fig. 22.21 (**a**) An OBFs filter bank can be used to generate the dynamics for an external dynamics approach in combination with any type of approximator $f(\cdot)$. (**b**) Individual OBF models can be used for each LLM in a local linear neuro-fuzzy model

$m = \text{int}(T_{95}/T_0)$, where T_{95} is the settling time of the process and T_0 is the sampling time. This is infeasible for most applications.

- *The process is well damped*: With the approximate knowledge about the dominant process pole a bank of orthonormal *Laguerre* filters [352, 575, 583] $L_i(q)$ can be designed. The required number of Laguerre filters m depends on the accuracy of the prior knowledge of the process pole and on the operating point dependency of the process dynamics. If the assumptions on the process dynamics are reasonable and the process dynamics vary only slightly with the operating point, then a few filters, say $m = 2$–6, are sufficient.
- *The process is resonant*: With rough knowledge about the dominant conjugate complex pole pair of the process, a bank of orthonormal *Kautz* filters [352, 575, 584] $L_i(q)$ can be designed. The choice of the number of Kautz filters m follows the same criteria as for the Laguerre filters.
- *The process possesses several dispersed poles*: If approximate knowledge about several process poles is available, then generalized orthonormal basis functions can be designed that include Kautz, Laguerre, and FIR filters as special cases [245, 246, 352, 575]. This is an excellent approach to identify models with several modes stretching over a wide frequency band that are approximately known, a common problem in machine dynamics and vibration analysis.

The assumption that *approximate* prior knowledge about the dominant process pole(s) is available is quite realistic. This knowledge may stem from first principles, operator experiences, step responses, correlation analysis, etc. It is shown in [575] for linear OBF models that even with rough information about the process poles, an OBF model can outperform an ARX and even an OE model, especially for higher-order and highly disturbed processes. In particular, the dynamic order of the process,

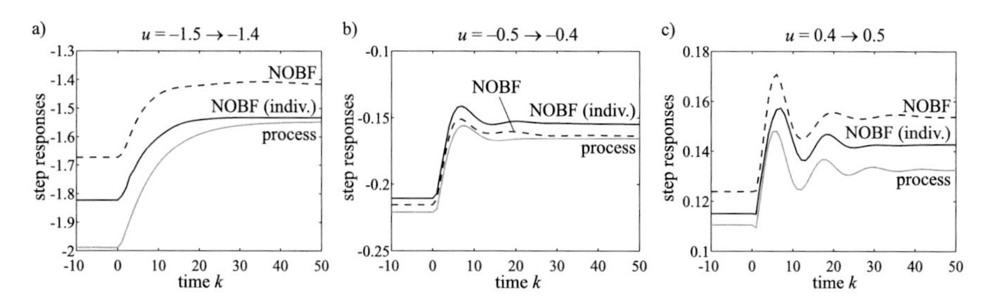

Fig. 22.22 Step responses of models with ten rules identified by LOLIMOT with a general NOBF approach and an NOBF approach with individual filter design for each LLM. Both NOBF models utilize generalized orthonormal filters of order six which means that compared with the NARX model, one parameter more has to be estimated for each LLM. Step responses for (**a**) low, (**b**) medium, (**c**) high input values

which typically is hard to determine, is less relevant for OBF models. Furthermore, a preceding identification of a NARX model can also yield the approximate process dynamics.

For nonlinear OBF models, however, the additional difficulty arises that the process dynamics must not depend strongly on the operating point; otherwise, the chosen OBF dynamics will deviate significantly from the process dynamics for some operating points. The consequence would be either a severe dynamic undermodeling or the choice of many filters m, which implies a large input space for the approximator $f(\cdot)$.

This dilemma can be overcome if a local linear modeling scheme is utilized for the approximator $f(\cdot)$. In [413] it is proposed to design *individual* OBF models for each local linear model instead of a single global OBF filter bank. This idea is illustrated in Fig. 22.21b. The major advantage of this approach is that the poles of the Laguerre or Kautz filters can be specified individually for each operating regime. This allows one to cope well with processes that possess extremely operating condition dependent dynamics at the price of an extra design effort. For example, it is possible to utilize OBFs of Laguerre type for well-damped operating regimes and Kautz filter banks for resonant regimes. A drawback of the individual design strategy besides the higher development effort is that the operating point (premise input space) still has to be defined conventionally in terms of previous inputs and outputs; see Fig. 22.21b. This diminishes the nice property of inherent stability for NOBF models if the premise space includes previous outputs.

For the dynamic nonlinearity test process, the NOBF model is a competitive choice only if the process is highly disturbed; otherwise, either the approximation error or the number of parameters is very high. For the noise standard deviation $\sigma_n = 0.04\,\sigma_y$, the step responses of the NOBF models with and without individually designed filters are shown in Fig. 22.22. The poles of the OBFs are calculated from the local ARX models obtained by a preceding identification with LOLIMOT. The NOBF model with LLM individual filter design utilizes generalized OBFs on the basis of Laguerre or Kautz filters, depending on the type of poles in the corresponding operating regime. The NOBF model with a single OBF filter bank is also based on a generalized OBF approach and incorporates information about the process dynamics in several operating regimes. This is necessary because it is not possible to characterize the dynamics of the test process with a single real pole or conjugate complex pole pair. Such a simple nonlinear Laguerre or Kautz filter-based model is not capable of describing this test process appropriately. However, note that for other processes that possess only slightly operating point dependent dynamics, a simple NOBF model based on Laguerre or Kautz filters is sufficient.

22.9 Structure Optimization of the Rule Consequents

One difficult and mainly unsolved problem for external dynamics approaches is the determination of the dynamic order m and dead time d. It can be beneficial to choose different orders for the input and the output, and for multivariable models, the number of possible variants even increases further. A signal-based method for order determination is proposed in [234] (see Sect. 19.9), but because of its shortcomings most commonly a trial-and-error approach is pursued, that is, models for some or all reasonable combinations of dynamic orders and dead times are identified and compared. Correlation analysis can give further hints. For linear system identification, such a trial-and-error approach is acceptable, but for nonlinear system identification, many other things have to be determined, such as the number of neurons or the strength of the regularization effect, so that the computational effort and the required user interaction can easily become overwhelming.

A combination of a linear subset selection technique such as the orthogonal least squares (OLS) algorithm with LOLIMOT proposed and applied in [414, 427, 429] allows one to partly solve the order determination problem by a structure optimization of the rule consequents. However, in addition to the static version presented in Sect. 14.3, for identification of dynamic systems, some difficulties arise. In principle, the OLS algorithm can be applied to any potential regression vector independently of whether it describes a static or dynamic relationship. When dealing with dynamic systems, however, only NARX, NFIR, or NOBF models can be estimated, since the OLS is based on linear regression; so the model must be linearly parameterized. For models with output feedback, this means that the structure selection is based on the one-step prediction (equation) error, not on the simulation (output) error. The difficulties caused by this fact will be illustrated with the following simple linear example.

Consider a third-order time-lag process with gain 1 and the time constants $T_1 = 10\,\text{s}$, $T_2 = 5\,\text{s}$, $T_3 = 3\,\text{s}$, sampled with $T_0 = 1\,\text{s}$. The true order of this process is assumed to be unknown. An OLS structure selection algorithm may be started with the assumption that the process is at most of fifth order, which implies the following vector of ten potential regressors:

$$\underline{x} = [u(k-1)\; u(k-2)\; u(k-3)\; u(k-4)\; u(k-5)$$
$$y(k-1)\; y(k-2)\; y(k-3)\; y(k-4)\; y(k-5)]^T\,.$$

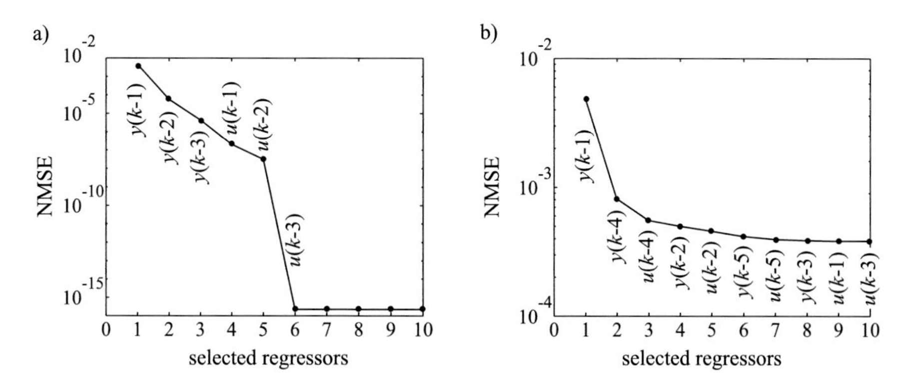

Fig. 22.23 Convergence curves for an OLS structure selection on data generated from a linear third-order process with the potential regressors $u(k-1), \ldots, u(k-5)$, $y(k-1), \ldots, y(k-5)$: (**a**) noise-free, (**b**) process output disturbed by noise

The process is excited with a pseudo random binary signal, and the data is utilized for structure optimization with an OLS forward selection scheme. Figure 22.23a shows the convergence curve of the structure selection for the noise-free case. As expected, the first six selected regressors correspond exactly to the third-order process structure. Furthermore, the significance of the seventh regressor drops by many orders of magnitude, indicating the irrelevance of all regressors selected in the sequel. (Theoretically, the selected regressors 1–6 should explain 100% of the output variance; 10^{-16} is the numerical accuracy of the Gram-Schmidt orthogonalization algorithm.) Thus, in principle an automatic order determination is possible with this algorithm.

Two remarkable effects can be observed in Fig. 22.23a:

- The first three selected regressors are the delayed process outputs $y(k-i)$, $i = 1, 2, 3$.
- The first selected regressor already explains 99.4% of the process output variance.

Both effects are due to the fact that the *one-step prediction* performance is the criterion for the OLS structure optimization. Both effects are undesirable when the intended use of the obtained model is *simulation*. For a simulation model, the most significant regressor clearly is a delayed input since the process output is unavailable during the simulation. In contrast, for a one-step prediction model, the previous process output $y(k-1)$ is very close to the actual process output and thus most significant. Consequently, $\hat{y}(k) = a_1 y(k-1)$ is usually quite a good one-step prediction model, and particularly if the sampling time is small compared with the major process time constant. For this reason, the regressor $y(k-1)$ is able to explain a huge fraction (usually more than 99%) of the process output variance.

As a consequence of these different goals for one-step prediction and simulation, the order determination becomes hard or even infeasible in practice when the model contains an autoregressive part. This can be seen from Fig. 22.23b, which illustrates the OLS convergence curve for the same process as in Fig. 22.23a disturbed by white noise with a signal-to-noise amplitude ratio of 50. The first selected regressor is still $y(k-1)$, but all others apparently change. In particular, irrelevant regressors such as $y(k-4)$ or $u(k-4)$ are chosen early. Furthermore, the convergence curve does not allow one to determine the order of the process. All this follows directly from the fact that for one-step prediction $y(k-1)$ already explains the process output almost fully, and thus all other regressors have a very small relevance (for one-step prediction accuracy). In the noisy case, this small relevance is negligible in comparison to the random effects caused by the stochastic disturbances, and no reasonable structure selection is possible. Currently, no satisfactory solution to the order selection problems for nonlinear dynamic processes is available.

It is important to note that these shortcomings are due solely to the autoregressive part of the model. They can be avoided with (N)FIR and (N)OBF models, which possess no output feedback. Furthermore, it is possible to fix the delays of the autoregressive model part a priori and to select only the numerator coefficients. Often additionally prior knowledge about dead times caused by transport delays might be available. It can be incorporated into the model as well by ensuring the selection of the corresponding regressors.

Although the above discussion shows that additional care must be taken when applying the OLS or any other linear subset selection technique to models with autoregressive parts, the LOLIMOT+OLS algorithm is a very powerful tool. It can well be applied for the selection of dead times and the input delays. This is particularly important for multivariable systems because a good selection of the numerator structure can compensate for deficiencies occurring from the identical denominator dynamics; see Sect. 19.6.

The features of the LOLIMOT+OLS algorithm will be demonstrated by a simulation example. The process depicted in Fig. 22.24a possesses first-order time-lag behavior with dead time. Both the dead time $d(n_{\text{eng}})$ and the time constant

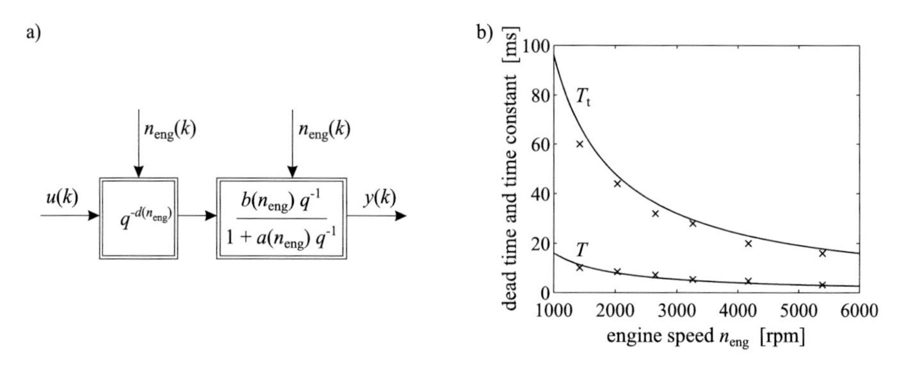

Fig. 22.24 (**a**) Process. (**b**) Process time constant T and dead time $T_t = dT_0$ as functions of the engine speed n_{eng}. The crosses mark the model parameters for the six local linear models at the centers of their validity functions

$T(n_{\text{eng}})$, and thus the parameters $b(n_{\text{eng}})$ and $a(n_{\text{eng}})$, depend on the external signal $n_{\text{eng}}(k)$; see Fig. 22.24b. Problems of this kind can occur, for example, in transport processes with energy storage (Sect. 25.3.2). They also occur in the modeling of combustion engines, where the delay between the time of fuel injection and maximum effect on the engine torque is proportional to the time required for a certain crankshaft angle. For a constant engine speed, this delay and thus the dead time $T_t = dT_0$ and the time constant T are constant. For varying engine speed n_{eng}, the model becomes nonlinear owing to its engine speed-dependent parameters. This difficulty has led to the development of angle-discrete instead of time-discrete models [319, 513, 533], which are sampled at fixed crankshaft angles instead of fixed time instants and thus are linear in the new domain.

LOLIMOT offers an elegant nonlinear dynamic modeling approach in discrete time. The local linear models are simple first-order dynamic models with dead time. This leads to rule consequent regressors that are not universal for the complete model but LLM specific. Since each LLM represents another operating point (engine speed), different dead times $d_i T_0$ are used in the rule consequent vectors, i.e., $\underline{x}_i = [u(k-d_i)\ y(k-1)]^T$. Note that for this example, no offset is required in the local linear models so that each LLM possesses only two parameters. These local linear models are scheduled by the engine speed, which therefore is the only variable for the rule premises, i.e., $\underline{z} = n_{\text{eng}}(k)$. This is a further nice example of the reduced premise input space approach discussed in Sect. 22.2.

The LOLIMOT+OLS algorithm is required to select the dead times $d_i T_0$ if they cannot be determined a priori. Then the rule consequent regression vector must contain all possible dead times $\underline{x} = [u(k-1)\ u(k-2)\ \cdots\ u(k-d_{\max})\ y(k-1)]^T$, where $d_{\max}$ has to be supplied by the user. The task of the OLS algorithm is to select the two most significant regressors from $\underline{x}$ individually for each LLM and thus for each engine speed operating point.

In this example the time constant T varies between 4, and 18 ms, and the dead time T_t varies from 18 to 95 ms. With a sampling time of $T_0 = 4$ ms, the time delays due to the dead time are in the range $d = T_t/T_0 = 4, \ldots, 23$. The data shown in Fig. 22.25a is used for training with the LOLIMOT+OLS algorithm. In the first iteration, it yields the following global linear model:

$$R_1\text{: IF } n_{\text{eng}}(k) = \textsf{don't care} \text{ THEN } \hat{y}(k) = 0.1541\, u(k-4) + 0.8481\, \hat{y}(k-1)$$

In the second iteration, the operating point dependent dead times and time constants can already be observed:

$$R_1\text{: IF } n_{\text{eng}}(k) = \textsf{small} \text{ THEN } \hat{y}(k) = 0.0922\, u(k-6) + 0.9090\, \hat{y}(k-1)$$

$$R_2\text{: IF } n_{\text{eng}}(k) = \textsf{large} \text{ THEN } \hat{y}(k) = 0.2995\, u(k-4) + 0.7035\, \hat{y}(k-1)$$

Finally, the LOLIMOT+OLS algorithm identified the process very accurately by constructing the following six rules as demonstrated in Fig. 22.25b (the argument "(k)" is omitted for n_{eng}):

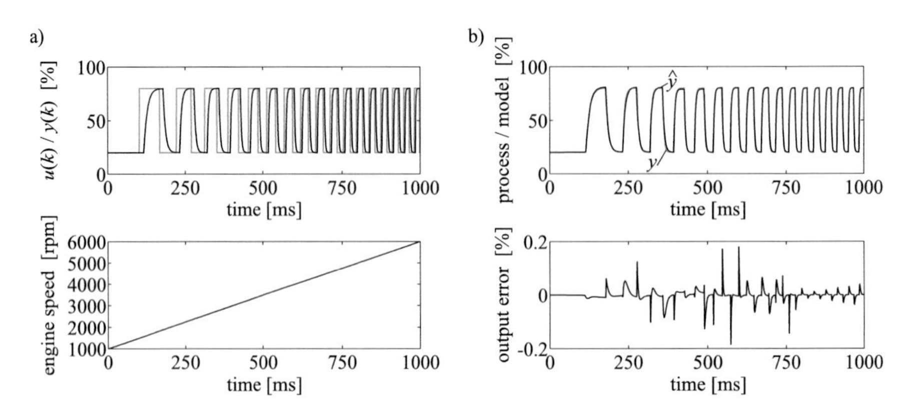

Fig. 22.25 (**a**) Training data. (**b**) Comparison between process and model output and the output (simulation) error

$$R_1\text{: IF } n_{\text{eng}} = \text{tiny} \quad \text{THEN } \hat{y}(k) = 0.0815\,u(k-15) + 0.9189\,\hat{y}(k-1)$$

$$R_2\text{: IF } n_{\text{eng}} = \text{very small} \quad \text{THEN } \hat{y}(k) = 0.1229\,u(k-11) + 0.8785\,\hat{y}(k-1)$$

$$R_3\text{: IF } n_{\text{eng}} = \text{small} \quad \text{THEN } \hat{y}(k) = 0.1684\,u(k-8) + 0.8331\,\hat{y}(k-1)$$

$$R_4\text{: IF } n_{\text{eng}} = \text{medium} \quad \text{THEN } \hat{y}(k) = 0.2620\,u(k-7) + 0.7347\,\hat{y}(k-1)$$

$$R_5\text{: IF } n_{\text{eng}} = \text{large} \quad \text{THEN } \hat{y}(k) = 0.3091\,u(k-6) + 0.6925\,\hat{y}(k-1)$$

$$R_6\text{: IF } n_{\text{eng}} = \text{very large} \quad \text{THEN } \hat{y}(k) = 0.4490\,u(k-4) + 0.5512\,\hat{y}(k-1)$$

The validity or normalized membership functions, respectively, are depicted in Fig. 22.26a. For a better understanding, the identified dead times and the time constants of the six rules are marked as crosses in Fig. 22.24b. Obviously, LOLIMOT+OLS can accurately describe the process, and the fuzzy rules are very transparent, giving insights into the process behavior.

Since different dead times are involved, the adjusted interpolation method according to (22.16) proposed in Sect. 22.3.3 should be used. Figure 22.26b illustrates the improvement achieved by the dead time adjusted interpolation method compared with the standard interpolation.

Two application examples of the dynamic LOLIMOT+OLS algorithm are presented in Sect. 25.3.

22.10 Summary and Conclusions

In this chapter, the local linear model tree algorithm was extended to dynamic systems. It was shown that local linear neuro-fuzzy models and in particular the LOLIMOT algorithm offer some important advantages over other approaches for

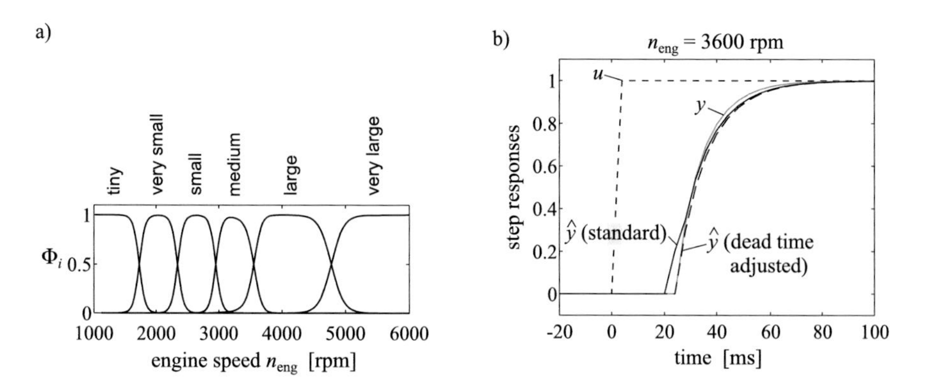

Fig. 22.26 (**a**) Validity functions constructed by LOLIMOT+OLS. (**b**) Step response of the process in comparison with the model with standard interpolation and dead time adjusted interpolation

identification of nonlinear dynamic processes. The equation error can be used for linear estimation of the local linear model parameters, while the output error can be utilized for structure optimization. The model structure represented by the rule premises allows one to incorporate prior knowledge by defining the variables that describe the operating point. Furthermore, the partitioning of the input space generated by LOLIMOT allows one to gain insights into the process structure. With different linearization strategies, a large variety of mature linear design techniques can be exploited for local linear neuro-fuzzy models in an elegant manner. For proving the stability of local linear neuro-fuzzy models, some powerful tools based on the solution of linear matrix inequalities are available, and when the parallel distributed compensation, controller design strategy is applied even closed-loop stability can be shown. Although these stability tests are conservative, they represent a distinct advantage over other model architectures where no such results are available. The simulation studies in Sect. 22.5 demonstrated the capability of LOLIMOT to identify processes of very different nonlinear structures. The consistency problem encountered in the equation error-based estimation of local ARX models can be overcome by the extension of the instrumental variables (IV) method and the output error (OE) and ARMAX models to local linear neuro-fuzzy models. With the introduction of orthonormal basis functions (OBF) to local linear neuro-fuzzy models, linear parameterized output error models can be designed if prior knowledge about the approximate process dynamics is available. Finally, a local orthonormal least squares (OLS) subset selection technique was proposed to automatically select the relevant inputs and the dynamic order of the local linear models in the rule consequents.

Additionally to the large number of benefits for *static* modeling (Sect. 13.4) obtained from local linear modeling schemes in general and the LOLIMOT algorithm and its extensions, in particular, the following particular advantages can be stated for *dynamic* models:

1. The *interpretability* of dynamic local linear neuro-fuzzy models is superior to that of many other model architectures since the local linear models can be understood as transfer functions with local gains, zeros, poles, and dead times, and thus analysis and synthesis methods for linear systems can be applied. The computational complexity increase from linear models to local linear neuro-fuzzy models is moderate. Local linear neuro-fuzzy models cover linear models, gain scheduled models, and parameter scheduled models as special cases.
2. The *distinction* between the rule *premises* and *consequents* on the one hand allows one to incorporate prior knowledge about the nonlinear structure of the process, and on the other hand it allows one to extract information about some unknown parts of the process structure with the LOLIMOT+OLS algorithm.
3. An *automatic order selection* is possible (with some limitations for the autoregressive regressors) by the combination of the LOLIMOT and the OLS algorithms. In contrast to all other sophisticated model structure optimization approaches available up to now, the LOLIMOT+OLS algorithm possesses the important advantage of *local* structure selection, i.e., the relevant inputs (regressors) are not determined globally for the overall model but rather in an operating regime dependent manner separately for each local linear model.
4. The capability of using *different loss functions for parameter and structure optimization* in the LOLIMOT algorithm allows one to combine two benefits. Linear optimization techniques can be exploited to estimate local ARX models, and nevertheless the simulation error, which in many applications is the actual measure of model quality, is utilized for model structure optimization. Furthermore, a generalization effect is involved by this strategy, which allows one to detect and avoid error accumulation and overfitting.
5. For dynamic local linear models, various *sophisticated and mature concepts known from linear system identification* literature can be applied, such as the instrumental variables method and orthonormal basis function approaches.
6. For *online learning* with recursive algorithms, many concepts that have been successfully developed and applied to linear adaptive control, such as excitation based control of the forgetting factor or design of a supervisory level [165, 323], can be extended to nonlinear adaptive models. First steps toward online learning of dynamic local linear neuro-fuzzy models can be found in [409]. They are significantly extended in [25, 145, 153, 158]; see also Sect. 28.2.
7. Powerful tools are available to prove the *stability* of local linear neuro-fuzzy models [555]. Moreover, closed-loop stability can be checked if a local linear neuro-fuzzy model-based controller is designed according to the parallel distributed compensation principle [554].
8. The *extrapolation behavior is dynamic,* and can be chosen by the user to be stable. This ensures high reliability of the model in practice where robust extrapolation is a very important issue.

Note that most of these features do not hold for other model architectures pursuing the external dynamics approach. For some internal dynamics approaches, the investigation of the model stability is much easier; see Chap. 23. This, however, does not extend to control design. All other properties do not hold for any internal dynamics model architecture either.

The dynamic LOLIMOT and the LOLIMOT+OLS algorithms have been and currently are utilized for the following applications:

- Nonlinear system identification of a cooling blast was carried out, which possesses strongly operating condition dependent dynamic characteristics [422].
- On the basis of this model, nonlinear PID controllers have been designed [151], and generalized predictive controllers suitable for fast sampled processes have been developed [152].
- A nonlinear dynamic model has been built for a truck Diesel engine turbocharger for a hardware-in-the-loop simulation [412, 434, 533].
- Concepts have been developed for nonlinear system identification and nonlinear predictive control of a tubular heat exchanger [236, 427, 429].
- Neural networks with internal and external dynamics are compared theoretically in [420]. In particular, the LOLIMOT+OLS algorithm is compared with an internal dynamics MLP network with locally recurrent globally feedforward structure for nonlinear system identification of a turbocharger and a tubular heat exchanger [284].
- Nonlinear gray box modeling and identification of a cross-flow heat exchanger can be found in [156, 424]. A complete gray box modeling and identification approach with the utilization of first principles, expert knowledge, and data is realized in [150].
- Based on these models, various predictive control concepts have been developed for temperature control of a cross-flow heat exchanger [148, 154]. Moreover, sophisticated online model adaptation strategies were designed in order to make the nonlinear predictive controller adaptive [145, 153, 155, 157, 158].
- Furthermore, the nonlinear dynamic models of the cross-flow heat exchanger were used as a basis for the development of fault detection and diagnosis schemes [24, 26, 27].
- A combination of online identification, nonlinear model-based predictive control, and fault detection and diagnosis methods have culminated in the integrated control, diagnosis, and reconfiguration of a heat exchanger [25].
- For modeling and identification of an electrically driven pump, a local linear neuro-fuzzy model trained with LOLIMOT complements a first principles model for improved accuracy.
- For modeling and identification of the dynamics of a vehicle, a hybrid model has been developed consisting of a first principles dynamic physical model in combination with a dynamic local linear neuro-fuzzy model trained with LOLIMOT [209, 257].
- A nonlinear dynamic model has been developed for the cylinder individual air/fuel ratio λ of a spark-ignition engine.
- For the exhaust gas recirculation (EGR) flow of a spark-ignition engine, a nonlinear dynamic model-based on the engine speed, the throttle angle, and the

EGR valve position has been implemented. This model was utilized by a newly designed EGR flow feedback controller.
- A nonlinear dynamic model has been developed for the load of a spark-ignition engine with and without exhaust gas recirculation.
- Modeling and identification of a variable nozzle turbocharger have been carried out. The nonlinear dynamic local linear neuro-fuzzy model is trained with both static and dynamic measurement data and describes the relationship between the charging pressure and the Diesel engine injection mass, speed, and the actuation signal for the turbine inlet guide vanes.
- Finally, LOLIMOT finds a number of applications in industry that are realized with the LOLIMOT MATLAB Toolbox [421]. These industrial applications are currently mainly in the field of automotive systems.

22.11 Problems

1. Explain the terms "static linearization" and "dynamic linearization."
2. In which way can the prediction error and the simulation error both be considered in the LOLIMOT algorithm?
3. Why is it often possible to get by with a premise input space of reduced dimensionality for dynamic models?
4. Consider two or more stable linear models whose parameters are interpolated. What can be concluded for the stability of the interpolated model if the models are of first, second, third, or higher order?
5. What can be concluded in Problem 3 if the interpolation depends on the state of the model itself as it is the case for a simulation model where previous outputs fed back?
6. How can the stability properties of the interpolated model be checked?
7. How does the (ideal) partitioning of the premise input space performed by LOLIMOT look like in the case of a Hammerstein, Wiener, and NDE system type? Explain why.
8. How can the structure, i.e., the relevant regressors, for the rule consequents be determined automatically in an efficient manner? What are the main obstacles for such an automatic structure selection in the context of dynamic systems?

Chapter 23
Neural Networks with Internal Dynamics

Abstract This chapter covers neural networks with internal dynamics. The traditional ways to incorporate dynamics into the networks are presented, and some of their properties analyzed. The breakthroughs in machine learning for sequence learning with long short-term memory (LSTM) networks motivate to look at this architecture in an engineering context. The network structure is visualized in a block scheme manner, which makes its characteristics more obvious from a control engineering point of view. It is also compared with a nonlinear state space formulation where similarities and differences are pointed out. Finally, internal and external dynamics approaches are compared with their advantages and drawbacks.

Keywords Internal dynamics · Recurrent network · LSTM · Nonlinear state space

The internal dynamics approach realizes a nonlinear state space model without information about the true process states. Consequently, the model's internal states can be seen as a somewhat artificial tool for the realization of the desired dynamic input/output behavior. Models with internal dynamics are most frequently based on an MLP network architecture. A systematic overview of internal dynamics neural networks can be found in [570]. Basically, four types can be distinguished: *fully recurrent* networks (Sect. 23.1) [601]; *partially recurrent* networks (Sect. 23.2) with the particular well-known Elman [134] and Jordan [308] architectures; *nonlinear state space* networks (Sect. 23.3) proposed by Schenker [508]; and *locally recurrent globally feedforward* networks (Sect. 23.4) systematized by Tsoi and Back [570]. Finally, the major differences between internal and external dynamics are analyzed in Sect. 23.6. An overview of and a comparison between the external and internal dynamics approaches on the basis of their fundamental properties are given in [139, 419, 420]. For additional case studies, refer to [284].

Because about 90% of the literature and applications of neural networks to nonlinear system identification focuses on external dynamics, the goal of this chapter is only to give a brief summary of the state of the art and the fundamental features of the internal dynamics approach.

O. Nelles, *Nonlinear System Identification*,
https://doi.org/10.1007/978-3-030-47439-3_23

For the training of these neural networks, the backpropagation-through-time or the real-time recurrent learning algorithms discussed in Sect. 19.5 can be applied. The gradient calculations have to be performed individually for each network structure. Although the gradients may look quite complicated, and their derivation can be tedious, they follow from a straightforward application of the chain rule.

23.1 Fully Recurrent Networks

Williams and Zipser [601] developed a fully recurrent neural network consisting of M fully connected neurons with sigmoidal activation functions (the same type of neurons as used in a multilayer perceptron), p inputs, and r outputs. Each link between two neurons represents an internal state of the model. With reference to Fig. 23.1a, the resulting structure is not organized in layers, and clearly such a network has no feedforward architecture. Originally, the fully recurrent network has been suggested for sequence recognition tasks, but because of their nonlinear dynamic behavior, these fully recurrent networks can be used for the identification of nonlinear dynamic systems, too. In [570] the disadvantages of this architecture are denoted by slow convergence of the training algorithms and stability problems. In general, this architecture seems to be too complex for a reliable practical implementation. Furthermore, the fixed relationship between the number of states and the number of neurons does not allow one to adjust separately the dynamic order of the model and the flexibility of the nonlinear behavior. Therefore, fully recurrent networks are rarely used for nonlinear system identification tasks.

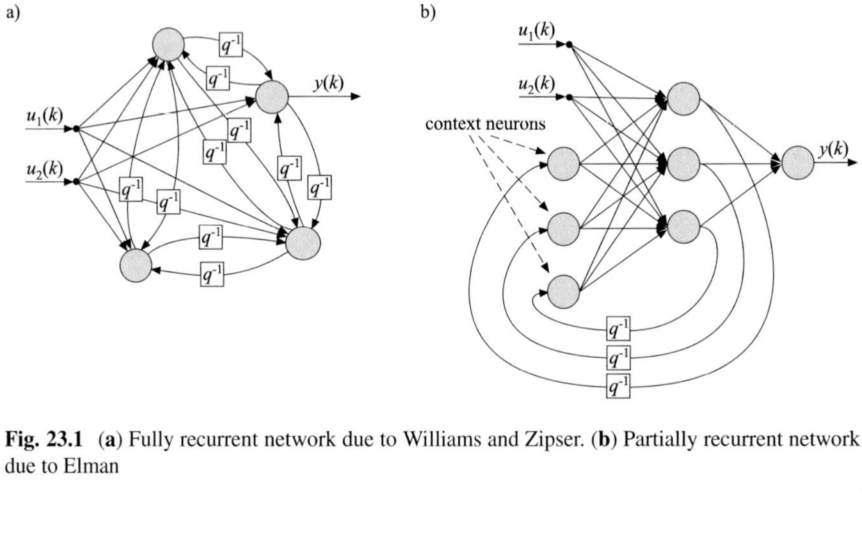

Fig. 23.1 (**a**) Fully recurrent network due to Williams and Zipser. (**b**) Partially recurrent network due to Elman

23.2 Partially Recurrent Networks

Unlike to the fully recurrent structures, the architecture of partially recurrent networks is based on feedforward multilayer perceptrons containing an additional so-called context layer; see Fig. 23.1b. The neurons of this context layer serve as internal states of the model. Elman [134] and Jordan [308] proposed partially recurrent networks where feedback connections from the hidden and the output layer, respectively, are fed to the context units. The partially recurrent architectures possess the important advantage over the fully recurrent ones that their recurrency is more structured, which leads to faster training and fewer stability problems. Nevertheless, the number of states (i.e., the dynamic order of the model) is still related to the number of hidden (for Elman) or output (for Jordan) neurons, which severely restricts their flexibility. Extended Elman and Jordan networks additionally implement recurrent connections from the context units to themselves. The outputs of the context neurons represent the states of the model, and they depend on previous states and previous hidden (for Elman) or output (for Jordan) neuron outputs. Partially recurrent networks have first been suggested for natural language processing tasks. In comparison with fully recurrent networks, they possess better convergence and stability problems. However, a lot of trial and error is still required to successfully train such an architecture.

23.3 State Recurrent Networks

The most straightforward realization of the internal dynamics approach is the direct implementation of a nonlinear state space model as proposed by Schenker [508]. Figure 23.2 shows that this approach looks similar to an external dynamics configuration. The important difference is, however, that for external dynamics the *outputs* are fed back, which are *known* during training, while for the structure in Fig. 23.2, the *states* are fed back, which are *unknown* during training.[1] As a consequence, a state recurrent structure can be trained only by minimizing the simulation error.

The major advantages of the state recurrent structure over the fully and partially recurrent structures discussed above are as follows:

- The number of states (dynamic model order) can be adjusted separately from the number of neurons. Although the networks tend to become more complex as the number of states increases, since each state is a network input and thus causes additional links, the number of hidden neurons can be separately determined by the user.

[1] If the full state is measurable, the problem simplifies to the approximation of the nonlinear state space mappings $\underline{f}(\cdot)$ and $g(\cdot)$ in (19.6b), and no feedback is required at all; see Sect. 19.1.

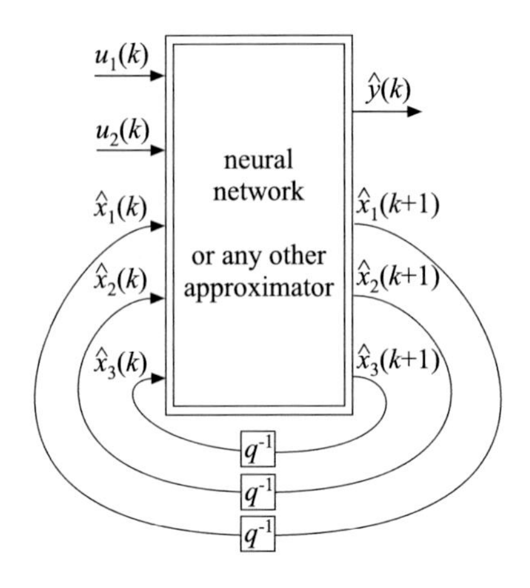

Fig. 23.2 State recurrent network

- The model states act as network inputs and thus are easily accessible from outside. This can be utilized when state measurements are available at some time instants, e.g., the initial conditions. This is particularly interesting and important for batch processes in the chemical industry, as pointed out by Schenker [508], but may be unrealistic for other domains.
- Owing to the state space structure, it may be possible, as proposed in [508], to incorporate state space models obtained by first principles within the neural network and let the neural network learn or compensate only for the unmodeled process characteristics.
- As indicated in Fig. 23.2, any type of nonlinear static approximator can be utilized instead of the multilayer perceptron used in [508]. Note, however, that linearly parameterized models offer no distinct advantages here because the state feedback makes the parameter optimization problem nonlinear anyway.

It should be underlined that the state recurrent approach seems to be more promising than the fully and partially recurrent structures. However, in practice many difficulties can be encountered, in particular, if no state measurements (and no initial conditions) are available:

- The states of the model do not approach (or even relate to) the true process states.
- Wrong initial conditions deteriorate the performance when only short training data sets are available and/or the dynamics are very slow.
- Training can become unstable.
- The trained model can be unstable.

These drawbacks are characteristic of the internal dynamics approach. They can be partly overcome by the architecture discussed in the next section.

23.4 Locally Recurrent Globally Feedforward Networks

The architecture of locally recurrent globally feedforward (LRGF) networks proposed by Tsoi and Back [570] and also extensively studied by Ayoubi in [14] is based on static feedforward networks that are extended with local recurrence. This means there are neither feedback connections between neurons of successive layers nor lateral connections between neurons of one layer. Recurrency is always restricted to the links (synapses) or to the neurons themselves. The dynamics can be introduced in the form of finite impulse response filters (FIR) or infinite impulse response filters (IIR). It is important to note here that the gain of the filters should be normalized to 1 in order to avoid redundant parameters because the neural network weights already realize a scaling factor (gain) with their weights. One motivation for the LRGF approach stems from the fact that these architectures include Hammerstein and Wiener model structures as special cases; see Sects. 20.6 and 20.7. As shown in Fig. 23.3, three types of local recurrency can be distinguished:

- *Local synapse feedback*: Instead of a constant weight, each synapse incorporates a linear filter.
- *Local activation feedback*: This kind of local recurrency represents a special case of local synapse feedback. If the transfer functions of all synapses leading to one neuron are identical, the resulting structure can be simplified. Then all filters in the synapses can be replaced by one filter with the same poles and zeros behind the summation of the neuron input.

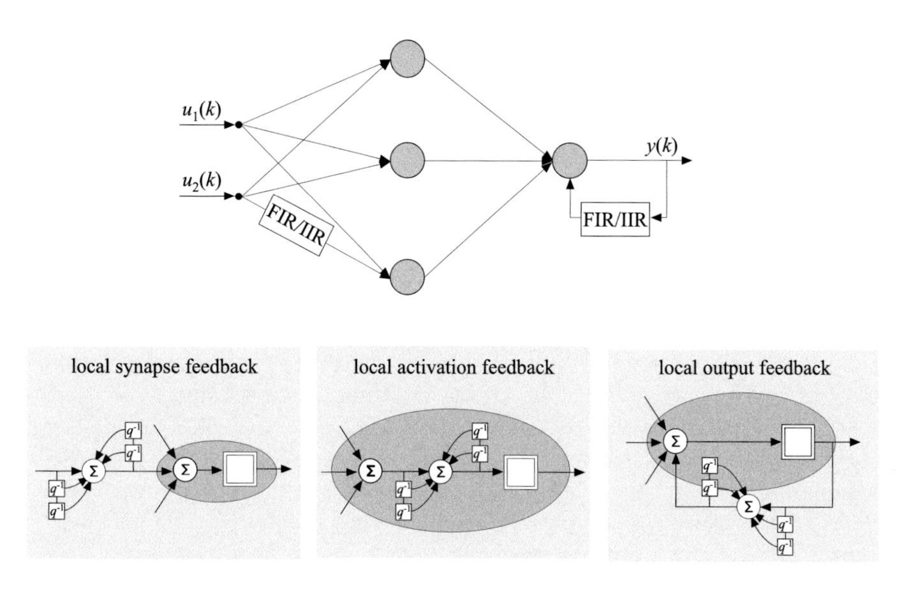

Fig. 23.3 Locally recurrent globally feedforward network with three alternative strategies for incorporation of the linear filters

- *Local output feedback*: Local output feedback models possess a linear transfer function from the neuron output to the neuron input.

An important difference between these different types of LRGF recurrency is the separation into nonlinear statics and linear dynamics for local synapse and activation feedback on the one hand and the non-separable nonlinear dynamics for local output feedback on the other hand.

Typically, multilayer perceptrons are extended to LRGF architectures; however, the idea of LRGF can also be applied to radial basis function networks. The only reasonable way to extend an RBF network (Sect. 11.3) to internal dynamics is to replace the constant weights in the output layer by linear filters. This internal dynamics RBF network architecture is, however, of limited flexibility since the nonlinear mapping is based solely on the static network inputs. It is shown in [15] that this internal RBF network is only able to identify Hammerstein model structures.

An advantage of LRGF networks, in general, is that their stability is much easier to check, owing to the merely local recurrency. If all filters within the network are stable, then the complete network is stable (except when local output feedback is applied, which requires one to take the gain of the activation function into account). On the other hand, the local recurrency restricts the class of nonlinear dynamic systems that can be represented. Another problem with the LRGF approach is that the resulting model will generally be of high dynamic order even when the filters are of low order since transfer functions with different denominator dynamics are additively combined. This fact may cause undesirable dynamic effects on the generalization behavior. Often non-minimum phase models are identified even if the process is minimum phase. The choice of the linear filters is generally based on previous experience that suggests IIR filters of second order are sufficient. With first-order filters, oscillatory behavior could not be modeled, and for higher than second order, the required additional parameters usually do not pay off [14].

23.5 Long Short-Term Memory (LSTM) Networks

Complex internal dynamics neural networks severely suffer from a handicap called *vanishing gradient problem*. During the gradient calculations with backpropagation-through-time (compare Sect. 19.5.1), the influence of "old" inputs, i.e., inputs many time steps into the past, decays exponentially. Therefore the gradients become exponentially smaller, and practically it is impossible to learn long-term relationships. The reason for this effect is that any stable (first-order) filter behaves like $x(k) = f \cdot x(k-1)$ with $f < 1$. In fact, it can also happen that the gradient "explodes" if $f > 1$ which corresponds to an unstable filter. Both effects make learning long-term relationships very difficult.

In 1997 Hochreiter and Schmidhuber proposed a new network architecture to circumvent this difficulty [252]. It is called *long short-term memory (LSTM)*.

The following intuitive overview of the basics of recurrent neural networks is recommended: [354]. A recent meta-study which compares various variants of LSTMs can be found in [195]. They have been applied with extreme success in areas like natural language translation or hybrid systems where discrete and continuous states matter. Their applicability or suitability to system identification problems as discussed in this book is a question for future research. First encouraging results for LSTMs in the system identification world are available [436, 520]. However, they are hampered by huge amounts of parameters and consequently huge amounts of required training data and time.

A hidden layer neuron is displayed in Fig. 23.4 in a manner known from block diagram schematics used in controls. This significantly deviates from the typical drawings shown in machine learning literature. The key idea of LSTMs is to build a memory cell into each neuron with a state $s(k)$ which allows to model first-order dynamics. With the input gate, the input into this memory cell can be controlled by multiplication. The gate activity is always between 0 and 1 since it is the output of a sigmoid. Correspondingly the output gate controls which fraction of the state is fed to the neuron output $h(k)$.

Note that this framework is slightly different from the standard controls formulation of the nonlinear state space description introduced in Sect. 19.1. In this section the state is denoted by $\underline{s}(k)$ (computer science notation) and $\underline{h}(k)$ rather than $\underline{x}(k)$ (controls notation). In LSTMs and other recurrent neural network architectures, it is common to influence the state $\underline{s}(k)$ and $\underline{h}(k)$ directly by the input $\underline{u}(k)$ (direct feedthrough), while in controls the state depends on the delayed input $\underline{u}(k-1)$; more exactly $\underline{x}(k+1)$ depends on $\underline{u}(k)$ [520]. In controls, the direct feedthrough is realized in the output equation where $\underline{u}(k)$ enters directly in cases where direct feedthrough is present. On the other hand, because LSTMs lack this direct feedthrough in the output equation, an additional artificial state might be required to represent $\underline{u}(k)$ inside $\underline{s}(k)$. For a more detailed discussion on the relationship between LSTM networks and state space representations refer to [519].

With respect to the internal neuron dynamics, the decisive part happens through the forget gate. It controls the pole of the first-order system inside. Figure 23.5 illustrated how this is typically presented in the standard literature. The output of the forget gate $f(k)$ is equivalent to the pole. Since $f(k)$ comes out of a sigmoid, only stable first-order dynamics result – or in the limit for $f(k) \approx 1$ marginally stable systems, i.e., integral behavior. The latter can be utilized to store information for an arbitrarily long time because $s(k) \approx s(k-1)$ which gives LSTMs their name.

It is very important to note, however, that the internal dynamics *inside* each neuron is not the only type of feedback in LSTMs. All hidden neuron outputs are fed back with one-time step delay to all hidden layer neurons of the same layer. This is illustrated in Fig. 23.6.

Typical applications of LSTMs deal with huge networks with 100,000 s to millions of weights. In order to keep training times in a realistic order of magnitude, parallel computing and graphics processing units (GPUs) are exploited heavily.

LSTMs correspond to an extended nonlinear state space representation (refer to Sects. 19.1 and 19.3) where each neuron realizes one classical state variable as it

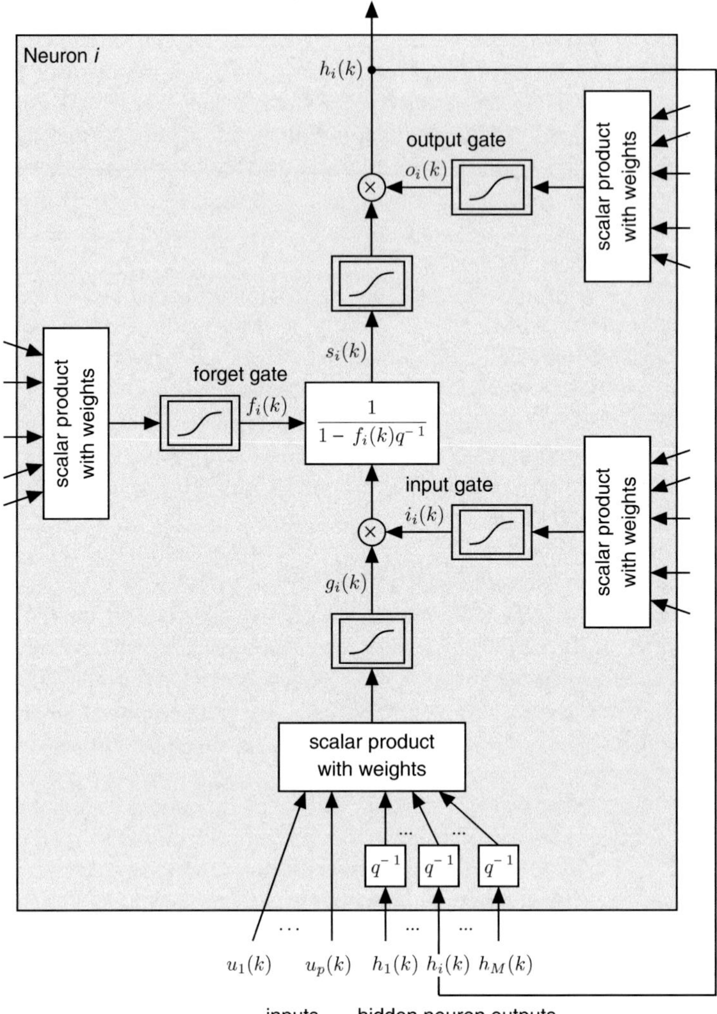

Fig. 23.4 One out of M LSTM neurons in a hidden layer. Each neuron processes the network inputs $u_1(k), \ldots, u_p(k)$ and the previous (i.e., delayed) neuron outputs from the same layer $h_1(k-1), \ldots, h_M(k-1)$. Like in an MLP network (ridge construction), these neuron input quantities are weighted and summed up (plus an additional offset) to form a scalar quantity indicated with the blocks "scalar product with weights." Four individual set of weights are used for the three gates and the input paths. The $p + M$ neuron inputs enter four times into the "scalar product with weights" blocks. The sigmoidal blocks at the gates are always chosen as sigmoids between 0 (inactive) and 1 (active), while for the computation of $g_i(k)$ and $h_i(k)$ in modern times, often tanh functions between -1 and 1 are chosen

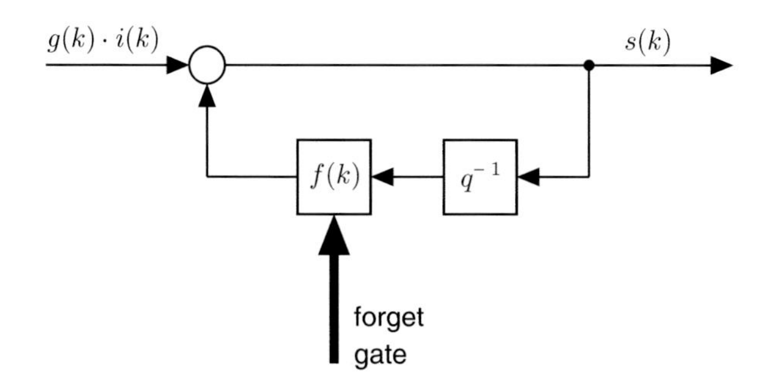

Fig. 23.5 Internal LSTM neuron dynamics which corresponds to a first-order filter with a pole at $f(k)$. For $f(k) = 1$ it is called a *memory carousel* which corresponds to an integrator in controls language

is fed back to the same layer's inputs over memory units (z^{-1} blocks). In addition, each neuron possesses its own individual memory unit which realizes a first-order dynamics. In summary, an LSTM network corresponds to a nonlinear state space model with as many states as neurons and each state equation; see Fig. 23.7. For more details and first benchmark, investigations, refer to [520].

23.6 Internal Versus External Dynamics

This section compares the external with the internal dynamics approach. Of course, such a comparison can only be quite rough, since many details depend on the specific type of architecture and training algorithm chosen. Nevertheless, some main differences between both approaches can be pointed out:

- *Training*: The feedback involved in the internal dynamics approach yields a nonlinear output error (NOE) model (for LRGF this is only true if IIR filters are used); see Sect. 19.2.3. This implies that nonlinear parameter optimization techniques have to be applied independently of whether the utilized function approximator is linear or nonlinear parameterized. Furthermore, the gradient calculations carried out with the real-time recurrent learning or backpropagation-through-time algorithm can become tedious and time-consuming. So some authors employ direct search methods to avoid any gradient calculation [508]; see Sect. 4.3. Additionally, the training procedure can become unstable.

  In contrast, with the external dynamics approach, the NARX model structure can be utilized to exploit linear relationships. Thus, the use of linear parameterized model architectures does make much more sense in combination with the external dynamics approach. The price to be paid for this advantage is that the one-step prediction error rather than the simulation error is optimized. This is not identical to the actual objective when the intended model use is a simulation.

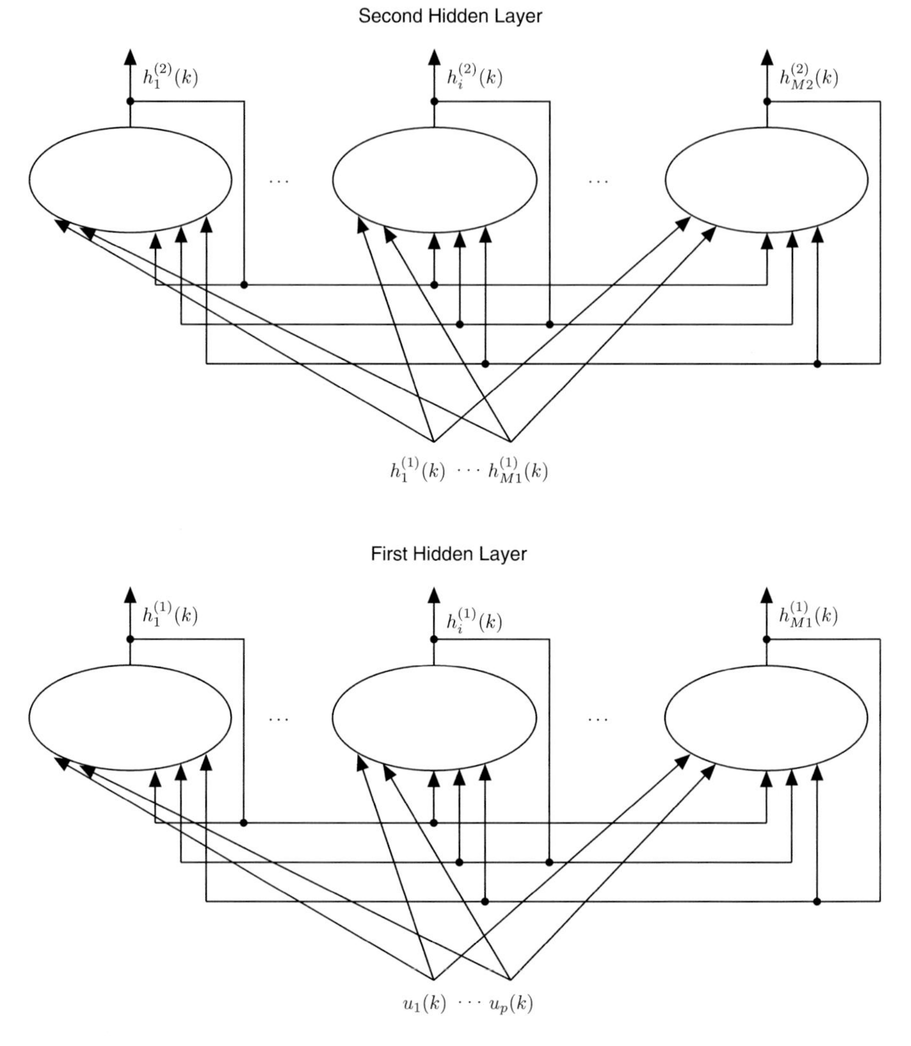

Fig. 23.6 Two hidden layers of an LSTM network. This first hidden layer has $M1$ neurons; the second hidden layer has $M2$ neurons. Often $M1 = M2$

- *Generalization*: Owing to the internal feedback, internal dynamics models inherently perform a simulation. They usually cannot be used for one-step prediction since the previous process states cannot be fed into the model (with the exception of the state recurrency network if it is able to reconstruct the process states within the network[2]). Internal dynamics models can only represent stable

[2]This can hardly be expected if no state measurements are available for training.

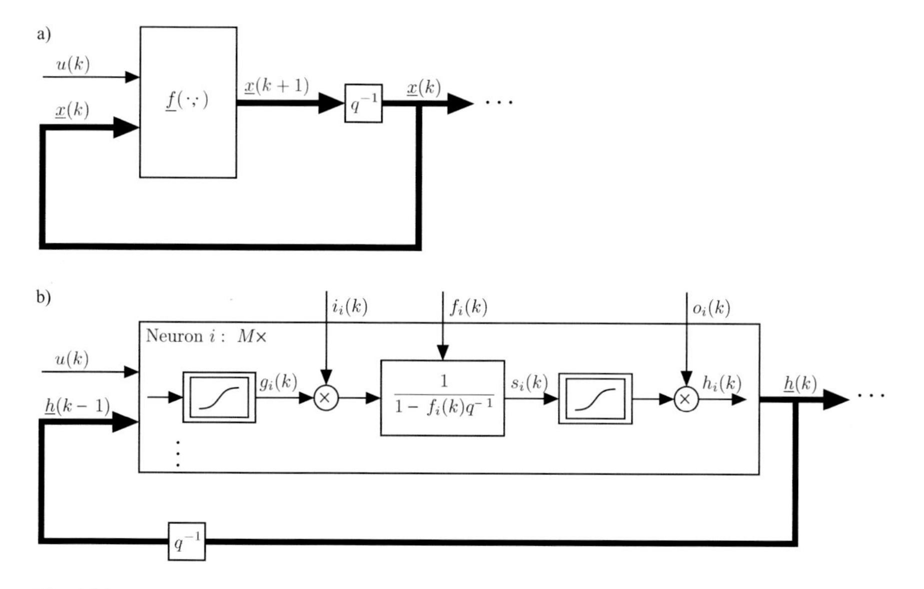

Fig. 23.7 (**a**) Nonlinear state equations: number of states is $\dim\{\underline{x}(k)\}$. (**b**) One layer LSTM network with M neurons in state-space-like representation: number of states is $2\dim\{\underline{h}(k)\} = 2M$ because each neuron i possesses one fed back state $h_i(k)$ and additionally one internal state $s_i(k)$ whose time constant is scheduled by the forget gate $f_i(k)$ and whose input and output are scheduled by the input gate $i_i(k)$ and output gate $o_i(k)$, respectively. While in (**a**) the mapping from $\{\underline{x}(k), u(k)\} \rightarrow \underline{x}(k+1)$ is static, in (**b**) the mapping from $\{\underline{h}(k-1), u(k)\} \rightarrow \underline{h}(k)$ is filtered by a first-order nonlinear filter with pole $f_i(k)$ for neuron i. While in (**a**) no direct feedthrough from $u(k)$ to $\underline{x}(k)$ exists in the state equations (if necessary it is realized by the subsequent output equation), in (**b**) there exists a direct feedthrough path from $u(k)$ to $\underline{h}(k)$

processes, while the external dynamics approach would allow unstable when trained in the NARX structure.

- *Dynamic order*: The choice of the dynamic order for internal dynamics models is not as explicit (the state recurrent networks are an exception in all that follows under this point) as for the external dynamics approach. Usually, the order of the model is somehow related to the network structure and the number of neurons. Internal dynamics approaches, therefore, possess a drawback if the true dynamic order of the process is known. If it is unknown, however, they are able to cope with a variety of process orders automatically. In other words, the internal dynamics approach is much more robust (or less sensitive) with respect to the information about the process order. So the internal dynamics approach does not necessarily require an assumption about the order of the process dynamics. It is more like a "black box" than the external dynamics approach and thus offers an advantage whenever knowledge about the process order does not exist and vice versa.
- *Stability*: For the fully and partially recurrent and also for the state feedback structures, stability is as hard to prove as for the external dynamics approach

(except that of local linear neuro-fuzzy models, which offer some advantages in this respect; see Sect. 22.4). The internal dynamics approaches with LRGF, however, allow an easy check for stability by simple investigation of the poles of their linear filters.

- *Curse of dimensionality*: The curse of dimensionality is much weaker for the internal dynamics approach because the network is fed only with the actual input signal and possibly with the actual states but not with previous inputs and outputs. Thus, the input space is of lower dimensionality. This is a clear advantage for the internal dynamics approach and in particular when higher-order systems and multivariable systems are handled.
- *Interpretability*: Owing to the higher complexity and the interaction between neurons and dynamics, any interpretation of internal dynamics neural networks is almost impossible. The LRGF architecture allows one to gain some insights into the process dynamics and structure for very simple cases [14]. For the external dynamics approach, interpretability depends strongly on the chosen network architecture. The local linear neuro-fuzzy models offer by far the best interpretation possibilities; see Chap. 22. The missing interpretability is a drawback for internal dynamics.
- *Prior knowledge*: The possibility for incorporation of prior knowledge is related to the interpretability of the model. Thus, few such possibilities exist for internal dynamics networks (with the partial exception of state recurrent networks).

23.7 Problems

1. Explain the fundamental differences between neural networks with external and internal dynamics.
2. Sketch a fully recurrent network with two inputs, one output, and three neurons that are connected by single delay units.
3. Sketch a partially recurrent network duc to Elman with two inputs, one output, three hidden layer neurons, and two context neurons.
4. Sketch a state recurrent network with two inputs, one output, and two states.
5. Sketch some alternatives for a locally recurrent globally feedforward network with two inputs, one output, and three hidden neurons.

Part IV
Applications

Chapter 24
Applications of Static Models

Abstract This chapter deals with static real-world applications. It covers the nonlinear smoothing of a driving cycle. The goal is to generate a smooth version that ensures a given tolerable deviation is not exceeded. The second application concerns modeling of exhaust gases of a combustion engine. Here static models are built, but an outlook to dynamic modeling is given as well. The models are used for the optimization of feedforward controllers for an engine control unit (ECU).

Keywords Application · Driving cycle · Look-up table · Combustion engine · Exhaust gas · ECU

This chapter is the first of three that present some real-world application examples to illustrate the practical usefulness and wide-ranging applicability of modern nonlinear system identification approaches. These chapters focus on local linear neuro-fuzzy models and the LOLIMOT training algorithm because this combination revealed very promising features in the preceding analysis. Clearly, a thorough comparison with other model architectures and algorithms is an interesting topic for future studies.

This chapter deals with static modeling problems. In Sect. 24.1 it is demonstrated how nonlinear models can be used for the adaptive filtering of a reference signal. The task is to smooth a driving cycle that is used for standardized exhaust gas tests while meeting given accuracy requirements in the form of constraints. Section 24.1.4 is concerned with modeling and optimization of exhaust gases for combustion engines. This is an important topic of current research in the automotive industry in order to fulfill the ever-increasing demand to reduce pollution and fuel consumption. The overall problem is emphasized from a system-wide perspective, and the important role that modeling and identification play for the systematic achievement of the ultimate goal is illustrated.

O. Nelles, *Nonlinear System Identification*,
https://doi.org/10.1007/978-3-030-47439-3_24

24.1 Driving Cycle

This section presents the application of local linear neuro-fuzzy models trained with LOLIMOT to a one-dimensional static approximation problem. The proposed methodology can also be seen as acausal nonlinear filtering of signals. This application underlines the following important features of LOLIMOT:

- efficient training of very complex models with more than 100 neurons;
- incorporation of constraints by means of an appropriate objective for structure optimization;
- interpretation and information extraction from the local linear models' parameters.

The presented results have been partly published in [515].

24.1.1 Process Description

Driving cycles are standardized testing procedures in automotive applications to evaluate certain properties of vehicles. For example, the fuel consumption and the amount of exhaust gases of different engines and vehicles can be compared. In this way, regulations by law enforce an upper bound on emissions. Here, the FTP 75 cycle (FTP = Federal Test Procedure) shown in Fig. 24.1, which is used for passenger cars worldwide, is considered [119].

At present, driving cycles are mostly driven by skilled human drivers, who are able to meet the tolerance band that is shown in Fig. 24.1 for the FTP 75. In order to automate this procedure and improve the reproducibility, the human driver has to be replaced by an automatic controller. The scheme in Fig. 24.2 illustrates how the driving cycle can be preprocessed and subsequently utilized by the automatic speed controller, which controls either a real or a simulated car. The preprocessing phase that maps the driving cycle to a smoothed speed profile is motivated and explained in the following.

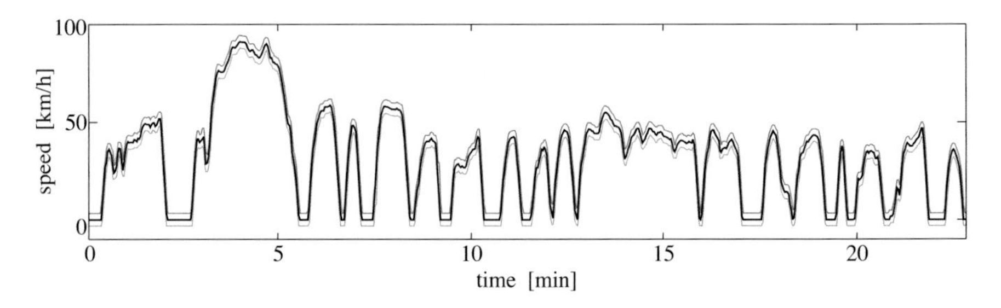

Fig. 24.1 FTP 75 driving cycle with tolerance band

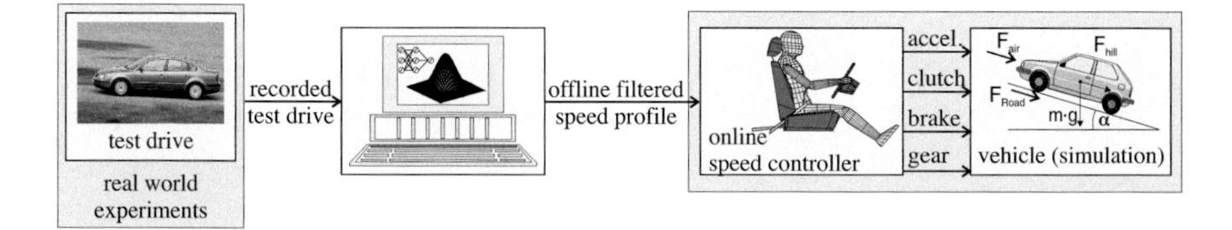

Fig. 24.2 A driver simulation requires a speed profile that is obtained from the driving cycle by offline filtering. (This figure was kindly provided by Martin Schmidt, Institute of Automatic Control, TU Darmstadt)

It is difficult to design a controller that solves this tracking task within the tolerance band. Clearly, the a priori knowledge about the speed reference signal should be exploited, and the derivative of the reference signal is required for a feedforward component in the controller. Consequently, the reference signal should be as smooth as possible in order to be able to design a fast controller that is able to meet the tolerances. Since the smoothed speed signal will be used as the reference signal for the control loop, the tolerance band for the smoothing has to be tightened in order to allow for some control errors. The smoothing of the driving cycle by a standard (linear and non-adaptive) filter is not successful since the tolerance band requires a time-dependent filter bandwidth. In the ideal case, the maximum smoothing effect should be achieved locally with the constraint that the tolerances are met. LOLIMOT is a well-suited tool for solving this task.

24.1.2 Smoothing of a Driving Cycle

A one-dimensional local linear neuro-fuzzy model

$$y = f(u) \tag{24.1}$$

with time as the input $u = t$ and speed as the output $y = v$ is constructed with LOLIMOT.

The iterations of LOLIMOT are terminated when the model meets the tolerance band. This ensures the realization of the minimal number of local linear models and therefore the maximum smoothing effect. As proposed in Sect. 13.3.2, the following objective function may be used for structure optimization in order to make the LOLIMOT construction more goal-oriented:

$$I_i^{(\mathrm{tol})} = \sum_{j=1}^{N} \mathrm{tol}_j \cdot \Phi_i(\underline{u}(j)) \tag{24.2}$$

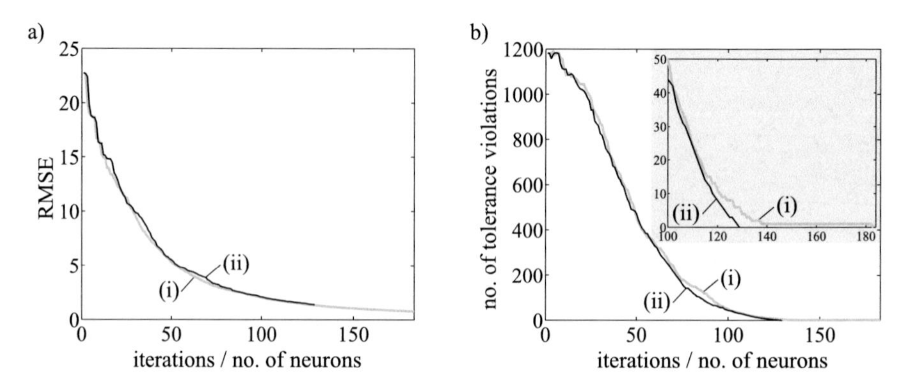

Fig. 24.3 Convergence behavior for driving cycle approximation: (i) standard LOLIMOT, (ii) LOLIMOT with (24.2) as objective for structure optimization. (**a**) RMS error (**b**) number of violations

where $\text{tol}_j = 1$ if the model violates the tolerance band for data sample j and $\text{tol}_j = 0$ otherwise. The use of (24.2) guarantees that new local linear models are generated only where the tolerance band is violated, independent of the model quality in terms of model errors. Figure 24.3 compares this approach with the standard LOLIMOT algorithm, which employs a sum of squared error loss function for structure optimization. As Fig. 24.3a shows, the root mean squared error of the fit is slightly better for the standard approach. However, the number of tolerance violations depicted in Fig. 24.3b demonstrates the advantage of using (24.2). While standard LOLIMOT (i) yields a model with $M = 183$ neurons, the algorithm with objective (24.2) for structure optimization (ii) leads in shorter training time to a significantly simpler model with $M = 128$, which represents a much smoother speed profile.

The performance of the obtained model (ii) is depicted in Fig. 24.4. The local linear neuro-fuzzy model successfully smooths the driving cycle while concurrently meeting the tolerance band. In the interval, $t \approx 1$–1.8 min, very few local linear models have been generated since the speed profile and the tolerance band basically allow one to draw a straight line in this area. In contrast, for the interval $t \approx 0.5$–1 min, more local linear models are required in order to describe the oscillations. The model smooths the driving cycle in a similar fashion as probably a human driver does in his mind. Note that the density of local linear models does depend only on the tolerance band and not on the approximation error.

24.1.3 Improvements and Extensions

The original driving cycle as shown in Fig. 24.1 is given in samples of $T_0 = 1$ s [119]. For the driver and vehicle simulation, however, much smaller sampling times are required. Since the local linear neuro-fuzzy model represents a continuous

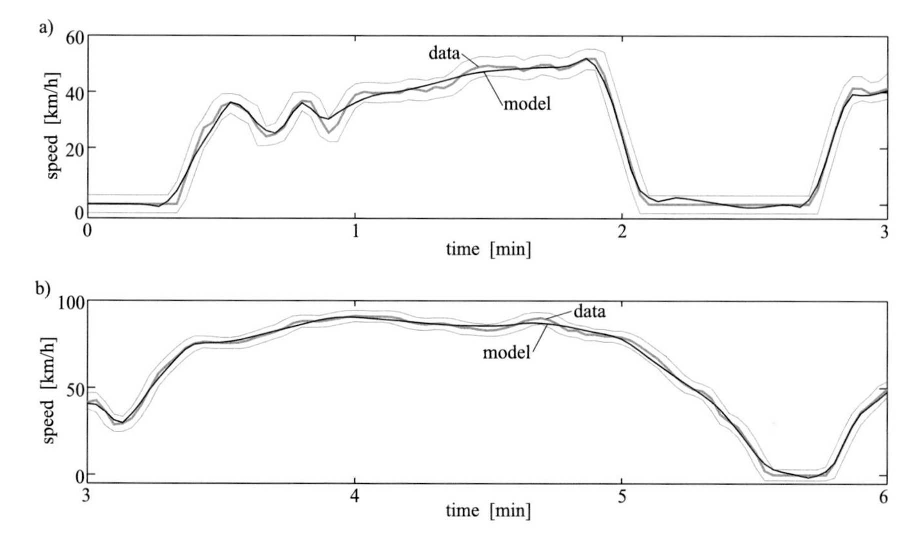

Fig. 24.4 Local linear neuro-fuzzy model (ii) for smoothing of the driving cycle. (**a**) first section, (**b**) second section of driving cycle

function, it can be generalized at any time instant t, and therefore a resampling with a higher sampling frequency can easily be performed. Note that this would not be possible with a digital adaptive filter, which would require an additional interpolation algorithm of a high order to generate data points between the original samples.

It can be observed in Fig. 24.4 that the model does not behave well in regions where the original driving cycle possesses idle phases with $v = 0\,\text{km/h}$, as in the intervals $t \approx 2.1$–$2.7\,\text{min}$ and $t \approx 5.6$–$5.8\,\text{min}$. The model tends to oscillate in these regions, and the model output even becomes partly negative ($<0\,\text{km/h}$). The reason for this undesirable behavior is the abrupt slope change of the original cycle. In order to avoid these unwanted effects, it is reasonable to extract the parts of the driving cycle that lie between idle phases and to model those parts separately. In a second step, these submodels are joined together by including the idle phases from the original cycle between them. This procedure ensures that the submodels are not deteriorated from the idle phases.

24.1.4 Differentiation

As has been mentioned above, the derivative of the driving cycle is required for the feedforward controller, which certainly has differential behavior since it (partly) inverts the vehicle dynamics that possess time-lag behavior. Figure 24.5 demonstrates the improvement achieved by differentiating the smoothed reference

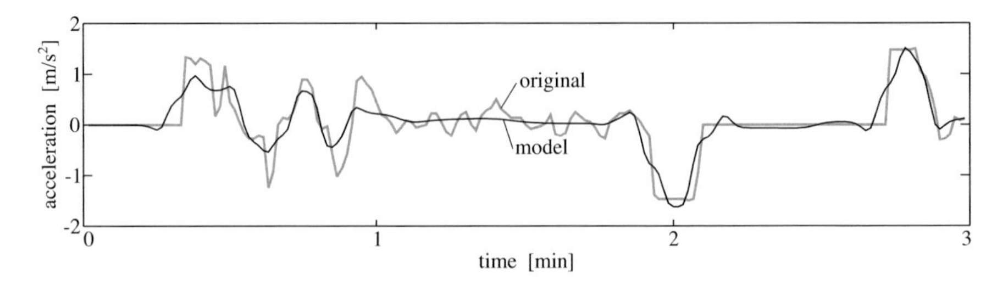

Fig. 24.5 Derivatives of the driving cycle (original) and the smoothed speed profile (model)

signal instead of the original one. Note that the derivative of the model can be evaluated according to the local linearization approach introduced in Sect. 14.5. This local derivative is equivalent to the slopes of the local linear models, which can be directly interpreted as accelerations. Figure 24.5 compares the derivative of the original driving cycle and the local derivative of the neuro-fuzzy model. The accelerations obtained from the model are much smoother, and their absolute value is smaller in the mean. This leads to more economical control in terms of fuel consumption and exhaust gases.

A similar strategy as demonstrated above is utilized online and in real time in the automotive application of nonlinear adaptive filtering of wheel speed sensor signals [522, 524].

tionModeling and Optimization of Combustion Engine Exhaust This section demonstrates how multivariable nonlinear static models can be utilized in automotive electronics to cope with the ever-increasing complexity of modern combustion engines. The purpose of this section is to illustrate the importance that models have for efficient design procedures. The following features of local linear neuro-fuzzy models and the LOLIMOT training algorithm are addressed:

- fast training of static approximators with many inputs;
- usefulness of user-defined extrapolation behavior;
- smoothness of the model, allowing it to be used successfully for optimization.

The presented results have been partly published in [203, 205].[1]

24.1.5 The Role of Look-Up Tables in Automotive Electronics

The automotive electronics industry is characterized by a huge number of units sold (hundred thousands to millions). Therefore, the saving of some cents by using

[1]Most of the results and figures presented in this section have been taken from the research work of Michael Hafner and Matthias Schüler, Institute of Automatic Control, TU Darmstadt.

cheaper hardware can be profitable even if it requires a man-year of engineering effort. Consequently, the computational possibilities are usually severely limited.

Today, grid-based look-up tables are by far the most common way to describe nonlinear relationships in automotive electronics. These look-up tables are restricted to one or two inputs,[2] and their parameters (i.e., the choice of the grid and the heights) are usually not optimized but directly obtained from measurement data or by manual manipulation after inspection. The reasons for the dominance of look-up tables in this application area are as follows (see Sect. 10.3):

- The measurement data can be directly stored in the look-up table without the application of structure or parameter optimization techniques.
- The evaluation speed (access time) of look-up tables is fast (short).
- The implementation effort on cheap micro-controllers without a floating-point unit is low.
- The development and application engineers are used to coping with look-up tables and utilize a number of specialized tools that are designed to support their tasks.
- The restriction to one or two inputs did not impose severe performance limitations on the engine in the past when the complexity of engine control systems was relatively low. Rather it corresponds to the visualization capability of humans, which is confined to three dimensions (2 inputs, 1 output).

If a quantity depends on more than two inputs, typically the following approach is taken to circumvent the restriction of look-up tables to two inputs. The two most important quantities are chosen as inputs for a look-up table that yields the desired output keeping the additional, not considered, input constant. Then the look-up table output is corrected by an additive or multiplicative correction model. This scheme is depicted in Fig. 24.6b and c for a problem with three inputs; see Sect. 7.6.2. This approach can be regarded as a special case of a hierarchical model as shown in Fig. 24.6d; see Sect. 7.6.5. Note that all schemes shown in Fig. 24.6b, c, and d are special cases of a general model with three inputs (Fig. 24.6a). These special cases can yield satisfactory performance only as long as the inputs used for the corrections (u_3) are either not significant or almost decoupled from the primary inputs (u_1, u_2).

The limitation of look-up tables to low-dimensional mappings calls for such approaches as a correction or hierarchical models. Two main difficulties arise with these conventional approaches as the complexity of automotive electronics increases. This complexity increase is a consequence of stricter laws and regulation with respect to the exhaust gases and the endeavor to improve performance and reduce fuel consumption. First, the number of required look-up tables increases dramatically with the dimensionality of the modeling problems. For example, in a modern electronic engine control unit, typically more than 100 look-up tables are

[2]Owing to the curse of dimensionality more than two inputs cannot really be handled except in cases where the grid resolution can be chosen very low; see Sect. 10.3. It is therefore assumed in the following that grid-based look-up tables have either one or two inputs only.

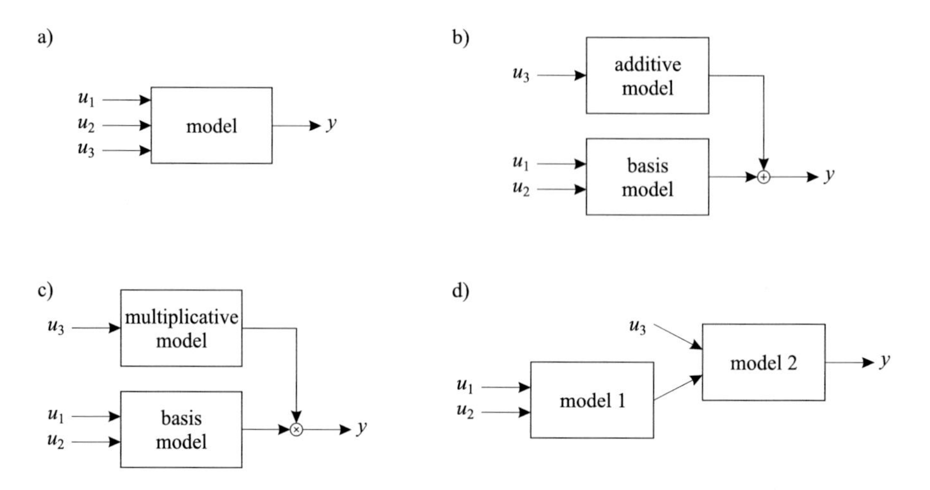

Fig. 24.6 Schemes for solving a modeling problem with three inputs shown in (**a**) if the models are restricted to two inputs: (**b**) additive, (**c**) multiplicative, (**d**) hierarchical. The inputs for the multiplicative correction model in reality may be, e.g., u_1 = engine charge (cylinder fill), u_2 = engine speed, u_3 = ignition angle efficiency, and the output y = engine torque. The output of the basis model is the "ideal" engine torque for optimal injection angle and optimal air/fuel ratio $\lambda = 1$. The input u_3 accounts for the deviations of the injection angle from its optimal value. A similar correction is necessary for the deviation of λ from 1 or for temperature dependencies that would require additional inputs u_4, u_5

implemented. Thus the sheer number of look-up tables becomes hardly manageable. Second, more manipulated variables are introduced to influence the characteristics of the engine in a favorable manner. In the past, merely two manipulated variables existed: the fuel injection mass and the injection angle for a diesel engine, or the ignition angle for a spark-ignition engine, respectively. The fuel injection mass is determined by the required engine torque, while the injection/ignition angle has to be optimized for each operating point of the engine given by the fuel injection mass and the engine speed. Additional manipulated variables in current and future engines allow one to control the following: exhaust gas recirculation, wastegate (with turbocharging), variable nozzle turbine (diesel with turbocharging), pilot injection (common rail diesel), and variable camshaft or variable valve timing (spark ignition engine). Thus, additional inputs arise for the models that cannot really be dealt with as depicted in Fig. 24.6b, c and d since their influence is significant and they are strongly coupled. The higher dimensionality of these problems calls for new solutions and the replacement of the dimensionality restricting look-up table models.

Besides the shortcomings of look-up tables, a further, more fundamental difficulty arises as a consequence of the higher dimensionality of the modeling problems: the question of data acquisition or experiment design. Currently, the inputs are usually varied according to a given grid, and the measured outputs are stored in look-up tables. Such grid-based measurement strategies underlie the curse of dimensionality, i.e., the measurement time increases exponentially with the number of inputs. Since for each measurement, the responses must settle to their stationary values, the time required cannot be arbitrarily shortened. The limited capacity of engine test stands, and the constraints on development time and cost call for new measurement strategies in the future. Although the solution to this problem is far from clear, some promising new ideas can be pointed out:

- Take measurements more densely where strongly nonlinear effects are expected and sparingly where mainly linear behavior is expected. Use prior knowledge obtained from similar engines to form these expectations.
- Use models of similar engines and only adapt some different characteristics with new measurements. Since fewer parameters have to be estimated, less data is required.
- Replace the step-like input changes with a slow (low frequency) ramp. Instead of utilizing only the stationary value of each measurement (and wasting all data describing transient behavior), take all measurements into account, neglecting the small dynamic error that is introduced by the fact that the ramp is not infinitely slow. Improve accuracy by sweeping through the input space in both directions and averaging out dynamic errors, i.e., change an input from 0% to 100% and back from 100% to 0% and take the mean between both measurements.
- Improve the strategy described above by identifying a nonlinear *dynamic* model. Subsequently, calculate the static characteristics of the nonlinear dynamic model. Since the dynamic model can represent the transient behavior, the ramp can be much faster for equivalent accuracy. A tradeoff exists between the measurement time and the expected static model accuracy because the accuracy of a dynamic model in a given frequency range (here around zero) depends on the proportion of this frequency range in the excitation signal; see Sect. 18.7.3. Such a tradeoff is very appealing because it allows one to adjust the measurement time with regard to the desired model accuracy.

All these ideas lead to a non-uniform sampling of the input space. None of them is compatible with look-up table models. Rather they require modern nonlinear approximators that can deal with high-dimensional input spaces.

The following benefits of modern nonlinear models such as neural networks compared with grid-based look-up tables in automotive electronics can be summarized:

- General models with more than two inputs are possible.
- Arbitrary distribution of measurement data is possible.
- Smoothness properties exist, i.e., the model output is differentiable several times. This is important if the models are to be used for optimization.
- A more compact description is possible because one high-dimensional model can replace a conglomerate of look-up tables.
- Trial and error and manual tuning can be reduced because of the unified approach.

- They lead one step further toward nonlinear *dynamic* models that promise to deliver the next performance and flexibility boost.

As the complexity of automotive electronics grows, the above advantages tend to outweigh the advantages of look-up tables. Probably, in a first phase, modern nonlinear models such as neural networks will be utilized mainly in the development process for modeling and optimization, while the micro-controller will still contain conventional look-up tables. In a second phase, more flexible and complex models will also be implemented at the micro-controller level.

24.1.6 Modeling of Exhaust Gases

Figure 24.7 shows a steady and rapid decrease of the exhaust gas limits for diesel engines enforced by law within the European Union. In order to be able to meet these limits in the future, improvements in various areas are required, including constructive mechanical changes of the motor, the introduction of additional manipulated variables for higher flexibility, filtering and catalytic conversion of the exhaust gases, and optimization of the engine management and control systems. Traditionally, many optimization problems have been solved by a trial-and-error approach with the knowledge and experience of application engineers. The reasons for this are manifold. They range from a lack of suitable models and tools to all the difficulties arising in interdisciplinary fields were mechanical and control engineers have to work together in a team. In particular, with the increasing number

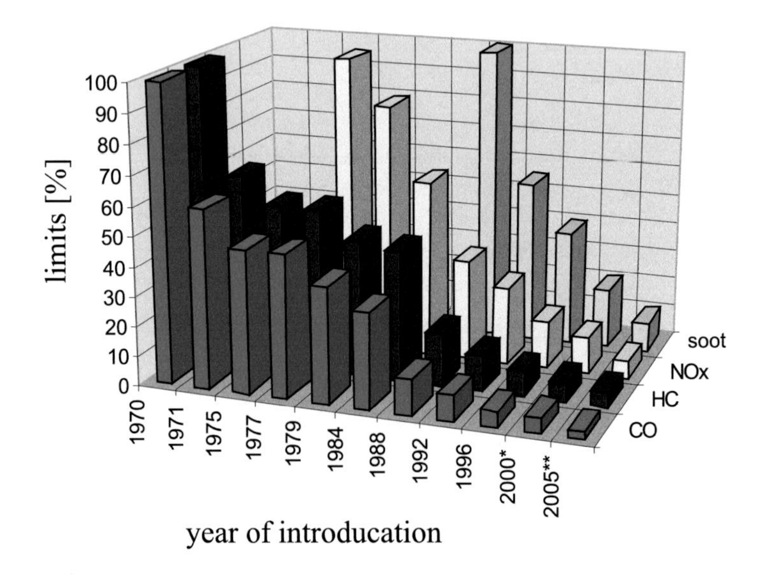

Fig. 24.7 Exhaust gas limits enforced by law in Europe

of manipulated variables, appropriate models for the exhaust gases are required in order to carry out an exhaust gas optimization that yields optimal engine control.

It is extremely difficult to derive exhaust gas models from first principles. The exhaust is very sensitive with respect to the fuel, temperature, and pressure distribution within the cylinders. Small local changes can influence the exhaust significantly, and these effects are still not well understood. Numerical finite element simulation studies require about 70 hours of computation time on the fastest available computers for a single combustion cycle [459]. Thus, this approach may be useful for supporting the mechanical construction and design changes, but a different strategy is required for modeling for optimization and control. Fortunately, the demand for accuracy is not very high because the engine characteristics even after production and clearly after wear can vary considerably.

The following quantities are interesting for diesel engine control optimization:

- *Fuel consumption* b_e *(or the proportional carbon dioxide* CO_2*)*: Consumes valuable, limited resources and is the major cause of the greenhouse effect, i.e., global warming.
- *Nitrogen oxides* (NO_x): Form acids, which in the form of acid rain damage nature and buildings and cause the death of forests.
- *Hydrocarbons (HC)*: Poisonous for humans; may cause cancer.
- *Carbon monoxide (CO)*: Only 0.3 volume percent in the air can cause suffocation for humans and animals.
- *Particles (soot)*: May promote cancer in humans.

A model should be able to predict these quantities from the relevant inputs, e.g., for a modern diesel engine these may be

- engine speed n_{eng},
- injection mass m_b,
- injection angle θ_{si},
- exhaust gas recirculation (EGR) rate,
- variable nozzle turbine (VNT) position,

and possibly injection pressure and pilot injection profile for a common rail diesel. Clearly, that many inputs cannot be reasonably processed with grid-based look-up tables. With neural networks these static relationships, however, can be easily learned provided that representative measurement data is available.

Figure 24.8 depicts the output of local linear neuro-fuzzy models for the four major exhaust components for an older diesel engine with just the first three manipulated variables from the above list. These models possess 15 neurons each and are trained with LOLIMOT in less than a minute.[3] Note that these three-dimensional plots represent just one cut through the four-dimensional mapping for a single engine speed. As the number of inputs grows further, the relationships between the inputs and outputs become increasingly difficult to visualize. This is another

[3]With a Pentium 100 MHz PC.

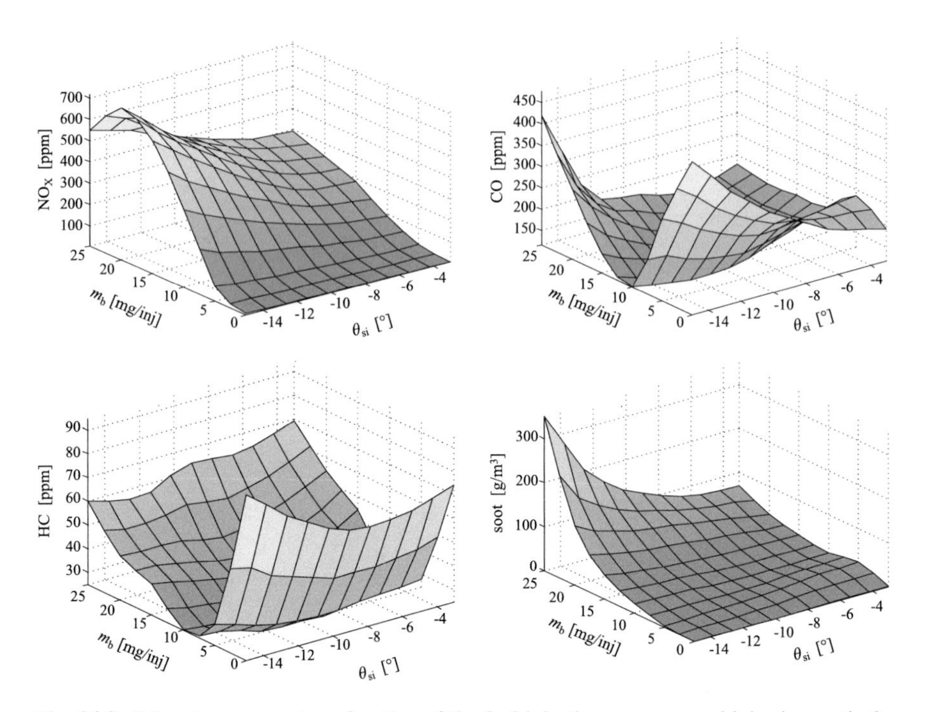

Fig. 24.8 Exhaust components as function of the fuel injection mass m_b and injection angle θ_{si} for a fixed engine speed of $n_{eng} = 2000$ rpm

manifestation of the curse of dimensionality. Smooth and reliable interpolation behavior of the model architecture is extremely important since for more than three[4] inputs, it is barely possible to detect "strange" model behavior visually. Note that these difficulties are caused by the exponential complexity increase with the number of *inputs*. The complexity grows only linearly with the number of *outputs*.

As the number of model inputs increases, the data acquisition problem becomes harder; see Sect. 24.1.5. At some point, it may no longer be possible to include measurements from all boundary conditions (all combinations of minimal and maximum input values) in the training data set. Then a reasonable model extrapolation becomes a very important issue. For the processes under investigation, the principle behavior is known from simple balance considerations and experience. This allows one to construct an extrapolation behavior that at least does not violate physics. For local linear neuro-fuzzy models, the user-defined extrapolation design procedure described in Sect. 14.4.2 can be used.

[4]Mappings with more than two inputs can be visualized by plotting several cuts through the higher-dimensional mapping, e.g., for $n_{eng} = 1000, 2000, \ldots, 5000$ rpm. However, the number of plots and the information about the overall mapping contained in these plots fades as the input dimensionality increases.

24.1.7 Optimization of Exhaust Gases

Once a model for the engine fuel consumption and the exhaust has been built, the question arises: How it should be utilized for optimization? Figure 24.9 illustrates a possible optimization scheme. Basically, the manipulated variables that are model inputs are varied by the optimization technique until their optimal values are found. Optimality is measured in terms of a loss function that depends on the fuel consumption and exhaust gases, i.e., the model outputs. When the operating point-dependent optimal input values are found, they can be stored in a static map that represents the feedforward controller, which can be realized either as a look-up table (since it has only two inputs) or as a neural network.

The two most straightforward choices for the loss function will be introduced in the following. One possibility is to incorporate all relevant model outputs directly into a loss function of the type

$$J\left(b_{\mathrm{e}}, \mathrm{NO_x}, \mathrm{HC}, \mathrm{CO}, \mathrm{soot}\right) \longrightarrow \min_{\theta_{\mathrm{si}}, \mathrm{EGR}, \mathrm{VNT}} . \tag{24.3}$$

More concretely the loss function can be chosen as a weighted sum of the fuel consumption and the exhaust gases

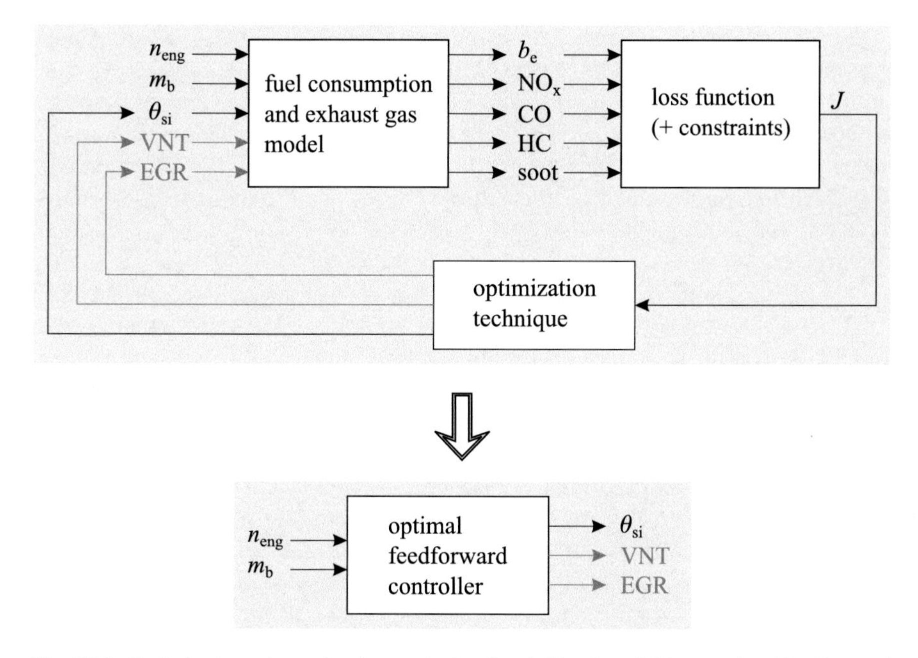

Fig. 24.9 Optimization scheme for the manipulated variables that yields an optimal feedforward controller. The gray lines are optional

$$J\left(b_{\mathrm{e}}, \mathrm{NO_x}, \mathrm{HC}, \mathrm{CO}, \mathrm{soot}\right) = w_1 b_{\mathrm{e}} + w_2 \mathrm{NO_x} + w_3 \mathrm{HC} + w_4 \mathrm{CO} + w_5 \mathrm{soot} \qquad (24.4)$$

with all weights summing up to 1, i.e., $\sum_{i=1}^{5} w_i = 1$. It can be extremely difficult to normalize the different fuel consumption and exhaust gas components in such a way that the weights w_i can be chosen in an intuitive fashion. Only with a reasonable normalization can the sensitivity of the optimum with respect to all weights be adjusted to the same order of magnitude; for more details refer to [204]. The advantage of the approach in (24.4) is that a good tradeoff can be found between the fuel consumption and the individual exhaust gases components, i.e., the optimum is generally well balanced. For example, a huge increase in one exhaust gas for a slight decrease in fuel consumption can be avoided unless the weight corresponding to this exhaust component is chosen (virtually) zero.

An alternative approach is to optimize only the fuel consumption and to take the exhaust gases into consideration by constraints, i.e.,

$$J\left(b_{\mathrm{e}}\right) \longrightarrow \min_{\theta_{\mathrm{si}}, \mathrm{EGR}, \mathrm{VNT}} \quad \text{with } \mathrm{NO_x} < \mathrm{NO_x^{(max)}}, \mathrm{HC} < \mathrm{HC^{(max)}}, \mathrm{CO} < \mathrm{CO^{(max)}}, \mathrm{soot} < \mathrm{soot^{(max)}} . \qquad (24.5)$$

This constrained version has the advantage being closer to the real requirements where the maximum exhaust gas values are given by law. Then (24.5) allows one to achieve minimum fuel consumption, while all exhaust gases are kept at their upper bound. Of course, the upper bounds in (24.5) must be chosen somewhat stricter than those given by the regulations to account for process/model mismatch. The difficulties involved in the decision for a specific loss function type show that this problem is a good candidate for the application of multi-objective optimization methods. This certainly is a promising undertaking for future research; see Sect. 7.3.4.

The optimization is carried out with respect to the manipulated variables that can be freely adjusted. Note that the injection mass is *not* free because it is determined in order to deliver the demanded engine torque. Thus, in the past, merely one variable had to be optimized, the injection angle θ_{si}. This could be done manually. However, optimization with respect to several variables cannot really be carried out by hand. Therefore, the application of an automated optimization technique as described in Part I is required. Figure 24.10a illustrates how the optimization is performed in the one-dimensional case where only the injection angle is optimized. The loss function depends on three quantities, the engine speed n_{eng}, the fuel injection mass m_{b}, and the injection angle θ_{si}, but only the last one is optimized. Thus, a cut through the loss function at a specific engine speed (not shown) and a specific fuel injection mass (shown) determines the one-dimensional curve that depends solely on the injection angle. The task of optimization is to find the optimal value of θ_{si}. In Fig. 24.10b the same procedure is shown for a different choice of the weights in (24.4), yielding a

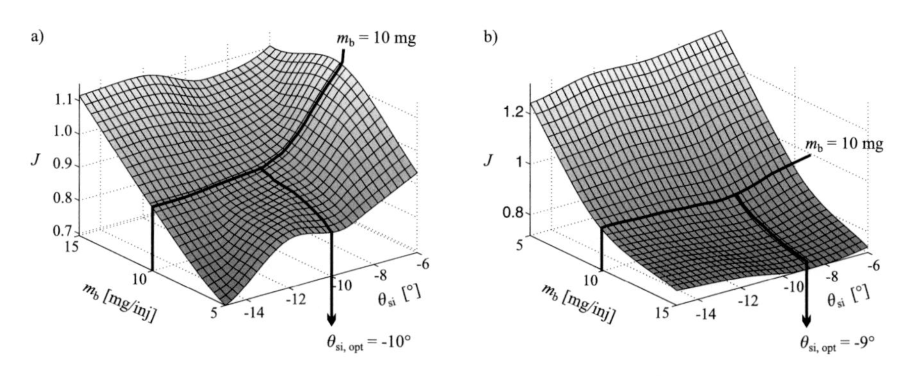

Fig. 24.10 Illustration of the one-dimensional optimization with respect to the injection angle θ_{si} with a fixed engine speed n_{eng} and injection mass m_b: (**a**) $w_1 = 0$, $w_2 = 0.2$, $w_3 = 0.8$; (**b**) $w_1 = 35$, $w_2 = 0.3$, $w_3 = 0.35$. Note the reverse scaling of the m_b-axis in (**b**)

totally different loss function shape (although here by chance the optimal injection angle is similar).

The optimization procedure illustrated in Fig. 24.10 relies on a certain smoothness of the loss function. Since the loss function here is just a linear combination of the fuel consumption and exhaust gas models, the smoothness requirement passes on to the models themselves. For local linear neuro-fuzzy models, this motivates the use of relatively large smoothness factors k_σ (see Sect. 13.3) to avoid local optima. Otherwise, local optima may be caused by a single (or very few) disturbed measurement since usually the total number of measurement data samples available for training is small. The application of local *quadratic* neuro-fuzzy models may also improve the reliability and accuracy of the optimization results. If look-up tables with linear interpolation are utilized, the gradients are piecewise constant, which can cause difficulties for the optimization technique owing to the abrupt gradient changes. Furthermore, local optima become much more likely since all measurement samples are stored in the look-up tables and no smoothing (averaging) takes place.

It should further be remarked that the manipulated variables cannot take any combination of values within their physical boundaries. Rather, large areas in the search space may not be allowed because they would damage the engine. This leads to additional constraints in the optimization. For example, for spark ignition engines, the ignition angle must not become too small, to avoid knocking of the engine. This knocking limit imposes constraints on the search space that vary with the operating point (n_{eng}, m_b).

One important aspect of the optimization scheme shown in Fig. 24.9 still has to be explained in more detail. The manipulated variables are in fact *functions* of the operating point (n_{eng}, m_b); see the feedforward controller in Fig. 24.9. So the question arises: How exactly are the operating point-dependent manipulated variables optimized? Basically, two strategies can be distinguished:

- *Local strategy*: The manipulated variables are optimized for each operating point *separately*, while an outer loop goes through all operating points lying on a grid. Then the optimal manipulated variables are stored in a look-up table at their associated operating points.
- *Global strategy*: The operating points are defined, e.g., by a grid, and the manipulated variables of all operating points are optimized *concurrently*. Note that the number of parameters to be optimized is equal to the number of manipulated variables (say 3) times the number of operating points (say 64 for an 8×8 grid), i.e., a large number (say 192). The loss function covers all operating points and accumulates their effects on a given *driving cycle*; see Sect. 24.1.

The global strategy is much more straightforward. A driving cycle with upper bounds for the various exhaust gases is given by law; thus these bounds can explicitly be taken into account. Furthermore, the driving cycle ideally should imitate realistic driving behavior (hopefully future driving cycles do so) that optimization with regard to this driving cycle should give the best exhaust emissions in practice. In future, dynamic models may make it possible to improve the exhaust emission further, which is of course only possible with a global strategy since the local strategy separates the operating points and thus does not allow the consideration of dynamics at all. Another advantage of the global strategy is that the different operating points will contribute with different relevance to the driving cycle, and this relevance can be accounted for and exploited by the optimization.

The major drawback of the global strategy is, of course, the huge number of parameters that have to be optimized. In addition, difficulties arise if the driving cycle does not include all operating points since the values of the manipulated variables associated with these non-covered operating points do not influence the global loss function and consequently cannot be optimized. Then the local strategy has to be applied. Basically, the local strategy possesses the corresponding disadvantages as opposed to all the advantages of the global strategy. The most severe drawback of the local strategy is that the constraints on the exhaust gases cannot be taken into account in any direct manner. Therefore, in case the global strategy yields a too highly dimensional optimization problem, it might be reasonable to pursue a compromise approach. Such a *compromise strategy* may extract the most important operating points covered by the driving cycle and then apply the global strategy with this problem of reduced size. Somehow the constraints have to be adjusted to this sub-cycle. All remaining operating points may be handled with the local strategy. For the results presented in the following, the local strategy together with the loss function in (24.4) was applied, which required some tuning with the weights.

In Fig. 24.11 the fuel consumption and the exhaust gases NO_x and soot achieved with the standard electronic diesel control (EDC) unit and the local optimization strategy are compared for a driving cycle. The other exhaust gases are not considered here for the sake of simplicity, but they all stayed within their allowable limits. For the optimization (after normalization), the weights are chosen identically, i.e., $w_1 = w_2 = w_5 = 1/3$. The comparison clearly shows that the settings in the standard EDC unit are suboptimal. Fuel consumption and all exhaust gas components can

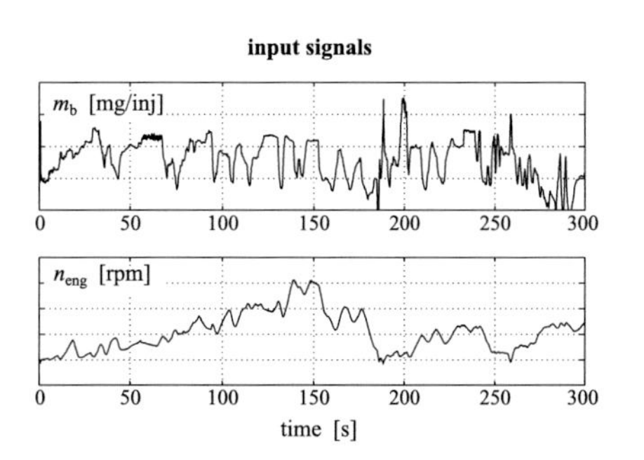

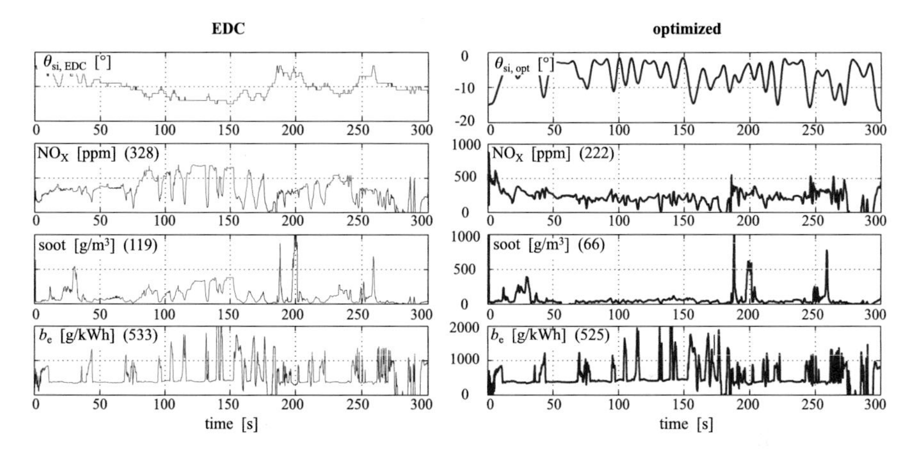

Fig. 24.11 Comparison between the fuel consumption b_e and the exhaust gases NO_x and soot for the standard EDC unit and the locally optimized approach with $w_1 = w_2 = w_5 = 1/3$. The optimized approach improves the standard solution in all respects. The average values over the whole cycle are given in parenthesis

be improved simultaneously. This underlines the potential improvements that can be realized by the consequent application of advanced modeling and optimization tools.

Figure 24.12 compares two optimization results with the standard EDC unit. Only fuel consumption and NO_x are considered in this example. In the left the weights are chosen as $w_1 = 2/3$, $w_2 = 1/3$, while in the right they are chosen vice versa, i.e., $w_1 = 1/3$, $w_2 = 2/3$. As expected, in the left the fuel consumption is reduced compared with the EDC, while in the right the fuel consumption yielded by the optimization and the EDC is indistinguishable, but the NO_x exhaust is even more improved. Obviously, the weights easily allow a tradeoff between the different objectives. Meeting of strict exhaust limits, however, requires some trial-and-error experimentation with the weights.

Table 24.1 demonstrates the effect of the weight choice in (24.4) on the fuel consumption and the exhaust gases NO_x and soot. Furthermore, it compares these results with the performance realized by the settings of the standard electronic diesel

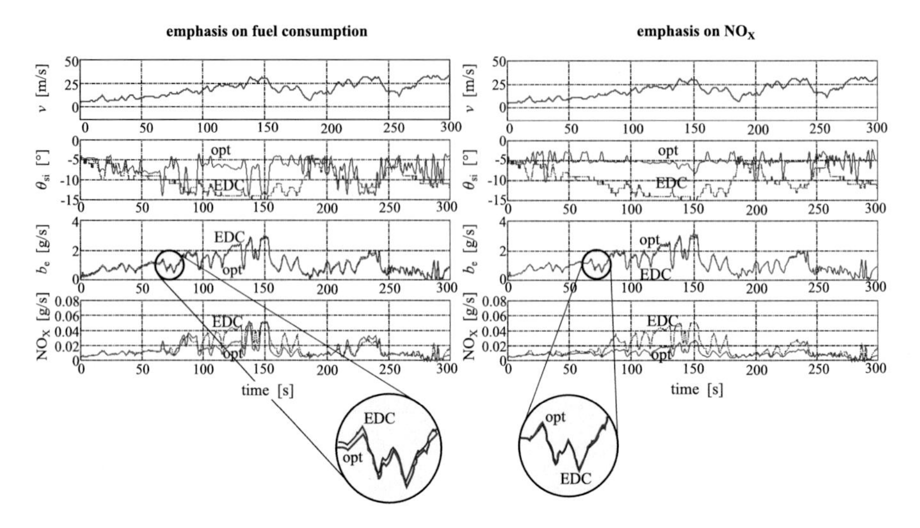

Fig. 24.12 Comparison between the fuel consumption b_e and the exhaust gas NO_x for the standard EDC unit and the locally optimized approach with (left) more emphasis on fuel consumption, i.e., $w_1 = 2/3$, $w_2 = 1/3$ and (right) more emphasis on NO_x, i.e., $w_1 = 1/3$, $w_2 = 2/3$

Table 24.1 Comparison of the influence of different loss function weights

w_1	w_2	w_5	b_e [%]	NO_x [%]	Soot [%]	Comments
-	-	-	100	100	100	Default values from standard EDC (suboptimal)
0.3	0.7	0	100	70	85	Improvement on exhaust gases with constant fuel
0.7	0.3	0	94	102	184	Small decrease in fuel yields strong increase in soot
0.5	0.25	0.25	95	88	68	Lower fuel with acceptable exhaust gases

EDC electronic diesel control

control unit. Obviously, as already shown in Fig. 24.11, the standard realization is suboptimal because fuel consumption and both exhaust gases can *simultaneously* be reduced. All results yielded by the optimization approach are pareto-optimal. This means that it is not possible to improve all outputs simultaneously; see Sect. 7.3.4. Table 24.1 shows that at some point small improvements in fuel consumption must be paid for by huge increases in the exhaust (third row compared with the last row). However, if the weights are chosen reasonably, i.e., all $w_i > 0$, then a good compromise between the different objectives can be obtained (last row).

All the discussion up to this point has assumed that the optimization is carried out offline and the resulting optimal feedforward controller shown in Fig. 24.9 is implemented during production and subsequently is kept fixed. If the optimization

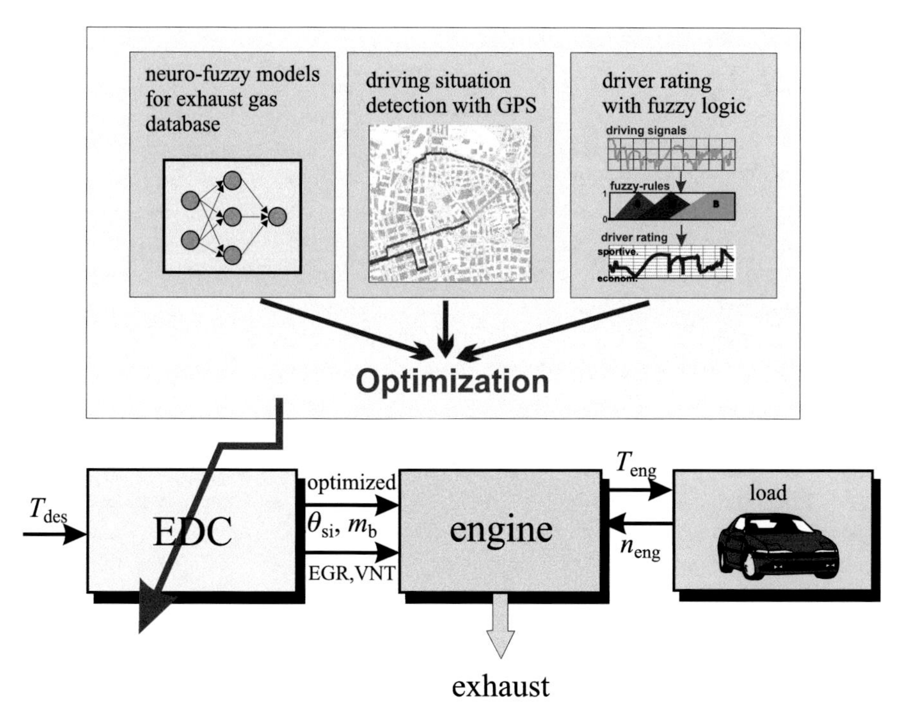

Fig. 24.13 Engine management and optimization system

can be performed in a reliable manner, and sufficiently fast hardware[5] is available in the car, then the optimization can also be done *online*. It then can adapt to different kinds of drivers (ecologic, moderate, sportive, etc.) and different traffic situations (urban, freeway, stop and go, etc.) by adjusting the weights of the loss function (24.4). For example, low exhaust emissions such as CO and NO_x may be more important than fuel consumption and CO_2 within cities, while the opposite applies on a freeway. For a successful adaptation to these situations, they have to be automatically detected. In [204] a strategy based on the information of a global positioning and navigation system and the evaluation of a fuzzy rule-based decision-making system is proposed; see Fig. 24.13.

[5]The hardware should be powerful enough to carry out an optimization within less than a minute or so.

24.1.8 Outlook: Dynamic Models

In the future, the modeling and optimization approaches described above will have to be extended to *dynamic models*. Although this topic is beyond the scope of this chapter, a brief outlook will be given in the following. For more details refer to [204]. The reason for the emphasis on static models in automotive electronics is *not* that dynamic effects can be neglected. Rather, it is that non-linear dynamic models have been much too difficult to derive. More commonly realized engine features like turbocharging and exhaust gas recirculation introduce significant dynamic effects into the engine characteristics that cannot be properly described with static models. With modern dynamic neural network architectures, the derivation of experimental nonlinear dynamic models will be feasible in the future.

The next chapter presents some nonlinear dynamic modeling and identification applications in much greater detail. The following figures will just give the reader an idea of what dynamic exhaust gas modeling may look like. A considerable difficulty with the *dynamic* experimental data of exhaust gases is that the sensors used are generally very slow. Thus, before identification can start, the raw data must be processed in order to eliminate most of the sensor characteristics. Otherwise, the major time constant (in the range of several seconds) will be caused by the sensor rather than the process. One possible way of eliminating the sensor distortion is to filter the measurement data by an inverse of a sensor model. Unfortunately, this procedure cannot compensate fully for the sensor behavior because the strongly differentiating characteristics of the inverse model would yield an amplification of high-frequency noise. However, sufficient compensation can be achieved with regard to the required accuracy of the exhaust model.

Figure 24.14 shows the training data used for the identification of a local linear neuro-fuzzy model for NO_X. A first-order model with 15 neurons is identified with LOLIMOT. Its performance is very good on both the training data and fresh generalization data, as depicted in Fig. 24.15. Such a dynamic model allows extracting the static relationship as a special case; see Fig. 24.16. This is an interesting feature for speeding up measurement times, i.e., measuring dynamically (fast) and extracting the static model behavior instead of taking static (slow) measurements; see Sect. 24.1.5. The reason for the quite inaccurate approximation of the static behavior in Fig. 24.16 lies in the relatively fast excitation signal which forces the identification to sacrifice static accuracy for good high-frequency accuracy. The results shown in Fig. 24.16 can only demonstrate the principal feasibility of the idea and can certainly be improved in the future.

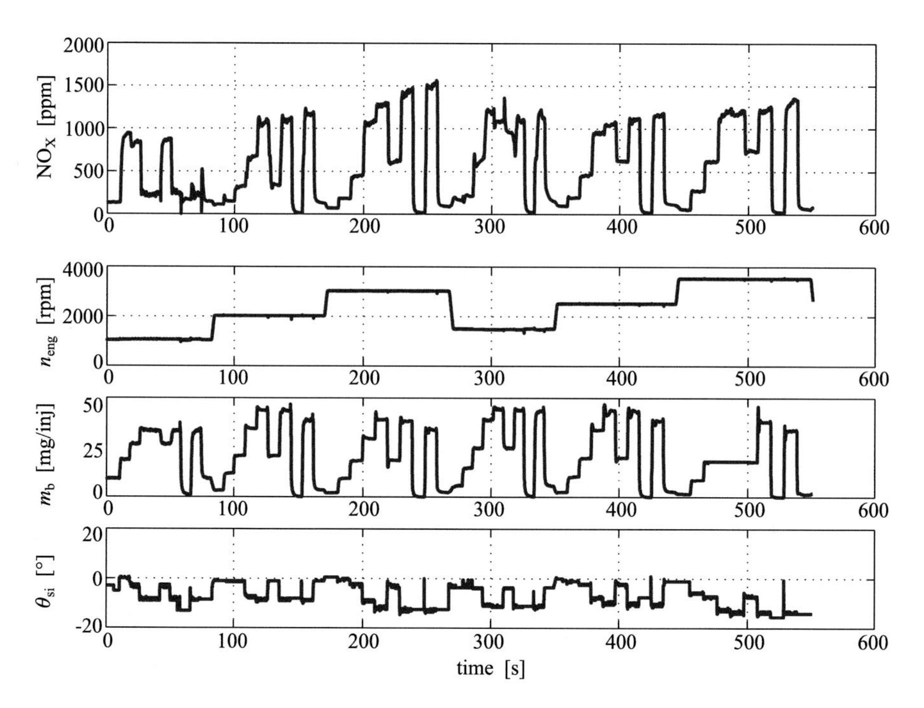

Fig. 24.14 Training data for identification of an NO_x model

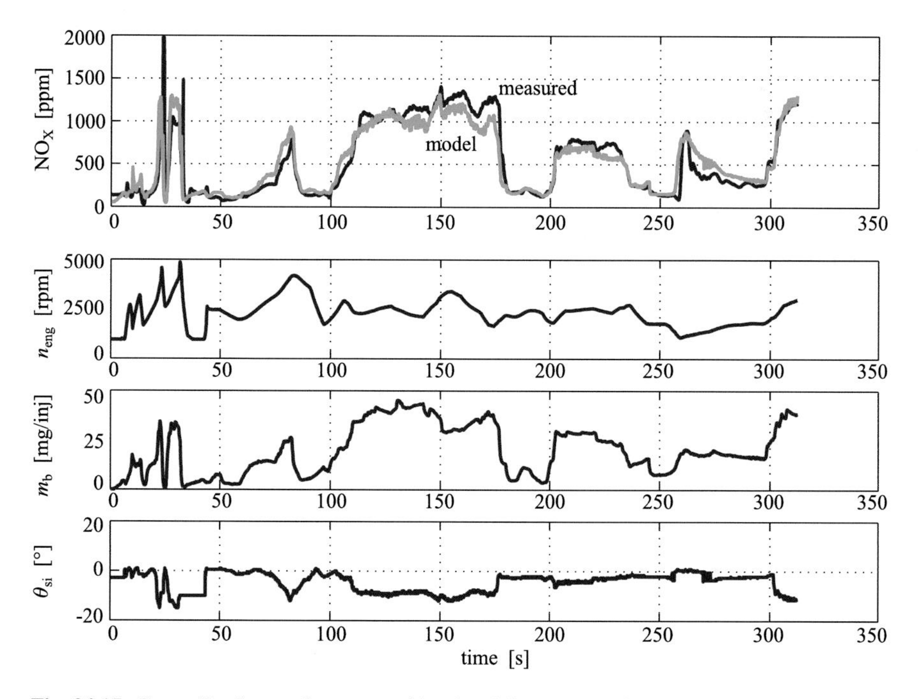

Fig. 24.15 Generalization performance with a local linear neuro-fuzzy model for NO_x with 15 neurons trained by LOLIMOT with the data shown in Fig. 24.14

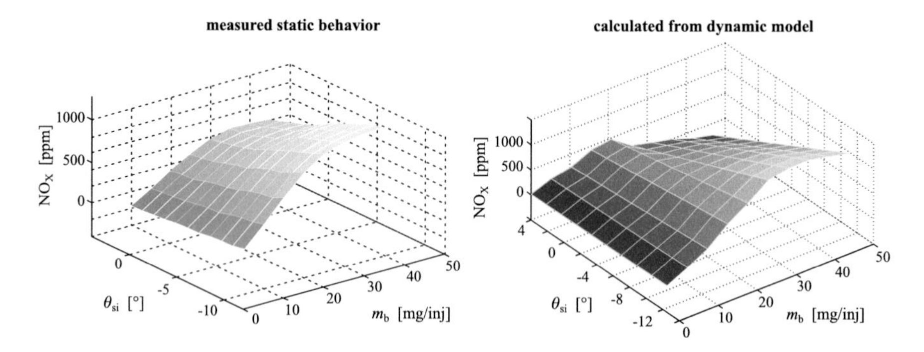

Fig. 24.16 Static NO_X characteristics measured (left) and calculated from the dynamic model (right) for engine speed $n_{eng} = 2500$ rpm

24.2 Summary

In this chapter, some static nonlinear system identification examples have been presented. In the driving cycle application, it was shown that local linear neuro-fuzzy models with a large number of neurons can be efficiently trained with the LOLIMOT algorithm. Also, constraints are taken into account in a very simple and computationally inexpensive way. The second example of the modeling of fuel consumption and exhaust gases of combustion engines illustrated how models can be utilized to systematize and optimize engine control systems. The restrictions caused by the limitations of grid-based look-up tables to low-dimensional mappings have been analyzed, and the potential improvements by utilizing neural networks have been pointed out. It was demonstrated that multivariable mappings can easily be realized by neural networks and that these models can be used for the multi-objective optimization of fuel consumption and exhaust gases. The success of this approach and the achieved improvements compared with the state-of-the-art industrial solution motivate to go even one step further by utilizing nonlinear dynamic models. This topic is pursued in the next chapter.

Chapter 25
Applications of Dynamic Models

Abstract This chapter deals with various dynamic modeling tasks. First, a cooling blast is identified. It has complicated dynamic behavior as the time constants are extremely direction-dependent. Second, a turbocharger for a Diesel engine is modeled with purely dynamic measurement data. Nevertheless, from the dynamic model, the static nonlinearity can be extracted, allowing a nice visualization that also can be used for model validation. Third, several sub-processes of a thermal plant are modeled. MISO models for a tubular and a cross-flow heat exchanger are built. With a simple SISO transport process, the capabilities of local model networks are nicely demonstrated to automatically identify flow rate dependent dead times via a local subset selection algorithm. This shows how easily powerful tools for linear regression can be transferred into the nonlinear world via local linear models.

Keywords Application · Cooling blast · Diesel engine · Turbocharger · Thermal plant · Subset selection · Dead times

This chapter presents various case studies for the modeling of nonlinear *dynamic* systems with local linear neuro-fuzzy models. In Sect. 25.1, LOLIMOT is used for nonlinear system identification of a cooling blast. Section 25.2 describes the experimental modeling of a turbocharger for hardware-in-the-loop simulation of a truck diesel engine. Several subprocesses of a thermal pilot plant are modeled in Sect. 25.3. Finally, Sect. 25.4 summarizes the results accomplished and the insights gained by the applications.

25.1 Cooling Blast

This section demonstrates the application of the LOLIMOT algorithm for identification of a single input, single output nonlinear dynamic process. This application underlines the following important features of LOLIMOT:

O. Nelles, *Nonlinear System Identification*,
https://doi.org/10.1007/978-3-030-47439-3_25

- identification of highly operating point-dependent static and dynamic behavior;
- identification of nonlinear dynamic behavior with structural changes in the dynamics;
- application of data weighting for compensation of non-uniformly distributed training data;
- utilization of a low-order premise input space;
- interpretation of the static and dynamic characteristics of the local linear models.

Some of the presented results have been published in [422]. For an extensive discussion and the application of nonlinear model-based control design methods to the cooling blast, refer to [149, 151, 152].

25.1.1 Process Description

The radial industrial cooling blast for ventilation and air conditioning of buildings shown in Fig. 25.1a is to be modeled. It is a pilot plant driven by a capacitor motor of 1.44 kW power, delivering up to 5000 m^3 of air per hour at a maximum speed of 1200 rpm (revolutions per minute). The power is supplied by phase control, and the speed is measured by a light-dependent resistor, which senses the light reflected from the rotor wings. The speed resolution of 4.7 rpm is quite rough owing to the low-cost 8-bit counter used for the speed measurement. The sampling time for this process is chosen as $T_0 = 0.164$ s.

Since the plain cooling blast without load is an almost linear process, it is additionally equipped with a throttle flap. This makes the system strongly nonlinear in its statics and dynamics and yields an extremely challenging identification task. The static nonlinearity of the process is depicted in Fig. 25.1b. The nonlinear static behavior can be explained as follows. As long as the throttle is closed, the gain of the process is very large since almost no mechanical work is required to speed up the fan because there is no air transported and only some turbulence losses occur. At

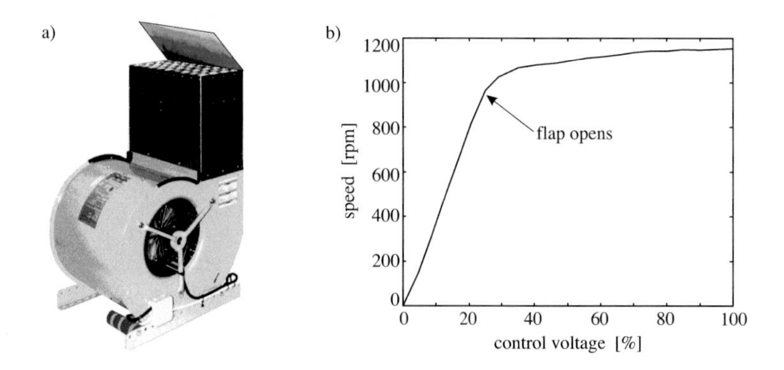

Fig. 25.1 (**a**) Cooling blast. (**b**) Measured static nonlinearity

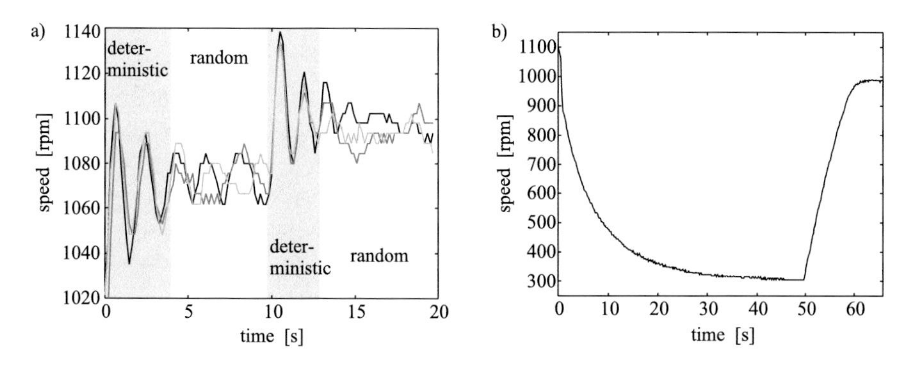

Fig. 25.2 Measured dynamics of the cooling blast: (**a**) flap opened (measured speed for three experiments with the same input signal), (**b**) flap closed (step responses u : 30% $\rightarrow$ 0% and u : 0% $\rightarrow$ 25%)

a speed of about 1000 rpm, the throttle opens and the gain abruptly decreases since the fan has to accelerate the air.

The system dynamics illustrated in Fig. 25.2, which is also severely nonlinear, can be explained as follows. For an opened throttle flap, the process exhibits weakly damped oscillatory behavior stemming from the acceleration of the flap. As Fig. 25.2a clearly demonstrates, the first two oscillations in a step response are reproducible, while the subsequent oscillations are obviously different for the three experiments with identical input signals. The reason for this stochastic behavior probably lies in the interaction of the airflow with the surroundings. Consequently, any model of the cooling blast can only be expected to be able to describe the first two oscillations. If the throttle flap is closed, the dynamic process characteristics change drastically to well-damped first-order time lag behavior; see Fig. 25.2b. Besides this change in characteristics, the time constants increase by more than one order of magnitude. Furthermore, the behavior becomes strongly direction-dependent since the time constants for decreasing speed are much larger than those for increasing speed. This effect is caused by the fact that on the one hand an increasing voltage *actively* accelerates the fan while on the other hand the fan is not actively braked after a voltage drop. In the latter case it rather *passively* slows down merely due to the air drag losses.

25.1.2 Experimental Results

For modeling and identification, a black box approach will be adopted. The prior knowledge discussed above, however, enters the excitation signal design procedure. This is of fundamental importance since for black box modeling the major information utilized is contained in the training data. In the following, the excitation signal design philosophy is described.

25.1.2.1 Excitation Signal Design

The high speed (open flap) and medium speed (flap about to open or close) regions are excited with an APRBS that covers different amplitude levels and a wide frequency range; see Sect. 19.7 and [430, 431]. Since the cooling blast changes its static and dynamic behavior strongly for fan speeds around 1000 rpm it is important to include enough training data within this region. Furthermore, enough training data should be generated to enable the identification method to average out the stochastic components in the speed signal. Therefore, this part of the training data set is chosen as a 2/3 fraction of the whole training data set.

The low speed region (closed flap) is excited by several steps of various heights, taking into account the different rise times of the process for acceleration and deceleration. The complete signal is shown in Fig. 25.3. A second experiment with a similar signal is made for the generation of a separate validation data set.

25.1.2.2 Modeling and Identification

From the prior knowledge discussed above, it is clear that the model has to be at least of second dynamic order to represent the oscillatory behavior of the cooling

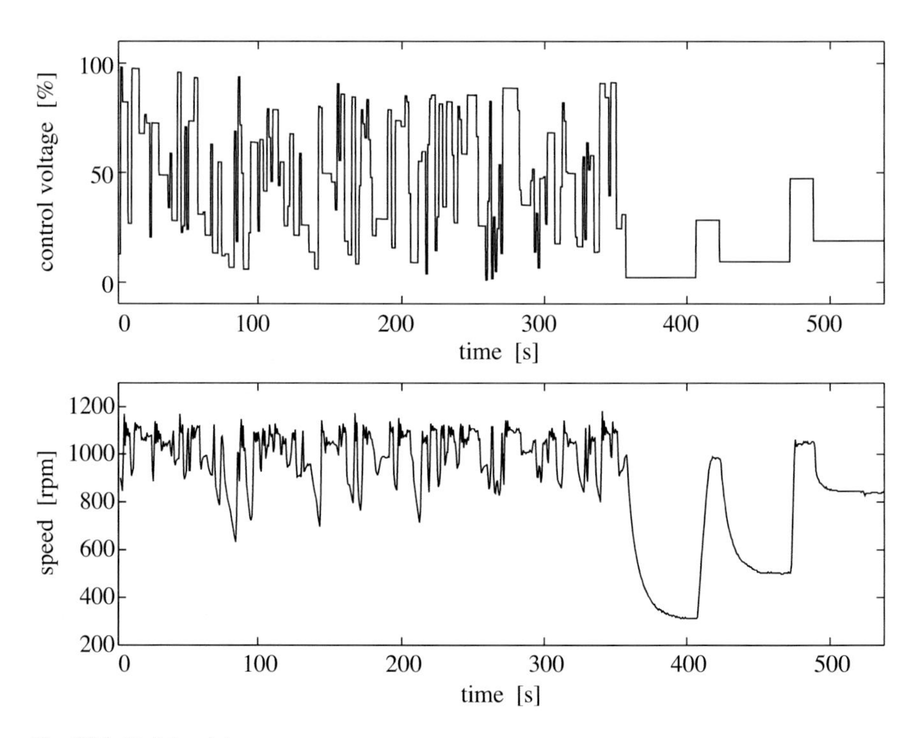

Fig. 25.3 Training data

blast for an open flap. In a trial-and-error procedure, models higher than second order did not yield a significant improvement in performance. Thus, the following approach is taken:

$$y(k) = f\left(u(k-1), u(k-2),\ y(k-1), y(k-2)\right) , \tag{25.1}$$

where $u(k)$ represents the control voltage and $y(k)$ represents the fan speed. In order to compensate for the over-representation of the high-speed regions in the training data set, a data weighting is introduced as follows; see Sect. 13.2.6. All data points within the first 360 s of the training data set are weighted with the factor 1/2, while the remaining data samples are weighted with 1. Investigations confirm the assumption that this data weighting improves the overall model quality. In particular, the model dynamics for low speeds improve, leading to an overall gain in model quality of 30% compared with the non-weighted case. The LOLIMOT algorithm is first run with full premise and consequent spaces, i.e., $\underline{z}(k) = \underline{x}(k) = [u(k-1)\ u(k-2)\ y(k-1)\ y(k-2)]^T$.

The convergence curve of LOLIMOT is depicted in Fig. 25.4a. The strong decrease in the model error compared with the linear model obtained in the first iteration reveals the highly nonlinear process characteristics. The convergence curve gives a clear indication of which model complexity might be appropriate. Because no significant further improvement can be obtained by choosing more than ten local linear models, a neuro-fuzzy model with $M = 10$ rules is selected.

Figure 25.4b shows the poles of the ten local linear second-order models identified by LOLIMOT. Obviously, four LLMs possess real poles, while six LLMs possess conjugate complex pole pairs. LOLIMOT starts with a global linear model, and already in the second iteration, the fundamental characteristics of the process dynamics are captured by the following two rules:

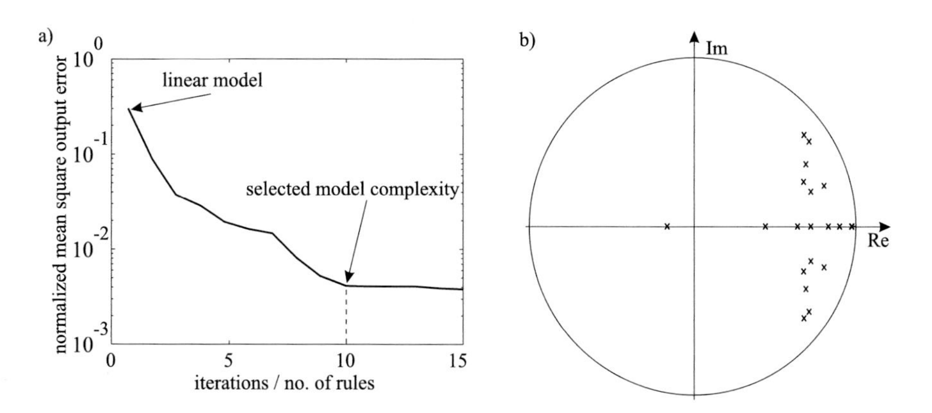

Fig. 25.4 (**a**) Convergence curve of LOLIMOT. (**b**) Poles of the ten local linear second-order models

R_1: IF $u(k-2) = \textsf{small}$ THEN $K = 10.2\,s^{-1}$, $T_1 = 1.4\,\text{s}$, $T_2 = 0.7\,\text{s}$

R_2: IF $u(k-2) = \textsf{large}$ THEN $K = 1.9\,s^{-1}$, $T = 0.3\,\text{s}^2$, $D = 0.33$,

where the LLMs are expressed by their gains and time constants in continuous time for the sake of a better understanding, i.e., the first and the second rule consequents represent the following continuous-time transfer functions

$$G_1(s) = \frac{K}{(1+T_1 s)(1+T_2 s)}, \qquad G_2(s) = \frac{K}{1+2DTs+T^2 s^2}. \tag{25.2}$$

As Fig. 25.4b shows, the final set of ten rules contains one rule with a consequent LLM has a negative real pole. This does not possess a reasonable correspondence in continuous time. Because this pole is very close to zero, the associated dynamics are several orders of magnitude faster than those associated with the other poles. This indicates that the process actually might be only of first dynamic order in the corresponding operating regime (as prior knowledge already suggested). Indeed, the corresponding LLM is active in the operating regime that represents the lowest fan speed and voltage. This second-order LLM can thus be replaced by a first-order LLM without a significant loss in model quality. Note that in this case, the interpolation algorithm, in fact, introduces a "dummy" pole at zero for this first-order LLM; see Sect. 22.3.3.

LOLIMOT partitioned the premise input space spanned by $\underline{z}(k)$ only along the $u(k-2)$- and $y(k-2)$-dimensions. The resulting decomposition is shown in Fig. 25.5a together with the training data distribution. Obviously, the LLMs 3, 5, 8, and 9 represent off-equilibrium regimes that are hardly excited by the training data. Nevertheless, owing to the regularization effect of the local parameter estimation, even these local linear models possess acceptable statics and dynamics, which also ensures a reasonable extrapolation behavior. Note that with a global parameter estimation, these barely excited LLMs could easily become unstable owing to the measurement noise.

Since LOLIMOT partitions the premise input space only in the two dimensions $u(k-2)$ and $y(k-2)$, the other inputs can be discarded from the rule premises:

$$\underline{z}(k) = [u(k-2)\ y(k-2)]^T, \tag{25.3}$$

$$\underline{x}(k) = [u(k-1)\ u(k-2)\ y(k-1)\ y(k-2)]^T. \tag{25.4}$$

This is an example of the practical relevance for a low-order premise input space in connection with higher-order rule consequents as discussed in Sect. 22.2. Because the premise input space is only two-dimensional, a visualization is possible allowing an easy understanding of the local models' validity regions.

Figure 25.5b shows some step responses illustrating the strongly nonlinear statics and dynamics of the model. Both gains and time constants vary by more than one order of magnitude. The behavior of the local linear models agrees well with the

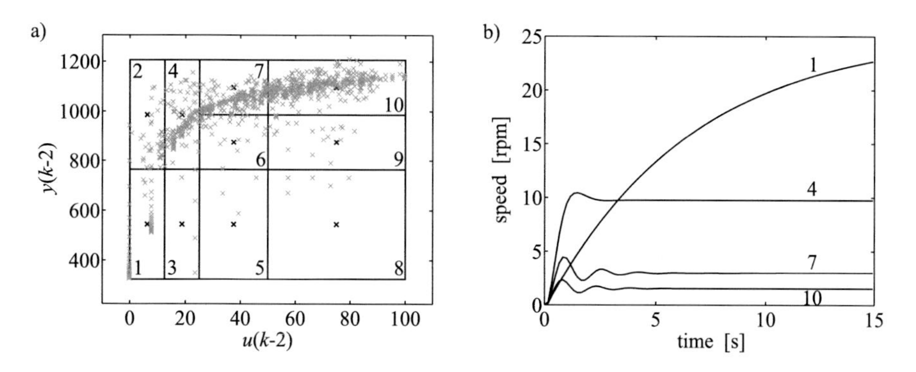

Fig. 25.5 (**a**) Premise input space decomposition created by LOLIMOT. The black crosses mark the centers of the validity functions; the gray crosses show the training data samples. (**b**) Examples for step responses of the four local linear models 1, 4, 7, and 10

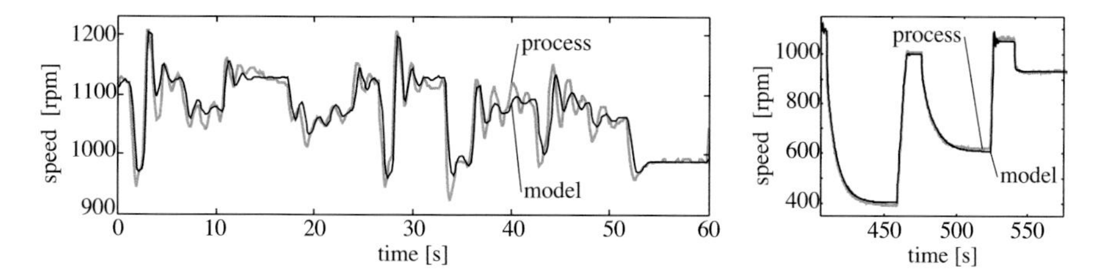

Fig. 25.6 Performance of the local linear neuro-fuzzy model with ten rules

observed process characteristics. Note that a corresponding interpretation of the off-equilibrium regimes is not straightforward.

Figure 25.6 illustrates the performance of the neuro-fuzzy model obtained with ten rules. The model quality is extremely good and could not be reproduced by any alternative modeling and identification technique so far. As expected above the model is not capable of describing more than the first two oscillation periods of a step response at high fan speeds.

25.2 Diesel Engine Turbocharger

This section presents the modeling and identification of an exhaust turbocharger. The need for this turbocharger model arose within an industrial project with the goal of a hardware-in-the-loop simulation of truck diesel engines. The objective of this project is to be able to develop, examine, and test new engine control systems and their electronic hardware realization with a real-time simulation rather than with a real engine; see Fig. 25.7. The use of hardware-in-the-loop simulations makes it possible to save a tremendous amount of time and money compared with experiments on expensive and often not readily available dynamic engine test

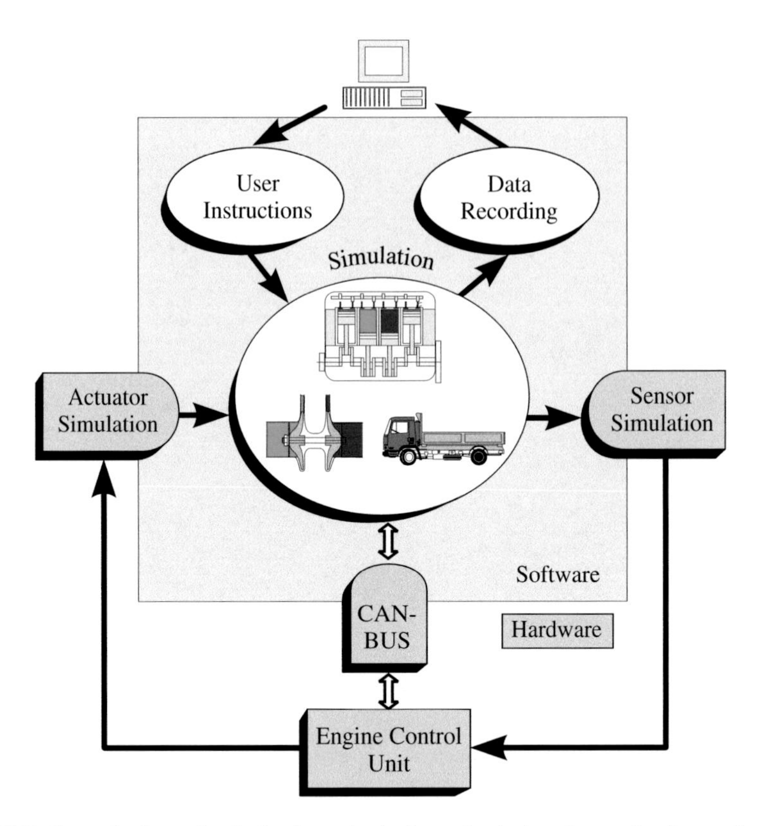

Fig. 25.7 General scheme for the hardware-in-the-loop simulation of a combustion engine with vehicle [283]. (This figure was kindly provided by Jochen Schaffnit, Institute of Automatic Control, TU Darmstadt)

stands. Obviously, a hardware-in-the-loop simulation of a truck Diesel engine relies strongly on accurate dynamic models that can be evaluated in real time on the given hardware platform (here an Alpha chip signal processing board). For more details on the concept and realization of the overall hardware-in-the-loop simulation, refer to [283, 533, 534]. This section demonstrates the modeling of the turbocharger, which is a decisive component for the engine's dynamics. This application underlines the following important features of LOLIMOT:

- utilization of a reduced and low-order premise input space;
- interpretation of the statics and dynamics of the local linear models;
- transformation of the identified model to different sampling time.

25.2.1 Process Description

Figure 25.8 shows a picture of an exhaust turbocharger, and Fig. 25.9 illustrates its role in the charging process of a diesel engine. The operation of the turbocharger is as follows. The exhaust enthalpy is utilized by the turbine to drive a compressor that aspirates and precompresses fresh air in the cylinder. Thus, the turbocharger allows a higher boost pressure, increasing the maximum power of the engine while its stroke volume remains the same. This is particularly important in the middle-speed range. The charging process possesses a nonlinear static input/output behavior as well as a strong dependency of the dynamic parameters on the operating point. This is known by physical insights and is confirmed by the poor quality of linear process models.

The static behavior of a turbocharger may be sufficiently described by characteristic maps of the compressor and turbine. However, if the dynamics of a turbocharger need to be considered, first principles modeling based on mechanical and thermodynamical laws is required [68, 471, 616]. Practical applications have shown that first principles models are capable of reproducing the characteristic dynamic behavior of a turbocharger. The model quality, however, depends crucially on the accurate knowledge of various process parameters. They typically have to be laboriously derived, estimated, or in most cases approximated by analogy considerations. The tedious determination of the physical parameters makes the development of a turbocharger model time-consuming and expensive. Another severe drawback of the first principles modeling approach is the considerable computational effort required for the simulation of these models owing to the high complexity of the equations obtained.

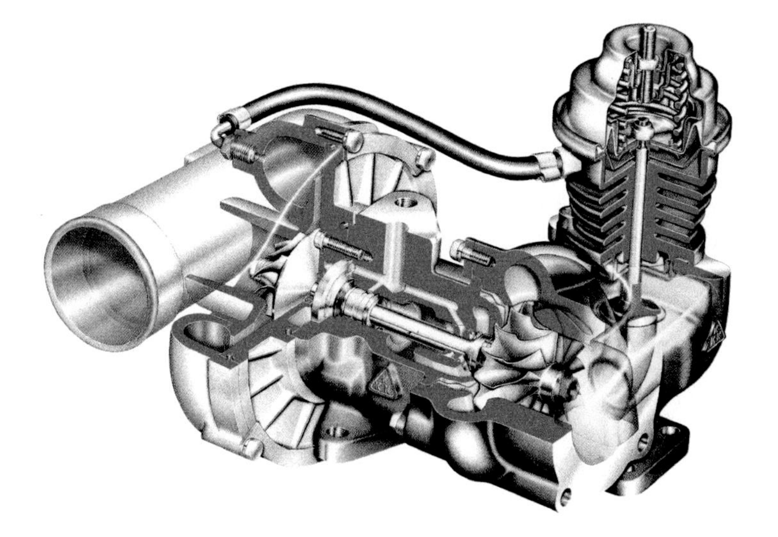

Fig. 25.8 Picture of a turbocharger

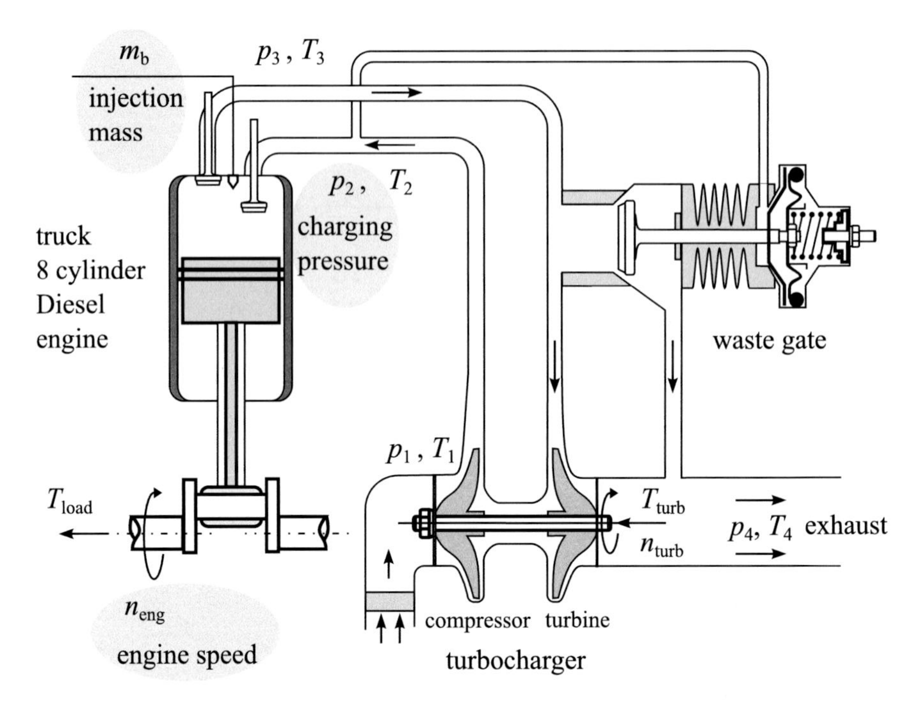

Fig. 25.9 Scheme of the charging process of a diesel engine by an exhaust turbocharger

For these reasons, such an approach is considered to be inconsistent with the usual requirements of control engineering applications such as controller design, fault diagnosis, and hardware-in-the-loop simulation. Here, simple and cost-effective models suitable for real-time simulation are needed. Consequently, an experimental modeling approach with LOLIMOT is presented in the following [434]. The nonlinear system identification is performed in a black box manner. However, the obtained local linear neuro-fuzzy model can be easily interpreted and reveals insights about the process.

The model will describe the charging pressure p_2 generated by the turbocharger. As input signals, the directly relevant quantities such as pressure and temperature at the input of the turbocharger and its speed are not available because they cannot be measured. Therefore, the injection mass m_b and the engine speed n_{eng} serve as input signals for the model. Thus, the model includes a part of the engine's behavior as well. The examined motor is an 8 cylinder diesel engine with 420 kW maximum power and 2200 Nm maximum torque. The sampling time is chosen with respect to the approximate process dynamics as $T_0 = 0.2$ s.

25.2.2 Experimental Results

For the turbocharger model, the charging pressure $y = p_2$ is used as output; the injection mass $u_1 = m_b$ and engine speed $u_2 = n_{eng}$ are the inputs.

25.2.2.1 Excitation and Validation Signals

The training data shown in Fig. 25.10 was generated by a special driving cycle to excite the process with all amplitudes and frequencies of interest. This driving cycle was designed by an experienced driver. The measurements were recorded on a flat test track. In order to be able to operate the engine in high load ranges, the truck was driven with the highest possible load. Besides the highly exciting training data, two additional data sets were recorded that reproduce realistic conditions in urban and interstate traffic. These urban and interstate traffic data sets can be used for model validation.

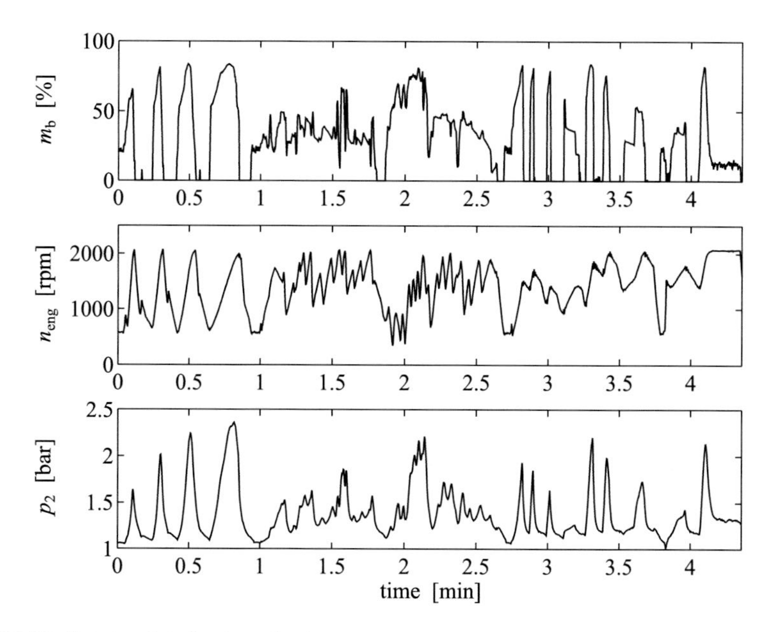

Fig. 25.10 Training data for the turbocharger

25.2.2.2 Modeling and Identification

In a trial-and-error procedure, the following second-order modeling approach with direct feedthrough yielded the best results:

$$y(k) = f\,(u_1(k), u_1(k-1), u_1(k-2),\ u_2(k), u_2(k-1), u_2(k-2),\\ y(k-1), y(k-2))\,. \quad (25.5)$$

Since the improvement compared with a first-order model is not very significant, for many purposes even the first-order model

$$y(k) = f\,(u_1(k), u_1(k-1), u_2(k), u_2(k-1), y(k-1)) \quad (25.6)$$

might be sufficiently accurate. Both approaches differ mainly only in the beginning of their step responses as demonstrated in the next paragraph. Interestingly, the first- and second-order modeling approaches lead to the same kind of input space decomposition by the LOLIMOT algorithm.

It is not quite clear, in terms of first principles, why the incorporation of direct feedthrough, i.e., the regressors $u_1(k)$ and $u_2(k)$, is advantageous with respect to the model quality. When $u_1(k)$ and $u_2(k)$ are discarded from the model, its accuracy drops by 20%. One possible explanation for this effect may lie in the relatively long sampling time and some small unknown and uncontrollable time delays in the measurement and signal processing units.

The convergence curve of LOLIMOT is depicted in Fig. 25.11. It does not suggest an optimal model complexity as unambiguously as the cooling blast application did in the previous section. The choice of $M = 10$ local linear models seems reasonable. This model complexity is justified because for a model with 11 rules, the number of data samples within the least activated local model falls below the threshold given in (13.39).

As Fig. 25.12 demonstrates, the second-order local linear neuro-fuzzy model with ten rules performs very well on all three data sets, the training data, and the validation data. Figure 25.12 compares the measured process output with the simulated model output. The maximum output error is below 0.1 bar.

25.2.2.3 Model Properties

Similar to the cooling blast application presented in the previous section, LOLIMOT partitioned the premise input space only along two input axes, namely, $u_1(k-1)$ and $u_2(k-1)$, which are the delayed injection mass m_b and the delayed engine speed n_{eng}. Obviously, the process is almost linear in its output, the charging pressure $y = p_2$. Consequently, it can be seen as an extension of a Hammerstein structure that also is only nonlinear in the inputs but, of course, does not allow for operating point-dependent dynamics as local linear neuro-fuzzy models do.

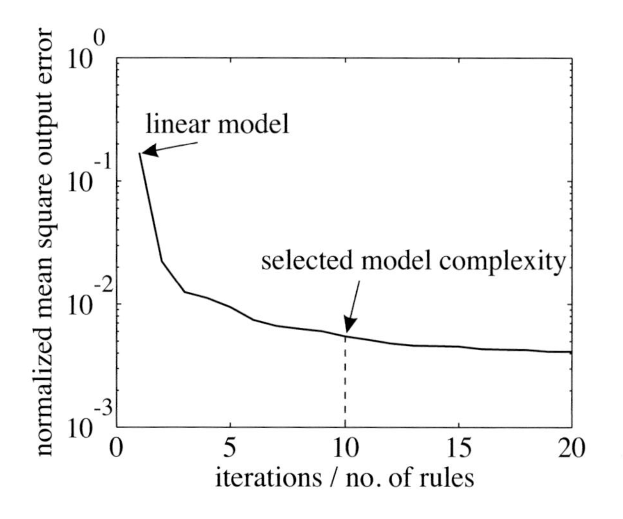

Fig. 25.11 Convergence curve of LOLIMOT

Once the relevance of the eight regressors in (25.5) with respect to the nonlinear process behavior is understood, the premise input space can be reduced to

$$\underline{z}(k) = [u_1(k-1)\; u_2(k-1)]^T, \tag{25.7}$$

while the regression vector for the local linear models remains

$$\underline{x}(k) = [u_1(k)\; u_1(k-1)\; u_1(k-2)\;\; u_2(k)\; u_2(k-1)\; u_2(k-2) \\ y(k-1)\; y(k-2)]^T. \tag{25.8}$$

This is an example of a reduced and low-order premise input space, as discussed in Sect. 22.2.

Figure 25.13a depicts the input space, which again can be easily visualized since it is only two dimensional. The injection mass seems to influence the charging pressure in a more strongly nonlinear manner than the engine speed because LOLIMOT performs more splits along m_b. The training data is also plotted in Fig. 25.13a. This shows that despite all the expertise and effort in its design, it is far from being perfectly distributed. In comparison, Fig. 25.13b shows the distribution of both validation data sets. Obviously, some extrapolation is required for large injection masses and for high engine speeds. These extreme cases have not been covered by the training data. This is a situation that typically cannot be completely avoided in real-world applications. It underlines the importance of reasonable extrapolation behavior of the nonlinear dynamic model; see Sects. 14.4.2 and 22.4.3.

Figure 25.14 shows the step responses of some local linear models in which the injection mass performs a 1% increase at time $k = 0$ and the engine speed is kept fixed. These types of step responses are realistic because in practice the

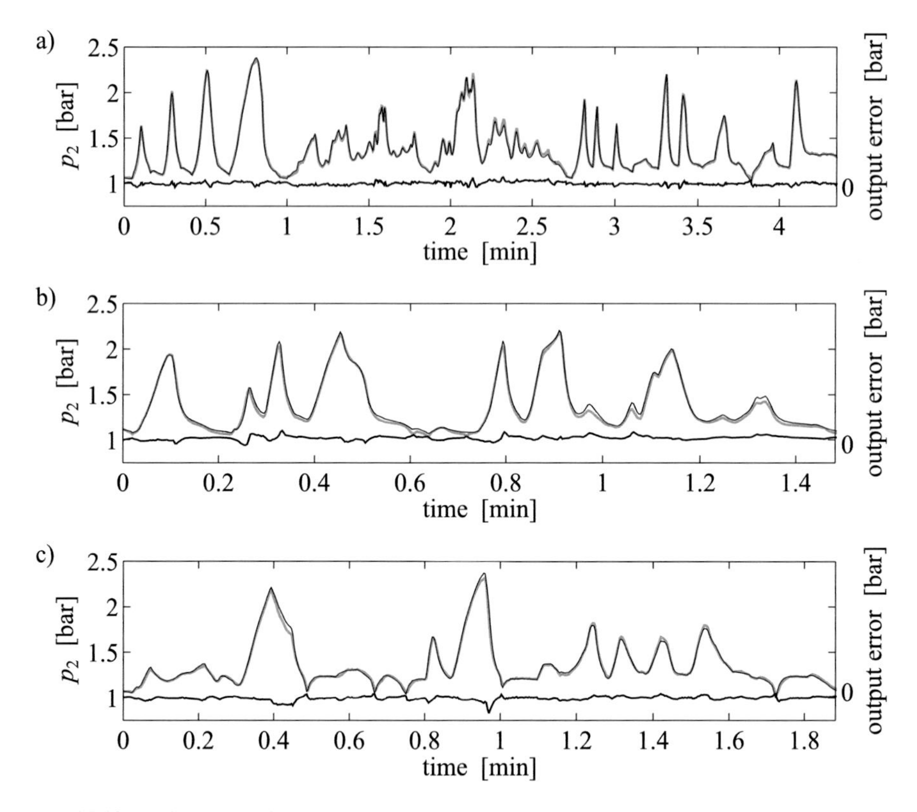

Fig. 25.12 Performance of the second-order model on (**a**) training, (**b**) urban traffic, and (**c**) interstate traffic data

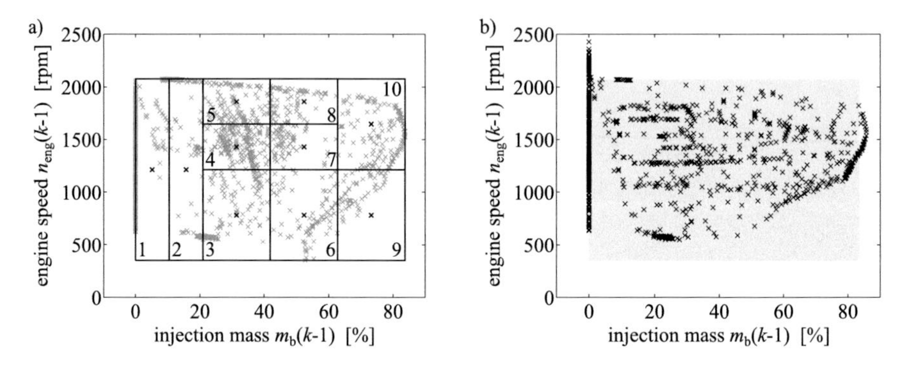

Fig. 25.13 (**a**) Premise input space decomposition performed by LOLIMOT and training data distribution. (**b**) Validation data distribution

injection mass m_b can change stepwise but the engine speed n_{eng} cannot. The step responses of the first-order local linear models are depicted in Fig. 25.14a, and those of the second-order models are plotted in Fig. 25.14b. The left-hand side of

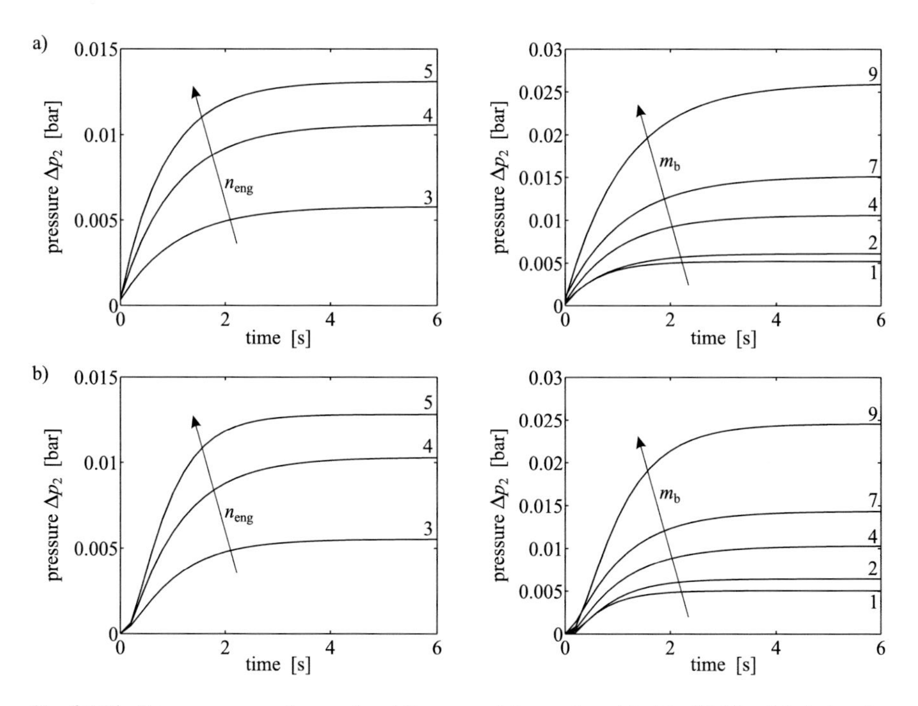

Fig. 25.14 Step responses of some local linear models numbered in Fig. 25.13a: (**a**) first-order model, (**b**) second-order model

Fig. 25.14 depicts three-step responses with increasing engine speed; the right-hand side shows five step responses with increasing injection mass. As assumed from the fine decomposition of the premise input space with respect to the injection mass, this quantity has a stronger influence on the local models' time constants and gains. The time constants vary by a factor of up to 3; the gains differ by a factor of up to 5.

The static behavior calculated from the nonlinear dynamic turbocharger model is shown in Fig. 25.15a. Unfortunately, static measurements from the turbocharger can only be gathered with large effort on a test stand, and they are currently not available for comparison. Nevertheless, static mapping looks very reasonable. However, there is one area for low injection mass and low engine speed in which the static model characteristic possesses a negative slope in the m_b-direction. This effect is clearly unrealistic. Basically, two explanations are possible: Either it is an artifact caused by inadequate data in this region or it is due to the undesired interpolation effects discussed in Sect. 14.4.1. The following analysis shows why the second assumption is correct. All local linear models possess a positive gain. Consequently, the data and the identified local linear models represent the process properly, and a negative gain can result only from undesired interpolation effects. Indeed, Fig. 25.15b, which shows the operating point-dependent gain with respect to the injection mass, reveals the negative gain region and other "oscillations" in the gain that are due to interpolation effects.

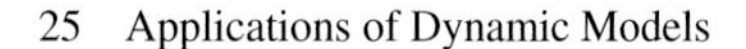

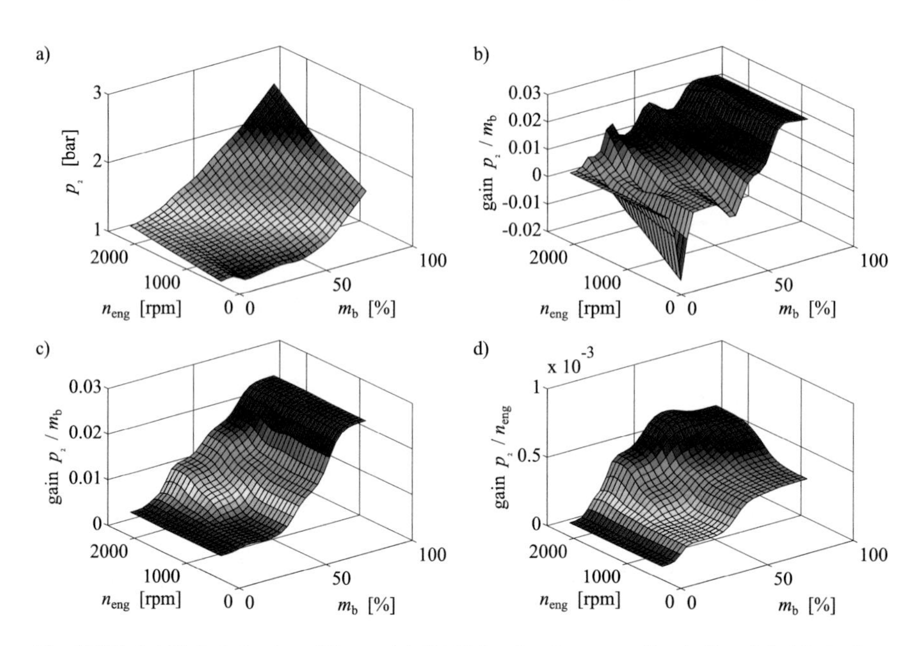

Fig. 25.15 (**a**) Static behavior of the model. (**b**) Gain of p_2/m_b according to the *global* derivative. (**c**) Gain of p_2/m_b according to the *local* derivative. (**d**) Gain of p_2/n_{eng} according to the *local* derivative

As proposed in Sect. 14.5, the method of local differentiation can overcome these undesired effects in the gain calculation. Note, however, that this method affects only the gains, not the original static mapping. The gain calculated as the local derivative of the static model output with respect to the injection mass is shown in Fig. 25.15c; it is a strictly positive function. For the sake of completeness, Fig. 25.15d shows the gain with respect to the engine speed, also calculated with the local derivative.

In practice, the local derivative can be used for the calculation of the gains and time constants in order to avoid interpolation effects. Controller design can also be based either on local linearization or on the parallel distributed compensation strategy; see Sect. 22.3. Nevertheless, the interpolation effects can still be a serious drawback of local linear modeling schemes. Since the suggested remedies for this problem (see Sect. 14.4.1) are still not general and convincing enough, this is left as a major issue for future research.

25.2.2.4 Choice of Sampling Time

The sampling time $T_0 = 0.2$ s for the identification data was chosen with respect to the process dynamics. However, the hardware-in-the-loop simulation has to be carried out 20 times faster with $T_0^* = 0.01$ s since it contains other significant components with much faster dynamics. There are several reasons why the identification should not be carried out on much faster-sampled data, e.g., with $T_0^* = 0.01$ s. First, the amount of data would increase by a factor of 20 which would make its processing tedious. Second, the local linear ARX models obtained by a least squares estimation of the parameters are very sensitive to the choice of the sampling time. ARX models are known to overemphasize high-frequency components in the model fit owing to their unrealistic noise model assumptions, and consequently, their quality deteriorates for overly short sampling times [278, 356]. The latter drawback can be overcome by an appropriate prefiltering of the data that deemphasizes high frequencies accordingly.

Thus, it is beneficial to identify the turbocharger model with data sampled at $T_0 = 0.2$ s and subsequently to transform the obtained model to the faster sampling time of the hardware-in-the-loop simulation $T_0^* = 0.01$ s. The goal of this procedure is merely to utilize the model directly within the hardware-in-the-loop simulation. The transformation is roughly possible for local linear neuro-fuzzy models since each local linear model can be transformed separately. This feature is another important advantage of local linear neuro-fuzzy models over other external dynamic model architectures. Note, however, that the sampling time transformation is strictly speaking incorrect since the signals entering the rule premises are also affected and thus the interpolation between the LLMs is influenced. In many practical cases, this effect is negligible, but special care has to be taken if the premise space contains previous outputs $y(k-i)$. Furthermore, the transformed model correctly describes only the low-frequency range up to half of the original sampling frequency $1/2T_0$ (Nyquist frequency). Effects faster than $1/2T_0$ cannot be reflected by the model.

25.3 Thermal Plant

In this section, a thermal plant with two heat exchangers of different types is considered for modeling. After a description of the complete plant in Sect. 25.3.1, three different parts of this process are modeled in what follows. The simple transport process of water through a pipe with different flow rates is utilized in Sect. 25.3.2 to illustrate the combined LOLIMOT+OLS algorithm for dynamic systems. Section 25.3.3 deals briefly with modeling and identification of the tubular heat exchanger and Sect. 25.3.4 discusses the cross-flow heat exchanger more extensively. These applications underline the following important features of LOLIMOT:

- automatic OLS structure selection for the dynamic local linear models;
- incorporation of prior knowledge into the structure of the local linear models;
- interpolation between local models with different dead times;
- combination of a static operating point with dynamic local linear models;
- Interpretation of the obtained input space partitioning;
- interpretation of the statics and dynamics of the local linear models;
- utilization of error bars for excitation signal design.

25.3.1 Process Description

The industrial-scale thermal plant as shown in Fig. 25.16 will be considered. It contains a steam generator SG, a tubular steam/water heat exchanger HE_1 and a cross-flow air/water heat exchanger HE_2. In the *primary circuit* of the tubular heat exchanger, the steam generator with a power of 54 kW produces saturated steam at a pressure of about 6 bar. The steam flow rate $\dot{V}_s$ can be changed by means of an electric valve. This valve is controlled by G_s in an inner loop by a three-level controller with integral behavior. The steam condenses in the tubular heat exchanger, and the liquid condensate is pumped back to the steam generator.

In the *secondary circuit*, water is heated in the tubular heat exchanger and is cooled again in the cross-flow heat exchanger. The water leaves HE_1 with temperature $\vartheta_{wo}^{(1)}$ and enters HE_2 with temperature $\vartheta_{wi}^{(2)}$. In this cross-flow heat exchanger, the water is cooled by the air stream $\dot{V}_a$ which is controlled by the fan speed[1] F. The incoming air stream has environmental temperature ϑ_{ai}, and

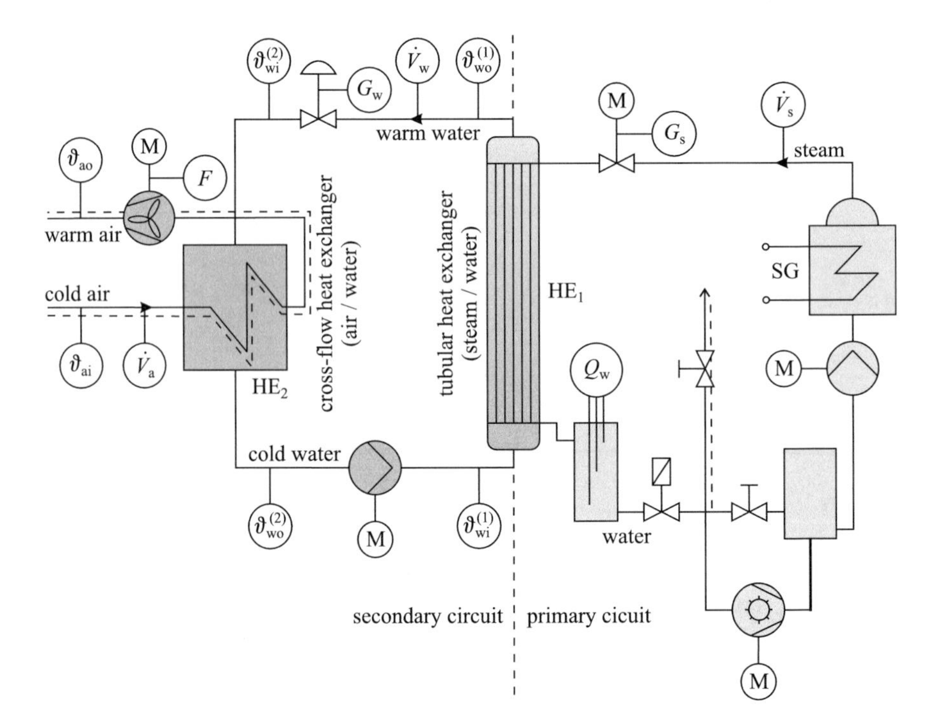

Fig. 25.16 Scheme of the thermal plant with the steam generator SG and two heat exchangers HE_1 and HE_2

[1]Note that the air stream cannot be directly measured, but the measurable fan speed is an almost proportional quantity.

Table 25.1 Overview of the inputs and outputs of the considered processes in the thermal plant

Process	u_1	u_2	u_3	y
Transport process	$\vartheta_{\text{wo}}^{(2)}$	$\dot{V}_{\text{w}}$	–	$\vartheta_{\text{wi}}^{(1)}$
Tubular heat exchanger	$G_{\text{s}} \to \dot{V}_{\text{s}}$	$\dot{V}_{\text{w}}$	$\vartheta_{\text{wo}}^{(2)}$	$\vartheta_{\text{wo}}^{(1)}$
Cross-flow heat exchanger	$F \to \dot{V}_{\text{a}}$	$\dot{V}_{\text{w}}$	$\vartheta_{\text{wi}}^{(2)}$	$\vartheta_{\text{wo}}^{(2)}$

"→" = "influences directly"

the outcoming heated air stream has temperature ϑ_{ao}. The water leaves HE_2 with temperature $\vartheta_{\text{wo}}^{(2)}$ and finally enters HE_1 again with temperature $\vartheta_{\text{wi}}^{(1)}$. The water flow rate $\dot{V}_{\text{w}}$ is changed by an electro-pneumatically driven valve controlled in an inner loop by a PI controller with the command signal G_{w}.

In the next three subsections, the following parts of the thermal plant are investigated: (1) the transport process from $\vartheta_{\text{wo}}^{(2)}$ to $\vartheta_{\text{wi}}^{(1)}$, (2) the tubular heat exchanger, and (3) the cross-flow heat exchanger. Table 25.1 summarizes the inputs and outputs for these processes.

25.3.2 *Transport Process*

This section illustrates the OLS structure selection for the rule consequents with a simple transport process. Modeling by first principles would be quite easy for this example when the geometry of the pipe system in the thermal plant is assumed to be known. Here, this information will not be exploited, to demonstrate the power of the experimental modeling approach. The water temperature $y = \vartheta_{\text{wi}}^{(1)}$ at the HE_1 input will be modeled in dependency on the temperature $u_1 = \vartheta_{\text{wo}}^{(2)}$ at the HE_2 output and the water flow rate $u_2 = \dot{V}_{\text{w}}$.

Two different measurement data sets, each 120 min long, were acquired for training and validation. Figure 25.17 shows the first half of the training data set. The sampling time was chosen as $T_0 = 1$ s.

25.3.2.1 Modeling and Identification

From simple physical considerations, the following basic relations are known:

- The temperature $y = \vartheta_{\text{wi}}^{(1)}$ is, with a certain time delay, about equal to the temperature $u_1 = \vartheta_{\text{wo}}^{(2)}$. Thus, the static gain of the process can be assumed to be around 1. In fact, a value slightly smaller than 1 has to be expected because the water temperature is higher than the environment temperature and some energy will be lost to the environment.
- The flow rate of the water $u_2 = \dot{V}_{\text{w}}$ determines the dead time (large dead time at low flow rates, small dead time at high flow rates). These relations are also obvious from Fig. 25.17.

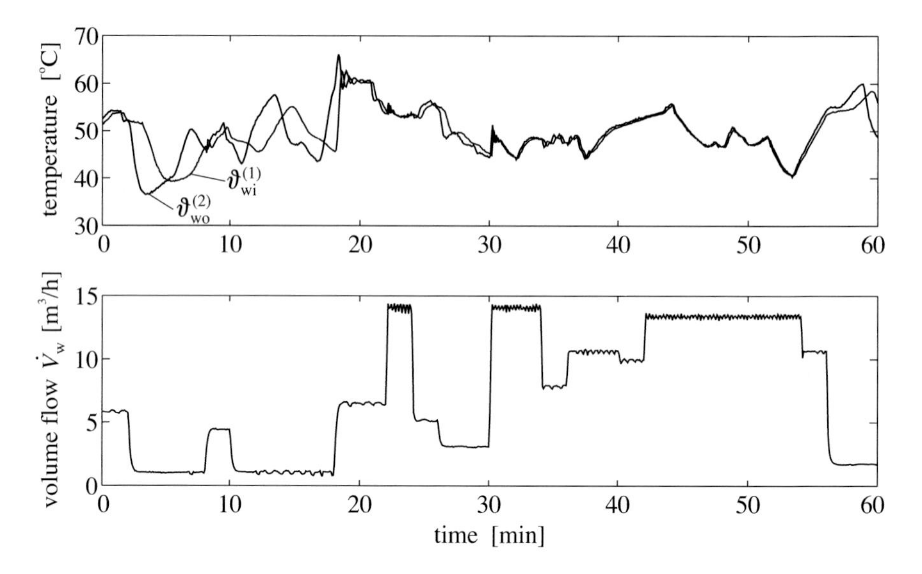

Fig. 25.17 First half of the training data set for the transport process

- Especially at low flow rates, deviations from a pure dead time behavior can be seen, which are probably due to energy storage in the water pipe. Thus, a first-order time-lag model is reasonable.

This analysis implies the following modeling structure:

$$y(k) = f\left(u_1(k-1), \ldots, u_1(k-d_{\max}),\ u_2(k-1),\ y(k-1)\right) \tag{25.9}$$

with $y = \vartheta_{\text{wi}}^{(1)}$, $u_1 = \vartheta_{\text{wo}}^{(2)}$, and $u_2 = \dot{V}_{\text{w}}$. The current input $u_1(k)$ is not used in (25.9) because the dead time is larger than the sampling time for all operating conditions $T_{\text{t}} > T_0$.

From Fig. 25.17, the maximum dead time $d_{\max}$ can be estimated to be smaller than 100 s. The nonlinear behavior of the function $f(\cdot)$ is assumed to be dependent only on the water flow rate. Therefore, a one-dimensional premise input space can be chosen:

$$\underline{z}(k) = u_2(k-1)\,. \tag{25.10}$$

For the rule conclusions, the following 16 terms are presented to the structure determination algorithm:

$$\underline{x}(k) = [u_1(k-1)\ u_1(k-2)\ u_2(k-4)\ u_2(k-8)\ u_1(k-16)$$
$$u_1(k-24)\ \cdots\ u_1(k-96)\ \ y(k-1)]^T\,. \tag{25.11}$$

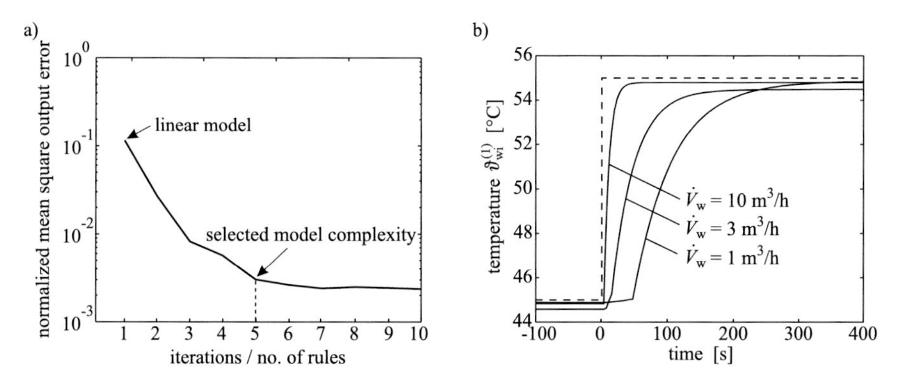

Fig. 25.18 (**a**) Convergence curve of LOLIMOT. (**b**) Step responses of the model for a change in the input temperature $\vartheta_{\mathrm{wo}}^{(2)}$ from 45 °C to 55 °C

In principle, all terms with dead times varying from 1 to 100 could be presented to the OLS structure determination algorithm. From previous experience, however, this leads only to very small improvements in the model accuracy at the cost of the significantly increased training effort. Out of the terms given in (25.11), the OLS algorithm was allowed to select two regressors for each rule consequent. Therefore only two parameters have to be estimated for each rule. Note that no offset is modeled because it is clear from physical insights that there exists no offset between $y = \vartheta_{\mathrm{wi}}^{(1)}$ and $u_1 = \vartheta_{\mathrm{wo}}^{(2)}$.

Figure 25.18a shows the convergence curve of LOLIMOT, indicating that a model with five rules is sufficiently accurate. Figure 25.19 summarizes the modeling strategy pursued.

25.3.2.2 Model Properties

Figure 25.18b depicts three step responses of the neuro-fuzzy model for a change of $u_1 = \vartheta_{\mathrm{wo}}^{(2)}$ from 45 °C to 55 °C. It can clearly be seen that the time constants and dead times decrease for increasing flow rates. In the static operating points, $y = \vartheta_{\mathrm{wi}}^{(1)}$ is slightly smaller than $u_1 = \vartheta_{\mathrm{wo}}^{(2)}$ as expected owing to a heat loss to the environment during the water transport. However, from physical considerations, the response for $\dot{V}_{\mathrm{w}} = 1\,\mathrm{m}^3/\mathrm{h}$ should be below that for $\dot{V}_{\mathrm{w}} = 3\,\mathrm{m}^3/\mathrm{h}$. This small model error is probably caused by insufficient static excitation for very low flow rates in the training data because the time constant and dead time are very large. Another interesting aspect can be noticed by examining the model behavior of the step response with $\dot{V}_{\mathrm{w}} = 1\,\mathrm{m}^3/\mathrm{h}$. From Fig. 25.17 it can be observed that the dead time is around 48 s. Because all rules in the model are active (at least to a small degree), the model's response is not exactly equal to zero for the first 48 sampling periods after the step input. Nevertheless, this effect is negligible since the validity values of the rules representing larger flow rates are very small. This effect can be

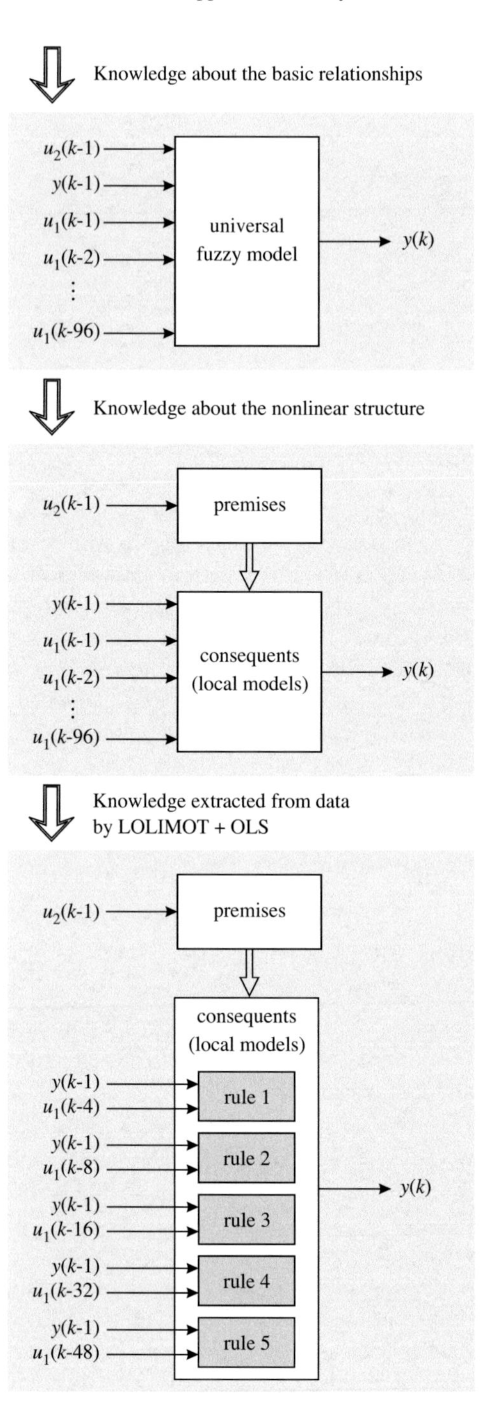

Fig. 25.19 Construction of the neuro-fuzzy model by prior knowledge and structure optimization

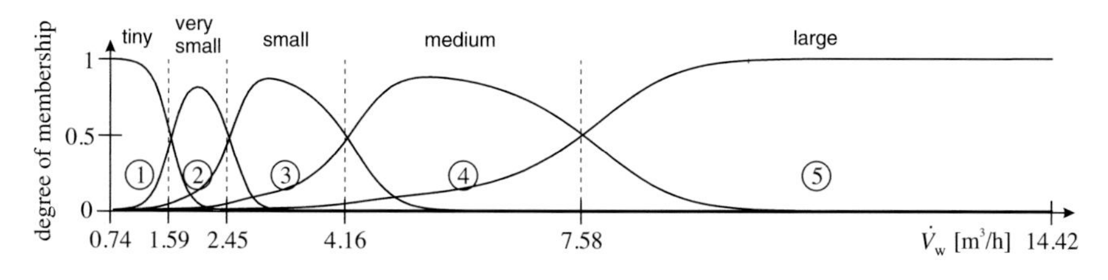

Fig. 25.20 Membership functions generated by LOLIMOT

overcome by a slight change in the interpolation law with respect to the dead times, as explained in Sect. 22.3.3.

The neuro-fuzzy model comprises the following five rules with the membership functions depicted in Fig. 25.20:

1. IF $u_2(k-1)$ is tiny THEN $y(k) = 0.017\, u_1(k-48) + 0.984\, y(k-1)$
2. IF $u_2(k-1)$ is very small THEN $y(k) = 0.023\, u_1(k-32) + 0.977\, y(k-1)$
3. IF $u_2(k-1)$ is small THEN $y(k) = 0.023\, u_1(k-16) + 0.977\, y(k-1)$
4. IF $u_2(k-1)$ is medium THEN $y(k) = 0.068\, u_1(k-8) + 0.932\, y(k-1)$
5. IF $u_2(k-1)$ is large THEN $y(k) = 0.121\, u_1(k-4) + 0.878\, y(k-1)$

This result can easily be interpreted. LOLIMOT has recognized that the dead times and time constants of the local linear models change much more strongly at low flow rates than at higher ones. This can be seen from the membership functions in Fig. 25.20 (fine divisions for low flow rates, coarse divisions for high flow rates). For all local linear models, the OLS algorithm selects one delayed input temperature and the previous output temperature. The dead times change in a wide interval and are, as expected, large for low flow rates and small for high flow rates. The same holds for the time constants (computed from the poles of the local linear models), which vary between 60 s (LLM 1) and 7.7 s (LLM 5). For all local linear models, a gain slightly smaller than 1 was identified, which also corresponds to the physical insights. Figure 25.21 shows the performance of the model simulated in parallel to the process.

Finally, it should be remarked that an automatic structure selection of the rule consequents is not necessary if the significant regressors can be chosen by prior knowledge. For example, the dead time can be calculated from the water flow rate if the length and the diameter of the pipe are known. However, in less trivial cases, such as those discussed in the following subsections, structure selection serves as an important tool.

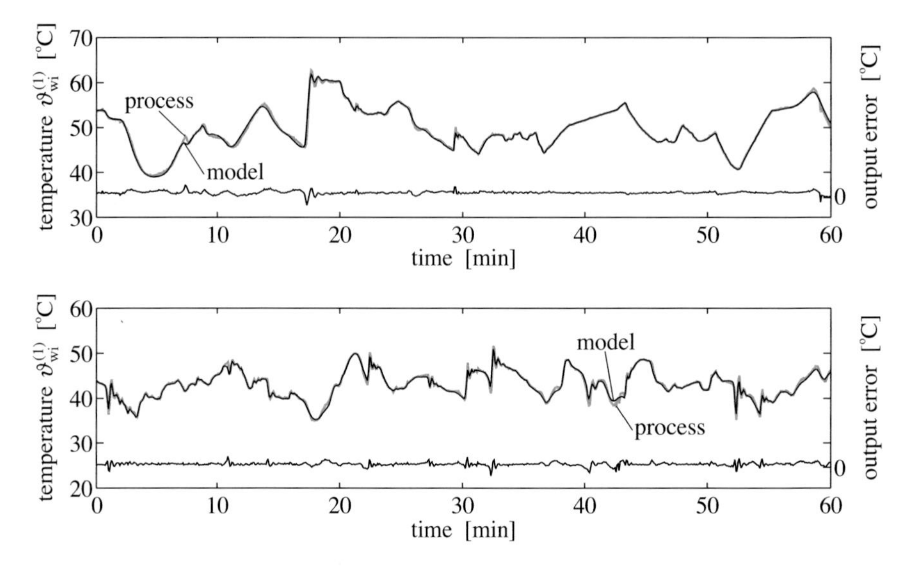

Fig. 25.21 Performance of the local linear neuro-fuzzy model on training (top) and validation (bottom) data

25.3.3 Tubular Heat Exchanger

In this subsection, the tubular steam/water heat exchanger HE_1 will be modeled. The output of the model is the water outlet temperature $y = \vartheta_{wo}^{(1)}$. It depends on the following inputs: the position $u_1 = G_s$ of the valve that controls the steam flow $\dot{V}_s$, the water flow rate $u_2 = \dot{V}_w$, and the inlet temperature $u_3 = \vartheta_{wo}^{(2)}$. Previous results in modeling and identification of the tubular heat exchanger can be found in [188, 284]. The obtained model can be used for controller design where the valve position is the manipulated variable, the outlet temperature is the controlled variable, and the water flow rate and the inlet temperature are measurable disturbances. For the design of a predictive controller, refer to [136, 236]. The utilization of the model for predictive control also clarifies the question why rather the outlet temperature of HE_2, $\vartheta_{wo}^{(2)}$, is used as model input instead of the much closer inlet temperature $\vartheta_{wi}^{(1)}$. The reason is that it is advantageous for the controller to "know" this disturbance earlier, in particular, if the significant dead time from the actuation signal G_s to the controlled variable is considered.

Two different measurement data sets, each about 120 min long, were acquired for training and validation. Figure 25.22 shows the highly exciting training data set. The sampling time was chosen as $T_0 = 1$ s.

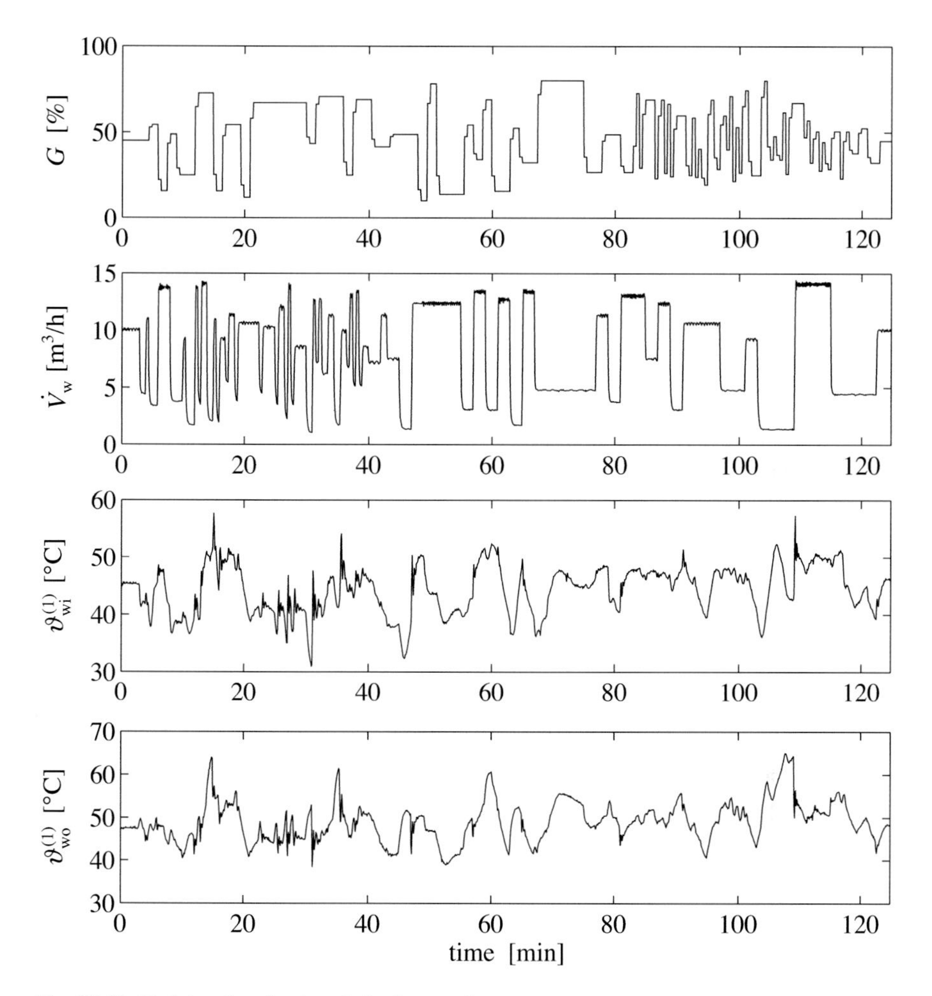

Fig. 25.22 Training data for the tubular heat exchanger

25.3.3.1 Modeling and Identification

For modeling the tubular heat exchanger, the operating condition-dependent dead times from the valve position $u_1 = G_s$ and the inlet temperature $u_3 = \vartheta_{wo}^{(2)}$ to the process output have to be taken into account. As demonstrated in the previous section, this can be done by supplying several delayed versions of these inputs as potential regressors for the local linear models to the LOLIMOT+OLS algorithm. By trial and error, it can be found that a model of first dynamic order can describe the process well. No significant improvement can be achieved by increasing the model order. In fact, for a second-order model, the $y(k-2)$ regressors are partly not selected by the OLS, indicating a low relevance of these terms. Those local linear models for which $y(k-2)$ is selected possess one dominant real pole (comparable to the poles of the first-order model), and the second pole is negative real. This also indicates that a second-order model is overparameterized. Therefore, the following first-order modeling approach is taken:

$$y(k) = f\,(u_1(k-1), \ldots, u_1(k-18),\ u_2(k-1),\\ u_3(k-1), \ldots, u_3(k-57),\ y(k-1))\ , \tag{25.12}$$

with $u_1 = G_S$, $u_2 = \dot{V}_W$, $u_3 = \vartheta_{WO}^{(2)}$, and $y = \vartheta_{WO}^{(1)}$. The maximum dead times from the valve position, 18 s, and from the inlet temperature, 57 s, to the process output are estimated from prior experiments. In order to reduce the computational demand of the OLS algorithm, some of the delayed inputs in (25.12) can be discarded, thereby reducing the number of potential regressors. For example, only the following terms are considered without a significant accuracy loss: $u_1(k-i)$ with $i = 1, 2, 3, 4, 6, 9, 12, 15, 18$ and $u_3(k-i)$ with $i = 1, 2, 3, 5, 7, 11, 17, 25, 38, 57$. This approach is pursued in the sequel.

The number of selected regressors is chosen equal to seven in order to allow the selection of the autoregressive term $y(k-1)$, an offset, one parameter for $u_2(k-1)$, and two parameters for $u_1(k-i)$ and $u_3(k-i)$ each. The selection of two regressors for the inputs u_1 and u_3 allows some compensation for the unavailability of some dead times in the list of potential regressors. Furthermore, it gives the model some flexibility to compensate for the common "denominator" dynamics for all inputs; see Sect. 19.6.

The premise input space is chosen as a first-order operating point that represents a subset of the regressors in (25.12):

$$\underline{z}(k) = [u_1(k-1)\ u_2(k-1)\ u_3(k-1)\ y(k-1)]^T\ . \tag{25.13}$$

Figure 25.23a depicts the convergence curve of LOLIMOT. A local linear neuro-fuzzy model with eight rules is selected. As Fig. 25.24 demonstrates, this local linear

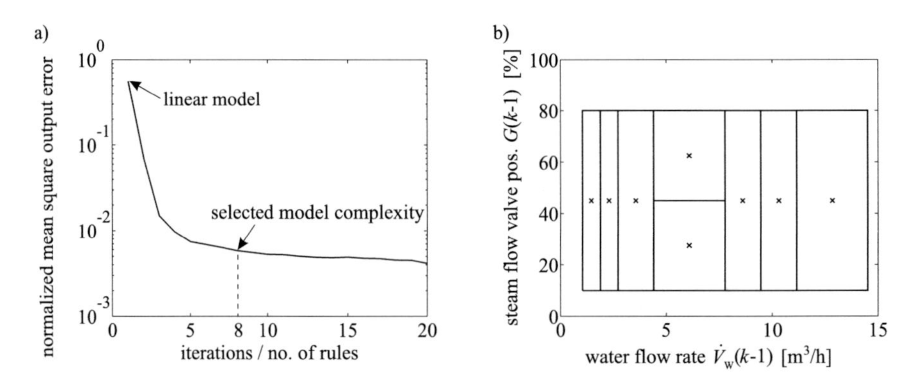

Fig. 25.23 (**a**) Convergence curve of LOLIMOT. (**b**) Premise input space decomposition performed by LOLIMOT

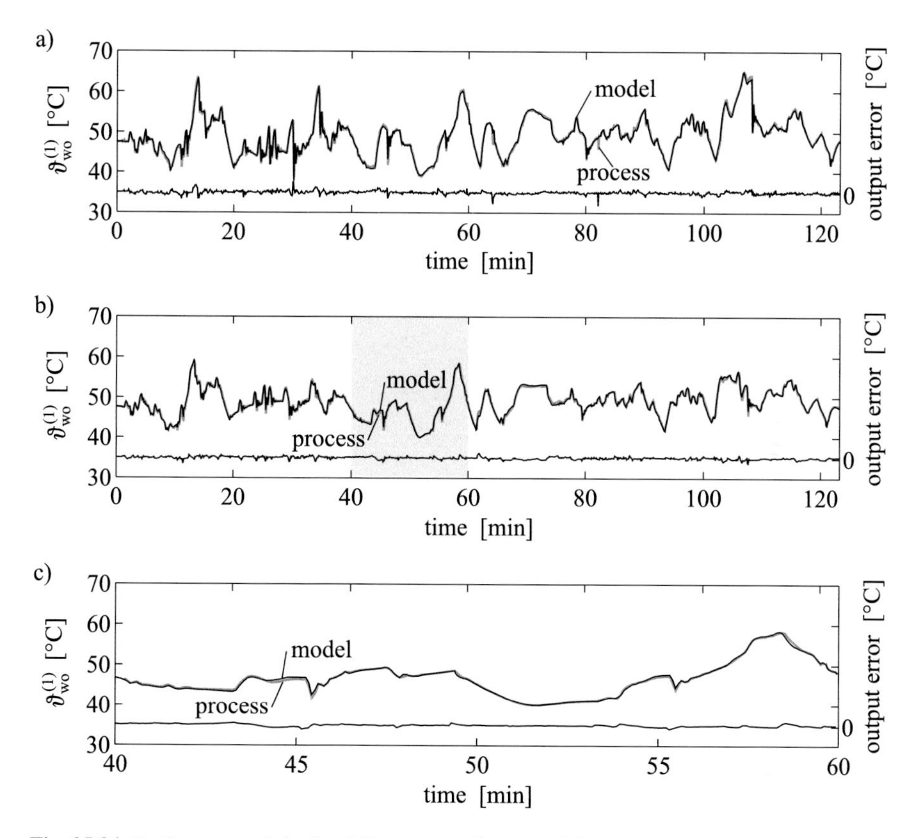

Fig. 25.24 Performance of the local linear neuro-fuzzy model on (**a**) training and (**b**) validation data, (**c**) validation data enlarged

neuro-fuzzy model can describe the process very accurately with a maximum error of 1.5 °C.

25.3.3.2 Model Properties

Figure 25.23b shows the premise input space partitioning performed by LOLIMOT. Obviously, the effect of the inlet temperature $u_3 = \vartheta_{\text{wo}}^{(2)}$ and the outlet temperature $y = \vartheta_{\text{wo}}^{(1)}$ is not significantly nonlinear since LOLIMOT does not split along these axes. Thus, the premise input space can be reduced to $\underline{z}(k) = [u_1(k-1)\ u_2(k-1)]^T$ without any accuracy loss. Furthermore, the water flow rate $u_2 = \dot{V}_{\text{w}}$ has a strongly nonlinear influence on the process model, which causes the fine decomposition of this input. The nonlinearity seems to increase with decreasing water flow rates since the partitioning becomes finer.

The static model behavior is shown in Fig. 25.25a. This plot verifies the assumption made from the premise input space partitioning that the nonlinearity

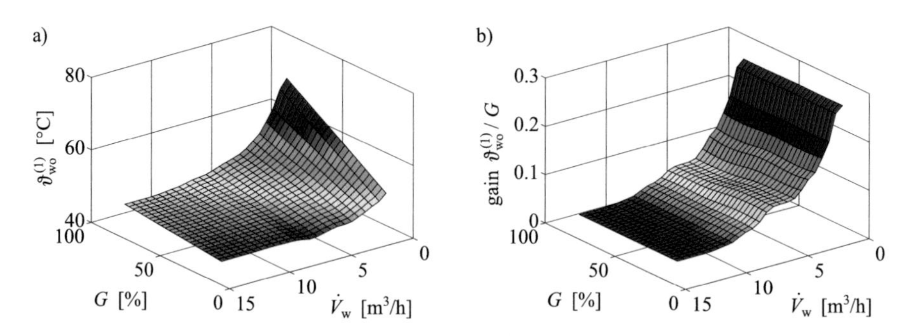

Fig. 25.25 (**a**) Static behavior of the model for inlet temperature $\vartheta_{\text{wo}}^{(2)} = 45\,°\text{C}$. (**b**) Gain of $\vartheta_{\text{wo}}^{(1)}/G_{\text{s}}$ for $\vartheta_{\text{wo}}^{(2)} = 45\,°\text{C}$ according to the local derivative

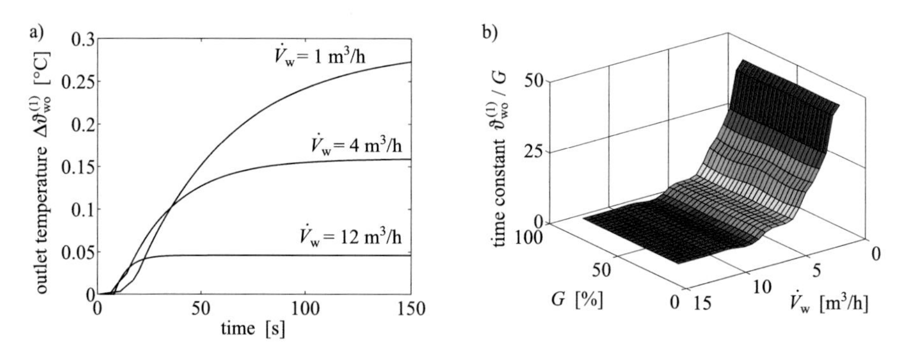

Fig. 25.26 (**a**) Step responses of the model for a change of the steam valve position G_{s} from 50% to 51% for the inlet temperature $\vartheta_{\text{wo}}^{(2)} = 45\,°\text{C}$. (**b**) Time constant of $\vartheta_{\text{wo}}^{(1)}/G_{\text{s}}$ for $\vartheta_{\text{wo}}^{(2)} = 45\,°\text{C}$ according to the local derivative

becomes increasingly stronger as the water flow rate decreases. The local derivative of the static mapping with respect to the input valve position G_{s} yields the gain from this input to the process output; see Fig. 25.25b. By first principles considerations, it can be shown that the gain should be approximately of hyperbolic shape. This is in good correspondence to the experimental model.

Figure 25.26a depicts three step responses of the model for small, medium, and large water flow rates to underline the dramatically different gains and to illustrate the strongly operating point-dependent dynamics. The inlet temperature is kept at $\vartheta_{\text{wo}}^{(2)} = 45\,°\text{C}$, and the valve position G_{s} is changed in a step from 50% to 51%. With decreasing water flow rate, it is not only the gain that grows, but the time constants and the dead times also increase. Owing to the selection of several $G_{\text{s}}(k - i)$-regressors by the OLS, the step responses for small water flow rates are very similar to those of a higher-order system. Thus, it is not easy to compare the dead times of the different local linear models directly. Note that with several "numerator" coefficients, i.e., with more than one selected regressor per input, the dynamics becomes some mixture between a first-order time-lag system and a finite impulse response model. This makes the local linear models less interpretable, but this effect is well known from linear MISO modeling by matrix polynomial models [279, 356]. Figure 25.26b shows the operating point dependency of the model's time constant.

25.3.4 Cross-Flow Heat Exchanger

In this section, the cross-flow air/water heat exchanger HE_2 will be modeled; see [424]. Some characteristic properties of this process are illustrated in Fig. 25.27. Obviously, the process is strongly nonlinear in its static and dynamic behavior with respect to the water flow rate $\dot{V}_w$. The fan speed F has a nonlinear influence mainly for low water flow rates. The output of the model is the water outlet temperature $y = \vartheta_{wo}^{(2)}$. It is influenced by the following inputs: the air stream $\dot{V}_a$, which is controlled by the fan speed $u_1 = F$; the water flow rate $u_2 = \dot{V}_w$; the water inlet temperature $u_3 = \vartheta_{wi}^{(2)}$; and the air inlet temperature ϑ_{ai}. Since the air temperature is equal to the environment temperature, it cannot be actively influenced by the user. It cannot be excited and thus is not included as a (fourth) model input. In what follows, it is assumed that the environment temperature stays about constant at $\vartheta_{ai} \approx 12\,°C$. Clearly, the accuracy of the experimental model will deteriorate when this assumption is violated. This is a fundamental restriction of experimental modeling. The drawback can be overcome by the incorporation of prior knowledge obtained by static first principles modeling. For details refer to [147], and see also Sect. 28.2.

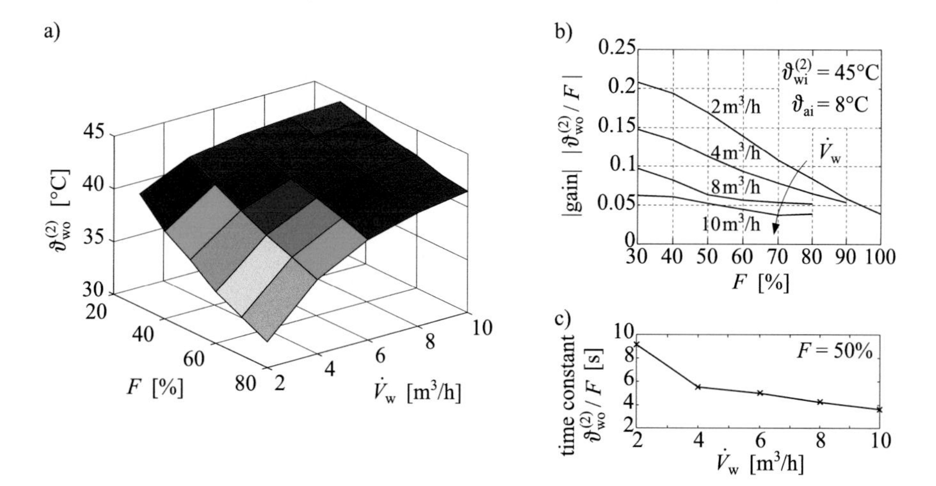

Fig. 25.27 Measured cross-flow HE characteristics: (**a**) static behavior, (**b**) gains, (**c**) time constants

25.3.4.1 Data

In the design of the excitation signals depicted in Fig. 25.28, the prior knowledge available from the measurements shown in Fig. 25.27 is used. Furthermore, the method of error bars is utilized; see Sect. 14.9 and [156]. Since the strength of the process nonlinearity increases with decreasing water flow rates, regions with low $\dot{V}_W$ are covered more densely with data than regions with high $\dot{V}_W$. The fan speed F is dynamically excited with different APRBS sequences; see Sect. 19.7 and [430, 431]. For low water flow rates where the fan speed possesses a significant nonlinear influence, several set points are covered, while for high water flow rates, one APRBS covers the whole operating range. The minimum hold time of the APRB signals is adjusted with respect to the approximate time constants of the process.

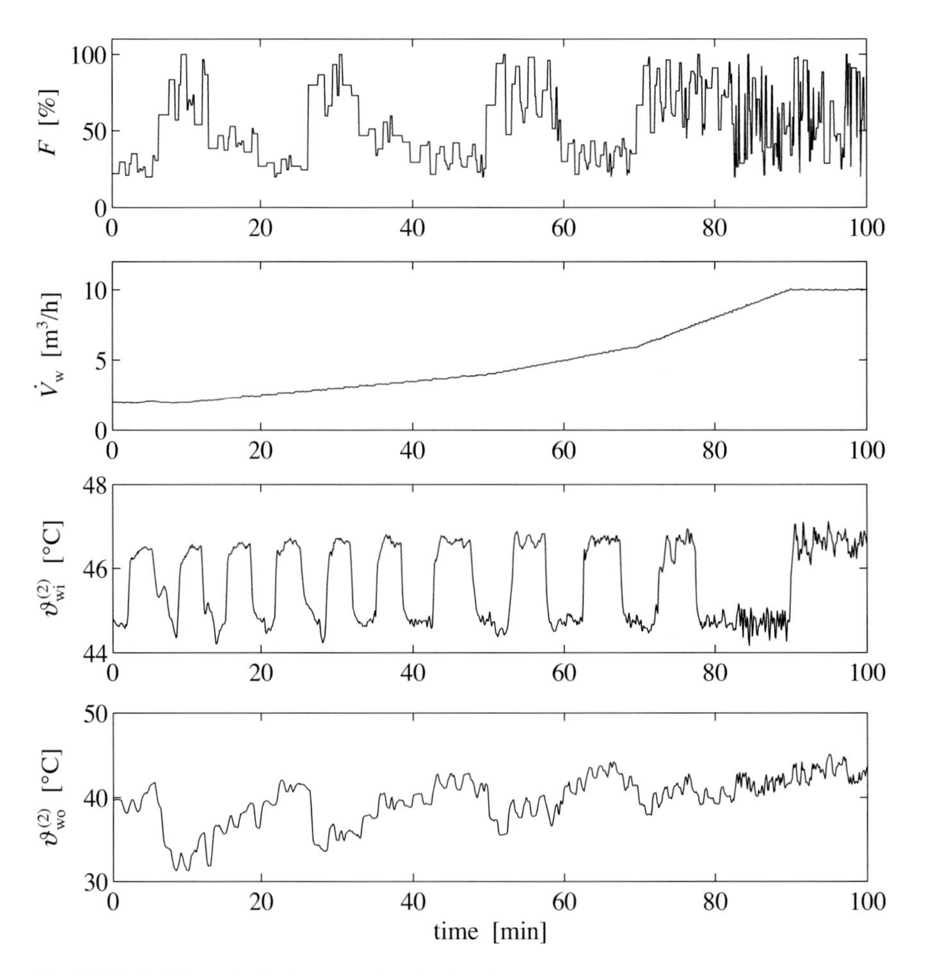

Fig. 25.28 Training data for the cross-flow heat exchanger

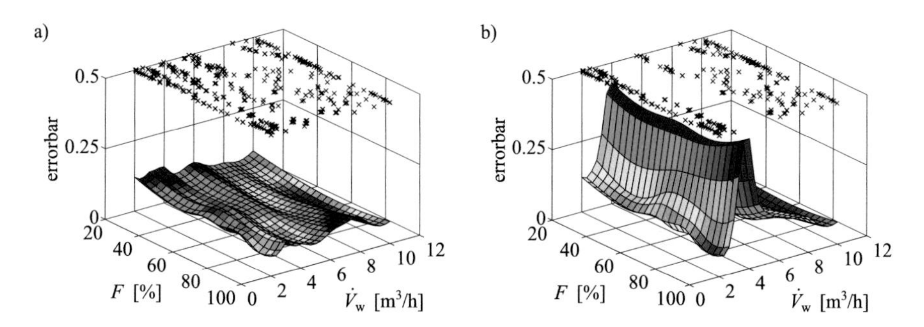

Fig. 25.29 Error bars of the model for (**a**) a well-designed excitation signal and (**b**) missing excitation around the water flow rate $\dot{V}_w \approx 4\,\text{m}^3/\text{h}$. The water inlet temperature is fixed at $\vartheta_{\text{wi}}^{(2)} = 45\,°\text{C}$. The crosses mark training data samples

Thus, the fan speed is excited with higher frequencies for higher water flow rates. The inlet temperature is changed between two operating points in order to be able to estimate the parameters associated with $\vartheta_{\text{wi}}^{(2)}$. For more details on the design of excitation signals for nonlinear dynamic processes, refer to [147, 156].

The method of error bars proposed in Sect. 14.9 can be used as a tool for excitation signal design. It allows one to detect regions with insufficient data density in the input space. For example, Fig. 25.29a depicts the error bars of the local linear neuro-fuzzy model as it is identified with LOLIMOT from the training data shown in Fig. 25.28. These error bars are calculated for the static operating points, which allows a three-dimensional representation. Note that error bars can also be calculated for dynamic operation. The flat shape of the error bar indicates an appropriate data distribution. If data with water flow rates around $\dot{V}_w \approx 4\,\text{m}^3/\text{h}$ is not included in the training data set, the error bar indicates this, as shown in Fig. 25.29b. Note that in this two-dimensional case, it would be possible for the user to assess the quality of the training data by comparing visually the density of the data distribution with the density of the local linear models. In higher-dimensional problems, however, tools such as the error bars are clearly most useful.

25.3.4.2 Modeling and Identification

For modeling the cross-flow heat exchanger HE_2, similar to HE_1, the dead times from the water inlet temperature and the fan speed to the outlet temperature have to be taken into account. The model is also chosen to be of first dynamic order because for a second-order model one pole of each estimated local linear model lies very close to zero ($|p_i| < 0.2$ for all $i = 1, 2, \ldots, M$), indicating negligible second-order dynamic effects. With the minimal and maximum dead times obtained from prior experiments, these considerations lead to the following modeling approach:

$$y(k) = f\,(u_1(k-3), \ldots, u_1(k-9),\ u_2(k-1),$$
$$u_3(k-1), \ldots, u_3(k-21),\ y(k-1))\ , \qquad (25.14)$$

where $u_1 = F$, $u_2 = \dot{V}_{\rm w}$, $u_3 = \vartheta^{(2)}_{\rm wi}$, and $y = \vartheta^{(2)}_{\rm wo}$.

In contrast to the tubular heat exchanger, here it will be assumed that the pipe system geometry is known. Thus, the dead time from the inlet to the outlet temperature can be calculated [147]. For each local linear model, the water flow rate at its center is used for this calculation, and the two regressors with the closest time delays are selected automatically. If, e.g., the calculated dead time is equal to 8.3 s, then the two regressors $u_3(k-8)$ and $u_3(k-9)$ will be included in the corresponding LLM. Therefore, only the significant regressors of the fan speed $u_1(k-3), \ldots, u_1(k-9)$ have to be selected by the OLS. The selection of an autoregressive term $y(k-1)$, an offset, the water flow rate $u_2(k-1)$, and two inlet temperature regressors $u_3(k-i)$ is thus enforced. Therefore, seven parameters have to be estimated for each LLM. The task of the OLS is merely to select two fan speed regressors. By this strategy, the computational demand for the structure selection is reduced significantly since its complexity grows strongly with the number of potential regressors, which here is reduced to the seven terms $u_1(k-3), \ldots, u_1(k-9)$. As an alternative modeling approach, no OLS structure selection is applied at all. Rather the inlet temperature regressors are chosen by prior knowledge, and all seven fan speed regressors are incorporated into the model. This strategy increases the number of parameters per LLM to 12. Both modeling approaches yield almost identical results. Only the early phases of the step responses differ slightly, as demonstrated below.

Figure 25.30a depicts the convergence curve of LOLIMOT. A local linear neuro-fuzzy model with nine rules is selected. Its accuracy on training and validation data is shown in Fig. 25.31. The maximum output error of the model is smaller than 1 °C.

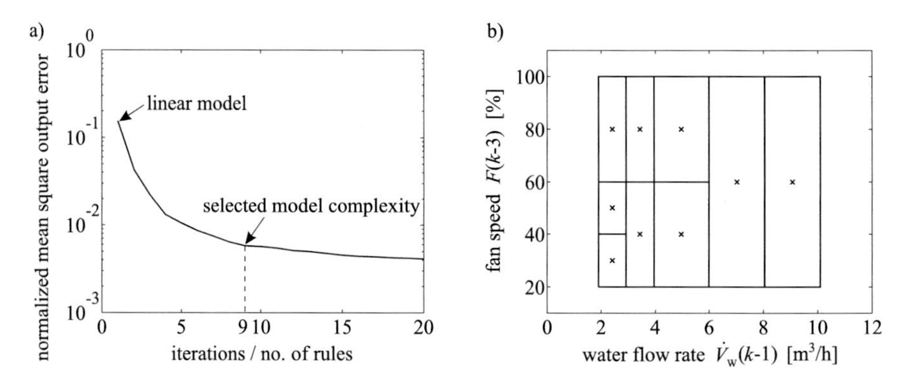

Fig. 25.30 (**a**) Convergence curve of LOLIMOT. (**b**) Premise input space decomposition performed by LOLIMOT

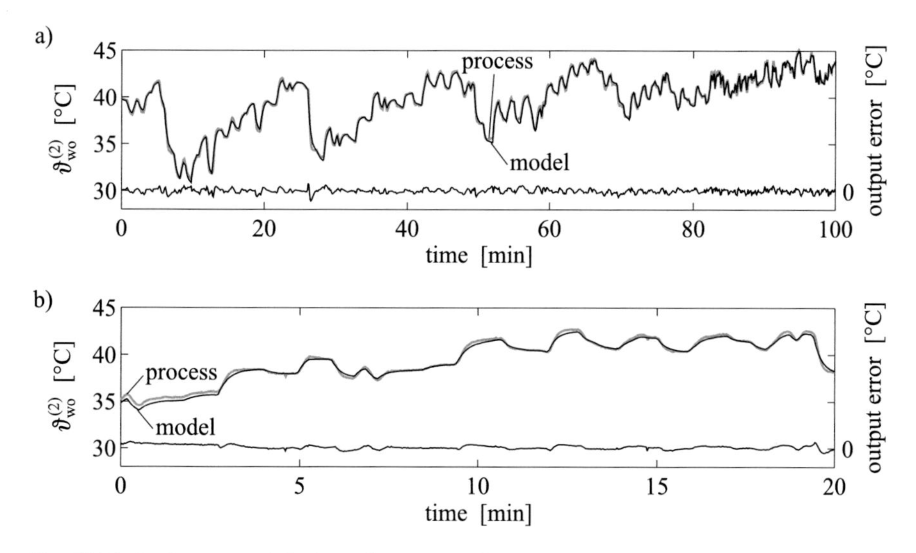

Fig. 25.31 Performance of the local linear neuro-fuzzy model on (**a**) training and (**b**) validation data

25.3.4.3 Model Properties

Figure 25.30b shows the premise input space partitioning performed by LOLIMOT. The results are quite similar to those for the tubular heat exchanger; see Fig. 25.23b. The process obviously is significantly nonlinear only in the water flow rate and the fan speed. Thus the premise input space can be restricted to the two dimensions $\underline{z}(k) = [u_1(k-3)\ u_2(k-1)]^T$. Since the nonlinear behavior is stronger for smaller water flow rates, the partitioning there becomes finer. Also, as expected, the fan speed affects the process in a significant nonlinear way only for small water flow rates. The partitioning represents well the shape of the measured static process nonlinearity shown in Fig. 25.27a.

The nonlinear static behavior of the model shown in Fig. 25.32a is also in agreement with the static measurements shown in Fig. 25.27a. The same is true for the operating point-dependent gains with respect to the fan speed. Figure 25.32b underlines the strong influence of the fan speed on the gain for small water flow rates.

Figure 25.33a, b compare some step responses of the neuro-fuzzy model for different water flow rates. Figure 25.33a shows the step responses for the structure selection modeling approach, where two fan speed regressors are selected by an OLS for each LLM. In contrast, Fig. 25.33b shows the step responses for the approach in which all seven fan speed regressors are incorporated into the model. Obviously, the two results are very similar. They demonstrate the strong dependency of gains and time constants on the water flow rate. The step responses in Fig. 25.33b are smoother and look like higher-order behavior in the beginning. This is a direct consequence of the higher model flexibility due to the larger number of parameters.

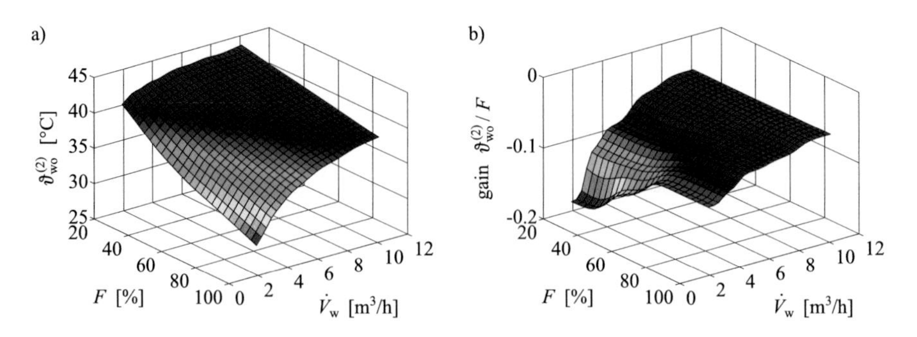

Fig. 25.32 (**a**) Static behavior of the model. (**b**) Gain of $\vartheta_{\rm wo}^{(2)}/F$ according to the local derivative. The water inlet temperature is fixed at $\vartheta_{\rm wi}^{(2)} = 45\,°\rm C$

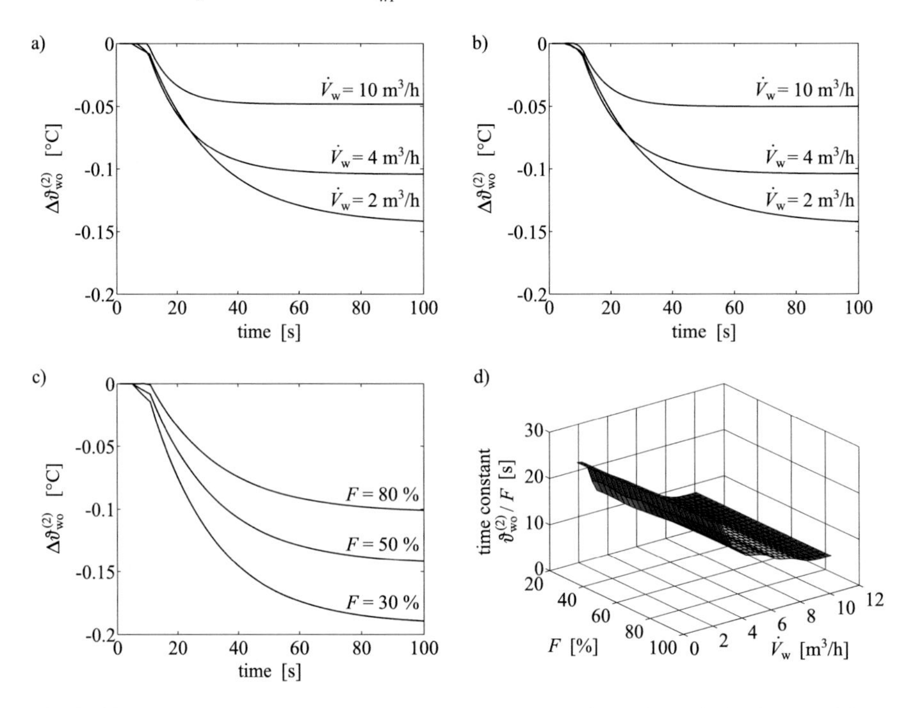

Fig. 25.33 Dynamic behavior of the model. (**a**) Step responses for a change in the fan speed from 50% to 51% for different water flow rates and for the modeling approach with OLS structure selection. (**b**) Same as a but with the modeling approach that includes all fan speed regressors. (**c**) Step responses for a change in the fan speed from 50% to 51% for different fan speeds and for the water flow rate $\dot{V}_{\rm W} = 2\,\rm m^3/h$. (**d**) Operating point-dependent time constants. The water inlet temperature for all figures is constant at $\vartheta_{\rm wi}^{(2)} = 45\,°\rm C$

The seven "numerator" coefficients give the model the same flexibility as an FIR filter in the first samples.

Figure 25.33c illustrates step responses for three different fan speeds, which show that the gain also depends on the fan speed but the time constants do not. Indeed,

Fig. 25.33d confirms the strong dependency of the model's time constants on the water flow rate but their independence from the fan speed.

25.4 Summary

This chapter demonstrated that local linear neuro-fuzzy models trained with the LOLIMOT algorithm are a universal tool for modeling and identification of nonlinear dynamic real-world processes. The following observations can be made from the application examples considered:

- Training times on a standard PC[2] are in the range between some seconds to some minutes for the cooling blast and turbocharger applications and one order of magnitude slower if OLS structure selection is used as for the thermal plant applications.
- The number of required local linear models for many applications is surprisingly low (around ten) compared with the typical complexity of other neural networks reported in the literature. This emphasizes that each neuron or rule with its local linear representation in the neuro-fuzzy model is a good description of reality.
- It is quite realistic for many applications to assume a lower order or a reduced operating point in the premise input space. This allows one to reduce the complexity of the problem significantly. The premise input space partitioning performed by LOLIMOT suggests which regressors may be relevant for the nonlinear process behavior.
- The considered applications indicate that low (first or second)-order models can often be sufficiently accurate. In many cases, high-order nonlinear models are obviously overparameterized. This can be seen, for instance, in the heat exchanger applications. The thermal plant is clearly a higher than first-order process, and in prior investigations, linear models for one operating point are chosen of third-order [282]. Nevertheless, it turned out in Sect. 25.3 that higher than first-order nonlinear models include local linear models with non-interpretable poles (either negative real or conjugate complex).
  The following reasons probably cause this effect. First, in context with the bias/variance dilemma (see Sect. 7.6.1), the maximum complexity of a nonlinear model is limited more severely than for a linear model, which is much simpler by nature. Second, the process/model mismatch due to unmodeled nonlinear effects as a consequence of approximation errors typically dominates those errors caused by unmodeled dynamics. Third, the operating regime covered by one local linear model still possesses nonlinear behavior, which can degrade the estimation of the local linear models' parameters.

[2]With a Pentium 100 MHz chip.

- An interpretation of the local linear neuro-fuzzy models identified by LOLIMOT is easy. The calculation of the step responses of the local linear models and the investigation of their poles and gains allow the user to gain insight into the model and thereby into the process. The good interpretability of the model also serves as a good tool for the selection of a suitable model structure and the detection of overfitting.
- The partitioning of the premise input space and the evaluation of the errorbars serve as valuable tools for the design or redesign of excitation signals.

Chapter 26
Design of Experiments

Abstract This chapter is devoted to the application of various design of experiment strategies. First, some fundamental questions are addressed in simulation studies, such as "In what order should measurements be carried out?" and "Should corners of the input space be measured?" Then, HILOMOT-DoE, an active learning strategy based on the HILOMOT algorithm, is applied to a structural health monitoring application. In a second application example, HILOMOT-DoE is utilized for efficient combustion engine measurement at a test stand. With this approach, half of the measurement time could be saved compared to conventional DoE strategies, still achieving comparable model quality. Finally, the excitation signal generator proposed in Chap. 19 is very successfully applied to a common rail fuel injection system. The benefits of such a generator-based DoE approach are demonstrated.

Keywords Application · DoE · Active learning · Clustering · Corners · Combustion engine · Test stand · Common rail

In the first section of this chapter, some practical, methodological aspects which are important for the design of experiments are discussed and analyzed. In order to draw quite general conclusions, a synthetic function generator is developed which allows varying important design aspects like dimensionality, number of points, complexity of the process/function, etc.

This chapter furthermore presents two real-world applications of HILO-MOT DoE which is methodically introduced and analyzed in Sect. 14.10.4. The first application is in the area of structural health monitoring in cooperation with the chair of Experimental Technical Mechanics, University of Siegen (Prof. Fritzen). The second application is in the field of combustion engine measurement in collaboration with industry (Daimler).

The third application discusses the design of nonlinear *dynamic* excitation signals for a common rail Diesel injection system. This is joint work together with industry (Bosch Engineering).

O. Nelles, *Nonlinear System Identification*,
https://doi.org/10.1007/978-3-030-47439-3_26

26.1 Practical DoE Aspects: Authored by Julian Belz

As already outlined in Sect. 14.10, design of experiments (DoE) plays a key role in experimental modeling. Independent of the model type, the characteristics of a process can only be represented by a black-box model if these characteristics are contained in the data used for the training. Section 26.1.2 deals with the order in which experiments should be conducted if the goal is to yield the best possible model performance with fractions of the whole experimental design. In black-box modeling tasks, active learners need at least some data for the generation of an initial model, on which the strategy can rely on. Through a clever order of experimentation, the amount of data needed for the initial model might be decreased. In Sect. 26.1.9, advice about the incorporation of corner measurements is given. The corners (vertices) of a hypercube are defined by the most extreme values of the input variables. This section focuses on the case where the amount of possible measurements is not much bigger than the number of corners.

26.1.1 Function Generator

During the development of methods and algorithms situated in the field of data-driven techniques, it is always difficult to test new ideas. For the demonstration of strengths and weaknesses of new algorithms, an arbitrary amount of synthetic examples would be of great help. In contrast to typical real-world or simulation data of specific applications, this offers the possibility to easily vary important factors such as:

- amount of data,
- dimensionality (number of inputs),
- strength of nonlinearity,
- data distribution,
- noise level, etc.

Furthermore, the generation of multiple data sets of arbitrary size, e.g., for training, validation, and testing, is no problem. DoE and active learning strategies can be investigated nicely which is completely impossible for fixed data sets.

Infinite possibilities exist for building a function generator, and there exists no overall "best" solution. Extremely few proposals for a function generator can be found in the literature. One very primitive approach has been made in the context of the massive online analysis (MOA) project [52] which is based on a random radial basis function network. A much more sophisticated approach proposed by Friedman [170] is based on additive Gaussians which gives an unfair advantage to all algorithms utilizing Gaussian kernels (typically used in support vector machines, radial basis function networks, Gaussian process models, etc.). The function generator used for investigations is proposed in [43] and is meant to mimic static nonlinear regression problems. In contrast to other proposals, it is able to mimic saturation effects that often naturally arise in physical systems. Additionally, a wide range of nonlinear characteristics can be generated and is controlled by very few parameters. However, the function generator is not intended to mimic a specific considered process. It is a generic way to generate test problems.

The function generator is based on randomly generated polynomials. For a p-dimensional input space, a polynomial arises from the sum of M_{poly} monomials according to the following equation:

$$g(u_1, u_2, \ldots, u_p) = \sum_{i=1}^{M_{poly}} c_i \cdot (u_1 - s_{i1})^{\gamma_{i1}} \cdot (u_2 - s_{i2})^{\gamma_{i2}} \cdot \ldots \cdot (u_p - s_{ip})^{\gamma_{ip}} \,. \tag{26.1}$$

Each monomial is a product of powers of variables with non-negative integer exponents and a coefficient c_i. Here, these variables are the physical inputs u_j, $j = 1, 2, \ldots, p$ shifted by the randomly generated values s_{ij} with $i = 1, 2, \cdots, M_{poly}$. The shifts s_{ij} are drawn from a uniform distribution $\mathcal{U}[0, 1]$, whereas the coefficients c_i originate from a normal distribution $\mathcal{N}(0, 1)$. The physical input values should be normalized to the interval [0, 1], such that all bases $u_j - s_{ij}$ lie in the interval $[-1, 1]$ after the shifts s_{ij} are subtracted. The powers γ_{ij} are non-negative integer values that are yielded by taking the floor of values coming from an exponential distribution with expected value μ. Therefore the strength of the nonlinearity is determined via two user-specified values M_{poly} and μ.

An optional extension is the transformation of the resulting polynomials with a sigmoid function:

$$h(u_1, u_2, \ldots, u_p) = \frac{1}{1 + \exp(-\alpha \cdot g(u_1, u_2, \ldots, u_p))} \,. \tag{26.2}$$

This yields functions that have the potential to possess saturation characteristics, which might be desired in order to mimic many typical real-world applications. The tuning parameter α determines the probability of saturation effects. Figure 26.1 shows exemplarily three two-dimensional polynomial functions originating from this function generator in the left column. The right column of Fig. 26.1 displays the sigmoidally saturated counterparts with $\alpha = 10$. As can be seen, the original polynomials are more or less deformed, depending on the output range of $g(\cdot)$. Note that for Fig. 26.1 all final output ranges have been scaled to the interval [0, 1]. A comparison to the function generator proposed by Friedman in 2001 [170] can be found in [42].

During the subsequent investigations, one weakness of the polynomial-based function generator described in this section was observed. For relatively high-dimensional input spaces, the way the polynomial functions are created is very likely to lead to almost constant functions. With an increasing input dimensionality, the

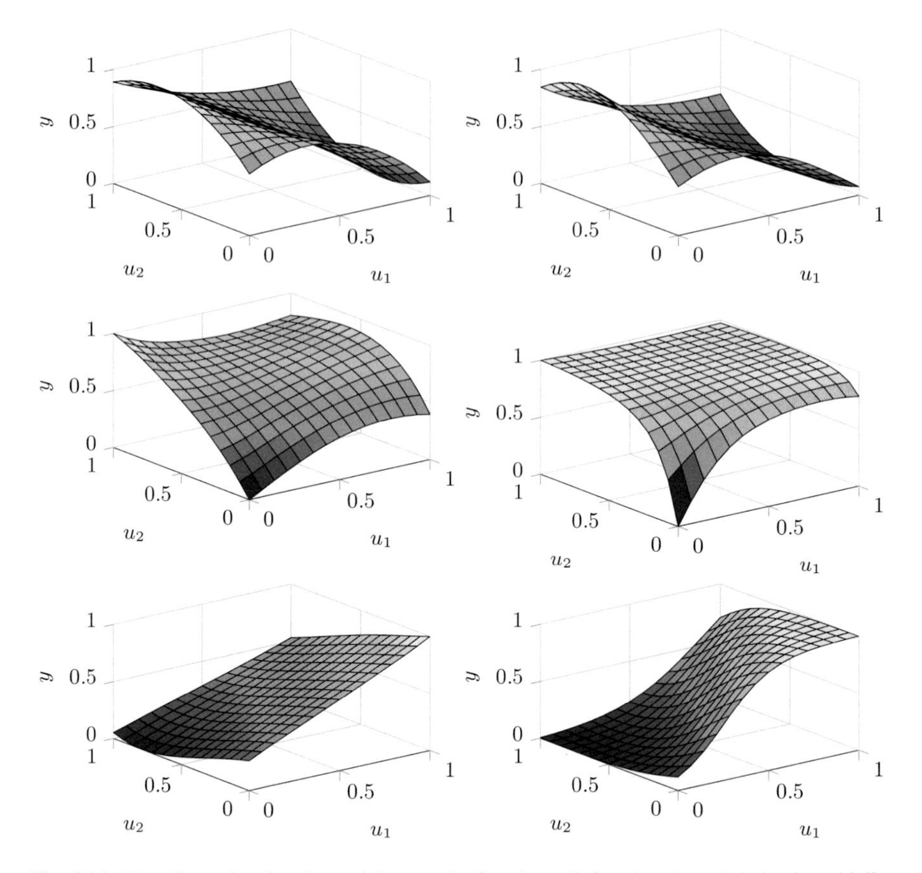

Fig. 26.1 Two-dimensional polynomial example functions (left column) and their sigmoidally saturated counterparts (right column)

number of bases in each of the M_{poly} monomials increases linearly, and therefore more exponents have to be drawn. This increases the chance to obtain a high value for at least some of the exponents in each of the M_{poly} monomials; see (26.1). High values of the exponents are critical, because all bases $u_j - s_{ij}$ lie in the interval $[-1, 1]$ and the expression $(u_j - s_{ij})^{\gamma_{ij}}$ tends to zero if $|(u_j - s_{ij})| < 1$. This leads to counter-intuitive results if the expected value of the exponential distribution is chosen to be $\mu > 1$ in order to create more complex functions. To circumvent this weakness, the number of monomials M_{poly} has been increased to at least 20, and the standard deviation of the test function's output is tested on a large data set ($N = 10^5$) originating from a uniform distribution. The latter step is utilized to detect and discard almost constant test functions. Through the increased number of terms, the probability that all M_{poly} terms become zero at the same time is decreased. In the case that the standard deviation of the output of a generated test function is below 0.01, an alternative test function is randomly generated.

An idea to improve the function generator in order to overcome the above-described weakness systematically is to use the following modification of (26.1):

$$g_2(u_1, u_2, \ldots, u_p) = \sum_{i=1}^{p} \left[\sum_{j=1}^{M_{poly}} c_{ij} \prod_{k=1}^{i} b_{ijk}^{\gamma_{ijk}} \right] . \tag{26.3}$$

Already known symbols from (26.1) maintain their meaning. The only new symbol is an abbreviation for the bases:

$$b_{ijk} = u_l - s_{ijk} , \tag{26.4}$$

where index l determines which of the p inputs should be used with equal probability for the particular basis. For one of the M_{poly} monomials or a fixed value of j, respectively, i different inputs are selected by drawing values for l without replacement. The number of bases raised to the powers γ that are multiplied within one monomial is determined by the outer-sum-variable i. Thus, the originally proposed function generator from (26.1) is obtained if the outer-sum would start not at $i = 1$ but at $i = p$. Compared to the original function generator, this incorporates guaranteed low-order interactions, which are not prone to the weakness described in the last paragraph. The way the modified function generator works is related to the structure of functional analysis of variance (ANOVA) models, where the prediction is composed of a sum of functions describing the main effects (functions of one variable) and interactions of two and more variables. More details about ANOVA models can be found in [172, 229]. Note that the modified function generator is just a proposal and has not been used for the following investigations.

26.1.2 Order of Experimentation

An often neglected aspect and the main topic of this section regards the order in which the measurements are conducted to yield the best possible model performance with fractions of the whole experimental design. Only little literature could be found addressing the order of experimentation aiming at this specific goal. For example, [116] and [249] consider the order of experimentation only in the context of neutralizing influences of undesirable factors on the experimentation or efforts needed to change factor levels. Additionally, these two publications deal only with factorial designs, whereas methods proposed in this section can be applied to arbitrary experimental designs. The method presented in [553] aims at a reduction of the training set size in order to decrease computational demands and to improve the convergence speed of the model training. In differentiation to that, the proposed methods here aim to improve the convergence with respect to the *data amount* and do not consider computational demands at all. Good models in the early stages of the measurement process yield several advantages, e.g., time can be saved because

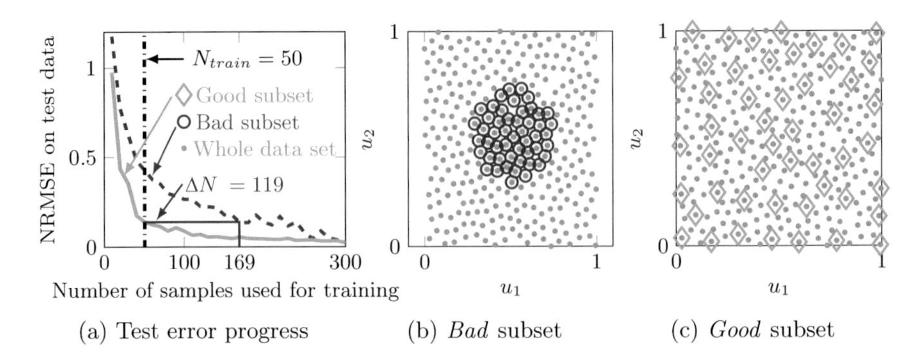

Fig. 26.2 Demonstration example for the importance of the order of experimentation. Test error progress (**a**) together with the chosen subsets ($N_{train} = 50$) in a *bad* (**b**) and a *good* way (**c**)

demanded model qualities can be reached with less data which is exemplarily demonstrated in Fig. 26.2. For this example, the experimental design is a maximin Latin Hypercube (LH) from which subsets are chosen in a *bad* and a *good* way. In order to reach the same model quality, $\Delta N = 119$ additional samples are needed following the *bad* ordering strategy; see Fig. 26.2a. The chosen subsets for $N_{train} = 50$ for both order determination strategies are visualized. Only points near the design space center are chosen for the subset designated as bad (red circles in Fig. 26.2b). The other subset, designated as good, originates from the intelligent k-means sequence (IKMS) explained in more detail in Sect. 26.1.5 (green diamonds in Fig. 26.2c). Of course, Fig. 26.2 contrasts worst- and best-case scenarios which are extremely unlikely to happen in practice through the standard approach of random selection. The point of the approach proposed here is to guarantee the best worst-case scenario. As a result, the model can be used earlier, while the measurement process is still in progress. For example, model-based optimization runs become more reliable with small data subsets. Active learning strategies enlarge the experimental design in an iterative and adaptive way, based on models trained with the currently available training data, such that the information gained by additional measurements is maximized. Typically, an initial experimental design is needed before the active learning phase can start [224]. Through a good order of experimentation, the necessary amount of data serving as an initial experimental design might be decreased. Note that throughout the discussions in this book, the term "measurement" is used synonymously for obtained data points, regardless of their origin. For example, the data might originate from real-world measurements as well as from computer simulations.

The proposed methods are based on the following considerations. Models relying only on small amounts of data are limited in their possible model accuracy. Through a clever placement of measurements, the possible model accuracy can be influenced. The aim of all order determination strategies is to find subsets of points from the overall experimental design that lead to the possibly best model quality. Therefore a wide and nearly uniform coverage of the whole input space with small subsets is the goal if no prior knowledge about the process is available. The biggest gap sequence (BGS) tries to tackle the described considerations straightforward, filling the biggest gap of the input space with each additional point. One possible weakness of BGS might be the exclusive concentration on points near the boundary at the beginning of the algorithm, especially in high-dimensional input spaces. This potential issue motivates the median distance sequence (MDS) that tries to avoid the concentration of boundary-near points at the beginning. Adding the next point corresponding to the median distance instead of the maximum distance should lead to a better balance between points at the boundary and inside the input space for very small subsets. The IKMS method exploits in a first step possibly existing structures in the experimental design by the intelligent k-means algorithm [379]. All resulting clusters represent a specific region in the input space. Starting with all points next to these cluster centers should obtain information from all input space regions and in addition avoids the concentration of points at the boundary for very small subsets. It is a mechanism to balance the exploration and exploitation of the whole experimental design. At the beginning, all areas of the input space covered by the experimental design are explored and then successively exploited in more detail.

The different methods to determine the order of experimentation for regression problems are explained in more detail in Sects. 26.1.3, 26.1.4, and 26.1.5. All methods are originally published in [39] and are based on the aforementioned considerations. Section 26.1.7 compares all methods for the determination of the order of experimentation on functions generated with the static function generator that is explained in Sect. 26.1.1. Section 26.1.8 summarizes the order of experimentation findings and gives some remarks about the applicability for real-world test benches.

For the following explanations, it is always necessary to distinguish between three sets of data points. Set $\mathbb{N}$ contains all points of a data set. $\mathbb{S}$, containing all already sorted points, and $\mathbb{F}$, containing all not yet sorted points, are non-overlapping subsets of $\mathbb{N}$. Here and throughout the rest of this section, "sorted" refers to the successional order of experimentation, i.e., the order in which the measurements should be conducted. The relationships between the previously defined sets can mathematically be expressed as follows:

$$\mathbb{N} = \mathbb{S} \cup \mathbb{F} \text{ and } \mathbb{S} \cap \mathbb{F} = \emptyset \,. \tag{26.5}$$

At the beginning of each method, $\mathbb{F}$ contains all data points ($\mathbb{F} = \mathbb{N}$). Then points are sequentially moved from $\mathbb{F}$ to $\mathbb{S}$ until $\mathbb{F}$ is empty ($\mathbb{F} = \emptyset$).

26.1.3 Biggest Gap Sequence

For the BGS, the first point to be added to $\mathbb{S}$ is the one closest to the center of gravity of all data points. In the following, one iteration of the BGS is explained and illustrated in Fig. 26.3. Here one iteration refers to all steps necessary to determine one data point that should be added to the sorted list next. In Fig. 26.3a, there are

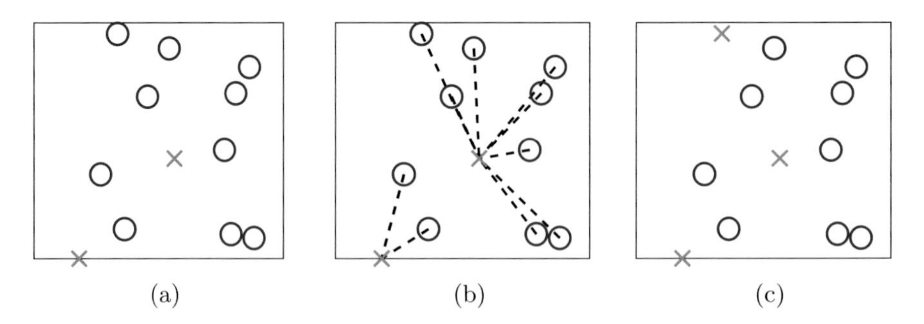

Fig. 26.3 Illustration of the procedure for the MDS and BGS methods. (**a**) Two already sorted points (x). (**b**) Assigning each point in $\mathbb{F}$ (o) to its NN in $\mathbb{S}$ (x). (**c**) Maximum NN is added to $\mathbb{S}$ (x)

two already sorted points in $\mathbb{S}$ (orange crosses), whereas all other points still belong to set $\mathbb{F}$ (blue circles). Each point in $\mathbb{F}$ is now assigned to its nearest neighbor (NN) from subset $\mathbb{S}$. The dashed lines in Fig. 26.3b connect each point in $\mathbb{F}$ with its NN in $\mathbb{S}$. The corresponding distances (lengths of the dashed lines) from all points in $\mathbb{F}$ to their NN in $\mathbb{S}$ are calculated. The point with the maximum distance to its NN is selected and is moved from $\mathbb{F}$ to $\mathbb{S}$; see Fig. 26.3c. After adding a point to $\mathbb{S}$, the next iteration starts, and the whole procedure continues until $\mathbb{F}$ is empty. Note that $\mathbb{S}$ is incremented in size only after the last step of each iteration.

26.1.4 Median Distance Sequence

The MDS starts similar to the BGS method, i.e., the first point added to $\mathbb{S}$ is the one closest to the center of all data points. After that, each point in $\mathbb{F}$ is assigned to its NN from subset $\mathbb{S}$, like in the BGS strategy; see Fig. 26.3b. Again, the distances from all points in $\mathbb{F}$ to their NN in $\mathbb{S}$ are calculated. Now the point that corresponds to the median value of all calculated distances is moved from $\mathbb{F}$ to $\mathbb{S}$, instead of the one with the maximum distance. An update of the NN relationships between $\mathbb{F}$ and $\mathbb{S}$ has to be performed. After that, the procedure continues until $\mathbb{F}$ is empty.

26.1.5 Intelligent k-Means Sequence

For the ordinary k-means algorithm, the number of clusters k as well as initial centroids (centers of the clusters) has to be specified by the user before the algorithm can proceed. In the *intelligent* k-means algorithm, the number of clusters k and the corresponding initial centroids are automatically determined, before the ordinary k-means algorithm is used [379]. Therefore, the intelligent k-means clustering contains the ordinary k-means clustering algorithm but adds a clever initialization for it as a preprocessing step [104].

26.1.5.1 Intelligent k-Means Initialization

1. Determine the center of gravity of all data points $\underline{c}_1$.
2. Determine the data point $\underline{c}_2$ farthest away from $\underline{c}_1$. Treat both points as cluster centers.
3. All data points are assigned to their nearest cluster center.
4. In contrast to the ordinary k-means algorithm, only $\underline{c}_2$ is updated to the center of gravity of all points assigned to $\underline{c}_2$. Cluster center $\underline{c}_1$ remains unchanged.
5. Repeat steps 3 and 4 until convergence, i.e., no further changes in $\underline{c}_2$.
6. The cluster center $\underline{c}_2$ is saved, and all data points belonging to it are removed from the data set for further calculations. If the number of points that have been assigned to cluster center $\underline{c}_2$ is greater than a user-specified number (typically 2), it is used later for the initialization of the ordinary k-means clustering.
7. If there are data points left, go to step 1; otherwise stop.

At the end of this procedure, there are k saved cluster centers for the initialization of the ordinary k-means algorithm leading to k clusters. Then all data points belonging to a cluster C_i, $i = 1 \ldots k$ are sorted according to the BGS strategy described in Sect. 26.1.3. As a result, there are k sorted data point lists $\mathbb{L}_i$, $i = 1, 2, \ldots, k$, or cluster lists, respectively. To get the final order of experimentation, the first element of each list is moved from $\mathbb{F}$ to $\mathbb{S}$ as described in more detail in Algorithm 1. In that way, the first k points to be measured are the ones closest to the cluster centers because these points are the first elements in the lists $\mathbb{L}_i$. After that, the second element of each cluster list is added until all data points are ordered according to the IKMS. Figure 26.4 demonstrates the IKMS procedure for 12 points. First, the intelligent k-means algorithm determines three clusters and assigns each point to one cluster; see Fig. 26.4a. The numbers shown in Fig. 26.4b represent the final ordering determined by the IKMS method.

26.1.6 *Other Determination Strategies*

The aforementioned order determination strategies are compared to one active learning strategy and simple randomization of the order of the measurements. As active learning strategy HILOMOT for design of experiments (HilomotDoE) [224] is utilized. For the selection of the point that should be moved next from $\mathbb{F}$ to $\mathbb{S}$, a local model network (LMN) is trained according to the HIerarchical LOcal MOdel Tree (HILOMOT) algorithm based on all points already present in $\mathbb{S}$. As a result, the whole input space is partitioned into subregions, and for each subregion, a local error measure is calculated. The point in $\mathbb{F}$ is chosen that fills the biggest gap within the

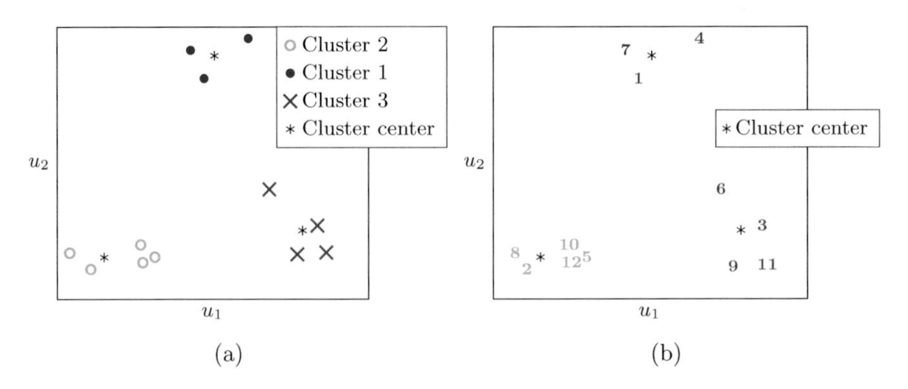

Fig. 26.4 Explanation of the IKMS with the help of a demonstration example. (**a**) The intelligent k-means determines three clusters. (**b**) The numbers represent the sequence determined by the IKMS procedure

Input: Set of sorted data point lists $\mathbb{L}_i$, $i = 1, 2, \dots, k$
Output: One ordered list $\mathbb{S}$ which specifies the order of experimentation
Step 1: Define the set of not yet sorted points $\mathbb{F}$ as the union of all points contained in the lists $\mathbb{L}_i$, $i = 1, 2, \dots, k$:
$\mathbb{F} := \mathbb{L}_1 \cup \mathbb{L}_2 \cup \dots \cup \mathbb{L}_k$;
Step 2: Move the first element of each list $\mathbb{L}_i$, $i = 1, 2, \dots, k$ to $\mathbb{S}$ until $\mathbb{F}$ is empty. The processing order of the lists is arbitrarily determined but fixed.

```
while F ≠ ∅ do
    /* Go through all non-empty lists                                   */
    for each non-empty list do
        move first element of the current list to S;
        if current list is empty then
            remove current list from the set of non-empty lists
        end
    end
    /* Update the set of all not yet sorted points                      */
    F := L1 ∪ L2 ∪ ... ∪ Lk
end
```

Algorithm 1: Specification of the order of experimentation once the clustering of the data is finished according to the intelligent k-means algorithm

subregion with the worst local error measure. If there are no points from $\mathbb{F}$ inside the worst-performing subregion, the second-worst subregion is considered and so on. After one point is moved from $\mathbb{F}$ to $\mathbb{S}$, the whole procedure starts again (including a new training) and continues until all measurements are ordered. HilomotDoE requires some initial points in $\mathbb{S}$ such that the parameters of the local model network can be estimated. Typically $N_{ini} = 2(p + 1)$ initial points are determined according to the BGS strategy. Further details about HilomotDoE can be found in Sect. 14.10.4 and in [228].

Optimality criteria are not considered here to determine the order of experimentation. For these criteria typically a real-valued summary statistics of the Fisher information matrix is optimized to achieve some desired properties, like minimum variance of the estimated parameters (D-optimal design) [164]. The Fisher information matrix of the used model type (here it would be a local model network) depends on parameters that are responsible for the input space partitioning. If this partitioning of the local model network is determined and fixed, the Fisher information matrix could easily be derived. However, this information matrix would only be valid in the case of the *global* (concurrent) estimation of all local model parameters. The used identification algorithm utilizes a *local* estimation scheme for these parameters in order to introduce a regularization effect, improve interpretability, and avoid overfitting; see [416] for further details. Therefore it is impossible to straightforwardly implement an optimality criterion-based strategy.

26.1.7 Comparison on Synthetic Functions

The experimental setup is described that is chosen to compare the different methods for the determination of the order of experimentation for synthetic test functions. These test functions are generated randomly with the help of the function generator described in Sect. 26.1.1. The dimensionality of the design space is varied from $p = 2$ to $p = 8$. Three training data sets with different input distributions are used for each random function and input dimension. Additionally, one huge test data set is generated for each random function and input dimension in order to assess the model quality. For each training data set, the order of experimentation is determined according to the methods defined in Sects. 26.1.3, 26.1.4, 26.1.5, and 26.1.6. Then, for training data increments of 10% (in the determined order), a model is trained, and its performance is assessed using the test data. Strategies for the order determination are compared based on how quick the model performance increases (corresponding to a decrease of the test error). Therefore, results are averaged over all random functions created with the function generator. Here the advantages of a function generator are exploited, i.e., designing synthetic functions with desired properties such as the input dimensionality, the amount of data, and the data distribution, to name only a few. In order to conduct the comparison, three different input distributions are used. Maximin LH designs, optimized according to the algorithm proposed in [131], data samples drawn from a uniform distribution, and data samples drawn from two normal distributions with different mean values and equal standard deviations are employed. The two normal distributions have equal standard deviations, but the centers differ, $\mathcal{N}(0.3 \cdot \underline{1}, 0.12 \cdot \underline{I})$ and $\mathcal{N}(0.7 \cdot \underline{1}, 0.12 \cdot \underline{I})$, with $\underline{1}$ being a vector consisting of ones and $\underline{I}$ being the identity matrix. Figure 26.5 shows the different data distributions exemplarily in a two-dimensional input space. It can be recognized that the input coverage decreases from the maximin LH design over data drawn from a uniform distribution down to the drawing from two normal distributions. Additionally the chance to repeat almost identical inputs twice or

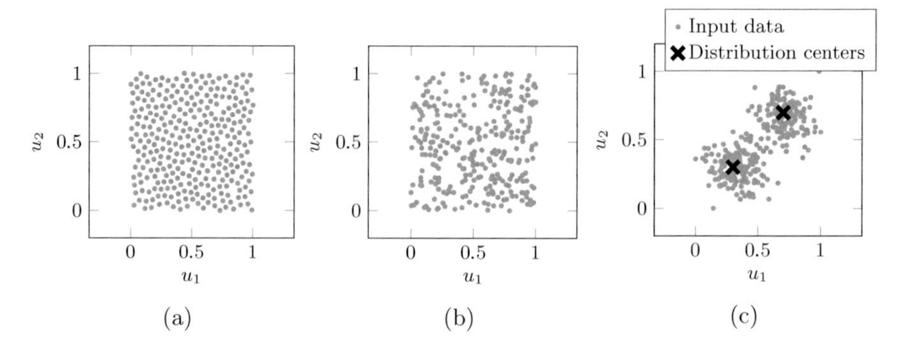

Fig. 26.5 Data used for the comparison of different order determination strategies. (**a**) Maximin optimized LH. (**b**) Data drawn from uniform distribution. (**c**) Data drawn from two normal distributions

more often grows from Fig. 26.5a–c. The number of training samples for all data distributions is kept constant at $N = 300$, while the number of inputs varies from $p = 2, \ldots, 8$. Due to the constant number of training samples and the increasing input dimension, different local data densities arise.

In the case of the uniform distribution and the 2 normal distributions, 20 random test functions are used, and 20 different realizations (per input dimensionality) are drawn from the probability density functions for the comparisons. Because there is only one deterministic LH design (per input dimensionality), the number of test functions for this input type is increased to 100. For each input dimension, the number of test samples is kept constant at $N_t = 10^5$. The location of the test data samples is determined by a Sobol sequence [435] for each input dimension and is fixed for all synthetic functions. HILOMOT is used as training algorithm; see Sect. 14.8 for details.

The achieved test errors for an increasing amount of training data are presented in Fig. 26.6 for all three data distributions and input dimensionalities $p = 2$, $p = 5$, and $p = 8$. Each column corresponds to one input dimensionality and each row to a data distribution. Most diverse results are yielded in the case of Fig. 26.6g, where the experimental design is drawn from two normal distributions and the input dimensionality is $p = 2$. In this case, the IKMS, BGS, and the HilomotDoE-based method perform equally well. MDS turns out to be the worst method on average, even worse than the randomly chosen order of experimentation. With an increasing input dimensionality, the benefit of IKMS, BGS, and HilomotDoE vanishes, and the test errors of all methods get closer to each other. For input distributions yielding a good coverage of the input space, all procedures perform equally well except for MDS, which at least for two-dimensional input spaces turns out to be the worst method. Altogether MDS performs worst and is on average even beaten by random orderings. Because of this MDS results are omitted in subsequent discussions.

In Fig. 26.7, the achieved model performances are plotted against the amount of used training data. The mean normalized root mean squared error (NRMSE) of the

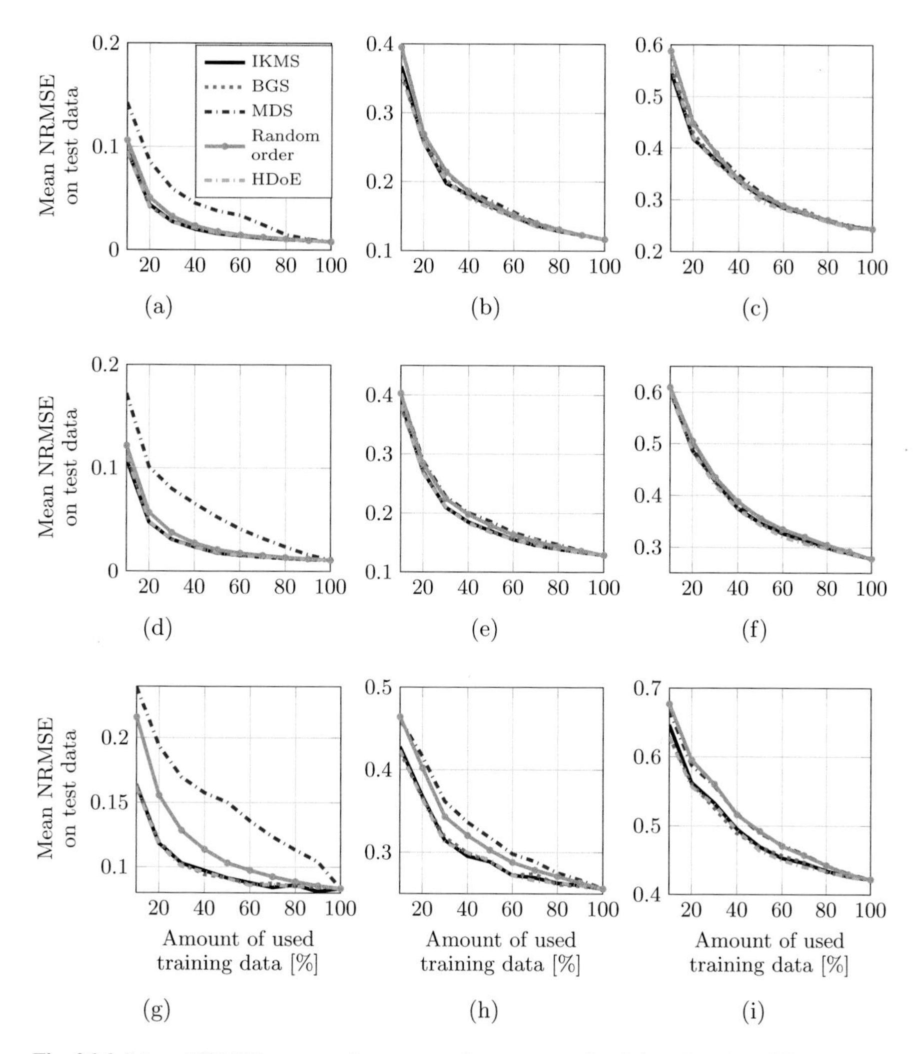

Fig. 26.6 Mean NRMSE on test data versus the amount of training data for different input dimensionalities and distributions. (**a**) 2 inputs, maximin LH design. (**b**) 5 inputs, maximin LH design. (**c**) 8 inputs, maximin LH design. (**d**) 2 inputs, uniform distribution. (**e**) 5 inputs, uniform distribution. (**f**) 8 inputs, uniform distribution. (**g**) 2 inputs, 2 normal distributions. (**h**) 5 inputs, 2 normal distributions. (**i**) 8 inputs, 2 normal distributions

randomly determined order of experimentation ($\overline{\text{NRMSE}}_{\text{rand}}$) is used to normalize each strategy's mean NRMSE value. All NRMSE values are calculated based on test data. The shown achieved model performances are averaged over all input distributions. Results are shown for input dimensionalities $p = 2$ and $p = 8$. The curves of all remaining input dimensions lay in between the shown ones. Compared to simple random ordering, the maximum benefit amounts to 19% and is obtained at

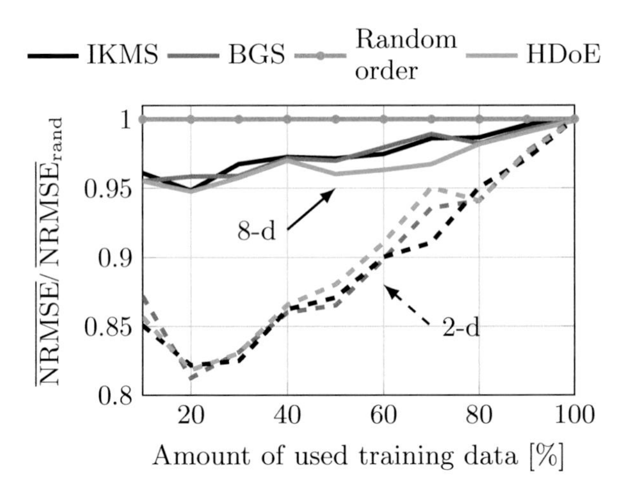

Fig. 26.7 Model performances vs. the amount of used training data for $p = 2$ and $p = 8$, averaged over all input distributions

usage of 20% of the training data. The advantage of IKMS, BGS, and HilomotDoE is more pronounced for low-dimensional input spaces (or higher data densities). For almost all cases, IKMS and BGS perform equally well and on a similar level as HilomotDoE (abbreviated as HDoE in the figure). Even though the active learning strategy is able to use more information, the performance is not significantly better. However, not the full potential of HilomotDoE is exploited since it can only pick data points present in the existing experimental design. Usually, far more potential points would be provided to the active learning strategy in order to cover the design space better.

Figure 26.8 shows the comparison for all order of experimentation strategies, except for MDS, in case of 10% and 50% used training data for all input dimensions and input distributions individually. Based on the shown results, no general recommendation for BGS or IKMS can be given. All proposed methods are significantly superior to a random ordering as it is typically carried out nowadays.

26.1.8 Summary

Models can already be used while the measurement process is still in progress. The accuracy of models in the early stages of the measurement process highly depends on the order, in which the measurements are carried out. Several newly invented strategies for the order of experimentation are compared given an already existing experimental design for regression problems. The main purpose is to gain information from all subregions of the input space as early as possible, such that certain reliability is ensured throughout the whole input space. With the help of a

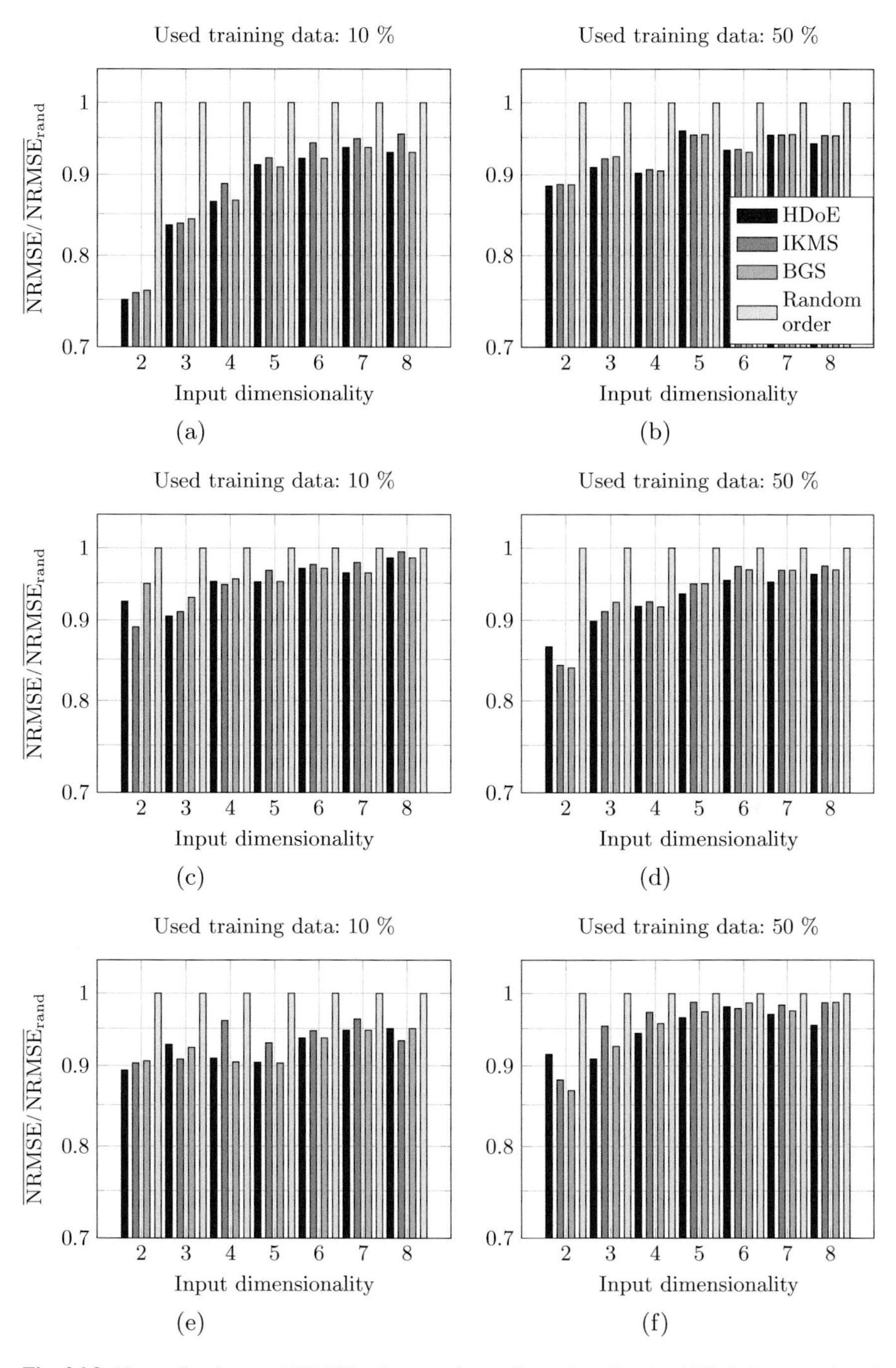

Fig. 26.8 Normalized mean NRMSE values vs. input dimensionality for 10% (left column) and 50% (right column) of used training data. (**a**) 2 normal distributions. (**b**) 2 normal distributions. (**c**) Uniform distribution. (**d**) Uniform distribution. (**e**) Maximin LH design. (**f**) Maximin LH design

function generator, the data distribution and the data density are varied for several randomly generated synthetic functions. It is shown that especially dense and structured input data benefits from well-ordered measurements. For sparsely and uniformly covered input spaces, almost all presented ordering strategies perform equally well. In summary, IKMS and BGS perform in almost all scenarios equally well and are pretty close to the active learning strategy. The yielded improvements of both model-free ordering strategies (IKMS and BGS) lie between 5% and 25% compared to the random approach. In most cases HilomotDoE outperforms the model-free approaches, but only slightly. For real-world applications, the proposed order determination strategies can be used in combination with active learning strategies. It might be a good idea to use one of the proposed order determination methods in very early stages of the measurement process since the model lacks reliability if only very few data are available. At some point, one may switch completely to the active learning strategy.

For the application on real-world test benches, additional considerations are necessary. The proposed strategies are all based on distance measures. If there are hard-to- or expensive-to-vary factors on the test bench, a suitable input weighting can be used to influence the methods. Through such an input weighting, points appear closer or farther away, such that sequences can be generated, where hard-to-vary factors are only changed slightly going from point to point of the ordered list. Experimental design principles for controlling noise and bias are (I) replication, (II) blocking, and (III) randomization.

Replication cannot be influenced by the order of experimentation strategies since it is assumed to determine the order of an already existing experimental design. If this DoE should contain replication points or not has to be decided in advance before the order of experimentation is determined.

Blocking is the division of the whole DoE into several but similar groups. The motivation behind blocking is to determine the effect of variables that are not considered in the experimental design and that clearly influence the output variable. Examples for such so-called nuisance variables can be the room temperature or air humidity. Each generated group of the DoE should then be measured, while the corresponding nuisance variables are at least approximately constant. Output variations between different groups can then be associated with the nuisance variables, whereas output variations within these groups can be attributed to the variables contained in the experimental design. If the a priori determined experimental design contains several groups as a result of blocking, the strategies to determine the order of experimentation should be applied to each group individually.

Randomization is somehow already included in the proposed methods, i.e., IKMS and BGS, since biggest gaps are filled. As a result, points that are close to each other in the input space will be far away in the ordered lists generated by IKMS and BGS. Thus, the time between the measurements of two similar measuring points is increased, and effects resulting from factors (inputs), which cannot be controlled, can average out.

The applicability to a real-world metamodeling task is demonstrated in Sect. 27.2. As the most promising methods, IKMS and BGS are used to determine the order of experimentation for the generation of a centrifugal fan metamodel, where data is yielded by computational fluid dynamics (CFD) simulations.

26.1.9 Corner Measurement

Typical human intuition favors the measurement of all inputs at their most extreme values, i.e., minima and maxima. If only box constraints exist, these combinations correspond to the corners of the input space. Measuring these corners in p dimensions is called a 2^p full factorial design in DoE language [386]. Other commonly used DoE strategies like D-optimality with polynomial model assumptions or partial fractional designs also incorporate many design space corners [386]. This section analyzes in which cases (number of data points, dimensionality of the problem) measuring corners is advantageous and in which cases it is counter-productive. Obviously, in a regression context, additional measurements *within* the design space are absolutely necessary to detect any nonlinear behavior. Intuitive benefits of measuring corners are:

- Extrapolation can be completely avoided if the physical minima and maxima of each input are measured. This should tighten the confidence intervals and improve reliability.
- Arbitrarily high-order interactions can be discovered in principle.
- Optima with respect to the inputs frequently occur at the boundary of the input space. Thus, it might be a good idea to measure there in order to improve the accuracy of the model in these regions.

On the other hand, the number of corners of an p-dimensional input space is 2^p. The following three cases can be distinguished (N = number of data points that shall be measured):

- Low-dimensional problems ($N \gg 2^p$): The number of corners is negligible to the overall number of points to be measured. Therefore it seems reasonable to include the corners to cover the most extreme combinations.
- Medium-dimensional problems ($N > 2^p$): This is the critical case mainly discussed in this section and probably the most frequently occurring one (at least in an engineering context). It will be shown empirically that measuring corners offers increasing advantages when the amount of extrapolation and function complexity grow.
- High-dimensional problems ($N \leq 2^p$): The number of corners is similar or larger than the overall number of points to be measured. Therefore measuring corners is infeasible.

A priori it is not clear for the second case whether the inclusion of all corner points yields benefits for the final model performance.

In order to bring some light in the above-addressed question, several design of experiments with and without explicitly added corner measurements are compared in terms of the achieved model accuracy. Therefore, the function generator described in Sect. 26.1.1 is used to generate a large number of synthetic examples. The results are statistically evaluated.

For the comparisons made here, LH designs [369], Sobol sequences [539] and data drawn from a uniform distribution is supplemented with additional corner points. In order to achieve good space-filling properties, the LH designs are optimized with the extended deterministic local search (EDLS) algorithm described in [131]. Sobol sequences are inherently space-filling, and the uniform distribution should also cover the design space in a more or less space-filling manner. In order to investigate the influence of corner points on the model quality, two data sets for each setting (input dimensionality p, number of points N, basic input design, i.e., LH, Sobol, and uniform) are created. One data set consists of $N_{wC} = N$ points, referred to as the designs without corners (DWC). The other data set consists of $N_C = N - 2^p$ points coming from one of the three basic input designs and is supplemented with 2^p corner points such that the overall number of samples is equal. The latter one is referred to as combined design (CD). Both the CD and DWC designs are compared in Fig. 26.9 exemplarily for a three-dimensional input space and $N = 12$ samples. The comparison of the DWC and CD design corresponds to the question for which settings the explicit incorporation of corner points should be favored.

For the comparison the number of inputs is varied from $p = 2, \ldots, 8$, while the number of samples is held constant at $N = 300$. The DWC and CD designs are used for the training. In the case of the data drawn from a uniform distribution and the Sobol sequences, 20 different realizations are generated, and results are averaged over these 20 input designs. Only one LH design per setting is utilized. For each

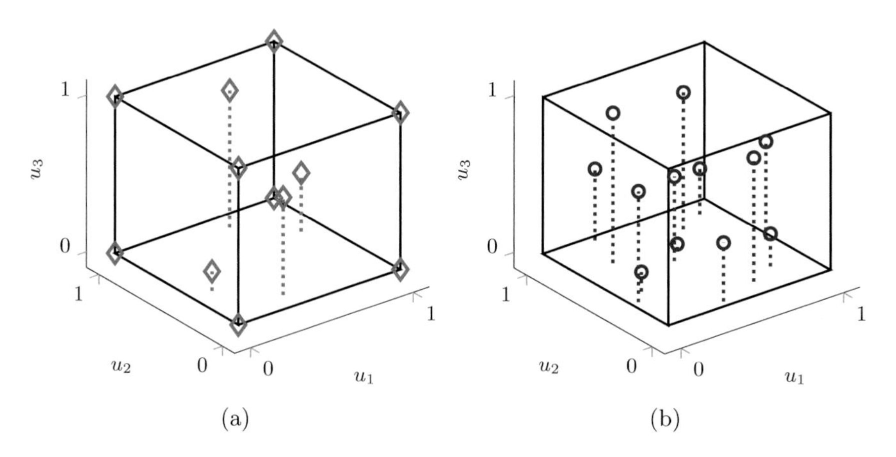

Fig. 26.9 Comparison of a combined design (CD) and a design without corners (DWC) maximin LH design in a three-dimensional input space. The number of points is $N = 12$. (**a**) CD design with 8 corner points. (**b**) DWC design

setting, 100 sigmoidally saturated random functions (see Sect. 26.1.1) are generated and statistically evaluated afterward. For these functions $M = 10$ polynomial terms are used, and the expected value of the exponential distribution is set to $\mu = 1$. Independent of the input dimensionality p and the case under consideration, the test data consists of $N_t = 10^5$ samples, and it is generated by Sobol sequences. It is assured that the Sobol sequences used for the test data and training data have no intersection. Therefore the models trained with data originating from the Sobol sequences should not have a significant unfair advantage compared to models trained with the other two input designs.

The training is done with the HILOMOT algorithm (see Sect. 14.8 and [417] for details), and three different extrapolation scenarios are investigated:

- No extrapolation: Design space for training is $[0, 1]^p$.
- Small extrapolation: Design space for training is $[0.1, 0.9]^p$.
- Large extrapolation: Design space for training is $[0.2, 0.8]^p$.

For all scenarios the test data points lie in the hypercube $[0, 1]^p$.

Results for the investigations with the help of the synthetic functions are shown in Fig. 26.10. The averaged NRMSE values are plotted against the input dimensionality for all extrapolation scenarios and all distributions of the training data. For the scenarios with no and small extrapolation, DWC designs yield smaller NRMSE values and are therefore superior to the CD designs. In the scenario, with large extrapolation, the DWC designs outperform the CD designs only for the highest input dimensionality. The reason why the DWC designs are better suited for the highest input dimensionality is probably due to the high fraction of corner points in relation to the overall number of training samples. Table 26.1 shows the ratios defined by the corners of the design space (2^p) divided by the number of training samples N. Since the number of training samples is kept constant at $N = 300$, the shown ratio tends to one as the input dimensionality increases. This means that almost no samples are left to gather information from the "inside," which is missing in order to approximate the given test function properly. The distribution of the training data seems to play no role, since the results are qualitatively very similar regardless if LH designs, Sobol sequences, or data coming from a uniform distribution is used.

Interestingly the models obtained by LH designs perform always better than the ones from Sobol sequences in case of no and small extrapolation. In order to visualize this, results of all training data distributions are shown altogether in one graph in Fig. 26.11a in case of no extrapolation and in Fig. 26.11b in case of small extrapolation. This observation is investigated in more detail in Sect. 26.1.10.

Another noticeable result that can be seen in Fig. 26.10 is the decrease in the mean test NRMSE value for large extrapolation demands at least for higher input dimensionalities ($p > 6$). This phenomenon is counter-intuitive and occurs only in the CD designs but regardless of the used training data distribution. It also appears for different model types. The same phenomenon was observed with multilayer perceptron networks (see [416] for details about this model type) and Gaussian process models as described in detail in [476]. To clarify this phenomenon, designs

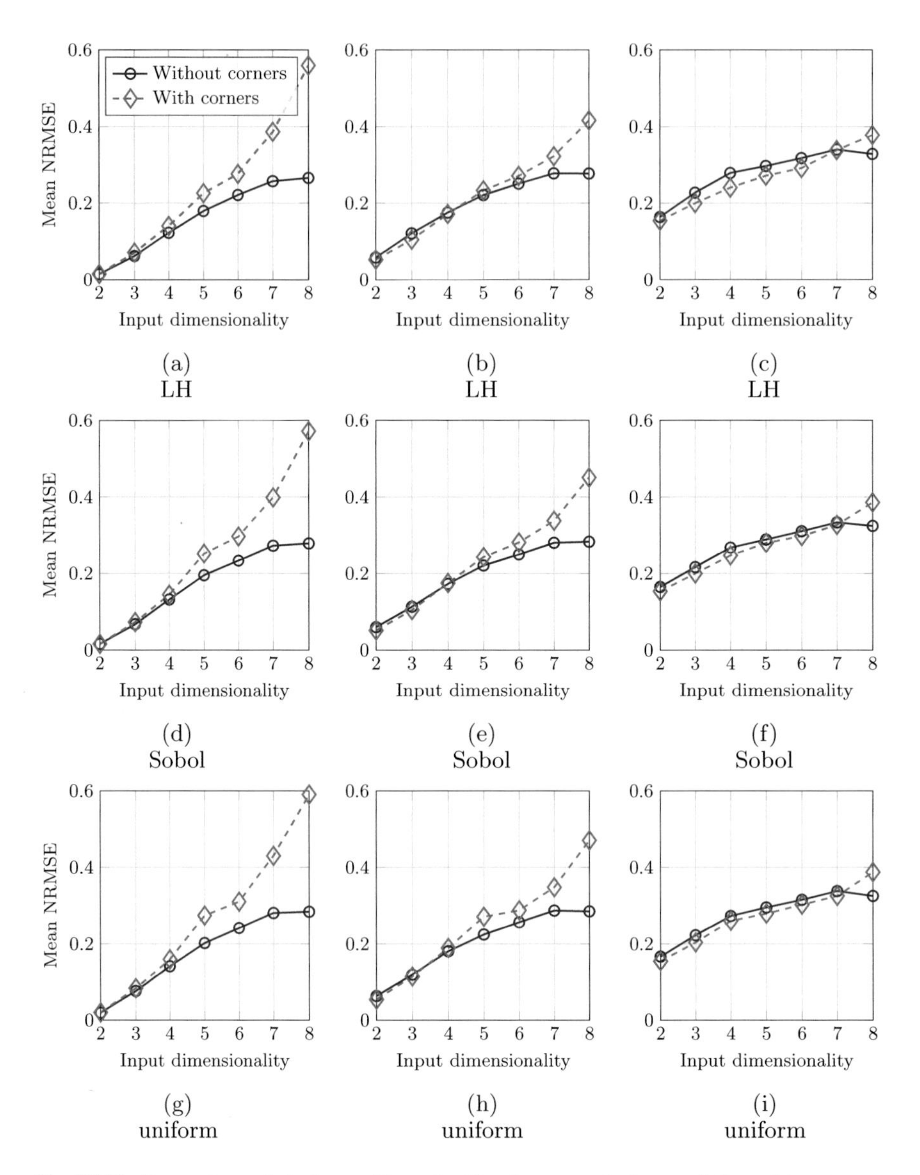

Fig. 26.10 Comparison of designs with and without corners in terms of the achieved mean NRMSE on test data for all extrapolation scenarios and all training data distributions. (**a**) No extrapolation, LH. (**b**) Small extrapolation, LH. (**c**) Large extrapolation, LH. (**d**) No extrapolation, Sobol. (**e**) Small extrapolation, Sobol. (**f**) Large extrapolation, Sobol. (**g**) No extrapolation, uniform. (**h**) Small extrapolation, uniform. (**i**) Large extrapolation, uniform

only consisting of corner points are investigated, since the number of non-corner points is negligible in cases where this phenomenon is observed. The objective function J_t is the error on the test data which always lies in the hypercube $[0, 1]^p$

Table 26.1 Ratios between the corners of the design space (2^p) and the number of training samples $N = 300$

Input dimensionality p	2	3	4	5	6	7	8
Ratio $2^p/N$	0.013	0.027	0.053	0.107	0.213	0.427	0.853

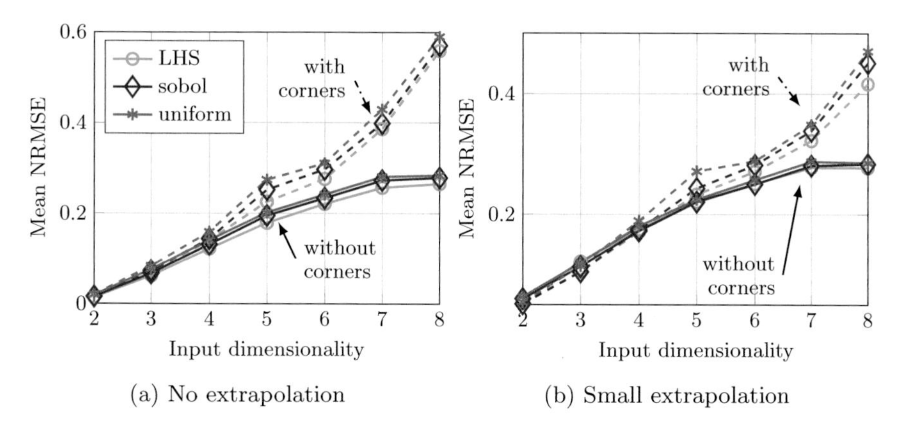

Fig. 26.11 Comparison of different distributions of the training data for no (**a**) and small (**b**) extrapolation

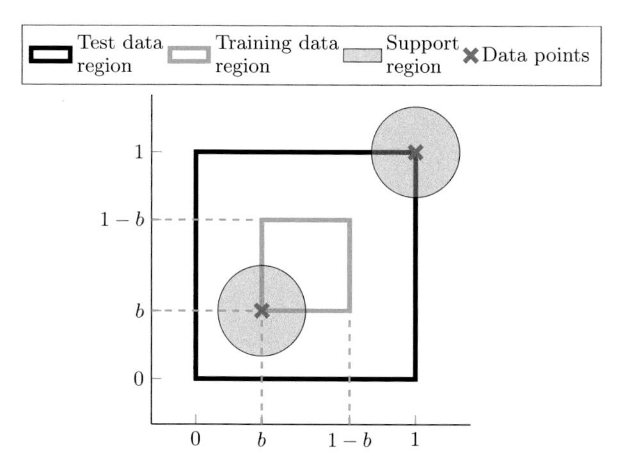

Fig. 26.12 Visualization of the test data region, the training data region depending on b, and exemplary support regions for two data points

and is optimized with respect to the region $[b, 1 - b]^p$ into which the training data is scaled down to; compare Fig. 26.12:

$$b_{opt} = \arg\min_{b} J_t \, .$$

The value b_{opt} leads to the minimal test error and corresponds to a specific region in which the training data is located. This optimization problem is solved for all 100 eight-dimensional test functions generated with the function generator. On average b_{opt} turned out to be 0.2 with a standard deviation of 0.03. Assuming that each training data point supports a model in a local region around it with information about the process that should be approximated, this is a reasonable result. The region supported by a training data point is visualized in Fig. 26.12 for two cases corresponding to (i) $b = 0$ (upper right corner of the test data region) and (ii) $b > 0$ (lower left corner of the training data region). As can be seen in case (i), most of the supported region lies outside the test data hypercube and does therefore not contribute to a lower error on test data. Case (ii) covers more hypervolume of the test data hypercube leading to better model performance on the test data. Note that the shape and size of the shown support regions in Fig. 26.12 are chosen arbitrarily to explain the mechanism responsible for the counter-intuitive observation that larger extrapolation demands lead to better model performance. Of course, the true shape and size of such regions are unknown and may vary with the location in the input space as well as with the specific process under consideration.

In summary, it can be recommended to measure corners only if "enough" data points remain inside the design space and large extrapolation is required. Another important outcome is the superiority of maximin LH designs compared to Sobol sequences in most cases, leading directly to the investigations of Sect. 26.1.10.

26.1.10 Comparison of Space-Filling Designs

Commonly used space-filling experimental designs are compared by means of the model quality achieved with them. These DoEs are typically chosen for computer simulation experiments, leading to so-called metamodels. In cases where there is relatively little prior knowledge about the process of interest, space-filling experimental designs have the advantage to explore the whole design space uniformly. This might be one reason for their popularity in the field of metamodel-based design optimization (metamodel-based design optimization (MBDO)), in which data is sequentially collected in regions of the design space, where an optimum is suspected. Note that the term *design* in MBDO refers to constructive design and not to the experimental design.

The experimental designs that are compared are several maximin LHs, originating from different optimization strategies, Sobol sequences, and data coming from uniform distributions. Examples where LH designs are chosen as DoE are, e.g., [314, 586–588], and [88]. Sobol sequences are increasingly used for computer experiments according to [469]. Examples for work in which they are used are [144, 585], and [108].

Most of the published related work, such as [196, 294, 295, 305, 306], and [97], relies on comparison criteria based on measures of the input space coverage or on the estimated prediction variance. In contrast to that, the comparison criterion used here is the achieved model quality. Again, the random function generator described in Sect. 26.1.1 is used to build several test functions. Through the randomness incorporated in the function generator, a variety of different characteristics can be imitated, and a favoring of any experimental design through a specific choice of functions is avoided. Once the test functions are generated, an arbitrary amount of data samples for training or testing purposes can be produced. Chen et al. [102] conducted a comparison of some experimental designs based on the resulting model quality, but only for far fewer test functions compared to this work.

Three types of maximin LH designs are used for the comparison. Two of the maximin LH designs are obtained by the optimization with the EDLS algorithm [131]. The third maximin design is obtained by the function implemented in the commercially available software MATLAB. As an additional reference, data drawn from a uniform distribution is also incorporated in the comparison.

The EDLS algorithm tries to maximize the distance between the point pair having the smallest nearest-neighbor distance in a given data set (maximin criterion) by coordinate exchanges. Therefore, the distances between all possible point pairs have to be determined, and coordinates of the points having the smallest distance to each other are tentatively exchanged with other points. If such a tentative exchange leads to an improvement of the maximin criterion, it is actually carried out. In case no exchange partner can be found that leads to an improvement of the maximin criterion, phase one of the EDLS algorithm is terminated. In phase two of the EDLS algorithm, additional points are considered for coordinate exchanges. These additional points are point pairs having the second (third, fourth, etc.) smallest distance to each other. For more details, the reader is referred to [131].

The input space dimensionality is varied from $p = 2, \ldots, 8$ while the number of training samples is held fixed at $N = 300$. The test data to assess the model quality consists of $N_t = 10^5$ samples coming from a Sobol sequence independent of the input dimensionality. The high number of test samples compared to the training data size should guarantee a good measure of the generalization performance of the obtained models. It is assumed that the Sobol sequence for the model quality assessment and the ones used for the training do not contain the same data points. For each investigated input dimensionality, 30 different realizations of each experimental design are generated and serve as training data for 30 different test functions created by the function generator. The realizations of the EDLS-optimized LH designs are obtained by varying the initial LH design. The design achieved after finishing phase one of the EDLS algorithm will be abbreviated as LH (EDLS I) in the following. Phase two takes the result of phase one and continues the optimization until convergence. The optimization result of phase two will be abbreviated as LH (EDLS II). For the Sobol sequences, different parts of a very long sequence are taken.

The aforementioned settings for the comparison are done for three different complexities of the test functions. Through the variation of the μ parameter (see Sect. 26.1.1), functions are likely to arise that are of rather low complexity ($\mu = 0.5$), of medium complexity ($\mu = 1$), and of high complexity ($\mu = 2$). All models are LMNs trained with HILOMOT as explained in Sect. 14.8. The model complexities are determined with the help of the Akaike's information criterion (AIC_C).

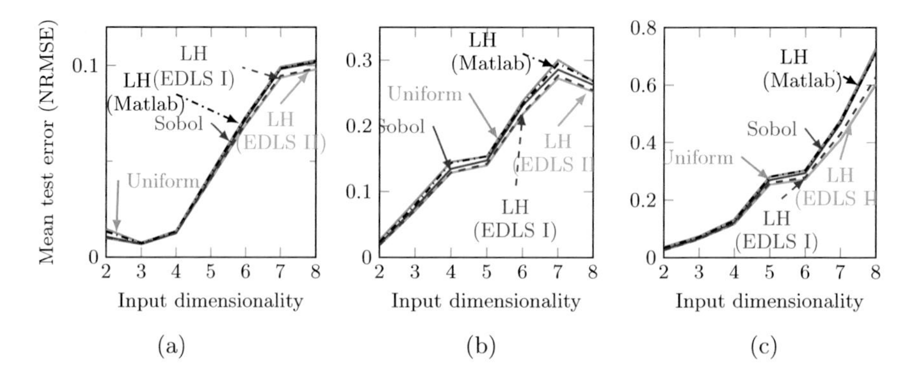

Fig. 26.13 Mean test errors of functions that are of low complexity ($\mu = 0.5$), medium complexity ($\mu = 1$), and high complexity ($\mu = 2$). (**a**) Low complexity ($\mu = 0.5$). (**b**) Medium complexity ($\mu = 1$). (**c**) High complexity ($\mu = 2$)

Figure 26.13 shows the mean test errors of all experimental designs versus the input dimensionality for all test function complexities. For each input dimensionality, the results of all 30 test functions and 30 input data realizations are averaged to obtain the mean values for each experimental design. The NRMSE is calculated and employed on $N_t = 10^5$ test data samples. Lower NRMSE values indicate better generalization performance. The difference between the model qualities achieved with the investigated experimental designs increases with an increasing complexity of the test functions and an increasing input dimensionality. In all cases, the maximin LH designs optimized with the EDLS algorithm yield the best results. The continuation of the EDLS optimization (phase two) after phase one has completed brings advantages especially for complex functions ($\mu = 2$) and high input dimensionalities; see Fig. 26.13c. The next best experimental design according to the achieved model quality is the Sobol sequence. The maximin LH designs obtained by the MATLAB function yield slightly better model performances compared to the data coming from the uniform distribution.

The variability due to varying test functions is far bigger than the variation due to different realizations of the experimental designs. Here, the standard deviations of the model errors resulting from the different realizations of the experimental designs are of interest. Therefore the standard deviations for each test function are compared. If an experimental design reveals a small standard deviation in the corresponding test errors, the danger of obtaining a bad DoE by picking just one realization for a specific task is small. This is an extremely appealing property in combination with a low mean value of the expected model error. In order to compare the best maximin LH design (EDLS II) with the Sobol sequences, the relative difference between the mean errors $\bar{e}_i$

$$\Delta e_r = \frac{\bar{e}_{Sobol} - \bar{e}_{LH}}{\bar{e}_{LH}} \cdot 100\% \,, \tag{26.6}$$

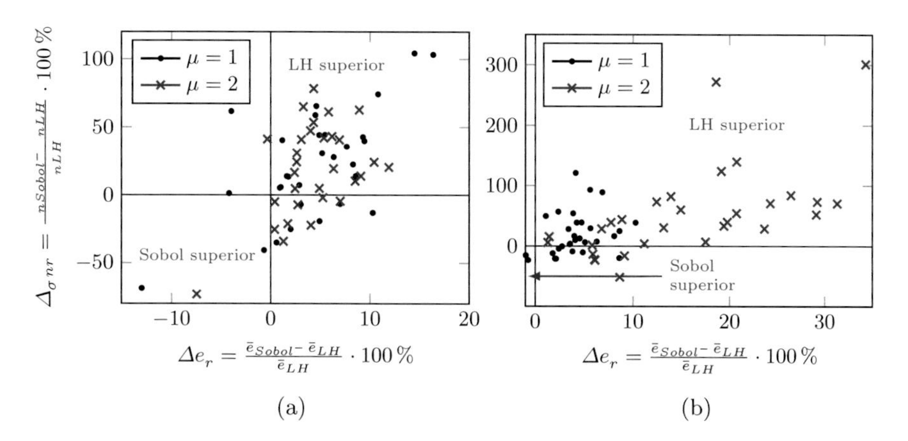

Fig. 26.14 Relative mean test errors and standard deviations for all 30 test functions of two complexities ($\mu = 1$, $\mu = 2$) for $p = 4$ and $p = 8$. (**a**) Input dimensionality $p = 4$. (**b**) Input dimensionality $p = 8$

and between the standard deviations σ_{ni}

$$\Delta\sigma_{nr} = \frac{\sigma_{nSobol} - \sigma_{nLH}}{\sigma_{nLH}} \cdot 100\%\,, \tag{26.7}$$

is used. According to the definitions of (26.6) and (26.7), positive values indicate the improvement of the EDLS-optimized maximin LH design in percent compared to the Sobol sequence. Figure 26.14 shows the values of Δe_r and $\Delta\sigma_{nr}$ for all 30 medium and complex test functions together in one graph for input dimensionalities $p = 4$ and $p = 8$. Each dot and each cross corresponds to one test function. The upper right corner of the graph contains test functions for which the maximin LH design leads to better mean test errors ($\Delta e_r > 0$) and lower standard deviations ($\Delta\sigma_{nr} > 0$), which is the case for most of the 60 shown test functions in each graph. The selected graphs, i.e., Fig. 26.14a, b, show the results for the input dimensionalities where the EDLS-optimized maximin LH designs come off worst and best. The improvements in the relative mean error difference and the standard deviation difference go up to 35% and 300%, respectively. The maximum decline in Δe_r is about 15% and 70% in $\Delta\sigma_r$. However, in far more cases, the EDLS-optimized LH designs are superior to Sobol sequences in both aspects.

Figure 26.15 shows the required mean computation time for the generation of all investigated experimental designs. In this category, the Sobol sequences are superior to all LH designs and are only outperformed by the uniform designs. On an absolute scale, the optimization of the EDLS I LH designs takes up to 1 minute on average, while the generation of EDLS II designs consumes up to 660 min on average. The creation of all other designs takes far less than 1 second.

Unfortunately it is not possible to clarify with certainty what the reasons are that in almost all cases the maximin LH designs outperform the other investi-

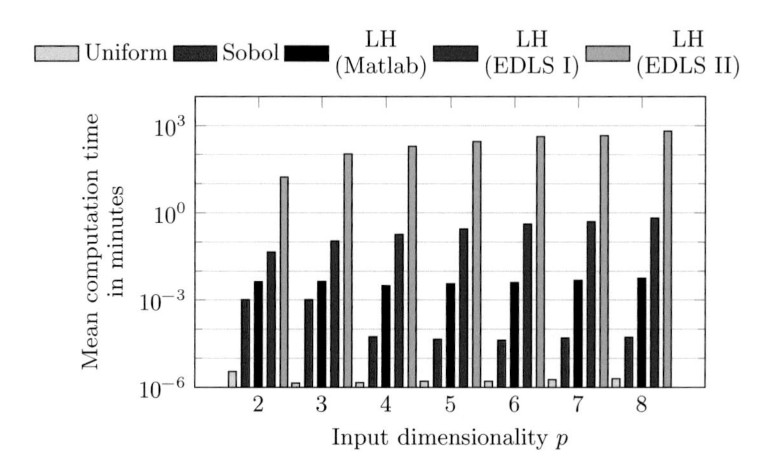

Fig. 26.15 Mean computation time for the generation of the experimental designs

gated experimental designs regarding the mean model quality. The superiority regarding the variance in the model qualities originates probably due to the use of an optimization algorithm in order to generate the training data. Through the optimization, the variation between different training data sets should be very low which finally results in a lower model quality variance. However, one property of the designs is investigated in some more detail in the following in order to find a hint that might explain the superiority regarding the better mean model quality. For the comparison of the space-filling designs carried out in this section, it is assumed that designs covering the input space uniformly are advantageous for the expected model quality. In the opinion of the author, one crucial point is to fulfill the uniformity requirement for reasonable few data samples. For example, data drawn from a uniform distribution should cover the input space uniformly, but quite a lot of samples are needed until clumps and empty spaces vanish. Figure 26.5b shows $N = 300$ samples drawn from a uniform distribution in a two-dimensional input space and clumps as well as empty spaces are apparently present. A commonly used measure to quantify the geometric non-uniformity of points is the L_∞ discrepancy [532]. For any data set containing N points, it is defined as

$$D_N^* = \sup_{K \subseteq \mathbb{R}^p} \left| \frac{N_{in}(K)}{N} - \frac{V_{in}(K)}{V} \right| . \tag{26.8}$$

In this equation K defines a hypercube contained within the p-dimensional input space $\mathbb{R}^p$ with one corner of it located at the origin $\underline{0}$. The number of points inside K is N_{in} and the hypervolume of K is V_{in}. V corresponds to the total hypervolume in which all data points N are contained. In other words, a hypercube is sought in which the discrepancy between the proportion of points inside the hypercube differs the most from the proportion of points that *should* be present according to the volume proportion of that hypercube. In Fig. 26.16 the proportion of points is

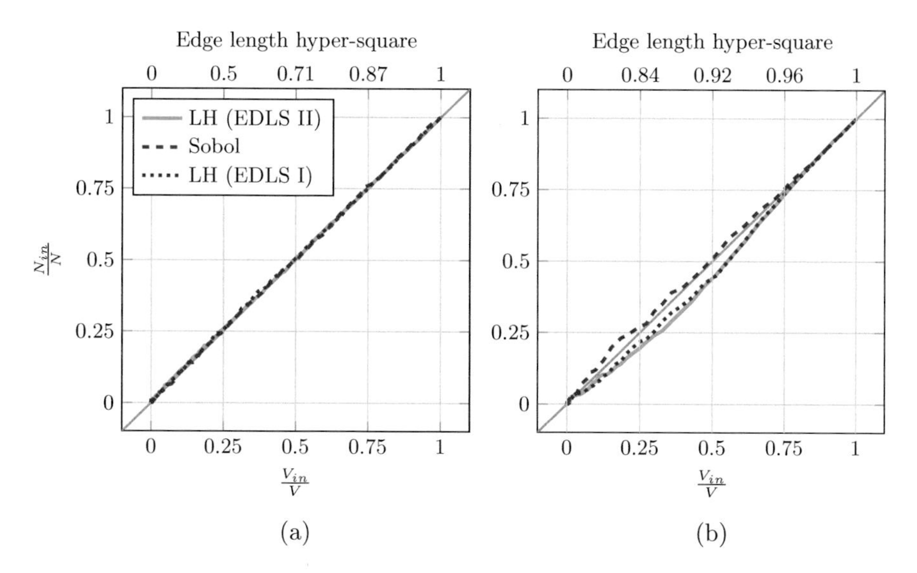

Fig. 26.16 Proportion of points inside a defined hypervolume to all points (N_{in}/N) versus proportion of that hypervolume to the total hypervolume (V_{in}/V) for input dimensionalities $p = 2$ and $p = 8$. (**a**) $p = 2$. (**b**) $p = 8$

plotted against the proportion of hypervolume of a hyper-square for the two- and eight-dimensional case. The shown curves correspond to the worst L_∞ discrepancy of all 30 input data realizations of the maximin LH designs, i.e., EDLS I and II, and the Sobol sequences. On top of each shown figure, there is an alternative x-axis showing the edge length of the hyper-square that corresponds to the hypervolume V_{in}. The closer the curves are to the diagonal, the lower is the discrepancy measure according to (26.8). Curves lying below the diagonal represent a too low data density and vice versa. For the two-dimensional case shown in Fig. 26.16a, all experimental designs lie upon the diagonal. With an increasing input dimensionality, the curves deviate more and more from the diagonal. On average the Sobol sequences tend to be a little bit above the diagonal, whereas the maximin LH designs are typically below it. The interpretation of this observation is that the LH designs possess more points very close to the boundary of the outer hypercube. In an eight-dimensional input space, the ratio $V_{in}/V = 0.5$ corresponds to an edge length of ≈ 0.92; see Fig. 26.16b. At this point the curves of the two LH designs lie clearly below the diagonal, meaning that in relation to the proportion of the hypervolume, too few data samples lie inside the hyper-square. Viewed from a different angle, the number of points in the remaining hypervolume ($V - V_{in}$) has to be relatively high, because the point ($V_{in}/V = 1$, $N_{in}/N = 1$) has to be met exactly. Since the advantage of the maximin LH designs with respect to model quality increases with the input dimensionality (see Fig. 26.13), this property of the training data set seems to be favorable w.r.t. to the model quality.

Table 26.2 Advantages and drawbacks of Sobol sequences and maximin LH designs

Maximin LH designs	Sobol sequences
+ Better model qualities on average + Lower deviations in the model qualities (even worst case delivers satisfactory results) – Higher computational effort and therefore more time-demanding to generate the experimental designs – Difficult to add further design points incrementally – Not suited for very high amounts of data samples ($N > 10^4$) because optimization takes too long • Higher L_∞ discrepancy on average (theoretically a drawback, but has a positive effect in the investigations of this section)	– Lower model qualities on average – Higher deviations in the model qualities (worst case might be very inferior) + Lower computational effort and therefore less time-demanding to generate the experimental designs + Easy to add further design points incrementally + Very high amounts of data samples not a problem • Lower L_∞ discrepancy on average (theoretically an advantage, but has no positive effect in the investigations of this section)

In summary, maximin LH designs created by the EDLS algorithm (and probably other optimization schemes as well) are superior to all other investigated experimental designs regarding both the achieved model qualities and the variation of these model qualities. These advantages have to be paid off by a higher computational effort for the DoE. However, the LH designs can be optimized and stored in advance to a specific task, such that this drawback can be weakened at least for some applications. Another drawback of the maximin LH designs regards their extensibility. If just a few points should be added to an already optimized LH design, it is not as easy as to append additional points to an existing Sobol sequence, because further optimization runs are necessary. The results are important for both the initial DoE and subsequent refinements of it in presumably near-optimum regions of the design space. Table 26.2 compares the advantages and drawbacks of the Sobol sequences and the maximin LH designs.

26.2 Active Learning for Structural Health Monitoring

In [225, 226, 384] and the German thesis [222], HILOMOT DoE is applied to a challenging structural health monitoring problem. In structural health monitoring (SHM), the goal is to continuously supervise or monitor technical structures or elements during normal operation in a nondestructive manner. Due to economic and ecological reasons, structures or elements increasingly are not designed for their whole life cycle. Rather it is planned that parts are substituted or serviced and maintained. In this context, monitoring systems play a very important role in order to *detect* and *localize* looming faults at a very early phase. Favorably, besides fault detection and localization, even the amount of damage occurring and the remaining lifetime shall be predicted by an SHM system. Important application areas are, e.g., the monitoring of bridges, pipelines, and wind turbines or in aerospace.

The main benefits of SHM compared to more traditional approaches are:

- Supervision of complex structures is possible.
- For monitoring, no disassembly is required which can save significant time and cost.
- Damages in/at the structure can be detected in the early stages.
- With structural health monitoring, structures can be continuously monitored during normal operation with the help of installed sensors.

Many approaches to carry out SHM are possible [23]. One promising alternative is based on ultrasound waves. The application considered here is particularly interesting for thin-walled structures as they abound in the aerospace industry. For experiments, many piezoelectric sensors which also can be utilized as actuators are glued on a thin aluminum plate. Ultrasonic waves are sent through the structure by one actuator and received by the other piezoelectric sensors. Reference measurements are made for the undamaged structure. The measured characteristics then can later be compared with new measurements of a possibly damaged structure. The change in characteristics can be exploited to detect and localize the damage. With ultrasonic waves, the characteristics are sensitive with respect to the type of damage (cracks, delamination in compound materials, or corrosion damage).

Figure 26.17 shows a snapshot of the wave propagation in an aluminum plate measured with a laser Doppler vibrometer. It can clearly be seen that damage causes reflections of the waves dependent on the orientation of the damage with respect to the actuator. With algorithms as described in, e.g., [383], it is possible to construct a so-called damage map for visualization of the damage likelihood. With

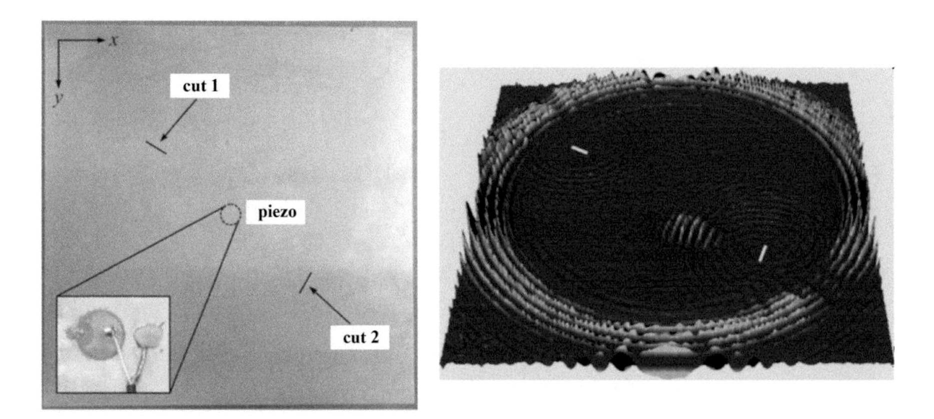

Fig. 26.17 Left: Aluminum plate with two cuts introduced which are oriented differently with respect to the piezo actuator. Right: Snapshot of the wave propagation measured with a laser Doppler vibrometer. (Taken from [383])

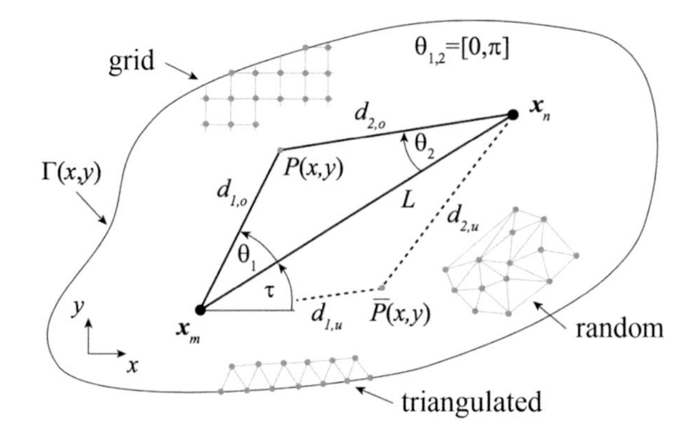

Fig. 26.18 The computation of the damage map can only be carried out inside a discretized design space $\Gamma(x, y)$. Standard types of meshes include grid, random, or triangulated partitioning [383]

these algorithms a damage intensity $P(x, y)$ or $\bar{P}(x, y)$ can be computed on the plate geometry at discrete points; compare Fig. 26.18. The point of highest intensity characterizes the estimated damage location (x^*, y^*).

Classical strategies place the points where the intensity is evaluated according to a geometric pattern. This is comparable to an unsupervised DoE method with similar drawbacks; compare Sect. 14.10.1. With active learning based on HILOMOT DoE, these drawbacks can be overcome, and the specific nonlinear shape of the intensity can be exploited. The placement of the points automatically focuses on the most promising regions with the highest intensities since there the nonlinearity is most severe and the worst local model will likely occur in these regions.

26.2.1 Simulation Results

According to the localization method proposed in [383], the discretization for the geometry shown in Fig. 26.19 first was carried out on the basis of simulation data. The discretization was performed by HILOMOT DoE for the two fault scenarios D1 and D2. Section 26.2.2 deals with one experimental study for fault D = D1.

Figure 26.20 shows the fault scenario D1 with the measurement points generated by HILOMOT DoE and the associated contours of the local model network for 5, 10, 15, 20, 25, and 30 points. The concentration of measurement points around the damage and the differentiation of the estimated damage intensity can clearly be observed. The precision of the damage location increases quickly. The position of the highest estimated intensity accurately coincides with the simulated damage location after 30 measurements.

Figure 26.21 shows the corresponding results for fault D2. Here, the point density and the intensity values close to the damage are slightly lower than in the example

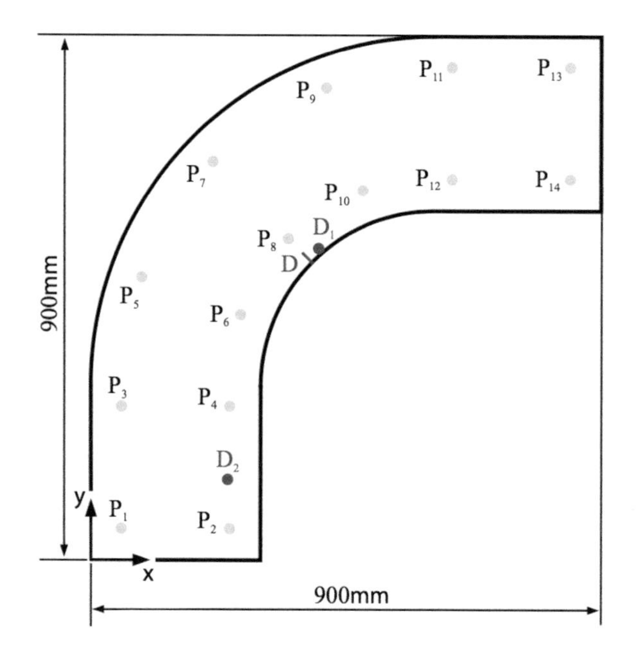

Fig. 26.19 Geometry of the investigated structure with the positions of 14 actuators P1 to P14. The thickness of the plate is 1.5 mm. For simulation, the damage positions D1 and D2 were chosen. In the experiment, the damage position D was selected [382]

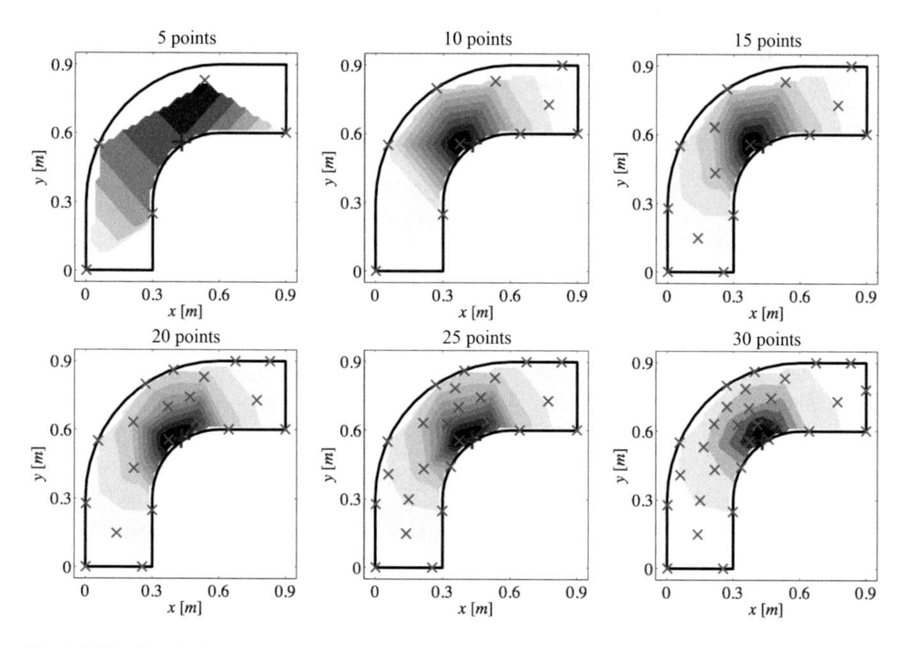

Fig. 26.20 Simulation results with HILOMOT DoE for the fault D1. The damage is marked with a "+"; the measurement points are marked with an "×." The contour lines illustrate the fault model [222]

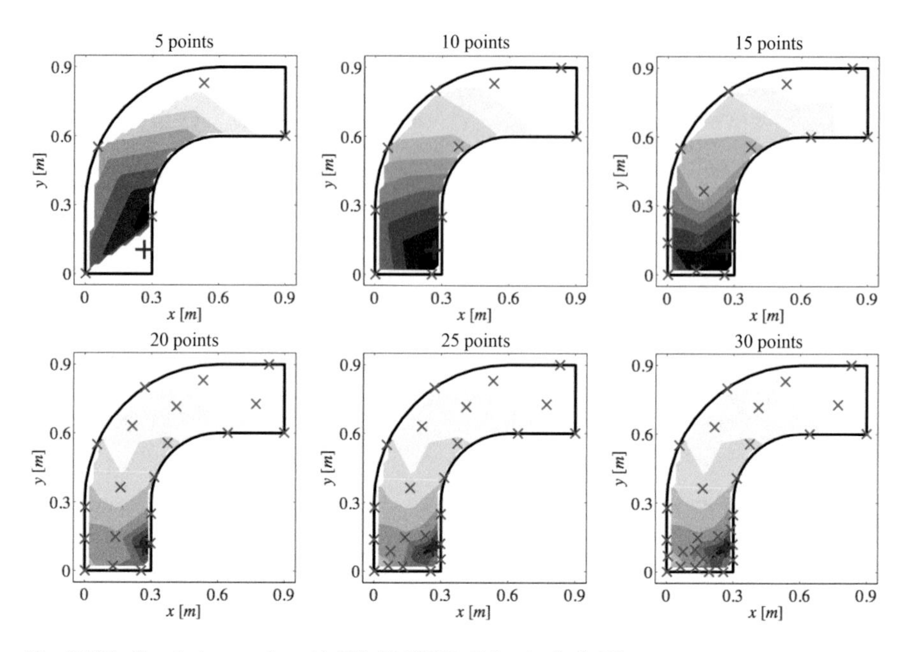

Fig. 26.21 Simulation results with HILOMOT DoE for the fault D2

in Fig. 26.20. Probably this is due to the more difficult damage location close to a corner of the structure with more dominant wave reflections. Nevertheless, the simulation results are more than satisfactory and highly encouraging.

26.2.2 Experimental Results

In contrast to the simulated wave propagation considered in the previous section, the experimental measurements include all kinds of disturbing effects like (i) measurement noise, (ii) nonlinear and dynamic sensor distortions, (iii) non-ideal geometric properties, (iv) non-ideal material properties, etc. The experimental setting is shown in Fig. 26.22 with 14 piezoelectric sensors/actuators and a close-up of the cut. The geometric shape is identical to the one depicted in Fig. 26.19.

Figure 26.23 compares the discretization on a grid with the one carried out by HILOMOT DoE for 25 data points. The latter one yields a significantly superior fault location and a much more differentiated intensity model. The grid evaluates many measurement points in uninteresting regions, thereby wasting lots of performance.

Figure 26.24 demonstrates that HILOMOT DoE is only negligibly affected by all real-world difficulties. The results are quite comparable to the simulation case. One key reason for this robustness of HILOMOT DoE certainly is due to the fact that its behavior is never based on only one (or very few) data samples but rather

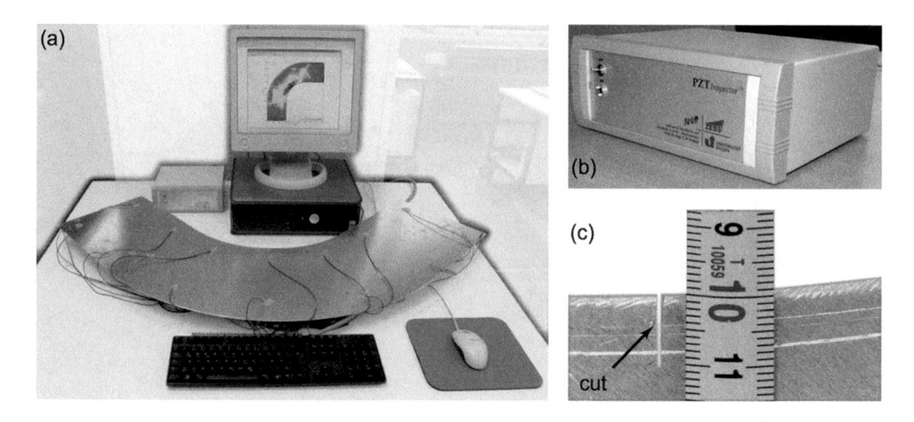

Fig. 26.22 (**a**) Experimental setting of the investigated, non-convex structure, (**b**) switch and amplifier for the measurement recording, (**c**) introduced cut in form of a slice of 10 mm length and 1 mm width [383]

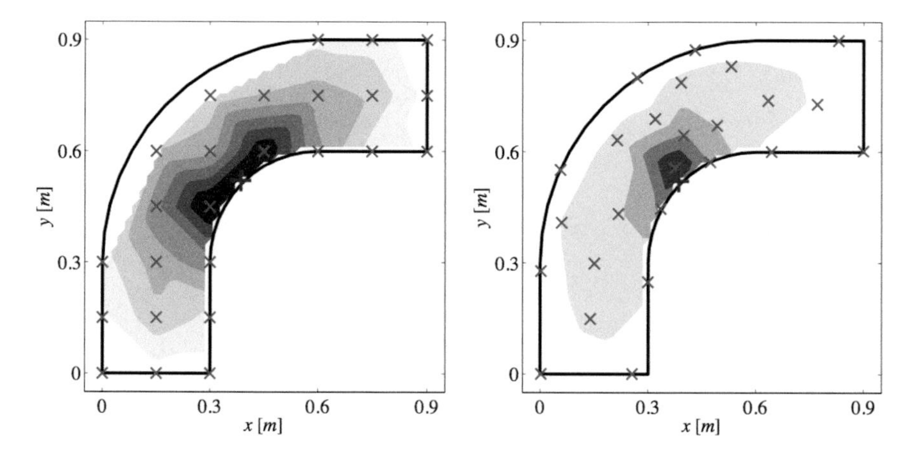

Fig. 26.23 Experimental results for 25 measurements on a regular grid (left) and with active learning by HILOMOT DoE (right). The plots demonstrate that the damage can be localized much more precisely with measurement points placed by HILOMOT DoE than by a grid with an identical number of measurements

considers averaging effects in all of its steps. Finally, Fig. 26.25 illustrates the model constructed with the HILOMOT DoE-generated data together with the measured intensity and the true damage location.

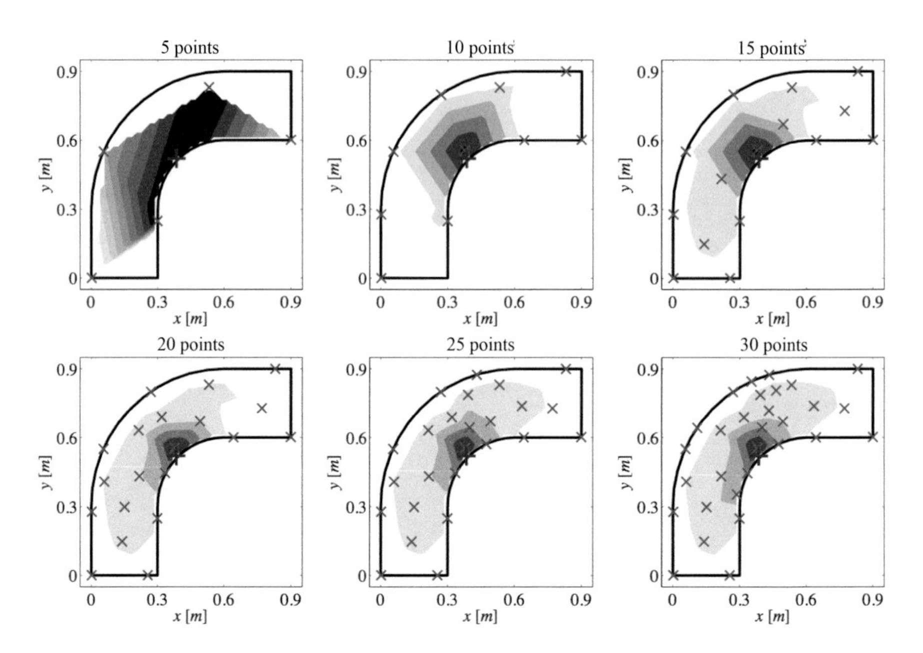

Fig. 26.24 Experimental results for fault D shown in Fig. 26.19. The measurement points placed by HILOMOT DoE are marked with an "×"; the damage is marked with a "+." The fault model is illustrated by its contours. With a growing number of measurements, the modeled damage localization steadily improves

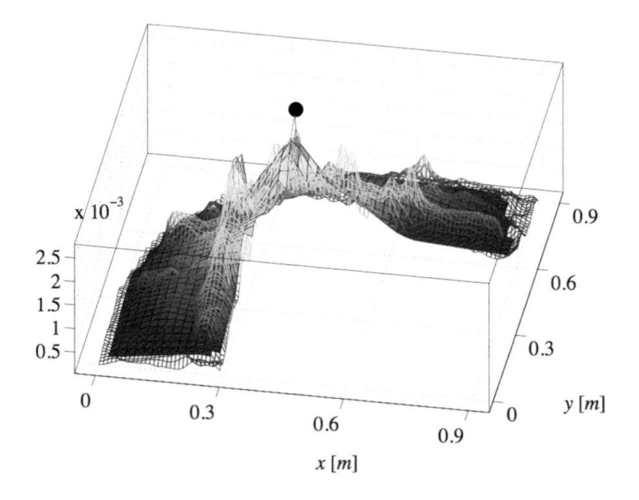

Fig. 26.25 3D plot of the local model network constructed with HILOMOT DoE (solid) compared to the noisy damage intensity which represents the process (light). The true damage is marked by the big dot

26.3 Active Learning for Engine Measurement

This section is mainly based on the publications [223, 228, 322] and discusses active learning with HILOMOT DoE for efficient combustion engine measurement on a test stand together with Daimler. The engine under test is the Mercedes-Benz V6 Diesel Engine (internal designation OM 642 LS) depicted in Fig. 26.26 with the characteristic data listed in Table 26.3.

26.3.1 Problem Setting

A modern combustion engine used on many means of transportation like cars, trucks, and ships requires an electronic control unit (ECU). Control is often feedforward since many appropriate sensors that are robust and cheap enough are lacking. However, the number of feedback control loops is steadily increasing

Fig. 26.26 Mercedes-Benz V6 Diesel Engine OM642 LS [322]

Table 26.3 Characteristic engine data [322]

Number of cylinders	6
Bank angle	72°
Displacement	2987 cm^3
Cylinder gap width	106 mm
Compression	15.5
Bore	83 mm
Stroke	92 mm
Maximum power at rpm	195 kW at 3800 rpm
Maximum torque at rpm	620 Nm at 1600–2400 rpm

and certainly will do so in the future. Often technological breakthroughs were enabled by the realization of feedback control. A prominent example is λ-control for gasoline engines which controls a stoichiometric fuel rate and thus made the utilization of a three-way catalytic converter possible.

In order to tune or optimize the control systems in an ECU engine, models are necessary that are capable of predicting important engine outputs from the control actuations and the operating point (load, speed). With the help of these models, the control can be optimized. Currently, it is typically represented as a look-up table in the ECU. With the increasing complexity of the engine, the number of actuation signals grows, intensifying the need for *high-dimensional models*.

Since the relationship between the actuation signals and the operating point as the inputs and the fuel consumption and emissions as the outputs can hardly be modeled by first principles, one must resort to data-driven modeling. Two key difficulties arise in the context of the curse of dimensionality that cannot be solved by traditional approaches anymore:

1. *Model architecture*: The high input dimensionality more and more rules out the traditional look-up table models. Rather neural network-type architectures like local model networks or Gaussian process models have to be employed.
2. *Data acquisition*: Measurement times and costs explode with the input dimensionality. It is not feasible anymore to vary each input with respect to all other inputs in all combinations (grid measurement). Rather sophisticated DoE strategies are called for that measure only where the most valuable information is expected.

In this section, Point 1 is solved with a local linear model network trained by HILOMOT. Point 2 is addressed with HILOMOT DoE. Figure 26.27 shows a scheme of the modeling task discussed in the following. The five actuation signals are [322]:

- Quantity of the pilot injection (PI): A small fraction of the fuel is injected before the main injection. This prolongs the combustion process and reduces vibrations and noise because the combustion is not as harsh.
- Crank angle of 50% accumulated heat release (CA50): The angle of 50% heat release could be easily varied since the engine has a closed-loop combustion control. Controlling this angle gives maximum influence over the combustion process.
- Charge motion control valve (CMCV): The engine is equipped with an electronic charge motion control valve system that provides increased air velocity at low engine speeds for improved emissions and low-rpm torque. "This is a variable intake system in order to optimize power and torque across the range of engine speed operation, as well as help provide better fuel efficiency."[1]

[1]https://en.wikipedia.org/wiki/Variable-length_intake_manifold

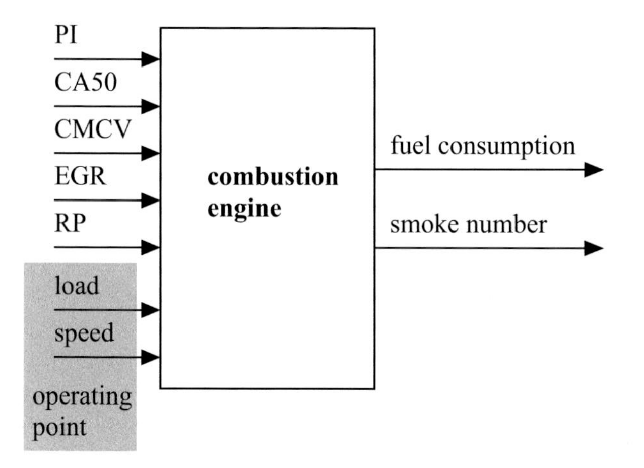

Fig. 26.27 Combustion engine process considered here: 2 outputs depend on 7 inputs of which 5 inputs are actuation signals and 2 inputs describe the operating point

- Exhaust gas recirculation rate (EGR): Exhaust gas is recirculated in order to lower the combustion chamber temperatures and thereby reduce the NO_X emissions.
- Rail pressure (RP): If fuel is injected with high pressure, its particles become smaller and better distributed creating favorable combustion.

Two types of modeling and DoE approaches are taken:

- *Operating point-specific or local approach*: For a number of operating points, typically on a grid of load-speed combinations, an operating point-specific local model network with 5 inputs and 2 outputs is identified. This also requires a five-dimensional DoE individually for each operation point. This route is taken in Sect. 26.3.2.
- *Global approach*: All 7 inputs including the operating point are varied by the DoE. This means just one big DoE and one big model exist. This approach is taken in Sect. 26.3.3.

In the subsequent discussion, the following terminology is used:

- The *coefficient of determination* R^2 as a measure for the goodness of fit. It is frequently used in statistics. It is related to the error measures used in the rest of this book as follows: $R^2 = 1 - \text{NMSE}$.
- *PRESS* stands for the prediction residual error sum of squares and is usually called leave-one-out (LOO) error in other parts of the book; compare Sect. 7.3.2. It is a cross-validation error that can be calculated in closed analytic form for linear regression models.
- *FNN* for Fast Neural Network is the name of a local model network trained by LOLIMOT (or something very similar) offered in the industrial tool CAMEO. This tool is promoted by the AVL LIST GmbH, one major supplier in engine

powertrain engineering. CAMEO handles the complete calibration process for combustion engines—from data collection to look-up tables. Experimental modeling is a key ingredient.

26.3.2 *Operating Point-Specific Engine Models*

An operating point-specific approach is closer to the traditional workflow in the automotive industry. It offers some important benefits because it is easier to understand and visualize. This is particularly advantageous if the number of inputs (besides the operating point) is ≤ 3. The main drawback is that no information sharing between operating points takes place. No exploitation of similar behaviors in different (but close-by) operating points is possible. Note that these similar characteristics always occur in physical relationships. Therefore potentially more measurement data needs to be acquired if the whole operating range shall be covered.

For each operating point, 80 training data points and 160 test data points were measured. On the training data, HILOMOT DoE was run for data acquisition which built up a local model network. For comparison sake, two alternative models were identified on training data with 80 points generated by a D-optimal design assuming a polynomial of degree 3. This data was used to build a polynomial of degree 3 as assumed in the DoE and a local model network trained with LOLIMOT. The polynomial possesses $n = 56$ parameters which allows for sufficient redundancy in the context of the $N = 80$ training data points. A factor $N/n \approx 1.5$ is a common choice in the automotive industry for global models like polynomials (without regularization).

26.3.2.1 Results

Figure 26.28 shows the performance of a local model network trained with data obtained by HILOMOT DoE for 80 data points. The criterion for HILOMOT DoE was the model performance on the output "fuel consumption." Nevertheless, the model for the other output also develops nicely. It is important to not assess the model quality by observing the training data error of the model since it includes overfitting effects. Rather it can roughly be assessed by an approximated leave-one-out error calculated for the local models and aggregated for the whole network

$$J_{\text{PRESS}} = \sum_{j=1}^{M} \sum_{i=1}^{N} e_{j,\text{LOO}}^2(i)\ \Phi_j(i) \tag{26.9}$$

where $e_{j,\text{LOO}}(i)$ denotes the leave-one-out error at data sample i for local model j which can be evaluated in closed form if the local model is linearly parameterized

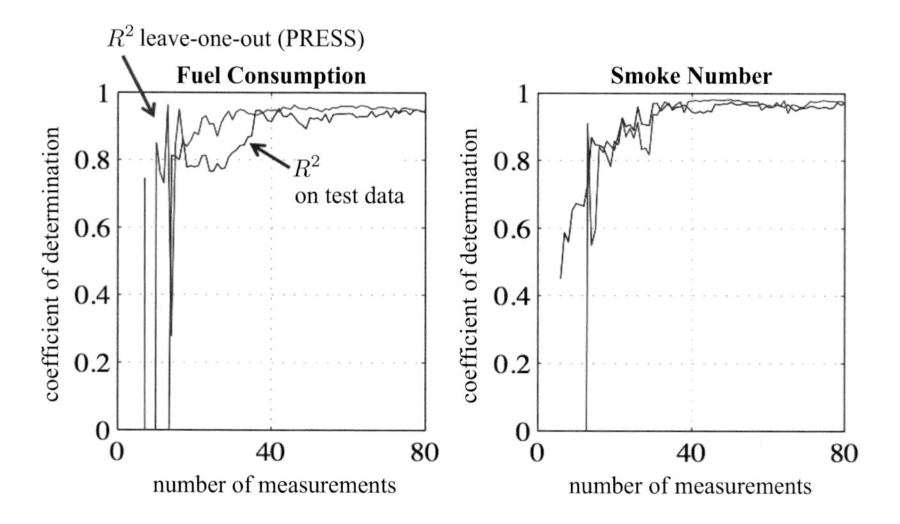

Fig. 26.28 HILOMOT DoE model performance on fuel consumption and smoke number where the second model output "smoke number" simultaneously improves although the DoE is purely optimized with respect to first output "fuel consumption"

and $\Phi_j(i)$ is the validity function of model j for data $\underline{z}(i)$. Normalized by $\sum_{i=1}^{N}(y(i) - \bar{y})^2$, this gives the NRMSE$_{\text{PRESS}}$ and $R^2_{\text{PRESS}} = 1 - \text{NRMSE}_{\text{PRESS}}$.

CAUTION: This evaluation of R^2_{PRESS} does not include the variance error induced by the partitioning but gives a quick orientation on the expected test performance. Figure 26.28 also shows the model quality on test data (which of course was not available at the time of measurement). It can clearly be observed that both quantities develop almost synchronously. This observation inspires a new algorithm for exploiting HILOMOT DoE for measurement time reduction. R^2_{PRESS} is slightly over-optimistic as expected but can be seen as a good approximation of the cross-validation error as long as the number of local models is not excessively large. This demand is met by utilizing the AIC$_\text{C}$ as HILOMOT termination criterion.

26.3.2.2 Reducing Measurement Time with HILOMOT DoE

One strength of most active learning strategies is that they operate incrementally, i.e., point by point. Instead of a pre-established list of points that has to be processed from start to finish in order to make any sense, active learning strategies can be halted anytime. For a discussion of this topic, refer to Sect. 14.10.4. The tradeoff between model performance (longer measurement times are better) and cost, time, etc. (shorter measurement times are better) can be made explicit. And it can be carried out process- and model-individually. In early phases of development where coarse models might suffice, another tradeoff is desired than in the final phase.

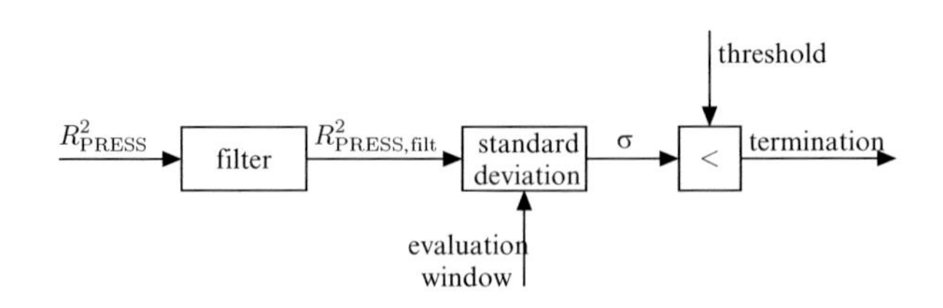

Fig. 26.29 Scheme for termination of HILOMOT DoE if the model is good enough, i.e., a filtered version of R^2_{PRESS} falls below a threshold

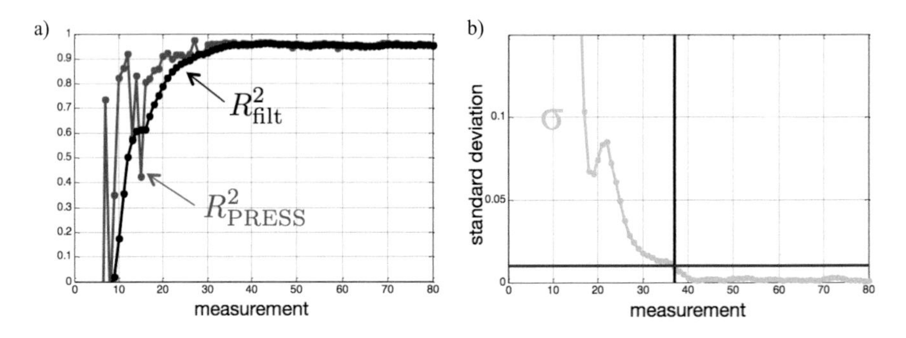

Fig. 26.30 Example for termination of HILOMOT DoE according to Fig. 26.29: (**a**) filtered and unfiltered model quality (1 is best), (**b**) variation of filtered model quality (0 is best)

Figure 26.29 outlines a simple scheme that can be used to automatize these tradeoffs. It is based on evaluating the standard deviation σ of the filtered R^2_{PRESS} over an evaluation window. If this value falls below a threshold, HILOMOT DoE is terminated. Three parameters can be adjusted in this scheme: (i) cutoff frequency of the filter, (ii) window length, and (iii) threshold. In principle, these parameters are redundant and could be replaced by a single, more complex FIR filter where the window length would correspond to the filter order. In this application, this scheme has proven easy to calibrate. Note that the threshold must be tuned to a performance variation that realistically can be achieved. Otherwise, the measurement would go on forever. In practice, typically the termination criterion will be a logical combination of different criteria where one will be the maximally allowed measurement time.

Figure 26.30 shows that the filtered version R^2_{filt} of the model performance R^2_{PRESS} behaves monotonously with appropriate filter settings. Also, it stabilizes and settles to its final value with little variation. Therefore comparing its standard deviation with the threshold can terminate the HILOMOT DoE procedure. It is important to understand that the *absolute* model performance is irrelevant for this step. Of course the *absolute* value of R^2_{PRESS} could also enter the termination condition. However, in practice, it is often unrealistic to ensure that certain model performances are achievable. Thus, it more reasonable to terminate if no significant progress can be made for a couple (i.e., the window length) of active learning steps. This is indicated by a small standard deviation below the given threshold.

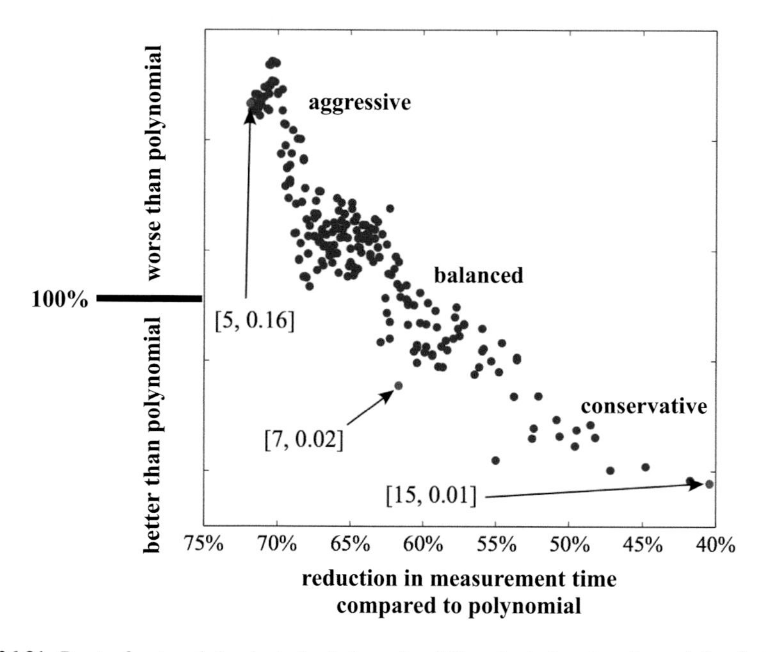

Fig. 26.31 Pareto front and dominated solutions for different window lengths and thresholds in Fig. 26.29. Lower thresholds yield more conservative models and less measurement time savings. For similar model performance compared to a D-optimal design with a polynomial model, 55–60% measurement time can be saved with HILOMOT DoE. The numbers in the brackets mean "[window length, threshold]"

Different values of the parameters window length and threshold yield different tradeoffs between measurement time and model performance. Figure 26.31 depicts a wide range of reasonable tradeoffs. For a model performance similar to the polynomial model with D-optimal data, about 60% of the measurement time could be saved. More conservative tradeoffs even allow for model improvements with less significant reductions in measurement time. More aggressive tradeoffs that even allow model deteriorations can boost time savings up to 70%. However, from the distribution of points in Fig. 26.31, it is obvious that significant model performance must be sacrificed for that reduction. At a measurement time reduction of 67%, there seems to be a steep decline in model performance.

26.3.2.3 Multiple Outputs

The focus on the above discussion was on the first output "fuel consumption" although the second output "smoke number" was also modeled well as a by-product. In practice, frequently even many more outputs (e.g., several types of emissions) arise. In this application, the local model network possesses two outputs with individual local model parameters. However, *one* structure, i.e., the partitioning, has to be determined for the whole network. An alternative would be to choose individual local model networks for each output; see also Sects. 13.3.7 and 9.1.

But independent of the taken modeling approach for multiple outputs, HILOMOT DoE (or any other supervised, bias-oriented strategy) requires a *single* criterion which it is based on. Several approaches are possible to solve this dilemma:

- *Most important output*: Oftentimes one output is the most important one. Or at least it is the one which is required with the highest precision or can be measured with the highest accuracy. Then the DoE should focus on this output. The other outputs can be modeled with the same data—only their DoE is suboptimal. In the engine example, this approach seems the most reasonable since emissions cannot be determined very accurately by measurements anyway.
- *Alternating between outputs*: If the DoE should reflect all outputs, HILOMOT DoE can alternate between these criteria. Since each measurement point is treated individually, this causes no computational overhead.
- *Weighted average of outputs*: All outputs can be included in the criterion by taking a weighted average where each weight reflects the importance of the corresponding output for the DoE.

The second idea is pursued successfully in [322].

It is obvious that multiple outputs dilute the effect of HILOMOT DoE. A worst-case scenario would be that all outputs exhibit completely different nonlinear characteristics and are sensitive in disjunct regions of the input space. However, if there exist common regions of benign and common regions of joint strongly nonlinear behavior for all outputs, then HILOMOT DoE can be very effective.

26.3.3 Global Engine Model

The engine was operated in a slightly different setting. The input CMCV was fixed, and a second pilot injection was introduced. Thus, the number of actuation signals remained being 5.

Operating point-specific or local DoE may be the first step toward modern high-performance DoE strategies. It basically splits the problem into two subproblems: (i) DoE for each operating point and (ii) decision on which operating points to cover. Although this is lower-dimensional and easier to understand from a human perspective, it is suboptimal. The consequent alternative is to apply a global DoE to an overall model that concurrently describes the influences of all actuation signals and the operating point.

This global DoE has to deal with seven input variables, two more than the local DoEs which is the major disadvantage. The advantages are basically the opposite of the drawbacks for local DoE. 800 measurements points were gathered for

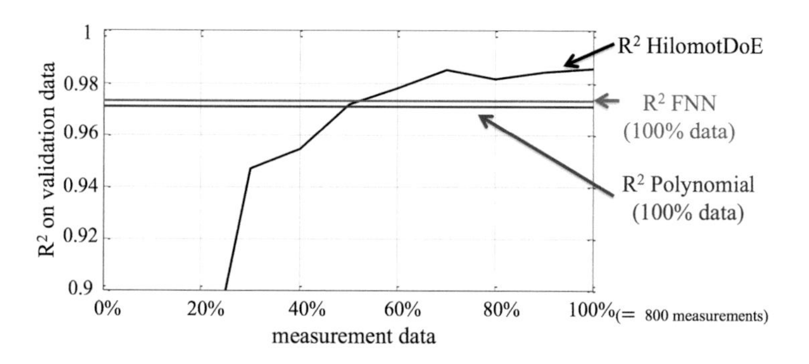

Fig. 26.32 Comparison of a local model network trained by active learning with HILOMOT DoE and a polynomial and a local model network trained with LOLIMOT (FNN) both on S-optimal data

training; 900 space-filling data points were measured for testing. For comparison, 40 operating points were covered with 800 data points in an S-optimal[2] manner.

Figure 26.32 illustrates the model performance of a local model network trained with LOLIMOT (FNN) and a polynomial of 4th[3] degree, both utilizing 800 S-optimally designed data. At around 50% of the data, the local model network with HILOMOT DoE achieved similar performance as the polynomial and the local model network trained with LOLIMOT with the traditional design. Thus, either half of the measurement time can be saved, or significantly better model performance can be accomplished.

Figure 26.33 (top) compares model output vs. measurement output for the test data for three modeling approaches. Since HILOMOT performs so well, it was also considered in a reduced training data variant. Figure 26.33 (bottom) shows the same results in terms of histograms of the model test errors. The plots "Hilomot" show the result of a local model network trained by HILOMOT using all 1200 measurements. "Hilomot reduced" shows the results using the first 800 measurements for modeling. For this comparison, the HILOMOT DoE measurements of every single operating point were reduced equally with a certain percentage value in order to get the same amount of data as it was measured with the S-optimal DoE approach. The plots "FNN" show the results using the Fast Neuronal Network approach implemented in CAMEO based on the S-optimal measurements. The plots "Polynomial" present the results of a polynomial approach of fourth degree.

[2]This is a special kind of optimal block design that is beyond the scope of this book. It was offered by the software and achieved good results in the past.

[3]Because the amount of data is much higher for the global DoE than for each local DoE, the degree of the polynomial is chosen higher as well.

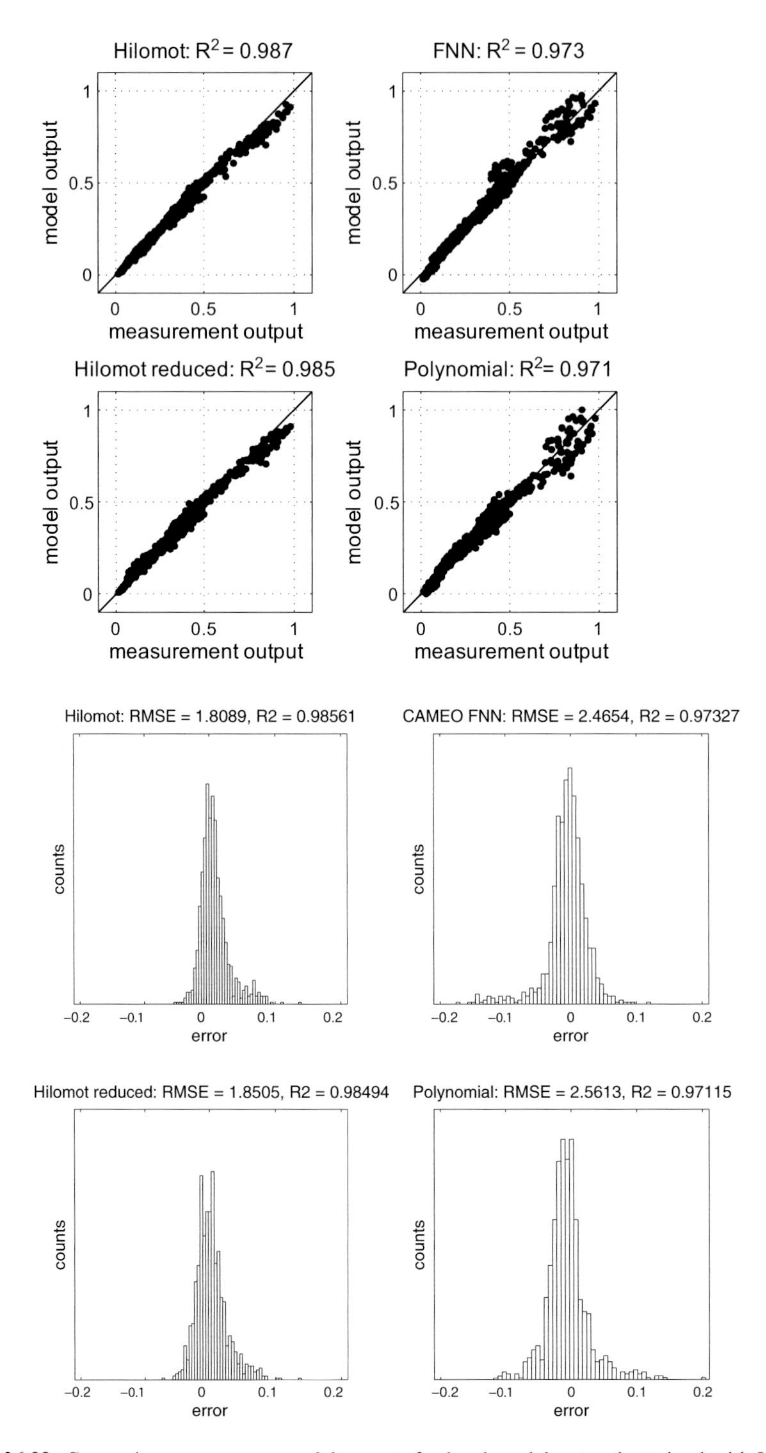

Fig. 26.33 Comparison process vs. model outputs for local model networks trained with HILOMOT (standard and a reduced data variant), LOLIMOT (FNN), and a third-degree polynomial (top). Corresponding histograms of the test errors (bottom)

26.4 Nonlinear Dynamic Excitation Signal Design for Common Rail Injection

The superiority of the new excitation signal OMNIPUS described in Sect. 19.8 in Part III is demonstrated on a real high-pressure fuel supply systems (HPFS) which is described in Sect. 26.4.1.[4] The identification procedure and the results are given in Sect. 26.4.2 and in Sect. 26.4.3, respectively. For more details on the methodological and algorithmic side of this example, refer to [510, 511]. For an illustration of the strengths of Gaussian process models and their uncertainty description in optimization, [509] is a nice example in the context of controller design.

The OMNIPUS excitation signal design applied in the following is introduced and analyzed in Sects. 19.8.2 for the SISO and 19.8.3 for the MISO case. Additionally, a methodological comparison between different nonlinear dynamic excitation signals can be found in [239]. The following results are taken from [242].

26.4.1 Example: High-Pressure Fuel Supply System

The main components of an HPFS system are the high-pressure rail, the high-pressure fuel pump, and the ECU (engine control unit; compare Fig. 26.34). The pump is actuated by the crankshaft of the engine. A demand control valve in the pump allows controlling the delivered volume per stroke. A pressure-relief valve is also included in the pump, but should never open, if possible. Hence, we want to limit the maximum pressure during the whole measurement process. The pump

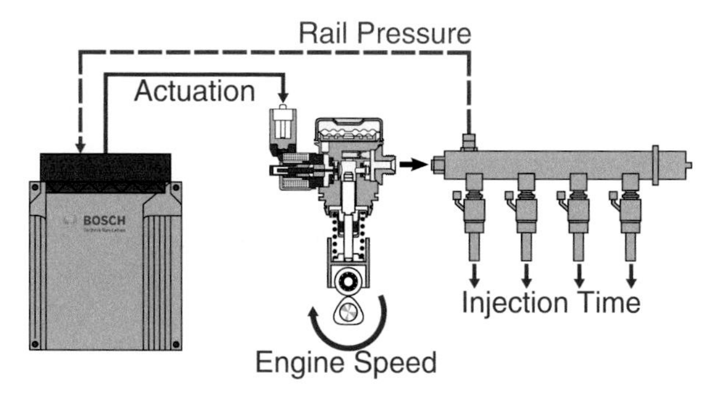

Fig. 26.34 Main components of the HPFS system, inputs (continuous lines), and output (dashed line). (The figure is taken from [563])

[4]Many thanks to my research assistant Tim O. Heinz and my external Ph.D. student Mark Schillinger, Bosch Engineering, for providing me with the above application example.

transports the fuel to the rail which contains the pressure sensor. From there, it is injected into the combustion chambers. See [489] for more details.

The system has three inputs and one output. The engine speed (*nmot*) affects the number of strokes per minute of the pump and the engine's fuel consumption. The fuel pump actuation (Mengensteuerventil, *MSV*) gives the fuel volume which is transported with every stroke of the pump. It is applied by opening and closing the demand control valve accordingly during one stroke of the pump. The injection time t_{inj} is a variable calculated by the ECU, which sums up the opening times of the single injectors and is, thus, related to the discharge of fuel from the rail.

During the measurement procedure, the injection time is not varied manually but set by the ECU. The permissible times depend on many factors, and a wrong choice could extinguish the combustion or even damage components. The engine load would have a major influence on the HPFS system via the injection time but is omitted to prevent the necessity of a vehicle test bench.

In summary, three physical inputs are relevant for modeling the rail pressure p_{r}, namely, engine speed $nmot$, pump actuation MSV, and injection timing t_{inj}, but the first two are actively controlled, while the third is determined by the ECU. Thus, an excitation signal for the first two inputs needs to be designed.

26.4.2 *Identifying the Rail Pressure System*

When comparing excitation signals for nonlinear system identification, the question of quantifying the signal quality arises. In general, good excitation signals lead to informative data, which lead to good models. Thus, in this section, the model quality is used to assess the quality of the excitation signal. The process is modeled with local model networks (LMN) and Gaussian process models (GPM) in order to also demonstrate the dependency on model architecture. Both models utilize the external dynamics approach in NARX structure to represent the dynamic behavior.

26.4.2.1 Local Model Networks

The first model architecture applied in this section is a local linear model network built with the HILOMOT algorithm; refer to Sect. 14.8 for details. The $\underline{x}$-input space consists of all three process inputs and the output delayed by 1, 2, and 3 time steps (third-order). The $\underline{z}$-input space is reduced to one time-delayed process inputs and output:

$$\underline{z}(k) = [u_1(k-1)\ u_2(k-1)\ u_3(k-1)\ y(k-1)]\ . \tag{26.10}$$

This choice takes into account that the actual operating point is defined by the level of the actual process inputs and output. The first (and higher)-order derivatives of the model inputs and output are assumed to be insignificant to describe the operating point.

26.4.2.2 Gaussian Process Models (GPMs)

The second type of model employed for this application is based on Gaussian processes; refer to Chap. 16. In order to reduce the computational complexity, it utilizes a sparse Gaussian process implementation instead of the classic algorithm presented in Chap. 16. It is quite typical for many *dynamic* systems that huge amounts of data are collected within a relatively short period of time. And this dynamic data usually is not as informative due to its worse distribution in the model input space. Therefore kernel methods often require some data or kernel reduction technique to be feasible. The software ASCMO by ETAS was used. See [198] for more information regarding dynamic modeling in ASCMO.

26.4.3 Results

After preprocessing the data, the results of the models are compared. Besides the quantitative analysis based on the error values on a test data set, the simulated data are visualized and compared to the measured data.

26.4.3.1 Operating Point Depending Constraints

For the design of an excitation signal for an unknown system, the operating range has to be explored. In the first instance, the amplitude constraints were determined based on (previous) stationary measurements. The feasible amplitude levels are given in (26.11) and (26.13). The maximum rate of change were chosen by expert knowledge as given in (26.12) and (26.14) based on the given actuator limits.

$$nmot \in [1200, 4000]\ \text{min}^{-1} \tag{26.11}$$

$$\frac{\text{d}}{\text{d}t} nmot \in [-1000, 4000]\ \text{min}^{-1}/\text{s} \tag{26.12}$$

$$MSV \in [0, 50]\ \text{mm}^3 \tag{26.13}$$

$$\frac{\text{d}}{\text{d}t} MSV \in [-500, 500]\ \text{mm}^3/\text{s} \tag{26.14}$$

To validate the constraints of the process, a combined signal consisting of ramp and chirp sequences (ramp-chirp) was generated according to (26.11)–(26.14). All the operating points defined by this input signal were feasible for the process.

The coverage of operating points depends strongly on the chosen excitation signal type. By changing the signal type for the inputs of the process, the coverage of operating points changes as well. Consequently, more extreme rail pressure values may occur compared to the ramp-chirp excitation. These operating point-dependent effects to the rail pressure call for a further tightening of the constraints.

A first test with an OMNIPUS optimized according to the constraints in (26.11)–(26.14) unveils a much higher dynamic characteristics than the ramp-chirp signal. Therefore the constraints given in (26.11)–(26.14) are not sufficient for all operating conditions excited by OMNIPUS. In consequence this forced the engine into infeasible operating conditions. By decreasing the maximum rate of the engine speed from $\frac{d}{dt}nmot_{max} = 4000\,\text{min}^{-1}/\text{s}$ to $\frac{d}{dt}nmot_{max} = 1000\,\text{min}^{-1}/\text{s}$, the rail pressure stayed inside an uncritical pressure range during operation. The more conservative constraints for

$$\frac{d}{dt}nmot \in [-1000, 1000]\ \text{min}^{-1}/\text{s} \tag{26.15}$$

are used in all further measurements.

26.4.3.2 Data Acquisition

The OMNIPUS is compared to a signal combination consisting of ramp and chirp sequences proposed in [562]. The measurement time for each signal is limited to $t_{max} = 10\,\text{min}$. The sampling frequency is $f_0 = 100\,\text{Hz}$; thus, a signal length of $N = 60{,}000$ results.

The OMNIPUS is optimized over the whole signal length. For the generation of the proxy output, two first-order systems are used. The poles in the z-plane are chosen equally to $p_1 = p_2 = 0.99$ which correspond to a time constant of $T \approx 1$ s. Within the constraints, all generated sequences are permissible; thus, no classification function is used. The transition between piecewise constant phases of the excitation signals is adjusted to the rate constraints using sine transitions.

This optimized signal is compared to a ramp-chirp signal. By generating the ramp-chirp combination, the first half of the signal consists of a ramp sequence; the second half consists of a chirp sequence. The high frequent chirp together with a low frequent ramp sequence seems to be a reasonable excitation of the process. All signals are generated according to the abovementioned constraints.

To test the achievable model quality for both excitation signals, an independent test signal is generated. For a fair comparison, the test signal consists of a chirp-ramp signal and an OMNIPUS one half each.

26.4.3.3 Accuracy of the Simulation Results

The dynamic order of the models is chosen to be three; thus, the model inputs are three time-delayed process inputs and outputs:

$$\hat{y}(k) = f\,(u_1(k-1), \ldots, u_1(k-3), u_2(k-1), \ldots, u_2(k-3),$$
$$u_3(k-1), \ldots, u_3(k-3), y(k-1), \ldots, y(k-3)) \qquad (26.16)$$

$$u_1 = nmot, \quad u_2 = MSV, \quad u_3 = t_{\text{inj}}, \quad y = p_{\text{r}} \qquad (26.17)$$

The function $f(\cdot)$ will be approximated using LMNs and GPMs.

26.4.3.4 Qualitative Analysis

For qualitative analysis, the model outputs of both models are compared to the measured data. Figures 26.35 and 26.36 show the simulated output on the test data set, based on the models which have been trained with either the ramp-chirp or the OMNIPUS data.

The GPMs do not show dramatical mismatches between the process and the identified models (Fig. 26.35). In the last half of the signal, the ramp-chirp-trained model seems to be slightly worse because of a static model error (too low gain). But overall the GPMs behave well in the whole range of operation.

For the LMN, the mismatch between model and process is small in most cases, e.g., Fig. 26.36a, d. But there are also some significant mismatches, for example, in Fig. 26.36c. The nonlinear behavior of the process is not well identified by the ramp-chirp-trained LMN. This major mismatch between process and model is most likely due to the lack of informative data in this area of operation. Figure 26.36b shows only a minor mismatch between the process and the OMNIPUS-trained LMN.

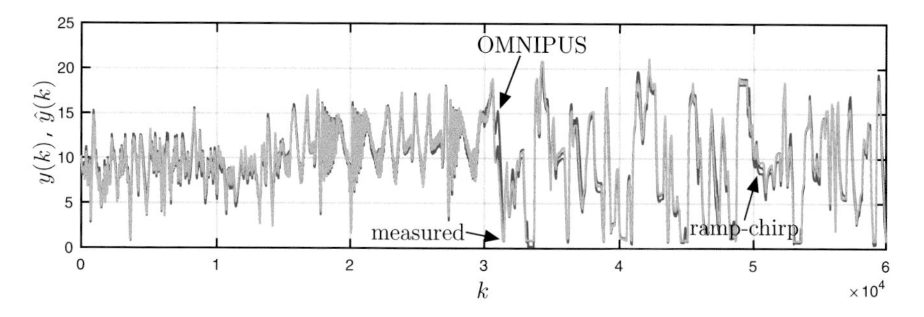

Fig. 26.35 Output signal of the ramp-chirp-trained GPM and the OMNIPUS-trained GPM on test data. The measured output is given as reference

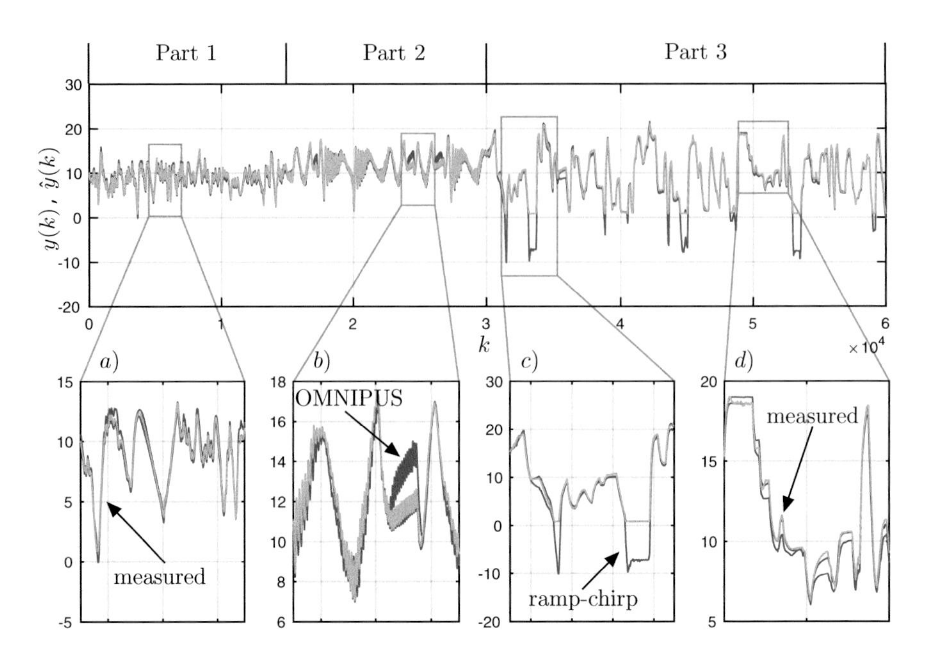

Fig. 26.36 Output signal of the ramp-chirp-trained and the OMNIPUS-trained HILOMOT model. The measured output is given as reference. (**a**) good model fit for both models on the ramp-chirp sequence (**b**) plant model mismatch of the OMNIPUS model on the ramp-chirp sequence (**c**) plant model mismatch of the ramp-chirp model on the OMNIPUS sequence (**d**) good model fit for both models on the OMNIPUS sequence

26.4.3.5 Quantitative Analysis

For comparison, the normalized root-mean-squared errors (NRMSE) based on the simulated data are used. For a thorough analysis, the errors were calculated for the following six different signals:

- u_{t}: Whole test signal—displayed in Figs. 26.35 and 26.36.
- $u_{\text{t,ramp}}$: Ramp sequence of the test data set—Fig. 26.36 part 1.
- $u_{\text{t,chirp}}$: Chirp sequence of the test data set—Fig. 26.36 part 2.
- $u_{\text{t,OMNIPUS}}$: OMNIPUS sequence of the test data set—Fig. 26.36 part 3.
- u_{OMNIPUS}: Training data of the OMNIPUS-trained model.
- $u_{\text{ramp-chirp}}$: Training data of the ramp-chirp-trained model.

In Table 26.4, the results of the LMN models are highlighted. Since the partitioning of a HILOMOT model depends on the data point distribution in the $\underline{z}$-input space, the excitation plays an important role in the achievable model quality. By comparing the simulation error on the training data sets, the model trained with the OMNIPUS unveils significant benefits compared to the ramp-chirp model. The simulation error of the ramp-chirp-trained model on the OMNIPUS data (u_{OMNIPUS}) is 15 times higher compared to the OMNIPUS-trained model. In contrast, the error

Table 26.4 Normalized root-mean-squared error on simulation data of the local model network constructed with HILOMOT

Model type	Test data				Training data	
HILOMOT	u_t	$u_{t,ramp}$	$u_{t,chirp}$	$u_{t,OMNIP.}$	$u_{OMNIP.}$	$u_{ramp\text{-}chip}$
OMNIPUS	0.0244	0.0282	0.0428	0.0198	0.0061	0.0396
Ramp-chirp	0.0781	0.0294	0.0153	0.1091	0.0931	0.0130

Table 26.5 Normalized root-mean-squared error on simulation data of the sparse GP model

Model type	Test data				Training data	
GPM	u_t	$u_{t,ramp}$	$u_{t,chirp}$	$u_{t,OMNIP.}$	$u_{OMNIP.}$	$u_{ramp\text{-}chirp}$
OMNIPUS	0.0343	0.0344	0.0268	0.0427	0.0349	0.0376
Ramp-chirp	0.0302	0.0258	0.0208	0.0389	0.0383	0.0197

of the OMNIPUS-trained model evaluated on the ramp-chirp data set ($u_{ramp\text{-}chirp}$) is only three times higher compared to the ramp-chirp-trained model.

On the test data set, the OMNIPUS-trained model is in most cases superior to the ramp-chirp model. Since the ramp-chirp-trained model has a strong mismatch for the OMNIPUS test sequence (see Fig. 26.36), the difference between the two models is maximal on the sequence $u_{t,OMNIPUS}$. The chirp sequence $u_{t,chirp}$, in contrast, is better described by the ramp-chirp-trained model, and the OMNIPUS-trained model performs worse.

The simulation errors of the GP models are shown in Table 26.5. The difference of the two GP models evaluated on the training data sets is much smaller compared to the HILOMOT models. Even the simulation errors on the whole test data set are comparable. Actually, the ramp-chirp-trained model performs slightly better on the test data set.

Surprisingly, the ramp-chirp-trained model shows better generalization abilities on the OMNIPUS sequence $u_{t,OMNIPUS}$ of the test data compared with the OMNIPUS-trained model. In general, this is an abnormal behavior that needs further investigation.

By comparing the test data results of Tables 26.4 and 26.5, it is noticeable that the HILOMOT model has the best quality with OMNIPUS excitation but the worst quality with the ramp-chirp excitation. The qualities of the two GP models are in between the HILOMOT models.

Obviously, the generalization ability depends (i) on the data-point distribution of the training data in the input space and (ii) on the chosen model structure. It seems that the GP model is less sensitive with respect to the data-point distribution. The utilized sparse GP approach is based on projections and works excellently for the discussed application. The exact root cause for this desirable behavior is an issue for future research. Thus it is unclear at the moment whether this very robust behavior generalized to other processes, dimensions, and data distributions. In contrast, the HILOMOT model is very sensitive to the data-point distribution. The optimization of the excitation signal yields the most benefit for the HILOMOT model.

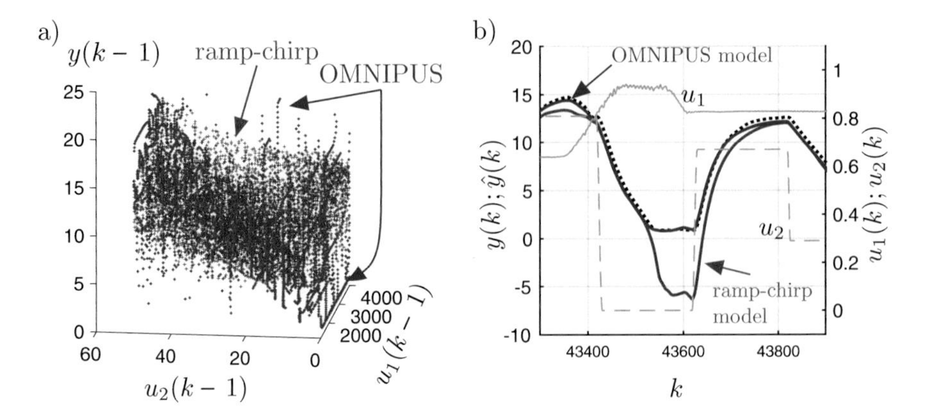

Fig. 26.37 (**a**) Subspace of the original input space consisting only the model inputs $u_1(k-1)$, $u_2(k-1)$, and $y(k-1)$ and (**b**) the excitation of the system (gray) scaled to $[0, 1]$. The corresponding outputs of the process (black dotted), the LMN based on the ramp-chirp data, and the LMN based on the OMNIPUS data

26.4.3.6 Data Coverage of the Input Space

The objective of the OMNIPUS is generating uniform distributed data in the input space. Figure 26.37 shows the data distribution with a reduced input space of the identification problem assuming a first-order MISO system. The input u_3 is fixed through the engine control and therefore irrelevant for the dynamic DOE. Thus, only projections to the remaining model inputs are considered. This figure reveals the data coverage of the different excitation signals in this subspace. The ramp-chirp signal has a uniform data distribution in the u_1-u_2-projection. The OMNIPUS operates on discrete amplitude values; thus, the data distribution concentrates on a few input levels. The benefit of the OMNIPUS becomes visible by taking the output y into account. The OMNIPUS generates more points at the extreme values of the output (outer regions in Fig. 26.37a). In contrast, the ramp-chirp signal generates no points in these areas. Thus, the ramp-chirp data set contains no information close to and at the boundary of the input space. This results in a significantly better performance of the OMNIPUS-trained model compared to the ramp-chirp-trained one; see Fig. 26.37b for an excerpt of the time signals for the local model network outputs.

A model based on such a data set has an increased risk of wrong extrapolation behavior. Why this issue has no observable negative consequences for the sparse GP model investigated here cannot be answered at this point in time. It is most likely, however, that the bad data coverage of the ramp-chirp signal in the input space is responsible for the strong model-plant mismatch for the ramp-chirp-trained HILOMOT model in Fig. 26.36c.

Chapter 27
Input Selection Applications

Abstract This chapter deals with applications of the input selection strategies introduced in Chapter 15. First, a static process whose characteristics are well-known is studied. Some variables influence the output in a linear way, others in nonlinear ways. The sophisticated local model network-based input selection strategy successfully assigned the linear variable to the consequent space only, while the nonlinear variables are assigned to the premise space as well. This is in perfect alignment with the best-case scenario. This theoretically optimal result was achieved based on the information in the data only. Second, the efficiencies of two fans were modeled by computational fluid dynamics (CFD) – an axial and a radial fan. With local model networks metamodels were built approximating the CFD simulation results. Input selection could automatically discover the inputs considered most relevant by experts. They were selected as the most important ones by the subset selection procedure. Finally, a dynamic process HVAC process has been modeled as a basis for predictive control design. Here the input selection automatically discovered not only the relevant physical inputs but also the orders and dead times. This corresponds to an extremely high-dimensional input selection problem since 6 physical inputs/outputs with a time delay between 1 and 10 acted as potential input variables for the premises and consequents of the local model network.

Keywords Application · Input selection · Partition · Wrapper · Embedded · Goal-oriented · HVAC

The potential of input selection methods introduced in Chap. 15 shall be demonstrated for three applications. First, a model for substitution of a number of nested look-up tables in an engine control unit of a combustion engine is investigated. This application nicely illustrates the capability of the proposed methods to discover linear relationships automatically from data without utilizing any prior knowledge.

The substantial contribution by Dr. Julian Belz for writing this chapter is gratefully acknowledged and appreciated.

O. Nelles, *Nonlinear System Identification*,
https://doi.org/10.1007/978-3-030-47439-3_27

The second application deals with fan metamodels depending on a large number of geometric parameters. It demonstrates that on the one hand, the input selection strategy can hardly be beneficial if only highly relevant (potential) inputs are provided to the algorithm and, on the other hand, when provided with a wealth of potential inputs, the selection is capable of automatically selecting the most relevant ones (geometric parameters) in agreement with the assessment of human experts before modeling. Additionally, some design of experiment aspects already discussed in previous chapters in an application-independent manner are applied to the efficiency optimization of centrifugal fans. Particularly, these are active learning approaches as introduced in Sect. 14.10.4 and issues concerning the order of experimentation as discussed in Sect. 26.1.1. An extension to the active learning procedure HILOMOT for design of experiments (HilomotDoE) is presented in Sect. 27.2.7. This extension incorporates a new strategy to pick more than one query point which can easily be expanded for multiple-input multiple-output (MIMO) systems. Additionally, an alternative for the generation of so-called candidate points is presented.

Thirdly, a *dynamic* process is analyzed. Since the dynamic orders are unknown, plenty of delayed regressors are provided to the model, and the input selection is capable of automatically selecting the most relevant ones. Particularly impressive is the fact that the separated $\underline{x}$-$\underline{z}$-strategy can reduce the premise input space $\underline{z}$ to only two dimensions (from originally 60). The two selected inputs for $\underline{z}$ correspond to both actuating variables which makes perfectly sense because they are most highly excited and most important for the model quality.

27.1 Air Mass Flow Prediction

In modern combustion engines, the engine control unit (ECU) is responsible for managing the engine performance. It is used to control, e.g., the air-fuel ratio, the idle speed, and the variable valve timing [247]. Therefore, the ECU has to model the air-mass flow (AMF) depending on the engine's operating point. The parameters needed to calculate the AMF are traditionally stored in look-up tables with supporting points upon a grid. This runs into difficulties due to the required storage, which grows exponentially with higher input dimensions. In order to overcome the exponential increase of required storage, Bänfer et. al. proposed, investigated, and realized the substitution of the look-up tables with local model networks (LMNs); see [30, 31] for details. The data set used in this section originates from the cooperation of the University of Siegen with Continental Automotive GmbH, in which the substitution of look-up tables by LMNs has been developed.

The AMF into the combustion chamber of a gasoline engine should be predicted in order to determine the optimal amount of fuel that has to be injected for a complete combustion. The AMF y depends on the following six inputs:

- Engine speed u_1
- Valve timing for the intake camshaft u_2
- Valve timing for the exhaust camshaft u_3
- Position of the swirl flap u_4
- Position of the variable intake manifold (VIM) flap u_5
- intake manifold air pressure (MAP) u_6

A schematic sketch of the technical system is shown in Fig. 27.1.

It is known that the system can be adequately described with scheduled affine models [247]. These affine models depend explicitly on the MAP u_6, while the dependency on the remaining inputs is implicitly contained in operating-point-dependent slopes m_a and offsets b. In particular, the slope m_a and offset b depend only on the inputs $u_1, u_2, \ldots, u_5$:

$$\hat{y}(u_1, u_2, u_3, u_4, u_5, u_6) = m_a(u_1, u_2, u_3, u_4, u_5) \cdot u_6 + b(u_1, u_2, u_3, u_4, u_5)\,. \tag{27.1}$$

The AMF prediction has traditionally been modeled by grid-based look-up tables. Recently it has been described by LMNs which is shown in Fig. 27.2. As can be

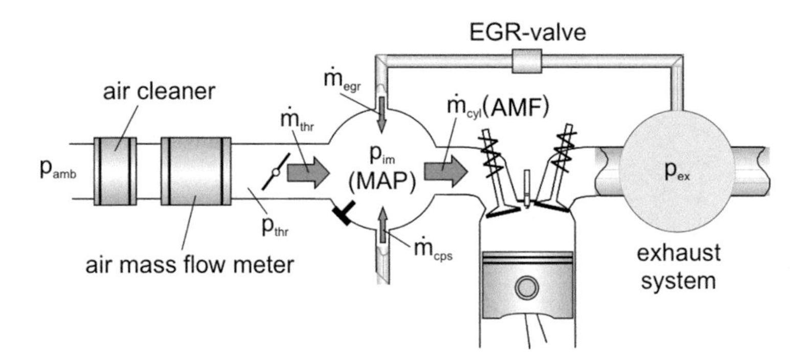

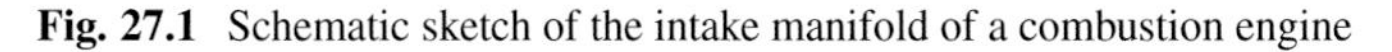

Fig. 27.1 Schematic sketch of the intake manifold of a combustion engine

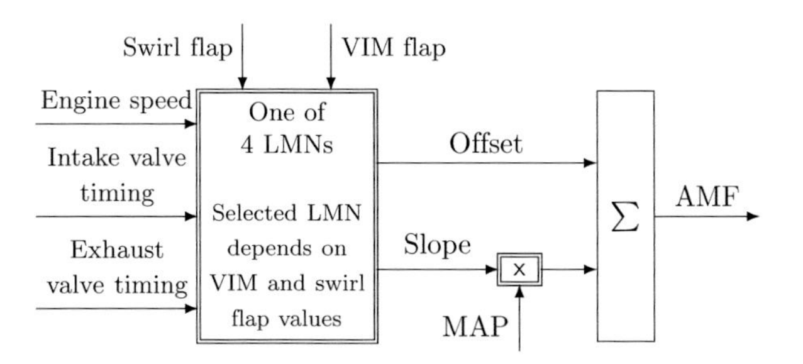

Fig. 27.2 Schematic sketch of the "traditional" model structure for the AMF prediction. Depending on the positions of the VIM and swirl flap, one of four LMNs is used to determine the offset and the slope for an affine model depending on the intake MAP

seen, the position of the swirl flap u_4 and the position of the VIM flap u_5 are treated in a special way. Both of these inputs are binary ones. As a result, there are four possible combinations of u_4 and u_5 values. For each combination an individual LMN is trained with three inputs, namely, the engine speed u_1 as well as the valve timings for the intake and exhaust camshaft u_2 and u_3, respectively. The outputs of each LMN are the slope m_a and the offset b, which are used to finally calculate the AMF under consideration of the MAP.

All of the previously described prior knowledge about the modeling of the AMF is not exploited for the investigations in this section. Instead, it should be analyzed if a separated $\underline{x}$-$\underline{z}$-input selection is able to reveal the model structure known from expert knowledge solely based on data. For the separated $\underline{x}$-$\underline{z}$-input selection, all physical inputs u_1 to u_6 are treated equally and are used as inputs for one LMN. Additionally, the partitioning of the resulting LMN is analyzed following the embedded input selection approach described in Sect. 15.4.1.

For the investigations, one data set is available with a total number of $N = 3045$ samples. This data set is split into one training data set consisting of $N_{\text{train}} = 2589$ samples and one test data set consisting of $N_{\text{test}} = 456$ samples.

27.1.1 Mixed Wrapper-Embedded Input Selection

The separated $\underline{x}$-$\underline{z}$-input selection is used with backward elimination (BE) as search strategy and the Akaike's information criterion (AIC_C) as evaluation criterion. Note that due to the used input selection scheme, the number of LMN inputs to choose from is 12. Only local affine models are used for LMNs that are trained with

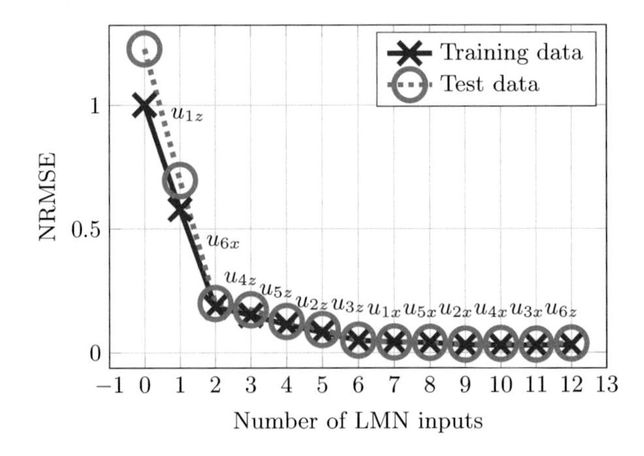

Fig. 27.3 NRMSE values on training and test data versus the number of LMN inputs for the AMF prediction

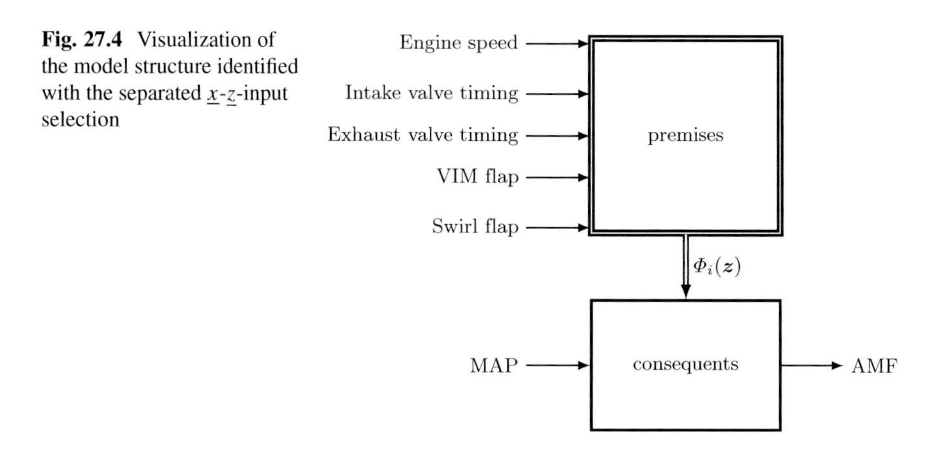

Fig. 27.4 Visualization of the model structure identified with the separated $\underline{x}$-$\underline{z}$-input selection

HIerarchical LOcal MOdel Tree (HILOMOT). The model complexity of the LMNs is also determined with the AIC_C.

Figure 27.3 shows errors on training and test data versus the number of LMN inputs. In between two subsequent error values, the input is denoted that is discarded in the corresponding BE step. Besides the information about the input number, the input space from which the input is removed is denoted, i.e., x or z. The minimum NRMSE value is obtained with 11 LMN inputs. If the number of LMN inputs is decreased to six, the NRMSE value does not change significantly. Therefore, having six LMN inputs can be seen as a good compromise between the model quality and the simplicity of the model.

The model structure identified solely based on the available data for six LMN inputs is shown in Fig. 27.4. It matches the "traditional" model structure which is based on expert knowledge; see Fig. 27.2. In other words, the six most important inputs found with the separated $\underline{x}$-$\underline{z}$-input selection are in accordance with expert knowledge and are assigned to the correct input space. Note that the likelihood of obtaining this result just by chance is 1/924, because the number of possible combinations of 6 inputs out of 12 is $\binom{12}{6} = 924$.

Finally, the model quality of the LMN with six inputs resulting from the separated $\underline{x}$-$\underline{z}$-input selection is compared to the existing model structure. The main difference between both models is the handling of the binary inputs u_4 (swirl flap) and u_5 (VIM flap). In the existing model structure, these inputs are used to decide which of four distinct LMNs is used. In contrast to that, the model structure identified with the separated $\underline{x}$-$\underline{z}$-input selection uses only one LMN with u_4 and u_5 only incorporated in the $\underline{z}$-input space. It turns out that the model quality of the existing model structure is slightly better in terms of the achieved NRMSE value on test data, namely, 0.03 opposed to 0.05.

In summary, the confidence in the separated $\underline{x}$-$\underline{z}$-input selection is increased through the AMF application example. Even though the model structure identified solely based on the available data is worse compared to the traditional structure in terms of the achieved prediction quality, it is very close to it. Additionally, the six most important inputs are revealed and are assigned correctly to the respective input space.

27.1.2 Partition Analysis

In order to rank all inputs according to their nonlinear influence, an LMN with local affine models is trained, and subsequently, the resulting input space partitioning is analyzed as described in Sect. 15.4.1. HILOMOT is used to train the LMN. The model complexity is determined with the $\mathrm{AIC_C}$. All physical inputs are used as inputs for the $\underline{x}$- and $\underline{z}$-input space.

The resulting relevance factors are shown in Fig. 27.5. The engine speed has the highest relevance factor while the swirl flap has the lowest relevance factor. Surprisingly, the relevance factor of the intake MAP is not the lowest even though it is the first $\underline{z}$-input that is removed in the separated $\underline{x}$-$\underline{z}$-input selection; see Fig. 27.3. Compared to the relevance factor of the engine speed, all other relevance factors are almost zero, and differences are only visible due to the logarithmic scaling of the y-axis. It seems that due to the small differences in the relevance factors, the ranking of inputs u_2 to u_6 is not trustworthy. Based on the obtained relevance factors, only the engine speed seems to be relevant for the $\underline{z}$-input space. From prior knowledge and from the separated $\underline{x}$-$\underline{z}$-input selection, this cannot be confirmed. However, the results of the separated $\underline{x}$-$\underline{z}$-input selection indicate that having the engine speed (u_1) in the $\underline{z}$-input space and the intake MAP (u_6) in the $\underline{x}$-input space already leads to quite a good model; see Fig. 27.3.

In summary, the partition analysis fails to fully convince. Even though the partition analysis is able to reveal the most relevant $\underline{z}$-input, which is the engine speed, the ranking of the remaining inputs seems to be completely random. In defense of the partition analysis, it can be said that the differences between the relevance factors ρ_2, ρ_3, ρ_4, ρ_5, and ρ_6 are too small to draw conclusions about

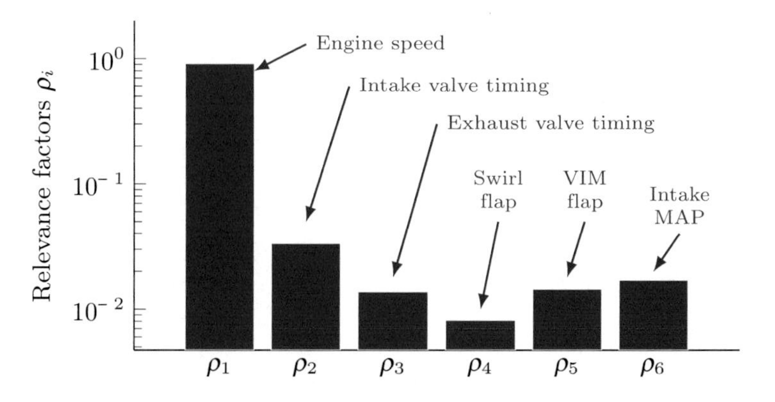

Fig. 27.5 Relevance factors ρ_i for the LMN that predicts the AMF

their relative ranking. However, it cannot be recommended to rely on the partition analysis results in general. Performing an additional mixed wrapper-embedded input selection seems to be mandatory in order to find good input subsets for the $\underline{z}$-input space.

27.2 Fan Metamodeling: Authored by Julian Belz

In this section the generation of computational fluid dynamics (CFD) metamodels serves as an example to prove the usefulness and applicability of some approaches discussed in Chap. 15 and Sect. 26.1. In particular, advantages obtained through the chosen order of experimentation, the goal-oriented active learning strategy with LMNs, and the mixed wrapper-embedded input selection are pointed out. Two different types of fans are considered, which are centrifugal and axial fans. Most of the information contained in this section about the axial fans is obtained from [28]. Information regarding the centrifugal fans originates from [29, 39], and [40].

The metamodels should capture the behavior of either centrifugal or axial fans. The general purpose of fans is to generate a gaseous fluid flow under the buildup of pressure. Typically, the design of a new fan comprises two main targets. Firstly, the design point (i.e., the desired flow rate V and pressure rise Δp) must be fulfilled. Secondly, the shaft power P_{shaft} shall be as low as possible. The achievability of the first design target mainly depends on the choice of the outer fan diameter D and the rotational speed n_R. Cordier [114] found that the specific fan diameter

$$\delta = \frac{D}{\left(\frac{8}{\pi^2}\right)^{1/4}\left(\frac{\Delta p}{\rho_f}\right)^{-1/4} V^{1/2}} \tag{27.2}$$

and the specific fan speed

$$\sigma = \frac{n_R}{(2\pi^2)^{-1/4}\left(\frac{\Delta p}{\rho_f}\right)^{3/4} V^{-1/2}} \tag{27.3}$$

of all built fans and pumps lie in a narrow band around the curve depicted in Fig. 27.6 which were later known as the Cordier curve and the Cordier diagram, respectively.

The original Cordier diagram is based on fan performance data stemming from the 1950s, but its validity was confirmed in numerous more recent studies; see, e.g., the work by Willinger et al. [602–604]. Figure 27.6 furthermore indicates the typical realm of centrifugal and axial fans. The rest of the Cordier band is associated with other fan types such as propellers or mixed flow. The second design target (the minimization of P_{shaft}) is equivalent to the maximization of the aerodynamic efficiency defined as

$$\eta = \frac{V \cdot \Delta p}{P_{shaft}}. \tag{27.4}$$

Independent of the fan type, i.e., centrifugal or axial, only the impeller as the key component with respect to aerodynamic efficiency is investigated.

27.2.1 Centrifugal Impeller Geometry

The centrifugal impeller geometry is described by nine geometrical parameters including, among others, the number of blades, the inner diameter, the inlet width, the outlet width, the inlet blade angle, and the outlet blade angle. Those parameters are supposed to be most relevant in order to adapt the impeller geometry to a large variety of potential design points [60]. Figure 27.7 shows a technical drawing in which almost all geometrical design parameters are explained. Note that the centrifugal fan metamodels have ten inputs, including the geometrical parameters as well as the flow rate.

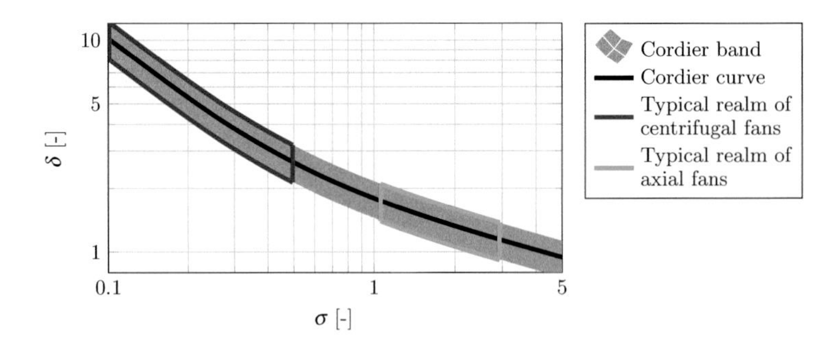

Fig. 27.6 Cordier diagram with indication of the typical realm of centrifugal and axial fans

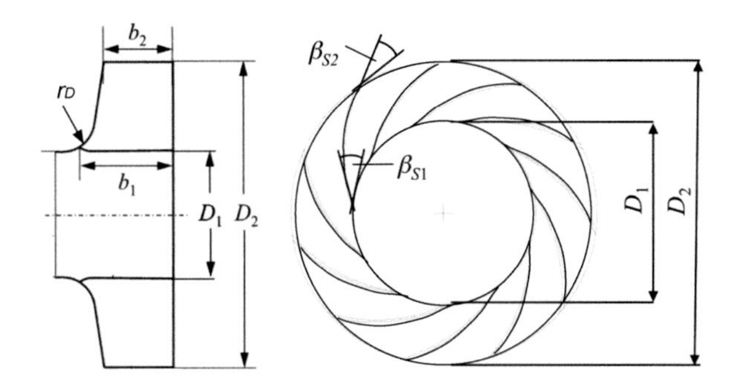

Fig. 27.7 Impeller design parameters of the centrifugal impeller

27.2.2 Axial Impeller Geometry

The axial impeller geometry is described by 26 geometrical parameters including, among others, the number of blades, the hub-to-tip ratio, the chord length, the angle of incidence, and the sweep angle. Figure 27.8 shows two technical drawings in which some of the geometrical parameters are depicted. Note that the axial fan metamodels have 28 inputs, including the geometrical parameters as well as the flow rate and the target flow rate.

27.2.3 Why Metamodels?

The generated CFD metamodels approximate, among other flow field quantities, the efficiency and the pressure rise of fans depending on the geometric parameters of the impeller. The CFD metamodels are used to aerodynamically optimize the design parameters of the impeller with the evolutionary algorithm described in [28]. In principle, CFD simulations could directly be coupled with optimization algorithms to optimize impeller designs. However, this would lead to a very high computational expense required for the optimization of each impeller design. Both ways to optimize the impeller geometry are depicted in Fig. 27.9. By using a metamodel, the computational demand that is necessary for the numerical CFD simulations

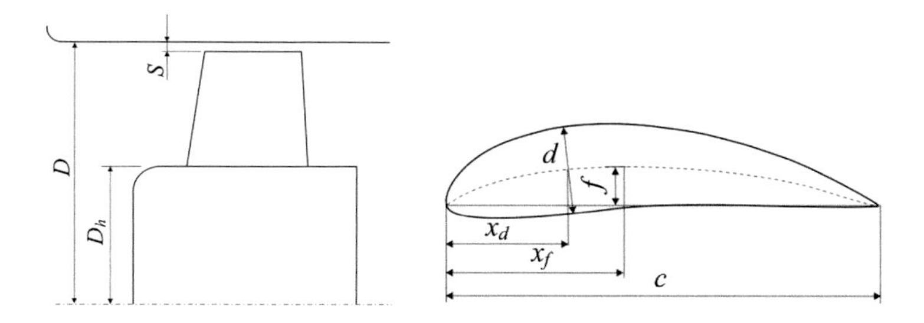

Fig. 27.8 Some impeller design parameters of the axial impeller

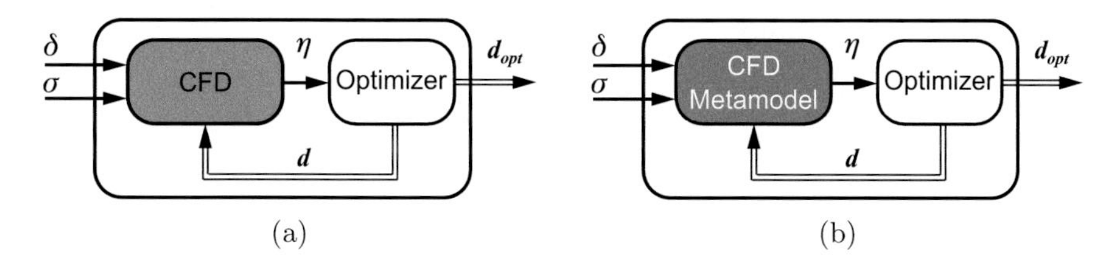

Fig. 27.9 Impeller geometry optimization with the help of CFD simulations (**a**) and a CFD metamodel (**b**)

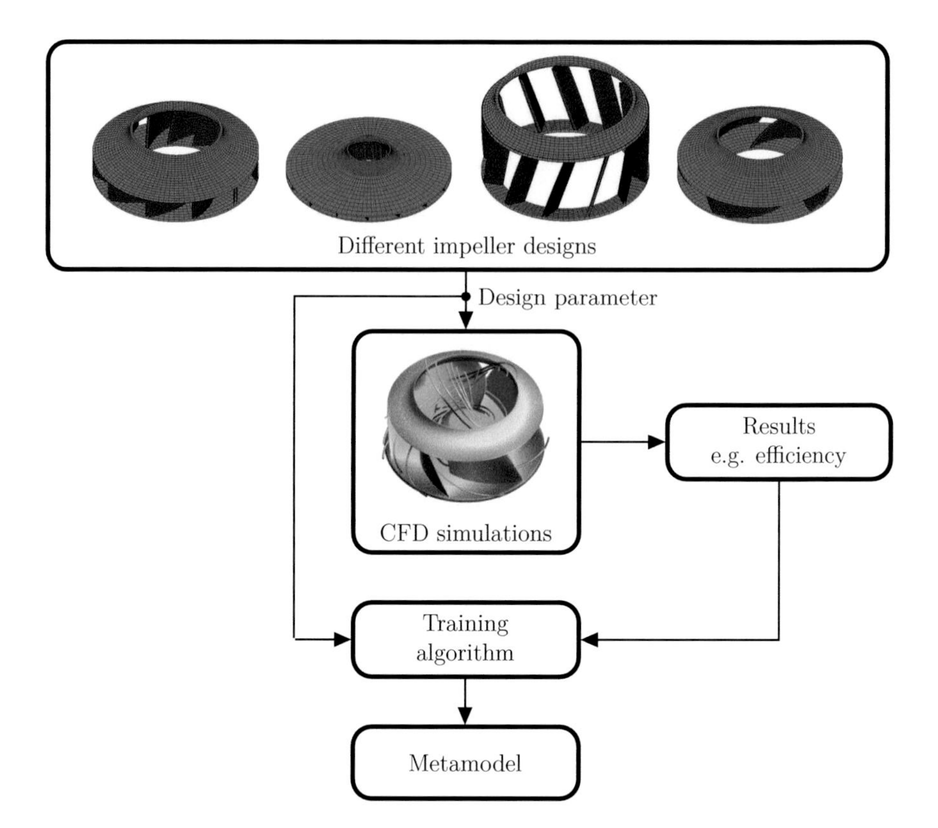

Fig. 27.10 Metamodeling procedure for the substitution of computationally expensive CFD simulations

reduces to the generation of data, which is needed for the metamodel training as shown in Fig. 27.10. Once the metamodel is trained, no further CFD simulations are necessary. The time needed to compute the efficiency and the pressure rise for one impeller geometry at one specific volume flow could be reduced from approximately 30 minutes to 0.02 seconds by substituting the CFD simulation with the metamodel. More information about the centrifugal impeller parameterization, the CFD model, and the optimization of the CFD grid resolution can be found in [29]. For the same information regarding the axial impeller, the reader is referred to [28]. A more detailed discussion about fans, in general, can be found in [59]. The fundamentals of fluid dynamics are covered in [398].

27.2.4 Design of Experiments: Centrifugal Fan Metamodel

The training data acquisition for the metamodel generation can be divided into two phases. The first one is a passive learning phase in which CFD simulations are carried out according to a predefined maximin optimized Latin Hypercube (LH) design. In this phase, the intelligent k-means sequence (IKMS) method is used to determine in which order the CFD simulations should be carried out that are contained in the design of experiments. Section 27.2.6 describes the effect of all order determination methods on the metamodel quality. The second phase follows an active learning scheme in which information about already simulated impeller designs is exploited to determine further CFD simulation queries. The goal-oriented active learning strategy with LMNs is used for the second phase. Section 27.2.7 reviews the influence on the model quality and additional improvements of the resulting metamodels. Roughly, 900 and 3000 different impeller geometries are CFD simulated in phases one and two, respectively.

By using this two-phase data acquisition approach, it is assured that the whole design space is explored in phase one such that metamodel predictions are reliable for a variety of different impeller geometries. Once the metamodel quality is saturated or considered to be sufficiently reliable, information from the metamodel is used to plan additional measurements. By exploiting information of the metamodel, these adaptively planned measurements aim to improve the metamodel in areas of the design space where the metamodel's performance is considered to be worst while simultaneously concentrating on near-optimum regions and geometrical designs that differ as much as possible from already measured geometrical designs. Since the quality of the metamodel was saturated after the CFD simulation of approximately 900 different impeller geometries in phase one, phase two was started and used for the rest of the project duration.

27.2.5 Design of Experiments: Axial Fan Metamodel

The training data acquisition for the metamodel generation consists only of a passive learning phase. It dates back to times where the design of experiments (DoE) know-how was less developed. At first, 2000 different geometrical designs are generated with the help of a Sobol sequence [539] and CFD simulated. Subsequently, 1000 random candidate points are generated from which the 10% are chosen that are farthest away from the already existing points. The second step is repeated several times until the DoE contains approximately 12,000 different geometrical designs.

27.2.6 Order of Experimentation

Although the CFD simulations have been carried out according to the IKMS method during the first data acquisition phase, this section tries to compare all order of experimentation methods except for the median distance sequence (MDS) method. Because the MDS method's results on the artificial test processes are the worst ones, it is completely neglected here. In addition, HilomotDoE and simple randomization are used to determine an ordering of all training data samples as references. For an explanation of how HilomotDoE is used to determine the order of experimentation, see Sect. 26.1.6. However, all calculations are done in retrospect, after all data for the CFD metamodel generation has been completely collected. Metamodels for both relevant outputs, i.e., the specific pressure rise ψ and the efficiency η, are considered.

In order to compare all order determination strategies on a fixed data set, all available data is randomly split into 85% for training and 15% for testing purposes. This is done 50 times, such that there are 50 different training and test data sets. For each of these data splittings, the order of experimentation for the training data is determined with each order determination strategy, i.e., biggest gap sequence (BGS), IKMS, HilomotDoE, and the simple randomization. Once each order determination strategy has determined the ordering for each of the 50 training data sets, LMNs are trained with fractions of these sorted training data sets. At first, an LMN with the first 10% of the training data is trained, then with the first 20% of the training data, and so on until all training data is used, as visualized in Fig. 27.11. HILOMOT is used as a training algorithm to generate the LMNs, and with the help of the test data, the model quality is assessed. As a result, one curve of the model quality versus the amount of used training data arises for the BGS, IKMS, and HilomotDoE strategy and each split of the whole data set. In the case of the randomized ordering, ten different random orderings are determined for each of the 50 training data sets, such that there are 500 of these curves. Figure 27.12a and b show the mean values of these curves for the prediction of ψ (specific pressure rise) and η (efficiency) in the case of the BGS, IKMS, and HilomotDoE strategies. In the

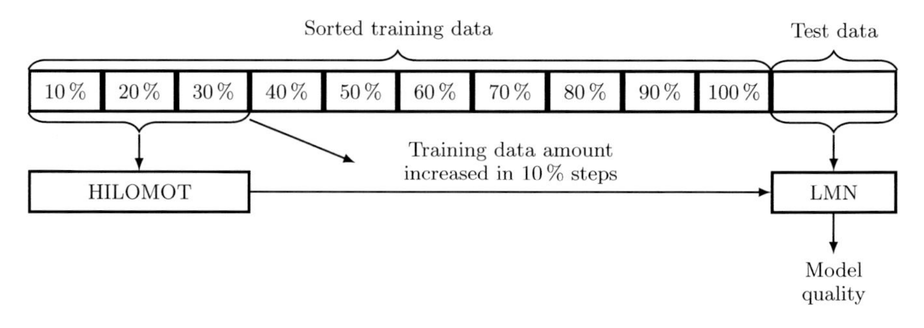

Fig. 27.11 Procedure to assess the model quality for 10% increments of the sorted training data. Here, only the model quality assessment for 30% of the training data is visualized

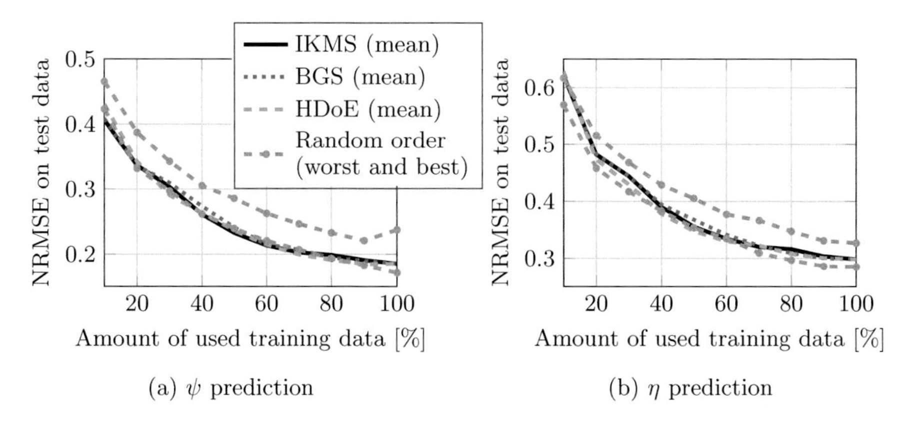

Fig. 27.12 Comparison of different order determination strategies for the prediction of ψ **(a)** and η **(b)**

case of the randomized orderings, only the best and worst curves (out of all 500) are shown. An order determination strategy is considered to be good if the resulting model quality quickly improves. This means a good model quality is reached with a rather low training data set size. Comparing the two model-free algorithms, IKMS and BGS perform equally well and on a similar level as the active learning strategy, which is abbreviated with HDoE. Showing the best and the worst case of the random orderings should provide a feeling of how good or bad the strategies perform on an absolute scale. Note that the error of the worst and best random order does not coincide with the error of the other determination strategies at 100% of the used training data. This is due to the fact that each of the two shown random sequence curves is generated with only one of the 50 resulting training data sets while the other curves represent mean values. The IKMS, BGS, and HDoE curves are very close to the best case of all random sequences for all amounts of used training data.

The advantage of using a specific order of experimentation is pointed out for the real-world example of generating a data base used to build a CFD metamodel. The achieved model qualities with the two model-free approaches, namely, the BGS and the IKMS, are even comparable to an active learning strategy (HilomotDoE) as can be seen in Fig. 27.12. The reason for the comparable model qualities might be the rather high-dimensional input space ($p = 9$), while the overall amount of training data even at the end of the first data acquisition phase is rather low ($N \approx 900$). In other words, HilomotDoE might still be exploring the whole input space instead of concentrating on highly nonlinear regions in it. Another disadvantage for HilomotDoE here is that the candidate points are restricted to the data points contained in the existing data set and cannot be chosen arbitrarily in the input space.

27.2.7 Goal-Oriented Active Learning

In the case of the active learning for the CFD metamodel, the goal-oriented procedure illustrated in Fig. 27.13 is applied with minor adjustments to meet problem-specific needs. A maximin LH design for an 11-dimensional input space is generated with $N_{LH} = 586$ samples. Two of the eleven inputs specify a design point defined by the desired flow rate and pressure rise. These two inputs always belong to the set of constrained optimization variables, since the efficiency is maximized for these design points. The remaining $p = 9$ inputs correspond to the geometric parameters that should be optimized. It follows that there are $n_c = 2^p = 512$ possible optimization constraints for each point contained in the LH design. As explained in detail in [40], all possible combinations of constrained optimization variables are applied to each point contained in the LH design, resulting in $N_c = 512 \cdot 586 = 300{,}032$ candidate points after all optimization runs are finished.

Here, in each loop of the active learning strategy, a list of 500 query points is determined. Each query point is intended to be CFD simulated. As soon as new CFD simulations are finished and the data set is updated with new information, the process of updating the list of query points starts again. The number of demanded queries is chosen quite high to prevent the current query list from becoming empty, i.e., all queries have been CFD simulated, before the new query list is readily determined. Until the update of the query point list is finished, the "old" list is further used. Once the optimization of all candidate points is accomplished and all-new 500 queries are calculated, the new list is used.

The influence of the proposed goal-oriented active learning strategy is evaluated based on three aspects. First, the model quality is assessed for increasing portions of training data. Second, the extension of achievable design points in the Cordier diagram is shown. And third, the improvements of the achievable total-to-static efficiencies are visualized for all achievable points in the Cordier diagram.

For the model quality assessment, 545 data points are chosen from the CFD data basis according to a specific data splitting procedure. All 545 chosen data points cover the area of possible design points spanned by the desired flow rate

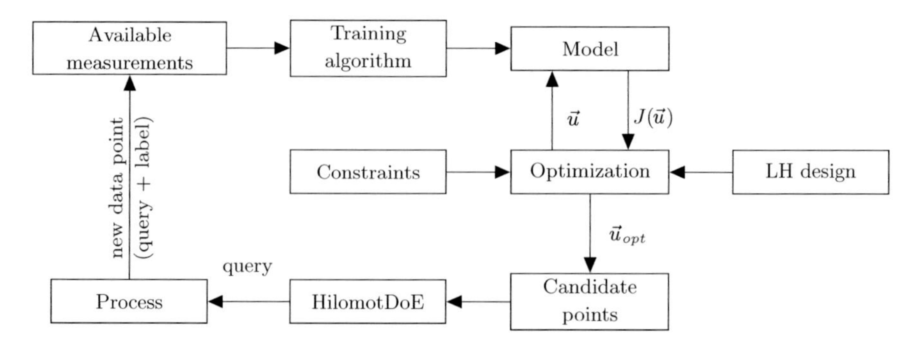

Fig. 27.13 Block diagram of the goal-oriented active learning procedure

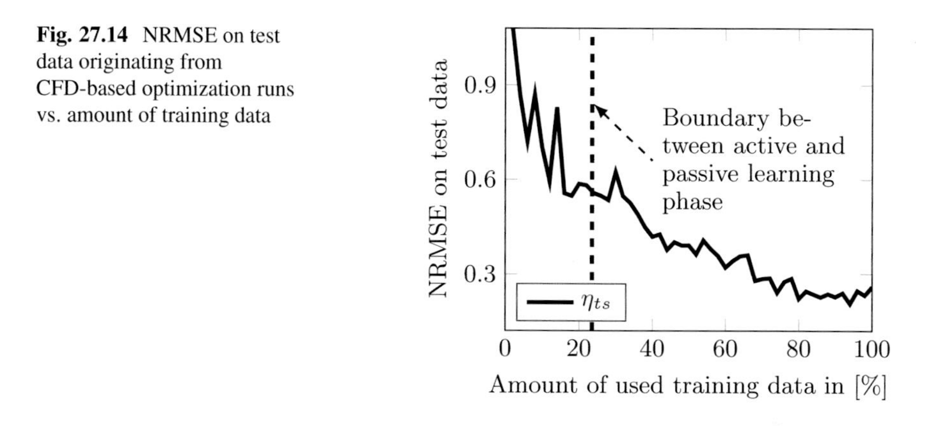

Fig. 27.14 NRMSE on test data originating from CFD-based optimization runs vs. amount of training data

and pressure rise uniformly. These designs together with the achieved total-to-static efficiencies serve as test data for the models generated with different amounts of training data. The target value of the model is the total-to-static efficiency η_{ts}. Figure 27.14 shows the curve for the model quality versus the training data amount.

The dashed line marks the point, where the passive learning phase has ended and the active learning strategy has started. It is observable that the model quality at the end of the passive learning phase started to saturate. A likely cause for the saturation is that the global characteristics of the centrifugal fans are already identified with the 900 different impeller designs. No new information seems to be added by further exploring the design space uniformly since the model quality does not improve. Hence, a more goal-orientated DoE strategy is needed that exploits already available information in the form of the metamodel to find new queries (impeller designs) that contain new information. Note that the test data used to create the shown curve here differs from the test data used in the previous section. Therefore the errors from Fig. 27.14 are not comparable to the ones shown in Fig. 27.12.

Figure 27.15 shows the extension of achievable regions in the Cordier diagram through the proposed goal-oriented active learning strategy. For almost each viable specific fan speed σ, higher specific fan diameters δ are possible. Additionally, lower and higher specific fan speeds could be achieved through the active learning phase compared to the passive one. The impact of the extension of achievable points cannot be evaluated without taking into account the corresponding efficiencies. Therefore, the discussion of the impact is postponed to the end of this section.

The absolute total-to-static efficiencies in the area of possible impeller designs after the active learning strategy and the corresponding improvements are shown in Fig. 27.16a and b, respectively. The efficiency could be improved at some points from $\eta_{ts} \approx 0.3$ (before the active learning took place) to $\eta_{ts} \approx 0.6$. The improvements are mostly achieved above the Cordier curve, where the specific fan diameter δ is relatively high. The efficiency improvements through active learning can only be shown in areas, where data from the passive learning phase is available because a reference value exists only there; compare Fig. 27.15. That is the reason

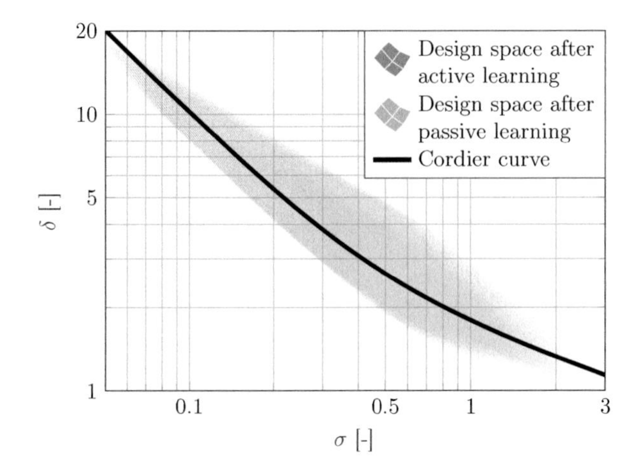

Fig. 27.15 Extension of achievable design point area in the Cordier diagram

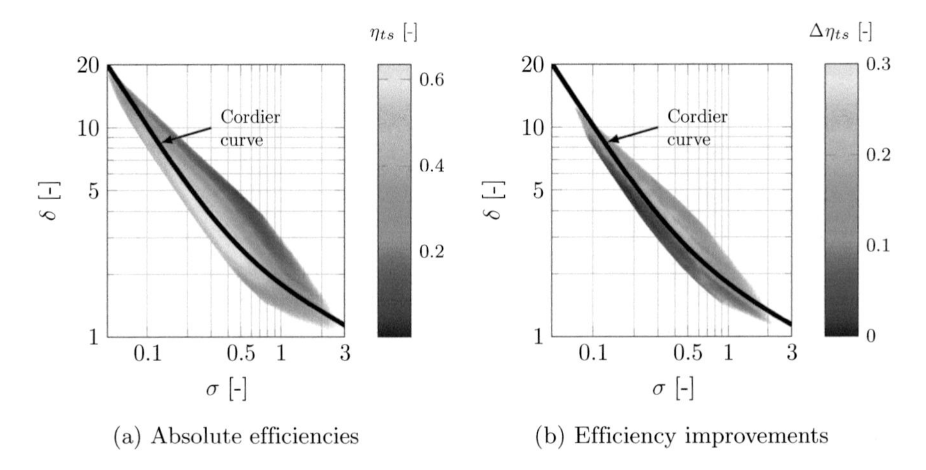

Fig. 27.16 Absolute total-to-static efficiencies after the active learning strategy (**a**) and improvements through the active learning strategy (**b**)

why the area covered by possible designs is smaller in Fig. 27.16b compared to Fig. 27.16a.

The extension of achievable regions in the Cordier diagram together with the absolute efficiencies shown in Fig. 27.16a indicates that designs in the extended regions are not attractive to be realized in practice. The achieved efficiencies in the extended regions are rather low and therefore not desirable. However, the efficiency improvements close to the Cordier curve are quite high and demonstrate the usefulness of the active learning strategy for one real-world application.

Through an extension of the already proposed HilomotDoE [224] algorithm, a goal orientation is introduced; see [40] for more details. There are three goals: (I) the concentration on possibly optimal geometries, (II) the focus on areas in the input space with inferior generalization performance, and (III) a high diversity of training data samples. In the application example shown here, all three goals are met. (I) The maximum efficiencies could be improved as shown in Fig. 27.16, (II) the overall model performance could be increased as can be seen in Fig. 27.14, and (III) the diversity increase of the training data samples leads to an extension of achievable design points; see Fig. 27.15.

27.2.8 Mixed Wrapper-Embedded Input Selection

The mixed wrapper-embedded input selection is applied to both the centrifugal and the axial fan metamodels. In this section the linked $\underline{x}$-$\underline{z}$-input selection is used with the AIC$_C$ as evaluation criterion. Two search strategies are employed and compared, which are BE and forward selection (FS). For both input selection approaches, 80% of all available data is used, while 20% of it is saved to subsequently assess the model quality for specific input subsets on fresh data.

27.2.9 Centrifugal Fan Metamodel

Figure 27.17 shows the results of the linked $\underline{x}$-$\underline{z}$-input selection for the centrifugal fan metamodel. Regardless of the used search strategy, it is observable that no better bias/variance tradeoff could be found by removing inputs; see Fig. 27.17a. This means that each of the inputs carries enough information to overcompensate the additionally introduced variance. It can be concluded that (I) only physically meaningful inputs have been chosen to describe the behavior of the centrifugal fans

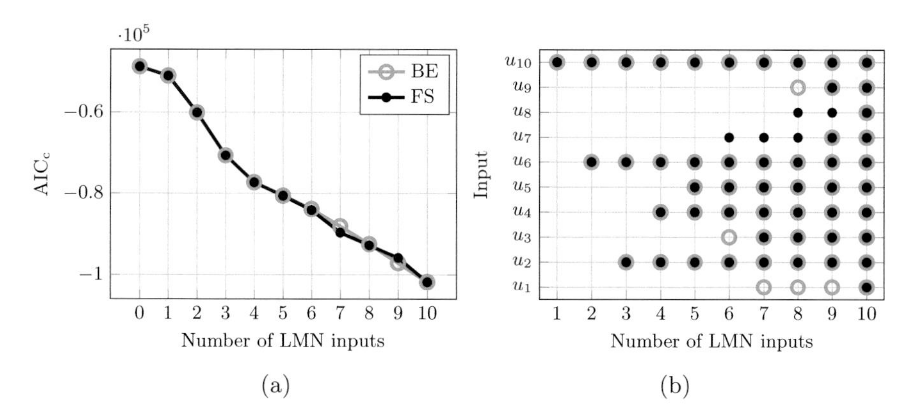

Fig. 27.17 Obtained results of the linked $\underline{x}$-$\underline{z}$-input selection for the metamodel of centrifugal fans. (**a**) AIC$_C$ vs. number of LMN inputs. (**b**) Selected inputs

Table 27.1 NRMSE values for models based on different LMN input subsets evaluated on the test data set of the centrifugal fan metamodel

Number of LMN inputs	6	8	10
Selected BE subset	**0.4598**	0.3773	0.3041
Selected FS subset	0.4622	**0.3691**	0.3041

and (II) the experimental design ensures a meaningful variation of these inputs. This outcome is reasonable and was expected since experts in the field of fans, including several experts from established companies, discussed and finally agreed on which inputs to use for the metamodel.

For this application, there is no noticeable difference between the two used search strategies. From one up to five LMN inputs, the subsets chosen by BE and FS are in fact identical; see Fig. 27.17b. But even with different subsets of inputs for the higher-dimensional cases, the model accuracies are comparable. Table 27.1 presents errors on the test data set for several input subsets, including six, eight, and ten LMN inputs. Of course, the error for the model incorporating all available inputs is identical. With six LMN inputs, the model based on the subset selected by BE performs insignificantly better. In the case of eight LMN inputs, the model based on the subset selected by FS is slightly better.

According to commonly used impeller design methodologies, as described, e.g., in [90], the most important geometrical parameters are those that directly affect the guidance of the gaseous fluid flow and how it enters and exits each flow channel. Therefore, the most important geometrical parameters for the CFD metamodel should be:

- The inner diameter u_2
- The inlet blade angle u_3
- The inlet width u_5
- The outlet width u_6
- The outlet blade angle u_4

Note that the order of the list above does not reflect the importance of the individual parameters, but rather the path from the gaseous fluid flow from the impeller inlet to the outlet. Since the rotational speed for all CFD simulations was fixed, the inner diameter together with the inlet blade angle determines the flow angle with which the gaseous fluid enters each flow channel. The inlet and outlet width has a big impact on how the cross-sectional area, through which the gaseous fluid flows, changes from the inlet to the outlet of the impeller. The change in the cross-sectional area has a significant influence on the deceleration of the gaseous fluid [135] which in turn affects the losses of the impeller and therefore the efficiency [90]. The outlet blade angle directly influences the deflection of the gaseous fluid flow at the impeller outlet.

The five most important geometrical parameters according to the linked $\underline{x}$-$\underline{z}$-input selection with BE as search strategy are in perfect accordance with expert knowledge, i.e., all five parameters from the above list are selected; see Fig. 27.17b.

The only metamodel input that is considered even more important is the volume flow rate u_{10}, which is also quite reasonable because it specifies the operating point of the centrifugal fan. When using the FS search strategy, the shroud radius u_7 is considered to be more important than the inlet blade angle u_3 which seems not reasonable according to expert knowledge. However, no significant influence on the model quality can be observed in Fig. 27.17a.

27.2.10 Axial Fan Metamodel

Figure 27.18 shows the results of the linked $\underline{x}$-$\underline{z}$-input selection for the axial fan metamodel. Both search strategies deliver for most input subset sizes comparable to AIC_C values. From 20 up to 28 LMN inputs, no more significant improvements are observable. The best achieved AIC_C values are obtained with 27 and 24 LMN inputs for BE and FS, respectively. The biggest difference in the achieved AIC_C values appears at eleven LMN inputs.

Both used search strategies lead to identical input subsets from a subset size of one up to three; see Fig. 27.18b. According to the achieved AIC_C values, both search strategies deliver comparable results for almost all subset sizes. Selected input subsets are compared regarding the error on the test data set in Table 27.2. The compared subsets include:

- 11 LMN inputs because the biggest difference in the achieved AIC_C value of both search strategies appears at this point; see Fig. 27.18a.
- 24 LMN inputs because the lowest AIC_C value is achieved in the case of FS.
- 27 LMN inputs because the lowest AIC_C value is achieved in the case of BE.
- 28 LMN inputs, corresponding to a model incorporating all available inputs.

Again, it is hard to decide whether one of the two search strategies delivers significantly better subsets. A noteworthy fact is that, according to the test errors, no model quality improvement can be achieved by removing any of the 28 LMN inputs. This could not be concluded from the obtained AIC_C values; see Fig. 27.18a. However, this indicates again that (I) only physically meaningful inputs have been chosen to describe the behavior of the axial fans and (II) the experimental design ensures a meaningful variation of all inputs.

Prior to the input selection of the axial fan metamodel, an expert in the field of axial fans made an assumption about the most important inputs according to his expertise. Seven inputs are stated as presumably most important for the efficiency of the axial fans, including:

- The volume flow u_1
- The target volume flow u_2
- The number of blades u_3
- The hub-to-tip ratio u_4
- The tip clearance u_5

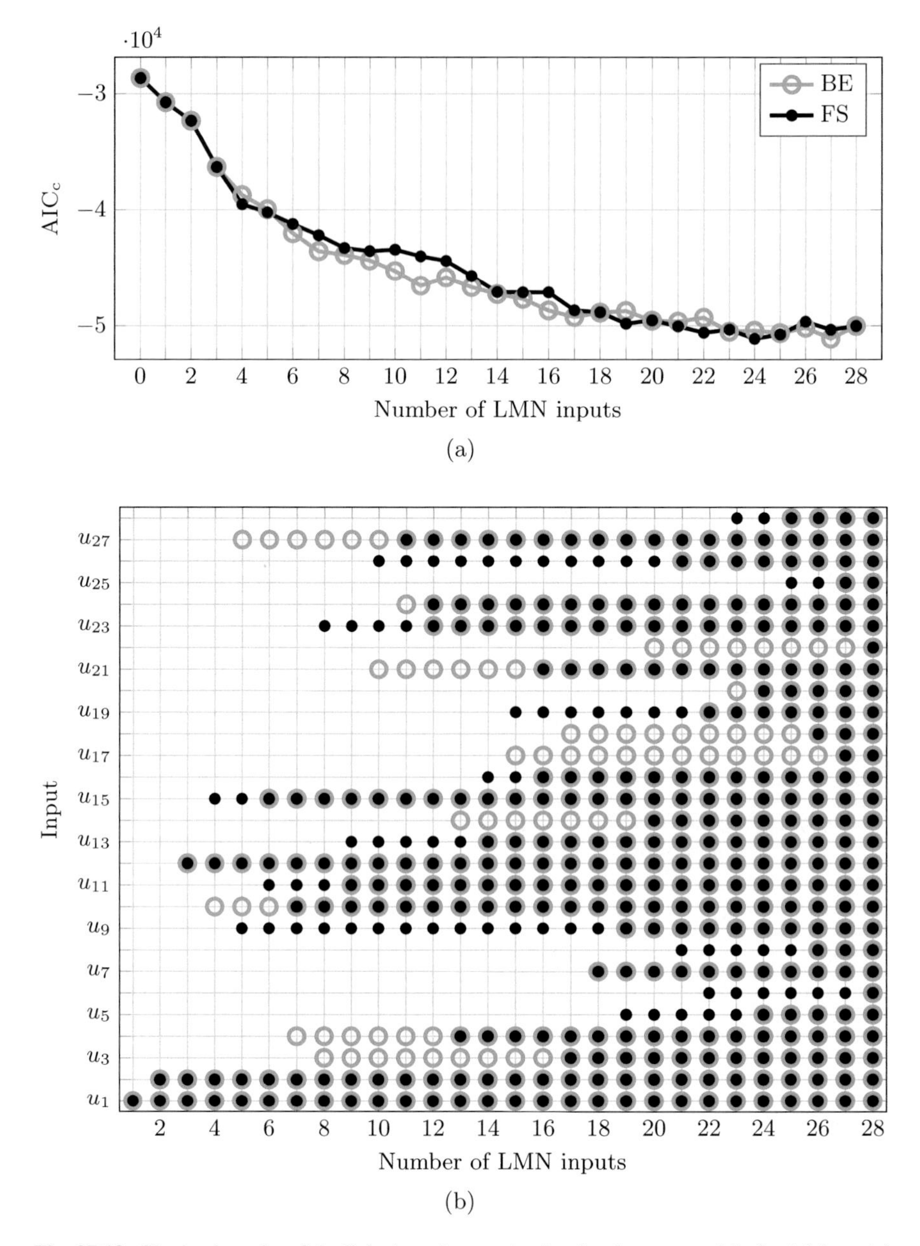

Fig. 27.18 Obtained results of the linked $\underline{x}$-$\underline{z}$-input selection for the metamodel of axial fans. (**a**) AIC$_C$ vs. number of LMN inputs. (**b**) Selected inputs

Table 27.2 NRMSE values for models based on different LMN input subsets evaluated on the test data set of the axial fan metamodel

Number of LMN inputs	11	24	27	28
Selected BE subset	**0.5934**	0.5168	**0.4814**	0.4339
Selected FS subset	0.6450	**0.4948**	0.5305	0.4339

- The chord length at midspan u_{12}
- The maximum camber at midspan u_{15}

Reasons for this assumption can be found, e.g., in [90]. It is remarkable that in the case of the BE search strategy, the subset with seven LMN inputs contains five of these inputs; see Fig. 27.18b. The subset containing seven inputs found with FS includes four of the assumed seven most important inputs.

27.2.11 Summary

For the two fan metamodels, the mixed wrapper-embedded input selection approach is not able to improve the model accuracy by removing inputs through the linked $\underline{x}$-$\underline{z}$-input selection. Both metamodels are a good example of reasonably selected inputs and a well-chosen DoE. However, with the trend of an ever-increasing input space dimensionality, it will get harder, even for domain experts, to select the right amount of inputs for the best bias/variance tradeoff. Even though the mixed wrapper-embedded input selection approach is not able to improve the accuracy of the metamodels, information about the most important inputs is gathered. In the case of the axial fan metamodel, expert knowledge confirms the results regarding the input importance obtained from the input selection. Input selection allows gaining insights and confidence in the data-driven models which is a very important factor for industrial acceptance of these abstract approaches.

27.3 Heating, Ventilating, and Air Conditioning System

A heating, ventilating, and air conditioning (heating, ventilating, and air conditioning (HVAC)) system that has already been investigated in [527] and [480] is the real-world application dealt within this section. The goal is to build a dynamic model that can be used for model predictive control [526]. The focus in this section lies in the generation of a good dynamic model and not on the model predictive control. The mixed wrapper-embedded input selection approach with the separated $\underline{x}$-$\underline{z}$-input selection is applied to the HVAC system in order to find the best subset of inputs. Three data sets are available that serve as training, validation, and test data, which are described in more detail at the end of this section. The training data is only used for the estimation of all LMN parameters during the training with LOcal LInear MOdel Tree (LOLIMOT). Here, LOLIMOT is used instead of HILOMOT due to the increased required computation time that results from the very high number of potential LMN inputs, which is 140; see Sect. 27.3.3 for details. With the help of the simulation error on validation data, the model complexity, i.e., the number of local models, is chosen. Additionally, the simulation error on validation data is used as evaluation criterion during the separated $\underline{x}$-$\underline{z}$-input selection. For this investigation, the results obtained with the validation error as an evaluation criterion for the separated $\underline{x}$-$\underline{z}$-input selection turned out to be superior compared to using the $\mathrm{AIC_C}$. The test data set is only used after the separated $\underline{x}$-$\underline{z}$-input selection is finished. The reason for the inferiority of the $\mathrm{AIC_C}$ as an evaluation criterion in this particular application is unknown. It is assumed that the $\mathrm{AIC_C}$ is not that well suited for the model complexity determination of dynamic models for short sampling times and therefore also struggles as evaluation criterion during the input selection.

27.3.1 Problem Configuration

The setup presented in Fig. 27.19 shows a typical application of a series connection of cooling and heating coils. The air is dehumidified by the cooling coil in order to adjust the air humidity. Since the air temperature is decreased by the cooling coil too, the air has to be reheated by the heating coil in order to meet the desired temperature. The power of the coils can be adjusted by valves. For the cooling coil, the water mass flow is varied via the valve position u_1, whereas for the heating coil, the mixing ratio of the returned cold water and the hot supply water is varied via the valve position u_2. For this configuration, a constant air mass flow of $\dot{m}_a = 1$ kg

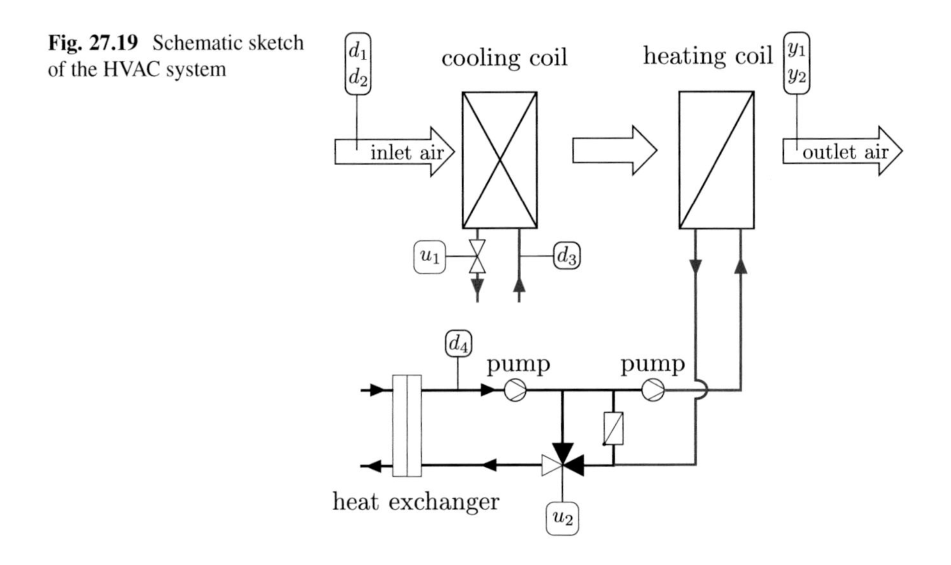

Fig. 27.19 Schematic sketch of the HVAC system

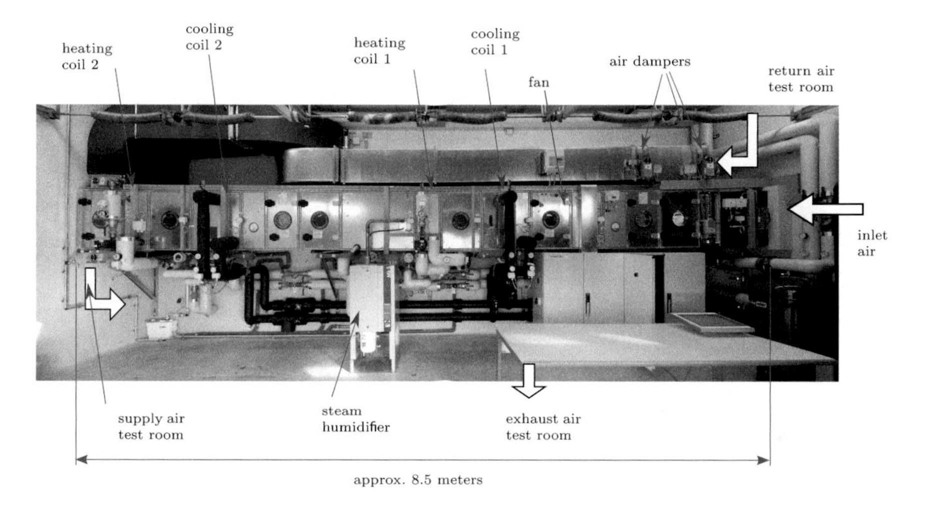

Fig. 27.20 Picture of the real-world HVAC system [527]

per second is assumed. The output signals of the system are the temperature y_1 and the relative humidity y_2 of the outlet air. The input signals of the system are the valve positions u_1 and u_2, the temperature d_1 and the relative humidity d_2 of the inlet air, the inlet water temperature d_3 of the cooling coil, and the water supply temperature d_4 of the heating circuit. The valve positions u_1 and u_2 are actuating variables which can be varied arbitrarily to excite the dynamic system as desired. In contrast to that, the LMN inputs d_1, d_2, d_3, and d_4 can be measured but not be manipulated freely. The units of the measured temperatures and humidities are °C and %, respectively. The data sets used in this work are generated on a real-world system shown in Fig. 27.20. This pilot plant is provided by the company Fischer&Co Luft- und Klimatechnik in Graz/Austria. Figures 27.19 and 27.20 are kindly provided by Daniel Schwingshackl (University of Klagenfurth), Jakob Rehrl (Graz University of Technology), and Martin Horn (Graz University of Technology).

27.3.2 Available Data Sets

The training data set consists of $N_{\text{train}} = 5490$ samples, the validation data set has $N_{\text{val}} = 224$ samples, and there are $N_{\text{test}} = 262$ test samples. The sampling frequency for all data sets is $f_0 = 1/16\,\text{Hz}$. As already mentioned, only two valve positions u_1 and u_2 can be manipulated. As can be seen in Fig. 27.21, the excitation signal for the valve positions consists of several steps around different operating points. The variation in the measurable disturbances is by far weaker and lower frequent compared to the manipulated variables. Figure 27.21 also shows the signals

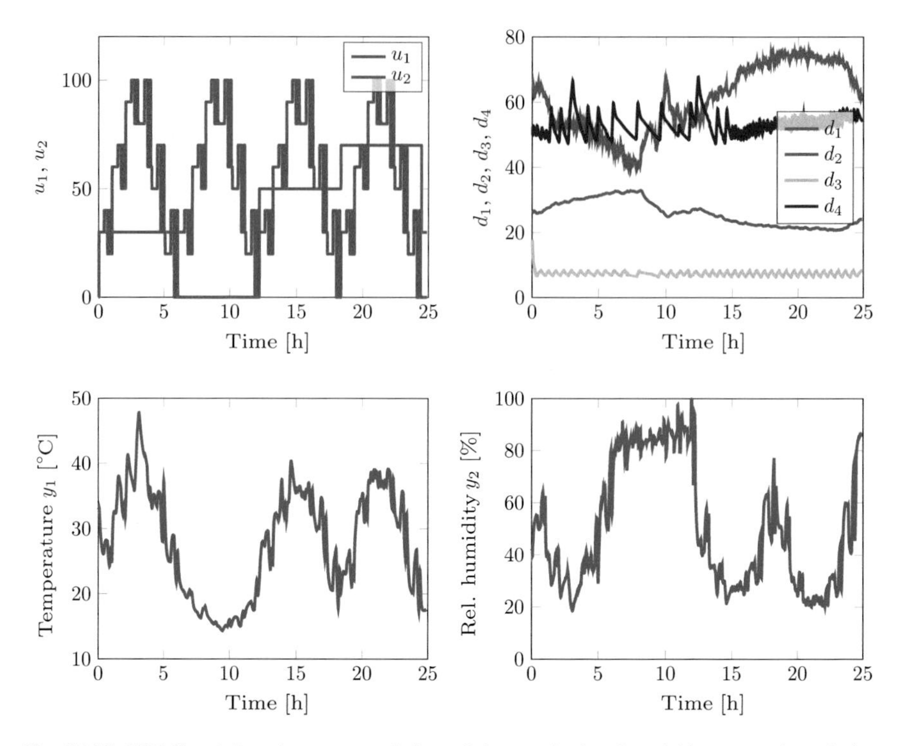

Fig. 27.21 HVAC training data set consisting of the manipulated variables u_1 and u_2 (valve positions); the measurable disturbances d_1 (inlet air temperature), d_2 (rel. humidity inlet air), d_3 (inlet water temperature cooling coil), d_4 (water supply temperature); and the control variables y_1 (outlet air temperature) and y_2 (rel. humidity outlet air)

of the control variables. Especially the dependency of both control variables on the second valve position u_2 is clearly observable.

27.3.3 Mixed Wrapper-Embedded Input Selection

The search strategies used for the separated $\underline{x}$-$\underline{z}$-input selection are backward elimination (BE) and forward selection (FS). In order to test the mixed wrapper-embedded input selection, we assume to have very limited insights into the physical background of the HVAC system. Nevertheless the maximum and minimum delay $n_{\max}$ and $n_{\min}$, respectively, has to be provided for the BE search strategy because it starts with all potential inputs. $n_{\min}$ and $n_{\max}$ are chosen to be 1 and 10, respectively, for all physical inputs as well as for all physical outputs. For each target value, i.e., the temperature y_1 and the relative humidity y_2, distinct models are built, and therefore distinct input selections are performed. Because of the chosen minimum and maximum delays, there are 70 possible LMN inputs for each of the two input spaces, i.e., the $\underline{x}$- and $\underline{z}$-input space. As a result, the total amount of potential LMN inputs is virtually doubled to 140. Following the separated $\underline{x}$-$\underline{z}$-input selection with the two search strategies BE and FS, the methods choose LMN inputs to be removed or added without paying attention from which input space it is eliminated from or added to.

27.3.4 Results

For the performed separated $\underline{x}$-$\underline{z}$-input selection, Fig. 27.22a shows results for the temperature y_1 and Fig. 27.22b the results for the relative humidity y_2. The validation error is plotted against the number of LMN inputs for both search

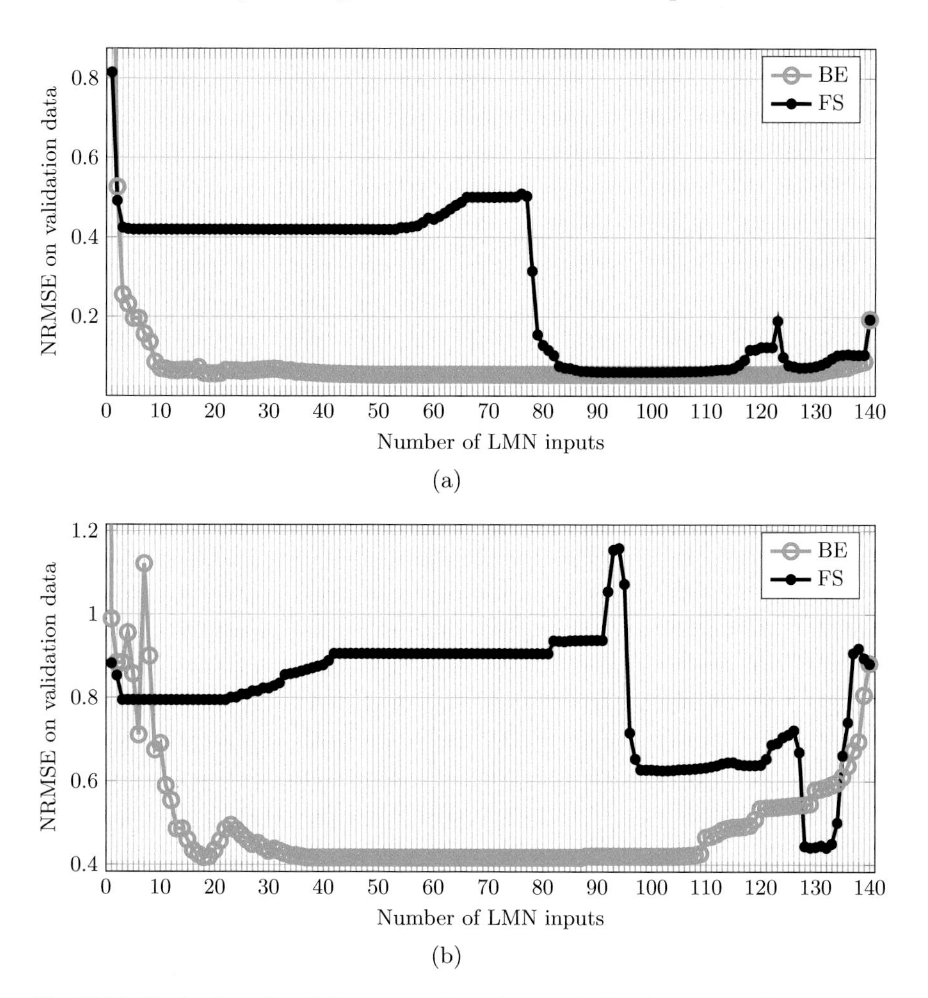

Fig. 27.22 Obtained results of the separated $\underline{x}$-$\underline{z}$-input selection for the HVAC system. (**a**) Temperature y_1. (**b**) Relative humidity y_2

strategies. It is evident that for both outputs y_1 and y_2, the subsets found by BE are far superior to the ones found by FS. Especially for input subsets containing less than 80 inputs, the difference in the model qualities achieved with the found subsets is huge. The main reason for this turns out to be a lack of delayed outputs in the $\underline{x}$-input space of the LMNs in the FS case. For the temperature y_1, the FS selects the first delayed output, in particular $y_1(k-3)$, as the 78th LMN input. Therefore, subsets with less than 78 LMN inputs lead to local finite impulse response (FIR) models. In the case of the relative humidity y_2, the same effect takes place, but the first delayed output, in particular $y_2(k-6)$, is selected as 94th LMN input. In this application, one huge disadvantage of FS becomes clearly visible. The interaction between several inputs are not discoverable by simply adding one input at a time. In contrast to that, when following a BE strategy, all interactions are present from the start. The deterioration of the model performance by removing one input that is important in combination with any other input is directly recognizable. In the following, only results of the BE are further discussed because of the far worse model accuracies obtained with subsets found by FS.

The performance improvement through the mixed wrapper-embedded input selection with BE as search strategy is significant for both outputs. In general, the achieved model performance for the temperature is by far better than for the relative humidity. One noticeable aspect is the long plateaus of the simulated validation error for both output variables. For the temperature the same validation error is achieved between 54 and 121 LMN inputs. In the case of the relative humidity, there are two of these plateaus, one ranging from 37 up to 87 and the other from 88 up to 104. We discovered that during these plateaus only inputs from the $\underline{z}$-input space are eliminated that are never chosen as split directions. The $\underline{x}$-input space is kept unchanged. Therefore discarding those LMN inputs does neither affect the $\underline{z}$-input space partitioning nor the parameter estimation of the local models. As a result, all LMNs corresponding to one plateau are in fact identical, which explains the constant validation errors. The calculations took 90 hours (17 for FS) for the temperature and 41 hours (14 for FS) for the relative humidity on a cluster node with 12 CPUs and 48 GB of RAM. The difference in the calculation times originates from the fact that generally a higher complexity (more local models) was chosen for the temperature models throughout the whole input selection procedure. This might also explain the much lower error rates for the temperature model.

For the temperature, the optimal subset size of LMN inputs according to the simulation error on validation data is 50; for the relative humidity, it is 37. Now several LMN input subsets are compared to each other in terms of their NRMSE values (J) on all three available data sets; see Table 27.3. Besides the best LMN input subsets, models including all LMN inputs are listed. In addition, local minima with far less LMN inputs are compared as well, in particular the temperature model relying on 19 LMN inputs and the relative humidity model relying on 18 LMN inputs. Highlighted loss function values indicate the best value for each data set and each output. Rows two to four of Table 27.3 show the loss function values for the model of the temperature. Especially the performance improvement on validation and test data through the separated $\underline{x}$-$\underline{z}$-input selection is impressive. Even with

Table 27.3 NRMSE values (J) on different LMN input subsets for the training, validation, and test data sets

Number of LMN inputs	J_{train}	J_{vali}	J_{test}
19 LMN inputs (y_1)	0.013	0.052	**0.230**
50 LMN inputs (y_1)	**0.013**	**0.049**	0.248
140 LMN inputs (y_1)	0.025	0.193	0.550
18 LMN inputs (y_2)	**0.046**	0.417	0.723
37 LMN inputs (y_2)	0.070	**0.417**	**0.408**
140 LMN inputs (y_2)	0.073	0.882	0.459

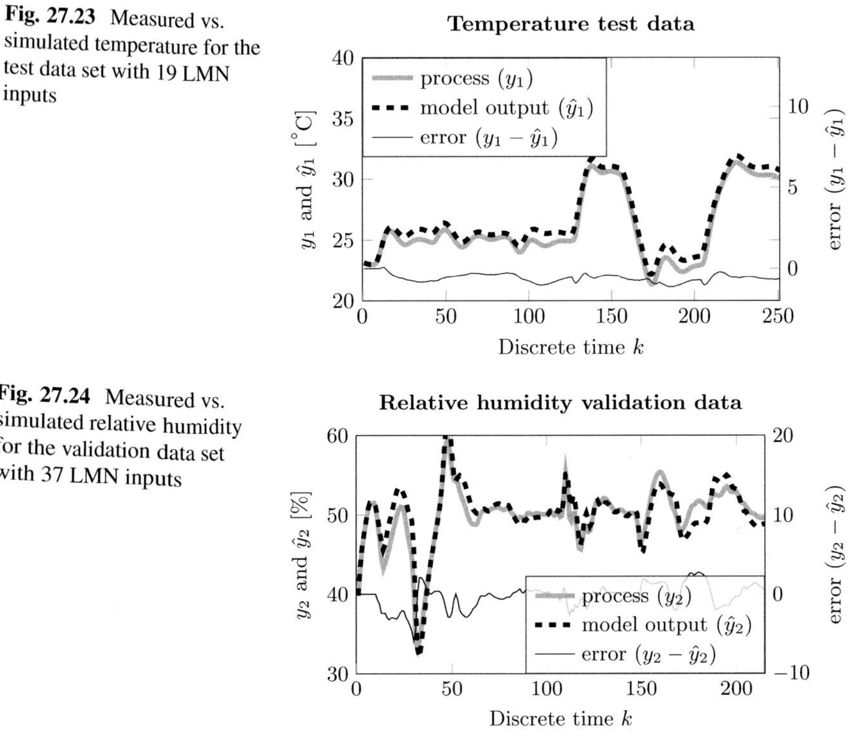

Fig. 27.23 Measured vs. simulated temperature for the test data set with 19 LMN inputs

Fig. 27.24 Measured vs. simulated relative humidity for the validation data set with 37 LMN inputs

only 19 LMN inputs, almost the same model quality is achieved compared to a model with 50 LMN inputs. Since the curves of the simulated model outputs for 19 and 50 LMN inputs are barely distinguishable, Fig. 27.23 shows only the case with 19 LMN inputs. In addition, the process output (here the temperature y_1) from the test data and the test error is shown.

Rows five to seven of Table 27.3 show the model performances for the relative humidity. Here the improvement on validation data is significant, but the improvement on test data is only moderate. Eventually, the generalization performance does not reach the same accuracy level as the temperature model, but the simulated model output follows the measured relative humidity quite well. Figure 27.24 demonstrates this for 37 LMN inputs on the validation data.

In addition to the gain in model performance, the separated $\underline{x}$-$\underline{z}$-input selection makes the final model more concise and simplifies its interpretation. After a closer look at the best LMN input subsets, it is remarkable that for both outputs there are only two LMN inputs left in the $\underline{z}$-input space (serving as operating point variables). In both cases, these are the same physical inputs, namely, the valve positions u_1 and u_2, but with different delays. Actually, nonlinearities originating from the valve characteristics are identified correctly to be important for the definition of an operating point. This is extremely appealing since these are the manipulated variables in control of the HVAC system. Therefore their influence is most important, and the accuracy of the model w.r.t. u_1 and u_2 is crucial. Since the excitation of u_1 and u_2 in the training data is most active, it is a natural outcome of the input and order selection procedure to keep both in the LMN inputs.

For the model of the temperature, the LMN inputs that are left in the $\underline{z}$-input space are $u_1(k-3)$ and $u_2(k-1)$. The LMN inputs $u_1(k-2)$ and $u_2(k-2)$ are in the relative humidity model's $\underline{z}$-input space. Since the dimensionality of both $\underline{z}$-input spaces is two-dimensional, we can visualize the partitioning of them. Figure 27.25 shows the partitioning for the temperature model exemplarily. More splits along the u_2-axis indicate a stronger nonlinearity in that direction. Each validity area contains a number that corresponds to the relevant local affine model.

The interpretation of the selected LMN inputs for the affine local models ($\underline{x}$-input space) is difficult since there are so many variables left. Easy interpretation, however, regards the relative humidity model, for which the physical input d_1 (inlet air temperature) is neglected completely, i.e., none of its delayed versions is used. In summary, it is shown that the mixed wrapper-embedded input selection approach,

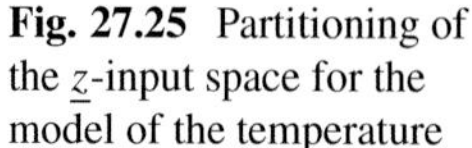
Fig. 27.25 Partitioning of the $\underline{z}$-input space for the model of the temperature

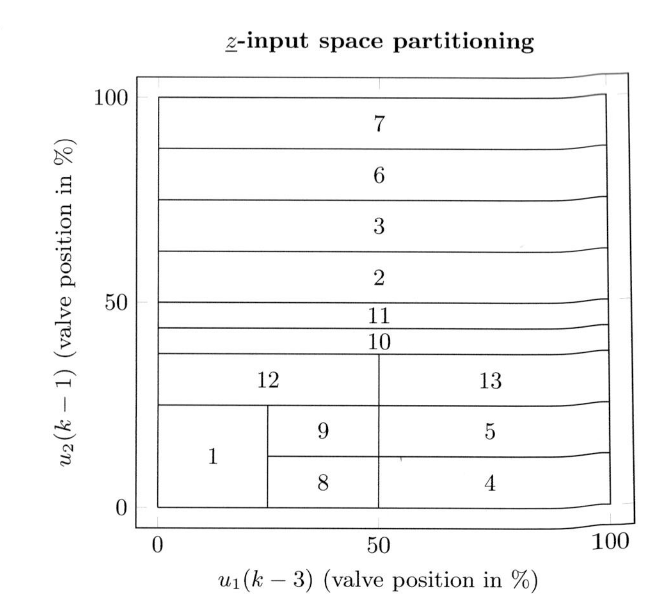

in particular, the separated $\underline{x}$-$\underline{z}$-input selection, for the modeling of the HVAC system leads to significant improvement and simplification of the model as shown in Table 27.3. Additionally, the interpretability is increased at least for the $\underline{z}$-input space that can be described for both outputs with only two operating point variables.

Chapter 28
Applications of Advanced Methods

Abstract This chapter deals with the application of the heat exchanger models from Chap. 25. Nonlinear predictive control design is discussed. In a simple approach, the linearized models can be used for a fast linear predictive control algorithm that is updated with the model linearized in each time step. In a more sophisticated and computationally demanding approach, a full nonlinear predictive control algorithm can be run. The simpler approach seems worth considering since it offers many advantages in implementation with just a slight sacrifice in performance. Robust online adaptation is also demonstrated were local model networks are adapted only locally, resulting in almost identical computational load compared to a linear model. Next, fault detection and diagnosis are carried out, running a nominal model and fault models in parallel to the process, calculating a residuum for each model. Finally, the information from the diagnosis can be exploited for the potential reconfiguration of the plant.

Keywords Application · Predictive control · Online · Adaptation · Forgetting · MISO · Fault detection · Fault diagnosis

This chapter gives an overview of the way in which nonlinear dynamic models can be utilized for control and fault detection. As an illustrative application example, the thermal plant presented in Sect. 25.3 is chosen. Section 28.1 discusses the design of a predictive controller based on a local linear neuro-fuzzy model. The online adaptation of this model yields a nonlinear adaptive controller. It allows an adjustment to time-variant behavior and changing environmental conditions – in this particular example to changing environment temperature. As pointed out in Sect. 28.2, some precautions must be taken to make adaptive control robust against insufficient excitation. Section 28.3 introduces the topic of model-based fault detection, and Sect. 28.4 briefly addresses the subject of fault diagnosis, i.e., the determination of the fault cause. Finally, a reconfiguration strategy for the controller is discussed, based on the fault diagnosis results obtained. This chapter can only offer a glimpse of the topics addressed. For a more detailed treatment, the reader is referred to the vast literature on nonlinear control and fault diagnosis.

O. Nelles, *Nonlinear System Identification*,
https://doi.org/10.1007/978-3-030-47439-3_28

For a more extensive treatment of integrated gray box modeling approaches to the thermal plant, refer to [147, 150, 156]. The predictive controller design is thoroughly discussed in [157], and various control strategies are compared in [154]. The adaptive predictive control of the cross-flow heat exchanger can be found in [148, 155]. In [145, 153, 158] the sophisticated supervisory level for the adaptive predictive controller is proposed, a local variable forgetting factor is introduced, and the adaptation mechanism is based on prior knowledge obtained from static first principles modeling. The tubular and cross-flow heat exchanger models are utilized for fault detection and diagnosis in [26, 27, 177, 426]. Finally, the adaptive predictive controller can be combined with fault detection and diagnosis strategy to the integrated control, diagnosis, and reconfiguration framework presented in [25].

28.1 Nonlinear Model Predictive Control

In this section,[1] a nonlinear model predictive controller (NMPC) based on a local linear neuro-fuzzy model is introduced. By utilizing the online adaptation scheme for the model as discussed in the subsequent section and depicted in Fig. 28.2, the controller becomes *adaptive* (ANMPC). The key idea of predictive control is illustrated in Fig. 28.1. The model is utilized to predict the process response (N_2 steps in advance) to a given sequence of process inputs generated by the controller. The goal is to find the sequence of process inputs that yields the optimal process response with respect to some criteria. No specific controller structure is chosen. Rather an optimizer searches for the best actuation signal sequence. In order to limit the complexity of the problem, the actuation signal sequence may be allowed to change only a few times, given by the number N_u. At each sampling instant, an optimization is carried out to determine the optimal next N_u actuation values. Then only the first value of this sequence is applied. This procedure is repeated at the next time instant by shifting everything by one time step $k \rightarrow k + 1$. This is called the *receding horizon strategy*. For more details on predictive control, refer to [483, 542, 557] and the references therein.

For the sake of simplicity, a SISO (single input, single output) process with the manipulated variable u and system output y is chosen. Disturbances can be distinguished into m measurable disturbances $\underline{n} = [n_1\ n_2\ \cdots\ n_m]^T$ and an arbitrary number of unmeasurable disturbances gathered in $\underline{v}$. The optimizer determines the new actuation signal $u(k) = u(k-1) + \Delta u(k)$ by finding the optimal control increment $\Delta u(k)$ in each sampling instant. At the current time instant k, the

[1]This section is based on research undertaken by Martin Fischer, Institute of Automatic Control, TU Darmstadt.

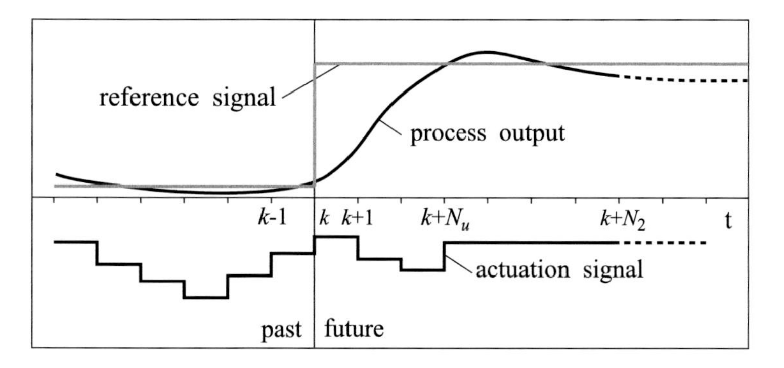

Fig. 28.1 Key idea of predictive control

sequence of N_u future control increments $\Delta u(k+j)$ is obtained by minimization of the following quadratic loss function:

$$J = \sum_{j=N_1}^{N_2} \left(r(k+j) - \hat{y}(k+j)\right)^2 + \beta(k) \sum_{j=0}^{N_u-1} (\Delta u(k+j))^2 . \tag{28.1}$$

Here N_1 and N_2 denote minimum and maximum prediction horizons, and $\beta(k)$ is the penalty factor for future changes of the manipulated variable $\Delta u(k+j)$. N_1 is typically chosen equal to the dead time of the process, which is the earliest time instant where the controller can influence the process output. N_2 is usually chosen according to the dominant time constants of the process and must not be chosen too small for non-minimum phase processes. The first sum in (28.1) penalizes the control error, which is calculated as the difference between the reference values $r(k+j)$ taken from the reference vector $\underline{r} = [r(k+N_1)\ r(k+N_1+1)\ \cdots\ r(k+N_2)]^T$ and the (corrected) predicted process outputs $\hat{y}(k+j)$. If the reference signal is not known in advance, r is assumed to keep its current value over the complete prediction horizon, i.e., $r(k+j) = r(k)$ for all $N_1 \leq j \leq N_2$. For calculating the (corrected) predicted values $\hat{y}(k+j)$, the fuzzy process model output $y^*(k+j)$ is utilized and additively corrected by an compensation term e^*. This correction procedure is described in more detail below. The fuzzy model is run in parallel to the process in order to predict the response of the process to a given input sequence gathered in the vector $\underline{u}^*$. The prediction vector $\hat{y}^* = [\hat{y}^*(k+N_1)\ \hat{y}^*(k+N_1+1)\ \cdots\ \hat{y}^*(k+N_2)]^T$ is calculated by the local linear neuro-fuzzy model with the following regression vector:

$$\begin{aligned}\underline{\varphi}(k) = [&u(k-d-1)\ \cdots\ u(k-d-nu)\\ &n_1(k-d_1-1)\ \cdots\ n_1(k-d_1-nn_1)\ \cdots\\ &n_m(k-d_m-1)\ \cdots\ n_m(k-d_m-nn_m)\\ &\hat{y}^*(k-1)\ \cdots\ \hat{y}^*(k-ny)]^T ,\end{aligned} \tag{28.2a}$$

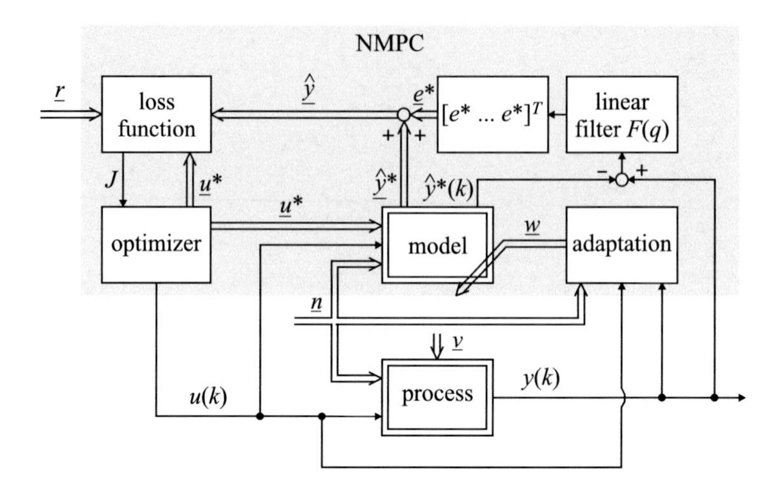

Fig. 28.2 Adaptive nonlinear model predictive control (ANMPC)

where d and nu are the dead time and dynamic order of the manipulated variable, d_i and nn_i ($i = 1, \ldots, m$) are the dead times and dynamic orders of the measurable disturbances, and ny is the dynamic order of the output. Note that from the modeling perspective, no distinction exists between the manipulated variable u and the measurable disturbances n_i. Typically, however, the manipulated variable should be excited with fast dynamics within the training data set to yield a high model accuracy in the case of tight (high performance) control. The disturbances usually have individual characteristic patterns in the frequency and amplitude range (Fig. 28.2).

For future time instants, the actuation signals

$$\underline{u}^* = [u^*(k) \ \cdots \ u^*(k + N_u - 1)]^T$$

provided by the optimizer are substituted in the regression vector (28.2a), and the measurable disturbances are assumed to keep their current values $\underline{n}(k + j) = \underline{n}(k)$. In contrast to series-parallel mode, where measured outputs y are used in the predictor, here previously predicted values $\hat{y}^*$ are fed back, that is, the model is used for *simulation* rather than prediction (in spite of the fact that it is called *predictive* control). In literature, both types, simulation, and prediction models, can be found. One disadvantage of using the prediction model is that the process/model mismatch causes a transient, which declines over the prediction horizon and disguises the actual model dynamics. For an illustration of this effect, it is helpful to assume that the process is in steady state. Thus, a fixed manipulated variable should keep the process in steady state. Owing to a process/model mismatch (steady-state levels of the model are not equivalent to those of the process), however, the model predicts a transient that leads to the model's steady state (different from the steady state of the process). This "artificial" transient causes wrong control action by the optimizer. Therefore, the use of a simulation model is recommended with this scheme. A simulation model avoids this problem since it possesses this process/model mismatch transients just once after initialization. Then the process states are not forced upon the model by feeding it with the measured process outputs (series-parallel mode). Rather the model feeds back its own predicted outputs (parallel mode).

The standard formulation of the loss function is extended by the time-dependent penalty factor $\beta(k)$; see (28.1). While for linear processes a constant $\beta(k) = \beta_0$ is usually sufficient, for nonlinear processes the changing process gain must be taken into consideration. In operating points of relatively low process gain, larger changes in the actuation signal can be accepted. By contrast, aggressive control actions are not desirable in regimes of high gain [277]. Therefore, the following equation is recommended for nonlinear processes:

$$\beta(k) = K_p^2(k)\beta_0 \,. \tag{28.3}$$

The current process gain K_p can be determined by dynamic linearization of the local linear neuro-fuzzy model; see Sect. 22.3. The gain of the linearized model can be easily calculated from the parameters of the local dynamic linearization.

For linear processes, the optimization problem can be solved analytically because then the loss function J depends linearly on the control increments Δu; see [542]. If nonlinear process models are utilized or nonlinear constraints on any of the variables are to be considered, nonlinear optimization routines must be applied; refer to Chaps. 4 and 5. For the experiments shown below, a combination of a one-dimensional grid search, a subsequent Hooke-Jeeves search, and an optimization with Newton's method is used. This staggered optimization procedure utilizes the advantages of the three different techniques. Grid search does not trap into local optima and yields a fairly good starting point for the Hooke-Jeeves search. The latter is easy to implement and does not require the calculation of gradients. Hence, it works independently from the underlying model structure. When the parameters are sufficiently close to the optimum, the Newton method guarantees second-order convergence because the loss function is approximately quadratic in this region. Alternatively, a Levenberg-Marquardt method may yield similar results.

The unmeasurable disturbances $\underline{v}$ affect only the real process and cause process/model mismatch and thereby deviations between process and model output signals. Because of this, the prediction vector $\underline{\hat{y}}^*$ should not be directly used for the calculation of the loss function. Instead, a feedback component is added to the predictive controller. The current model error

$$e(k) = y(k) - \hat{y}^*(k) \tag{28.4}$$

is calculated, low-pass filtered with a linear filter $F(q)$, and used for the correction of the simulated outputs $\underline{\hat{y}}^*$

$$\hat{y}(k+j) = \hat{y}^*(k+j) + F(q)e(k) = \hat{y}^*(k+j) + e^*(k), \tag{28.5}$$

with $j = N_1, \ldots, N_2$. This corresponds to the internal model control (IMC) scheme where the difference between the measured process output and the simulated model output is fed back through a robustness filter [388]. The most obvious improvement is the cancellation of steady-state control errors in the case of offset errors in the fuzzy model. The linear filter attenuates measurement noise, and it can be heuristically tuned to ensure the stability of the feedback loop. Note that *without* such a correction, the controller possesses only feedforward action (if the model is run in parallel mode) since no information about the measured process output would enter the optimization. It is also possible to introduce feedback indirectly by adapting the model online.

In the ANMPC scheme, a further component is introduced that counteracts the adverse effects of process/model mismatch. The online parameter estimator serves to abolish model errors. The local RLS adaptation strategy described in Sect. 14.6.1 operates in series-parallel mode. The model adaptation acts as an additional feedback component. Therefore, the linear filter in (28.5) can be omitted for the special case where the premise input vector $\underline{z}$ does not contain previous process outputs $y(k-i)$. Then the adaptation updates the local models and compensates for process/model mismatch. In the general case, however, $\underline{z}$ depends on outputs $y(k-i)$. If the model error is large, i.e., $y(k-i)$ are significantly different from $\hat{y}^*(k-i)$, different local models are active in the series-parallel estimation and the parallel simulation. Thus, the local model that is updated does not necessarily contribute to the actual process simulation. Consequently, in this general case, steady-state control errors may occur if the linear feedback filter $F(q)$ is not implemented.

28.2 Online Adaptation

In Sect. 14.6.1 online adaptation schemes for local linear neuro-fuzzy models have already been discussed. It was mentioned that the weighted recursive least squares (RWLS) algorithm requires concepts for the supervision of the adaptation, in particular when applied to dynamic systems; see Sect. 18.8. The goal of the supervisory level concept proposed in the sequel is to ensure fast parameter tracking while avoiding parameter divergence and thus ensuring safe operation of the control loop when applied in the context of adaptive control.[2]

In the subsequent section, a variable forgetting factor is computed with regard to the information content of the excitation signal. Section 28.2.2 introduces an additional adaptation model that prevents the steadily changing parameters from degrading the control performance. In Sect. 28.2.3 a method for the bumpless transfer of parameters from the adaptation to the control model is proposed.

[2]This section is based on research done together with Alexander Fink and Martin Fischer, Institute of Automatic Control, TU Darmstadt.

The extension to multiple-input systems (with regard to the model, that does not necessarily mean that the control is multivariable since some model inputs can be measurable disturbances; see (28.2a)) is discussed in Sect. 28.2.4. Finally, some experimental results illustrate this approach in Sect. 28.2.5.

28.2.1 Variable Forgetting Factor

It was shown in Sect. 14.6.1 how a forgetting factor $\lambda \leq 1$ can be implemented in the RWLS algorithm (14.25a)–(14.25c). Usually, a constant forgetting factor λ is chosen in the range from 0.95 to 0.99. The choice of λ represents a tradeoff between fast parameter tracking (small λ) and good noise attenuation (large λ). A problem of the exponential forgetting arises during phases of little dynamic excitation; see Sect. 18.8. Then the entries of the covariance matrices $\underline{P}_i$ can increase exponentially. This so-called "blow-up" effect is undesirable because the parameter estimates become sensitive to measurement noise; see Sect. 18.8 and [278, 356] for more details.

This drawback can be avoided by the introduction of a variable forgetting factor. In [165] a method is proposed that performs an automatic tradeoff between fast parameter tracking and good noise attenuation based on the current excitation of the system. This approach is modified for use with local linear neuro-fuzzy models. Since the online adaptation is performed locally, for each local linear model i, a local forgetting factor is calculated by

$$\lambda_i(k) = 1 - \left(1 - \underline{x}_i(k)^T \underline{\gamma}_i(k)\right) \frac{e_i^2(k)}{\Sigma_0} \Phi_i(k)\,, \tag{28.6}$$

where $\underline{x}_i(k)$ is the regression vector of the rule consequents, $\underline{\gamma}_i$ is the correction factor, $e_i(k) = y(k) - \hat{y}_i(k)$ is the a priori model error of the ith LLM, Σ_0 is proportional to the assumed noise variance, and Φ_i is the value of the validity function of the ith LLM. The choice of the tuning factor Σ_0 influences the noise sensitivity of the forgetting factor. It can either be determined by a trial-and-error approach, or the heuristic guidelines given in [324] can be followed.

The idea of variable forgetting is as follows. When the process is not excited, the forgetting factor becomes equal to 1, and old data is not forgotten. In the case of good process excitation, λ decreases, and the impact of the current data becomes stronger. Including the a priori model error $e_i(k)$ in (28.6) slows down the adaptation if the model error is small. In (28.6) the quantity $(1 - \underline{x}_i^T \underline{\gamma}_i)$ is the only non-zero eigenvalue of the estimator. It is a measure for the current excitation of the plant; see [165]. In comparison to the original formulation in [165], the value of the validity function Φ_i has been incorporated. Since $\Phi_i \approx 0$ holds for inactive rules, the forgetting factor of the inactive local models is close to 1.

A lower bound λ_{min} is imposed on the forgetting factor in order to prevent too fast a forgetting of the past measurements:

$$\lambda_j(k) > \lambda_{\text{min}} \,. \tag{28.7}$$

In experiments $\lambda_{\text{min}} = 0.7$ turned out to be a suitable value. Besides the lower bound, an upper bound is introduced above which the adaptation is frozen. This is necessary because even for a forgetting factor of $\lambda = 1$ non-exciting measurement data influence the estimates and might degrade the model performance. An upper bound of $\lambda_{\text{freeze}} = 0.9999$ is recommended.

28.2.2 Control and Adaptation Models

During online model adaptation, the following problem can arise even if the variable forgetting factor discussed in the previous subsection is applied. As long as the parameters have not converged to their final values, they can be unreliable, and the local model may become unstable. The introduction of an additional adaptation model provides a solution to this problem. The adaptation model is steadily adapted with variable forgetting factors, and its parameters are transferred to the model utilized for control (control model) only at distinct time instants. This parameter transfer is triggered when the adaptation model (AM) performs better than the control model (CM). Moreover, model properties such as local or global stability can be checked as a transfer condition. The performance comparison is carried out when dynamic excitation is detected in the control loop. Clearly, this is the case when the reference signal or disturbances change.

Figure 28.3a explains the model comparison in detail. Two reference changes can be observed at $t = 50$ s and $t = 100$ s, respectively. The closed-loop response

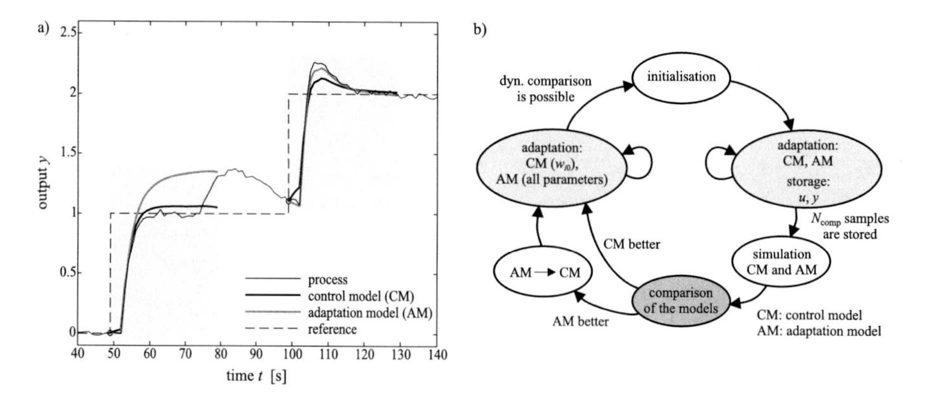

Fig. 28.3 (**a**) Comparison of the control model (CM) and the adaptation model (AM). (**b**) State transition diagram for operation of the supervisory level

of the process is depicted in comparison with the simulated outputs of the CM and the AM. The latter are obtained by feeding the control input into the models in parallel configuration. During the first reference step, the AM is worse than the CM, and consequently, the controller keeps the original CM. In the second case, the AM performs better than the CM, and thus the parameters are transferred.

The state transition diagram in Fig. 28.3b depicts the operation of the supervisory level in detail. As long as dynamic excitation in the form of reference steps has not been detected, the AM is continuously updated (light gray ellipse on the left). On reference steps the comparison of the models is prepared in the state "Initialization." Memory is allocated to store the actuation signal u and the process output y over the next N_{comp} samples. N_{comp} is chosen large enough that settling of the process output in closed loop is guaranteed. Over these N_{comp} samples, adaptation is still active, and u, y are recorded (light gray ellipse on the right). Afterward, within one time instant, the CM and AM simulations are performed with the current parameter sets. Then the "comparison of the models" is carried out by evaluating the following expression:

$$R_{\text{AM/CM}} = \frac{\sum\limits_{i=0}^{N_{\text{comp}}-1} \left(y(k-i) - \hat{y}_{\text{AM}}(k-i)\right)^2}{\sum\limits_{i=0}^{N_{\text{comp}}-1} \left(y(k-i) - \hat{y}_{\text{CM}}(k-i)\right)^2}, \tag{28.8}$$

where $\hat{y}_{\text{AM}}$ and $\hat{y}_{\text{CM}}$ denote the outputs of the AM and CM simulations, respectively. If $R_{\text{AM/CM}} < 1$, the AM parameters are transferred to the CM in state "AM $\longrightarrow$ CM." Otherwise, CM remains unchanged. Typically, the AM is better than the CM. The bad AM performance in Fig. 28.3a from 50 to 80 s is caused by a 50% change in the process gain at $t = 75$ s. The AM has already grasped the process change during adaptation between $t = 75, \ldots, 80$ s. The simulation shown in the left gray area runs with the latest available parameter sets from $t = 80$ s.

So far, the CM has been changed solely by the parameter transfer from the AM. The performance of the controller can be significantly enhanced if a limited adaptation is performed on the CM. Only the offset parameters $w_{i,0}$ of the local models are updated. In this case, there is no need for persistent excitation because the "blow-up" effect cannot occur. The major benefit of the offset adaptation for the CM is that time-variant static behavior of the process can be partially tracked.

28.2.3 Parameter Transfer

In the following, the realization of the parameters' transfer from the control model (CM) to the adaptation model (AM) will be discussed. Three different approaches are compared: hard switching without further adjustments, continuous fading over

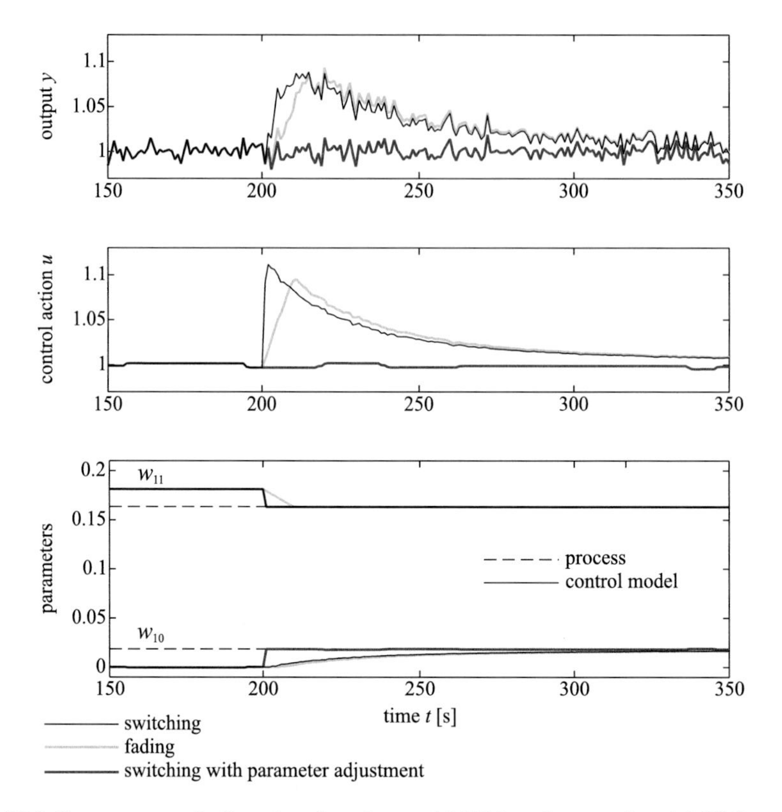

Fig. 28.4 Parameter transfer from the adaptation model (AM) to the control model (CM)

a given number of samples, and bumpless switching with model adjustment. The example in Fig. 28.4 demonstrates the properties of these approaches. A linear first-order system

$$y(k) = w_{11}u(k-1) - w_{12}y(k-1) + w_{10} \tag{28.9}$$

is closed-loop controlled to the reference value $r = 1$. At $t = 200$ s, the parameters of the adaptation model are transferred to the control model. The process gain of the adaptation model is 10% less than the gain of the control model. The hard switching approach can be simply described by

$$\underline{w}_i^{\mathrm{CM}}(k) = \underline{w}_i^{\mathrm{AM}}(k)\,. \tag{28.10}$$

All parameters of the AM are transferred to the CM at the current time instant. It can be seen in Fig. 28.4 that there is a large deviation in the system output. Gradually, the offset parameter w_{10} is adapted, and the steady-state error vanishes.

A straightforward modification is to fade in N_{fade} steps from one set of parameters to the other:

$$\Delta \underline{w}_i = \frac{1}{N_{\text{fade}}} \left(\underline{w}_i^{\text{AM}}(k) - \underline{w}_i^{\text{CM}}(k) \right) . \tag{28.11}$$

For the next N_{fade} samples, the parameter set of the CM is modified by linearly blending over from the AM to the CM:

$$\underline{w}_i^{\text{CM}}(k) = \underline{w}_i^{\text{CM}}(k-1) + \Delta \underline{w}_i . \tag{28.12}$$

The impact of the parameter transfer is still unsatisfactory. Even if N_{fade} is increased, the adverse effects on the control performance cannot be abolished.

An alternative approach is switching with parameter adjustment. The dynamic parameters are simply transferred from the AM to the new CM:

$$\left[w_{i,1}^{\text{CM,new}} \ w_{i,2}^{\text{CM,new}} \ \cdots \ w_{i,nx}^{\text{CM,new}} \right] = \left[w_{i,1}^{\text{AM}} \ w_{i,2}^{\text{AM}} \ \cdots \ w_{i,nx}^{\text{AM}} \right] . \tag{28.13}$$

The offset parameters $w_{i,0}^{\text{CM,new}}$ of the new CM are calculated such that the new CM model output equals the old CM model output:

$$w_{i,0}^{\text{CM,new}} + [x_1 \ \cdots x_{nx}] \begin{bmatrix} w_{i,1}^{\text{CM,new}} \\ \vdots \\ w_{i,nx}^{\text{CM,new}} \end{bmatrix} = w_{i,0}^{\text{CM,old}} + [x_1 \ \cdots x_{nx}] \begin{bmatrix} w_{i,1}^{\text{CM,old}} \\ \vdots \\ w_{i,nx}^{\text{CM,old}} \end{bmatrix} . \tag{28.14}$$

This guarantees a bumpless transfer from the AM to the CM, because in the stationary case, the predicted plant outputs do not change abruptly. The adjustment is performed in each local model separately following the local adaptation strategy discussed in Sect. 14.6.1. A global estimation of the adjustment would not even be possible since the resulting set of linear equations would be underdetermined (M unknown offsets but only one data sample).

28.2.4 Systems with Multiple Inputs

When systems with multiple inputs are adapted, all inputs including the measurable disturbances $\underline{n}$ must dynamically excite the process. If inputs do not excite the system, the estimation problem is ill-conditioned since the local RLS algorithm cannot distinguish between the offset parameter and the parameters associated with the non-exciting inputs. In practice, this occurs for quasi-static disturbances or inputs, respectively. Therefore, the excitation of each model input must be supervised individually. According to the current excitation, only a subset of $\underline{w}_i$ corresponding to the exciting model inputs is adapted. The excitation σ_s^2 of all model inputs $s(k) \in \{u(k),\ n_1(k),\ \ldots,\ n_m(k)\}$ will be defined as the windowed variance

$$\sigma_s^2(k) = \frac{K_{p,s}^2}{N_{\text{excite}}} \sum_{j=0}^{N_{\text{excite}}-1} (s(k-j) - \bar{s}(k))^2 \tag{28.15}$$

with

$$\bar{s}(k) = \frac{1}{N_{\text{excite}}} \sum_{j=0}^{N_{\text{excite}}-1} s(k-j)\,, \tag{28.16}$$

where N_{excite} is the length of the observation window. The expression (28.15) allows one to assess the degree of excitation from each input $s(k) \in \{u(k),\ n_1(k),\ \ldots,\ n_m(k)\}$. For the same reasons as in the control loss function (28.1) in Sect. 28.1, the variance has to be weighted with the gain $K_{p,s}$ from the input s to the model output $\hat{y}$. This gain can be obtained from a local dynamic linearization. The parameters associated with the most exciting model input $s(k)$ with the excitation $\max(\sigma_s^2(k))$ are always adapted. Furthermore, all other model inputs whose excitations are "sufficiently large" defined by exceeding a threshold $S_{\text{th}} = K_{\text{th}} \max(\sigma_s^2(k))$ are adapted, too. A typical choice is $K_{\text{th}} = 0.5$.

The strategy described above means that only a subset of the parameters of a local linear model $\underline{w}_i$ are adapted except in the rare case where all inputs are sufficiently exciting. If different subsets of the parameters are adapted, only the corresponding entries in the covariance matrices $\underline{P}_i(k)$ $(i = 1, \ldots, M)$ must be manipulated. When the adaptation is initialized, full covariance matrices $\underline{P}_{i,\text{all}}$ representing all parameters are generated. Every time the adaptation parameter sets are switched with regard to the excitation of the different inputs, new (smaller) covariance matrices $\underline{P}_{i,\text{curr}}$ are initialized, representing only the currently adapted parameter subset. For this purpose, the main diagonal entries corresponding to the currently, adapted parameters are transferred from $\underline{P}_{i,\text{all}}$ to $\underline{P}_{i,\text{curr}}$. The other entries of $\underline{P}_{i,\text{curr}}$ are set to zero. This procedure saves the variance information about the parameters and ensures positive definite matrices $\underline{P}_{i,\text{curr}}$. If $\underline{P}_{i,\text{curr}}$ becomes obsolete because a new excitation condition is detected, the parameter variances that are no longer needed in $\underline{P}_{i,\text{curr}}$ are transferred back to $\underline{P}_{i,\text{all}}$.

28.2.5 Experimental Results

The thermal plan shown in Fig. 25.16 is considered. The goal is to control the water outlet temperature $\vartheta_{\text{wo}}^{(2)}$ of the cross-flow heat exchanger by manipulating the fan speed F. The model built in Sect. 25.3.4 is used for adaptive nonlinear model predictive control. The model inputs are the fan speed $u = F$ (actuation signal) which commands the air stream $\dot{V}_a$, the water flow rate $n_1 = \dot{V}_w$ (first measurable disturbance), and the water inlet temperature $n_2 = \vartheta_{wi}^{(2)}$ (second measurable disturbance). Furthermore, the controlled variable depends on the air inlet temperature ϑ_{ai}, but as explained in Sect. 25.3.4, this quantity cannot be incorporated into a black box model because it cannot be actively excited. Variations of the environment temperature ϑ_{ai} thus influence the process behavior but are not reflected by the model. If high performance is to be achieved, this calls for control that adapts to these changes.

Figure 28.5 shows closed-loop control with the online adaptation of the fuzzy model. The process model was identified offline for an environmental temperature

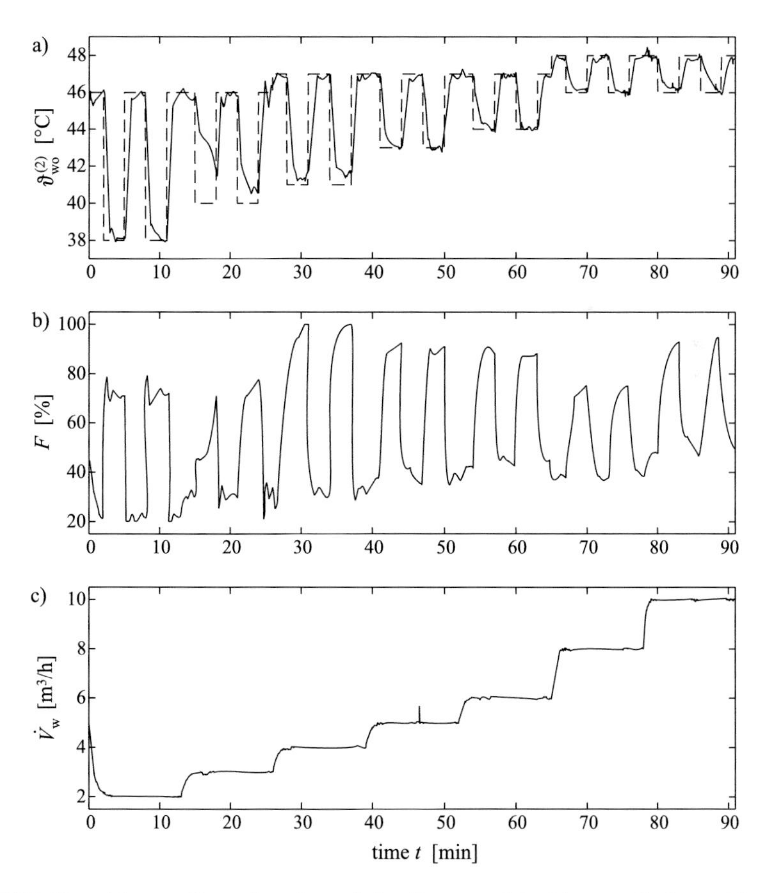

Fig. 28.5 Closed-loop online adaptation of the fuzzy model: (**a**) controlled variable water outlet temperature $\vartheta_{wo}^{(2)}$, (**b**) actuation signal fan speed F (which commands the air flow $\dot{V}_a$), (**c**) disturbance water flow rate $\dot{V}_w$

of $\vartheta_{ai} = 5\,°C$. The air temperature for the experiment in Fig. 28.5, however, is $\vartheta_{ai} = 15\,°C$.

In Fig. 28.5a the controlled variable $\vartheta_{wo}^{(2)}$ and the reference trajectory are plotted. The corresponding actuation signal F is shown in Fig. 28.5b. The control performance during the experiment is not satisfactory since the parameters are to be updated. The disturbance $\dot{V}_w$ shown in Fig. 28.5c covers the complete operating range from 2 to $10\,m^3/h$. The maximum power of the heat exchanger limits the range of the reference steps in Fig. 28.5a depending on the water flow rate. The flow profiles imply that all local linear models are activated during the adaptation process.

The adaptation of three of these LLMs is illustrated in Fig. 28.6. The first column of Fig. 28.6 depicts the validity function values Φ_i for the LLMs 1, 3, and 4, respectively. As can easily be seen in the second column, the forgetting factor λ_i drops to values less than 1 only if the corresponding local model is active. The third column shows the gains of the LLMs from the actuation signal $u = F$ to the output $y = \vartheta_{wo}^{(2)}$, which depend on the adapted parameters. (The gains are shown because this is more illustrative than the actual parameter values themselves.) They change significantly only if the forgetting factor is smaller than 1. In LLM 1 the adaptation is completely switched off during $t = 30-90$ min; the same is true for LLM 3 during $t = 50-90$ min. Note that during the first 30 minutes in LLM 4, the forgetting factor is smaller than 1 although this can be hardly discovered from the plot. Figure 28.7 shows offline simulation runs for the original model identified offline and the online adapted model obtained from the experiment described in Fig. 28.5. The input signals F and $\dot{V}_w$ are identical to the ones in Fig. 28.5b.

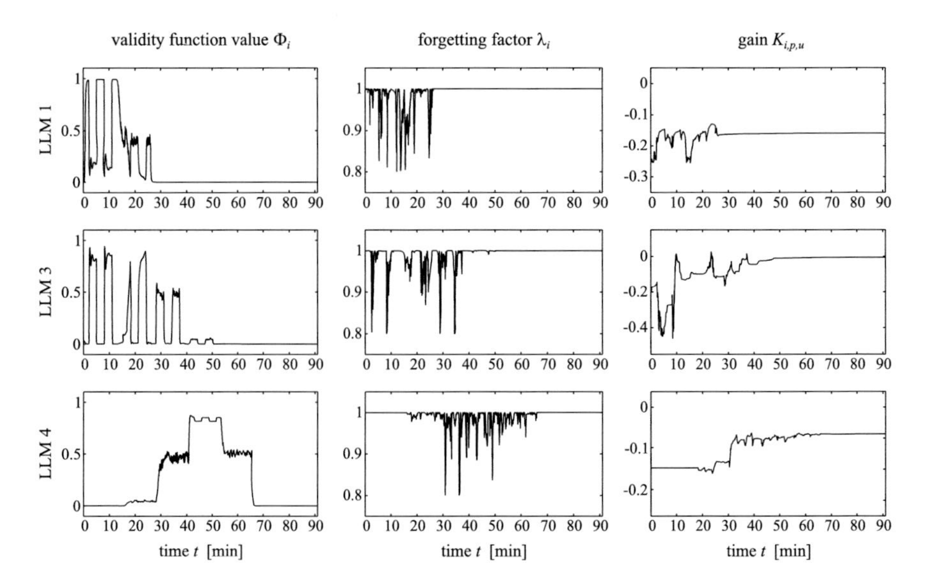

Fig. 28.6 Characteristic quantities during the local adaptation

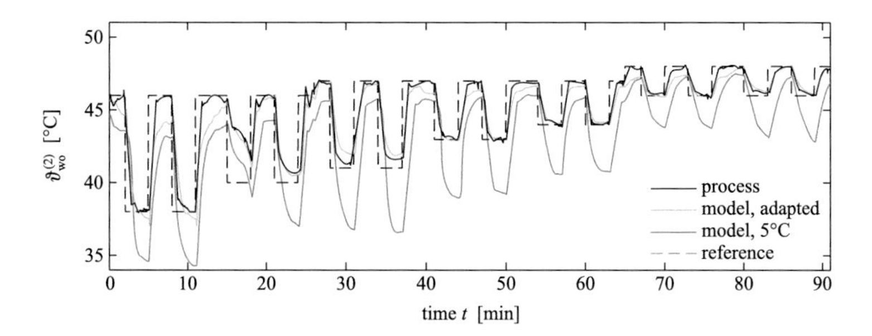

Fig. 28.7 Simulation of the original and the adapted model

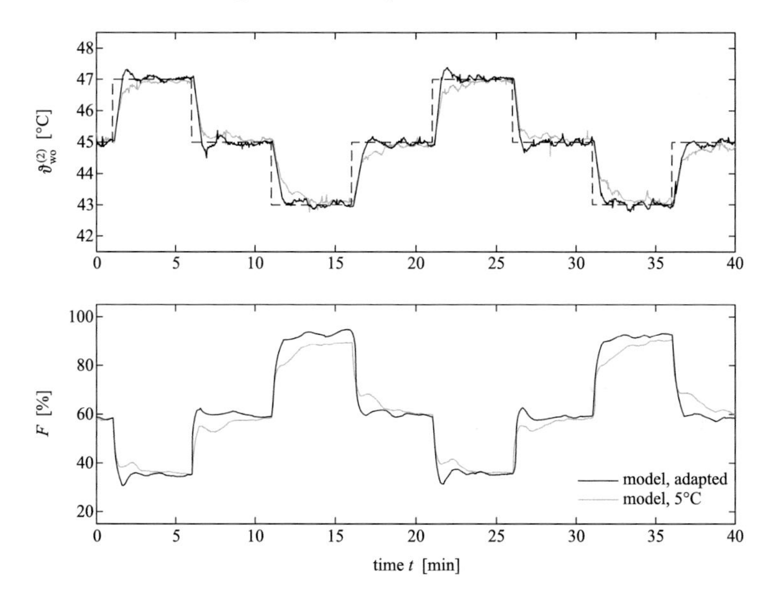

Fig. 28.8 Closed-loop control with the original and the adapted models: (**a**) controlled variable water outlet temperature $\vartheta_{wo}^{(2)}$, (**b**) actuation signal fan speed F

The simulation results for the offline and online estimated models are compared with the measured process output. It can be observed that the online adapted model performs significantly better than the offline counterpart. Since the reference signal in Fig. 28.5 contains only about two or three steps per local linear model, the adaptation obviously is very fast. Further improvement can be expected if the online adaptation is continued with more data.

In Fig. 28.8 the original and the adapted models are utilized for predictive temperature control. Obviously, the control performance with the online adapted model is better than with the offline estimated model. The process gain has decreased because the difference between the temperatures $\vartheta_{wi}^{(2)}$ and ϑ_{ai} has dropped (recall that ϑ_{ai} is now 15 °C compared with the 5 °C during training data acquisition). The original model pretends a gain that is too large, and therefore the controller acts weakly. Consequently, the controlled variable approaches the setpoint slowly. The adapted model is closer to the true process gain, which results in a faster closed-loop response. However, the discrepancy is not as dramatic in closed loop (Fig. 28.8) as in the process/model simulation comparison (Fig. 28.7). The reason for this is that the controller is robust against the process/model mismatch and can abolish steady-state errors.

28.3 Fault Detection

The task of each process fault detection and isolation (FDI) scheme is to monitor the process state, to decide whether the process is healthy and if not to identify the source of the fault.[3] Most schemes consist of two levels, a symptom generation part and a diagnostic part (Fig. 28.9). In the first one, symptoms are generated, which indicate the state of the process. It is a major challenge to generate significant symptoms that are robust against noise and disturbances. Modern approaches exploit the physical relationships between different process signals for the generation of significant symptoms. Compared with signal-based approaches, they can be derived from a process model (model-based approaches). Therefore deeper insights and understanding of the process are required, which in turn provides a higher depth of diagnosis, because more process knowledge can be exploited. One typical symptom type is the deviation between measured signals and predicted ones, the so-called output residuals $r(k)$ [107, 183]. They can be calculated independently from the process excitation and have the property of being close to zero in the fault-free case and significantly deviate from zero if a fault affects the process or the measurements used:

- $r(k) \approx 0$: fault-free case (noise effects, uncertainty).
- $|r(k)| \gg 0$: a fault has occurred.

In the case of MISO processes, a set of structured residuals can be designed, each dependent on different of inputs. Therefore, some inputs have no impact on specific residuals (decoupled residuals), and in the case of a faulty measurement, the decoupled residuals remain small, while all others are affected. The pattern of deflected and undeflected residuals indicates the possible source of the fault. Besides signal deviations, symptoms can be defined as time constants, static gains, or direct physical parameters of the process, which can be calculated only if the process is sufficiently excited.

[3] This section is based on research carried out by Peter Ballé, Institute of Automatic Control, TU Darmstadt.

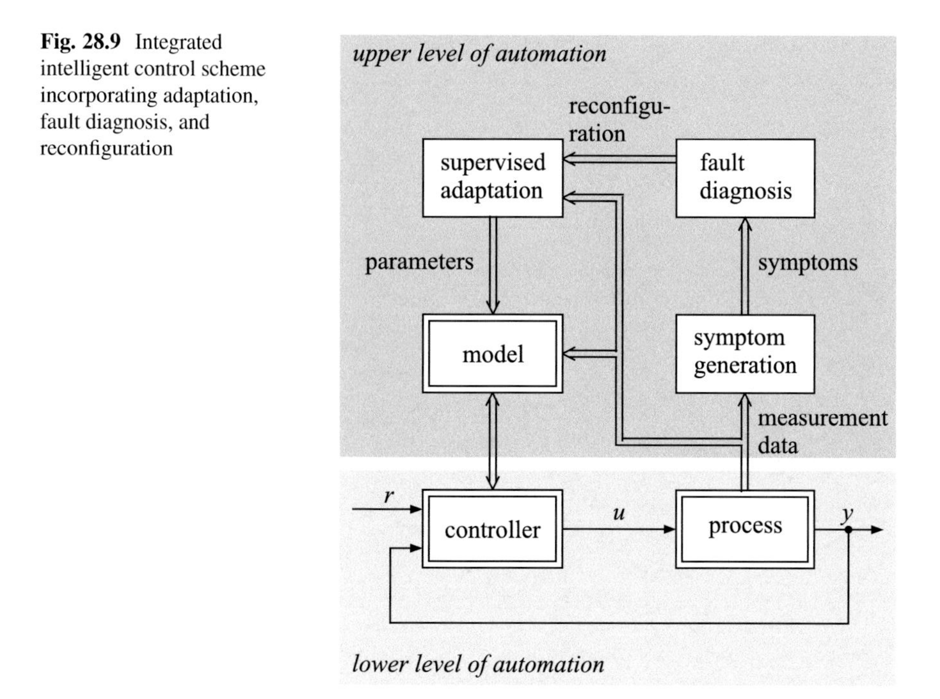

Fig. 28.9 Integrated intelligent control scheme incorporating adaptation, fault diagnosis, and reconfiguration

28.3.1 Methodology

The supervision of a nonlinear process is often very difficult to achieve owing to their complexity. In particular, model-based approaches that provide a high depth of diagnosis are difficult to implement and often require laborious modeling. Here, a multi-model approach is implemented based on black box local linear neuro-fuzzy models. This approach is suitable for all classes of processes that can be decomposed into several MISO subprocesses. A separate model for each subprocess is identified and used for the generation of symptoms. In this approach only the output residuals (difference between measured and simulated or predicted output) for each subprocess are used; see Fig. 1.1 and Sect. 1.1.7.

'The task is to generate a set of structured residuals in such a way that each residual depends on a different set of inputs and components of the process. Therefore, the pattern of deflected and undeflected residuals indicates the location of the fault. For the decomposition into different subprocesses, expert knowledge can be integrated that provides information about the general relations between the different process signals in the form of the following: $\hat{y}_1$ depends on (u_1, u_2, u_3), $\hat{y}_2$ depends on (u_2, u_3, u_4), $\hat{y}_3$ depends on (u_1, u_3, u_4), etc. (assuming a system with four inputs). Depending on the specific process, these relationships can be either static or dynamic and linear or nonlinear. In the following, nonlinear dynamic processes and subprocesses are investigated. The information can be used for the design of an incidence matrix containing all relationships between residuals and faults as shown in Table 28.1.

Table 28.1 Example for an incidence matrix

Fault in →	u_1	u_2	u_3	u_4
$r_1 = y_1 - \hat{y}_1$	1	1	1	0
$r_2 = y_2 - \hat{y}_2$	0	1	1	1
$r_3 = y_3 - \hat{y}_3$	1	0	1	1

1 = dependent, 0 = independent

The interpretable structure of the neuro-fuzzy model provides information about the sensitivity of the residuals to different faults. For example, consider an additive constant sensor fault Δu_j on an input signal u_j that affects a residual

$$r_f = r + \Delta r = y - (\hat{y} + \Delta \hat{y}) = \underbrace{y - \hat{y}}_{r} \underbrace{-\Delta \hat{y}}_{\Delta r} . \tag{28.17}$$

The deflection of the residual Δr caused by the fault is determined by deviations in the validity functions (rule premise part) and by the faulty input itself:

$$\Delta \hat{y} = f\left(\Delta \Phi_i, \Delta u_j\right) . \tag{28.18}$$

The deviation $\Delta \Phi_i$ in the validity functions is similar to a change of the model operating point that then deviates from the process operating point. This means that the fault causes different local linear models to be activated than actually should be. This is similar to a process fault in addition to a single sensor fault. The possible residual deflection can be evaluated by comparing the parameters of different rule consequents. This allows one to determine the maximum possible difference between model operating point and process operating point; refer to [26] for a more detailed description.

If the physical input u_j is contained only in the rule consequent vector $\underline{x}$ but not in the rule premise vector $\underline{z}$, then $\Delta \Phi_i = 0$. In spite of a fault, model and process are then run in the same operating regime. The residual deflection is determined by the fault size multiplied by the weighted parameters of the rule consequent. For an additive constant fault, the deflection in steady state is proportional to u_j multiplied by the coefficients associated with the $u_j(k-l)$ regressors. The proportionality factor will be called the *fault gain factor* K_f. This fault gain factor can be computed by

$$K_f = \sum_{i=1}^{M} \left(w_{i,a} + w_{i,a+1} + \ldots + w_{i,b}\right) u_j \Phi_i(\underline{z}) , \tag{28.19}$$

where $w_{i,a}, w_{i,a+1}, \ldots, w_{i,b}$ are the weights associated with the $u(k-l)$ regressors. Note that (28.19) may hold approximately even if u_j enters the premise input space spanned by $\underline{z}$ when the change of operating point is negligible. However, (28.19) is in any case valid only in steady state.

Table 28.2 Modified incidence matrix from Table 28.1

Fault in →	u_1	u_2	u_3	u_4
$r_1 = y_1 - \hat{y}_1$	1	1	1	0
$r_2 = y_2 - \hat{y}_2$	0	1	1	1
$r_3 = y_3 - \hat{y}_3$	–	0	1	1

1 = dependent, 0 = independent

Evaluating the fault gain factors for each residual provides information on how sensitive the residuals are. Large K_f lead to sensitive residuals that allow the detection of small faults. Small K_f indicate a low sensitivity of the residual with respect to the considered fault and thus only allow the detection of large faults. Note that the factors can be evaluated for each local linear model separately:

$$K_f^{(i)} = \left(w_{i,a} + w_{i,a+1} + \ldots + w_{i,b}\right) u_j , \quad K_f = \sum_{i=1}^{M} K_f^{(i)} \Phi_i(\underline{z}) \, . \tag{28.20}$$

Strongly different $K_f^{(i)}$ for the same fault indicate that the sensitivity is highly dependent on the operating point, i.e., some faults may be easily detectable in one operating regime but not in another.

This information about the $K_f^{(i)}$ and K_f can be exploited to modify the incidence matrix. Basically, the information content of a residual for the detection of a specific fault depends on the ratio between its K_f and the standard deviation of the residual in the fault-free case. If this ratio becomes too small, a reliable detection of the fault cannot be based on this residual. Thus, only the most sensitive residuals are used for FDI. If the structure is no longer isolating, less sensitive residuals are also evaluated. For example, assume the residual r_3 in the above example is not very sensitive to faults in u_1. Then this residual should not be used for fault isolation, and the modified incidence matrix is shown in Table 28.2. The structure in Table 28.2 would still be suitable for the isolation of each sensor fault.

28.3.2 Experimental Results

The task is to supervise the temperature and flow rate sensors of the secondary circuit (cross-flow heat exchanger) of the thermal plant shown in Fig. 25.16, the valve position, and the fan speed signal. If a fault is detected and isolated, the faulty signal is no longer used, and the control level is reconfigured; see Sect. 28.5.

The following subprocesses have been selected for the multi-model approach and have been identified with LOLIMOT as local linear neuro-fuzzy models: fan → r_1, pipe 1 → r_2, pipe 1 + valve → r_3, pipe 2 → r_4, and valve → r_5. Tables 28.3 and 28.4 summarize the specification of these five submodels.

The five output residuals generated by the difference between the measured signals and the output of the corresponding submodels provide sufficient analytical

Table 28.3 Summary of the neuro-fuzzy submodels identified for fault detection with LOLIMOT

Model	Physical inputs	Physical output	Train/test error
1	F / $\dot{V}_{\text{w}}$ / $\vartheta_{\text{wi}}^{(2)}$	$\vartheta_{\text{wo}}^{(2)}$	0.006/0.020
2	$\dot{V}_w$ / $\vartheta_{\text{wo}}^{(2)}$	$\vartheta_{\text{wo}}^{(1)}$	0.002/0.006
3	G_{w} / $\vartheta_{\text{wo}}^{(2)}$	$\vartheta_{\text{wo}}^{(1)}$	0.002/0.006
4	$\dot{V}_w$ / $\vartheta_{\text{wi}}^{(1)}$	$\vartheta_{\text{wi}}^{(2)}$	0.002/0.008
5	G_{w}	$\dot{V}_{\text{w}}$	0.001/0.001

Table 28.4 Rule premise and consequent regressors for models in Table 28.3

Model	Premises	Consequents
1	1 / 1 / 1 / –	1, 2, 3 / 3, 7, 11, 15 / 2, 4, ..., 22 (OLS) / 1
2	1 / – / –	– / 2, 4, ..., 34 (OLS) / 1
3	1 / – / –	– / 2, 4, ..., 34 (OLS) / 1
4	1 / – / –	– / 2, 4, ..., 26 (OLS) / 1
5	1, 2 / –	1 / 1

The numbers represent delays used for the inputs and outputs given in Table 28.3. The last entry "/ .." corresponds to the delays of a possibly fed back output. "(OLS)" indicates that structure selection was used for dead time determination

Table 28.5 Incidence matrix for heat exchanger

	G_{w}	$\dot{V}_{\text{w}}$	F	$\vartheta_{\text{wi}}^{(2)}$	$\vartheta_{\text{wi}}^{(1)}$	$\vartheta_{\text{wo}}^{(2)}$	$\vartheta_{\text{wo}}^{(1)}$	Valve	Pipe 1	HE_2	Pipe 2	Threshold
r_1	0	1	1	1	0	1	0	0	0	1	0	1.5 °C
r_2	0	1/–	0	0	0	1	1	0	0	0	1	1.1 °C
r_3	1	0	0	0	0	1/–	1/–	1	0	0	1	1.1 °C
r_4	0	1/–	0	1	1	0	0	0	1	0	0	0.9 °C
r_5	1	1	0	0	0	0	0	1	1	0	1	0.5 m^3/h

"/–" = excluded for the modified incidence matrix

redundancy for fault isolation. They are sensitive to faults in the sensor signals and also to faults in the respective subprocesses. If, for example, the behavior of the valve changes, the residuals r_3 and r_5 are deflected. But it is not possible to make a further decision on whether this deflection is caused by corrosion of the valve plug, by partial clogging in the pipe, or by a leak in the pneumatic air supply. Hence, process faults can be detected, but only the faulty components of the process can be isolated. The structure of the residuals is shown in the incidence matrix in Table 28.5.

Faults 4–9 affect only the rule consequent part, while faults 1–3 also affect the rule premise part of the fuzzy model. Note that the residuals r_3 and r_5 both provide information to isolate flow rate sensor faults and valve position faults, and therefore the consideration of one of these two residuals is sufficient. Preferably r_3 would be omitted because fault gain factors K_f are lower, as can be seen from the unused entries in the modified incidence matrix "/–" in Table 28.5.

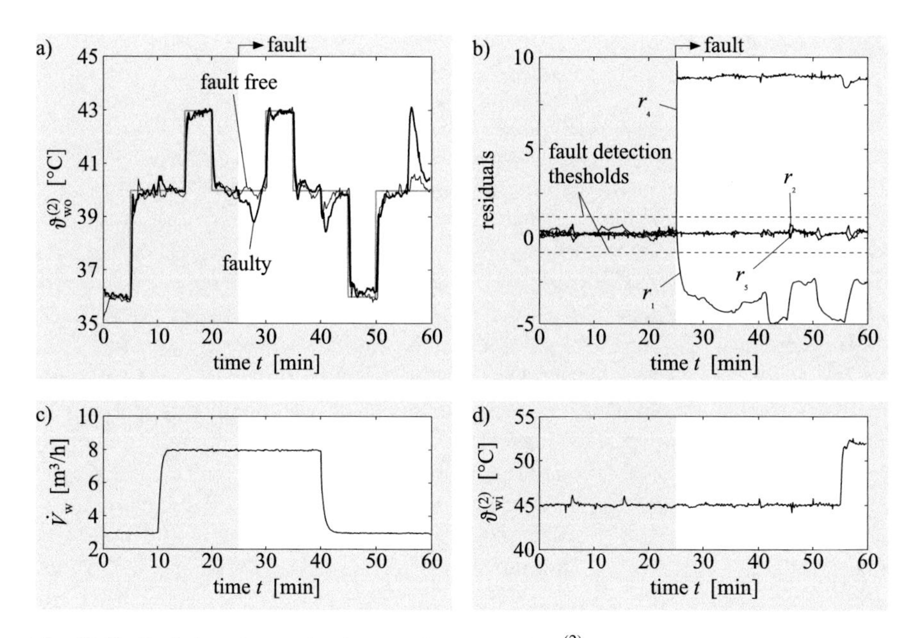

Fig. 28.10 Fault detection results. A sensor fault occurs in $\vartheta_{wi}^{(2)}$ at $t = 25$ min. The residuals r_1 and r_4 are deflected. Thus, the correct fault cause can be concluded from Table 28.5. The plots show the true (fault free) signals: (**a**) control performance, (**b**) residuals, (**c**) water flow rate, (**d**) true inlet temperature (different from the faulty sensor signal)

After evaluating the sensitivity and also during the training of the classification algorithm, it turned out that the residual r_5 is needed for isolation of faults $\dot{V}_w$ and F. Therefore, r_3 is not used in the final FDI scheme. Based on calculations for different operating points using (28.20), the smallest detectable sensor fault size can be determined in all regimes of operation. The sensitivity for faults on the temperature sensors is high, and faults as low as 2 °C can be detected. The sensitivity for flow rate, fan speed, and valve position signal is lower. Faults in F must be larger than 10%, faults in G_w higher than 15%, and faults in $\dot{V}_w$ higher than 2 m³/h to be isolated in all regimes of operation. Nevertheless, the detection of smaller faults and also isolation in certain regimes of operation are still possible.

Figure 28.10 illustrates the results for a larger fault in $\vartheta_{wi}^{(2)}$. Table 28.5 allows one to detect correctly a fault in the $\vartheta_{wi}^{(2)}$ temperature sensor because the residuals r_1 and r_4 are deflected. The residual r_4 reacts very quickly (after a dead time corresponding to the transport from $\vartheta_{wi}^{(1)}$ to $\vartheta_{wi}^{(2)}$) because the output of the submodel 4 is directly the "faulty" temperature $\vartheta_{wi}^{(2)}$. The residual r_1 reacts slower because one input of the submodel 1 is affected by the fault. The effect of this fault thus has to pass the dynamics of the cross-flow heat exchanger until it can be discovered by a comparison between the measurement of $\vartheta_{wo}^{(2)}$ and the output of the submodel 1.

28.4 Fault Diagnosis

When supervising technical processes, it is common to include a fault diagnosis that evaluates the generated symptoms; see Fig. 28.9.[4] In some cases, an explicit symptom generation is actually omitted, and process measurements act directly as the inputs for the diagnosis. A typical output of the diagnosis comprises information about which fault has occurred as well as possible alternative fault causes. The diagnostic system serves the following purposes: In the existence of noise and model inaccuracies, it identifies the optimal decision boundaries between the different fault situations. Additional external information sources (such as observations from an operator) can be exploited. A diagnosis can provide insight into the relevance and performance of symptoms that are used. Finally, the system should make the fault decision transparent. This increases the acceptance of the system and allows later changes to be easily applied. Common approaches are fault symptom trees and fuzzy rule bases that are constructed manually. Systems based on measurement data use multilayer perceptron networks in most cases. Occasionally self-organizing maps, ART networks, and simple clustering techniques are implemented. Radial basis function networks, regression trees, and neuro-fuzzy approaches have also gained more attention. Each of the methods has special advantages and disadvantages associated with it, and a selection of the appropriate method has to be carefully made. For comparison and overview, refer to [544].

28.4.1 Methodology

An approach that combines the learning ability of neural networks with the transparency of a fuzzy system is the SElf-LEarning Classification Tree (SELECT) proposed by Füssel [176]. This procedure relies on measured data to create a diagnostic tree consisting of neural decision nodes. The resulting tree can be augmented with a priori knowledge, and finally, some parameters allow fine-tuning by optimization algorithms. The procedure consists of the following steps:

1. Creation of membership functions, e.g., by unsupervised clustering.
2. Selection of a fuzzy rule for the easiest separable fault.
3. All data belonging to the fault situation that can be diagnosed with the rule from Step 2 is removed from the training set. Thereby, a new data set is created containing one class less than before. The algorithm now returns to Step 2 and iteratively selects new rules for the other fault classes. This procedure creates a sequential classification tree where different fault possibilities are tested one by one.

[4] This section is based on research undertaken by Dominik Füssel, Institute of Automatic Control, TU Darmstadt.

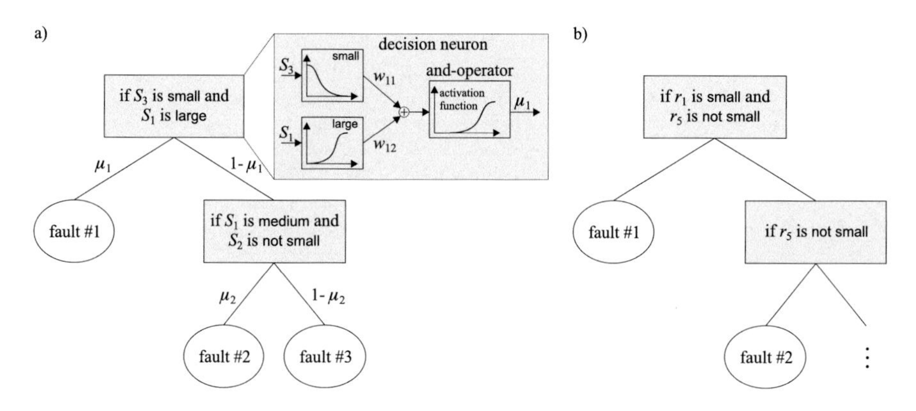

Fig. 28.11 (**a**) Structure of the self-learning classification tree (SELECT). (**b**) Upper part of the tree built by SELECT for fault diagnosis

4. Adding a priori known rules to the top of the tree.
5. Optimization of parameters yielding optimal classification accuracy.

Figure 28.11a shows an example of the tree structure that evolves from the SELECT procedure. The nodes of the tree are conjunctional operators (AND operators) that are implemented as artificial neurons. The operator performs a weighted sum of the fuzzy membership values $\widetilde{\mu}_j$ $(j = 1, \ldots, p)$ and uses a sigmoidal function to generate the output of the neuron

$$\mu_i = \frac{1}{1 + \exp\left(-\frac{2}{p} \sum_{j=1}^{p} w_{ij}\widetilde{\mu}_j + \alpha_i\right)}, \tag{28.21}$$

where w_{ij} represents the relative importance of the corresponding rule premise, p is the number of inputs to the rule, and α_i is an offset parameter. Note that (28.21) can be understood as a weighted and "soft" (the max-function is replaced by a sigmoid) version of the bounded difference operator; see Sect. 12.1.2. Successive outputs down the tree are multiplied by $1-\mu_i$. One easily realizes that the AND operator can only perform a linear separation in the space of the membership values. This might appear to be a strong drawback of the method. One has to consider, nevertheless, that this translates into a nonlinear separation in the original feature space, and furthermore the sequence of linear decision boundaries in the tree leads to an overall nonlinear boundary. The optimization normally involves the relevance weights w_{ip} only, but the membership functions or other parameters of the operator can also be optimized. For that purpose a nonlinear least squares approach is chosen (Levenberg-Marquardt); see Sect. 4.5.2. The optimization and rule building phases are usually performed offline. The advantages of this system are as follows:

1. *Transparency*: The decision neuron is implemented in such a way that the individual decisions can be interpreted. The use of fuzzy membership functions also supports this interpretation.
2. *Speed*: The speed of the rule building as well as of the optimization is very high. A decreasing data set allows an acceleration of the rule generation. Furthermore, the optimization can be implemented in a sequential manner because the weights of each neuron can be trained individually instead of in parallel as in most other neural network structures.

Thereby, the dimensionality of the search space is greatly reduced, and also problems associated with the curse of dimensionality are lessened; see Sect. 7.6.1. For a detailed discussion of the SELECT procedure, refer to [176].

28.4.2 Experimental Results

Different fault events were simulated by adding wrong sensor information of varying magnitudes to experimental data. Table 28.6 gives an overview of the faults and fault sizes considered. The diagnostic system is activated when at least one of the residuals has exceeded its threshold, i.e., a fault has occurred. It remains the task to identify which of the fault causes is most likely. The classification tree was originally constructed using the residuals r_1 through r_4. A mean value of these residuals over a certain range of operations was used as inputs for the diagnosis. The system did nevertheless fail to reliably distinguish the faults at the fan speed sensor F and flow rate sensor $\dot{V}_{\mathrm{w}}$. The overall achievable classification rate was therefore just below 90%. The reason was found to be the low sensitivity of r_2 and r_4 to the fault at the flow rate sensor, which made the residual deflections of the two faults look very similar. While in simulation runs the deflection was still visible, it was hidden under the normal measurement noise when using real measured data. Perceiving that problem leads to the introduction of the additional residual r_5. The new classification tree was trained with r_1, r_2, r_4, and r_5. Figure 28.11b pictures the top of the tree.

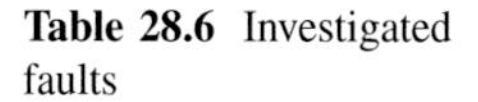

Table 28.6 Investigated faults

Faults	Fault size
G_{w}	$\pm 20, \ldots, 50\%$
$\dot{V}_{\mathrm{w}}$	$\pm 2, \ldots, 5\,\mathrm{m^3/h}$
F	$\pm 20, \ldots, 50\%$
$\vartheta_{\mathrm{wi}}^{(2)}$	$\pm 6, \ldots, 15\,°\mathrm{C}$
$\vartheta_{\mathrm{wi}}^{(1)}$	$\pm 6, \ldots, 15\,°\mathrm{C}$
$\vartheta_{\mathrm{wo}}^{(2)}$	$\pm 6, \ldots, 15\,°\mathrm{C}$
$\vartheta_{\mathrm{wo}}^{(1)}$	$\pm 6, \ldots, 15\,°\mathrm{C}$

Considering the incidence matrix in Table 28.5 makes the rules being picked by SELECT understandable. It is also interesting to note that not all residuals are used in every decision node. Those chosen seem to be sufficient and provide a good separation of the faults. The small rule set, on the other hand, makes the system comprehensible. By examining the individual rules, it can further be seen that the residual r_1 was most often used. It obviously has the most discriminatory power.

The resulting tree was able to classify all generalization data correctly. It must be said, however, that smaller fault magnitudes than those considered will increase the misclassification rate.

28.5 Reconfiguration

Autonomous control systems require a reconfiguration module in order to cope with undesirable effects such as disturbances, altering process behavior, or faults. Reference [477] defines control reconfiguration to apply in the following three situations: establishment of the system's operating regime, performance improvement during operation, and control reconfiguration as part of fault accommodation. The reconfiguration of the controller is typically performed by adaptation in the broad sense. It can be divided into direct and indirect adaptation [282]. While in the former case the controller parameters are directly manipulated, in the latter case, a process model is updated. Besides the continual adaptation of a single model, multiple-model control is a common approach [405]. Here, the purpose of applying reconfiguration is to maintain safe operation and reasonable control performance in the presence of faults. In particular, it is investigated how the fuzzy-model-based predictive controller from the previous section can be adjusted to sensor faults. Distorted measurements enter the controller in two different ways (see Fig. 28.2): via the parameter estimator and via the corrected fuzzy model prediction. A fault in the sensor of the process output y influences the online parameter estimation of the fuzzy model. Hence, the feedback component of the controller will not function properly anymore. This also happens if the sensors of the measurable disturbances $\underline{n}$ are faulty.

Once the fault detection and isolation (FDI) scheme has isolated a sensor fault, appropriate reconfiguration actions must be launched. The most obvious approach is the reconfiguration of the respective sensor signal by means of an observer-based estimation. Here, the outputs of nonlinear observers replace the sensor information and are fed into the adaptive NMPC. This method is highly suitable if observers are already used for fault detection purposes. In the sequel, an approach tailored to the local linear neuro-fuzzy models and the adaptive NMPC is presented. In the case of an incorrectly sensed process output, only the parameter estimator has to be switched off. Since the neuro-fuzzy model is run in parallel to the process, the measured output y is not used for prediction. Hence, the NMPC works as a pure feedforward controller. If the diagnosis level recognizes a fault in one of the disturbance sensors, the measured value is not reliable anymore, and therefore it is fixed to a value that had been measured before the error occurred. The effect of freezing the input depends on whether it is only an entry of the consequents' input vector $\underline{x}$ or whether it is also an entry of the premises' input vector $\underline{z}$. In the parameter estimator, the characteristic of the entries of $\underline{x}$ changes from a measurable to a non-measurable disturbance, which is from then on compensated by the adaptation of the offset parameters $w_{i,0}$. In contrast, neglecting changes in entries of $\underline{z}$ degrades the nonlinear approximation capabilities of the model. This can be easily understood from the extreme case when the complete vector $\underline{z}$ is fixed. Then the validity functions Φ_i have constant values, and the fuzzy model acts as a purely linear model whose offset is adapted by the parameter estimator. Because of this one might argue that it is advantageous to adapt not only the offset parameters of the local linear models but also the dynamic parameters. This can clearly be done if persistent excitation is guaranteed or the forgetting factors of the estimators are adjusted to the current degree of excitation as proposed in Sect. 28.2.

Chapter 29
LMN Toolbox

Abstract This chapter introduces a MATLAB-based toolbox for local model network, which has been implemented in an object-oriented manner. The LOLIMOT and HILOMOT algorithms are available. Simple settings are discussed, such as the choice of the termination criterion, specifying the polynomial degree of the local models, defining the input spaces for rule premises and consequents (local models), and the smoothness parameter. Furthermore, data weighting and visualization of the model performance are explained.

Keywords Toolbox · Implementation · Local model network · MATLAB · Simplicity · Download

This chapter briefly introduces a local model network (LMN) toolbox in MATLAB. It can be downloaded from

http://www.mb.uni-siegen.de/mrt/lmn-tool/

for non-commercial use. The two algorithms for building local model networks with axis-orthogonal partitioning (LOLIMOT) and with axis-oblique partitioning (HILOMOT) introduced in Chaps. 13 and 14 are implemented.

Both algorithms incrementally construct a tree with local models as their leaf nodes. However, the obtained model structures are flat with LOLIMOT and hierarchical with HILOMOT. While LOLIMOT ensures the partition of unity by normalizing the membership functions (defuzzification step), HILOMOT hierarchically ensures that each split maintains the partition of unity by utilizing complementary splitting functions Ψ and $1 - \Psi$. LOLIMOT uses local membership functions (Gaussians), while HILOMOT uses S-shaped splitting functions (sigmoids). Note that the exact functional from of these functions is not very relevant; just their principal shape and their smoothness is important.

O. Nelles, *Nonlinear System Identification*,
https://doi.org/10.1007/978-3-030-47439-3_29

The toolbox is implemented in MATLAB in an object-oriented manner. First an object corresponding to a tree of local model networks needs to be established for either axis-orthogonal or -oblique partitioning by

$$\texttt{lmn} = \texttt{lolimot}$$

or

$$\texttt{lmn} = \texttt{hilomot}\,.$$

Then the networks can be trained with the default settings by

$$\texttt{lmnTrained} = \texttt{lmn.train}$$

resulting in a tree of trained networks `lmnTrained`. Afterward, the output of the trained model can be calculated by

$$\texttt{ModelOutput} = \texttt{lmnTrained.calculateModelOutput(input)}$$

or

$$\texttt{ModelOutput} = \texttt{lmnTrained.calculateModelOutput(input, output)}$$

where `output` is optional and only used for one-step prediction in dynamic models (if `kStepPrediction=0`) and initialization in case of simulation (if `kStepPrediction=inf`).

In the following, the most important settings to influence the training are explained. These settings can be altered from their default value by

$$\texttt{lmn.property} = \texttt{new value}$$

before training.

29.1 Termination Criteria

One decisive factor for the quality of a model is an appropriate bias/variance tradeoff. The toolbox offers a variety of criteria for finding a good model complexity. In particular, these are information criteria (AIC_C and BIC_C) and performance on validation data. In addition, the model complexity can be directly controlled by the maximal number of local models or effective number of parameters or indirectly controlled by choosing a maximal training time.

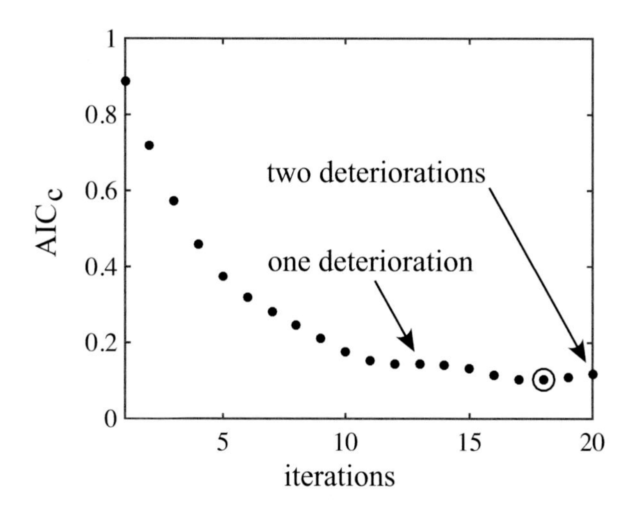

Fig. 29.1 Convergence of the AIC_C criterion. The model corresponding to the global optimum (here at $M = 18$) shall be selected. Training is terminated after the criterion increases two times (by default)

29.1.1 Corrected AIC

The Akaike information criterion (AIC) corrected for finite data sets called AIC_C is the most popular choice; compare Sect. 7.3.3 and [3, 87]. For the number of parameters, the *effective* not the nominal number is used; compare Sect. 29.1.5. The goal is finding the network associated with the global minimum of the information criterion. This model is suggested by the toolbox.

At some point, the incremental training has to be terminated to keep the computational demand low. Sometimes random fluctuations on the convergence curve can be observed that are due to noise and suboptimalities in the training algorithm. Therefore it is advisable to NOT terminate training after the *first* deterioration of the criterion. Rather a couple of subsequent deteriorations are required for termination; see Fig. 29.1. The default number is

$$\texttt{maxValidationDeterioration} = 2.$$

29.1.2 Corrected BIC

The Bayesian information criterion (BIC) corrected for finite data sets called BIC_C exhibits a larger parameter penalty than the AIC; compare Sect. 7.3.3 and [3, 87]. Therefore it yields simpler models. This may be advisable for scenarios with huge training data sets or whenever the motivation for simple models is higher than normal. These circumstances are often fulfilled for dynamic models since typically much more data can be gathered in the same amount of time than with static measurements.

Note that many motivations exist for simpler models than the best bias/variance tradeoff which can be seen as the upper threshold for model complexity. These additional motivations include:

- Computational demand (computing time and memory) during training and use
- Interpretability
- Robustness

In order to force even simpler models than those suggested by the BIC_C, the following property allows to increase a factor multiplied with the parameter penalty of the AIC_C or BIC_C criterions, i.e.,

$$\texttt{complexityPenalty} = 1\,.$$

29.1.3 Validation

If plenty of data is available, it may be the best choice to generate a separate representative validation data set. Instead of the information criteria discussed above, then the performance on validation data is monitored in exactly the same manner.

Note that the user possesses two alternative procedures

1. Validation data: Splitting the data into training and validation data. Both should be representative. Determining the model complexity on the validation data performance.
2. No validation data: Using all data for training. Determining the model complexity on an information criterion.

Procedure 2 offers the potential for better models (more training data), but this comes with a worse choice of model complexity (no validation data).

29.1.4 Maximum Number of Local Models

The number of local models or neurons or rules M can be used as termination criterion. Note that a minimum number of data is requested for each local model. By default it is required that the number of data points within each local models is at least as big as the number of parameters to be estimated. Because the validity functions overlap this requirement is formulated in a fuzzy way

$$n_i \leq \sum_{j=1}^{N} \Phi_i(j) \tag{29.1}$$

where n_i is the number of parameters in local model i and $\Phi_i(j)$ is the validity of data point j in the local model i under investigation. If (29.1) is violated then this split of the local model is not allowed.

In case of LOLIMOT, other splits of the same model may be possible. If none of the splits of the worst local model is possible, then the next worse local model is considered for splitting. Therefore a reasonable limit on the choice of $M_{\max}$ is given by the data density and distribution.

In case of HILOMOT, the split optimization is constrained with (29.1); thus it is always met. However, local models which violate (29.1) for all initial splits even *before* splitting optimization starts are locked, and the next worse local model is split.

29.1.5 Effective Number of Parameters

The effective number of parameters gives a good indication of the flexibility of the model. For LOLIMOT just the effective number of parameters of the local models are summed up; see also (13.27) in Sect. 13.2.2

$$n_{\mathrm{eff}} = \sum_{i=1}^{M} n_{\mathrm{eff},i} \tag{29.2}$$

with

$$n_{\mathrm{eff},i} = \mathrm{tr}\{\underline{S}_i\} \tag{29.3}$$

where $\underline{S}_i$ is the smoothness matrix of local model i. There is some discussion whether the effective number of parameter should be calculated as $\mathrm{tr}\{\underline{S}_i\}$ or $\mathrm{tr}\{\underline{S}_i^T\,\underline{S}_i\}$ but the statistics literature seems to agree on $\mathrm{tr}\{\underline{S}_i\}$.

The degrees of freedom contained in the flexibility of the partitioning determined by the centers and standard deviations of the Gaussians are neglected. This can be justified by the coarse LOLIMOT algorithm which delivers far from optimal partitioning.

For HILOMOT the partitioning is much more flexible, and the splitting parameters are numerically optimized. Thus, in addition to the local model parameters in (29.2), the number of splitting parameters needs to be considered. For each split there is one splitting parameter per dimension (in the space of the validity functions spanned by $\underline{z}$) plus the offset. However, the smoothness of the sigmoid is not optimized but rather normalized and determined by a heuristic smoothness adjustment. Thus each split possesses n_z independent splitting parameters; compare Sect. 14.7.1. Since for an LMN with M local models there are $M - 1$ splits, this yields

$$n_{\mathrm{split}} = n_z(M-1)\,. \tag{29.4}$$

Thus a HILOMOT trained LMN with M local models is considered more flexible than a LOLIMOT trained one. This discrepancy increases if the validity function input space dimensionality n_z grows.

These effective number of parameters also enter the AIC_C and BIC_C criteria mentioned above. Note that these are very rough estimates. The splitting parameters are not optimized concurrently which would make the model much more flexible. Thus, (29.2)+(29.4) is an overestimation of HILOMOTs flexibility.

29.1.6 Maximum Training Time

For huge models that may be generated for large training data sets, it can be reasonable to specify the maximum training time [in seconds] which can be utilized as termination criterion. In practice, this feature can be used for generating the best possible results over night, for example.

29.2 Polynominal Degree of Local Models

The local models in the LMN toolbox can be polynomials of any order. If the order of the polynomials and/or input space dimensionality nx grows, then the number of parameters increases rapidly. Therefore, good tradeoffs typically are low-order polynomials. The most frequently chosen local model types certainly are

- *Linear* models: The default choice. The number of parameters per local model is $nx + 1$. They offer particular advantages for dynamic models due to their extrapolation behavior which maintains the dynamics in extrapolation.
- *Sparse quadratic* models: This refers to quadratic models without cross-product (or interaction) terms, i.e., in the two-dimensional case

$$y = w_0 + w_1 u_1 + w_2 u_2 + w_3 u_1^2 + w_4 u_2^2 \tag{29.5}$$

  without $w_5 u_1 u_2$. The number of parameters per local model is $2nx + 1$, i.e., grows only linearly with nx. Nevertheless they are able to describe a minimum or maximum without the support of neighboring local models which is a clear benefit for non-monotonous models. In contrast, local linear models need to change sign in each dimension to describe a minimum or maximum. This is difficult to achieve, particularly in high dimensions. Furthermore, it is sensitive with respect to the exact shape of the validity functions which makes the model less robust. Therefore sparse quadratic models are a good choice in these cases. Contrary to full quadratic models they can be used even for large nx.
- *Full quadratic* models: The number of parameters per local model is $(nx + 2)(nx + 1)/2$. Certainly the best choice if optima of the model are of interest. They can be interpreted as a nonlinear extension of Newton's method for optimization where local quadratic models are estimated in certain areas

of the input space; compare Sect. 4.4.4. However, the number of parameters grows quadratically with nx which limits their applicability to low-dimensional problems.

Higher-order polynomials usually are recommended only if motivated by prior knowledge about the process. Of course, subset selection techniques like orthogonal least squares or lasso allow to extend the limits by local regressor selection; compare Sect. 14.3.

29.3 Dynamic Models

In order to build dynamic models, the delays of the inputs and possibly outputs must be specified. To allow for a maximum of flexibility, not just the dynamic order (and dead time(s)) can be chosen but specifically every delay for every network input. This also be done individually for the regressors of the local models (or rule consequents) gathered in $\underline{x}$ and those for the validity functions (or rule premises) gathered in $\underline{z}$; compare Sect. 14.1.

Pure feedforward structures of NFIR type just require delays of inputs. They are specified in the properties `xInputDelay` for $\underline{x}$ and `zInputDelay` for $\underline{z}$.

Example 29.3.1 NFIR of Fourth Order
The model is

$$\hat{y}(k) = f\left(u(k-1), u(k-2), u(k-3), u(k-4)\right) . \tag{29.6}$$

This require in the general form $\underline{x} = \underline{z}$ and thus

$$\texttt{xInputDelay} = \texttt{zInputDelay} = \{[1\ 2\ 3\ 4]\} .$$

The "{ }" indicates a cell array.

The delayed output does not enter either $\underline{x}$ or $\underline{z}$ and thus

$$\texttt{xOutputDelay} = \texttt{zOutputDelay} = \{[\ ]\} .$$

It may be sufficient to provide the validity functions (rule premises) just with the level of $u(k)$, not with any dynamics (like first and/or second derivate of $u(k)$). Then it would be sufficient to choose

$$\texttt{zInputDelay} = \{[1]\}$$

which makes the whole problem significantly simpler going from 4D to 1D still maintaining the fourth order in the local models.

□

Example 29.3.2 Dead Time
A fourth-order model with dead time d is

$$\hat{y}(k) = f\,(u(k-1-d), u(k-2-d), u(k-3-d), u(k-4-d)) \tag{29.7}$$

which translates in the following delays

$$\texttt{xInputDelay} = \texttt{zInputDelay} = \{[1+d \quad 2+d \quad 3+d \quad 4+d]\}\,.$$

□

Example 29.3.3 MISO NFIR
For system with multiple inputs, it is possible to specify the delays individually because the properties are cell arrays and thus can have individual lengths for each input. Assume the following model:

$$\hat{y}(k) = f\,(u_1(k-1), u_1(k-2), u_1(k-3),\; u_2(k-1), u_2(k-2))\,. \tag{29.8}$$

For $\underline{x} = \underline{z}$ this is specified by

$$\texttt{xInputDelay} = \texttt{zInputDelay} = \begin{Bmatrix} [1\ \ 2\ \ 3] \\ [1\ 2] \end{Bmatrix}.$$

□

Example 29.3.4 MISO NARX
For NARX models the delayed outputs also enter the network. Assume

$$\hat{y}(k) = f\,(u_1(k),\; u_2(k-1), u_2(k-2),\; y(k-1), y(k-2))\,. \tag{29.9}$$

For $\underline{x} = \underline{z}$, this is specified by the following input delays

$$\texttt{xInputDelay} = \texttt{zInputDelay} = \begin{Bmatrix} [0] \\ [1\ \ 2] \end{Bmatrix}$$

and the following output delays

$$\texttt{xOutputDelay} = \texttt{zOutputDelay} = \{[1\ 2]\}\,.$$

□

Example 29.3.5 MIMO NARX
For a first-order NARX model with two inputs and two outputs, different possibilities exist. Two separate models of the following type can be used

$$\hat{y}_1(k) = f_1\,(u_1(k-1),\; u_2(k-1),\; y_1(k-1)) \tag{29.10}$$

$$\hat{y}_2(k) = f_2\,(u_1(k-1),\; u_2(k-1),\; y_2(k-1)) \tag{29.11}$$

where each model describes one output without relating to the other output. Here two models – one for $y_1(k)$, the other for $y_2(k)$ – are set up. For $\underline{x} = \underline{z}$ with

$$\texttt{xInputDelay} = \texttt{zInputDelay} = \begin{Bmatrix} [1] \\ [1] \end{Bmatrix}$$

and

$$\texttt{xOutputDelay} = \texttt{zOutputDelay} = \{[1]\}\,.$$

Alternatively, both delayed outputs can be used in both models

$$\hat{y}_1(k) = f_1\left(u_1(k-1),\ u_2(k-1),\ y_1(k-1),\ y_2(k-1)\right) \tag{29.12}$$

$$\hat{y}_2(k) = f_2\left(u_1(k-1),\ u_2(k-1),\ y_1(k-1),\ y_2(k-1)\right) \tag{29.13}$$

which not only requires feedback from each model's output to its input in simulation but also from each model's output to the other model's input. This certainly will increase potential stability problems.

WARNING: Such an approach can only be used for one-step prediction with the discussed toolbox. For simulation it would require two simulate two models in parallel and feed back the outputs in a crosswise manner.

For prediction the "other" output can be treated as an additional (third) input which could be realized by

$$\texttt{xInputDelay} = \texttt{zInputDelay} = \begin{Bmatrix} [1] \\ [1] \\ [1] \end{Bmatrix}.$$

and

$$\texttt{xOutputDelay} = \texttt{zOutputDelay} = \{[1]\}$$

with $u_3(k) = y_2(k)$ for model 1 and $u_3(k) = y_1(k)$ for model 2.

Finally, a real MIMO model can be established:

$$\begin{bmatrix} \hat{y}_1(k) \\ \hat{y}_1(k) \end{bmatrix} = \underline{f}\left(u_1(k-1),\ u_2(k-1),\ y_1(k-1),\ y_2(k-1)\right)\,. \tag{29.14}$$

□

Often significant improvements can be achieved by keeping $\underline{z}$ low-dimensional. Especially for high-order dynamics, it is very beneficial to choose only $\underline{x}$ high-dimensional where it only increases the estimation variance and computational demand moderately while keeping the dimensionality of the $\underline{z}$ input space to a minimum which is essential for combating the curse of dimensionality.

29.3.1 Nonlinear Orthonormal Basis Function Models

This dynamic realization is not specifically supported. It can be realized by filtering the inputs appropriately and utilizing them with a "static" model.

29.4 Different Input Spaces $\underline{x}$ and $\underline{z}$

Static models do not require any delays. Therefore always

$$\texttt{xOutputDelay} = \texttt{zOutputDelay} = \{[\,]\}\,.$$

The entries for `xInputDelay` and `zInputDelay` can either be 0 or empty dependent on whether the inputs do exist in $\underline{x}$ and $\underline{z}$ or not.

Example 29.4.1 Scheduling
The following relationship shall be modeled where the parameters depend on u_3. Such models are called *linear parameter varying (LPV)*:

$$\hat{y}(k) = w_0(u_3) + w_1(u_3)u_1 + w_2(u_3)u_2\,. \tag{29.15}$$

The local model network describing this relationship has three inputs: u_1 and u_2 enter $\underline{x}$ and u_3 enters $\underline{z}$. Thus

$$\texttt{xInputDelay} = \begin{Bmatrix} [0] \\ [0] \\ [\,] \end{Bmatrix}$$

and

$$\texttt{zInputDelay} = \begin{Bmatrix} [\,] \\ [\,] \\ [0] \end{Bmatrix}\,.$$

□

29.5 Smoothness

The smoothness of the validity functions is determined by the standard deviations of the Gaussians in the LOLIMOT case and by the absolute values of the direction weights of the sigmoids in the HILOMOT case, respectively. The property `smoothness` determines the proportionality factor between the partitions' extensions and the standard deviations in the LOLIMOT case and determines a minimum validity value for data points close to the split or close to the center of the local model in the HILOMOT case.

Since the smoothness is not optimized but heuristically determined, it can be increased to make the model smoother or decreased to make the model crisper. Its default value is 1.

$$\texttt{smoothness} = \texttt{value}.$$

Even after training this value can be changed. Although, strictly speaking, it changes the validity values and thus the outcome of the weighted LS estimations of the local models' parameters, this effect typically is small. If the smoothness is altered significantly, a re-estimation of the local models' parameters is recommended. Slight suboptimalities with respect to the partitioning usually can be tolerated.

29.6 Data Weighting

By default all data points are weighted equally with weight 1. It can be set by the property

$$\texttt{dataWeighting} = [w(1)\ w(2)\ \cdots\ w(N)]'.$$

By choosing an appropriate weighting "badly" distributed data which is undesirably dense in some regions and sparse in others can be "normalized" in its importance on the model fit. Alternatively, sometimes it is desirable to over-pronounce certain regions compared to others because their performance it more important. Data weighting is an easy and effective method to deal with these issues.

29.7 Visualization and Simplified Tool

For first steps that do not require (and allow) delving into the details of the toolbox and dealing with the settings/properties explained above a very simplified approach is as follows. Local model networks can be trained by the trivial function call

```
LMNTrain(data)
```

where the training data

```
data = [u_1 u_2 ... u_p y]
```

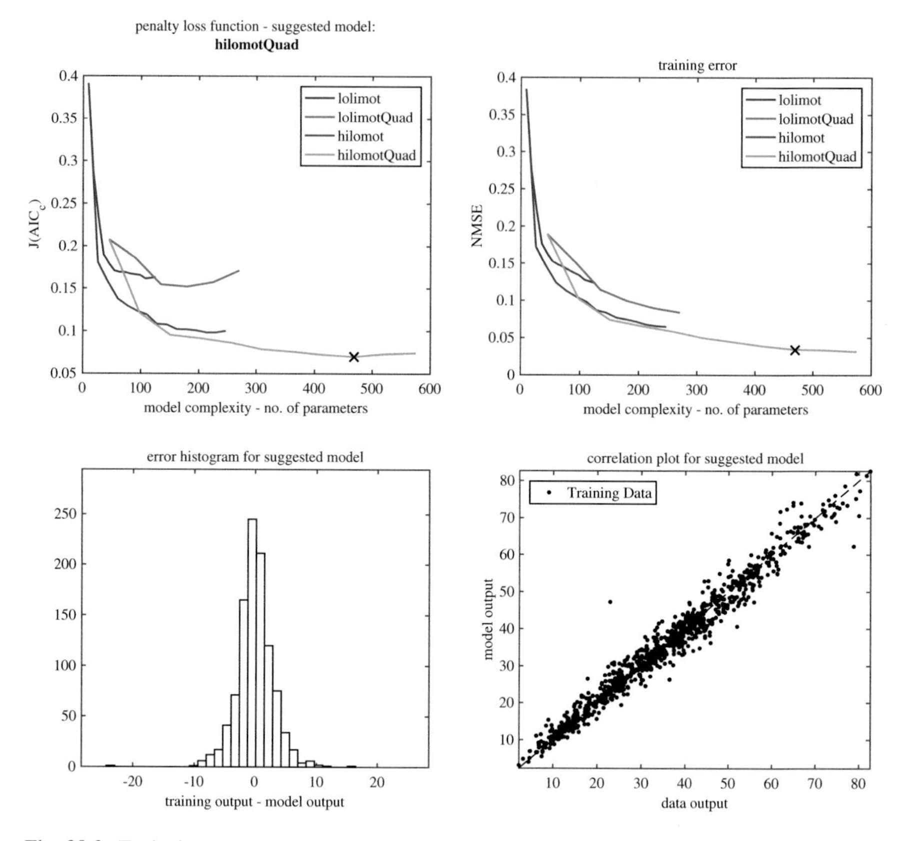

Fig. 29.2 Typical outcome of the LMN toolbox. Default setting is the training of four networks with local linear and quadratic models with axis-orthogonal (LOLIMOT) and -oblique (HILOMOT) partitioning

contains the inputs in the first columns and the output in the last column. The model with the best penalty loss function value (AIC_C) is recommended [3, 87]. Figure 29.2 shows the plot that is generated by this function call. The data set is the benchmark "Concrete Compressive Strength" from the UCI data repository.[1]

In extension, this simplified tool can be called with two or three inputs `traindata` and *optionally* `valdata` and/or `testdata`, respectively, and delivers two outputs `LMNBest` and `AllLMN`:

```
[LMNBest, AllLMN] = LMNTrain(traindata, valdata, testdata)
```

Then the model is assessed and the complexity recommendation is made according to this separate validation data set which is more reliable (if representative).

[1]https://archive.ics.uci.edu/ml/datasets.html

Commonly, no separate validation data set is available. Then the function is just called as `LMNTrain(traindata)` without test data or `LMNTrain(traindata, [], testdata)` with test data. The complexity then is determined with the help of the AIC_C criterion [3, 87].

If the user provides a separate test data set, then it is used purely for testing the model. This allows comparability to other models. It is important that no actions are based on the test data performance, e.g., any further subsequent selection step by comparing different model architectures. Then the test data would be used as some kind of validation data whose performance is over-optimistic.

Correction to: Nonlinear System Identification

Correction to:
O. Nelles, *Nonlinear System Identification*, https://doi.org/10.1007/978-3-030-47439-3

Chapter 14:

The original version of the book was inadvertently published with an error in equation 14.40 in page 502 of the fourteenth chapter in Part II. This has now been corrected.

Chapter 16:

The original version of the chapter was inadvertently published with the incorrect abbreviation for advanced simulation for calibration, modeling and optimization (ACSMO) which has now been corrected to "ASCMO" in page 640 of the sixteenth chapter in Part II.

The updated online version of these chapters can be found at
https://doi.org/10.1007/978-3-030-47439-3_14
https://doi.org/10.1007/978-3-030-47439-3_16

O. Nelles, *Nonlinear System Identification*,
https://doi.org/10.1007/978-3-030-47439-3_30

Appendix A
Vectors and Matrices

Abstract This appendix gathers some simple fundamentals for matrix and vector calculation, in particular when calculating derivatives.

Keywords Matrix · Vector · Derivative

This appendix summarizes some of the basics of derivatives involving vectors and matrices.

A.1 Vector and Matrix Derivatives

The derivative of an m-dimensional column vector $\underline{x} = [x_1\ x_2\ \cdots\ x_m]^T$ with respect to a scalar θ is defined as

$$\frac{\partial \underline{x}}{\partial \theta} = \begin{bmatrix} \partial x_1/\partial \theta \\ \partial x_2/\partial \theta \\ \vdots \\ \partial x_m/\partial \theta \end{bmatrix} . \qquad \text{(A.1)}$$

The derivative of an $l \times m$-dimensional matrix $\underline{X}$ with respect to a scalar θ is defined as

$$\frac{\partial \underline{X}}{\partial \theta} = \begin{bmatrix} \partial X_{11}/\partial \theta & \partial X_{12}/\partial \theta & \cdots & \partial X_{1m}/\partial \theta \\ \partial X_{21}/\partial \theta & \partial X_{22}/\partial \theta & \cdots & \partial X_{2m}/\partial \theta \\ \vdots & \vdots & \ddots & \vdots \\ \partial X_{l1}/\partial \theta & \partial X_{l2}/\partial \theta & \cdots & \partial X_{lm}/\partial \theta \end{bmatrix} . \qquad \text{(A.2)}$$

O. Nelles, *Nonlinear System Identification*,
https://doi.org/10.1007/978-3-030-47439-3

When the derivative of a quantity with respect to an n-dimensional vector $\underline{\theta} = [\theta_1\ \theta_2\ \cdots\ \theta_n]^T$ is to be calculated, it is helpful to first define the following derivative operator vector:

$$\frac{\partial}{\partial \underline{\theta}} = \begin{bmatrix} \partial/\partial\theta_1 \\ \partial/\partial\theta_2 \\ \vdots \\ \partial/\partial\theta_n \end{bmatrix}. \tag{A.3}$$

With (A.3) the derivative of a scalar x with respect to an n-dimensional column vector $\underline{\theta}$ is defined as

$$\frac{\partial x}{\partial \underline{\theta}} = \begin{bmatrix} \partial x/\partial\theta_1 \\ \partial x/\partial\theta_2 \\ \vdots \\ \partial x/\partial\theta_n \end{bmatrix}. \tag{A.4}$$

The derivative of an m-dimensional row vector $\underline{x}^T$ with respect to an n-dimensional column vector $\underline{\theta}$ is defined via the outer product as

$$\frac{\partial \underline{x}^T}{\partial \underline{\theta}} = \frac{\partial}{\partial \underline{\theta}} \cdot \underline{x}^T = \begin{bmatrix} \partial x_1/\partial\theta_1 & \partial x_2/\partial\theta_1 & \cdots & \partial x_m/\partial\theta_1 \\ \partial x_1/\partial\theta_2 & \partial x_2/\partial\theta_2 & \cdots & \partial x_m/\partial\theta_2 \\ \cdots & \vdots & & \vdots \\ \partial x_1/\partial\theta_n & \partial x_2/\partial\theta_n & \cdots & \partial x_m/\partial\theta_n \end{bmatrix}. \tag{A.5}$$

The chain rule can be extended to the vector case by defining the vector derivative of a scalar product:

$$\frac{\partial}{\partial \underline{\theta}} \left(\underline{x}^T \underline{y} \right) = \frac{\partial \underline{x}^T}{\partial \underline{\theta}} \cdot \underline{y} + \frac{\partial \underline{y}^T}{\partial \underline{\theta}} \cdot \underline{x}, \tag{A.6}$$

where $\underline{x}^T$ is an m-dimensional row vector, $\underline{y}$ is an m-dimensional column vector, and $\underline{\theta}$ is an n-dimensional column vector.

An interesting special case of (A.6) occurs if one vector does not depend on $\underline{\theta}$ and the other one is equivalent to $\underline{\theta}$ (this of course requires $m = n$):

$$\frac{\partial}{\partial \underline{\theta}} \left(\underline{z}^T \underline{\theta} \right) = \underline{z} \tag{A.7}$$

and

$$\frac{\partial}{\partial \underline{\theta}} \left(\underline{\theta}^T \underline{z} \right) = \underline{z}\,, \tag{A.8}$$

where the row and column vectors $\underline{z}^T$ and $\underline{z}$ do not depend on $\underline{\theta}$. Corresponding expressions hold if the vector $\underline{z}$ is replaced by a matrix $\underline{Z}$:

$$\frac{\partial}{\partial \underline{\theta}} \left(\underline{Z}^T \underline{\theta} \right) = \underline{Z}\,, \tag{A.9}$$

$$\frac{\partial}{\partial \underline{\theta}} \left(\underline{\theta}^T \underline{Z} \right) = \underline{Z}\,. \tag{A.10}$$

Finally, the derivative of the quadratic form $\underline{\theta}^T \underline{Z}\,\underline{\theta}$ is important where $\underline{Z}$ is a square $n \times n$ matrix:

$$\begin{aligned} \frac{\partial}{\partial \underline{\theta}} \left(\underbrace{\underline{\theta}^T \underline{Z}}_{\underline{x}^T} \underbrace{\underline{\theta}}_{\underline{y}} \right) &= \frac{\partial}{\partial \underline{\theta}} \left(\underline{\theta}^T \underline{Z} \right) \cdot \underline{\theta} + \frac{\partial}{\partial \underline{\theta}} \left(\underline{\theta}^T \right) \cdot \left(\underline{\theta}^T \underline{Z} \right)^T \\ &= \underline{Z}\,\underline{\theta} + \underline{I}\,\underline{Z}^T \underline{\theta} = \left(\underline{Z} + \underline{Z}^T \right) \underline{\theta}\,. \end{aligned} \tag{A.11}$$

In the first step, (A.6) is applied as indicated. In the second step, $\partial \underline{\theta}^T / \partial \underline{\theta}$ is equal to the identity matrix $\underline{I}$ according to (A.5). The last term in (A.11) can be further simplified to $2\underline{Z}\,\underline{\theta}$ if $\underline{Z}$ is symmetric.

A.2 Gradient, Hessian, and Jacobian

The *gradient* $\underline{g}$ is the first derivative of a scalar function with respect to a vector. The gradient of the function $f(\underline{\theta})$ dependent on the n-dimensional parameter vector $\underline{\theta} = [\theta_1\ \theta_2\ \cdots\ \theta_n]^T$ with respect to these parameters is defined as the following n-dimensional vector:

$$\underline{g} = \frac{\partial f(\underline{\theta})}{\partial \underline{\theta}} = \begin{bmatrix} g_1 \\ g_2 \\ \vdots \\ g_n \end{bmatrix} = \begin{bmatrix} \partial f(\underline{\theta}) \,/\, \partial \theta_1 \\ \partial f(\underline{\theta}) \,/\, \partial \theta_2 \\ \vdots \\ \partial f(\underline{\theta}) \,/\, \partial \theta_n \end{bmatrix}. \tag{A.12}$$

The geometric interpretation of the gradient is that it points into the direction of the steepest ascent of function $f(\underline{\theta})$. The gradient is zero where $f(\underline{\theta})$ possesses an optimum.

The *Hessian* $\underline{H}$ is the second derivative of a scalar function with respect to a vector. The Hessian of the function $f(\underline{\theta})$ dependent on the n-dimensional parameter vector $\underline{\theta} = [\theta_1\ \theta_2\ \cdots\ \theta_n]^T$ with respect to these parameters is defined as the following $n \times n$-dimensional matrix:

$$\underline{H} = \frac{\partial^2 f(\underline{\theta})}{\partial \underline{\theta}^2} = \begin{bmatrix} H_{11} & H_{12} & \cdots & H_{1n} \\ H_{21} & H_{22} & \cdots & H_{2n} \\ \vdots & \vdots & \ddots & \vdots \\ H_{n1} & H_{n2} & \cdots & H_{nn} \end{bmatrix}$$

$$= \begin{bmatrix} \partial^2 f(\underline{\theta}) \,/\, \partial\theta_1^2 & \partial^2 f(\underline{\theta}) \,/\, \partial\theta_1\partial\theta_2 & \cdots & \partial^2 f(\underline{\theta}) \,/\, \partial\theta_1\partial\theta_n \\ \partial^2 f(\underline{\theta}) \,/\, \partial\theta_2\partial\theta_1 & \partial^2 f(\underline{\theta}) \,/\, \partial\theta_2^2 & \cdots & \partial^2 f(\underline{\theta}) \,/\, \partial\theta_2\partial\theta_n \\ \vdots & \vdots & \ddots & \vdots \\ \partial^2 f(\underline{\theta}) \,/\, \partial\theta_n\partial\theta_1 & \partial^2 f(\underline{\theta}) \,/\, \partial\theta_n\partial\theta_2 & \cdots & \partial^2 f(\underline{\theta}) \,/\, \partial\theta_n^2 \end{bmatrix}. \quad \text{(A.13a)}$$

The geometric interpretation of the Hessian is that it gives the curvature of the function $f(\underline{\theta})$, e.g., for a linear $f(\underline{\theta})$, the Hessian is zero, and for a quadratic $f(\underline{\theta})$, the Hessian is constant. The Hessian is symmetric. It is positive definite at a minimum of $f(\underline{\theta})$ and negative definite at a maximum of $f(\underline{\theta})$.

The *Jacobian* is the first derivative of a vector function with respect to a vector. The Jacobian of the N-dimensional vector function $\underline{f}(\underline{\theta}) = [f_1(\underline{\theta})\ f_2(\underline{\theta})\ \cdots\ f_N(\underline{\theta})]^T$ dependent on the n-dimensional parameter vector $\underline{\theta} = [\theta_1\ \theta_2\ \cdots\ \theta_n]^T$ with respect to these parameters is defined as the following $N \times n$-dimensional matrix:

$$\underline{J} = \frac{\partial \underline{f}(\underline{\theta})}{\partial \underline{\theta}} = \begin{bmatrix} \left(\partial f_1(\underline{\theta}) \,/\, \partial\underline{\theta}\right)^T \\ \left(\partial f_2(\underline{\theta}) \,/\, \partial\underline{\theta}\right)^T \\ \vdots \\ \left(\partial f_N(\underline{\theta}) \,/\, \partial\underline{\theta}\right)^T \end{bmatrix} = \begin{bmatrix} J_{11} & J_{12} & \cdots & J_{1n} \\ J_{21} & J_{22} & \cdots & J_{2n} \\ \vdots & \vdots & & \vdots \\ J_{N1} & J_{N2} & \cdots & J_{Nn} \end{bmatrix}$$

$$= \begin{bmatrix} \partial f_1(\underline{\theta}) \,/\, \partial\theta_1 & \partial f_1(\underline{\theta}) \,/\, \partial\theta_2 & \cdots & \partial f_1(\underline{\theta}) \,/\, \partial\theta_n \\ \partial f_2(\underline{\theta}) \,/\, \partial\theta_1 & \partial f_2(\underline{\theta}) \,/\, \partial\theta_2 & \cdots & \partial f_2(\underline{\theta}) \,/\, \partial\theta_n \\ \vdots & \vdots & & \vdots \\ \partial f_N(\underline{\theta}) \,/\, \partial\theta_1 & \partial f_N(\underline{\theta}) \,/\, \partial\theta_2 & \cdots & \partial f_N(\underline{\theta}) \,/\, \partial\theta_n \end{bmatrix}. \quad \text{(A.14a)}$$

The Jacobian is, for example, used to calculate the entry-wise gradient of an error vector $\underline{e} = [e_1\ e_2\ \cdots\ e_N]^T$, where N is the number of measurements. Then the Jacobian contains the transposed gradients of each entry of $\underline{e}$ with respect to $\underline{\theta}$, i.e., $J = [\underline{g}_1\ \underline{g}_2\ \cdots\ \underline{g}_N]^T$, where $\underline{g}_i$ is the gradient of e_i.

Appendix B
Statistics

Abstract This appendix summarizes some elementary facts on random variables, probabilities, and stochastic processes.

Keywords Random variable · Probability · Pdf · Expectation · Variance · Correlation · Covariance · Estimator · Bias

This appendix summarizes some useful statistical basics. For a deeper discussion, refer to [443]. After a brief introduction to deterministic and random variables, the elementary tools for dealing with random variables, such as the expectation operator, auto- and cross-correlation, and the variance and covariance, are introduced. Finally, some important statistical properties of estimates and estimators, such as bias, consistency, and efficiency, are discussed.

B.1 Deterministic and Random Variables

If an experiment, such as feeding data into a process and measuring the resulting process output, is carried out several times and the outcome always is identical, then the process and the signals involved can be properly modeled deterministically. It can be predicted that a specific input value will result in a specific output value. However, in practice, the outcome of an experiment often will not be identical each time it is performed. It is not possible to exactly predict the result for a given input. A common example is tossing a dice. In the context of feeding input data into a process and measuring the output, the reason for non-reproducibility is often described as noise. Under "noise" all undesirable effects are summarized that do not have an exactly predictable nature. This non-deterministic behavior is described by random variables, and the whole non-deterministic signal over time is described by a stochastic process; see below.

Although classical physics is purely deterministic, random variables are very important for modeling the real world. Many deterministic effects seem to be of

O. Nelles, *Nonlinear System Identification*,
https://doi.org/10.1007/978-3-030-47439-3

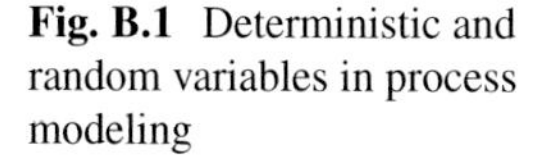
Fig. B.1 Deterministic and random variables in process modeling

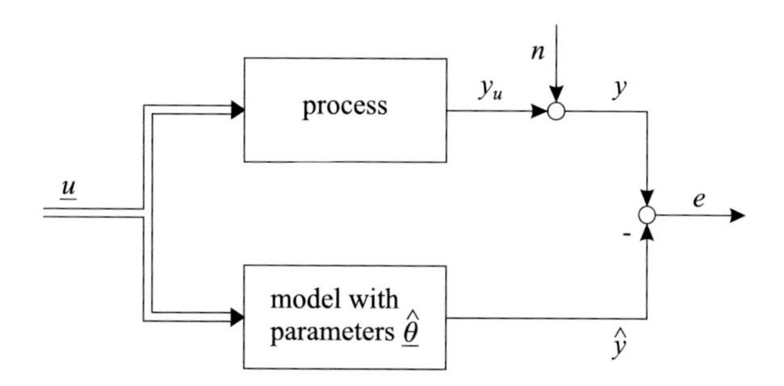

random nature to an observer who lacks the amount of necessary information in order to gain insights about the truly deterministic origin of the effects. These effects are usually caused by many unmeasurable (or at least not measured) variables. Thus, the notion of noise depends highly on knowledge about the deterministic effects and the measured variables. The effects of all non-modeled behavior are often summarized by incorporating a stochastic component into a model.

Owing to the stochastic character of noise, it cannot be properly described by the signal itself. Rather, random variables and stochastic processes are characterized by their stochastic properties, such as expected value, variance, etc. To clarify the concept of deterministic and random variables, the example in Fig. B.1 will be examined. The (non-measurable) process output y_u is assumed to be additively corrupted by noise n, resulting in the measurable y. No other stochastic effects are assumed. Such a model of reality is very simple and common, although in reality certainly, the noise will corrupt variables inside the process as well. These effects, however, can often be thought to be transformed into the process output. The input vector is usually generated by the user, and the model is calculated on a digital computer. Therefore, $\underline{u}$ and $\hat{y}$ are not noisy. If the model of the process was perfect in the sense that $\hat{y} = y_u$, then the error e would be equal to the noise n. The model parameters $\underline{\hat{\theta}}$ are estimated in order to minimize some loss function depending on the error.

Which signals in Fig. B.1 should be described in a deterministic and which in a stochastic way? It is clear that the input $\underline{u}$ and the process output y_u have no relation to the noise n; therefore they are seen as deterministic variables. The noise n certainly is a stochastic process, i.e., a random variable dependent on time. The measurable process output y and the error e depend on the noise and consequently are also described stochastically. At first sight, the model output, which may be simulated in a computer, has nothing to do with noise n. But this statement is wrong! Before the model can be simulated, its parameters have to be estimated. And these parameters $\underline{\hat{\theta}}$ are estimated by minimizing a loss function that depends on the error (a stochastic process). Thus, the loss function itself, and consequently the estimated parameters, are random variables. Therefore, the model output $\hat{y}$ is a stochastic process, because it is computed with the estimated parameters. All this means that it is possible (or necessary) to apply statistical methods in order to analyze the properties of parameter estimates and the model output. The expected value and the covariance matrix of the estimated parameters and the model output reveal interesting properties that allow an assessment of the quality of the parameter estimates and the model output.

B.2 Probability Density Function (pdf)

In contrast to deterministic variables, random variables are described in terms of probabilities. The *probability density function (pdf)* is the most complete type of knowledge that can be derived through the application of statistical techniques [141]. As an example, Fig. B.2 depicts a *Gaussian* pdf $p(x)$ for a random variable x. A random variable with a Gaussian pdf is also said to have a *normal* distribution. The interpretation is as follows. The probability that x lies in the interval $[a, b]$ is equal to the gray-shaded area below the pdf. Note that because x is a continuous variable, each specific exact value of x has probability zero. Since the probability of x being in the interval $[-\infty, \infty]$ is 1, the area under the pdf must be equal to 1.

If the width of the Gaussian in Fig. B.2 or of any other continuous pdf approaches zero, the interval of highly probable values of the random variable becomes very small. In the limit of an infinitesimally narrow pdf, it becomes a Dirac impulse. Then the value where the Dirac is positioned has a probability of 1, while all other realizations of the random variable have a probability of zero. Therefore, in this limit a random variable becomes deterministic.

The pdf determines the probability of each realization of the random variable if an experiment is performed. Figure B.3 shows this relationship for a Gaussian pdf and 100 experiments. Each experiment forces the random variable to a realization. The values around zero are the most frequent since they have the highest probability

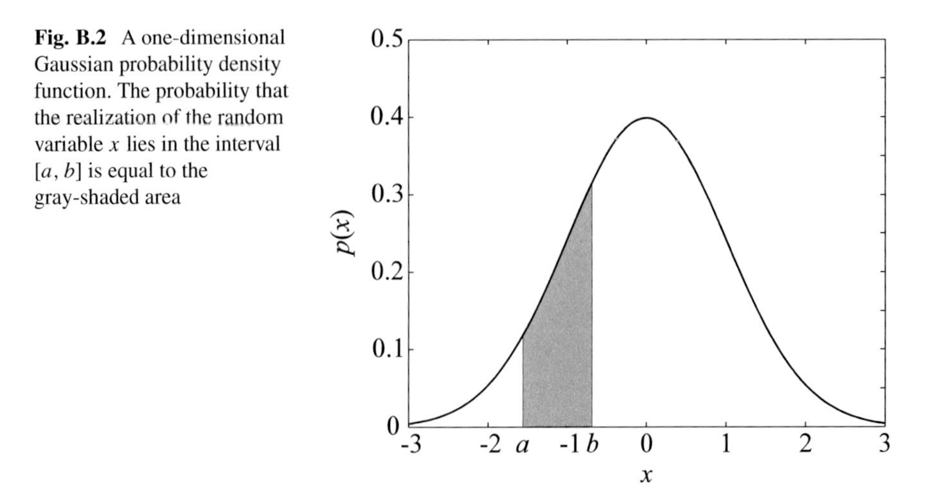

Fig. B.2 A one-dimensional Gaussian probability density function. The probability that the realization of the random variable x lies in the interval $[a, b]$ is equal to the gray-shaded area

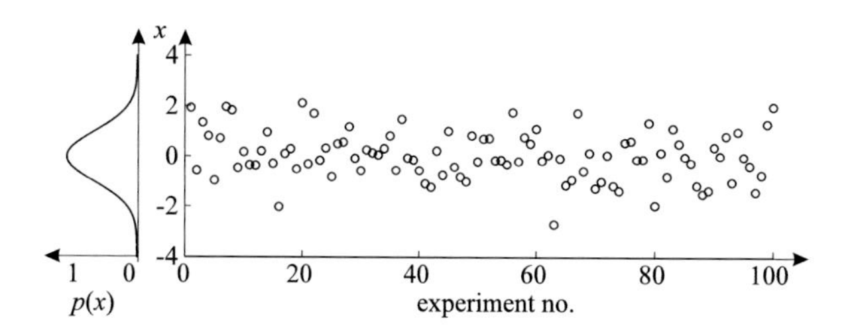

Fig. B.3 A random variable with Gaussian pdf yields a (different) realization for each experiment. The probability of each realization is described by its pdf. Note that the experiment number has nothing to do with time

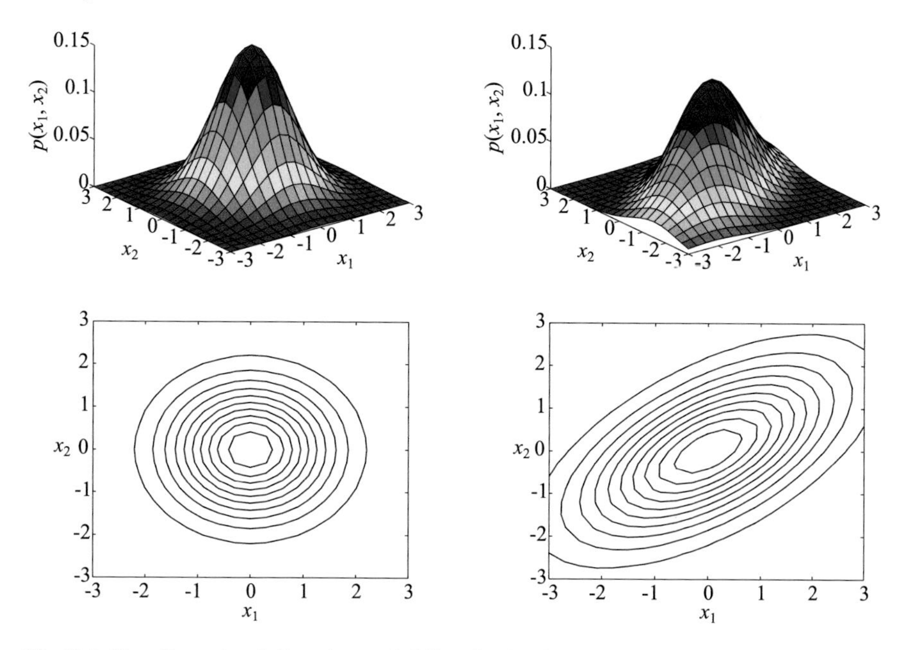

Fig. B.4 Two-dimensional Gaussian probability density functions. The left pdf describes two independent random variables. Knowledge about the realization of one random variable does not yield any information about the distribution of the other random variable. The right pdf describes two dependent random variables. Knowledge about the realization of one random variable yields information about the other random variable

with the given pdf. Note that the experiment number has nothing to do with time, e.g., all experiments could be performed at once.

In the case of more than one random variable, a vector of random variables $\underline{x} = [x_1\ x_2\ \cdots\ x_n]^T$ is constructed, and the pdf $p(\underline{x})$ becomes multidimensional. Such a multidimensional pdf is called a *joint pdf*. As an example, Fig. B.4 shows two different two-dimensional Gaussian pdfs with their contour lines. Obviously, the pdf on the left is axis-orthogonal, that is, it can be constructed by the multiplication of two one-dimensional (Gaussian) pdfs $p_{x_1}(x_1)$ and $p_{x_2}(x_2)$. This property does not hold for the pdf on the right. The one-dimensional pdfs are called the *marginal densities* of $p(\underline{x})$. They can be calculated by $p_{x_1}(x_1) = \int_{-\infty}^{\infty} p(\underline{x})\, dx_2$ and $p_{x_2}(x_2) = \int_{-\infty}^{\infty} p(\underline{x})\, dx_1$, respectively.

It is interesting to compare the pdfs in Fig. B.4 in more detail and to analyze the consequences of their different shapes for the two random variables x_1 and x_2. First, the left pdf is considered. If the realization of one random variable, say $x_1 = c$, is known, what information does this yield on the other random variable x_2? In other words, how does the one-dimensional pdf change that is obtained by cutting a slice through $p(\underline{x})$ at $x_1 = c$? Obviously, all these "sliced" pdfs are identical in shape (not in scaling); they do not depend on c. Knowing the realization of one random variable does not yield any information about the other. Two random variables that have this property are called independent.

Two random variables x_1 and x_2 are *independent* if a multiplication of their marginal densities yields their joint probability density:

$$p(x_1, x_2) = p_{x_1}(x_1) \cdot p_{x_2}(x_2) . \tag{B.1}$$

This implies that independent random variables can be described completely by their one-dimensional pdfs $p_{x_i}(x_i)$.

The pdf in Fig. B.4(right) describes two dependent random variables. If the realization of x_1 is known, one gains information about x_2. For small values of x_1, small values of x_2 are more likely than large ones and vice versa. The pdf "slices" along $x_1 = c$ move toward higher values of x_2 for increasing $x_1 = c$.

B.3 Stochastic Processes and Ergodicity

A random variable is described by its pdf and yields a realization if an experiment is performed. Often stochastic signals (such as noise) have to be modeled. Signals evolve over time. Thus, it is reasonable to make a random variable x dependent not only on the experiment but also on time. Such a random variable $x(k)$, which is a function of time k, is called a *stochastic process*. A stochastic process $x(k)$ is described by its pdf $p(x, k)$, which clearly also depends on time k. Thus, a stochastic process can be seen as an infinite number of random variables, each for one time instant k [443].

Figure B.5 shows several different realizations of a stochastic process. It is important to note that it depends on both the experiment realization and time. As an illustrative example, assume Fig. B.5 shows measurement noise. Then typically in one measurement, one realization (one row in Fig. B.5) of the stochastic process is measured. If the statistical properties do not change from one measurement to

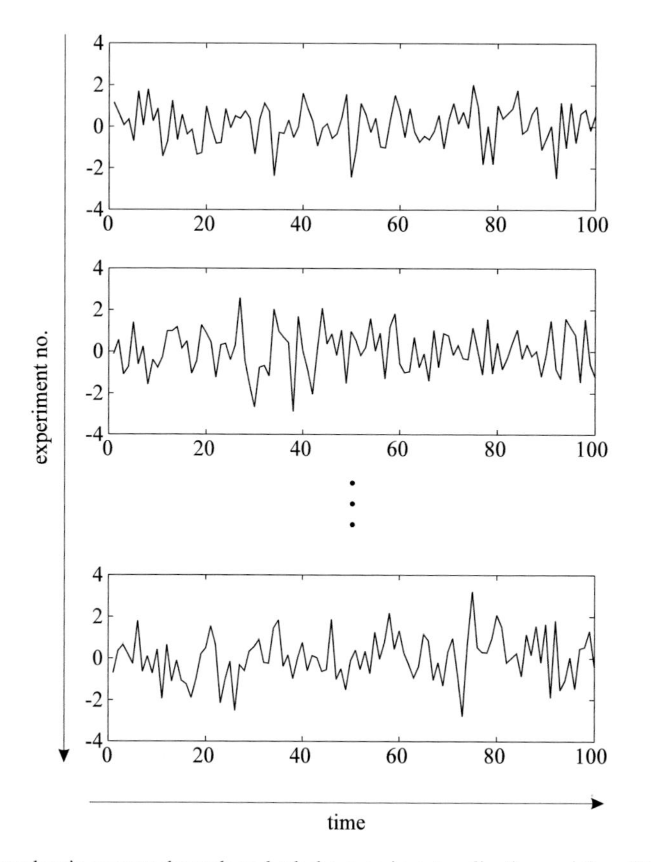

Fig. B.5 A stochastic process depends on both the experiment realization and time [217]

another, in a second measurement, a different realization of the same stochastic process will be observed. This is the stochastic character of random variables. Deterministic variables yield the same values in each measurement; they just change as time progresses.

A typical and widely used type of noise is "Gaussian white" noise. "Gaussian" tells something about the amplitude distribution of the stochastic process at each time instant k_0, i.e., it describes the pdf $p(x, k_0)$. It determines how probable small and large values $x(k_0)$ are. In contrast, "white" tells us something about the relationship of the stochastic process between two time instants, that is, between the pdfs $p(x, k_1)$ and $p(x, k_2)$. For more details on the property "white," see below. Thus, it is important to distinguish between the properties related to amplitudes and the properties related to time (or equivalently frequency).

In general, a stochastic process can change its statistical properties with time; for this reason, it explicitly depends on k. This means that the kind of pdf $p(x, k)$ may change with k, e.g., from a Gaussian distribution to a uniform distribution (where within an interval all values are equally probable). Or the kind of distribution may stay constant over time, but some of its characteristic properties, such as mean or variance (see below), may change with time. However, a special class of stochastic processes does *not* depend on time. These stochastic processes are called stationary.

> A stochastic is called *stationary* if its statistical properties do not depend on time, i.e., if
>
> $$p(x, k_1) = p(x, k_2) \qquad \text{for all } k_1 \text{ and } k_2 . \tag{B.2}$$

In most applications noise is modeled as a stationary process since there is often no reason to assume that the noise characteristics depend on time. A counterexample is a sensor that yields more or less noisy results depending on its temperature. The temperature may be related to time, e.g., monotonically increasing from the start of operation. In such a case, the sensor noise can be modeled by a non-stationary stochastic process with time-varying properties.

In most applications, even more than stationarity is assumed about the process. It is usually assumed that the expectation (see below) over *all* different realizations, that is, from top to bottom of Fig. B.5, is equivalent to the expectation of a *single* realization over time, that is, from left to right of Fig. B.5. The main reason for this assumption is practicability. In most applications, just a single realization over time of the stochastic process is known from measurements. Thus, averaging different realizations is impossible. Because the expectation operator is of fundamental importance for all statistical calculations, it is of great relevance to have some practical method for its calculation. If a stochastic process allows one to compute the expectation over the realizations as the expectation over time, it is called ergodic.[1]

> A stochastic process is called *ergodic* if the expectation over its realizations can be calculated as the time average of one realization.

Ergodicity implies that one realization of the stochastic process contains all information about the statistical properties as $k \to \infty$. Thus, for ergodic processes, one can reach conclusions about typical statistical characteristic properties such as mean and variance (see below) by observation of one process realization over time.

An ergodic process is stationary. This becomes obvious considering that by averaging over time, any time dependency is lost. Therefore, the statistical properties of an ergodic process cannot be time-dependent. Because ergodicity is usually assumed without stating this assumption explicitly, it may be helpful to consider stochastic processes that are not ergodic. First of all, any non-stationary process automatically is non-ergodic. See Fig. B.6 for an example of a non-stationary and consequently a

[1]More exactly speaking: mean ergodic [443].

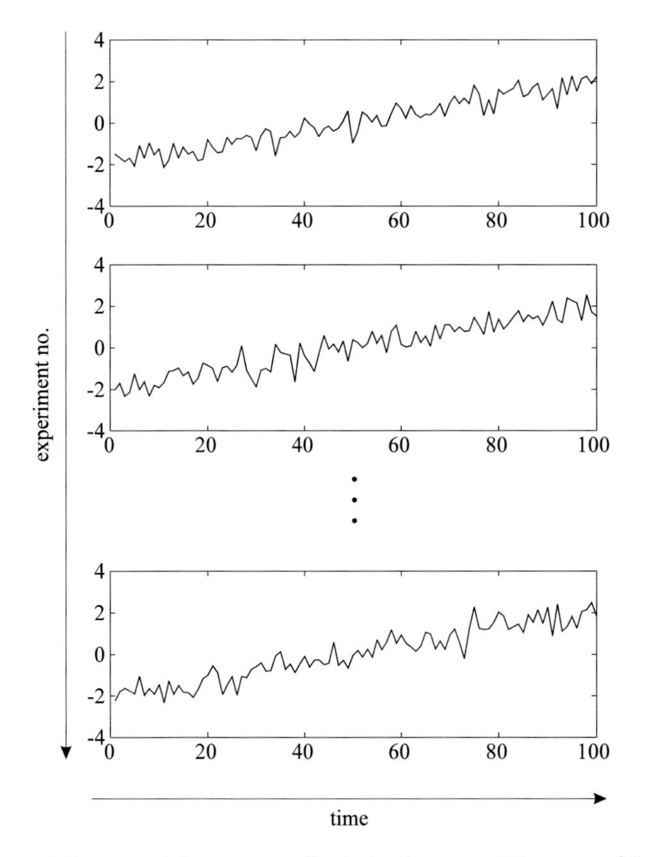

Fig. B.6 A non-stationary and thus non-ergodic stochastic process. The mean of the stochastic process increases with time. Therefore, the statistical properties depend on the time, and the process is non-stationary. Obviously, taking the average over time (left to right) is totally different from taking the average over the realizations (top to bottom)

non-ergodic process. However, a simple process such as $x(k) = c$ with some $p(x, k)$ is also not ergodic [443]. This stochastic process shown in Fig. B.7 is constant (over time), that is, $x(k) = c$ for each experiment (i.e., in each row). This value $x(k)$ is different in each experiment according to its distribution $p(x, k)$. By picking a single realization and calculating the time average, one ends up with the corresponding value of $x(k)$. This, however, is not (necessarily) equivalent to the expected value of the pdf $p(x, k)$.

B.4 Expectation

The two most important quantities in order to characterize a random variable and a stochastic process are its *expected value* or *mean* and its *variance*. They are introduced in this and the following section. Other important characteristics, the

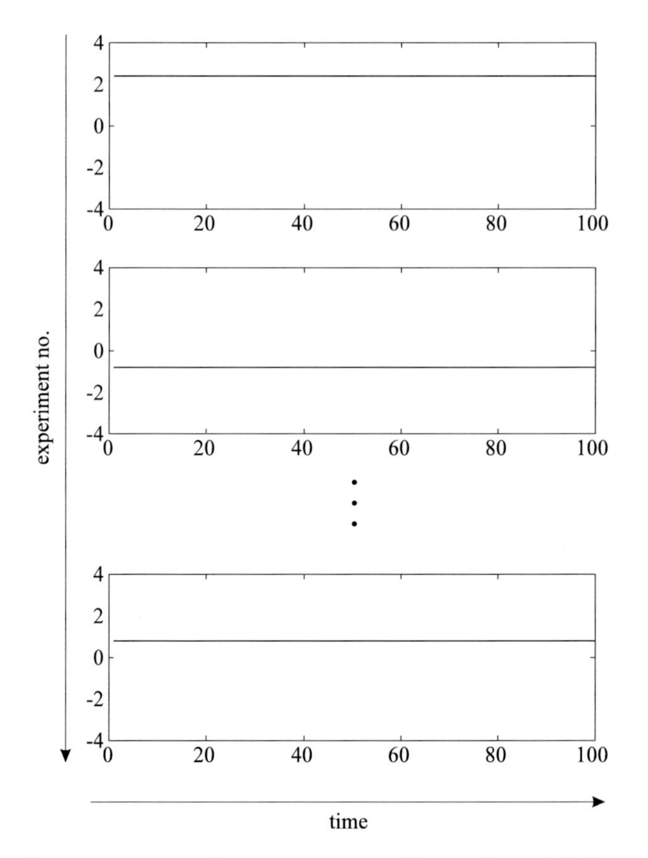

Fig. B.7 A stationary but non-ergodic stochastic process. Because the statistical properties of this process do not depend on time, it is stationary. However, this process is not ergodic since one realization does not reveal the statistical properties of the process

correlation and covariance, are defined for two random variables or stochastic processes. They are discussed in Sect. B.6.

The basis of all statistical calculations is the *expectation* operator. Its argument usually is a random variable or a stochastic process, and its result is the expected value of this random variable or stochastic process, respectively. If the argument of the expectation operator is deterministic $\mathrm{E}\{x\} = x$.

In order to calculate the expected value of a random variable, all realizations have to be weighted with their corresponding probability and have to be summed up. Thus, the expectation of a random variable x becomes

$$\text{mean}\{x\} = \mu_x = \mathrm{E}\{x\} = \int_{-\infty}^{\infty} x\, p(x)\, dx\,. \tag{B.3}$$

Correspondingly, the expectation of a stochastic process is

$$\text{mean}\{x(k)\} = \mu_x(k) = \mathrm{E}\{x(k)\} = \int_{-\infty}^{\infty} x(k)\, p(x,k)\, dx\,. \tag{B.4}$$

The equations for calculation of the expected value given above require knowledge about the pdf. Usually in practice this knowledge is not available. Therefore, the stochastic process is assumed to be ergodic, and averaging over all realizations in (B.4) is replaced by averaging over time. If $x(k)$ denotes the measurement (i.e., one realization) of the stochastic process for time instant $k = 1, \ldots, N$, the expectation is estimated by

$$\mathrm{E}\{x\} \approx \frac{1}{N} \sum_{k=1}^{N} x(k) = \bar{x}\,. \tag{B.5}$$

Sometimes this estimation of the expected value of x is written as $\bar{x}$ for shortness and in order to explicitly express the experimental determination of the mean. Note that in (B.5) $\mathrm{E}\{x\}$ does not depend on time k since this is the variable that is averaged over. Here it becomes clear again that a stochastic process must be stationary (not depend on time) in order to be able to apply (B.5). If a stochastic process is ergodic, the estimation in (B.5) converges to the true expectation as $N \to \infty$. Although it is not explicitly noted throughout the book, ergodicity is supposed. For all expectations, it is assumed that they can be calculated according to (B.5). It can be shown that (B.5) is an unbiased estimate of the expected value of x if the stochastic process is ergodic [443].

The definition of the expectation operator can be extended to vectors and matrices as follows:

$$\mathrm{E}\{\underline{x}\} = \mathrm{E}\left\{\begin{bmatrix} x_1 \\ x_2 \\ \vdots \\ x_n \end{bmatrix}\right\} = \begin{bmatrix} \mathrm{E}\{x_1\} \\ \mathrm{E}\{x_2\} \\ \vdots \\ \mathrm{E}\{x_n\} \end{bmatrix}, \tag{B.6}$$

$$\mathrm{E}\{\underline{X}\} = \mathrm{E}\left\{\begin{bmatrix} x_{11} & x_{12} & \cdots & x_{1n} \\ x_{21} & x_{22} & \cdots & x_{2n} \\ \vdots & \vdots & \ddots & \vdots \\ x_{m1} & x_{m2} & \cdots & x_{mn} \end{bmatrix}\right\}$$

$$= \begin{bmatrix} \mathrm{E}\{x_{11}\} & \mathrm{E}\{x_{12}\} & \cdots & \mathrm{E}\{x_{1n}\} \\ \mathrm{E}\{x_{21}\} & \mathrm{E}\{x_{22}\} & \cdots & \mathrm{E}\{x_{2n}\} \\ \vdots & \vdots & \ddots & \vdots \\ \mathrm{E}\{x_{m1}\} & \mathrm{E}\{x_{m2}\} & \cdots & \mathrm{E}\{x_{mn}\} \end{bmatrix} . \tag{B.7}$$

Since all other statistical expressions are based on the expectation operator, these vector and matrix definitions can be applied to them as well.

B.5 Variance

Roughly speaking, the variance of a random variable measures the width of its pdf. A low variance means that realizations close to the expected value are highly probable, while a high variance implies high probabilities for realization far away from the mean.[2] The variance of a random variable can be calculated as the sum over the squared distance from the expected value weighted with the corresponding probability:

$$\mathrm{var}\{x\} = \sigma_x^2 = \mathrm{E}\left\{(x - \mu_x)^2\right\} = \int_{-\infty}^{\infty} (x - \mu_x)^2\, p(x)\, dx\,. \tag{B.8}$$

Correspondingly, the variance of a stochastic process is

$$\begin{aligned} \mathrm{var}\{x(k)\} &= \sigma_x^2(k) = \mathrm{E}\left\{(x(k) - \mu_x(k))^2\right\} \\ &= \int_{-\infty}^{\infty} (x(k) - \mu_x(k))^2\, p(x, k)\, dx\,. \end{aligned} \tag{B.9}$$

In practice, as discussed in the previous section, ergodicity is assumed, and the time average replaces the average over the realizations. Two cases have to be distinguished in this context: (i) the mean of the stochastic process is known, and (ii) the mean of the stochastic process is estimated by (B.5).

The somewhat unrealistic case (i) with the known (not estimated) mean μ_x leads to the following unbiased estimate of the variance:

$$\mathrm{E}\{(x - \mu_x)^2\} \approx \frac{1}{N} \sum_{i=1}^{N} (x(i) - \mu_x)^2\,. \tag{B.10}$$

[2]Counterexamples to these qualitative statements can be constructed. However, they involve pdfs of strange shapes, e.g., multimodal, which are not very relevant in practice.

For case (ii) with an experimentally estimated mean $\bar{x}$, the unbiased variance estimate becomes

$$\mathrm{E}\{(x - \bar{x})^2\} \approx \frac{1}{N-1} \sum_{i=1}^{N} (x(i) - \bar{x})^2 = s^2 . \tag{B.11}$$

Because the variance σ^2 is a quadratic distance measure, it is often easier to think in terms of the *standard deviation* σ, that is, the square root of the variance. For example, with a Gaussian pdf with mean μ and standard deviation σ, the following intervals around its mean $[\mu \pm \sigma]$, $[\mu \pm 2\sigma]$, and $[\mu \pm 3\sigma]$, which are called the one-, two-, and three-sigma intervals, represent 68.3%, 95.4%, and 99.7% of the whole area under the pdf. In other words only 31.7%, 4.6%, and 0.3% of all realizations of the random variable lie outside these intervals. Thus, the standard deviation gives the user intuitively clear information about the width of the pdf. For example, the errorbars introduced in Sect. 3.1.2 represent the $[\mu - \sigma, \mu + \sigma]$ interval, where σ is the standard deviation of the noise. Also, signal to noise amplitude (power) ratios are expressed by the quotient of the standard deviation (variance) of the noise and of the signal.

B.6 Correlation and Covariance

The properties mean and variance are defined for single random variables or stochastic processes. In contrast, the *correlation* and *covariance* are defined for *two* random variables or stochastic processes, say x and y or $x(k)$ and $y(k)$, respectively. The interpretation of correlation and covariance is basically the same; in fact they are identical for random variables or stochastic processes with zero mean. Both, correlation and covariance, measure the similarity between two random variables or stochastic processes. For two random variables x and y, the correlation is defined by

$$\mathrm{corr}\{x, y\} = \mathrm{corr}_{xy} = \mathrm{E}\{x \cdot y\} , \tag{B.12}$$

and the covariance is defined by

$$\mathrm{cov}\{x, y\} = \mathrm{cov}_{xy} = \mathrm{E}\{(x - \mu_x) \cdot (y - \mu_y)\} . \tag{B.13}$$

Obviously, if the means μ_x and μ_y are equal to zero, correlation and covariance are identical. Those definitions are quite intuitive because correlation and covariance are large (x and y are highly correlated) if the random variables are closely related and they are small (x and y are weakly correlated) if the random variables have little in common. An extreme case arises if both random variables are identical, i.e., $x = y$. Then the covariance is equivalent to the variance of x: $\text{cov}\{x, x\} = \text{var}\{x\}$. The correlation $\text{corr}\{x, x\}$ is called *auto*-correlation because it correlates a variable with itself, while $\text{corr}\{x, y\}$ for $x \neq y$ is called *cross*-correlation because it correlates two different variables.

Two random variables are called *uncorrelated* if

$$\text{E}\{x \cdot y\} = \text{E}\{x\} \cdot \text{E}\{y\} = \mu_x \cdot \mu_y \,. \tag{B.14}$$

If furthermore the mean of one of the uncorrelated random variables is zero, they are called *orthogonal*, that is,

$$\text{E}\{x \cdot y\} = 0 \,. \tag{B.15}$$

For stochastic processes the correlation and covariance are defined as

$$\text{corr}\{x, y, k_1, k_2\} = \text{corr}_{xy}(k_1, k_2) = \text{E}\{x(k_1) \cdot y(k_2)\} \,, \tag{B.16}$$

$$\begin{aligned} \text{cov}\{x, y, k_1, k_2\} &= \text{cov}_{xy}(k_1, k_2) \\ &= \text{E}\{(x(k_1) - \mu_x(k_1)) \cdot \big(y(k_2) - \mu_y(k_2)\big)\} \,. \end{aligned} \tag{B.17}$$

Obviously, the correlation and covariance of stochastic processes depend on time k_1 and k_2. If the stochastic process is stationary, the statistical properties do not depend on time. Consequently, only the time shift $\kappa = k_2 - k_1$ between $x(k_1)$ and $y(k_2)$ matters, and the correlation and covariance simplify to (note that owing to stationarity the means no longer depend on time)

$$\text{corr}\{x, y, \kappa\} = \text{corr}_{xy}(\kappa) = \text{E}\{x(k) \cdot y(k + \kappa)\} \,, \tag{B.18}$$

$$\text{cov}\{x, y, \kappa\} = \text{cov}_{xy}(\kappa) = \text{E}\{(x(k) - \mu_x) \cdot \big(y(k + \kappa) - \mu_y\big)\} \,. \tag{B.19}$$

In practice, the pdfs for calculation of the expectations are usually unknown. Therefore, ergodicity is assumed, and the correlation and covariance are computed as time averages instead of realization averages. If $x(k)$ and $y(k)$ denote the measurements (i.e., one realization) of the stochastic processes for time instant $k = 1, \ldots, N$, the correlation and covariance can be estimated by

$$\mathrm{corr}_{xy}(\kappa) \approx \frac{1}{N - |\kappa|} \sum_{k=1}^{N-|\kappa|} x(k)\, y(k+\kappa)\,, \tag{B.20}$$

$$\mathrm{cov}_{xy}(\kappa) \approx \frac{1}{N - |\kappa| - 1} \sum_{k=1}^{N-|\kappa|} (x(k) - \bar{x})\; (y(k+\kappa) - \bar{y})\,. \tag{B.21}$$

The above estimates are unbiased [278] because the factor before the sum divides by the number of terms within the sum. (For the covariance estimate, the denominator is $N - |\kappa| - 1$ since it contains the *estimated* mean values $\bar{x}$ and $\bar{y}$; compare the variance estimation in the previous section.) Because only N data samples are available, $|\kappa| = N - 1$ is the largest possible time shift for this formula. Thus, for example, for the estimation of $\mathrm{corr}_{xy}(N-1)$, only a single term appears under the sum.

Although the above estimates are unbiased, their estimation variance becomes very large for $\kappa \to N$ because the sum averages only over a small number of terms. This often makes the above estimates practically useless. Therefore, the following biased estimates with a lower estimation variance are much more common:

$$\mathrm{corr}_{xy}(\kappa) \approx \frac{1}{N} \sum_{k=1}^{N-|\kappa|} x(k)\, y(k+\kappa)\,, \tag{B.22}$$

$$\mathrm{cov}_{xy}(\kappa) \approx \frac{1}{N - 1} \sum_{k=1}^{N-|\kappa|} (x(k) - \bar{x})\; (y(k+\kappa) - \bar{y})\,. \tag{B.23}$$

For a more general discussion about the tradeoffs between the estimator bias and variance, refer to Sect. B.7. Note that again the autocorrelation is obtained by setting $x = y$. This autocorrelation $\mathrm{corr}_{xx}(\kappa)$ tells us something about the correlation of a signal $x(k)$ with its time-shifted version $x(k - \kappa)$. Certainly, the correlation without any time shift, that is, $\mathrm{corr}_{xx}(0)$, takes the largest value. Typically, the more the signal is shifted, i.e., the higher κ is, the weaker the correlation becomes. The autocorrelation is always symmetric.

An extreme case is realized by the so-called white noise. This stochastic process has a correlation equal to zero for all $\kappa \neq 0$ and $\mathrm{corr}_{xx}(0) = 1$. Consequently, for one realization of white noise, a measurement at one time instant does not correlate with a measurement at any other time instant. This implies that measuring a white noise source at one time instant reveals no information about the next. White noise contains no systematical part; it is totally unpredictable. Therefore, an error between process and model should be as close to white noise as possible because this guarantees that all information is incorporated into the model and only "pure randomness" remains. White noise does not exist in practice since its power would be infinity. However, white noise is of great theoretical interest since it represents the case that is simplest to analyze. Furthermore, any kind of correlated and more realistic noise can be generated by feeding white noise through a dynamic filter.

Because the so-called covariance matrix is frequently used in the context of parameter estimation, this matrix will be defined here. If $\underline{x}$ is a vector of random variables $[x_1\ x_2\ \cdots\ x_n]$, the covariance matrix of $\underline{x}$ is defined as

$$\text{cov}\{\underline{x}\} = \text{cov}\{\underline{x}, \underline{x}\} = \begin{bmatrix} \text{cov}\{x_1, x_1\} & \text{cov}\{x_1, x_2\} & \cdots & \text{cov}\{x_1, x_n\} \\ \text{cov}\{x_2, x_1\} & \text{cov}\{x_2, x_2\} & \cdots & \text{cov}\{x_2, x_n\} \\ \vdots & \vdots & \ddots & \vdots \\ \text{cov}\{x_n, x_1\} & \text{cov}\{x_n, x_2\} & \cdots & \text{cov}\{x_n, x_n\} \end{bmatrix}. \tag{B.24}$$

Strictly speaking, the main diagonal entries of the covariance matrix contain variances. Therefore, this matrix is sometimes referred to as the variance-covariance matrix. Here, the shorter terminology is used for simplicity. One example for the use of this covariance matrix is for a quality assessment of the parameter estimates $\underline{\hat{\theta}} = [\hat{\theta}_1\ \hat{\theta}_2\ \cdots\ \hat{\theta}_n]^T$. Then $\text{cov}\{\underline{\hat{\theta}}\}$ contains the variances and covariances of the estimated parameter vector; see (3.34). Another example is the noise covariance matrix $\text{cov}\{\underline{n}\}$, where $\underline{n} = [n(1)\ n(2)\ \cdots\ n(N)]^T$ is the vector of noise realizations for the measured data samples $i = 1, \ldots, N$; see (3.66).

A covariance vector is obtained by the following definition of the covariance between a vector of random variables $\underline{x} = [x_1\ x_2\ \cdots\ x_n]$ and a scalar random variable y:

$$\text{cov}\{\underline{x}, y\} = \begin{bmatrix} \text{cov}\{x_1, y\} \\ \text{cov}\{x_2, y\} \\ \vdots \\ \text{cov}\{x_n, y\} \end{bmatrix}. \tag{B.25}$$

Correspondingly to these covariance matrices and vectors, correlation matrices and vectors can also be defined in a straightforward manner.

B.7 Properties of Estimators

It is very interesting to study the properties of an estimator, and consequently, of the estimates, it yields, in order to assess its quality. An estimator takes a (usually large) number of measurements denoted by N and maps them to a (smaller)

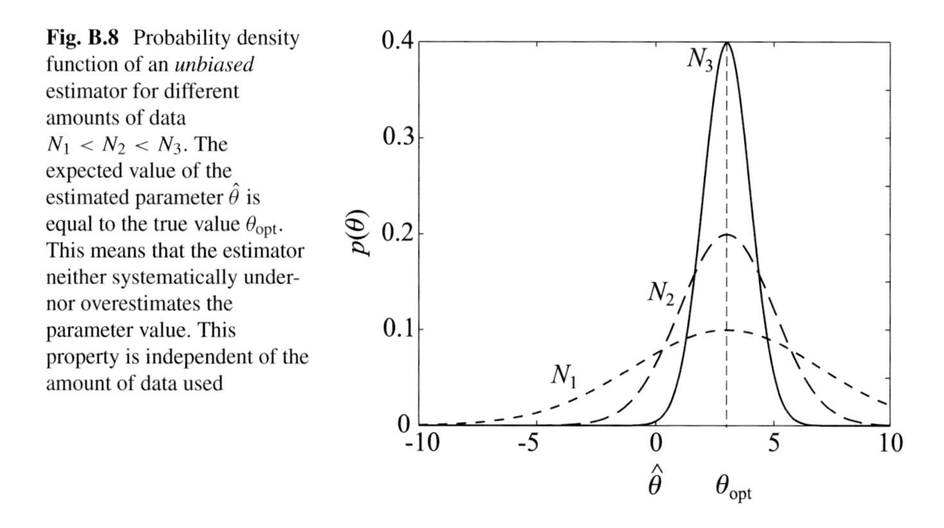

Fig. B.8 Probability density function of an *unbiased* estimator for different amounts of data $N_1 < N_2 < N_3$. The expected value of the estimated parameter $\hat{\theta}$ is equal to the true value θ_{opt}. This means that the estimator neither systematically under- nor overestimates the parameter value. This property is independent of the amount of data used

number of parameters $\underline{\theta} = [\theta_1\ \theta_2\ \cdots\ \theta_n]^T$. It can statistically be described in terms of a probability density function $p(\underline{\hat{\theta}}, N)$ which depends on the number of measurements N. This pdf describes what parameters are estimated with which probability given the data set. Figures B.8 and B.9 depict three examples. Since usually more than just one parameter is estimated, these pdfs are multivariate and cannot be visualized easily. Therefore, the typical properties of these pdfs are of interest. The most significant characteristics are the expected value (mean) $\text{E}\{\underline{\hat{\theta}}\}$ and the covariance matrix (variance in the univariate case) $\text{cov}\{\underline{\hat{\theta}}\}$. Many estimators' pdfs approach a Gaussian distribution as $N \to \infty$. If the pdf is Gaussian, the mean and covariance matrix characterize the distribution completely.

To interpret and analyze the pdf $p(\underline{\hat{\theta}}, N)$ and its characteristics, first Fig. B.8 is considered, which shows three pdfs for the univariate case $n = 1$ (one parameter). The true but in practice unknown parameter to be estimated is θ_{opt}. Now, N_1 input/output measurements are taken from the process in order to estimate the parameter. The output is assumed to be disturbed by stationary noise, i.e., noise that does not change its statistical characteristics (mean, covariance, etc.) over time. The estimator will yield some estimated parameter $\hat{\theta}^{(1)}$. Next, for exactly the same input data, N_1 measurements are taken again. Although the noise does not change its statistical characteristics, it yields a different realization of disturbances, and consequently, different output values are measured. Therefore, the estimator (generally) yields a different parameter estimate $\hat{\theta}^{(2)}$. Each time this experiment is repeated, different parameters are estimated, simply due to different disturbance realizations by the noise source. The pdf $p(\underline{\hat{\theta}}, N_1)$ shown in Fig. B.8 describes the distribution of these estimated parameters.

A good estimator should yield parameters that are close to the true value with a higher probability than parameters that are far away from the true value. This means that the maximum of the pdf should be close to the true value. Furthermore, the pdf should be "sharp" in order to give poor parameter estimates a low probability. The

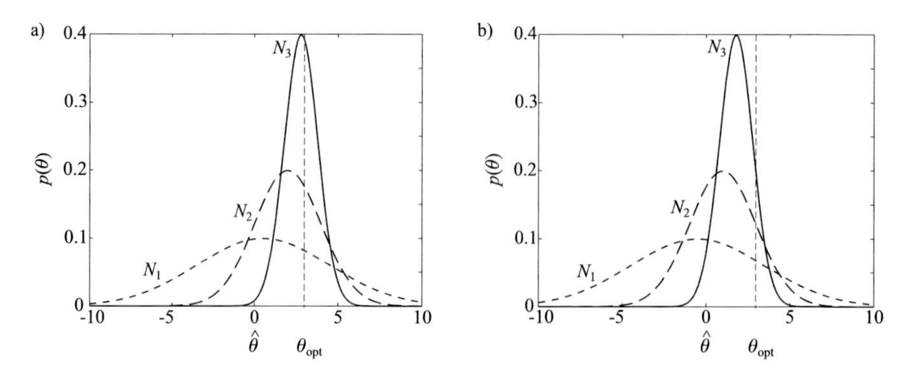

Fig. B.9 (**a**) Probability density function of a *biased* but *consistent* estimator for different amounts of data $N_1 < N_2 < N_3$. The expected value of the estimated parameter $\hat{\theta}$ is *not* equal to the true value θ_{opt}. However, the expected value of the estimated parameter approaches the true value as the amount of data used goes to infinity. This means the estimator systematically under- or overestimates the parameter value. However, this effect can be neglected if a large amount of data is used. (**b**) Probability density function of a *non-consistent* (and therefore also *biased*) estimator for different amounts of data $N_1 < N_2 < N_3$. The expected value of the estimated parameter $\hat{\theta}$ is *not* equal to the true value θ_{opt}. Even if the amount of data used approaches infinity, there is a systematical deviation between the expected value of the estimate and the true parameter value. This property is undesirable, since even for an optimal environment (low noise level, vast amount of data) one can expect a wrong estimate

optimal pdf would be a Dirac positioned at θ_{opt}. Another desirable feature for an estimator is to become "better" as the amount of data increases.

The three pdfs shown in Fig. B.8 correspond to different amounts of measurement data N_1, N_2, and N_3 with $N_1 < N_2 < N_3$. Figure B.8 illustrates that the pdf $p(\hat{\underline{\theta}}, N)$ becomes "sharper," i.e., its variance decreases, as the amount of data N increases. This property is quite natural because more data should allow one to better average out the disturbances and gather more information about the true parameters. For $N \to \infty$ the variance should approach zero. For most estimators the variance decreases with $1/N$ because the disturbances average out, and consequently the standard deviation decreases with $1/\sqrt{N}$. This relationship is encountered in almost any estimation task.

A comparison of the pdfs in Figs. B.8 and B.9 shows that the variance decrease for an increasing amount of data is similar. However, the behavior of the mean value differs. In Fig. B.8 the mean of the estimated parameter is equal to the true parameter θ_{opt} independent of the amount of data used. Such an estimator is called unbiased.

An estimator is called *unbiased* if for each amount of data N

$$E\{\hat{\underline{\theta}}\} = \underline{\theta}_{\text{opt}} \,. \tag{B.26}$$

Otherwise the estimator is called *biased*, and the systematical deviation between the expected value of the estimated parameters $E\{\hat{\underline{\theta}}\}$ and the true parameters $\underline{\theta}_{\text{opt}}$ is called the *bias*:

$$\underline{b} = E\{\hat{\underline{\theta}}\} - \underline{\theta}_{\text{opt}} \,. \tag{B.27}$$

In contrast to the estimator described by Figs. B.8, B.9a results from a biased estimator. The means of the pdfs reveal a systematical deviation from the true parameter value. However, this deviation decreases for an increasing amount of data: $\underline{b}(N_1) > \underline{b}(N_2) > \underline{b}(N_3)$. Moreover, the bias approaches zero as N approaches infinity. Such an estimator is called consistent or asymptotically unbiased (meaning that the bias vanishes as $N \to \infty$). Clearly, all unbiased estimators are consistent but not vice versa. Thus, consistency is a weaker property than unbiasedness.

An estimator is called *consistent* if for an increasing amount of data $N \to \infty$

$$\hat{\underline{\theta}} \to \underline{\theta}_{\text{opt}} \,. \tag{B.28}$$

Although an unbiased estimator is highly desirable for many applications, a biased but consistent estimator might be sufficient because a large amount of data is available, and the resulting bias can be neglected for large N. It can be shown that all maximum likelihood estimators are consistent. Figure B.9b shows the pdfs of a non-consistent estimator. This means that the estimated parameters do not converge to the true value as $N \to \infty$. Non-consistent estimators often are not acceptable.

The above discussion focused on the mean of the estimator's pdf. Now the variance (or covariance matrix in the multivariate case) is analyzed in more detail. A good estimator should be unbiased or at least consistent. However, this property is practically useless if the variance is very large. Then an estimator may yield parameter estimates far away from the true value with a high probability. Therefore, the estimator should have a small variance. Clearly, from all unbiased estimators, the one with the lowest variance is the best.

From all unbiased estimators, the one with the lowest variance is called *efficient*. In the multivariate case, this is the estimator with the smallest covariance matrix

$$\mathrm{cov}\{\hat{\underline{\theta}}\} \preceq \mathrm{cov}\{\hat{\underline{q}}\}\,, \tag{B.29}$$

where $\hat{\underline{q}}$ is the estimate of any unbiased estimator. Equation (B.29) means that the matrix $(\mathrm{cov}\{\hat{\underline{\theta}}\} - \mathrm{cov}\{\hat{\underline{q}}\})$ is negative semi-definite, i.e., has only non-positive eigenvalues. This relationship can also be defined via the determinants:

$$\det\left(\mathrm{cov}\{\hat{\underline{\theta}}\}\right) \le \det\left(\mathrm{cov}\{\hat{\underline{q}}\}\right)\,. \tag{B.30}$$

Note that although an efficient estimator has the smallest variance among all unbiased estimators, there exist *biased* estimators with lower variance. The quality of an estimator is given by its bias *and* its variance. This is demonstrated for the biased and unbiased correlation and covariance estimators discussed in Sect. B.6. Figure B.10 illustrates why a biased estimator with low variance can be more accurate than an unbiased estimator with high variance. In that case, the biased estimator is preferred over the unbiased one because it produces more reliable estimates. This bias/variance tradeoff is a fundamental and very general issue in statistics. In the context of modeling, it is addressed in Sect. 7.2. When many parameters have to be estimated from small, noisy data sets, it often pays to sacrifice unbiasedness in order to further reduce the variance. Methods that perform such a tradeoff between bias and variance are called *regularization techniques*; see Sect. 7.5.

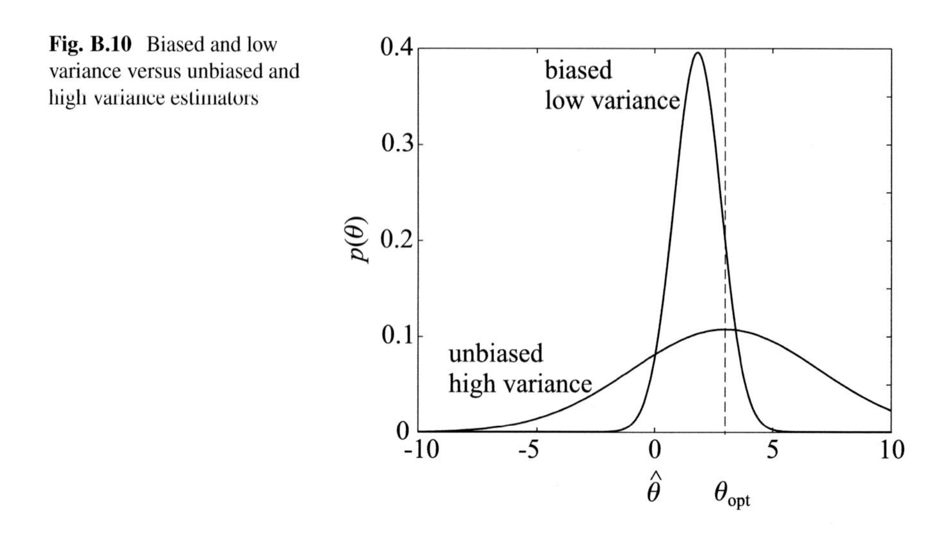

Fig. B.10 Biased and low variance versus unbiased and high variance estimators

References

1. Abou El Ela, A.: Sensorarme Methoden zur Bearbeitung komplexer Werkstücke mit Industrierobotern. Ph.D. Thesis, TU Darmstadt, Reihe 8: Mess-, Steuerungs- und Regelungstechnik, Nr. 824. VDI-Verlag, Düsseldorf (2000)
2. Aggarwal, C.C., Hinneburg, A., Keim, D.A.: On the surprising behavior of distance metrics in high dimensional space. In: International Conference on Database Theory, pp. 420–434. Springer (2001)
3. Akaike, H.: A new look at the statistical model identification. IEEE Trans. Autom. Control **19**(6), 716–723 (1974)
4. Alander, J.T.: An indexed bibliography of genetic algorithms and fuzzy systems. In: Pedrycz, W. (ed.) Fuzzy Evolutionary Computation, chapter 3.1, pp. 299–318. Kluwer Academic Publishers, Boston (1997). ftp: ftp.uwasa.fi, directory: /cs/report94-1, file: gaFUZZYbib.ps.Z
5. Albus, J.S.: Theoretical and Experimental Aspects of a Cerebellar Model. Ph.D. thesis, University of Maryland, Maryland, USA (1972)
6. Albus, J.S.: A new approach to manipulator control: the cerebellar model articulation controller (CMAC). Trans. ASME. J. Dyn. Syst. Meas. Control **63**(3), 220–227 (1975)
7. Aleksovski, D., Kocijan, J., Dzeroski, S.: Model tree ensembles for the identification of multiple-output systems. In: Control Conference (ECC), 2014 European, pp. 750–755. IEEE (2014)
8. Alessio, A., Bemporad, A.: A survey on explicit model predictive control. In: Nonlinear Model Predictive Control, pp. 345–369. Springer (2009)
9. Anderberg, M.R.: Cluster Analysis for Applications. Probability and Mathematical Statistics. Academic Press, New York (1971)
10. Anjyo, K., Lewis, J.P.: RBF interpolation and gaussian process regression through an RKHS formulation. J. Math. Ind. **3**(6), 63–71 (2011)
11. Aronszajn, N.: Theory of reproducing kernels. Trans. Am. Math. Soc. **68**(3), 337–404 (1950)
12. Atkeson, C.G., Moore, A.W., Schaal, S.: Locally weighted learning. Artif. Intell. Rev. **11**(1–5), 11–73 (1997)
13. Auer, P.: Using confidence bounds for exploitation-exploration trade-offs. J. Mach. Learn. Res. **3**, 397–422 (2002)
14. Ayoubi, M.: Nonlinear System Identification Based on Neural Networks with Locally Distributed Dynamics and Application to Technical Processes. Ph.D. Thesis, TU Darmstadt, Reihe 8: Mess-, Steuerungs- und Regelungstechnik, Nr. 591. VDI-Verlag, Düsseldorf (1996)

O. Nelles, *Nonlinear System Identification*,
https://doi.org/10.1007/978-3-030-47439-3

15. Ayoubi, M., Isermann, R.: Radial basis function networks with distributed dynamics for nonlinear dynamic system identification. In: European Congress on Intelligent Techniques and Soft Computing (EUFIT), pp. 631–635, Aachen, Germany (1995)
16. Babuška, R.: Fuzzy Modeling and Identification. Ph.D. thesis, Dept. of Control Engineering, Delft University of Technology, Delft, The Netherlands (1996)
17. Babuška, R., Fantuzzi, C., Kaymak, U., Verbruggen, H.B.: Improved inference for Takagi-Sugeno models. In: IEEE International Conference on Fuzzy Systems (FUZZ-IEEE), pp. 701–706, New Orleans, USA (1996)
18. Babuška, R., Setnes, M., Kaymak, U., van Naute Lemke, H.R.: Simplification of fuzzy rule bases. In: European Congress on Intelligent Techniques and Soft Computing (EUFIT), pp. 1115–1119, Aachen, Germany (1996)
19. Babuška, R., Verbruggen, H.B.: Comparing different methods for premise identification in Sugeno-Takagi models. In: European Congress on Intelligent Techniques and Soft Computing (EUFIT), pp. 1188–1192, Aachen, Germany (1994)
20. Babuška, R., Verbruggen, H.B.: An overview of fuzzy modeling for control. Control Eng. Pract. **4**(11), 1593–1606 (1996)
21. Babuška, R., Verbruggen, H.B.: Fuzzy set methods for local modelling and identification. In: Murray-Smith, R., Johansen, T.A. (eds.) Multiple Model Approaches to Modelling and Control, chapter 2, pp. 75–100. Taylor & Francis, London (1997)
22. Bäck, T., Hammel, U., Schwefel, H.-P.: Evolutionary computation: comments on the history and current state. IEEE Trans. Evol. Comput. **1**(1), 3–17 (1997)
23. Balageas, D., Fritzen, C.-P., Güemes, A.: Structural health monitoring, vol. 493. Wiley Online Library (2006)
24. Ballé, P.: Methoden zur Fehlerdiagnose für nichtlineare Prozesse mit linearen parameterveränderlichen Modellen. Ph.D. Thesis, TU Darmstadt, Reihe 8: Mess-, Steuerungs- und Regelungstechnik. VDI-Verlag, Düsseldorf (2000)
25. Ballé, P., Fischer, M., Füssel, D., Nelles, O., Isermann, R.: Integrated control, diagnosis, and reconfiguration of a heat exchanger. IEEE Control Syst. Mag. **18**(3), 52–63 (1998)
26. Ballé, P., Nelles, O.: Fault detection for a heat exchanger based on a bank of local linear fuzzy models of subprocesses. In: IFAC Symposium on Artificial Intelligence in Real-Time Control (AIRTC), pp. 606–613, Kuala Lumpur, Malaysia (1997)
27. Ballé, P., Nelles, O., Füssel, D.: Fault detection on nonlinear processes based on local linear fuzzy models in parallel- and series-parallel mode. In: IFAC Symposium on Fault Detection, Supervision and Safety for Technical Processes (SAFEPROCESS), pp. 1138–1143, Kingston upon Hull, UK (1997)
28. Bamberger, K.: Aerodynamic Optimization of Low-Pressure Axial Fans. Ph.D. thesis, University of Siegen (2015)
29. Bamberger, K., Belz, J., Carolus, T., Nelles, O.: Aerodynamic optimization of centrifugal fans using CFD-trained meta-models. In: 16th International Symposium on Transport Phenomena and Dynamics of Rotating Machinery (ISROMAC), Hawaii, USA (2016)
30. Bänfer, O., Nelles, O., Kainz, J., Beer, J.: Local model networks – the prospective method for modeling in electronic control units? ATZelektronik **3**(6), 36–39 (2008)
31. Bänfer, O., Nelles, O., Kainz, J., Beer, J.: Local model networks with modified parabolic membership functions. In: IEEE International Conference on Artificial Intelligence and Computational Intelligence (AICI), pp. 179–183, Shanghai, China (2009)
32. Barnard, E.: Optimization for training neural nets. IEEE Trans. Neural Netw. **3**(2), 322–325 (1992)
33. Battiti, R.: First and second order methods for learning: between stepest descent and Newton's method. Neural Comput. **4**(2), 141–166 (1992)
34. Battiti, R., Bertossi, A.A.: Greedy, prohibition, and reactive heuristics for graph partitioning. IEEE Trans. Comput. **48**(4), 361–385 (1999)
35. Baumann, W., Schaum, S., Roepke, K., Knaak, M.: Excitation signals for nonlinear dynamic modeling of combustion engines. In: Proceedings of the 17th World Congress, The International Federation of Automatic Control, Seoul, Korea (2008)

36. Baur, U.: On-line Parameterschätzverfahren zur Identifikation linearer, dynamischer Prozesse mit Prozeßrechner – Entwicklung, Vergleich, Erprobung. Ph.D. Thesis, Universität Stuttgart. KfK-PDV-Bericht-65, Kernforschungszentrum Karlsruhe (1976)
37. Bellman, R.E.: Adaptive Control Processes. Princeton University Press, New Jersey (1961)
38. Belz, J., Nelles, O.: Honda project: input selection with local linear models. Technical report, University of Siegen (2015)
39. Belz, J., Bamberger, K., Nelles, O.: Order of experimentation for metamodeling tasks. In: International Joint Conference on Neural Networks (IJCNN), pp. 4843–4849, Vancouver, Canada (2016)
40. Belz, J., Bamberger, K., Nelles, O., Carolus, T.: Goal-oriented active learning with local model networks. Int. J. Comput. Methods Exp. Measur. **6**(4), 785–796 (2018)
41. Belz, J., Münker, T., Heinz, T.O., Kampmann, G., Nelles, O.: Automatic modeling with local model networks for benchmark processes. In: Proceedings of the 20th IFAC World Congress, pp. 472–477, Toulouse, France (2017)
42. Belz, J., Nelles, O.: Function generator application: shall corners be measured? In: Proceedings 25. Workshop Computational Intelligence, Dortmund. KIT Scientific Publishing (2015)
43. Belz, J., Nelles, O.: Proposal for a function generator and extrapolation analysis. In: 2015 International Symposium on Innovations in Intelligent Systems and Applications (INISTA), pp. 1–6, Madrid, Spain. IEEE (2015)
44. Belz, J., Nelles, O.: Normalized L1 regularization for axis-oblique tree construction algorithms. In: IEEE Symposium Series on Computational Intelligence (SSCI), pp. 1–7, Hawaii, USA. IEEE (2017)
45. Bemporad, A., Borrelli, F., Morari, M., et al.: Model predictive control based on linear programming˜ the explicit solution. IEEE Trans. Autom. Control **47**(12), 1974–1985 (2002)
46. Benner, M.J., Tushman, M.L.: Exploitation, exploration, and process management: the productivity dilemma revisited. Acad. Manag Rev. **28**(2), 238–256 (2003)
47. Bergstra, J., Bengio, Y.: Random search for hyper-parameter optimization. J. Mach. Learn. Res. **13**(1), 281–305 (2012)
48. Bernd, T., Kroll, A., Schwarz, H.: Approximation nichtlinearer dynamischer Prozesse mit optimierten funktionalen Fuzzy-Modellen. In: 7. Workshop "Fuzzy Control" des GMA-UA 1.4.2, Dortmund, Germany (1997)
49. Bertram, T., Svaricek, F., Böhm, T., Kiendl, H., Pfeiffer, B.-M., Weber, M.: Fuzzy-Control. Zusammenstellung und Beschreibung wichtiger Begriffe. Automatisierungstechnik **42**(7), 322–326 (1994)
50. Bettenhausen, K.D.: Automatische Struktursuche für Regler und Strecke: Beiträge zur datengetriebenen Analyse und optimierenden Führung komplexer Prozesse mit Hilfe evolutionärer Methoden und lernfähiger Fuzzy-Systeme. Ph.D. Thesis, TU Darmstadt, Reihe 8: Mess-, Steuerungs- und Regelungstechnik, Nr. 574. VDI-Verlag, Düsseldorf (1996)
51. Bezdek, J.C.: Pattern Recognition with Fuzzy Objective Function. Plenum Press, New York (1981)
52. Bifet, A., Holmes, G., Kirkby, R., Pfahringer, B.: Moa: massive online analysis. J. Mach. Learn. Res. **11**, 1601–1604 (2010)
53. Billings, S.A., Fakhouri, S.Y.: Identification of systems containing linear dynamic and static nonlinear elements. Automatica **18**(1), 15–26 (1982)
54. Billings, S.A., Voon, W.S.F.: Correlation based validity tests for nonlinear models. Int. J. Control **44**(1), 235–244 (1986)
55. Billings, S.A., Voon, W.S.F.: Piecewise linear identification of non-linear systems. Int. J. Control **46**(1), 215–235 (1987)
56. Billings, S.A., Zhu, Q.M.: Nonlinear model validation using correlation tests. Int. J. Control **60**(6), 1107–1120 (1994)
57. Bishop, C.M.: Neural Networks for Pattern Recognition. Clarendon Press, Oxford (1995)
58. Biswas, P., Grieder, P., Löfberg, J., Morari, M.: A survey on stability analysis of discrete-time piecewise affine systems. IFAC Proc. **38**(1), 283–294 (2005)

59. Bleier, F.P.: Fan Handbook: Selection, Application, and Design. McGraw-Hill, New York (1998)
60. Bommes, L., Fricke, J., Grundmann, R.: Ventilatoren. Vulkan Verlag, Essen (2003)
61. Bossley, K.M.: Neurofuzzy Construction Algorithms. ISIS Technical Report ISIS-TR1, Department of Electronics and Computer Science, University of Southampton, Southampton, UK (1995)
62. Bossley, K.M.: Neurofuzzy Modelling Approaches in System Identification. Ph.D. thesis, Department of Electronics and Computer Science, University of Southampton, Southampton, UK (1997)
63. Bossley, K.M., Brown, M., Gunn, S.R., Harris, C.J.: Intelligent data modelling using neurofuzzy algorithms. In: IEE Colloquium on Industrial Applications of Intelligent Control, pp. 1–6, London, UK (1997)
64. Bossley, K.M., Brown, M., Harris, C.J.: Bayesian regularisation applied to neurofuzzy models. In: European Congress on Intelligent Techniques and Soft Computing (EUFIT), pp. 757–761, Aachen, Germany (1996)
65. Bothe, H.-H.: Fuzzy Logic: Einführung in Theorie und Anwendungen. Springer, Berlin (1993)
66. Box, G.E.P., Meyer, R.D.: An analysis for unreplicated fractional factorials. Technometrics **28**(1), 11–18 (1986)
67. Box, G.E.P., Jenkins, G.M.: Times Series Analysis, Forecasting and Control. Holden-Day, San Francisco (1970)
68. Boy, P.: Beitrag zur Berechnung des instationären Betriebsverhaltens von mittelschnelllaufenden Schiffsdieselmotoren. Ph.D. thesis, TU Hannover, Hannover, Germany (1980)
69. Boyd, S., Ghaoui, L.E., Feron, E., Balakrishnan, V.: Linear Matrix Inequalities in System and Control Theory. Studies in Applied Mathematics. SIAM, Philadelphia (1994)
70. Branch, M.A., Grace, A.: MATLAB Optimization Toolbox User's Guide, Version 1.5. The MATHWORKS Inc., Natick, MA (1998)
71. Brandimarte, P.: Low-discrepancy sequences. In: Handbook in Monte Carlo Simulation: Applications in Financial Engineering, Risk Management, and Economics, pp. 379–401 (2014)
72. Braun, H., Zagorski, P.: ENZO-II: a powerful design tool to evolve multilayer feed forward networks. In: IEEE Conference on Evolutionary Computation, pp. 278–283, Orlando, USA (1994)
73. Breiman, L.: Hinging hyperplanes for regression, classification, and function approximation. IEEE Trans. Inf. Theory **39**(3), 999–1013 (1993)
74. Breiman, L., Stone, C.J., Friedman, J.H., Olshen, R.: Classification and Regression Trees. Chapman & Hall, New York (1984)
75. Breiman, L., Spector, P.: Submodel Selection and Evaluation in Regression. The X-Random Case. International Statistical Review/Revue Internationale de Statistique, pp. 291–319 (1992)
76. Breiman, L.: Bagging predictors. Mach. Learn. **24**(2), 123–140 (1996)
77. Breiman, L.: Out-of-bag estimation. Technical report, Citeseer (1996)
78. Breiman, L.: Random forests. Mach. Learn. **45**(1), 5–32 (2001)
79. Breiman, L., et al.: Arcing classifier (with discussion and a rejoinder by the author). Ann. Stat. **26**(3), 801–849 (1998)
80. Brockwell, P.J., Davis, R.A.: Introduction to Time Series and Forecasting. Springer, New York (1996)
81. Broomhead, D.S., Lowe, D.: Multivariable functional interpolation and adaptive networks. Complex Syst. **2**, 321–355 (1988)
82. Brovelli, M.A., Sanso, F., Venuti, G.: A discussion on the Wiener–Kolmogorov prediction principle with easy-to-compute and robust variants. J. Geod. **76**(11–12), 673–683 (2003)
83. Brown, M., Bossley, K.M., Mills, D.J., Harris, C.J.: High dimensional neurofuzzy systems: overcoming the curse of dimensionality. In: IEEE International Conference on Fuzzy Systems (FUZZ-IEEE), pp. 2139–2146, New Orleans, USA (1995)

84. Brown, M., Harris, C.J.: Neurofuzzy Adaptive Modelling and Control. Prentice Hall, New York (1994)
85. Brown, M.D., Lightbody, D.G., Irwin, G.W.: Nonlinear internal model control using local model networks. IEE Proc. **144**(6), 505–514 (1997)
86. Bühlmann, P., Hothorn, T.: Boosting algorithms: regularization, prediction and model fitting. Stat. Sci. **22**(4), 477–505 (2007)
87. Burnham, K.P., Anderson, D.R.: Model selection and multimodel inference: a practical information-theoretic approach. Springer Science & Business Media (2002)
88. Cai, X., Qiu, H., Gao, L., Shao, X.: Metamodeling for high dimensional design problems by multi-fidelity simulations. Struct. Multidiscip. Optim. **56**(1), 151–166 (2017)
89. Cao, S.G., Rees, N.W., Feng, G.: Analysis and design for a class of complex control systems – part I: fuzzy modelling and identification. Automatica **33**(6), 1017–1028 (1997)
90. Carolus, T.: Ventilatoren. Vieweg+Teubner Verlag (2013)
91. Carpenter, G., Grossberg, S.: The ART of adaptive pattern recognition by a self-organizing neural network. IEEE Comput. **21**(3), 77–88 (1988)
92. Carpenter, G., Grossberg, S.: ART2: an adaptive resonance algorithm for rapid category learning and recognition. Neural Netw. **4**(4), 493–504 (1991)
93. Cawley, G.C., Talbot, N.L.C.: On over-fitting in model selection and subsequent selection bias in performance evaluation. J. Mach. Learn. Res. **11**, 2079–2107 (2010)
94. Chapelle, O., Vapnik, V., Bousquet, O., Mukherjee, S.: Choosing multiple parameters for support vector machines. Mach. Learn. **46**(1–3), 131–159 (2002)
95. Charalambous, C.: Conjugate gradient algorithm for efficient training of artificial neural networks. IEE Proc.-G **139**(3), 301–310 (1992)
96. Chen, P.-W., Wang, J.-Y., Lee, H.-M.: Model selection of SVMs using GA approach. In: 2004 IEEE International Joint Conference on Neural Networks. Proceedings. vol. 3, pp. 2035–2040. IEEE (2004)
97. Chen, R.-B., Hsieh, D.-N., Hung, Y., Wang, W.: Optimizing Latin hypercube designs by particle swarm. Stat. Comput. **23**(5), 663–676 (2013)
98. Chen, S., Billings, S.A.: Representations of nonlinear systems: the NARMAX model. Int. J. Control **49**(3), 1013–1032 (1983)
99. Chen, S., Billings, S.A., Luo, W.: Orthogonal least squares methods and their application to nonlinear system identification. Int. J. Control **50**(5), 1873–1896 (1989)
100. Chen, S., Cowan, C.F.N., Grant, P.M.: Orthogonal least-squares learning algorithm for radial basis function networks. IEEE Trans. Neural Netw. **2**(2) (1991)
101. Chen, T., Ljung, L.: Implementation of algorithms for tuning parameters in regularized least squares problems in system identification. Automatica **49**(7), 2213–2220 (2013)
102. Chen, V.C.P., Tsui, K.-L., Barton, R.R., Meckesheimer, M.: A review on design, modeling and applications of computer experiments. IIE Trans. **38**(4), 273–291 (2006)
103. Chen, X., Bushnell, M.: Efficient Branch and Bound Search with Applications to Computer-Aided Design. Kluwer Academic Publishers, Boston (1996)
104. Chiang, M.M.-T., Mirkin, B.: Intelligent choice of the number of clusters in k-means clustering: an experimental study with different cluster spreads. J. Classif. **27**(1), 3–40 (2010)
105. Chikkula, Y., Lee, J.H.: Input sequence design for parametric identification of nonlinear systems. In: American Control Conference (ACC), vol. 5, pp. 3037–3041, Albuquerque, USA (1997)
106. Chow, C.-M., Kuznetsov, A.G., Clarke, D.W.: Using multiple models in predictive control. In: European Control Conference (ECC), pp. 1732–1737, Rome, Italy (1995)
107. Chow, E.Y., Willsky, A.S.: Analytical redundancy and the design of robust detection systems. IEEE Trans. Autom. Control **29**(7), 603–614 (1984)
108. Ciuffo, B., Casas, J., Montanino, M., Perarnau, J., Punzo, V.: Gaussian process metamodels for sensitivity analysis of traffic simulation models: case study of AIMSUN mesoscopic model. Transp. Res. Rec. J. Transp. Res. Board **2390**, 87–98 (2013)
109. Cleveland, W.S., Devlin, S.J., Grosse, E.: Regression by local fitting: methods, properties, and computational algorithms. J. Econ. **37**, 87–114 (1996)

110. Cohn, D.: Neural network exploration using optimal experiment design. In: Cowan, J.D., Tesauro, G., Alspector, J. (eds.) Advances in Neural Information Processing Systems, vol. 6, pp. 1071–1083. Morgan Kaufmann, San Francisco (1994)
111. Cohn, D., Atlas, L., Ladner, R.: Training connectionist networks with queries and selective sampling. In: Touretzky, D.S. (ed.) Advances in Neural Information Processing Systems, vol. 2. Morgan Kaufmann, San Mateo (1990)
112. Cohn, D., Ghahramani, Z., Jordan, M.I.: Active learning with mixture models. In: Murray-Smith, R., Johansen, T.A. (eds.) Multiple Model Approaches to Modelling and Control, chapter 6, pp. 167–184. Taylor & Francis, London (1997)
113. Cohn, K.H., Holstein-Rathlou, N.H., Marsh, D.J., Marmarelis, V.Z.: Comparative nonlinear modeling of renal autoregulation in rats: volterra approach versus artificial neural networks. IEEE Trans. Neural Netw. **9**(3), 430–435 (1998)
114. Cordier, O.: Ähnlichkeitsbedingungen für Strömungsmaschinen. Brennstoff-Wärme-Kraft (BWK) **5**(10), 337–340 (1953)
115. Cordón, O., Herrera, F., Lozano, M.: A Classified Review on the Combination Fuzzy Logic: Genetic Algorithms Bibliography. Technical report, Department of Computer Science and Artificial Intelligence, Spain (1996)
116. Correa, A.A., Grima, P., Tort-Martorell, X.: Experimentation order in factorial designs: new findings. J. Appl. Stat. **39**(7), 1577–1591 (2012)
117. Davis, L.: Handbook of Genetic Algorithms. van Nostrand Reinhold, New York (1991)
118. Deflorian, M., et al.: Versuchsplanung und Methoden zur Identifikation zeitkontinuierlicher Zustandsraummodelle am Beispiel des Verbrennungsmotors. Ph.D. thesis, Technische Universität München (2011)
119. Delphi Automotive Systems: Emissions Standards Passennger Cars Worldwide. Luxembourg (1997)
120. Demuth, H., Beale, M.: MATLAB Neural Network Toolbox User's Guide, Version 3.0. The MATHWORKS Inc., Natick, MA (1998)
121. Didcock, N., Jakubek, S., Kögeler, H.-M.: Regularisation methods for neural network model averaging. Eng. Appl. Artif. Intell. **41**, 128–138 (2015)
122. Dietterich, T.G.: Ensemble methods in machine learning. In: International Workshop on Multiple Classifier Systems, pp. 1–15. Springer (2000)
123. Donoho, D.L.: Compressed sensing. IEEE Trans. Inf. Theory **52**(4), 1289–1306 (2006)
124. Dotsenko, V.I., Faradzhev, R.G., Chakartisvhili, G.S.: Properties of maximum length sequences with p-levels. Automatika i Telemechanika **8**, 189–194 (1971)
125. Draper, N.R., Smith, H.: Applied Regression Analysis. Probability and Mathematical Statistics. John Wiley & Sons, New York (1981)
126. Driankov, D., Hellendoorn, H., Reinfrank, M.: An Introduction to Fuzzy Control. Springer, Berlin (1993)
127. Duan, K., Keerthi, S.S., Poo, A.N.: Evaluation of simple performance measures for tuning SVM hyperparameters. Neurocomputing **51**, 41–59 (2003)
128. Dubios, D., Prade, H.: Fuzzy Sets and Systems. Theory and Applications. Academic Press, New York (1980)
129. Duvenaud, D.: Automatic Model Construction with Gaussian Processes. Ph.D. thesis, University of Cambridge (2014)
130. Ebert, T., Nelles, O.: A Note on Analytical Gradient Calculation for Hilomot. Technical report, University of Siegen (2013)
131. Ebert, T., Fischer, T., Belz, J., Heinz, T.O., Kampmann, G., Nelles, O.: Extended deterministic local search algorithm for maximin Latin hypercube designs. In: 2015 IEEE Symposium Series on Computational Intelligence: IEEE Symposium on Computational Intelligence in Control and Automation (2015 IEEE CICA), Cape Town, South Africa (2015)
132. Efron, B., Efron, B.: The Jackknife, the Bootstrap and Other Resampling Plans, vol. 38. SIAM (1982)
133. Efron, B., Hastie, T., Johnstone, I., Tibshirani, R., et al.: Least angle regression. Ann. Stat. **32**(2), 407–499 (2004)

134. Elman, J.L.: Finding structure in time. Cognit. Sci. **14**, 179–211 (1990)
135. Engel, C.: Untersuchung der Laufradströmung in einem Radialventilator mittels Particle Image Velocimetry (PIV). Ph.D. thesis, Universität Duisburg-Essen, Fakultät für Ingenieurwissenschaften, Maschinenbau und Verfahrenstechnik, Institut für Energie-und Umweltverfahrenstechnik (2007)
136. Ernst, E.-J., Hecker, O.: Predictive control of a heat exchanger. In: Seminar of ESPRIT Working Group CIDIC on Theory and Applications of Model Based Predictive Control (MBPC), Brussels, Belgium (1996)
137. Ernst, S.: Hinging hyperplane trees for approximation and identification. In: IEEE Conference on Decision and Control (CDC), pp. 1261–1277, Tampa, USA (1998)
138. Ernst, S.: Nonlinear system identification with hinging hyperplane trees. In: European Congress on Intelligent Techniques and Soft Computing (EUFIT), pp. 659–663, Aachen, Germany (1998)
139. Ernst, S., Nelles, O., Isermann, R.: Neuronale Netze für die Identifikation nichtlinearer, dynamischer Systeme: Ein Überblick. In: GMA-Kongreß, pp. 713–722, Baden-Baden, Germany (1996)
140. Evgeniou, T., Pontil, M., Poggio, T.: Regularization networks and support vector machines. Adv. Comput Math. **13**(1), 1–50 (2000)
141. Eykhoff, P.: System Identification. John Wiley & Sons, London (1974)
142. Fabri, S., Kadirkamanathan, V.: Neural control of nonlinear systems with composite adaptation for improved convergence of gaussian networks. In: European Control Conference (ECC), pp. 231–236, Brussels, Belgium (1997)
143. Fahlman, S.E., Lebiere, C.: The cascade-correlation learning architecture. In: Touretzky, D.S. (ed.) Advances in Neural Information Processing Systems, vol. 2, pp. 524–532. Morgan Kaufmann, San Mateo (1990)
144. Fang, K.-T., Li, R.: Uniform design for computer experiments and its optimal properties. Int. J. Mater. Prod. Technol. **25**(1–3), 198–210 (2005)
145. Fink, A., Fischer, M., Nelles, O.: Supervison of nonlinear adaptive controllers based on fuzzy models. In: IFAC World Congress, vol. Q, pp. 335–340, Beijing, China (1999)
146. Fink, A., Nelles, O., Fischer, M.: Linearization based and local model based controller design. In: European Control Conference (ECC), Karlsruhe, Germany (1999)
147. Fischer, M.: Fuzzy-modellbasierte Regelung nichtlinearer Prozesse. Ph.D. Thesis, TU Darmstadt, Reihe 8: Mess-, Steuerungs- und Regelungstechnik, Nr. 750. VDI-Verlag, Düsseldorf (1999)
148. Fischer, M., Fink, A., Hecker, O., Pfeiffer, B.-M., Kofahl, R., Schumann, R.: Prädiktive Regelung eines Wärmetauschers auf der Basis on-line adaptierter Fuzzy-Modelle. Automatisierungstechnik **46**(9), 426–434 (1998)
149. Fischer, M., Isermann, R.: Inverse fuzzy process models for robust hybrid control. In: Driankov, D., Palm, R. (eds.) Advances in Fuzzy Control, Studies in Fuzziness and Soft Computing, pp. 103–128. Physica-Verlag, Heidelberg (1998)
150. Fischer, M., Isermann, R.: Neuro-Fuzzy-Modelle zur Integration theoretischer und experimenteller Modellbildung am Beispiel eines Kreuzstrom-Wärmeaustauschers. Automatisierungstechnik **47**(5), 224–232 (1999)
151. Fischer, M., Nelles, O.: Tuning of PID-controllers for nonlinear processes based on local linear fuzzy models. In: European Congress on Intelligent Techniques and Soft Computing (EUFIT), pp. 1891–1895, Aachen, Germany (1996)
152. Fischer, M., Nelles, O.: Fuzzy model-based predictive control of nonlinear processes with fast dynamics. In: International ICSC Symposium on Fuzzy Logic and Applications, pp. 57–63, Zürich, Switzerland (1997)
153. Fischer, M., Nelles, O., Fink, A.: Adaptive fuzzy model-based control. Journal A **39**(3), 22–28 (1998)
154. Fischer, M., Nelles, O., Fink, A.: Predictive control based on local linear fuzzy models. Int. J. Syst. Sci. **28**(7), 679–697 (1998)

155. Fischer, M., Nelles, O., Isermann, R.: Adaptive predictive control based on local linear fuzzy models. In: IFAC Symposium on System Identification (SYSID), pp. 859–864, Kitakyushu, Fukuoka, Japan (1997)
156. Fischer, M., Nelles, O., Isermann, R.: Exploiting prior knowledge in fuzzy model identification of a heat exchanger. In: IFAC Symposium on Artificial Intelligence in Real-Time Control (AIRTC), pp. 445–450, Kuala Lumpur, Malaysia (1997)
157. Fischer, M., Nelles, O., Isermann, R.: Fuzzy model-based predictive control of a heat exchanger. In: IFAC Symposium on Artificial Intelligence in Real-Time Control (AIRTC), pp. 463–468, Kuala Lumpur, Malaysia (1997)
158. Fischer, M., Nelles, O., Isermann, R.: Adaptive predictive control of a heat exchanger based on a fuzzy model. Control Eng. Pract. (CEP) **6**(6), 259–269 (1998)
159. Fischer, T., Hartmann, B., Nelles, O.: Increasing the performance of a training algorithm for local model networks. In: World Congress of Engineering and Computer Science (WCECS), pp. 1104–1109, San Francisco, USA (2012)
160. Fischer, T., Nelles, O.: Merging strategy for local model networks based on the Lolimot algorithm. In: International Conference on Artificial Neural Networks, pp. 153–160. Springer (2014)
161. Föllinger, O.: Nichtlineare Regelungen I, 7th edn. Oldenbourg-Verlag, München (1993)
162. Fonseca, C.M., Fleming, R.J.: Multiobjective optimization and multiple contraint handling with evolutionary algorithms – part I: a unified formulation. IEEE Trans. Syst. Man Cybern. Part A: Syst. Humans **28**(1), 38–47 (1998)
163. Fonseca, C.M., Fleming, R.J.: Multiobjective optimization and multiple contraint handling with evolutionary algorithms – part II: application example. IEEE Trans. Syst. Man Cybern. Part A: Syst. Humans **28**(1), 26–37 (1998)
164. Ford, I., Titterington, D.M., Kitsos, C.P.: Recent advances in nonlinear experimental design. Technometrics **31**(1), 49–60x (1989)
165. Fortescue, T.R., Kershenbaum, L.S., Ydstie, B.E.: Implementation of self-tuning regulators with variable forgetting factor. Automatica **17**, 831–835 (1981)
166. Francois, D., Wertz, V., Verleysen, M., et al.: About the locality of kernels in high-dimensional spaces. In: International Symposium on Applied Stochastic Models and Data Analysis, pp. 238–245 (2005)
167. Frean, M.: The upstart algorithm: a method for constructing and training feedforward neural networks. Neural Comput. **2**(2), 198–209 (1990)
168. Freeman, J.A.S., Saad, D.: Learning and generalization in radial basis function networks. Neural Comput. **7**(5), 1000–1020 (1995)
169. Freeman, J.A.S., Saad, D.: Online learning in radial basis function networks. Neural Comput. **9**(7), 1601–1622 (1997)
170. Freeman, J.A.S.: Greedy function approximation: a gradient boosting machine. Ann. Stat. **29**, 1189–1232 (2001)
171. Friedman, J.H.: Stochastic gradient boosting. Comput. Stat. Data Anal. **38**(4), 367–378 (2002)
172. Friedman, J.H.: Multivariate adaptive regression splines (with discussion). Ann. Stat. **19**(1), 1–141 (1991)
173. Friedman, J.H., Stuetzle, W.: Projection pursuit regression. J. Am. Stat. Assoc. **76**, 817–823 (1981)
174. Fritzke, B.: Fast learning with incremental radial basis function networks. Neural Process. Lett. **1**(1), 2–5 (1994)
175. Fritzke, B.: Growing cell structures: a self-organizing network for unsupervised and supervised learning. Neural Netw. **7**(9), 1441–1460 (1994)
176. Fuessel, D.: Self-learning classification tree (select) – a human-like approach to fault diagnosis. In: European Congress on Intelligent Techniques and Soft Computing (EUFIT), pp. 659–663, Aachen, Germany (1997)
177. Fuessel, D., Ballé, P., Isermann, R.: Closed loop fault diagnosis on a nonlinear process model and automatic fuzzy rule generation. In: IFAC Symposium on Fault Detection, Supervision and Safety for Technical Processes (SAFEPROCESS), Kingston upon Hull, UK (1997)

178. Gahinet, P., Nemirovski, A., Laub, A.J., Chilali, M.: MATLAB LMI Control Toolbox User's Guide, Version 1.0. The MATHWORKS Inc., Natick, MA (1995)
179. Gallant, S.I.: Neural Networks Learning and Expert Systems. MIT Press, Cambridge (1987)
180. Garulli, A., Paoletti, S., Vicino, A.: A survey on switched and piecewise affine system identification. IFAC Proc. **45**(16), 344–355 (2012)
181. Geman, S., Bienenstock, E., Doursat, R.: Neural networks and the bias/variance dilemma. Neural Comput. **4**(1), 1–58 (1992)
182. Gen, M.: Genetic Algorithms and Engineering Design. Wiley, New York (1997)
183. Gertler, J.: Analytical redundancy methods in failure detection and isolation. In: IFAC Symposium on Fault Detection, Supervision and Safety for Technical Processes (SAFEPROCESS), Baden-Baden, Germany (1991)
184. Gevers, M., Ljung, L.: Optimal experiment design with respect to the intended model application. Automatica **22**(5), 543–554 (1986)
185. Gill, P.E., Murray, W., Wright, M.H.: Practical Optimization. Academic Press, London (1981)
186. Girosi, F., Jones, M., Poggio, T.: Regularization theory and neural networks architectures. Neural Comput. **7**(2), 219–269 (1995)
187. Girosi, F., Poggio, T.: Representation properties of networks: Kolmogorov's theorem is irrelevant. Neural Comput. **1**(4), 465–469 (1989)
188. Goedicke, W.: Fehlererkennung an einem technischen Prozeß mit Methoden der Parameterschätzung. Ph.D. Thesis, TU Darmstadt, Reihe 8: Mess-, Steuerungs- und Regelungstechnik, Nr. 130. VDI-Verlag, Düsseldorf (1987)
189. Goldberg, D.E.: Genetic Algorithms in Search, Optimization & Machine Learning. Addison-Wesley, Reading (1989)
190. Golub, G., Pereyra, V.: Separable nonlinear least squares: the variable projection method and its applications. Inverse Prob. **19**(2), R1 (2003)
191. Golub, G.H., Heath, M., Wahba, G.: Generalized cross-validation as a method for choosing a good ridge parameter. Technometrics **21**(2), 215–223 (1979)
192. Golub, G.H., Pereyra, V.: The differentiation of pseudo-inverses and nonlinear least squares problems whose variables separate. SIAM J. Numer. Anal. **10**(2), 413–432 (1973)
193. Golub, G.H., Van Loan, C.F.: Matrix Computations. Mathematical Sciences. The Johns Hopkins University Press, Baltimore (1987)
194. Goodfellow, I., Bengio, Y., Courville, A.: Deep Learning. MIT Press (2016). http://www.deeplearningbook.org
195. Greff, K., Srivastava, R.K., Koutník, J., Steunebrink, B.R., Schmidhuber, J.: LSTM: a search space odyssey. IEEE Trans. Neural Netw. Learn. Syst. (2016)
196. Grosso, A., Jamali, A., Locatelli, M.: Finding maximin Latin hypercube designs by iterated local search heuristics. Eur. J. Oper. Res. **197**(2), 541–547 (2009)
197. Guo, X.C., Yang, J.H., Wu, C.G., Wang, C.Y., Liang, Y.C.: A novel LS-SVMS hyper-parameter selection based on particle swarm optimization. Neurocomputing **71**(16), 3211–3215 (2008)
198. Gutjahr, T., Kleinegräber, H., Ulmer, H., Kruse, T., Eckstein, C.: New approaches for modeling dynamic engine behavior with gaussian processes. In: Röpke, K. (ed.) Design of Experiments (DoE) in Engine Development. Expert Verlag (2013)
199. Haber, R.: Nonlinearity tests for dynamic processes. In: IFAC Symposium on Identification and System Parameter Estimation, pp. 409–414, York, UK (1985)
200. Haber, R., Keviczky, L.: Identification of 'linear' systems having state-dependent parameters. Int. J. Syst. Sci. **16**(7), 869–884 (1985)
201. Haber, R., Unbehauen, H.: Structure identification of nonlinear dynamic systems: a survey on input/output approaches. Automatica **26**(4), 651–677 (1990)
202. Häck, M., Köhne, M.: Internal Model Control mit neuronalen Netzen zur Regelung eines Prozessanalysators. Automatisierungstechnik **45**(1) (1997)
203. Hafner, M., Schüler, M., Isermann, R.: Fast neural networks for diesel engine control design. In: IFAC World Congress, Beijing, China (to appear in 1999)

204. Hafner, M., Schüler, M., Nelles, O.: Dynamical identification and control of combustion engine exhaust. In: American Control Conference (ACC), pp. 222–226, San Diego, USA (1999)
205. Hafner, M., Schüler, M., Nelles, O.: Neural net models for Diesel engines – simulation and exhaust optimization. In: European Congress on Intelligent Techniques and Soft Computing (EUFIT), vol. 1, pp. 215–219, Aachen, Germany (1998)
206. Hafner, S.: Neuronale Netze in der Automatisierungstechnik. Oldenbourg, München (1994)
207. Hagan, M.T., Menhaj, M.B.: Training feedforward networks with the Marquardt algorithm. IEEE Trans. Neural Netw. **5**(6), 989–993 (1994)
208. Hainmueller, J., Hazlett, C.: Kernel regularized least squares: reducing misspecification bias with a flexible and interpretable machine learning approach. Polit. Anal. mpt019 (2013)
209. Halfmann, C., Nelles, O., Holzmann, H.: Semi-physical modeling of the vertical vehicle dynamics. In: American Control Conference (ACC), pp. 1707–1711, San Diego, USA (1999)
210. Halgamuge, S.: Advanced Methods for Fusion of Fuzzy Systems and Neural Networks in Intelligent Data Processing. Reihe 10, Nr. 401. VDI-Verlag, Düsseldorf (1995)
211. Halgamuge, S., Glesner, M.: Fuzzy neural fusion techniques for industrial applications. In: ACM Symposium on Applied Computing, Phoenix, USA (1994)
212. Halme, A., Visala, A., Zhang, X.-C.: Process modelling using the functional state approach. In: Murray-Smith, R., Johansen, T.A. (eds.) Multiple Model Approaches to Modelling and Control, chapter 4, pp. 121–144. Taylor & Francis, London (1997)
213. Hametner, C., Jakubek, S.: Neuro-fuzzy modelling using a logistic discriminant tree. In: 2007 American Control Conference, pp. 864–869. IEEE (2007)
214. Hametner, C., Jakubek, S.: Nonlinear system identification through local model approaches: partitioning strategies and parameter estimation. INTECH Open Access Publisher (2010)
215. Hametner, C., Jakubek, S.: Local model network identification for online engine modelling. Inf. Sci. **220**, 210–225 (2013)
216. Hametner, C., Stadlbauer, M., Deregnaucourt, M., Jakubek, S., Winsel, T.: Optimal experiment design based on local model networks and multilayer perceptron networks. Eng. Appl. Artif. Intell. **26**(1), 251–261 (2013)
217. Hänsler, E.: Statistische Signale. Springer, Berlin (1991)
218. Hardt, M., Recht, B., Singer, Y.: Train faster, generalize better: stability of stochastic gradient descent. In: International Conference on Machine Learning, pp. 1225–1234 (2016)
219. Hartman, E., Keeler, J.D.: Predicting the future: advantages of semilocal units. Neural Comput. **3**(4), 566–578 (1991)
220. Hartman, E., Keeler, J.D., Kowalski, J.M.: Layered neural networks with gaussian hidden units as universal approximators. Neural Comput. **2**(2), 210–215 (1990)
221. Hartmann, B.: Lokale Modellnetze zur Identifikation und Versuchsplanung nichtlinearer Systeme. Ph.D. thesis, Universität Siegen (2014)
222. Hartmann, B.: Lokale Modellnetze zur Identifikation und Versuchsplanung nichtlinearer Systeme. Ph.D. thesis, Universitätsbibliothek der Universität Siegen (2014)
223. Hartmann, B., Baumann, W., Nelles, O.: Axes-oblique partitioning of local model networks for engine calibration. In: Design of Experiments (DoE) in Engine Development, pp. 92–106, Berlin, Germany. Expert Verlag (2013)
224. Hartmann, B., Ebert, T., Nelles, O.: Model-based design of experiments based on local model networks for nonlinear processes with low noise levels. In: American Control Conference (ACC), pp. 5306–5311, 29 2011–July 1 2011 (2011)
225. Hartmann, B., Moll, J., Nelles, O., Fritzen, C.-P.: Modeling of nonlinear wave velocity characteristics in a structural health monitoring system. In: IEEE International Conference on Control Applications (CCA), Yokohama, Japan (2010)
226. Hartmann, B., Moll, J., Nelles, O., Fritzen, C.-P.: Hierarchical local model trees for design of experiments in the framework of ultrasonic structural health monitoring. In: IEEE International Conference on Control Applications (CCA), pp. 1163–1170. IEEE (2011)
227. Hartmann, B., Nelles, O.: Automatic adjustment of the transition between local models in a hierarchical structure identification algorithm. In: European Control Conference (ECC), Budapest, Hungary (2009)

228. Hartmann, B., Nelles, O.: Adaptive test planning for the calibration of combustion engines – methodology. In: Design of Experiments (DoE) in Engine Development, pp. 1–16, Berlin, Germany. Expert Verlag (2013)
229. Hastie, T., Tibshirani, R.: Generalized Additive Models. Wiley Online Library (1990)
230. Hastie, T., Tibshirani, R., Friedman, J.: The Elements of Statistical Learning. Springer Series in Statistics, 2nd edn. Springer, Berlin (2009)
231. Hathaway, R.J., Bezdek, J.C.: Switching regression models and fuzzy clustering. IEEE Trans. Fuzzy Syst. **1**(3), 195–204 (1993)
232. Haykin, S.: Adaptive Filter Theory. Prentice Hall, Oxford (1991)
233. Haykin, S.: Neural Networks. A Comprehensive Foundation. Macmillan, New York (1994)
234. He, X., Asada, H.: A new method for identifying orders of input-output models for nonlinear dynamic systems. In: American Control Conference (ACC), pp. 2520–2524, San Francisco, USA (1993)
235. Hecht-Nielsen, R.: Neurocomputing. Addison-Wesley, Reading (1990)
236. Hecker, O., Nelles, O., Isermann, R.: Nonlinear system identification and predictive control of a heat exchanger based on local linear model trees. In: American Control Conference (ACC), pp. 3294–3299, Albuquerque, USA (1997)
237. Heinz, T.O., Belz, J., Nelles, O.: Design of experiments – combining linear and nonlinear inputs. In: Hoffman, F., Hüllermeier, E., Mikut, R. (eds.) Proceedings 27. Workshop Computational Intelligence, pp. 211–226. KIT Scientific Publishing (2017)
238. Heinz, T.O., Nelles, O.: Efficient pole optimization of nonlinear Laguerre filter models. In: 2015 IEEE Symposium Series on Computational Intelligence, pp. 1–7. IEEE (2016)
239. Heinz, T.O., Nelles, O.: Vergleich von anregungssignalen für nichtlineare identifikationsaufgaben. In: Hoffman, F., Hüllermeier, E., Mikut, R. (eds.) Proceedings 26. Workshop Computational Intelligence, pp. 139–158. KIT Scientific Publishing (2016)
240. Heinz, T.O., Nelles, O.: Iterative excitation signal design for nonlinear dynamic black-box models. Proc. Comput. Sci. **112**, 1054–1061 (2017)
241. Heinz, T.O., Nelles, O.: Excitation signal design for nonlinear dynamic systems with multiple inputs – a data distribution approach. at-Automatisierungstechnik **66**(9), 714–724 (2018)
242. Heinz, T.O., Schillinger, M., Hartmann, B., Nelles, O.: Excitation signal design for nonlinear dynamic systems. In: Röpke, K., Gühmann, C. (eds.) International Calibration Conference – Automotive Data Analytics, Methods, DoE, pp. 191–208. expertVerlag (2017)
243. Heitkoetter, J., Beasley, D.: The hitch-hiker's guide to evolutionary computation: a list of frequently asked questions (FAQ), usenet: comp. ai. genetic. available via anonymous FTP from RTFM (1996)
244. Herrera, F., Lozano, M., Verdegay, J.L.: A Learning for Fuzzy Control Rules Using Genetic Algorithms. Technical Report DECSAI-95108, Department of Computer Science and Artificial Intelligence, Spain (1995)
245. Heuberger, P.S.C., Van den Hof, P.M.J., Bosgra, O.H.: Modelling linear dynamical systems through generalized orthonormal basis functions. In: IFAC World Congress, vol. 5, pp. 19–22, Sydney, Australia (1993)
246. Heuberger, P.S.C., Van den Hof, P.M.J., Bosgra, O.H.: A generalized orthonormal basis for linear dynamical systems. IEEE Trans. Autom. Control **40**(3), 451–465 (1995)
247. Heywood, J.B.: Internal Combustion Engine Fundamentals. McGraw-Hill, Inc. (1988)
248. Hilhorst, R.A.: Supervisory Control of Mode-Switch Processes. Ph.D. thesis, Electrical Engineering Department, University of Twente, Twente, The Netherlands (1992)
249. Hilow, H.: Comparison among run order algorithms for sequential factorial experiments. Comput. Stat. Data Anal. **58**, 397–406 (2013)
250. Hinton G.E., Osindero, S., Teh, Y.-W.: A fast learning algorithm for deep belief nets. Neural Comput. **18**(7), 1527–1554 (2006)
251. Hjalmarsson, H., Gevers, M., de Bruyne, F.: For model-based control design, closed-loop identification gives better performance. Automatica **32**(12), 1659–1673 (1996)
252. Hochreiter, S., Schmidhuber, J.: Long short-term memory. Neural Comput. **9**(8), 1735–1780 (1997)

253. Hoffmann, S., Schrott, M., Huber, T., Kruse, T.: Model-based methods for the calibration of modern internal combustion engines. MTZ Worldwide **76**(4), 24–29 (2015)
254. Hofmann, T., Schölkopf, B., Smola, A.J.: A tutorial review of RKHS methods in machine learning (2005)
255. Hohensohn, J., Mendel, J.M.: Two-pass orthogonal least-squares algorithm to train and reduce fuzzy logic systems. In: IEEE International Conference on Fuzzy Systems (FUZZ-IEEE), pp. 696–700, Orlando, USA (1994)
256. Holland, J.H.: Adaptation in Natural and Artificial Systems. University of Michigan Press, Ann Arbor (1975)
257. Holzmann, H., Nelles, O., Halfmann, C., Isermann, R.: Vehicle dynamics simulation based on hybrid modeling. In: IEEE/ASME Conference on Advanced Intelligent Mechatronics, pp. 19–23, Atlanta, USA (1999)
258. Homaifar, A., McCormick, E.: Simultaneous design of membership functions and rule sets for fuzzy controllers using genetic algorithms. IEEE Trans. Fuzzy Syst. **3**(2), 129–139 (1995)
259. Hopfield, J.J.: Neural networks and physical systems with emergent collective computational abilities. Proc. Natl. Acad. Sci. USA **79** (1982)
260. Höppner, F., Klawonn, F., Kruse, R.: Fuzzy-Clusterananlyse. Vieweg, Braunschweig (1997)
261. Hornik, K., Stinchombe, M., White, H.: Multilayer feedforward networks are universal approximators. Neural Comput. **2**, 359–366 (1989)
262. Hsu, C.-W., Lin, C.-J.: A comparison of methods for multiclass support vector machines. IEEE Trans. Neural Netw. **13**(2), 415–425 (2002)
263. Hu, C., Wan, F.: Input selection in learning systems: a brief review of some important issues and recent developments. In: IEEE International Conference on Fuzzy Systems. FUZZ-IEEE 2009, pp. 530–535. IEEE (2009)
264. Huang, G.-B., Zhu, Q.-Y., Siew, C.-K.: Extreme learning machine: theory and applications. Neurocomputing **70**(1–3), 489–501 (2006)
265. Huber, P.: Projection pursuit (with discussion). Ann. Stat. **13**, 435–475 (1985)
266. Hunt, K.J., Haas, R., Murray-Smith, R.: Extending the functional equivalence of radial basis functions networks and fuzzy inference systems. IEEE Trans. Neural Netw. **7**(3), 776–781 (1996)
267. Hunt, K.J., Kalkkuhl, J.C., Fritz, H., Johansen, T.A.: Constructive empirical modelling of longitudinal vehicle dynamics with local model networks. Control Eng. Pract. **4**, 167–178 (1996)
268. Hunt, K.J., Sbarbaro, D., Zbikowski, R., Gawthrop, P.J.: Neural networks for control systems: a survey. Automatica **28**(6), 1083–1112 (1992)
269. Hwang, J., Lay, S., Maechler, M., Douglas, R., Schimet, J.: Regression modeling in back-propagation and projection pursuit learning. IEEE Trans. Neural Netw. **5**(3), 342–352 (1994)
270. Ingber, L.: Very fast simulated re-annealing. Math. Comput. Model. **12**, 967–973 (1989)
271. Ingber, L.: Simulated annealing: practice versus theory. Math. Comput. Model. **18**(11), 29–57 (1993)
272. Ingber, L.: Adaptive simulated annealing (ASA): lessons learned. Math. Comput. Model. **25**(1), 33–54 (1995)
273. Ingber, L., Rosen, B.: Genetic algorithms and very fast simulated reannealing: a comparison. Math. Comput. Model. **16**(11), 87–100 (1992)
274. Intrator, N.: Combining exploratory projection pursuit and projection pursuit regression with application to neural networks. Neural Comput. **5**(3), 443–457 (1993)
275. Isaksson, A.J., Ljung, L., Strömberg, J.-E.: On recursive construction of trees as models of dynamic systems. In: IEEE International Conference on Decision and Control, pp. 1686–1687, Brighton, UK (1991)
276. Isermann, R.: Required accuracy of mathematical models of linear time invariant controlled elements. Automatica **7**, 333–341 (1971)
277. Isermann, R.: Digitale Regelsysteme – Band 1, 2. ed. Springer, Berlin (1987)
278. Isermann, R.: Identifikation dynamischer Syteme – Band 1, 2. ed. Springer, Berlin (1992)
279. Isermann, R.: Identifikation dynamischer Syteme – Band 2, 2. ed. Springer, Berlin (1992)

280. Isermann, R.: Fault diagnosis of machines via parameter estimation and knowledge processing: tutorial paper. Automatica **29**(4), 815–835 (1993)
281. Isermann, R., Baur, U., Bamberger, W., Kneppo, P., Siebert, H.: Comparision of six on-line identification and parameter estimation methods. Automatica **10**, 81–103 (1974)
282. Isermann, R., Lachmann, K.-H., Matko, D.: Adaptive Control Systems. Prentice Hall, New York (1992)
283. Isermann, R., Sinsel, S., Schaffnit, J.: Hardwar-in-the-loop simulation of Diesel engines for the development of engine control systems. In: IFAC Workshop on Algorithms and Architectures for Real-Time Control (AARTC), pp. 90–92, Algarve, Portugal (1997)
284. Isermann, R., Ernst, S., Nelles, O.: Identification with dynamic neural networks – architectures, comparisons, applications –, plenary. In: IFAC Symposium on System Identification (SYSID), pp. 997–1022, Kitakyushu, Fukuoka, Japan (1997)
285. Ishibuchi, H., Nozaki, K., Yamamoto, N., Tanaka, H.: Construction of fuzzy classification systems with rectangular fuzzy rules using genetic algorithms. Fuzzy Sets Syst. **65**, 237–253 (1994)
286. Jacobs, R., Jordan, M.I., Nowlan, S.J., Hinton, G.E.: Adaptive mixture of local experts. Neural Comput. **3**, 79–87 (1991)
287. Jager, R.: Fuzzy Logic in Control. Ph.D. thesis, Deft University of Technology, Delft, The Netherlands (1995)
288. Jakubek, S., Keuth, N.: A new training algorithm for neuro-fuzzy networks. In: ANNIIP, pp. 23–34 (2005)
289. Jang, J.-S.R.: Neuro-Fuzzy Modeling: Architectures, Analyses, and Applications. Ph.D. thesis, EECS Department, Univ. of California at Berkeley, Berkeley, USA (1992)
290. Jang, J.-S.R.: ANFIS: adaptive-network-based fuzzy inference systems. IEEE Trans. Syst. Man Cybern. **23**(3), 665–685 (1993)
291. Jang, J.-S.R., Sun, C.-T.: Functional equivalence between radial basis function networks and fuzzy inference systems. IEEE Trans. Neural Netw. **4**(1), 156–159 (1993)
292. Jang, J.-S.R., Sun, C.T., Mizutani, E.: Neuro-Fuzzy and Soft Computing: A Computational Approach to Learning and Machine Intelligence. Prentice Hall, Englewood Cliffs (1997)
293. Jeng, J.-T.: Hybrid approach of selecting hyperparameters of support vector machine for regression. IEEE Trans. Syst. Man Cybern. Part B Cybern. **36**(3), 699–709 (2005)
294. Jin, R., Chen, W., Sudjianto, A.: On sequential sampling for global metamodeling in engineering design. In: ASME 2002 International Design Engineering Technical Conferences and Computers and Information in Engineering Conference, pp. 539–548. American Society of Mechanical Engineers (2002)
295. Jin, R., Chen, W., Sudjianto, A.: An efficient algorithm for constructing optimal design of computer experiments. J. Stat. Plan. Inference **134**(1), 268–287 (2005)
296. Johansen, T.A.: Fuzzy model based control: stability, robustness, and performance issues. IEEE Trans. Fuzzy Syst. **2**, 221–234 (1994)
297. Johansen, T.A.: Operating Regime Based Process Modeling and Identification. Ph.D. thesis, Dept. of Engineering Cybernetics, Norwegian Institute of Technology, Trondheim, Norway (1994)
298. Johansen, T.A.: Identification of non-linear system structure and parameters using regime decomposition. Automatica **31**(2), 321–326 (1995)
299. Johansen, T.A.: On Tikhonov regularization, bias and variance in nonlinear system identification. Automatica **33**(3), 441–446 (1997)
300. Johansen, T.A., Foss, B.A.: Constructing NARMAX models using ARMAX models. Int. J. Control **58**(5), 1125–1153 (1993)
301. Johansen, T.A., Hunt, K., Gawthrop, P.J., Fritz, H.: Off-equilibrium linearisation and design of gain scheduled control with application to vehicle speed control. Eng. Pract. **6**, 167–180 (1998)
302. Johansen, T.A., Hunt, K.J., Gawthrop, P.J.: Transient performance, robustness and off-equilibrium linearization in fuzzy gain scheduled control. In: Driankov, D., Palm, R. (eds.) Advances in Fuzzy Control, Studies in Fuzziness and Soft Computing, pp. 357–376. Physica-Verlag, Heidelberg (1998)

303. Johansen, T.A., Murray-Smith, R.: The operating regime approach to nonlinear modelling and control. In: Murray-Smith, R., Johansen, T.A. (eds.) Multiple Model Approaches to Modelling and Control, chapter 1, pp. 3–72. Taylor & Francis, London (1997)
304. Johansson, R.: System Modeling & Identification. Prentice Hall, Englewood Cliffs (1993)
305. Johnson, M.E., Moore, L.M., Ylvisaker, D.: Minimax and maximin distance designs. J. Stat. Plan. Inference **26**(2), 131–148 (1990)
306. Johnson, R.T., Montgomery, D.C., Jones, B., Fowler, J.W.: Comparing designs for computer simulation experiments. In: Proceedings of the 40th Conference on Winter Simulation, pp. 463–470. Winter Simulation Conference (2008)
307. Jones, R.D., Coworkers: Nonlinear Adaptive Networks: A Little Theory, a Few Applications. Technical Report 91-273, Los Alamos National Lab., New Mexico (1991)
308. Jordan, M.I.: Attractor dynamics and parallelism in a connectionist sequential machine. In: Conference of the Cognitive Science Society (1986)
309. Jordan, M.I., Jacobs, R.A.: Hierarchical mixtures of experts and the EM algorithm. Neural Comput. **6**(2), 181–214 (1994)
310. Junge, T.F., Unbehauen, H.: Real time learning control of an emergency turbo-generator plant using structurally adaptive neural networks. In: IEEE Conference on Industrial Electronics (IECON), pp. 2403–2408, Aachen, Germany (1998)
311. Kadirkamanathan, V., Fabri, S.: Stable nonlinear adaptive control with growing radial basis function networks. In: IFAC Symposium on Adaptive Systems for Control and Signal Processing (ASCAP), pp. 231–236, Budapest, Hungary (1995)
312. Kadirkamanathan, V., Niranjan, M.: A function estimation approach to sequential learning with neural networks. Neural Comput. **5**, 954–975 (1993)
313. Kahlert, J., Frank, H.: Fuzzy-Logik und Fuzzy-Control. Vieweg, Braunschweig (1993)
314. Kajero, O.T., Chen, T., Yao, Y., Chuang, Y.-C., Shan Hill Wong, D.: Meta-modelling in chemical process system engineering. J. Taiwan Inst. Chem. Eng. (2016)
315. Karnin, E.D.: A simple procedure for pruning back-propagation trained neural networks. IEEE Trans. Neural Netw. **1**(2), 239–242 (1990)
316. Kavli, T.: ASMOD: an algorithm for adaptive spline modeling of observation data. Int. J. Control **58**(4), 947–967 (1993)
317. Kbiob, D.: A statistical approach to some basic mine valuation problems on the witwatersrand. J. Chem. Metall. Min. Soc. S. Afr. (1951)
318. Kecman, V., Pfeiffer, B.-M.: Exploiting the structural equivalence of learning fuzzy systems and radial basis function networks. In: European Congress on Intelligent Techniques and Soft Computing (EUFIT), pp. 58–66, Aachen, Germany (1994)
319. Kessel, J.A., Schmidt, M., Isermann, R.: Model-based control and monitoring of combustion engines. Motortechnische Zeitschrift MTZ Worldwide 7–9 (1998)
320. Kiendl, H.: Fuzzy-Control methodenorientiert. Oldenbourg, München (1997)
321. Kirkpatrick, S., Gelatt, C.D., Vecchi, M.P.: Optimization by simulated annealing. Science **220**(4598), 671–680 (1983)
322. Klein, P., Kirschbaum, F., Hartmann, B., Bogachik, J., Nelles, O.: Adaptive test planning for the calibration of combustion engines – application. In: Design of Experiments (DoE) in Engine Development, pp. 17–30, Berlin, Germany. Expert Verlag (2013)
323. Kofahl, R.: Parameteradaptive Regelungen mit robusten Eigenschaften. Ph.D. Thesis, TU Darmstadt, FB MSR 39. Springer, Berlin (1988)
324. Kofahl, R.: Robuste parameteradaptive Regelungen. Fachbericht Nr. 19. Messen, Steuern, Regeln. Springer, Berlin (1988)
325. Kohavi, R., et al.: A study of cross-validation and bootstrap for accuracy estimation and model selection. In: International Joint Conference on Artificial Intelligence, vol. 14, pp. 1137–1145 (1995)
326. Kohonen, T.: Self-Organizing Maps, 2nd edn. Springer, Berlin (1997)
327. Kortmann, M.: Die Identifikation nichtlinearer Ein- und Mehrgrössensysteme auf der Basis nichtlinearer Modellansätze. Reihe 8: Mess-, Steuerungs- und Regelungstechnik, Nr. 177. VDI-Verlag, Düsseldorf (1989)

328. Kortmann, P.: Fuzzy-Modelle zur Systemidentifikation. Reihe 8: Mess-, Steuerungs- und Regelungstechnik, Nr. 647. VDI-Verlag, Düsseldorf (1997)
329. Kosko, B.: Fuzzy systems as universal approximators. IEEE Trans. Comput. **43**, 1329–1333 (1994)
330. Koza, J.R.: Genetic Programming: On the Programming of Computers by Means of Natural Selection. MIT Press, Cambridge, MA (1992)
331. Koza, J.R.: Genetic Programming II: Automatic Discovery of Reusable Programs. MIT Press, Cambridge, MA (1994)
332. Koza, J.R.: Genetic Programming III: Darwinian Invention and Problem Solving. Morgan Kaufmann, San Francisco, CA (1999)
333. Kramer, M.A.: Diagnosing dynamic faults using modular neural nets. IEEE Expert (1993)
334. Krizhevsky, A., Sutskever, I., Hinton, G.E.: Imagenet classification with deep convolutional neural networks. In: Advances in Neural Information Processing Systems, pp. 1097–1105 (2012)
335. Kroll, A.: Fuzzy-Systeme zur Modellierung und Regelung komplexer technischer Systeme. Reihe 8: Mess-, Steuerungs- und Regelungstechnik, Nr. 612. VDI-Verlag, Düsseldorf (1997)
336. Kruse, R., Gebhardt, J., Klawonn, F.: Foundations of Fuzzy Systems. John Wiley & Sons, Chichester (1994)
337. Kursawe, F., Schwefel, H.-P.: Optimierung mit Evolutionären Algorithmen. Automatisierungstechnische Praxis **39**(9), 10–17 (1997)
338. Kurth, J.: Identifikation nichtlinearer Systeme mit komprimierten Volterra-Reihen. Reihe 8: Mess-, Steuerungs- und Regelungstechnik, Nr. 459. VDI-Verlag, Düsseldorf (1995)
339. Kurth, J.: Komprimierte Volterra-Reihen: ein neuer Modellansatz zur Identifikation nichtlinearer Systeme. Automatisierungstechnik **44**(6), 265–273 (1996)
340. Lachmann, K.-H.: Parameteradaptive Regelalgorithmen für bestimmte Klassen nichtlinearer Prozesse mit eindeutigen Nichtlinearitäten. Ph.D. Thesis, TU Darmstadt, Reihe 8: Mess-, Steuerungs- und Regelungstechnik, Nr. 66. VDI-Verlag, Düsseldorf (1982)
341. Lawler, E.L., Wood, E.D.: Branch-and-bound methods: a survey. J. Oper. Res. **14**, 699–719 (1966)
342. Lee, C.C.: Fuzzy logic in control systems: fuzzy logic controller – part I. IEEE Trans. Syst. Man Cybern. **20**(2), 404–418 (1990)
343. Lee, C.C.: Fuzzy logic in control systems: fuzzy logic controller – part II. IEEE Trans. Syst. Man Cybern. **20**(2), 419–435 (1990)
344. Lee, Y., Oh, S., Kim, M.: The effect of initial weights on premature saturation in backpropagation learning. In: International Joint Conference on Neural Networks, vol. 1, pp. 765–770, Seattle, USA (1991)
345. Leonhard, J.A., Kramer, M.A., Ungar, L.H.: A neural network architecture that computes its own reliability. Comput. Chem. Eng. **16**(9), 818–835 (1992)
346. Leontaritis, I.J., Billings, S.A.: Input-output parametric models for nonlinear systems, part 1: deterministic nonlinear systems. Int. J. Control **41**, 303–344 (1985)
347. Levin, A.U., Narendra, K.S.: Identification using feedforward networks. Neural Comput. **7**(2), 349–357 (1995)
348. Lewis, J.M., Lakshmivarahan, S., Dhall, S.: Dynamic data assimilation: a least squares approach, vol. 13. Cambridge University Press, Cambridge (2006)
349. Lichman, M.: UCI Machine Learning Repository (2013)
350. Lim, R.K., Phan, M.Q.: Identification of a multistep-ahead observer and its application to predictive control. J. Guid. Control. Dyn. **20**(6), 1200–1206 (1997)
351. Lin, S.-W., Lee, Z.-J., Chen, S.-C., Tseng, T.-Y.: Parameter determination of support vector machine and feature selection using simulated annealing approach. Appl. Soft Comput. **8**(4), 1505–1512 (2008)
352. Lindskog, P.: Methods, Algorithms and Tools for System Identification Based on Prior Knowledge. Ph.D. thesis, Linköping University, Linköping, Sweden (1996)
353. Linkens, D.A., Nyongesa, H.O.: A hierachical multivariable fuzzy controller for learning with genetic algorithms. Int. J. Control **63**(5), 865–883 (1996)

354. Lipton, Z.C., Berkowitz, J., Elkan, C.: A critical review of recurrent neural networks for sequence learning. arXiv preprint arXiv:1506.00019 (2015)
355. Liu, H., Motoda, H.: Computational Methods of Feature Selection. Chapman & Hall (2007)
356. Ljung, L.: System Identification: Theory for the User, 2nd edn. Prentice Hall, Englewood Cliffs (1999)
357. Ljung, L.: System identification toolbox for MATLAB user's guide. The Matlab User's Guide (1988)
358. Lowe, D.: Adaptive radial basis function nonlinearities, and the problem of generalization. In: IEE International Conference on Artificial Neural Networks, pp. 171–175, London, UK (1989)
359. Lughofer, E.: Evolving Fuzzy Systems-Methodologies, Advanced Concepts and Applications, vol. 53. Springer (2011)
360. Mamdani, E.H.: Application of fuzzy logic to approximate reasoning using linguistic systems. Fuzzy Sets Syst. **26**, 1182–1191 (1977)
361. Marconato, A., Schoukens, M.: Tuning the hyperparameters of the filter-based regularization method for impulse response estimation. IFAC-PapersOnLine **50**(1), 12841–12846 (2017)
362. Marconato, A., Schoukens, M., Schoukens, J.: Filter-based regularisation for impulse response modelling. IET Control Theory Appl. (2016)
363. Marconato, A., Sjöberg, J., Suykens, J.A.K., Schoukens, J.: Improved initialization for nonlinear state-space modeling. IEEE Trans. Instrum. Meas. **63**(4), 972–980 (2014)
364. Marenbach, P., Bettenhausen, K.D., Cuno, B.: Selbstorganisierende Generierung strukturierter Prozessmodelle. Automatisierungstechnik **43**(6), 277–288 (1995)
365. Marenbach, P., Bettenhausen, K.D., Freyer, S.: Signal path-oriented approach for generation of dynamic process models. In: Conference on Genetic Programming, pp. 327–332, Stanford, USA (1996)
366. Martin, J.D., Simpson, T.W.: Use of kriging models to approximate deterministic computer models. AIAA J. **43**(4), 853–863 (2005)
367. Martin, J.P.: Fuzzy stability analysis of fuzzy systems: a Lyapunov approach. In: Driankov, D., Palm, R. (eds.) Advances in Fuzzy Control, Studies in Fuzziness and Soft Computing, pp. 67–101. Physica-Verlag, Heidelberg (1998)
368. Martinetz, T.M., Berkovich, S.G., Schulten, K.J.: "Neural Gas" Network for vector quantization and its application to time-series prediction. IEEE Trans. Neural Netw. **4**(4), 558–569 (1993)
369. McKay, D.M., Beckman, R.J., Conover, W.J.: A comparison of three methods for selecting values of input variables in the analysis of output from a computer code. Technometrics **21**(2), 239–245 (1979)
370. McKelvey, T.: Identification of State-Space Models from Time and Frequency Data. Ph.D. thesis, Linköping University, Linköping, Sweden (1995)
371. Mead, C.: Analog VLSI and Neural Systems. Addison Wesley, Reading (1989)
372. Meila, M., Jordan, M.I.: Markov mixtures of experts. In: Murray-Smith, R., Johansen, T.A. (eds.) Multiple Model Approaches to Modelling and Control, chapter 5, pp. 145–166. Taylor & Francis, London (1997)
373. Mendel, J.M.: Fuzzy logic systems for engineering: a tutorial. Proc. IEEE **83**(3), 345–377 (1995)
374. Michalewicz, Z.: Genetic Algorithms + Data Structures = Evolution Programs. Springer, Berlin (1992)
375. Miller, A.J.: Subset Selection in Regression. Statistics and Applied Probability. Chapman & Hall, New York (1990)
376. Miller, W.T., Glanz, F.H., Kraft, L.G.: CMAC: an associative neural network alternative to backpropagation. Proc. IEEE **78**(10), 1561–1567 (1990)
377. Miller, W.T., Sutton, R.S., Werbos, P.J.: Neural Networks for Control. MIT Press, Cambridge (1992)
378. Minsky, M.L., Pappert, S.A.: Perceptrons. MIT Press, Cambridge, MA (1969)

379. Mirkin, B.: Clustering for Data Mining: A Data Recovery Approach. Chapman & Hall/CRC, London (2005)
380. Mischo, W.S.: Ein neuronaler Interpolationsspeicher für die lernende Regelung: Konzeptwahl und mikroelektronischer Entwurf. Ph.D. Thesis, TU Darmstadt, Reihe 9: Elektronik, Nr. 249. VDI-Verlag, Düsseldorf (1997)
381. Mitten, L.G.: Branch-and-bound methods: general formulation and properties. J. Oper. Res. **18**, 24–34 (1970)
382. Moll, J., Hartmann, B., Chaaban, R., Fritzen, C.-P., Nelles, O.: A novel online learning approach for ultrasonic imaging applied to a non-convex structure. Technical report, Department of Mechanical Engineering (2011)
383. Moll, J.: Strukturdiagnose mit Ultraschallwellen durch Verwendung von piezoelektrischen Sensoren und Aktoren. Ph.D. thesis, Univ. of Siegen (2011)
384. Moll, J., Schulte, R.T., Hartmann, B., Fritzen, C.-P., Nelles, O.: Multi-site damage localization in anisotropic plate-like structures using an active guided wave structural health monitoring system. Smart Mater. Struct. **19**(4), 045022 (2010)
385. Monaghan, J.J., Gingold, R.A.: Shock simulation by the particle method SPH. J. Comput. Phys. **52**(2), 374–389 (1983)
386. Montgomery, D.C.: Design and Analysis of Experiments. John Wiley & Sons, Hoboken (2008)
387. Moody, J., Darken, C.J.: Fast learning in networks of locally-tuned processing units. Neural Comput. **1**(2), 281–294 (1989)
388. Morari, M., Zafiriou, E.: Robust Process Control. Prentice Hall, Englewood Cliffs, NJ (1989)
389. Morris, M.D., Mitchell, T.J.: Exploratory designs for computational experiments. J. Stat. Plan. Inference **43**(3), 381–402 (1995)
390. Müller, N., Nelles, O.: Closed-loop ignition control using on-line learning of locally-tuned radial basis function networks. In: American Control Conference (ACC), pp. 1356–1360, San Diego, USA (1999)
391. Münker, T., Belz, J., Nelles, O.: Improved incorporation of prior knowledge for regularized FIR model identification. In: 2018 Annual American Control Conference (ACC), pp. 1090–1095, Milwaukee, WI. IEEE (2018)
392. Münker, T., Heinz, T.O., Nelles, O.: Regularized local FIR model networks for a Bouc-Wen and a Wiener-Hammerstein system. In: Workshop on Nonlinear System Identification Benchmarks, p. 33 (2017)
393. Münker, T., Nelles, O.: Local model network with regularized MISO finite impulse response models. In: 2016 IEEE International Conference on Fuzzy Systems (FUZZ-IEEE), pp. 1–8. IEEE (2016)
394. Münker, T., Nelles, O.: Nonlinear system identification with regularized local FIR model networks. In: 4th IFAC International Conference on Intelligent Control and Automation Science, pp. 61–66, Reims, France (2016)
395. Münker, T., Nelles, O.: Nonlinear system identification with regularized local fir model networks. Eng. Appl. Artif. Intell. **67**, 345–354 (2018)
396. Münker, T., Nelles, O.: Sensitive order selection via identification of regularized fir models with impulse response preservation. In: 2018 IFAC Symposium on System Identification (SYSID), vol. 51, pp. 197–202, Stockholm, Sweden. Elsevier (2018)
397. Münker, T., Peter, T., Nelles, O.: Gray-box identification with regularized FIR models. at-Automatisierungstechnik **66**(9), 704–713 (2018)
398. Munson, B.R., Young, D.F., Okiishi, T.H.: Fundamentals of Fluid Mechanics. New York (1990)
399. Murray-Smith, R.: A Local Model Network Approach to Nonlinear Modeling. Ph.D. thesis, University of Strathclyde, Strathclyde, UK (1994)
400. Murray-Smith, R., Johansen, T.A. (eds.): Multiple Model Approaches to Modelling and Control. Taylor & Francis, London (1997)
401. Murray-Smith, R., Johansen, T.A.: Local learning in local model networks. In: IEE International Conference on Artificial Neural Networks, pp. 40–46 (1995)

402. Murray-Smith, R., Johansen, T.A.: Local learning in local model networks. In: Murray-Smith, R., Johansen, T.A. (eds.) Multiple Model Approaches to Modelling and Control, chapter 7, pp. 185–210. Taylor & Francis, London (1997)
403. Nadaraya, É.A.: On estimating regression. Theory Probab. Appl. **9**(1), 141–142 (1964)
404. Nakamori, Y., Ryoke, M.: Identification of fuzzy prediction models through hyperellipsoidal clustering. IEEE Trans. Syst. Man Cybern **24**(8), 1153–1173 (1994)
405. Narendra, K.S., Balakrishnan, J., Ciliz, M.K.: Adaptation and learning using multiple models, switching, and tuning. IEEE Control Syst. **15**(3), 37–51 (1995)
406. Narendra, K.S., Parthasarathy, K.: Identification and control of dynamical systems using neural networks. IEEE Trans. Neural Netw. **1**(1), 4–27 (1990)
407. Nelles, O.: On the identification with neural networks as series-parallel and parallel models. In: International Conference on Artificial Neural Networks (ICANN), pp. 255–260, Paris, France (1995)
408. Nelles, O.: Training neuronaler Netze als seriell-parallele oder parallele Modelle zur Identifikation nichtlinearer, dynamischer Systeme. In: 3. GI Fuzzy-Neuro-Systeme Workshop, pp. 333–339, Darmstadt, Germany (1995)
409. Nelles, O.: Local linear model tree for on-line identification of time-variant nonlinear dynamic systems. In: International Conference on Artificial Neural Networks (ICANN), pp. 115–120, Bochum, Germany (1996)
410. Nelles, O.: GA-based generation of fuzzy rules. In: Pedrycz, W. (ed.) Fuzzy Evolutionary Computation, chapter 2.9, pp. 269–295. Kluwer Academic Publishers, Boston, USA (1997)
411. Nelles, O.: LOLIMOT – Lokale, lineare Modelle zur Identifikation nichtlinearer, dynamischer Systeme. Automatisierungstechnik (at) **45**(4), 163–174 (1997)
412. Nelles, O.: Nonlinear system identification with neurofuzzy methods. In: Ruan, D. (ed.) Intelligent Hybrid Systems – Fuzzy Logic, Neural Networks, and Genetic Algorithms, chapter 12, pp. 283–310. Kluwer Academic Publishers, Boston, USA (1997)
413. Nelles, O.: Orthonormal basis functions for nonlinear system identification with local linear model trees (LOLIMOT). In: IFAC Symposium on System Identification (SYSID), pp. 667–672, Kitakyushu, Fukuoka, Japan (1997)
414. Nelles, O.: Structure optimization of Takagi-Sugeno fuzzy models. Int. J. Uncertainty Fuzziness Knowledge Based Syst. **5**(2), 161–170 (1998). Special Issue on Applications of New Functional Principles of Fuzzy Systems and Neural Networks within Computational Intelligence
415. Nelles, O.: Nonlinear System Identification with Local Linear Neuro-Fuzzy Models. Ph.D. Thesis, TU Darmstadt, Automatisierungstechnik series. Shaker Verlag, Aachen, Germany (1999)
416. Nelles, O.: Nonlinear System Identification. Springer, Berlin, Germany (2001)
417. Nelles, O.: Axes-oblique partitioning strategies for local model networks. In: IEEE International Symposium on Intelligent Control, pp. 2378–2383, Munich, Germany (2006)
418. Nelles, O., Ayoubi, M.: Learning the relevance of symtoms in fault trees. In: European Congress on Intelligent Techniques and Soft Computing (EUFIT), pp. 1148–1152, Aachen, Germany (1994)
419. Nelles, O., Ernst, S., Isermann, R.: Neural networks for identification of nonlinear dynamic systems – an overview. In: European Congress on Intelligent Techniques and Soft Computing (EUFIT), pp. 282–286, Aachen, Germany (1996)
420. Nelles, O., Ernst, S., Isermann, R.: Neuronale Netze zur Identifikation nichtlinearer, dynamischer Systeme: Ein Überblick. Automatisierungstechnik (at) **45**(6), 251–262 (1997)
421. Nelles, O., Fink, A., Isermann, R.: Local linear model trees (LOLIMOT) toolbox for nonlinear system identification. In: IFAC Symposium on System Identification (SYSID), Santa Barbara, USA (2000)
422. Nelles, O., Fischer, M.: Local linear model trees (LOLIMOT) for nonlinear system identification of a cooling blast. In: European Congress on Intelligent Techniques and Soft Computing (EUFIT), pp. 1187–1191, Aachen, Germany (1996)

423. Nelles, O., Fischer, M.: Lokale Linearisierung von Fuzzy-Modellen. Automatisierungstechnik (at) **47**(5), 217–223 (1999)
424. Nelles, O., Fischer, M., Isermann, R.: Identification of fuzzy models for predictive control of nonlinear time-variant processes. In: IFAC Symposium on Artificial Intelligence in Real-Time Control (AIRTC), pp. 457–462, Kuala Lumpur, Malaysia (1997)
425. Nelles, O., Fischer, M., Müller, B.: Fuzzy rule extraction by a genetic algorithm and constrained nonlinear optimization of membership functions. In: International IEEE Conference on Fuzzy Systems (FUZZ-IEEE), pp. 213–219, New Orleans, USA (1996)
426. Nelles, O., Füssel, D., Ballé, P., Isermann, R.: Model-based fault detection and diagnosis part B: applications. In: International Conference on Probability, Safety, Assessment, and Management, vol. 3, pp. 1875–1880, New York, USA (1998)
427. Nelles, O., Hecker, O., Isermann, R.: Automatic model selection in local linear model trees for nonlinear system identification of a transport delay process. In: IFAC Symposium on System Identification (SYSID), pp. 727–732, Kitakyushu, Fukuoka, Japan (1997)
428. Nelles, O., Hecker, O., Isermann, R.: Identifikation nichtlinearer, dynamischer Prozesse mit Takagi-Sugeno Fuzzy-Modellen variabler Struktur. In: 4. GI Fuzzy-Neuro-Systeme Workshop, pp. 388–395, Soest, Germany (1997)
429. Nelles, O., Hecker, O., Isermann, R.: Automatische Strukturselektion für Fuzzy-Modelle zur Identifikation nichtlinearer, dynamischer Prozesse. Automatisierungstechnik (at) **46**(6), 302–312 (1998)
430. Nelles, O., Isermann, R.: A comparison between radial basis function networks and classical methods for identification of nonlinear dynamic systems. In: IFAC Symposium on Adaptive Systems in Control and Signal Processing (ASCSP), pp. 233–238, Budapest, Hungary (1995)
431. Nelles, O., Isermann, R.: Identification of nonlinear dynamic systems – classical methods versus radial basis function networks. In: American Control Conference (ACC), pp. 3786–3790, Seattle, USA (1995)
432. Nelles, O., Isermann, R.: Basis function networks for interpolation of local linear models. In: IEEE Conference on Decision and Control (CDC), pp. 470–475, Kobe, Japan (1996)
433. Nelles, O., Isermann, R.: A new technique for determination of hidden layer parameters in RBF networks. In: IFAC World Congress, pp. 453–457, San Francisco, USA (1996)
434. Nelles, O., Sinsel, S., Isermann, R.: Local basis function networks for identification of a turbocharger. In: IEE UKACC International Conference on Control, pp. 7–12, Exeter, UK (1996)
435. Niederreiter, H.: Low-discrepancy and low-dispersion sequences. J. Number Theory **30**(1), 51–70 (1988)
436. Ogunmolu, O., Gu, X., Jiang, S., Gans, N.: Nonlinear systems identification using deep dynamic neural networks. arXiv preprint arXiv:1610.01439 (2016)
437. Omohundro, S.M.: Efficient algorithms with neural network behavior. J. Complex Syst. **1**, 273–347 (1987)
438. Omohundro, S.M.: Bumptrees for efficient function, constraint and classification learning. In: Lippmann, R.P., Moody, J.E., Touretzky, D.S. (eds.) Advances in Neural Information Processing Systems, pp. 693–699. Morgan Kaufmann, San Francisco (1991)
439. Ong, C.S., Williamson, R.C., Smola, A.J.: Learning the kernel with hyperkernels. J. Mach. Learn. Res. **6**(1), 1043–1071 (2005)
440. Orr, M.J.L.: Regularization in the selection of radial basis function centers. Neural Comput. **7**(3), 606–623 (1995)
441. Paduart, J., Lauwers, L., Swevers, J., Smolders, K., Schoukens, J., Pintelon, R.: Identification of nonlinear systems using polynomial nonlinear state space models. Automatica **46**(4), 647–656 (2010)
442. Paoletti, S., Juloski, A.L., Ferrari-Trecate, G., Vidal, R.: Identification of hybrid systems a tutorial. Eur. J. Control **13**(2), 242–260 (2007)
443. Papoulis, A.: Probability, Random Variables, and Stochastic Processes. Electrical & Electronic Engineering. McGraw-Hill, New York (1991)

444. Park, D., Kandel, A., Langholz, G.: Genetic-based new fuzzy reasoning models with application to fuzzy control. IEEE Trans. Syst. Man Cybern. **24**(1), 39–47 (1994)
445. Park, J., Sandberg, I.W.: Universal approximation using radial-basis-function networks. Neural Comput. **3**(2), 246–257 (1991)
446. Park, J., Sandberg, I.W.: Approximation and radial-basis-function networks. Neural Comput. **5**(3), 305–316 (1993)
447. Park, M.-K., Ji, S.-H., Kim, E.-T., Park, M.: Identification of Takagi-Sugeno fuzzy models via clustering and hough transform. In: Hellendoorn, H., Driankov, D. (eds.) Fuzzy Model Identification: Selected Approaches, chapter 3, pp. 91–161. Springer, Berlin (1997)
448. Parzen, E.: On estimation of probability density function and mode. Ann. Math. Stat. **33**, 1065–1076 (1962)
449. Pearlmutter, B.A.: Learning state space trajectories in recurrent neural networks. Neural Comput. **1**(2), 263–269 (1989)
450. Pearlmutter, B.A.: Gradient calculations for dynamic recurrent neural networks: a survey. IEEE Trans. Neural Netw. **6**(5), 1212–1228 (1995)
451. Pedrycz, W.: Fuzzy Control and Fuzzy Systems, 2nd edn. John Wiley & Sons, New York (1993)
452. Peter, T., Nelles, O.: Gray-box regularized fir modeling for linear system identification. In: Proceedings, 28th Workshop on Computational Intelligence, pp. 113–128, Dortmund, Germany (2018)
453. Peterka, V.: Bayesian approach to system identification. In: Eykhoff, P. (ed.) Trends and Progress in System Identification. Pergamon Press, Oxford (1991)
454. Pfeiffer, B.-M.: Identification von Fuzzy-Regeln aus Lerndaten. In: Workshop des GMA-UA 1.4.2 "Fuzzy-Control", pp. 238–250, Dortmund, Germany (1995)
455. Pfeiffer, B.-M., Isermann, R.: Criteria for successful applications of fuzzy control. Eng. Appl. Artif. Intell. **7**(3), 245–253 (1994)
456. Pickhardt, R.: Adaptive Regelung nach einem Multi-Modell-Verfahren. Reihe 8: Mess-, Steuerungs- und Regelungstechnik, Nr. 499. VDI-Verlag, Düsseldorf (1995)
457. Pickhardt, R.: Adaptive Regelung auf der Basis eines Multi-Modells bei einer Transportregelstrecke für Schüttgüter. Automatisierungstechnik **3**, 113–120 (1997)
458. Pillonetto, G., Dinuzzo, F., Chen, T., De Nicolao, G., Ljung, L.: Kernel methods in system identification, machine learning and function estimation: a survey. Automatica **50**(3), 657–682 (2014)
459. Pitsch, H., Barths, H., Peters, N.: Modellierung der Schadstoffbildung bei dieselmotorischen Verbrennungen. In: 2. Dresdner Motorenkolloquium, pp. 163–169, Dresden, Germany (1997)
460. Plate, T.: Re: Kangaroos (Was Re: BackProp without Calculus?). Usenet article <93Sep8.162519edt.997@neuron.ai.toronto.edu> in comp.ai.neural-nets (1993)
461. Platt, J.: A resource allocating network for function interpolation. Neural Comput. **3**, 213–225 (1991)
462. Plutowski, M.: Selecting Training Examplars for Neural Network Learning. Ph.D. thesis, University of California, San Diego, USA (1994)
463. Poggio, T., Girosi, F.: Networks for approximation and learning. Proc. IEEE **78**(9), 1481–1497 (1990)
464. Poggio, T., Girosi, F.: Regularization algorithms that are equivalent to multilayer networks. Science **247**, 978–982 (1990)
465. Pottmann, M., Unbehauen, H., Seborg, D.E.: Application of a general multi-model approach for identification of highly nonlinear processes: a case study. Int. J. Control **57**(1), 97–120 (1993)
466. Powell, M.J.D.: Radial basis functions for multivariable interpolation: a review. In: IMA Conference on Algorithms for the Approximation of Functions and Data, pp. 143–167, Shrivenham, UK (1985)
467. Preuss, H.P.: Fuzzy-Control: Heuristische Regelung mittels unscharfer Logik. Automatisierungstechnik **34**(5), 239–246 (1992)
468. Preuss, H.P., Tresp, V.: Neuro-Fuzzy. Automatisierungstechnische Praxis **36**(5), 10–24 (1994)

469. Pronzato, L., Müller, W.G.: Design of computer experiments: space filling and beyond. Stat. Comput. **22**(3), 681–701 (2012)
470. Pucar, P., Millnert, M.: Smooth hinging hyperplanes: a alternative to neural nets. In: European Control Conference (ECC), pp. 1173–1178, Rome, Italy (1995)
471. Pucher, H.: Aufladung von Verbrennungsmotoren. Expert-Verlag, Sindelfingen (1985)
472. Quinlan, J.: C4.5 Programs for Machine Learning. Morgan Kaufmann Publishers, San Francisico (1993)
473. Quinlan, R.: Combining instance-based and model-based learning. In: International Conference of Machine Learning, pp. 236–243, Amherst, MA (1993)
474. Quinonero-Candela, J., Rasmussen, C.E.: A unifying view of sparse approximate gaussian process regression. J. Mach. Learn. Res. **6**, 1939–1959 (2005)
475. Raju, G.V.S., Zhou, J.: Adaptive hierarchical fuzzy controller. IEEE Trans. Syst. Man Cybern. **23**(4), 973–980 (1993)
476. Rasmussen, C.E., Williams, C.K.I.: Gaussian processes for machine learning. MIT Press, Cambridge, MA (2006)
477. Rauch, H.E.: Autonomous control reconfiguration. IEEE Control Syst. Mag. **15**(6), 37–48 (1995)
478. Rechenberg, I.: Evolutionsstrategie: Optimierung technischer Systeme nach Prinzipien der biologischen Evolution. Frommann-Holzboog-Verlag, Stuttgart (1975)
479. Reed, R.: Pruning algorithms: a survey. IEEE Trans. Neural Netw. **4**(5), 740–747 (1993)
480. Rehrl, J., Schwingshackl, D., Horn, M.: A modeling approach for HVAC systems based on the LoLiMoT algorithm. In: 19th IFAC World Congress, pp. 10862–10868 (2014)
481. Reklaitis, G.V., Ravindran, A., Ragsdell, K.M.: Engineering Optimization – Methods and Applications. John Wiley & Sons, London (1983)
482. Reunanen, J.: Overfitting in making comparisons between variable selection methods. J. Mach. Learn. Res. **3**, 1371–1382 (2003)
483. Richalet, J.: Industrial applications of model based predictive control. Automatica **29**(5), 1251–1274 (1993)
484. Rifkin, R.M.: Everything Old Is New Again: A Fresh Look at Historical Approaches in Machine Learning. Ph.D. thesis, Massachusetts Institute of Technology (2002)
485. Rifkin, R., Klautau, A.: In defense of one-vs-all classification. J. Mach. Learn. Res. **5**, 101–141 (2004)
486. Rifkin, R.M., Lippert, R.A.: Notes on Regularized Least Squares (2007)
487. Ripley, B.D.: Pattern Recognition and Neural Networks. Cambridge University Press, Cambridge (1996)
488. Ritter, H., Martinetz, T., Schulten, K.: Neuronale Netze. Addison-Wesley, Bonn (1991)
489. Bosch GmbH, R. (ed.): Ottomotor-Management, 3rd edn. Friedr. Vieweg & Sohn Verlag (2005)
490. Ronco, E., Gawthrop, P.J.: Incremental model reference adaptive polynomial controller network. In: IEEE Conference on Decision and Control, pp. 4171–4172, New York, USA (1997)
491. Rumelhart, D.E., Hinton, G.E., Williams, R.J.: Learning internal representations by error propagation. In: Rumelhart, D.E., McClelland, J.L. (eds.) Parallel Distributed Processing: Explorations in the Mircostructure of Cognition, vol. 1, chapter 8. MIT Press, Cambridge (1986)
492. Runkler, T.A.: Automatische Selektion signifikanter scharfer Werte in unscharfen regelbasierten Systemen der Informations- und Automatisierungstechnik. Reihe 10, Nr. 417. VDI-Verlag, Düsseldorf (1996)
493. Runkler, T.A., Bezdek, J.C.: Polynomial membership functions for smooth first order Takagi-Sugeno systems. In: GI-Workshop Fuzzy-Neuro-Systeme: Computaional Intelligence, pp. 382–387, Soest, Germany (1997)
494. Sacks, J., Welch, W.J., Mitchell, T.J., Wynn, H.P.: Design and analysis of computer experiments. Stat. Sci. **4**(4), 409–423 (1989)

495. Sanger, T.D.: A tree-structured adaptive network for function approximation in high-dimensional spaces. IEEE Trans. Neural Netw. **2**(2), 285–293 (1991)
496. Sanger, T.D.: A tree-structured algorithm for reducing computation in networks with separable basis functions. Neural Comput. **3**(1), 67–78 (1991)
497. Sanner, R.M., Slotine, J.-J.E.: Gaussian networks for direct adaptive control. IEEE Trans. Neural Netw. **3**, 837–863 (1992)
498. Saridis, G.N.: Comparision of six on-line identification algorithms. Automatica **10**, 69–79 (1974)
499. Sarle, W.S.: Frequently asked questions (FAQ) of the neural network newsgroup
500. Sarle, W.S.: Why statisticians should not FART (1995). ftp://ftp.sys.com/pub/neural/fart.doc
501. Saunders, C., Gammerman, A., Vovk, V.: Ridge regression learning algorithm in dual variables. In: Proceedings of the 15th International Conference on Machine Learning (ICML-1998), pp. 515–521. Morgan Kaufmann (1998)
502. Sbarbaro, D.: Context sensitive networks for modelling nonlinear dynamic systems. In: European Control Conference (ECC), pp. 2420–2425, Rome, Italy (1995)
503. Sbarbaro, D.: Local Laguerre models. In: Murray-Smith, R., Johansen, T.A. (eds.) Multiple Model Approaches to Modelling and Control, chapter 10, pp. 257–268. Taylor & Francis, London (1997)
504. Scales, L.E.: Introduction to Non-Linear Optimization. Computer and Science Series. Macmillan, London (1985)
505. Schapire, R.E.: The strength of weak learnability. Mach. Learn. **5**(2), 197–227 (1990)
506. Schenker, B., Agarwal, M.: Long-range prediction for poorly-known systems. Int. J. Control **62**(1), 227–238 (1995)
507. Schenker, B., Agarwal, M.: Dynamic modelling using neural networks. Int. J. Syst. Sci. **28**(12), 1285–1298 (1997)
508. Schenker, B.G.E.: Prediction and Control Using Feedback Neural Networks and Partial Models. Ph.D. thesis, Swiss Federal Institute of Technology Zürich, Zürich, Switzerland (1996)
509. Schillinger, M., Hartmann, B., Skalecki, P., Meister, M., Nguyen-Tuong, D., Nelles, O.: Safe active learning and safe bayesian optimization for tuning a PI-controller. In: IFAC World Congress, pp. 5967–5972 (2017)
510. Schillinger, M., Mourat, K., Hartmann, B., Eckstein, C., Jacob, M., Kloppenburg, E., Nelles, O.: Modern online doe methods for calibration – constraint modeling, continuous boundary estimation, and active learning. In: Röpke, K., Gühmann, C. (eds.) Automotive Data Analytics, Methods, DoE. Expert Verlag (2017)
511. Schillinger, M., Ortelt, B., Hartmann, B., Schreiter, J., Meister, M., Nguyen-Tuong, D., Nelles, O.: Safe active learning of a high pressure fuel supply system. In: 9th EUROSIM Congress on Modelling and Simulation, pp. 286–292 (2016)
512. Schiøler, H., Hartmann, U.: Mapping neural network derived from the parzen window estimator. Neural Netw. **5**(6), 903–909 (1992)
513. Schmidt, C.: Digitale kurbelwinkelsynchrone Modellbildung und Drehschwingungsdämpfung eines Dieselmotors mit Last. Ph.D. Thesis, TU Darmstadt, Reihe 12: Verkehrstechnik/Fahrzeugtechnik, Nr. 253. VDI-Verlag, Düsseldorf (1995)
514. Schmidt, M., Fung, G., Rosales, R.: Fast optimization methods for L1 regularization: a comparative study and two new approaches. In: Machine Learning: ECML 2007, pp. 286–297. Springer (2007)
515. Schmidt, M., Nelles, O.: Filtering and deriving signals using neural networks. In: American Control Conference (ACC), pp. 2730–2731, Philadelphia, USA (1998)
516. Schmitt, M.: Untersuchungen zur Realisierung mehrdimensionaler lernfähiger Kennfelder in Großserien-Steuergeräten. Number 246 in Ph.D. Thesis, TU Darmstadt, Reihe 12: Verkehrstechnik/Fahrzeugtechnik. VDI-Verlag, Düsseldorf (1994)
517. Schoukens, J., Vaes, M., Pintelon, R.: Linear system identification in a nonlinear setting: nonparametric analysis of the nonlinear distortions and their impact on the best linear approximation. IEEE Control Syst. **36**(3), 38–69 (2016)

518. Schram, G., Krijgsman, A., Verhaegen, M.H.G.: System identification with orthogonal basis functions and neural networks. In: IFAC World Congress, pp. 221–226, San Francisco, USA (1996)
519. Schüssler, M., Münker, T., Nelles, O.: Deep recurrent neural networks for nonlinear system identification. In: IEEE Symposium Series on Computational Intelligence (SSCI), Xiamen, China (2019)
520. Schüssler, M., Münker, T., Nelles, O.: Local model networks for the identification of nonlinear state space models. In: IEEE Conference on Decision and Control (CDC), New Orleans, Lousiana, USA (2019)
521. Schwarz, R.: Rekonstruktion der Bremskraft bei Fahrzeugen mit elektromechanisch betätigten Radbremsen. Ph.D. Thesis, TU Darmstadt, Reihe 12: Verkehrstechnik/Fahrzeugtechnik, Nr. 393. VDI-Verlag, Düsseldorf (1999)
522. Schwarz, R., Nelles, O., Isermann, R.: Verbesserung der Signalgenauigkeit von Raddrehzahlsensoren mittels Online-Kompensation der Impulsradfehler. Automatisierungstechnische Praxis (atp) **41**(3), 35–42 (1999)
523. Schwarz, R., Nelles, O., Isermann, R., Scheerer, P.: Verfahren zum Ausgleich von Abweichungen eines Raddrehzahlsensors. Patent DE 197 21 488 A1 (1999)
524. Schwarz, R., Nelles, O., Scheerer, P., Isermann, R.: Increasing signal accuracy of automatic wheel-speed sensors by on-line learning. In: American Control Conference (ACC), pp. 1131–1135, Albuquerque, USA (1997)
525. Schwefel, H.-P.: Numerical Optimization of Computer Models. John Wiley & Sons, Chichester (1988)
526. Schwingshackl, D., Rehrl, J., Horn, M.: Model predictive control of a HVAC system based on the LoLiMoT algorithm. In: European Control Conference (ECC), pp. 4328–4333 (2013)
527. Schwingshackl, D., Rehrl, J., Horn, M.: LoLiMoT based MPC for air handling units in HVAC Systems. Build. Environ. **96**, 250–259 (2016)
528. Setnes, M., Babuška, R., Verbruggen, H.B.: Rule-based modeling: precision and transparency. IEEE Trans. Syst. Man Cybern. Part C **28**(1), 165–169 (1998)
529. Settles, B.: Active learning literature survey. Univ. Wisconsin Madison **52**(55–66), 11 (2010)
530. Shorten, R., Murray-Smith, R.: Side-effects of normalising basis functions in local model networks. In: Murray-Smith, R., Johansen, T.A. (eds.) Multiple Model Approaches to Modelling and Control, chapter 8, pp. 211–229. Taylor & Francis, London (1997)
531. Shorten, R., Murray-Smith, R., Bjørgan, R., Gollee, G.: On the interpretation of local models in blended multiple model structures. Int. J. Control **72**(7–8), 620–628 (1998). Special issue on Multiple Model approaches to modelling and control
532. Singhee, A., Rutenbar, R.A.: Why Quasi-Monte Carlo is better than Monte Carlo or Latin hypercube sampling for statistical circuit analysis. IEEE Trans. Comput. Aided Des. Integr. Circuits Syst. **29**(11), 1763–1776 (2010)
533. Sinsel, S.: Echtzeitsimulation von Nutzfahrzeug-Dieselmotoren mit Turbolader zur Entwicklung von Motormanagementsystemen. Ph.D. Thesis, TU Darmstadt. Logos-Verlag, Berlin (2000)
534. Sinsel, S., Schaffnit, J., Isermann, R.: Hardwar-in-the-Loop Simulation von Dieselmotoren für die Entwicklung moderner Motormanagementsysteme. In: VDI-Tagung Mechatronik im Maschinen- und Fahrzeugbau, Moers, Germany (1997)
535. Sjöberg, J.: Non-Linear System Identification with Neural Networks. Ph.D. thesis, Linköping University, Linköping, Sweden (1995)
536. Sjöberg, J., Zhang, Q., Ljung, L., Benveniste, A., Delyon, B., Glorennec, P.-Y., Hjalmarsson, H., Juditsky, A.: Nonlinear black-box modeling in system identification: a unified overview. Automatica **31**(12), 1691–1724 (1995)
537. Škrjanc, I.: Evolving fuzzy-model-based design of experiments with supervised hierarchical clustering. IEEE Trans. Fuzzy Syst. **23**(4), 861–871 (2015)
538. Snelson, E., Ghahramani, Z.: Local and global sparse gaussian process approximations. In: AISTATS, vol. 11, pp. 524–531 (2007)

539. Sobol', I.M.: On the distribution of points in a cube and the approximate evaluation of integrals. USSR Comput. Math. Math. Phys. **7**(4), 86–112 (1967)
540. Söderström, T., Stoica, P.: System Identification. Prentice Hall, New York (1989)
541. Söderström, T., Stoica, P.: On convariance function tests used in system identification. Automatica **26**(1), 125–133 (1990)
542. Soeterboek, R.: Predictive Control: A Unified Approach. Prentice Hall, Englewood Cliffs, USA (1992)
543. Sollich, P., Williams, C.K.I.: Understanding gaussian process regression using the equivalent kernel. In: Deterministic and statistical methods in machine learning, pp. 211–228. Springer (2005)
544. Sorsa, T., Koivo, H.: Application of artificial neural networks in process fault diagnosis. Automatica **29**(4), 834–849 (1993)
545. Specht, D.F.: Probabilistic neural networks. Neural Netw. **3**(1), 109–118 (1990)
546. Specht, D.F.: General regression neural network. IEEE Trans. Neural Netw. **2**(6), 568–576 (1995)
547. Srivastava, N., Hinton, G., Krizhevsky, A., Sutskever, I., Salakhutdinov, R.: Dropout: a simple way to prevent neural networks from overfitting. J. Mach. Learn. Res. **15**(1), 1929–1958 (2014)
548. Strokbro, K., Umberger, D.K., Hertz, J.A.: Exploiting neurons with localized receptive fields to learn chaos. J. Complex Syst. **4**(3), 603–622 (1990)
549. Sugeno, M., Kang, G.T.: Structure identification of fuzzy model. Fuzzy Sets Syst. **28**(1), 15–33 (1988)
550. Sugeno, M., Tanaka, K.: Successive identification of a fuzzy model and its application to prediction of a complex system. Fuzzy Sets Syst. **42**, 315–334 (1991)
551. Suykens, J.A.K., Gestel, T.V., Brabanter, J., Moor, B., Vandewalle, J.: Least Squares Support Vector Machines. World Scientific Publishing, New Jersey (2003)
552. Takagi, T., Sugeno, M.: Fuzzy identification of systems and its application to modeling and control. IEEE Trans. Syst. Man Cybern. **15**(1), 116–132 (1985)
553. Tambouratzis, T.: Counter-clustering for training pattern selection. Comput. J. **43**(3), 177–190 (2000)
554. Tanaka, K., Ikeda, T., Wang, H.O.: Fuzzy regulators and fuzzy observers: relaxed stability conditions and LMI-based design. IEEE Trans. Fuzzy Syst. **6**(2), 250–265 (1998)
555. Tanaka, K., Sugeno, M.: Stability analysis and design of fuzzy control systems. Fuzzy Sets Syst. **45**(2), 135–156 (1992)
556. Tanaka, M., Ye, J., Tanino, T.: Identification of nonlinear systems using fuzzy logic and genetic algorithms. In: IFAC Symposium on System Identification, pp. 301–306, Copenhagen, Denmark (1994)
557. Te Braake, H.A.B., Babuška, R., Van Can, H.J.L.: Fuzzy and neural models in predictive control. Journal A **35**(3), 44–51 (1994)
558. Te Braake, H.A.B., Van Can, H.J.L., Van Straten, G., Verbruggen, H.B.: Two-step approach in the training of regulated activation weight neural networks (RAWN). Eng. Appl. Artif. Intell. **10**(2), 157–170 (1997)
559. Thrun, S.B.: The role of exploration in learning control. In: Handbook of Intelligent Control: Neural Fuzzy and Adaptive Approaches. Van Nostrand Reinhold (1992)
560. Tibshirani, R.: Regression shrinkage and selection via the lasso. J. R. Stat. Soc. Ser. B (Methodolog.) **58**(1), 267–288 (1996)
561. Tibshirani, R.J.: Degrees of freedom and model search. arXiv preprint arXiv:1402.1920 (2014)
562. Tietze, N.: Model-Based Calibration of Engine Control Units Using Gaussian Process Regression. Ph.D. thesis, Technische Universität (2015)
563. Tietze, N., Konigorski, U., Fleck, C., Nguyen-Tuong, D.: Model-based calibration of engine controller using automated transient design of experiment. In: 14th Stuttgart International Symposium, Wiesbaden. Springer Fachmedien (2014)
564. Tikhonov, A.N., Arsenin, V.Y.: Solutions of Ill-Posed Problems. Wiley, New York (1977)

565. Titterington, D., Smith, A.F.M., Markov, U.E.: Statistical Analysis of Finite Mixture Distributions. Wiley, Chichester (1985)
566. Tolle, H., Ersü, E.: Neurocontrol: Learning Control Systems Inspired by Neural Architectures and Human Problem Solving. Lecture Notes in Control and Information Science, No. 172. Springer, Berlin (1992)
567. Tolle, H., Gehlen, S., Schmitt, M.: On interpolating memories for learning control. In: Hunt, K.J., Irwin, G.R., Warwick, K. (eds.) Neural Network Engineering in Dynamic Control Systems, Advances in Industrial Control, pp. 127–152. Springer, Berlin (1996)
568. Töpfer, S.: Realisation of hierarchical look-up tables on low-cost hardware. In: IFAC Symposium on System Identification (SYSID), Santa Barbara, USA (2000)
569. Töpfer, S., Nelles, O., Isermann, R.: Comparison of a hierarchically constructed neural network and a hierarchical look-up table. In: IFAC Symposium on System Identification (SYSID), Santa Barbara, USA (2000)
570. Tsoi, A., Back, A.D.: Locally recurrent globally feedforward networks: a critical review of architectures. IEEE Trans. Neural Netw. **5**(2), 229–239 (1994)
571. Tuis, L.: Anwendung von mehrwertigen pseudozufälligen Signalen zur Identifikation von nichtlinearen Regelungssystemen. Ph.D. thesis, Lehrstuhl für Mess- und Regelungstechnik, Abt. Maschinenbau, Ruhr-Universität Bochum, Bochum, Germany (1975)
572. Tulleken, H.J.A.F.: Grey-box modelling and identification using prior knowledge and Bayesian techniques. Automatica **29**(2), 285–308 (1993)
573. Ullrich, T.: Untersuchungen zur effizienten interpolierenden Speicherung von nichtlinearen Prozeßmodellen und Vorsteuerungsstrategien: Methodik und Anwendungen in der Automobilelektronik. Automatisierungstechnik series. Shaker Verlag, Aachen. Ph.D. Thesis, TU Darmstadt (1998)
574. Van den Hof, P.M.J.: Closed-loop issues in system identification. In: IFAC Symposium on System Identification, pp. 1651–1664, Fukuoka, Japan (1997)
575. Van den Hof, P.M.J., Heuberger, P.S.C., Bokor, J.: System identification with generalized orthonormal basis functions. Automatica **31**(12), 1821–1834 (1995)
576. Van Overschee, P., De Moor, B.: N4SID: subspace algorithms for the identification of combined deterministic-stochastic systems. Automatica **30**(1), 75–93 (1994)
577. Vanderplaats, G.N.: Numerical Optimization Techniques for Engineering Design. Series in Mechanical Engineering. McGraw-Hill, New York (1984)
578. Verhagen, M.: Identification of the deterministic part of MIMO state space models given in innovations form from input-output data. Automatica **30**(1), 61–74 (1993)
579. Viberg, M.: Subspace-based methods for the identification of linear time-invariant systems. Automatica **31**(12), 1835–1851 (1995)
580. Voigtländer, K., Wilfert, H.-H.: Systematische Strukturierung und Parametrierung neuronaler Netze mit interner Filterkettendynamik. In: 42. Internationales Wissenschaftliches Kolloquium, pp. 100–106, Ilmenau, Germany (1997)
581. Wahba, G.: A comparison of GCV and GML for choosing the smoothing parameter in the generalized spline smoothing problem. Ann. Stat. **13**(4), 1378–1402 (1985)
582. Wahba, G.: Spline Models for Observational data, vol. 59. SIAM (1990)
583. Wahlberg, B.: System identification using Laguerre models. IEEE Trans. Autom. Control **36**(5), 551–562 (1991)
584. Wahlberg, B.: System identification using Kautz models. IEEE Trans. Autom. Control **39**(6), 1276–1282 (1994)
585. Wan, H.-P., Ren, W.-X.: Parameter selection in finite-element-model updating by global sensitivity analysis using gaussian process metamodel. J. Struct. Eng. **141**(6), 04014164 (2014)
586. Wang, G.G.: Adaptive response surface method using inherited Latin hypercube design points. J. Mech. Des. **125**(2), 210–220 (2003)
587. Wang, G.G., Dong, Z., Aitchison, P.: Adaptive response surface method – a global optimization scheme for approximation-based design problems. Eng. Optim. **33**(6), 707–734 (2001)

588. Wang, G.G., Shan, S.: Review of metamodeling techniques in support of engineering design optimization. J. Mech. Des. **129**(4), 370–380 (2007)
589. Wang, L.-X.: Adaptive fuzzy systems and control. design and stability analysis. Prentice Hall, Englewood Cliffs (1994)
590. Wang, L.-X., Mendel, J.M.: Fuzzy basis functions, universal approximation, and orthogonal least-squares learning. IEEE Trans. Neural Netw. **3**(5), 807–814 (1992)
591. Watson, G.S.: Smooth regression analysis. Indian J. Stat. Ser. A **26**, 359–372 (1964)
592. Weigend, A.S., Gershenfeld, N.A.: Time series prediction: forecasting the future and understanding the past. Addison-Wesley, Reading (1994)
593. Weiss, S.M., Kulikowski, C.A.: Computer Systems that Learn. Morgan Kaufmann Publishers, San Francisico (1991)
594. Wellers, M., Kositza, N.: Identifikation nichtlinearer Systeme mit Wiener- und Hammerstein-Modellansätzen auf Basis der Volterra-Reihe. Automatisierungstechnik **47**(5), 209–216 (1999)
595. Welling, M.: Kernel ridge regression. Max Welling's Classnotes in Machine Learning (http://www.ics.uci.edu/welling/classnotes/classnotes.html), pp. 1–3 (2013)
596. Werbos, P.J.: Beyond Regression: New Tools for Prediction and Analysis in the Behavioural Sciences. Ph.D. thesis, Harvard University, Boston, USA (1974)
597. Werntges, H.W.: Partitions of unity improve neural function approximators. In: IEEE International Conference on Neural Networks (ICNN), vol. 2, pp. 914–918, San Francisco, USA (1993)
598. Wettschereck, D., Dietterich, T.: Improving the performance of radial basis function networks by learning center locations. In: Moody, J.E., Hanson, S.J., Lippmann, R.P. (eds.) Advances in Neural Information Processing Systems, vol. 4, pp. 1133–1140. Morgan Kaufmann, San Mateo (1992)
599. Whitely, D., Bogar, C.: The evolution of connectivity: pruning neural networks using genetic algorithms. In: IEEE International Conference on Neural Networks (ICNN), vol. 1, Washington, USA (1990)
600. Williams, R.J.: Adaptive state representation and estimation using recurrent connectionist. In: Miller, W.T., Sutton, R.S., Werbos, P.J. (eds.) Neural Networks for Control, pp. 97–114. MIT Press, Cambridge (1990)
601. Williams, R.J., Zipser, D.: A learning algorithm for continually running fully recurrent neural networks. Neural Comput. **1**(2), 270–280 (1989)
602. Willinger, R.: Das CORDIER-Diagramm für Strömungsarbeitsmaschinen: Eine theoretische Begründung mittels Stufenkennlinien. In: VDI-Berichte, number 2112, pp. 17–28 (2010)
603. Willinger, R.: Theoretical Interpretation of the CORDIER-Lines for Squirrel-Cage and Cross-Flow Fans. In: Proceedings of the ASME TurboExpo, pp. 675–684, Copenhagen, Denmark (2012)
604. Willinger, R., Köhler, M.: Influence of Blade Loading Criteria and Design Limits on the Cordier-Line for Axial Flow Fans. In: Proceedings of the ASME TurboExpo, Düsseldorf, Germany (2014)
605. Wilson, A.C., Roelofs, R., Stern, M., Srebro, N., Recht, B.: The marginal value of adaptive gradient methods in machine learning. In: Advances in Neural Information Processing Systems, pp. 4151–4161 (2017)
606. Yingwei, L., Sundararajan, N., Saratchandran, P.: Performance evaluation of a sequential minimal radial basis function (RBF) neural network learning algorithm. IEEE Trans. Neural Netw. **9**(2), 308–318 (1998)
607. Yoshinari, Y., Pedrycz, W., Hirota, K.: Construction of fuzzy models through clustering techniques. Fuzzy Sets Syst. **54**, 157–165 (1993)
608. Zadeh, L.A.: Fuzzy sets. Inf. Control **8**, 338–353 (1965)
609. Zadeh, L.A.: Outline of a new approach to the analysis of complex systems and decision processes. IEEE Trans. Syst. Man Cybern. **1**, 28–44 (1973)
610. Zhan, J., Ishida, M.: The multi-step predictive control of nonlinear siso processes with a neural model predictive control (NMPC) method. Comput. Chem. Eng. **21**(2), 201–210 (1997)

611. Zhang, C., Bengio, S., Hardt, M., Recht, B., Vinyals, O.: Understanding deep learning requires rethinking generalization. arXiv preprint arXiv:1611.03530 (2016)
612. Zhang, C., Liao, Q., Rakhlin, A., Sridharan, K., Miranda, B., Golowich, N., Poggio, T.: Theory of deep learning III: generalization properties of SGD. Technical report, Center for Brains, Minds and Machines (CBMM) (2017)
613. Zhang, X.-C., Visala, A., Halme, A., Linko, P.: Functional state modelling approach for bioprocesses: local models for aerobic yeast growth processes. J. Process Control **4**, 127–134 (1994)
614. Zhou, Z.-H., Wu, J., Tang, W.: Ensembling neural networks: many could be better than all. Artif. Intell. **137**(1), 239–263 (2002)
615. Zimmermann, H.-J.: Fuzzy Set Theory and its Application, 2nd edn. Kluwer Academic Publishers, Boston (1991)
616. Zinner, K.A.: Aufladung von Verbrennungsmotoren. Springer, Berlin (1985)

Index

O. Nelles, *Nonlinear System Identification*,
https://doi.org/10.1007/978-3-030-47439-3